Springer Collected Works in Mathematics

More information about this series at http://www.springer.com/series/11104

1938

Marston Morse.

Marston Morse

Selected Papers

Editor
Raoul Bott

Reprint of the 1981 Edition

Springer

Author
Marston Morse (1892 – 1977)
The Institute for Advanced Study
Princeton
USA

Editor
Raoul Bott (1926 – 2002)
Harvard University
Cambridge
USA

ISSN 2194-9875
Springer Collected Works in Mathematics
ISBN 978-1-4939-3415-7 (Softcover)
978-0-387-90532-7 (Hardcover)
DOI 10.1007/978-1-4939-3417-1

Library of Congress Control Number: 2012954381

Springer New York Heidelberg Dordrecht London

Printed on acid-free paper

Springer Science+Business Media LLC New York is part of Springer Science+Business Media (www.springer.com)

Marston Morse
Selected Papers

Edited by Raoul Bott

Springer-Verlag
New York Heidelberg Berlin

Marston Morse
The Institute for Advanced Study
(1935 – 1977)
Princeton, New Jersey 08540
USA

Editor
Raoul Bott
Department of Mathematics
Harvard University
Cambridge, Massachusetts 02138
USA

AMS Classification (1980): 01A75, 58-XX

With 22 illustrations

Library of Congress Cataloging in Publication Data

Morse, Marston, 1892 – 1977
Selected papers.

Bibliography: p.
1. Mathematics—Collected works. 2. Global analysis (Mathematics)—Collected works. I. Bott, Raoul, 1924–
QA3.M714 510 81-100
AACR1

9 8 7 6 5 4 3 2 1

ISBN 978-0-387-90532-7 Springer-Verlag New York Heidelberg Berlin
ISBN 978-0-387-90532-7 Springer-Verlag Berlin Heidelberg New York

Contents

Permissions

Springer-Verlag would like to thank the original publishers of M. Morse's papers for granting permissions to reprint specific papers in this volume. The following list contains the credit lines for those articles that are under copyright.

[2] Reprinted by permission from *Am. J. Math.* **43,** 33–51, © 1921 by Johns Hopkins University Press.

[3] Reprinted by permission from *Trans. Amer. Math. Soc.* **22,** 84–110, © 1921 by American Mathematical Society.

[4] Reprinted by permission from *Trans. Amer. Math. Soc.* **26,** 25–61, © 1924 by American Mathematical Society.

[5] Reprinted by permission from *Trans. Amer. Math. Soc.* **27,** 345–396, © 1925 by American Mathematical Society.

[7] Reprinted by permission from *Trans. Amer. Math. Soc.* **30,** 213–274, © 1928 by American Mathematical Society.

[17] Reprinted by permission from *Proc. Amer. Acad. Arts Sci.* **66,** 235–253, © 1931 by American Academy of Arts and Sciences.

[18] Reprinted by permission from *Ann. Math.* **32,** 549–566, © 1931 by Princeton University Press.

[21] Reprinted by permission from *Trans. Amer. Math. Soc.* **32,** 599–631, © 1930 by American Mathematical Society.

[35] Reprinted by permission from *Bull. Amer. Math. Soc.* **42,** 915–922, © 1936 by American Mathematical Society.

[36] Reprinted by permission from *Ann. Math.* **38,** 386–449, © 1937 by Princeton University Press.

[39] Reprinted by permission from *Amer. J. Math.* **60,** 815–866, © 1938 by Johns Hopkins University Press.

[44] Reprinted by permission from *Ann. Math.* **41,** 419–454, © 1940 by Princeton University Press.

[49] Reprinted by permission from *Ann. Math.* **42,** 62–72, © 1941 by Princeton University Press.

[60] Reprinted by permission from *Duke J. Math.* **11,** 1–7, © 1944 by Duke University Press, Durham (North Carolina).

[65] Reprinted by permission from *Acta Math.* **79,** 51–103, © 1947 by Institut Mittag-Leffler.

[73] Reprinted by permission from *Ann. Math.* **51,** 576–614, © 1950 by Princeton University Press.

[87] Reprinted by permission from *Yale Rev.* **40,** 604–612, © 1951 by Yale University Press.

[102] Reprinted by permission from *Ann. Math.* **64,** 480–504, © 1956 by Princeton University Press.

[113] Reprinted by permission from *Ann. Math.* **71,** 352–383, © 1960 by Princeton University Press.

Permissions

[114] Reprinted by permission from *Bull. Amer. Math. Soc.* **66,** 113–115, © 1960 by American Mathematical Society.

[139] Reprinted by permission from Differential and Combinatorial Topology, 81–103, © 1965 by Princeton University Press.

BULLETIN (New Series) OF THE
AMERICAN MATHEMATICAL SOCIETY
Volume 3, Number 3, November 1980

MARSTON MORSE AND HIS MATHEMATICAL WORKS

BY RAOUL BOTT[1]

1. Introduction. Marston Morse was born in 1892, so that he was 33 years old when in 1925 his paper *Relations between the critical points of a real-valued function of n independent variables* appeared in the Transactions of the American Mathematical Society. Thus Morse grew to maturity just at the time when the subject of Analysis Situs was being shaped by such masters[2] as Poincaré, Veblen, L. E. J. Brouwer, G. D. Birkhoff, Lefschetz and Alexander, and it was Morse's genius and destiny to discover one of the most beautiful and far-reaching relations between this fledgling and Analysis; a relation which is now known as *Morse Theory*.

In retrospect all great ideas take on a certain simplicity and inevitability, partly because they shape the whole subsequent development of the subject. And so to us, today, Morse Theory seems natural and inevitable. However one only has to glance at these early papers to see what a tour de force it was in the 1920's to go from the mini-max principle of Birkhoff to the Morse inequalities, let alone extend these inequalities to function spaces, so that by the early 30's Morse could establish the theorem that for any Riemann structure on the n-sphere, there must be an infinite number of geodesics joining any two points.

This whole flight of ideas was of course acclaimed by the mathematical world. It brought him to the Institute for Advanced Study in 1935, when, at 43, he also delivered the Colloquium Lectures of the Mathematical Society and wrote his monumental book on the Calculus of Variations in the Large; it eventually earned him practically every honor of the mathematical community, over twenty honorary degrees, the National Science Medal, the Legion of Honor of France, ...

Nevertheless, when I first met Marston in 1949 he was in a sense a solitary figure, battling the *algebraic topology*, into which his beloved Analysis Situs had grown. For Marston always saw topology from the side of Analysis, Mechanics, and Differential geometry. The unsolved problems he proposed had to do with dynamics–the three body problem, the billiard ball problem, and so on. The development of the algebraic tools of topology, or the project of bringing order into the vast number of homology theories which had sprung up in the thirties–and which was eventually accomplished by the Eilenberg-Steenrod axioms–these had little interest for him. "*The battle between algebra and geometry has been waged from antiquity to the present*" he wrote in his address *Mathematics and the Arts* at Kenyon College in 1949, and

Received by the editors April 15, 1980.

[1]This work was supported in part through funds provided by the National Science Foundation under the grant 33-966-7566-2.

[2]Poincaré was born in 1854, the others all in the 1880's.

0002-9904/80/0000-0500/$12.00

again, a few lines later: "Forever the foundation and never the Cathedral."

Poincaré's interest in Analysis Situs had two distinct sources: Dynamics and Algebraic Geometry. And these two mainsprings of Topology were magnificently represented by G. D. Birkhoff and S. Lefschetz respectively, in the early part of this century. Through his teacher, G. D. Birkhoff, Morse had inherited the dynamical tradition–in no uncertain terms– and it was in this framework that he understood Analysis Situs.

There was thus a natural ambivalence to him to the sibling algebraic branch which had evolved so strongly around Lefschetz in Princeton. I distinctly recall an afternoon in 1949 when he reminisced about his appointment at the I. A. S. and spoke wistfully about how differently things would look if he had gone to the University and Lefschetz had taken on the post at the Institute. His wrath, I should hasten to add, was never personal–and certainly not directed at Lefschetz's achievements in algebraic geometry. Rather, he resented the omnipresence of algebra in the topological scene at that time. And certainly it was true that in 1949 the Geometric tradition of topology was not nearly as well represented as it is, say, today. And it is also true that at decisive moments it was precisely the Morse Theory–in the highly geometric setting of Smale–which overcame the greatest obstacles. On the other hand the late forties and early fifties were exciting years for homotopy theory; for with the advent of the algebraic tools of the Steenrod Algebra and Serre's application of the Leray Spectral Sequence homotopy theory had become tamed, and it is not surprising that this development temporarily eclipsed all others.

The next sections will be devoted to a more detailed account of the Morse theory and other aspects of Morse's work, but before going on, I find the urge to reminisce–once indulged in–too strong to be denied. It was my good fortune to come to the Institute for Advanced Study in Princeton in 1949, largely through the good offices of my teacher R. Duffin at Carnegie Institute of Technology and of H. Weyl, who had befriended me there in Pittsburgh while visiting Carnegie on a lecture tour. The general plan of my appointment as I understood it, was that I was to write a book on network theory at the Institute.

I suppose that a young prospective knight approaching King Arthur's table for the first time must have felt as I did when I first walked into Fuld Hall and took possession of my small office on the third floor. The professors of Mathematics at that time were Oswald Veblen, Hermann Weyl, John Von Neumann, Carl Ludwig Siegel, Marston Morse and James W. Alexander. On the ground floor you passed Einstein's office and you were welcomed upon arrival by J. Robert Oppenheimer. The permanent members were Kurt Gödel, Deane Montgomery, and Atlè Selberg. The officer in charge of the temporary members that year was Marston Morse and it was in that capacity that I first met him. His office was also on the third floor and I had seen him bounding up the stairs several times before my official "reporting to work", so to speak, occurred.

The overwhelming impression which remains with me to this day is the vast energy which Marston somehow radiated. I compute now—with amazement–that he must have been fifty-seven at that time, and recall with some humiliation that a few weeks later at a party at our house, he easily beat me in a hundred yard dash to which he had–characteristically–challenged me. But to return to my "reporting to work". I approached this interview with some nervousness, because in the few weeks which I had spent at Princeton before it took place, the whole mysterious world of pure mathematics had burst upon me and all I wanted to do was explore it. In no way did I want to write the book.

Well, after five minutes with Marston all my uneasiness had vanished. First of all I found that I really did not have to say very much! I think it is a fair statement that in all conversations with Marston, one only had to do twenty percent of the talking. His energy was such that it just naturally took over. He immediately dismissed my fears of having to write a book. It was a matter of course to him that at the Institute a young man should only do what he wanted to do; that this was the place where a young man should find himself, and the last place in the world for performing a chore. And once this technical part of our interview was over he immediately, again characteristically I think, started to speak about the subject that absorbed his interest at the time. Actually, in 1949 this subject had nothing to do with critical point theory. Rather, he was deeply involved in his work with Transue, on functions of bounded variations. In any case, I remember leaving this interview with a light heart, newly liberated and buoyed by the energy and optimism I had just encountered. I was also elated by the directness of Marston's manner. There was not the slightest condescension in it. Although he dominated–I expect–all encounters, he treated everyone as an equal, with complete honesty, and in personal matters he showed great kindness and generosity.

Mathematically, Marston and I did not communicate too well, and I don't think we could have collaborated. I also recall really only one private lecture on critical point theory from him. His primary interests were elsewhere at the time. But on the personal side we got on right from the start, even though we often disagreed.

Marston loved music and played the piano beautifully and effortlessly. He was devoted to Bach and very knowledgeable about all aspects of music, and so music was our first and quite natural bond. But beyond that and quite apart from certain affinities of taste, I think I immediately sensed and revered his spiritual nature. Marston was a deeply religious man, yet I never heard him "preach". One was conscious of this aspect of his life only indirectly and quite marvelously. His daughter-in-law, Terry Morse, put it better than I every could.

"His personality had a light and a force which was very spiritual and mysterious. I think it was because he welcomed the ultimate mystery of life, embraced it, and took great joy in it, that we always came away from being with him feeling a heightened sense of awareness of the beauty and the possibilities in life," she wrote to Louise Morse after his death.

And then, there was his wonderful wife Louise, to whom we–indeed all of us new green Ph.D.'s–were immediately drawn, and who was such a natural complement to Marston. The Morses took their stewardship of the mathematical community very seriously and it was to them we turned in times of need. Of course Marston's advice could be disheartening. I remember calling on him for suggestions concerning a summer job in 1950. After some reflection he recommended mowing lawns or baby sitting! When it came time to move on in 1951, he would recommend the mathematical wasteland. "Get away from Princeton and the mathematical centers. Have your own thoughts in peace and quiet," he would say. "Look what Lefschetz did in Kansas."

But it is time to hear different memories of the past than mine. *Maurice Heins*, who knew Marston as a student and collaborator and friend, writes as follows:

"My first encounter with Marston was as a sophomore in what was then Mathematics 5 at Harvard. It was his last year at Harvard. Characteristic of his lectures were lucidity, simplicity and eloquence. One felt that one was in the presence of a master. He shared with Heinz Hopf a strongly tactile presentation. It was as if they were sculptors in their ateliers. In this connection, I recall that Marston thought that his experience with carpentry contributed to his geometric sense greatly. In 1934 the curriculum of a course like Math 5 was extremely rigid. Marston conveyed very beautifully and intuitively without belaboring the ε's the import and the essential ideas of the proofs of such theorems as the implicit function theorem, the Sturm separation theorem (curriculum not withstanding), the Euler-Lagrange condition. One has to keep in mind the quasitotal absence of mathematical sophistication of the part of his hearers."

"What was very impressive about Marston in the 20's and 30's was that he was devoid of the bigotry that paraded in the guise of gentility during that period. He valued one's worth and integrity, not the accidentals of birth. He had enormous drive and the physical capacity for many hours of work. Twenty hour work days were common."

"The intensity of his devotion to scholarship was coupled with uncompromising standards of rigor imposed both on himself and his collaborators. Working with him was a very intense experience."

"Our collaborative work was proposed by Marston who wanted to bring critical point theory to bear on the theory of harmonic functions and analytic functions, or, more generally, the pseudoharmonic functions and light interior transformations. The physical circumstances of the collaboration may appear surprising but exemplify some of the things said above about Marston's temperament. At the tme he was in Washington as a scientific consultant to the then War Department, Office Chief of Ordnance. I was a P-4 mathematician working with him and W. R. Transue. After the normal work day Marston and I worked on the joint papers from 7 to 11, Monday through Thursday. He wrote drafts of the work on the weekends back in Princeton . . . "

I quote next from *William Transue's* reminiscences:

"In the fall of 1942 I went to the Institute as Marston's assistant. I was a fresh Ph.D., with dissertation on subharmonic functions, but with no set direction except for the broad field of analysis. At that time Marston had the idea of applying topological methods to obtain information on the 3-body problem, and he set me to work reading F. R. Moulton's *Celestial mechanics* and a paper of G. D. Birkhoff in this area. However, Marston was at this time a consultant to the Office of the Chief of Ordnance and spent more and more time in Washington. In the spring of 1943 I moved to Washington and all of my attention and most of his was devoted to military problems. These were "applied" mathematics of the dirtiest sort, applied to whatever difficulties the Ordnance Department was encountering, but came to center principally on the area known as terminal ballistics–the study of the destructive effects of bombs and shells. The mathematics involved were usually of the most elementary sort, generally numerical integrations, and we were delighted when we could bring in something as sophisticated as the icosahedron. This was before the day of the computer, and our computations were done by a small staff using keyboard machines to add and subtract, multiply and divide. Programming computations for such a staff was roughly equivalent to programming a present-day computer. Marston's capacity for sheer hard, and usually dull, work on these problems was unbelievable. After the offices were closed in the evening we would go on for a couple of hours. I remember particularly one set of data on land mine explosions which seemed to make no good sense. The rest of us were ready to give up, but Marston returned to the attack again and again, determined to beat some kind of order into the data. and we did finally put out some kind of analysis."

"Several years later we undertook the study of integral representation of bilinear functionals and allied topics which continued for some years. I spent one full year at the Institute and many summers with him at various places–Princeton, Cape Cod, Maine. Working with Marston (for me at least) meant being completely taken over, spending almost all waking hours with him, talking mathematics all day, including during many meals taken with the Morse household, and continuing late into the evening. He was a real bear for work. In Princeton we usually worked in his bedroom (Richard Arens once remarked that he couldn't concentrate with someone else's pants hanging in the closet!), and in Maine sometimes in the car. Louise was always extremely good about finding him a quiet place to work, but with the number of children about, this was not always easy. For his part, he was a very considerate husband and father, and the Morse household, which I got to know pretty intimately, was a very happy and harmonious one."

"The stamp of his Maine upbringing was pretty heavy on him, and he retained not only the industry, but the frugality which characterizes the Maine citizen. Although he gave his money generously, he spent it carefully. He was equally generous with his time. I recall one evening in Princeton, when Louise and the children were out of town and we were doing our own cooking. In the midst of warming our soup, the telephone rang and a reporter from *Time* magazine wanted Marston to explain the theory of relativity. Well, he did! He could never refuse to teach anyone."

"He was a great Francophile and liked most to think of himself as a mathematical descendant of Poincaré. His conversion to Catholicism brought him also particularly close to Italy. Of course, his position made him widely known in the international community of mathematics, but his feelings for France and Italy were special . . . "

Finally I quote from a letter by *S. Cairns*, a life long friend and collabora-

Marston Morse with Dryden and Meroe
Waterville, Maine 1929

Louise and Marston
Louise II, Peter, Julia, Elizabeth, and William
Princeton, New Jersey 1950
(Photo by Meroe Morse)

tor, in which he gives a brief chronological account of Morse's career, as well as some personal remembrances.

"In the fall of 1926, after the completion of undergraduate work at Harvard, I made the acquaintance of Marston Morse. He had just joined the faculty of Harvard University. During the ensuing academic year, he conducted a lively seminar in topology. This subject, then in a relatively primitive stage, was essential to the research in global analysis on which he had embarked. In the following year, Morse suggested to me, as subject for a doctoral dissertation, the triangulation of the differentiable manifold, with the enticing comment that I could start on it at once because there was no literature to read. Thus commenced more than fifty years of professional association and personal friendship, culminating in collaboration during the last twelve years of Morse's life."

"As teacher and research worker for nine years at Harvard, Morse was a stimulating source of enlightenment and inspiration. He saw mathematics as a challenging thing of beauty and imparted his enthusiasm to others."

"Morse was born in 1892 in Waterville, Maine, where he had his early education and where he completed his undergraduate work in 1914 at Colby College. Three years later he received the degree of Ph.D. from Harvard, having meanwhile published his first research paper in 1916."

"World War I interrupted his career. He served with distinction in the American Expeditionary Force and was awarded the Croix de Guerre with Silver Star for bravery under fire."

"Resuming the academic life, Morse taught at Cornell, 1920–1925, and at Brown, 1925–1926, before his appointment to Harvard University."

"In 1935, Morse accepted a professorship at the Institute for Advanced Study, which had just been established in Princeton to provide an ideal environment for the most active and distinguished scholars in a broad spectrum of disciplines. Morse retired in 1962 as professor emeritus, but, for the rest of his eighty-five years, continued his research with extraordinary vigor and creativity."

"In World War II, Morse was a consultant in the Office of the Chief of Ordnance. His invaluable work on military applications of mathematics was recognized by a Meritorious Service Award, conferred in 1944 by President Roosevelt."

"After the war, he was a prime mover in the creation of the National Science Foundation. President Truman appointed him to serve on its first board from 1950 to 1954."

"In 1952, Morse was a representative of the Vatican at the Atoms for Peace Conference of the United Nations."

"He was president of the American Mathematical Society, 1940–1942; a vice president of the International Mathematical Union, starting in 1958; chairman of the Division of Mathematics of the National Research Council, 1951–1952. The list of such offices could be continued. In his community he served on the boards of two private schools and of the Princeton Chapter of Recording for the Blind."

"Many honors were bestowed on Morse, among them honorary degrees from twenty institutions in the U.S.A., Austria, France and Italy. These included the University of Paris (1946), Pisa (1948), Vienna (1952), Harvard (1965) and Modena (1975). In 1952, he became a Chevalier of the French Legion of Honor. He was elected in 1932 to the National Academy of Sciences and in 1956, as an associate member, to the French Academy of Sciences. His affinity for France made the honors from that country particularly gratifying. He also cherished his election as a corresponding member of the Italian National Academy Lincei. A National Medal of Science was awarded to him in 1964 and presented by President Johnson at the White House in 1965. Again the list could be prolonged."

"A year or two after his retirement, in a conversation, Morse outlined to me the problems he hoped to solve, if only he could live twenty years more and keep on doing research. It has been gratifying to see a substantial part of his hope fulfilled and to collaborate with him in its implementation."

"Essential to the remarkable prolongation of his long and brilliant career was the devoted care and sympathetic understanding of Mrs. Morse. She was the mistress of their home, the mother of their five children and a charming hostess to countless colleagues and friends. Also, with tender skill, she helped conserve her husband's health in his advancing years."

"Besides Mrs. Morse and their children Julia, William, Elizabeth, Peter and Louise there survives one of the two offspring of an earlier marriage, Dr. Dryden Morse. Meroë, his eldest daughter, died in 1969."

"Morse's cultural interests extended far beyond the boundaries of his profession. Music, in particular, was a lifelong avocation. He played the piano with consummate skill and sensitivity. His repertory of classical music was large and was increasing up to the end . . . ."

"Science, philosophy, religion and the arts were objects of Morse's inquiring mind. He recognized the fundamental unity of creativity in all these areas. A stimulating treatment of this subject is to be found in his paper, *Mathematics and the Arts*, Bulletin of the Atomic Scientists **9** (1959), 55–59, based on his lecture at Kenyon College in 1949, during a conference honoring Robert Frost . . . ."

A moment I will always cherish occurred upon Marston's return from his lecture at Kenyon. He met me as he came bounding up the stairs to our third floor, and immediately took me into his office to tell me what a marvelous speech he had given: "The speaker before me had a terrible voice," he remarked, "And really didn't have much to say, so when I finished, I brought the house down."

And of course he is quite right; his essay is a masterpiece, and had indeed brought the house down.–There was no modesty in Marston; he told it as he saw it.

> His life was gentle, and the elements
> So mixed in him that nature might stand up
> And say to all the world, "This was a man".

2. The works of Marston Morse. It would be impossible to comment in detail on the bulk of Morse's work. His bibliography has 180 entries, and includes seven books. Rather, I have selected eight topics which played a central role at various stages of his career and I will attempt to explain some aspects of each of these in detail.

Morse had many collaborators, and I have therefore included the names of his principal collaborators in the various subjects as well as a rough indication of the time when this work was done in the following list of the topics.

(3) Dynamics, geodesic flow (Hedlund, (1917–1940's))
(4) Morse theory (1921–1978)
(5) Minimal Surfaces (Tompkins, 1940's)
(6) Topological methods in a single complex variable (M. Heins, 1940's)
(7) Integral representations (Transue, 1940's, 1950's)
(8) Pseudoharmonic functions (Jenkins, 1950's)
(9) Differential topology (Hubsch, Cairns, 1960's–1970's)

Finally a word of thanks to the many colleagues who have helped in this enterprise. I am especially indebted to the collaborators of Morse already mentioned as well as: G. Mackey, D. Gromoll, J. Mather, D. Kazhdan, M. Brin, T. Goodwilly and N. Hingston.

3. Dynamics–geodesic flow. There are two outstanding results in Morse's work on this subject, and I will report on them under two headings–*The Morse Trajectory*, and *Instability implies transitivity*.

3.1. *The Morse Trajectory*. This discovery goes back to Morse's thesis of 1917 and was published under the title, *Recurrent geodesic on a surface of negative curvature* [3].

In 1944, Hedlund and Morse collaborated on a paper [**60**] where this same construction solves a problem in unending chess and in the theory of semigroups. This "Morse trajectory" also occurs in Novikov's disproof of the Frobenius-Burnside conjecture [**N1**] and was there attributed to a Russian mathematician writing in 1939. The trajectory in question occurs in the paper cited above dealing with the behavior of geodesics on surfaces of negative curvature imbedded in three-space. In this context this trajectory solves a problem posed by Birkhoff [**B1**] in 1912.

The surfaces which Morse considers are of the type indicated in Figure 1,

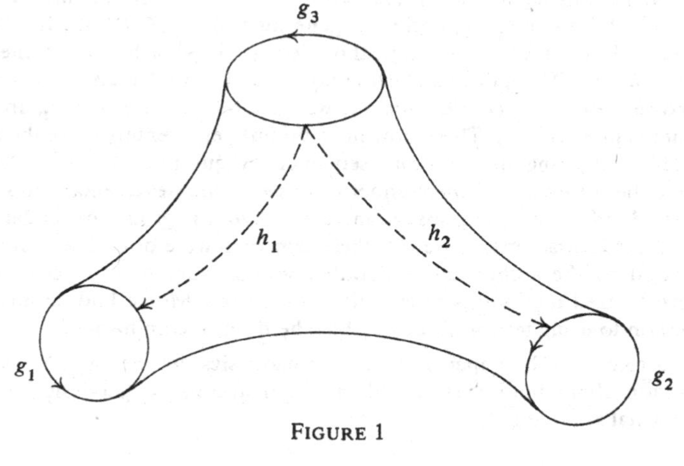

FIGURE 1

and in contemporary language are 2-manifolds with boundary, smoothly embedded in a compact region of $\mathbf{R}^3$ so as to inherit negative curvature except possibly at a finite number of points. The boundary circles are taken to be closed geodesics, and the genus of S is assumed to be not less than 2

$$\chi(S) > 2.$$

Using the basic existence theorems of Hadamard **[H]**, Morse gives a completely combinatorial description of those geodesics g *which lie entirely on* S. Precisely, such a geodesic is then a map

$$g\colon \mathbf{R} \to S$$

satisfying the geodesic differential equation.

To every such "completely extended" geodesic Morse now assigns an object which nowadays we might call an infinite word in the generators of the fundamental group $\pi_1(S)$. Morse's strategy for doing this is to introduce geodesic segments h_1, h_2 as indicated above. Cut along these and S becomes simply connected. Now any geodesic g must cut these segments transversally at all points of intersection, and using this fact Morse assigns to g a sequence of symbols

$$\cdots C_{-2}C_{-1}C_0C_1C_2 \cdots$$

indexed by the integers, with each C ranging over the set g_1, $g_2g_1^{-1}$, g_2^{-1}, with the understanding that a g_i is never followed by a g_i^{-1} and a g_i^{-1} is never followed by a g_i. Morse calls such a sequence a *Normal set* and shows that modulo translation the Normal set gives a one-to-one *representation of the geodesics entirely on S*.

A slightly different, but quite equivalent description of the geodesic flow on S had already been given by Morse in **[2]**, but this normal set representation is the most useful for the purposes of this paper.

In modern language, this representation can be thought of this way: We first interpret the symbols g_1 and g_2 as generators of $\pi_1(S; P)$, the fundamental group of S based at P; thus think of g_1 as shorthand for the loop $h_1^{-1} \circ g_1 \circ h_1$, etc. Then, once a starting point has been selected, g gives rise to a nested sequence of geodesic segments with endpoints on h_1 and h_2, and thus to elements in $\pi_1(S^1; P)$. These now have unique representations in the g_i and so result in the unending normal sequence in question. I say in modern language, because in the papers under discussion one never encounters terms such as: Manifold, covering space, universal covering, group, or fundamental group, Euler Characteristic, etc. All these concepts were of course profoundly understood by the author and his contemporaries, but they do not seem to have crystallized into words at that time. Of course Morse had, in any case, an aversion to using technical terms which he did not coin himself.

But to return to the paper on recurrent geodesics, let me say one word of explanation about the inverse problem of constructing a geodesic g from a given normal sequence.

The proof of this fact depends on the following fundamental result of Hadamard: "Given a curve $c(t)$, $0 \leq t \leq 1$, on S there is one and only one geodesic joining the endpoints of c on S, which is homotopic to c (with end points fixed). Furthermore this geodesic is of minimal length in its homotopy class". At this point the negative curvature of course enters vitally; this result is patently incorrect, say on the sphere.

The description of the geodesic flow on S furnished by the Normal Sequences is the starting point of a discipline called "symbolic dynamics". It seems to have been independently discovered by various authors–but as far as I can see, always considerably later and usually in connection with surfaces of *constant* negative curvature. Marston Morse and Hedlund took up this subject in considerable detail much later in **[39]** and **[43]**, where they developed delicate combinatorial criteria for symbolic trajectories; and of course by now this subject is well established in all aspects of the study of dynamical systems, see for instance **[S4]**, **[S5]**.

Some of the early consequences which Morse deduced from his representation are the following.

THEOREM. *Every geodesic wholly on S is the limit of closed geodesics.*

DEFINITION. A set R of geodesics wholly on S is called minimal if every element of R has every other element of R but no other geodesic as a limit geodesic.

Thus a closed geodesic is the simplest example of a minimal set. On the other hand Morse shows that

THEOREM. *Every minimal set other than a closed geodesic has the power of the continuum.*

And he then goes on to construct a nontrivial minimal set. By the way, he *defines* a geodesic to be *recurrent* if it is a member of such a minimal set. Thus in this terminology he constructs a *recurrent geodesic* which is not *closed*.

His method is of course to translate all these concepts into combinatorial form first, and then to construct a *normal sequence* with the desired combinatorial properties. The pertinent definitions are as follows.

A sequence

$$\cdots R_{-2}R_{-1}R_0R_1R_2 \cdots \tag{3.1}$$

indexed by the integers is called recurrent if for every positive integer r there is a positive integer s such that every "segment" of length r

$$R_m R_{m+1} \cdots R_{m+r}$$

in (3.1) is contained in every segment of length s

$$R_n R_{n+1} \cdots R_{n+s}$$

of (3.1).

With this concept understood, Morse then constructs his ingenious *Morse Trajectory* on two symbols which he denotes 1 and 2–by the following inductive procedure (I quote verbatim): "Set

$$\begin{aligned} a_0 &= 1\\ b_0 &= 2\\ a_1 &= a_0b_0\\ b_1 &= b_0a_0\\ &\vdots\\ a_{n+1} &= a_n\cdot b_n\\ b_{n+2} &= b_n\cdot a_n. \end{aligned}$$

We introduce the set of symbols

$$\cdots d_{-2}d_{-1}d_0d_1d_2\cdots \tag{3.2}$$

of which $d_0d_1 \cdots d_{2^n}$ are defined respectively as the 2^n integers of a_n; further if m is any positive integer, d_{-m} is defined as equal to d_{m-1}. The set (2.2) so defined will be proved to be *recurrent* without being *periodic*".

Explicitly

$$a_0 = 1,\quad a_1 = 12,\quad a_2 = 1221,\quad a_3 = 1221\,2112,$$
$$b_0 = 2,\quad b_1 = 21,\quad b_2 = 2112,\quad b_3 = 2112\,1221.$$

Note that if we replace 1 and 2 by g_1 and g_2 respectively, this Morse trajectory corresponds to a geodesic on the surface S of Figure 1 and thus constitutes the first example of a set of "recurrent motions of discontinuous typè" in the sense of Birkhoff **[B2]**.

A final explicit result of the paper under discussion is the following.

THEOREM. *On a surface S of genus $\geqslant 2$ of the type we are considering, every geodesic wholly on S is the limit of a recurrent but not periodic geodesic.*

3.2. *Instability and transitivity*. On a Riemann manifold M, the "geodesic flow" defines an action of the real numbers $\mathbf{R}$ on the unit tangent bundle T_1M of M: The real number r assigns to a tangent vector X_m the unit tangent vector at the "final" end of a geodesic segment s of length r starting at m and with initial direction X_m. The transitivity properties of this flow have been a subject of great interest throughout the last hundred years; in the period around 1930 especially, great advances were made with contributions from a veritable who's who of mathematicians of the period: Birkhoff, Hopf, Koopman, von Neumann, P. A. Smith. In 1934 Hedlund **[H2]** proved that this flow was *metrically transitive* (*Ergodic*) for any compact Riemann Surface M of *constant* negative curvature. At the same time Morse in his paper **[28]** bearing the title of our heading, gave very much more general conditions on a surface M, for this flow to be *topologically transitive*.

Recall that in context, *metrically transitive* means that all L^2-functions on T_1M invariant under the flow are constant (up to a set of measure 0) while *topological transitivity* means that some orbit of the flow be dense in T_1M. By the Ergodic theorems of Birkhoff and von Neumann metric transitivity implies topological transitivity but not vice versa.

Morse proves that a motion is *topologically transitive* provided it satisfies a *uniform instability*[3] criterion, which he informally describes as follows: "... it may be roughly regarded as the hypothesis that the first conjugate point of any point p on M be beyond the point at ∞ ...". More precisely it is a hypothesis which ensures that neighboring geodesics diverge in a uniform manner. Technically it means the following: Suppose that $g\colon \mathbf{R} \to M$ is any geodesic parametrized by arc length and that $w(x)$ is any solution of the Jacobi equations along g

$$d^2 w/dx^2 + k(x)w(x) = 0,$$

which satisfies the condition

$$(dw/dx)^2|_{x=0} + w(0)^2 = 1.$$

The geodesics on M are then *uniformly unstable* if there are no pairs of conjugate points on any g and if there exists a function $A(x)$, $x \in \mathbf{R}$ subject to the following conditions:

(α) A is positive continuous exceeding some constant λ and becomes infinite with x.

(β) $|w(x_1)| + |w(x_2)| > A(x)$ for $x > \lambda$ and $-x_1$ and $x_2 > x$.

(c) The function A is independent of g.

In particular then, this hypothesis is fulfilled on all surfaces of *negative curvature* and can also hold on surfaces having regions of *positive curvature*.

The methods of this paper depend heavily on the paper **[4]** of 1924, which in turn explores the ideas we encountered in the Morse Trajectory paper, but now in the context of *closed* surfaces of genus > 1. On an arbitrary such surface one of course cannot expect a purely combinatorial description of all geodesics. On the other hand Morse shows that certain geodesics–which he calls of class A–behave very much like the geodesics in the constant negative curvature case.

A geodesic g is of class A if any segment of g, i.e. $g|[a, b]$, minimizes the distance between its endpoints, amongst curves in its homotopy class.

To every such geodesic g on M Morse assigns a geodesic g^* in the constant curvature model of M by the following strategem.

Because the genus of M is > 1, its universal cover $\tilde{M}$ can be identified with the unit disc $|z| < 1$ in the complex plane. Thus we may think of $\tilde{M}$ as this *disc* in *some Riemann structure* which is invariant under the action of $\pi_1(M)$ acting as a subgroup of $SL(2, \mathbf{R})$ on $|z| < 1$. Now then every g on M lifts to a curve $\tilde{g}$ on $\tilde{M}$, and Morse shows that there is at *least one non-Euclidean straight line g^* on $|z| < 1$, such that $\tilde{g}$ is in a finite neighborhood of g^* and vice versa*. Further he shows that this correspondence is *one-to-one under the hypothesis of uniform instability*.

The ideas and insights of these papers are clearly precursors of the great new developments in this field in the 60's, due in large part to Anosov, Arnold, Sinai, Smale and others. In particular the condition of uniform instability is a precursor of the notion of an Anosov flow. See **[S4]** and **[S5]** for instance.

[3] *Warning*: This notion should not be confused with structural instability of a flow!

4. The Morse theory. In 1925 Marston Morse published his first paper concerned with the distribution of critical points of a function [5]. He returned to this subject in one form or another for the rest of his career, and there are at least fifty papers in his bibliography presenting different settings of this theory.

The paper [5] discusses the finite-dimensional case, but right away for manifolds with boundary, so that the famous "Morse inequalities" appear here less symmetrically than in later versions. Because these inequalities play such a natural role in contemporary mathematics, let me formulate them and outline their proof here in contemporary language, and only then return to the subject of Morse's methods in these early papers.

4.1. *The Morse inequalities.* Let f be a smooth function on a smooth manifold M. A critical point of f is then a point p at which

$$\left.\frac{\partial f}{\partial x^i}\right|_p = 0 \tag{4.1}$$

relative to some local system of coordinates $(x^1, \ldots, x^n)$. At such a point the matrix

$$H_p f = \left.\frac{\partial^2 f}{\partial x^i \partial x^j}\right|_p \tag{4.2}$$

has two intrinsic invariants which can be taken to be the rank and the number of negative eigenvalues. Morse calls the corank of $H_p f$ the *nullity* $n(p)$ of p (as a critical point of f) and the number of negative eigenvalues the *index* of p. This index is usually denoted by λ_p. Clearly it represents the dimension of the *largest subspace on which the quadratic form* $H_p f$ *is negative definite*.

With these concepts understood, Morse calls f a *nondegenerate* function if all of its critical points are nondegenerate, and a first quite general result concerning such functions is that *on a compact* M *a nondegenerate function can only have a finite number of critical points*. One may therefore count the critical points of a fixed index k, to obtain an integer $m_k(f)$ or simply m_k, which Morse calls the kth *type number of* f.

Some of us also like to introduce the polynomial

$$\mathfrak{M}_t(f) \equiv \sum t^k m_k \equiv \sum_p t^{\lambda_p} \tag{4.3}$$

and refer to it as the "Morse polynomial" of f.

This terminology recalls another famous polynomial in topology, the Poincaré polynomial

$$P_t(M) = \sum \dim H_i(M) t^i \tag{4.4}$$

where H_i denotes the ith homology group of M with coefficients in some field, and Morse's insight is now expressed by the following theorem, which essentially determines a *lower bound for* $\mathfrak{M}_t(f)$ *in terms of* $P_t(M)$.

THEOREM. *If f is any nondegenerate function on the compact n-manifold M, then its type numbers m_k and the Betti numbers* $b_k = \dim H^k(M)$ *of M satisfy* *the inequalities*

$$
\begin{aligned}
m_0 &\geq b_0,\\
m_0 - m_1 &\leq b_0 - b_1,\\
m_0 - m_1 + m_2 &\geq b_0 - b_1 + b_2, \qquad (4.5)\\
&\cdots\cdots\\
m_0 - m_1 + \cdots + (-1)^n m_n &= b_0 - b_1 + \cdots + (-1)^n b_n.
\end{aligned}
$$

These then are the famous "*Morse inequalities.*" They clearly set bounds for the m_k's in terms of the b_k's. Indeed subtracting successive lines of (4.5) yields

$$m_k \geq b_k. \tag{4.6}$$

But (4.5) is of course stronger. Note also the equality of the last line.

All in all one should think of these inequalities as a beautiful extension of the *minimum principle*. Indeed in the present context this principle simply asserts that $m_0 \geq b_0$.

The inspiration for Morse's work was the minimax principle of G. D. Birkhoff, which occurs in the middle of Birkhoff's famous paper on dynamical systems of two degrees of freedom, **[B1]** and there amounted to the inequality

$$m_0 - m_1 \leq b_0 - b_1 \tag{4.7}$$

on 2-surfaces.

Indeed on p. 346 of the paper **[5]** under discussion, Morse quotes the previous inequality and then writes: "Upon reading Birkhoff's paper it occurred to the author that inasmuch as there are $(n + 1)$ different kinds of critical points possible (in a sense to be defined later) there ought to be relations analogous to (4.7) . . . ."

Morse's method however is quite different from Birkhoff's, and it was only in the 30's that Birkhoff–in collaboration with Hestenes **[B3]**–produced a proof of the Morse inequalities along the lines of the minimax principle.

In his book **[C.V.]**, Morse also mentions that the equality in the last line of (4.5) already occurs in Poincaré. The general equality seems to have been discovered quite independently–and indeed from quite different points of view–at about the same time by H. Hopf, M. Morse and S. Lefschetz.

The method of proof which led Morse to the result is the one we still use today, and it naturally falls into two quite distinct parts. Part 1 is purely geometric, and is essentially based on the principle of deforming M along the directions of steepest descent for f. These arguments culminate in the following two theorems.

THEOREM A. *Suppose that our nondegenerate f has no critical points in the region $a \leq f \leq b$. Then if M_t denotes the "half space" where $f \leq t$ on M, there is a diffeomorphism of M_a with M_b*

$$M_b \simeq M_a. \tag{4.8}$$

This result (Morse's Lemma 6, p. 359) follows pretty directly from "pushing M_b into M_a along the gradient of f" and of course uses the compactness of the M_t's.

The next result on the other hand, really goes to the heart of the matter of what happens to M_t as t passes a critical value. To explain it in contemporary language let me remind the reader of the concept of attaching a *thickened cell* to a manifold. The underlying geometric idea is best gleaned from the following diagram

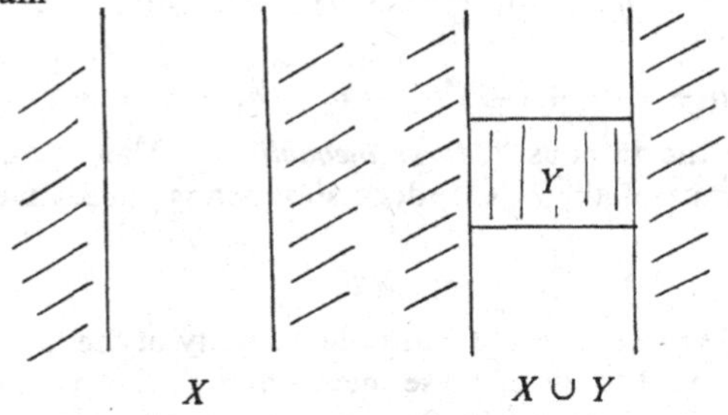

FIGURE 2

where we have attached the "thickened 1-cell Y" to X. Here $X = \{(x, y) | |x| > 1\}$; Y is the square $I \times I$, given by $|x| \leq 1$, $|y| \leq 1$, and the terminology arises from the fact that homotopically $X \cup Y$ is quite equivalent to the space

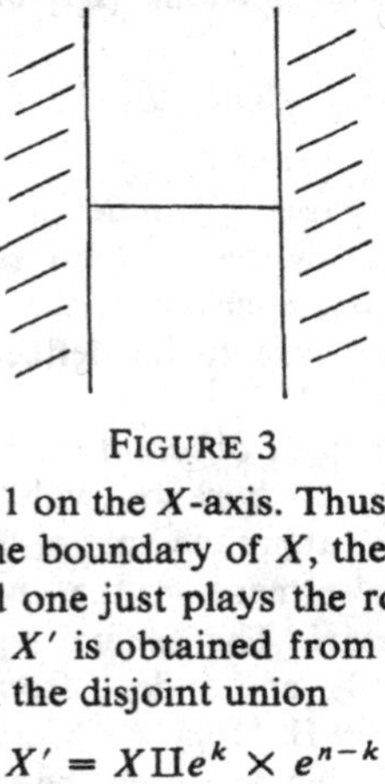

FIGURE 3

where I is the interval $|x| \leq 1$ on the X-axis. Thus as far as the glueing of part of the boundary of Y into the boundary of X, the two factors of Y play quite distinct roles, i.e. the second one just plays the role of a "thickening". Quite generally one now says that X' is obtained from X by attaching a thickened k-cell, if X' is obtained from the disjoint union

$$X' = X \amalg e^k \times e^{n-k} \tag{4.9}$$

by glueing "half" the boundary, $\partial e^k \times e^{n-k}$, of the cell $e^k \times e^{n-k}$ into the boundary ∂X by a diffeomorphism

$$\alpha: \partial e^k \times e^{n-k} \to \partial X. \tag{4.10}$$

(The resulting manifold is then also often denoted by $X \cup_\alpha e^k \times e^{n-k}$.)

With all this understood we come now to the fundamental theorem of the Morse theory.

THEOREM B. *Suppose that f has only one nondegenerate critical point p, of index λ in the range $a \leq f \leq b$ and that $a < f(p) < b$. Then*

$$M_b \simeq M_a \cup_\alpha e^\lambda \times e^{n-\lambda}; \tag{4.11}$$

that is, M_b is diffeomorphic to M_a with a thickened λ-cell attached.

This theorem summarizes–in modern terminology–the Lemmas 7, 8, 9, 10 of [5], and the figure which follows should make the theorem plausible. Here we have drawn the behavior of M_b relative to M_a in a 2-dimensional example, near a critical point p of f, in terms of a coordinate system x, y, such that, near p,

$$f = f(p) + y^2 - x^2. \tag{4.12}$$

The so-called *Morse Lemma* assures one that *such coordinates always exist near a nondegenerate critical point of index* 1. In fact this lemma asserts that near a nondegenerate critical point of index λ, one can always find coordinates $(x_1 \ldots x_\lambda, y_1 \ldots y_{n-\lambda})$ such that

$$f = f(p) + \sum_{i=1}^{n-\lambda} y_i^2 - \sum_{j=1}^{\lambda} x_j^2 \quad \text{near } p. \tag{4.13}$$

Now then, in such a coordinate patch U, and taking a and b equal to $f(p) - \varepsilon$ and $f(p) + \varepsilon$ respectively with ε small and positive, one clearly finds that ∂M_b–the boundary of M_b–intersects U in the upper and lower arcs of the hyperbola

$$y^2 - x^2 = \varepsilon > 0$$

while M_a appears in U as the shaded region. Finally consider the cross hatched region Y. It is clearly a thickened 1-handle attached to the closure of $M_b - Y$. On the other hand, the gradient flow can be used to "push" this $M_b - Y$ into M_a and in fact to construct a diffeomorphism of $M_b - Y$ and M_a. Q.E.D.

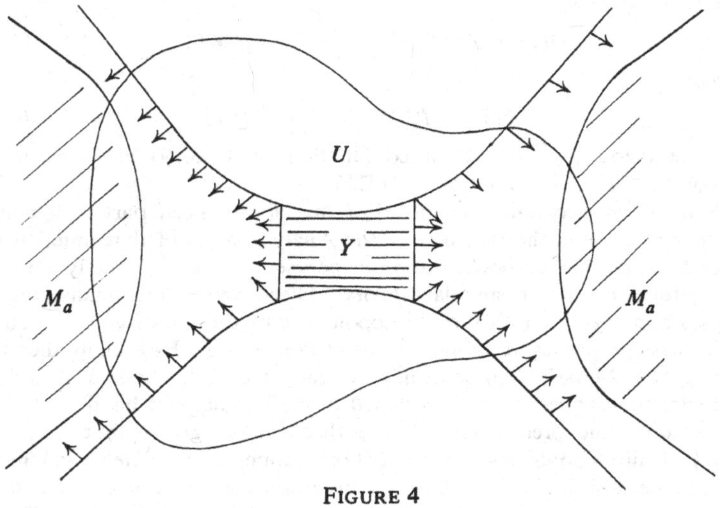

FIGURE 4

Now in 1980, the Morse inequalities follow from Theorems A and B by absolutely standard procedures of algebraic topology.

Precisely, one considers the sequence of maps

$$M_a \hookrightarrow M_b \to M_p / M_a \tag{4.14}$$

where in the last place we mean the space obtained from M_b by collapsing M_a to a point. From our thickened *handle construction*, that is *Theorem* B, it follows that

$$M_b/M_a \simeq S^\lambda$$

in the sense of homotopy. Here S^λ of course denotes a λ-sphere. On the other hand, two of the basic properties of any homology theory are that

(A) They are invariant under homotopy equivalences.

(B) For any inclusion $X \subset Y$ of reasonable spaces there is induced a "long exact sequence"

$$\rightarrow H_{k+1}(Y/X) \xrightarrow{\partial} H_k(X) \rightarrow H_k(Y) \rightarrow H_k(X/Y) \xrightarrow{\partial}. \tag{4.15}$$

I will not describe this notion in detail here; however in our situation this exactness immediately implies that

$$\Delta P_t \equiv P_t(M_b) - P_t(M_a) = \begin{cases} t^\lambda \\ \text{or} \\ -t^{\lambda-1}. \end{cases} \tag{4.16}$$

On the other hand the corresponding change $\Delta \mathfrak{M}_t$ in the Morse-series is clearly given by t^λ. Thus

$$\Delta \mathfrak{M}_t - \Delta P_t = \begin{cases} 0 \\ \text{or} \\ t^{\lambda-1}(1+t). \end{cases} \tag{4.17}$$

Proceeding inductively we see that *there exists a polynomial* $Q(t)$ *with nonnegative coefficients*

$$Q(t) = q_0 + q_1 t + \cdots, \qquad q_i \geqslant 0, \tag{4.18}$$

such that

$$\mathfrak{M}_t(M) - P_t(M) = (1+t)Q(t). \tag{4.19}$$

But the inequalities $q_i \geqslant 0$ in conjunction with (3.19) are seen to be precisely the "Morse inequalities". Q.E.D.

REMARKS. The account given in [5] of this homological part is of course couched by Morse in the language of the Analysis Situs of that time; that is of Veblen's fundamental book. From our perspective this essentially amounts to an informal sort of singular theory. There was no formalization of exactness about yet, but all the fine topologists of the era of course used all the exactness properties at various stages of their proofs. Essentially then it is fair to say that Morse's arguments in [5] already establish Theorem A and the homological consequences of Theorem B in the language of his time. On the other hand in the precise formulation that I have given, that is, in the category of diffeomophisms, Theorem B only appears in Smale's work in the 60's, and to establish Theorem B in this precision considerable care has to be taken with concepts such as smoothing corners, etc. In the 50's René Thom, E. Pitcher and I used to formulate this Theorem B in purely homotopy-theoretic terms–that is in the form

$$M_b = M_a \underset{\alpha}{\cup} e_\lambda, \tag{4.20}$$

expressing the fact that as far as *homotopy theory* is concerned one simply attaches a λ cell as one passes a critical point of index λ.

Smale's proper understanding of the Morse theory in the diffeomorphism category of course was the first step in his handle-body theory, and finally in his fundamental contributions to differential topology. (See [S1], [S3] for instance.)

Before proceeding to the infinite-dimensional settings of the theory, which really was the driving force in Morse's work throughout, let me show off the power of these inequalities in one or two examples.

First of all let me record the "*Lacunary Principle*" of Morse, which follows directly from the $(1 + t)$ factor on the right to (3.19).

LACUNARY PRINCIPLE. *If in* $\mathfrak{M}_t(f)$ *all products of two consecutive coefficients vanish, i.e.* $m_j \cdot m_{j+1} \equiv 0$, *then*

$$\mathfrak{M}_t(f) = P_t(M) \tag{4.21}$$

for any coefficients field.

Thus for such a function $\mathfrak{M}_t(f)$ computes the Poincaré Polynomial of M–and as a consequence the space M is even "torsion-free". We call functions satisfying (4.21) *perfect Morse functions*, and they clearly furnish us with a convenient way of computing $P_t(M)$. As an example consider the complex projective space $P(V)$ of one-dimensional subspaces of a finite-dimensional vector space.

Given a hermitian form (Hx, x) on V, it can be used to define the function

$$\tilde{f}(x) = (Hx, x)/(x, x) \tag{4.22}$$

on the unit sphere $S(V)$, of V, and as $f(\lambda x) = f(x)$ for all $\lambda \neq 0$ in the complex number field $\mathbf{C}$, this $\tilde{f}$ induces a function f on $P(V)$. The critical points of f are now easily seen to be given by the eigenspaces of H

$$Hx_i = \mu_i x_i. \tag{4.23}$$

Hence if we assume, as we may, that the eigenvalues of H are distinct, and that they are ordered in ascending order,

$$\mu_1 < \mu_2 < \cdots < \mu_n, \qquad n = \dim V, \tag{4.24}$$

we see that f will have precisely n-critical points. Furthermore they turn out to be *nondegenerate* and their indices are easily computed to be

$$\lambda\{x_i\} = 2(i - 1). \tag{4.25}$$

(Indeed moving x_i in the direction of the earlier x_j clearly decreases f quadratically, and as we are dealing in complex directions each of these contributes 2 to the index.) In short then

$$\mathfrak{M}_t(f) = 1 + t^2 + \cdots + t^{2(n-1)}. \tag{4.26}$$

This is a lacunary series and therefore f is perfect. Thus

$$P_t(\mathbf{C}P_n) = 1 + t^2 + \cdots + t^{2(n-1)} \tag{4.27}$$

and $\mathbf{C}P_n$ *is torsion-free.* Q.E.D.

In this and similar examples one may thus turn the Morse theory around so to speak and use it as a computational tool. This was often done by Morse in his subsequent work. For instance in his book written in 1934 **[C.V.]** he computes the mod 2 cohomology of the symmetric product of two spheres in this manner. Still, he left many examples where the Morse theory, used similarly, easily produces results which seem quite inaccessible by other means. Let me just mention two. The first, which I noticed in 1951 (see **[B6]**) is in a sense a generalization of the complex projective space example we just discussed. The theorem asserts that wherever *M is the orbit of a point under the adjoint representative of a compact Lie group G on its Lie algebra* $\mathfrak{g}$, *then the distance function f_p from a generic $p \in \mathfrak{g}$ to M is perfect because its Morse series is a function of t^2 alone and hence lacunary*. Thus these orbits are all free of torsion and the Morse theory leads to a description of their homology in terms of the usual paraphernalia of Lie group theory, i.e. diagrams, Weyl groups, etc.

A second–and in my mind maybe the most striking application of this procedure–is to the Lefschetz Hyperplane Theorem in algebraic geometry. Various ways of deducing this fundamental result were found in the 50's and finally a beautifully simple proof of the Lefschetz theorem was given by Andreotti and Frankel **[A]**. Here, as really also in my examples, the complex variable situation rather naturally forces special properties on the indices of suitably natural functions on the spaces in question. In my case they all turned out to be even; in the Lefschetz theorem they all could be bounded. Via the Morse theory these easy *local computations* then turn out to have global topological implications.

4.2. *The calculus of variation setting of the Morse Theory*. Beautiful as the considerations of the previous sections are, it is clear that to Morse they were mainly "results along the way" to his real goal–a corresponding theory in the calculus of variations. For instance while he was writing the paper discussed previously, he had already lectured the Society on his results in this infinite-dimensional context.

The situation envisaged by Morse is the following one. There is first of all an underlying manifold M, and on it a fixed variational form $F(q, \dot{q})$, which defines a "functional"

$$J(u) = \int_a^b F(u, \dot{u})\, dt \tag{4.28}$$

on the space of piecewise differentiable paths u on M. In modern terminology F is of course a function on the tangent bundle of M, and it is understood that it satisfies the usual nondegeneracy condition

$$\left\{ \frac{\partial^2 F}{\partial \dot{q}^i \partial \dot{q}^j} \right\} > 0. \tag{4.29}$$

Now, given a set of "admissible boundary conditions" Morse restricts himself to the space Ω of those paths which satisfy the condition in question and develops the "Morse-theory" for the function J on Ω. That is, he succeeds in defining the notion of *index* and *nullity* for any extremal in Ω, and in proving that for any nondegenerate J (i.e. all extremals of J in Ω have nullity zero) the Morse inequalities persist.

With hindsight, and fifty years of experience, it is of course now pretty clear what to expect. The "tangent space" to an extremal, s, is the set of vector fields along it, subject to certain boundary conditions, so that the Hessian of J at such a critical segment should be the quadratic form on this space furnished by the second variation of J.

Once a framing of the tangent-space to M along s is chosen, the space of vector-fields along s is simply the space of $\mathbf{R}^n$-valued functions of t–the parameter along s–satisfying certain boundary conditions at the endpoints. Finally after integration by parts, this Hessian should take the form

$$H_s J(x, x) = \int_a^b (Lx, x)\, dt, \qquad a < t < b, \tag{4.30}$$

where $x(t)$ represents the tangent field, and L is a linear second order differential operator

$$L = A(t)d^2/dt^2 + B(t)d/dt + C(t). \tag{4.31}$$

It follows that the eigenvalue problem

$$Lx = \lambda x, \qquad x \in \beta(s), \tag{4.32}$$

with $x(t)$ subject to the boundary conditions $\beta(s)$, is well posed and hence has only a *finite number* of independent solutions with $\lambda <$ some constant.

Thus Morse defines the index and nullity of an extremal s by

$$\begin{aligned} \textit{nullity}(s) &= \text{dim of solutions of } Lx = 0, \qquad x \in \beta(s), \\ \textit{index}(s) &= \text{dim of solutions to } Lx = \lambda x, \qquad x \in \beta(s) \text{ with } \lambda < 0 \end{aligned} \tag{4.33}$$

and with this definition every nondegenerate J should have a well-defined Morse-series

$$\mathfrak{M}_t(J) = \sum_s t^{\lambda(s)} \tag{4.34}$$

where s runs over the extremals of J in Ω and $\lambda(s)$ denotes the index.

I say *series*, because now there is of course no a priori bound on $\lambda(s)$ as there was in the finite-dimensional case. There remains the question however, whether this series is a well-defined formal power series; for conceivably there might be an infinite number of extremals of fixed index. However, as Morse shows, under appropriate completeness conditions this cannot happen on any component of Ω_0 of Ω, and then *the coefficients of the restricted Morse series*

$$\mathfrak{M}_t^0(J) = \sum t^{\lambda(s)}, \qquad s \in \Omega_0, \tag{4.35}$$

satisfy the Morse inequalities relative to the Poincaré Series of Ω_0

$$P_t(\Omega_0) = \sum t^i \dim H^i(\Omega_0). \tag{4.36}$$

Let me illustrate the situation with an example which is certainly of the greatest *geometric* significance. For our functional J we take the energy-function on a complete Riemann manifold M

$$J(u) = \int_0^1 |\dot{u}|^2\, dt. \tag{4.37}$$

For simplicity our space Ω will be taken to be the space of piecewise differentiable curves parametrized by a parameter $0 \leq t \leq 1$ proportional to arc-length, and the boundary conditon we impose on Ω is the *fixed endpoint condition*

$$u(0) = p, \qquad u(1) = q, \qquad p, q \in M. \tag{4.38}$$

The eigenvalue problem associated to a given extremal s, in this situation is then invariably described by

$$-\left\{\nabla_X^2 \cdot Y + R(X, Y)X\right\} = \lambda Y, \qquad Y_0 = Y_1 = 0, \tag{4.39}$$

where Y is a normal vector field along s, X is the tangent field along s, ∇_X is the Levi-Civita invariant derivative along X and $R(X, Y)$ the curvature of the Riemann structure.

In Riemannian geometry one calls the vector-fields Y along s subject to

$$\nabla_X^2 Y + R(X, Y)X = 0 \tag{4.40}$$

the space of *Jacobi-fields* along s, and in terms of them two points a and b on s are called *conjugate along s, if and only if there is a Jacobi field $Y \not\equiv 0$, which vanishes at a* and b. The multiplicity of such a conjugate pair is then the dimension of the subspace of Jacobi-fields which vanish at a and b.

In view of (4.39), we see then that the critical *segment s in Ω has nullity* 0 *if and only if the endpoints of s are not conjugate along s*. Conjugate points are of course of great geometric interest and also often intuitively apparent. Thus if s can be embedded in a "k-parameter family" $s(\alpha)$ of geodesics, all having the same endpoints as s, then these endpoints are conjugate along s to order at least k. This follows from Jacobi's principle that the *vector fields* $\partial s_\alpha / \partial\alpha|_{\alpha=0}$ *along s satisfy the Jacobi equation whenever* s_α *is a parameter family of extremals reducing to s at* $\alpha = 0$.

For instance on the unit sphere S^n in $\mathbf{R}^{n+1}$, only antipodal points are conjugate and the *multiplicity of an antipode pair* $(p, \bar{p})$ *is* $(n-1)$, *independently of the segment s joining them.*

This follows from the great symmetry of the sphere. Indeed, every geodesic on S^n must lie in the two-plane spanned by its initial point and direction, and so is simply an arc along a great circle S^n. Furthermore, as the curvature of S^n is 1, the Jacobi equations simply take the form

$$(d^2/dt^2)X(t) + k^2X(t) = 0, \tag{4.41}$$

k being the length of the segment s under consideration. (Recall our parametrization convention in Ω.)

Thus the nullity of a segment s is always 0 unless s has length a multiple of π, and then its endpoints are clearly antipodal or equal.

Finally the nullity of an s with antipodal endpoints $(p, \bar{p})$ is $(n-1)$ in conformity with the fact that all the geodesics through p pass through $\bar{p}$.

It is now a simple matter to compute the Morse series appropriate to this example. Indeed the number of negative eigenvalues of (4.41) are also quite easily estimated, and one then finds the following state of affairs. First of all let us survey the totality of geodesics joining the north pole p say, to the point

q, one quarter of the great circle away. Thus we start with the figure

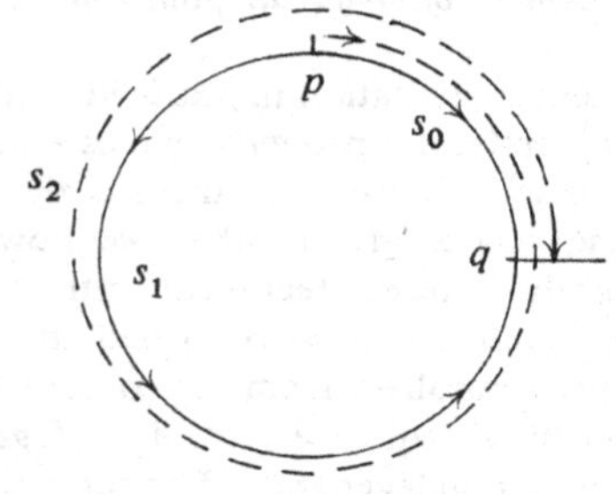

FIGURE 5

where we have also indicated the first three extremals joining p to q. Notice that if one "unfurls" the circle, i.e. passes to its universal cover **R**, one obtains the following overview of the situation

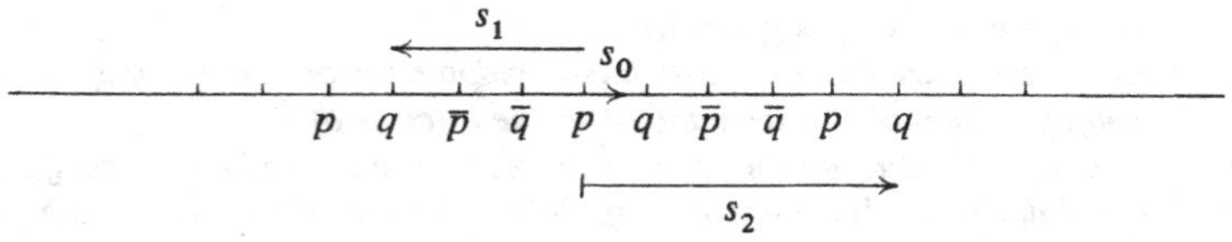

FIGURE 6

where the bar denotes antipodes.

In short then the geodesic segments in Ω correspond to the line-segments in **R** joining the origin to the *lattice of integer points n on* **R** *which are congruent to* 1 mod 4.

Now then if we estimate the index of each s_k we find that

$$\textit{index } s_k = |k|(n-1) \tag{4.42}$$

so that the Morse Series for this situation takes the form

$$\mathfrak{M}_t(J) = \sum_{k=0}^{\infty} t^{k(n-1)} \equiv \frac{1}{1-t^{n-1}}. \tag{4.43}$$

Note that this series is lacunary for $n > 2$, so that the Morse inequalities become equalities and yield the formula

$$P_t(\Omega) = \frac{1}{1-t^{n-1}} \tag{4.44}$$

in that range. Actually this equality also holds for $n = 2$, but for that one needs a more geometric argument.

Morse was in possession of this formula in the early thirties and most probably in the late twenties. (It certainly occurs in his book [C.V.],–rather buried–as Theorem 15.1 on p. 247, but I find no reference to it earlier.)

Morse did not seem to attach particular interest to this computation except insofar as it enabled him to prove the theorem I already mentioned in the introduction. That is: *For any Riemann structure on* S^n, "*there must be an infinite number of geodesics* $g_1, g_2, \ldots$ *joining any two fixed points* A_1 *and* A_2. *The length of* g_n *and the number of conjugate points of* A_1 *on* g_n *become infinite with* n".

The part of this statement in quotation marks is verbatim from the *corollary* on p. 248 of [C.V.]. To understand it properly two essentially new points have to be added to our discussion. The first of these is the *generic nature of the nondegnerate case.* Morse was aware of what we nowadays call the *Sard Lemma,* although in slightly different technical settings, and it is an understatement to say that he preferred it to be called the *Morse-Sard Lemma.* Thus, to proceed to the prior corollary from the computations of $P_t(\Omega)$ he first approximates A_1 and A_2 by a sequence (A_1^n, A_2^n) of generic points–i.e. for which the energy function is nondegenerate. For these the Morse inequalities clearly imply the existence of geodesics $\{g_n\}$ with index tending to ∞, and as nondegenerate critical points are isolated, with length also tending to ∞. Morse then argues by continuity to establish that a sequence $\{g_n\}$ with index and length tending to ∞ exists in the limiting case also.

The second point to be added is now the beautiful *index-theorem* of Morse which in the present context is given by

The index of an extremal s *in the fixed endpoint problem is equal to the number of conjugate points of one endpoint in the interior of* s.

Thus if the index of g_n tends to infinity so do the number of conjugate points of *one endpoint* in the interior of g_n. With the aid of this result the path to the corollary under discussion should now be clear.

Note that this index theorem also enables one to read off the index formula (4.42) from Figure 5: The index of s_1 is equal to $(n - 1)$ times the number of $\bar{p}$ points on it.

For a general symmetric space there is, by the way, a completely analogous description of the geodesics joining two points. Again they correspond to segments joining the origin to a lattice–this time in $\mathbf{R}^k$ where k is the rank of the space, and the index of a segment s can be read off from the number of times s pierces a certain family of hyperplanes. See [**B8**].

This *index theorem* fits, properly speaking, into a natural extension of the classical Sturm theory for the differential equation

$$-(x'' + qx) = \lambda x \tag{4.45}$$

and coming at it from the calculus of variations as he did led Morse to redo the classical theory as well as extend it in a variety of ways along quite new lines.

A thorough account of this work is to be found in his book [V.A.] called Variational Analysis.

The point is that in all properly posed variational problems there is some type of index theorem; however for some boundary conditions the answer is more difficult and less satisfying. Let me illustrate. If we consider the J of our discussion on the space $\Omega(p, N)$ of paths starting on a submanifold N and ending at a point p, then the index theorem in this context becomes what Morse calls the *focal-point* theorem:

The index of an extremal s *is equal to the number of focal points of* N *in the interior of* s.

The following figure illustrates this theorem clearly

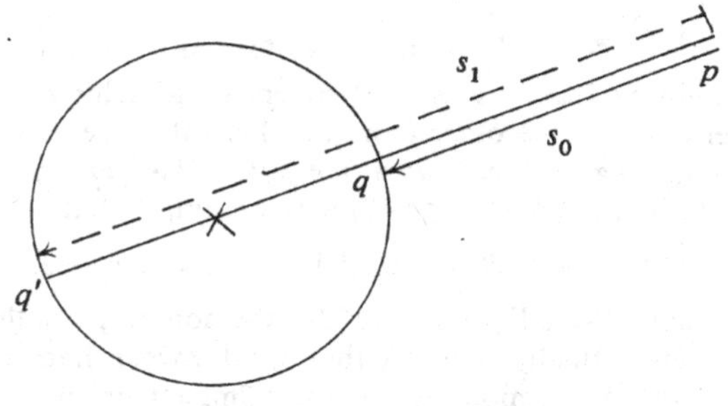

FIGURE 7

Here the underlying manifold is the plane and N is the unit circle, while p is taken as a point not on the circle. The extremals of J on $\Omega(N, p)$ are geodesic segments starting perpendicularly on N and joining N to p. Here they are therefore given by the two straight lines s_0 and s_1 joining q and q' to P. The only focal point of N is the center of the circle and it has multiplicity 1. Thus index $s_1 = 1$. Note that the same picture is valid for the n-space in $\mathbf{R}^n$ but that then s_1 would have index $(n - 1)$ corresponding to the fact that the origin is a focal point of multiplicity $(n - 1)$.

However, for more complicated situations, for instance when we deal with two end manifolds N_1 and N_2 both of dimensions greater than 0, or with periodic boundary conditions

$$u(1) = u(a), \qquad \dot{u}(1) = \dot{u}(0), \tag{4.46}$$

the index of an extremal cannot be read off so simply from the behavior of the solutions to the Jacobi equation along the extremal.

This whole subject is still of great interest and has been taken up by a variety of authors (see [K]). Basic in all of them is however Morse's insight that this index has a topological meaning which can be computed by an intersection-number.

REMARKS. Inadequate as the previous account of what one might call the concrete "Morse Theory" is, it will have to do in the present context. It is meant more as an excursion into his work; and to a certain extent as a biased excursion into that part of the theory which I fell in love with thirty years ago. It is also the part of the theory which has had the most direct bearing on homotopy theory. My first remark is therefore devoted to this connection.

(1) The space Ω of paths joining two points on M is nowadays called the *Loop Space* of S and denoted by ΩM. Its homotopy type is independent of the points chosen (as was already shown by Morse), and it plays a vital role in all of homotopy theory. In fact it appeared right away in the first papers of Hurewitz on the higher homotopy groups through its characteristic property

$$\pi_k(\Omega M) = \pi_{k+1}(M). \tag{4.47}$$

Of course the homotopy theorists and Morse each went their own way–essentially until R. Thom and I realized what a powerful tool the Morse theory could be in this context. For instance our discussion of the critical points of J on ΩS^n immediately implies a formula of the type

$$\Omega S^n = S^{n-1} \cup_\alpha e_{2(n-1)} \cup_\alpha e_{3(n-1)} \cdots . \tag{4.48}$$

That is, ΩS^n is built up from the $(n-1)$ sphere by attaching cells of higher and higher dimension. On the other hand an elementary result of homotopy theory is that *attaching a k-cell does not effect the homotopy groups in dimension* $< k - 2$. Thus (4.47) and (4.48) together imply that

$$\pi_{k+1}(S^n) = \pi_k(\Omega S^n) = \pi_k(S^{n-1}), \qquad k < n-3.$$

This is the "Freudenthal stability theorem" for the homotopy of the spheres.

In the later fifties I finally realized that what Morse had done for the spheres worked even more smoothly for the compact groups and some of their homogeneous spaces. For instance, if U_n is the nth Unitary group, the correct analogue of (4.48) is given by two formulas

$$\begin{cases} \Omega U_{2n} = U_{2n}/U_n \times U_n \cup_\alpha e_1 \cup_\alpha e_2 \ldots, & \dim e_i > n, \\ \Omega U_{2n}/U_n \times U_n = U_n \cup_\alpha e_1 \cup e_2 \cup \ldots, & \dim e_i > n. \end{cases}$$

Combined with the known fact that $\pi_k(U_n)$ is also "stable" in the sense that it does not depend on n for n large compared to k, these relations immediately lead to the periodicity $\pi_k(U) = \pi_{k+2}(U)$ with $\pi_k(U)$ the stable value of $\pi_k(U_n)$. In this framework **[B5]**, eight such formulas finally lead to the 8-fold periodicity $\pi_k(0) \simeq \pi_{k+8}(0)$ of the stable orthogonal group.

(2) My next remark deals with the techniques Morse uses in his papers and his book. Basically he deals with the infinite-dimensional function space by deforming compact sets in it into a finite-dimensional space of broken extremals, and then applying his finite-dimensional results.

Apart from technicalities–which actually can become quite formidable–this is an approach some of us still favor. Others, like Smale-Palais **[P]**, **[S3]**, have cast the whole theory in terms of infinite-dimensional Hilbert-manifolds from the very outset. In such an account the space Ω of course has to be topologized in a slightly different way than in Morse's approach. Morse essentially endowed Ω with a metric topology, with metric

$$d(g_1, g_2) = \sup_t |g_1(t), g_2(t)| + J(g_1) - J(g_2).$$

Topologists think of Ω as simply the space of continuous maps in the compact open topology. Luckily all these topologies yield the same "weak homotopy type" for Ω, so that the same Morse inequalities emerge at the end. I should confess here also that Morse did not work with the energy integral, but always with the parameter-free length integral

$$l(u) = \int_a^b |\dot{u}|\, dt$$

or its more general analogue. This involved him in considerable technical difficulties which, I should add, he overcomes with tremendous ingenuity and geometric insight.

Altogether Morse's geometric power is phenomenal and he seems to have been able to compute anything he set his mind to. For instance, he computes the homology of the symmetric product of the n-sphere by hand, so to speak.

That is, he constructs the explicit cycles and thus in 1930 he really anticipates the work of P. A. Smith and Richardson on the one hand, and that of Steenrod in the 50's on the other.

Morse's sheer power is however best exhibited in his attack on the *closed geodesics* problem. These closed geodesics can be thought of as extremals of our J considered on the space of all piecewise smooth paths Γ subject to $u(0) = u(1)$. However for this problem there is no "nondegenerate case" for, with a given u, all its translates are also critical. Thus in the "most nondegenerate situation" one can hope for one still has to deal with *critical sets of circles*. Furthermore each genuine geometric closed geodesic gives rise to an infinite number of such critical circles corresponding to the number of times one circumnavigates it. Thus it is hard to estimate the number of closed geodesics on S^n say, in terms of the topology of ΓS^n. Still one can, in this manner, easily deduce that there is at least one nontrivial closed geodesic on S^n in any Riemann structure. On the other hand Morse was shooting for more, and so, to my knowledge, he never bothered to compute the homology of ΓS^n. Rather, he felt that because J was naturally invariant under translations, the correct space on which J was to be considered was in some sense the *quotient* of ΓS^n *by the action of* S^1 on Γ, given by $u(x) \to u(x + \theta)$, $0 \leqslant \theta \leqslant 1$.

Thus he made polygonal approximations to ΛS^n, divided them by the appropriate cyclic group actions and then attempted to pass to the direct limit (!) in homology. The Betti-numbers he obtained that way he called the "*circular connectivities of* S^n", and they were supposed to play the same role vis-à-vis the closed extremal problem which the Betti numbers of ΩM play vis-à-vis the fixed endpoint problem.

Unfortunately an error crept into this computation–an error, which by the way, I fell into myself in my paper **[B4]** trying to "modernize" Morse's computations–which was caught by L. Schwartz in 1957. This whole subject has had a lively history since then with many contributors–and also many subsequent mistakes. (See **[K]** for an excellent bibliography and account of this general topic. However, Klingenberg's main result is still in doubt at this time.) All in all the last word has not yet been written on it and I cannot help feeling that there remain many clues to be found in Morse's amazing last chapter to his Colloquium volume [C.V.].

4.3. *Abstract settings of the Morse theory*. In a series of papers **[36]**, **[44]**, **[59]**, in the thirties, forties, and fifties, Morse explores various *general formulations* of his theory. In these papers he usually works on an abstract metric space M, with a function F not necessarily continuous, and to obtain some analogue of the Morse inequalities in these situations he uses the Vietoris homology theory.

This is then the study of the homology behavior of the set $F \leqslant a$ under very general conditions. More precisely let

$$H(a, b) = H(F \leqslant b, F \leqslant a), \qquad a \leqslant b,$$

denote the relative homology of the sets indicated. A critical value of F is then defined by Morse as any number c, *such that*

$$H(c, c^-) = \lim_{a \to c^-} H(c, a)$$

is nontrivial, and the problem is now to relate the "sizes" of these critical $H(c, c^-)$ to the "size" of $H(M)$. In a sense Morse was here grappling with ideas which were properly understood only 15 years later by Leray.

Indeed the general scheme of the Leray spectral sequence can be thought of in the following context. One is given a filtering

$$M_0 \subset M_1 \subset \cdots M_n$$

and asks how the groups $E_1 = \bigoplus H(M_i, M_{i-1})$ are related to $H(M)$. Leray's answer is of course that one can proceed in an orderly manner from E_1 to $E_2 \ldots$ the passage from E_k to E_{k+1} always being a passage from a chain complex to its homology. Finally at E_∞ one is close to $H(M)$. Now it is a fact that if all objects in a chain complex A are finite dimensional, then the *Morse inequalities pertain between the dimensions of the* A_i *and those of* $H(A_i)$. Thus the Morse inequalities already reflect a certain part of the "Spectral Sequence magic", and a modern and tremendously general account of Morse's work on rank and span in the framework of Leray's theory was developed by Deheuvels **[D]** in the 50's.

Unfortunately both Morse's and Deheuvel's papers are not easy reading. On the other hand there is no question in my mind that the papers **[36]** and **[44]** constitute another tour de force by Morse. Let me therefore illustrate rather than explain some of the ideas of the rank and span theory in a very simple and tame example.

In the figure which follows I have drawn a homeomorph of $M = S^1$ in the plane, and I will be studying the height function $F = y$ on M.

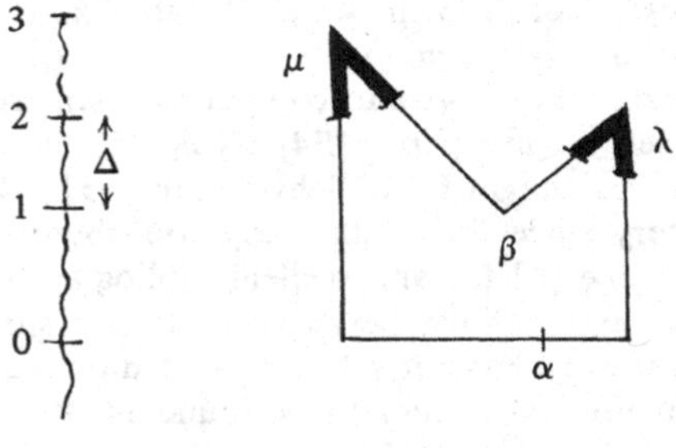

FIGURE 8

The values a where $H(a, a^-) \neq 0$ are indicated on the left, and corresponding to each of these *critical values* a generator of $H(a, a^-)$ is drawn on M, using the singular theory for simplicity. Morse calls such generators "caps". Thus α and β are two "0-caps" and μ and λ two "1-caps". Notice that every cap u defines a definite boundary element ∂u in

$$H(a^-) = \lim_{\varepsilon \to 0^+} H(F < a - \varepsilon);$$

Morse calls a cap u linkable iff $\partial u = 0$. Otherwise it is called *nonlinkable*.

In our example, α, β and μ are linkable while λ is *not*.

Next Morse defines the *span* of a cap u associated to the critical level a in the following manner.

Case I. u linkable. Consider the inclusion homomorphism

$$H(a, a^-) \xrightarrow{\iota_b} H(b, a^-), \qquad b > a,$$

and let $s(u)$ be the least upper bound of all b with $\iota_b u \neq 0$. (If the set of such b's is unbounded $s(u) = \infty$.) The *span* of u is now defined as

$$span(u) = s(u) - a.$$

In our example α and μ have infinite span, while β has span Δ. Indeed the moment b has passed (2) the cycle α can be deformed in $F < b$ into the region $F < F(\beta)$.

Case II. u is *not linkable*. In this case consider $\partial u \in H(a^-)$ and let $t(u)$ be the greatest lower bound of b's such that ∂u is in the image of

$$\iota_b \colon H(b) \to H(a^-), \qquad b < a.$$

Then set *span* $u = a - t(u)$.

In our example λ is the only nonlinkable cap, and its span is now clearly again the number Δ.

With these definitions understood Morse sets m_k^e equal to the *number of k-caps of span* $> e$ and proves that under very general conditions these numbers will *be finite for every* $e > 0$, *and then obey the Morse inequalities*.

For instance, in our example this amounts to $m_0^e \geq 1$ and $m_0^e = m_1^e$–for all $e > 0$. And indeed, for $0 < e < \Delta$ we have $m_0^e = 2$, $m_1^e = 2$, while for $e \geq \Delta$, $m_0^e = m_1^e = 1$.

Our example is of course a very tame one; however, it should be clear that as one complicates it–for instance by introducing an infinite number of critical values clustering at some point, that as long as some sort of continuity is preserved, *the number of k-caps of span* $> e > 0$ *will be finite*. On the other hand one then also sees that the Vietoris or Čech theory is the correct one in this framework.

5. Minimal surfaces. Marston Morse had developed the abstract setting of the variational theory which we just described in large part because he hoped to make it applicable to minimal surface theory and other variational problems. Unfortunately however a direct extension of the Morse Theory just does not work for variational problems in more than one variable. That is, if in analogy to the preceding discussion one attempts to study the area function, or its "energy", on the space $\Omega_2(M)$ of piecewise smooth maps of a disc into a Riemann manifold M then the Morse theory will not work. In the context of the Palais-Smale theory **[P]**, one understands this phenomenon in terms of the Sobolev inequalities, which show that the conditions on a map μ in $\Omega_2(M)$ to have finite area are far from forcing μ to be continuous. Thus, roughly, the space of continuous maps–which carries the topology–is too sparse in the space of admissible maps for a proper Morse theory to exist. (See **[P]** in particular 16.)

Nevertheless in a series of papers, **[40]**, **[45]**, **[47]**, **[49]** Morse and Tompkins do obtain applications of the Morse theory in this context. For instance in **[40]** they prove the existence of a minimal surface of *nonminimum type*, by applying the Morse inequalities to the following situation. Let D be the unit disc $|z| \leq 1$, and let $S = \partial D$ be its bounding circle.

Now let $g\colon S \to \mathbf{R}^n$ be a simple closed curve in $\mathbf{R}^n$, satisfying a Lipschitz condition and parametrized by arc-length, i.e. defining an isometry of S into $\mathbf{R}^n$. Next Morse and Tompkins introduced a space, Ω, of "reparametrizations" of g

$$g_\varphi(s) = g(\varphi(s))$$

where φ ranges over the homeomorphisms of S^1 which move points in only one direction, and keep three distinct points fixed. On this space they now consider the "Douglas-function"

$$A(\varphi) = \frac{1}{16\pi}\int\int_{S^1\times S^1} \frac{|g_\varphi(\alpha) - g_\varphi(\beta)|^2}{\sin^2\left(\frac{\alpha-\beta}{2}\right)}\, d\alpha\, d\beta$$

and show that it satisfies the abstract conditions laid down in **[32]** by Morse. The geometric pertinence of A is of course that $A(\varphi)$ is equal to the Dirichlet integral of the *harmonic* map $h_\varphi\colon D \to \mathbf{R}^n$ *extending* g_φ.

Now the space Ω is clearly contractible, so that in particular the Morse inequalities read

$$m_1 \geqslant m_0 - 1. \tag{5.2}$$

Morse and Tompkins cite an explicit g with this property and so achieve their result.

In their other papers they explore this framework of questions in various directions, increasing the number of bounding circles, etc.

The paper **[45]** is by Morse alone, and is essentially analytic in character. There he gives an explicit form for the first variation of A in the case of n-bounding contours, and shows that, if this variation vanishes, then the harmonic surface defined by the boundary conditions is indeed minimal.

6. Topological methods in the theory of a complex variable. In the middle forties Marston Morse and Maurice Heins collaborated in a series of papers on the before-mentioned subject. Their aim was two-fold. On the one hand they generalized many theorems from the classical theory of one complex variable to the domain of *light interior* maps of one Riemann surface to another. Thus they often found topological proofs of topological properties of holomorphic mappings. At the same time they also discovered some remarkable new properties of such mappings, and I would like to describe one of these here. For a more complete survey of their collaboration the reader is referred to Morse's own account **[64]**.

Consider the space M_α of all meromorphic maps $f\colon D \to S^2$ of the disc, $|z| < 1$, into the Gauss Sphere subject to the boundary condition

(α): f has a prescribed number of zeros, poles and branch-points of prescribed multiplicity at prescribed points of D.

Morse and Heins ask for the components, $\pi_0(M_\alpha)$, of M_α; or, put differently, for the *deformation classes of such maps*. They then go on to solve this question by constructing a complete set of invariants "J_α" characterizing these components. When the multiplicities of all the zeros $a_0, \ldots, a_k$, and the poles $a_{k+1}, \ldots, a_n$, and the branch-points $b_1, \ldots, b_\mu$ of α are taken to be 1 the cardinality of the $\{J_\alpha\}$ is n, and the invariants $J_\alpha(f)$ can be computed for a map $f \in M_\alpha$ by the following recipe: One first selects regular arcs h_i joining a_0 to a_i, $i = 1, \ldots, n$, which, except for their endpoints, lie in the complement of the a's and b's of α. It follows that the $f(h_i)$ are regular arcs on S^2 which are either closed curves containing the origin, or joining 0 to ∞ on S^2. The recipe for $J^i_\alpha(f)$–in the first case for instance–is as follows

$$J^i_\alpha(f) = \Delta \frac{1}{2\pi} \arg h'_i(z) - \Delta \frac{1}{2\pi} \arg h_i(z) - \Delta \frac{1}{2\pi} \arg C_i(z) \tag{6.1}$$

where Δ denotes the algebraic change as z moves on h_i, and $C_i(z)$ is given by

$$C_i(z) = \left[\frac{(z-b_1)\cdots(z-b_n)}{(z-a_1)\cdots(z-a_n)} \right] (z-a_i).$$

The b's here denote the location of the branch-points of f, and the virtue of the last term is that it makes $J^i_\alpha(f)$ independent of the choice of h_i joining a_0 to a_i.

In the *light interior* category the same result holds; however there the derivative $h'_i(z)$ is replaced by a sufficiently short chord. In an earlier paper **[61]**, the authors had already generalized the Whitney theory of regular curves to the domain of locally simple ones, and this classification plays an important role throughout their work.

It seems to me that their result fits into the general framework of the Morse Theory in a way which is not made explicit in their papers, and which would be of contemporary interest. My question is the following one. Recall that any $f \in M_\alpha$ is an immersion at all points other than the branch-points, so that its differential df, defines a bundle map

$$df\colon T_\alpha D \to TS^2$$

of the unit tangent bundle of D (with the branch-points deleted) to the unit tangent bundle of S^2. One may therefore construct a space $\tilde{M}_\alpha$ of all *continuous bundle-maps* from $T_\alpha D$ to TS^2, subject to the boundary conditions α, and one then has a natural inclusion

$$M_\alpha \hookrightarrow \tilde{M}_\alpha. \tag{6.2}$$

The Morse-Heins results seem to be equivalent to the assertion that M_α and $\tilde{M}_\alpha$ have the "same" components–and so the conjecture which comes to mind is that the *inclusion* (6.2) *is a homotopy equivalence*. Similar questions seem to be of interest at this time not only to physicists but also to workers in control theory.

7. Integral representations. The very active collaboration of Morse and Transue extended over a period of 10 years or so, starting in 1949. Altogether they produced over 20 papers of considerable complexity, all dealing in one form or another with an extension to bilinear forms of the Riesz representation theorem

$$F(x) = \int_0^1 x(t)\, df \tag{7.1}$$

for a continuous linear functional on the space of continuous functions C on $[0, 1]$.

Thus they sought representation for a bilinear function $B(x, y)$ of the form

$$B(x, y) = \int_0^1 x(s)\, d_s \int_0^1 y(t)\, d_t k(s, t) \tag{7.2}$$

where the integrals are taken in the Stieltjes sense.

The original motivation seems to have come from the quadratic form which describes the second variation of an extremal, but I think it is fair to say that after the initial impetus Morse's and Transue's interests became aroused and they explored this territory for its own sake.

The starting point of this work is a (1915) representation theorem for bilinear functions on $C \times C$ due to Fréchet which introduces the notion of Fréchet variation of functions of two variables. In the course of their work Morse and Transue extended this notion to n-dimensions and undertook a detailed study of the limit properties of functions of bounded variation **[71]**, **[72]**, **[73]**. In the process they discovered that in many of the test for convergence of a multiple Fourier series the condition of bounded Vitali variation could be replaced by the weaker condition of bounded Fréchet variation (see **[100]**, **[101]**), and in the paper **[55]** they even make a delicate contribution to the convergence theory of Fourier series in one variable. Among other results they show that the Young-Pollard conditions imply the L_2 conditions of Lebesgue.

8. Pseudo-harmonic-functions. "Among the characteristics of a function U which is harmonic on a Riemann surface G^* are the topological interrelations of the level lines of U. One merely has to look at the level lines of Rz, Re^z, $R \log z$ to sense both complexity and order. The dual level lines of a conjugate V to U add to this order complexity . . . "

This is the beginning sentence of the paper **[92]**, in which Morse and Jenkins solve the difficult problem of showing that on a simply connected Riemann surface every pseudo-harmonic function has a pseudo-conjugate. Thus in particular they show that on such a surface any pseudo-harmonic function can be made harmonic by a change of the conformal structure.

Recall here that pseudo-harmonic means "harmonic after a suitable homeomorphism" so that the topological properties of harmonic functions automatically carry over to pseudo-harmonic ones. In this context V is a pseudo-conjugate to U if there is a homeomorphism of the domain of these functions φ such that $(U + iV) \circ \varphi$ is analytic.

The work of Morse and Jenkins extending over the early fifties is devoted to exploring the "order" in the "complexity" mentioned in their "Fundamenta" paper. After essentially setting the simply connected case, where they extended and completed earlier work of Kaplan, Boothby and others, they go on in **[94]** to discuss these problems on doubly connected surfaces. In particular they there give a very complete analysis of the structure of the level sets of a pseudo-harmonic function.

9. Differential topology. In 1953 M. Morse and E. Baiada wrote a paper **[90]** which was an assault on the Schoenflies problem by Morse-theoretic means, but it was not until 1959, after B. Mazur's remarkable contribution, that

Morse became more and more intrigued by the Schoenflies question in all its ramifications.

Most impressive is his 1959 paper [**114**] where he removes the "unnatural" part of Mazur's hypothesis. Recall that the Schoenflies problem deals with the question of whether a homeomorphism

$$\varphi: S^{n-1} \to \mathbf{R}^n$$

of the standard $(n-1)$ sphere into $\mathbf{R}^n$ admits an extension to the unit disc D^n interior to S^{n-1}. The "Horned Sphere" of Alexander shows that in general this is not possible without some tameness assumption on φ. Essentially untouched for 30 years, Mazur finally decided this question in the affirmative under the following two tameness assumptions:

The first is the natural "shell condition", that φ admit a local extension on both sides. Precisely there should exist a homeomorphism

$$\tilde{\varphi}: S^n \times I \to \mathbf{R}^n, \qquad I = [0, 1],$$

reducing to φ on $S^n \times (1/2)$.

His second assumption was the less natural one, that $\tilde{\varphi}$ *be semilinear near some point* $p \in S^n$.

To set these conditions in perspective note that if φ is smooth, both conditions–with p any point on S^n–follow immediately. In any case Mazur's inspired construction [**M1**] led to an easy proof that any φ subject to these conditions does admit an extension.

Morse's very fine contribution in [**114**] was now to remove the second hypothesis, thus achieving the ultimate solution of the Schoenflies problem in the topological setting. By quite different methods the same result was achieved at around the same time by M. Brown [**B7**].

In his subsequent work on this question–mostly in collaboration with Huebsch, Morse takes up the same question in the smooth and analytic categories, and also considers parameter families of extensions. Thus in [**120**] they prove the following theorem.

Corresponding to a real analytic mapping

$$\varphi: S \times \Gamma \to E$$

each of whose restrictions to $S \times p$, $p \in \Gamma$, is an analytic diffeomorphism, there exists an extension

$$\Lambda: D \times \Gamma \to E$$

each of whose restrictions to $D \times p$ is real analytic except possibly for one point.

When Γ reduces to a point, this theorem was first established by Royden [**R**], and to extend it to this parameter form, Huebsch and Morse reproved the Royden result along different lines. Note here that, just as in the C^m category, $m > 0$, one cannot expect the extension to have smoothness properties at *all points* of D. Indeed the constructions of Milnor produce counterexamples in dimension > 6. On the other hand the iterative constructions of the extension make it fairly clear that the bad behavior of an extension can be concentrated at one point. In [**123**] Huebsch and Morse introduce the concept of a conical singular point and show how to make extensions whose singular behavior is no worse than conical. The article **[126]** surveys this whole question in considerable detail.

The period 1958–1960 was an exciting one in differential topology. Not only was the Schoenflies problem solved, but Smale **[S2]** had by that time announced his fundamental results: These included proofs of the *Poincaré Conjecture* in dimensions > 4 and of the *h-cobordism theorem*. Both depended on his refined version of the Morse theory, and the resulting *handle-body decomposition of manifolds*. Morse was of course very pleased with these great successes of the Morse theory. Still, I think he must also have felt a little scooped. In his book with Cairns **[CP]**, he recalls how such handle-bodies were introduced (but not named–or used as spectacularly) by him in an earlier paper. In any case starting in about 1965, and continuing until the late 70's, he undertook to redo this part of differential topology in his own way. Apart from the before mentioned book Morse and Cairns collaborated in many papers on this project. Let me close my account of this phase of Morse's work with a theorem on the removal of critical points which Morse presented in 1965–at a symposium held in his honor–under the title: *Bowls of a nondegenerate function on a compact differentiable manifold* **[139]**. This theorem will also serve to introduce us to the ideas of Thom and Smale of the fifties.

The situation envisaged here is that of a nondegenerate function f on a compact manifold M. Let the necessarily finite number of critical points of f be denoted by $\mathrm{Cr}(f)$ and consider the complement $\hat{M} = M - \mathrm{Cr}(f)$.

Once a Riemann structure (,) is selected on M, the differential of f uniquely specifies a vector field $X = \hat{df}$, the gradient of f, by the formula

$$Y \cdot f = (X, Y).$$

This vector field therefore points orthogonally to the level surfaces of f at each point of M and is nonvanishing on $\hat{M}$.

It follows now that the trajectories of X on $\hat{M}$, i.e. the maximal integral manifolds, l of X, will be diffeomorphic to the open interval (0, 1), and their closures will be curves on M joining two critical points of f at distinct f-levels. These are referred to as the lower and upper endpoints of l. Now let p be a critical point of M. Then Morse calls the point-set consisting *of p and all points of trajectories of X with upper endpoint p, the "descending Bowl" of p, and denotes it by* $B^-(p)$. Similarly the ascending Bowl, $B^+(p)$, is defined by replacing f with $-f$. A first step of the paper under discussion is the theorem that

A Bowl $B^-(p)$ *is diffeomorphic to* $\mathbf{R}^k$ *with k the index* $\lambda(p)$ *of* p.

Thus the Bowls define two decompositions of M into disjoint cells:

$$M = \bigcup_p B^-(p); \tag{9.1}$$

$$M = \bigcup_p B^+(p) \tag{9.2}$$

with cell dimensions

$$\dim B^-(p) = \lambda(p), \tag{9.3}$$

$$\dim B^{+}(p) = n - \lambda(p). \tag{9.4}$$

REMARKS. The terminology and treatment of this cell-decomposition is somewhat anachronistic. This cell-decomposition already appears in a C-R. Note of Thom's in (1949) **[T]**. It also plays an essential role in Smale's work of the late 1950s, **[S1]** and his terminology of *stable* and *unstable manifolds* for $B^{\pm}(p)$ is the generally accepted one. In my own recollection I heard Thom discuss this decomposition in 1951 but its implications were largely lost on me and my contemporaries. Roughly we felt that the flow X became much too complicated "far" from the critical set to be of further use. And locally it, of course, had already been put to use in Theorems A and B. I still remember the surprise therefore when in the late 1950s Smale explained the next step in his program to me. Coming at the Morse Theory from his individual point of view he immediately realized that by *perturbing* the flow X a little–if necessary–it could be arranged that the two cell decompositions (9.1) and (9.2) be *as transversal to each other as possible.*

Note in this connection that as both B^{-} and B^{+} are stable under our flow X, the intersection $B^{-} \cap B^{+}$ of a descending Bowl from p and an ascending Bowl from q, $p \neq q$, consists of a family of trajectories of X. Hence the proper way of defining transversality is to demand that *at any point* $p \in B^{-} \cap B^{+}$ with $f(p) = \alpha$, the intersections of B^{-} and B^{+} with the level set $M_{\alpha} = \{f = \alpha\}$ in M intersect transversally on M_{α}.

These concepts are maybe best understood by studying the two examples of Figure 9.

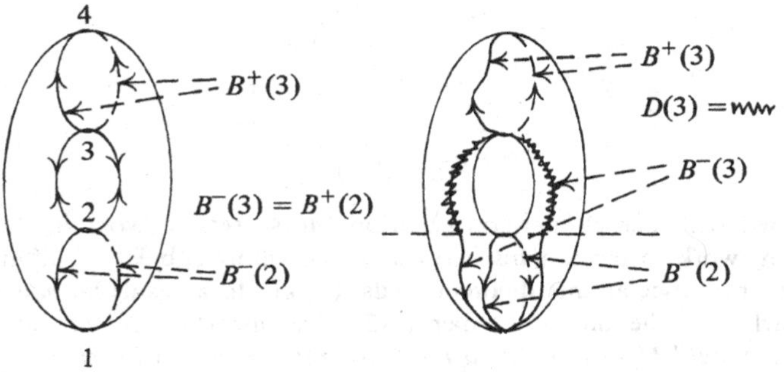

FIGURE 9

On the left we are dealing with the gradient of the z-coordinate of a regular torus in $\mathbf{R}^3$. The flow it generates is *not transversal*; for instance $B^{-}(3)$ and $B^{+}(2)$ have no business intersecting at all. The right-hand picture is generic and there they do not.

Note also that $B^{+}(1) \cap B^{-}(3)$ consists of the *two trajectories* $B^{-}(3) - (3)$.

In his paper **[139]**, Morse constructs a transversal decomposition of M according to a flow X in his own way and then in terms of it formulates and proves a criterion for the elimination of critical points. We still need one additional concept to make this criterion intelligible. Morse calls the "*dome*" *of a "descending Bowl"* $B^{-}(p)$, *that part of* $B^{-}(p)$ *where* f *takes values* $> f(c)$ *as* c *ranges over the critical points in the closure of* $B^{-}(p)$ *other than* p. In Figure 7 we have indicated the dome of $B^-(3)$ by $D(3)$ on the right-hand figure. The dome of $B^-(3)$ is all of $B^-(3)$ in the left-hand example.

With this terminology and transversally understood the *Bowl-theorem* of Morse asserts:

Suppose that $B^-(p)$ is a descending k Bowl and $B^+(q)$ an ascending $(n - k + 1)$ bowl such that $B^-(p) \cap B^+(q)$ intersect transversally in a *single trajectory* and such that the closure of $B^-(p)$ *meets no critical point between the f levels of p and q.* Also let N be a prescribed open neighborhood of the closure of the *dome* of $B^-(p)$ which contains p and q and such that $\bar{N}$ contains no other critical points of f.

Then there exists a nondegenerate function $\hat{f}$ on M without critical points on $\bar{N}$ and such that $\hat{f} = f$ on $M - N$.

For instance in the one-dimensional example of Figure 10 we see that the conditions are satisfied for $B^-(1)$ and $B^+(2)$ and it should also be clear how to eliminate the critical points 1 and 3. In the generic example of Figure 9 the theorem does not apply–indeed all the pertinent intersections consist of *two trajectories*. And of course none of these critical points are removable.

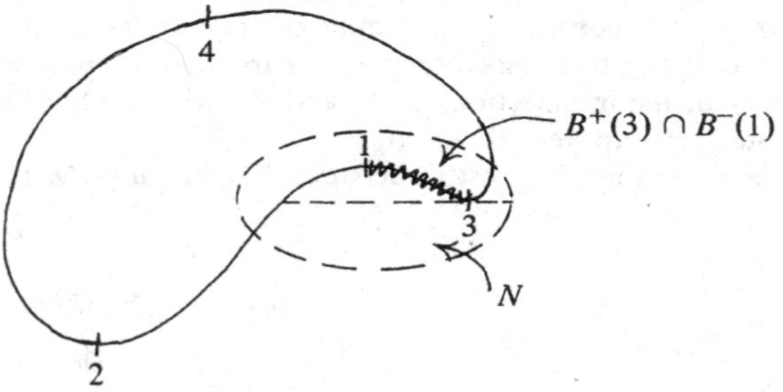

FIGURE 10

Equipped with this elimination criterion Morse sets to work in this and subsequent work to fashion his version of the whole subject of differential topology. In particular this theorem leads quickly to a result he had established earlier in the difficult paper **[113]**. The assertion is that *on every compact manifold M there exists a nondegenerate function with only one local minimum and one local maximum.*

10. Concluding remarks. Alas, there remain many areas of Morse's work which are not accounted for in the previous sections. First of all, there is a wealth of material on the more technical aspects of the calculus of variations. These papers deal with such matters as sufficient conditions for the problem of Meyer and Lagrange under different boundary assumptions, theorems concerning envelopes of extremals, the analytic continuation of closed geodesics, etc. There are also papers on singular quadratic functionals e.g. **[34]** and related topics. Morse had many collaborators in addition to the ones we have already mentioned: G. Ewing, D. Lander, W. Leighton, G. B. Van Schaack, E. Pitcher–and others. There are also three books (one with S. S. Cairns) and four sets of lecture notes. All in all this magnificent "oeuvre" bespeaks Morse's attitude towards his work. Morse was first and foremost a mathematical "craftsman" who did mathematics every day of the year, naturally, and–like Bach–under all conditions; with children on the lap, in the car . . .

Unfortunately under such a regime there remained little time for non-mathematical writing. Indeed I find only four entries **[46]**, **[54]**, **[87]**, **[108]** which are not mathematical research or mathematical exposition.

Characteristically all of these deal with the relationship of mathematics to the arts and the practical world. Morse rejoiced in the affinity of mathematics to both, and his views were at once eloquent and optimistic. One meets in these articles a quality not easily gleaned from his purely mathematical writing; a quality which he really showed only to his closer friends and family. Let me conclude therefore with three quotes from these sources.

His Kenyon address **[87]** starts as follows:

> "To talk about art other than in the impersonal sense of history, is to talk about the moments when one has been confronted with beauty. Every essay on art that lights a hidden niche has its source in the life of the writer. You will then perhaps understand why I start with the mood of my childhood. One hundred miles northeast of Derry, New Hampshire, lie the Belgrade Lakes, and out of the last and longest of these lakes flows the Messalonskee. I was born in its valley, 'north of Boston' in the land of Robert Frost."

He then continues with a beautiful account of this countryside which he loved so much, and of his early love of music. Then he turns to the art of cabinet making, and in his article on *Mathematics, the Arts and Freedom* he describes this encounter as follows:

> "Turning to another art I recall the shop of an old cabinet-maker on the banks of the Kennebec. It was a place where much could be learned of the relation between mathematics and the arts. Around the room were scattered Sheraton chairs and tables, broken pieces of beauty. Fluted columns, capped by acanthus leaves, made ancient Greece seem near. The cabinet-maker was a provincial Socrates who discoursed with his hands and tools. I wanted to learn more of his art. In a nearby library there was a copy of Sheraton's *Cabinet-maker*. It started with descriptive geometry and ended with designs of cornices with ruler and compass. Beauty and perfection reigned throughout. Mathematics was the handmaiden of art, faithful if not creative. I did not then realize the difference between mathematics as a servant and as a sister of the arts" (p. 17).

And now listen to how he starts this address:

> "Mathematics is an art, and as an art chooses beauty and freedom. It is an aid to technology, but is not a part of technology. It is a handmaiden of the arts, but it is not for this reason an art. Mathematics is an art because its mode of discovery and its inner life are like those of the arts" (p. 16).

One needs first hand knowledge of all these three; the Arts, Mathematics and Freedom, to speak with such authority.

BIBLIOGRAPHY

[A] A. Andreotti and T. Frankel, *The Lefschetz theorem on hyperplane sections*, Ann. of Math. (2) **69** (1959), 713–717.

[B1] G. D. Birkhoff, *Dynamical systems with two degrees of freedom*, Trans. Amer. Math. Soc. **18** (1917), 240.

[B2] ______, *Quelques théorèmes sur le mouvement des systèmes dynamiques*, Bull. Soc. Math. France **40** (1912), 305–323.

[B3] G. D. Birkhoff and M. R. Hestenes, *Generalized minimax principle in the calculus of variations*, Duke Math. J. **1** (1935).

[B4] R. Bott, *Nondegenerate critical manifolds*, Ann. of Math. (2) **60** (1954), 248–261.

[B5] ______, *The stable homotopy of the classical groups*, Ann. of Math. (2) **70** (1959), 179–203.

[B6] R. Bott and H. Samelson, *Applications of the theory of Morse to symmetric spaces*, Amer. J. Math. **80** (1958), 964–1029.

[B7] Morton Brown, *A proof of the generalized Schoenflies theorem*, Bull. Amer. Math. Soc. **66** (1960), 74–76.

[D1] R. Deheuvels, *Topologie d'une fonctionelle*, Ann. of Math. (2) **61** (1955), 13–72.

[H1] J. Hadamard, *Les surfaces à courbure opposées et leur lignes géodesiques*, J. Math. Pures Appl. (5) **4** (1898).

[H2] G. Hedlund, *On the metrical transitivity of geodesics on closed surfaces of constant negative curvature*, Ann. of Math. (2) **35** (1934), 787–808.

[K] W. Klingenberg, *Lectures on closed geodesics*, Grundlehren der Mathematischen Wissenschaften, Springer-Verlag.

[M1] B. Mazur, *On embeddings of spheres*, Bull. Amer. Math. Soc. **65** (1959), 59–65.

[M2] J. Milnor, *Morse theory*, Ann. of Math. Studies no. 51, Princeton Univ. Press, Princeton, N. J., 1963.

[N1] P. Novikov, *One periodic groups*, Dokl. Akad. Nauk SSSR **127** (1959), 749–752; English transl., Amer. Math. Soc. Transl. (2) **45** (1965), 19–22.

[P1] R. Palais, *Morse theory on Hilbert manifolds*, Topology **2** (1963), 299–340.

[R] H. L. Royden, *The analytic approximation of differentiable mappings*, Math. Ann. **139** (1960), 171–179.

[S1] Seifert-Threlfall, *Variationsrechnung im Grossen*, Teubner, Leipzig, 1938.

[S2] S. Smale, *Generalized Poincaré conjecture in higher dimensions*, Bull. Amer. Math. Soc. **66** (1960), 373–375.

[S3] ______, *A survey of some recent developments in differential topology*, Bull. Amer. Math. Soc. **69** (1963), 131–145.

[S4] ______, *Differentiable dynamical systems*, Bull. Amer. Math. Soc. **73** (1967), 747–817.

[S5] G. Series, *Symbolic dynamics for geodesic flows*, Univ. Warwick, 1979.

[T1] R. Thom, *Sur une partition en cellules associée á une fonction sur une variété*, C. R. Acad. Sci. Paris **228** (1949), 973–975.

[Reprinted from AMERICAN JOURNAL OF MATHEMATICS, Vol. XLIII, No. 1, Jan., 1921.]

A ONE-TO-ONE REPRESENTATION OF GEODESICS ON A SURFACE OF NEGATIVE CURVATURE.

BY HAROLD MARSTON MORSE.

INTRODUCTION.

Surfaces of negative curvature and geodesics upon such surfaces have been considered by J. Hadamard in the paper,* cited below. The paper by Hadamard is in two parts, in the first of which he establishes the existence of a very general class of surfaces of negative curvature; the remainder of Hadamard's paper is devoted to considerations of geodesics upon such surfaces.

In the present paper only those geodesics on the given surfaces of negative curvature are considered which, if continued indefinitely in either sense, lie wholly in a finite portion of space. A class of curves is introduced each of which consists of an unending succession of the curve segments by which the given surface, when rendered simply connected, is bounded. It is shown how a curve of this class can be chosen so as to uniquely characterize some geodesic lying wholly in a finite portion of space. Conversely, it is shown that every geodesic lying wholly in a finite portion of space, is uniquely characterized by some curve of the above class. Use is made of this representation of the geodesics considered to prove several fundamental theorems.

The results of this paper and the representation of geodesics obtained here, will be used in a later paper to establish the existence of a class of geodesics called recurrent geodesics of the discontinuous type.† This class of geodesics offers the first proof that has been given in the general theory of dynamical systems, of the existence of recurrent motions of the discontinuous type.

PART I.

The Surface.

§1. We will consider surfaces without singularities in finite space. We will suppose the surface divisible into overlapping regions such that every point of the surface lying in a finite portion of space is contained as an interior point in some one of a finite number of these regions, and such that

*"Les surfaces à courbures opposées et leur lignes geodesiques," Liouville, *Journal de Mathématique*, 5 Sér., t. 4, p. 27, 1898.

† G. D. Birkhoff, "Quelques théorèmes sur le mouvement des systèmes dynamiques," *Bull. de la Soc. Math. de France*, Vol. 40, p. 303, 1912.

the Cartesian coördinates x, y, z, of the points of any one of these regions can be expressed in terms of two parameters, u and v, by means of functions with continuous derivatives up to a convenient order, at least the third, and such that

$$\left(\frac{D(xy)}{D(uv)}\right)^2+\left(\frac{D(xz)}{D(uv)}\right)^2+\left(\frac{D(yz)}{D(uv)}\right)^2 \neq 0.$$

By a *curve* on the surface we will understand any set of points on the surface in continuous correspondence with the points of an interval on a straight line, including one, both, or neither of its end points.

We will suppose the Gaussian curvature of the surface to be negative at every point, with the possible exception of a finite number of points, at which points the curvature will necessarily be zero. A first result, given by Hadamard in the paper already referred to, is that a surface of negative curvature cannot be contained in any finite portion of space.

§ 2. By a *funnel* of a surface will be meant a portion of a surface topographically equivalent to either one of the two surfaces obtained by cutting an unbounded circular cylinder by a plane perpendicular to its axis. We will consider surfaces of negative curvature whose points outside of a sufficiently large sphere with center at the origin consist of a finite number of funnels. Each of these funnels will be cut off from the rest of the surface along a simple closed curve. These curves will be taken sufficiently remote on the funnels to be entirely distinct from one another.

An unparted hyperboloid of revolution is an example of a surface of negative curvature with two funnels.

From the definition of a funnel, it follows that by a continuous deformation of the closed curve forming the boundary of the funnel, the funnel may be swept out in such a way that every point of the funnel is reached once and only once. Hadamard considers two classes of funnels: those which can be swept out by closed curves which remain less in length than some fixed quantity, and those which do not possess this property. Surfaces with funnels of the first sort are for several reasons of less general interest than those with funnels of the second sort. In the present paper surfaces with funnels only of the second sort will be considered. Hadamard showed that there exist surfaces of negative curvature possessing funnels all of the second sort of any arbitrary number exceeding one, and such that the surface obtained by cutting off these funnels is of an arbitrary genus.

§ 3. We shall consider surfaces which possess at least two funnels of the second sort, and of the surfaces with just two funnels of the second sort we will exclude those surfaces that are topographically equivalent to an unbounded circular cylinder. Hadamard proves that on the surfaces we are considering there exists one and only one closed geodesic that is deformable into the boundary of a given funnel, and that this geodesic possesses no multiple points, and no points in common with the other closed geodesics that are deformable into the boundaries of the other funnels.

We shall denote these closed geodesics, say v in number, by

(1) $$g_1, g_2, \cdots, g_v.$$

They will form the complete boundary of a part of the surface, contained in a finite part of space. We denote this bounded surface by S. It may be proved as a consequence of the assumptions concerning the surface, made in § 1, that S is of finite connectivity. According to the well known theory, S may be rendered simply connected as follows: We first cut S along a system of curves,

$$h_1, h_2, \cdots, h_{v-1},$$

each of which has one end point on an arbitrarily chosen point, P on g_v, and the other, respectively, on the geodesic,

$$g_1, g_2, \cdots, g_{v-1}$$

with the same subscript; and no two of which have a point other than P in common, and no points other than their end points in common with the geodesics of the set (1). There then results a surface with a single boundary. This surface can accordingly be rendered simply connected by $2p$ curves,

$$c_1, c_2, \cdots, c_{2p},$$

each of which curves can be taken as beginning and ending at P, and having no other points than P in common with any of the other curves or geodesics.

We denote by T, the simply connected piece of surface obtained by cutting S along the above curves. It may readily be proved as a consequence of the assumptions made concerning the representation of the given surface, that T is topographically equivalent to a plane region consisting of the interior and boundary points of a circle.

The Infinitely Leaved Surface M.

§ 4. We suppose an infinite number of distinct copies of T spread out above the surface S, in such a manner that points arising from a given point of S overhang the given point of S. We denote by

$$A_i^+, \qquad i = 1, 2, \cdots, 2p + v - 1.$$

that boundary piece of T which arises from an arbitrarily chosen side of the ith one of the $2p + v - 1$ cuts,

(1) $$c_1, c_2, \cdots, c_{2p}, \qquad h_1, h_2, \cdots, h_{v-1},$$

which we have made in S, and by A_i^- that boundary piece of T that arises from the opposite side of the same cut.

Consider now a first copy of T, say M_1, and a particular boundary piece of M_1, say A_i^+. We join M_1 to a second copy of T, joining A_i^+ of M_1 to A_i^- of the second copy of T. In the same manner, we join to M_1 at each different piece of the remaining boundary pieces, a different copy of T, joining each two copies of T along two boundary pieces arising from opposite sides of the same cut. We denote the bounded surface so obtained by M_2.

Proceeding in the same manner, we join to each different boundary piece of M_2 a different copy of T, and one not already a part of M_2, obtaining thereby a new surface M_3. From M_3 we form a new surface M_4, as we formed M_3 from M_2, and continue this process indefinitely, obtaining an enumerable infinity of surfaces each of which contains the preceding. We will understand two different copies of T making up M_n, to have a point in common, if and only if we have expressly joined the two copies together at this point.

§ 5. DEFINITION. *By the surface M we will understand that infinitely leaved surface spread over S that contains each of the surfaces M_n, and every point of which is contained in some one of the surfaces M_n.*

The boundaries of M will consist of those boundary pieces, of the copies of T that make M up, that arise from the geodesics,

$$g_1, g_2, \cdots, g_v.$$

We have made the cuts in S so that on each of the boundary pieces,

$$g_1, g_2, \cdots, g_{v-1}, \tag{2}$$

lies one end point of just one cut. The remaining $4p + v - 1$ end points of the cuts lie at a single point on g_v. On M a sufficiently restricted neighborhood of any one of the points on the geodesics (2), at which lies an end point of a cut, is, when severed along this cut, best described as topographically equivalent to the neighborhood of a point on the boundary of a half plane, when that half plane is severed along a ray distinct from the boundary of the half plane, and issuing from the given point. Similarly a sufficiently restricted neighborhood of that point on g_v at which lie the remaining $4p + v - 1$ end points of cuts, is, when severed along these cuts, topographically equivalent to the neighborhood of a point on the boundary of a half plane, when that half plane is severed along $4p + v - 1$ rays, distinct from each other and from the boundary of the half plane, and issuing from the given point. A sufficiently restricted neighborhood of any other point of M, is topographically equivalent to the complete neighborhood of any point in the plane.

We may conclude that *a sufficiently restricted neighborhood of any point on M is a one-leaved copy of the neighborhood on S of the point overhung on S.*

§ 6. It will be convenient to denote by R a plane region consisting of the points on and interior to a circle.

Any single copy of T is topographically equivalent to a region R. From the method of formation of the surfaces M_n of § 4, it follows that any one of the surfaces M_n is topographically equivalent to a region R. Now from the definition of M in § 5, it follows that any set of points on M that lie in only a finite number of copies of T making up M, will be contained in some one of the surfaces M_n forming a part of M. *Hence any set of points that lie in only a finite number of the copies of T that make up M, will be contained in some region of M, topographically equivalent to a region R.*

§ 7. An infinite set of points on M will be said to have a *limit point* P, if there exists an infinite number of points of the set in every neighborhood of P, on M.

We will prove that a set of points every infinite subset of which has at least one limit point on M, cannot have points in more than a finite number of different copies of T making up M.

If there were points of the set in more than a finite number of the copies of T, there would exist a subset of the given set, say Q, containing an infinite number of points of the given set, and such that each different point would lie in a different copy of T. By the hypothesis made concerning the given set, Q has at least one limit point on M, say P. There are accordingly an infinite number of points of Q in every neighborhood of P, and since no two points of Q lie in the same copy of T, there must be an infinite number of copies of T with points in the neighborhood of P. This is impossible; for all the points in a sufficiently restricted neighborhood of a point on M lie on at most $4p + v$ copies of T, that are joined together at that point.

Hence the result required is proved. Combining this result with the conclusion of the preceding section, we have that *a set of points on M every infinite subset of which has at least one limit point on M, can be included in a region of M, made up of a finite number of copies of T, and topographically equivalent to the plane region consisting of the interior and boundary points of a circle.*

Curves on M Overhanging Curves on S.

§ 8. DEFINITION. Let there be given on S a curve k, and on k a point P. Let P' be any point on M, overhanging P. Starting from a particular point of k, say Q, let P trace out k, first in one sense, and then starting again at Q, in the other sense. As P moves continuously on S, let P' move continuously on M, always overhanging P, and starting at both the times that P starts at Q, at the same point, a point, say Q', overhanging Q. The continuous curve traced out by P' on M, will be said to *correspond* to k on S.

It follows from the result of § 5, that the curve traced out by P', is uniquely determined by k and the choice of the points Q and Q .

DEFINITION. Let h be a curve on S that is closed. Let h be cut at some point so as to form a curve segment k. Let h' be any curve segment on M corresponding, in the sense of the preceding definition, to k. h and $h,$ will be said to *correspond.*

Every curve on M obviously corresponds to one and only one curve on S, while a given curve on S will correspond to an infinite number of different curves on M.

§ 9. *If a curve on S is both closed and deformable into a point, it corresponds on M only to curves that are closed.*

Let h be a curve on S that is both closed and deformable into a point. Let Q' and P' be the end points of a curve segment on M corresponding in the sense of the preceding definition, to h. Since h is closed, Q' and P' overhang a single point on h, say Q.

Suppose now that the above lemma were not true, and that Q' and P' were distinct points on M. Let h be continuously deformed into a point. At the same time let the curve segment joining Q' to P', say h', be continuously deformed on M in such a manner that h' corresponds at each stage of the deformation to the curve into which h has been deformed at that stage, while Q' and P' overhang Q. When, at the end of this process, h reduces to a point, h' must also reduce to a point.

On the other hand P' and Q', the end points of h', in their initial positions were supposed to be distinct points. Since P' and Q' move continuously throughout the deformation, there must then be a first position on M, say P'', at which they are coincident. P' and Q', throughout this deformation, overhung one point of S, namely Q. Hence the point P'' will have in every neighborhood on M pairs of points overhanging the same point of S. This, however, is contrary to the result of § 5, from which contradiction we infer the truth of the above lemma.

§ 10. The following lemma serves as the converse of the preceding.

If a curve on M is closed, it corresponds on S to a curve that is both closed and deformable into a point.

Denote by h the given closed curve on M. It follows from § 7 that there exists a region of M, say M', that contains all the points of h as interior points, and is topographically equivalent to a plane region consisting of the interior and boundary points of a circle. On M', h can be deformed into a point, and hence on M.

Denote by F a continuous family of closed curves on M through the mediation of which h may be deformed into a point on M. On S the closed curve corresponding to h, may be deformed into a point through the mediation of that continuous family of closed curves which correspond to the family F on M.

§ 11. DEFINITION. Let r be any integer, positive, negative, or zero. Let T_r denote a particular copy of T of M. By a *linear set* of copies of T of M, will be understood a region of M consisting of a set of the copies of T in M of the form,

$$(1) \qquad \cdots\ T_{-2},\ T_{-1},\ T_0,\ T_1,\ T_2,\ \cdots,$$

or of the form of any subset of consecutive symbols of (1), in which any three successive copies of T are three different copies of T of M, and in which any two successive copies of T are copies of T that in M are joined along a common boundary piece. A linear set which has no first or last copy of T, will be termed an *unending* linear set.

A linear set is of the nature of an unclosed ribbon of surface. That a linear set does not form a multiple-covered region of M will now be proved.

(*A*) *No copy of T of M appears more than once in any linear set.*

Starting with any copy of T, say T^0, of a given linear set, if possible, let T' be the first successor of T^0 that is identical with some copy of T of the given linear set between T^0 and T', say T''. That linear subset of the given linear set that consists of T'' and its successors up to, but not including T', forms a simply covered, connected region of M, in which the two sides of any boundary piece common to two of its successive copies of T, can be joined by a curve that does not cross that boundary piece. This is impossible, since each boundary piece common to two copies of T of M joins two boundary points of M and divides M into two regions. Cf. § 5 and § 6.

(*B*) *If T' and T'' are any two copies of T of the copies of T that make up M, there exists one and only one linear set of which T' is the first, and T'' the last member.*

We form a linear set in which the first copy of T is T' and in which the successor of T', say T_1, is that copy of T that adjoins T along that boundary piece of T' that separates T' from that part of the surface M that contains T''. If T_1 is not T'', its successor shall be that copy of T that adjoins T_1 along that boundary piece of T_1 that separates T_1 from that part of the surface in which T'' lies. We continue this process indefinitely, or until T'' appears in the linear set formed.

Join any interior point of T' to an interior point of T'' by a curve, say k, lying wholly on M. Each copy of T determined in the process of the preceding paragraph contains at least a point of k. But from the result of § 7 we infer that there are only a finite number of copies of T of M that contain any point of k. It follows that the process of the preceding paragraph will lead after a finite number of steps to T''.

There is but one linear set of which T' is the first and T'' the last copy of T. For no such linear set can contain as a part of its boundary that boundary piece of T' that separates T' from that part of the surface that contains T''. Hence, in any such linear set T' and T_1 must appear at least once as successive copies of T. But according to the result (A), no copy of T of M appears more than once in any linear set; it follows that T_1 is the successor of T' in any linear set joining T' to T''.

In a similar manner it follows that any linear set joining T' to T'', contains T_2 as the successor to T_1. Continuing this method of proof, it is seen that every linear set joining T' to T'', is identical with the first such linear set that we have formed.

§ 12. We denote by H, the set of all curve segments on M each of which corresponds on the original surface to some one of the segments,

$$c_1, c_2, \cdots, c_{2p}; \qquad h_1, h_2, \cdots, h_{v-1}; \qquad g_1, g_2, \cdots, g_v.$$

Every boundary piece of the copies of T of M, is included in the set H.

DEFINITION. Let r be any integer, positive, negative, or zero. Let k_r be a particular piece of the set H of M. By a *reduced curve* of the set H will be understood a continuous curve composed of a set of pieces of the set H of the form,

$$(1) \qquad \cdots, k_{-2}, k_{-1}, k_0, k_1, k_2, \cdots,$$

or of the form of any subset of consecutive symbols of (1), in which no two consecutive pieces are copies of the same piece of the set H, and which contains no pieces corresponding to g_v on the original surface. A reduced curve without end points will be called an *unending* reduced curve.

(A) *No reduced curve can begin and end at the same point.*

Starting with any point, say P, of a given reduced curve, if possible, let P' be the first point following P that is identical with some point of the reduced curve between P and P', say P''. The segment $P'P''$ of the given reduced curve forms a closed curve without multiple points, which closed curve we denote by C.

According to the result of section 7, C can be included in some simply connected region of M consisting of a finite number of copies of T. C, accordingly, forms the boundary of a simply connected region of M, say R, all of whose points are contained in a finite number of copies of T of M. Since C contains no points interior to any copy of T, R contains each copy of T of which it contains any interior points.

Taking the copies of T in the order in which they were added in section 4 to form M, let T'' be the last copy of T of R. T'', like every other copy of T of M, is joined along a common boundary piece to but one copy of T of

M, say T''', that was added to M prior to T'. R accordingly contains none of the copies of T that are adjoined to T' along a common boundary piece except T'''. Hence all of the boundary of T', except the piece common to T' and T''', must form part of the boundary of R, and hence a part of C. In particular, C contains that piece of the boundary of T' that corresponds on the original surface to the boundary geodesic g_v, contrary to the definition of a reduced curve. From this contradiction we infer the truth of the lemma.

The following result is an immediate consequence of the result (A).

(B) *A reduced curve can have no multiple points.*

(C) *There is one and only one reduced curve joining any two given points on pieces of the set H.*

Denote the two given points by P and Q. Among the curves consisting of any continuous succession of pieces of the set H, there obviously exists a variety of curves joining P to Q. If in any such curve, each piece that corresponds on the original surface to g_v, be replaced by the remainder of the boundary of the copy of T containing that piece, there will then result a curve containing no piece corresponding to g_v on the original surface. Of all curves of the latter class joining P to Q, any one that contains the minimum number of pieces of the set H, will be such that no two successive pieces are copies of the same piece of the set H, and will accordingly be a reduced curve.

If possible, let h' and h'' be two different reduced curves joining P to Q. Tracing out h' and h'' respectively, starting from P, let k' and k'' be the first pieces on h' and h'' respectively, that are different. That portion of h' that begins at Q, and ends with k', followed by that portion of h'' that begins with k'', and ends with Q, is a reduced curve with initial and final point at the same point, namely Q. This is contrary to the result (A) of this section, from which contradiction we infer the truth of the lemma (C).

(D) *A reduced curve never returns to a copy of T which it has once left.*

If this assertion be false, there exists some continuous segment of a reduced curve, say k, with end points, say P and Q, belonging to the same copy of T of M, and with no further points in common with that copy of T. A reduced curve cannot begin and end at the same point; hence P and Q are different points. However, let k' be that part of the boundary of the given copy of T that joins P to Q, and does not contain any piece of the set H corresponding to g_v on the original surface. k', taken with k, forms a segment of a reduced curve that may be considered as beginning and ending at P. We conclude that no such segment as k exists, and the lemma is proved.

§ 13. The following is the first of two theorems showing the relation of reduced curves to linear sets.

(*A*) *An unending linear set of copies of T contains one and only one unending reduced curve.*

Let T^0, T' and T'', be any three successive copies of T in the given linear set. There exist two distinct and continuous portions of the boundary of T', of which one may reduce to a point such that each joins some boundary point of T^0 to a boundary point of T'', without containing any other points of T^0 and T''. We associate with T' that one of these portions that does not contain a piece of the set H corresponding to g_v on the original surface. Denote this portion by k. The set of all curve segments such as k, taken in the order in which they arise from the given linear set, will not necessarily form a continuous curve. However by a proper addition of the boundary pieces common to successive copies of T of the given linear set, the resulting curve will be continuous. If these additions are made only when necessary to make the curve continuous, the resulting curve will also be a reduced curve.

If possible, let r be a second reduced curve contained in the given linear set. As proved in the preceding section, a reduced curve never returns to a copy of T which it has once left; it follows that r contains at least a point of each of the pieces of the set H that separate members of the given linear set, and hence contains each of the points or segments k, associated in the first paragraph of this proof with each copy of T of the given linear set. But according to the result (C) of section 12, there is but one segment of a reduced curve joining any two points of M. Hence any reduced curve that contains each of the points or curve segments k, associated with each copy of T of the given linear set, is thereby uniquely determined. There is thus but one reduced curve contained in the given linear set.

The preceding result has its converse in the following:

(*B*) *An unending reduced curve is contained in one and only one linear set, which set is an unending linear set.*

Let

$$\cdots,\ P_{-2},\ P_{-1},\ P_0,\ P_1,\ P_2,\ \cdots,$$

be a set of points which lie on the given reduced curve in the order of their subscripts, and which are such that any point of the given curve lies between P_{-n} and P_n for n a sufficiently large positive integer, and where P_{-1} and P_1 are supposed to be taken so as not to lie on a common copy of T of M. Let C_n be the segment joining P_{-n} to P_n of the given reduced curve. Let L_n be a linear set containing P_{-n} and P_n, and made up of the smallest possible number of copies of T.

C_n lies wholly in L_n. If this assertion be false, let k be a piece of the set H which forms part of the boundary of L_n, and contains a point at which C_n leaves L_n. The end points of C_n lie on L_n; hence C_n returns to L_n. But k divides the whole of M into two regions; hence to return to L_n, C_n would have to return to k, which is impossible, since a reduced curve never returns to a copy of T which it has once left.

We will now prove that if r is any positive integer greater than n, the linear set L_r contains L_n. Let h be any one of the boundary pieces common to two adjacent copies of T of L_n. C_n contains points, not on h, in each of the two regions into which M is divided by h, for otherwise one of the two linear sets into which L_n is divided by h, would contain all the points of C_n and be a linear set containing fewer copies of T than does L_n, contrary to the hypothesis concerning L_n. Hence any connected region that contains all the points of C_n, must contain the two copies of T of L_n that are adjoined along h. Hence any connected region that contains all of the points of C_n, must contain each copy of T of L_n. In particular, L_r contains all the points of C_n, and hence contains L_n.

The set of all copies of T of all regions L_n, for all positive integers n, will accordingly constitute a linear set, which we denote by L. Any point of the given curve is contained on every curve C_n for sufficiently large values of n, and hence is contained in L. We will now prove that L is an unending linear set.

As seen earlier in the proof, the two linear sets into which L_n and hence L is divided by any boundary piece h, common to copies of T of L, each contain points of the given reduced curve that are not on h. Since a reduced curve never returns to a copy of T which it has once left, the removal of h divides the given reduced curve into just two continuous portions, which lie respectively in the two linear sets into which L is divided by h. But each of these continuous portions of the given reduced curve will contain an infinite number of different pieces of the set H, so that each of the two linear sets into which L is divided by h, must contain an infinite number of different copies of T. We conclude that L is an unending linear set.

Any second linear set that contains L, would have to contain each of the linear sets L_n, hence every copy of T of L. Since there is but one linear set joining any two copies of T of M, any linear set containing all the copies of T of L, is identical with L.

The two results of this section may be combined in the following fundamental theorem:

THEOREM I. *There is a one to one correspondence between the set of all unending reduced curves, and the set of all unending linear sets of M, in which each reduced curve corresponds to that linear set in which it is contained.*

PART II.

Some General Properties of Geodesics on M.

§ 14. Hadamard proved that there is on S one and only one geodesic joining two given points, and deformable into an arbitrary curve joining the two given points, and that this geodesic is shorter than any other rectifiable curve deformable into the given curve joining the two given points. With the aid of the results of sections 9 and 10, Hadamard's result becomes the following theorem.

THEOREM II. *There exists on M one and only one geodesic joining two given points, and this geodesic is shorter than any other rectifiable curve joining the two given points.*

COROLLARY I. *On M two geodesics can intersect in but one point.*

COROLLARY II. *On M a geodesic can have no multiple points.*

§ 15. On a surface representable in the manner in which the given surface is representable, there exists one and only one geodesic through a given point, and tangent to a given direction.

DEFINITION. A point on the surface, and a direction tangent to the surface will be called an *element*, and will be said to *define that sensed geodesic* that passes through the initial point of the given element, and is such that its positive tangent direction at that point agrees with the direction of the given element.

If u and v are parameters in any representation of a part of the surface, and if θ' is the angle which a given tangent direction makes at the point $(u'v')$ with the positive tangent to the curve $u = u'$, then $(u'v'\theta')$ will represent an element of the given surface. We shall understand by each statement of metric relations between elements, the same statement of metric relations between the points in space of three dimensions, obtained by considering the complex $(u'v'\theta')$ as the Cartesian coördinates of a point.

§ 16. Let G be any geodesic segment lying on the original uncut surface. G is an extremal in the Calculus of Variations problem of minimizing the arc length, from which theory we can readily obtain the following theorem that describes the nature of the variation of G with variation of its initial element.*

Corresponding to any positive constants, e and h, there exists a positive constant d, so small, that if any two elements, with initial points on the bounded surface S, lie within d of each other, and if a second pair of elements lie respectively on the two geodesics defined by the first two elements, and if further the initial points of this second pair of elements lie respectively at a distance, measured along the given geodesics from the geodesics' initial

* Cf. Bolza, *Vorlesungen über Variationsrechnung*, 1909, p. 219.

points, that is the same in both cases and that does not exceed h, the second pair of elements will lie within e of each other.

§ 17. According to Theorem II, § 14, any given geodesic on M is shorter than any other rectifiable curve joining its end points. According to the theory of the Calculus of Variations, this is a case where there exists on the given geodesic on M no point conjugate to a given point on the given geodesic. A particular consequence* of this result, as given in the theory of the Calculus of Variations, is that, if we vary the end points of a given geodesic segment, there exists further geodesic segments joining these end points and varying continuously with the end points, both in position and in length. We have seen that there is no more than one geodesic segment joining any two points on M. Whence we may say that the length of the geodesic segment, joining two points on the surface M, is a single-valued continuous function of the position of these points. In particular we have the result:

There exists a finite upper limit to the lengths of all geodesic segments joining pairs of points lying on a closed set of points on M.

§ 18. The following statement, of importance in the developments that are to follow, may readily be proved as a consequence of the development given by Hadamard.

Corresponding to two arbitrary positive constants, e and d, there exists a positive constant h, so large, that if on M each point of a first geodesic segment G, of length h, lies within a geodesic distance d of some point of a second geodesic segment, then this second geodesic segment has at least one element within an e of that element of G which lies at the mid point of G.

§ 19. *The Network H Replaced by a Network H′ of Geodesic Segments.* We denote by H', the set of all geodesic segments on M each of which joins two end points of a piece of the set H, defined in § 12.

We shall prove the following lemma:

Two pieces of the set H′ arising from two different pieces of H, can have no points in common other than one end point.

Let P and Q, P' and Q', be respectively the end points of two different pieces of the set H. Denote by h the geodesic segment that joins P to Q, and by h' the geodesic segment that joins P' to Q'. All cases are included in the three following cases:

CASE I. The two given pieces of the set H have an end point in common. In particular suppose $P = P'$; if $P = P'$, it follows from the nature of the construction of the surface M, that $Q \neq Q'$. Hence, if $P = P'$, the geodesic segments, h and h', have P as a common end point, and are not identical, since $Q \neq Q'$. With the aid of Corollary II, § 14, we conclude that h and h' can have no other point than P in common.

* Cf. Bolza, loc. cit., p. 307.

CASE II. $P \neq P'$, and $Q \neq Q'$, and one or both of the geodesic segments, h and h', form a piece of the geodesic boundary of the surface M.

h and h' are not identical, since $P \neq P'$, and $Q \neq Q'$. In case h were a piece of the geodesic boundary of M, h' could not meet h in other than an end point without passing off from the surface at that point, contrary to the result of Theorem II, § 14.

CASE III. $P \neq P'$, and $Q \neq Q'$, while neither h nor h' form a piece of the geodesic boundary of the surface M.

According to the result of § 7, there exists on M, a simply connected region, bounded by a simple closed curve, and containing the two given pieces of the set H and the geodesic segments h and h'. This region, which we denote by R, we will suppose taken so large that no points of the two given pieces of the set H, nor of the geodesic segments h and h', are boundary points of R unless they are boundary points of M.

P, Q, P', *and* Q', all lie on the boundary of M, and hence on the boundary of R. Since $P \neq P'$, and $Q \neq Q'$, the two given pieces of the set H have no points in common; it follows that P, Q, P', and Q', lie on the boundary of R in the circular order named.

On the other hand, it is an hypothesis of this case, that neither h nor h' are a piece of the geodesic boundary of M; it follows that neither h nor h' have points other than their end points on the boundary of M, and hence of R A particular consequence is that h divides R into two regions. If now h' crossed h, its end point P' would lie in one of the two regions into which R is divided by h, while its remaining end point, Q', would lie in the other such region. P, P', Q, Q', would thus lie on the boundary of R in the circular order named. From this contradiction we infer the truth of the lemma in this case. The proof is thus given in general.

§ 20. We seek now to render the original surface simply connected by means of cuts made along geodesics.

In accordance with the result given in § 14, the end points of each one of the cuts by means of which the original surface was rendered simply connected, can be joined on the original surface by a geodesic segment deformable into the given cut. With the aid of the result of § 9, it appears that these geodesic segments correspond on M to the geodesic segments of H', and further that two of these geodesic segments arising from different cuts correspond on M only to different members of the set H'. From the result of the preceding section we conclude that no two of these geodesic segments have any points in common other than their end points.

If now the original surface be cut along these geodesic segments, the resulting surface, which we denote by U, will be simply connected. The infinitely leaved surface M can be formed by joining together copies of U in the same manner as it was formed from copies of T.

We define a reduced curve of the set H', and a linear set of copies of U, by replacing H and T in the definitions of sections 11 and 12, by H' and U, respectively. In the same manner, we obtain from the results of sections 11 and 13, the following results in terms of H' and U.

(*A*) *If U' and U'' are any two copies of U of the copies of U that make up M, there exists one and only one linear set of which U' is the first and U'' the last member.*

(*B*) *There is a one to one correspondence between the set of all unending reduced curves of the set H', and the set of all unending linear sets of copies of U, in which each reduced curve corresponds to that linear set in which it is contained.*

Geodesics Lying Wholly on M.

§ 21. Let G be a geodesic lying wholly on M. G cannot become infinite in length in any one copy of U, as follows from the result of section 17. In leaving a copy of U of M, G cannot be tangent to any one of the geodesic segments that separate that copy of U from the remainder of M. For in such a case G would coincide with the geodesic segment, and pass off from M at its end points. Further, it follows from Corollary I, section 14, that G can have but one point of intersection with any of the geodesic segments that separate the different copies of U of M. Hence G never returns to a copy of U of M which it has once left. Since G cannot become infinite in length in any one copy of U, it follows that either of the two portions into which G may be divided by an arbitrary one of its points, has points in common with an infinite number of different copies of U.

From these last two results, it follows by a proof, which except for terminology, may be given as a repetition of the proof of (*B*), section 13, that *a geodesic lying wholly on M, is contained in one and only one linear set, which set must be an unending linear set.*

§ 22. As a converse to the preceding result, we have the following:

If there be given any unending linear set, there exists one, and only one geodesic contained wholly in the given linear set, and this geodesic has at least one point in each copy of U of the given linear set.

Let an unending linear set of copies of U be given as follows:

$$\cdots, U_{-2}, U_{-1}, U_0, U_1, U_2, \cdots.$$

Let g_n be a geodesic segment joining any interior point of U_{-n} to an interior point of U_n. The set of copies of U which g_n passes through, is seen to form a linear set joining U_{-n} to U_n. But from the result of section 11, it follows that there is but one such linear set joining U_{-n} to U_n. The linear set,

$$U_{-n}, U_{-n+1}, \cdots, U_0, \cdots, U_{n-1} U_n,$$

is one such set; we conclude that this is the set c. copies of U through which g_n passes.

Denote by E_n an element on that part of g_n that lies in U_0. The set of all elements E_n will have a limit element. Let G be the geodesic defined by this element. We will first show that G has a point in each copy of U of the given linear set.

Let r be any positive integer. For integers $n > r$ that portion of g_n in

$$(1) \qquad U_{-r},\ U_{-r+1},\ \cdots,\ U_0,\ \cdots,\ U_{r-1}U_r,$$

is according to the result of section 17, less in length than some fixed quantity independent of n. Now a finite segment of a geodesic varies continuously with its initial element. It follows that G possesses a finite segment, say G_r, which has a point in each copy of U of (1), and which is wholly contained in the set (1). From the fact that G_r has a point in each copy of U of (1), we may conclude that G has a point in each copy of the given linear set.

We seek to prove that G is wholly contained in the given linear set. Any finite segment of any geodesic on M has points in not more than a finite number of copies of U. Cf. section 7. We may conclude that, for r sufficiently large, any given segment of G that begins with a point of U_0, is included in one of the two portions into which G_r is divided by that point. Thus every point of G lies on some segment G_r. But every point of G_r, and hence every point of G lies in the given linear set.

If there were a second geodesic, say G', contained in the same linear set as G, it would follow from the result of section 18, that every element of G would be a limit element of elements on G'. This is impossible, since G' can never return to a copy of U which it has once left.

The results of this section and the preceding, are summed up in the following:

THEOREM III. *There is a one-to-one correspondence between the set of all geodesics lying wholly on M, and the set of all unending linear sets, in which each geodesic corresponds to that linear set in which it is contained.*

A similar theorem is now obtained from (B) in § 20.

THEOREM IV. *There is a one-to-one correspondence between the set of all geodesics lying wholly on M, and the set of all reduced curves of the set H', in which each geodesic corresponds to that reduced curve that is contained in the same linear set.*

Congruent Curves.

§ 23. DEFINITION. Two curves on M that correspond to the same curve on S will be said to be *congruent.*

The first statement of the following theorem serves as an existence proof for closed geodesics on the original surface.

(*A*) *If a reduced curve*[4] *on M consists of successive mutually congruent portions, the geodesic that lies in the same linear set as the given reduced curve, consists also of successive mutually congruent portions. Conversely, if a geodesic lying wholly on M consists of successive mutually congruent portions, the reduced curve that lies in the same linear set, consists also of successive mutually congruent portions.*

Suppose the given reduced curve consists of successive mutually congruent portions. Let the configuration, consisting of the given reduced curve and the geodesic lying in the same linear set, be carried as a whole into that congruent configuration in which one of the successive mutually congruent portions of the given reduced curve is carried into the succeeding congruent portion. Denote this transformation by T.

The given geodesic is carried into itself by T. For otherwise there would exist two different geodesics, namely the given geodesic and the geodesic into which it is carried by T, both lying in the same linear set as the given reduced curve. This is contrary to the result of Theorem IV, section 22.

Now let P' be any point on the given geodesic. Let P'' be the point in which P' is carried by T. P'' lies on the given geodesic, since the given geodesic is carried into itself by T. The successive images of the geodesic segment $P'P''$ which result through repetition of T and its inverse, will be successive mutually congruent portions of the given geodesic, all of whose points can be reached by a sufficient number of such transformations.

The converse statement of the theorem can be proved in a similar manner.

In the same manner also, the following can be established.

(*B*) *Let there be given a first reduced curve and geodesic contained in a single linear set, and a second reduced curve and geodesic that are also contained in a single linear set. If the two given reduced curves are congruent, the two geodesics must also be congruent; and conversely, if the two given geodesics are congruent, the two reduced curves must also be congruent.*

Variation of Geodesics with Their Initial Elements.

§ 24. The following statement depends upon the result given in section 18.

(*A*) Corresponding to any positive quantity e, there exists a positive integer n, so large, that if any two unending linear sets have in common a region U, say U', together with the first n regions U succeeding U' in either sense in the given linear set, then there exists in U', and on each of the two geodesics that pass respectively through the two given linear sets, at least one element that lies within e of some element on that part of the other geodesic that lies in U'.

The converse statement depends upon the property of continuous variation of a geodesic with its initial element, as given in section 16.

(*B*) Corresponding to any positive integer n, there exists a positive constant e, so small, that if on each of two geodesics there exists some element within e of some element on the other, then the two linear sets through which the two given geodesics respectively pass, possess in common that linear set that contains the region or regions U, in which the two given elements lie, together with n regions U succeeding these regions in either sense.

By virtue of the relations between unending reduced curves and corresponding unending linear sets, as given in the proofs of section 13, the statements of (*A*) and (*B*) of this section become the following:

THEOREM VI. *Corresponding to any positive constant e, there exists a positive constant k, so large, that if two unending reduced curves possess in common a continuous segment of length exceeding k, the two corresponding geodesics each have at least one element within e of some element on the other, and with initial point in the same copy of U as the mid point of the common reduced segment. Conversely, corresponding to any positive constant k, there exists a positive constant d, so small, that if on each of two geodesics there exists some element within d of some element on the other, the two corresponding reduced curves possess in common a segment of length k, with mid point in the same copy of U as the initial point of either of the two elements.*

§ 25. DEFINITION. Two elements E' and E'' of a set of elements E on M, will be said to be *mutually accessible* in E, if corresponding to any positive constant e, there exists in the set E a finite ordered subset of elements of which the first element is E', and the last element E'', while each element of the subset, excepting the last, lies within e of the following element.

THEOREM VI. *In the set of all elements on a closed region of M, and on geodesics lying wholly on M, no two elements lying on different geodesics are mutually accessible.*

Let G' and G'' be two different geodesics lying wholly on M. Let E' and E'' be two elements lying respectively on G' and G''. Let R be any closed region of M, in which the initial points of E' and E'' lie. We will show that E' and E'' are not mutually accessible in the set of all elements on R, and on geodesics lying wholly on M.

G' and G'' are different geodesics, and are accordingly not contained in the same linear set of copies of U. There therefore exists some geodesic segment, of the geodesic segments that separate different copies of U of M, that G' crosses and that G'' does not cross. Denote this geodesic segment by h. Denote by B' the set of all elements on R, and on geodesics that lie wholly on M and cross h. Denote by B'' the set of all elements on R, and that lie on geodesics that lie wholly on M and do not cross h.

B' is a closed set. For it follows from the result of sections 16 and 17, that a limit element of elements of the set B', defines a geodesic, say G, lying wholly on M, and that is either tangent to h or else crosses h. But G cannot be tangent to h without passing off from M at the end points of h. Hence G crosses h. Thus B' is a closed set. In a similar manner it follows that B'' is a closed set.

From their definitions it follows that B' and B'' can have no elements in common. There accordingly exists a positive constant d, such that no element of B' lies within d of any element of B'' Hence E', which belongs to B', and E'', which belongs to B'', cannot be mutually accessible in the set of elements comprising the elements of B' and B''. But the elements of B' and B'' comprise all the elements on geodesics lying wholly on M, and with initial points in the region R.

The theorem thus is proved.

HARVARD UNIVERSITY,
June, 1917.

Reprinted from the
TRANSACTIONS OF THE AMERICAN MATHEMATICAL SOCIETY
Vol. XXII., No. 1, pp. 84–100
January, 1921

RECURRENT GEODESICS ON A SURFACE OF NEGATIVE CURVATURE*

BY

HAROLD MARSTON MORSE

Introduction

The results necessary for the development of this paper are contained in a paper by G. D. Birkhoff,† in a paper by J. Hadamard,‡ and in an earlier paper by the present writer.§

In this earlier paper, as in the present paper, only those geodesics on the given surfaces of negative curvature are considered which, if continued indefinitely in either sense, lie wholly in a finite portion of space. A class of curves is introduced, each of which consists of an unending succession of the curve segments by which the given surface, when rendered simply connected, is bounded. It is shown how a curve of this class can be chosen so as to uniquely characterize some geodesic lying wholly in a finite portion of space. Conversely, it is shown that every geodesic lying wholly in a finite portion of space, is uniquely characterized by some curve of the above class.

The results of the earlier paper on geodesics, and the representation obtained there, will be used in the present paper to establish various theorems concerning sets of geodesics and their limit geodesics. In particular, the existence of a class of geodesics called *recurrent geodesics of the discontinuous type*,‖ will be established. This class of geodesics offers the first proof that has been given in the general theory of dynamical systems, of the existence of recurrent motions of the discontinuous type.

For a more complete treatment of the questions of the existence of surfaces of negative curvature, the reader is referred to the paper by Hadamard, already cited.

---

* Presented to the Society, Dec. 28, 1920.

† *Quelques théorèmes sur le mouvement des systèmes dynamiques*, Bulletin de la Société Mathématique de France, vol. 40 (1912), p. 303.

‡ *Les surfaces à courbures opposées et leur lignes géodésiques*, Journal de Mathématiques pures et appliquées, (5), vol. 4 (1898), p. 27.

§ *A one to one representation of geodesics on a surface of negative curvature*, American Journal of Mathematics, vol. 42 (1920).

‖ G. D. Birkhoff, loc. cit.

The surface

§ 1. We will consider surfaces without singularities in finite space. We will suppose the surface divisible into overlapping regions, such that every point of the surface lying in a finite portion of space is contained as an interior point in some one of a finite number of these regions, and such that the Cartesian coördinates x, y, z of the points of any one of these regions can be expressed in terms of two parameters, u and v, by means of functions with continuous derivatives up to a convenient order, at least the third, and such that

$$\left(\frac{D(xy)}{D(uv)}\right)^2 + \left(\frac{D(xz)}{D(uv)}\right)^2 + \left(\frac{D(yz)}{D(uv)}\right)^2 \neq 0.$$

By a *curve* on the surface we will understand any set of points on the surface in continuous correspondence with the points of an interval on a straight line, including one, both, or neither of its end points.

We will suppose the Gaussian curvature of the surface to be negative at every point, with the possible exception of a finite number of points, at which points the curvature will necessarily be zero. A first result, given by Hadamard in the paper already referred to, is that a surface of negative curvature cannot be contained in any finite portion of space.

§ 2. By a *funnel* of a surface will be meant a portion of a surface topographically equivalent to either one of the two surfaces obtained by cutting an unbounded circular cylinder by a plane perpendicular to its axis. We will consider surfaces of negative curvature whose points, outside of a sufficiently large sphere with center at the origin, consist of a finite number of funnels. Each of these funnels will be cut off from the rest of the surface along a simple closed curve. These curves will be taken sufficiently remote on the funnels to be entirely distinct from one another.

An unparted hyperboloid of revolution is an example of a surface of negative curvature with two funnels.

From the definition of a funnel it follows that, by a continuous deformation of the closed curve forming the boundary of the funnel, the funnel may be swept out in such a way that every point of the funnel is reached once and only once. Hadamard considers two classes of funnels: those which can be swept out by closed curves which remain less in length than some fixed quantity, and those which do not possess this property. Surfaces with funnels of the first sort are for several reasons of less general interest than those with funnels of the second sort. In the present paper surfaces with funnels only of the second sort will be considered. Hadamard showed that there exist surfaces of negative curvature possessing funnels all of the second sort, of any arbitrary number exceeding one, and such that the surface obtained by cutting off these funnels is of an arbitrary genus.

§ 3. We shall consider surfaces which possess at least two funnels of the second sort, and of the surfaces with just two funnels of the second sort, we will exclude those surfaces that are topographically equivalent to an unbounded circular cylinder. Hadamard proves that on such surfaces there exists one and only one closed geodesic that is deformable into the boundary of a given funnel, and that this geodesic possesses no multiple points, and no points in common with the other closed geodesics that are deformable into the boundaries of the other funnels.

We shall denote these closed geodesics, say v in number, by

$$g_1 g_2 \cdots g_v . \tag{1}$$

They will form the complete boundary of a part of the surface, contained in a finite part of space. We denote this bounded surface by S. As shown in § 18 and § 19 of my earlier paper on geodesics, S may be rendered simply connected as follows: S is first cut along a system of geodesics,

$$h_1 h_2 \cdots h_{v-1},$$

each of which has one end point on an arbitrarily chosen point, P, on g_v, and the other, respectively, on the geodesic of the set

$$g_1 g_2 \cdots g_{v-1},$$

with the same subscript, and no two of which have a point other than P in common, and no points other than their end points in common with the geodesics, of the set (1). There then results a surface with a single boundary. This surface can be rendered simply connected by $2p$ geodesics,

$$c_1 c_2 \cdots c_{2p},$$

which can be taken as beginning and ending at P, and which will have no other points than P in common with any of the other geodesics or with each other.

We denote by T, the simply connected piece of surface obtained by cutting S along the above geodesics. It may be proved as a consequence of the assumptions made concerning the representation of the given surface, that T is topographically equivalent to a plane region consisting of the interior and boundary points of a circle.

Representation of geodesics by linear sets or by reduced curves

§ 4. We suppose that we have at our disposal an unlimited number of copies of the simply connected surface T, and that each of these copies of T is entirely distinct from every other copy of T.

Definition. Let r be any integer, positive, negative or zero. Let T_r

denote a particular copy of T. By a *linear set* of copies of T will be understood a surface consisting of a set of copies of T of the form

$$\cdots\ T_{-2}\ T_{-3}\ T_0\ T_1\ T_2 \cdots, \tag{1}$$

or of the form of any subset of successive symbols of (1), in which no one copy of T appears twice, and in which each copy of T is joined along some one of its boundary pieces to that boundary piece of the succeeding copy of T that arises from the opposite side of the same cut, while no copy of T is joined to its predecessor and successor along the same boundary piece. A linear set which has no first or last copy of T will be called an *unending* linear set.

Two linear sets will be considered the *same* if the two sets of their copies of T can both be expressed by the same form (1), in such a manner that successive symbols represent successive copies of T in the respective linear sets, joined along copies of the *same* cut.

A linear set in which the number of copies of T is finite is seen to be a multiple-leaved, simply connected surface, bounded by a single closed curve.

Let the set of geodesic segments,

$$g_1\, g_2 \cdots g_v;\ h_1\, h_2 \cdots h_{v-1};\ c_1\, c_2 \cdots c_{2p},$$

described in § 3, be denoted by H.

Definition. Let r be any integer, positive, negative, or zero. Let k_r be any member of the set H. By a *reduced curve* we shall understand any *continuous* curve that consists of a set of members of the set H, excluding g_v, of the form

$$\cdots\ k_{-2}\ k_{-1}\ k_0\ k_1\ k_2 \cdots, \tag{1}$$

or of the form of any subset of consecutive symbols of (1). In the special* case where a k_r and a k_{r+1} of (1) are copies of the same member of the set H, say l, we require that the end point of k_r and the end point of k_{r+1} which are joined, be points which on l would be considered as opposite end points. A reduced curve without end points will be termed an *unending* reduced curve.

If a given reduced curve be traced out in an arbitrary sense, it follows from the last condition of the definition of a reduced curve that no two consecutive pieces of the given reduced curve will thereby appear as copies of the same piece of H taken in opposite senses.

§ 5. The results of this section are established in § 12 and § 13, of my earlier paper on geodesics, already referred to.

A given unending reduced curve is contained in one and only one linear

* We admit the possibility of two symbols in (1) representing the same member of the set H, but as parts of the reduced curve we shall consider two such copies as distinct, in a manner analogous to the convention ordinarily made in the construction of a Riemann surface.

set, which set is an unending linear set. Conversely, every unending linear set contains one and only one unending reduced curve. Each copy of T of an unending linear set that contains an unending reduced curve contains either a point or a single continuous segment of the given reduced curve, and no other points of the given reduced curve. The results necessary for the developments of this paper are summed up in the following:

THEOREM 1. *There is a one to one correspondence between the set of all unending reduced curves on S, and the set of all unending linear sets, in which each reduced curve corresponds to that linear set in which it is contained.*

§ 6. The results of this section follow from the results of §§ 21, 22, and 23, of the earlier paper on geodesics. For the purpose of representing geodesics that lie wholly on S, it will be convenient to suppose each closed geodesic replaced by that geodesic obtained by tracing out the given closed geodesic an infinite number of times in either sense.

Every geodesic lying wholly on S is contained on one and only one linear set, which set must be an unending linear set. Conversely, every unending linear set contains one and only one of the geodesics lying wholly on S. Every copy of T of a linear set that contains a geodesic lying wholly on S, contains either a point or a single continuous segment of this geodesic, and no other point on this geodesic.

THEOREM 2. *There is a one to one correspondence between the set of all geodesics lying wholly on S, and the set of all unending linear sets, in which each geodesic corresponds to that linear set in which it is contained.*

The results of Theorems 1 and 2 can be combined in the following:

THEOREM 3. *There is a one to one correspondence between the set of all geodesics lying wholly on S, and the set of all unending reduced curves, in which each geodesic corresponds to that unending reduced curve that is contained in the same linear set.*

THEOREM 4. *If an unending reduced curve consists wholly of repetitions of a closed curve, the geodesic that passes through the same linear set consists wholly of successive repetitions of a closed geodesic. Conversely, if a geodesic consists wholly of successive repetitions of a closed curve, the unending reduced curve that passes through the same linear set, consists wholly of successive repetitions of a closed curve.*

VARIATION OF GEODESICS WITH INITIAL ELEMENTS

§ 7. On a surface representable in the manner in which the given surface is representable, there is one and only one geodesic through a given point, and tangent to a given direction.

DEFINITION. A point on the surface and a direction tangent to the surface will be called an *element*, and will be said to define that sensed geodesic that passes through the initial point of the given element, and is such that its positive tangent direction at that point agrees with the direction of the given element.

If u and v are parameters in any representation of a part of the surface, and if θ' is the angle which a given tangent direction makes at the point $(u'\,v')$ with the positive tangent to the curve, $u = u'$, then $(u'\,v'\,\theta')$ will represent an element of the given surface. We shall understand by each statement of metric relations between elements, the same statement of metric relations between the points in space of three dimensions obtained by considering the complex $(u'\,v'\,\theta')$ as the Cartesian coördinates of a point.

Let G be any geodesic segment lying on the original uncut surface. G is an extremal in the Calculus of Variations problem of minimizing the arc length, from which theory we can readily obtain the following theorem that describes the nature of the variation of G with variation of its initial element.*

THEOREM 5. *Corresponding to any positive constants e and h, there exists a positive constant d so small, that if any two elements, with initial points on the bounded surface S, lie within d of each other, and if a second pair of elements lie respectively on the two geodesics defined by the first two elements, and if further the initial points of this second pair of elements lie respectively at a distance, measured along the given geodesics from the geodesics' initial points, that is the same in both cases and that does not exceed h, the second pair of elements will lie within e of each other.*

The following theorem describes the manner in which a geodesic varies with the reduced curve contained in the same linear set. It is given in § 24 of the earlier paper on geodesics.

THEOREM 6. *Corresponding to any positive constant e, there exists a positive constant k, so large, that if two unending reduced curves possess in common a continuous segment of length exceeding k, the two corresponding geodesics each have at least one element within e of some element on the other, and with initial point in the same copy of T, in the geodesic's linear set, as the mid point of the common reduced curve segment.*

Conversely, corresponding to any positive constant k, there exists a positive constant e, so small, that if on each of two geodesics there exists some element within e of some element on the other, the two corresponding reduced curves possess in common a segment of length k, with mid point in the same copy of T in the reduced curve's linear set, as the initial point of either of the two elements.

REPRESENTATION OF GEODESICS BY SETS OF NORMAL CURVES

§ 8. The previous representation of geodesics by means of linear sets and reduced curves can now be replaced by another representation which will be

* Cf. Bolza, *Vorlesungen über Variationsrechnung* (1909), p. 219.

fundamental in the work of this paper. This representation will be in terms of the geodesic segments,

(1) $$c_1\, c_2 \cdots c_{2p},$$

(2) $$g_1\, g_2 \cdots g_{v-1},$$

which form a subset of the boundary pieces of each copy of T.

Definition. I. Each one of the geodesic segments of (1) and (2) will be called a *normal segment.*

II. Let m be any integer, positive, negative, or zero. Let C_m represent any sensed normal segment. By a *normal set* C, will be understood an unending ordered set of sensed normal segments, in the form

(3) $$\cdots C_{-2}\, C_{-1}\, C_0\, C_1\, C_2 \cdots,$$

in which no two successive members are the same normal segment taken in opposite senses.

III. Two normal sets C will be considered the *same* if they contain the same normal segments in the same order with the same senses.

A normal set C will not in general constitute a reduced curve. For a reduced curve may include any normal segment, and in addition any geodesic segment of the set,

(4) $$h_1\, h_2 \cdots h_{v-1}.$$

However, it is readily seen that, with the aid of the members of the set (4), there can be formed from a given normal set C one and only one sensed reduced curve whose normal segments taken in the order and with the senses in which they appear on the given reduced curve constitute the given normal set C. Conversely, if there be given any unending sensed reduced curve, its normal segments taken in the order and with the senses in which they appear on the given unending sensed reduced curve, constitute a normal set C.

Thus there is a one to one correspondence between the set of all unending sensed reduced curves and the set of all normal sets C, in which each normal set C corresponds to that unending sensed reduced curve whose normal segments, taken in the order and with the senses in which they appear on the given unending sensed reduced curve, constitute the given normal set C.

Definition. If an unending sensed reduced curve and a normal set C correspond in the sense of the preceding statement, the normal set C will be said to *represent* the given unending sensed reduced curve, and also that sensed geodesic that passes through the same linear set of copies of T in the same sense as does the given unending sensed reduced curve.

By virtue of Theorem 3, § 6, every sensed geodesic lying wholly on the surface S, is represented by one and only one normal set C, while every normal set C represents one and only one sensed geodesic.

Closed geodesics

§ 9. If a normal set C of the from (3) of the preceding section represents a closed geodesic it follows from Theorem 4, § 6, that there exists a positive integer p, such that in the set C

$$C_m = C_{m+p},$$

where m is any integer, positive, negative, or zero. The given normal set will then be said to be *periodic,* and to have the period p. Theorem 4, § 6 now becomes the following:

Theorem 7. *A necessary and sufficient condition that a geodesic be closed, is that the normal set C representing that geodesic be periodic.*

Let q be the smallest period of a periodic set C. Then any other period p must either equal q, or else be a multiple of q. For if p were not equal to q or a multiple of q, it follows from Euclid's Algorithm that there exist three integers, A, B, and r, of which r is less than q, and is greater than zero, and which are such that

$$Aq + Bp = r.$$

It follows from this equation that r is also a period of the given periodic set C, contrary to the assumption that q was the smallest period of the given periodic set C.

Definition. If q is the smallest period of a periodic normal set C, then any q successive sensed normal segments of the given set C will be called a *generating set* of the given set C, and also of the closed geodesic represented by the given set C.

If B is a generating set of a normal set C, this set C consists merely of an unending succession of sets B, which we will write in the form,

$$\cdots B\,B\,B\,B\,B\,B\,B\,B \cdots$$

All generating sets of a periodic set C, can evidently be obtained from any one such generating set by a circular permutation of the sensed normal segments composing the given generating set.

§ 10. We consider now the question of the arbitrary formation of sets that may serve as generating sets of some geodesic. To that end we form a finite ordered set of sensed normal segments, in which neither the first and last members, nor any two successive members are the same normal segment taken in opposite senses, and which cannot be obtained through repetitions of a similar set containing fewer sensed normal segments. Denote the set so obtained by D. The set

(1) $$\cdots D\,D\,D\,D\,D \cdots$$

is, in the first place, a normal set C. For D is made up of sensed normal segments in which neither the first and last members, nor any two successive members are the same normal segment taken in opposite senses. Further D is a generating set of the set (1), for otherwise (1) would have a period smaller than the number of successive segments in D, and hence a period that is a divisor of the number of successive segments in D. D could then be obtained by a finite number of repetitions of a similar set containing fewer sensed normal segments, contrary to the last hypothesis made concerning D.

The number of different periodic normal sets C is seen to equal the number of generating sets not obtainable one from the other by a circular permutation of their normal segments. The number of such generating sets is readily seen to be an enumerable infinity. From this result, together with the theorem of the preceding section, we have the result given by Hadamard:

There are an enumerable infinity of distinct closed geodesics on the surface S.

Limit geodesics of sets of geodesics

§ 11. Definition. A geodesic G will be said to be a *limit geodesic of a set of geodesics* if a set of elements, M, lying on the given set of geodesics, have as a limit an element E, on G, while all the initial points of those elements of the set M that lie on G, are at distances, measured along G from the initial point of E, exceeding a fixed positive quantity.

From the property of continuous variation of a geodesic with its initial element, as given in Theorem 5, § 7, it follows that if one element on G is a limit element of elements on a given set of geodesics, then every element on G is a limit element of elements on the given set of geodesics.

If a closed geodesic should be considered as replaced by an unclosed geodesic that traces out the given closed geodesic an infinite number of times in either sense, the latter geodesic would be a limit geodesic of itself. In this sense any closed geodesic will be considered a limit geodesic of itself.

From Theorem 6, § 7, it follows that a necessary and sufficient condition that a geodesic G be a limit geodesic of a set of geodesics J, not including G, is that every finite segment of the unending reduced curve corresponding to G be contained in the unending reduced curve corresponding to some geodesic of the set J. In terms of normal sets C, this result becomes the following:

Theorem 8. *A necessary and sufficient condition that a geodesic G be a limit geodesic of a set of geodesics J, not including G, is that every subset of consecutive normal segments of the normal set representing G, be a subset of consecutive normal segments of some normal set representing a geodesic of the set J.*

The following theorem is given by Hadamard, with a proof, however, that is different from the following.

Theorem 9. *Every geodesic lying wholly on a surface of negative curvature for which* $2p + v - 1 \geqq 2$ (cf. section 3), *is a limit geodesic of the set of all closed geodesics on that surface.*

The number of different normal segments equals $2p + v - 1$. Hence on any of the surfaces considered, there are at least two different normal segments. Since any closed geodesic is a limit geodesic of itself, we need only consider the case of a geodesic not a closed geodesic. Let G be any geodesic lying wholly on S, and not a closed geodesic. Let there be given an arbitrary finite subset of consecutive normal segments of the normal set C representing G. If this subset does not begin and end with the same normal segment taken in opposite senses, we denote the subset by D; in the other case we add to the given subset a normal segment different from the first and last normal segment, and denote this set also by D.

In either case,

$$\cdots D\,D\,D\,D\,D\,D\,D \cdots$$

will be a normal set C. This normal set is periodic; according to Theorem 7, § 9, it then represents a closed geodesic. Further this normal set contains as a subset of successive normal segments the given arbitrary subset of the normal set representing G. From Theorem 8, it accordingly follows that G is a limit geodesic of the set of all closed geodesic on S, and the theorem is proved.

THEOREM 10. *On a surface of negative curvature for which* $2p + v - 1 \geqq 2$, *there exists at least one geodesic which has for a limit geodesic every geodesic lying wholly on* S.

The set of all possible finite subsets of consecutive normal segments of normal sets C, form an enumerable set which may accordingly be put into one to one correspondence with the set of all integers, positive, negative, or zero. In this correspondence that one of these subsets that corresponds to the integer n, we denote by B_n. The set,

$$\cdots B_{-2}\, B_{-1}\, B_0\, B_1\, B_2 \cdots$$

will be a normal set C, unless for some integer n, the last sensed normal segment of B_{n-1} and the first sensed normal segment of B_n are the same normal segments taken in opposite senses. In every such case we insert between B_{n-1} and B_n a normal segment different from the normal segment in question. The resulting set will be a normal set, which we denote by C'.

C' contains each subset B_n as a subset of consecutive normal segments. It follows from Theorem 8, that every geodesic lying wholly on S, with the possible exception of the geodesic represented by C', is a limit geodesic of the geodesic represented by C'. That the geodesic represented by C' is a limit geodesic of itself, follows from the fact that every *closed* geodesic is a limit geodesic of the geodesic represented by C', while *every* geodesic lying wholly on S is a limit geodesic of the set of all closed geodesics on S.

Recurrent geodesics

§ 12. The following definition, and Theorems 11, 12, and 13, are restatements for the case of geodesics of what is given by Professor Birkhoff for a dynamical system, in the paper referred to in the introduction.

Definition. By a *minimal set* of geodesics we shall understand any set of geodesics lying wholly on S, each of which has every other geodesic of the set, and no other geodesic, as a limit geodesic. Any geodesic of a minimal set will be called a *recurrent* geodesic.

A closed geodesic constitutes a minimal set in which it is the only geodesic.

The following theorem serves as an existence proof for recurrent geodesics.

Theorem 11. *Every geodesic lying wholly on S contains among its limit geodesics at least one minimal set of geodesics.*

Concerning the number of recurrent geodesics in a minimal set, we have

Theorem 12. *The power of any minimal set not simply a closed geodesic, is that of the continuum.*

The characteristic property of a recurrent geodesic is given by the following:

Theorem 13. *A necessary and sufficient condition that a geodesic lying wholly on S be a recurrent geodesic is that, corresponding to any arbitrary positive constant e, there exist a positive constant h, so large, that if L be any segment of the given geodesic of length at least equal to h, any element of the given geodesic lies within e of some element of L.*

§ 13. Let there be given a set of symbols of the form,

$$\cdots R_{-2}\, R_{-1}\, R_0\, R_1\, R_2 \cdots . \tag{1}$$

Let m and n be any integers, positive, negative, or zero.

Definition. I. A set of symbols of the form (1) will be said to be *recurrent*, if corresponding to any positive integer r, there exists a positive integer s, so large that any subset of (1) of the form,

$$R_m\, R_{m+1} \cdots R_{m+r} \tag{2}$$

is contained in every subset of (1) of the form

$$R_n\, R_{n+1} \cdots R_{n+s}. \tag{3}$$

II. The set (1) will be said to be *periodic*, if there exists a positive integer p, such that

$$R_n = R_{n+p},$$

whatever integer n may be, and p will be said to be a *period* of the set (1).

It appears at once that a set (1) that is periodic, is also recurrent.

Theorem 13, § 12, interpreted in terms of normal sets C by means of Theorem 6, § 7, becomes the following

THEOREM 14. *A necessary and sufficient condition that a geodesic lying wholly on S be recurrent, is that the set C representing the given geodesic be recurrent.*

EXISTENCE OF RECURRENT GEODESICS, NOT PERIODIC.

§ 14. We come now to the question of the existence of recurrent geodesics that are not closed geodesics.

On a surface of negative curvature topographically equivalent to an unbounded circular cylinder, the only possible recurrent geodesic is a single closed geodesic. On a surface of negative curvature topographically equivalent to an unbounded plane, there are no recurrent geodesics whatever. The surfaces of negative curvature which we have been considering include neither of these two types of surfaces (cf. § 3).

We have seen in Theorem 7, § 9, that a geodesic that is periodic is represented by a normal set C that is periodic; while Theorem 14, § 13, states that a normal set C that is recurrent represents a geodesic that is recurrent. Hence, to prove the existence of a geodesic that is recurrent without being periodic, it is sufficient to prove the existence of a normal set C that is recurrent without being periodic.

Now there are just $2p + v - 1$ normal segments (cf. § 8). We are considering surfaces for which $2p + v - 1 \geqq 2$. Hence any of the surfaces of negative curvature considered will possess at least two normal segments. We will seek a normal set C that is composed solely of two normal segments. For that purpose the following lemma is introduced.

LEMMA. *There exists an unending set of symbols each of which is either* 1 *or* 2, *which forms a set that is recurrent without being periodic.*

By the juxtaposition of two or more symbols representing ordered sets of symbols, we shall mean here, as elsewhere, the ordered set obtained by taking the symbols of the given sets in the order in which the sets are written.

Let n be any positive integer. We introduce the following definitions:

$$
\begin{aligned}
a_0 &= 1, \\
b_0 &= 2, \\
a_1 &= a_0 \, b_0, \\
b_1 &= b_0 \, a_0, \\
&\cdots \\
a_{n+1} &= a_n \, b_n, \\
b_{n+1} &= b_n \, a_n .
\end{aligned}
\tag{1}
$$

We introduce the set of symbols,

$$\cdots d_{-2}\, d_{-1}\, d_0\, d_1\, d_2 \cdots, \tag{2}$$

of which

$$d_0\, d_1 \cdots d_{2^n}$$

are defined respectively as the 2^n integers of a_n; further, if m is any positive integer, d_{-m} is defined as equal to d_{m-1}. The set (2), so defined, will be proved to be recurrent without being periodic.

For definiteness we write out (2) in part, beginning with d_0:

$$1221\ 2112 \quad 2112\ 1221 \quad 2112\ 1221 \quad \cdots. \tag{3}$$

It follows from the definitions (1), that if the integers of (2) be grouped in groups of 2^n integers, then the set (2) can be expressed, beginning with d_0, by a succession of the sets a_n and b_n, obtained by replacing the integers 1 and 2 in the set (2), respectively by a_n and b_n. Thus beginning with d_0, (2) is given in part as

$$a_n\, b_n\, b_n\, a_n \quad b_n\, a_n\, a_n\, b_n \quad b_n\, a_n\, a_n\, b_n \quad \cdots. \tag{4}$$

The symbols of the set (2) that have negative subscripts, can be obtained, according to their definition, by taking the symbols of (2) with positive or zero subscripts in reverse order. It follows from the definitions (1) that the integers of a_n and b_n, taken respectively in their reverse orders, give a_n and b_n when n is even, and b_n and a_n when n is odd. We have the result:

Whatever integer n may be, the set (2) *can be expressed by a properly chosen succession of the sets, a_n and b_n. Thus, if r be any integer such that*

$$r \equiv 0 \text{ modulo } 2^n,$$

then any subset of (2) *of the form*

$$d_r\, d_{r+1} \cdots d_{r+2^n}$$

is either a set a_n or a set b_n.

We will now prove that the set (2) is recurrent.

Let there be given any subset of (2) of the form

$$d_s\, d_{s+1} \cdots d_{s+m}, \tag{5}$$

where s is any integer, positive, negative, or zero, and m is any positive integer. Let r' be the largest integer less than s such that

$$r' \equiv 0 \text{ modulo } 2^m.$$

From the choice of r', we have,

$$r' < s < s + m < r' + 2^{m+1}.$$

Hence the set (5) is a subset of the set,

$$d_{r'}\, d_{r'+1} \cdots d_{r'+2^{m+1}}. \tag{6}$$

From the result of the preceding paragraph, it appears that (6) must be one of the four possible ordered combinations of a_m and b_m, that is, one of the four sets,

$$a_m\, b_m, \quad b_m\, b_m, \quad b_m\, a_m, \quad a_m\, a_m.$$

Each of these four sets is a subset of a_{m+3} and b_{m+3}; for from the equations (1), we have,

$$a_{m+3} = a_{m+2}\, b_{m+2} = a_{m+1}\, b_{m+1}\, b_{m+1}\, a_{m+1} = a_m\, b_m\, b_m\, a_m\, b_m\, a_m\, a_m\, b_m,$$

$$b_{m+3} = b_{m+2}\, a_{m+2} = b_{m+1}\, a_{m+1}\, a_{m+1}\, b_{m+1} = b_m\, a_m\, a_m\, b_m\, a_m\, b_m\, b_m\, a_m.$$

Since the set (2) can be expressed as a succession of the sets a_{m+3} and b_{m+3}, each of which contains 2^{m+3} integers of the set (2), it appears that any subset of at least 2^{m+4} successive integers of (2), say R, contains at least one of the sets a_{m+3} and b_{m+3}. Retracing the steps it is seen that R contains a subset identical with the given set (5). The set (2) is thus recurrent.

We will now show that the set (2) is not periodic. Suppose that the set (2) had a period prime to 2. Since p is prime to 2, there exists an integer m, greater than one, such that

$$2 \equiv 2^m, \qquad \text{modulo } p. \tag{7}$$

Since the set (2) has the period p, it follows from (7) that the set

$$d_2\, d_3\, d_4 \cdots \tag{8}$$

must be identical with the set

$$d_{2^m}\, d_{2^m+1}\, d_{2^m+2} \cdots. \tag{9}$$

The set (8) commences with the integers

$$2\,1\,2\,1\,1\,2 \cdots, \tag{10}$$

while the set (9) commences as does b_m, which is seen from the equations (1) to commence with the integers

$$2\,1\,1\,2\,1\,2 \cdots. \tag{11}$$

The sets (10) and (11) are not identical. The set (2) can thus have no period prime to p.

Finally suppose that the set (2) had a period $2^r\, p$, where r is any positive integer, and p is prime to 2. Let the set (2), commencing with d_0, be written in terms of a_r and b_r:

$$a_r\, b_r\, b_r\, a_r\, b_r\, a_r \cdots. \tag{12}$$

Considered as a succession of symbols a_r and b_r, (12) has the period p. But the original expression for (2) in terms of its integers, and commencing with d_0, is obtained from (12) by replacing the symbols a_r and b_r respectively by 1 and 2. Thus the expression for the set (2), in terms of its integers, and commencing with d_0, would have a period prime to 2. We have seen this to be impossible.

Thus the set (2) is recurrent without being periodic, and the lemma is proved.

§ 15. THEOREM 15. *On a surface of negative curvature for which*

$$2p + v - 1 \geqq 2,$$

there exists a set of geodesics that are recurrent without being periodic, and this set has the power of the continuum.

The number of different normal segments equals $2p + v - 1$. We are considering surfaces of negative curvature for which $2p + v - 1 \geqq 2$. Hence on any of the surfaces considered, there are at least two different normal segments. Let N_1 and N_2 be two different normal segments, each taken in an arbitrary sense.

In the preceding lemma we have established the existence of a set,

$$\cdots d_{-2}\, d_{-1}\, d_0\, d_1\, d_2 \cdots,$$

that is composed entirely of the integers one and two, and which is a set that is recurrent without being periodic. The set

$$\cdots N_{d_{-2}}\, N_{d_{-1}}\, N_{d_0}\, N_{d_1}\, N_{d_2} \cdots \tag{1}$$

is accordingly recurrent; from Theorem 14, § 13, it follows that the geodesic represented by the set (1) is recurrent. The set (1) is not periodic; it follows from Theorem 7, § 9, that the geodesic represented by (1) is not periodic. We have thus established the existence of a geodesic that is recurrent without being periodic.

According to Theorem 12, § 12, the existence of one geodesic that is recurrent without being periodic, is sufficient to establish that the power of the complete set of geodesics that are recurrent without being periodic is that of the continuum.

§ 16. THEOREM 16. *On a surface of negative curvature for which*

$$2p + v - 1 \geqq 2,$$

the set of all geodesics that are recurrent without being periodic, has as a limit geodesic every geodesic lying wholly on S.

Let there be given an arbitrary closed geodesic lying on S. Let B be any finite subset of successive normal segments of the normal set C representing the given closed geodesic. Let N_1 and N_2 be the two sensed normal segments used in the proof of the preceding theorem, and $-N_1$ and $-N_2$ be, respectively, the same normal segments taken in opposite senses.

If now the set B does not begin or end with $-N_1$, or $-N_2$, we denote the set B, by D. If the set B begins with $-N_1$ or $-N_2$, we prefix N_2 or N_1, respectively, to the set B, while if the set B ends with $-N_1$ or $-N_2$, we add N_2 or N_1, respectively to the set B, and in either case denote the resulting set by D. We interpose this set D between each two successive sensed normal segments of the normal set C, given by (1) in the proof of the preceding theorem, and denote the resulting set by C'.

It is a consequence of the nature of the construction of the set C', that no two of its successive sensed normal segments are the same normal segment taken in opposite senses. The set C' is thus a normal set. The normal set (1) of the proof of the preceding theorem, is recurrent without being periodic; it follows that the set C' is recurrent without being periodic. The geodesic represented by C' is accordingly recurrent without being periodic. The set C' contains B as a subset of successive normal segments. It follows from Theorem 8, § 11, that the given closed geodesic is a limit geodesic of the set of all recurrent geodesics that are not periodic.

That every geodesic lying wholly on S is a limit geodesic of the set of all geodesic that are recurrent without being periodic, follows now from the fact that *every* geodesic lying wholly on S is a limit geodesic of the set of all closed geodesics on S.

Distribution of elements on recurrent geodesics

§ 17. Definition. Two elements E' and E'' of a set of elements M on a region R of S, will be said to be *mutually accessible in M and on R*, if corresponding to any positive constant e, there exists in the set M, a finite ordered subset of elements of which the first is E', and the last E'', while each element of the subset, excepting the last, lies within a geodesic distance e, measured on R, of the following element.

The following theorem is established in § 25 of the earlier paper on geodesics.

Theorem 17. *On any simply connected region R of S, and in the set of all elements on R, and on geodesics lying wholly on S, no two elements on different geodesics are mutually accessible.*

A particular consequence of the preceding theorem is that, on any simply connected region R of S, and in the set of all elements on R, and on geodesics that are *recurrent*, no two elements on different geodesics are mutually accessible. A set of recurrent geodesics with this property are of a type called *discontinuous* recurrent motions by Professor Birkhoff. Thus:

Theorem 18. *The set of all recurrent geodesics on S constitutes a set of recurrent motions of the discontinuous type.*

The proof, given in this paper, of the existence of a set of this type, is the first proof of the existence of a discontinuous set of recurrent motions.

§ 18. A recurrent geodesic was defined as a member of a minimal set,—a set in which every geodesic has every geodesic of the set, and no other geodesic, as a limit geodesic. In case a given recurrent geodesic is a closed geodesic, the minimal set containing the given geodesic consists merely of the given closed geodesic. In case a recurrent geodesic is not a closed geodesic, the power of the minimal set that contains the given recurrent geodesic, is, according to Theorem 12, § 12, that of the continuum.

From the definition of a minimal set, it appears that no two minimal sets that are not identical, have any geodesic in common. Each recurrent geodesic thus belongs to one and only one minimal set. The question arises as to how many different minimal sets there are on the given surface. That there are at least an enumerable infinity, follows at once from the fact that there are an enumerable infinity of closed geodesics. The number of minimal sets that do not consist simply of one closed geodesic still remains to be determined.

It has been seen that any geodesic lying wholly on S can be completely characterized by means of normal sets of sensed normal segments. It may be inquired whether or not any minimal set may not be characterized in terms of sensed normal segments, and if so, what is the explicit nature of the characterization. These questions seem to indicate the opening to an interesting field of inquiry.

Harvard University,
June, 1917

A FUNDAMENTAL CLASS OF GEODESICS ON ANY CLOSED SURFACE OF GENUS GREATER THAN ONE*

BY

HAROLD MARSTON MORSE

Introduction

1. The study of geodesics on closed surfaces is influenced largely by the genus of the surface. Of closed surfaces of genus zero an important class has been studied by H. Poincaré,† namely, those closed surfaces which are everywhere convex. Of the surfaces with genus one, the torus has been studied by G. A. Bliss,‡ and it seems probable that the types of geodesics found there will be found among the geodesics on any closed surfaces of genus one. It is the object of this paper to consider geodesics on any closed surface of genus greater than one.

In studying geodesics on a surface of negative curvature, the author§ found that, of those geodesics which if extended indefinitely in either sense remained in a finite part of space, any particular one could be characterized in a manner which depended only upon a succession of fundamental contours of the surface. Now these surfaces of negative curvature are never closed. The question accordingly arose, is it possible also on closed surfaces, to characterize the geodesics in terms of the topographical elements of the surface, that is to characterize the geodesics of any particular closed surface in terms which would serve likewise for any other closed surface of the same genus? The answer to this question was readily seen to be no. However, on surfaces of genus $p > 1$, there appeared a fundamental class of geodesics which could be identified once and for all for all closed surfaces of the same genus. This paper is devoted to the definition and study of such a fundamental class of geodesics.

The surface is first mapped upon a hyperbolic non-euclidean plane. In the theory as developed each non-euclidean straight line represents a different

* Presented to the Society, September 8, 1921.

† *Sur les lignes géodésiques des surfaces convexes,* these Transactions, vol. 6 (1905), p. 237.

‡ *The geodesic lines on the anchor ring,* Annals of Mathematics, ser. 2, vol. 4 (1903), p. 1.

§ *Recurrent geodesics on a surface of negative curvature,* these Transactions, vol. 22 (1921), p. 84.

type of fundamental geodesic and it is with the aid of this representation that the fundamental class of geodesics is studied.

THE SURFACE

2. The surface defined. The surfaces to be considered are to be two-sided, closed surfaces of finite genus p, p greater than unity. They are to be without singularities. More specifically we will suppose that the points in the neighborhood of any point of the surface can be put into one to one continuous correspondence with the points in the neighborhood of some point in a plane, in such a manner, that for the neighborhoods considered, the cartesian coördinates, x, y, and z, of a point of the surface, be continuous functions of the cartesian coördinates, u, v, of the plane, provided with continuous partial derivatives up to the fourth order, while further

$$\left[\frac{D(xy)}{D(uv)}\right]^2+\left[\frac{D(xz)}{D(uv)}\right]^2+\left[\frac{D(yz)}{D(uv)}\right]^2 \neq 0.$$

The given surface will eventually be mapped upon the interior of a unit circle, in a manner that will be one to one and continuous as far as the neighborhoods of any two corresponding points are concerned, but by a correspondence that will be one to infinity as a whole. With that end in view there will now be defined a special group of linear transformations of a complex variable carrying the interior of the unit circle into itself. The method of definition will be similar to the method used by Poincaré in the article cited below.* From the "fundamental domain" of this group will be formed a canonical surface which can be mapped on the interior of the unit circle in the desired manner, and with the aid of which the given surface can also be so mapped.

3. The polygon S_0. Let there be given a unit circle. In the plane of this circle let there be drawn a second circle concentric with the unit circle and with a radius r greater than one. Let p be a positive integer greater than one. On the circle of radius r let there be placed $4p$ equidistant points. With these points as centers let there be drawn $4p$ circles orthogonal to the unit circle. Now let r and the $4p$ circles vary, the $4p$ circles still remaining orthogonal to the unit circle and still having their centers equidistant on the circle of radius r. As r becomes arbitrarily large the $4p$ circles will approach straight lines passing through the center of the unit circle, and such that each makes an angle of $\pi - 2\pi/4p$ with its successor or predecessor. For very large values of r there will thus be formed a curvilinear polygon, lying

* *Théorie des groupes fuchsiens*, Acta Mathematica, vol. 1 (1882), p. 1.

within the unit circle, containing the center of the unit circle, and absolutely symmetrical with respect to the center of the unit circle (cf. Fig. 1). If r now decrease from very large values to small values, for some value of r the $4p$ circles will become tangent to each other. That is, the interior angles of the curvilinear polygon will diminish from $\pi - 2\pi/4p$ to zero. But for $p > 1$

$$\pi - \frac{2\pi}{4p} > \frac{2\pi}{4p} = \frac{\pi}{2p}$$

so that for a properly chosen value of r the curvilinear polygon will have interior angles of the magnitude of $\pi/2p$. The sides of this polygon will all be segments of circles orthogonal to the unit circle and will be equal in length. The polygon will contain the center of the unit circle and be absolutely symmetrical with respect to that center. The sum of the interior angles of the polygon will be 2π. This polygon will be denoted by S_0 and the corresponding value of r by r_0. (Cf. Fig. 1, for the case $p = 2$.)

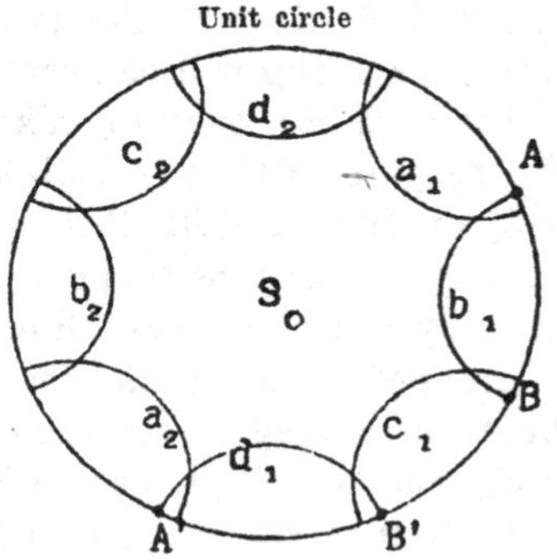

Fig. 1.

A property of S_0 to be used later is that *among the* $4p$ *circles bounding* S_0, *alternating circles do not meet.* Suppose they did meet. If r and the common radius of the $4p$ circles should then be increased, the $4p$ circles still remaining orthogonal to the unit circle, the alternating circles would still continue to meet. Now the interior angles of S_0 have a magnitude $\pi/2p$ which is less than $\pi/2$ for $p > 1$. On the other hand if r should be increased beyond all limit the interior angles of the polygon thereby formed, say S', would approach $\pi - 2\pi/4p$ which is greater than $\pi/2$ for $p > 1$. Thus for some value of $r > r_0$ the interior angles of S' would be right angles. Thus of any three successive circles bounding S', the first and last would be orthogonal to the second and at the same time would be orthogonal to the unit circle. This is impossible. For two circles orthogonal to two non-tangent circles which meet, cannot themselves meet.

4. **A group with principal circle.** The polygon S_0 will be used to define a "group with principal circle."* The $4p$ sides of S_0 will be taken in one of their two circular orders and labelled as follows:

$$a_1\, b_1\, c_1\, d_1\, a_2\, b_2\, c_2\, d_2, \cdots, a_p\, b_p\, c_p\, d_p.$$

* H. Poincaré, *Théorie des groupes fuchsiens*, loc. cit. Also L. R. Ford, *An introduction to the theory of automorphic functions*, Edinburgh Mathematical Tracts.

3*

No confusion need arise if the circles bearing these sides be indicated by the same letters. We will term a_k conjugate to c_k, and b_k conjugate to d_k, where k ranges over the integers from 1 to p. Now a reflection of the plane in the radical axis of a_k and c_k will carry S_0 into itself and carry a_k into its conjugate side c_k. On the other hand a reflection of the plane in c_k will carry c_k into itself and S_0 into an adjacent polygon bordering S_0 along c_k. The product of these two reflections will be a transformation expressible as a linear transformation of a complex variable, and will carry the side a_k into the side c_k, and S_0 into a polygon bordering S_0 along c_k. Similarly there exist linear transformations of a complex variable which will carry any one of the sides of S_0 into its conjugate, and S_0 into a polygon bordering S_0 along the second of these conjugate sides. These transformations and all possible combinations of them will form a group G. All transformations of this group will carry the unit circle into itself. Points and regions which are the images of each other under this group will be termed *congruent*.

It is here necessary to make a convention, that just one side of each pair of conjugate sides of S_0 shall be considered as belonging to S_0, and just one of S_0's vertices. With this understood it may be stated that the set of all curvilinear polygons congruent to S_0 cover the interior of the unit circle once and only once. About each of the vertices of each of the polygons there will be grouped $4p$ polygons each with one corner at the given vertex. Each of these corner regions at such vertices will be congruent to just one corner in S_0, and conversely each corner of S_0 will be congruent to just one corner at each such vertex. For the proofs of the results of this paragraph the reader is referred to the literature on the subject.

Nature of the fundamental group

5. Distances between congruent points. The circles orthogonal to the unit circle may be looked upon as representing the straight lines of a hyperbolic non-euclidean (written N E) geometry in which the points of the hyperbolic plane consist of the points interior to the unit circle. If we suppose the unit circle has its center at the origin of the euclidean plane, the N E length of any rectifiable curve interior to the unit circle is taken as the value of the integral

$$\int \frac{2\,ds}{1-x^2-y^2},$$

taken along the given curve, and the N E distance between any two points within the unit circle as the N E length of the N E straight line joining the two given points. It is possible to prove directly that the N E length and distance as defined by the above integral remain unchanged under any linear transformation of the complex variable that carries the interior of the unit circle into itself.

The N E *distance between any two congruent points has a lower limit not zero.*

It follows from the explicit nature (cf. § 4) of the transformations which carry S_0 into the polygons which are contiguous to S_0, that in a sufficiently small neighborhood of any point of S_0, or of its boundary, there are no pairs of mutually congruent points. Now if the lemma were false, it would be possible to pick out an enumerable infinity of pairs of congruent points whose N E distances would approach zero as the enumerating integer n became infinite. Each of these pairs could be replaced by a congruent pair of which one member at least might be taken so as to lie in S_0 or on its boundary. N E distances between the two points of any pair would be the same as before. These points in S_0 would have at least one limit point in S_0 or on its boundary. In every neighborhood of this limit point there could be found points which would be mutually congruent. This, as just stated, is impossible and the lemma is proved.

6. Nature of the group. *The group can contain no elliptic transformations.* An elliptic transformation carrying the unit circle into itself would have one fixed point within the unit circle and one without. Now there are in every neighborhood of a fixed point of an elliptic transformation points which are mutually congruent under the transformation. This is here impossible according to the result of the preceding section.

The group can contain no parabolic transformations.

To establish this fact it will be convenient to carry the interior of the unit circle into the upper half plane by means of a linear transformation of the complex variable. The integral

$$\int \frac{2\,ds}{1-x^2-y^2}$$

will go over into the integral

$$\int \frac{ds}{y}.$$

If now the points above the axis of reals be considered as the points of a hyperbolic N E plane, the circles orthogonal to the axis of reals as the straight lines of this geometry, and the value of the integral $\int (1/y)\,ds$ taken along any curve segment in the upper half plane as the N E length of that curve segment, then any curve segment within the unit circle may be considered as unchanged in N E length by the transformation from the unit circle to the upper half plane.

Corresponding to the fundamental group carrying the unit circle into itself there will exist a transformed group carrying the upper half plane into itself. If the first group had a parabolic transformation the transformed group would have one also. Now any parabolic transformation, say T, which carried the axis of reals into itself would have its fixed point on the axis of reals. Suppose now the upper half plane be carried into itself and thereby the group be still further transformed in such a manner that the fixed point of the parabolic transformation T' goes into the point at infinity. N E distances would thereby be unchanged but the original parabolic transformation would now be transformed into a translation parallel to the axis of reals through a euclidean distance, say d. A point, no matter how remote from the axis of reals, would still be carried by this parabolic transformation through this euclidean distance d. If sufficiently remote, however, from the axis of reals, the N E distance between two such congruent points is seen with the aid of the integral

$$\int \frac{ds}{y}$$

to be arbitrarily small. This is contrary to the result of the preceding section. Thus the group can have no parabolic transformations.

Now the only linear transformations of a complex variable which carry the interior of a circle into itself are either elliptic, parabolic, or hyperbolic. *Hence all the transformations of the group are of the hyperbolic type.*

7. A lemma on polygons near the unit circle. The N E distance between any two points of S_0 has an upper limit. The same upper limit necessarily holds for distances between any two points on any one of the congruent polygons. Accordingly a polygon which has any point sufficiently near the unit circle will be arbitrarily small in its euclidean dimensions, for otherwise its N E dimensions as measured by the N E integral

$$\int \frac{2\,ds}{1-x^2-y^2}$$

would become arbitrarily large. *The polygons accordingly cluster about each point of the unit circle.*

The following lemma will now be proved.

LEMMA. *If a polygon lie sufficiently near the unit circle, all of its bounding circles, with the possible exception of one of them, will be arbitrarily small.*

Denote by O the center of the unit circle. Let A and B be two points on the unit circle at the extremities of a diameter. Let d be a small positive constant. Consider all the radii through O, except those that make an angle with OA numerically less than d. Now let O move along AB towards B. At the same time let each of the radii not excepted be replaced by an arc of a circle orthogonal to the unit circle, and drawn from O to the unit circle so as to make the same angle with OA as the radius to be replaced. Denote this pencil of circles through O by C. If O move along AB to a point sufficiently near B, it is obvious that the circles of the pencil C just described will become arbitrarily small.

To prove the lemma consider the $8p$ radii which pass from the center O to the points on the unit circle where the circles bounding the polygon S_0 (cf. § 3) meet the unit circle. Denote this set of radii by R. Let d, the constant of the preceding paragraph, here be a positive quantity less than half the magnitude of the angle between any two successive radii of the set R. From the result of the preceding paragraph it follows that if O and S_0 be carried by a transformation of the group respectively into a point O' and a new polygon sufficiently near the unit circle, then the set of circular segments, say C', into which the radii of the set R are transformed will all be arbitrarily small *except* any of them which make an angle with $O'O$ of magnitude less than d. But from the choice of d there will be at most one circle of C' making an angle with $O'O$ of magnitude less than d. Now the circles bounding the transformed polygon and the circles of the set C' will intersect the unit circle in the same points, and accordingly, with the possible exception of one of them, each of the circles bounding the transformed polygon will meet the unit circle in two arbitrarily near points, and being orthogonal to the unit circle will consequently be arbitrarily small.

8. **Fixed points of the transformations of the group.** The distribution of the fixed points of the transformations of the group will now be considered.

(*a*). Let T be a transformation of the group. T is hyperbolic, and accordingly has two fixed points on the unit circle. Let A and B be two points on the unit circle not the fixed points of T, and let A' and B' be respectively the images of A and B under T. It follows from the nature of a hyperbolic transformation that in case the fixed points of T separate A and B, $ABA'B'$ will appear on the unit circle *in the circular order* $ABB'A'$, and that in the contrary case they will appear *in the order* $ABA'B'$ or else *in the order* $AA'BB'$.

(*b*). Consider any two conjugate circles of the circles bounding the polygon S_0. Suppose the first of these two circles meets the unit circle in the points A and B and the second of the two circles meets the unit circle in the points A' and B'

where A is congruent to A' and B to B' (cf. Fig. 1). According to the result of § 3 no two conjugate circles intersect. A reference to the explicit description of the transformations which carry any circle bounding S_0 into its conjugate (cf. § 4) will now show that *$ABB'A'$ appear on the unit circle in the order $ABB'A'$*. After subjection to any other transformation of the group the circular order of these four points will still be the same.

With the aid of the results of the preceding paragraphs (a) and (b) the following lemma can be proved. This lemma will be used to show that among the geodesics considered the types that correspond to closed geodesics are everywhere dense in a sense to be explained later.

LEMMA. *There exists a transformation of the group which has fixed points arbitrarily near the end points of any preassigned arc of the unit circle.*

Let M_1 and M_2 be the end points of a preassigned arc of the unit circle. Let e be an arbitrarily small positive constant less than half the distance between M_1 and M_2. According to the lemma of § 7, it is possible to choose a polygon S_1 so near M_1 that all of its bounding circles with the possible exception of one of them will lie within e of M_1. Let S_2 be a second polygon similarly chosen relative to M_2. Since $p > 1$ and the number of pairs of conjugate sides of a polygon is $2p$, S_1 and S_2 each have at least four pairs of conjugate sides, and on both S_1 and S_2 at least three of these pairs of conjugate sides lie entirely within e of M_1 and M_2 respectively. It is accordingly possible to choose a side of S_1 such that its intersections with the unit circle, say A_1 and B_1, and the intersections of its conjugate side with the unit circle, say A_1' and B_1', lie within e of M_1, while at the same time the four points $A_2 B_2 A_2' B_2'$ corresponding to $A_1 B_1 A_1' B_1'$ under the transformation of the group that carries S_1 into S_2 also lie within e of M_2.

According to the choice of e, the e neighborhood of M_1 is entirely distinct from the e neighborhood of M_2. Accordingly $A_1 B_1 A_2 B_2$ have either the circular order $A_1 B_1 B_2 A_2$ or else the circular order $A_1 B_1 A_2 B_2$. In the first case it follows from the result of paragraph (a) that the transformation of the group that carries $A_1 B_1$ into $A_2 B_2$ will have one of its fixed points on the arbitrarily small arc between A_1 and B_1 and the other fixed point on the small arc between A_2 and B_2, and the theorem is proved for that case.

Consider the second case. Here $A_1 B_1 A_2 B_2$ appear in the order $A_1 B_1 A_2 B_2$. According to the result of paragraph (b) $A_1 B_1 B_1' A_1'$ always appear in the circular order $A_1 B_1 B_1' A_1'$. It follows that in this second case $A_2 B_2 B_1' A_1'$ appear in the order $A_2 B_2 B_1' A_1'$. Using again the result of paragraph (a), the transformation of the group which carries $A_2 B_2$ into $A_1' B_1'$ is seen to be one which has one fixed point on the small arc between A_2 and B_2 and the other fixed point on the small arc between A_1' and B_1', and the lemma is proved for the second case as well.

THE GIVEN SURFACE MAPPED ON THE UNIT CIRCLE

9. Definition of the correspondence. Let a closed surface be formed from S_0 by continuously deforming S_0 so as to bring congruent points of conjugate sides into coincidence. S_0 has $2p$ pairs of conjugate sides. Hence the resulting closed surface, say T, will be of genus p. Each point P of T will arise from one point P_0 of S_0. The correspondence between T and S_0 can be extended over the interior of the unit circle by requiring that P shall not only correspond to P_0, but also to all points of the unit circle congruent to P_0. The correspondence thereby established between the interior of the unit circle and T is everywhere continuous and one to one as far as the neighborhoods of corresponding points are concerned, but one to infinity as a whole.

Now by a fundamental theorem of analysis situs any two closed two-sided surfaces of the same genus can be put into one to one continuous correspondence. The original surface, say Σ (cf. § 2), can then be put into one to one continuous correspondence with that one of the above surfaces T which has the same genus as does Σ. The surface Σ will thereby be mapped on the interior of the unit circle in a manner essentially the same as the manner in which T has just been mapped upon the interior of the unit circle.

10. Surface distances and N E distances. In contradistinction to NE lengths and distances, surface lengths and distances will now be defined.

DEFINITION. If a curve segment h within the unit circle corresponds on the original surface to a curve segment k that is rectifiable, the ordinary length of k will be termed the *surface length of* h. Before defining the surface distance between two points within the unit circle the following lemma is needed. *Of the curves within the unit circle joining any two non-coincident points, there is at least one which corresponds on the original surface to a geodesic segment whose surface length is at least as small as that of any other curve joining the two given points.* This lemma is essentially equivalent for the case of geodesics to Hilbert's *a priori* existence theorem of the calculus of variations.* A method which can be used without any essential change to prove this lemma is given by G. D. Birkhoff† for a similar problem in the article cited below.

DEFINITION. The lower limit of the surface lengths of curves joining two given points within the unit circle will be termed the *surface distance* between these two points. The surface distance between two given points will be denoted by D_g while the N E distance between the same two points will be denoted by D_n.

---

* Cf. Bolza, *Vorlesungen über Variationsrechnung,* 1909, p. 428.

† *Dynamical systems with two degrees of freedom,* these Transactions, vol. 18 (1917), § 9, p. 219.

LEMMA 1. *Corresponding to a positive constant a there exists a positive constant b such that any pair of points interior to the unit circle for which*

$$D_g \leqq a$$

will be such that

$$D_n < b.$$

Any pair of points whose surface distance is at most a can be replaced by a congruent pair of which one member lies on S_0 and the other on the closed region consisting of all points whose surface distance from points of S_0 and its boundary is at most a. These new pairs will moreover have the same surface distances as the original pairs. But the N E distances between any two points lying on a closed region are all less than a properly chosen constant which we may take for b. The lemma thus is proved.

LEMMA 2. *There exist constants A and m such that for any pair of points interior to the unit circle, if either*

$$D_n > A,$$

or

$$D_g > A,$$

then

$$D_n < m D_g.$$

Let there be given two points whose surface distance is D_g. Starting from one end of a geodesic of surface length D_g joining the two given points, let there be marked off segments of surface length equal to the constant a of Lemma 1. The line will thereby be divided into not more than

$$\frac{D_g}{a} + 1$$

parts. According to the preceding lemma there is between the end points of each of these parts an N E distance less than the constant b. The N E distance between the end points of the geodesic is accordingly less than

$$b\left[\frac{D_g}{a} + 1\right].$$

That is,

$$D_n < \frac{b}{a} \cdot D_g + b. \tag{1}$$

If then

$$D_g > b \tag{2}$$

(1) becomes

$$D_n < \frac{b}{a} \cdot D_g + D_g$$

or

$$D_n < \left(\frac{b}{a} + 1\right) D_g. \tag{3}$$

Setting

$$\frac{b}{a} + 1 = m$$

we have

$$D_n < m \cdot D_g \tag{4}$$

subject to the condition (2). Now (1) holds unconditionally, and shows that D_g becomes infinite with D_n, so that the condition (2) will be fulfilled if for a sufficiently large constant c

$$D_n > c. \tag{5}$$

As seen now (4) holds if either (2) or (5) holds. Taking A as the larger of the constants b and c, and m as the constant m of (4) the inequalities of the lemma accordingly follow.

It may be observed that there is a large degree of reciprocity between D_n and D_g. In fact the two preceding lemmas and their proofs will remain true if the terms N E distance and surface distance be interchanged together with D_n and D_g. Lemma 2 and the lemma obtained by interchanging D_n and D_g in Lemma 2 may be combined into the following fundamental lemma:

LEMMA 3. *If either the* NE *distance or the surface distance between two points exceeds a properly chosen constant, then the* NE *distance lies between two positive constant multiples of the surface distance, and the surface distance lies between two positive constant multiples of the* NE *distance, where the constants involved are fixed once and for all for the whole unit circle and surface.*

11. Proof of Lemma 7. The question arises, do there exist geodesic segments which recede arbitrarily far from the NE straight lines joining their end points and which still have a surface length as small as that of any other geodesics joining their end points? This question is answered in Lemma 8. This section is devoted to proving Lemma 7 which is the principal aid in proving Lemma 8.

A slight modification of the proof of Lemma 2 will serve to establish the following lemma. The constants b and m of the following lemma are the constants b and m used in Lemmas 1 and 2.

LEMMA 4. *There exist constants b and m such that, if a geodesic segment has a surface length g for which $g > b$, then the end points of this segment can be joined by a curve all of whose points lie within an* NE *distance b of the given geodesic segment, and whose* NE *length, v, is such that*

$$v < mg.$$

LEMMA 5. *Corresponding to a positive constant μ there exists a positive constant R so large that if all the points of a curve segment h are at an* NE *distance R from some* NE *straight line, L, and the end points of h project through* NE *perpendiculars into the end points of a segment of L, say k, of more than unit length, then h has an* NE *length exceeding μ times the* NE *distance between its end points.*

The locus of points at a constant NE distance from L is a circle meeting the unit circle in the same points as does L. The segment h lies on this circle.

To simplify the proof let the interior of the unit circle be carried into the upper half plane by a linear transformation which carries the end points of L respectively into the origin, O, and the point at infinity. This can be done, as in § 6, without altering any NE distances. L will go into an ordinary straight line, L', perpendicular to the axis of reals at the origin (see Fig. 2). The transform of k, say k', will lie on this line. The circle on which h lies will go into an ordinary straight line through the origin making an angle with the axis of reals which we denote by α, measuring the angle from the axis of reals in the counterclockwise direction. We may suppose further that $\alpha < \pi/2$, since that could be brought about without altering any NE distances by a reflection of the upper half plane in L'. The transform of h, say h', will lie on this line, making an angle α with the axis of reals. The NE perpendiculars from the end points of h to those of k will go respectively into two arcs, say a' and b', of two circles both with center at the origin and leading from the end points of h' to those of k'. To compute the NE length of a', b', h', and k', suppose the

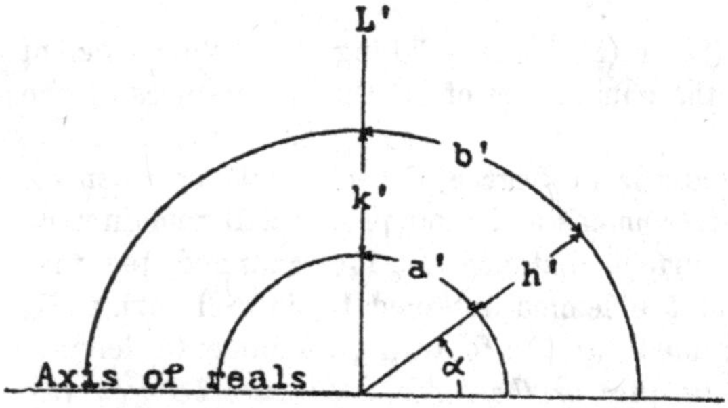

Fig. 2.

plane referred to polar coördinates ϱ and φ with pole at the origin. Denote by ϱ_1 the value of ϱ at the lower end point of k', and by ϱ_2 its value at the upper end point. The polar coördinates of the end points of h' will be $(\varrho_1\ \alpha)$ and $(\varrho_2\ \alpha)$. The NE length of k' as evaluated through the integral $\int (ds/y)$ is

$$(1)\qquad \int_{\rho_1}^{\rho_2} \frac{d\varrho}{\varrho} = \log\frac{\varrho_2}{\varrho_1}.$$

The NE length of h', along which φ equals the constant α, is

$$(2)\qquad \int_{\rho_1}^{\rho_2} \frac{d\varrho}{\varrho \sin\alpha} = \log\frac{\varrho_2}{\varrho_1}\cdot\frac{1}{\sin\alpha}.$$

The NE lengths of a' and b' are the same, namely

$$(3)\qquad \int_{\alpha}^{\frac{\pi}{2}} \frac{\varrho_1\, d\varphi}{\varrho_1 \sin\varphi} = \log\cot\frac{\alpha}{2}, \qquad 0<\alpha<\frac{\pi}{2}.$$

It is desired to show that if the end points of h' be joined by the broken line $a'\ k'\ b'$, there results a curve whose NE length has a ratio to the NE length of h' that approaches zero as R becomes infinite. The ratio of the NE length of the broken line $a'\ k'\ b'$ to that of h' is

$$\frac{2\log\cot\frac{\alpha}{2}+\log\frac{\varrho_2}{\varrho_1}}{\log\frac{\varrho_2}{\varrho_1}}\sin\alpha = \left[\frac{2\log\cot\frac{\alpha}{2}}{\log\frac{\varrho_2}{\varrho_1}}+1\right]\sin\alpha.$$

Now by hypothesis the NE length of k', namely $\log(\varrho_2/\varrho_1)$, is greater than one, so that the above ratio is less than

$$(4)\qquad \left(2\log\cot\frac{\alpha}{2}+1\right)\cdot\sin\alpha.$$

R, the length of a' or b', is given by (3) and becomes infinite only if α approaches zero. But as α approaches zero the expression (4) approaches zero, so that for R sufficiently large the N E length of h' will certainly exceed μ times the N E length of the broken line $a'k'b'$ joining the end points of h', and hence the N E length of h' will exceed μ times the shortest N E distance between the end points of h', as was to be proved.

DEFINITION. Let k be any curve segment within the unit circle. *The* N E *distance* between the end points of k will be denoted by $D_n(k)$, and the *surface distance* between the end points of k will be denoted by $D_g(k)$.

LEMMA 6. *Corresponding to positive constants b and m there exists a positive constant r so large that if a curve segment, v, has its end points at an* N E *distance r from some* N E *straight line L, if further all of v's points are at an* N E *distance from L exceeding $r - b$, and if v projects into a segment of L whose* N E *length exceeds unity, then the* N E *length of v exceeds m times the* N E *distance between v's end points.* (See Fig. 3.)

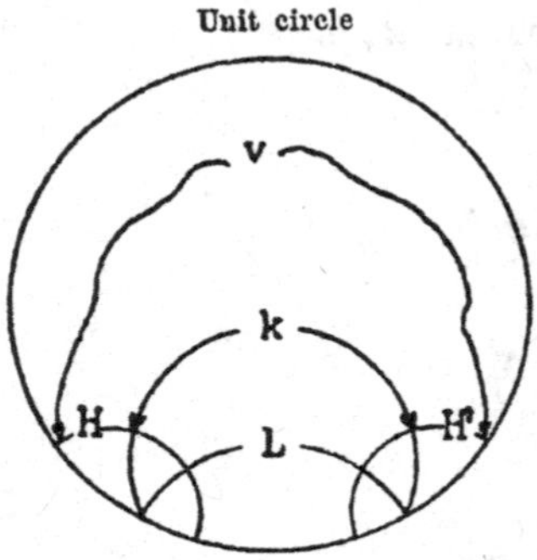

Fig. 3.

Denote by k a curve segment projecting through N E perpendiculars into the same segment on L as does v, and consisting further of points all at an N E distance $r - b$ from L. Let the N E length of v and k be denoted also by v and k respectively. We seek to prove that, for r sufficiently large,

$$v > m \cdot D_n(v). \tag{1}$$

To that end we shall obtain inequalities connecting v with k, k with $D_n(k)$, $D_n(k)$ with $D_n(v)$, and by combining these inequalities obtain the inequality (1).

By the aid of the construction of the preceding proof it is seen that

$$v > k.$$

Taking the constant μ of the preceding lemma as here equal to $m + 1$ it follows from that lemma that if $r - b$, and hence r, be sufficiently large,

$$k > D_n(k) \cdot (m + 1).$$

The two ends of v project into the two ends of k through N E perpendiculars to L, say H and H' respectively. Now the end points of v can be joined through the mediation of k by joining the first end point of v to the first end point of k, tracing out that segment of H that lies between these two points, following this segment by tracing out the N E line joining the end points of k and proceeding finally to the second end point of v by tracing out that portion of H' which lies between the second end points of v and k. In accordance with one of the hypotheses of this lemma the portions of H and H' thereby traced out are each of N E length b, so that the whole broken curve thus traced out will have an N E length $2b + D_n(k)$. This length will be at the least $D_n(v)$, since $D_n(v)$ stands for the shortest N E distance between the end points of v. Thus

$$2b + D_n(k) \geqq D_n(v) \text{ or } D_n(k) \geqq D_n(v) - 2b.$$

Combining the preceding inequalities we have

$$v > (m+1)[D_n(v) - 2b] = (m+1) D_n(v) - (m+1) 2b. \tag{2}$$

Now as r becomes infinite $D_n(v)$ will do likewise, and, for all cases in which the segment v projects on L into a segment of at least unit N E length, $D_n(v)$ will become infinite uniformly with r. Suppose r so large therefore that

$$D_n(v) > (m+1) 2b. \tag{3}$$

Adding (2) and (3) we have

$$v > m \cdot D_n(v)$$

with the condition simply that r be sufficiently large. The proof is accordingly complete.

LEMMA 7. *There exists a constant r such that if all the points of a segment of a geodesic g are at an* N E *distance at least r from some* N E *straight line L, if further the end points of the geodesic segment g are exactly at an* N E *distance r from L, and if g projects in the* N E *sense into a segment of L of more than unit* N E *length, then the surface length g of the given geodesic exceeds the surface distance between its end points.*

It is required to prove that for r sufficiently large

$$g > D_g(g). \tag{1}$$

Now as r becomes infinite $D_n(g)$ will also become infinite. It follows from Lemma 3 that $D_g(g)$ will do likewise, so that according to Lemma 4 there exist positive constants b and $\overline{m}$ such that if r exceed a properly chosen positive constant the end points of g can be joined by a curve v whose points are each within an N E distance b of g and whose N E length v is such that

$$\overline{m} g > v.$$

We now have a relation between g and v. We shall obtain a relation between v and $D_n(v)$, and a relation between $D_n(v)$ and $D_g(v)$. Combining these relations (1) will be obtained.

If r exceed a second constant so large that $D_n(v)$ exceeds the constant required for the application of Lemma 3, we will have

$$D_n(v) > A\, D_g(v),$$

where A is a positive constant. If further r exceed the positive constant required for the application of Lemma 6 for the case where the m of Lemma 6 equals $\overline{m}/A$, then in addition to the preceding inequalities there will also result the inequality

$$v > \frac{\overline{m}}{A} D_n(v).$$

If the above inequalities be combined there results the inequality

$$g > D_g(v),$$

subject simply to the condition that r exceed a properly chosen constant.

A FUNDAMENTAL CLASS OF GEODESICS

12. Geodesics of Class A. DEFINITION. A geodesic segment h will be said to be of *Class* A, if, on the original surface, h is at least as short as any other rectifiable curve joining h's end points and capable of being continously deformed on the surface into h without moving its end points. An image of a geodesic segment of Class A on the hyperbolic plane for simplicity also will be termed a geodesic segment of Class A. The definition of geodesic segments of Class A reduces in the hyperbolic plane to the following. An image of a geodesic segment within the unit circle is of Class A if its surface length is at least as small as that of any curve joining its end points and possessing a surface length. *An unending geodesic* will be said to be of Class A if each of its finite segments is of Class A.

LEMMA 8. *There exists a positive constant R, so large, that within the unit circle, no segment of a geodesic of Class* A *can recede an* NE *distance greater than R from the* NE *straight line joining its end points.*

Denote the given geodesic segment by g, and the NE straight line joining its end points by L. Consider the constant r whose existence is affirmed in Lemma 7. If g has points at more than r NE units from L, let P' be such a point. Let P and Q be the first points following and preceding P', respectively, that are at an NE distance r from L. If the segment PQ projects through NE perpendiculars into a segment on L of more than unit NE length, it follows from Lemma 7 that the surface length of the segment PQ exceeds the surface distance between P and Q contrary to the hypothesis that the given geodesic segment is of Class A.

If on the other hand the segment PQ projects into a segment on L of unit NE length or less than unit NE length, while P and Q are at an NE distance r from L, the NE distance between P and Q will be at most $2r+1$, and hence the surface distance between P and Q will in all cases be inferior to some positive constant, say d. Since g is of Class A, the surface distance between P' and P is at most that between P and Q, or at most d. It follows from Lemma 3 that the NE distance of P' from P is always less than some constant, say q. Hence P' is in all cases at most at NE distance $r+q$ from L, and the theorem is proved.

13. Unending curves of the same type. A definition will presently be given of what is meant by two unending curves being of the same type. This definition will be an extension of the definition of what is meant by two closed curves being mutually deformable. We begin accordingly by considering closed curves. A curve continuously deformable into a point either on the given surface or on the unit circle, will have an image deformable into a point on the other. A closed curve, say h, not deformable into a point on the given surface, if traced out just once will have as images on the unit circle an infinite number of curve segments whose end points are congruent, non-coincident points. Let A and B be respectively the end points of one of these images. A is congruent to B under a certain transformation T. The set of all points within the unit circle congruent to A under T and its inverse all lie on a circle through the fixed points of T and cluster in the neighborhood of these fixed points. The set of all segments congruent to AB under T and its inverse join up these points congruent to A so as to form a single curve passing from one fixed point of T to the other. An unending curve such as this, consisting of an unending succession of segments all congruent under the same transformation of the group, will be called a *periodic curve*. Traced out in either sense such a curve corresponds on the original surface to a closed curve traced out an infinite number of times.

It follows from the fact that the correspondence between the surface and unit circle is one to one and continuous as far as the neighborhood of corresponding points is concerned, that if the given closed curve, h, be continuously deformed on the surface, then the image AB will vary continuously through corresponding points, A and B always remaining congruent under the same transformation. At the same time the variation of the complete set of points congruent to A under T and its inverse is one in which the circle on which these points lie turns about the fixed points of T while the points themselves vary continuously on the circle in such a manner as to remain always congruent under T. Thus two closed curves which on the original surface are not deformable into points, but which are mutually deformable into each other, will correspond within the unit circle, among other images, to pairs of periodic curves with end points at the same points on the unit circle. For the sake of completeness the following is added. Two curves which correspond to two periodic curves within the unit circle with the same end points on the unit circle and with successive portions congruent under the *same* transformation of the fundamental group, are mutually deformable on the original surface.

DEFINITION. Two unending curves within the unit circle will be said to be of the *same type* if there exists a positive constant C such that every point of either curve has an N E distance from some point of the other less than C. Two unending curves *on the original surface* will be said to be of the *same type* if among their respective images on the unit circle there are at least two curves of the same type.

Two circular arcs within the unit circle are of the same type if their end points on the unit circle are the same, and of different types if either of their end points are different. Every circular arc with two end points on the unit circle is thus of the type of the N E straight line with the same two end points on the unit circle. *Only those curves will be considered which are of the types of* N E *straight lines.* There is thereby included a type for every periodic curve, that is, curves which correspond on the original surface to closed curves. Future developments will still further show the inclusiveness of types of curves as restricted to N E straight lines. From Lemma 3, it follows that N E distance and surface distances become infinite together uniformly. Accordingly if two unending curves are of the same type there exists a constant D such that every point of either curve has a surface distance less than D from some point on the other curve. It is not difficult to show that if on the original surface two curves are closed and mutually deformable they have for their images on the unit circle an infinite number of pairs of periodic curves of the same type. It is in this sense that *the property of two curves being of the same type is an extension of the property of two curves being mutually deformable.* The definition of what is meant by two curves being of the same type will be given in the following paragraph more in the spirit of analysis situs. The statements of the following paragraph will not be proved as they will not be used later.

Let the given surface, $p > 1$, be cut so as to form a single simply-connected piece S. Let there be provided an enumerable infinity of copies of S each spread over the original surface in exactly the position in which they were before the surface was cut. Without altering the position of any of these copies of S, these copies of S can be successively joined in a manner analogous to the way in which the congruent polygons of the unit circle were joined, so as to "heal" up all of the cuts and to form an infinitely-sheeted, simply-connected, unbounded surface, say M, the neighborhood of each point of which will consist on M of an exact copy of the neighborhood of the point which is overhung on the original uncut surface. Two curves of M could now be defined as being of the same type if there exists a positive integer n, so large, that any point of either curve could be joined to some point of the other curve by a curve which has points in at most n of the copies of S making up M. From this definition it could be shown that *two curves of the same type on the given surface will correspond, on any surface which corresponds to the given surface in a one to one continuous manner, to two curves again of the same type.*

14. Existence of geodesics of Class A. An image on the unit circle of a geodesic on the original surface will be spoken of as a geodesic on the unit circle. By an *element* is meant a point and a direction through the point. By an element on a geodesic on the surface is meant a point on the geodesic together with a direction tangent to the given geodesic at the given point. By an *element on a geodesic on the unit circle* is meant a point P within the unit circle on such a geodesic, together with the direction of a tangent to the original geodesic at the point on the original surface corresponding to P. In either case an element may be represented by the coördinates x, y of its initial point in the unit circle, together with the direction cosines a, b, c of the direction tangent to the original geodesic on the surface. The element may be considered as represented by a point $(xyabc)$ in five dimensions. A set of elements will be said to have a *limit element* $(x_0\, y_0\, a_0\, b_0\, c_0)$ if the corresponding points in five dimensions have a limit point $(x_0\, y_0\, a_0\, b_0\, c_0)$. With these conventions the following definition is given.

Definition. A geodesic G will be said to be a *limit geodesic* of a set of geodesics M not containing G, if every element on G is a limit element of elements on geodesics of the set M.

An element will be said to be on an NE straight line in the unit circle if its initial point P is on the NE straight line, and if its direction is the direction of a tangent to the NE straight line at P.

4*

LEMMA 9. *If a geodesic is a limit geodesic of a set of geodesics of Class* A, *it is itself of Class* A.

Let H be an image of the limit geodesic within the unit circle. If H were not of Class A, it follows from the definition of geodesics of Class A, that H has some segment, h, whose end points can be joined by a curve k whose surface length k is less than the surface length h of the segment h. Set

$$k = h - e.$$

On the other hand, if the given geodesic is a limit geodesic of geodesics of Class A, it follows, from the property of continuous variation of a finite geodesic segment with its initial point and direction, that there exists a geodesic segment m, of Class A, with end points within $e/4$ of those of h, and with a surface length m, such that

$$m > h - \frac{e}{2}.$$

Now the end points of m can be joined through the mediation of k by first passing from one end point of m to the nearer end point of k, passing thereby along a curve which can be taken so as to have a surface length less than $e/4$; thence passing along k to k's other end point, thereby passing over an additional surface length of $h-e$; finally passing from the second end point of k to the second end point of m along a curve taken so as to have a surface length less than $e/4$; in all passing over a curve of surface length less than $h-(e/2)$, that is, over a length less then that of m. This is impossible if m be of Class A. Thus the lemma is proved.

THEOREM 1. *Corresponding to any* NE *straight line, there exists at least one geodesic of Class* A *of the same type. Conversely, every geodesic of Class* A *is of the type of some one* NE *straight line.*

Let there be given an NE straight line L. Let

$$\cdots P_{-2}\, P_{-1}\, P_0\, P_1\, P_2 \cdots$$

be an unending succession of points on L which follow each other on L in the order of their subscripts, and which with increasing subscripts approach one end point of L on the unit circle, and with decreasing subscripts approach the other end point of L. Within the unit circle let P_{-n} be joined to P_n by a geodesic of Class A, say g_n. According to Lemma 8, g_n will have at least one point, say Q_n, within an NE distance R of P_0. Let E_n be the geodesic element on g_n at Q_n. The elements E_n will have at least one limit element, E. Let G be the geodesic within the unit circle bearing the element E. G will be shown to be of Class A, and to be a curve of the type of the given NE straight line.

If any finite segment of G be given, it follows from the definition of G, that there can be found among the segments g_n, one which possesses a sub-segment lying arbitrarily near the given segment of G, not only in position but also in direction and length. A repetition of the proof of the preceding lemma will suffice to show that the given segment of G is of Class A, as was to be proved.

Every point of the segments g_n lies within an NE distance R of some point of L (cf. Lemma 8). All the limit points of the points of the segments g_n, including the points of G, accordingly lie at most at an NE distance R from points of L. G thus lies in a region bounded by the two circles at an NE distance of R from L. Any two such circles and L meet the unit circle in the same two points. G passes from one of these points to the other. From this it is obvious that each point of L, also, lies at most at a distance R from some point of G, and G is thereby proved to be of the type of L.

To prove the converse proposition let G be a geodesic of Class A, within the unit circle, and let

$$\cdots P_{-2}\, P_{-1}\, P_0\, P_1\, P_2 \cdots$$

be a set of points appearing on G in the order of their subscripts, and so chosen that their surface distances from P_0 increase beyond all limit as n becomes infinite in absolute value. Denote by h_n an NE straight line joining P_n to P_{-n}. At least one point of h_n, say Q_n, will be at an NE distance from P_0 not greater than the constant R of Lemma 8. Denote by E_n, the element on h_n at the point Q_n. The elements E_n will have at least one limit element, say E. The NE straight line, L, through E, will be shown to be of the type of G.

Let P be any point on G. Let m be the smallest value of n such that all segments $P_{-n}\, P_n$ for which $n > m$ include the point P. For $n > m$ there is a point A_n on each h_n which is within an NE distance R (cf. Lemma 8) of the point P. But as n becomes infinite the points A_n will have at least one cluster point which, according to the definition of L, will lie on L. This cluster point will be at most at an NE distance R from the given point P of G. G thus lies within the band consisting of all points within an NE distance R of L. G and L will accordingly be of the same type, and the theorem is completely proved.

As just stated, a geodesic of Class A lies within a band consisting of all points within an NE distance R of the NE straight line of the same type. This constant R is the constant of Lemma 8, and is entirely independent of the particular geodesic of Class A that is considered. Combining this fact with the fact given in Lemma 3, that N E distances and surface distances become infinite together uniformly for the whole unit circle, we have the following theorem:

THEOREM 2. *There exists a positive constant D, fixed for all types of geodesics of Class* A, *such that no geodesic of Class* A, *and* N E *straight line of the same type, can recede either a surface distance or an* N E *distance exceeding D from each other.*

15. Intersecting and asymptotic geodesics of Class A.

THEOREM 3. *Two geodesics of Class* A *can intersect at most once within the unit circle.*

If two such curves, G and H, intersected twice in two points, A and B, the portions, g and h, respectively, of G and H, between the points A and B, would have the same surface length. For if, for example, h were shorter than g, G could not be of Class A, as follows directly from the definition of curves of Class A. But g and h cannot have the same surface length. For consider a segment k of G large enough to contain the segment g as an interior segment. If the surface length of g equaled that of h, k would be unaltered in surface length if its subsegment g be replaced by the segment h. But the curve so obtained would have corners at A and B, and so could be shortened, contrary to the fact that k is of Class A, and hence could not be shortened. Thus the theorem is proved.

A geodesic of Class A within the unit circle passes from one end point of the N E straight line of the same type, to the other such end. Hence the theorem:

THEOREM 4. *If two* N E *straight lines have both end points distinct, the corresponding geodesics of Class* A *of the same type cross or do not cross each other within the unit circle, according to whether or not the given* N E *lines cross or do not cross each other.*

If two N E straight lines have one end point on the unit circle in common, the minimum N E distance from a point P on either to a point on the other approaches zero as the point P approaches the common end point. The N E straight lines are termed *asymptotic*. A direct result of Theorem 2 is accordingly

THEOREM 5. *If two geodesics of Class* A *are respectively of the types of two asymptotic* N E *straight lines, as they approach the common end point on the unit circle of the given* N E *straight lines, they come and remain within a finite surface distance of one another.*

THEOREM 6. *No two geodesics of Class* A *which are asymptotic can intersect within the unit circle.*

Let a and b be two geodesics of Class A which are asymptotic.

Choose as the positive senses of a and b the senses in which a and b are asymptotic. Suppose the theorem were false and that a and b did intersect in a point O (cf. Fig. 4). Now any broken geodesic which contained any one of the corners formed by a and b together with a segment of a or b on either side of O of at least unit surface length could be replaced by a curve joining its end points and with a surface length smaller by at least a positive constant e.

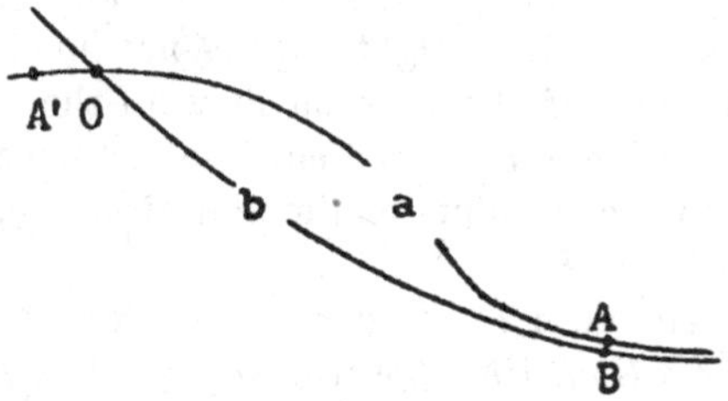

Fig. 4.

By hypothesis a and b are asymptotic. Accordingly it will be possible to find two points A and B, on a and b, so remote in the positive senses of a and b that they can be joined by a curve whose surface length AB is less than $e/4$:

$$AB < \frac{e}{4}.$$

Observe that the surface length of the segment AO of a cannot differ from that of BO on b by $e/4$ or more, for otherwise the larger of the two segments AO and BO, say AO, could be replaced by a broken curve of shorter surface length, namely ABO. Hence, in particular,

$$BO < AO + \frac{e}{4}.$$

Now let A' be any point on a, at least a unit surface distance beyond O in the negative sense. The segment AOA' of a can be replaced by a segment of shorter surface length. For adding to the two preceding inequalities the equality

$$OA' = OA',$$

we have

$$ABOA' < AOA' + \frac{e}{2}.$$

But the broken curve, $ABOA'$, contains a corner formed by a and b at O, so that it can be shortened by at least the constant e without altering its end points. The new curve will still join the end points of AOA', and will now in addition have a surface length less than that of AOA'. This is impossible if a is of Class A. The theorem thus is proved.

We have already defined what is meant by a limit geodesic of a class of geodesics (cf. § 14).

DEFINITION. We here define an NE straight line L to be a *limit line* of a set of NE straight lines M not including L, if every element of L is a limit element of elements on lines of the set M. A necessary and sufficient condition that L be a limit line of the set M is that there be in the set M lines whose end points on the unit circle lie, in the euclidean sense, arbitrarily near those of L.

THEOREM 7. *If a set M of* NE *straight lines have an* NE *straight line L for a limit line, the geodesics of Class* A *of the type of the lines of M will have at least one geodesic of the type of L as a limit geodesic.*

Let A be any element on L. It follows from the hypothesis of the theorem hat there exists a set of elements,

$$A_1 A_2 A_3 \cdots,$$

all on lines of M, and such that

$$\lim_{n\to\infty} A_n = A.$$

Let L_n be the NE straight line of M bearing A_n. Let C_n be a geodesic within the unit circle of Class A and of the type of L_n. According to Lemma 8, there exists on C_n at least one point, say P_n, within an NE distance R of the initial point of the element A_n. Let E_n be the geodesic element on C_n at P_n. The elements E_n will have at least one limit element, E. It follows from the result of Lemma 9, § 14, that the geodesic G, defined by E, is of Class A. It will now be proved that G is of the type of the NE straight line L.

It follows, from the property of continuous variation of a finite segment of a geodesic with its initial element, that every point P of G is a limit point, as n becomes infinite, of a properly chosen set of points on the geodesics C_n. But the latter points are each at an NE distance less than the constant R of Lemma 8 from some point of the corresponding NE straight line L_n. These points on the lines L_n will, however, as n becomes infinite, approach some point of L, say H, and this point H will be at most at an NE distance R from the given point P on G. Thus G is of the type of L as was to be proved.

THEOREM 8. *If there is more than one geodesic of Class* A *of a given type within the unit circle, there always exist two particular non-intersecting geodesics of Class* A *of the given type bounding a region outside of which no geodesics of Class* A *of the given type can pass.*

Let L be an NE straight line of the given type. Let M be a set of NE straight lines which have L as a limit line, which have no points in common with L either within or on the unit circle, and which all lie on one side of L. Let N be a similar set of NE straight lines all on the other side of L. Let M' and N' be two sets of geodesics consisting of one geodesic of Class A of the type of each NE straight line of the sets M and N respectively. According to Theorem 7 the geodesics of the sets M' and N' each have at least one limit geodesic, say M_0 and N_0, respectively, that is of Class A, and of the type of L. The NE straight lines of the sets M and N do not cross L or have any end points in common with L. It follows from Theorem 4 that all geodesics of Class A of the type of L will lie in the region between any geodesic of the set M', and any second geodesic of the set N'. It follows further from the property of continuous variation of any finite segment of a geodesic with its initial element that the geodesics M_0 and N_0, by virtue of the fact that they are respectively limit geodesics of the sets M' and N', will, in case they are non-coincident, be non-intersecting, and have between them all geodesics, other than M_0 and N_0, of Class A and of the type of L.

Closed geodesics

16. The variation of the distance of an arbitrary geodesic V of Class A from a given type of periodic curve h. Let there be given an NE straight line, h, made up of successive segments all congruent under a transformation, T, of the fundamental group. We have previously termed h a periodic curve. Let k be a second NE straight line crossing h. Let K be the geodesic of Class A of the type of k. The transformation T carries h into itself. Accordingly the successive images of k,

$$\cdots k_{-2}\, k_{-1}\, k_0\, k_1\, k_2 \cdots,$$

under T and its inverse, will consist of a set of non-intersecting circles, all meeting h and shrinking down about the two ends of h, the fixed points of T. The successive images of K under T and its inverse, say

$$\cdots K_{-2}\, K_{-1}\, K_0\, K_1\, K_2 \cdots,$$

will accordingly not meet each other (cf. Theorem 4), and will, like the k_n, shrink down about the two fixed points of T. Now let there be given any NE straight line, v, and let V be a geodesic of Class A of the type of v. If v crosses any one of the k_n, V will cross the geodesic K_n with the same subscript.

Three cases are distinguished here. The end points of v on the unit circle may be distinct from those of the given periodic curve h. In this case v

crosses at most a finite number of the k_n. Or v may be asymptotic to h, in which case v will cross all of the k_n with subscripts exceeding a certain integer, or else less than a certain integer. Finally v may coincide with h, in which case v will cross all of the k_n. A study will now be made of the positions of the points of intersection of V with the different K_n.

Let A_0 be any point on K_0. Let A_n on K_n be the image of A_0 under T, or its inverse. N E distances will be measured along each K_n from the corresponding A_n. The sense that corresponds to positive distances on each of the K_n will be chosen so as to be unchanged through application of T. Now if V intersects K_n, denote this point of intersection by P_n and let s_n be the surface distance of P_n from A_n, measured along K_n. A proof will be given that the s_n behave in one of the following ways: (1) *remain constant,* (2) *increase with their subscripts,* (3) *decrease with their subscripts,* (4) *increase to a maximum and thereafter decrease, or* (5) *decrease to a minimum and thereafter increase. In the latter two cases the maximum or minimum may be taken on by a single s_n or by two successive s_n.* The proof will be given under the heads A, B, C, D, E.

A. *If three successive s_n are equal all of the s_n are equal.* An application of the transformation T to V will carry V into a new geodesic of Class A, say V'. If in particular $s_1 = s_2 = s_3$, P_1 and P_2 on V would go into P_2 and P_3 on V' so that V and V' would intersect in P_2 and P_3. According to Theorem 3, V must then be identical with V'. Thus each P_n will be carried by T into P_{n+1}, so that s_n equals s_{n+1}.

B. *If just two successive s_n are equal, the next preceding and the next following s_n, if they exist, are either both greater or both less than the equal s_n.*

Of the two regions into which V divides the unit circle, that region which a point enters upon being projected from V along any of the curves K_n in the positive sense of that particular K_n will be termed the region above V, and the other region the region below V. As before let V' be the image of V under the transformation T. Suppose now that $s_2 = s_3$ and that $s_1 > s_2$. To prove $s_3 < s_4$: V' will meet K_2 above V since $s_1 > s_2$. V' will meet K_3 in the same point as does V since $s_2 = s_3$, and at this point will pass from above to below V. Thereafter V' will remain below V, since it cannot cross V a second time. In particular V' will meet K_4 below V, so that $s_3 < s_4$ as was to be proved. Similarly it could be shown that if $s_2 = s_3$ and $s_1 < s_2$, then $s_3 > s_4$.

The preceding proof includes a proof of the following.

C. *If the s_n are equal for $n = m$ and $n = m-1$, V and V' meet on K_m.*

D. *If the s_n have a relative maximum or minimum for $n = m$, then between K_m and K_{m+1}, V and V' cross each other.* More explicitly suppose that

s_2 is greater than both s_1 and s_3. Then V' meets K_2 below V since $s_1 < s_2$, and meets K_3 above V since $s_2 > s_3$. Thus V' must cross V between K_2 and K_3. The case of a minimum among the s_n is similarly treated.

E. *The s_n cannot have more than one extremum of any sort.* For if the s_n had more than one extremum it follows from C and D that V and V' would have more than one point of intersection. V would then have to be identical with V'. All the s_n would be equal in that case.

A proof can now be given that the s_n behave only in the ways enumerated. Any sequence of numbers such as the s_n are either (1) all equal, or (2) increase or decrease monotonically with n, or (3) have at least one extremum. The first case is admittedly possible. The second case can be achieved, according to A and B, only when no two of the s_n are equal, and then is admittedly possible.

The third case, according to E, can occur only when there is a single extremum among the s_n. In this case the extremum will be taken on by not more than two successive s_n, as follows from A. Further, no two of the remaining s_n are equal, for if just two of them were equal, it would follow from B that they would constitute another extremum among the s_n, contrary to E; and if three or more of the remaining s_n were equal all of the s_n would be equal and we would have the first case.

Thus all three cases reduce to the cases originally stated to be possible.

17. Periodic geodesics of Class A. DEFINITION. Of the transformations of the fundamental group which have the same fixed points there are two which will be called the corresponding *primitive transformations.* They are inverses of each other. Each is such that every other transformation of the group with the same fixed points is either its multiple or the multiple of its inverse. Such transformation exists. For every transformation of the group which has the two given points for fixed points carries the NE straight line through these points into itself. With the aid of NE distances measured along this line the existence of the primitive transformations can be proved in exactly the same way as the existence of a primitive period is proved in the theory of simply periodic functions.

THEOREM 9. *A geodesic of Class A which is carried into itself by a transformation T' is also carried into itself by the corresponding primitive transformations.*

Denote the given geodesic by V. We shall make use of the notation of the preceding section, identifying the V and T of this section with the V and T of that section. Now according to the result of the preceding paragraph, there exists a positive integer m, such that

$$T' = T^m$$

where T is one of the two primitive transformations. Accordingly, upon applying $T\,m$ times to V, V will be carried into itself. Hence among the s_n used in the preceding section those whose subscripts differ by m must be equal. But if all of the s_n were not then equal, the s_n would have an infinite number of maxima and minima, which has been shown to be impossible. The s_n are then all equal. The result of applying T to V must carry V into itself. For the geodesic into which V is carried by T would otherwise intersect V in each of the points at which the s_n were equal; according to Theorem 3, this is impossible. Thus T carries V into itself, as was to be proved.

THEOREM 10. *A geodesic, V, of Class* A, *which if traced out in one of its senses, say the positive sense, comes and remains at an* NE *distance less than an assignable positive constant from a periodic* NE *straight line L, is either a periodic geodesic of the type of L, or else is asymptotic in its positive sense to a geodesic that is itself periodic, and is of the type of L, and of Class* A.

L is made up of successive portions congruent under a primitive transformation, say T. We will again make use of the notation of the preceding section identifying the V, T and L of the preceding section with the V, T and L of this theorem. We suppose that the subscripts n of the s_n of the preceding section are here so chosen that they increase as V is traced out in its positive sense. If the s_n are all equal, V must be periodic. The case where the s_n are not all equal will be considered now.

According to the result of the preceding section, if the s_n are not all equal, then for values of n exceeding a properly chosen integer, say m, the s_n must either all increase with n, or else all decrease with n. It follows from the hypothesis that V traced out in its positive sense comes and remains within an assignable positive NE distance of L, that the s_n are bounded, for $n > m$. Hence as n becomes infinite the quantities s_n approach a limit, say b.

Let k be any positive integer greater than m. The transformation which carries the geodesics K_k of the preceding section into K_0, carries L into itself, and V into a curve of Class A, say V_k, for which the corresponding s_n will be obtained from the s_n for V by advancing the subscripts by k. Thus the point of intersection of V_k with K_0, say Q_k, will be at an NE distance s_{0+k} from the point A_0 of K_0, from which NE distances along K_0 are being measured. As k becomes infinite the points Q_k will approach a point on K_0, say Q, whose surface distance from A_0 will equal $\lim_{k\to\infty} s_k = b$. Let E_k be the element on V_k, at the point Q_k. The elements E_k will have one or more limit elements. Let E be one such limit element. The initial point of E will be Q. The geodesic, say G, defined by E will be of Class A.

The point of intersection of G with any particular K_n, say K_3, for example, will be the limit of the points of intersection of the geodesics V_k with that particular geodesic K_3, as k becomes infinite; so that G will intersect K_3 at a surface distance from A_3 on K_3 equal to $\lim_{k\to\infty} s_{3+k} = b$. Similarly G will meet all other K_n at a surface distance from the corresponding A_n equal to b. It follows that the transformation T will carry G into itself, so that G is a periodic geodesic.

It remains to be proved that V is asymptotic to G. Now E was defined as a limit element of the elements E_k. But the elements E_k can have no other limit element than E. For if the elements had another limit element E', then E', like E, would serve to define a second periodic geodesic, say G', passing through the point Q. G' and G would intersect not only in Q but also in all the points congruent to Q under the transformation T. This is impossible for two geodesics of Class A. Hence $G' = G$, and $E' = E$, and the set E_k has no other limit element than E. This may be written

$$\lim_{k\to\infty} E_k = E. \tag{1}$$

Now E_k is the initial element of the portion of V_k, say h_k, between K_0 and K_1, while E is the initial element of the portion of G between K_0 and K_1. Hence for k sufficiently large, h_k, according to (1), will lie arbitrarily near the portion of G between K_0 and K_1. But the portion of V between K_k and K_{k+1} is, by definition of V_k, congruent to h_k, and accordingly for k sufficiently large lies arbitrarily near the portion of G between K_k and K_{k+1}. Thus V is asymptotic to G in its positive sense. The proof of the theorem is complete.

DEFINITION. The two non-intersecting geodesics of Class A of a given type which form the two boundary curves of the region outside of which no geodesics of Class A of the given type can pass, will be termed the *boundary geodesics* of that type (cf. Theorem 8). The name will be applied even in the special case where the two bounding geodesics are identical.

THEOREM 11. *The boundary geodesics of the type of a periodic* NE *straight line L are themselves periodic.*

Let T be a transformation of the group that carries L into itself. T carries L into itself, and hence every curve of the type of L into another such curve. Further, T, as well as every other transformation of the group, preserves surface distances, and hence carries every geodesic of Class A into a geodesic of Class A. Thus T carries a geodesic of Class A of the type of L into another such geodesic.

To prove the theorem, observe first that the two given boundary geodesics, say G_1 and G_2, divide the unit circle into three non-overlapping regions, say S_1, S_2, and S, of which S lies between G_1 and G_2, S_1 lies between G_1 and the unit circle, and S_2 lies between G_2 and the unit circle. S, in case G_1 is identical with G_2, will reduce simply to the geodesic $G_1 = G_2$. Now S_1 contains in its interior no geodesic of Class A of the type of L, but is, however, bounded within the unit circle by one such geodesic. These two properties of S_1 must be preserved in the region, say S', into which S_1 is carried by the transformation T. For otherwise, upon applying the inverse of T to S', we could conclude that the two properties attributed to S_1 did not really belong to S_1. Now the only regions of the unit circle which have these properties of S_1, and like S_1 are bounded in part by the unit circle, are the regions S_1 and S_2. Hence the transformation T must carry S_1 either into S_1, or else into S_2. But T cannot carry S_1 into S_2; for that portion of the unit circle that bounds S_1 lies between the two fixed points of T, and under a hyperbolic transformation, such as T, is carried into the same portion of the unit circle. Thus S_1 goes into S_1. Hence G_1, as the boundary of S_1 within the unit circle, goes into G_1. It follows similarly that G_2 is carried into G_2 by T, and the theorem is proved.

THEOREM 12. *If two closed geodesics of Class* A *are of the same type they are also of the same length.*

Since the two given closed geodesics are of the same type they will be represented on the unit circle by an infinite number of pairs of periodic curves, the members of each pair of the same type. Let C and D be the members of one such pair. Let T be a primitive transformation which carries C and D into themselves. Let c be a segment of C of which one end point A is carried into the other end point by T. Let d be a segment of D of which one end point B is carried into the other end point by T. Denote the surface lengths of c and d also by c and d, respectively.

Suppose now that the theorem were false, and that

$$d - c = e > 0.$$

Let h be the surface distance between A and B. Let n be a positive integer such that

$$n > \frac{2h}{e}.$$

Consider now a continuous portion of C, say c', consisting of all of the images of c obtained by applying $T\,n$ times to c. Let d' be a similar piece of D obtained by applying $T\,n$ times to d'. The end points of d' can be joined by a curve whose surface length is shorter than that of d', by passing from the end point B of d' to the end point A of c' through a surface distance h, passing along c through a surface distance nc, thence returning to the second end point of d' through a surface distance h, in all over a path whose surface length would be $2h+nc$ which according to the preceding inequalities would be less than nd, or the surface length of d'. This is impossible if D is of Class A. The theorem thus is proved.

18. Geodesics of Class A asymptotic to periodic geodesics of Class A.

THEOREM 13. *A geodesic a of Class* A *which is asymptotic to a periodic geodesic b of Class* A *within the unit circle cannot cross any periodic geodesic of Class* A *of the type of b.*

Suppose that a did cross a periodic geodesic c of Class A of the type of b, at a point P. (See Fig. 5.) Let there be assigned to a, b, and c, positive senses that lead to their common end point on the unit circle. Let m be the length of the closed geodesic which on the original surface corresponds to b.

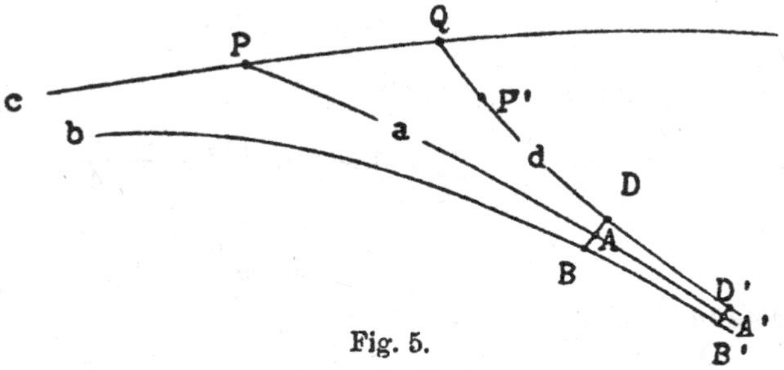

Fig. 5.

According to Theorem 12, the length of the closed geodesic which corresponds to c is also m. On c let there be laid off from P in c's positive sense a segment whose surface length is m; denote the second end point of this segment by Q. The transformation T which carries P into Q, and b and c into themselves, carries a into a geodesic d of Class A that is again asymptotic to b.

Let P' be any point on d, following Q, and neighboring Q. Any curve which contains the corner PQP', formed at Q by c and d, can be shortened in surface length, without disturbing P and P', by a positive constant, say e. Let B and B' be two points on b which are the two end points of a segment of b of surface length m, and which are so remote on b that two geodesics perpendicular to b respectively at B and B' meet the asymptotes a and d in points within a surface distance $e/4$ of B and B' respectively, and cut out of a and d respectively two segments whose surface lengths differ from m by less than $e/4$. Denote the point in which the geodesic perpendicular to b at B meets a by A, and the point at which it meets d by D, and denote the point in which the congruent geodesic perpendicular to b at B' meets a by A', and d by D'.

We will now show that the segment PA on a can be replaced by a curve joining its end points, which has a smaller surface length. To that end we compare the surface length of PA with that of the broken curve $PQDA$. We shall use the letters PA to denote the surface length of PA measured along the geodesic a, and will extend this convention to all other segments to be considered. Now since the segment PA is congruent to the segment QD' we have

$$PA = QD', \tag{1}$$

or

$$PA = QD + DD'. \tag{2}$$

Now, from the choice of B and B',

$$DD' > BB' - \frac{e}{4}. \tag{3}$$

Adding (2) and (3),

$$PA > QD + BB' - \frac{e}{4}. \tag{4}$$

Now

$$BB' = m = PQ.$$

Hence (4) becomes

$$PA > QD + PQ - \frac{e}{4}. \tag{5}$$

Further, from the choice of B,

$$\frac{e}{4} > DA. \tag{6}$$

Adding (5) und (6)

$$PA > PQ + QD + DA - \frac{e}{2},$$

or

$$PA > PQDA - \frac{e}{2}.$$

Now the broken curve $PQDA$ contains the above mentioned corner at Q, and can be replaced by a curve joining its end points whose surface length will, according to the choice of e, be e less than that of $PQDA$. This new curve will accordingly join the end points of PA, and according to the last inequality be less in surface length than PA contrary to the fact that PA is a segment of Class A. Thus our assumption that a crosses a periodic geodesic of Class A of the type of b is impossible, and the theorem is proved.

One of the hypotheses of Theorem 10 requires that the geodesic V of that theorem be of Class A. However, what is essential in the proof of that theorem can be carried over to the case where every segment on V from a certain point on is of Class A. In this way we get the following lemma:

LEMMA 10. *A portion of a geodesic which in one sense, say its positive sense, is unending, every segment of which portion is of Class* A, *and which in its positive sense comes and remains at an* NE *distance less than an assignable positive constant from some periodic geodesic b of Class* A *is either asymptotic in its positive sense to a periodic geodesic of the type of b, and of Class* A, *or is itself periodic of the type of b.*

There is a similar difference between Theorem 13 and the following lemma:

LEMMA 11. *A portion of a geodesic which in one sense from a certain point on is of Class* A, *and which is asymptotic in the given sense to a periodic geodesic b of Class* A, *cannot cross any periodic geodesic of Class* A *of the type of b.*

THEOREM 14. *If there exist two different periodic geodesics of Class* A, *namely, b and c, of the same type and between which there are no other periodic geodesics of Class* A, *then if to b and c there be assigned positive senses that lead on b and c to a common end point of b and c on the unit circle, there exist at least two geodesics of Class* A *lying in the region between b and c, of which one geodesic is asymptotic to b in b's positive sense and to c in c's negative sense, while the other geodesic is asymptotic to b in b's negative sense and to c in c's positive sense.*

Let $B_1\, B_2\, B_3 \cdots$ be a set of points which follow one another on b in b's positive sense, and whose surface distances from B_1 become infinite with the subscript n. Let $C_1\, C_2\, C_3 \cdots$ be a set of points on c which follow each other other on c in c's negative sense and whose surface distances from C_1 become infinite with the subscript n. Let h_n be a segment of a geodesic of Class A joining B_n to C_n. Let P_n be a point on h_n whose surface distances from b and c are equal.

Let k be a segment of a geodesic of Class A joining some point of b to some point of c. Let T be a transformation of the fundamental group that carries b and c into themselves. (Cf. Theorem 9.) Under T, k will be carried into a second geodesic segment, say k', also of Class A, and also joining a point of b to one of c. Between k and k', b and c, there will be a region, say S, contained in the region, say S', between b and c, and such that every point of S' is congruent under some multiple of T or its inverse to some point of S.

A proper multiple of T or its inverse will carry P_n into a point P_n' of S. The same transformation will carry h_n into a second geodesic segment of Class A, say h_n', again joining a point of b to a point of c. Let E_n be the element on h_n' at the point P_n'. The elements E_n will have at least one limit element E. Denote the initial point of E by P. Let d be the geodesic defined by E. We

5

will show that d is asymptotic to b in b's positive sense, and to c in c's negative sense.

The h'_n all become infinite in surface length with n. (Cf. Lemma 3.) Accordingly d will become infinite in surface length, in at least one sense, without passing out of the region between b and c. That portion of d that lies between b and c, and contains the point P, will certainly be of Class A. For any of its finite segments between b and c permit an arbitrarily close approximation, both in position and in surface length, by a properly chosen segment of some one of the segments h'_n, which are themselves of Class A. (Cf. proof of Lemma 9, § 14.) According to Lemma 10, d traced out in that sense in which it is already proved to remain between b and c, becomes asymptotic to some periodic curve of Class A of the type of b and c. According to Lemma 11, d cannot then cross any periodic curve of Class A, of the type of b or c. In particular d cannot cross b or c, and hence remains between b and c. As stated before every segment of d which contains the point P, and remains between b and c, is of Class A. We can conclude therefore that every segment of d is of Class A.

Finally it would be impossible for d, as defined, to be asymptotic to b in b's negative sense, and to c in c's positive sense. For the methods of § 16, applied to any geodesic segment h_n, will show that if h_n be traced from its end point B_n to its end point C_n, its successive intersections with the geodesic segments congruent to k under T will recede from b and progress toward c, in the sense that their surface distances measured from b along the segments congruent to k will increase. Now according to the definition of d, there exists a geodesic segment h'_n, which, starting near the point P on d, will follow along d arbitrarily near d, for an arbitrarily long surface distance, measured on d from P, in either sense. Hence d can only be asymptotic to b in b's positive sense, and to c in c's negative sense.

The existence of a geodesic of Class A asymptotic to b in b's negative sense and to c in c's positive sense is similarly proved.

THEOREM 15. *It is impossible within the unit circle for a geodesic of Class* A *to be asymptotic in both of its senses to the same periodic geodesic of Class* A.

Suppose the theorem were false and that a geodesic b of Class A were asymptotic in both of its senses to a periodic geodesic c of Class A. Upon applying any transformation of the fundamental group which carries c into itself b would be carried into a second geodesic, say b', of Class A also asymptotic in both of its senses to c. Now b and b' would intersect. For if b, for example, lay entirely within the region between b' and c the maximum N E distance of any of b's points from c would be less than that of b'. This maximum, however, should be the same for both b and b' since they are congruent. Thus b would intersect b' contrary to the fact, as stated in Theorem 6, that no two geodesics which are asymptotic can intersect. Thus the initial assumption of this proof is false, and the theorem is true.

19. Completion of results on periodic geodesics. Let L be any periodic N E straight line. Denote the set of all geodesics of Class A and of the type of L by M. The properties of the geodesics of the set M may be summed up and completed as follows:

If there is but one geodesic in the set M it is periodic. If there is more than one geodesic in M there are two non-intersecting geodesics of M, say b_1 and b_2, called boundary geodesics of M, delimiting a region outside of which no geodesics of M can pass. The boundary geodesics are periodic. No periodic geodesics of M intersect. Between b_1 and b_2 there may be a finite or infinite number of periodic geodesics of M or none. If there exists a single point in the region between b_1 and b_2 through which there does not pass a periodic geodesic of M, there is at least one region bounded by two periodic geodesics of M, say p_1 and p_2, between which there are no other periodic geodesics of M. Denote one end point of L on the unit circle by A, and the other end point by B. In the region between p_1 and p_2 there is at least one geodesic of M asymptotic to p_1 as A is approached, and asymptotic to p_2 as B is approached, and at least one geodesic of M asymptotic to p_1 as B is approached and to p_2 as A is approached. Any geodesic of M which is not periodic lies in a region between two periodic geodesics of M, between which there are no periodic geodesics of M. Any such non-periodic geodesic of M is always asymptotic in one sense to one of the periodic geodesics bounding the region in which it lies, and to the other such periodic geodesic in the other sense.

Geodesics of Class A, of the type of N E straight lines that cross L, cross all geodesics of the set M and recede indefinitely from all points of geodesics of M. Any geodesic, say c, of Class A, of the type of any one of the infinite number of N E straight lines asymptotic to L, lies wholly outside the region between b_1 and b_2, and as A is approached becomes asymptotic to that one of the two geodesics, b_1 or b_2, which separates c from the region between b_1 and b_2. A similar statement may be made replacing the point A by the point B.

We conclude with a proof of the following theorem.

THEOREM 16. *The set of all periodic geodesics of Class* A *include among their limit geodesics all of the "boundary" geodesics.* (Cf. § 17.)

Let L be a given N E straight line. According to the lemma of § 8 there exists a transformation of the fundamental group, having its fixed points arbitrarily near the end points of any preassigned arc of the unit circle. The N E straight line passing through the fixed points of a transformation of the group is a periodic straight line. Accordingly it will be possible to choose a set of periodic N E straight lines, say M, which have L as a limit line, which have no points in common with L either within or on the unit circle, and which all lie on one side of L. Let N be a similar set of periodic N E straight lines all on the other side of L. According to Theorem 11, § 17, the boundary geodesics of the type of any periodic N E straight line are themselves periodic. Accordingly, corresponding to each periodic N E straight line of the set M there exists at least one geodesic of Class A of the same type that is itself periodic; denote the set of such periodic geodesics by M'. Let N' be a similar set of periodic geodesics consisting of one periodic geodesic of Class A of the type of each N E straight line of the set N. Now exactly as was shown in the proof of Theorem 8, § 15, the geodesics of the sets M' and N' will have, the one, a limit geodesic M_0, and the other, a limit geodesic N_0, that are the two boundary geodesics of the geodesics of Class A of the type of L. Thus the theorem is proved.

Cornell University,
Ithaca, N. Y.

RELATIONS BETWEEN THE CRITICAL POINTS OF A REAL FUNCTION OF n INDEPENDENT VARIABLES*

BY

MARSTON MORSE

Introduction

In the main body of the paper the results and proofs are given for the general case of a real function of n independent variables, and for a domain of definition of the given function that is very general from the point of view of analysis situs. For the purposes of an introduction one of the principal results will be given for the case where $n = 2$. It must be stated, however, that the results for the general case differ tremendously from those for the case $n = 2$. The results for the case $n \leq 2$ can conveniently be given in geometric language as follows:

Let R consist of the interior points of a finite, connected region of the x, y plane, bounded by a finite number of distinct closed curves without multiple points of any sort, possessing at each point a tangent which turns continuously with movement of its point of contact. Denote the boundary of R by B. Let $f(x, y)$ be a real function of x and y defined and continuous in R and on B, and possessing continuous third order partial derivatives in R taking on continuous boundary values on B. By the positive direction of the normal to B will be understood the direction that leads across B from points in R to points not in R. It is assumed that the unilateral, directional derivative of $f(x, y)$ on the side of the inner half of the normal and in the positive direction along the normal is everywhere positive on B.

Let $z = f(x, y)$ be interpreted as a surface in ordinary euclidean (x, y, z) space. By a critical point of $f(x, y)$ is meant a point in the x, y plane corresponding to which the tangent plane to the surface $z = f(x, y)$ is parallel to the x, y plane. In the neighborhood of such a critical point let $f(x, y)$ be expanded according to Taylor's formula in powers of differences of the x's and y's which vanish at the critical point, and with the remainder as a term of the third order. We hereby assume that in this expansion the discriminant of the terms of the second order is not zero.

From these assumptions it follows that the critical points are isolated and finite in number, and that they are of the two general types called

* Presented to the Society, December 28, 1923.

in differential geometry elliptic and hyperbolic points. The elliptic critical points can be further subdivided into points at which $f(x, y)$ takes on a relative minimum, and those at which $f(x, y)$ takes on a relative maximum. Suppose that $f(x, y)$ has a relative minimum at M_0 different points, and a relative maximum at M_2 different points, while the hyperbolic critical points are in number M_1. Suppose that R and its boundary form a region of linear connectivity R_1.* With this understood we have the equality fundamental for the case $n = 2$,

$$\text{(a)} \qquad M_0 - M_1 + M_2 = 2 - R_1 .$$

From this it follows immediately that

$$\text{(b)} \qquad M_1 \geqq M_0 + R_1 - 2 .$$

It should be noted that (a) does not follow from (b). The inequality (b) has been obtained by G. D. Birkhoff.† Enunciated in slightly different form for the general case, he has called his result the "minimax principle" and he uses it to infer the existence of at least $M_0 + R_1 - 2$ critical points of the hyperbolic type when the existence of M_0 points of relative minimum are known. He has carried out his work for the general case of real, analytic function of n real variables and applied his result to the theory of dynamical systems. Upon reading Birkhoff's paper it occurred to the author that inasmuch as there are $n + 1$ different kinds of critical points possible (in a sense to be defined later) there ought to be relations analogous to (b) involving the numbers of critical points of all kinds and not simply of just two kinds. With a proper interpretation of M_1 and R_1, (b) however will serve to state Birkhoff's "minimax principle" in any number of dimensions.

The author has replaced the inequality (b) by an equality involving all of the $n + 1$ possible types of critical points, and in addition all of the connectivity numbers of the n-dimensional region of definition of the given function (Theorems 8 and 9). Further, a set of n additional inequality relations have been discovered. The author ventures without proof the opinion that these relations, or relations derivable from them, are all the the relations there are between the numbers involved.

* Oswald Veblen, *Analysis Situs, The Cambridge Colloquium*, Part 2, p. 50, § 28.
We shall have occasion in the future as well as here to make use of the terms of analysis situs. They will be used in the senses in which they are defined in the work just cited. In the remainder of this paper reference to this work will be indicated by the letter V.

† *Dynamical systems with two degrees of freedom*, these Transactions, vol. 18 (1917), p. 240.

Note. For the case $n = 2$ the author has been able to outline an alternative proof of one of the relations between the critical points by making use of the Kronecker characteristic of the given function and its partial derivatives. That the Kronecker theory as developed by Kronecker (*Werke*, vol. I, pp. 175–226, and vol. II, pp. 71–82) will not suffice in general to obtain the results of this paper is evident from the fact that Kronecker distinguishes between the critical points according to the sign of the hessian of the function, while in this paper both in the proofs and in the results there is an essential distinction between the $n+1$ types of critical points. Further, in Kronecker no reference is made to anything like the connectivity numbers of the regions in which the critical points are being studied.*

THE FUNCTION $f(x_1, x_2, \cdots, x_n)$

1. The function $f(x_1, x_2, \cdots, x_n)$ and its critical points. Let $(x_1, x_2, \cdots, x_n)$ be the coördinates of any point (X) in a euclidean space of n dimensions. In a finite part of this space let there be given an n-dimensional region Σ. The set of interior points of Σ will be denoted by R. At every point of R let there be defined a real, single-valued, continuous function of the x's, namely $f(x_1, x_2, \cdots, x_n)$. Suppose further that in R this function possesses continuous partial derivatives with respect to each of the x's of at least the first and second orders.

A point of R at which all of the first partial derivatives of f vanish will be called a *critical point* of f. Any other point of R will be called an *ordinary point* of f. The value of f at a critical point will be termed a *critical value* of f. As a further assumption regarding f we will suppose that in the neighborhood of each critical point f possesses continuous third partial derivatives. Let there be given a critical point of f at which f takes on the value c. For simplicity suppose this critical point at the origin. In the neighborhood of the origin, f can be represented with the aid of Taylor's formula in the form

$$f(x_1, x_2, \cdots, x_n) - c = \sum_{ij} a_{ij}\, x_i x_j + r(x_1, x_2, \cdots, x_n) \tag{1}$$

$$(i, j = 1, 2, \cdots, n; \quad a_{ij} = a_{ji}),$$

where the terms of first order are missing because the origin is a critical point, where the terms of second order are represented as a symmetric quadratic form in the n x's with constant coefficients, and where the remainder $r(x_1, x_2, \cdots, x_n)$ is included as a term of the third order. In

* For closely connected results in the case $n = 2$ see a paper by H. Poincaré, Journal de Mathématiques, ser. 3, vol. 7 (1881), pp. 375–421.

22*

any representation such as this of the function f in the neighborhood of a critical point, it is hereby assumed that the terms of the second order constitute a non-singular quadratic form. Analytically this means that the determinant

$$|a_{ij}| \neq 0. \tag{2}$$

It is a consequence of this last restriction upon f that *in a sufficiently small neighborhood of any critical point there is no other critical point.* For, if the partial derivative of f with respect to x_i be denoted by f_i $(i = 1, 2, \cdots, n)$, each critical point of f is a solution of the equations

$$f_1 = f_2 = \cdots = f_n = 0. \tag{3}$$

Now the jacobian of the n functions f_i with respect to the n variables x_i at the critical point we are here considering, namely the origin, vanishes only with the determinant $|a_{ij}|$ which we have assumed not zero. It follows from the theory of implicit functions that the system of equations (3) has no other solution in the neighborhood of the origin than the origin itself. Thus there are no other critical points in the neighborhood of the origin.

A consequence of the result just obtained is that *in any closed region S all of whose points are points of R, there are at most a finite number of critical points of the given function.* For if there were an infinite number of critical points in S, by the Weierstrass cluster point theorem, these critical points would have at least one cluster point, which, since S is closed, would belong to S, and hence would be a point of R. At each of these critical points all of the first partial derivatives of f would vanish, and since these partial derivatives are assumed continuous at every point in R, they would all be zero at the cluster point. Thus the cluster point would also be a critical point contrary to the result of the preceding paragraph that the critical points of f are isolated.

2. Certain preliminary lemmas. Lemmas 1 and 2 of this section are important aids in later proofs. They refer to a function $F(x_1, x_2, \cdots, x_n)$, real, single-valued, and continuous in the neighborhood of the origin (written N). To aid in proving Lemma 1, results (A), (B), and (C) will first be established.

(A) *If $F(x_1, x_2, \cdots, x_n)$ possesses continuous first partial derivatives in N, then in N*

$$F(x_1, x_2, \cdots, x_n) = F(0, x_2, x_3, \cdots, x_n) + a(x_1, x_2, \cdots, x_n)\, x_1 \tag{4}$$

where $a(x_1, x_2, \cdots, x_n)$ is single-valued and continuous in N.

A function $a(x_1, x_2, \cdots, x_n)$ readily seen to satisfy the requirements of (A) is defined by the equations

$$(5) \qquad a(x_1, x_2, \cdots, x_n) = \frac{F(x_1, x_2, \cdots, x_n) - F(0, x_2, \cdots, x_n)}{x_1}, \quad x_1 \neq 0,$$

$$(6) \qquad a(0, x_2, x_3, \cdots, x_n) = F_1(0, x_2, \cdots, x_n),$$

where F_i will be taken to mean the partial derivative of F with respect to x_i $(i = 1, 2, \cdots, n)$.

(B) *If $F(x_1, x_2, \cdots, x_n)$ possesses continuous second partial derivatives in N, then in N*

$$(7) \qquad F(x_1, x_2, \cdots, x_n) = F(0, x_2, \cdots, x_n) + F_1(0, x_2, \cdots, x_n)\, x_1 + b(x_1, x_2, \cdots, x_n)\, x_1^2$$

where $b(x_1, x_2, \cdots, x_n)$ is single-valued and continuous in N.

A function $b(x_1, x_2, \cdots, x_n)$ readily seen to satisfy the requirements of (B) is defined by the equations

$$(8) \qquad b(x_1, x_2, \cdots, x_n) = \frac{F(x_1, x_2, \cdots, x_n) - F(0, x_2, \cdots, x_n) - F_1(0, x_2, \cdots, x_n)\, x_1}{x_1^2}, \quad x_1 \neq 0,$$

$$(9) \qquad b(0, x_2, \cdots, x_n) = \frac{F_{11}(0, x_2, \cdots, x_n)}{2},$$

where F_{ij} will be taken to mean the second partial derivative of F with respect to x_i and x_j $(i, j = 1, 2, \cdots, n)$.

(C) *If $F(x_1, x_2, \cdots, x_n)$ possesses continuous first partial derivatives in N, then in N*

$$(10) \qquad F(x_1, x_2, \cdots, x_n) = F(0, 0, \cdots, 0) + \sum_i a_i(x_1, x_2, \cdots, x_n)\, x_i \quad (i = 1, 2, \cdots, n),$$

where each a_i can be so chosen as to be continuous in N.

From (A) it follows that (C) is true for $n = 1$. Following the method of mathematical induction the truth of (C) is now assumed for the case of $n-1$ independent variables. Whence

$$(11) \qquad F(0, x_2, \cdots, x_n) = F(0, 0, \cdots, 0) + \sum_i b_i(x_2, \cdots, x_n)\, x_i \quad (i = 2, \cdots, n)$$

where each b_i is continuous in N. If $F(0, x_2, \cdots, x_n)$ in (4) be replaced by the right hand member of (11) the result (C) follows directly.

LEMMA 1. *If $F(x_1, x_2, \cdots, x_n)$ possesses continuous second partial derivatives in N, and vanishes with its first partial derivatives at the origin, then in N*

$$F(x_1, x_2, \cdots, x_n) = \sum_{ij} a_{ij}(x_1, x_2, \cdots, x_n) x_i x_j \qquad (i, j = 1, 2, \cdots, n) \tag{12}$$

where each function a_{ij} can be so chosen as to be continuous in N.

For $n = 1$ (B) shows that Lemma 1 is true. If Lemma 1 be assumed to be true for the case of $n-1$ independent variables, then

$$F(0, x_2, \cdots, x_n) = \sum_{ij} b_{ij}(x_2, \cdots, x_n) x_i x_j \qquad (i, j = 2, 3, \cdots, n) \tag{13}$$

where each b_{ij} is continuous in N. If $F(x_1, x_2, \cdots, x_n)$ satisfies the hypotheses of this lemma, $F_1(0, x_2, \cdots, x_n)$ will satisfy the hypotheses (C) so that

$$F_1(0, x_2, \cdots, x_n) = \sum_{i} c_i(x_2, \cdots, x_n) x_i \qquad (i = 2, 3, \cdots, n) \tag{14}$$

where each c_i is continuous in N. If now in (7) $F(0, x_2, \cdots, x_n)$ and $F_1(0, x_2, \cdots, x_n)$ be replaced by their values from (13) and (14) respectively, then Lemma 1 follows as stated.

To prove Lemma 2, the following results, (A′), (B′), and (C′), will first be proved.

(A′) *If F possesses continuous second partial derivatives in N then the function $a(x_1, x_2, \cdots, x_n)$ of* (A) *as defined by* (5) *and* (6) *has continuous first partial derivatives in N.*

It follows from (5) that $a(x_1, x_2, \cdots, x_n)$ has continuous first partial derivatives in N if $x_1 \neq 0$. For $x_1 = 0$ the partial derivatives of $a(x_1, x_2, \cdots, x_n)$ with respect to $(x_2, x_3, \cdots, x_n)$ exist, as follows from (6). For $x_1 \neq 0$ it follows from Taylor's formula that

$$a(x_1, x_2, \cdots, x_n) = F_1(0, x_2, \cdots, x_n) + \frac{F_{11}(\theta x_1, x_2, \cdots, x_n) x_1}{2}, \quad 0 < \theta < 1. \tag{15}$$

From (15) and (6) and the definition of a derivative it follows that the partial derivative

$$a_{x_1}(0, x_2, \cdots, x_n) = \frac{F_{11}(0, x_2, \cdots, x_n)}{2}.$$

To prove $a_{x_1}(x_1, x_2, \cdots, x_n)$ continuous for $x_1 = 0$, observe that from (5) for $x_1 \neq 0$ it follows that

$$a_{x_1} = \frac{x_1 F_1(x_1, x_2, \cdots, x_n) - [F(x_1, x_2, \cdots, x_n) - F(0, x_2, \cdots, x_n)]}{x_1^2},$$

which by two applications of Taylor's formula becomes

$$\frac{x_1[F_1(0, x_2, \cdots, x_n) + F_{11}(\theta x_1, x_2, \cdots, x_n)x_1]}{x_1^2} - \frac{F_1(0, x_2, \cdots, x_n)x_1 + \dfrac{F_{11}(\theta' x_1, x_2, \cdots, x_n)x_1^2}{2}}{x_1^2}$$

$$= F_{11}(\theta x_1, x_2, \cdots, x_n) - \frac{F_{11}(\theta' x_1, x_2, \cdots, x_n)}{2}, \quad 0<\theta<1, \ 0<\theta'<1,$$

whence

$$\lim_{x_1 \to 0} a_{x_1}(x_1, x_2, \cdots, x_n) = \frac{F_{11}(0, x_2, \cdots, x_n)}{2}$$

and a_{x_1} is thereby proved continuous in N without exception.

To show that $a_{x_2}(x_1, x_2, \cdots, x_n)$ is continuous in N for $x_1 = 0$ observe that from (5) for $x_1 \neq 0$ it follows that

$$a_{x_2}(x_1, x_2, \cdots, x_n) = \frac{F_2(x_1, x_2, \cdots, x_n) - F_2(0, x_2, \cdots, x_n)}{x_1},$$

which by Taylor's formula gives

(16) $$a_{x_2}(x_1, x_2, \cdots, x_n) = F_{21}(\theta x_1, x_2, \cdots, x_n), \qquad 0<\theta<1, \ x_1 \neq 0,$$

while from (6)

(17) $$a_{x_2}(0, x_2, \cdots, x_n) = F_{12}(0, x_2, \cdots, x_n).$$

From (16) and (17) it follows that a_{x_2} is continuous in N even when $x_1 = 0$. The remaining partial derivatives of $a(x_1, x_2, \cdots, x_n)$ are similarly seen to be continuous in N.

(B′) *If F possesses continuous third partial derivatives in N, then the function $b(x_1, x_2, \cdots, x_n)$ of* (B) *as defined by* (8) *and* (9) *possesses continuous first partial derivatives in N.*

The proof of (B′) is similar to that of (A′).

(C′) *If F possesses continuous second partial derivatives in N then the functions $a_i(x_1, x_2, \cdots, x_n)$ of* (C) *can be so chosen as to possess continuous first partial derivatives in N.*

A review of the proof of (C) with (A′) in mind will show the truth of (C′).

LEMMA 2. *If F possesses continuous third partial derivatives in N, and vanishes with its first partial derivatives at the origin, then F can be represented in the form*

(18) $$F(x_1, x_2, \cdots, x_n) = \sum_{ij} a_{ij} x_i x_j \qquad (i, j = 1, 2, \cdots, n)$$

where each a_{ij} possesses continuous first partial derivatives in N.

A review of the proof of Lemma 1 with (A′), (B′) and (C′) in mind will show the truth of Lemma 2.

LEMMA 3. *The functions* $a_{ij}(x_1, x_2, \cdots, x_n)$ *of Lemmas* 1 *and* 2 *are such that at the origin*

$$a_{ij} + a_{ji} = F_{ij}. \tag{19}$$

Under the hypotheses of Lemmas 1 and 2

$$F(x_1, x_2, \cdots, x_n) = \frac{\sum_{ij} F_{ij}(\theta_1 x_1, \theta_2 x_2, \cdots, \theta_n x_n)\, x_i x_j}{2}, \quad 0 < \theta_i < 1. \tag{20}$$

Hence from (20) and (18)

$$\sum_{ij} a_{ij} x_i x_j = \frac{\sum_{ij} F_{ij}(\theta_1 x_1, \theta_2 x_2, \cdots, \theta_n x_n)\, x_i x_j}{2}. \tag{21}$$

Let $(y_1, y_2, \cdots, y_n)$ be any point in N, and let r be a positive constant so small that $(ry_1, ry_2, \cdots, ry_n)$ is a point in N for any choice of $(y_1, y_2, \cdots, y_n)$ in N. In (21) set $x_i = ry_i$ and in the resulting equation cancel the factor r^2 from both members. There will then result the equation

$$\sum_{ij} a_{ij}(ry_1, ry_2, \cdots, ry_n)\, y_i y_j = \frac{\sum_{ij} F_{ij}(r\theta_1 y_1, r\theta_2 y_2, \cdots, r\theta_n y_n)\, y_i y_j}{2}. \tag{22}$$

The two members of (22) possess limits as r approaches zero, and these limits are equal, giving

$$\sum_{ij} a_{ij}(0, 0, \cdots, 0)\, y_i y_j = \frac{\sum_{ij} F_{ij}(0, 0, \cdots, 0)\, y_i y_j}{2}. \tag{23}$$

Equation (23) is an identity in the y's. Hence the coefficients of corresponding powers are equal. The equation (19) follows directly and the lemma is proved.

3. Different types of critical points. Let there be given a critical point of f. At this critical point let c be the value of f. If this critical point be supposed to lie at the origin, it follows from Lemma 2 that in the neighborhood of the origin the difference $f - c$ can be represented in the form

$$f(x_1, x_2, \cdots, x_n) - c = \sum_{ij} a_{ij}(x_1, x_2, \cdots, x_n)\, x_i x_j \quad (i, j = 1, 2, \cdots, n) \tag{24}$$

where $a_{ij}(x_1, x_2, \cdots, x_n)$ is a single-valued, continuous function of the x's in a sufficiently small neighborhood of the origin, provided with continuous first partial derivatives with respect to each of the x's. Further without any loss of generality, we can suppose that

$$a_{ij} = a_{ji}. \tag{25}$$

If we set $a_{ij}(0, 0, \cdots, 0) = \alpha_{ij}$, according to Lemma 3, we will have at the origin

$$2\alpha_{ij} = f_{x_i x_j}. \tag{26}$$

Equation (26) shows that the constants α_{ij} appearing here are the same as the constants α_{ij} used in (1). In § 1 we assumed that $|\alpha_{ij}| \neq 0$. We can conclude that

$$|a_{ij}| \neq 0, \tag{27}$$

in a sufficiently small neighborhood of the origin.

According to the well known theory of quadratic forms a non-singular quadratic form in n variables $(x_1, x_2, \cdots, x_n)$, by a real, non-singular, linear transformation of $(x_1, x_2, \cdots, x_n)$ into n new variables $(y_1, y_2, \cdots, y_n)$, can be reduced to a quadratic form

$$-y_1^2 - y_2^2 - \cdots - y_k^2 + y_{k+1}^2 + \cdots + y_n^2 \tag{28}$$

where k is one of the integers from 0 to n, inclusive.* An examination of the Lagrange method of making this reduction shows that it can be carried through in the case where the coefficients a_{ij} are not constants, by formally following the method used for the case where the coefficients a_{ij} are constants, writing down the Lagrange transformation at each stage with the variable coefficients a_{ij} substituted for the constant coefficients a_{ij} of the ordinary theory, provided at each stage at least one of the coefficients a_{ii} is not zero at the origin. If at any stage all of the coefficients a_{ii} are zero at the origin, at least one of the remaining coefficients, say a_{rs}, will not be zero at the origin since $|\alpha_{ij}| \neq 0$. If the preliminary transformation

$$\begin{aligned} x_r &= z_r - z_s, \\ x_s &= z_r + z_s, \\ x_i &= z_i \qquad\qquad (i \neq r, s) \end{aligned}$$

be made the resulting coefficients of z_{rr} and z_{ss} will not be zero at the origin, and the Lagrange type of transformation can then be used. The transformation will of course be restricted to a neighborhood of the origin for which $|a_{ij}| \neq 0$ and such that each a_{ij} by which it has been necessary

* Bôcher, *Introduction to Higher Algebra*, p. 131.

to divide shall not be zero. If the given critical point is not at the origin a proper translation of the space of the n x's will bring the critical point to the origin. The above method of reduction of the function f will then give the following lemma.

LEMMA 4. *If $f(x_1, x_2, \cdots, x_n)$ takes on the value c at a critical point $(x_1^0, x_2^0, \cdots, x_n^0) = (X_0)$ then there exists a real transformation of the form*

$$y_i = \sum_j c_{ij}(x_j - x_j^0), \quad |c_{ij}| \neq 0 \quad (i, j = 1, 2, \cdots, n) \tag{29}$$

under which

$$f(x_1, x_2, \cdots, x_n) - c \equiv -y_1^2 - y_2^2 - \cdots - y_k^2 + y_{k+1}^2 + \cdots + y_n^2 \tag{30}$$

where each c_{ij} is a continuous function of the x's provided with continuous first partial derivatives in a sufficiently small neighborhood of (X_0), and where k is some integer from 0 to n, inclusive.

The integer k is called the type of the given critical point. The number of critical points of the kth type under boundary conditions to be introduced presently will be shown to be finite and will be denoted by m_k. It is the purpose of this paper to obtain the relations between the numbers m_k under various types of boundary conditions.

The jacobian of the functions $(y_1, y_2, \cdots, y_n)$ of (29) with respect to the variables $(x_1, x_2, \cdots, x_n)$ reduces at (X_0) to the value of $|c_{ij}|$ at (X_0). Thus this jacobian is not zero at (X_0). It follows that *the transformation* (29) *is one-to-one in the neighborhood of the given critical point.*

4. **The boundary conditions α.** In order to proceed further it is necessary to make some assumptions regarding boundary conditions. The first boundary conditions that we introduce are, strictly speaking, not boundary conditions alone. They will be shown later to lead to results depending upon boundary conditions in a strict sense. These first boundary conditions which we consider are introduced because of their simplicity.

The boundary conditions α will be considered as fulfilled if among the $(n-1)$-dimensional spreads

$$f(x_1, x_2, \cdots, x_n) = \text{const.}$$

there exists one, say A, with the following properties:

I. *A is a closed manifold;*

II. *A forms the complete boundary of a portion of space, S, including A, all of whose points are points of R;*

III. *Each critical point of f is an interior point of S;*

IV. *The value of f on A is greater than the value of f at any interior point of S.*

The region S as defined in the boundary conditions α is a closed region containing all of the critical points of f. It follows from the results of § 1 that there are at most a finite number of critical points of f. According to condition III there is no critical point of f on A.

5. The complex of points satisfying $f(x_1, x_2, \cdots, x_n) \leqq c$. According to the boundary condition IV the absolute maximum of f in S is taken on at each point on the boundary of S and at no other point of S. If any absolute minimum point of f in S be joined by a continuous curve to some boundary point of S, every constant c between the absolute minimum m of f in S and the absolute maximum M of f in S will be taken on by f at some point in S on this curve. Let c be any such constant between m and M, not a critical value of f. The set of points $(x_1, x_2, \cdots, x_n) = (X)$, lying in S and satisfying

$$f(x_1, x_2, \cdots, x_n) = c \tag{31}$$

according to the choice of c, are all ordinary points of f. This set of points is closed since f is continuous. Its points are all interior points of S since the value of f on S's boundary is M. If (X_0) be any point on (31), at least one of the partial derivatives of f, say f_{x_1}, will not be zero at (X_0). It follows that the points $(x_1, x_2, \cdots, x_n)$ which satisfy (31) in the neighborhood of (X_0) can be represented in the form

$$x_1 = h(x_2, x_3, \cdots, x_n)$$

where $h(x_2, x_3, \cdots, x_n)$ is a single-valued function of its arguments provided with continuous partial derivatives up to the second order all in the neighborhood of (X_0). Geometrically this means that the manifold of points satisfying (31) possesses at each point of (31) a hypertangent plane which turns continuously as the point of tangency varies continuously on (31). The manifold (31) thus has no singular points.

The set of points (X) in S satisfying

$$f \leqq c \tag{32}$$

will make up an n-dimensional region, say C_n. In particular C_n will include all of the points neighboring any absolute minimum of f. Its points are all interior points of S, and its boundary obviously consists of the points (X) satisfying (31). It can be shown that C_n is an n-dimensional complex in the technical sense of analysis situs.* In a proof of this the same sort of considerations would enter as enter in connection with the so called "regular regions" in the plane.†

* See V, loc. cit., Chap. 3.

† W. F. Osgood, *Lehrbuch der Funktionentheorie*, p. 179.

The variation of the connectivity numbers $R_0, R_1, \cdots, R_n$ of C_n with variations of c will now be studied. To that end it will be convenient to consider first the trajectories orthogonal to the manifolds $f =$ const.

6. The trajectories orthogonal to the manifolds $f =$ const. In studying the trajectories orthogonal to the manifolds $f = c$ the neighborhoods of critical points of f will be excluded from consideration. The trajectories orthogonal to the manifolds $f = c$ at ordinary points of f might be defined as those curves along which the differentials $dx_1, dx_2, \cdots, dx_n$ at any point are proportional to the partial derivatives $f_1, f_2, \cdots, f_n$ at that point, where f_i is used to denote the partial derivative of f with respect to x_i $(i = 1, 2, \cdots, n)$. *Two trajectories are considered as the same* if they consist of the same points $(x_1, x_2, \cdots, x_n)$. It is thus explicitly pointed out that a difference of parametric representative alone does not, according to this convention, constitute a difference in the trajectories.

At any ordinary point of f at least one of the partial derivatives of f is not zero. Let (P) be an ordinary point of f and suppose that at (P) the partial derivative $f_m \neq 0$. The trajectories passing through the points neighboring (P) will be solutions of the differential equations

$$\frac{dx_i}{dx_m} = \frac{f_i}{f_m}, \tag{33}$$

where i takes on all the integers from 1 to n except m. According to the fundamental existence theorems for ordinary differential equations there is one and only one trajectory through each point of a sufficiently small neighborhood of (P).

A set of differential equations will now be introduced in which the independent variable is not x_m but a parameter τ, but whose solutions also represent orthogonal trajectories. The equations are

$$\frac{dx_i}{d\tau} = rf_i \qquad (i = 1, 2, \cdots, n) \tag{34}$$

where

$$r = \frac{1}{f_1^2 + f_2^2 + \cdots + f_n^2}.$$

The particular form of dependence of the x's upon τ which is required by the above differential equations (34) is chosen in preference to any other because for a solution $x_1(\tau), x_2(\tau), \cdots, x_n(\tau)$ of (34) the following identity in τ holds:

$$f[x_1(\tau), x_2(\tau), \cdots, x_n(\tau)] \equiv \tau + \text{const.} \tag{35}$$

For if the left hand member of (35) be differentiated with respect to τ, making use of (34) it is found that the resulting derivative is unity. Whence (35) follows directly.

Let there be given an ordinary point (B) with coördinates $(b_1, b_2, \cdots, b_n)$. In a sufficiently small neighborhood of (B) the denominator of the quotient in (34) giving r is not zero, and in such a neighborhood the right hand members of the differential equations (34) possess continuous partial derivatives with respect to each of the x's. Further these right hand members do not involve τ at all. It follows from the fundamental existence theorems of ordinary differential equations that there exists a positive constant e so small that for any point $(A) = (a_1, a_2, \cdots, a_n)$ in the e neighborhood of (B) and for any constant τ_0 whatsoever, and range of τ within e of τ_0, there exist functions

$$x_i = h_i(a_1, a_2, \cdots, a_n, \tau_0, \tau) \qquad (i = 1, 2, \cdots, n) \tag{35'}$$

which give for constant values of $(a_1, a_2, \cdots, a_n, \tau_0)$ and for a variation of τ, a solution of (34) that passes through (A) when $\tau = \tau_0$, and is the only such solution of (34). Further, for the above prescribed domain of $(a_1, a_2, \cdots, a_n, \tau_0)$ the right hand members of (35′) possess continuous partial derivatives of at least the first order with respect to all of their arguments.

Because the right hand members of (34) do not involve τ it follows that two solutions of (34), for which the points $(a_1, a_2, \cdots, a_n)$ are the same but for which τ_0 differs, will give the same set of points $(x_1, x_2, \cdots, x_n)$, assigned however by (35′) to different values of τ. What is here desired is to obtain all of the orthogonal trajectories in a form which will be the simplest possible for our purposes. This end will be served by setting

$$\tau_0 = f(a_1, a_2, \cdots, a_n).$$

Upon writing

$$h_i[a_1, a_2, \cdots, a_n, f(a_1, a_2, \cdots, a_n), \tau] = H_i(a_1, a_2, \cdots, a_n, \tau) \; (i = 1, 2, \cdots, n)$$

(35′) reduces to

$$x_i = H_i(a_1, a_2, \cdots, a_n, \tau). \tag{36}$$

For initial conditions we now have

$$a_i = H_i\,[a_1, a_2, \cdots, a_n, f(a_1, a_2, \cdots, a_n)], \tag{37}$$

an identity in all of the a_i's in the neighborhood of (B). If the solution of (34), namely $x_1(\tau), x_2(\tau), \cdots, x_n(\tau)$, appearing in (35) be now understood to be a solution given by (36), a substitution in (35) of $\tau = f(a_1, a_2, \cdots, a_n)$ will show that the constant in (35) is zero. Thus for all solutions of (34) given by (36) there results the identity in τ

$$(38) \qquad f[x_1(\tau), x_2(\tau), \cdots, x_n(\tau)] \equiv \tau .$$

Thus we have proved the following lemma:

LEMMA 5. *The trajectories orthogonal to the manifolds $f = c$ can be so represented that the parameter τ at any one of their points equals the value of f at that point.*

7. Invariance of the connectivity numbers of the complex $f \leqq c$ with variation of c through ordinary values of f. The following lemma will now be proved.

LEMMA 6. *Let a and b $(a < b)$ be two ordinary values of f in S such that the interval between a and b contains only ordinary values of f; then the set of points in S satisfying*

$$(39) \qquad f \leqq a$$

can be put into one-to-one continuous correspondence with the set of points in S satisfying

$$(40) \qquad f \leqq b .$$

The constant a is by hypothesis not a critical value of f. Since there are altogether at most a finite number of critical values of f there can be chosen a constant α less than a and differing from a by so little that between an a and α there is no constant equal to a critical value of f. Denote by H a set of orthogonal trajectories in which there is just one trajectory passing through each point of

$$(41) \qquad f = \alpha .$$

Let these trajectories be represented in the form described in Lemma 5 in which the value of the parameter τ at each point equals the value of f at that point. The point on each trajectory that lies on (41) will correspond to the parameter value $\tau = \alpha$.

Since there is no critical value of f equal to a constant between α and b, each trajectory that starts from a point of (41) can be continued through an arc corresponding to a range of values of τ from α to b. On these arcs the parameter values $\tau = a$ and $\tau = b$, respectively, will correspond to points on $f = a$ and $f = b$. To the complete set of values of τ satisfying

$$(42) \qquad \alpha \leqq \tau \leqq b$$

there will correspond points satisfying

$$\alpha \leqq f \leqq b. \tag{43}$$

Further, the set of trajectories H, including as it does a trajectory through each point of (41), if continued through the interval (42) will include at least one trajectory through each point $(x_1, x_2, \cdots, x_n)$ satisfying (43). For if

$$(a_1, a_2, \cdots, a_n) = (A)$$

be any point such that

$$\alpha \leqq f(a_1, a_2, \cdots, a_n) \leqq b \tag{44}$$

a trajectory through (A), represented as described in Lemma 5, will pass through (A) for a parameter value $\tau = f(a_1, a_2, \cdots, a_n)$ and if continued in the sense of decreasing τ will reach a point on (41) when τ reaches the value $\tau = \alpha$. Finally, it follows from the fundamental existence theorem stated in § 6 that there is not more than one of these trajectories through each point satisfying (43).

The correspondence whose existence is affirmed in the lemma can now be set up. First observe that the points satisfying (39) consist of the sum of the points satisfying

$$f \leqq \alpha \tag{45}$$

and those satisfying

$$\alpha < f \leqq a, \tag{46}$$

while the points satisfying (40) consist of the points satisfying (45) together with the points satisfying

$$\alpha < f \leqq b. \tag{47}$$

Thus to put the points satisfying (39) into correspondence with the points satisfying (40) the points satisfying (45) are first made to correspond to themselves. The points satisfying (46) are then put into correspondence with those satisfying (47) by requiring that any point satisfying (46) that arises on the trajectory through it from a value of τ that divides the interval from α to a in a certain ratio, shall correspond to that point satisfying (47) that lies on the same trajectory and arises from a value of τ that divides the interval from α to b in the same ratio.

That the latter correspondence is one-to-one follows from the fact that the trajectories concerned pass through each of the points concerned with one and only one trajectory through each point. That this correspondence is continuous follows from the nature of the dependence of the points of the trajectories on τ and on their initial points on (41). Further, the correspondence of the points satisfying (45) with themselves obviously joins up in a continuous manner with the correspondence of the points satisfying (46) with those satisfying (47). Thus the lemma is proved.

Critical points of the type zero

8. **The connectivity numbers of the complex $f \leqq m + e^2$ where m is the absolute minimum of f.** Let e be a positive constant so small that between m and $m + e^2$ there is no constant equal to a critical value of f. Under the boundary conditions α, § 4, f will take on its absolute minimum at one or more points all interior to S. Let P be one such point. It follows from Lemma 4 of § 3 that the points $(x_1, x_2, \cdots, x_n) = (X)$ neighboring P can be transformed in a one-to-one continuous manner into the points $(y_1, y_2, \cdots, y_n) = (Y)$ neighboring the origin in such a manner that

$$f(x_1, x_2, \cdots, x_n) - m = y_1^2 + y_2^2 + \cdots + y_n^2 .$$

Thus the points in the neighborhood of P satisfying

$$f(x_1, x_2, \cdots, x_n) \leqq m + e^2,$$

for a sufficiently small positive constant e, will correspond in a one-to-one continuous manner to the points satisfying

$$y_1^2 + y_2^2 + \cdots + y_n^2 \leqq e^2$$

or what might be described as the points interior to and on an $(n-1)$-dimensional hypersphere in an n-dimensional space. Such a set of points would constitute, in the terms of analysis situs, an n-cell and its boundary. With this understood we can state the following lemma.

LEMMA 7. *If f takes on its absolute minimum m in S at s distinct points of S, then for a sufficiently small positive constant e the set of points in S satisfying*

$$f(x_1, x_2, \cdots, x_n) \leqq m + e^2$$

constitutes s distinct n-cells and their boundaries.

From this lemma and the definitions of the connectivity numbers* we have the following theorem.

* V, loc. cit.

THEOREM 1. *If f takes on its absolute minimum m in S at s distinct points of S then for a sufficiently small positive constant e the set of points in S satisfying*

$$f(x_1, x_2, \cdots, x_n) \leqq m + e^2$$

makes up an n-dimensional complex with the connectivity numbers $R_0 = s$, $R_1 = R_2 = \cdots = R_n = 1$.

9. A critical point of type zero at which f is not equal to its absolute minimum.

THEOREM 2. *If a and b $(a < b)$ are any two ordinary values of f in S between which there is just one critical value of f, say c, taken on at p_0 critical points all of the zeroth type, then the connectivity numbers R_i, of the set of points in S satisfying*

$$f(x_1, x_2, \cdots, x_n) \leqq b, \tag{48}$$

differ from the connectivity numbers R'_i of the set of points in S satisfying

$$f(x_1, x_2, \cdots, x_n) \leqq a \tag{49}$$

only in that

$$R_0 = R'_0 + p_0 .$$

Let P be any one of the p_0 critical points of the zeroth type. If e be a sufficiently small positive constant it follows from Lemma 4, § 3, that the points $(x_1, x_2, \cdots, x_n)$ neighboring P and satisfying

$$f \leqq c + e^2 \tag{50}$$

can be made to correspond in a one-to-one continuous manner to a set of points $(y_1, y_2, \cdots, y_n)$ satisfying

$$y_1^2 + y_2^2 + \cdots + y_n^2 \leqq e^2 . \tag{51}$$

Now the points satisfying (51) make up an n-cell and its boundary. There will thus be associated with each of the p_0 critical points an n-cell and its boundary consisting of points satisfying (50). For e sufficiently small these n-cells will be distinct from each other and from the set of remaining points, say C_n, satisfying (50).

The set C_n is bounded by points at which $f = c + e^2$, and contains only those critical points which are already contained among the points satisfying (49). With the aid of the orthogonal trajectories as used in the proof of Lemma 6, it is easy to show that C_n can be put into one-to-one continuous

28

correspondence with the set of points satisfying (49). The connectivity numbers of C_n are accordingly equal to the connectivity numbers R_i' of the complex (49).

If now the p_0 n-cells and their boundaries be added to the complex C_n there will result the entire set of points satisfying (50) and the only connectivity number to be thereby changed will be R_0', which will be increased by p_0. If finally we suppose e originally chosen so small that $c+e^2 < b$, it follows from Lemma 6 that the set of points satisfying (50) can be put into one-to-one correspondence with the set of points satisfying (48). Hence (48) defines a complex with the same connectivity numbers as the complex defined by (50). Retracing the steps it is seen that the relation between the connectivity numbers of the complexes (48) and (49) is as stated.

CRITICAL POINTS OF TYPES $1, 2, \cdots, n-1$

10. Outline of the method to be followed. Throughout this chapter we shall assume that we are dealing with a critical value c taken on by f at just one critical point of the kth type, where k may be any one of the integers $1, 2, \cdots, n-1$. We can and will suppose e to be so small a positive constant that there is no critical value of f between $c-e^2$ and $c+e^2$ or equal to $c-e^2$ or $c+e^2$. It is the aim of this chapter to determine the difference between the connectivity numbers of the complex of points (X) satisfying

$$f \leqq c-e^2 \tag{52}$$

and the complex of points (X) satisfying

$$f \leqq c+e^2. \tag{53}$$

To accomplish this we will first show that the complex (52) can be put into one-to-one continuous correspondence with the complex (53) provided we exclude a properly chosen neighborhood of the given critical point. This will be accomplished with the aid of the orthogonal trajectories already used in § 7. The problem then will be resolved into one of determining the difference between the connectivities of the complex (53) and of the complex (53) with this neighborhood excluded.

11. Simultaneous reduction of $f(x_1, x_2, \cdots, x_n)$ and the differential equations of the orthogonal trajectories to a simple form. It follows from Lemma 2, § 2, that, in the neighborhood of a critical point $(x_1^0, x_2^0, \cdots, x_n^0)$ at which $f = c$,

(54) $$f - c = \sum_{ij} b_{ij}(x_i - x_i^0)(x_j - x_j^0) \qquad (i, j = 1, 2, \cdots, n),$$

where each b_{ij} is a continuous function of the x's possessing continuous first partial derivatives, and where we can suppose that $b_{ij} = b_{ji}$. In the following the matrix whose element in the ith row and jth column is given by a symbol such as b_{ij} will itself be denoted be the same symbol b without the subscripts. Conjugates of matrices will be indicated by adding a prime.

LEMMA A. *In the neighborhood of a critical point* $(x_1^0, x_2^0, \cdots, x_n^0) = (X_0)$ *there exists a real transformation of the form*

(55) $$x_i - x_i^0 = \sum_j a_{ij} z_j \qquad (i, j = 1, 2, \cdots, n)$$

such that (54) *becomes*

(56) $$\sum_{ij} b_{ij}(x_i - x_i^0)(x_j - x_j^0) = r_1 z_1^2 + \cdots + r_n z_n^2,$$

where $r_1, r_2, \cdots, r_k$ *are negative constants and* $r_{k+1}, r_{k+2}, \cdots, r_n$ *are positive constants, while in* (55) *each* a_{ij} *is a function of the* z*'s possessing continuous first partial derivatives in the neighborhood of the origin of the space of the* z*'s, and at the origin the matrix equation*

(57) $$a(0, 0, \cdots, 0)\, a'(0, 0, \cdots, 0) = \mathrm{I}$$

holds where I *is the unit matrix.*

Proof. From the results of § 3 it follows that in the neighborhood of the given critical point there exists a transformation

(58) $$x_i - x_i^0 = \sum_j e_{ij} y_j \qquad (i, j = 1, 2, \cdots, n)$$

under which

(59) $$\sum_{ij} b_{ij}(x_i - x_i^0)(x_j - x_j^0) = -y_1^2 - y_2^2 - \cdots - y_k^2 + y_{k+1}^2 + y_{k+2}^2 + \cdots + y_n^2$$

where each e_{ij} is a function of the y's possessing continuous first partial derivatives in the neighborhood of the origin of the space of the y's. Denote by η_{ij} the value of e_{ij} at the origin of the y's. Suppose that under the transformation

(60) $$x_i = \sum_j \eta_{ij} y_j \qquad (i, j = 1, 2, \cdots, n),$$

(61) $$\sum_i x_i^2 = \sum_{ij} \gamma_{ij} y_i y_j.$$

23*

According to the theory of pairs of quadratic forms the quadratic forms in the right hand members of (59) and (61), by a real non-singular transformation with constant coefficients of the form

$$y_i = \sum_j \delta_{ij} z_j \qquad (i, j = 1, 2, \cdots, n) \tag{62}$$

can be reduced to a pair of forms

$$r_1 z_1^2 + r_2 z_2^2 + \cdots + r_n z_n^2,$$
$$z_1^2 + z_2^2 + \cdots + z_n^2.$$

The transformation obtained by expressing the x's in (58) in terms of the z's in (62) is the one whose existence is affirmed in the lemma. For the matrix of this combined transformation, namely $e\delta$, reduces at the origin to $\eta\delta$, and $\eta\delta$ is the matrix of a transformation under which

$$\sum_{i=1}^{n} x_i^2 = \sum_{i=1}^{n} z_i^2,$$

so that $\eta\delta$ is orthogonal as (57) requires. The remaining assertions of Lemma A follow directly.

LEMMA B. *In the neighborhood of the critical point* (X_0) *there exists a real transformation*

$$x_i - x_i^0 = \sum_j g_{ij} y_j \qquad (i, j = 1, 2, \cdots, n) \tag{63}$$

such that

$$\sum_{ij} b_{ij}(x_i - x_i^0)(x_j - x_j^0) = -y_1^2 - y_2^2 - \cdots - y_k^2 + y_{k+1}^2 + \cdots + y_n^2, \tag{64}$$

while the differential equations of the orthogonal trajectories to the manifolds $f =$ const. *if taken in the form*

$$\frac{dx_h}{dt} = f_h \qquad (h = 1, 2, \cdots, n) \tag{65}$$

are transformed into the equations

$$\frac{dy_h}{dt} = 2r_h y_h + \sum_{ij} A_{hij} y_i y_j \quad (h, i, j = 1, 2, \cdots, n) \tag{66}$$

where $r_1, r_2, \cdots, r_k$ *are positive constants and* $r_{k+1}, r_{k+2}, \cdots, r_n$ *are negative constants, where each* g_{ij} *and* A_{hij} *is a continuous function of the* y's *in the neighborhood of the origin and the* g_{ij} *there possess continuous first partial derivatives.*

Proof. With the aid of the representation of f given in (54), (65) becomes

$$\frac{dx_h}{dt} = 2\sum_i b_{hi}(x_i - x_i^0) + \sum_{ij}\frac{\partial b_{ij}}{\partial x_h}(x_i - x_i^0)(x_j - x_j^0). \tag{67}$$

It follows from (C), § 2, that each b_{hi} equals its value, say β_{hi} at (X_0), plus a sum of terms each a product of a difference $(x_j - x_j^0)$ multiplied by a function of the x's continuous in the neighborhood of (X_0). Thus (67) becomes

$$\frac{dx_h}{dt} = 2\sum_i \beta_{hi}(x_i - x_i^0) + \sum_{ij} B_{hij}(x_i - x_i^0)(x_j - x_j^0) \tag{68}$$

where each B_{hij} is a continuous function of the x's in the neighborhood of the origin. The transformation (55) of Lemma A will be applied to (68). Remembering that the value of a_{ij} at the origin is written as α_{ij}, it is seen that (55) may be written

$$x_h - x_h^0 = \sum_i \alpha_{hi} z_i + \sum_{ij} C_{hij} z_i z_j \tag{69}$$

where each C_{hij} is a continuous function of the z's in the neighborhood of their origin.

The transformation (55) applied to (68) gives

$$\frac{dz_h}{dt} = 2\sum_i p_{hi} z_i + \sum_{ij} D_{hij} z_i z_j, \tag{70}$$

where the elements p_{hi} are constants such that, in matrix notation,

$$p = \alpha^{-1}\beta\alpha \tag{71}$$

and each D_{hij} is a continuous function of the z's in the neighborhood of their origin. But according to (57) $\alpha\alpha' = 1$ so that $\alpha' = \alpha^{-1}$. Hence (71) may be written as

$$p = \alpha'\beta\alpha. \tag{72}$$

But from (55) and (56) it follows that

$$a'ba = r. \tag{73}$$

Matrix relation (73) becomes at the origin

$$\alpha'\beta\alpha = r \tag{74}$$

so that from (74) and (72) we have

$$p = \bar{r}.$$

Thus (70) takes the form

$$\frac{dz_h}{dt} = 2 r_h z_h + \sum_{ij} D_{hij} z_i z_j. \tag{75}$$

Finally the transformation

$$y_h = \sqrt{|r_h|}\, z_h \qquad (h = 1, 2, \cdots, n) \tag{76}$$

changes the right hand member of (56) to the right hand member of (64) and takes (75) into the form (66). Thus the transformation obtained by combining (55) and (76) so as to eliminate the z's is of the nature required of (63) and the lemma is proved.

12. **The correspondence between the complex $f \leqq c - e^2$ and the complex $f \leqq c + e^2$ with the points (79) excluded.** We are concerned with a critical value c taken on by f at just one critical point (X_0) of the kth type $(0 < k < n)$. Let the points (X) in the neighborhood of (X_0) be subjected to the transformation (63) of Lemma B, § 11, so that in terms of the new variables $(y_1, y_2, \cdots, y_n) = (Y)$

$$f - c = q^2 - p^2 \tag{77}$$

where

$$\begin{aligned} p^2 &= y_1^2 + y_2^2 + \cdots + y_k^2, \\ q^2 &= y_{k+1}^2 \quad + \cdots + y_n^2. \end{aligned} \tag{78}$$

It is the purpose of this section to show that for a proper choice of constants a and $e\,(a > e > 0)$, the complex of points satisfying $f \leqq c - e^2$ can be put into one-to-one continuous correspondence with the complex of points satisfying $f \leqq c + e^2$ provided we exclude from the latter complex those points which satisfy

$$c - e^2 < f \leqq c + e^2, \qquad p^2 < a^2. \tag{79}$$

The existence of this correspondence is affirmed in slightly different terms in the following lemma.

LEMMA 8. *If c be a critical value of f taken on by f at just one critical point (X_0), and if (X_0) is of the type for which $0 < k < n$, then for a proper choice of a and e, $a > e > 0$, the points (X) satisfying*

$$f \leqq c - e^2 \tag{80}$$

can be put into one-to-one continuous correspondence with the set of points consisting of the points (80) *together with the points satisfying*

$$c - e^2 < f \leqq c + e^2, \qquad p^2 \geqq a^2.* \tag{81}$$

The latter inequality $p^2 \geqq a^2$ has a meaning and is intended to have a meaning only where the variables $y_1, y_2, \cdots, y_n$ are defined (Lemma B), that is, in the neighborhood of the given critical point. To prove this lemma, Lemmas C, D, E, F, and G will first be established. For this purpose the boundary points of (81) will be grouped into the following sets of points: the points on $f = c + e^2$ satisfying

$$f = c + e^2, \qquad p^2 > a^2,* \tag{82}$$

the points on $f = c - e^2$ satisfying

$$f = c - e^2, \qquad p^2 > a^2,* \tag{83}$$

the points on $p^2 = a^2$ satisfying

$$p^2 = a^2, \qquad c - e^2 < f \leqq c + e^2, \tag{84}$$

and the points

$$p^2 = a^2, \qquad f = c - e^2. \tag{85}$$

LEMMA C. *There exists a positive constant d so small that when a is chosen less than d, e can then be chosen so small that none of the trajectories orthogonal to the manifolds f* = const. *passes through more than one point of* (84) *or its boundary.*

Proof of Lemma C. For e sufficiently small each point of (84) lies arbitrarily near some point satisfying

$$p^2 = q^2 = a^2. \tag{86}$$

Let $(B) = (b_1, b_2, \cdots, b_n)$ be a point satisfying (86). Let $(U) = (u_1, u_2, \cdots, u_n)$ be a point in the space of the y's near (B). The solutions of (66) which pass through (U) when $t = 0$ are representable for t near 0 in the form

$$y_h = Y_h(u_1, u_2, \cdots, u_n, t) \qquad (h = 1, 2, \cdots, n), \tag{87}$$

for which

$$u_h \equiv Y_h(u_1, u_2, \cdots, u_n, 0), \tag{88}$$

* Where the latter condition has no meaning through lack of definition of p it will be supposed satisfied vacuously.

where each Y_h possesses continuous partial derivatives with respect to all of its arguments for (U) neighboring (B) and t neighboring 0.

If now any point (U) neighboring (B) be given, the trajectory passing through (U) for $t = 0$ meets the manifold $p^2 = a^2$ in as many points as there are solutions $(y_1, y_2, \cdots, y_n, t)$ of the equations

$$
\begin{aligned}
y_h &= Y_h(u_1, u_2, \cdots, u_n, t), \\
p^2 &= a^2.
\end{aligned}
\tag{89}
$$

Equations (89) will be written in the form

$$
\begin{aligned}
y_h - Y_h(u_1, u_2, \cdots, u_n, t) &= 0, \\
y_1^2 + y_2^2 + \cdots + y_k^2 - a^2 &= 0.
\end{aligned}
\tag{90}
$$

The jacobian J of the left hand members of (90) with respect to $y_1, y_2, \cdots, y_n$ and t for $t = 0$ is obtained by bordering an n-square unit matrix I as follows (Lemma B, § 11):

$$
J = \left| \begin{array}{c|c}
 & -2r_1 y_1 + \cdots \\
 & -2r_2 y_2 + \cdots \\
\mathrm{I} & \cdot \quad \cdot \quad \cdot \quad \cdot \quad \cdot \quad \cdot \\
 & \cdot \quad \cdot \quad \cdot \quad \cdot \quad \cdot \quad \cdot \\
 & \cdot \quad \cdot \quad \cdot \quad \cdot \quad \cdot \quad \cdot \\
 & -2r_n y_n + \cdots \\
\hline
2y_1, \quad 2y_2, \cdots, 2y_k, \quad 0, \cdots, 0 & 0
\end{array} \right|
$$

where the terms omitted in the last column are of the form $A_{ij}\, y_i\, y_j$, where each A_{ij} is a continuous function of the y's in the neighborhood of the origin. The value of this jacobian when $(U) = (Y) = (b_1, b_2, \cdots, b_n)$ is

$$
J = 4 \sum_h r_h b_h^2 + \sum_{mij} B_{mij}\, b_m\, b_i\, b_j \qquad \begin{pmatrix} m, i, j = 1, 2, \cdots, n; \\ h \quad = 1, 2, \cdots, k \end{pmatrix}
\tag{91}
$$

where each B_{mij} is a continuous function of the b's in the neighborhood of the origin.

Recalling that $r_1, r_2, \cdots, r_k$ are all negative constants, let m and M be two positive constants such that

$$
m < -4r_h < M \qquad (h = 1, 2, \cdots, k).
$$

Then

$$
m \sum b_h^2 < -4 \sum r_h b_h^2 < M \sum b_h^2.
\tag{92}
$$

Since the point (B) satisfies (86),

$$a^2 = \sum_h b_h^2.$$

Hence from (92)

$$m a^2 < -4 \sum_h r_h b_h^2 < M a^2 \qquad (h = 1, 2, \cdots, k).$$

That is

(93) $$4 \sum_h r_h b_h^2 = R a^2,$$

where R is a function of $(b_1, b_2, \cdots, b_n)$ such that

(94) $$m < |R| < M.$$

Now the point (B) satisfies $p^2 = q^2 = a^2$, from which it follows that

(95) $$|b_i| \leqq a.$$

Moreover since the functions B_{mij} in (91) are continuous functions of the point (B) for (B) in the neighborhood of the origin, there exists a constant H such that for that neighborhood

(96) $$|B_{mij}| < H.$$

For this neighborhood it follows from (95) and (96) that

$$\sum_{mij} B_{mij} b_m b_i b_j < n^3 H a^3,$$

where n^3 appears since there are not more than n^3 terms $b_m b_i b_j$. Thus

$$\sum_{mij} B_{mij} b_m b_i b_j = T a^3,$$

where T is a function of the point (B) such that in the neighborhood of the origin

(97) $$|T| < n^3 H.$$

From (91), (93), and (97) it follows that

$$J = R a^2 + T a^3.$$

We can and do hereby choose a positive constant d so small that for $0 < a < d$

$$J \neq 0.$$

For this choice of a none of the segments of the trajectories with points all lying sufficiently near (B) pass through more than one point of $p^2 = a^2$ neighboring (B). To show that none of these trajectories return to points of (84) it will be useful to take again that representation of the trajectories in which the parameter τ at any point on a trajectory equals the value of f at that point. For e sufficiently small the points on the trajectories passing through some point in a sufficiently small neighborhood of (B) and for which

$$c - e^2 \leqq \tau \leqq c + e^2$$

will all lie arbitrarily near (B) and hence have at most one intersection with $p^2 = a^2$. But these trajectories no matter how extended will not return to points of (84) or its boundary, for points on the extended trajectories will be points for which

$$\tau > c + e^2 \quad \text{or} \quad \tau < c - e^2$$

and hence will be points at which

$$f > c + e^2 \quad \text{or} \quad f < c - e^2,$$

and hence will not be points satisfying (84) or its boundary.

The following lemma follows from the above proof, and in particular from the non-vanishing of the above jacobian.

LEMMA D. *For a choice of a such that the above jacobian is not zero and for any point (U) sufficiently near some point of*

$$p^2 = q^2 = a^2,$$

the trajectory passing through (U) for $t = 0$ meets the manifold

$$p^2 = a^2$$

in a point (V) whose coördinates and parameter value t are continuous functions of the coördinates of (U) and are provided with continuous first partial derivatives.

The proof of the following lemma could be given along the same lines as the proof of Lemmas C and D.

LEMMA E. *For a sufficiently small positive constant a, and corresponding to this a, a sufficiently small positive constant e, and for any point (U) sufficiently near some point of*

$$p^2 = q^2 = a^2,$$

the trajectory which passes through (U) *when* $t = 0$ *meets the manifold*

$$p^2 - q^2 = e^2$$

in a point (Y) *whose coördinates and parameter value* t *are continuous functions of the coördinates of* (U) *and are provided with continuous first partial derivatives.*

The following lemma will now be proved:

LEMMA F. *For a sufficiently small positive constant* a, *and corresponding to this* a, *a sufficiently small positive constant* e, *all of the trajectories passing through points of* (84) *pass through points of* (83).

The equations (84) may be written in terms of the point (Y) as

$$-e^2 \leqq p^2 - q^2 < e^2, \qquad p^2 = a^2, \tag{98}$$

or again as

$$p^2 = a^2, \qquad a^2 - e^2 < q^2 \leqq a^2 + e^2, \tag{99}$$

while (83) may be written as

$$p^2 - q^2 = e^2, \qquad p^2 > a^2. \tag{100}$$

To distinguish between a point (Y) on (99) and a point (Y) on (100), any point (Y) on (99) will be written as $(Y) = (U) = (u_1, u_2, \cdots, u_n)$. Now (99) and (100) have as common boundary points the points satisfying

$$p^2 = a^2, \qquad q^2 = a^2 - e^2. \tag{101}$$

Let $(B) = (b_1, b_2, \cdots, b_n)$ be any point on (101). The complete set of points on (99) consists of the points on the following family of straight lines:

$$\begin{aligned} u_1 &= b_1, \\ u_2 &= b_2, \\ &\cdots\cdots, \\ u_k &= b_k, \\ u_{k+1} &= m\, b_{k+1}, \\ &\cdots\cdots\cdots, \\ u_n &= m\, b_n, \end{aligned} \tag{102}$$

where m is a parameter varying on each straight line on the interval

$$1 < m \leqq \sqrt{\frac{a^2 + e^2}{a^2 - e^2}}. \tag{103}$$

For $m = 1$, $(U) = (B)$. If m, considered as the single independent variable, vary on its interval (103), (U) will then vary on the straight line (102), and the trajectory through (U), according to Lemma E, will meet $p^2 - q^2 = e^2$ in a point (Y) whose coördinates and parameter value t will be continuous functions of m provided with continuous derivatives with respect to m. (U), (Y), and t, thus depending upon m, satisfy (87) making (87) an indentity in m. The lemma will follow readily if it can be shown that as m increases on its interval, and (Y) depends on m as just described, $p^2 = y_1^2 + y_2^2 + \cdots + y_k^2$, considered as a function of m, increases.

To this end let (87) be differentiated with respect to m. At $m = 1$ and hence $t = 0$ (cf. (66)),

$$\frac{dy_h}{dm} = \left[2r_h y_h + \sum_{ij} A_{hij} y_i y_j\right]\frac{dt}{dm} \qquad \begin{pmatrix} h = 1, 2, \cdots, k; \\ i, j = 1, 2, \cdots, n \end{pmatrix}, \tag{104}$$

$$\frac{dy_h}{dm} = b_h + \left[2r_h y_h + \sum_{ij} A_{hij} y_i y_j\right]\frac{dt}{dm} \qquad (h = k+1, \cdots, n). \tag{105}$$

If the two members of the equation in (104) which gives dy_h/dm be multiplied by $2y_h$ $(h = 1, 2, \cdots, k)$ and the resulting k equations added, the left hand sum will become dp^2/dm, and if in the right hand sum we set $(Y) = (U) = (B)$ we will have

$$\frac{dp^2}{dm} = 4\left[\sum_h r_h b_h^2 + \cdots\right]\frac{dt}{dm} \qquad (h = 1, 2, \cdots, k), \tag{106}$$

where the terms omitted here and in the subsequent equations of this proof are sums of terms of the form

$$B_{hij} b_h b_i b_j,$$

where B_{hij} is a continuous function of all of the b's in the neighborhood of the origin.

Similarly, upon multiplying the two members of the equation in (105) that gives dy_h/dm by $2y_h$ $(h = k+1, \cdots, n)$ and adding the resulting $n-k$ equations we will have at $(Y) = (U) = (B)$, using (101),

$$\frac{dq^2}{dm} = 2(a^2 - e^2) + 4\left[\sum_h r_h b_h^2 + \cdots\right]\frac{dt}{dm} \qquad (h = k+1, \cdots, n). \tag{107}$$

Now the coördinates of (Y) considered as the above functions of m satisfy $p^2 - q^2 = e^2$ identically. Thus

$$\frac{dq^2}{dm} = \frac{dp^2}{dm}. \tag{108}$$

Upon eliminating $\frac{dq^2}{dm}$ and $\frac{dt}{dm}$ from (106), (107), and (108), we have

$$\frac{dp^2}{dm} = \frac{2\left[\sum_h r_h b_h^2 + \cdots\right](a^2-e^2)}{\sum_h r_h b_h^2 - \sum_i r_i b_i^2 + \cdots} \quad \begin{pmatrix} h = 1, 2, & \cdots, k; \\ i = k+1, & \cdots, n \end{pmatrix}$$

(109)

$$= \frac{2\left[\sum_h r_h b_h^2 + \cdots\right](a^2-e^2)}{-\sum_j |r_j| b_j^2 + \cdots} \quad (j = 1, 2, \cdots, n).$$

Now if a be a sufficiently small positive constant and e be less, say, than $a/2$, it can be shown by the methods used at the end of the proof of Lemma C that the expression (109) has the sign of

$$\frac{2\sum_h r_h b_h^2}{-\sum_j |r_j| b_j^2},$$

and hence is positive.

For this choice of a and e, dp^2/dm is then positive at (B), that is, when $m = 1$. For a fixed $(a>0)$ and for e sufficiently small, the interval (103) will be arbitrarily small. Hence for e sufficiently small dp^2/dm will be positive throughout the interval (103). But for $m = 1$, $p^2 = a^2$. Thus on the interval (103) $p^2 > a^2$, and the lemma follows directly.

The preceding proof requires only a formal alteration to serve as the proof of the following lemma.

LEMMA G. *For a sufficiently small positive constant a and, corresponding to this a, a sufficiently small positive constant e, none of the trajectories passing through points of* (84) *or its boundary will pass through a point of* (82).

Proof of Lemma 8. Let us return to that form of representation of the orthogonal trajectories in which the parameter τ at any point on a trajectory equals the value of f at that point. Let (X) be any interior point of (81). On the trajectory through (X) there will be a first point P preceding (X) in the sense of a decreasing τ which is on the boundary of (81), and a first point Q following (X) in the sense of an increasing τ which is on the boundary of (81). It follows from Lemmas C and G that neither P nor Q can lie on (85). P will lie then either on (83) or else on (84), while Q will lie on (82) or else on (84). There are four possibilities:

(a) P on (84), Q on (84),

(b) P on (84), Q on (82),

(c) P on (83), Q on (84),
(d) P on (83), Q on (82).

The combination (a) is impossible according to Lemma C, while combination (b) is impossible according to Lemma G.

Thus P always lies on (83). Further, the points P will include each point P_1 on (83). For there are no boundary points of (81) other than points $f = c - e^2$ neighboring P_1. Hence all of the points neighboring P_1 for which $f > c - e^2$ will be interior points of (81). Hence on a trajectory through P_1 the points neighboring P_1 for which $\tau > c - e^2$ will be interior points of (81). Thus P_1 will be one of the points P.

According to (c) and (d) all of the points Q lie either on (82) or else on (84). Further, the points Q include all of the points of (82), a fact which can be proved after the manner of proof in the preceding paragraph. The points Q also include all the points Q_1 of (84). For according to Lemma F a trajectory which passes through a point Q_1 on (84) in the sense of a decreasing τ will meet (83) in a point, say P_1. Now any point of (83) such as P_1 has just been proved to be a point P. Since there can be no point on (82) or (84) between P_1 and Q_1, Q_1 must be a point Q. Thus the points Q include all of the points of (82) and (84). To sum up, the segments PQ include all of the points of (81) and the boundary of (81) save the points of (85). At each of the points of (85) it is convenient to suppose a point P coincident with a point Q.

Let τ_1 and τ_2 be respectively the values of τ at a point P and corresponding Q. The value of τ_1 is $c - e^2$. The position of Q and value of τ_2 are readily seen to be continuous functions of the position of P. To define the correspondence whose existence is affirmed in Lemma 8 let each trajectory PQ be continued from P in the sense of a decreasing τ to a point R at which $\tau = \tau_0$, where

$$\tau_1 - \tau_0 = \tau_2 - \tau_1 .$$

The new trajectory segments RP will consist of points satisfying (80). Now let every point of (80) not on a segment RP correspond to itself, and let every point of (80) with parameter value τ on a segment RP correspond to that point on RPQ with parameter value τ', at which

$$(\tau' - \tau_0) = 2\,(\tau - \tau_0).$$

The correspondence thereby established is readily seen to be of the nature required in the lemma to be proved.

13. **Incidence relations between the boundary of D_n and the remainder of the complex $f \leq c + e^2$.** Following the general outline

of the method given in § 10 this section is devoted first to proving that the set of points D_n satisfying

$$-e^2 \leqq q^2 - p^2 \leqq e^2, \qquad p^2 \leqq a^2, \qquad 0 < e < a, \tag{110}$$

make up an n-cell a_n and its boundary, and secondly to investigating the incidence relations between D_n and the remainder of the complex of points satisfying $f \leqq c + e^2$.

For the moment let p and q in (110) be interpreted as the rectangular cartesian coördinates (pq) of a point in a p, q plane. So interpreted the points (pq) which satisfy (110) consist of all those points lying between the two conjugate hyperbolas

$$p^2 - q^2 = e^2, \qquad q^2 - p^2 = e^2$$

which also lie between the two straight lines

$$p = a, \qquad p = -a.$$

Since $a > e$ these two straight lines meet the two conjugate hyperbolas altogether in eight real points. The two-dimensional set of points in the (pq) plane satisfying (110) will be denoted by D_2. D_2 contains the origin and is symmetrical with respect to it.

We will now consider a transformation T of the coördinates (pq) into rectangular coördinates (uv) which carries D_2 into a rectangle E_2 with vertices at the intersections of the straight lines $p = \pm a$ with the hyperbola $q^2 - p^2 = e^2$. By the methods of elementary analytic geometry it is seen that a straight line segment issuing from the origin in the (pq) plane in any direction whatsoever meets the boundary of D_2 in just one point whose distance from the origin varies continuously with the angle the given line segment makes with the p axis. The same is true of the boundary of the rectangle E_2. D_2 and E_2 can be put into a one-to-one continuous correspondence T by requiring that a point (pq) on D_2 that divides the straight line segment from the origin to the boundary of D_2 in a particular ratio shall correspond on E_2 to a point on the same straight line and so placed as to divide the line segment from the origin to the boundary of E_2 in the same ratio.

The following properties of the transformation T will be used.

(A) Under the transformation T the point (pq) is related to its correspondent (uv) as follows:

$$\begin{aligned} u &= r(p^2 q^2)\, p, \\ v &= r(p^2 q^2)\, q, \end{aligned} \qquad r(00) = 0, \tag{111}$$

where $r(p^2 q^2)$ is a positive, single-valued, continuous function of p^2 and q^2 for each pair (pq) in D_2 not equal to (00). The pair (pq) is obviously also related to its correspondent (uv) as follows:

$$(112)\qquad \begin{aligned} p &= \sigma(u^2 v^2)\, u, \\ q &= \sigma(u^2 v^2)\, v, \end{aligned} \qquad \sigma(00) = 0,$$

where $\sigma(u^2 v^2)$ is a positive, single-valued, continuous function of u^2 and v^2 for each pair (uv) in E_2 not equal to (00).

(B) Under T the points (pq) on D_2, that is the points satisfying

$$(113)\qquad -e^2 \leqq q^2 - p^2 \leqq e^2, \qquad p^2 \leqq a^2,$$

correspond to the points (uv) on E_2, that is the points satisfying

$$(114)\qquad u^2 \leqq a^2, \qquad v^2 \leqq b^2,$$

where we have set $\qquad a^2 + e^2 = b^2.$

(C) The points on the boundary of D_2 satisfying

$$(115)\qquad p^2 - q^2 = e^2, \qquad q^2 \leqq a^2 - e^2,$$

taken with the points on the boundary of D_2 which satisfy

$$(116)\qquad p^2 = a^2, \qquad a^2 - e^2 < q^2 \leqq a^2 + e^2,$$

form two continuous curve segments which correspond under T to those two sides of the rectangle E_2 which satisfy

$$(117)\qquad u^2 = a^2, \qquad v^2 \leqq b^2.$$

The points (pq) satisfying (115) are the two finite curve segments cut off from the hyperbola $p^2 - q^2 = e^2$ by the straight lines $p = \pm a$, while the points (pq) satisfying (116) make up the four finite segments of the straight lines $p = \pm a$ which lie between the two conjugate hyperbolas.

(D) The points on the boundary of D_2 which satisfy

$$(118)\qquad q^2 - p^2 = e^2, \qquad p^2 < a^2,$$

correspond under T to those two sides of the rectangle E_2 which satisfy

$$(119)\qquad u^2 = b^2, \qquad v^2 < a^2.$$

The points (pq) satisfying (118) obviously make up those two finite segments of the hyperbola $q^2 - p^2 = e^2$ which lie between the two straight lines $p = \pm a$.

The results just obtained properly interpreted will give us the following lemma. In this lemma we will again understand by p^2 and q^2 the expressions

$$(120) \qquad \begin{aligned} p^2 &= y_1^2 + y_2^2 + \cdots + y_k^2, \\ q^2 &= y_{k+1}^2 + y_{k+2}^2 + \cdots + y_n^2, \end{aligned}$$

and in addition, in terms of the variables $(z_1, z_2, \cdots, z_n) = (Z)$ we set

$$(121) \qquad \begin{aligned} u^2 &= z_1^2 + z_2^2 + \cdots + z_k^2, \\ v^2 &= z_{k+1}^2 + z_{k+2}^2 + \cdots + z_n^2. \end{aligned}$$

LEMMA 9. *There exists a one-to-one continuous correspondence between a set of points* $(y_1, y_2, \cdots, y_n) = (Y)$ *and a set of points* $(z_1, z_2, \cdots, z_n) = (Z)$ (a) *under which the set of points* (Y) *satisfying*

$$(122) \qquad -e^2 \leqq q^2 - p^2 \leqq e^2, \qquad p^2 \leqq a^2$$

corresponds to the points (Z) *satisfying*

$$(123) \qquad u^2 \leqq a^2, \qquad v^2 \leqq b^2,$$

(b) *while that part of the boundary of* (122) *that consists of points* (Y) *satisfying*

$$(124) \qquad p^2 - q^2 = e^2, \qquad q^2 \leqq a^2 - e^2$$

and of points (Y) *satisfying*

$$(125) \qquad p^2 = a^2, \qquad a^2 - e^2 < q^2 \leqq a^2 + e^2$$

corresponds to those boundary points (Z) *of* (123) *which satisfy*

$$(126) \qquad u^2 = a^2, \qquad v^2 \leqq b^2,$$

(c) *while the boundary points* (Y) *of* (122) *satisfying*

$$(127) \qquad q^2 - p^2 = e^2, \qquad p^2 < a^2$$

24

correspond to those boundary points (Z) *of* (123) *which satisfy*

$$v^2 = b^2, \qquad u^2 < a^2. \tag{128}$$

The correspondence whose existence is affirmed in this lemma may be given in terms of the function $r(p^2 q^2)$ of (111) as follows:

$$z_i = r(p^2 q^2) y_i \qquad (i = 1, 2, \cdots, n), \tag{129}$$

where p^2 and q^2 here as in the future stand for the functions of the y's to which they are set equal in (120).

From (129) we have at once that

$$\begin{aligned} [z_1^2 + z_2^2 + \cdots + z_k^2] &= r^2 (p^2 q^2)[y_1^2 + y_2^2 + \cdots + y_k^2], \\ [z_{k+1}^2 \quad + \cdots + z_n^2] &= r^2 (p^2 q^2)[y_{k+1}^2 \quad + \cdots + y_n^2]. \end{aligned} \tag{130}$$

Equations (130) show that the functions p^2 and q^2 are related under the transformation (129) to the functions u^2 and v^2 by equations obtained by squaring the different members of (111). But from equations (111) so squared follow those parts of the results of (B), (C) and (D) which describe in detail how p^2 and q^2 are transformed into u^2 and v^2, and the analytical results there contained abstracted from their geometric setting lead directly to the results (a), (b) and (c) of this lemma.

The correspondence is obviously continuous. That it is one-to-one will now be proved by giving its inverse. From (111) and (112) it follows that if neither (pq) nor $(uv) = (00)$

$$\frac{1}{r(p^2 q^2)} = \sigma(u^2 v^2).$$

Thus (129) may be written

$$y_i = \sigma(u^2 v^2) z_i \qquad (i = 1, 2, \cdots, n), \tag{131}$$

where here u^2 and v^2 stand for those functions of the z's to which they are set equal in (121). Further, (131) obviously also holds for the points just excepted, namely $(Y) = (Z) = (0, 0, \cdots, 0)$. Thus the correspondence is one-to-one, and the lemma is completely proved.

The set of points (Z) satisfying (123) can readily be shown to make up an n-cell and its boundary. Hence the corresponding set of points (Y) satisfying (122) make up an n-cell and its boundary. *The set of points* (Y) *satisfying* (122) *has been termed the set* D_n. D_n will also be used to designate the set of points $(X) = (x_1, x_2, \cdots, x_n)$ corresponding to the points (Y) of D_n.

Let c be the value of f at the given critical point. All of the points (X) of D_n satisfy

$$f \leqq c + e^2, \tag{132}$$

and are separated from the remaining points (X) satisfying (132) by those boundary points of D_n satisfying (124) and (125). These boundary points of D_n satisfying (124) and (125) will be termed *the old boundary points of* D_n while the remaining boundary points of D_n, namely those satisfying (127), will be termed *the new boundary points of* D_n. The set of all points of D_n other than the old boundary points of D_n will be called *the new points of* D_n.

The points (Z) satisfying (123) may be represented by a pair of points (PQ) of which P is a point in a k-dimensional space with coördinates $(z_1, z_2, \cdots, z_k)$ satisfying

$$u^2 \leqq a^2,$$

and Q is a point in an $(n-k)$-dimensional space with coördinates $(z_{k+1}, z_{k+2}, \cdots, z_n)$ satisfying

$$v^2 \leqq b^2.$$

Let the $(k-1)$-dimensional hypersphere $u^2 = a^2$ in the space of the points P be denoted by S_{k-1} and the $(n-k-1)$-dimensional hypersphere $v^2 = b^2$ in the space of the points Q be denoted by S_{n-k-1}. With the above convention the results of Lemma 9 may be translated into the following lemma.

LEMMA 10. *The set D_n of points (Y) satisfying* (122) *make up an n-cell and its boundary. It can be put into one-to-one continuous correspondence with the set of pairs of points (PQ) obtained by combining an arbitrary point P interior to or on a $(k-1)$-dimensional hypersphere S_{k-1} with an arbitrary point Q interior to or on an $(n-k-1)$-dimensional hypersphere S_{n-k-1}. In this correspondence, points on the old boundary of D_n correspond to those pairs (PQ) that are obtained by combining an arbitrary point P on S_{k-1} with an arbitrary point Q interior to or on S_{n-k-1}, while points on the new boundary of D_n correspond to those pairs of (PQ) that are obtained by combining an arbitrary point P interior to S_{k-1} with an arbitrary point Q on S_{n-k-1}.*

If we make use of the terms of the preceding lemma, Lemma 8 may be restated and completed as follows.

24*

LEMMA 11. *The complex of points satisfying $f \leqq c - e^2$ can be put into one-to-one continuous correspondence with the complex of points, say C_n', consisting of the points satisfying $f \geqq c + e^2$ with the new points of D_n excluded. The remaining points of D_n, namely the points on the old boundary of D_n, are on the boundary of C_n'.*

14. A representation of D_n in terms of the cells of S_{k-1} and S_{n-k-1}. Suppose the hypersphere S_{k-1} "covered just once" by a non-singular complex whose matrices of incidence H_i are given by the equations

$$\text{H}_i = \begin{vmatrix} 1 & 1 \\ 1 & 1 \end{vmatrix} \qquad (i = 1, 2, \cdots, k-1).$$

Denote the two i-cells of S_{k-1} by a_i' and a_i'' $(i = 0, 1, \cdots, k-1)$. Denote the k-cell consisting of the points interior to S_{k-1} by a_k. Similarly suppose S_{n-k-1} covered just once by a non-singular complex whose matrices of incidence H_j $(j = 1, 2, \cdots, n-k-1)$ are each identical with the preceding matrices H_i. Denote the two j-cells of S_{n-k-1} by b_j' and b_j'' $(j = 0, 1, \cdots, n-k-1)$. Denote the $(n-k)$-cell consisting of the points interior to S_{n-k-1} by b_{n-k}.

Let α be any one of the cells a_i', a_i'' or a_k, and β be any one of the cells b_j', b_j'', or b_{n-k}. If an arbitrary point P of a cell α be combined with an arbitrary point Q of a cell β there will result a set of pairs of points (PQ) which may be considered as the elements of an $(i+j)$-cell. Denote such an $(i+j)$-cell by $(\alpha\beta)$. According to Lemma 10 the complex consisting of all such $(i+j)$-cells will be a non-singular complex covering D_n just once. This complex of $(i+j)$-cells will be adopted as the future representation of D_n.

It follows from Lemma 10 that in the representation of D_n just adopted the interior points of D_n will be represented by the n-cell $(a_k b_{n-k})$ while the new boundary of D_n will be represented by the $(k+j)$-cells $(a_k b_j')$ and $(a_k b_j'')$. Here k is the number specifying the type of critical point being considered, and j takes on the values $0, 1, \cdots, n-k-1$. It should be noted that for each value of j there are just two of these $(k+j)$-cells. The cells $(a_k b_j')$, $(a_k b_j'')$, and $(a_k b_{n-k})$ will be termed *the new cells* of D_n and the remaining cells of D_n *the old cells* of D_n.

The above representation of D_n shows that *there are no new cells of dimensionality less than k.* The points on the boundary of a cell $(\alpha\beta)$ will be represented by the pairs (PQ) obtained by combining a point P on the boundary of α with a point Q on β or its boundary, or by combining a point Q on the boundary of β with a point P on α or its boundary. For the work to follow it is important to determine the $(m-1)$-cells on the boundary of each new m-cell. We begin with the new cells of lowest dimensionality.

The only $(k-1)$-cells on the boundary of the new k-cell $(a_k b'_0)$ will be obtained by combining an arbitrary point of a boundary $(k-1)$-cell of a_k, namely a'_{k-1} or a''_{k-1}, with b'_0 giving thus the two $(k-1)$-cells

$$(a'_{k-1} b'_0), \ (a''_{k-1} b'_0) \quad \text{on} \quad (a_k b'_0),$$

both of which are old cells of D_n. Similarly the only $(k-1)$-cells on the boundary of $(a_k b''_0)$ are the old cells

$$(a'_{k-1} b''_0), \ (a''_{k-1} b''_0) \quad \text{on} \quad (a_k b''_0).$$

For $j = 1, 2, \cdots, n-k-1$ the $(k+j-1)$-cells on the boundary of the new $(k+j)$-cell $(a_k b'_j)$ are obtained by combining a boundary $(k-1)$-cell of a_k, namely a'_{k-1} or a''_{k-1}, with b'_j, or by combining a boundary $(j-1)$-cell of b'_j, namely b'_{j-1} or b''_{j-1}, with a_k, giving thus the $(k+j-1)$-cells

$$(a'_{k-1} b'_j), \ (a''_{k-1} b'_j), \ (a_k b'_{j-1}), \ (a_k b''_{j-1}) \ \text{on} \ (a_k b'_j) \quad (j = 1, 2, \cdots, n-k-1),$$

of which the first two cells are old cells and the latter two cells are new cells. Similarly the $(k+j-1)$-cells on the boundary of the new $(k+j)$-cell $(a_k b''_j)$ are the cells

$$(a'_{k-1} b''_j), \ (a''_{k-1} b''_j), \ (a_k b'_{j-1}), \ (a_k b''_{j-1}) \ \text{on} \ (a_k b''_j) \quad (j = 1, 2, \cdots, n-k-1),$$

of which the first two are old cells and the latter two are new cells. Finally the $(n-1)$-cells on the boundary of the new n-cell $(a_k b_{n-k})$ are

$$(a'_{k-1} b_{n-k}), \quad (a''_{k-1} b_{n-k}), \quad (a_k b'_{n-k-1}), \quad (a_k b''_{n-k-1}) \quad \text{on} \quad (a_k b_{n-k}),$$

of which the first two are old cells and the latter two are new cells.

15. Differences between the connectivity numbers of the complex $f \leqq c - e^2$ and the complex $f \leqq c + e^2$. It follows from Lemma 11 that the complex $f \leqq c - e^2$ has the same connectivity numbers as the complex C'_n consisting of the points satisfying $f \leqq c + e^2$ with the new points of D_n excluded. The problem of this section then resolves itself into one of determining the relations between the connectivity numbers of the complex C'_n and those of the complex $f \leqq c + e^2$. Denote the latter complex by C_n.

If the new points of D_n be added to C'_n there results C_n. The question then is what is the effect upon the connectivity numbers of adding the new points of D_n to C'_n? According to Lemma 11 the only points of D_n coinciding with points of C'_n are the old boundary points of D_n. Because of the simple nature of the functions defining the boundaries of these complexes we can and will suppose C'_n broken up into its component cells in such a manner that points of C'_n that coincide with points of D_n all lie on cells of C'_n that coincide with cells of D_n.

Let H_i $(i = 1, 2, \cdots, n)$ be the ith matrix of incidence of C_n. Let ϱ_i be the rank (mod 2) of H_i. For the same values of i let H'_i be the ith matrix of incidence of C'_n. Let ϱ'_i be the rank (mod 2) of H'_i. Let $R_j (j = 0, 1, \cdots, n)$ be the jth connectivity number of C_n and R'_j that of C'_n. For the same values of j let α_j be the number of j-cells on C_n and α'_j the corresponding number for C'_n. The following well known equations will be used:

$$
\begin{aligned}
R_0 - 1 &= \alpha_0 - \varrho_1 - 1, \\
R_1 - 1 &= \alpha_1 - \varrho_1 - \varrho_2, \\
R_2 - 1 &= \alpha_2 - \varrho_2 - \varrho_3, \\
&\cdots\cdots\cdots\cdots, \\
R_{n-1} - 1 &= \alpha_{n-1} - \varrho_{n-1} - \varrho_n, \\
R_n - 1 &= \alpha_n - \varrho_n.
\end{aligned} \tag{133}
$$

Equations (133) refer to C_n. Similar equations hold for C'_n and are obtained by priming each letter in (133). From equations (133) and equations (133) with the primes added there can be obtained by a subtraction of the corresponding members of the respective equalities the equations

$$
\begin{aligned}
R_0 - R'_0 &= \alpha_0 - \alpha'_0 - (\varrho_1 - \varrho'_1), \\
R_1 - R'_1 &= \alpha_1 - \alpha'_1 - (\varrho_1 - \varrho'_1) - (\varrho_2 - \varrho'_2), \\
R_2 - R'_2 &= \alpha_2 - \alpha'_2 - (\varrho_2 - \varrho'_2) - (\varrho_3 - \varrho'_3), \\
&\cdots\cdots\cdots\cdots\cdots\cdots, \\
R_{n-1} - R'_{n-1} &= \alpha_{n-1} - \alpha'_{n-1} - (\varrho_{n-1} - \varrho'_{n-1}) - (\varrho_n - \varrho'_n), \\
R_n - R'_n &= \alpha_n - \alpha'_n - (\varrho_n - \varrho'_n).
\end{aligned} \tag{134}
$$

To apply these equations the relations between the matrices H_i and H'_i will now be determined. In the first place we have

$$\mathrm{H}_i = \mathrm{H}'_i \qquad (i = 1, 2, \cdots, k-1),$$

since for these values of i there are no new i-cells (§ 14) so that for these values of i, C_n and C'_n contain the same i-cells. According to the results of § 14 the table of incidence of the $(k-1)$-cells with the k-cells of C_n will take the following form:

	x	x	$\cdots$	x	$(a_k\, b_0')$	$(a_k\, b_0'')$
y					0	0
y					0	0
$\cdot$					$\cdot$	$\cdot$
$\cdot$					$\cdot$	$\cdot$
$\cdot$					$\cdot$	$\cdot$
y					0	0
$(a_{k-1}'\, b_0')$		H_k'			1	0
$(a_{k-1}''\, b_0')$					1	0
$(a_{k-1}'\, b_0'')$					0	1
$(a_{k-1}''\, b_0'')$					0	1

where the x's represent k-cells which it is not desired to give more explicitly and the y's represent $(k-1)$-cells. The rectangle containing H_k' is supposed filled out with the elements of the matrix H_k'. The matrix H_k of course consists of the elements of the above two rectangles combined. It can be simplified as follows. The two new k-cells written above the last two columns are on at least one new $(k+1)$-cell (§ 14). The boundary of this $(k+1)$-cell is a k-circuit. Let the columns, other than the last column, which in the table are written under k-cells that are on this k-circuit, be added to the last column. The last column will thereby be reduced to zero (mod 2) without altering the rank of H_k. Let the third from the last row be added to fourth from the last row. As a result there will be but one element not zero in the last two columns. The rank ϱ_k' of the matrix of elements of H_k' has by these processes not been altered (mod 2). It is seen that for ϱ_k there are two possibilities depending upon the nature of the resulting matrix in the left hand rectangle, namely either

Case 1, $\qquad \varrho_k = \varrho_k'$

or $\qquad\qquad\qquad\qquad (0<k<n)$.

Case 2, $\qquad \varrho_k = \varrho_k'+1$

It may be shown by examples that both of these cases actually occur.

The tables of incidence corresponding to the matrices H_{k+j} for $j = 1, 2, \cdots, n-k-1$ (§ 14) are given by the first table on page 384. In the special case where $k = n-1$ there is no table of this sort.

In this table conventions similar to those of the previous table are made. The matrix H_{k+j} is of course the matrix of all the elements included in the four rectangles. It can be simplified as follows. First, exactly as in the case of the preceding table there is a set of columns which added to the last column (mod 2) reduce that column to zeros. If now the next to the last row be added to the three other rows that have

	x	x	$\cdots$	x	$(a_k b'_j)$	$(a_k b''_j)$
y					0	0
y					0	0
.					.	.
.					.	.
.		H'_{k+j}			.	.
y					0	0
$(a'_{k-1} b'_j)$					1	0
$(a''_{k-1} b'_j)$					1	0
$(a'_{k-1} b''_j)$					0	1
$(a''_{k-1} b''_j)$					0	1
$(a_k b'_{j-1})$	0	0	$\cdots$	0	1	1
$(a_k b''_{j-1})$	0	0	$\cdots$	0	1	1

unity in the next to the last column, there will be left (mod 2) in the last two rows and columns only one element not zero and that a 1 in the next to the last row and column. It follows that

$$\varrho_{k+j} = \varrho'_{k+j} + 1 \qquad (j = 1, 2, \cdots, n-k-1).$$

Finally, corresponding to the matrix H_n there is the table

	x	x	$\cdots$	x	$(a_k b_{n-k})$
y					0
y					0
.					.
.					.
.		H'_n			.
y					0
$(a'_{k-1} b_{n-k})$					1
$(a''_{k-1} b_{n-k})$					1
$(a_k b'_{n-k-1})$	0	0	$\cdots$	0	1
$(a_k b''_{n-k-1})$	0	0	$\cdots$	0	1

from which it follows that

$$\varrho_n = \varrho'_n + 1.$$

The number of cells of any particular rank which are contained in C_n but not in C'_n can be read off from the preceding tables. We tabulate the following results:

$$\alpha_i - \alpha'_i = 0, \quad \varrho_i - \varrho'_i = 0 \qquad (i = 1, 2, \cdots, k-1);$$
$$\alpha_i - \alpha'_i = 2, \quad \varrho_i - \varrho'_i = 0 \qquad (i = k \quad \text{Case 1});$$
$$\alpha_i - \alpha'_i = 2, \quad \varrho_i - \varrho'_i = 1 \qquad (i = k \quad \text{Case 2});$$
$$\alpha_i - \alpha'_i = 2, \quad \varrho_i - \varrho'_i = 1 \qquad (i = k+1, k+2, \cdots, n-1);$$
$$\alpha_i - \alpha'_i = 1, \quad \varrho_i - \varrho'_i = 1 \qquad (i = n).$$

From these equations and equations (134) it follows for a critical point of the kth type $(0<k<n)$ that

$$R_i = R'_i \qquad (i = 0, 1, \cdots, k-2, k+1, k+2, \cdots, n)$$

while either

$$R_{k-1} = R'_{k-1} \qquad \text{and} \qquad R_k = R'_k + 1 \qquad \text{(Case 1)}$$

or

$$R_{k-1} = R'_{k-1} - 1 \qquad \text{and} \qquad R_k = R'_k \qquad \text{(Case 2).}$$

That both Case 1 and Case 2 are possible may be shown by simple examples. If we recall that the complex C'_n has the same connectivity numbers as the complex $f \leqq c - e^2$ (Lemma 11, § 13) we have the following fundamental theorem.

THEOREM 3. *Let c be a critical value assumed by f at just one critical point and that a critical point of a type for which $0<k<n$; then for a sufficiently small positive constant e the connectivity numbers R_i of the complex of points satisfying $f \leqq c+e^2$ differ from the connectivity numbers R'_i of the complex of points satisfying $f \leqq c-e^2$ only in that either*

$$R_k = R'_k + 1$$

or else

$$R_{k-1} = R'_{k-1} - 1.$$

CRITICAL POINTS OF THE nTH TYPE

16. Let there be given a critical value of the given function, say c, taken on by f at a critical point of the nth type, and at no other critical point. Let e be a positive constant so small that there is no critical value, other than c, between $c-e^2$ and $c+e^2$, or equal to $c-e^2$ or $c+e^2$. This section is devoted to proving the following theorem.

THEOREM 4. *If c be a critical value of f taken on at just one critical point, and that of the nth type, then for a sufficiently small positive constant e, the connectivity numbers R_i of the complex of points in S satisfying*

$$f \leqq c + e^2 \tag{135}$$

differ from the connectivity numbers R'_i of the complex of points in S satisfying

$$f \leqq c - e^2 \tag{136}$$

only in that

$$R_{n-1} = R'_{n-1} - 1.$$

Before proving this theorem the following lemma will be established.

LEMMA. *For any complex C_n lying in a finite part of euclidean n-space, and bounded by a finite number of distinct closed manifolds on each of* *which f takes on a non-critical constant value the connectivity number R_n equals unity.*

While a theorem similar to this could be proved under much more general hypotheses this particular lemma is all that we need here. To prove the lemma observe that in the given euclidean n-space we can choose an $(n-1)$-dimensional hypersphere S_{n-1} so large that it contains C_n entirely in its interior. Let A_n be the complex consisting of the points interior to and on S_{n-1}. A_n can be broken up into component cells in such a manner as to contain C_n as a sub-complex of its cells. Now as is well known, for A_n, $R_n = 1$, or in equivalent terms A_n contains no n-circuits. Hence C_n contains no n-circuits, and its connectivity number R_n must also be unity. Thus the lemma is proved.

In terms of the y's of Lemma 4, § 3, f can be represented in the neighborhood of the given critical point in the form

$$f-c = -y_1^2 - y_2^2 - \cdots - y_n^2.$$

If the points (Y) satisfy

$$0 \leq y_1^2 + y_2^2 + \cdots + y_n^2 < e^2, \tag{137}$$

the corresponding points (X) satisfy the relations

$$c - e^2 < f \leq c$$

and thus satisfy (135) but not (136).

Denote the complex (135) by C_n and the complex (136) by C_n'. The points (X) corresponding to the points (Y) satisfying (137) obviously make up an n-cell, say a_n, which belongs to C_n. The boundary of a_n consists of points at which the sum of the squares of the y's equals e^2 and so consists of points at which $f = c-e^2$. The boundary of a_n thus lies on the boundary of C_n'. Let a_n be added to C_n', thereby forming a complex denoted by C_n''. If use be made of the orthogonal trajectories employed for a similar purpose in § 7, it is easy to prove that C_n and C_n'' can be put into one-to-one continuous correspondence and hence have the same connectivity numbers R^i.

The question remains as to what is the relation between the connectivity numbers R_i of C_n'' and those R_i' of C_n'. C_n' and C_n'' differ only in that a_n belongs to C_n'' and not to C_n'. If then C_n'' be supposed represented by the cells of C_n' together with the n-cell a_n, it appears that all the matrices of incidence of C_n' and of C_n'' will be identical except the nth ones. From formulas (134) it follows that the connectivity numbers of C_n' and C_n'' are the same with the possible exception of the nth and $(n-1)$th connectivity numbers.

Now it follows from the preceding lemma that both R_n and R'_n equal unity so that

$$R_n - R'_n = 0. \tag{138}$$

Let α'_i and α''_i be respectively the numbers of i-cells in C'_n and C''_n, and ϱ'_j and ϱ''_j be respectively the ranks of the jth matrix of incidence of C'_n and C''_n. Since a_n is the only n-cell in C''_n that is not in C'_n we have

$$\alpha''_n - \alpha'_n = 1. \tag{139}$$

The last two equations of (134) become under the present notation

$$R_{n-1} - R'_{n-1} = \alpha''_{n-1} - \alpha'_{n-1} - (\varrho''_{n-1} - \varrho'_{n-1}) - (\varrho''_n - \varrho'_n), \tag{140}$$

$$R_n - R'_n = \alpha''_n - \alpha'_n - (\varrho''_n - \varrho'_n). \tag{141}$$

From (138), (139) and (141) it follows that

$$\varrho''_n - \varrho'_n = 1. \tag{142}$$

Since $\alpha''_{n-1} = \alpha'_{n-1}$ and $\varrho''_{n-1} = \varrho'_{n-1}$ it follows from (140) that

$$R_{n-1} = R'_{n-1} - 1.$$

Thus the theorem is completely proved.

GENERAL THEOREMS CONCERNING THE CRITICAL POINTS

17. The case of a critical value of f taken on at several different critical points. Up to this point only those critical values of f have been considered which are taken on at just one critical point. A first theorem in the general case is the following:

THEOREM 5. *Let c be any critical value of f that is not the absolute minimum of f in S, and m_i be the number of critical points of the ith type at which $f = c$. Let e be a small positive constant. Let R_i and R'_i be respectively the ith connectivity numbers of the complex of points in S satisfying*

$$f \leq c + e^2 \tag{143}$$

and

$$f \leq c - e^2. \tag{144}$$

Then for e sufficiently small there exist integers $p_0, p_1, \cdots, p_{n-1}$ and $q_1, q_2, \cdots, q_n$, all positive or zero, such that

$$
(145) \qquad
\begin{aligned}
m_0 &= p_0, \\
m_1 &= p_1 + q_1, \\
m_2 &= p_2 + q_2, \\
&\cdots\cdots\cdots, \\
m_{n-1} &= p_{n-1} + q_{n-1}, \\
m_n &= q_n,
\end{aligned}
$$

and such that

$$
(146) \qquad
\begin{aligned}
R_0 - R_0' &= p_0 - q_1, \\
R_1 - R_1' &= p_1 - q_2, \\
&\cdots\cdots\cdots, \\
R_{n-1} - R_{n-1}' &= p_{n-1} - q_n,
\end{aligned}
$$

while, as ever,

$$
(147) \qquad R_n = R_n' = 1.
$$

Denote the complex defined by (143) by C_n, and that defined by (144) by C_n'. The critical points at which $f = c$ are divided for convenience into those of type 0, those of type n, and those of the remaining types. Corresponding to this division of the critical points into classes the proof of the above theorem is divided into parts A, B, and C.

A. *Critical points of type* 0 *at which* $f = c$. For e sufficiently small it follows exactly as in the proof of Theorem 2 of § 9 that the points in C_n neighboring the m_0 critical points of type 0 make up m_0 n-cells and their boundaries, distinct from each other and from any other points of C_n, and not included at all among the points of C_n'. If these m_0 n-cells and their boundaries be added to C_n' there will result a complex, say C_n'', whose connectivity numbers R_n'' will differ from those of C_n' only in that

$$
(148) \qquad R_0'' - R_0' = m_0.
$$

B. *Critical points of type n at which* $f = c$. For e sufficiently small it follows exactly as in the proof of Theorem 4, § 16, that the points in the neighborhood of the m_n critical points of type n make up m_n distinct n-cells whose boundaries belong to C_n', but which themselves belong only to C_n. If these m_n n-cells be added to C_n'' there will result a complex C_n''' whose connectivity numbers R_i''' will differ from those of C_n'' only in that

$$
(149) \qquad R_{n-1}''' - R_{n-1}'' = -m_n.
$$

C. *Critical points not of type* 0 *or* n *at which* $f = c$. With each critical point of this class there can be associated a complex of points D_n, exactly as is done in Lemmas 10 and 11 for the case where the given critical point is the only critical point at which $f = c$. For e and a sufficiently small the different complexes D_n will all be distinct from each other. The so called "new points" of each complex D_n will belong to C_n. Suppose that there are just r critical points, at which $f = c$ and which are of a type neither 0 or n. Let A_j be the set of new points in the complex D_n associated with the jth one of these critical points ($j = 1, 2, \cdots, r$). Let all of the points of C_n not already in C_n''' and not among the points of any set A_j be added to C_n''', thereby forming a complex C_n''''. C_n'''' and C_n''' can be put into one-to-one continuous correspondence by the methods of Lemma 11, § 13, so that the complexes C_n'''' and C_n''' will have the same connectivity numbers.

Finally to obtain C_n we have only to add to C_n'''' the different sets A_j. We will suppose that these sets A_j are added in the order of their subscripts. Suppose that A_j was associated with a critical point of type k_j. The addition of A_j to what we shall call the *old complex*, consisting of C_n'''' and the points of $A_1, A_2, \cdots, A_{j-1}$, will give a *new complex* whose connectivity numbers will differ from the connectivity numbers of the old complex, either (Case 1) in that the k_jth connectivity number of the new complex will be one greater than that of the old complex, or else (Case 2) the (k_j-1)th connectivity number of the new complex will be one less than that of the old complex. We now define p_i, for $i = 1, 2, \cdots, n-1$, as the number of critical points of type i, at which $f = c$, which come under Case 1, and define q_i for the same values of i as the number of critical points of the ith type at which $f = c$ which come under Case 2. The connectivity numbers R_i'''' of C_n'''' are obviously then related to the connectivity numbers R_i of C_n as follows:

$$
(150)\qquad
\begin{aligned}
R_0 - R_0'''' &= - q_1,\\
R_1 - R_1'''' &= p_1 - q_2,\\
\cdots&\cdots,\\
R_{n-2} - R_{n-2}'''' &= p_{n-2} - q_{n-1},\\
R_{n-1} - R_{n-1}'''' &= p_{n-1}.
\end{aligned}
$$

We now define p_0 as equal to m_0 and define q_n as equal to m_n. If use be made of the fact that the connectivity numbers of C_n' and C_n'' have no differences other than that given by (148), and that the connectivity numbers of C_n'' and C_n''' have no differences other than that given by (149), while $R_i'''' = R_i'''$ without exception, then from equations (150) and the definitions of the p's and the q's the theorem follows.

18. Relations between the connectivity numbers of the complex $f \leq c_1$ and the complex $f \leq c_2$.

THEOREM 6. *Let c_1 and c_2 ($c_1 < c_2$) be two non-critical values taken on by f in S. Let m_i be the number of critical points of the ith type at which f equals some constant between c_1 and c_2. Let R_i and R_i' be the ith connectivity numbers of the complexes of points in S respectively satisfying $f \leqq c_2$ and $f \leqq c_1$; then there exist integers, $p_0, p_1, \cdots, p_{n-1}$, and $q_1, q_2, \cdots, q_n$, all positive or zero, such that*

$$m_0 = p_0,$$
$$m_i = p_i + q_i \qquad (i = 1, 2, \cdots, n-1),$$
$$m_n = q_n,$$

and such that

$$R_i - R_i' = p_i - q_{i+1} \qquad (i = 0, 1, \cdots, n-1)$$

while as ever

$$R_n = R_n' = 1.$$

Let $a_1, a_2, \cdots, a_r$ be the critical values of f between c_1 and c_2, taken in the order of increasing magnitude. With each of these critical values of f, say a_j, let there be associated a positive constant e_j taken so small that for this choice of e_j Theorem 5 holds if c and e in Theorem 5 are here taken to be a_j and e_j respectively. Let the constants e_j be also taken so small that the constants

$$(151) \quad c_1, \quad a_1 - e_1, \quad a_1 + e_1, \quad a_2 - e_2, \quad a_2 + e_2, \quad \cdots, \quad a_r - e_r, \quad a_r + e_r, \quad c_2$$

are none of them equal and appear in (151) in the order of increasing magnitude. Now there are no critical values of f between c_1 and $a_1 - e_1$. It follows from Lemma 6, § 7, that the complex $f \leqq c_1$ has the same connectivity numbers as the complex $f \leqq a_1 - e_1$. For the same reason the complexes $f \leqq a_i + e_i$ and $f \leqq a_{i+1} - e_{i+1}$, for $i = 1, 2, \cdots, r-1$, have the same connectivity numbers, as well as the complexes $f \leqq a_r + e_r$ and $f \leqq c_2$. The relations between the connectivity numbers of the complex $f \leqq a_j + e_j$ and the complex $f \leqq a_j - e_j$ are of the nature of those given by the preceding Theorem 5.

If then there be considered in succession the complexes of points in S satisfying $f \leqq c$, where c takes on, in succession, the constants in (151) in their order from left to right, the relations between these successive complexes will combine into the relations of which this theorem affirms the existence.

19. Relations between all of the critical points in S and the connectivities of S. The following theorem is proved under the boundary conditions α of § 4.

THEOREM 7. *Let M_i be the total number of critical points of f of the ith type ($i = 0, 1, \cdots, n$). Let R_j be the jth connectivity number of S.*

Under the boundary conditions α there exist integers $P_0, P_1, \cdots, P_{n-1}$ and $Q_1, Q_2, \cdots, Q_n$, all positive or zero, such that

$$
\begin{aligned}
M_0 &= P_0 + 1, \\
M_i &= P_i + Q_i \qquad (i = 1, 2, \cdots, n-1), \qquad (152)\\
M_n &= \phantom{P_i +{}} Q_n
\end{aligned}
$$

and such that

$$R_i - 1 = P_i - Q_{i+1} \qquad (i = 0, 1, \cdots, n-1) \qquad (153)$$

while as ever

$$R_n - 1 = 0. \qquad (154)$$

Let s be the number of critical points at which f takes on its absolute minimum m. According to Theorem 2, § 9, for a sufficiently small positive constant e, the connectivity numbers R_i' of the complex of points satisfying $f \leqq m + e^2$ are given by the relations

$$R_0' = s, \qquad R_1' = R_2' = \cdots = R_n' = 1. \qquad (155)$$

The proof of these results in Theorem 2 presupposed that e was so small that there were no other critical points than the s absolute minimum points of f among the points in S satisfying $f \leqq m + e^2$. This supposition is again made here. For this choice of e, an application can now be made of Theorem 6. In this application the notation of Theorem 6 will be taken over. The constants c_1 and c_2 of Theorem 6 will here be taken respectively as $m + e^2$ and the value M which f takes on upon the boundary of S. Observe now that M_0, the total number of critical points of type 0 in S, equals $s + m_0$. This we write

$$m_0 = M_0 - s. \qquad (156)$$

Otherwise

$$m_i = M_i \qquad (i = 1, 2, \cdots, n). \qquad (157)$$

We now give the following definitions of $P_0, P_1, \cdots, P_{n-1}$ and of $Q_1, Q_2, \cdots, Q_n$:

$$
\begin{aligned}
P_0 &= p_0 + s - 1, \\
P_i &= p_i \qquad (i = 1, 2, \cdots, n-1), \qquad (158)\\
Q_j &= q_j \qquad (j = 1, 2, \cdots, n),
\end{aligned}
$$

and note that P_0, in particular, as defined, is either positive or zero, since p_0 is at least zero, and s at least one.

If in the relations of Theorem 6, for the constants R_i' there be substituted their values as given by (155), and for m_0 and m_i their values from (156) and (157) and for p_0 and the p_i and q_j their values as given by (158), there will result the equations which are to be proved.

The following theorem gives the answer to one of the fundamental questions with which this paper is concerned.

THEOREM 8. *Under the boundary conditions* α *the following relations exist between the set of numbers* M_i, *where* M_i *is the number of critical points of the ith type of* f *in* S, *and the set of connectivity numbers* R_j *of* S $(i,j = 0, 1, \cdots, n)$:

$$\begin{aligned}
1 &\leqq (M_0 - R_0 + 1),\\
1 &\geqq (M_0 - R_0 + 1) - (M_1 - R_1 + 1),\\
1 &\leqq (M_0 - R_0 + 1) - (M_1 - R_1 + 1) + (M_2 - R_2 + 1),\\
1 &\geqq (M_0 - R_0 + 1) - (M_1 - R_1 + 1) + (M_2 - R_2 + 1) - (M_3 - R_3 + 1),\\
&\cdots\cdots\cdots\cdots\cdots\cdots\cdots\cdots,\\
1 &\leqq [(M_0 - R_0 + 1) - + \cdots + (-1)^{n-1}(M_{n-1} - R_{n-1} + 1)](-1)^{n-1},\\
1 &= (M_0 - R_0 + 1) - + \cdots + (-1)^n \ (M_n \ - R_n \ + 1).
\end{aligned}$$

To prove the theorem consider equations (152) together with equations (153), without the last equation of (152). These equations together form the system

$$\begin{aligned}
M_0 &= P_0 + 1, &&\\
M_i &= P_i + Q_i &&(i = 1, 2, \cdots, n-1),\\
R_j - 1 &= P_j - Q_{j+1} &&(j = 0, 1, \cdots, n-1).
\end{aligned} \tag{159}$$

Suppose the left hand members of (159) known constants, and the P's and Q's in the right hand members unknown constants or variables. The determinant of the coefficients of these unknown variables is readily seen to be different from zero, so that if the M's and the connectivity numbers are known, the equations (159) uniquely determine the P's and Q's as constants satisfying (159). In particular the equations (159) can be solved successively for the Q's as follows:

$$\begin{aligned}
1 + Q_1 &= (M_0 - R_0 + 1),\\
1 - Q_2 &= (M_0 - R_0 + 1) - (M_1 - R_1 + 1),\\
1 + Q_3 &= (M_0 - R_0 + 1) - (M_1 - R_1 + 1) + (M_2 - R_2 + 1),\\
1 - Q_4 &= (M_0 - R_0 + 1) - (M_1 - R_1 + 1) + (M_2 - R_2 + 1) - (M_3 - R_3 + 1),\\
&\cdots\cdots\cdots\cdots\cdots\cdots\cdots\cdots,\\
1 + (-1)^{n-1}Q_n &= (M_0 - R_0 + 1) - + \cdots + (-1)^{n-1}(M_{n-1} - R_{n-1} + 1).
\end{aligned} \tag{160}$$

Of the equations of Theorem 7 which are not included in (159) there remain the equations

$$\begin{aligned}
R_n - 1 &= 0,\\
M_n &= Q_n.
\end{aligned} \tag{161}$$

With the aid of (161) we can write

$$Q_n = M_n - R_n + 1.$$

If this expression for Q_n be substituted for Q_n in the last of equations (160) there will result the equality given by the last of the equations of Theorem 8 The remaining relations of Theorem 8 result immediately from (160) upon recalling that the Q's are all either positive or zero. Thus the theorem is proved.

If the domain S be one which can in particular be put into one-to-one continuous correspondence with the points in n-space interior to and on an $(n-1)$-dimensional hypersphere, then the connectivity numbers R_i of S are all unity. Hence we have the following corollary to Theorem 8.

COROLLARY. *If S be an n-dimensional region homeomorphic with the points in n-space interior to and on an $(n-1)$-dimensional hypersphere, and M_i be the number of critical points of the ith type of f in S $(i = 0, 1, \cdots, n)$ then under the boundary conditions α the following relations hold true*:

$$\begin{aligned}
1 &\leqq M_0, \\
1 &\geqq M_0 - M_1, \\
1 &\leqq M_0 - M_1 + M_2, \\
1 &\geqq M_0 - M_1 + M_2 - M_3, \\
&\cdots\cdots\cdots\cdots, \\
1 &\leqq [M_0 - M_1 + M_2 - M_3 + - \cdots + (-1)^{n-1} M_{n-1}] (-1)^{n-1}, \\
1 &= M_0 - M_1 + M_2 - M_3 + - \cdots + (-1)^n M_n.
\end{aligned}$$

20. The boundary conditions β defined. The boundary conditions β are the following.

I. The boundary B of the domain R of § 1 shall consist of a closed set of points lying in a finite part of the space of the variables $(x_1, x_2, \cdots, x_n)$.

II. The points on B in the neighborhood of any particular point (X_0) on B shall satisfy a relation of the sort

$$F(x_1, x_2, \cdots, x_n) = 0,$$

where $F(x_1, x_2, \cdots, x_n)$ is a single-valued continuous function of $(x_1, x_2, \cdots, x_n)$ possessing continuous first, second, and third partial derivatives in the neighborhood of each point of B, of which not all of the first partial derivatives are to be zero at any point of B.

III. The function $f(x_1, x_2, \cdots, x_n)$ in addition to satisfying the conditions of § 1 shall be continuous on B, and its first partial derivatives shall take on continuous boundary values on B.

IV. At each point of B the unilateral directional derivative of $f(x_1, x_2, \cdots, x_n)$ along the normal to B in the sense that leads from points of R to points not in R, and on the side that lies in R, shall be positive.

21. The boundary conditions β reduced to the boundary conditions α (§ 4). By a redefinition of $f(x_1, x_2, \cdots, x_n)$ in the neighborhood of B there can be obtained a function $L(x_1, x_2, \cdots, x_n)$ with critical points identical in position and in type with those of $f(x_1, x_2, \cdots, x_n)$, but such that one of the manifolds

$$L(x_1, x_2, \cdots, x_n) = \text{const.}$$

will serve as a boundary A satisfying the earlier boundary conditions.

Under condition II of the boundary conditions β there is a definite normal to B at each point $(x_1^0, x_2^0, \cdots, x_n^0) = (X_0)$ of B whose equations may be given in the form

(162) $$x_i - x_i^0 = A_i(x_1^0, x_2^0, \cdots, x_n^0)s \qquad (i = 1, 2, \cdots, n),$$

where $A_i(x_1^0, x_2^0, \cdots, x_n^0)$ together with its first and second partial derivatives are continuous functions of their arguments in the neighborhood of any given point on B, where s is the distance along the normal, being zero on B, and increasing as the point on the normal crosses B from points of R to points not in R. We can and will suppose s_1 to be a negative constant chosen so small that the points $(x_1, x_2, \cdots, x_n) = (X)$ on the normal (162) at the points at which

(163) $$s_1 \leqq s < 0$$

include once and only once every point of R, not including B, in the neighborhood of B.

Because of the boundary conditions IV we can and will suppose that s_1 is so small that the points (X) given by (162) for values of s satisfying (163) contain no critical points of f. It follows that there is one and only one of the trajectories orthogonal to the manifolds $f =$ const., through each point of (162) for which (163) holds. Let s_2, s_3, and s_4 be three constants such that

$$s_1 < s_2 < s_3 < s_4 < 0.$$

Let T_1, T_2, T_3, and T_4 be respectively the $(n-1)$-dimensional manifolds of points (X) obtained by putting $s = s_1$, s_2, s_3, and s_4 in (162) and letting (X_0) vary on B. It is a consequence of boundary conditions IV, as the calculation of the appropriate jacobian will show, that if s_1 be a sufficiently small negative constant, any trajectory orthogonal to the manifolds $f =$ const. passing through a point P_1 of T_1 also passes through uniquely determined points P_2, P_3, and P_4 on T_2, T_3, and T_4, respectively, such that the positions of P_2, P_3, and P_4 are continuous functions of the position of P_1 on T_1.

We again make use of that representation of the orthogonal trajectories in which the parameter at any point equals the value of f at that point, and we let τ_1, τ_2, τ_3, and τ_4 respectively be the values of τ at P_1, P_2, P_3 and P_4.

We now are in a position to replace $f(x_1, x_2, \cdots, x_n)$ by a new function $L(x_1, x_2, \cdots, x_n)$ defined as follows. At each point of R not a point given by (162) for a value of $s > s_2$, we define

$$L(x_1, x_2, \cdots, x_n) \equiv f(x_1, x_2, \cdots, x_n). \tag{164}$$

To define $L(x_1, x_2, \cdots, x_n)$ further it will be convenient, for those points (X) given by (162) for which

$$s_1 \leqq s \leqq s_4, \tag{165}$$

to replace $(x_1, x_2, \cdots, x_n)$ by a new set of n independent variables, namely $(n-1)$ independent parameters $(u_1, u_2, \cdots, u_{n-1})$, determining a point P_1 on T_1 and thereby a trajectory through P_1, and an nth independent parameter, namely the above parameter τ determining a point (X) on the trajectory through P_1. To represent a set of points (X) satisfying (165) and neighboring a particular trajectory, the parameters $(u_1, u_2, \cdots, u_{n-1})$ can be so chosen that the coördinates of (X) are single-valued functions of the variables $(u_1, u_2, \cdots, u_{n-1}, \tau)$ provided with continuous second partial derivatives. Now in terms of the new variables the points satisfying (165) are the points at which

$$\tau_1 \leqq \tau \leqq \tau_4. \tag{166}$$

Of these points the points at which

$$\tau_1 \leqq \tau \leqq \tau_2 \tag{167}$$

are points at which $L(x_1, x_2, \cdots, x_n)$ has already been defined. Setting

$$L(x_1, x_2, \cdots, x_n) = \varphi(\tau, u_1, u_2, \cdots, u_{n-1})$$

the previous definition of L, namely (164), reduces for the points of (167) to

$$\varphi(\tau, u_1, u_2, \cdots, u_{n-1}) \equiv \tau. \tag{168}$$

To define $L(x_1, x_2, \cdots, x_n)$ further note that τ_1, τ_2, τ_3, and τ_4 are functions of $(u_1, u_2, \cdots, u_{n-1})$. Let G be a constant greater than the value of f in R. For those points of (166) for which

(169) $$\tau_2 \leqq \tau \leqq \tau_4$$

we now define

(170) $$\varphi(\tau, u_1, u_2, \cdots, u_{n-1}) \equiv \tau + H(u_1, \cdots, u_{n-1})\,[\tau - \tau_2(u_1, \cdots, u_{n-1})]^4$$

where H is defined by the equation

(171) $$G = \tau_3 + H[\tau_3 - \tau_2]^4$$

as a positive function of $(u_1, u_2, \cdots, u_{n-1})$. From (170) and (171) it follows that

(172) $$\varphi(\tau_3, u_1, u_2, \cdots, u_{n-1}) = G.$$

For any point corresponding to a value of τ satisfying (169) we now define

$$L(x_1, x_2, \cdots, x_n) \equiv \varphi(\tau, u_1, \cdots, u_{n-1})$$

at that point.

Let R' be the domain in which L has been defined. $L(x_1, x_2, \cdots, x_n)$ and its second partial derivatives are readily seen to be continuous in R'. No critical points are introduced by the definition (170), as follows upon verification of the fact that the partial derivative $\varphi_\tau(\tau, u_1, \cdots, u_{n-1}) > 0$ for points for which (170) and (171) hold. Thus in the neighborhood of *critical points* of f or of L, (164) holds. From (172) it follows that on the manifold T_3

$$L(x_1, x_2, \cdots, x_n) = G.$$

Boundary conditions α of § 4 will obviously be satisfied if we take for the manifold A used in boundary conditions α the manifold T_3. Further, with the aid of the normals (162) it is easy to see that the complex of points satisfying $L(x_1, x_2, \cdots, x_n) \leqq G$ can be put into one-to-one continuous correspondence with the complex consisting of the domain R and the above boundary B. Thus these two complexes have the same connectivity numbers. From Theorem 8 we obtain accordingly the following theorem.

THEOREM 9. *Under the boundary conditions β the relations between the numbers M_i, where M_i $(i = 0, 1, \cdots, n)$ is the number of critical points of f of type i in R, and the numbers R_j, where R_j $(j = 0, 1, \cdots, n)$ is the jth connectivity number of the complex consisting of R and its boundary B, are the same as those given in Theorem 8.*

CORNELL UNIVERSITY,
ITHACA, N. Y.

Reprinted from the
TRANSACTIONS OF THE AMERICAN MATHEMATICAL SOCIETY
Vol. 30, No. 2, pp. 213–274

THE FOUNDATIONS OF A THEORY IN THE CALCULUS OF VARIATIONS IN THE LARGE*

BY
MARSTON MORSE

INTRODUCTION

The conventional object in a paper on the calculus of variations is the investigation of the conditions under which a maximum or minimum of a given integral occurs. Writers have accordingly done little with extremal segments that have contained more than one point conjugate to a given point. An extended theory is needed for several reasons.

One reason is that in applying the calculus of variations to geodesics, or to that very general class of dynamical systems or differential equations which may be put in the form of the Euler equations, *it is by no means a minimum or a maximum that is always sought.* For example, if in the problem of two bodies we make use of the corresponding Jacobi principle of least action† the ellipses which thereby appear as extremals always have pairs of conjugate points on them, and do not accordingly give a minimum to the integral relative to neighboring closed curves, so that *no example of periodic motion would be found by a search for a minimum* of the Jacobi integral. In general if one is looking for extremals joining two points or periodic extremals deformable into a given closed curve, the a priori expectation, as justified by the results of this paper, in general problems, would seem to be that more solutions would not give an extremum than would give an extremum.

A second reason for the study of extremal segments and periodic extremals that do not furnish an extremum for the integral is that *if the ultimate object sought is an extremum, the existence of such extrema is tied up "in the large" with the existence of extremals which do not furnish extrema.* It is one of the purposes of this paper to show the relations in the large between all sorts of extremals joining two fixed points, or deformable into a given closed curve.

A first type of a priori existence theorem is Hilbert's theorem concerning

---

* Presented to the Society in part December 30, 1924, under the title *Relations in the large between the numbers of extremals of different types joining two fixed points*, and in part December 29, 1926, under the title *The type number and rank of a closed extremal, and the consequent theory in the large*; received by the editors January 15, 1927.

† See P. Appell, *Traité de Mécanique Rationnelle*, Paris, 1919, vol. 1, pp. 547–548.

the absolute minimum.* Reference should be given to the more recent work of Tonelli,† also on the absolute minimum. Other references to studies of this sort will be found in the standard treatises. Birkhoff‡ effectively departed from the study of the absolute minimum alone, when he stated his "minimax principle," and applied it to find closed trajectories that do not give a minimum to his "least action" integral. The present writer§ has shown that Birkhoff's points of minimum and minimax appear as two types of critical points among $n+1$ such types, and has replaced Birkhoff's inequality relation by n inequality relations and one equality relation. One of the objects of the present paper is to show how these $n+1$ relations between critical points can be translated into relations between different types of extremals.

It was necessary to develop for the first time a complete parallelism between types of critical points and types of extremals. It was found that the type of an extremal segment whose ends were not conjugate was completely determined by the number of mutual conjugate points on the segment. In the case where the end points were conjugate it was necessary to bring in the envelope theory.

Turning to periodic extremals it was found, even in the most general cases, that conjugate points would not serve to determine the type of a periodic extremal g, but that other relations of g to neighboring extremals had to be brought in. A periodic extremal was called *degenerate* if the corresponding Jacobi differential equation possessed periodic solutions not identically zero. It is shown for the first time how a parameter may be introduced into the integrand in such a fashion that the degenerate periodic extremal disappears and non-degenerate periodic extremals, or no periodic extremals, take its place.

This theory is brought to a head in two applications, one "in the large," giving relations between different types of extremals, and one "in the small," showing how the type of a given extremal may be characterized in a third way, in terms of the possibility or impossibility of deformations of m-parameter families of neighboring extremals into families of lower dimensionality. This deformation theory was made possible by the application

* Bolza, *Vorlesungen über Variationsrechnung*, 1909, pp. 419–437. Further references to Bolza will be indicated by the letter B.

† Tonelli, *Fondamenti di Calcolo delle Variazioni*, vol. 2.

‡ Birkhoff, *Dynamical systems with two degrees of freedom*, these Transactions, vol. 18 (1917), p. 240.

§ Marston Morse, *Relations between the critical points of a real function of n independent variables*, these Transactions, vol. 27 (1925), pp. 345–396.

of certain powerful theorems of analysis situs. It is believed that this deformation theory will serve as the basis for an even more extended theory "in the large."

PART I. THE TYPE NUMBER OF AN EXTREMAL SEGMENT

1. **The integrand $F(x, y, \dot{x}, \dot{y})$ and the function $J(v_1, \cdots, v_n)$.** Let (x, y) be any point in an open two-dimensional region S of the x, y plane. Let there be given a function $F(x, y, x, \dot{y})$ of class C''' (B, loc. cit., p. 193), and positively homogeneous in $\dot{x}$ and $\dot{y}$ of dimension one, for (x, y) in S and $\dot{x}$ and $\dot{y}$ any two numbers not both zero. Corresponding to the calculus of variations problem in the parametric form with integral J, and with $F(x, y, \dot{x}, \dot{y})$ as the integrand, let there be given an extremal g of class C''' (B, p. 191), without multiple points, passing from a point A to a point B, and such that along g we have (B, p. 196)

$$F_1(x, y, \dot{x}, \dot{y}) > 0. \tag{1}$$

Let the arc length along g, measured from A toward B, be denoted by u. Let

$$u_1, u_2, \cdots, u_n \tag{2}$$

be n values of u increasing with their subscripts, and corresponding to points on g between A and B. Denote the value of u at A by u_0, and its value at B by u_{n+1}. We suppose the points (2) so chosen on g that no one of the closed segments of g bounded by successive points of the set

$$u_0, u_1, \cdots, u_{n+1} \tag{3}$$

contains a conjugate point of either end point of that segment. Let there be given n short arcs of class C''', say $h_1, h_2, \cdots, h_n$, passing respectively through the points of g at which u takes on the values (2), arcs not tangent to g at these points. Let positive senses be assigned to $h_1, h_2, \cdots, h_n$ in such a fashion that the positive tangent to g at $u = u_i$ has to be turned through a positive angle less than π to coincide with the positive tangent to h_i at the same point. Let v_i be the arc length measured along h_i in h_i's positive sense from the point $u = u_i$ on g. We regard (u_i, v_i) as representing the point on h_i at the distance v_i from g.

If each v_i be sufficiently small in absolute value the points

$$A, (u_1, v_1), (u_2, v_2), \cdots, (u_n, v_n), B \tag{4}$$

can be successively joined by unique extremal segments neighboring g.

Let the integral J taken along this broken extremal joining A to B be denoted by

$$J(v_1, \cdots, v_n). \tag{5}$$

Corresponding to the extremal g the function (5) will be shown to have a *critical point* for $(v_1, \cdots, v_n)=(0, 0, \cdots, 0)$, that is *a point at which all of its partial derivatives are zero*. We will also investigate the terms of second order in the expansion of the function (5) about $(0, 0, \cdots, 0)$.

2. **A transformation of the problem.** We introduce the following lemma which simplifies the problem.

Lemma. *It is possible to make a transformation T from the (x, y) plane to the (u, v) plane with the following properties:*

(A) *Under T the extremal g is carried into a portion γ of the u axis bounded by $u=u_0$ and $u=u_{n+1}$.*

(B) *The transformation establishes a one-to-one correspondence between a suitably chosen region of the (x, y) plane enclosing g, and a region R_1 of the (u, v) plane enclosing γ.*

(C) *The transformation carries the arcs $h_1, h_2, \cdots, h_n$ into straight line segments on which $u=u_1, u_2, \cdots, u_n$, respectively.*

(D) *The transformation is representable in the form*

$$x = x(u, v), \quad y = y(u, v) \tag{1}$$

where $x(u,v), y(u,v)$ are of class C''' in R_1 and possess there a positive jacobian.

(E) *The transformation preserves arc lengths along g and $h_1, h_2, \cdots, h_n$.*

In the first preparation of this paper the complete details of the proof of this lemma were given, but second thought makes it seem not too much to leave to the reader.

3. **The integral in the (u, v) plane.** Under the transformation of the preceding lemma the integrand $F(x, y, \dot{x}, \dot{y})$ can be replaced by a new integrand $G(u, v, \dot{u}, \dot{v})$, where $\dot{x}, \dot{y}, \dot{u}, \dot{v}$ stand for derivatives of x, y, u, v, respectively, with respect to a parameter t (B, p. 344). We are concerned with properties of the extremal segment γ, namely,

$$v \equiv 0, \quad u_0 \leqq u \leqq u_{n+1}, \tag{1}$$

which corresponds under the preceding transformation T to the extremal g in the (x, y) plane. For the present we need only consider curves on which $\dot{u}>0$, and set

$$G(u, v, 1, v') = f(u, v, v'), \quad v' = \frac{dv}{du}, \tag{2}$$

thus defining $f(u, v, v')$ for all points (u, v) neighboring γ, and all numbers v'. For this domain $f(u, v, v')$ is of class C'''. The points which are enumerated in §1, (4), here correspond to points whose actual coördinates are

$$(u_0, 0),\ (u_1, v_1),\ \cdots,\ (u_n, v_n),\ (u_{n+1}, 0). \tag{3}$$

Subject to the limitation that we deal here only with curves on which $\dot{u} > 0$, the integral J of §1 becomes here

$$J = \int_{u_0}^{u_{n+1}} f(u, v, v')du, \tag{4}$$

and the function

$$J(v_1, \cdots, v_n) \tag{5}$$

is the value of the integral (4) taken along the successive extremals joining the successive points of (3), varying the coördinates v_i, but holding all coördinates u_i fast. It should be expressly noted that the function (5) *is not a new function set up for the first time in the* (u, v) *plane, but that it is identical with the function* $J(v_1, \cdots, v_n)$ *defined in* §1.

In terms of $f(u, v, v')$, and for the extremal γ given by (1), we define the functions $P(u)$, $Q(u)$, and $R(u)$ in the usual way (B, p. 55). These functions are of class C' along γ. Because of the assumption that $F_1 > 0$ along g, it follows that

$$R(u) > 0 \tag{6}$$

along γ. The Jacobi differential equation corresponding to the extremal segment γ will be written in the well known form

$$Rw'' + R'w' + (Q' - P)w = 0, \tag{7}$$

with w the dependent variable, and u the independent variable. *We are always going to write J. D. E. for* (7).

4. **The second partial derivatives of** $J(v_1, \cdots, v_n)$. We shall see presently that $J(v_1, \cdots, v_n)$ has a critical point when $(v_1, \cdots, v_n) = (0, \cdots, 0)$. To determine the nature of this critical point we proceed to the determination of the second partial derivatives of $J(v_1, \cdots, v_n)$. To that end we represent the family of extremals which join the points

$$(u_i, v_i),\ (u_{i+1}, v_{i+1}) \qquad (i = 0, 1, \cdots, n) \tag{1}$$

in the form

$$v = r^i(u, v_i, v_{i+1}),\quad u_i \leqq u \leqq u_{i+1}, \tag{2}$$

where it is understood that the coördinates u_i and u_{i+1} are held fast. The functions

$$r^i,\ r^i_u,\ r^i_{uu} \tag{3}$$

will be of class C' in all of their arguments for u on the interval in (2), and for v_i and v_{i+1} neighboring zero (B, p. 73 and p. 307).

We shall understand by $w_{\mu\nu}(u)$, a solution of the J. D. E., such that

$$w_{\mu\nu}(u_\mu) = 0,\quad w_{\mu\nu}(u_\nu) = 1 \quad (|\mu - \nu| = 1,\ \mu, \nu = 0, 1, 2, \cdots, n+1). \tag{4}$$

Because of the fact that the functions (2) satisfy the identities

$$\begin{aligned} v_i &\equiv r^i(u_i, v_i, v_{i+1}), \\ v_{i+1} &\equiv r^i(u_{i+1}, v_i, v_{i+1}), \end{aligned} \tag{5}$$

it follows that

$$\begin{aligned} w_{i+1,i}(u) &\equiv r^i_{v_i}(u, 0, 0), \\ w_{i,i+1}(u) &\equiv r^i_{v_{i+1}}(u, 0, 0). \end{aligned} \tag{6}$$

Differentiation of the integral J, and integration by parts in the usual way will now give

$$J_{v_i}(v_1, \cdots, v_n) = f_{v'}[u_i, v_i, r_u^{i-1}(u_i, v_{i-1}, v_i)] - f_{v'}[u_i, v_i, r^i_u(u_i, v_i, v_{i+1})]. \tag{7}$$

Here it is understood that

$$i = 1, 2, \cdots, n,\quad v_0 = 0,\quad v_{n+1} = 0. \tag{8}$$

From (7) we see that all the partial derivatives of $J(v_1, \cdots, v_n)$ are zero at $(0, \cdots, 0)$, and note that $J(v_1, \cdots, v_n)$ is of class C' in the neighborhood of $(0, \cdots, 0)$.

From (7) we see that

$$J_{v_i, v_j} = 0,\quad |i - j| \neq 1, 0. \tag{9}$$

The remaining second partial derivatives will be evaluated at $(0, \cdots, 0)$. Evaluation at the latter point will be indicated by a subscript zero preceding the partial derivative. We obtain the following results:*

$$\begin{aligned} {}_0J_{v_i, v_i} &= R(u_i)[r^{i-1}_{u, v_i}(u_i, 0, 0) - r^i_{u, v_i}(u_i, 0, 0)] \\ &= R(u_i)[w'_{i-1,i}(u_i) - w'_{i+1,i}(u_i)] \qquad (i = 1, 2, \cdots, n); \end{aligned} \tag{10}$$

* A. Dresden has given complete formulas for the second partial derivatives of the extremal integral (Bolza, p. 310). Use has not been made of this work, however, because the formulas given in the present paper need not have the general form given by Dresden, and do need a different notation.

$$(11)\qquad {}_0J_{v_i,v_{i+1}} = -R(u_i)w'_{i,i+1}(u_i) \qquad (i = 1, 2, \cdots, n-1);$$

$$(12)\qquad {}_0J_{v_{i+1}\,v_i} = R(u_{i+1})w'_{i+1,i}(u_{i+1}) \qquad (i = 1, 2, \cdots, n-1).$$

The last two partial derivatives are necessarily equal, as can be proved directly from the properties of the J. D. E.

The matrix whose elements are

$$(13)\qquad a_{i,j} = {}_0J_{v_i,v_j} \qquad (i, j = 1, 2, \cdots, n)$$

has now been determined. Note first that all elements in the ith row of this matrix have the factor $R(u_i)$. Let this factor be removed from the ith row ($i=1, 2, \cdots, n$). We will write down the resulting matrix for a typical case, $n=5$.

$$\left\|\begin{matrix} w'_{01}(u_1)-w'_{21}(u_1), & -w'_{12}(u_1) & , & 0 & , & 0 & , & 0 \\ w'_{21}(u_2) & , & w'_{12}(u_2)-w'_{32}(u_2), & -w'_{23}(u_2) & , & 0 & , & 0 \\ 0 & , & w'_{32}(u_3) & , & w'_{23}(u_3)-w'_{43}(u_3), & -w'_{34}(u_3) & , & 0 \\ 0 & , & 0 & , & w'_{43}(u_4) & , & w'_{34}(u_4)-w'_{54}(u_4), & -w'_{45}(u_4) \\ 0 & , & 0 & , & 0 & , & w'_{54}(u_5) & , & w'_{45}(u_5)-w'_{65}(u_5) \end{matrix}\right\|$$

5. **The rank of the extremal segment** g. We prove the following theorem:

THEOREM 1. *Let the function* $J(v_1, \cdots, v_n)$ *be set up for the extremal segment* g, *as described in* §1. *The matrix* a, *whose elements are*

$$a_{ij} = {}_0J_{v_iv_j},$$

is of rank n *if the final point of* g *is not among the conjugate points of the initial point of* g. *Otherwise* a *is of rank* $n-1$.

To prove this theorem let us turn to the (u, w) plane of the J. D. E., and in that plane join the successive points

$$(u_0, 0)(u_1, c_1), \cdots, (u_n, c_n)(u_{n+1}, 0),$$

by curve segments representing solutions of the J. D. E. The successive segments are of the form

$$(1)\qquad w = c_iw_{i+1,i}(u) + c_{i+1}w_{i,i+1}(u), \qquad u_i \leqq u \leqq u_{i+1},$$

where

$$(2)\qquad i = 0, 1, \cdots, n, \quad c_0 = 0, \quad c_{n+1} = 0.$$

Let the curve obtained by combining the successive segments (1) be denoted by λ. A necessary and sufficient condition that the point $(u_0, 0)$ be conjugate to $(u_{n+1}, 0)$ is that among the curves λ there exist at least one, not

$w \equiv 0$, that has no corners at the junctions of its successive segments. The n conditions that λ have no corners at the junctions of these segments are

$$(3)\quad [c_{i-1}w'_{i,i-1}(u_i) + c_i w'_{i-1,i}(u_i)] - [c_i w'_{i+1,i}(u_i) + c_{i+1}w'_{i,i+1}(u_i) = 0],$$

where

$$(4)\qquad i = 1, 2, \cdots, n, \qquad c_0 = c_{n+1} = 0.$$

Equations (3) may be written in the form

$$(5)\quad c_{i-1}w'_{i,i-1}(u_i) + c_i[w'_{i-1,i}(u_i) - w'_{i+1,i}(u_i)] - c_{i+1}w'_{i,i+1}(u_i) = 0,$$

subject again to the conditions (4).

Let the matrix of the coefficients of $(c_1, \cdots, c_n)$ in (5) be denoted by w and the value of the corresponding determinant by $|w|$. With the aid of (9), (10), (11), and (12) of §4 we obtain the following equation, giving the determinant $|a|$ of the theorem to be proved:

$$(6)\qquad |a| = R(u_1),\ R(u_2), \cdots, R(u_n) \mid w \mid .$$

For $|a|$, and accordingly $|w|$, to be zero, it is necessary and sufficient that the equations (5), subject to the conditions (4), admit a solution $(c_1, \cdots, c_n)$ in which the c_i's are not all zero. That is, for $|w|$ and $|a|$ to be zero, it is necessary and sufficient that $(u_0, 0)$ be conjugate to $(u_{n+1}, 0)$.

It remains to prove that if a is of rank less than n, its rank is exactly $n-1$. Suppose $|a|=0$. Then $(u_0, 0)$ is conjugate to $(u_{n+1}, 0)$. The point $(u_0, 0)$ cannot also be conjugate to $(u_n, 0)$, for otherwise $(u_n, 0)$ would be conjugate to $(u_{n+1}, 0)$, contrary to the original choice of $(u_1, \cdots, u_n)$. Now a consideration of the form of the minor A_{n-1} of the element a_{nn} of a shows that the vanishing of A_{n-1} is the condition that $(u_0, 0)$ be conjugate to $(u_n, 0)$. Hence $A_{n-1} \neq 0$, and a is of rank $n-1$. Thus the theorem is completely proved.

We denote by A_i $(i=1, 2, \cdots, n)$ the determinant obtained from a by striking out the last $n-i$ rows and columns of a, and set $A_0 = 1$.

COROLLARY 1. *A necessary and sufficient condition that* $A_i = 0$ $(i=1, 2, \cdots, n)$ *is that* $(u_0, 0)$ *be conjugate to* $(u_{i+1}, 0)$.

This follows at once from the form of A_i and from Theorem I.

COROLLARY 2. *The matrix* a *is in normal form.**

For if A_r and A_{r+1} $(0<r<n)$ were both zero, $(u_0, 0)$ would be conjugate

* Bôcher, *Introduction to Higher Algebra*, p. 59.

to both $(u_r, 0)$ and $(u_{r+1}, 0)$, contrary to the original restrictions on $(u_1, u_2, \cdots, u_n)$.

6. **The type number of the extremal segment g.** We prove the following theorem:

THEOREM 2.* *If the final end point B of the extremal segment g of §1 is not conjugate to its initial point A, but there are k points $(0 \leqq k \leqq n)$ on g between A and B conjugate to A, then the symmetric quadratic form*

$$\sum_{ij} {}_0J_{v_i v_j} v_i v_j \qquad (i, j = 1, 2, \cdots, n), \tag{1}$$

when reduced by a real non-singular linear transformation to squared terms only, will have k coefficients that are negative, and $n-k$ that are positive.

The number k is called the type number of the critical point $(0, \cdots, 0)$ *of $J(v_1, \cdots, v_n)$, and also the type number of g.*

To prove this theorem we shall make use of the well known fact (Bôcher, loc. cit., p. 147) that in a regularly arranged quadratic form the type number k equals the number of changes of sign in the sequence $A_0, A_1, \cdots, A_n$ (§5) where an A_i which is zero is counted as positive or negative at pleasure. To continue we adopt the method of mathematical induction. In the case $n=1$ we are concerned simply with the fixed end points $(u_0, 0)$ and $(u_2, 0)$, and an intermediate point $(u_1, 0)$. Here

$$A_1 = [w'_{0,1}(u_1) - w'_{2,1}(u_1)]R(u_1). \tag{2}$$

It is readily seen that this difference is negative or positive according to whether or not $(u_0, 0)$ possesses a conjugate point prior to $(u_2, 0)$. We now distinguish between the cases in which $A_{n-1}=0$, and $A_{n-1}\neq 0$.

Case I. $A_{n-1}\neq 0$. If we assume the validity of the theorem for the case where there are $n-1$ coördinates $v_1, \cdots, v_{n-1}$, we can conclude that there are as many changes in sign in the sequence $A_0, \cdots, A_{n-1}$ as there are conjugate points prior to $(u_n, 0)$. To determine whether there is a change of sign in passing from A_{n-1} to A_n, it is useful to consider again equations (5) of §5, unaltered except that the zero in the right hand member of the last equation is here replaced by

$$A_{n-1}/A_n \tag{3}$$

and the left hand member of the ith one of these equations is here to be multiplied by $R(u_i)$ $(i=1, 2, \cdots, n)$.

---

* The corresponding theorem in m dimensions has recently been discovered by the author.

The resulting equations can now be solved for the constants $c_1, \cdots, c_n$. In particular we obtain, by Cramer's rule,

$$c_n = \frac{A_{n-1}^2}{A_n^2} > 0.$$

The curve λ of §5 corresponding to these constants $(c_1, \cdots, c_n)$ has just one corner, and that at (u_n, c_n). The last equation of (5), §5, altered as we have said, now tells us that at this corner the slope of the first segment of λ exceeds or is less than the slope of the second segment of λ, according as the right hand member (3) is positive or negative. This result can be interpreted in terms of the J. D. E. to mean that there is, or is not, a conjugate point of $(u_0, 0)$ between $(u_n, 0)$ and $(u_{n+1}, 0)$, according as the quotient (3) is negative or positive. Thus, in case $A_{n-1} \neq 0$, the total number of changes of sign in the sequence $A_0, A_1, \cdots, A_n$ equals the total number of conjugate points of $(u_0, 0)$ prior to $(u_{n+1}, 0)$.

Case II. $A_{n-1} = 0$. In this case we make use of the fact that, if $A_{n-1} = 0$ in a regularly arranged non-singular quadratic form, then $A_{n-2} \neq 0$. Assuming then the validity of the theorem for the case where there are $n-2$ points between the end points of a given extremal we can conclude that there are as many changes of sign, say h, in the sequence $A_0, A_1, \cdots, A_{n-2}$ as there are conjugate points of $(u_0, 0)$ prior to $(u_{n-1}, 0)$. But since $A_{n-1} = 0$ it follows from Corollary 1, §5, that $(u_n, 0)$ is conjugate to $(u_0, 0)$ so that there are altogether $h+1$ conjugate points of $(u_0, 0)$ prior to $(u_{n+1}, 0)$. On the other hand it follows from the theory of regularly arranged quadratic forms that if $A_{n-1} = 0$, then A_{n-2} and A_n have opposite signs. Thus in the sequence A_0, $A_1, \cdots, A_n$, there are $h+1$ changes of sign, and the theorem is proved in Case II as well as in Case I.

Part II. Relations in the large between extremals joining A to B

7. The integrand and region S. In the developement of this chapter we do not wish to exclude the case where the end points of the extremal segment are conjugate. The complete treatment of an extremal segment whose ends are conjugate, both in our theory and in the classical theory, depends upon the nature of the singularity at B of the envelope of the extremals passing through A. Such a treatment, at least as developed so far, requires the assumption that the functions used be analytic.

I. *We therefore assume that the function $F(x, y, \dot{x}, \dot{y})$ is positively homogeneous of the first degree in $\dot{x}$ and $\dot{y}$, and analytic in all of its arguments, for x, y any point interior to an open region S of the (x, y) plane, and $\dot{x}$ and $\dot{y}$ any* *two numbers not both zero. We shall also assume that for the same arguments*

$$F_1(x, y, \dot{x}, \dot{y}) > 0.$$

In S the differential equations of the extremals can be put in the Bliss* form,

$$\frac{dx}{ds} = \cos\theta, \quad \frac{dy}{ds} = \sin\theta, \quad \frac{d\theta}{ds} = H(x, y, \cos\theta, \sin\theta), \tag{1}$$

where

$$H(x, y, \dot{x}, \dot{y}) = \frac{F_{y\dot{x}} - F_{x\dot{y}}}{F_1[\dot{x}^2 + \dot{y}^2]^{3/2}}. \tag{2}$$

A necessary and sufficient condition that a solution g of (1) on which $(x, y, \theta) = (x_0, y_0, \theta_0)$ for some value of s, be identical as a set of points (x, y), with the solution on which $(x, y, \theta) = (x_0, y_0, \theta_0 + \pi)$ for some value of s, is that, on g,

$$H(x, y, \cos\theta, \sin\theta) + H(x, y, -\cos\theta, -\sin\theta) = 0. \tag{3}$$

II.† *We assume that* (3) *holds identically for every point* (x, y) *in* S, *and for every* θ. *This type of problem is termed reversible.*

8. **The regions** S_1, **and** R. We can choose between a great variety of assumptions that will serve as boundary conditions. The following perhaps are as simple as any.

III. *Let there be given a closed region* S_1, *consisting of points interior to* S, *bounded by a simple closed curve* β *consisting of a finite number of analytic arcs. We assume that this boundary* β *is extremal-convex*‡, *in the sense that the interior angles at the vertices of* β *shall be between* 0 *and* π, *and that an extremal tangent to an analytic arc* β' *of* β *at a point* P, *shall in the neighborhood of* P, *except for* P, *lie wholly on that side of* β' *which is exterior to* S_1.

IV.† *We assume further that the region* S_1 *is covered in a one-to-one manner by a proper field of extremals of the form*

$$x = h(u, v), \quad y = k(u, v), \tag{1}$$

where u *is the parameter, and* v *is the arc length measured along the extremals, where at every point* (u, v) *that corresponds to a point* (x, y) *in* S_1 *the functions* (1) *are single-valued and analytic in* u *and* v, *and*

* Bliss, these Transactions, vol. 7 (1906), p. 180.

† The author has recently removed hypotheses II and IV, assuming then that F is positive definite. Powerful results obtain, but the proofs are necessarily more difficult.

‡ Compare Bolza, loc. cit., pp. 276–278, and also Birkhoff, loc. cit., pp. 216–219.

$$D = \begin{vmatrix} h_u & h_v \\ k_u & k_v \end{vmatrix} \neq 0.$$

The region S_1 in the (x, y) plane will correspond in the (u, v) plane to a closed region R, bounded by a simple closed curve γ consisting of a finite number of analytic arcs, and again extremal-convex. When u is the independent variable, and v the dependent variable, the integral J can be put in the non-parametric form as in §3 with $f(u, v, v')$ the integrand. Because of the assumption of reversibility the non-parametric problem, with integrand $f(u, v, v')$, will here include all the extremals of the parametric form except those of the family $u=$constant. From our assumption regarding F_1, we have here that

$$f_{v'v'}(u, v, v') = 0,$$

for all points (u, v) in R, and any number v'. Concerning the region R, we can now establish the following lemma:

LEMMA. *Let a and b be, respectively, the minimum and maximum values of u on γ. Then γ consists of two arcs of the form*

$$v = A(u), \quad v = B(u), \qquad a \leqq u \leqq b,$$

where $A(u)$ and $B(u)$ are of class C, and analytic in u on the interval $a \leqq u \leqq b$, except for a finite set of values of u, while

$$A(u) < B(u), \qquad a < u < b, \tag{2}$$

$$A(a) = B(a), \quad A(b) = B(b). \tag{3}$$

Finally the interior points (u, v) of R are the points satisfying

$$A(u) < v < B(u), \qquad a < u < b. \tag{4}$$

To prove these statements let (u_0, v_0) be any interior point of R. There exist two numbers, v_1 and v_2, of such sort that the points (u, v) for which

$$u = u_0, \qquad v_1 < v < v_2, \tag{5}$$

include the point (u_0, v_0), but no points other than interior points of R, while the points (u_0, v_1) and (u_0, v_2) are on the boundary of R. Now the extremal-segment (5) is not tangent to γ at either of its ends, since γ is extremal convex. If γ has a vertex at (u_0, v_2), the two analytic arcs adjoining (u_0, v_2) must lie on opposite sides of $u=u_0$, at least in the neighborhood of (u_0, v_2), for otherwise the points of (5) in the neighborhood of (u_0, v_2) would not lie in R. A similar statement applies to (u_0, v_1). Thus, in any case, there must exist functions $h(u)$ and $k(u)$ of class C, and constants a_1 and b_1, differing from u_0 by so little, that the points (u, v) satisfying

$$h(u) < v < k(u), \qquad a_1 < u < b_1, \tag{6}$$

are all interior points of R, while the curves

$$\begin{aligned} v &= h(u), \quad v_1 = h(u_0), \qquad a_1 \leqq u \leqq b_1, \\ v &= k(u), \quad v_2 = k(u_0), \qquad a_1 < u_0 < b_1, \end{aligned} \tag{7}$$

lie on the boundary of R.

We wish to show that these functions $h(u)$ and $k(u)$ can be extended in definition, as functions of class C, so that the preceding statements regarding (6) and (7) still hold true when $a_1 = a$ and $b_1 = b$. Whether this is true or not there certainly exist constants a_2 and b_2, such that

$$a \leqq a_2 < u_0 < b_2 \leqq b,$$

and functions $h(u)$ and $k(u)$ of class C, such that the curves

$$\begin{aligned} v &= h(u), \quad v_1 = h(u_0), \\ v &= k(u), \quad v_2 = k(u_0), \qquad a_2 \leqq u \leqq b_2, \end{aligned} \tag{8}$$

lie on the boundary of R, and the points (u, v) which satisfy

$$h(u) < v < k(u), \qquad a_2 < u < b_2, \tag{9}$$

are all interior points of R, while further, a_2 and b_2 are constants such that the interval in (8), and the corresponding interval in (9), cannot be expanded at either or both ends, and the preceding statements about (8) and (9) still hold true for these expanded intervals.

The following division into cases is exhaustive:

Case I $\quad a_2 = a, \quad b_2 = b;$
Case II $\quad a_2 = a, \quad b_2 < b;$
Case III $\quad a < a_2, \quad b_2 = b;$
Case IV $\quad a < a_2, \quad b_2 < b.$

Case I. In this case the theorem follows readily.

Case II. In this case we will prove that the inequality $b_2 < b$ is impossible. In the first place if we had $h(b_2) = k(b_2)$, the curve (8) would completely bound the region (9), a sub-region of R. This is impossible, since R consists of a single connected region. Hence $h(b_2) < k(b_2)$. The points (u, v) which satisfy

$$h(b_2) < v < k(b_2), \qquad u = b_2, \tag{10}$$

are limit points of points of (9), and are accordingly points of R. We distinguish again between two cases, Cases IIa and IIb.

In Case IIa we suppose the points of (10) are all interior points of R. In this case (8) and (9) will hold as stated for a slightly larger b_2, and thus we have a contradiction.

In Case IIb we suppose the points of (10) include at least one boundary point P of R. Such a point P cannot be a vertex of the boundary without the interior angle at the vertex being greater than π, contrary to an hypothesis. Neither can P be an ordinary point of the boundary, for in that case the boundary would have to be tangent to the extremal segment (10), a result which is again impossible, since the boundary is extremal-convex. Thus Case IIb leads to a contradiction under all circumstances, and is impossible.

Similarly, it can be proved that Cases III and IV are impossible. Case I alone is possible, and the theorem in italics follows readily.

9. **Further properties of the region R.** In R no extremals other than the extremals u = constant can be tangent to the extremals u = constant. Hence every extremal segment other than the extremal segments u = constant is representable in the form $v = M(u)$, where $M(u)$ is an analytic function of u, for u on the interval (§8, Lemma)

$$a \leqq u \leqq b,$$

or some sub-interval of that interval.

Let A and B be two distinct points of R, either interior to R, or on the boundary of R. *Any extremal joining A to B in R will have no points on the boundary of R, except possibly its end points A or B.* This follows readily from the extremal convex nature of the boundary (§8).

We will now prove *that two points A and B in R can be joined in R by at most a finite number of extremal segments.* If A and B are both on the same extremal segment $u = u_0$, that extremal segment is the only extremal segment in R which can join A to B. Suppose then that (u_0, v_0) and (u_1, v_1) are two points of R for which $u_0 < u_1$, and which can be joined by an infinite set of extremal segments in R. Denote by m the angles in the (u, v) plane which the tangents to these extremals at (u_0, v_0) make with a parallel to the positive u axis. Take these angles between $-\pi/2$ and $\pi/2$.

Now there will be at least one limit angle m_0 of these angles m. The v coördinate of an extremal E_0 issuing from (u_0, v_0) with the angle m_0 can be continued as an analytic function of u until E_0 passes out of R or through (u_1, v_1). The extremal E_0 cannot pass out of R before passing through (u_1, v_1), because if E_0 did so pass out of R, extremals with angles m sufficiently near m_0 would also pass out of R before passing through (u_1, v_1), which is impossible.

It would follow then from the envelope theory of extremals that all extremals, without exception, that issue from (u_0, v_0) with angles *neighboring* m_0 would pass through (u_1, v_1). Moreover, we could then prove that *all* extremals issuing from (u_0, v_0) with angles m for which

$$m_0 \leqq m < \pi/2, \tag{1}$$

would remain in R, at least until they passed through (u_1, v_1). For otherwise there would be a least upper bound $m_1 < \pi/2$, of angles m at which extremals in R issue from (u_0, v_0) and pass through (u_1, v_1). But the extremal issuing from (u_0, v_0) with the angle m_1 would lie entirely within R, except possibly for its end points, so that extremals issuing from (u_0, v_0) with angles slightly larger than m_1 would also lie in R, and pass through (u_1, v_1). Therefore no such least upper bound, m_1, for which $m < \pi/2$ exists. Thus all extremals which issue from (u_0, v_0) with angles m satisfying (1) must pass through (u_1, v_1) before passing out of R.

But an extremal issuing from (u_0, v_0) with an angle sufficiently near $\pi/2$ will pass out of R without passing through (u_1, v_1). From this contradiction we infer the truth of the statement to be proved.

10. **The function** $J(v_1, \cdots, v_n)$. *None of the extremal segments* $u = u_0$ *in* R *have a point on them conjugate to either end point.* To prove this let the J. D. E. in the Weierstrass form with independent variable $t = v$, be set up corresponding to the extremal $u = u_0$ in the (x, y) plane. This differential equation has as a solution the determinant D of §8, provided we set $u = u_0$ and let v vary. The absence of any conjugate points on $u = u_0$ follows from the non-vanishing of this determinant. From the absence of conjugate points in R on the extremals $u = u_0$, the "regularity" $(F_1 > 0)$, and analyticity of the problem, and the extremal convex nature of the boundary, it follows that *a positive constant* e *can be determined small enough to have the following properties. Any point of* R *for which* $u = u_0$, *can be joined to any point of* R *for which* u *differs from* u_0 *in absolute value by less than* e, *by a unique analytic extremal* h *lying entirely within* R, *excepting possibly its end points, and such that on* h *there are no pairs of conjugate points* (B, p. 307). The questions of uniformity arising can be met by the reader without too much difficulty.

Now let there be given two points A and B of R, not on the same extremal segment $u = u_0$. We are concerned with the extremals joining A to B, if any such exist. Let $u_0, u_1, \cdots, u_{n+1}$ be a set of increasing values of u of which successive values differ at most by the constant e of the preceding paragraph, and which are such that A lies on $u = u_0$, and B on $u = u_{n+1}$. Consistent with this, let the coördinates of A and B be respectively (u_0, v_0) and (u_{n+1}, v_{n+1}), and $(v_1, \cdots, v_n)$ be variables such that the points

$$(1) \qquad (u_1, v_1), \cdots, (u_n, v_n)$$

are all in R. Let the successive points of the set

$$(2) \qquad (u_0, v_0), (u_1, v_1), \cdots, (u_{n+1}, v_{n+1})$$

be joined by the unique extremal segments of the preceding paragraph, and let the integral J be evaluated along the resulting curve. If we hold the end points A and B fast, as well as the u coördinates of the intermediate points, the value of J will be a function, $J(v_1, \cdots, v_n)$, that will be analytic in $(v_1, \cdots, v_n)$ in the domain (Lemma §8)

$$(3) \qquad A(u_i) \leqq v_i \leqq B(u_i) \qquad (i = 1, 2, \cdots, n).$$

The partial derivatives of the function J are given by the formulas

$$(4) \quad J_{v_i}(v_1, \cdots, v_n) = f_{v'}(u_i, v_i, p_i) - f_{v'}(u_i, v_i, q_i) \qquad (i = 1, 2, \cdots, n),$$

where p_i is the slope at (u_i, v_i) of the extremal joining (u_{i-1}, v_{i-1}) to (u_i, v_i), and q_i is the slope at (u_i, v_i) of the extremal joining (u_i, v_i) to (u_{i+1}, v_{i+1}). Since

$$f_{v'v'}(u, v, v') > 0$$

for every (u, v) in R and every v', the partial derivative (4) is zero, when and only when $p_i = q_i$. We have the result that *the function* $J(v_1, \cdots, v_n)$ *has a critical point* $(v_1, \cdots, v_n)$, *when and only when the points* (2) *all lie on a single analytic extremal.*

11. Relations between critical points. A lemma fundamental for our present purposes is derived from the Corollary to Theorem 8, page 392 of the paper of the author's already cited. Before stating the lemma let it be agreed that the positive normal to an analytic boundary of a region Σ will be understood to be that sensed normal that leads from points in Σ to points without Σ.

LEMMA 1. *Let there be given a closed region Σ in the space of the n variables $(x_1, \cdots, x_n)$, bounded by a closed analytic manifold, without singularity and homeomorphic with the interior and boundary of an $(n-1)$-dimensional hypersphere. In Σ let there be given a function $f(x_1, \cdots, x_n)$, of class C''' at each point of Σ, and possessing on the boundary of Σ a normal directional derivative that is positive. Suppose the critical points of $f(x_1, \cdots, x_n)$ are all of rank n. Of the type numbers k of these critical points, let m be the maximum. Let M_k be the number of critical points of type k. Then between the integers M_k the following relations hold:*

$$
\text{(R)}\quad
\begin{aligned}
1 &\leqq M_0,\\
1 &\geqq M_0 - M_1,\\
1 &\leqq M_0 - M_1 + M_2,\\
1 &\geqq M_0 - M_1 + M_2 - M_3,\\
&\cdots\cdots\cdots\cdots\cdots\cdots\\
1 &\leqq [M_0 - M_1 + M_2 - M_3 + - \cdots + (-1)^{m-1}M_{m-1}](-1)^{m-1},\\
1 &= M_0 - M_1 + M_2 - M_3 + - \cdots + (-1)^m M_m.
\end{aligned}
$$

In terms of the functions of the Lemma, §8, we can say that the domain of the points $(v_1, \cdots, v_n) = (V)$ for which the function $J(v_1, \cdots, v_n)$ of §10 is defined, is a rectangular hyperspace made up of points (V) satisfying the inequalities

$$
(1)\qquad A(u_i) \leqq v_i \leqq B(u_i) \qquad (i = 1, 2, \cdots, n),
$$

where the numbers $(u_1, \cdots, u_n)$ are constants specified in §10. The boundary of the domain obviously consists of hyperrectangles on $2n$ hyperplanes

$$
v_i = A(u_i), \quad v_i = B(u_i) \qquad (i = 1, 2, \cdots, n),
$$

taken separately.

To apply Lemma 1 of this section we need to find the normal directional derivative of $J(v_1, \cdots, v_n)$ at points on the boundary of (1).

Let $(a_1, a_2, \cdots, a_n)$ be a point (V) of (1) for which $a_1 = A(u_1)$, that is, a point on the bounding hyperplane, $v_1 = A(u_1)$. If we hold the variables $(v_2, \cdots, v_n)$ constantly equal to $(a_2, \cdots, a_n)$, respectively, and let v_1 decrease so as to pass through $v_1 = a_1$, the corresponding point $(v_1, a_2, \cdots, a_n)$ will vary on a normal to the hyperplane $v_1 = A(u_1)$, and pass out of the domain (1). But the point $(v_1, \cdots, v_n) = (a_1, \cdots, a_n)$ will correspond (§10) to a succession of $n+1$ extremal segments joining A to B in R. Of these segments the first two will adjoin each other in R at the point $(u, v) = (u_1, a_1)$. The partial derivative of $J(v_1, \cdots, v_n)$ with respect to v_1 at the point $(a_1, \cdots, a_n)$ can be obtained from (4), §10, by putting $i = 1$. Using the law of the mean we then obtain the equation

$$
(2)\qquad J_{v_1}(a_1, a_2, \cdots, a_n) = f_{v'v'}[u_1, a_1, \bar{p}][p_1 - q_1],
$$

where $\bar{p}$ is a number between p_1 and q_1. From the extremal-convex nature of the boundary it follows that $p_1 < q_1$. Thus this derivative (2) is negative.

The directional derivative of J along the normal to the hyperplane $v_1 = A(u_1)$ at the point $(a_1, a_2, \cdots, a_n)$ is to be taken in the sense that leads out of (1), that is in the sense of decreasing v_1. This directional derivative is thus equal to the left hand member of (2) multiplied by (-1), and is accordingly positive. Similarly the directional derivative of J along a positive normal to any of the other hyperplanes bounding (1) may be seen to be positive.

But we cannot as yet apply the lemma on critical points because the boundary of (1) is made up of portions of $2n$ hyperplanes instead of one analytic manifold. To meet this difficulty let O be any interior point of (1), and let straight line rays be drawn from O to each point of the boundary of (1). It follows from the results of the preceding paragraph that the directional derivative of $J(v_1, \cdots, v_n)$, at a point P on the boundary of (1), taken along the ray joining O to P in the sense that leads from O to P is always positive. Now the domain (1) can be projectively transformed into an n-dimensional hypercube that lies in the space of $(y_1, \cdots, y_n)$, and that is bounded by the $(n-1)$-dimensional hyperplanes

$$y_i = \pm 1 \qquad (i = 1, 2, \cdots, n). \tag{3}$$

This hypercube can be approximated to by the analytic manifold

$$y_1^{2r} + y_2^{2r} + \cdots + y_n^{2r} = 1, \tag{4}$$

where r is a positive integer. In fact, if e be a positive constant, it is easy to show that for r sufficiently large, points on the above hypercube and the manifold (4) that lie on the same ray issuing from the origin will be within a distance e of one another.

Let now the hypercube be projectively transformed back into (1), and suppose the manifold (4) goes into a manifold M. Let the point O from which rays were drawn in (1) be the image of the origin in the hypercube. The rays in (1) will each meet the manifold M in a single point. If r be sufficiently large, M will approximate the domain (1) so closely that the directional derivatives of J at points of M on the rays issuing from O in the sense that leads away from O, will all be positive. It follows that the outer normal directional derivatives at points of M will be positive.

A second requirement on M is that it approximate the domain (1) so closely that it contain in its interior all the critical points of $J(v_1, \cdots, v_n)$ that (1) contains. The manifold M will serve as the manifold Σ of Lemma 1 of this section.

The critical points of Lemma 1 are of rank n. For the moment, then, we restrict ourselves to extremals joining A to B on which A is not conjugate to B. Reference to Theorem 2 of §6 and Lemma 1 of this section gives the lemma.

LEMMA 2. *Let there be given regions S and S_1, and integrand F, satisfying the hypotheses* I, II, III, *and* IV, *of* §7 *and* §8. *In S_1 let there be given two* *points A and B which are joined by no extremals on which A is conjugate to B. Let the number of extremals joining A to B in S_1 on which there are k conjugate points of A prior to B be denoted by M_k. Let m be an integer equal to the maximum of these integers k. Then between the numbers M_k the relations* (R) *of Lemma* 1 *hold.*

12. **A theorem in the large.** We seek now to remove from Lemma 2, §11, the restriction that on no extremal joining A to B in S_1 is A conjugate to B. Before doing this it will be necessary to recall certain results obtained by Lindeberg.*

Let there be given in R an analytic extremal E_0 of the form

$$v = E(u), \qquad c \leqq u \leqq d,$$

joining A to B, and on which the point at which $u=c$ has for its $(m+1)$st conjugate point the point at which $u=d$. Let α be the slope at A of any extremal through A. In particular let $\alpha=\alpha_0$ be the slope of E_0 at A. That part of the envelope of the family of extremals through A which lies in the neighborhood of B, will not here consist merely of the point B. For according to the envelope theory this could only happen if all the extremals through A with slopes α near α_0 should pass through B, contrary to results in §9.

According to Lindeberg, the envelope T of the family of extremals passing through A, with a slope α near α_0, will then be of the form

$$v - E(u) = (\alpha - \alpha_0)^{r+1}K(\alpha), \quad r > 0, \qquad K(\alpha_0) \neq 0,$$
$$u - d = (\alpha - \alpha_0)^{r}H(\alpha), \qquad H(\alpha_0) \neq 0,$$

where $K(\alpha)$ and $H(\alpha)$ are analytic in α at $\alpha=\alpha_0$, and r is a positive integer.

Three classes of envelopes can be distinguished:

CLASS I r is odd;

CLASS II r is even and $H(\alpha_0)<0$;

CLASS III r is even and $H(\alpha_0)>0$.

CLASS I. Here the envelope T is tangent to the extremal E_0 at B, has no cusp there, and lies wholly on one side of E_0 near B. If the class of extremals through A be restricted to extremals E for which $|\alpha-\alpha_0|$ is sufficiently small, we can say that through each point P, not on T, but sufficiently near B, and on the same side of T as E_0, there will pass just two extremals of the set E. On these two extremals the type number k, that is, the number of points which are conjugate to A and prior to P, will equal m and $m+1$,

* Lindeberg, Mathematische Annalen, vol. 59 (1904), p. 321.

respectively. Through any point P, not on T, but sufficiently near B and on the opposite side of T from E_0, there will pass no extremals of the set E.

CLASS II. Here T is tangent to the extremal E_0 at B, but has a cusp there. On the envelope near B, $u<d$, except for the point B. The two branches of the cusp lie on opposite sides of E_0. A point P sufficiently near B, within the cusp, but not on T, can be joined to A by three extremals neighboring E_0. On these three extremals the type numbers k will have the values $m+1$, $m+1$, and m, respectively. Through any point P sufficiently near B, without the cusp, but not on T, there passes just one extremal issuing from A and neighboring E_0. On this extremal $k=m+1$.

CLASS III. Here T is tangent to the extremal E_0 at B, but has a cusp there. On the envelope near B, $u>d$, except for the point B. The two branches of the cusp lie on opposite sides of E_0. A point P, sufficiently near B, within the cusp, but not on T, can be joined to A by three extremals neighboring E_0. On these three extremals k has the values m, m, and $m+1$, respectively. Through each point P sufficiently near B, without the cusp, and not on T, there passes just one extremal issuing from A and neighboring E_0. On this extremal $k=m$.

In the following theorem there are a number of conventions to be adopted. Let g be an extremal joining A to B on which there are m points conjugate to A prior to B. If B is not conjugate to A, g is to be counted as one extremal of type $k=m$. *If B is conjugate to A, and g belongs to Class* I, *of the preceding classification, g is to be counted as two extremals, of types $k=m+1$ and $k=m$ respectively. If B is conjugate to A, and g belongs to Class* II, *or to Class* III, *then g is to be counted as one extremal of type $k=m+1$, or of type $k=m$, respectively.* We can now prove the following theorem "in the large."

THEOREM 3. *Let there be given regions S and S_1 and an integrand F satisfying hypotheses* I, II, III, *and* IV, *of* §7 *and* §8. *Let A and B be any two points of S_1. A first conclusion is that there are at most a finite number of extremals g joining A and B and lying in S_1. Let M_k be the number of these extremals g of type k, counted according to the conventions preceding this theorem, and let m be the maximum of these integers k. Then between the numbers M_k the relations* (R) *of Lemma* 1, §11, *hold true.*

That there are at most a finite number of extremals g joining A to B in S_1, was proved in §9. If on no extremal g, A is conjugate to B, the remainder of the theorem follows from Lemma 2, §11.

If there are a number of extremals g on which A is conjugate to B, each of these extremals g will be tangent at B to an envelope T of extremals issuing from A and neighboring that g. Let L be a short straight line segment passing through B, lying in S_1, and tangent to none of these envelopes. Any point P on L, not B, but sufficiently near B, will lie on none of these envelopes. Concerning the possibility of joining A to P by extremals g' in S_1, we can state the following:

Corresponding to any extremal g on which A is not conjugate to B there will exist one extremal g' neighboring g, joining A to P, and of the same type as g.

If g be an extremal on which A is conjugate to B, and g is of Class II, the point P if sufficiently near B, will lie without the corresponding cusp. According to the preceding description of Class II there will then be just one extremal g' joining A to P and neighboring g. On g', A will not be conjugate to P, and the number of conjugate points of A prior to P will equal the type number k which we have agreed to assign to g. The facts are similar for extremals g of Class III.

If g be any extremal of Class I, joining A to B, let T be the corresponding envelope neighboring B. We consider two cases. In Case I we suppose P lies on the same side of T as g. In this case A and P can be joined by two extremals neighboring g, on which A is not conjugate to P, and whose type numbers are the numbers we have agreed to assign to g. In Case II we suppose P lies on the opposite side of T from g. In this case there will exist no extremals neighboring g' passing from A to P. Relative to this case we observe that if the relations (R) of Lemma 1, §11, hold between any given set of integers M_k, these relations will also hold if we replace M_i and M_{i-1} by M_i+1 and $M_{i-1}+1$, for any particular i $(i=1, 2, \cdots, n)$. Finally, if P be sufficiently near B, there will be no other extremals g' joining A to P than those just enumerated. For if there were more, we could prove by a limiting process that there would be other extremals g joining A to B besides those first supposed to exist.

The extremals which we have just proved join A and P are all extremals on which A is not conjugate to P. Concerning them Lemma 1 of §11 holds. If there are no extremals g joining A to B, of Class I, Case II, the theorem follows directly. But if there are extremals g of Class I, Case II, the relations of Lemma 1, §11, hold if all extremals g be counted except those of Class I, Case II. As we have already noted, the counting of extremals of Class I, Case II, as if they were in Class I, Case I, will not alter the validity of the relations (1) of §11, if these relations hold true prior to such a change. Thus the theorem is proved.

NOTE. *We could prove, by the methods of the paper on critical points, that extremals g of Class* I *could be omitted from the count entirely, and the relations* (R), §11, *still hold true. Such a proof would, however, lead us too far astray.*

Another important question is whether there are relations between the numbers M_k other than those affirmed in the Theorem. For the case of functions in general, apart from the calculus of variations, the answer is that there are no other relations without further hypotheses. More specifically the author has proved that *if there be given any set of integers* M_i *satisfying the relations* (R) *of Lemma* 1, §11, *there exists a function f of class* C'' *within and on a unit* $(n-1)$*-sphere, which on this* $(n-1)$*-sphere satisfies the boundary conditions of Lemma* 1, *and which possesses for each i,* M_i *critical points of type i, and no other critical points of any sort.**

For the case of the calculus of variations the author has shown that if under the hypotheses of Theorem 3 there is more than one extremal joining A and B, then there are at least two such extremals of minimum type. Except for this, Dr. Richmond, National Research Fellow at Harvard, has recently shown, *for the case where* $m<3$ *in the Theorem, that an example can be set up in the calculus of variations corresponding to any set of integers* M_i *satisfying the relations* (R).

13. **Existence of extremal-convex boundaries.** Let there be given in the (x, y) plane an open or closed curve γ, without multiple points, and made up of a finite number of arcs of class C'. If γ is closed, the points neighboring γ and not on γ make up two disconnected regions. If γ is open, the points neighboring γ, not on γ, and lying on short perpendiculars to the component arcs of γ, slightly extended at the vertices, again make up two distinct regions. *These two disconnected regions will be called the sides of* γ.

The curve γ *will be said to be extremal-convex relative to one of its sides S, if the angles at its vertices on the side S are between* 0 *and* π, *and if an extremal tangent to* γ *at any point P has no other point than P in common with* γ *or S in the neighborhood of P.*

Let g be any open or closed curve of class C' and without multiple points. A curve g' of class C' will be said to lie *arbitrarily near g, in position and direction,* if corresponding to an arbitrarily small positive constant e, a one-to-one continuous correspondence can be set up between g and g', in such a fashion that corresponding points are within a distance e of one another, and direction cosines of tangents at corresponding points differ by at most e. We can now prove the following lemma:

LEMMA 1. *Let g be any extremal segment satisfying the hypothesis in* §1, *and in addition derived from a reversible problem* (§7).

(A) *Then there can be found a curve* g_1 *of class* C''', *arbitrarily near to g*

* Morse, *The analysis and analysis situs of regular n-spreads in* $(n+r)$*-space*, Proceedings of the National Academy of Sciences, vol. 13 (1927), pp. 813–817.

in position and direction, lying wholly on an arbitrary side of g, and such that g_1 will be extremal-convex relative to the side of g_1 that does not contain g.

(B) *If g contains no conjugate point to its end points, then in addition to g_1 there can also be found a second curve g_2 with the same properties as g_1, except that g_2 will be extremal-convex relative to the side of g_2 that contains g.*

To prove (A) we take the problem into the (u, v) plane as in §2, so that g becomes a segment of the u axis. The differential equation of the extremals near g can be put in the form

$$v'' = A(u, v, v') \tag{1}$$

where $A(u, v, v')$ is of class C''' for (u, v) neighboring g, and any number v' sufficiently small in absolute value. Alongside of (1) let us consider the differential equation

$$v'' = A(u, v, v') + Mv, \tag{2}$$

where M is a positive constant such that on g

$$M > A_v(u, 0, 0). \tag{3}$$

Now $v \equiv 0$ will represent a solution of (2), as well as of (1). The differential equation of the first variation, corresponding to a solution of (2), set up in particular for g, on which $v \equiv 0$, will be

$$w'' = A_{v'}(u, 0, 0)w' + [A_v(u, 0, 0) + M]w. \tag{4}$$

Let us compare (4), by Sturm's method, with

$$w'' = A_{v'}(u, 0, 0)w'. \tag{5}$$

Since (5) has a solution $w \equiv \text{constant} \neq 0$, (4) has no solution except $w \equiv 0$ which vanishes twice. Hence there is no conjugate point on g to either end point of g, relative to the differential equation (2).

Accordingly there exists a curve segment g_1, which represents a solution of (2), which lies arbitrarily near g, and on which $v > 0$. A comparison of the value of v'', say v_2'', given by (2) for a v and v' on g_1, with the value, say v_1'', of v'' given by (1) for the same v and v', shows that $v_2'' > v_1''$. The proof is similar for the side of g where $v < 0$.

To prove (B), compare (1) with

$$v'' = A(u, v, v') - ev^3, \tag{6}$$

where e is a positive constant. The equations of first variation corresponding to a solution of (6), or a solution of (1), respectively, are identical when set up for g, that is for $v \equiv 0$. Accordingly there is no conjugate point on g

to any point on g, when g is considered as a solution of (6). Hence a curve segment g_1 exists that represents a solution of (6), that lies arbitrarily near g, and on which $v>0$. Part (B) follows on comparing the v'' given by (6) with that given by (1) for the same v and v'. Lemma 1 leads to the following lemma.

LEMMA 2. *Let there be given in the region S of* §1 *a region S' bounded by a simple closed curve γ made up of a finite number of extremal segments g of* §1, *at least two in number, and making interior angles between* 0 *and* π. *Suppose further that the problem is reversible.*

(A) *Then each arc g may be replaced by an arc g_1 of class C''', arbitrarily near g_1, within S', and such that the set of arcs g_1 form a simple closed curve extremal-convex relative to its interior.*

(B) *If the end points of each arc g have no conjugate points on that arc g, each arc g may be replaced by an arc g_2 of class C''', arbitrarily near g_2, exterior to S', and such that the set of arcs g_2 form a simple closed curve extremal-convex relative to its interior.*

NOTE: The proof shows that when the integrand is analytic the preceding curves g_1 and g_2 may be taken as analytic.

14. **An example.** Let a surface of revolution be defined by revolving an analytic open or closed curve G, without singularities, about an axis lying in a plane with G_1, but not intersecting G. Let the surface be referred to parameters (u, v), of which u measures the angle through which G has been revolved from an initial position, and of which v represents arc lengths measured along G from a point chosen on G. The meridians $u=\text{constant}$ are all geodesics. The parallels $v=v_0$ are geodesics, if at the point $v=v_0$ the tangent to G is parallel to the axis of revolution. These facts follow at once from the differential equations of the geodesics.

Let us represent the surface in the (u, v) plane. Let there be given in the (u, v) plane a rectangle S, bounded by any two curves $u=\text{constant}$, and by any two curves $v=\text{constant}$ which represent geodesics. The rectangle S can be replaced by a region S_1, interior to the rectangle, but differing from the rectangle arbitrarily little, and bounded by a curve extremal-convex relative to its interior (Lemma 2, §13). The curves $u=\text{constant}$ will form the field of extremals contemplated in §8, so that Theorem 3, §12, applies to S_1.

If the curves $v=\text{constant}$ bounding the rectangle S represent parallels on which the distance to the axis has a proper or improper minimum relative to neighboring parallels, then these curves $v=\text{constant}$ have no conjugate points on them, as can be readily proved. The curves $u=\text{constant}$ never have conjugate points on them. According to Lemma 2, §13, the rectangle S can in this case be replaced by a region S_1 *slightly larger* than S, bounded by a curve again extremal-convex relative to its interior. To S_1 Theorem 3, §12, will then apply. In particular the torus presents several different types of regions S_1 to which Theorem 3, §12, applies.

PART III. THE TYPE NUMBER OF A NON-DEGENERATE PERIODIC EXTREMAL

15. The function $J(v_1, \cdots, v_n)$. We start here with the same assumptions regarding the integrand $F(x, y, \dot{x}, \dot{y})$ as we made in §1. We suppose here, however, that we have given a periodic extremal g of length ω and of class C'''. We again suppose $F_1(x, y, \dot{x}, \dot{y})>0$ along g.

Let u be the arc length measured along g in the given sense. Let d be any positive constant less than the minimum distance between any two successive conjugate points on g. By an *admissible integer n, and constants $u_1, u_2, \cdots, u_n$*, will be meant an integer $n>1$, and constants u_i increasing with their subscripts, all less than $u_1+\omega$, and such that no one of the closed intervals on g consisting of points corresponding to a segment of the u axis bounded by successive points of the set $u_1, u_2, \cdots, u_n, u_1+\omega$, exceeds d in length. Let $h_1, h_2, \cdots, h_n$ be n short arcs of class C''', crossing g respectively at $u_1, u_2, \cdots, u_n$, but not tangent to g, and let v_i be the arc length measured along h_i as in §1.

Let the point on h_i at the distance v_i from g be denoted by (u_i, v_i). If v_i be sufficiently small in absolute value, the successive points of the set

$$(u_1, v_1), \ (u_2, v_2), \cdots, (u_n, v_n), \ (u_1, v_1)$$

can be joined by unique extremals arbitrarily near segments of g between successive points on g at which u takes on respectively the values $u_1, u_2, \cdots, u_n, u_1+\omega$. The value of the integral J taken along this succession of extremal segments will be again denoted by $J(v_1, \cdots, v_n)$.

16. The second partial derivatives of $J(v_1, \cdots, v_n)$. As in §2, g can be mapped onto the u axis in the (u, v) plane, with the additional fact that here the transformation from the (x, y) to the (u, v) plane can be taken as one in which x and y will be functions of u and v with a period ω in u. Points in the (u, v) plane whose u coördinates differ by ω will be termed *congruent*. The periodic extremal g will be represented in the (u, v) plane by any segment of the u axis of length ω. The function $f(u, v, v')$ derived from $F(x, y, \dot{x}, \dot{y})$ as in §3 will have a period ω in u, as will the coefficients of the J. D. E. corresponding to the extremal $v\equiv 0$. As in §2, so here, the transformation from the (x, y) to the (u, v) plane may be made to preserve distances along g and the arcs h_i, which become respectively in the (u, v) plane the curves

$v \equiv 0$ and $u = u_i$. Thus the function $J(v_1, \cdots, v_n)$, set up in the preceding section, will here equal the value of the integral in the (u, v) plane taken along extremal segments joining the successive points

$$(u_1, v_1), \ (u_2, v_2), \ \cdots, \ (u_n, v_n), \ (u_1 + \omega, v_1) \tag{1}$$

of the (u, v) plane.

$J(v_1, \cdots, v_n)$ will have a critical point when $(v_1, \cdots, v_n) = (0, \cdots, 0)$. To determine the nature of that critical point we shall examine the second partial derivatives of J.

In §4 the points (u_1, v_1) and (u_n, v_n) played a special rôle because they were adjacent to the end points of the given extremal. Here a set of formulas giving the second partial derivatives should still hold after a circular permutation of the points $(u_1, v_1), \cdots, (u_n, v_n)$. Such a circular permutation would be equivalent to advancing all the subscripts by the same integer, provided for any integer i we set

$$u_{i+n} = u_i + \omega, \qquad v_{i+n} = v_i. \tag{2}$$

With (2) understood we now repeat the definitions and assertions of §4 associated with (1), (2), (3), (4), (5) and (6) of §4.

The values of J_{v_i} are again given by (7) of §4 subject not to (8) of §4, but to the limitations (3) as follows:

$$i = 1, 2, \cdots, n, \qquad v_0 = v_n, \qquad v_{n+1} = v_1. \tag{3}$$

Instead of (9) in §4 we have here

$$J_{v_i v_j} \equiv 0, \qquad i \neq j, \qquad |\, i - j \,| \neq 1, \text{ or } n - 1, \tag{4}$$

subject also to (3). Equations (10), (11), and (12), of §4, hold here exactly as given in §4. From (7) of §4, as qualified by (3) above, we obtain finally

$$\begin{aligned} {}_0J_{v_n v_1} &= - R(u_n) w'_{n,n+1}(u_n), \\ {}_0J_{v_1 v_n} &= R(u_1) w'_{1,0}(u_1). \end{aligned} \tag{5}$$

The two partial derivatives of J just given in (5) were given in §4 by (9) and were there zero; *they are the only two second partial derivatives of J for which the formulas of this section differ from those of* §4.

17. **Periodic extremals classified: non-degenerate, simply-degenerate, and doubly-degenerate.** Before going further it is necessary to consider more in detail the J. D. E. as set up in §3 for the extremal $v \equiv 0$ in the (u, v) plane. We note here however that $P(u)$, $Q(u)$, and $R(u)$, have the period ω in u. In terms of the solutions of this J. D. E. we distinguish three different kinds of periodic extremals E.

I. *There are no solutions of the J. D. E. with the period* ω *other than* $w(u)\equiv 0$.

II. *There is a set of solutions with period* ω *of the form* $Cw(u)$, *where* $w(u)\not\equiv 0$, *and* C *is any constant, but no other solutions with period* ω.

III. *Every solution of the J. D. E. has the period* ω.

In these three cases we shall say, respectively, that the periodic extremal is I, *non-degenerate*, II, *simply-degenerate*, and III, *doubly-degenerate*.

Let $p(u)$ and $q(u)$ be two solutions of the J. D. E. that satisfy the initial conditions

$$\begin{aligned} p(0) &= 1, & q(0) &= 0, \\ p'(0) &= 0, & q'(0) &= 1. \end{aligned} \tag{1}$$

Abel's integral for these solutions becomes

$$R(u)\left[p(u)q'(u) - p'(u)q(u)\right] \equiv \text{constant}. \tag{2}$$

A substitution of $u=0$ and of $u=\omega$ in (2), gives the well known result

$$p(\omega)q'(\omega) - p'(\omega)q(\omega) = 1. \tag{3}$$

We state without proof the following known result.

The given periodic extremal will be non-degenerate, simply-degenerate, or doubly-degenerate, according as the rank of the matrix

$$\left\|\begin{matrix} p(\omega) - 1, & q(\omega) \\ p'(\omega) \quad , & q'(\omega) - 1 \end{matrix}\right\| \tag{4}$$

is 2, 1, *or* 0.

18. **Non-degenerate periodic extremal segments: convex, concave, or conjugate.** Let us represent solutions $w(u)$ of the J. D. E. as curves in the (u, w) plane. We will now prove the following lemma.

LEMMA 1. *If the given periodic extremal is non-degenerate, a necessary and sufficient condition that in the* (u, w) *plane every point* $(0, b)$ *be capable of being joined to its congruent point* (ω, b) *by a solution of the J. D. E., is that* $u=0$ *be not conjugate to* $u=\omega$.

Suppose a point $(0, b)$, not $(0, 0)$, can be joined to the point (ω, b) by a solution of the J. D. E. Such a solution will be of the form

$$bp(u) + cq(u), \qquad b \neq 0, \tag{1}$$

where c is a suitably chosen constant. Since the solution passes through (ω, b) we have

$$bp(\omega) + cq(\omega) = b. \tag{2}$$

Now if $q(\omega)$ were zero, from (2) it would follow that $p(\omega)=1$, whence the rank of the matrix (4), §17, would be less than 2, contrary to the fact that the J. D. E. has no periodic solutions other than $w(u)\equiv 0$. Thus the condition $q(\omega)\neq 0$ is a necessary consequence of the existence of solutions of the J. D. E. joining any point $(0, b)$ to its congruent point (ω, b).

To prove that if $q(\omega)\neq 0$, any point $(0, b)$ can be joined to (ω, b) by a solution of the J. D. E., say $w(u, b)$, we exhibit such a solution, namely,

$$w(u, b) = \frac{b}{q(\omega)}\begin{vmatrix} p(u) & q(u) \\ p(\omega)-1, & q(\omega) \end{vmatrix} \tag{3}$$

Thus the lemma is proved.

From equation (3) we obtain the following:

$$w_u(\omega,b) - w_u(0,b) = \frac{-b}{q(\omega)}\begin{vmatrix} p(\omega)-1, & q(\omega) \\ p'(\omega) & q'(\omega)-1 \end{vmatrix}. \tag{4}$$

Now the determinant in (4) is not zero if the given periodic extremal is non-degenerate. The usual methods of the calculus of variations serve with the aid of (4) to furnish a ready proof of the following lemma.*

LEMMA 2. *If there be given a non-degenerate periodic extremal g on which the point $u=u_0$ is not conjugate to $u=u_0+\omega$, then in the (u, v) plane any point (u_0, a), $a\neq 0$, neighboring $(u_0, 0)$, can be joined to $(u_0+\omega, a)$ by an extremal segment g'. If congruent points be regarded as identical, these extremal segments g' will make an angle α with themselves at (u_0, a) measured on the side of g' towards g, which in magnitude will be either (Case I) always less than π, or else (Case II) always greater than π. If $u_0=0$, Cases I or II will occur according as the sign of*

$$M = -\frac{1}{q(\omega)}\begin{vmatrix} p(\omega)-1, & q(\omega) \\ p'(\omega) & q'(\omega)-1 \end{vmatrix} \tag{5}$$

is positive or negative.

The extremal segment g taken from $u=u_0$ to $u=u_0+\omega$, will be said to be convex or concave according as $\alpha<\pi$ or $\alpha>\pi$. We shall term M the test quotient.

In case a point $u=u_0$ on a non-degenerate periodic extremal g is conjugate to its congruent point, we shall say that the segment of g from $u=u_0$ to $u_0+\omega$ is a *conjugate segment*. It may happen in the case of a non-degenerate periodic extremal segment that every point is conjugate to its congruent point, as examples would show. In any case we see that a non-

---

* Compare Hadamard, *Leçons sur le Calcul des Variations*, Paris, 1910, pp. 434–435.

degenerate periodic extremal segment from $u=u_0$ to $u=u_0+\omega$, is either *convex*, *concave*, or *conjugate*.

19. **Conjugate points on simply-degenerate and doubly-degenerate periodic extremals.** We prove the following lemma.

LEMMA 1. *Let there be given a simply-degenerate, periodic extremal. Then if $w(u)$ is any one of the 1-parameter family of periodic solutions of the J. D. E. which is not identically zero, the only points which are conjugate to their congruent points are points at which $w(u)$ is zero.*

Suppose the lemma false. In particular suppose that $u=a$ is a point conjugate to $u=a+\omega$, while $w(a)\neq 0$.

Let $w_1(u)$ be a solution of the J. D. E. which vanishes at a, but is not identically zero. Since a is conjugate to $a+\omega$ we have

$$w_1(a+\omega) = w_1(a) = 0. \tag{1}$$

Abel's integral gives

$$R(u)\left[w(u)w_1'(u) - w'(u)w_1(u)\right] = \text{constant}. \tag{2}$$

Upon successively substituting $u=a$, and $u=a+\omega$ in this integral we obtain

$$w_1'(a+\omega) = w_1'(a). \tag{3}$$

Because of (1) and (3) we infer that $w_1(u)$ is periodic. Since $w_1(u)$ and $w(u)$ are linearly independent, every solution of the J. D. E. is periodic, contrary to the fact that we are dealing with the simply-degenerate case, and not the doubly-degenerate case. Thus the lemma is proved.

Concerning conjugate points *on doubly-degenerate periodic extremals, we say simply that every point is conjugate to its first congruent point.* It would be a mistake to believe that this property is characteristic of doubly-degenerate periodic extremals, for it may occur, in particular, in the case of a non-degenerate periodic extremal, as examples would show.

20. **The rank and form of the matrix of second partial derivatives of J.** We prove the following

THEOREM 4. *Corresponding to the given periodic extremal g of §15, the symmetric matrix* a *of elements*

$$a_{ij} = {}_0J_{v_iv_j}$$

is of rank n, $n-1$, or $n-2$, according as g is non-degenerate, simply-degenerate, or doubly-degenerate. If g is non-degenerate a *is always in normal form.**

* Bôcher, loc. cit., p. 59.

The case where g is non-degenerate. To prove the theorem in this case we turn again to the (u, w) plane of the J. D. E. of §18, and in that plane consider the points

$$(1) \qquad (u_0, c_0), \ (u_1, c_1), \cdots, (u_n, c_n),$$

of which (u_0, c_0) is supposed congruent to (u_n, c_n). We join the successive points of (1) by curve segments representing solutions of the J. D. E., and denote the resulting curve by λ. The curve λ will represent a periodic solution of the J. D. E., if and only if the slope of λ at (u_0, c_0) equals its slope at (u_n, c_n), and λ has no corners at the remaining points of (1). These conditions will be fulfilled if equations (5) of §5 are satisfied for $i=1, 2, \cdots, n$, and for c_0 and c_{n+1} respectively replaced by c_n and c_1. Let equations (5) of §5, so altered and understood, be denoted by (5a). The J. D. E. will have a periodic solution, not identically zero, if and only if equations (5a) are satisfied by a set of constants $(c_1, \cdots, c_n)$ not all zero. Such a periodic solution is possible if, and only if, the matrix of the coefficients of $(c_1, \cdots, c_n)$ in (5a) is of rank less than n. With the aid of the results of §16 this matrix is seen to be identical with the matrix $\mathbf{a}$ of the present theorem, provided the factor $R(u_i)$ be removed, for each i, from the ith row of $\mathbf{a}$. Thus the J. D. E. has no periodic solution except $w\equiv 0$, if and only if the rank of $\mathbf{a}$ is n.

That the matrix $\mathbf{a}$ is always arranged in normal form when g is a non-degenerate periodic extremal, can now be proved after the manner of proof that the matrix $\mathbf{a}$ of §5 is in normal form.

The case where g is simply-degenerate. According to the proof already given the rank of $\mathbf{a}$ here is less than n. It remains to prove that the rank of $\mathbf{a}$ is $n-1$.

Of the points $(u_1, 0), \cdots, (u_n, 0)$ at least one is not conjugate to its congruent point, for otherwise it would follow from the lemma of §19 that the set of points $(u_1, 0), \cdots, (u_n, 0)$ are mutually conjugate, contrary to the original choice of $(u_1, u_2, \cdots, u_n)$. We can then suppose, without loss of generality, that $u=u_n$ has been so chosen as not to be conjugate to $u=u_n-\omega=u_0$.

Now the minor A_{n-1}, obtained from $\mathbf{a}$ by striking out the last row and column, is one identical in form with the minor A_{n-1}, considered in §5. According to Corollary 1, §5, A_{n-1} will not be zero, since $(u_0, 0)$ is not conjugate to $(u_n, 0)$. Thus in the simply-degenerate case the rank of $\mathbf{a}$ is $n-1$.

The case where g is doubly-degenerate. As in the preceding case the rank of $\mathbf{a}$ is less than n. Further, every point on $v=0$ is here conjugate to its congruent point. In particular the point $(u_0, 0)$ is conjugate to its congruent point $(u_n, 0)$. It follows from Corollary 1, §5, that $A_{n-1}=0$. But if we understand that $u_i+\omega=u_{i+n}$ for each integer i, then by advancing the subscripts we can bring it to pass that $(u_i, 0)$ is denoted by $(u_n, 0)$. In the matrix a the principal minor obtained by deleting the ith row and column becomes the principal minor A_{n-1} when $(u_i, 0)$ becomes $(u_n, 0)$. Thus all the principal $(n-1)$-rowed minors of a are zero. But according to Corollary 1, §5, A_{n-2} is not zero, since $(u_0, 0)$ is conjugate to $(u_n, 0)$ and cannot at the same time be conjugate to $(u_{n-1}, 0)$. Thus a is in this case of rank $n-2$, and the proof of the theorem is complete.

21. The type number of a non-degenerate periodic extremal. We prove the following

THEOREM 5. *If the periodic extremal g of* §15 *is non-degenerate, the type number k of the corresponding critical point of the function $J(v_1, \cdots, v_n)$ of* §15 *will be independent of the choice of n among admissible integers n, and of the points $(u_1, \cdots, u_n)$ on g among admissible points $(u_1, \cdots, u_n)$, and may be determined as follows. Setting $u_n - \omega = u_0$, let m be the number of conjugate points to $u = u_0$, preceding $u = u_n$. If $u = u_0$ is conjugate to $u = u_n$, $k = m+1$. If $u = u_0$ is not conjugate to $u = u_n$, then $k = m$, or $m+1$, according as the segment of g from $u = u_0$ to $u = u_n$ is convex or concave* (§18).

The number k will be called the type number of g.

As previously we concern ourselves here with the symmetric matrix a of which the elements are

$$a_{ij} = {}_0J_{v_iv_j}.$$

According to Theorem 4, §20, the matrix a in the case of a non-degenerate periodic extremal, is of rank n, and arranged in normal form. The type number k desired is then simply the number of changes in sign of the principal minors $A_0, A_1, \cdots, A_n$ (cf. §5, §6). To proceed further we agree again to set $u_{i+n} = u_i + \omega$, and $v_{i+n} = v_i$ for all integers i.

Now the matrix whose elements are those in A_{n-1} would be identical with the matrix a of §5 and §6, if in §5 and §6 we were dealing with the segment of g for which u lies between $u = u_0$ and $u = u_n$, and if we had chosen $n-1$ intermediate points $(u_1, \cdots, u_{n-1})$ instead of n. To proceed further we distinguish between two cases.

Case I. *The point $u = u_0$ is not conjugate to $u = u_n = u_0 + \omega$.* In this case it follows from Theorem 2, §6, that the segment of g for which u lies between $u = u_0$ and $u = u_n$ is of type m, and that $A_{n-1} \neq 0$. Thus there are m changes of sign in the sequence $A_0, A_1, \cdots, A_{n-1}$. Hence in Case I the number k of the present theorem is m or $m+1$ according to whether or not A_{n-1} and A_n have the same sign.

Now consider the n equations (5) of §5, altered as follows. We here replace c_0 by c_n, c_{n+1} by c_1, and the zero of the right hand member of the nth equation by

$$A_{n-1}/A_n, \tag{1}$$

while finally we here multiply the left hand member of the ith one of these equations ($i=1, 2, \cdots, n$) by $R(u_i)$. We denote the resulting set of equations by (5b). Equations (5b) can be solved for $(c_1, \cdots, c_n)$, giving in particular for c_n a positive value

$$c_n = \frac{A_{n-1}^2}{A_n^2}.$$

The interpretation of this solution is that there is a solution of the J. D. E. which in the (u, w) plane passes through the points

$$(u_0, c_n), \ (u_1, c_1), \ (u_2, c_2), \cdots, (u_n, c_n),$$

whose slope at the point (u_n, c_n) minus its slope at (u_0, c_n) equals the fraction (1) divided by $R(u_n)$.

This difference of slopes, and hence the sign of (1), is readily seen to be positive or negative according as the segment of g taken from $u=u_0$ to $u=u_n$ is convex or concave in the sense of §18. Thus in Case I, k equals m or $m+1$ according as the segment of g from $u=u_0$ to $u=u_n$ is convex or concave.

Case II. *The point $u=u_0$ is conjugate to $u=u_n=u_0+\omega$.* In this case $A_{n-1}=0$, according to Corollary I, §5. Since $u=u_0$ is conjugate to $u=u_n$, it cannot be conjugate to $u=u_{n-1}$. If then Theorem 2, §6, be applied to the segment of g from $u=u_0$ to $u=u_{n-1}$, we find that that segment is of type m, since there are m points conjugate to u_0 preceding u_{n-1}. Hence there are m changes of sign in $A_0, A_1, \cdots, A_{n-2}$. But according to the theory of regularly arranged quadratic forms, when $A_{n-1}=0$, A_{n-2} and A_n have opposite signs. Thus there are $m+1$ changes of sign in $A_0, A_1, \cdots, A_n$. Hence $k=m+1$ in this case.

We can now prove that the type number k is independent of the choise of n, and $(U)=(u_1, u_2, \cdots, u_n)$, among admissible integers n and points (U).

The formulas of §16 and §4 show that the partial derivatives

$${}_0J_{v_i v_j}$$

are continuous functions of the position of admissible points (U). Now any admissible point (U) can be varied continuously through admissible points (U) into any other admissible point (U) for which n is the same. During such a variation the terms of the sequence

$$A_0, A_1, \cdots, A_n \tag{2}$$

will vary continuously, and A_n will never be zero. We can now prove that there can be no variation in the total count of changes of sign in (2). This will certainly be true if no member of the sequence (2) is ever zero. If A_i becomes zero (where i cannot be zero or n) then A_{i-1} and A_{i+1} will have opposite signs at that stage of the variation.* Thus there can be no variation in the total count of changes of sign in (2), and the statement in italics is proved, except for possible changes of n.

That permissible changes in n do not alter the number k follows from the fact that, if we hold u_0 fast and introduce or remove any set of points u_i so as to have an admissible set u_i left, the number k, as determined under Cases I and II, will be the same before and after the change. Thus the theorem is completely proved.

PART IV. THE TYPE-NUMBER OF A DEGENERATE PERIODIC EXTREMAL

22. **Simply-degenerate, isolated, analytic, periodic, extremals.** By an *isolated* periodic extremal is understood a periodic extremal g, in the neighborhood of which there is no other periodic extremal with a length neighboring that of g. In this section and the two following we assume that *we are dealing with a periodic extremal g given in the region S of* §1. *Concerning $F(x, y, \dot{x}, \dot{y})$ we make the same assumptions as in* §1 *together with the assumption that $F(x, y, \dot{x}, \dot{y})$ be analytic for the values of $(x, y, \dot{x}, \dot{y})$ admitted. The extremal g is to be without multiple points, isolated, simply-degenerate, and analytic. Along g, $F_1(x, y, \dot{x}, \dot{y})$ is to be positive. We denote g's length by ω.*

The region neighboring g can be mapped conformally on the region neighboring the u axis in the (u, v) plane in such a fashion that g corresponds to any segment of the u axis of length ω, and so that congruent points (u, v) and $(u+\omega, v)$ correspond to the same point (x, y) in S, but that otherwise the transformation is one-to-one. As in §3 we can derive from $F(x, y, \dot{x}, \dot{y})$ an integrand $f(u, v, v')$ corresponding to which $v=0$ is an extremal. According to the lemma of §19 only a finite number of points of g are conjugate to their congruent points in the simply-degenerate case. Let $(u, v) = (0, 0)$ be one of the points of g not congruent to its conjugate point. As is well known, for a sufficiently small constant $a \neq 0$, any point $(0, a)$ can be joined to its congruent point (ω, a) by an extremal segment g' neighboring g. Further, if *congruent points be considered as identical these extremal segments g' make an angle with themselves at (ω, a) measured on the side of g' toward g which in magnitude will now be shown to be either*

* Bôcher, loc. cit., p. 147.

(I) *Less than* π *for all vertices* (ω, a) *on one side of* g, *and greater than* π *for* (ω, a) *on the other side of* g;

(II) *Less than* π *for all vertices* (ω, a) *on either side of* g; *or*

(III) *Greater than* π *for all vertices* (ω, a) *on either side of* g.

Further, whether I, II, *or* III *occurs depends only upon* g *and the choice of the point* $(u, v) = (0, 0)$.

The family of extremals passing through the point $(u, v) = (0, a)$ with a slope b can be represented in the form

$$v = A(u, a, b), \qquad a^2 + b^2 \leqq e, \tag{1}$$

where $A(u, a, b)$ is a real analytic function of (u, a, b) for u on any finite segment of the u axis and for a corresponding positive constant e sufficiently small, where further

$$a \equiv A(0, a, b), \tag{2}$$

$$b \equiv A_u(0, a, b). \tag{3}$$

The condition that an extremal (a, b) pass from a point $(0, a)$ to (ω, a) is that

$$A(\omega, a, b) - a = 0. \tag{4}$$

Now we have

$$A_b(\omega, 0, 0) \neq 0, \tag{5}$$

since $(0, 0)$ has been chosen so as not to be conjugate to $(\omega, 0)$. Hence (4) possesses a solution

$$b = \beta(a), \qquad \beta(0) = 0, \tag{6}$$

where $\beta(a)$ is analytic in a at $a=0$. The slope of an extremal (a, b) given by (6) at (ω, a), minus its slope at $(0, a)$, is

$$A_u(\omega, a, \beta(a)) - \beta(a). \tag{7}$$

This function is not identically zero in a. For otherwise extremals (a, b) given by (6) would make up a continuous family of periodic extremals, a case which has been barred. Hence we can write

$$A_u(\omega, a, \beta(a)) - \beta(a) = \overline{A}(a) = a^r k(a), \qquad k(0) \neq 0, \quad r > 1, \tag{8}$$

where $k(a)$ is analytic in a at $a=0$. That r is an integer exceeding one is seen from the fact that

$$\overline{A}'(0) = -\frac{1}{A_b(\omega, 0, 0)} \begin{vmatrix} A_a(\omega, 0, 0) - 1, & A_b(\omega, 0, 0) \\ A_{u,a}(\omega, 0, 0), & A_{u,b}(\omega, 0, 0) - 1 \end{vmatrix} \tag{9}$$

In terms of $p(u)$ and $q(u)$ of §17 this gives

$$(10)\qquad \bar{A}'(0) = -\frac{1}{q(\omega)}\begin{vmatrix} p(\omega)-1, & q(\omega) \\ p'(\omega) \quad , & q'(\omega)-1 \end{vmatrix}.$$

Now the determinant in (10) is always zero if the periodic extremal is degenerate (§17). Thus here $\bar{A}'(0)=0$ and $r>0$. The statements in italics follow at once from (8).

23. **A simply-degenerate periodic extremal. Cases: concave-convex, concave, convex.** A simply-degenerate, isolated, analytic, periodic, extremal g, taken from $u=0$ to $u=\omega$ will be said to be *concave-convex, convex, or concave, according as* I, II, *or* III *of the preceding section holds.* We have previously given similar definitions for a non-degenerate periodic extremal (§18) involving the terms convex and concave. *The concave-convex case did not present itself in the non-degenerate case.* In the present case we state in terms of r and $k(0)$ in (8) of the preceding section that the segment of g from $u=0$ to $u=\omega$ is

(I) *Concave-convex when r is even,*

(II) *Convex when r is odd and $k(0)>0$,*

(III) *Concave when r is odd and $k(0)<0$.*

24. **The type number of a simply-degenerate periodic extremal.** To define the type number of g we are going to modify the integrand $f(u, v, v')$ in such a fashion that we shall no longer have to do with a degenerate periodic extremal. Let $h(u)$ be any real analytic function of u with a period ω, and μ be a parameter. An integrand of the form

$$(1)\qquad f(u, v, v') + \mu v h(u)$$

reduces for $\mu=0$ to the original integrand. We will now prove the following theorem.

THEOREM 6. *Concerning the simply-degenerate periodic extremal g of* §22, *on which* $(0, 0)$ *can and will be chosen so as not to be conjugate to* $(\omega, 0)$, *we can say the following. It is possible to choose a function $h(u)$ real and analytic in u with a period ω, and values of μ arbitrarily near $\mu=0$ in such a fashion, that if $f(u, v, v')$ be replaced by the integrand* (1) *then we have the following*:

(A) *If g, taken from $u=0$ to $u=\omega$, is convex-concave, the modified problem will possess no extremals neighboring g of period ω in u.*

(B_1) *If g, taken from $u=0$ to $u=\omega$, is convex or concave, the modified problem will possess just one extremal E neighboring g of period ω in u.*

(B_2) *The extremal E of (B_1) will be non-degenerate, and if there are on g just m points conjugate to $u=0$ and preceding $u=\omega$, the type number of E as determined in Theorem 5, §21, will be m or $m+1$ according as g, taken from $u=0$ to $u=\omega$, is convex or concave.*

The choice of $h(u)$. The extremals which correspond to the integrand (1) for values of μ neighboring $\mu=0$, and which lie in the neighborhood of the extremal $\mu=0$, can be represented in the form

$$v = B(u, a, b, \mu), \qquad a^2 + b^2 + \mu^2 \leqq e, \tag{2}$$

with

$$a \equiv B(0, a, b, \mu), \tag{3}$$

$$b \equiv B_u(0, a, b, \mu), \tag{4}$$

where $B(u, a, b, \mu)$ is analytic in its arguments, for u on any closed interval, and e a correspondingly sufficiently small positive constant.

We will show that we can choose $h(u)$ so that

$$B_\mu(\omega, 0, 0, 0) \neq 0, \tag{5}$$

$$B_{u\mu}(\omega, 0, 0, 0) \neq 0. \tag{6}$$

Differentiation of the Euler equation with respect to μ will show that the function $B_\mu(u, 0, 0, 0)$, set equal to $w(u)$, satisfies

$$Rw'' + R'w' + (Q' - P)w = h(u) \tag{7}$$

where $R(u)$, $P(u)$, and $Q(u)$ are the functions already used in the J. D. E. set up for the extremal $v\equiv 0$, when $\mu=0$.

It follows from (3) and (4) that we have initially

$$\begin{aligned} w(0) &= B_\mu(0, 0, 0, 0) = 0, \\ w'(0) &= B_{u\mu}(0, 0, 0, 0) = 0. \end{aligned} \tag{8}$$

If we make use of the functions $p(u)$ and $q(u)$ of §17, it is readily seen that this solution $w(u)$ of (7) is given by

$$w(u) = B_\mu(u, 0, 0, 0) = \int_0^u \frac{h(t)}{R(0)} \begin{vmatrix} p(t), & q(t) \\ p(u), & q(u) \end{vmatrix} dt.$$

In particular

$$B_\mu(\omega, 0, 0, 0) = \int_0^\omega \frac{h(t)}{R(0)} \begin{vmatrix} p(t), & q(t) \\ p(\omega), & q(\omega) \end{vmatrix} dt, \tag{9}$$

$$B_{\mu u}(\omega, 0, 0, 0) = \int_0^\omega \frac{h(t)}{R(0)} \begin{vmatrix} p(t), & q(t) \\ p'(\omega), & q'(\omega) \end{vmatrix} dt. \tag{10}$$

Now the coefficients of $h(t)$ in the integrands of (9) and (10) are respectively 0 and $R_{(0)}^{-1}$ for $t=\omega$. They are both positive for t in an interval

$$\omega - e < t < \omega, \tag{11}$$

if e be a sufficiently small positive constant. Let $h_1(u)$ be a function which is identically zero as t ranges from $t=0$ to $t=\omega$, except that in the interval (11), $v=h_1(u)$ shall be equal to the v ordinate on a semicircle with end points at $(\omega-e, 0)$ and $(\omega, 0)$ and on which $v>0$. If in (9) and (10), $h(t)$ be replaced by $h_1(t)$, the resulting integrals would both be positive. Another fact of importance is that if e in (11) be a sufficiently small positive constant the corresponding function $h_1(t)$ will not only make the integrals (9) and (10) positive but will cause the ratio of the integral (9) to the integral (10) to be arbitrarily small. Now $h_1(t)$ can be approximated by a number of terms of a Fourier series with period ω. The resulting function, which will serve as our definition of $h(t)$, can be taken as a function approximating $h_1(t)$ so closely that *for this choice of* $h(t)$ *in* (9) *and* (10) *the ratio of* $B_\mu(\omega, 0, 0, 0)$ *to* $B_{u\mu}(\omega, 0, 0, 0)$ *will be arbitrarily small, and both* $B_\mu(\omega, 0, 0, 0)$ *and* $B_{u\mu}(\omega, 0, 0, 0)$ *will be positive.*

We can now settle *the question of the existence of periodic extremals near the extremal* $v\equiv 0$ *corresponding to the integrand* (1) *when* μ *is near* $\mu=0$. The conditions for a periodic extremal neighboring $v\equiv 0$ for μ near $\mu=0$ are

$$B(\omega, a, b, \mu) - a = 0, \tag{12}$$

$$B_u(\omega, a, b, \mu) - b = 0. \tag{13}$$

Now equation (12) reduces for $\mu=0$ to equation (4) of §22. But (4) of §22 is satisfied by the function $b=\beta(a)$ of (6), §22. Hence (12) is satisfied by $b=\beta(a)$ and $\mu=0$. Further

$$B_b(\omega, 0, 0, 0) \neq 0, \tag{14}$$

since on the extremal $v\equiv 0$, the point $(0, 0)$ was chosen not conjugate to $(\omega, 0)$. Hence (12) admits a solution of the form

$$b = b(a, \mu) = \beta(a) + \mu R(a, \mu), \tag{15}$$

where $R(a, \mu)$ is analytic in a and μ, at $a=0$ and $\mu=0$.

We are next concerned with solving (13) subject to (12), that is with solving

$$B_u[\omega, a, b(a, \mu), \mu] - b(a, \mu) = 0. \tag{16}$$

Using the fact that $b(a, 0)=\beta(a)$ we obtain the identity

$$B_u[\omega, a, b(a, 0), 0] - b(a, 0) \equiv A_u[\omega, a, \beta(a)] - \beta(a),$$

which becomes with the aid of (8), §22,

$$\equiv a^r k(a), \qquad k(0) \neq 0.$$

Hence (16) becomes

$$(17) \qquad B_u[\omega, a, b(\alpha, \mu), \mu] - b(a, \mu) = a^r k(a) + \mu S(a, \mu) = 0,$$

where $S(a, \mu)$ is analytic at $a=0$ and $\mu=0$. We can solve (17) for μ if $S(0,0)\neq 0$. But $S(0, 0)$ is the partial derivative with respect to μ, at $(a, \mu) = (0, 0)$, of the left hand member of (17). Thus

$$(18) \quad S(0, 0) = [1 - B_{ub}(\omega, 0, 0, 0)]\frac{B_\mu(\omega, 0, 0, 0)}{B_b(\omega, 0, 0, 0)} + B_{\mu u}(\omega, 0, 0, 0).$$

Now as we have seen we can choose $h(t)$ so that $B_{\mu u}(\omega, 0, 0, 0)$ will be positive at the same time that the ratio of $B_\mu(\omega, 0, 0, 0)$ to $B_{\mu u}(\omega, 0, 0, 0)$ is arbitrarily small, so that the term $B_{\mu u}(\omega, 0, 0, 0)$ will dominate the sign in (18). Thus $S(0, 0)\neq 0$ for a proper choice of $h(t)$. Hence, (17) admits a solution of the form

$$19) \qquad \mu = \mu(a) = a^r d(a), \qquad d(0) \neq 0,$$

where $d(a)$ is analytic at $a=0$. The solution of (12) and (13) is now given by (19), taken with (15).

Final proof of (A). Now the integer r in (19) is the integer r in (8), §22. If the given periodic extremal is convex-concave, r is even. In this case let μ be chosen arbitrarily small, different from zero, and opposite in sign to $d(0)$ in (19). For this μ, (19) will admit no real solution a neighboring $a=0$, and corresponding to this μ the problem will possess no periodic extremal neighboring the extremal $v\equiv 0$.

Proof of (B_1). If the given periodic extremal is convex or concave, r is odd, and corresponding to any value of μ say $\mu_1\neq 0$, neighboring $\mu=0$, (19) gives a value of a, say a_1, and (15) then a value of b, say b_1, corresponding to which the modified problem for which $\mu=\mu_1$ will possess a *single* periodic extremal E, with initial point $(0, a_1)$ and initial slope b_1.

Proof of (B_2). *The sign of the test quotient M of* (5), §18, *determined for E.* To show that E is non-degenerate it will be sufficient to prove that the test quotient M of Lemma 2, §18, evaluated for E, is not zero. According to the results of §18, E, taken from $u=0$ to $u=\omega$, will be convex or concave according as M is positive or negative. By thus determining the sign of M we intend to prove that the non-degenerate extremal E is convex or concave according as the given simply-degenerate extremal $v\equiv 0$ is convex or concave, taking both extremals from $u=0$ to $u=\omega$.

More generally the quotient M of §18 set up for any extremal with parameters (a, b, μ) in (2), will be denoted by $M(a, b, \mu)$. We have

$$\frac{-1}{B_b}\begin{vmatrix} B_a - 1, & B_b \\ B_{ua}\quad , & B_{ub} - 1 \end{vmatrix} = M(a, b, \mu), \tag{20}$$

where in each of the partial derivatives we set $u=\omega$, but where (a, b, μ) may take on any values neighboring (0, 0, 0). The form (20) shows that in terms of the function $b(a, \mu)$ of (15)

$$M[a, b(a, \mu), \mu] \equiv \frac{\partial}{\partial a}[B_u\{\omega, a, b(a, \mu), \mu\} - b(a, \mu)].$$

With the aid of (17) we therefore have

$$M[a, b(a, \mu), \mu] \equiv ra^{r-1}k(a) + a^r k'(a) + \mu S_a(a, \mu). \tag{21}$$

We shal evaluate (21) for values of μ given by (19), thereby evaluating $M(a, b, \mu)$ for values of (a, b, μ) which correspond to periodic extremals. Thus

$$M[a, b\{a, \mu(a)\}, \mu(a)] \equiv ra^{r-1}k(0) + \cdots, \qquad k(0) \neq 0, \tag{22}$$

where the terms omitted are of higher order than $r-1$ in a.

Now according to the results of §23 the given simply-degenerate extremal $v\equiv 0$, taken from $u=0$ to $u=\omega$, is convex or concave when r is odd, and more particularly is convex or concave according as $k(0)$ is positive or negative. It follows from (22) and (19) that if E corresponds to a parameter $\mu\neq 0$, sufficiently small in absolute value, then E will be convex or concave according as the extremal $v\equiv 0$ is convex or concave (§ 18).

Finally the type of the non-degenerate extremal E, determined according to Theorem 5, §21, equals the number m of points on E conjugate to $u=0$ and preceding $u=\omega$, plus one when E is concave, and exactly m when E is convex. But this number m will equal the corresponding number determined on the extremal $v=0$, provided only that μ be sufficiently small in absolute value. Thus part (B_2) of the theorem is proved.

In accordance with the results of the preceding theorem, the given simply-degenerate periodic extremal g, in case it is convex or concave, will be said to be equivalent in type to the non-degenerate extremal E whose existence is affirmed by the preceding theorem, and in case it is convex-concave will be said to be equivalent to a null set of extremals and to be neutral in type.

25. **A doubly-degenerate, isolated, analytic, periodic, extremal.** We make the same assumptions here as we made in the case of a simply-degenerate periodic extremal (§22), except that here the given extremal g

is to be doubly-degenerate (§17). We will reduce the determination of the type of g to the cases already considered, that is, the non-degenerate and simply-degenerate cases.

THEOREM 7. *Concerning the preceding doubly-degenerate, isolated, periodic extremal g, and the corresponding simplified integrand $f(u, v, v')$ for which g becomes the extremal $v\equiv 0$, we can say the following. It is possible to choose a function $h(u)$ analytic and periodic in u with the period ω, such that the extremals corresponding to the modified integrand*

$$f(u, v, v') + \mu v h(u),$$

for properly chosen values of the parameter μ, neighboring $\mu=0$, will include at most a finite set σ of extremals with the period ω in u that lie in the neighborhood of $v\equiv 0$, and such that none of these extremals will be doubly-degenerate.

To prove Theorem 7 we proceed as in the proof of Theorem 6, §24. We represent the extremals neighboring $v\equiv 0$ as in §24 and choose $h(u)$ in the same way, so that we may regard equations (1) to (13) of §24 as holding here. At this point the two proofs diverge, since (14) of §24 does not hold here. In terms of the function $A(u, a, b)$ of §22 of the unmodified problem, equations (12) and (13) of §24 become

$$B(\omega, a, b, \mu) - a \equiv A(\omega, a, b) - a + \mu D(a, b, \mu) = 0, \tag{1}$$

$$B_u(\omega, a, b, \mu) - b \equiv A_u(\omega, a, b) - b + \mu E(a, b, \mu) = 0, \tag{2}$$

where $D(a, b, \mu)$ and $E(a, b, \mu)$ are analytic in (a, b, μ) at $(a, b, \mu) = (0, 0, 0)$. With the aid of (5) and (6), §24, we see that

$$D(0, 0, 0) = B_\mu(\omega, 0, 0, 0) \neq 0, \tag{3}$$

$$E(0, 0, 0) = B_{u\mu}(\omega, 0, 0, 0) \neq 0. \tag{4}$$

Because of (3) and (4), (1) and (2) can be solved for μ. These solutions take the forms, respectively,

$$\mu = [A(\omega, a, b) - a]G(a, b), \tag{5}$$

$$\mu = [A_u(\omega, a, b) - b]H(a, b), \tag{6}$$

where $G(a, b)$ and $H(a, b)$ are analytic at $(a, b) = (0, 0)$ and do not vanish there. To solve (5) and (6) simultaneously we are led to the equation

$$[A(\omega, a, b) - a]G(a, b) = [A_u(\omega, a, b) - b]H(a, b). \tag{7}$$

We distinguish between two cases:

Case I: Equation (7) holds identically.

Case II: Equation (7) does not hold identically.

The proof in Case I. In this case neither of the differences

$$A(\omega, a, b) - a, \quad A_u(\omega, a, b) - b$$

can vanish along any real analytic arcs in the (a, b) plane neighboring $(a, b) = (0, 0)$. For if either of these two differences vanished along such real arcs, according to (7) the other difference would also so vanish, and the extremals for the unmodified problem $\mu=0$ would include a family of periodic extremals contrary to the assumption that g is isolated. Except at $(0, 0)$, the two members of (7) must then be of one sign, say positive, throughout the neighborhood of $(a, b)=(0, 0)$. If then μ be chosen negative and sufficiently near $\mu=0$, there will be no solution of (5) and (6) and hence no solution of (11) and (12). For each such choice of μ the modified problem will possess no periodic extremals neighboring g.

Proof in Case II. It may happen in this case that (7) possesses no real solutions neighboring $(a, b)=(0, 0)$ other than $(0, 0)$. If this occurs, then for any choice of $\mu \neq 0$, sufficiently near $\mu=0$, (1) and (2) admit no real solutions (a, b, μ) and the modified problem possesses no periodic extremals neighboring g.

If, on the other hand, (7) possesses real solutions $(a, b) \neq (0, 0)$, arbitrarily near $(a, b)=(0, 0)$, these real solutions will make up a finite number of real analytic arcs representable in a one-to-one manner by function pairs of the form

$$\begin{aligned} a &= a(t), & a(0) &= 0, \\ b &= b(t), & b(0) &= 0, \end{aligned} \tag{8}$$

where $a(t)$ and $b(t)$ are real analytic functions of t at $t=0$, and are not both identically zero. The differences

$$\begin{aligned} &A\,[\omega, a(t), b(t)] - a(t), \\ &A_u[\omega, a(t), b(t)] - b(t) \end{aligned} \tag{9}$$

cannot both be identically zero, for otherwise $a(t)$, $b(t)$ would correspond in the unmodified problem to a continuous family of periodic extremals. On the other hand neither of the functions in (9) can be identically zero in t without the other being identically zero in t, as follows from (7). We can then set

$$A[\omega, a(t), b(t)] - a(t) \equiv t^s F(t), \qquad F(0) \neq 0, \qquad s > 0, \tag{10}$$

where $F(t)$ is analytic at $t=0$. From (5) it follows that the values of μ that go with (8) to give a solution of (1) and (2) are representable in the form

$$\mu = \mu(t) \equiv t^s K(t), \qquad K(0) \neq 0, \tag{11}$$

where $K(t)$ is analytic at $t=0$. For any particular t, neighboring $t=0$, and corresponding values of $a(t)$, $b(t)$, and $\mu(t)$, the corresponding extremal g' will be periodic.

Among the conditions that g' be doubly-degenerate are that for the value of t that gives g' (§17),

$$B_a[\omega, a(t), b(t), \mu(t)] = 1, \tag{12}$$

$$B_b[\omega, a(t), b(t), \mu(t)] = 0. \tag{13}$$

We proceed to show that (12) and (13) do not hold identically in t. For if (12) and (13) did hold identically in t, a performance of the following indicated differentiation would show that

$$\frac{\partial}{\partial t}\{B[\omega, a(t), b(t), \mu] - a(t)\} \equiv 0, \quad \mu = \mu(t), \tag{14}$$

was an identity in t, where, as indicated, μ is to be held fast during the differentiation and set equal to $\mu(t)$ thereafter. If use be made of (1) and (10) we obtain the identity

$$B[\omega, a(t), b(t), \mu] - a(t) \equiv t^s F(t) + \mu D[a(t), b(t), \mu], \tag{15}$$

and the partial derivative (14) becomes, upon using (11) and (15),

$$st^{s-1} F(t) + t^s F'(t) + t^s K(t)[D_a \cdot a'(t) + D_b \cdot b'(t)]. \tag{16}$$

This function does not vanish identically since $s>0$ and $F(0) \neq 0$. Hence (12) and (13) do not hold identically, and accordingly hold simultaneously for no value of t, neighboring $t=0$, other than $t=0$.

Now to any real value of μ not zero, but sufficiently near zero, there will correspond, under (11), either no real value, one real value, or two real values of t, according to the evenness, or oddness of s, and the sign of $K(0)$. To such a μ there will then correspond, by virtue of (8), no periodic extremal, one periodic extremal, or two periodic extremals, and none of these extremals will be doubly-degenerate, since (12) and (13) will not both hold.

Similarly there may arise a finite number of other periodic extremals from real solutions of (7) other than $a(t)$, $b(t)$, but in any case for a $\mu \neq 0$, and sufficiently near $\mu=0$, these periodic extremals will not be doubly-degenerate. Thus the theorem is proved.

Let the set σ of periodic extremals appearing in Theorem 7 be modified by replacing each simply-degenerate periodic extremal g_1 of σ by an equivalent non-degenerate extremal, or a null set of extremals, according to the conventions at the end of §24. The set σ_1 of non-degenerate extremals thereby obtained will be *said to be equivalent in type to the given doubly-degenerate extremal g. If neither σ nor σ_1 contains any extremals, g will be said to be equivalent to a null set of extremals, and to be of neutral type.*

PART V. RELATIONS IN THE LARGE BETWEEN PERIODIC EXTREMALS

26. The integrand and regions S and S_1. We make here again the assumptions I and II of §7 and §8, qualifying F and S. We replace III by III′, and IV by IV′ as follows:

III′. *Let there be given a closed region S_1 consisting of points interior to S bounded by two simple closed curves β_1 and β_2, of which β_2 lies within β_1. We suppose both β_1 and β_2 consist of a finite number of analytic arcs without singularities and are extremal-convex relative to S_1 in the sense of* III §8.

IV′. *We assume that we have in S a proper field of extremals representable in the form*

$$x = h(u, v), \qquad y = k(u, v), \tag{1}$$

where u is the parameter and v is the arc length measured along the extremals, and $h(u, v)$ and $k(u, v)$ have a period ω in u. We assume further that for any constant a and interval

$$a \leqq u < a + \omega, \tag{2}$$

the field (1) *covers S_1 in a one-to-one manner, and that at each point (u, v) that corresponds to a point (x, y) in S_1 the functions* (1) *are single-valued and analytic in u and v, and*

$$\begin{vmatrix} h_u, & h_v \\ k_u, & k_v \end{vmatrix} \neq 0.$$

The lemma of §8 will here be replaced by the following lemma of which the method of proof is very similar to that used in §8. No proof need be given.

LEMMA 1. *The region S_1 of the (x, y) plane will correspond under* (1) *to a region R of the (u, v) plane bounded by two unending arcs of the form*

$$\begin{aligned} &v = A(u), \qquad A(u + \omega) = A(u), \\ &v = B(u), \qquad B(u + \omega) = B(u), \qquad A(u) < B(u), \end{aligned} \tag{3}$$

where $A(u)$ and $B(u)$ are of class C for all values of u, and analytic in u except for a finite set of values of u, on any interval of the form (2). *The interior points of R are the points (u, v) such that*

$$A(u) < v < B(u). \tag{4}$$

The correspondence between S_1 and any set of points of R limited as in (2) *will be one-to-one.*

As in §9, so here, it follows that all extremals, except the extremals u = constant, are representable in the form

$$v = M(u), \tag{5}$$

where $M(u)$ is an analytic function of u for all values of u that give points (u, v) in R. Further any extremal joining two points in R will have no points on the boundary of R, with the possible exception of its end points.

We shall concern ourselves at first with periodic extremals continuously deformable in S_1, in the (x, y) plane, into either one of the two closed boundary curves of S_1, taken just once. In R, in the (u, v) plane, these extremals will be representable in the form (5), with $M(u)$ possessing the period ω. A point of difference between the developments for periodic extremals and those for extremals joining two fixed points A and B, is that in the latter case we were able to prove, under hypotheses I, II, III, and IV, of §7 and §8, that there were at most a finite number of extremals joining A to B, while in the case of periodic extremals the hypotheses I, II, III′, and IV′ are not sufficient to bar the existence of analytic families of periodic extremals lying in S_1 and deformable into a boundary of S_1. Simple examples can be given to show the truth of this statement. However, cases where families of periodic extremals exist are certainly specialized in that the differential equation of the first variation, that is the J. D. E., set up for each member of such families, must possess a periodic solution not identically zero. Not only is this true but the differential equations of the variations of orders higher than the first must all possess periodic solutions. In excluding such families we are therefore excluding exceptional cases. We are the more justified in such exclusion by the fact that we are developing a theory that will serve to prove the existence of additional periodic extremals when a finite set of such extremals is given. If infinite sets of mutually deformable periodic extremals existed, then more than we hoped to prove would be granted. We state here the fundamental lemma which replaces Lemma 2 of §11.

LEMMA 2. *Let there be given regions S and S_1 satisfying* I *and* II *of* §7, *and* III′ *and* IV′ *of this section. In S_1 suppose there are at most a finite number of periodic extremals continuously deformable into either boundary of S_1, taken just once, and that all of these extremals are non-degenerate* (§17). *Let the number of these extremals which are of type k*(§21) *be denoted by M_k. Let m be the maximum of the integers k. Then between these numbers M_k the relations* (R) *of* §11 *hold.*

This lemma is proved with the aid of the fundamental lemma on critical points, §11, and Theorem 5, §21, in a manner similar to the manner of proof of Lemma 2, §11.

27. **The theorem in the large.** In order to remove the restriction from the above lemma that the periodic extremals appearing there be non-degenerate we cannot rely on the envelope theory as in the case of extremals joining two fixed points. The following lemma will serve our purpose. In this lemma let the $(n-1)$-dimensional hypersphere with radius r and center (A) in the space of $(v_1, \cdots, v_n)$ be denoted by $S^{A,r}$.

LEMMA 1. *Let there be given a function $J(v_1, \cdots, v_n)$ of class C''' at each point $(V)=(v_1, \cdots, v_n)$ of an open n-dimensional region Σ. Let $(A)=(a_1, \cdots, a_n)$ be a point of Σ at which $J(v_1, \cdots, v_n)$ has an isolated critical point. Let $J(v_1, \cdots, v_n, \mu)$ be a function of class C''' for (V) in Σ, and μ in the neighborhood of $\mu=0$, and such that in Σ*

$$J(v_1, \cdots, v_n) \equiv J(v_1, \cdots, v_n, 0). \tag{1}$$

Let e be a positive constant so small that $J(v_1, \cdots, v_n)$ has no critical point other than (A) within or on $S^{A,2e}$. Then for any fixed value of $\mu \neq 0$ and sufficiently small in absolute value, it is possible to replace $J(v_1, \cdots, v_n)$ by a function $\phi(v_1, \cdots, v_n)$ of class C''' throughout Σ, and of such sort that in $S^{A,e}$

$$\phi(v_1, \cdots, v_n) \equiv J(v_1, \cdots, v_n, \mu), \tag{2}$$

while in Σ but outside of $S^{A,2e}$

$$\phi(v_1, \cdots, v_n) \equiv J(v_1, \cdots, v_n), \tag{3}$$

and finally in the closed domain between $S^{A,2e}$ and $S^{A,e}$, $\phi(v_1, \cdots, v_n)$ has no critical points.

To prove this lemma we are going to use a function $h(x)$ that is of class C''' for all values of x, and is such that for the preceding constant e

$$\begin{aligned} h(x) &\equiv 1, \qquad |x| \leqq e, \\ h(x) &\equiv 0, \qquad |x| \geqq 2e. \end{aligned} \tag{4}$$

Such a function as $h(x)$ can readily be set up in terms of the elementary functions.

For points $(v_1, \cdots, v_n)$ of the closed domain between $S^{A,2e}$ and $S^{A,e}$ set

$$(5)\qquad D(v_1, \cdots, v_n, \mu) \equiv J(v_1, \cdots, v_n, \mu) - J(v_1, \cdots, v_n),$$

and now define ϕ for the same points $(v_1, \cdots, v_n)$ as follows:

$$(6)\ \phi(v_1, \cdots, v_n) \equiv J(v_1, \cdots, v_n) + D(v_1, \cdots, v_n, \mu)h([v_1^2 + \cdots + v_n^2]^{1/2}).$$

Now the constant μ can be chosen so small in absolute value that the function D of (6), as well as each of the partial derivatives D_{v_i}, is less in absolute value than a preassigned positive constant d throughout the whole region between $S^{A,2e}$ and $S^{A,e}$. For the same domain the sum of the squares of the first partial derivatives of $J(v_1, \cdots, v_n)$ exceeds some positive constant. It follows readily from (6), that for a $\mu \neq 0$, sufficiently small in absolute value, the function ϕ of (6) will have no critical point between $S^{A,2e}$ and $S^{A,e}$. We accordingly understand such a value of μ chosen, and hereafter held fast.

If now the function $\phi(v_1, \cdots, v_n)$ be defined at the remaining points of Σ by (2) and (3), it is readily seen that $\phi(v_1, \cdots, v_n)$ has the properties affirmed in the lemma.

We are now in a position to state the following theorem.

THEOREM 8. *Let there be given regions S and S_1 and integrand $F(x, y, \dot{x}, \dot{y})$ satisfying* I, II, II′, IV′ *of* §7 *and* §26. *In S_1 suppose there are at most a finite set T, of periodic extremals g, continuously deformable into a boundary curve of S_1. In the set T let each degenerate extremal be replaced by an equivalent set of non-degenerate extremals in accordance with the conventions at the ends of* §24 *and* §25. *Let M_k be the number of periodic extremals of type k in the resulting set, and m be the maximum of the numbers k. Between the numbers M_k the relations* (R) *of* §11 *hold.*

If the above periodic extremals g are all non-degenerate, the present theorem is identical with Lemma 2, §26.

Suppose on the other hand that not all of the above periodic extremals g are non-degenerate. To be specific, suppose the above set includes a simply-degenerate periodic extremal g_1 which taken from $u=0$ to $u=\omega$ is convex or concave. Then according to Theorem 6 of §24 it will be possible to modify the integrand by the introduction of a parameter μ in such a fashion, that for a suitably chosen value of μ, say μ_1, arbitrarily small in absolute value, there will appear, instead of g_1, in the neighborhood of g_1, a non-degenerate extremal E of what we have agreed to take as the type equivalent to g_1. Corresponding to this introduction of a parameter μ into the integrand, the function $J(v_1, \cdots, v_n)$ whose critical points correspond to the given periodic extremals will be replaced by a function $J(v_1, \cdots,$

$v_n, \mu)$. In setting up $J(v_1, \cdots, v_n, \mu)$ we can and will use the same field of extremals (IV′, §26) and corresponding parametric system (u, v) as in setting up $J(v_1, \cdots, v_n)$. The broken extremals along which $J(v_1, \cdots, v_n, \mu_1)$ is to be the integral will, however, be taken as the extremals corresponding to the integrand in which $\mu = \mu_1$. Let $(A) = (a_1, \cdots, a_n)$ be the point at which the unmodified function $J(v_1, \cdots, v_n)$ has the isolated critical point corresponding to g_1. By virtue of our modification of the integrand of the problem, $J(v_1, \cdots, v_n, \mu_1)$ will have, in the neighborhood of the point (A), a critical point (B) of the type termed equivalent to that of g_1.

It follows from the lemma of this section that we can set up a function $\phi(v_1, \cdots, v_n)$ that will have in a suitably chosen neighborhood of (A) no other critical point than (B), and that will be identical with $J(v_1, \cdots, v_n)$ without this neighborhood of (A).

We can similarly modify $J(v_1, \cdots, v_n)$ in the neighborhood of each other critical point (A) that corresponds to a simply-degenerate periodic extremal g_1, so that the modified function has in the neighborhood of (A) a critical point (B) of a type equivalent to that of g_1, in case g_1 is convex or concave, or has no critical point at all in the neighborhood of (A) in case g_1 is convex-concave, while except for the neighborhoods of these points (A), the modified function is identical with the original function $J(v_1, \cdots, v_n)$.

In case the set T includes doubly-degenerate periodic extremals, it follows from Theorem 7, §25, and Lemma 1 of this section, that we can first modify $J(v_1, \cdots, v_n)$ in such a fashion that the resulting function no longer has critical points corresponding to doubly-degenerate extremals. We can then, by additional successive modifications, obtain finally a function replacing $J(v_1, \cdots, v_n)$ whose critical points all correspond to non-degenerate extremals, including thereby all of the non-degenerate extremals of the original set T, together with the complete set of non-degenerate extremals equivalent to the degenerate extremals of T. The theorem then follows upon applying the lemma on critical points of §11 to the function finally evolved from $J(v_1, \cdots, v_n)$.

Part VI. Deformation theory

28. **Families of curves joining A to B.** In this part of the paper we shall show how the type of a given extremal segment can be characterized in terms of the possibility or impossibility of making certain deformations of families of curves joining A to B. The methods will be extended to the case of closed extremals in a later section. The results obtained, it is hoped by

the author, will serve as a part of a necessary basis for a "theory in the large" more extensive than any already developed.

By an m-family of curves, Z_m, will be understood a set of ordinary curves (B, p. 192) lying in the (x, y) plane and passing from an initial point A to a final point B; where, further, the set of all points of Z_m make up a single-valued continuous point function of the points on a product complex C_{m+1} obtained by combining an arbitrary point t on a closed interval of the t axis with an arbitrary point P on some m-dimensional manifold M_m; and where, finally, the dependence of the points of Z_m upon the points of C_{m+1} is such that to hold P fast and vary t gives a representation of the individual curves of Z_m by virtue of which they may be termed ordinary.*

The point P will be called the *parametric* point and the manifold M_m the *parametric* manifold.

Let there be given two families Z_m' and Z_m'' consisting of curves which join the same two points A and B. If the parametric manifolds of Z_m' and Z_m'' are *homeomorphic* the two m-families can be represented by the aid of the parametric manifold of either Z_m' or Z_m'', in particular, say, by the manifold M_m. *In such a case Z_m' and Z_m'' will be said to be mutually deformable if corresponding to each value of a parameter μ on the interval*

$$0 \leqq \mu \leqq 1$$

there exists an m-family $Z_m(\mu)$ of which $Z_m(0)$ is Z_m' and $Z_m(1)$ is Z_m'', while each m-family $Z_m(\mu)$ is representable in terms of the same parametric manifold M_m, and the same interval for t, and joins the same two points, A and B; and if further the complete set of points (x, y) on these m-families, $Z_m(\mu)$, by virtue of their dependence upon μ, t and P, make up a single-valued continuous point function of an arbitrary point on the product complex C_{m+2} obtained by combining an arbitrary point μ on its interval, an arbitrary point t on its interval, and an arbitrary point P on M_m.

In a deformation D such as the one just defined Z_m' will be called the *initial* family and Z_m'' the *final* family. The m-families $Z_m(\mu)$, for values of μ between 0 and 1, will be called the *intermediate families.* A point $P=P_0$ on M_m, and a value $\mu=\mu_0$ held fast while t varies, determine a curve in $Z_m(\mu_0)$. The curve $P=P_0$ of Z_m' will be said to be *replaced* in the deformation D when $\mu=\mu_0$, by the curve $P=P_0$ of $Z_m(\mu_0)$. Points on a curve P of Z_m', and points on any curve replacing P under D will be said to *correspond* if they are given by the same values of t. If we should hold P and t fast

* Cf. Veblen, The Cambridge Colloquium, 1916, Part II, *Analysis Situs*, p. 88.

in $Z_m(\mu)$, and vary μ, the resulting set of points would be the locus of points *corresponding* under D to a single point of Z_m'.

We hereby understand that all of the definitions of this section have been given in terms of (u, v) as well as of (x, y).

29. **Hypotheses. Fundamental lemmas on deformations.** We make here concerning $F(x, y, \dot{x}, \dot{y})$, and the given extremal g, the same hypothesis as in §1, except that here we suppose that F and g are of class C'''' instead of class C''', and that $F_1(x, y, \dot{x}, \dot{y})$ is positive, not only along g, but also for (x, y) on g and for $(\dot{x}, \dot{y})$ any two numbers not both zero. A consequence of the assumption that F be of class C'''' instead of class C''', is that the function $J(v_1, \cdots, v_n)$ set up in §1 is here of class C''' for $(v_1, \cdots, v_n)$ in the neighborhood of $(0, \cdots, 0)$. As in §2 we transfer the problem to the (u, v) plane, carrying g into a segment γ of the u axis.

By a *canonical* curve will be understood a succession of extremal segments joining the successive points of the set

$$(u_0, 0)(u_1, v_1), \cdots, (u_n, v_n)(u_{n+1}, 0), \tag{1}$$

determined as in §1. An m-family of canonical curves will be called a *canonical m-family*.

Let z stand for any positive constant a, b, c, d, e, etc. Let R_z denote the set of points (u, v) within a distance z of γ.

LEMMA 1. *Let R_1 be any region in the (u, v) plane enclosing γ in its interior and in which the problem is "regular." If a be a sufficiently small positive constant, the region R_a, consisting of the points (u, v) within a distance a of γ, will possess the following property. Any m-family Z_m consisting of curves that lie in R_a, join γ's end points, and give to J a value such that*

$$J \leqq J_0 - e^2, \tag{2}$$

where e is a positive constant, and J_0 is the value of J along γ, can be deformed within R_1, through the mediation of curves that always satisfy (2), into an m-family of canonical curves. This deformation can be so made that if any of the curves of Z_m are canonical curves they are replaced in the deformation only by curves along which v is a function of u of at least class C.

Before coming to the proof proper we make a number of preliminary statements and definitions.

(A) We can and will choose a positive constant r so small that of the intervals I_i:

$$u_i - r \leqq u \leqq u_i + r \qquad (i = 0, 1, \cdots, n+1); \tag{3}$$

no two successive intervals I_i and I_{i+1} have any points in common or contain any conjugate points of each other.

(B) If in the results obtained by Lindeberg* in a paper cited below we set

$$G(x, y, x', y') = (x'^2 + y'^2)^{1/2},$$

we readily obtain the following. Corresponding to the positive constant r just chosen in (A) there can be found a region R_b so small that if $\bar{\gamma}$ be any "ordinary curve" (B, p. 192) joining γ's end points within R_b and satisfying (2), if u and $\bar{u}$ are u coördinates of points Q and Q on $\bar{\gamma}$ and γ respectively, and if $\bar{Q}$ and Q lie at the same distance measured along $\bar{\gamma}$ and γ respectively from $(u_0, 0)$ or from $(u_{n+1}, 0)$, then

$$|\bar{u} - u| \leqq r.$$

(C) We will now choose a region R_c as follows. We first require the region R_c to lie within the region R_b of the preceding paragraph (B). Further we can and will choose c so small (B, pp. 275 and 307) that any two *distinct* points Q and Q' which both lie in one of the regions

$$u_{i-1} - r \leqq u \leqq u_i + r, \qquad v \leqq c \quad (i = 1, 2, \cdots, n+1) \tag{4}$$

can be joined in the order Q, Q' by an extremal segment E with the following properties:

(a) The extremal E gives a minimum to J relative to all "ordinary curves" joining Q to Q' and lying in (4).

(b) The coördinates (u, v) of points of E are functions of at least class C' of the coördinates of Q and Q' and of the distance s of the points (u, v) from Q measured along E.

(c) In case Q and Q' lie in two successive intervals I_i and I_{i+1} of (A), the coördinate v of a point (u, v) on E will also be a function of class C' of u as well as of the coördinates of Q and Q'.

(D) Finally we choose a region R_d which we will prove can serve as the region R_a of the lemma. We first require that d be so small that if the end points of any of the extremal segments E described in (C) lie in R_d the whole of E will lie in R_c. A second requirement on d will be added later.

Now let Z_m be any m-family each curve of which lies within R_d and satisfies the relation (2). In accordance with the conventions of §28, let P be the parametric point of the family Z_m, and M_m the parametric manifold on which P lies. We come now to the deformation of Z_m into an m-family of canonical curves. This deformation will be given as the resultant of two deformations,

* J. W. Lindeberg, *Über einige Fragen der Variationsrechnung*, Mathematische Annalen, vol. 67 (1909), p. 351, §8.

D' and D'', which obviously can be combined into a single deformation if the final m-family of D' is identical with the initial m-family of D''.

Let each curve of Z_m be divided into $n+1$ successive segments of which the ith, say g_i, consists of points for which the arc length s measured from $(u_0, 0)$ satisfies

$$u_{i-1} \leqq s \leqq u_i \qquad (i = 1, 2, \cdots, n),$$

and when $i=n+1$ consists of the remaining points on the given curve. In accordance with the preceding results, (A) and (B), the coördinates of the points of g_i will satisfy

$$u_{i-1} - r \leqq u \leqq u_i + r \qquad (i = 1, 2, \cdots, n+1). \tag{5}$$

The deformation D' will be defined as follows: For any value of the deforming parameter μ for which

$$0 < \mu \leqq 1$$

let each arc g_i on each curve of Z_m be divided into two successive segments whose arc lengths are in the ratio of μ to $1-\mu$. Corresponding to the given value of μ let the second of these two segments of g_i *be replaced* by itself but let the first of these two segments, say k_i, be replaced by that extremal segment, say h_i, that belongs to the class of extremals described in (C) and that joins k_i's end points. Points of h_i and k_i which divide h_i and k_i respectively in the same ratio as measured by arc lengths, shall be made to *correspond*, and shall accordingly be assigned the same values of t, namely the values already assigned to the points of k_i in the representation of Z_m in terms of t and P.

That we have here actually defined a deformation of Z_m readily follows from (C). Let Z_m' denote the final m-family of the deformation D'. According to paragraph (D) the curves of Z_m' will all lie in R_e, and according to paragraph (C), (c), their v coördinates will be functions of their u coördinates of class C at least. Further, the curves used in the deformation D' give to J at most the value which the corresponding curves of Z_m give, so that the curves used in D' satisfy (2).

We can now define a second deformation D'' under which Z_m' will be deformed into a family of canonical curves satisfying (2). The definition of D'' is similar to that of D' except that the segments g_i into which we divide the curves of Z_m' can here be defined in terms of u instead of s as follows:

$$u_{i-1} \leqq u \leqq u_i \qquad (i = 1, 2, \cdots, n+1).$$

The combination of the deformations D' and D'' will obviously give a deformation of the desired sort and the lemma is proved except for its concluding sentence. This concluding sentence is also readily seen to be true if the constant d of (D) be further restricted in magnitude so that the slopes of the canonical curves in Z_m will be less in absolute value than a sufficiently small positive constant.

In the deformation D'' a curve which initially was canonical is for each value of the deforming parameter μ replaced by itself. If in the original m-family the variable t had been u, then of the deformations D' and D'' only D'' would have been needed. We thus have the following lemma.

LEMMA 2. *If in Lemma* 1 *an m-family* Z_m *is given in which the variable* $t=u$, *then the deformation whose existence is affirmed in Lemma* 1 *can be defined in such a fashion that any curve of* Z_m *which is a canonical curve remains unaltered throughout the deformation.*

The following Lemma is important.

LEMMA 3. *Let* R_1 *be a region of the* (u, v) *plane enclosing* γ *in its interior, and in which the problem is regular. If* e' *be any sufficiently small positive constant, the region* $R_{e'}$ *consisting of points* (u, v) *within a distance* e' *of* γ *has the following property. If a canonical m-family* Z_m' *can be deformed within* $R_{e'}$ *into a canonical m-family* Z_m'' *through a deformation D all of whose curves give to J a value such that*

$$J \leqq J_0 - e^2, \tag{6}$$

where e is a positive constant, and J_0 *the value of J along* γ, *then* Z_m' *can be deformed into* Z_m'', *within* R_1, *through the mediation of curves all of which are canonical and which satisfy* (6).

For a moment suppose the region R_1 given in Lemma 1 is the region R_1 given in the present lemma. Corresponding to this choice of R_1 let $R_{a'}$ be a particular choice of the region R_a such that Lemmas 1 and 2 hold true. Now let us apply Lemma 1 again, this time taking for the region R_1 of Lemma 1 the region $R_{a'}$ just determined. Corresponding to this second choice of R_1, Lemma 1 will hold true for a second proper choice of R_a which we now denote by $R_{a''}$. We will prove Lemma 3 is true if $R_{e'}$ in Lemma 3 be taken as $R_{a''}$.

We first note that the complete set of curves used in the deformation D can be considered as an $(m+1)$-family Z_{m+1}. For if M_m is the common parametric manifold of Z_m' and Z_m'', each curve of D is specified by a pair (P, μ), where P is on M_m and μ on its interval $(0, 1)$. These curves of D can then be regarded as specified by a point $P_1=(P, \mu)$ the totality of which points form an $(m+1)$-dimensional manifold M_{m+1}.

Now the $(m+1)$-family Z_{m+1} can be deformed subject to (6), and within $R_{a'}$, into a canonical $(m+1)$-family $\bar{Z}_{m+1}$ using thereby the deformation, say $\bar{D}$, of Lemma 1. Under $\bar{D}$ the given canonical m-families Z_m' and Z_m'' appear initially in Z_{m+1}, and will be replaced finally in $\bar{Z}_{m+1}$ by canonical families, say $\bar{Z}_m'$ and $\bar{Z}_m''$ respectively. We will prove the lemma by showing how each of the first three canonical m-families of the set

$$Z_m', \bar{Z}_m', \bar{Z}_m'', Z_m'' \tag{7}$$

can be deformed, subject to (6), into its successor in (7) by a deformation in which only canonical curves are used.

(a) *The deformation of* Z_m' *into* $\bar{Z}_m'$. We may fix our attention upon those curves of $\bar{D}$ which serve to deform Z_m' into $\bar{Z}_m'$ and let Z_{m+1}' be the $(m+1)$-family of curves in $\bar{D}$ replacing Z_m', including also Z_m' and $\bar{Z}_m'$.

According to the concluding sentence of Lemma 1, Z_{m+1}' consists of curves along which v is a function of u of class C. According to Lemma 2, Z_{m+1} can then be carried by a deformation D' which does not alter Z_m' and $\bar{Z}_m'$ into a set of canonical curves. The final canonical $(m+1)$-family of curves of D' may be considered as the curves of the desired deformation, say D_1, of Z_m' into $\bar{Z}_m'$.

(b) *The deformation of* $\bar{Z}_m'$ *into* $\bar{Z}_m''$. The deformation, say D_2, required here is furnished by the set of canonical curves which make up $\bar{Z}_{m+1}$.

(c) *The deformation of* $\bar{Z}_m''$ *into* Z_m''. This deformation, say D_3, is set up as D_1 was set up in (a).

The combination of D_1, D_2, and D_3, in the order written, will give the required deformation of Z_m' into Z_m'' and the lemma is proved.

30. The analysis situs of the deformation problem. We have been denoting by $J(v_1, \cdots, v_n)$ the value of the integral J along the canonical curves joining the points (1) of §29. If $(u_0, 0)$ is not conjugate to $(u_{n+1}, 0)$ but if there are k conjugate points of $(u_0, 0)$ preceding $(u_{n+1}, 0)$, it follows from Theorem 2 of §6 that $J(v_1, \cdots, v_n)$ has a critical point of rank n and type k at the point $(v_1, \cdots, v_n) = (0, \cdots, 0)$. It follows from a lemma proved by the author* that there exists a one-to-one continuous transformation of the variables $(v_1, \cdots, v_n)$ into variables $(y_1, \cdots, y_n)$ in which the point $(v_1, \cdots, v_n) = (0, \cdots, 0)$ corresponds to $(y_1, \cdots, y_n) = (0, \cdots, 0)$, and under which

$$J(v_1, \cdots, v_n) - J(0, \cdots, 0) = -y_1^2 - y_2^2 - \cdots - y_k^2 + y_{k+1}^2 + \cdots + y_n^2 \tag{1}$$

* Marston Morse, these Transactions, loc. cit., p. 354.

for $(y_1, \cdots, y_n)$ in the neighborhood of $(0, \cdots, 0)$. We exclude for the present the case where $k=0$.

For the sake of brevity we set

$$p^2 = y_1^2 + y_2^2 + \cdots + y_k^2, \qquad 0 < k \leqq n, \tag{2}$$

$$q^2 = y_{k+1}^2 + y_{k+2}^2 + \cdots + y_n^2, \tag{3}$$

understanding q^2 to be identically zero if $k=n$. Now let a be a positive constant so small that the domain

$$p^2 + q^2 \leqq a^2 \tag{4}$$

is one for which (1) holds as described.

In the present section we shall concern ourselves with deformations, in the neighborhood of the extremal segment γ, of m-families Z_m composed of canonical curves subject to a condition of the form

$$J(v_1, \cdots, v_n) \leqq J(0, \cdots, 0) - e^2, \quad e < a, \tag{5}$$

where e is an arbitrarily small positive constant. We shall confine ourselves to m-families corresponding to which the parametric manifold M_m is closed. Corresponding to each point P on M_m there is a curve on the given m-family Z_m. In the present case each such curve is canonical, and the points at which it crosses the straight lines $u=u_i$ have for their v coördinates v_i. Thus each point P on M_m determines a point $(v_1, \cdots, v_n)=(V)$. If we regard the points (V) as distinct when they arise from distinct points P on M_m, the set of all such points (V) form a closed m-dimensional manifold M_m' homeomorphic with the manifold M_m. If the given canonical m-family be deformed in the sense of §28, the manifold M_m' arising at each stage of the deformation from the m-family which replaces Z_m at that stage, will itself be "homotopically deformed" in the sense of analysis situs.*

We turn next to the space of the points $(Y)=(y_1, \cdots, y_n)$ and confine ourselves to a spherical neighborhood of the origin in the form (4), and to points in the space of the points (V) that correspond to the neighborhood in the space of the points (Y). Corresponding to a deformation of M_m' in the space of the points (V) we shall have to deal with a deformation of a corresponding manifold M_m'' in the space of the points (Y). Our requirement that the curves of the given m-family satisfy (5) becomes in terms of the points (Y) on the manifold M_m'' the condition

$$p^2 - q^2 \geqq e^2, \qquad 0 < k \leqq n. \tag{6}$$

* Veblen, loc. cit., pp. 125–126.

31. **Homotopic deformations in the space (Y).** The preceding condition (6) and the restriction of the points (Y) to points in a spherical neighborhood of $(Y)=(0, \cdots, 0)$ combine to give us the relations

$$(1) \qquad \begin{aligned} p^2 + q^2 &\leqq a^2, \qquad & 0 < e < a, \\ p^2 - q^2 &\geqq e^2, \qquad & 0 < k \leqq n. \end{aligned}$$

The relations (1) define a finite closed region in the space of the points (Y). The deformation of m-families thus leads us to study deformations of closed manifolds in the space (1).

We could get some but not all of the properties of the space (1) by breaking it up into cells, and showing that of its connectivity numbers

$$R_0, R_1, \cdots, R_n,$$

all are unity except one, namely

$$R_{k-1} = 2.$$

Note that the $(k-1)$-dimensional sphere

$$(2) \qquad p^2 = e^2, \qquad q^2 = 0$$

is among the points of (1). Denote this sphere by S_{k-1}. A very important fact is that the region (1) can be deformed through the mediation of its own points into a singular complex on S_{k-1}. In terms of a deforming parameter μ the set of intermediate and final positions, $(y_1, \cdots, y_n)$, corresponding to any initial point $(a_1, \cdots, a_n)$ of (1), in a deformation D of (1) into a complex on S_{k-1} can be given as follows:

$$(3) \qquad \begin{aligned} y_i^2 &= (1-\mu)a_i^2 + \frac{\mu e^2 a_i^2}{a_1^2 + \cdots + a_k^2} \qquad (i = 1, 2, \cdots, k), \\ y_i^2 &= (1-\mu)a_i^2 \qquad (i = k+1, k+2, \cdots, n), \end{aligned}$$

where

$$(4) \qquad 0 < \mu \leqq 1,$$

and where the coördinates $(y_1, \cdots, y_n)$ are respectively required to have the sign of the coördinates $(a_1, \cdots, a_n)$ from which they are deformed.

We can now prove that any *m-dimensional closed manifold M_m, singular or non-singular, lying on* (1) *and such that m is less than $k-1$, can be deformed on* (1) *into a point.* For under the above deformation (3), M_m can be deformed into a manifold M_m' on S_{k-1}. Regardless of whether M_m' is singular or non-singular on S_{k-1}, there is a fundamental theorem on homotopic deformations* to the effect that M_m' can be deformed on S_{k-1} into a manifold

* Veblen, loc. cit., p. 131.

M_m'' on S_{k-1} each of whose cells "covers" a cell of S_{k-1}. Hence M_m'' can be deformed into a point on S_{k-1}; and the statement in italics follows at once.

The sphere S_{k-1} *cannot itself be deformed on* (1) *into a point on* (1). For if there were such a deformation, say D', let the family of intermediate manifolds homeomorphic with S_{k-1} under D' be deformed onto S_{k-1} under the deformation (3). The resulting family of manifolds on S_{k-1} would constitute a deformation on S_{k-1} in which the initial manifold would cover S_{k-1} just once and the final manifold coincide with a point. This, however, is impossible according to a fundamental theorem of analysis situs.*

We can now add the statement that S_{k-1} *cannot be deformed on* (1) *into any manifold on a manifold* M_r *on* (1) *of lower dimension than* $k-1$. For according to the preceding paragraph, M_r, and hence S_{k-1}, could be deformed on (1) into a point on (1). By making use of product manifolds analogous to surfaces of revolution in 3-space it is easy to set up for each $m>k-1$ an example of a manifold on (1) which cannot be deformed on (1) into any manifold of dimensionality less than $k-1$.

Finally with the aid of the deformation (3) we obtain the following result. *Any closed manifold* M_m *on* (1) *for which* $m \geqq k-1$ *can always be deformed on* (1) *into a manifold on* S_{k-1}, *and in special cases can be further deformed on* S_{k-1} *into a point.*

32. The theorem on deformations of m-families joining A to B. We can now translate the results of the preceding section into terms of deformations of canonical m-families. We shall understand that an *m-family shall be considered as "on" an r-family if every curve of the m-family coincides with some curve of the r-family.* Let it also be understood that a *closed m-family shall mean an m-family whose parametric manifold is closed.* In particular it should be noted that *a closed 0-family means a pair of ordinary curves joining the same two points.* The results of the preceding section now give the following lemma.

LEMMA. *Let γ be the extremal segment $v=0$ of §29. Suppose there are k points $(k>0)$, conjugate to γ's initial point A and preceding γ's final point B, but that A is not conjugate to B. Then within any region R enclosing γ in its interior there exists a region R_1 enclosing γ in its interior, and corresponding to R_1 an arbitrarily small positive constant such that the closed m-families of canonical curves that satisfy*

$$J \leqq J_0 - e^2 \tag{1}$$

and lie in R_1 are conditioned as follows:

* Veblen, p. 131, §14.

(a) *Those for which* $m<k-1$ *can be deformed among canonical curves, within* R_1, *and subject to* (1), *into a single canonical curve.*

(b) *Those for which* $m\geqq k-1$ *include for each* m *at least one* m*-family that cannot be deformed among canonical curves, within* R_1, *and subject to* (1), *into a single canonical curve, or even into a* $(k-1)$*-family on an* r*-family for which* $r<k-1$.

(c) *Those for which* $m\geqq k-1$ *can always be deformed among canonical curves, within* R_1, *and subject to* (1), *into an* m*-family of canonical curves on a* $(k-1)$*-family* Z_{k-1}, *and, in special cases, can be further deformed into a single canonical curve.*

This lemma follows from the italicized statements of the preceding section, taking for Z_{k-1} the m-family determined by the points (V) that correspond to the points (Y) on S_{k-1}.

Lemmas 1, 2, 3 of §29 tell how and when any m-family of ordinary curves in the (u, v) plane can be deformed into a closed m-family of canonical curves. The preceding lemma describes the limitations on the deformations of closed m-families of canonical curves. The combination of the results of these lemmas carried over into the (x, y) plane gives the following fundamental theorem. (Here it will be convenient to call a region enclosing g in its interior, and consisting of points (x, y) within an arbitrarily small positive constant distance of g, *an arbitrarily small neighborhood of* g. The term *a sufficiently small neighborhood of* g will be similarly defined.)

THEOREM 9. *Let there be given in the* (x, y) *plane the extremal* g *of* §29.

On g *let there be* k *points conjugate to* A, $k>0$, *but suppose* B *is not conjugate to* A. *Let* J_0 *be the value of the integral* J *along* g.

Then corresponding to any sufficiently small neighborhood R *of* g, *there exists within* R *an arbitrarily small neighborhood* R_1 *of* g, *and an arbitrarily small positive constant* e *with the following properties. Closed* m*-families of curves which lie in* R_1, *which join* g*'s end points, and give to the integral* J *a value for which* (1) *holds, are conditioned as follows:*

(a) *Those for which* $m<k-1$ *can be deformed in* R *and subject to* (1) *into a single curve.*

(b) *Those for which* $m\geqq k-1$ *include for each* m *at least one* m*-family that cannot be deformed, within* R *and subject to* (1), *into any* $(k-1)$*-family on an* r*-family for which* $r<k-1$, *or into a single curve.*

(c) *Those for which* $m\geqq k-1$ *can always be deformed, within* R *and subject to* (1), *into an* m*-family on a* $(k-1)$*-family, and in special cases can be further deformed into a single curve.*

There remains the case where A is conjugate to B. Here we need to assume that the problem is analytic, as well as regular, in the neighborhood of g. We suppose that not every extremal through A with slope at A near that of g passes through B. In accordance with the conventions preceding Theorem 3, §12, g will have a type number $k=s$, or $k=s+1$, or type numbers $k=s$ and $s+1$, where s is the number of points on g conjugate to A and preceding B. *It can be shown that if $k=s$, or $k=s+1$, the preceding theorem holds if the sentence concerning conjugate points be omitted, while if g is of the composite type where $k=s$ and $k=s+1$, the preceding theorem should be further altered by replacing* (a), (b), *and* (c) *by the statement that any m-family whatsoever lying in R_1, joining A to B, and satisfying* (1) *can be deformed within R and subject to* (1), *into a single curve joining A to B.*

A proof of these last facts would lead us too far astray. The writer hopes to return to a discussion of critical points of rank less than n in a separate paper.

33. **An illustrative example.** Consider the equator of a unit sphere. Refer the sphere near C to coördinates (u, v), giving respectively the latitude and longitude of the point. Consider the sphere near C as now covered an infinite number of times by an unending strip S which overhangs the neighborhood of C after the manner of a Riemann surface. We may suppose S represented in the (u, v) plane by an unending strip R containing the u axis. We suppose our integral corresponds to the arc length on S.

The case $k=1$. If we take a segment of the u axis whose length g lies between π and 2π we will have the case $k=1$ of the preceding theorem. In the strip R we can clearly take two curves which join g's end points but otherwise lie on opposite sides of g, and whose lengths are equal but both less than g. These two curves form a 0-family which cannot be deformed into a single curve without passing out of R or increasing the lengths on S.

The case $k=2$. Suppose now that the length g of g is between 2π and 3π. Here we have the case $k=2$. Let u_0 and u_3 correspond to the end points of g, and let u_0, u_1, u_2, u_3 correspond to four successive points on the u axis such that no one of the resulting three successive intervals is as great as π in length. We consider a one-dimensional closed manifold of pairs (v_1, v_2), namely the pairs

$$v_1(a) = b \cos a, \quad v_2(a) = b \sin a,$$

where a is any real number, and is to be our parameter, and b is a positive constant. If b be sufficiently small, the following will hold true. On S the images of the points

$$[u_0, 0], \quad [u_1, v_1(a)], \quad [u_2, v_2(a)], \quad [(u_3, 0]$$

can be successively joined by arcs of great circles which will give the shortest paths between their end points. The images on R of all such paths for all values of a will give a closed 1-family. The lengths of the different paths of the family considered as a function of a will have a maximum M which is readily seen to be less than g. If our choice of b be sufficiently small, it follows from the general theory that this 1-family cannot be deformed into a single curve joining g's end points without increasing the lengths beyond M or passing out of R.

If however we join the end points of g by any two ordinary curves whose lengths on S are less than a constant $g-e<g$, and which lie in a sufficiently small neighborhood of g, then, according to the preceding theorem, these two curves can be deformed into each other without passing out of R, or using curves whose lengths exceed $g-e$.

34. **Osgood's Theorem.** It has doubtless been noted that the case where $k=0$ has been omitted in the preceding discussion for the reason that there are in this case no curves neighboring the given extremal segment which satisfy the inequality (1) of the preceding section. For the case $k=0$ it seems to the author that the theorem that is due to Osgood* represents the nearest approach to the preceding developments. The questions at issue are sufficiently similar to have suggested to the author *a proof of Osgood's Theorem in the form stated by Hahn* (*B*, *p.* 281) *without, however, the hypothesis, used by Hahn and by recent writers on the calculus of variations, that at the ends of the extremal segment the problem be regular.* Such a proof seems desirable because it reduces the hypotheses under which Osgood's Theorem can be proved to a set identical with the least hypotheses under which the proper strong minimum is ordinarily established. This proof will be published later.

35. **Deformations of periodic extremals.** Although the results here are different from those of the preceding sections yet the methods may be carried over, and afford a proof of the desired results once the problem is well defined.

By an m-family of closed curves Z_m will be understood a set of ordinary closed curves in the (x, y) plane, the set of all of whose points make up a single-valued, continuous, point function of the point on a product complex C_{m+1}, obtained by combining an arbitrary point Q on a unit circle with an arbitrary point P on some m-dimensional manifold M_m. The dependence of the points of Z_m upon the points of C_{m+1} is to be such that each individual closed curve is obtained by holding P fast and varying Q on the unit circle;

* Osgood, these Transactions, vol. 2 (1901), p. 273.

and the dependence of the variable point of each individual curve upon Q, or more particularly upon the arc length measured from a fixed point on the unit circle to the point Q, is to be one by virtue of which the curve may be termed ordinary.

The remaining definitions and results of §§ 28, 29, 30, 31, and 32, may be carried over here either unchanged or with the changes obvious at each point. We have the following central theorem:

THEOREM 10. *Let there be given an integrand* $F(x, y, \dot{x}, y)$ *satisfying the hypotheses of* §1 *except that here* F *is to be of class* C''''. *Let there be given in* S *a non-degenerate* (§17) *periodic extremal* g, *at every point* (x, y) *of which, and for every* θ, $F_1(x, y, \cos\theta, \sin\theta)$ *is positive. Let* k *be the type number of* g, *determined as in Theorem* 5, §21. *Let* J_0 *be the value of* J *along* g.

Then corresponding to any sufficiently small neighborhood R *of* g, *there exists within* R *an arbitrarily small neighborhood* R_1 *of* g, *and an arbitrarily small positive constant* e *with the following properties. Closed* m*-families of periodic curves which are deformable within* R_1 *into* g, *and which give to the integral* J *a value such that* $J < J_0 - e^2$ *satisfy the statements* (a), (b), *and* (c) *of Theorem* 9.

36. **Birkhoff's Theorem.*** In dealing with deformations of one closed trajectory into another Birkhoff has given a theorem which for the case of dynamical systems is essentially equivalent to that special part of the preceding theorem which has to do with the mutual deformability of two closed trajectories. Birkhoff does not consider such entities as m-families in general. However his pair of closed trajectories comes under the head of what the author has called closed 0-families. Birkhoff's theorem is stated in terms of the Poincaré rotation number, and in terms of that rotation number tells when a pair of closed trajectories can be deformed into each other. *The author's theorem shows that periodic extremals for which* $k > 1$ *cannot be distinguished in type simply by a consideration of the mutual deformability of two closed trajectories.* It should be stated in explanation that Birkhoff was not seeking to distinguish between all types of periodic extremals by means of deformations, but rather to determine to what types his "minimax principle" applied, and for that purpose his theorem was sufficient.

Speaking generally, the results of Part VI show that *the number of conjugate points on extremals joining two fixed points, or the type number of periodic extremals, is the least integer* m *for which some* $(m-1)$*-family of curves*

* Birkhoff, loc. cit., p. 249.

cannot be deformed into a single curve subject to the conditions described in the text. This result suggests another new and powerful way of approaching a theory "in the large."

Part VII. Variable end points

37. A geometric theorem. No general development for this case will be presented. However, a suggestive and intrinsically complete geometric theorem has been obtained in a problem that comes under this head, namely the problem of the number and nature of the normals which can be drawn from a fixed point P in a euclidean $(n+1)$-space E_{n+1} to an n-spread S_n in that space. Proofs of the following results will be published later.

The n-spread S_n is to be a manifold without boundary in the sense of Veblen (loc. cit.). Moreover we assume that the points on S_n in the neighborhood of any given point of S_n are representable by an equation which gives one of the rectangular coördinates as a function of class C'' of the others. A preliminary result is that the loci of centers of principal normal curvature* C_n corresponding to S_n are not space filling in E_{n+1}, that is, there are in the neighborhood of each point of E_{n+1} points not on C_n. Let P be any point not on C_n or S_n. A second preliminary result is that any straight line L through P will be normal to S_n at no more than a finite number of points Q_i $(i=1, 2, \cdots, m)$. Each of the m segments PQ_i of L will be counted as a different normal from P to S_n. *A normal PQ_i will be said to be of type k if there are k centers of principal normal curvature corresponding to Q_i, on L, between P and Q_i.* We have the following theorem.

Theorem. *Let P be any point not on the manifold S_n, nor on the loci of centers of principal normal curvature of S_n, and let M_k be the number of different normals from P to S_n of type k. Then between the numbers M_k, and the connectivities R_i of S_n, the following relations hold true:*

$$M_0 \geqq 1 + (R_0 - 1),$$

$$M_0 - M_1 \leqq 1 + (R_0 - 1) - (R_1 - 1),$$

$$M_0 - M_1 + M_2 \geqq 1 + (R_0 - 1) - (R_1 - 1) + (R_2 - 1),$$

$$\cdots\cdots\cdots\cdots\cdots\cdots\cdots$$

$$M_0 - M_1 + \cdots + (-1)^n M_n = 1 + (R_0 - 1) - + \cdots + (-1)^n (R_n - 1),$$

where the inequality signs alternate until the final equality is reached.

---

* Eisenhart, *Riemannian Geometry*, p. 153.

To this theorem* we add the result that points Q on S_n such that the straight line segment PQ gives a relative minimum or maximum to the distance from P to S_n, yield normals PQ, respectively of types 0 or n. We now make the following specific applications of the theorem.

In 3-space let S_2 be homeomorphic to a sphere. Here

$$R_0 = 1, \quad R_1 = 1, \quad R_2 = 2.$$

If there are r normals from P to S_2 which give a relative minimum to the distance from P to S_2, and s normals that give a relative maximum, *there are* $r+s-2$ *other normals of type* 1. The total number of normals is always even.

In 3-space let S_2 be homeomorphic to a torus. Here

$$R_0 = 1, \qquad R_1 = 3, \qquad R_2 = 2,$$

$$M_0 \geqq 1, \qquad M_1 \geqq 2, \qquad M_2 \geqq 1.$$

Thus there are always four normals at least. The total number of normals is always even.

In 4-space let S_3 be homeomorphic to a manifold obtained by identifying the opposite faces of an ordinary cube. Here

$$R_0 = 1, \qquad R_1 = 4, \qquad R_2 = 4, \qquad R_3 = 2,$$

$$M_0 \geqq 1, \qquad M_1 \geqq 3, \qquad M_2 \geqq 3, \qquad M_3 \geqq 1.$$

There are thus at least eight normals, while the total number of normals is always even.

---

* A theorem essentially the same as the above has been proved to hold in the general problem one variable end point in m dimensions.

Harvard University,
Cambridge, Mass.

Reprinted from the
TRANSACTIONS OF THE AMERICAN MATHEMATICAL SOCIETY
Vol. 31, No. 3, pp. 379–404

THE FOUNDATIONS OF THE CALCULUS OF VARIATIONS IN THE LARGE IN m-SPACE (FIRST PAPER)*

BY
MARSTON MORSE

1. Introduction. The calculus of variations in the large as developed by the author may be roughly divided into two parts.

The first part involves the characterization of the number of conjugate points or focal points on a given extremal in terms of the type number of a basic quadratic form, and furnishes a connecting link between the theory of such extremals and the theory of critical points of functions.

The second part goes beyond the properties of one extremal alone; it involves families of extremals, families both closed and open. It necessarily leads to considerations which belong to analysis situs. It seeks to prove the existence of extremals on open or closed manifolds regardless of whether or not the extremals give minima.

The present paper consists of three parts of which the first two set up and study the fundamental forms associated respectively with an extremal joining two points, and an extremal cut transversally by a given manifold.

The third part presents a relatively complete picture of the theory in the large of extremals from a given point cut transversally by a given closed manifold. It gives strong existence theorems immediately applicable, for example, to the question of how many straight lines one can draw from a point normal to a given manifold. For the case where the manifold is homeomorphic to an $(m-1)$-sphere it can be shown that the relations found in this part are exhaustive.

The first two parts taken with the *deformation theory* previously developed by the author† will culminate in a second paper which will give the central existence theorems in the large for extremals joining two fixed points.

The first two parts taken in quite another way will lead to a generalization of the famous Sturm separation and comparison theorems for the case of a system of n linear, second-order, homogeneous, self-adjoint differential

* Presented to the Society as part of an address given at the request of the program committee April 6, 1928, under the title *The critical points of functions and the calculus of variations in the large*; see also Bulletin of the American Mathematical Society, vol. 35 (1929), pp. 38–54.

Received by the editors in April, 1929.

† Marston Morse, *The foundations of a theory in the calculus of variations in the large*, these Transactions, vol. 30 (1928), pp. 213–274.

equations. The methods and the results of this paper are also extensible to the case of two variable end points.

I. Extremals through fixed points

2. The integrand* $F(x, r)$. Let R be a closed region in the space of the variables $(x_1, \cdots, x_m) = (x)$. Let

$$F(x_1, \cdots, x_m, r_1, \cdots, r_m) = F(x, r)$$

be a function of class C''' for (x) in R and (r) any set not (0). We suppose further that F is homogeneous in that

$$F(x_1, \cdots, x_m, kr_1, \cdots, kr_m) = kF(x, r) \tag{2.1}$$

for any positive constant k. We shall start with the integral in the usual parametric form

$$J = \int_{t_0}^{t_1} F(x, \dot{x}) dt \tag{2.2}$$

where $(\dot{x})$ stands for the set of derivatives of (x) with respect to t.

Let g be an extremal segment lying in R. We suppose that g is an ordinary curve of class C'. Denote the second partial derivative of $F(x, r)$ with respect to r_i and r_j by F_{ij}. Along g we assume that F is positively regular, that is, that†

$$F_{ij}(x, \dot{x}) \eta_i \eta_j > 0 \qquad (i, j = 1, 2, \cdots, m) \tag{2.3}$$

for $(x, \dot{x})$ taken along g, and for (η) any set not (0) nor proportional to $(\dot{x})$.

Because of the homogeneity relation (2.1) we have the m identities

$$r_i F_{ij}(x, r) \equiv 0 \qquad (j = 1, 2, \cdots, m) \tag{2.4}$$

from which it appears that

$$|F_{ij}| \equiv 0. \tag{2.5}$$

---

* For that which concerns the classical theory of the calculus of variations in space the reader may refer to the following articles:

Mason and Bliss, *The properties of curves in space which minimize a definite integral*, these Transactions, vol. 9 (1908), pp. 440–466.

Bliss, *The Weierstrass E-function for problems of the calculus of variations in space*, these Transactions, vol. 15 (1914), pp. 369–378.

Bliss, *Jacobi's condition for problems of the calculus of variations in parametric form*, these Transactions, vol. 17 (1916), pp. 195–206.

Hadamard, *Leçons sur le Calcul des Variations*, vol. 1, Paris, 1910.

Carathéodory, *Die Methode der geodätischen Äquidistanten und das Problem von Lagrange*, Acta Mathematica, vol. 47 (1926), pp. 199–236.

† Here and elsewhere we adopt the convention that the repetition of a subscript indicates a summation with respect to that subscript.

The generalized Weierstrassian function $F_1(x, r)$ can here be conveniently defined as the sum of the principal minors of order $m-1$ of the determinant (2.5) divided by the sum of the squares of the r_i's.*

For $(x, \dot{x})$ on g we now form the characteristic determinant $D(\lambda)$ of the determinant (2.5). We see that one of the roots of $D(\lambda)=0$ must be zero on account of (2.5). The remaining roots must be positive on account of (2.3). A consequence is that the coefficient of λ in $D(\lambda)$ must be negative. But this coefficient is

$$-(\dot{x}_1^2 + \cdots + \dot{x}_m^2)F_1(x, \dot{x})$$

so that F_1 is thereby proved to be positive along g.

3. **The invariance of our hypotheses under admissible transformations.** By an admissible transformation from the variables (x) to a new set of variables (z) will be understood a transformation of the form

$$x_i = \phi_i(z_1, \cdots, z_m) \qquad (i = 1, 2, \cdots, m), \tag{3.1}$$

where the functions ϕ_i are of class C''' in their arguments and where the jacobian of the ϕ_i's with respect to the z_i's is never zero. We obtain a new integrand $G(z, r)$ by setting

$$G(z_1, \cdots, z_m, r_1, \cdots, r_m) = F\left(\phi_1, \cdots, \phi_m, \frac{\partial \phi_1}{\partial z_i} r_i, \cdots, \frac{\partial \phi_m}{\partial z_i} r_i\right). \tag{3.2}$$

The positive regularity of F along g implies the positive regularity of G along the transform of g, say γ. That is, it follows from (2.3) and (3.2) that

$$G_{ij}(z, \dot{z})\zeta_i\zeta_j > 0 \qquad (i, j = 1, 2, \cdots, m) \tag{3.3}$$

for $(z, \dot{z})$ on γ, and for (ζ) any set not (0) nor proportional to $(\dot{z})$. The proof of this statement appears readily if one obtains the second partial derivatives of G in terms of those of F.

That G possesses the same homogeneity property as does F, follows from (3.2).

4. **A transformation carrying g into a straight line.** The following lemma will greatly simplify our future work.

LEMMA. *There exists an admissible transformation T of the variables (x) into variables (z) under which the image of g is a segment of the z_1 axis.*

Suppose g is given in terms of its arc length in the form

$$x_i = h_i(s) \qquad (i = 1, 2, \cdots, m). \tag{4.1}$$

* See Bliss, second paper cited, p. 378, for a differently phrased but equivalent definition.

We now state the following preliminary proposition.

(A). Let the direction cosines of the direction tangent to g be represented by a point P_1 on a unit $(m-1)$-sphere S with center at the origin O. As P_1 moves continuously on S there can be chosen $m-1$ other points $P_2, \cdots, P_m$ on S which vary continuously with s, and are such that the points

$$O, P_1, \cdots, P_m$$

for no value of s lie on the same $(m-1)$-plane.

The proof of this proposition may be given in so many ways that it can very well be left to the reader.

Granting the truth of (A) consider the determinant b whose ith column gives the coördinates of the point P_i. We now replace the elements of the last $m-1$ columns of b by analytic approximations taken so close that the new determinant $a(s)$ is never zero. Let $a_{ij}(s)$ be the ijth element of $a(s)$. Our transformation T can now be defined as follows:

$$x_i = h_i(z_1) + a_{ik}(z_1)z_k \qquad (k = 2,3, \cdots, m; \; i = 1,2, \cdots, m) \tag{4.2}$$

where z_k is taken in the neighborhood of $z_k = 0$, and z_1 may take on the same values as s. We have for $z_k = 0$

$$\frac{D(x_1, \cdots, x_m)}{D(z_1, \cdots, z_m)} = a(z_1) \neq 0, \tag{4.3}$$

so that the lemma follows readily.

5. The integral in non-parametric form. According to the result of the last section there exists an admissible transformation T under which g is carried into a segment γ of the z_1 axis. We suppose that the positive sense along g corresponds to the positive sense of the z_1 axis. Suppose that under T the new integrand G is given by (3. 2). We now set

$$(z_1, \cdots, z_m) = (x, y_1, \cdots, y_n) = (x, y), \qquad n = m - 1,$$

and further set

$$\begin{aligned} f(x,y,p) &= f(x, y_1, \cdots, y_n, p_1, \cdots, p_n) \\ &= G(x, y_1, \cdots, y_n, 1, p_1, \cdots, p_n) \end{aligned} \tag{5.1}$$

for (x, y) neighboring γ and for any set (p).

For ordinary curves neighboring γ for which $\dot{z}_1 > 0$, our integral J now takes the non-parametric form

$$J = \int_a^b f(x, y, y')dx. \tag{5.2}$$

It will be convenient to set

$$\overset{0}{f}_{p_ip_j} = A_{ij}(x), \quad \overset{0}{f}_{p_iv_j} = B_{ij}(x), \quad \overset{0}{f}_{v_iv_j} = C_{ij}(x),$$

where the partial derivatives involved are evaluated for $(x, y, p) = (x, 0, 0)$, as indicated by the superscript zero.

We shall now reduce the regularity condition (3.3) *to the condition that*

$$A_{ij}(x)\zeta_i\zeta_j > 0 \qquad (i,j = 1, \cdots, n) \tag{5.3}$$

for every set $(\zeta) \neq (0)$.

Because G is homogeneous in (r) we have

$$G_{ij}(z,\dot{z})\dot{z}_j = 0. \tag{5.4}$$

In particular along γ we have

$$\dot{z}_2 = \cdots = \dot{z}_m = 0,$$

so that (5.4) reduces along γ to

$$G_{i1}(z,\dot{z})\dot{z}_1 = 0. \tag{5.5}$$

Since $\dot{z}_1 \neq 0$ along γ we infer from (5.5) that the elements in the first row and column of the matrix of the form (3.3) are zero along γ. With this understood the condition (3.3) taken along γ reduces with the aid of (5.1) to (5.3).

A particular consequence of (5.3) is that

$$|A_{ij}| \neq 0 \tag{5.6}$$

along γ.

6. **The invariance of the order of a conjugate point.** Let O be a point of g and (λ) the set of the m direction cosines of the direction of g at O. Let (λ) represent a point P on a unit $(m-1)$-sphere S with center at the origin. Let (α) be a set of parameters in an admissible* representation of S neighboring P, or, in other words, of the directions neighboring (λ). Suppose that $(\alpha) = (\alpha^0)$ corresponds to P. Let s be the arc length on the extremals issuing from O, measuring s from O in the positive senses of the extremals, and confining s to the set of values it has on g. It is well known that the extremals that issue from O with the directions determined by (α) can be represented in the form

$$x_i = x_i(s,\alpha) \qquad (i = 1,2, \cdots, m), \tag{6.1}$$

* That is, a representation in which the coördinates (x) are to be functions of class C''' of $m-1$ parameters (α) of such sort that not all of the jacobians of $m-1$ of the coördinates with respect to the parameters (α) are zero.

where the functions (6.1) and their first partial derivatives as to s are of class C''.

The jacobian of the functions (6.1), namely

$$(6.2)\qquad \Delta(s) = \frac{D(x_1, \cdots, x_m)}{D(s, \alpha_1, \cdots, \alpha_n)}, \quad (\alpha) = (\alpha^0), \quad n = m - 1,$$

is clearly zero for $s=0$. That the zero of $\Delta(s)$ at $s=0$ is isolated can also be readily proved.* We come now to an important definition.

The points on g, not O, at which $\Delta(s)$ vanishes are called the conjugate points of O, and the order of the vanishing of $\Delta(s)$ at each of these points the order of that conjugate point.

We turn next to the non-parametric integral (5.2), and suppose the J. D. E.† set up in the non-parametric form corresponding to the extremal γ. Suppose that under the transformation T the point O on g corresponds to the point $x=a$ on γ. Let $D(x)$ be an n-square determinant whose n columns give respectively n linearly independent solutions of the J. D. E. which vanish at $x=a$. Any other such determinant will be a non-vanishing constant times $D(x)$. We shall first prove the following

LEMMA. *Under the transformation T the zeros of $\Delta(s)$ on g correspond to the zeros of $D(x)$ on γ, and the orders of corresponding zeros are the same.*

By virtue of the transformation T the extremals given by (6.1) can be represented in the space of the new variables (x, y) in the form

$$(6.3)\qquad x = h(s, \alpha),$$

$$(6.4)\qquad y_i = k_i(s, \alpha) \qquad (i = 1, 2, \cdots, n = m - 1),$$

where it appears that the jacobian

$$(6.5)\qquad \Delta_1(s) = \frac{D(h, k_1, \cdots, k_n)}{D(s, \alpha_1, \cdots, \alpha_n)}, \qquad (\alpha) = (\alpha^0),$$

equals the jacobian $\Delta(s)$ divided by the non-vanishing jacobian of the transformation T. Thus $\Delta(s)$ and $\Delta_1(s)$ vanish together and to the same orders.

Now γ coincides with the x axis so that $h_s(s, \alpha^0) \neq 0$ on γ. We can thus solve (6.3) for s as a function of x and (α), substitute the result in (6.4), and so represent the extremals (6.1) in the form

* See Mason and Bliss, loc. cit., p. 447. It is easy to see that the determinant which multiplies U on p. 447 is not zero when (u, v) are a pair of parameters admissible in our sense.

† We write J.D.E. for the Jacobi differential equations.

$$y_i = w_i(x, \alpha) \qquad (i = 1, 2, \cdots, n). \tag{6.6}$$

The jacobian

$$\Delta_2(x) = \frac{D(w_1, \cdots, w_n)}{D(\alpha_1, \cdots, \alpha_n)}, \qquad (\alpha) = (\alpha^0), \tag{6.7}$$

is related to the jacobian $\Delta_1(s)$ in that

$$h_s(s, \alpha^0)\Delta_2(x) \equiv \Delta_1(s), \tag{6.8}$$

where x and s correspond under (6.3). Hence $\Delta_2(x)$ and $\Delta(s)$ vanish at corresponding points and to the same orders.

Finally, as is well known, the jacobian $\Delta_2(x)$ consists of columns which satisfy the J. D. E. The elements of these columns all vanish at $x = a$, and lastly are linearly independent since $\Delta_2(x) \not\equiv 0$. Hence

$$\Delta_2(x) \equiv cD(x), \tag{6.9}$$

where c is a non-vanishing constant. The lemma follows directly.

The following statement is an easy corollary.

Under admissible space transformations conjugate points correspond to conjugate points and the orders of conjugate points are preserved.

7. A lemma on the order of a conjugate point. In terms of the determinant $D(x)$ of §6 we have the following lemma.

LEMMA. *If $D(x)$ vanishes to the rth order at $x = a$ the rank of $D(a)$ is $n - r$.*

For simplicity suppose $a = 0$. Let $a_{ij}(x)$ be the general element of $D(x)$. Let $\mathbf{b}(x)$ be a matrix whose general element is

$$b_{ij}(x) = a_{ij}(0) + a_{ij}'(0)x. \tag{7.1}$$

According to the theory of λ matrices (here $\lambda = x$) there exist n-square non-singular matrices $\mathbf{c}$ and $\mathbf{d}$ of constant elements such that*

$$\mathbf{c}\mathbf{b}(x)\mathbf{d} = \mathbf{e}(x) \tag{7.2}$$

where $\mathbf{e}(x)$ is a matrix all of whose elements are identically zero except those in the principal diagonal.

With $\mathbf{c}$ and $\mathbf{d}$ so determined let us transform the dependent variables (η) of the J. D. E. into variables (w) by setting

$$w_i = c_{ij}\eta_j \qquad (i, j = 1, 2, \cdots, n), \tag{7.3}$$

* See Bôcher, *Introduction to Higher Algebra*, Chapter XX.

where c_{ij} is the general element of $\mathbf{c}$. The transformed J. D. E. will have for solutions the columns of the matrix

$$\mathbf{c}\mathbf{a}(x). \tag{7.4}$$

It will also have for solutions the columns of the matrix

$$\mathbf{c}\mathbf{a}(x)\mathbf{d} = \mathbf{g}(x). \tag{7.5}$$

It follows from (7.1), (7.2), and (7.5), that

$$e_{ij}(x) = g_{ij}(0) + g_{ij}'(0)x. \tag{7.6}$$

Further I say that none of the elements in the principal diagonal of $\mathbf{e}(x)$ are identically zero. For if so the corresponding column of $\mathbf{g}(x)$ would represent a solution of the transformed J. D. E. whose coördinates as well as their derivatives would all be zero at $x=0$, and so would be identically zero. Hence $|g(x)|$ would be identically zero contrary to (7.5).

Thus the elements in the principal diagonal of $\mathbf{e}(x)$ if suitably reordered must be of the form

$$a_1x, \cdots, a_kx, \quad a_{k+1}x + b_{k+1}, \cdots, a_nx + b_n$$

where k is an integer from 1 to n inclusive, and none of the constants

$$a_1, \cdots, a_k, b_{k+1}, \cdots, b_n$$

are zero.

By virtue of the relations (7.6) between $\mathbf{e}(x)$ and $\mathbf{g}(x)$ it appears then that $|g(x)|$ vanishes to the order k at $x=0$, and has the rank $n-k$ when $x=0$. From (7.5) it follows that $|g(x)|$ and $|a(x)|$ vanish to the same order at $x=0$, and have the same rank when $x=0$. Since $D(x)$ equals $|a(x)|$ it vanishes to the order k at $x=0$, and has the rank $n-k$ when $x=0$. Thus the lemma is proved.

8. Two additional lemmas. The preceding lemma leads to the following.

LEMMA 1. *If $x=b$ is a conjugate point of $x=a$ of the rth order there will be just r linearly independent solutions of the J. D. E. which vanish at both $x=b$ and $x=a$.*

Any solution which vanishes at a must be linearly dependent on the columns of $D(x)$, and conversely any such linear combination will vanish at a. If now we turn to b and write down the n conditions that a linear combination of the columns of $D(x)$ vanish at b we see from the preceding lemma that there are just r linearly independent combinations that satisfy the n conditions. The lemma follows directly.

We note the following consequence.

If $x=b$ is a conjugate point of $x=a$ of the rth order, then $x=a$ is a conjugate point of $x=b$ of the rth order.

For future use Lemma 1 will be generalized as follows. The preceding proofs hold for the generalization.

LEMMA 2. *Let $D(x)$ be a determinant whose n columns are n linearly independent solutions of the J. D. E. If $D(x)$ vanishes to the rth order at $x=c$ the number of linearly independent solutions dependent on the columns of $D(x)$ which vanish at $x=c$ equals r.*

9. **The fundamental quadratic form.** Our extremal g has been transformed into a segment γ of the x axis in the space of the variables

$$(x, y) = (x, y_1, \cdots, y_n).$$

We suppose γ bounded by the points $x=a$ and $x=b$ $(a<b)$.

Let us cut across the portion of γ for which $a<x<b$ with p successive n-planes t_i, perpendicular to γ, cutting γ respectively in points at which $x=x_i$. These n-planes divide γ into $p+1$ successive segments. Suppose these n-planes are placed so near together that no one of these $p+1$ segments contains a conjugate point of its initial end point. Let P_i be any point on t_i near γ. Let the points on γ at which $x=a$ and $x=b$, respectively, be denoted by A and B.

If the points

$$A, P_1, \cdots, P_p, B \tag{9.1}$$

are sufficiently near γ they can be successively joined by extremal segments neighboring γ. Let the resulting broken extremal be denoted by E. Let (u) be the set of $\mu=pn$ variables of which the first n are the coördinates (y) of P_1, the second those of P_2, and so on, until finally the last n are the coördinates (y) of P_p. The value of the integral J taken along E will be a function of the variables (u), and will be denoted by

$$J(u_1, \cdots, u_\mu) = J(u).$$

When $(u)=(0)$, E will become γ. Considerations similar to these taken up by the author* for the case $n=1$ show that $J(u)$ is of class C'' for (u) near (0). It clearly has a critical point when $(u)=(0)$. We come next to the terms of second order in $J(u)$, that is, to the fundamental form

$$_0J_{ij}u_iu_j \qquad (i,j = 1,2, \cdots, \mu), \tag{9.2}$$

* See Morse, loc. cit., pp. 217–219.

where ${}_0J_{ij}$ is the partial derivative of J with respect to u_i and u_j, and is evaluated for $(u)=(0)$, as indicated by the subscript 0.

Our first problem is to determine the rank and signature of this form.

10. The fundamental form and the J. D. E. As is conventional we set

$$(10.1)\qquad \Omega(x,\eta,\eta') = A_{ij}\eta_i'\eta_j' + 2B_{ij}\eta_i'\eta_j + C_{ij}\eta_i\eta_j \qquad (i,j = 1, \cdots, n)$$

where the arguments (x, y, p) in the partial derivatives of $f(x, y, p)$ are $(x, 0, 0)$. The J. D. E. corresponding to the extremal γ are

$$(10.2)\qquad \Omega_{\eta_i} - \frac{d\Omega_{\eta_i'}}{dx} = 0 \qquad (i = 1, \cdots, n).$$

It will be convenient to suppose that the points (9.1) lie not only in the space (x, y) but also in the space (x, η) of the J. D. E. A curve representing a solution of the J. D. E. will be called a *secondary extremal*. We see that successive points of (9.1) can be joined by secondary extremals. The set of these secondary extremals will form a broken secondary extremal which we may regard as *determined* by the set (u) of §9. With this understood we can prove the following

LEMMA. *The fundamental form may be represented as follows:*

$$(10.3)\qquad {}_0J_{ij}u_iu_j = \int_a^b \Omega(x,\eta,\eta')dx \qquad (i,j = 1, \cdots, \mu),$$

where (η) stands for the set of n functions of x which give the coördinates other than x of a point on the broken secondary extremal determined by (u).

To prove this let e be a variable neighboring $e=0$, and suppose the integral J evaluated along the broken extremal E *determined* when the set (u) is replaced by

$$(10.4)\qquad (eu_1, \cdots, eu_\mu).$$

This value of J is given by the function

$$(10.5)\qquad J(eu_1, \cdots, eu_\mu).$$

If now we differentiate the function (10.5) twice with respect to e, and then put $e=0$, we get the left hand member of (10.3).

If on the other hand we take the integral J along the curve E, differentiate under the integral sign twice with respect to e, and then set $e=0$, we will get the right hand member of (10.3). Here the functions η_i will be the partial derivatives with respect to e of the coördinates y_i of the point on E, and will thus represent a broken secondary extremal. If the conditions that E be the broken extremal determined by (10.4) be differentiated with respect to e it will follow that the functions (η) in (10.3) give that particular broken secondary extremal which is determined by (u).

The identity (10.3) follows at once.

11. The rank of the fundamental form. The theorem here is the following.

THEOREM 1. *If $x=b$ is a conjugate point of $x=a$ of the rth order the rank of the matrix of the form*

$$ {}_0J_{ij}u_iu_j = Q \qquad (i,j = 1,2,\cdots,\mu) \tag{11.1} $$

is $\mu - r$. If $x=b$ is not conjugate to $x=a$ the rank is μ.

The proof depends upon the following statements, (A), (B), and (C).

(A). The rank of Q will be $\mu - r$ if the linear equations

$$ Q_{u_\alpha} = 0 \qquad (\alpha = 1,2,\cdots,\mu) \tag{11.2} $$

have just r linearly independent solutions, and conversely.

(B). The equations (11.2) are necessary and sufficient conditions on a set (u) for such a set to determine a broken secondary extremal E' without corners at the points (9.1).

To prove (B) suppose (11.2) holds for some given set (u). Suppose u_α in (11.2) is the coördinate y_h of the point P_k in the set (9.1). We differentiate the integral in (10.3) with respect to u_α, integrate by parts, and obtain the following result, holding for the given set (u):

$$ Q_{u_\alpha} = \lim_{x=x_k^-} \Omega_{\eta_h'} - \lim_{x=x_k^+} \Omega_{\eta_h'} = 0. \tag{11.3} $$

In the limiting process the arguments in $\Omega_{\eta_h'}$ are those on E'. If now we hold k fast and let $h=1,2,\cdots,n$, (11.3) gives us n equations corresponding to the point $x=x_k$. It follows from (5.6) that these n equations are compatible only if E' has no corner at $x=x_k$.

Conversely equations (11.3) clearly hold for a given set (u) if the secondary extremal determined by (u) has no corners.

Thus the statement (B) is proved.

(C). A set of broken secondary extremals E' are linearly dependent or not according as the corresponding sets (u) are linearly dependent or not.

For if the coördinates (η) on q such curves E' satisfy a linear relation identically in x, then upon setting x successively equal to

$$ x_1,\cdots,x_p \tag{11.4} $$

we see that the corresponding sets (u) satisfy the same linear relation.

Conversely if a linear combination of q of the sets (u) gives the null set (0), and if these q sets determine q broken secondary extremals E' then the same linear combination of the coördinates (η) on the q curves E' will be zero at each point at which x has the values

$$a, x_1, \cdots, x_p, b, \tag{11.5}$$

and hence be identically zero, since no point of (11.5) is conjugate to its predecessor.

Thus (C) is proved.

It follows from Lemma 1 of § 8 that $x=b$ is a conjugate point of $x=a$ of the rth order, when and only when there are just r linearly independent secondary extremals which pass through the points $x=a$ and $x=b$ on the x axis. The theorem now follows from statements (C), (B), and (A).

12. The type number of the fundamental form. A real, symmetric, non-singular quadratic form,

$$Q = a_{\alpha\beta}u_\alpha u_\beta \qquad (\alpha, \beta = 1, 2, \cdots, \mu), \tag{12.1}$$

can be transformed by a suitable real, linear, non-singular transformation into a form

$$-z_1^2 - \cdots - z_q^2 + z_{q+1}^2 + \cdots + z_\mu^2.$$

The integer q is called the *type number* of the given form.

LEMMA 1. *If the form Q is negative at each point of an r-plane π through the origin, excepting the origin, then the type number of Q is at least r.*

By a suitable non-singular linear transformation into variables (w) we first take the r-plane π into the r-plane

$$w_{r+1} = w_{r+2} = \cdots = w_\mu = 0. \tag{12.2}$$

After this transformation suppose Q takes the form

$$b_{\alpha\beta}w_\alpha w_\beta \qquad (\alpha, \beta = 1, \cdots, \mu). \tag{12.3}$$

On (12.2) Q will become a negative definite form

$$b_{\alpha\beta}w_\alpha w_\beta \qquad (\alpha, \beta = 1, \cdots, r). \tag{12.4}$$

Let A_m be that principal minor of the determinant of the form (12.3) which is obtained by striking out all but the first m rows and columns. Let $A_0=1$.

Note that A_r cannot be zero since the form (12.4) is definite. According to the well known theory of quadratic forms the variables $w_1, \cdots, w_r$ can be reordered so that no two consecutive members of the sequence

$$A_0, A_1, \cdots, A_r \tag{12.5}$$

are zero. So ordered, the number of changes of sign in (12.5) is r. The remaining variables of the set (w) can now be so reordered among themselves that no two consecutive numbers of the sequence

$$A_0, A_1, \cdots, A_\mu \tag{12.6}$$

are zero. So ordered, the number of changes of sign in (12.6) gives the type number of Q. Thus this type number is at least r and the lemma is proved.

LEMMA 2. *Suppose the coefficients of the form Q are continuous functions of a parameter t. If for $t=0$ the matrix a is of rank $\mu-r$, but is of rank μ for other values of t, then as t passes through $t=0$ the type number of Q can change in absolute value by at most r.*

According to the theory of symmetric quadratic forms the type number of Q will equal the number of λ roots of the characteristic matrix of a which are negative. These λ roots are continuous functions of the elements $a_{\alpha\beta}$ and hence of t. If a is of rank $\mu-r$ at $t=0$, just r of these roots will be zero. Hence r at most of these roots will change sign. The lemma follows directly.

LEMMA 3. *If $x=b$ is not conjugate to $x=a$, the type number of the form,*

$$Q = {}_0J_{ij}u_iu_j \qquad (i,j = 1, \cdots, \mu) \tag{12.7}$$

is at most equal to the sum of the orders of the conjugate points of $x=a$ preceding $x=b$.

Consider the set of x coördinates

$$a, x_1, \cdots, x_p, b \tag{12.8}$$

of §9. To prove the lemma we shall hold a fast but vary the remaining coördinates in (12.8), using, however, only sets (12.8) which are admissible in the sense of §9.

Suppose the first conjugate point of $x=a$ is $x=a_1$. We start our variation of the coördinates (12.8) with a choice of $b<a_1$. For this choice of b and under the condition (3.3) the x axis from $x=a$ to $x=b$ furnishes a proper minimum* for the integral J, so that the type number q will be zero.

Suppose that a_1 is a conjugate point of a of the rth order. When $b=a_1$ the rank of the matrix of Q, according to Theorem 1, will be $\mu-r$. According to Lemma 2 of this section an increase of b through a_1 will be accompanied by an increase of the type number of Q at most r.

It is clear that a set of coördinates (12.8) for which $b<a_1$ can be varied into any other admissible set with the same a and the same p. One way of

* See Bliss, loc. cit., second paper.

doing this is first to increase x_1 to its final position keeping the following coördinates in (12.8) so near x_1 as to be admissible, then while holding a and x_1 fast, to increase x_2 to its final position, keeping the following coördinates in (12.8) so near x_2 as to be admissible, and so on.

As b thereby increases through each point conjugate to a prior to b's final position, there will be an accompanying increase of the type number of Q at most equal to the order of the conjugate point thereby passed. The lemma follows at once.

13. The theorem on the type number of Q. By an *auxiliary curve* g will be understood a finite succession of secondary extremals joining $(a, 0)$ to $(b, 0)$, whose points of junction are all on the x axis, and not all of which are segments of the x axis. An auxiliary curve g will intersect the successive hyperplanes

$$x = x_1, \cdots, x = x_p$$

in successive points P_i which we may suppose given in (9.1). But the points (9.1) determine the set (u) of §9. A set (u) thereby determined by an auxiliary curve g will be termed an *auxiliary set* (u).

No auxiliary set (u) *is null.*

This follows from the fact that each point of the set (11.5) precedes its predecessor's conjugate point. Further a set of auxiliary curves g will be linearly dependent if and only if the corresponding sets (u) are linearly dependent.

The following theorem is fundamental.

THEOREM 2. *In case $x = b$ is not conjugate to $x = a$ the type number q of the form*

$$Q = {}_0J_{ij}u_iu_j \qquad (i,j = 1, \cdots, \mu) \tag{13.1}$$

will equal the sum of the orders of the conjugate points of a between $x = a$ and $x = b$.

Suppose the conjugate points of $x = a$ prior to $x = b$ have x coördinates

$$a_1 < a_2 < \cdots < a_s, \tag{13.2}$$

and their respective orders are

$$r_1, r_2, \cdots, r_s. \tag{13.3}$$

According to Lemma 3, §12, the type number q is such that

$$q \leqq r_1 + r_2 + \cdots + r_s. \tag{13.4}$$

We shall use Lemma 1 of §12 to prove that (13.4) must be an equality.

Corresponding to the ith conjugate point a_i, Lemma 1, §8, affirms the existence of r_i linearly independent secondary extremals,

$$h_j^{(i)} \qquad (j = 1, \cdots, r_i; \quad i = 1,2, \cdots, s), \tag{13.5}$$

which pass from $(a, 0)$ to $(a_i, 0)$. From the curves (13.5) we can now form r_i auxiliary curves,

$$g_j^{(i)}, \tag{13.6}$$

which are identical with the curves (13.5) as x varies from a to a_i, and are identical with the x axis as x varies from a_i to b.

Let

$$(u)_j^{(i)} \qquad (j = 1, \cdots, r_i; \quad i = 1,2, \cdots, s) \tag{13.7}$$

be the auxiliary sets (u) determined respectively by the curves (13.6). Our theorem will follow once we have proved the two following statements.

(A). *The* $r_1+r_2+ \cdots +r_s$ *sets* (u) *in* (13.7) *are linearly independent.*

(B). *Any linear combination* $(u)\neq 0$ *of the sets* (13.7) *will make the fundamental form* Q *negative.*

To prove (A) it will be sufficient to prove the auxiliary curves (13.6) are linearly independent.

If there were any linear relation between the curves (13.6) which in particular involved those for which $i=s$. this would entail a linear relation among the curves for which $i=s$ alone, since the remaining curves of the set (13.6) coincide with the x axis between $x=a_{s-1}$ and $x=a_s$, and the curves for which $i=s$ do not. Similarly there can be no linear relation between the curves (13.6) which involves the curves for which $i=s-1$, and so on down to the curves for which $i=1$.

Thus (A) is proved.

We come to the proof of (B). Let (u) be any linear combination of the sets (13.7) not $(u)=(0)$. Let g be the auxiliary curve obtained by taking the same linear combination of the η-coördinates of the curves of the set (13.6). Let E' be the broken secondary extremal determined by (u) as in §10.

Without loss of generality we can suppose that none of the points at which x equals $x_1, \cdots, x_p$ are conjugate to $x=a$. For if such were not the case a slight displacement of these points would move them from positions conjugate to $x=a$, and would not alter the type number of Q. With this understood we see that the corners of g cannot coincide with any of the corners of E', since the corners of g are at the conjugate points of $x=a$, and the corners of E' are not.

I say now that E' will make the integral in (10.3) negative.

For g will make that integral zero since g is composed of secondary extremals with corners on the x axis.* But there will be some component secondary extremal of E', say E'', without any corner, which joins the end points of a portion g' of g with a corner. Under the condition (5.3), E'' will give a smaller value to the integral than g' so that E' must make the integral negative. Statement (B) now follows from (10.3).

To return to the theorem we note that the set of linear combinations (u) of the sets (13.7) may be regarded as the set of points on an

$$r_1 + r_2 + \cdots + r_s$$

plane through the origin in a space of the points (u). On this plane Q is negative except for the origin. The theorem follows from Lemma 1 of §12.

We saw in §8 that if $x=b$ is not conjugate to $x=a$, then $x=a$ is not conjugate to $x=b$. By interchanging the rôles of b and a in the preceding proof we can prove the following:

The type number of the form Q of Theorem 2 is also equal to the sum of the orders of the conjugate points of $x=b$, between $x=a$ and $x=b$.

We have thus proved the following:

If the point $x=b$ on γ is not conjugate to $x=a$, then $x=a$ is not conjugate to $x=b$, and the sum of the orders of the conjugate points of $x=a$ between $x=a$ and $x=b$ equals the sum of the orders of the conjugate points of $x=b$ between $x=a$ and $x=b$.

II. The fundamental form for the case of one variable end point

14. The hypotheses and definition of the form. Let O be any point in the region R of §2. Except for O suppose that R is covered in a one-to-one manner by the field of extremals through O, and that each extremal through O passes out of R before a first conjugate point of O is reached. We assume that F is *positively regular* along each extremal through O. In R let Σ be any closed regular† $(m-1)$-manifold of class C'''. We assume that $F(x, r)$ is positive for each point (x) on Σ, and for the direction (r) of the field at that point, or in particular if O lies on Σ, for all directions r at O. This concludes our statement of hypotheses.

We shall next set up the fundamental quadratic form. Let

$$x_i = h_i(v_1, \cdots, v_n) \qquad (i = 1, 2, \cdots, m)$$

* See Bolza, *Variationsrechnung*, p. 62, first formula.

† That is, a manifold the neighborhood of each point of which can be admissibly represented. See footnote §6.

be an admissible representation of Σ in the neighborhood of a point P of Σ. At P suppose $(v)=(v^0)$. Let the value of the integral J taken along the field extremal from O to the point (v) on Σ be denoted by

$$I(v_1, \cdots, v_n) = I(v) \qquad (n = m-1). \tag{14.1}$$

The function $I(v)$ will be of class* C'' in (v) for (v) neighboring (v^0). Suppose $I(v)$ has a critical point when $(v)=(v^0)$. The conditions that $I(v)$ have a critical point are the conditions that Σ cut the extremal g from O to P transversally. They are that

$$I_{v_j} = F_{r_i}(x,\dot{x})\frac{\partial h_i}{\partial v_j} = 0 \quad (j = 1,2,\cdots,m-1;\ i = 1,2,\cdots,m), \tag{14.2}$$

where (x) gives the coördinates of P, and $(\dot{x})$ is the direction of the field at P.

If then Σ cuts the extremal g transversally the terms of the first order in $I(v)$ at (v^0) all vanish. We next turn to the terms of the second order in $I(v)$, and in particular are concerned with the rank and type of the fundamental quadratic form,

$$Q = {}_0I_{ij}(v_i - v_i^0)(v_j - v_j^0) \qquad (i,j = 1,2,\cdots,n), \tag{14.3}$$

where the subscript zero indicates that the second partial derivatives I_{ij} are evaluated at $(v)=(v^0)$. This leads us to a study of focal points.

15. **Focal† points and their orders.** It is known that under our hypotheses‡ there exists a family of extremals neighboring g, passing through the points Q on Σ neighboring P in the senses of the extremals from O to Q, and cut transversally by Σ. They may be represented in the form

$$x_i = w_i(s,v) \qquad (i = 1,2,\cdots,m) \tag{15.1}$$

where s is the arc length taken as positive when measured along these extremals from Σ in their positive senses, and where the functions w_i and their first partial derivatives with respect to s are of class C'' for (v) neighboring (v^0), and for such values of s as give points neighboring g. The jacobian

$$\Delta(s) = \frac{D(w_1,\cdots,w_m)}{D(s,v_1,\cdots,v_n)}, \quad n = m-1, \quad (v) = (v^0) \tag{15.2}$$

does not vanish at $s=0$.

* This follows readily upon computing the first partial derivatives of $I(v)$ as in (14.2).

† An interesting paper on focal points has been written by M. B. White, *The dependence of focal points upon curvature for problems of the calculus of variations in space*, these Transactions, vol. 13 (1912), pp. 175–198.

‡ See Mason and Bliss, loc. cit., p. 448.

The points on g at which $\Delta(s)=0$ are called the focal points of P on Σ, and the order of vanishing of $\Delta(s)$ at each of these points the order of that point.

A necessary fact in what follows is that g cannot be tangent to Σ at P, as well as cut transversally by Σ at P. For the transversality conditions would then lead to the equation

$$F_{r_j}(x,\dot{x})\dot{x}_j = 0. \tag{15.3}$$

where (x) gives the coördinates of P, and $(\dot{x})$ the direction of g at P. But the left member of (15.3) equals F. Thus F would be zero contrary to our hypotheses on Σ.

16. The reduction to non-parametric form. As in §5, so here, the extremal g can be carried by an admissible transformation T into a segment γ of the x axis of a space of coördinates $(x, y_1, \cdots, y_n)$. Without loss of generality we can also suppose that T carries Σ into a manifold S that is orthogonal to γ at γ's final end point. Suppose g's end points O and P correspond to $x=a$ and $x=b$ on γ, respectively, with $a<b$. In the neighborhood of its intersection with γ, S can be represented in the form

$$x = \bar{x}(u_1, \cdots, u_n), \qquad y_i = u_i \qquad (i = 1,2, \cdots, n), \tag{16.1}$$

where $\bar{x}(u)$ is of class C''' for (u) neighboring (0).

The transformation T establishes a one-to-one correspondence between the points determined by (v) on Σ, and the points determined by (u) on S, at least for (u) neighboring (0). This correspondence will take the form

$$v_i = h_i(u_1, \cdots, u_n)$$

where the functions $h_i(u)$ are of class C''' for (u) neighboring $(u)=(0)$.

By virtue of the transformation T the family of extremals neighboring γ, cut transversally by S, can be represented in terms of the variables s and (v) of (15.1) in the form

$$x = h(s,v), \quad y_i = k_i(s,v) \qquad (i = 1,2, \cdots, n). \tag{16.2}$$

As in the proof of the lemma in §6, so here, we can take x instead of s as a parameter along the extremals. We shall also replace the v_i's by their values $h_i(u)$ in term of the u_i's. The family (16.2) will then take the form

$$y_i = a^{(i)}(x,u) \qquad (i = 1,2, \cdots, n). \tag{16.3}$$

As in the proof of the lemma of §6, so here, a consideration of our successive transformations leads to the following results.

The focal points of Σ relative to the end point of g on Σ correspond under T to the zeros on γ of the jacobian

$$\frac{D(a^{(1)}, \cdots, a^{(n)})}{D(u_1, \cdots, u_n)} = \Delta_1(x), \quad (u) = (0), \tag{16.4}$$

and the orders of these focal points equal the orders of the corresponding zeros of $\Delta_1(x)$.

Under admissible space transformations focal points correspond to focal points, and their orders are preserved.

The columns of the jacobian (16.4) are solutions of the J.D.E. We shall find what conditions these columns satisfy at $x=b$.

The manifold $x=\bar{x}(u)$, $y_i=u_i$, cuts the family (16.3) transversally. We accordingly have the following identities in (u):

$$(f - p_h f_{p_h})\bar{x}_{u_i} + f_{p_i} \equiv 0 \qquad (i = 1,2, \cdots, n), \tag{16.5}$$

in which we must set

$$x = \bar{x}(u), \ y_i = u_i, \ p_h = a_x^{(h)}[\bar{x}(u), u] \qquad (h = 1, \cdots, n).$$

Denote the partial derivatives of $\bar{x}(u)$ with respect to u_i and u_j by $\bar{x}_{ij}$. Now let $a_{ij}(x)$ be the general element of the jacobian (16.4). If we differentiate (16.5) with respect to u_j, and set $(u)=(0)$, we find that

$$f^0\bar{x}_{ij}(0) + B_{ik}(b)a_{kj}(b) + A_{ik}(b)a'_{kj}(b) = 0 \qquad (i,j,k = 1, \cdots, n), \tag{16.6}$$

where the superscript zero means that (x, y, p) is to be replaced by $(b, 0, 0)$. Further, differentiation of the identity

$$u_i \equiv a^{(i)}[\bar{x}(u), u]$$

with respect to u_i and u_j respectively yields the result

$$a_{ij}(b) = 1, \qquad i = j,$$
$$a_{ij}(b) = 0, \qquad i \neq j.$$

Consider now any linear combination of the columns of (16.4) of the form

$$\eta_k(x) = u_j a_{kj}(x) \qquad (k,j = 1,2, \cdots, n)$$

where the u_j's are constant. Such a set takes on the values $(u_1, \cdots, u_n)$ when $x=b$. If we multiply (16.6) by u_j and sum with respect to j, we find that (η) satisfies the n relations,

$$f^0\bar{x}_{ij}(0)u_j + B_{ik}(b)\eta_k(b) + A_{ik}(b)\eta_k'(b) = 0 \qquad (i = 1, \cdots, n). \tag{16.7}$$

We conclude with the following lemma.

LEMMA. *Any linear combination of the columns of* (16.4) *which takes on the values* $(u_1, \cdots, u_n)$ *when* $x=b$, *will satisfy the* n *relations* (16.7) *at* $x=b$, *and as a solution of the J.D.E. be uniquely determined by these values* (u) *at* $x=b$, *and by the relations* (16.7).

17. The rank of the fundamental form. Consider again the function $I(v)$ of §14 which gives the value of the integral along the extremals from O to Σ. In terms of the parameters (u) which represent S in the neighborhood of the end point of γ, $I(v)$ becomes a function

$$J(u_1, \cdots, u_n) = J(u).$$

The function $J(u)$ gives the value of the transformed integral along the extremals from the point $x=a$ on γ to the point on S at which $(y)=(u)$. The form

$$Q = {}_0J_{ij}u_iu_j \qquad (i,j = 1,2, \cdots, n),$$

in which the subscript 0 indicates that the partial derivatives are evaluated at $(u)=(0)$, will have the same rank and type number as the form (14.3).

We shall now prove the following lemma.

LEMMA. *The fundamental form can be expressed as follows:*

$$(17.1)\qquad {}_0J_{ij}u_iu_j = f^0\bar{x}_{ij}(0)u_iu_j + \int_a^b \Omega(x,\eta,\eta')dx \qquad (i,j = 1, \cdots, n)$$

where (η) *stands for the coördinates* $[\eta_1(x), \cdots, \eta_n(x)]$ *on the secondary extremal joining* $x=a$ *on* γ *to the point* $(\eta)=(u)$ *on the* n*-plane* $x=b$.

Let the family of extremals which join the point $x=a$ on γ to the point at which $(y)=(u)$ on S, be represented in the form

$$y_i = m^{(i)}(x,u) \qquad (i = 1,2, \cdots, n),$$

where (u) lies in the neighborhood of $(u)=(0)$. As in §10, so here, we consider the function

$$(17.2)\qquad J(eu_1, \cdots, eu_n) = \int_a^{\bar{x}(eu)} f[x,m(x,eu),m_x(x,eu)]dx,$$

where e is a variable neighboring $e=0$.

If the two members of (17.2) be differentiated twice with respect to e, and then e be set equal to zero, (17.1) will result with

$$\eta_i(x) = u_jm_{u_j}{}^{(i)}(x,0).$$

That $\eta_i(b)=u_i$ follows upon differentiating the identity

$$u_i = m^{(i)}[\bar{x}(u), u]$$

with respect to u_j, and setting $(u) = (0)$.

The lemma is thereby proved.

The rank of Q is given by the following theorem.

THEOREM 3. *If $x=a$ on γ is a focal point of the rth order of S relative to the point $x=b$ on γ, then the rank of the matrix of the form Q is $n-r$. If $x=a$ is not a focal point of S the form is non-singular.*

As in the proof of Theorem 1, so here, we see that the rank of Q is $n-r$ if the n linear equations

$$Q_{u_i} = 0 \qquad (i = 1, 2, \cdots, n) \tag{17.3}$$

have just r linearly independent solutions (u) and conversely.

If we differentiate the right hand member of (17.1) with respect to u_i, and perform the usual integration by parts, we find that when (17.3) holds for a set (u), the functions (η) of (17.1) which take on at $x=b$ these values (u), satisfy (16.7). According to the lemma of §16 these functions (η) must be dependent on the columns of the jacobian of (16.4). But the functions (η) of (17.1) vanish when $x=a$.

We see that the number of linearly independent solutions (u) of the equations (17.3) equals the number of linearly independent solutions of the J.D.E. which are dependent on the columns of Δ_1, and which vanish at $x=a$. This number equals the order of vanishing of $\Delta_1(x)$ at $x=a$ and thus equals the order of the focal point $x=a$. Thus the theorem is proved.

18. **The type number of the form Q.** We shall now prove the following lemma.

LEMMA. *If $x=a$ on γ is not a focal point of S relative to the point $x=b$ on γ, the type number of the form Q is at most equal to the sum of the orders of the focal points of S on γ between $x=a$, and $x=b$.*

To prove this lemma we vary a from a value near b, $a<b$, to its given value. For $x=a$ nearer $x=b$ than any focal point of S, the form Q will be positively definite, since (5.3) holds along γ. As $x=a$ decreases through a focal point of order r there will be an accompanying increase of the type number of Q at most r. This follows from Lemma 2 §12 and Theorem 3. The present lemma follows directly.

The following theorem will now be proved.

THEOREM 4. *If $x=a$ on γ is not a focal point of S relative to $x=b$ on γ, the type number q of the form Q will equal the sum of the orders of the focal points of S between $x=a$ and $x=b$.*

The proof is similar to the proof of Theorem 2. The following differences should however be noted.

By an *auxiliary curve* $y_i = \eta_i(x)$ we shall here mean a curve which is made of two secondary extremals as follows. The first shall consist of the x axis from $x = a$ to a point $x = a_1$ between $x = a$ and $x = b$. The second shall pass from the point $x = a_1$ on the x axis to a point on S at which $(y) = (u) \neq (0)$, and shall be dependent on the columns of Δ_1 in (16.4).

If in (17.1) (η) be taken as the coördinates along an auxiliary curve, and if u_i be taken as $\eta_i(b)$, the right hand member of (17.1) is zero. This is readily seen if the integral in (17.1) be written in the form

$$\frac{1}{2}\int_a^b (\eta_i \Omega_{\eta_i'} + \eta_{i'}\Omega_{\eta_{i'}})dx,$$

and the second sum be integrated by parts. The resulting integral vanishes. The right member of (17.1) then becomes

$$f^0 x_{ij}(0) u_i u_j + \tfrac{1}{2} u_i \Omega_{\eta_i'}[b, \eta(b), \eta'(b)]. \tag{18.1}$$

This expression may also be obtained by multiplying the left member of (16.7) by u_i, and summing with respect to i, and is accordingly zero.

It follows that any secondary extremal E that joins the end points of an auxiliary curve will make the right member of (17.1) negative, since E will give the integral in (17.1) a value that is smaller than that given by the auxiliary curve.

If $x = a_1$ is a focal point of order r there will be r linearly independent auxiliary curves with corners at $x = a_1$ on γ. We now continue as in the proof of Theorem 2, and complete the proof of the present theorem.

III. Theorems in the large on extremals through a fixed point

19. The function $I(v)$. We return to the manifold Σ of §14 and to the field of extremals through O. We assume *that O is not a focal point of* Σ. The locus of focal points at a finite distance from Σ will form in general a finite number of manifolds of dimensionality at most $m-1$. To take O therefore as a point which is not a focal point is to take the general point O.

The integral J taken along the extremals of the field from O to a point Q on Σ will be a function I of the point Q on Σ. The critical points of I are to be determined by expressing I in the neighborhood of each point P on Σ in terms of admissible local parameters (v). Since O is not a focal point on Σ these critical points will be non-degenerate, that is, points at which the fundamental quadratic form (14.3) will be of rank n. It follows that these critical points are isolated on Σ and therefore finite in number. Their type numbers will be given by Theorem 4 of §18, and will be called the type numbers of the corresponding extremals of the field.

We shall now prove the following theorem.

THEOREM 5. *Corresponding to a continuous variation of a point O' in the neighborhood of O, the end points of the extremals from O' to Σ which are cut transversally by Σ will vary continuously on Σ while the corresponding type numbers will be unchanged.*

In §14 we imposed conditions on the field of extremals through O. If O' be restricted to a sufficiently small neighborhood of O, the field of extremals through O' will satisfy these same conditions. Let $(x)=(a)$ be the coördinates of O', and $(x)=(a^0)$ the coördinates of O. Let (v) be a set of parameters in an admissible representation of Σ in the neighborhood of the point P. The integral taken along the extremal from O' to a point (v) on Σ will be a function $I(v, a)$ of (v) and (a).

The conditions that $I(v, a)$ have a critical point when considered as a function of (v) alone are

$$I_{v_i}(v,a) = 0 \qquad (i = 1,2, \cdots, n). \tag{19.1}$$

Suppose the extremal from O to P is cut transversally by Σ at P. Let $(v)=(v^0)$ correspond to P. The equations (19.1) will have an initial solution (v^0, a^0). The jacobian of the functions I_{v_j} with respect to the parameters (v) is

$$| I_{v_i v_j} | .$$

According to Theorem 3 of §17 this jacobian will not be zero when $(v, a) = (v^0, a^0)$, since O is not a focal point of Σ.

The equations (19.1) accordingly determine the parameters (v) as single-valued continuous functions of the point (a) at least for positions of (a) neighboring (a^0). Thus the end points of the extremals from O' to Σ which are cut transversally by Σ vary continuously on Σ as required.

That the type numbers of these extremals remain unchanged follows from the fact that the coefficients of the fundamental form (14.3) can now be regarded as continuous functions of (a).

20. **The fundamental theorem in the large.** We need the following lemma.

LEMMA. *At a point $(x)=(a)$ at which both $F(a, r)$ and $F_1(a, r)$ vanish for no direction (r), there will exist, corresponding to any regular $(m-1)$-dimensional manifold S through (a), at least one direction cut transversally by S at (a).*

Reference to the transversality conditions shows that a direction $(\dot{x})$ will be cut transversally by S at $(x)=(a)$, if it is cut transversally by the $(m-1)$-plane π tangent to S at $(x)=(a)$.

Let (λ) be any direction. The conditions of transversality show that there exists at least one $(m-1)$-plane ω through (a) which cuts (λ) transversally. Now let ω be continuously turned about (a) so as to coincide with π. That we can solve the transversality conditions for the directions (λ) in terms of the parameters which determine the variable position of ω follows upon consideration of a jacobian set up by Mason and Bliss.* Although this jacobian is not used by Mason and Bliss for exactly our purpose, still they show that it does not vanish if F and F_1 are not zero.

Thus as ω is continuously turned into the plane π the direction (λ) will continuously turn to a direction that is cut transversally by π. The lemma is thereby proved.

If O is not a focal point of Σ the critical points of the function I of §14 are non-degenerate. Let M_k be the number of critical points of type k. Let R_i be the ith connectivity of Σ. According to a theorem proved by the author the following relations hold:†

$$
\begin{aligned}
M_0 &\geqq 1 + (R_0 - 1), \\
M_0 - M_1 &\leqq 1 + (R_0 - 1) - (R_1 - 1), \\
M_0 - M_1 + M_2 &\geqq 1 + (R_0 - 1) - (R_1 - 1) + (R_2 - 1), \\
&\cdots\cdots\cdots\cdots\cdots\cdots \\
M_0 - M_1 + \cdots + (-1)^n M_n &= 1 + (R_0 - 1) - + \cdots + (-1)^n (R_n - 1),
\end{aligned}
\tag{20.1}
$$

where the inequality signs alternate until the final equality is reached.

By virtue of Theorem 4, §18, the preceding statement may be put into the following form.

THEOREM 6. *Suppose O is not a focal point of Σ nor on Σ. Of the extremal segments from O to Σ which are cut transversally by Σ at their ends P on Σ let M_k be the number upon which the sum of the orders of the focal points of P between O and P is k. Then between the numbers M_k and the connectivities R_i of Σ the relations* (20.1) *hold.*

If O is on Σ the above relations still hold if we count O as if it were an extremal of type zero.

---

* Loc. cit., p. 447.

† Marston Morse, *Relations between the critical points of a real function of n independent variables*, these Transactions, vol. 27 (1925), p. 392. The theorem proved in this reference admits an obvious generalization to closed manifolds such as Σ.

The theorem is already established except in the special case where O is at a point (a) on Σ. In this case it follows from the lemma of this section that there is at least one extremal E issuing from Σ at (a) which is cut transversally by Σ at (a). If we now take O in a position on E not (a) but sufficiently near (a), the extremal from O to (a) will be of type zero. According to the proof of Theorem 5, §19, the other extremals from O to Σ which are cut transversally by Σ will be changed at most slightly in position, but will be unchanged in type and number. Thus the theorem follows as stated when O is on Σ.

The following corollary furnishes a strong existence theorem.

COROLLARY. *If O is not a focal point of Σ, the number of extremals from O to Σ cut transversally by Σ on which the sum of the orders of the focal points is k is at least R_k-1. The total number of extremals is at least*

$$1+(R_0-1)+\cdots+(R_n-1).$$

Even when O is a focal point of Σ one can use a limiting process to get strong existence theorems. Considerations of this sort will be presented in a later paper.

21. **Normals from a point to a manifold.** A beautiful geometric application of these results obtains if we consider the integral which gives the arc length. The hypotheses of §14 are here automatically fulfilled for any point O. It remains to consider the focal points.

Let P be an arbitrary point on Σ. Let P be taken as the origin, and the $(m-1)$-plane tangent to Σ at P be taken as the $(m-1)$-plane $x_m=0$. After a suitable rotation of the remaining axes in the $(m-1)$-plane $x_m=0$, Σ may be represented as follows:

$$2x_m = b_i x_i^2 + H \qquad (i=1,2,\cdots,m-1), \tag{21.1}$$

where b_i is a constant, and H is of the third order in $(x_1,\cdots,x_{m-1})$.

For such of the constants b_i as are not zero we set

$$R_i = \frac{1}{b_i},$$

and call the point P_i on the x_m axis at which $x_m=R_i$, the ith center of a principal normal curvature of Σ corresponding to P.

If a constant $b_i=0$, we say that the corresponding center P_i is at infinity. We define the centers P_i for any axes obtained from our specialized set by a rotation or translation, by imagining each normal to Σ rigidly fixed to Σ, and each center P_i rigidly fixed to its normal.

To determine the focal points on the normal to Σ at P we take the parameters (v) of §14 as the set $(x_1,\cdots,x_{m-1})$, and measure s along the normals in the sense of increasing x_m. We make use of the representation

(21.1) to obtain the family of normals to Σ at points neighboring P. The jacobian $\Delta(s)$ of §15 evaluated for $(v)=(v^0)$ is here seen to have the form

$$\Delta(s) = (1 - b_1 s)(1 - b_2 s) \cdots (1 - b_{m-1} s).$$

Thus the sum of the orders of the focal points on a normal to Σ, between the foot P of that normal on Σ and a point O on the normal, equals the number of centers of principal normal curvature between P and O.

It is understood that for the purpose of counting the number of centers P_i between P and O, two centers P_i and P_j are to be counted separately if $i \neq j$, even if P_i and P_j are identical in position. In case O is on Σ it is also understood that O is to be counted as a single normal on which there are, of course, no centers P_i between O and P. Finally two normals from O to Σ which have the same direction are to be counted as distinct if they have different feet on Σ. With this understood Theorem 6 gives us the following:

THEOREM 7. *Let O be a point that is not a center of principal normal curvature of Σ. Of the normals from O to Σ let M_i be the number of those on which there are i centers of principal normal curvature of Σ between O and the feet of the normals on Σ. Then between these numbers M_i and the connectivities R_i of Σ the relations* (20.1) *hold.*

COROLLARY. *If O is not a center of principal normal curvature of Σ, the number of distinct normals from O to Σ on which there are k centers of principal normal curvature is at least $R_k - 1$. The total number of distinct normals is at least*

$$1 + (R_0 - 1) + \cdots + (R_n - 1).$$

22. **Examples.** Let Σ be a regular surface of genus p in three-space. Corresponding to a point O which is not a center of principal normal curvature of Σ we have

$$M_0 \geqq 1, \qquad M_1 \geqq 2p, \qquad M_2 \geqq 1.$$

Thus there will be at least $2p+2$ distinct normals from O to Σ. In addition we have the relation

$$M_0 - M_1 + M_2 = 2 - 2p.$$

Or again let Σ be a manifold in 4-space homeomorphic to a manifold obtained by identifying the opposite faces of an ordinary cube. If O is not a center of principal normal curvature of Σ we have

$$M_0 \geqq 1, \qquad M_1 \geqq 3, \qquad M_2 \geqq 3, \qquad M_3 \geqq 1.$$

There are thus at least eight normals from O to Σ.

HARVARD UNIVERSITY,
CAMBRIDGE, MASS.

Reprinted from the Proceedings of the NATIONAL ACADEMY OF SCIENCES,
Vol. 15, No. 11, pp. 856–859. November, 1929.

CLOSED EXTREMALS

BY MARSTON MORSE

DEPARTMENT OF MATHEMATICS, HARVARD UNIVERSITY

Communicated September 26, 1929

The object of this paper is to present briefly some of the principal results of a theory of closed extremals.

We are concerned with a calculus of variations problem in the ordinary parametric form with an integrand which is positive, analytic, homogeneous in the usual way, and positively regular. See reference 3, ¶2. The domain of the coördinates is to be a regular, analytic, orientable $(m\text{-}1)$-dimensional manifold R in m-space.

We classify closed extremals g according to their type numbers as follows.

We start by mapping g and its neighborhood on an x-axis and its neighborhood in a space $(x, y_1, \ldots, y_n)$, using thereby a transformation which has a period in x equal to the length ω of g. Let γ denote the portion of the x axis for which $o \leqq x \leqq \omega$.

We cut across γ by $p + 1$ successive n-planes, $t_0, \ldots, t_p$, of which the first is $x = o$, and the last $x = \omega$, and which are placed so near together that there are no pairs of conjugate points on the successive segments into which γ is divided. Let P_i be any point on t_i near γ, except that P_o and P_p shall differ only in their x coördinates. The points $P_o, \ldots, P_p$, can be successively joined by extremal segments neighboring γ. Let the resulting broken extremal be denoted by E. Let (u) be the set of $\mu = pn$ variables, of which the first n are the coördinates (y) of P_o, the second n those of P_e, and so on, until finally the last n are the coördinates (y) of P_{p-e}. The value of the transformed integral taken along E will be a function of (u) and will be denoted by $J(u)$.

The function $J(u)$ will have a critical point when $(u) = (o)$. We consider next the form

$$Q = \sum_{ij} J_{u_i u_j} u_i u_j \qquad (i, j = l, \ldots, \mu), \qquad (1)$$

where the partial derivatives are evaluated for $(u) = (o)$. We classify our closed extremals according to the type number and the nullity of the form Q.

THEOREM 1. *The nullity of the form Q equals the number of linearly independent solutions of the Jacobi differential which have the period ω.*

To aid in determining the type number of Q we first define the order of concavity of γ.

If $x = o$ on γ is not conjugate to $x = \omega$ on γ, any of the above points

P_o can be joined to its congruent point P_p by a secondary extremal.* Let P_o be restricted to points on $x = o$ for which

$$y_1^2 + \ldots + y_n^2 = 1.$$

The resulting set of secondary extremals joining P_o to P_p will form a sort of tubular manifold M roughly enclosing γ. The exterior normals to M at P_o and P_p will be in a 2-plane through the x axis. Let θ be the angle between these normals, measuring θ from the normal at P_o, and counting that sense of rotation positive which leads from the positive x axis to either of these normals. We have

$$r(y) \sin \theta = \sum_{ij} a_{ij}\, y_i y_j \qquad (i, j = 1, \ldots, n), \tag{2}$$

where (y) gives the coördinates of P_o, $r(y)$ is a positive continuous function of (y), and the coefficients a_{ij} are constants.

The type number of the right-hand form in (2) will be called the order of concavity of γ.

THEOREM 2. *If $x = o$ is not conjugate to $x = \omega$ on γ, the type number of Q will equal the sum of the orders of the conjugate points of $x = o$ preceding $x = \omega$, plus the order of concavity of g.*

As an example consider the (m-1)-dimensional ellipsoid

$$\frac{x_1^2}{a_1^2} + \ldots + \frac{x_m^2}{a_m^2} = 1 \qquad o < a_1 < \ldots < a_m. \tag{3}$$

The ellipse lying in the coördinate 2-plane in which all the x's are zero except x_i and x_j, $i \neq j$ will be called the ijth principal ellipse.

THEOREM 3. *The type number of the ijth principal ellipse is $m + i + j - 5$.*

Thus in 3-space the type numbers are 1, 2 and 3. For a general m the total number of principal ellipses is ${}_mC_2$, the number of combinations of m objects two at a time.

The author has developed a general deformation theory in m-space, extending and completing the theory already developed in the small in the paper 2 cited below. In this theory the following is a fundamental theorem.

THEOREM 4. *Suppose g is non-singular of type $k > o$, and gives to J the value c.*

Then corresponding to a sufficiently small neighborhood N of g, there exists within N an arbitrarily small neighborhood N' of g, such that closed m-families of curves which are deformable into g on N, and which satisfy

$$J \leqq c - e, \tag{4}$$

where e is a sufficiently small positive constant, are conditioned as follows.

(a) *Those for which* $m \neq k\text{-}1$ *can be deformed on* N *without increasing* J *into a family of lower dimensionality.*

(b) *Those for which* $m = k\text{-}1$ *include a* $(k\text{-}1)$*-family* Z *which is non-bounding on* N *and* (4), *and which is such that every other* $(k\text{-}1)$*-family can be deformed on* N *without increasing* J *either into* Z *or else a single closed curve.*

More generally still one replaces each sensed closed curve on R on which J is less than a constant J_0, by a succession of q short extremals (q large), joining successive vertices on the given closed curve. The set of vertices on the given curve furnish a point (π) in an auxiliary domain Σ. Two points (π) are to be regarded as identical if the corresponding sets of vertices define the same sensed broken extremal. Particular attention is paid to the reversible problem, that is, the problem in which a point and initial direction, and the same point and opposite direction define the same extremal. In the reversible problem two points (π) are regarded as identical if they define the same broken extremal regardless of sense.

The relation between the connectivities of the domain Σ and the type numbers of closed extremals for which $J < J_0$ is illustrated by the following typical theorem.

THEOREM 5. *Corresponding to an increase of the constant* J_0 *through a value of* J *taken on by a single, non-singular, closed extremal of type* k *in a reversible problem, either the* k*th connectivity of the domain* Σ *increases by one, or else the* $(k\text{-}1)$*st connectivity decreases by one.*

To come to a particular case of great importance let R be the topological image of an $(m\text{-}1)$-sphere under a correspondence in which great circles are carried into closed curves of length J less than J_l.

THEOREM 6.[5] *On a manifold* R *which is the topological image of an* $(m\text{-}1)$*-sphere, and under a proper count of multiplicities, there will always exist at least* ${}_mC_2$ *closed geodesics on which* $J < J_l$. *If these closed geodesics are non-singular they will be distinct, and of the respective types of the co-ordinate ellipses on an* $(m\text{-}1)$*-dimensional ellipsoid.*

A specialization of the preceding methods leads to interesting results in the theory of diameters, that is chords which are orthogonal to a manifold at both ends.

In the first place the length of the chord may be regarded as a function F of the parameters which specify its end-points. The function F will have a critical point if and only if the chord is a diameter. At a critical point the diameter will be said to have the type number of F at that point, and to be non-singular if that critical point is non-singular.

For example the ellipsoid has a diameter on each axis, and the type number of the diameter on the ith axis equals $m + i - 2$. Corresponding to these facts we have the following existence theorem in general.

THEOREM 7. *A regular manifold which is a topological image of an*

$(m$-1)-sphere possesses at least m diameters, counting multiplicities properly. If these diameters are non-singular they will be distinct, and of types m-1 to 2(m-1), *respectively.*

The method of proof is to set up explicitly the non-bounding circuits in the domain of the variables in F, and to prove the non-existence of homologies between them.

* A curve representing a solution of the Jacobi differential equations.

[1] "Relations Between the Critical Points of a Real Function of n Independent Variables," *Trans. Amer. Math. Soc.*, **29** (1927), pp. 429–463.

[2] "The Foundations of a Theory in the Calculus of Variations in the Large," *Ibid.*, **30** (1928), pp. 213–274.

[3] "The Foundations of the Calculus of Variations in the Large in m-Space," (first paper) *Ibid.*, **31** (1929), pp. 379–404.

[4] "The Critical Points of Functions and the Calculus of Variations in the Large," *Bull. Amer. Math. Soc.*, **35** (1929), pp. 38–54.

[5] For a proof of the existence of three closed geodesics on a convex surface, and of at least one closed geodesic on R in the m-dimensional case see Birkhoff, "Dynamical Systems," *Amer. Math. Soc., Colloquium Publications*, **9**, 1927.

Reprinted from the
MATHEMATISCHE ANNALEN
Vol. 103, No. 1, pp. 52–69

A generalization of the Sturm[1]) Separation and Comparison Theorems in n-space.

Von

Marston Morse in Cambridge (Mass., U.S.A.).

§ 1.
Introduction.

The author considers a system of n ordinary, second-order, linear, homogeneous differential equations. The system is supposed self-adjoint and regular. Starting with the theorem that any such system may be considered as the Euler equations of an integral of a quadratic form, he shows that the Sturm Separation and Comparison Theorems naturally generalize into theorems concerning the interrelation of conjugate and focal points. Care is taken at every stage to state the final results in a purely analytical form belonging to the theory of differential equations proper.

The results obtained were made possible by certain other discoveries already made in connection with the author's *theory of the calculus of variations in the large*[2]).

§ 2.
Self-adjoint systems.

Let there be given a system of n differential equations of the form[3])

$$a_{ij} z_j'' + b_{ij} z_j' + c_{ij} z_j = 0 \qquad (i, j = 1, \ldots, n), \; |a_{ij}| \neq 0 \tag{2.1}$$

[1]) See Bôcher, Leçons sur les Méthodes de Sturm, Paris (1917), Gauthier-Villars et Cie.

[2]) Marston Morse, The foundations of the calculus of variations in the large in m-space (first paper). Transactions of the American Mathematical Society **31**, No. 3 (1929). See also, The critical points of functions and the calculus of variations in the large, Bulletin of the American Mathematical Society **35** (1929), pp. 38–54. The first paper will hereafter be referred to as Morse.

[3]) We adopt the convention that a repetition of a subscript in a term shall indicate a sum with respect to that subscript.

where x is the independent variable, and a_{ij}, b_{ij}, and c_{ij}, are of class C'' on a segment γ of the x axis. The system adjoint to (2. 1) is

$$(2.2)\qquad (a_{ij} y_i)'' - (b_{ij} y_i)' + (c_{ij} y_i) = 0.$$

A set of necessary and sufficient conditions that (2. 1) be self-adjoint is readily seen to be[4])

$$(2.3)'\qquad a_{ij} \equiv a_{ji},$$

$$(2.3)''\qquad 2a'_{ij} \equiv b_{ij} + b_{ji},$$

$$(2.3)'''\qquad a''_{i} - b'_{ij} \equiv c_{ji} - c_{ij}.$$

On the other hand let us turn to the integral

$$\mathrm{I} = \int_{x_0}^{x_1} (A_{ij} z'_i z'_j + 2 B_{ij} z_i z'_j + C_{ij} z_i z_j)\, dx$$

where A_{ij}, and B_{ij}, and C_{ij} are functions of class C' on γ. No loss of generality results if we assume

$$(2.4)\qquad A_{ij} \equiv A_{ji}, \qquad C_{ij} \equiv C_{ji}.$$

The Euler equations corresponding to the integral I are

$$(2.5)\qquad \frac{d}{dx}(A_{ij} z'_j + B_{ji} z_j) - (B_{ij} z'_j + C_{ij} z_j) = 0.$$

We see that a set of necessary and sufficient conditions for the system (2. 5), when expanded, to be identical with the system (2. 1), is the following

$$(2.6)'\qquad a_{ij} \equiv A_{ij},$$

$$(2.6)''\qquad b_{ij} \equiv A'_{ij} + B_{ji} - B_{ij},$$

$$(2.6)'''\qquad c_{ij} \equiv B'_{ji} - C_{ij}.$$

We shall prove the following theorem.

Theorem 1. *A necessary and sufficient condition that the system* (2. 1) *admit the form* (2. 5) *is that the system* (2. 1) *be self-adjoint.*

That it is necessary that (2. 1) be self-adjoint is a well known fact[5]) whose proof we need not give.

To prove that it is sufficient that (2. 1) be self-adjoint for (2. 1) to admit the form (2. 5), we assume that the equations (2. 3) hold, and

---

[4]) See Davis R. D., The inverse problem of the calculus of variations in higher space, Transactions of the American Mathematical Society **30** (1928), p. 711–713.

[5]) See Hadamard, Leçons sur le calcul des variations, Paris 1910, 1, p. 319.

then show that it is possible to choose A_{ij}, B_{ij}, and C_{ij}, so as to satisfy (2.4) and (2.6).

We first choose A_{ij} as a_{ij}. We then choose the functions C_{ij} as arbitrary functions of class C' except for the conditions $C_{ij} = C_{ji}$. We next choose B_{ji} and B_{ij} so as to satisfy (2.6)″ as an initial condition at some point $x = x_0$. Equations (2.6)″ give apparently two conditions on $B_{ji} - B_{ij}$, obtained by interchanging i and j, namely,

$$B_{ji} - B_{ij} \equiv b_{ij} - A'_{ij},$$
$$B_{ij} - B_{ji} \equiv b_{ji} - A'_{ji}.$$

But these two conditions are really the same as is seen from the fact that A_{ij} has been chosen as a_{ij}, and from the fact that (2.3)″ is assumed to hold. Subject to this condition we now choose B_{ji} and B_{ij} so as to satisfy (2.6)‴.

We have made our choices of A_{ij}, B_{ij}, and C_{ij}. It remains to show that (2.6)″ holds not only at $x = x_0$ but identically.

According to the choices of B_{ij} and B_{ji} as satisfying (2.6)‴, we have

$$B'_{ji} - B'_{ij} \equiv c_{ij} - c_{ji}.$$

By virtue of (2.3) and of our choice of A_{ij} as a_{ij} this becomes

$$A''_{ij} + B'_{ji} - B'_{ij} \equiv b'_{ij}. \tag{2.7}$$

Equation (2.7) is also obtainable by differentiating (2.6)″. Since B_{ij} and B_{ji} have been chosen so that (2.6)″ holds for at least one x, by virtue of (2.7) it holds identically.

Thus the system (2.6) is satisfied by our choices of A_{ij}, B_{ij}, and C_{ij}, and the theorem is proved.

We note the following.

We can choose $C_{ij} = C_{ji}$ arbitrarily, and $A_{ij} = a_{ij}$. The functions B_{ii} are then uniquely determined to an arbitrary constant. The functions B_{ij} for $i \neq j$, are uniquely determined, except that an arbitrary constant may be added to an admissible choice of B_{ij} if substracted from the corresponding admissible choice of B_{ji}.

The second variation of the integral I takes the form of $2I$ with (z) replaced by (η) as is conventional. Thus if a system (2.1) is self-adjoint it may be considered as the system of Jacobi differential equations of a problem of the calculus of variations. Conversely, as is well known, the Jacobi differential equations of an ordinary problem in non-parametric form will always reduce to a system (2.1) that is self-adjoint.

§ 3.

A canonical form for the differential equations.

We now make a transformation from the variables (z) to variables (y) of the form

$$z_i = u_{ik}(x)\, y_k \qquad (i, k = 1, \ldots, n). \tag{3.1}$$

We seek to determine the functions $u_{ij}(x)$ so that they shall be of class C', their determinant shall not be zero, and so that after (3.1) the terms in $y_h y'_k$ shall disappear from the integral I.

We can also write (3.1) in the form

$$z_j = u_{jh} y_h \qquad (j, h = 1, \ldots, n). \tag{3.2}$$

If we make these substitutions in the integral I the terms in the integrand involving both the y's and their derivatives take the form

$$A_{ij} u'_{ik} u_{jh} y_k y'_h + A_{ij} u_{ik} u'_{jh} y'_k y_h + 2 B_{ij} u_{ik} u_{jh} y_k y'_h. \tag{3.3}$$

The second sum in (3.3) will not be changed if we interchange h with k and i with j. The sum (3.3) then becomes

$$2\,[A_{ij} u'_{ik} + B_{ij} u_{ik}]\, u_{jh} y_k y'_h. \tag{3.4}$$

We can make this sum vanish identically by choosing u_{ik} so that

$$A_{ij} u'_{ik} + B_{ij} u_{ik} = 0. \tag{3.5}$$

We choose the kth column of $|u_{ik}|$ as that solution of the n differential equations

$$A_{ij} u'_i + B_{ij} u_i = 0 \qquad (i, j = 1, 2, \ldots, n),$$

which takes on the values

$$u_k(a) = 1, \quad u_i(a) = 0 \qquad (i \neq k).$$

For this choice of the elements u_{ij} the determinant $|u_{ik}|$ is not zero at $x = a$, and is accordingly never zero.

The transformation (3.1) so determined reduces the integral I to the form

$$I = \int_{x_0}^{x_1} \frac{1}{2}\,[r_{ij}(x) y'_i y'_j + p_{ij}(x) y_i y_j]\, dx \tag{3.6}$$

where

$$r_{ij} = r_{ji}, \quad p_{ij} = p_{ji}, \qquad (|r_{ij}| \neq 0).$$

Our differential equations have thus been reduced to the canonical form,

$$\frac{d}{dx}(r_{ij} y'_j) - p_{ij} y_j = 0. \tag{3.7}$$

We shall assume that the calculus of variations problem arising from I is regular. That is we assume that

$$r_{ij}(x)\,\eta_i\eta_j > 0 \tag{3.8}$$

for every set $(\eta) \neq 0$, *and for every value of* x *on the given interval.*

§ 4.

Conjugate families of solutions.

From this point on we shall take our system in the form (3.7). If (y) and (z) stand for any two solutions of (3.7) it follows readily from (3.7) that

$$r_{ij}(y_i z_j' - z_i y_j') \equiv \text{const.} \tag{4.1}$$

If the constant in (4.1) is zero (y) and (z) are called *mutually conjugate* according to von Escherich[6]).

A system of n linearly independent mutually conjugate solutions will be called a *conjugate base.* The set of all solutions linearly dependent on the solutions of a conjugate base will be called a *conjugate family.*

By a determinant $D(x)$ of a conjugate base is meant the determinant whose columns are the respective solutions of the base. A determinant $D(x)$ cannot vanish identically. In fact one can easily show that it vanishes at a point $x = a$ at most to a finite order equal to the nullity[7]) of $D(a)$. (Morse § 7.) It is also clear that the determinants of two different conjugate bases of the same conjugate family are non-vanishing constant multiples of each other.

A zero of the rth order of a determinant $D(x)$ will be called a *focal point* of the rth order of the corresponding family.

If $x = a$ is not a focal point of a given conjugate family, one can always select a conjugate base of elements $y_{ij}(x)$ which at a point $x = a$ take on the values

$$y_{ii}(a) = 1, \quad y_{ij}(a) = 0 \qquad (i \neq j). \tag{4.2}$$

Such a base will be termed *unitary* at $x = a$.

§ 5.

The initial elements $g_{ij}(a)$ of a conjugate family.

We shall now investigate with what freedom one can determine a conjugate family by prescribing initial conditions at $x = a$.

---

[6]) See Bolza, Vorlesungen über Variationsrechnung, p. 626.

[7]) The nullity of a determinant is its order minus its rank.

We shall restrict ourselves in this section to conjugate families for which $x = a$ is not a focal point. Such a family will have a conjugate base $y_{ij}(x)$ that is unitary at $x = a$. Let us set

(5.1) $$g_{ij}(a) = r_{ik}(a)\, y'_{kj}(a).$$

We see that $g_{ij} = g_{ji}$. For the condition that the ith and jth column of the base $y_{ij}(x)$ be conjugate is that

(5.2) $$r_{hk}(y_{hi}\, y'_{kj} - y_{hj}\, y'_{ki}) \equiv 0$$

which reduces at $x = a$ with the aid of (4.2) to the condition $g_{ij} = g_{ji}$.

Conversely if the g_{ij}'s are arbitrary elements of a symmetric matrix one sees at once that a matrix of functions $y_{ij}(x)$ whose columns satisfy the differential equations, whose values at $x = a$ are given by (4.2), and which satisfy (5.1), will also satisfy (5.2), and thus be a conjugate base that is unitary at $x = a$.

We have thus proved the following.

Lemma. *The most general conjugate family without focal point at $x = a$ may be determined by giving an arbitrary symmetric matrix of constants $g_{ij}(a)$, and determining a conjugate base that is unitary at $x = a$ and satisfies* (5.1).

When the constants $g_{ij}(a)$ are related to a conjugate family as in the preceding lemma the constants $g_{ij}(a)$ will be called *the initial elements of the family at $x = a$*.

§ 6.

Transverse manifolds.

We need to develop the theory of conjugate families in connection with the theory of transverse manifolds. But the latter theory proceeds with difficulty if the integral be zero along an extremal. Now the integrand H of I is zero along γ. We can avoid any difficulty by taking a new integrand

(6.1) $$L(x, y, y') = H(x, y, y') + 1.$$

The corresponding integral

(6.2) $$J = \int_{x_0}^{x_1} L(x, y, y')\, dx$$

will have the same extremals as the previous one.

Relative to our new integral J we shall now briefly indicate a proof of the following lemma.

Lemma. *Through each point P of γ at which $D(x) \neq 0$ there passes an n-manifold S which in the neighborhood of P is regular*[8]) *and of class C'', and which cuts the extremals of the given conjugate family transversally.*

Let $y_{ij}(x)$ be the elements of a base of the given conjugate family. The extremals of the conjugate family may be represented in the form

$$y_i = c_j y_{ij}(x) \qquad (i, j = 1, 2, \ldots, n). \tag{6.3}$$

In $L(x, y, y')$ let y_i and y_i' be replaced by the right members of (6.3) and their derivatives respectively. The condition that the hth and kth columns of our base be mutually conjugate may now be written in the form

$$\frac{\partial y_i}{\partial c_h}\frac{\partial L_{y_i'}}{\partial c_k} - \frac{\partial y_i}{\partial c_k}\frac{\partial L_{y_i'}}{\partial c_h} = 0 \qquad (i, h, k = 1, 2, \ldots, n). \tag{6.4}$$

In this form the condition (6.4) may be identified with the usual condition appearing in the literature[9]) that the Hilbert invariant integral I^0, set up for the family (6.3), be independent of the path, at least in the neighborhood of any point $x = a$ on γ at which $|y_{ij}| \neq 0$. In the neighborhood of $x = a$ on γ the Hilbert integral then becomes a function $I^0(x, y_1, \ldots, y_n)$ of class C''. The equation

$$I^0(x, y_1, \ldots, y_n) = I^0(a, 0, \ldots, 0) \tag{6.5}$$

is solvable for x as a function $\bar{x}(y)$ of the y_i's, since

$$I_x^0(a, 0) = L(a, 0, 0) = 1 \neq 0.$$

Equation (6.5) thus defines a manifold S which in the neighborhood of $x = a$ on γ satisfies the lemma.

Reference to the form of I^0 shows that all the partial derivatives of I^0 with respect to the y_i's vanish when $(y) = (0)$, from which it follows that S is orthogonal to the x axis at $x = a$.

If now we suppose the base $y_{ij}(x)$ is unitary at $a = x$ we may apply the lemma of Morse § 16 to the jth column of y_{ij}, and thereby obtain the following.

$$\bar{x}_{y_i y_j}(0) + r_{ik}(a)\, y'_{kj}(a) = 0 \qquad (i, j, k = 1, \ldots, n).$$

If we compare this with (5.1) *we see that*

$$\bar{x}_{y_i y_j}(0) = -g_{ij}, \tag{6.6}$$

a relation that will be useful presently.

---

[8]) An n-manifold S is called regular if on it the coordinates of points neighboring a given point admit a representation in terms of n parameters in which not all of the jacobians of the coordinates with respect to the n parameters are zero.

[9]) Bliss, The transformations of Clebsch in the calculus of variations, Transactions of the American Mathematical Society 17 (1916), p. 595.

§ 7.

The fundamental quadratic form of a conjugate family.

We shall now review certain theorems which show how to characterize focal points in terms of a fundamental quadratic form. Consider a conjugate family *without focal point* at $x = a$ on γ. Let S be the manifold which cuts the family transversally at $x = a$.

Let us cut across the portion of γ for which $a < x < b$ with m successive n-planes t_i, perpendicular to γ, cutting respectively in points at which $x = x_i$. These n-planes divide γ into $m+1$ successive segments. Suppose these n-planes are placed so near together that no one of these $m+1$ segments contains a conjugate point of its initial end point. Let P_i be any point on t_i near γ. Let the points on γ at which $x = a$ and $x = b$ respectively be denoted by A and B. Let Q denote any point near A on the manifold S.

The points

$$Q, P_1, P_2, \ldots, P_m, B \tag{7.1}$$

can be successively joined by extremal segments neighboring γ. Let the resulting broken extremal be denoted by E. Let (v) be a set of $\mu = (m+1)n$ variables of which the first n are the coordinates (y) of Q, the second n are those of P_1, and so on, until finally the last n are the coordinates (y) of P_m. The value of the integral I taken along E will be a function of the variables (v) of class C'' at least, and will be denoted by $J(v)$.

The function $J(v)$ will have a critical point when $(v) = (0)$. We set

$$H(u) = J_{v_h v_k}(0)\, u_h u_k \qquad (h, k = 1, 2, \ldots, \mu). \tag{7.2}$$

The form $H(u)$ will be called the *fundamental form* of the given conjugate family taken from $x = a$ to $x = b$.

Consider now the points

$$P_0, P_1, \ldots, P_m, B, \tag{7.3}$$

where Q in (7.1) is here replaced by a point P_0 in the plane $x = a$. Let (u) be a complex of μ variables composed of the sets of coordinates (y) of the points (7.3) omitting B, and taking these sets in the order of the points (7.3). A curve of class C' which passes from P_0 to B through points (7.3) will be said to *determine* the corresponding set (u).

If we differentiate the function

$$J(e u_1, \ldots, e u_\mu)$$

twice with respect to e, and set $e = 0$, making use of (6.6), we obtain, as in the proof of the lemma, Morse § 17, the following result.

Lemma. *The fundamental form* $H(u)$ *of the given conjugate family is given as follows:*

$$H(u) = g_{ij} u_i u_j + \int_a^b (r_{ij} \eta_i' \eta_j' + p_{ij} \eta_i \eta_j) dx \qquad (i, j = 1, \ldots, n) \tag{7.4}$$

where g_{ij} *is the* ij*th initial element of the family at* $x = a$, *and* $[\eta(x)]$ *gives the coordinates* (y) *along the broken extremal from* P_0 *to* B *which determines the set* (u).

We come to the following theorem.

Theorem 2. *If the point* $x = b$ *is a focal point of the given family of the* r*th order, the rank of* $H(u)$ *is* $\mu - r$. *If* $x = b$ *is not a focal point of the family the rank of* $H(u)$ *is* μ, *and the negative type number*[10]) *of* $H(u)$ *equals the number of focal points of the given family between* $x = a$ *and* $x = b$, *always counting focal points according to their orders.*

For the reader who has examined the proofs of Theorems 1 to 4 of the earlier paper it will suffice here to enumerate the essential steps as follows.

a) The nullity of the matrix of $H(u)$ will equal the number of linearly independent solutions (u) of the μ equations $H_{u_i} = 0$. The latter equations when expressed in the terms of the right member of (7.4), are seen to be necessary and sufficient conditions on a set (u) for such a set to determine a member of the conjugate family passing through B. If B is a focal point of the family of the rth order there will be exactly r such sets which are linearly independent. The statements of the theorem about the rank of $H(u)$ follow.

b) To determine the type numbers of $H(u)$ we vary the position of the point b from a point nearer $x = a$ than any focal point of the family, to its final position as given. At the start of this variation we see that $H(u)$ will be of negative type zero. As b increases through each focal point the negative type number of $H(u)$ can change by at most the nullity of $H(u)$, that is by at most the order of the focal point. Hence the negative type number is at most the sum q of the orders of the focal points.

c) A lemma on quadratic forms could be stated as follows. A quadratic form which is negative on a q-plane through the origin, excepting the origin, will be of negative type at least q.

---

[10]) By the positive and negative type numbers of a real quadratic form, will be understood the number of positive and negative terms respectively in the form when reduced by a real non-singular linear transformation to squared terms only.

Let $x = c$ be any focal point of the family. We now join B to a point P_0 on the n-plane $x = a$ as follows. We pass along the x axis from B to $x = c$, and then along any curve of the conjugate family to a point P_0 on $x = a$. Let (u) be the set determined by this curve. If c is of order r we can get r such sets which are linearly independent, and from all the focal points we can get q sets (u) which are linearly independent.

Each of the curves just described and their linear combinations will make the right member of (7.4) zero if (η) in that member be taken along such a curve, and (u) be the set which that curve determines. This is proved by suitably integrating the first sum in the integral by parts. If now each of those curves be replaced by the broken extremal which determines the same (u), the integral will be decreased, at least unless some of the points $x = x_i$ are focal points, or unless $(\eta) \equiv (0)$.

Without changing the type numbers of $H(u)$ we can, however, always displace the points $x = x_i$ so that they are not focal points. We conclude that $H(u)$ is negative at each point of a q-plane π_q through the origin, excepting the origin, so that the negative type number of $H(u)$ must be exactly q.

Thus the theorem is proved.

The preceding proof also includes a proof of the following lemma.

Lemma A. *Suppose none of the points x_i are focal points but that $x = b$ is a focal point of the rth order. Let q equal the sum of the orders of the focal points on the interval $a < x \leqq b$. Then there exists a q-plane π_q through the origin, at each point of which $H(u) < 0$ except at the points of a sub r-plane π_r of points (u) determined by the members of the conjugate family through B.*

At no point of π_q excepting the origin are the first n of the u_i's all zero.

The last statement follows from the fact that the only member of the conjugate family which vanishes at $x = a$ is coincident with the x axis.

§ 8.

The special quadratic form for conjugate points.

Suppose now that we have a conjugate family every member of which vanishes at $x = a$. The focal points of such a family determine what are called the conjugate points of $x = a$. We proceed with this case much as in § 7, defining the n-planes t_i, the points x_i, the points P_i, A, and B.

The points

(8.1) $$A, P_1, \ldots, P_m, B$$

determine a broken extremal E along which I is evaluated.

By the set (u) is here meant the $\nu = mn$ variables of which the first n are the y-coordinates of P_1, the second n those of P_2, etc., and the last those of P_m. The integral I taken along E thus becomes a function $J(u)$. We set

$$Q(u) = J_{u_h u_k}(0)\, u_h u_k \qquad (h, k = 1, 2, \ldots, \nu)$$

and term $Q(u)$ *the special form associated with* γ taken from $x = a$ to $x = b$.

The following lemma and theorem are established in Morse § 10 — § 13.

Lemma. *The special form $Q(u)$ associated with γ taken from $x = a$ to $x = b$ is given by the formula*

$$Q(u) = \int_a^b (r_{ij}\eta_i'\eta_j' + p_{ij}\eta_i\eta_j)\, dx \qquad (i, j = 1, 2, \ldots, n)$$

where $[\eta(x)]$ gives the y-coordinates along the broken extremal joining the points (8.1) determined by (u).

Theorem 3. *If $x = b$ is a conjugate point of $x = a$ of the rth order, the rank of $Q(u)$ is $\nu - r$. If $x = b$ is not conjugate to $x = a$ the rank of $Q(u)$ is ν, and its negative type number is the sum of the orders of the conjugate points of $x = a$ preceding $x = b$.*

§ 9.

The continuous variation of conjugate points and focal points.

By the kth conjugate point of a point $x = a$ on γ is meant the kth conjugate point of $x = a$ following $x = a$, counting conjugate points according to their orders. We shall prove the following lemma.

Lemma. *The kth conjugate point $x = c$ of a point $x = a$ on γ varies continuously with $x = a$ as long as $x = c$ remains on γ.*

Let $x = b$ and $x = b'$ be any two points on γ or on a slight extension of γ, not conjugate to $x = a$, and such that

$$a < b' < c < b.$$

The special form $Q(u)$ of Theorem 3 set up for the segment of γ between $x = a$ and $x = b$ will be non-singular, and of negative type at least k. For a sufficiently small variation of $x = a$, Q will remain non-singular and hence unchanged in type, so that the kth conjugate point will exist and preceed $x = b$.

On the other hand the form $Q(u)$ set up for the segment of γ between $x = a$ and $x = b'$ will be non-singular and of negative type less than k. For a sufficiently small variation of $x = a$, $Q(u)$ will remain non-singular, and hence of type less than k, so that the kth conjugate point of $x = a$ will follow $x = b'$.

The lemma follows from the fact that b and b' may be taken arbitrarily near the given value of c.

We shall now prove the following theorem.

Theorem 4. *The kth conjugate point of a point $x = a$ on γ advances or regresses continuously with $x = a$ as long as it remains on γ.*

According to the preceding lemma the kth conjugate point of $x = a$ varies continuously with $x = a$. To prove that it advances with $x = a$, suppose the theorem false and that when $x = a$ advances to a nearby point $x = a_1$ the kth conjugate point of $x = a$ regresses from $x = c$ to $x = c_1$. Let $x = b$ be a point not conjugate to $x = a$, and between c_1 and c. We are supposing that

$$a < a_1 < c_1 < b < c.$$

The form $Q(u)$ set up for the segment of γ from $x = a$ to $x = b$ will be non-singular, and of negative type less than k, since b precedes c. On the other hand if we choose the point of division x_1 of § 7 as a_1, and set the first n of the variables (u) in $Q(u)$ equal zero, $Q(u)$ will reduce to a non-singular form which will have a negative type number equal to the number of conjugate points of $x = a_1$ preceding $x = b$, or since c_1 precedes b, a negative type number at least k. According to the theory of quadratic forms $Q(u)$ must then be of negative type number at least k. From this contradiction we infer that the kth conjugate point of $x = a$ advances with $x = a$.

Similarly it follows that the kth conjugate point of $x = a$ regresses with $x = a$ and the theorem is proved.

We state the following theorem.

Theorem 5. *The kth focal point of a conjugate family following a point $x = a$ not a focal point, varies continuously with the initial elements g_{ij} at $x = a$, and with the coefficients of the differential equations.*

That the coefficients of the form $H(u)$ vary continuously with the elements g_{ij}, r_{ij} and p_{ij}, follows with the aid of (7.4). One can then repeat the proof of the lemma in this section, referring to focal points where conjugate points are there referred to, and using $H(u)$ in place of $Q(u)$. In this manner one arrives at a proof of the present theorem.

§ 10.

Separation Theorems.

We commence with the following theorem.

Theorem 6. *Let g_{ij} and $\overset{0}{g}_{ij}$ be respectively the initial elements at $x = a$ of two conjugate families F and F^0. Let P and N be respectively the positive and negative type numbers of the form*

$$D(u) = (g_{ij} - \overset{0}{g}_{ij})\, u_i u_j \qquad (i, j = 1, \ldots, n).$$

If q and q^0 are respectively the numbers of focal points of F and F^0 on the interval $a \leqq x \leqq b$, we have

$$q^0 - P \leqq q \leqq q^0 + N. \tag{10.0}$$

It will be sufficient to give the proof for the case that $x = b$ is not a focal point of F or F^0. For if for the given a the theorem is true for every $b > a$ for which $x = b$ is not a focal point, it is clearly true for the special values of b for which $x = b$ is a focal point.

Let the fundamental forms corresponding to F and F^0 taken from $x = a$ to $x = b$, be denoted by $H(u)$ and $H^0(u)$ respectively. According to the lemma of § 7 we have

$$H(u) - H^0(u) = D(u) \qquad (i, j = 1, \ldots, n). \tag{10.1}$$

Now $D(u)$ may be written as a form $D(u)$ of rank P and positive type P, minus a form $N(u)$ of rank N and positive type N. We can therefore write (10.1) in the form

$$H(u) - P(u) = H^0(u) - N(u). \tag{10.2}$$

The form $H^0(u)$ is of negative type q^0, and hence will be negatively definite on a suitably chosen q^0-plane π through the origin in the space of the μ variables (u). Further $P(u)$ will be zero on a suitably chosen $(\mu - P)$-plane π_1 also through the origin. If $q^0 - P > 0$, π will have at least a $(q^0 - P)$-plane π_2 in common with π_1. We see then from (10.2) that $H(u)$ will be negatively definite on π_2. Thus $q \geqq q^0 - P$.

On reversing the roles of H and H^0 one proves that q^0 is at least $q - N$. Hence $q \leqq q^0 + N$. Thus the theorem is proved.

Since $P + N \leqq n$ we have the following corollary.

Corollary 1. *The number of focal points of any conjugate family on a given interval differs from that of any other conjugate family by at most n.*

If we are dealing with a second order differential equation in the plane, any solution of the differential equation not identically zero gives the base of a conjugate family. Here $n = 1$ and the corollary becomes the famous Sturm Separation Theorem.

Corollary 1 applied to that particular conjugate family F which determines the conjugate points of a given point, leads to Corollary 2.

Corollary 2. *If there are q conjugate points of the point $x = a$ on the interval $a < x \leqq b$, there will be at most $q + n$ focal points of any conjugate family on that interval.*

To still further extend Theorem 6 we need the following lemma.

Lemma. *The nullity of the difference form $D(u)$ equals the number of linearly independent extremals common to the two families F and F^0.*

The nullity of the form $D(u)$ is the number of linearly independent solutions $(u_1, \ldots, u_n)$ of the equations

$$D_{u_i} = 2(g_{ij} - g_{ij}^0)\, u_j = 0 \qquad (i, j = 1, \ldots, n). \tag{10.3}$$

If we refer to the definition (5.1) of the initial elements g_{ij}, we see that (10.3) may be written in the form

$$r_{ik}(a)\,[y'_{kj}(a) - y^{0\prime}_{kj}(a)]\, u_j = 0 \tag{10.4}$$

where $y_{ij}(x)$ and $y^0_{ij}(x)$ are respectively elements of bases of F and F^0 which are unitary at $x = a$. But (10.4) is equivalent to the equations

$$y'_{kj}(a)\, u_j = y^{0\prime}_{kj}(a)\, u_j \qquad (k, j = 1, \ldots, n). \tag{10.5}$$

Since (10.5) gives the conditions that the members of F and F^0 which take on the values $(u_1, \ldots, u_n)$ at $x = a$ be identical with each other, the lemma follows at once.

This leads to the following theorem.

Theorem 7. *If two conjugate families have r linearly independent solutions in common, the number of focal points of the one family on a given interval differs from that of the other family by at most $n - r$.*

According to the lemma the nullity of $D(u)$ is here r, so that in Theorem 6, $P + N = n - r$. The present theorem now follows from (10.0).

To illustrate the general content of Theorem 6 and its corollaries, let us consider an extremal segment g of a regular problem in parametric form in the calculus of variations. (See Morse § 1.) We can transform g into a segment γ of the x axis (Morse § 4), and obtain the following theorem.

Theorem 8. *Suppose g is an extremal in a regular problem in parametric form in the calculus of variations. Let S_1 and S_2 be two regular manifolds of class C''' which cut g transversally at points at which the integrand F is positive. Then the number of focal points of S_1*

on any segment of g differs from the corresponding number for S_2 by at most n.

The proof will be at once clear to the student of the calculus of variations, and will not be given here.

§ 11.

A comparison of coefficients.

Let there be given two self-adjoint differential systems

$$\frac{d}{dx}(r_{ij}\, y_j') - p_{ij}\, y_j = 0, \tag{11.1}$$

$$\frac{d}{dx}(\bar{r}_{ij}\, y_j') - \bar{p}_{ij}\, y_j = 0 \tag{11.2}$$

satisfying the requirements of § 3. We shall prove the following theorem.

Theorem 9. *Let F and $\bar{F}$ be respectively conjugate families of* (11.1) *and* (11.2). *If F and $\bar{F}$ have the same initial elements g_{ij} at $x = a$, and if*

$$\bar{r}_{ij}\, \eta_i\, \eta_j < r_{ij}\, \eta_i\, \eta_j \qquad (\eta) \neq 0, \tag{11.3}$$

$$\bar{p}_{ij}\, \eta_i\, \eta_j < p_{ij}\, \eta_i\, \eta_j \qquad (\eta) \neq 0 \tag{11.4}$$

then the qth focal point of F, if it exists, must be preceded by the qth focal point of $\bar{F}$. Moreover the same result holds if either (11.3) *or* (11.4), *but not both* (11.3) *and* (11.4) *become equalities.*

Let $H(u)$ and $\bar{H}(u)$ be respectively the fundamental forms for F and $\bar{F}$, taken from $x = a$ to $x = b$. Let us suppose $H(u)$ expressed by (7.4). For the same functions $[\eta(x)]$ it follows from the minimizing properties of the integral in (7.4) that

$$\bar{H}(u) \leqq g_{ij}\, u_i\, u_j + \int_a^b [\bar{r}_{ij}\, \eta_i'\, \eta_j' + \bar{p}_{ij}\, \eta_i\, \eta_j]\, dx.$$

From this relation and (7.4), together with (11.3) and (11.4) we see that

$$\bar{H}(u) < H(u) \qquad (u) \neq 0. \tag{11.5}$$

Now holding a fast, place $x = b$ on γ at a focal point of F of the rth order. Suppose b is then preceded by k focal points, counting focal points from $x = a$. It follows from Lemma A of § 7 that $H(u) \leqq 0$ along some $(k + r)$-plane t through the origin. According to (11.5), $\bar{H}(u) < 0$ on t, except at most at the origin.

The negative type number of $\bar{H}(u)$ must then be at least $k + r$, so that the $(k + r)$th focal point of $\bar{F}$ must precede b, and the theorem is proved.

§ 12.

Further comparison of initial conditions.

Theorem 10. *Let F and F^0 be respectively conjugate families of* (11.1) *determined at $x = a$ by initial elements g_{ij} and g^0_{ij}. If*

$$(12.1) \qquad g_{ij} z_i z_j < g^0_{ij} z_i z_j \qquad (i, j = 1, 2, \ldots, n), \quad (z) \neq (0)$$

then the kth focal point of F^0 following $x = a$, if it exists, must be preceded by the kth focal point of F.

Let $H(u)$ and $H^0(u)$ be respectively the fundamental forms for F and F^0 taken from $x = a$ to $x = b$. Suppose b is a focal point of F^0. With the aid of (7.4) we see that

$$(12.2) \qquad H(u) - H^0(u) = (g_{ij} - g^0_{ij}) u_i u_j \quad (i, j = 1, 2, \ldots, n).$$

It follows from the final lemma of § 7 that $H^0(u) \leqq 0$ at each point of a q-plane π_q, where q is the sum of the orders of the conjugate points of F^0 on the interval from a to b inclusive, and that at no point on π_q except the origin are the first n of the u_i's zero. Thus on π_q the right member of (12.2) is negative except when $(u) = (0)$. Hence $H(u) < 0$ on π_q except when $(u) = (0)$.

The type number of $H(u)$ must then be at least q, and hence the qth focal point of F must precede $x = b$. The theorem follows at once.

This theorem can be given a geometric form in terms of manifolds S and S_0 which cut the x axis transversally at $x = a$. If we recall that transversality reduces to orthogonality at a point of the x axis, and recall also that

$$g_{ij} = -\bar{x}_{y_i y_j}(0),$$

where $x = \bar{x}(y)$ was our representation of S, we have the following result.

If the difference between the x coordinate of a point on S, and a point on S_0 with the same (y) is a positive definite form in the set (y), then the kth focal point of S_0 following $x = a$, if it exists, must be preceded by the kth focal point of S.

If we have a single second order differential equation

$$\frac{d}{dx}(ry') - py = 0 \qquad (r > 0)$$

the condition (12.1) becomes

$$(12.3) \qquad g_{11} < g^0_{11}.$$

If we refer to the definition of the elements g_{ij} we see that (12.3) takes the form

$$y'_{11}(a) < y^{0\prime}_{11}(a)$$

where

$$y_{11}(a) = y^0_{11}(a) = 1.$$

The theorem then becomes a well known comparison theorem.

5*

§ 13.

Principal planes.

In order to obtain a more intimate knowledge of the variation of the focal points with the initial elements of a conjugate family F we introduce a generalization of the principal directions on a surface as follows.

Suppose $x = a$ is not a focal point of F. Let $x = c$ be a focal point of F of the r th order. The curves of F which pass through $x = c$ on γ intersect the n-plane $x = a$ in an r-plane π_r which will be called the *principal r-plane* corresponding to the focal point c.

Suppose now that the initial elements g_{ij} at $x = a$ are functions $\bar{g}_{ij}(\alpha)$ of class C' of a parameter α for α neighboring α_0. For each α we obtain then a conjugate family, say F_α. We need the following lemma.

Lemma. *If as α approaches α_0, certain focal points approach the point $x = c$ as a limit point, the corresponding principal planes on $x = a$ will have all their limit points on the principal plane corresponding to $x = c$.*

Let $y_{ij}(x, \alpha)$ be a base of F_α that is unitary at $x = a$. From the equations (5.1) defining the elements g_{ij} it follows that $y_{ij}(x, \alpha)$ is continuous in both x and α. A curve of F which meets the n-plane $x = a$ in a point $(y) = (d)$ will be given by

$$y_i = d_j y_{ij}(x, \alpha). \tag{13.1}$$

Now the right hand members of (13.1) are continuous in (d), x, and α. The lemma follows readily.

We state now our final comparison theorem.

Theorem 11. *Let $x = c$ be a focal point of the r th order of the family F_{α_0}, and let t be the corresponding principal r-plane on $x = a$. If on t the form*

$$\bar{g}'_{ij}(\alpha_0)\, y_i y_j \qquad (i, j = 1, 2, \ldots, n) \tag{13.2}$$

is definite, then an increase of α in any sufficiently small neighborhood of α_0 will cause an advance or regression of the r focal points at $x = c$ according as the form (13.2) is positively or negatively definite on t.

Let $H_0(u)$ and $H(u)$ be respectively the fundamental forms corresponding to the families F_{α_0} and F_α, taken on the interval $a \leqq x \leqq c$. From (7.4) we see that if we set (y) equal to the first n variables in the set (u) we have

$$H(u) - H_0(u) = (\alpha - \alpha_0)\, \bar{g}'_{ij}(\bar{\alpha})\, y_i y_j \qquad (i, j = 1, \ldots, n) \tag{13.3}$$

where $\bar{\alpha}$ lies between α and α_0.

To be specific suppose the form (13.2) is negatively definite on t. For $(\alpha - \alpha_0)$ positive, and sufficiently small, the right member of (10.3) will still be negatively definite on t.

It will be convenient to call a point (u) *unitary* if $u_i u_i = 1$.

Let us apply Lemma A of § 7 to the family F_{α_0}. We shall prove that $H(u)$ is negatively definite on the q-plane π_q of this lemma, where q is the number of focal points of F_{α_0} on the interval $a \leqq x \leqq c$. To prove this it will suffice to prove that $H(u) < 0$ at all unitary points on π_q.

Now, essentially by hypothesis the right member of (13.3) is negatively definite at the points (y) on t, and accordingly is negative at the unitary points (u) on the r-plane π_r of Lemma A, since the first n coordinates of (u) on π_r give points (y) on t. Hence $H(u) < 0$ at unitary points (u) within a sufficiently small positive distance e of the unitary points on π_r.

At the remaining unitary points on π_q, Lemma A affirms that $H_0(u)$ is negative. Hence at these points $H(u) < 0$, if $\alpha - \alpha_0$ be sufficiently small.

Thus after a sufficiently small increase of α from α_0, $H(u) < 0$ for all unitary points on π_q. Hence $H(u)$ is negatively definite on π_q. Thus the negative type number of $H(u)$ will be at least q. Hence as α increases slightly from α_0, the r focal points originally at $x = c$ regress.

Not only will the focal points at $x = c$ regress when α increases from α_0, but also any small increase of α in the neighborhood of α_0 will correspond to a regression of the focal points in the neighborhood of $x = c$.

This follows from the lemma of this section, because with the aid of that lemma, and with the use of unitary points we see that the form

$$\bar{g}'_{ij}(\alpha)\, y_i y_j \qquad (i, j = 1, \ldots, n)$$

will be negatively definite on each of the principal planes that correspond to focal points near $x = c$, provided $\alpha - \alpha_0$ be sufficiently small. Thus in the proof already given we can replace α_0 by any other value of α, say α_1, sufficiently near α_0, and then show that a small increase of α from α_1 also corresponds to a regression of the focal points near $x = c$.

The case where the form (13.2) is positively definite is treated by first supposing that α decreases, and repeating in essence the preceding proof. The results for an increasing α then follow.

Thus the theorem is proved.

(Eingegangen am 10. 7. 1929.)

Proceedings of the American Academy of Arts and Sciences.
Vol. 66. No. 6.—March, 1931.

THE PROBLEMS OF LAGRANGE AND MAYER WITH VARIABLE END POINTS.

BY MARSTON MORSE AND SUMNER BYRON MYERS.

1. **Introduction.** In developing sufficient conditions[1] in this problem, Morse found it convenient to give the finite conditions on the end points of the curves by representing these end points as functions of parameters rather than by expressing these conditions in the usual implicit form. This eliminated much of the algebra involved in the second variation and brought to the fore the essentials of the problem.

From the point of view of simplicity and elegance this was compatible only with a treatment of the Euler equations and the transversality conditions in corresponding form.

This was the first objective of the present paper. Incidentally the transversality condition is given a form that has apparently never appeared in the general problem, a form that reduces to the classic condition in the ordinary problem.

The paper also contains a new criterion for "normalcy" for the case of variable end points. This criterion has the advantage that it is near to the criterion given in the fixed end point problem, and has an immediate geometric interpretation which leads to various new theorems.

The form of the problem is an extension of the one used by Bolza.[2] The expression to be minimized is the sum of an integral and a function of a set of parameters which determine the end points, but which are not themselves in general uniquely determined by the end points. It is more general than any form hitherto treated at length, unless one resorts to devices such as introducing new differential conditions or terminal conditions. Such devices often change normal extremals to anormal extremals, or vice versa, and since sufficient conditions usually require normalcy, such changes lead to new difficulties. Furthermore, the generality of the form of the problem used here corresponds to an actual need in the treatment of the second variation.

The authors have made extensive use of the lectures of Bliss.[3]

2. **The problem.** In the space of the variables x and

$$(y) = (y_1, \cdots, y_n)$$

let there be given a curve E

$$y_i = \bar{y}_i(x) \qquad a^1 \leqq x \leqq a^2 \qquad (i = 1, \cdots, n) \tag{2.1}$$

of class D'*. Points neighboring the initial and final end points of E will be denoted respectively by

$$(x^s, y_1^s, \cdots, y_n^s) = (x^s, y^s), \tag{2.2}$$

where $s = 2$ at the final end point and $s = 1$ at the initial end point.

We consider curves of class D' neighboring E. Such curves will be called *differentially admissible* if they satisfy m differential equations of the form

$$\phi_\beta(x, y, y') = 0 \qquad (\beta = 1, \cdots, m). \qquad m < n† \tag{2.3}$$

We suppose that E is differentially admissible and that along E the functional matrix of the functions (2.3) with respect to the variables y_i' is of rank m.

A curve neighboring E will be said to be *terminally admissible* if its end points are given for some value of (α) by the functions

$$x^s = x^s(\alpha_1, \cdots, \alpha_r) \qquad y_i^s = y_i^s(\alpha_1, \cdots, \alpha_r), \tag{2.4}$$

where these functions of (α) are defined for (α) near (0), and reduce to the end points of E for $(\alpha) = (0)$.

A curve that is both differentially and terminally admissible will be called *admissible*. A set (α) neighboring $(\alpha) = (0)$ will be called *admissible* if it determines through (2.4) the end points of some admissible curve. We note that infinitely many sets (α) might determine the same admissible curve.

We seek conditions under which E and the set $(\alpha) = (0)$ afford a minimum for the expression

$$J = \int_{x^1}^{x^2} f(x, y, y')dx + \theta(\alpha) \tag{2.5}$$

among admissible sets (α) and corresponding admissible curves of class D'. The functions f and ϕ_β are to be of class C'' while the function $\theta(\alpha)$ and the end point functions in (2.4) need be of no more than class C'.

We note that $\theta(\alpha)$ *may contain variables which do not explicitly occur in the terminal conditions (2.4).*

---

* See Bolza, "Vorlesungen über Variationsrechnung," p. 63.

† Here m could equal n as well, and the following developments would hold. One would not have to "extend" the differential equations $\phi_\beta = 0$ nor prove any of the "multipliers" zero.

3. **The variations.** Suppose that we have a family of differentially admissible curves

$$y_i = y_i(x, t) \qquad x_0^1(t) \leqq x \leqq x_0^2(t) \quad (i = 1, \cdots, n) \tag{3.1}$$

containing E for $t = 0$. The *variations* of the family (3.1) along E with respect to t are defined to be the following derivatives:

$$\gamma^s = x_{0t}^s(0) \qquad \eta_i(x) = y_{it}(x, 0) \tag{3.2}$$

where the subscript t denotes differentiation with respect to t. These variations satisfy the *differential equations of variation**

$$\Phi_\beta \equiv \phi_{\beta y_i}\eta_i + \phi_{\beta y_i'}\eta_i' = 0 \tag{3.3}$$

obtained by differentiating (2.3) with respect to t and setting $t = 0$. Here (x, y, y') are to be taken on E.

A set (α) in (2.4) determines a set of end points. Let us take a set of points (α) of the form

$$\alpha_h = \alpha_h(t) \qquad \alpha_h(0) = 0, \qquad (h = 1, \cdots, r) \tag{3.4}$$

where the functions $\alpha_h(t)$ are of class C' for t near 0. Suppose that the family (3.1) takes on the end points determined by (α) in (3.4). That is, we suppose we have, subject to (3.4), the following identities in t:

$$\begin{aligned} x_0^s(t) &= x^s(\alpha) \\ y_i[x_0^s(t), t] &= y_i^s(\alpha). \end{aligned} \tag{3.5}$$

The curves of the family (3.1) are now also terminally admissible, and we include among the variations the following derivatives:

$$u_h = \alpha_h'(0).$$

These, together with the variations (3.2) satisfy what we shall call the *terminal equations of variation*

$$\begin{aligned} \gamma^s &= x_h^s u_h \quad (s = 1, 2;\ h = 1, \cdots, r;\ i = 1, \cdots, n) \\ \eta_i^s &= (y_{ih}^s - \bar{y}_i'^s x_h^s)u_h \qquad\qquad u_h = \alpha_h'(0) \end{aligned} \tag{3.6}$$

obtained by differentiation of (3.5) with respect to t and setting $t = 0$. In (3.6) the superscript s means evaluation at the corresponding end of E and the subscript h attached to x^s or y_i^s means differentiation with respect to α_h.

---

* The summation convention of tensor analysis is used throughout.

A set γ^s, $\eta_i(x)$, where the set γ^s consists of two arbitrary constants and the functions $\eta_i(x)$ are n functions of class D' which satisfy (3.3), will be called a set of *differentially admissible variations.*

A set of numbers γ^s, η_i^s which satisfy (3.6) for some set $(u) = (u_1 \cdots, u_r)$ will be called a set of *terminally admissible variations*, and will be said to be determined by the set (u).

With these definitions, we can now prove the following theorem:

THEOREM 1. *For every set of differentially admissible variations γ^s, $\eta_i(x)$ there exists a one-parameter family*

$$(3.7) \qquad y_i = y_i(x, t) \qquad x_0^1(t) \leqq x \leqq x_0^2(t)$$

of differentially admissible arcs containing E for $t = 0$ and having γ^s, $\eta_i(x)$ as its variations along E. The functions $y_i(x, t)$ are continuous and have continuous derivatives with respect to t for all values (x, t) near those defining E, and the derivatives $y_{ix}(x, t)$ have the same properties except possibly at the values of x defining corners of E, while the functions $x_0^s(t)$ are of class C' for t near 0.

This theorem and its proof are due to Bliss.* For purposes of reference we will reproduce the essentials of the proof here.

First we enlarge the system $\phi_\beta = 0$ to the extended system of differential equations

$$(3.8) \qquad \phi_\beta = 0 \qquad \phi_\tau = z_\tau(x) \qquad (\beta = 1, \cdots, m; \tau = m+1, \cdots, n)$$

where ϕ_τ are new functions of (x, y, y') such that the functional determinant of the extended system with respect to the variables y_i' is different from 0 along E† and the functions $z_\tau(x)$ are defined by (3.8) upon substituting in ϕ_τ the functions $\bar{y}_i(x)$ defining E.

From this extended system we obtain an extended system of differential equations of variation

$$(3.9) \qquad \Phi_\beta = 0 \qquad \Phi_\tau = \zeta_\tau(x) \qquad (\beta = 1, \cdots, m; \tau = m+1, \cdots, n)$$

along E where the functions $\zeta_\tau(x)$ are defined by these equations corresponding to a set of functions $\eta_i(x)$ of any set of differentially admissible variations.

---

* See Bliss,[3] "The Problem of Lagrange in the Calculus of Variations," loc. cit., p. 4–6, and p. 20.

† For a proof of the possibility of this enlargement see Bliss, "The Problem of Mayer with Variable End Points," Transactions of the American Mathematical Society, vol. 19 (1918), p. 312.

Let the set γ^s, $\eta_i(x)$ be a set of differentially admissible variations for E, and $\zeta_r(x)$ the corresponding set defined by (3.9). The existence theorems for differential equations applied to the system

$$\phi_\beta = 0 \qquad \phi_r = z_r(x) + t\zeta_r(x) \tag{3.10}$$

furnish a one-parameter family of continuous solutions

$$y_i = y_i(x, t) \tag{3.11}$$

where

$$x_0^1(t) \leqq x \leqq x_0^2(t) \qquad x_0^s(t) = a^s + t\gamma^s$$

and where the initial values are given by

$$y_i(a^1, t) = \bar{y}_i(a^1) + t\eta_i(a^1)$$

The solutions are required to be continuous at the corner values of x. This family contains the arc E for $t = 0$, and has the prescribed variations γ^s. Its η variations have the initial values $\eta_i(a^1)$ and satisfy (3.9) with the functions $\zeta_r(x)$, and hence coincide with the prescribed functions $\eta_i(x)$.

In a similar manner the following corollary can be proved:

For a matrix

$$\begin{vmatrix} \gamma_1^1 & \cdots\cdots & \gamma_q^1 \\ \gamma_1^2 & \cdots\cdots & \gamma_q^2 \\ \eta_{11} & \cdots\cdots & \eta_{1q} \\ \cdots & \cdots\cdots & \cdots \\ \eta_{n1} & \cdots\cdots & \eta_{nq} \end{vmatrix}$$

where each column is a set of differentially admissible variations, there exists a q-parameter family of differentially admissible arcs

$$y_i = y_i(x, e_1, \cdots, e_q) \quad x_0^1(e_1, \cdots, e_q) \leqq x \leqq x_0^2(e_1, \cdots, e_q) \tag{3.12}$$

containing the arc E for $(e_1, \cdots, e_q) = (0, \cdots 0)$ *and having* γ_p^s, η_{ip} *as its variations along E with respect to* e_p. *The continuity properties are similar to those of the preceding theorem.*

4. **The Euler equations and the transversality condition.** Let us substitute in the integral in J the functions (3.7) defining a one-parameter family of differentially admissible arcs containing E for $t = 0$. If we take the parameters (α) thus,

$$\alpha_h = tu_h$$

where u_h is a set of r arbitrary constants, and substitute them in $\theta(\alpha)$, then J becomes a function of t defined by the formula

$$J(t) = \int_{x_0^1(t)}^{x_0^2(t)} f[x, y(x, t), y'(x, t)]dx + \theta(tu). \tag{4.1}$$

The derivative of this function with respect to t at the value $t = 0$ has the value*

$$V(\gamma, \eta, u) = \int_{a^1}^{a^2} (f_{y_i}\eta_i + f_{y_i'}\eta_i')dx + (f^s\gamma^s)_1^2 + \theta_h u_h \tag{4.2}$$

where f^1 and f^2 are the values of f at the respective end points of E, and the arguments of the derivatives are those defining E.

Let us obtain a more useful form of $V(\gamma, \eta, u)$. Let F be defined by the equation

$$F(x, y, y', \lambda) = \lambda_0 f + \lambda_1\phi_1 + \cdots + \lambda_n\phi_n$$

where λ_0 is a constant and $\lambda_i(x)$ are functions of x in the interval $(x^1\ x^2)$. The value of $\lambda_0 V(\gamma, \eta, u)$ is not changed if we add the sum

$$\lambda_\beta\Phi_\beta + \lambda_r(\Phi_r - \zeta_r)$$

to its integrand, since the variations $\eta_i(x)$ satisfy (3.9). We obtain the equation

$$\lambda_0 V(\gamma, \eta, u) = \int_{a^1}^{a^2} (F_{y_i}\eta_i + F_{y_i'}\eta_i' - \lambda_r\zeta_r)dx + (\lambda_0 f^s\gamma^s)_1^2 + \lambda_0\theta_h u_h. \tag{4.3}$$

So far the functions $\lambda_i(x)$ have been arbitrary. With the admissible curve E we now associate a constant λ_0 and multipliers $\lambda_i(x)$ so that the equations

$$F_{y_i'} = \int_{a^1}^{x} F_{y_i}dx + c_i \qquad (i = 1, \cdots, n) \tag{4.4}$$

are satisfied for an arbitrarily selected set of constants λ_0, c_i and (x, y, y') on E. This is possible since if we introduce the new variables.

$$v_i = F_{y_i'} = \lambda_0 f_{y_i'} + \lambda_1\phi_{1y_i'} + \cdots + \lambda_n\phi_{ny_i'} \quad (i = 1, \cdots, n) \tag{4.4a}$$

* Here $(\ )_1^2$ means the value of $(\ )$ for $s = 2$, minus its value for $s = 1$.

equations (4.4) are equivalent to the following equations:

$$\frac{dv_i}{dx} = F_{y_i} = A_{i1}v_1 + \cdots + A_{in}v_n + B_i \quad (i = 1, \cdots, n) \tag{4.4b}$$
$$v_i(a^1) = c_i$$

The coefficients A_{ij}, B_i are determined by solving (4.4a) for $\lambda_1, \cdots, \lambda_n$ and substituting in F_{y_i}, and they are continuous except possibly at the values of x defining corners of E. Hence equations (4.4b) have unique continuous solutions $v_i(x)$ which are of class C' except possibly at the corner values of x. Equations (4.4a) then determine the functions $\lambda_i(x)$, continuous except possibly at the corner values of x.

By means of equations (4.3), after integrating by parts we find that

$$\lambda_0 V(\gamma, \eta, u) = -\int_{a^1}^{a^2} \lambda_r \zeta_r dx + (\lambda_0 f^s \gamma^s + \eta_i^s F^s{}_{y_i'})_1^2 + \lambda_0 \theta_h u_h. \tag{4.5}$$

This equation holds for arbitrary values of λ_0 and $F_{y_i'}{}^1$, for any set of differentially admissible variations γ^s, $\eta_i(x)$, and for any set of numbers u_h.

Consider the family (3.12) of the corollary of § 3, letting $q = 2n + 3$. Let

$$u_{hj} \quad (h = 1, \cdots, r; j = 1, \cdots, 2n + 3)$$

be $2n + 3$ sets of arbitrary constants. Take the parameters α_h of (2.4) as follows:

$$\alpha_h = e_j u_{hj} \quad (h = 1, \cdots, r; j = 1, \cdots, 2n+3). \tag{4.6}$$

If we substitute the functions defining the family (3.12) in the integral in J, and the parameters (4.6) in $\theta(\alpha)$, then J becomes a function $J(e_1, \cdots, e_{2n+3})$, which when $(e_1, \cdots, e_{2n+3}) = (0, \cdots, 0)$ takes on the value J_0, its value along E. The $2n + 3$ equations

$$\begin{aligned} &J(e_1, \cdots, e_{2n+3}) - J_0 = w \\ &x_0^s(e_1, \cdots, e_{2n+3}) - x^s(e_j u_{1j}, \cdots, e_j u_{rj}) = 0 \\ &y_i[x_0^s(e_1, \cdots, e_{2n+3}), e_1, \cdots, e_{2n+3}] - y_i^s(e_j u_{1j}, \cdots, e_j u_{rj}) = 0 \end{aligned} \tag{4.7}$$

all except the first of which are conditions for terminal admissibility, have the initial solution

$$(w, e_1, \cdots, e_{2n+3}) = (0, 0, \cdots, 0).$$

If the jacobian of the left-hand sides of (4.7) with respect to $e_1, \cdots, e_{2n+3}$ were different from zero at this solution, then the implicit function theorems show that (4.7) would have solutions $(e_1, \cdots, e_{2n+3})$ neighboring $(0, \cdots, 0)$ for all values of w in the neighborhood of $w = 0$. In that case, for negative values of w there would be admissible arcs in the family (3.12) giving J a value smaller than J_0, which is impossible if E is a minimizing arc.

Hence, if E is a minimizing arc, we conclude that the jacobian must be zero.

Let us set

$$\frac{\partial J}{\partial e_j} = V_j \qquad (e_j = 0; j = 1, \cdots, 2n+3)$$

where V_j is $V(\gamma, \eta, u)$ evaluated for the variations (γ, η) arising from the family (3.12) for variation of e_j, and (u) stands for the set u_{hj}, $(h = 1, \cdots, r)$. Then the jacobian in question has as the jth of its $2n + 3$ columns:

$$\begin{array}{llll} & V_j & \text{1 row} & \\ (4.8) & \gamma_j^s - x_h^s u_{hj} & \text{2 rows} & (s = 1, 2) \\ & \bar{y}_i'^s \gamma_j^s + \eta^s_{ij} - y^s_{ih} u_{hj} & 2n \text{ rows} & (s = 1, 2; i = 1, \cdots, n). \end{array}$$

Let us use the notation $\bar{\gamma}_j^s$, $\bar{\eta}_{ij}^s$ for the jth set of terminally admissible variations determined, in the sense of § 3, by the jth set of constants u_{hj}, in contradistinction to the notation γ_j^s, η_{ij}^s for the set of end values of the jth set of differentially admissible variations. Then by means of (3.6), the column (4.8) may be written as

$$\begin{array}{lll} & V_j & \text{1 row} \\ (4.8a) & \gamma_j^s - \bar{\gamma}_j^s & \text{2 rows} \\ & \bar{y}_i'^s(\gamma_j^s - \bar{\gamma}_j^s) + (\eta^s_{ij} - \bar{\eta}^s_{ij}). & 2n \text{ rows} \end{array}$$

We can simplify these columns by multiplying the second and third rows respectively by $\bar{y}_i'^s$ and substracting from the last $2n$ rows.

We thus obtain the necessary condition

$$
(4.9)\qquad
\begin{vmatrix}
V_2 \cdots\cdots\cdots\cdots\cdots\cdots V_{2n+3} \\
\gamma_1{}^1 - \bar{\gamma}_1{}^1 \cdots\cdots \gamma^1{}_{2n+3} - \bar{\gamma}^1{}_{2n+3} \\
\gamma_1{}^2 - \bar{\gamma}_1{}^2 \cdots\cdots \gamma^2{}_{2n+3} - \bar{\gamma}^2{}_{2n+3} \\
\eta^1{}_{11} - \eta^1{}_{11} \cdots\cdots \eta^1{}_{1\,2n+3} - \bar{\eta}^1{}_{1\,2n+3} \\
\cdots\cdots\cdots\cdots\cdots\cdots\cdots\cdots \\
\eta^1{}_{n1} - \bar{\eta}^1{}_{n1} \cdots\cdots \eta^1{}_{n\,2n+3} - \bar{\eta}^1{}_{n\,2n+3} \\
\eta^2{}_{11} - \bar{\eta}^2{}_{11} \cdots\cdots \eta^2{}_{1\,2n+3} - \bar{\eta}^2{}_{1\,2n+3} \\
\cdots\cdots\cdots\cdots\cdots\cdots\cdots\cdots \\
\eta^2{}_{n1} - \bar{\eta}^2{}_{n1} \cdots\cdots \eta^2{}_{n\,2n+3} - \bar{\eta}^2{}_{n\,2n+3}
\end{vmatrix} = 0.
$$

Here $\gamma_j{}^s$, $\eta_{ij}{}^s$ are the end values of a set of differentially admissible variations, otherwise unrestricted, and $\bar{\gamma}_j{}^s$, $\bar{\eta}_{ij}{}^s$ is any set of terminally admissible variations.

Hence for any particular choice of $2n + 3$ sets of differentially admissible variations and of $2n + 3$ sets of constants u_{hj} $(j = 1, \cdots, 2n + 3)$, there must exist $2n + 3$ constants $(\lambda_0, b^s, k_i{}^s)$ not all zero such that the equation

$$
(4.10)\qquad \lambda_0 V(\gamma, \eta, u) + b^s(\gamma^s - \bar{\gamma}^s) + k_i{}^s(\eta_i{}^s - \bar{\eta}_i{}^s) = 0
$$

is satisfied for every set γ^s, $\eta_i(x)$ of the chosen sets of differentially admissible variations, and for every set u_h of the chosen sets of constants u_{hj}. Here $\bar{\gamma}^s$, $\bar{\eta}_i{}^s$ is the set of terminally admissible variations determined by u_h.

In fact, if we choose the $2n + 3$ constants $(\lambda_0, b^s, k_i{}^s)$ corresponding to the choice of $\gamma_j{}^s$, $\eta_{ij}(x)$, u_{hj} which gives the determinant (4.9) its highest rank ρ, then (4.10) must be satisfied for any differentially admissible variations and any set of numbers u_h whatever, $\bar{\gamma}^s$, $\bar{\eta}_i{}^s$ being the terminally admissible variations determined by u_h. For otherwise one of the columns of (4.9) could be replaced by another making the determinant have the rank $\rho + 1$.

(a) *We shall now obtain* (4.17) *and* (4.19) *as consequences of* (4.10). *We call attention to the fact that the derivation of* (4.17) *and* (4.19) *from* (4.10) *will hold even if E is not a minimizing arc, in fact if E is simply an arbitrary admissible arc for which* (4.10) *holds as stated.*

If we substitute in (4.10) the value of $\lambda_0 V(\gamma, \eta, u)$ given by (4.5), we obtain the equation

$$(4.11)\quad -\int_{a^1}^{a^2} \lambda_r\zeta_r dx + (\lambda_0 f^s\gamma^s + \eta_i{}^s F_{y_i'}{}^s)_1^2 + \lambda_0\theta_h u_h + b^s(\gamma^s - \bar\gamma^s) + k_i{}^s(\eta_i{}^s - \bar\eta_i{}^s) = 0.$$

This equation must be satisfied for every choice of differentially admissible variations γ^s, $\eta_i(x)$; that is, for arbitrary γ^s, ζ_r, and $\eta_i{}^2$. The values of $F_{y_i'}{}^1$ are also arbitrary. Furthermore, u_h is any set of r numbers, and $\bar\gamma^s$, $\bar\eta_i{}^s$ is the set of terminally admissible variations determined by u_h.

We can rewrite (4.11) as

$$(4.12)\quad -\int_{a^1}^{a^2} \lambda_r\zeta_r dx + \gamma^s[(-1)^s\lambda_0 f^s + b^s] + \eta_i{}^s[(-1)^s F_{y_i'}{}^s + k_i{}^s] - b^s\bar\gamma^s - k_i{}^s\bar\eta_i{}^s + \lambda_0\theta_h u_h = 0.$$

Since the value of $F_{y'_i}{}^1$ is arbitrary, we can choose

$$(4.13)\quad F_{y_i'}{}^1 = k_i{}^1$$

If we choose

$$\gamma^s = \eta_i{}^2 = u_h = 0,$$

thus determining $\bar\gamma^s$ and $\bar\eta_i{}^s$ as zero, we find that since ζ_r is arbitrary, we must have the identities $\lambda_{m+1} \equiv \cdots \equiv \lambda_n \equiv 0$. Since γ^2, γ^1, and $\eta_i{}^2$ are arbitrary, we find that

$$(4.14)\quad b^2 = -\lambda_0 f^2 = -F^2 \qquad b^1 = \lambda_0 f^1 = F^1 \qquad k_i{}^2 = -F_{y_i'}{}^2.$$

Substituting these values for b^s, $k_i{}^s$ and setting $\lambda_r = 0$, (4.12) becomes

$$(4.15)\quad (F^s\bar\gamma^s + F_{y_i'}{}^s\bar\eta_i{}^s)_1^2 + \lambda_0\theta_h u_h = 0.$$

Recalling (3.6), we can write

$$\bar\gamma^s = dx^s \qquad \bar\eta_i{}^s = dy_i{}^s - \bar y_i'{}^s dx^s \qquad \theta_h u_h = d\theta$$

From (4.15) we now obtain the "transversality" condition* in the form of the following identity in the differentials $d\alpha_h$:

$$(4.16)\quad [F^s dx^s + F_{y_i'}{}^s(dy_i{}^s - \bar y_i'{}^s dx^s)]_1^2 + \lambda_0 d\theta = 0$$

* The transversality condition given by Bliss in his lectures, loc. cit. p. 21, can be simply derived from this condition, but this condition, if derivable at all from the Bliss form, would clearly be so derivable only after complicated transformations. This is partly due to the presence of θ, and partly to the greater generality of the end conditions.

where dx^s, dy_i^s, and $d\theta$ are understood as evaluated for $(\alpha) = (0)$, and expressed in terms of the differentials $d\alpha_h$.

The multipliers λ_0, $\lambda_\beta(x)$ cannot all vanish at any one point of (a^1a^2). For if $\lambda_0 = 0$, the differential equations which the v's satisfy are homogeneous, and hence if the v's are zero for any one value of x, they are identically zero on the interval between corners containing that value of x. Thus they are identically zero on (a^1a^2). But this makes the constants λ_0, b^s, k_i^s all zero, as follows from (4.13) and (4.14), and so is impossible.

A set of multipliers λ_0, $\lambda_\beta(x)$, where λ_0 is a constant and $\lambda_1(x)$, $\cdots$, $\lambda_m(x)$ are functions of x which are continuous except possibly at the corners, and where λ_0, $\lambda_\beta(x)$ are not all zero at any one point of (a^1a^2), will be called an *admissible* set of multipliers.

The results just attained may be summarized in the following theorem:

THEOREM 2. *If E affords a minimum in the problem, there exists a set of constants $c_1, \cdots, c_n$ and an admissible set of multipliers $\lambda_0, \lambda_1(x), \cdots, \lambda_m(x)$ such that the equations*

$$F_{y_i'} = \int_{a^1}^{x} F_{y_i}dx + c_i \tag{4.17}$$

A

are satisfied at every point of E, where

$$F(x, y, y', \lambda) = \lambda_0 f + \lambda_\beta \phi_\beta, \tag{4.18}$$

and such that the following identity

$$[(F^s - \bar{y}_i'^s F_{y_i'}^s)dx^s + F_{y_i'}^s dy_i^s]_1^2 + \lambda_0 d\theta = 0 \tag{4.19}$$

B

in $d\alpha_h$ is satisfied when dx^s, dy_i^s, and $d\theta$ are evaluated for $(\alpha) = (0)$ and expressed in terms of the differentials $d\alpha_h$.

An immediate consequence of this theorem is the so-called Euler-Lagrange Multiplier Rule:

On every sub-arc between corners of a minimizing arc the equations

$$\frac{d}{dx}F_{y_i'} = F_{y_i} \qquad \phi_\beta(x, y, y') = 0 \tag{4.20}$$

must be satisfied, where F is the function (4.18).

5. **Relative normalcy. A new criterion.** An admissible arc E shall be said to be *normal relative to conditions A and B of Theorem*

2, or, more shortly, *normal* (AB),* if it does not satisfy these conditions with any admissible set of multipliers in which $\lambda_0 = 0$.

A differentially admissible arc with an admissible set of multipliers is called an *extremaloid* if it satisfies equations (4.17) with some set of constants $c_1, \cdots, c_n$.

THEOREM 3. *For an extremaloid E which is normal (AB) and which satisfies A and B with an admissible set of multipliers, there always exist admissible multipliers in the form* $\lambda_0 = 1, \lambda_\beta(x)$ *with which E satisfies conditions A and B. In this form the multipliers are unique.*

We have multipliers in the form λ_0, $\lambda_\beta(x)$ where $\lambda_0 \neq 0$, and we can evidently divide them by λ_0 and obtain a set with $\lambda_0 = 1$. If we had two sets with $\lambda_0 = 1$, then the difference of the sets would be a set of multipliers with $\lambda_0 = 0$. This is impossible unless the two sets are identical, if the extremaloid is normal (AB). Thus the multipliers $\lambda_0 = 1$, $\lambda_\beta(x)$ are unique in this form.

Let us now consider the determinant

$$
\begin{vmatrix}
\gamma_1^1 - \bar{\gamma}_1^1 \cdots\cdots \gamma^1_{2n+2} - \bar{\gamma}^1_{2n+2} \\
\gamma_1^2 - \bar{\gamma}_1^2 \cdots\cdots \gamma^2_{2n+2} - \bar{\gamma}^2_{2n+2} \\
\eta^1_{11} - \bar{\eta}^1_{11} \cdots\cdots \eta^1_{1\,2n+2} - \bar{\eta}^1_{1\,2n+2} \\
\cdots\cdots\cdots\cdots\cdots\cdots \\
\eta^1_{n1} - \bar{\eta}^1_{n1} \cdots\cdots \eta^1_{n\,2n+2} - \bar{\eta}^1_{n\,2n+2} \\
\eta^2_{11} - \bar{\eta}^2_{11} \cdots\cdots \eta^2_{1\,2n+2} - \bar{\eta}^2_{1\,2n+2} \\
\cdots\cdots\cdots\cdots\cdots\cdots \\
\eta^2_{n1} - \bar{\eta}^2_{n1} \cdots\cdots \eta^2_{n\,2n+2} - \bar{\eta}^2_{n\,2n+2}
\end{vmatrix}
\tag{5.1}
$$

formed from $2n + 2$ sets of differentially admissible variations

$$\gamma_j^s, \eta_{ij}(x) \qquad (i = 1, \cdots, n) \quad (j = 1, \cdots, 2n + 2; s = 1, 2)$$

and $2n + 2$ sets of terminally admissable variations

$$\bar{\gamma}_j^s, \bar{\eta}^s_{ij} \qquad (i = 1, \cdots, n; j = 1, \cdots, 2n + 2; s = 1, 2)$$

* We say *normal* (AB) rather than simply *normal* for the following reason. Morse has shown that Theorem 2 holds with the following condition C added at each corner $x = c$,

$$[F - y_i' F_{y_i'}]_{c+}^{c-} = 0 \qquad C$$

and that moreover the multipliers in this new theorem do not always agree with those in the old theorem. He has shown that an admissible arc may be normal (ABC)—that is, relative to conditions A, B, and C—while anormal (AB).

and the determinant

$$
(5.2) \qquad \begin{vmatrix}
\eta^1{}_{11} - \bar{\eta}^1{}_{11} \cdots \eta^1{}_{1\,2n} - \bar{\eta}^1{}_{1\,2n} \\
\cdots\cdots\cdots\cdots\cdots\cdots \\
\eta^1{}_{n1} - \bar{\eta}^1{}_{n1} \cdots \eta^1{}_{n\,2n} - \bar{\eta}^1{}_{n\,2n} \\
\eta^2{}_{11} - \bar{\eta}^2{}_{11} \cdots \eta^2{}_{1\,2n} - \bar{\eta}^2{}_{1\,2n} \\
\cdots\cdots\cdots\cdots\cdots\cdots \\
\eta^2{}_{n1} - \bar{\eta}^2{}_{n1} \cdots \eta^2{}_{n\,2n} - \bar{\eta}^2{}_{n\,2n}
\end{vmatrix}
$$

formed from $2n$ sets of differentially admissible variations

$$\gamma_k{}^s, \eta_{ik}(x) \qquad (i = 2, \cdots, n;\ k = 1, \cdots, 2n;\ s = 1, 2)$$

and $2n$ sets of terminally admissible variations

$$\bar{\gamma}_k{}^s, \bar{\eta}^s{}_{ik} \qquad (i = 1, \cdots, n;\ k = 1, \cdots, 2n;\ s = 1, 2).$$

We will now prove the following lemma:

LEMMA. *A necessary and sufficient condition that there exist a determinant of form* (5.1) *different from zero is that there exist a determinant of form* (5.2) *different from zero.*

The sufficiency of this condition is proved by showing that if there exists a determinant of form (5.2) different from zero, then we can construct a determinant (5.1) different from zero.

In fact, if a_{pq} is the general element of the determinant (5.2), by choosing the arbitrary constants $\gamma_i{}^s$ of (5.1) suitably we obtain a determinant of the form

$$
(5.3) \qquad \left|\begin{array}{c|cc}
0 \cdots\cdots 0 & 0 & a \\
0 \cdots\cdots 0 & b & 0 \\
\hline
 & & \\
a_{pq} & & \\
 & &
\end{array}\right| \qquad a \neq 0,\ b \neq 0
$$

different from zero.

Conversely, if there exists a determinant of form (5.1) not zero, then determinants of form (5.2) obviously cannot all be zero. Thus the condition is also necessary.

The following theorem relates to the differential equations (3.3) alone, and is independent of whether E is or is not a minimizing arc.

THEOREM 4. *A necessary and sufficient condition that an admissible arc E be normal (AB) is that there exist for it a determinant of form* (5.1) *different from zero.*

If all determinants of form (5.1) vanished, there would clearly be a set of constants $\lambda_0 = 0$, b^s, k_i^s satisfying (4.10) for any sets of differentially and terminally admissible variations. But as pointed out in statement (a) of § 4, conditions A and B on E would then follow as consequences of (4.10), λ_0 being zero in this case. Thus E is anormal (AB) if all determinants of form (5.1) vanish, and the condition is proved necessary.

Now suppose that the arc E were anormal (AB). We shall show that there exists no determinant of form (5.1) different from zero.

Suppose that there were a determinant of form (5.1) *different from zero.*

Then there would also be a determinant of form (5.2) different from zero. By the definition of normalcy, conditions A and B would be satisfied by E with a set of multipliers in which $\lambda_0 = 0$. The function F in (4.17) and (4.19) would then have the form

$$F = \lambda_\beta \phi_\beta \tag{5.4}$$

Every set of differentially admissible variations would satisfy the equations

$$0 = \int_{a^1}^{a^2} \lambda_\beta \Phi_\beta dx = \int_{a^1}^{a^2} (F_{y_i}\eta_i + F_{y_i'}\eta_i')dx. \tag{5.5}$$

Upon integration by parts, and application of (4.17), we have

$$(F_{y_i'}{}^s\eta_i{}^s)_1^2 = 0. \tag{5.6}$$

Since we are assuming $\lambda_0 = 0$, and since $\phi_\beta = 0$ along E, we have $F^s = 0$. We then see from the transversality condition (4.16) that

$$(F_{y_i'}{}^s\bar{\eta}_i{}^s)_1^2 = 0 \tag{5.7}$$

and this would hold for any terminally admissible variations $\bar{\eta}_i{}^s$. Combining (5.6) and (5.7), we would obtain the equation

$$[F_{y_i'}{}^s(\eta_i{}^s - \bar{\eta}_i{}^s)]_1^2 = 0 \tag{5.8}$$

But we have a determinant of form (5.2) different from zero, so that from (5.8) we deduce the equations

$$F_{y_i'}{}^s = 0 \qquad (s = 1, 2;\ i = 1, \cdots, n). \tag{5.9}$$

But equations (5.9) would imply that $\lambda_\beta(a^s) = 0$, which is contrary to our hypothesis that λ_0, $\lambda_\beta(x)$ form an admissible set of multipliers. Thus there cannot be a determinant of form (5.1) different from zero if E is anormal (AB). This proves the sufficiency of the condition.

COROLLARY. *A necessary and sufficient condition that an admissible arc E be normal (AB) is that there exist for it a determinant of form* (5.2) *different from zero.*

THEOREM 5. *If a set of differentially admissible variations γ^s, $\eta_i(x)$ for an admissible arc E which is normal (AB) satisfies the terminal equations of variation with a set (u), then there exists a one-parameter family of admissible arcs of the form* (3.7) *containing E for the parameter value $t = 0$ and having γ^s, $\eta_i(x)$, u_h as the variations of x^s, y_i, and α_h.*

According to the preceding corollary we have a determinant of form (5.1) different from zero, with its $2n + 2$ sets of differentially admissible variations

$$\gamma_j^s,\ \eta^s_{ij} \qquad (j = 1, \cdots, 2n + 2)$$

and its $2n + 2$ sets of terminally admissible variations

$$\bar{\gamma}_j^s,\ \bar{\eta}^s_{ij} \qquad (j = 1, \cdots, 2n + 2) \tag{5.10}$$

Let

$$u_{hj} \qquad (h = 1, \cdots, r;\ j = 1, \cdots, 2n + 2)$$

be constants of which the jth set determines the jth set in (5.10), in the sense of § 3.

There exists a $(2n + 3)$-parameter family of differentially admissible arcs

$$y_i = y_i(x, t, e) \qquad x_0^1(t, e) \leqq x \leqq x_0^2(t, e) \tag{5.11}$$

with parameters t and $(e) = (e_1, \cdots, e_{2n+2})$, containing E for the parameter values $(t, e) = (0, 0)$ and having the sets γ^s, $\eta_i(x)$ and γ_j^s, $\eta_{ij}(x)$ as its variations along E with respect to t and e_j respectively.

Let us take the parameters α_h of (2.4) as follows:

$$\alpha_h = tu_h + e_j u_{hj} \qquad (h = 1, \cdots, r;\ j = 1, \cdots, 2n + 2).$$

Then the equations

$$\begin{aligned} x_0^s(t, e) - x^s(tu_1 + e_j u_{1j}, \cdots, tu_r + e_j u_{rj}) &= 0 \\ y_i[x_0^s(t, e), t, e] - y_i^s(tu_1 + e_j u_{1j}, \cdots, tu_r + e_j u_{rj}) &= 0, \end{aligned} \tag{5.12}$$

which are conditions for terminal admissibility, have the initial solution $(t, e) = (0, 0)$, at which the jacobian with respect to (e) is the non-vanishing determinant (5.1), as is seen by comparison with (4.7), (4.8) and (4.9). Hence equations (5.12) have solutions

$$e_j = e_j(t) \qquad (j = 1, \cdots, 2n + 2) \tag{5.13}$$

with initial values $e_j(0) = 0$. The one-parameter family

$$y_i = y_i[x, t, e(t)] \qquad x_0^1[t, e(t)] \leqq x \leqq x_0^2[t, e(t)] \tag{5.14}$$

consists of admissible arcs, and contains E for $t = 0$.

Let us compute the derivatives $e_j'(0)$. From the choice of γ^s, $\eta_i(x)$ and u_h we see that the partial derivatives of the left hand sides of (5.12) with respect to t are zero for $t = 0$, and therefore $e_j'(0) = 0$. We see then that the variations of the family (5.14) with respect to t are γ^s, $\eta_i(x)$, u_h, as was to be proved.

6. **Geometric criteria for normalcy (AB).** Let γ^s, η_i^s, be the end-values of a set of differentially admissible variations. Then we will consider the $2n + 2$ numbers γ^s, η_i^s, as the coördinates of a point in $(2n + 2)$-dimensional space. Such a point will be called a *differentially admissible point*. Similarly, consider any set of terminally admissible variations $\bar{\gamma}^s$, $\bar{\eta}_i^s$ as the coördinates of a point in $(2n + 2)$-space. Such a point will be called a *terminally admissible point*. The following theorem furnishes a useful geometrical interpretation of normalcy (AB).

THEOREM 6. *A necessary and sufficient condition that an admissible arc E be normal (AB) is that the family of $(2n + 2)$-dimensional points linearly dependent on the differentially admissible points and the terminally admissible points fill out the whole of $(2n + 2)$-space.*

For if an admissible arc is normal (AB) there exists a determinant of form (5.1) different from zero. The columns of this determinant are $2n + 2$ linearly independent points of the family; hence the family fills out $(2n + 2)$-space.

Now suppose that the family fills out $(2n + 2)$-space. Then there exist $2n + 2$ linearly independent points in the family. But any point of the family may be expressed as the difference of a differentially admissible point and a terminally admissible point because of the linear homogeneity of the differential and terminal equations of variation. Therefore there exists a determinant of form (5.1) not zero, and the arc E is normal AB.

In a similar manner the following theorem can be proved:

THEOREM 7. *A necessary and sufficient condition that an admissible arc be normal (AB) is that the family of 2n-dimensional points linearly dependent on the points in 2n-space with coördinates η_i^s and the points in 2n-space with coördinates $\bar{\eta}_i^s$ fill out the whole of 2n-space. Here the set η_i^s consists of the end values of differentially admissible variations η_i, and the set $\bar{\eta}_i^s$ consists of terminally admissible variations.*

The following test for normalcy (AB) is useful:

THEOREM 8. *Let ρ be the highest rank attained by any determinant consisting of $2n + 2$ columns of differentially admissible points in $(2n + 2)$-space. The number of parameters (α) in the terminal conditions (2.4) cannot be less than $2n + 2 - \rho$ without E being anormal (AB). One can, however, always choose terminal conditions of the form (2.4) with no more than $2n + 2 - \rho$ parameters such that E is normal (AB).*

The differentially admissible points fill out a ρ-plane, and the terminally admissible points must fill out at least the $(2n+2-\rho)$-plane orthogonal to the ρ-plane at the origin if E is to be normal (AB). This latter plane is certainly not filled out if the number of parameters is less than $2n + 2 - \rho$. This proves the first part of the theorem.

Now the $(2n + 2 - \rho)$-plane which must be filled out by the differentially admissible points if the arc E is to be normal (AB) may be represented in the following form, using $(\bar{\gamma}^s, \bar{\eta}_i^s)$ as coördinates in $(2n + 2)$-space:

$$\bar{\gamma}^s = c_h{}^s u_h \tag{6.1}$$
$$\bar{\eta}_i{}^s = c_{ih}{}^s u_h.$$

The matrix $\left\| \begin{matrix} c_h{}^s \\ c_{ih}{}^s \end{matrix} \right\|$ is to be of rank $2n+2-\rho$. A possible choice of terminal conditions which will produce (6.1) as the terminally admissible points is the following: (cf. (3.6))

$$x^s - a^s = c_h{}^s \alpha_h \qquad (h = 1, \cdots, 2n + 2 - \rho) \tag{6.2}$$
$$y_i{}^s - b_i{}^s = (\bar{y}_i{}'^{s} c_h{}^s + c^s{}_{ih})\alpha_h$$

where $(a^s, b_i{}^s)$ are the coördinates of the end points of E.

A theorem similar to the last one can be obtained by replacing $2n + 2 - \rho$ by $2n - \sigma$, where σ is the highest rank attained by any determinant of the form

$$
(6.3) \qquad \begin{vmatrix} \eta^1{}_{11} \cdots\cdots \eta^1{}_{1\,2n} \\ \cdots\cdots\cdots\cdots \\ \eta^1{}_{n1} \cdots\cdots \eta^1{}_{n\,2n} \\ \eta^2{}_{11} \cdots\cdots \eta^2{}_{1\,2n} \\ \cdots\cdots\cdots\cdots \\ \eta^2{}_{n1} \cdots\cdots \eta^2{}_{n\,2n} \end{vmatrix} .
$$

A differentially admissible arc shall be said to be *normal* (A) if it does not satisfy condition A of Theorem 2 with any admissible set of multipliers in which $\lambda_0 = 0$.

THEOREM 10.* *If a differentially admissible arc E is normal (A), there exists a family of differentially admissible arcs joining the points (x^1, y^1) and (x^2, y^2) neighboring the end points of E and representable in the form*

$$
(6.4) \qquad y_i = Y_i(x, x^1, y^1, x^2, y^2)
$$

where the functions Y_i are of class C'.

Since E is normal (A), it is normal (AB) if we choose the end-conditions so as to make the end points coincide with those of E. Hence, according to the corollary to Theorem 4, there exists a determinant of end values of differentially admissible variations of the form

$$
(6.5) \qquad |\eta^s{}_{ik}| \qquad (i = 1, \cdots, n;\ k = 1, \cdots, 2n)
$$

different from zero.

Now let us use the corollary of § 3 to set up a 2n-parameter family of differentially admissible arcs

$$
(6.6) \qquad y_i = y_i(x, e) \qquad (e) = (e_1, \cdots, e_{2n})
$$

whose variations with respect to e_k are precisely the variations in the k-th column of (6.5) and which contains the arc E for $(e) = (0)$. We then seek to satisfy the conditions

$$
(6.7) \qquad y_i^s = y_i(x^s, e).
$$

These conditions are satisfied by the end points of E and $(e) = (0)$, and the jacobian of the right hand members with respect to (e) is the determinant (6.5), different from zero. We can accordingly solve

* cf. Bliss, Lectures, loc. cit. p. 50.

(6.7) for (e), substitute the result in (6.6), and obtain the desired family.

References

[1] Morse. For an abstract see "The Problems of Lagrange and Mayer under General End Conditions," Proceedings of the National Academy of Sciences, vol. 16 (1930), pp. 229–233.

[2] Bolza. "Uber den Anormalen Fall beim Lagrangeschen und Mayerschen Problem mit gemischten Bedingungen und Variabeln Endpunkten," Mathematische Annalen, vol. 74 (1913), pp. 430–446.

[3] Bliss. "The Problem of Lagrange in the Calculus of Variations," lectures given at the University of Chicago, summer quarter, 1925. These lectures were prepared by O. E. Brown, Northwestern University, Evanston, Illinois.

Further references follow:

Radon. "Zum Problem von Lagrange," Abhandlungen aus dem Mathematischen Seminar, Hamburg, 1928.

Bliss. "The Problem of Mayer with Variable End Points," Transactions of the American Mathematical Society," vol. 19 (1918), p. 312.

Bolza. "Vorlesungen über Variationsrechnung."

Bliss. "The Problem of Lagrange in the Calculus of Variations," American Journal of Mathematics, vol. 52 (1930), pp. 673–744.

Reprinted from the
ANNALS OF MATHEMATICS
Second Series, Vol. 32, No. 3, pp. 549–566
June, 1931
PRINCETON, N.J.

CLOSED EXTREMALS.*

(FIRST PAPER.)

By Marston Morse.[1]

1. Introduction. The problem is one of the characterization and existence of closed extremals. It belongs both to the Calculus of Variations and to Differential Topology.

In problems of this sort it is important to distinguish between the general case and the special case. A closed extremal g is called general or "*non-degenerate*" if the corresponding equations of variation possess no periodic solutions except the null solution. To a non-degenerate extremal g the author attaches an *index* which characterizes g in the present theory in the same way that the number of conjugate points on an extremal segment characterizes the extremal segment in the theory of extremals joining two fixed points.[2] (See Trans. III and IV.) This index is shown to be not only a geometric invariant in the ordinary sense but also a semi-topological invariant. It is a geometric invariant in that it is independent of the parametric system or choice of coordinates. It is a semi-topological invariant in that it is definable by means of deformations of closed extremals of a restricted class. Cf. § 35 Trans. II.

This index of g equals the index of an associated quadratic form and depends upon an associated linear boundary value problem.[3] It is determined explicitly for the "principal ellipses" on an r-dimensional ellipsoid E_r whose semi-axes are unequal and have lengths near unity. It is shown that the closed geodesics covering these principal ellipses are the only closed geodesics on E_r with lengths less than a prescribed constant N, provided the semi-axes of E_r have lengths sufficiently near unity.

In the second paper we shall show that on a regular, analytic image S of an r-sphere there will always exist closed extremals with indices (suitably counted) such as the principal ellipses on E_r possess. If all the closed geodesics on S with lengths less than a sufficiently large constant N

* Received March 13, 1931.

[1] Morse, Proceedings of the National Academy of Sciences, vol. 15 (1929), pp. 856–859. The principal results of the present paper were outlined here.

[2] Morse, Transactions of the American Mathematical Society I, vol. 27 (1925), pp. 345–396; II, vol. 30 (1928), pp. 213–274; III, vol. 31 (1929), pp. 379–404; IV, vol. 32 (1930), pp 599–631; V, vol. 33 (1931), pp. 72–92.

These papers will be referred to as Trans. I, II, etc.

[3] Morse, American Journal of Mathematics, vol. 53 (1931).

are non-degenerate, the preceding statement is true without qualifications as to the count of indices. The final theorem of the preceding paragraph now assumes a peculiar importance because it shows that the result of the present paragraph cannot be made any stronger in the non-degenerate case. For E_r furnishes an example in which no more closed extremals exist than are affirmed to exist on S in general.

In the second paper the characterization of closed extremals will be made to depend on the author's theory of critical sets of functions regardless of whether the extremal is degenerate or not. (See Trans. I and V.) It is found that in the special case, where two closed extremals on S which are affirmed to exist and possess different indices, give the same value to the integral, there will appear a p-dimensional family of closed extremals, where p is the difference between the two indices.

The characterization of closed extremals is the authors. Except in the 2-dimensional reversible case where special methods[4] have been used, and except for the existence of at least one closed geodesic on S, as proved by Birkhoff,[5] the author's results on the existence of closed extremals are new. In particular the general theorem on the existence of extremals in the non-degenerate case is new without exception. In this case the author proves the existence of at least as many closed extremals on S as there are combinations of $r+1$ objects taken 2 at a time.

The results of the present paper will also be used in determining certain topological invariants to be called *circular connectivities* in a theory in which the closed curve replaces the point.

2. The r-spread S and the integral. Let

$$(w) = (w_1, \cdots, w_m) \qquad m > 2$$

be rectangular coördinates in an Euclidean m-space. Let $(v) = (v_1, \cdots, v_r)$ be rectangular coördinates in an auxiliary r-space, $(1 < r < m)$. By an *r-element* in the space (w) will be meant a set of points (w) homeomorphic with a set of points (v) within an $(r-1)$-sphere in the space (v). An r-element will be said to be of class $C^{(n)}$ if the w's are functions of the v's of class $C^{(n)}$, that is possess continuous nth order partial derivatives. An r-element will be called *regular* if it is of at least class C', and on it at least one of the jacobians of r of the w's with respect to the r v's is not zero.

We shall be concerned with a regular r-spread S of the following nature.

The r-spread S shall consist of a bounded, connected, closed set of

[4] M. M. Lusternik et Schnirelmann, Topological Methods in Variational Problems. Research Institute of Mathematics and Mechanics. Gosizdat Moscow (1930).

[5] Birkhoff, Dynamical Systems, American Mathematical Society Colloquium Publications, vol. 9.

points (w) the neighborhood of every point of which consists of a regular element of class C'''. We shall *admit* no representation of neighborhoods on S other than by regular elements of class C'''.

For each point (w) on S and for each set of direction numbers (σ) of a direction, tangent at (w) to S, let there be given a single-valued function,

$$F(w_1, \cdots, w_m, \sigma_1, \cdots, \sigma_m) = F(w, \sigma),$$

positive and *continuous* in (w) and (σ), for (w) on S and $(\sigma) \neq (0)$. Suppose further that

$$F(w, k\sigma) \equiv kF(w, \sigma)$$

for any positive constant k.

We shall suppose S *orientable.* For our purposes the following definition is convenient.

Let $D(p)$ be an ordered set of r mutually orthogonal directions each tangent to S at a variable point p of S. Let p_0 be any point on S. Suppose p traverses a closed curve h on S, starting and ending at p_0, and suppose that $D(p)$ varies continuously with p on h. If at the end of any such variation the final set D can be continuously turned about p_0 in its own r-plane so as to coincide with the original set D, then S will be termed *orientable.*

The hypothesis of orientability could readily be dispensed with. It is made for the sake of simplicity.

Corresponding to any admissible element $w_i = \bar{w}(v)$ of S we set

$$F\left[\bar{w}_1, \cdots, \bar{w}_m, \frac{\partial \bar{w}_1}{\partial v_j}\varrho_j, \cdots, \frac{\partial \bar{w}_m}{\partial v_j}\varrho_j\right] \equiv G(v, \varrho) \qquad (j = 1, \cdots, r)$$

for (v) on the element and for $(\varrho) \neq (0)$. Here j is to be summed following the usual conventions of tensor analysis to which we adhere.

We assume that the functions $G(v, \varrho)$ are of class C''' for (v) on the corresponding element and for $(\varrho) \neq (0)$.

The curves $w_i = w_i(t)$ to be admitted on S shall be of the "ordinary" type[6] with continuously turning tangents except at most at a finite number of corners.

Along any such curve g our integrand is to have the form

$$J = \int_{t^1}^{t^2} F(w, \dot{w})\, dt \tag{2.1}$$

[6] See Bolza, Vorlesungen über Variationsrechnung (1909), p. 192.

where the dot indicates differentiation with respect to t. If g lies on an element of parameters (v) then g can be represented in the form $v_j = v_j(t)$ and along g our integral will take the form

$$J = \int_{t^1}^{t^2} G(v, \dot{v})\, dt. \tag{2.2}$$

We assume that the integrand G is positively regular.

That is we assume that on each element the corresponding function G is such that

$$G_{\varrho_h \varrho_k}(v, \varrho)\, z_h z_k > 0 \qquad (h, k = 1, \cdots, r)$$

for $(\varrho) \neq (0)$, and (z) any set not (0) nor proportional to (ϱ).

An extremal shall be a regular curve of class C''' on S which satisfies the Euler equations set up for integrals (2.2). One proves readily that a curve which satisfies the Euler equations corresponding to any one admissible representation of an element will also satisfy the Euler equations arising from any other admissible representation of the whole or part of that element.

3. **A representation of the neighborhood of a closed curve.** Let g be a regular closed curve of class C''' and of length ω. Concerning g we shall prove the following lemma.

LEMMA. *The part of S near g can be admissibly represented in the form*

$$w_i = w_i(x, y_1, \cdots, y_n) \qquad (n = r - 1;\ i = 1, \cdots, m)$$

in such a fashion that g corresponds to the x axis in the space (x, y) and the functions $w_i(x, y)$ have a period ω in x.

The proof will be made to depend upon the following statements.

Let s be the arc length along g. For each value of s for which $0 \leqq s \leqq \omega$ there exists a set of m directions with direction cosines given by the columns of a matrix

$$\| a_{ij}(s) \| \qquad (i, j = 1, \cdots, m)$$

with the following properties.

A. The functions $a_{ij}(s)$ are continuous in s and $|a_{ij}(s)| > 0$. The first direction of the set is positively tangent to g at the point s, while the first r directions of the set are tangent to S at this point s. The sets $a_{ij}(0)$ and $a_{ij}(\omega)$ each consist of orthogonal directions.

B. The functions $a_{ij}(s)$ have a period ω in s.

To prove these statements we assume that there exists a set $\bar{a}_{ij}(s)$ which satisfies A but not necessarily B. The proof of this assumption can be given by a process of continuation along g and will be left to the reader.

We shall now show how to replace the set $\bar{a}_{ij}(s)$ by a set $a_{ij}(s)$ which satisfies B as well as A. That this can be done depends upon the orientability of S.

Let π_s be the r-plane tangent to S at the point s on g, and let L_s be the straight line tangent to g at the same point. As the time t varies from 0 to 1 let T represent a continuous rigid motion of our m-space which turns π_0 in itself, leaves the points of L_0 fixed, and which turns the directions of the set $\bar{a}_{ij}(0)$ into the corresponding directions of $\bar{a}_{ij}(\omega)$. Such a motion is possible because of our hypothesis of orientability of S.

We now return to the set $\bar{a}_{ij}(s)$ as originally defined.

Let e be a small positive constant. As the time t varies from 0 to 1 let each direction of the set $\bar{a}_{ij}(s)$ for which s lies between 0 and e be turned under T until t reaches that time t_s which divides the interval (0, 1) in the ratio inverse to that in which s divides the interval $(0, e)$. Of the directions into which the directions $\bar{a}_{ij}(s)$ are thereby turned we now replace the first direction by the direction of L_s and the 2nd to the rth by their orthogonal projections on π_s. The set $\bar{a}_{ij}(0)$ will thereby be replaced by the set $\bar{a}_{ij}(\omega)$, and if e be sufficiently small the final sets $a_{ij}(s)$ will satisfy both A and B.

To return to the proof of the lemma we now approximate the elements $a_{ij}(s)$ by elements $b_{ij}(s)$ which are analytic in s and have the period ω, except that we take the first column of $b_{ij}(s)$ as the first column of $a_{ij}(s)$. We make the approximations so close that $|b_{ij}(s)|$ is not zero.

Suppose that g is given in the form $w_i = w_i(s)$. Near g we shall make a transformation from the variables (w) to variables

$$(x, y_1, \cdots, y_{m-1}) \tag{3.1}$$

of the form

$$w_i = w_i(x) + b_{i\,p+1}(x) y_p \qquad (i = 1, \cdots, m;\ p = 1, \cdots, m-1) \tag{3.2}$$

where x is any number, and the variables y_p are near zero.

Under this transformation the x axis corresponds to g. The Jacobian of the right hand number of (3.2) with respect to the variables (3.1) evaluated for $y_p = 0$, is seen to be $|b_{ij}(x)| \neq 0$.

Suppose $w_i = \bar{w}_i(v)$ is an admissible representation of S near a point P on g. We shall see that the variables

$$(x, y_1, \cdots, y_n) \qquad n = r - 1 \tag{3.3}$$

can serve as parameters in an admissible representation of S near P.

To establish this fact we note that the equations

$$\bar{w}_i(v) = w_i(x) + b_{i\,p+1}(x) y_p \tag{3.4}$$

39*

have an initial solution corresponding to the point P on g. In the neighborhood of this initial solution these equations can be solved for the parameters (v) and the variables

$$y_r, \cdots, y_{m-1}$$

in terms of the variables (3.3). In fact the pertinent Jacobian D is not zero if the elements $b_{ij}(x)$ are sufficiently good approximations of the elements $a_{ij}(x)$. For one notes that in D the columns of partial derivatives of $\overline{w}_i(v)$ define a set of r independent directions in π_s, while the remaining columns of D define $m-r$ additional independent directions if our approximations are sufficiently close.

Thus the variables (3.3) form a set of admissible parameters in a representation of S near g, and the lemma follows directly.

In terms of the parameters (x, y) of the preceding lemma our integral J will now be given the form (cf. § 5, Trans. III)

$$J = \int_a^b f(x, y, y')\,dx, \qquad y'_k = \frac{dy_k}{dx} \tag{3.5}$$

where we restrict ourselves to ordinary curves near the x axis composed of a finite succession of closed segments on each of which y_k is a function of x of class C'.

4. **The fundamental form Q of a closed extremal.** Let g be a closed extremal on S of length ω. With g we shall now associate a quadratic form Q which brings to light certain remarkable characteristics of g and enables one to define in a very simple way certain integral geometric invariants of g.

We represent S near g as in the preceding lemma so that g is mapped on the x axis. In particular g will correspond to a segment γ of the x axis from $x = 0$ to $x = \omega$.

We cut orthogonally across γ with $p+1$ successive n-planes

$$t_0, \cdots, t_p \tag{4.1}$$

of which the first is $x = 0$ and the last $x = \omega$ and which are placed so near together that there are no pairs of conjugate points on the successive segments into which γ is divided. (See § 9, Trans. III). Let P_i be any point on t_i near γ arbitrarily placed on t_i except that P_0 and P_p shall have the same coördinates (y).

The points $P_0, \cdots, P_p$ can be successively joined by extremal segments neighboring γ. Let the resulting broken extremal be denoted by E. Let (u) be a set of $\mu = pn$ variables of which the first n are the coördinates (y) of P_1, the second n the coördinates (y) of P_2, and so on until finally the last are the coördinates (y) of P_p. The value of the integral J taken along E will be denoted by $J(u)$.

The function $J(u)$ will have a critical point at $(u) = (0)$, that is a point at which all of its first partial derivatives are zero.

We come next to our *fundamental form*

$$Q(z) = J^0_{u_h u_k} z_h z_k \qquad (h, k = 1, \cdots, \mu)$$

where the superscript 0 calls for the evaluation of the partial derivatives at $(u) = (0)$.

By the *index* of a quadratic form is meant the number of negative terms that appear in the form after the form has been reduced by a real, non-singular, linear transformation to squared terms alone. By the *nullity* of the form is meant the order of its matrix minus its rank.

We shall classify our closed extremals according to the index and nullity of the form Q.

A closed extremal whose nullity is zero will be called *non-degenerate.*

We set $y'_i = p_i$ and further set

$$2\,\Omega(x, \eta, \eta') = f^0_{p_i p_j} \eta'_i \eta'_j + 2 f^0_{p_i y_j} \eta'_i \eta_j + f^0_{y_i y_j} \eta_i \eta_j$$

where $i, j = 1, \cdots, n$, and where the partial derivatives are evaluated for $(x, y, p) = (x, 0, 0)$ as indicated by the superscript zero. Curves in the space (x, η) which are represented by functions $\eta_i = \eta_i(x)$ of class C'' and which satisfy the Jacobi differential equations

$$\Omega_{\eta_i} - \frac{d}{dx}\,\Omega_{\eta'_i} = 0 \qquad (i = 1, \cdots, n) \tag{4.2}$$

will be called *secondary* extremals.

It will convenient to suppose the spaces (x, η) and (x, y) are identical. With this understood we see that the successive points $P_0, \cdots, P_p$ can be joined by secondary extremals. The set of these secondary extremals form a broken secondary extremal which we shall say is *determined* by and *determines* the set (u) previously defined.

With this understood we state the following lemma.

LEMMA. *The fundamental form Q can be represented as follows,*

$$Q(z) = J^0_{u_h u_k} z_h z_k = 2\int_0^\omega \Omega(x, \eta, \eta')\,dx, \qquad (h, k = 1, \cdots, \mu)$$

where $\eta_j = \eta_j(x)$ represents the broken secondary extremal determined by (z).

The proof is essentially that in § 10, Trans. III.

We next consider the integral

$$2\int_0^\omega [\Omega(x, \eta, \eta') - \sigma \eta_i \eta_i]\, dx \qquad (i = 1, \cdots, n) \tag{4.3}$$

in which the parameter σ may have any value independent of x.

Let the system of n Euler equations corresponding to the integral (4.3) be represented by

$$\Omega^{\sigma i} = 0 \qquad (i = 1, \cdots, n).$$

We state the following theorem.

THEOREM I. *The nullity of the fundamental form $Q(z)$ equals the number of linearly independent solutions of the equations $\Omega^{\sigma i} = 0$ which have the period ω for $\sigma = 0$.*

The index of the form $Q(z)$ equals the number of linearly independent solutions of the equations $\Omega^{\sigma i} = 0$ which have a period ω and for which $\sigma < 0$.

This theorem is proved in § 15, Amer. Journ. loc. cit.

5. **The non-degenerate case when $n = 1$.** In this case, in § 18, Trans. II we have further analyzed the preceding segments γ as follows.

If $x = 0$ is not conjugate to $x = \omega$ on γ, there exists a secondary extremal E on which $\eta(0) = \eta(\omega) = 1$. Let E' be a secondary extremal obtained from E by replacing each point (x, η) on E by the point $(x + \omega, \eta)$. We see that E and E′ intersect at $(\omega, 1)$. Moreover, E and E' will not be tangent at $(\omega, 1)$ unless E and E' are identical and hence periodic. This could happen only in case the nullity of $Q(z)$ is positive, that is in the so called "*degenerate*" case. We exclude the degenerate case in this section.

As x increases through ω, E crosses E' at $(\omega, 1)$. If E then enters the region between E' and the x axis, we say that γ is *relatively concave*, in the contrary case *relatively convex*.

We have the following theorem from § 21, Trans. II.

THEOREM 2. *In the case $n = 1$ the index of a non-degenerate periodic extremal g may be determined as follows.*

If on the x axis representing g, $x = 0$ is not conjugate to $x = \omega$, the index of g equals the number of conjugate points of $x = 0$ which precede $x = \omega$ plus 0 or 1 according as the x axis between 0 and ω is relatively convex or concave.

If a point $x = a$ is conjugate to $x = a + \omega$, the index of g is the number of conjugate points of $x = a$ between a and $a + \omega$ including $a + \omega$.

We shall now establish the following theorem.

THEOREM 3. ($n = 1$). *If to each point x on the x axis representing the closed extremal there corresponds k conjugate points ($k > 0$) between x and $x + \omega$, while the point x is never conjugate to $x + \omega$, then the index of the closed extremal is k if k is odd and $k + 1$ if k is even.*

By virtue of the preceding theorem we have merely to show that γ is relatively convex if k is odd and relatively concave if k is even. The proof is largely an application of the Sturm separation theorem.

Suppose $k = 2r - 1$ where $r > 0$.

I say that E cannot pass below E' at the point $(\omega, 1)$, that is E cannot there enter the region between E' and the x axis.

A comparison of E with a secondary extremal which vanishes at $x = 0$ shows that E must vanish $2r$ times on the interval $(0, \omega)$. Let $x = a$ be the first zero of E following $x = 0$.

The first conjugate point of $x = 0$ following $x = 0$ must follow $x = a$. Hence the first conjugate point of $x = \omega$ following $x = \omega$ must follow $x = a + \omega$. Accordingly E cannot intersect E' except at $x = \omega$ on the closed interval $(\omega, a + \omega)$.

Moreover, E cannot intersect the x axis on the interval $(\omega, a + \omega)$ since that would mean that the point $x = a$ had $2r$ conjugate points on the interval $(a, a + \omega)$ contrary to our hypothesis.

It follows topologically that E cannot pass below E' at the point $(\omega, 1)$.

Thus if k is odd, γ is convex, and the index is k.

One can treat the case where k is even in a similar fashion.

6. Invariance of the index and degeneracy of a closed extremal. Let us call a set of parameters (x, y) of the type whose existence is affirmed in the lemma of § 3 a *canonical* set of parameters neighboring g. In terms of these parameters let the fundamental form Q be set up as in § 4. If Q is degenerate g will be momentarily termed *degenerate relative to the parameters* (x, y). The index of Q will be momentarily called the *index* of g *relative to the parameters* (x, y).

We shall prove the following theorem.

THEOREM 4. *The property of degeneracy of a closed extremal g and value of the index of a non-degenerate closed extremal are independent of the particular system of canonical parameters with the aid of which the fundamental form is defined.*

Let (x, y) be a set of canonical parameters and $f(x, y, y')$ the corresponding integrand of § 3. Let k be the index of g relative to the parameters (x, y).

If g is non-degenerate its index k can be characterized in a way clearly independent of the canonical parameters used. To proceed we need a definition.

Let e be a small positive constant. An ordinary closed curve h on S representable in the space (x, y) in the form $y_i = y_i(x)$ will be said to lie in the *neighborhood* R_e of g if y_i and y'_i are in absolute value less than e along h. $0 \leq x \leq \omega$.

The case $k = 0$ can now be characterized as follows.

(a) *In case g is non-degenerate relative to the parameters (x, y) a necessary and sufficient condition that its index k be zero relative to (x, y), is that g afford a proper minimum to J relative to all ordinary closed curves which lie in a sufficiently small neighborhood R_e of g.*

In characterizing the case $k \neq 0$ we shall *admit* only those ordinary closed curves which give J a value less than J_g. This characterization is an obvious generalization of a part of Theorem 10 in Trans. II.

(b) *Suppose g is non-degenerate relative to (x, y) and has an index $k \neq 0$. Corresponding to any sufficiently small neighborhood R_e of g there exists an arbitrarily small neighborhood $R_{e'}$ with the following property. The index k of g relative to (x, y) is one more than the minimum order m of closed m-families of admissible curves on $R_{e'}$ which cannot be deformed among admissible curves on R_e into a single admissible curve.*

To return to the proof of Theorem 4 we consider the following integral:

$$I^\sigma = \int_0^\omega [f(x, y, y') - \sigma y_i y_i]\, dx, \qquad (i = 1, \cdots, n). \tag{6.1}$$

The integrand has a period ω in x, and the x axis is still an extremal. If the integral be represented in terms of the original parameters of S, g will still appear as a closed extremal.

The second variation of the above integral will take the form (4.3) and the corresponding Euler equations will be represented by the set of equations,

$$\Omega^{\sigma i} = 0 \qquad (i = 1, \cdots, n). \tag{6.2}$$

Values of σ for which (6.2) has solutions (not null) of period ω are here called characteristic roots.

(c) *Relative to the parameters (x, y), and for $\sigma = 0$ the extremal g is now assumed degenerate.*

According to Theorem 1, $\sigma = 0$ must then be a characteristic root. Let $k(\sigma)$ be the index of g relative to the parameters (x, y) and the integral I^σ. According to Theorem 1 $k(\sigma)$ will suffer an increase as σ increases through 0 equal to the number of linearly independent periodic solutions of (6.2) for $\sigma = 0$. For σ near 0 but not 0, g will be non-degenerate, and the preceding characterizations of k in (a) and (b) are at our disposal.

Let $(\bar{x}, \bar{y})$ be a second set of canonical parameters. Restricting ourselves to ordinary curves in the (x, y) space in a sufficiently small neighborhood R_e of the x axis we see that the integral I^σ can be represented by an integral $\bar{I}^\sigma$ in terms of the parameters $(\bar{x}, \bar{y})$ with $\bar{x}$ as the independent variable.

(d) *I say that under the assumption* (c) g *must be degenerate for* $\sigma = 0$, *relative to the parameters* $(\bar{x}, \bar{y})$.

For in the contrary case the index $\bar{k}(\sigma)$ of g relative to $(\bar{x}, \bar{y})$ and $\bar{I}^{\sigma}$ would be constant for σ sufficiently near zero. But from (a) and (b) we see that $k(\sigma) = \bar{k}(\sigma)$ for σ near 0 but not 0. Thus $k(\sigma)$ would not change as σ increases through $\sigma = 0$. From this contradiction we infer the truth of (d).

The part of the theorem concerning the index now follows from (a) and (b).

7. The indices of the principal ellipses on an $(m-1)$-ellipsoid. We shall consider an $(m-1)$-ellipsoid E_{m-1} of the form

$$(7.1) \qquad a_1^2 w_1^2 + \cdots + a_m^2 w_m^2 = 1, \quad a_1 > a_2 > \cdots > a_m > 0$$

for which the constants a_i are near 1.

By *principal ellipse* g_{ij} of E_{m-1} $(i \neq j)$ will be meant the ellipse on E_{m-1} which lies on E_{m-1} and in the 2-plane of the w_i and w_j axes. The number of principal ellipses is $m(m-1)/2$. The principal ellipses are closed geodesics.

The determination of the indices of the principal ellipses on E_{m-1} can be reduced to a determination of the indices of the principal ellipses on E_2. We now proceed with the latter determination.

8. The ellipsoid E_2. First recall that if s is the arc length along any closed geodesic g on a 2-dimensional spread S, the conjugate points of the point $s = s_0$ on g are the zeros other than $s = s_0$ of a non-null solution of the differential equation

$$(8.1) \qquad \frac{d^2 u}{d s^2} + K(s) u = 0$$

where $K(s)$ is the total curvature of S at the point s on g, and u vanishes at $s = s_0$. (See Bolza, p. 231).

As in Trans. II we shall call g doubly-degenerate, simply-degenerate, or non-degenerate according as (8.1) possesses 2, 1, or 0 linearly independent, non-null solutions with a period ω.

We shall use the following lemma. (See § 19, Trans. II).

LEMMA 8.1. *If g is a simply-degenerate closed extremal and $u(s)$ is a non-null solution of* (8.1) *with a period ω, the only points s on g for which s is conjugate to $s + \omega$ are the points at which* $u(s) = 0$.

We shall now prove the following lemma.

LEMMA 8.2. *If the semi-axes of the ellipsoid E_2 of § 7 have lengths sufficiently near unity the principal ellipses of E_2 have the following properties.*

(a) *To each point s on g_{12} there corresponds just one conjugate point prior to $s+\omega$, while s is never conjugate to $s+\omega$.*

(b) *To each point s on g_{23} there correspond just two conjugate points prior to $s+\omega$ while s is never conjugate to $s+\omega$.*

(c) *On g_{13} opposite umbilical points are conjugate to each other and to no other points. The geodesic g_{13} is non-degenerate.*

Let the ellipsoid

$$b_1^2 x_1^2 + b_2^2 x_2^2 + b_3^2 x_3^2 = 1, \qquad b_i > 0$$

be denoted by $E(b_1, b_2, b_3)$.

We need the fact that the total curvature $K(s)$ of $E(b_1, b_2, b_3)$ along its principal ellipse g_{12} will be increased if b_3 is replaced by a larger positive constant. In fact $K(s)$ is the product of the curvature k_1 of g_{12} at the point s and the curvature k_2 of the ellipse π in which a plane orthogonal to g_{12} at the point s, cuts $E(b_1, b_2, b_3)$. An increase of b_3 will not alter k_1 but it will diminish the axis of the ellipse orthogonal to the plane g_{12}. It will accordingly increase k_2 and hence $K(s)$.

We now return to the constants $a_1 > a_2 > a_3$ of § 7, and consider the spheroid $E(a_1, a_2, a_2)$. On the ellipse g_{12} of this spheroid we shall measure s from the point $(a_1, 0, 0)$. We shall prove the following statement.

(A) *On the ellipse g_{12} of $E(a_1, a_2, a_2)$ the distance Δs from a point s to its first following conjugate point exceeds $\omega/2$ for all points except the points conjugate to $s = 0$, for which $\Delta s = \omega/2$.*

First note that $s = 0$ is conjugate to $\omega/2$ and ω, and that the corresponding solution of (8.1) has the period ω.

On the other hand on the ellipse g_{12} of $E(a_1, a_2, a_2)$ $K(s)$ is less than the total curvature at the same point on $E(a_1, a_2, a_1)$. But on $E(a_1, a_2, a_1)$ the point $s = \omega/4$ is conjugate to the opposite point on g_{12}. An application of the Sturm comparison theorem to (8.1) now shows that the distance from $s = \omega/4$ to its first conjugate point on g_{12} exceeds $\omega/2$ on $E(a_1, a_2, a_2)$. It follows from Lemma 8.1 that the same is true for all points of g_{12} on $E(a_1, a_2, a_2)$, except the points conjugate to $s = 0$.

Thus the statement (A) is proved.

We can now prove statement (a) of Lemma 8.2.

The curvature $K(s)$ on the ellipse g_{12} of $E(a_1, a_2, a_3)$ is less than that at the same points on $E(a_1, a_2, a_2)$. It follows from (A) and the Sturm comparison theorem that to each point s on the ellipse g_{12} of $E(a_1, a_2, a_3)$ there corresponds at most one conjugate point prior to or including $s+\omega$. If we bear in mind what happens on the unit sphere, we see that if the constants a_i are sufficiently near unity there will be exactly one conjugate point of the point s prior to $s+\omega$, and $s+\omega$ will not be conjugate to s.

Thus (a) is proved.

We can prove (b) similarly, first proving the following.

(B) *On the ellipse* g_{23} *of* $E(a_2, a_2, a_3)$ *the distance* Δs *from a point* s *to the first following conjugate point is less than* $\omega/2$ *for all points except the points conjugate to the point* $(w) = (0, 0, a_3)$, *for which* $\Delta s = \omega/2$.

To prove (B) we compare g_{23} on $E(a_2, a_2, a_3)$ with g_{23} on $E(a_3, a_2, a_3)$, and then use Lemma 8.1 as in the proof of (A).

To prove (b) we make use of (B) comparing g_{23} on $E(a_1, a_2, a_3)$ with g_{23} on $E(a_2, a_2, a_3)$.

To prove (c) we recall that the geodesics through an umbilical point pass through the opposite umbilical point but form a field otherwise. Hence, each umbilical point on g_{13} is conjugate to the opposite umbilical point.

Further, g_{13} cannot be doubly-degenerate. For if so $s + \omega$ would be the second conjugate point, not only of the umbilical points s, but also of all points s. After a slight increase of a_2 no point $s + \omega$ on g_{13} would be a conjugate point of s, contrary to the fact that there would still be umbilical points on g_{13}. Thus g_{13} cannot be doubly-degenerate.

Finally g_{13} cannot be simply-degenerate because it would then follow from Lemma 8.1 that the four umbilical points would be mutually conjugate, which is false.

Thus (c) is proved.

From Theorems 2 and 3 and the preceding lemma we now have the following theorem.

THEOREM 5. *If the semi-axes of the ellipsoid*

$$a_1^2 x_1^2 + a_2^2 x_2^2 + a_3^2 x_3^2 = 1 \qquad a_1 > a_2 > a_3 \tag{8.2}$$

are sufficiently near unity the indices of the principal ellipses g_{12}, g_{13}, *and* g_{23} *are* 1, 2, *and* 3, *respectively.*

9. **The index formula on E_{m-1}.** We shall now determine the indices of the principal ellipses g_{ij} on the $(m-1)$-ellipsoid E_{m-1} of (7.1).

By the *principal ellipsoids* of E_{m-1} we shall mean those ellipsoids E_2 which are obtained from E_{m-1} by setting all of the coördinates (w) equal to zero excepting three.

We note that the geodesics on any principal ellipsoid are geodesics on E_{m-1}. This is seen most simply by putting the equations of the geodesics in the form

$$\frac{d^2 w_i}{ds^2} = \lambda w_i a_i^2 \qquad (i = 1, \cdots, m;\ i \text{ not summed})$$

where (w) lies on E_{m-1}, and λ is a function of s.

We shall now fix our attention on a particular principal ellipse, $g_{m-1, m}$, *which we denote by* $\bar{g}$.

We can represent E_{m-1} near $\bar{g}$ in terms of parameters (x, y) as in the Lemma of § 3. In fact if we set

$$r^2(y) = 1 - a_h^2 y_h^2 \qquad (h = 1, \cdots, n = m - 2)$$

we can represent E_{m-1} near $\bar{g}$ in the form

$$
\begin{aligned}
w_h &= y_h & (h = 1, \cdots, n), \\
a_{m-1} w_{m-1} &= r(y) \cos x & r > 0, \\
a_m w_m &= r(y) \sin x. &
\end{aligned}
\tag{9.1}
$$

The integral of arc length on E_{m-1} near $\bar{g}$ becomes the integral (3.5). We continue with the notation of sections 3, 4, and 5.

The principal ellipsoids E_2^k. The principal ellipsoid which lies in the 3-space of the w_k, w_{m-1}, w_m axes $(k < m - 1)$ will be denoted by E_2^k.

If we set all of the parameters (y) except y_k equal to zero, (9.1) will give a representation of the neighborhood of $\bar{g}$ on E_2^k. The geodesics of E_{m-1} near $\bar{g}$ on E_2^k will be represented by extremals in the 2-plane of x and y_k.

One readily verifies the fact that the present integral of arc length in the form (3.5) will be unchanged if we replace y_h by $-y_h$ and y_h' by $-y_h'$ for any particular integer h. It follows that under this substitution each extremal arc is replaced by an extremal arc.

Now consider the fundamental form $Q(z)$ set up for $\bar{g}$ as in § 4.

Recall that (z) is composed of the successive sets of coördinates (y) of the points $P_1, \cdots, P_p$ of § 4. Let the coördinates y_h of P_k now be denoted by

$$y_h^k \qquad (k = 1, \cdots, p;\ h = 1, \cdots, n).$$

It follows from the preceding symmetry considerations that $Q(z)$ is unchanged if for any h we replace

$$(y_h^1, \cdots, y_h^p) \quad \text{by} \quad (-y_h^1, \cdots, -y_h^p).$$

As a matter of quadratic forms it then follows that we can write

$$Q(z) = Q_1(y_1^1, \cdots, y_1^p) + \cdots + Q_n(y_n^1, \cdots, y_n^p)$$

where Q_k is a quadratic form in its arguments alone.

Hence the index of $Q(z)$ will be the sum of the indices of the separate forms Q_k.

But if we set all of the variables (z) equal to zero except those in Q_k we see that Q_k is the fundamental form (except for the symbols for its arguments) which would be associated with $\bar{g}$ if $\bar{g}$ were regarded as a closed geodesic on the principal ellipsoid E_2^k.

We have made the preceding analysis for $\bar{g} = g_{m-1,m}$. It is not essentially different for g_{ij} in general. We therefore have the following theorem.

THEOREM 6. *The index of the principal ellipse g_{ij} on the ellipsoid E_{m-1} in m-space is the sum of the indices of g_{ij} regarded as an ellipse on each of the $m-2$ principal 2-dimensional ellipsoids on which g_{ij} lies.*

When the semi-axes of E_{m-1} have lengths sufficiently near unity we can evaluate the index k of g_{ij} by using Theorems 5 and 6.

Suppose $i<j$. Then the index k is three times the number of integers between i and 0, plus two times the number of integers between i and j, plus one times the number of integers between j and $m+1$. Thus

$$k = 3(i-1)+2(j-i-1)+(m-j) = m+i+j-5.$$

We accordingly have the following theorem.

THEOREM 7. *If the semi-axes of the $(m-1)$-ellipsoid* (8.1) *have lengths sufficiently near unity, the index of the principal ellipse g_{ij} is $m+i+j-5$.*

The number of distinct indices of principal ellipses is $2m-3$.

The preceding theorem can be immediately extended as follows.

THEOREM 8. *If the semi-axes of the $(m-1)$-ellipsoid* (7.1) *have lengths sufficiently near unity, the index k of the closed geodesic which covers a principal ellipse g_{ij} $r+1$ times is given by the formula*:

$$k = m+i+j-5+2r(m-2).$$

To prove this extension we review the proof of Theorem 7 and successively verify the following statements.

If the semi-axes of the ellipsoid E_2 have lengths sufficiently near unity the indices of the closed geodesics which cover the principal ellipses g_{12}, g_{13}, and g_{23}, $r+1$ times are $2r+1$, $2r+2$, and $2r+3$ respectively.

The index of a closed geodesic g on E_{m-1} which covers a principal ellipse is the sum of the indices of g regarded as a closed geodesic on each of the $m-2$ principal 2-dimensional ellipsoids on which it lies.

The index k of a closed geodesic g which covers g_{ij} on E_{m-1} $r+1$ times $(i<j)$ is accordingly

$$\begin{aligned} k &= (2r+3)(i-1)+(2r+2)(j-i-1)+(2r+1)(m-j) \\ &= m+i+j-5+2r(m-2) \end{aligned}$$

and the theorem is proved.

10. The exclusiveness of the principal ellipses as closed geodesics. In this section we shall prove the following theorem.

THEOREM 9. *Let N be an arbitrarily large positive constant. Upon any $(m-1)$-ellipsoid whose semi-axes are unequal and sufficiently near unity in* *length there are no closed geodesics with lengths less than N other than multiples of the principal ellipses.*

We state the following lemma without proof. See § 19, Trans. II and Lemma 8.1.

LEMMA. *Let there be given a differential equation*

$$w'' + q(s)\, w = 0 \tag{10.1}$$

in which $q(s)$ is continuous and has a period ω. If $u(s)$ is a solution ($\not\equiv 0$) of period ω upon which all periodic solutions of (10.1) *are dependent, the only solutions whose zeros have the period ω are dependent on $u(s)$.*

Let the $(m-1)$-ellipsoid have the form (7.1). The equations of the geodesics then take the form

$$w_j'' + \lambda w_j a_j^2 = 0 \qquad (j \text{ not summed}), \tag{10.2}$$

$$a_i^2 w_i^2 = 1 \qquad (i, j = 1, \cdots, m) \tag{10.3}$$

where the independent variable is the arc length s, and where λ is an analytic function of s which we now determine. If we differentiate (10.3) twice with respect to s and use (10.2) we find that

$$\lambda = \frac{a_i^2 w_i' w_i'}{a_i^4 w_i w_i} \qquad w_i' w_i' = 1. \tag{10.4}$$

When the constants (a) of (10.3) $= (1)$, $\lambda \equiv 1$. Accordingly for (a) sufficiently near (1), λ will be uniformly near 1 for any point (w) on (10.3) and $w_i' w_i' = 1$.

Let g be any geodesic on E_{m-1} and $\lambda(s)$ the corresponding function λ. Let $w_j(s, c)$ be a solution of the jth equation (10.2) such that

$$w_j(s, c) = 0, \quad w_j'(s, c) = 1.$$

We shall now restrict the constants (a) to a neighborhood R of the set $(a) = (1)$ so small that any variation of (a) in R will cause the successive zeros of $w_j(s, c)$ on the interval

$$c \leqq s \leqq c + N$$

to vary by less than $\pi/4$ regardless of the choice of g, of c, or of j.

Suppose now that g is a closed geodesic $w_i = u_i(s)$, with a length $\omega < N$, and that $\lambda(s)$ is the corresponding function λ. Suppose that the theorem is false. For definiteness assume that

$$u_k(s) \not\equiv 0 \qquad (k = 1, 2, 3). \tag{10.5}$$

We shall make repeated use of the Sturm comparison theorem which we denote by (S).

An obvious use of (S) shows that $u_k(s)$ must vanish at least once, say at s_k. Because of our restrictions on the constants (a) it follows that the zeros of $u_k(s)$ on the interval

$$s_k \leqq s \leqq s_k + \omega$$

lie respectively within $\pi/4$ of the points

$$s_k,\ s_k + \pi,\ s_k + 2\pi,\ \cdots,\ s_k + 2r\pi$$

where r is a positive integer dependent on ω.

Moreover the solution $u_k(s)$ can have no zero, say $s = a$, in common with

$$u_h(s) \qquad (h \neq k;\ h, k = 1, 2, 3).$$

For the conditions

$$u_k(a) = u_h(a) = u_k(a+\omega) = 0$$

taken with (S), imply that $u_h(a+\omega) \neq 0$ contrary to the periodicity of $u_h(s)$.

The equation (k not summed)

$$(k) \qquad w_k'' + \lambda(s)\, a_k^2\, w_k = 0 \qquad (k = 1, 2, 3)$$

possesses no solution ($\not\equiv 0$) of period ω, independent of $u_k(s)$. For if all solutions of (k) had the period ω a use of (S) would show that equations (h), $h \neq k$, could have no solutions of period ω not identically zero.

Let $D_k(s)$ be the distance along the s axis from a zero of a solution of (k) which is not identically zero and which vanishes at s, to the $2r$th following zero. It follows from the lemma that the only points at which $D_k(s) = \omega$ are the zeros of $u_k(s)$.

A use of (S) shows that $D_2(s) > \omega$ at the zeros of $u_1(s)$. But there is at least one zero of $u_1(s)$ between each two consecutive zeros of $u_2(s)$. Hence, $D_2(s) \geqq \omega$ without exception. This fact and a use of (S) now show that $D_3(s) > \omega$ at each point s.

From this contradiction we infer that (10.5) is impossible.

The theorem follows readily.

11. The set of all closed extremals in the analytic case. In the paper which is to follow we shall restrict ourselves to the analytic case, that is to the case where the functions defining the elements of S are analytic as well as the integrands $G(v, \varrho)$. In the analytic case certain very definite statements can be made about the set of all closed extremals neighboring a given closed extremal g. It is convenient to make these statements in this place.

In the lemma of § 3 we showed how to map the neighborhood of a closed curve g on S onto the neighborhood of the x axis in a space of variables $(x, y_1, \cdots, y_n)$. A review of the proof of this lemma shows that in the analytic case and in case g is analytic the mapping functions $w_i = w_i(x, y)$ there defined are also analytic.

We represent S neighboring g in terms of these parameters (x, y). The extremals neighboring g can then be represented by giving their coördinates y_i as functions $\varphi_i(x, \alpha)$ of x and $2n$ parameters (α) giving the initial values of (y) and (y') when $x = 0$. The functions φ_i will be analytic in their arguments for (α) near the set (0). The conditions that one of these extremals have a period ω are that

$$\varphi_i(0, \alpha) = \varphi_i(\omega, \alpha), \quad \varphi_{ix}(0, \alpha) = \varphi_{ix}(\omega, \alpha) \quad (i = 1, \cdots, n). \tag{11.1}$$

These equations may be satisfied for real sets (α) only when $(\alpha) = (0)$. Apart from this case the real solutions (α) of (11.1) will be representable as functions "in general" analytic on one or more suitably chosen "Gebilde" G[7] of r independent variables with $0 < r \leqq 2n$, each G including the point $(\alpha) = (0)$. To each such set (α) corresponds a periodic extremal.

These periodic extremals neighboring g all give the same value to J.

To see this we consider any regular curve h on one of the above "Gebilde" G, a curve h along which the parameters (α) are analytic on G. We evaluate J along the corresponding extremals from $x = 0$ to $x = \omega$. Upon differentiating J with respect to the arc length along h and making the usual reduction of the first variation we readily obtain the result zero. It follows that J is constant on G, and in fact takes on the value afforded by g.

We shall regard a continuous family of closed curves as *connected* if any closed curve of the family can be continuously deformed into any other closed curve of the family through the mediation of curves of the family.

From the preceding analysis "in the small" we now infer the following result "in the large".

In the analytic case the set of all closed extremals on S on which J is less than a prescribed positive constant make up at most a finite set of connected and continuous families of closed extremals on each of which J is constant, and which are either 0-dimensional or else analytically representable neighboring a given closed extremal in terms of a variable x and parameters (α) on one or more of the above "Gebilde".

[7] See Osgood, Funktionentheorie II. To Osgood's "Gebilde" we add the space of the variables (α) as a special "Gebilde".

Reprinted from the
TRANSACTIONS OF THE AMERICAN MATHEMATICAL SOCIETY
Vol. 32, No. 4, pp. 599–631

THE FOUNDATIONS OF A THEORY OF THE CALCULUS OF VARIATIONS IN THE LARGE IN m-SPACE (SECOND PAPER)*

BY
MARSTON MORSE

1. Introduction. The general problem† is the classification and existence of extremals under all kinds of conditions, a non-linear boundary value problem in the large. This paper deals primarily with the fixed end point problem.

The first paper by the author dealt with the fixed end point problem in the small, and with the problem with one end point variable in general. The latter problem could be readily studied in the large because it corresponds to the problem of finding the critical points of a function whose critical points are in general isolated.

Such is not the case in the fixed end point problem. The critical sets appearing there take the form of n-dimensional loci. This extreme difficulty is met by the aid of deformations. The author's characterization of the number of conjugate points on an extremal by means of deformations is essential. See Morse II.

A second difficulty which long appeared insurmountable was the fact

---

* Presented to the Society as part of an address given at the request of the program committee April 6, 1928, under the title *The critical points of functions and the calculus of variations in the large*; see also Bulletin of the American Mathematical Society, vol. 35 (1929), pp. 38–54; received by the editors May 8, 1930.

† The following references should be made.

For work on the absolute minimum including work of Hilbert see Bolza, *Vorlesungen über Variationsrechnung*, 1909, pp. 419–437, and Tonelli, *Fondamenti di Calcolo delle Variazioni*, vol. 2. Further references will be found in these works.

Signorini, Rendiconti del Circolo Matematico di Palermo, vol. 33 (1912), pp. 187–193. Here broken geodesics are used in the quest for minimizing periodic orbits.

Birkhoff, *Dynamical Systems*, American Mathematical Society Colloquium Publications, vol. 9. In chapter V, Birkhoff makes new and effective use of deformations of broken geodesics to obtain basic periodic motions. His coupling of the minimum and minimax methods was undoubtedly the first step towards a general theory.

The following papers by the author will be referred to: I These Transactions, vol. 27 (1925), pp. 345–396. II These Transactions, vol. 30 (1928), pp. 213–274. III These Transactions, vol. 31 (1929), pp. 379–404. IV Mathematische Annalen, vol. 103 (1930), pp. 52–69.

See also the following abstract: MM. Lusternik et Schnirelmann, *Existence de trois géodésiques fermées sur toute surface de genre* 0, Comptes Rendus, vol. 188 (1929), No. 8, p. 534.

For references to Bliss, Mason, Carathéodory, and Hadamard, see the papers Morse II and III.

that the domains of definition of the functions whose critical points were sought were not complexes, or at least could not readily be reduced to finite complexes.

This difficulty was again surmounted with the aid of deformations. The regions involved were shown to be deformable on themselves into subregions which were complexes. The existence of these deformations was a matter of the calculus of variations, their use a matter of analysis situs. Both aspects were indispensable.

A final difficulty as well as source of great interest was the fact that the domains usually have infinite connectivities. From the point of view of the calculus of variations this involved a study of the density of infinite sets of conjugate points of a fixed point. Over against the infinite set of connectivities is set a conjugate number sequence belonging to the calculus of variations, and the interrelations of these two infinite sequences appear central.

As a particular example, one can prove the existence of infinitely many geodesics joining any two points on any regular topological image of an m-sphere.

The author wishes to state in passing that a considerable part of the analysis of this problem has been carried over to the problem of periodic extremals.* The latter problem, however, presents additional difficulties both from the point of view of the calculus of variations and that of analysis situs. It involves the study of homologies on a given complex among complexes which are restricted to product complexes. The simplest complexes present decidedly new aspects from this point of view.

I. The problem and its deformations

2. The hypotheses. Let S be a closed region representable as an m-dimensional complex† A_m in the space of the variables $(x_1, \cdots, x_m) = (x)$. Let

$$(2.1)\qquad F(x_1, \cdots, x_m, r_1, \cdots, r_m) = F(x, r)$$

be a *positive, analytic* function of its arguments for (x) on a region slightly larger than S, and for (r) any set not (0). Suppose further that F is positively homogeneous of order one in the variables (r). We employ the parametric form, taking $F(x, \dot{x})$ as our integrand and

$$J = \int_{t_1}^{t_2} F(x, \dot{x})dt$$

* See Marston Morse, *Closed extremals*, Proceedings of the National Academy of Sciences, vol. 15 (1929), p. 856.

† See Veblen, The Cambridge Colloquium, 1916, Part II, *Analysis Situs*. Terms in analysis situs will be used in the sense of Veblen unless otherwise defined.

as our integral, where $(\dot{x})$ stands for the set of derivatives of (x) with respect to the parameter t. We suppose that the problem is ***positively regular***, that is, that

$$\sum_{ij} F_{r_i r_j}(x, r)\eta_i\eta_j > 0 \qquad (i, j = 1, 2, \cdots, m) \tag{2.2}$$

for (x) and (r) as before, and (η) any set not (0) nor proportional to (r).

*We suppose that S's boundary is extremal-convex.**

That is, we suppose there exists a positive constant e so small that any extremal segment on which $J<e$ and which joins two boundary points of S will lie interior to S except at most for its end points.

Let P and Q be any two distinct points on S. Let there be given an extremal segment g which joins P to Q on S. Let λ be the direction of g at P. Let the direction cosines of the directions neighboring λ be regularly† represented as functions of $m-1$ parameters (α). Let s be the value of J taken along the extremal through P in the direction determined by (α). On the extremals issuing from P with the directions determined by (α) the coordinates (x) will be analytic functions of s and (α).

All of the extremal segments through P neighboring g cannot go through Q.

This could happen if S were a closed manifold, but it cannot happen if S is an extremal-convex region. For if all such extremals did go through Q one could prove first that s would be a constant at Q, and secondly, one could show by a process of analytic continuation that all extremals whatsoever through P would reach Q at least as soon as they reach the boundary.

This cannot occur. For in particular the class of extremals which join P to an arbitrary point $R\neq Q$ of the boundary, and give an absolute minimum to J between P and R, cannot all pass through Q prior to R without violating their minimizing property. Thus the statement in italics is proved.

If in terms of the parameters s and (α) one writes down the m conditions that the extremal through P pass through Q, one sees from the theory of analytic functions that the extremals which join P to Q on S with J less than a finite constant J_0, are conditioned as follows:

These extremals are either finite in number or else make up a finite set of families on which (x) is representable by means of the parameters s and (α),

---

* Instead of assuming that S was an extremal-convex region in m-space, we could have assumed that S was a regular $(m-1)$-dimensional manifold in m-space. Such a change would entail at most obvious changes in the following. The results for this case will be reviewed at the end of the paper.

† That is, the direction cosines are to be analytic functions of the parameters (α) of such sort that not all of the jacobians of $m-1$ of the direction cosines with respect to the parameter (α) are zero.

with s a constant for each family, and (α) *analytic in general for each family on a suitably chosen "Gebilde"* of r independent variables with* $0<r<m-1$.

We note the following.

In the plane there are at most a finite number of extremals on S through P and Q on which $J<J_0$.

We wish to call attention to the case where Q is not conjugate to P on any of the extremals joining P to Q on which $J<J_0$. This is the general case. In this case there are at most a finite number of extremals joining P to Q on which $J<J_0$, as is readily seen from the fact that each such extremal is isolated.

3. The region Σ and function $J(\pi)$. Instead of considering the set of all curves joining P to Q our purposes will be equally well served by considering a particular class of broken extremals joining P to Q. To proceed we shall give successively a number of definitions.

We shall call the value of J taken along a curve γ the *J-length* of γ.

The constant J_0. We shall restrict ourselves to curves of J-length less than J_0, where J_0 is a positive constant larger than the absolute minimum of J along curves joining P to Q, and where further J_0 is not equal to the value of J along any extremals joining P to Q.

The constant ρ. It is well known that there exists a positive constant e_1 small enough to have the following properties. Any extremal E lying on S and with a J-length less than e_1, will give an absolute minimum to J relative to all curves of class D' joining its end points. The coördinates (x) of any point on E will be analytic functions of the coördinates of the end points of E and of the distance of (x) from the initial end point O of E, at least as long as E does not reduce to a point. The set of all extremal segments issuing from O with J-lengths equal to e_1 will form a field covering a neighborhood of O in a one-to-one manner, O alone excepted. With e_1 thus chosen we now choose the positive constant ρ so that

$$\rho < e_1, \quad \rho < e, \tag{3.1}$$

where e is the constant used in the definition of the term extremal-convex.

Any extremal segment on S whose J-length is at most ρ will be called an *elementary extremal*.

Admissible broken extremals. Let n be a positive integer so large that

$$J_0 < (n+1)\rho. \tag{3.2}$$

Consider now the class of broken extremals g which satisfy the following conditions:

* See Osgood, *Funktionentheorie*, II.

I. The broken extremal joins P to Q on S.

II. It consists of $n+1$ elementary extremals.

III. Its J-length is less than J_0.

Such a broken extremal g will be termed *admissible*.

Admissible points (π). Let

$$P, P_1, \cdots, P_n, Q \tag{3.3}$$

be the successive ends of the elementary extremals of g. We admit the possibility that two or more of these points be coincident. Let (π) represent the set of mn variables which give the coördinates of the points

$$P_1, P_2, \cdots, P_n, \tag{3.4}$$

taking first the coördinates of P_1, then those of P_2, and so on. The points (3.4) will be called the vertices of g. A point (π) derived from the vertices of an admissible broken extremal will itself be called *admissible*.

There are infinitely many admissible points (π). In particular let g_1 be an extremal joining P to Q with a J-length less than J_0. If we divide g_1 into $n+1$ successive extremals of equal J-length, these extremals turn out to be elementary extremals by virtue of (3.2), so that the resulting n points of division of g_1 combine into an admissible point (π). This point (π) has the further property that any point (π) in its neighborhood will also be admissible.

The set of all admissible points (π) *will form an mn-dimensional domain* Σ.

The value of the integral J *taken along the broken extremal determined by an admissible point* (π) *from the point* P *to the point* Q *will be denoted by* $J(\pi)$.

4. The boundary of Σ. Corresponding to an admissible point (π), let $M(\pi)$ be the maximum of the J-lengths of the elementary extremals that make up the broken extremal determined by (π).

The boundary of Σ will consist of points (π) of one or more of the following types:

Type I: Points at which $J(\pi)=J_0$;

Type II: Points at which $M(\pi)=\rho$;

Type III: Points corresponding to which at least one vertex P_i lies on the boundary of S.

The function $J(\pi)$ is analytic on Σ at least as long as successive vertices remain distinct. A point (π) corresponding to which the successive vertices are distinct will be called a *critical point* of $J(\pi)$ if all of the first partial derivatives of $J(\pi)$ are zero at that point. We shall prove for the case of distinct vertices that $J(\pi)$ will have a critical point (π) when and only when the corresponding broken extremal reduces to an unbroken extremal γ joining P to Q.

For the partial derivative of $J(\pi)$ with respect to the ith coördinate of a vertex (x), in the case of distinct vertices, is seen to be

$$F_{r_i}(x, p) - F_{r_i}(x, q) \qquad (i = 1, 2, \cdots, m) \tag{4.1}$$

where (p) and (q) give the directions at (x) of the elementary extremals preceding and following (x) respectively. If the differences (4.1) all vanished we would have

$$\sum_i [p_i F_{r_i}(x, p) - p_i F_{p_i}(x, q)] = 0. \tag{4.2}$$

But the sum (4.2) equals the Weierstrass E-function $E(x, p, q)$ which is known not to vanish* for $(p) \neq (q)$ whenever the hypothesis of regularity of §2 is granted. Hence $(p) = (q)$ if $J(\pi)$ has a critical point. The corresponding extremal can therefore have no corners.

If account be taken of the freedom in moving vertices on an extremal γ joining P to Q, it is seen that each such extremal corresponds to an n-dimensional set of critical points (π).

5. The deformation T. The boundary of Σ is exceedingly complex. We can avoid examining its structure more closely by showing how to deform Σ continuously as a whole into a part of Σ which contains none of the boundary points of Σ of types I and II. These deformations amount to deformations of broken extremals. For future use they need to have the following properties.

(α) *They do not increase $J(\pi)$ or $M(\pi)$ beyond their initial values.*

(β) *They deform admissible broken extremals into admissible broken extremals.*

(γ) *They deform continuous families of broken extremals continuously.*

Such deformations will be called J-deformations.

The deformation D_1. On an admissible broken extremal g let a point U be given. Let the value of J taken along g from P to U be termed the *J-coördinate*, u, of the point U on g. In case u is a function of the time t it will be convenient to term du/dt the *J-rate* of U on g.

As the time t varies from 0 to 1 let the n vertices P_i on g move along g from their initial positions to a set of positions which divide g into $n+1$ successive segments of equal J-lengths, each vertex moving at a constant J-rate.

We term this the deformation D_1. It is readily seen that D_1 is a J-deformation. We also note the following.

* See Bliss, *The Weierstrass E-function for problems of the calculus of variations in space*, these Transactions, vol. 15 (1914), pp. 369–378.

LEMMA 5.1. *The deformation D_1 carries each admissible broken extremal into one for which*

$$M(\pi)(n+1) < J_0.$$

The deformation D_2. Let h_i be the ith elementary extremal of g. As the time t increases from 0 to 1 let points $H_i(t)$ and $K_{i+1}(t)$ start from P_i and move away from P_i, respectively on h_i and h_{i+1}, at J-rates equal to one half the J-lengths of h_i and h_{i+1}. Let $H_i(t)$ and $K_{i+1}(t)$ be joined by an elementary extremal $E_i(t)$. The deformation D_2 is hereby defined as one in which P_i is replaced for each t by the point $P_i(t)$ which divides the elementary extremal $E_i(t)$ into two segments of equal J-lengths.

To show that D_2 is a J-deformation we shall first show that $M(\pi)$ is not increased beyond its initial value.

To that end let h_i also represent the J-length of h_i. Let $g(t)$ be the broken extremal replacing g at the time t. Note that the end points of the ith elementary extremal of $g(t)$ are also connected by a set of three elementary extremals joining successively the four points

$$P_{i-1}(t), \quad K_i(t), \quad H_i(t), \quad P_i(t), \tag{5.1}$$

the values $i=1$ and $n+1$ excepted. The J-lengths of these three elementary arcs are respectively at most

$$\frac{t}{4}(h_{i-1} + h_i), \quad (h_i - th_i), \quad \frac{t}{4}(h_i + h_{i+1}). \tag{5.2}$$

If the three constant h_{i-1}, h_i, and h_{i+1} be replaced by their maximum, the sum of the three quantities in (5.2) is readily seen to be at most that maximum. Thus the deformation D_2 does not increase $M(\pi)$ beyond its initial value.

With this established one sees readily that the remaining properties of J-deformations as previously listed are satisfied by D_2.

The deformations D_1 and D_2 combine to give us a deformation T concerning which we have the following lemma.

LEMMA 5.2. *If σ be a closed set of admissible points (π) at a positive distance from the set of critical points of $J(\pi)$, the product deformation $T = D_1D_2$ will carry each point (π) of σ into an admissible point (π') for which*

$$J(\pi') < J(\pi) - d,$$

where d is a positive constant.

In the region Σ let N be a closed neighborhood of the critical points of $J(\pi)$ so small that N and σ have no points in common. We divide the points (π) of Σ into three classes according to the nature of their images $(\pi)_1$ under D_1 at the time $t=1$.

Class I shall contain (π) if $(\pi)_1$ determines a broken extremal with at least one elementary extremal k of zero J-length.

Class II shall contain (π) if $(\pi)_1$ is in N and (π) is not in Class I.

Class III shall contain those points of σ which are neither in Class I nor in Class II.

The proof of the lemma will be given separately for each class.

Class I. To define D_1 we divided the given broken extremal into segments of equal J-length. When (π) is in Class I, one of these segments, say h, is replaced under D_1 by the elementary extremal k of zero J-length. Since the J-length of h is not zero, $J(\pi)$ must be lessened under D_1.

Class II. In general the only points (π) for which $J(\pi)$ is not lessened under D_1 are those for which the broken extremal determined by (π) is identical with the one determined by $(\pi)_1$. Hence $J(\pi)$ is lessened by D_1 for every point (π) of Class II.

Class III. Here the broken extremal determined by $(\pi)_1$ has no elementary extremals of zero J-length, and has at least two successive elementary extremals which intersect at an angle which is not straight. The application of D_2 to $(\pi)_1$ will then lessen $J(\pi)_1$.

Thus under $T = D_1D_2$, $J(\pi)$ will be lessened for each point of σ. Since σ is closed, and T is continuous on σ, $J(\pi)$ will be lessened on σ by at least some definite positive constant d, independent of (π) on σ. Thus the lemma is proved.

6. The domain (a, ρ) and its connectivities. Let a be a non-critical value of $J(\pi)$, and r any positive constant at most the constant ρ of §3.

The set of points (π) *which satisfy*

$$J(\pi) < a, \quad M(\pi) < r, \quad a < J_0 \tag{6.1}$$

will be called a domain (a, r).

We shall be more particularly concerned with domains (a, r) for which $r = \rho$. Because of the complicated nature of the boundary of a domain (a, ρ) it is not feasible to try to break such a domain up into cells. Because of our J-deformations it is fortunately possible to study the topological properties of the domain (a, ρ) by means of approximating complexes.

By hypothesis the region S of the space of the m variables (x) can be broken up into m-cells so as to form an m-complex A_m. We provide n copies of A_m, namely

$$A_m^1, A_m^2, \cdots, A_m^n, \tag{6.2}$$

upon which, in particular, we suppose the n vertices, respectively, of an admissible point (π) lie. The product complex A_{mn} formed by taking an arbitrary point from each of the complexes (6.2) will represent a portion of the space of the points (π) of which each domain (a, r) will be a subset.

By virtue of our choice of the integer n in (3.2) we can choose a positive constant e so small that

$$J_0 < (\rho - e)(n + 1), \tag{6.3}$$

choosing e also so small that no constant on the interval between a and $a-e$, including $a-e$, is a critical value of $J(\pi)$. If the deformation T be applied to the domain (a, ρ) it follows from (6.3) and Lemma 5.1 that the resulting points (π) will be such that

$$M(\pi) < \rho - e.$$

It follows then from Lemma 5.2 that a sufficient number of iterations of T will J-deform the domain (a, ρ) on itself into a set of points on the domain $(a-e, \rho-e)$.

Let C_{mn} be a sub-complex of cells of A_{mn} that contains all of the points of $(a-e, \rho-e)$. If A_{mn} be sufficiently finely divided, C_{mn} may be chosen so as to consist only of points on (a, ρ). It will serve as our e-approximation of (a, ρ).

As we have seen, any complex of A_{mn} on (a, ρ) can be J-deformed on (a, ρ) into a complex on C_{mn}. It follows that any cycle* on (a, ρ) is homologous (always mod 2) on (a, ρ) to a cycle of cells of C_{mn}. Therefore all cycles on (a, ρ) are homologous to a finite number of such cycles.

By a complete *j-set of non-bounding* j-cycles on (a, ρ) will be meant a set of j-cycles $(K)_j$ on (a, ρ) with the following properties.

I. Every j-cycle on (a, ρ) is homologous to a linear combination of members of $(K)_j$.

II. There are no proper† homologies on (a, ρ) between the members of $(K)_j$.

The following is an almost immediate consequence of the definition.

The number of j-cycles in a complete j-set for (a, ρ) is the same for all such complete j-sets.

The jth connectivity number R_j of the domain (a, ρ) will now be defined as the number of j-cycles in a complete j-set.

We come now to a first theorem.

THEOREM 1. *If a and b, $a<b$, are two non-critical values of $J(\pi)$ between which there are no critical values of $J(\pi)$, the connectivities of the domains $(a, \rho)=A$ and $(b, \rho)=B$ are the same.*

* By an i-cycle $(i>0)$ is meant a set of i-circuits. A 0-cycle shall mean any finite set of points.

† That is, homologies which are not empty (mod 2).

It will be sufficient to prove that a complete j-set for A, say $(K)_j$, is a complete j-set for B.

That every j-cycle on B is homologous to a linear combination of j-cycles of $(K)_j$ follows from the fact that the domain B can be J-deformed on itself into a set of points on A. It remains to prove that there are no homologies on B between the members of $(K)_j$.

If now a j-cycle, say H_j, composed of j-cycles of $(K)_j$ bounded a complex H_{j+1} on B, it would follow from Lemma 5.2 that we could J-deform H_{j+1} on B into a complex H'_{j+1} on A. The boundary of H'_{j+1} would be homologous on A to H_j, so that H_j would be bounding on A contrary to the choice of $(K)_j$.

Thus $(K)_j$ is a complete j-set for B. The theorem follows at once.

7. The preliminary homologies. Let a and b be two non-critical values of $J(\pi)$ between which $J(\pi)$ takes on just one critical value $J(\pi)=c$. Let us suppose that this critical value corresponds to a single extremal g, of type k, joining P to Q. We are going to show that the J-connectivities of the domains $A=(a,\rho)$ and $B=(b,\rho)$ differ only when $j=k$ or $k-1$.

Let ω denote the set of critical points of $J(\pi)$ on B at which $J(\pi)$ takes on the value $J(\pi)=c$.

Each point (π) of ω will determine the above extremal g. We come now to the following lemma.

LEMMA 7.1. *If there be given on B an arbitrary neighborhood R of the critical points ω, then in any sub-neighborhood R' sufficiently small every j-circuit is homologous to zero on R.*

It will be sufficient to show that, when suitably chosen, R' can be continuously deformed on R into a point on R. We shall prove that the required deformation can be effected by the product of three deformations $D_1D_3D_1$ of which D_3 will now be defined.

The deformation D_3. Let the extremal g be slightly extended at its ends to form an extremal g'. From each of the n vertices of points (π) defining a broken extremal near g we drop a perpendicular on S to g'. We let the vertices of (π) move along these perpendiculars towards g' with velocities equal to the lengths of the perpendiculars. As the time t varies from 0 to 1 the points (π) with vertices near g' will be deformed into points (π) with vertices on g'. This is the deformation D_3.

Observe now that the deformation D_1 will carry each point of ω into that point $(\pi)_0$ of ω whose vertices divide g into $n+1$ segments of equal J-length. Let N be a neighborhood of the point $(\pi)_0$ so small that the deformation D_3 will carry points (π) on N into points (π) defining g, rather than g multiply covered in part.

If now R' be sufficiently small, D_1, being continuous relative to different points (π), will deform R' on R into points near $(\pi)_0$, or in particular into N. Under D_3 the points (π) on N will be deformed on R into points (π) defining g, while the final application of D_1 will carry the resultant points (π) into the point $(\pi)_0$. Thus $D_1D_3D_1$ will serve as the required deformation.

In terms of the critical value c we have the following lemma.

LEMMA 7.2. *If R' be any neighborhood on $B-A$ of the points of the critical set ω, then any j-circuit K_j on B satisfies an homology,*

$$K_j \sim C_j^1 + C_j^2, \text{ on } B,$$

where C_j^1 is a complex on R' and C_j^2 satisfies $J(\pi) < c - e_1$, where e_1 is a positive constant which depends only on B.

To prove the theorem we shall first show that a sufficiently large number of iterations of the deformation T of Lemma 5.2 will carry each point of B either into R', or else into a point at which $J(\pi)$ is less than c.

Note first that T carries each point of ω into a point of ω and recall that T is continuous. Let R be a sub-neighborhood of R' so small that T will carry no point of R outside of R'.

The set of points $B-A-R$ has a positive distance from the critical points of $J(\pi)$, and according to Lemma 5.2 will then be carried by T into points (π') at which

$$J(\pi') < J(\pi) - d, \qquad d > 0,$$

where it will be convenient to suppose $d < c - a$.

Now a sufficiently large number of iterations of T, say T^r, will carry B into a set of points (π) which satisfy

$$J(\pi) < c + \frac{d}{2} \cdot$$

It follows that T^rT will carry all points of B whose rth images are not on R into points at which

$$J(\pi) < c + \frac{d}{2} - d = c - \frac{d}{2} > a, \tag{7.1}$$

while the remaining points of B will be carried into R'.

Let K_j^1 be the image of K_j under T^rT. We now take C_j^2 as the sum of the j-cells on K_j^1 which satisfy (7.1) at some one point at least, and take for C_j^1 the sum of the remaining j-cells of K_j^1. Thus C_j^1 will lie on R'. If K_j^1 be sufficiently finely divided it appears from (7.1) that on C_j^2, $J(\pi)$ will be less than $c-d/4$. Thus the lemma is proved.

We state the obvious extension.

LEMMA 7.3. *If R' be a neighborhood on $B-A$ of the points ω, then any complex K_j on B whose boundary is on A, is homologous on B to a complex $C_j^1+C_j^2$, which has the same boundary as K_j, and where C_j^1 and C_j^2 are complexes as in Lemma* 7.2.

8. **Deformations neighboring critical points.** Let us return to the space of the m variables (x). In an earlier paper we have proved for the plane essentially the following. (See Theorem 9 Morse II.*)

LEMMA 8.1. *On the extremal g let there be k points conjugate to P, with $k>0$, and Q not conjugate to A.*

Then corresponding to any sufficiently small neighborhood N of g, there exists within N an arbitrarily small neighborhood N' of g, such that closed m-families of curves which join g's end points on N', and satisfy

$$J \leqq c - e, \tag{8.1}$$

where e is a sufficiently small positive constant, are conditioned as follows:

(a) *Those for which $m \neq k-1$ can be deformed on N, without increasing J, into a family of lower dimensionality.*

(b) *Those for which $m=k-1$ include a $(k-1)$-family Z_{k-1} composed of admissible broken extremals, which is non-bounding† on N and* (8.1), *and which is such that every other $(k-1)$-family can be deformed on N without increasing J, either into Z_{k-1} or else into a single curve.*

The proof of this theorem in the plane depended in the first instance upon Theorem 2 of Morse II. The later theorem has been generalized for m-space and appears as Theorem 2 of Morse III.

We note the following differences between the present theorem and Theorem 9 of Morse II.

The deformations here are affirmed not to increase J, instead of simply satisfying (8.1). The deformations used in Morse II are products of the deformation D' of §29, Morse II, which as defined does not increase J, and of the deformation (3) of §31, Morse II. Now the latter deformation does increase J, but if in §31, (2) and (3), we replace e^2 by a^2, this deformation does not increase J, while the remainder of §31 holds as before. The remaining deformations of §31 are on S_{k-1} of Morse II, and hence will not increase J.

---

* As far as the developments of the paper Morse II are concerned, and also for the present paper, we need not restrict the nature of m-families by requiring that the domain of the parametric point be a manifold, but simply that it be a complex, and this extension we suppose made.

† That is, whose "parametric complex" is non-bounding relative to other admissible parametric complexes.

Returning to the space of the points (π) we have the following theorem.

THEOREM 2. *On the extremal g suppose there are k points conjugate to P, with $k>0$, and P not conjugate to Q.*

Then corresponding to any sufficiently small neighborhood R of the critical set ω, there exists within R an arbitrarily small neighborhood R' of ω, such that the circuits K_m on R' which satisfy

$$J(\pi) \leqq c - e, \tag{8.2}$$

where e is a sufficiently small positive constant, are conditioned as follows:

(a) *Those for which $m \neq k-1$ are homologous to zero on R and* (8.2).

(b) *Those for which $m=k-1$ include a $(k-1)$-circuit D_{k-1} that is not homologous to zero on R and* (8.2).

(c) *Those for which $m=k-1$ are either homologous to zero or to D_{k-1}, on R and* (8.2).

This theorem will follow from Theorem 1 once we have examined the relations between neighborhoods R of the set ω in the space (π) and neighborhoods N of g in the space (x).

Corresponding to a variable neighborhood R of the set ω there exists a variable neighborhood N of g that contains all the broken extremals determined by points (π) on R and which approaches g as R approaches ω.

Corresponding to a variable neighborhood N of g there exists a variable neighborhood R of ω that contains all the points (π) determined by those broken extremals on N which satisfy $J<c$, and which approaches ω as N approaches g. This follows from the generalization in m-space of (B) §29, Morse II.

With this understood the theorem follows readily.

II. THE ANALYSIS SITUS

9. Homologies among j-complexes when j is not the type number. For future use we now choose on the domain B distinct from A two closed neighborhoods R and R' of the critical points ω on B.

(a) *We choose R so small that any circuit on R is homologous to zero on B. See Lemma* 7.1. *We also choose R so as to be admissible in Theorem* 2.

(b) *We choose R' so small that any circuit on R' is homologous to zero on some region interior to R.*

(c) *If $k\neq 0$ we still further restrict R' so that R, R', and a positive constant e, taken less than $c-a$, satisfy Theorem* 2.

(d) *If $k=0$ we restrict R' so that on R', $J(\pi)>c$ except at the critical points ω and their limit points.*

Let C_j be any complex on B whose boundary C_{j-1} lies on A. An application of Lemma 7.3 gives the following homology and congruence:

$$C_j \sim C_j^1 + C_j^2 \quad \text{on } B, \tag{9.1}$$

$$C_{j-1} \equiv C_j^1 + C_j^2 \quad \text{on } B, \tag{9.2}$$

where C_j^1 is on R' and C_j^2 satisfies

$$J(\pi) < c - e_1, \quad e_1 < e, \tag{9.3}$$

where e_1 is any sufficiently small positive constant. We have taken $e_1<e$. Suppose C_{j-1}^1 is the common boundary* of C_j^1 and C_j^2. We have then

$$C_j^1 \equiv C_{j-1}^1 \quad \text{on } R', \tag{9.4}$$

$$C_j^2 \equiv C_{j-1}^1 + C_{j-1} \quad \text{on } B \text{ and } (9.3). \tag{9.5}$$

Lemma 9.1. *If $j \neq k$, C_j is homologous on B to some j-complex on A.*

We divide the proof into three cases.

Case 1. $j\neq 0$, $k\neq 0$. Because R, R', and e satisfy Theorem 2, and because $j\neq k$, there exists a complex C_j^3 such that [see Theorem 2, (a)]

$$C_{j-1}^1 \equiv C_j^3 \quad \text{on } R \text{ and } (9.3). \tag{9.6}$$

From (9.4) and (9.6) we see that $C_j^1 + C_j^3$ is without boundary and on R. Because R satisfies the restriction (a) we have then

$$C_j^1 + C_1^3 \sim 0 \quad \text{on } B. \tag{9.7}$$

From (9.1) and (9.7) we have

$$C_j \sim C_j^2 + C_j^3 \quad \text{on } B. \tag{9.8}$$

Now both C_j^2 and C_j^3 satisfy (9.3). With the aid of J-deformations we see that $C_j^2 + C_j^3$ is homologous to a complex K_j on A, with the same boundary on A. Hence

$$C_j \sim K_j \quad \text{on } B$$

and the proof is complete in Case 1.

Case 2. $j=0$, $k\neq 0$. For $j=0$, (9.1) still holds. We next define C_j^3 as any set of points on R and on (9.3) equal in number to the points of C_j^1. The proof may now be started with (9.7), and proceeds therefrom as before.

Case 3. $j\neq 0$, $k=0$. Here C_{j-1}^1 lies on R' and satisfies (9.3), which offers a contradiction to (d) unless C_{j-1}^1 is null. Thus C_{j-1}^1 is null. We now proceed exactly as in Case 1, understanding, however, that both members of (9.6) are now null.

---

* We admit the possibility that C_{j-1} or C_{j-1}^1 be null. For $j=0$ we regard them as meaningless.

10. **Homologies among k-circuits when k is the type number ($k \neq 0$).** We understand that R and R' are chosen as in the preceding section. According to Theorem 2 there then exists a $(k-1)$-cycle D_{k-1} on R' and (9.3) which bounds no complex on R and (9.3). Since R' satisfies (b) of §9 we have, however,

$$D_{k-1} \equiv D_k^1 \quad \text{on } R, \tag{10.1}$$

where D_k^1 is a k-complex interior to R.

We now come to a division into two cases according to the nature of B.

Case α. There exists a complex D_k^2 such that

$$D_{k-1} \equiv D_k^2 \quad \textit{on } (9.3) \textit{ and } B. \tag{10.2}$$

Case β. No such complex as D_k^2 exists.

LEMMA 10.1. *In Case α, the k-cycle D_k defined as*

$$D_k^1 + D_k^2 = D_k \quad \textit{on } B \tag{10.3}$$

is not homologous on B to any cycle on A.*

Suppose the lemma false and that D_k were homologous on B to a k-cycle D_k^3 lying on A. We would have then

$$D_k^1 + D_k^2 + D_k^3 \equiv D_{k+1} \quad \text{on } B \tag{10.4}$$

where D_{k+1} is a complex on B.

We can further suppose that D_{k+1} consists of points which either satisfy (9.3), or else are interior to R. For as in §7 we could J-deform D_{k+1} into such a complex. Such a deformation, say T', would alter the boundary of D_{k+1}, but it could be replaced by a deformation which did not alter this boundary. One would need simply to hold the boundary fast during T' and hold points near this boundary fast during a portion of the deformation starting from a suitable time depending continuously on the distance to the boundary.

With this understood, let D_k^4 be the boundary of the sum of those $(k+1)$-cells of D_{k+1} which do not wholly satisfy (9.3), or else whose boundaries do not wholly satisfy (9.3). We see that D_k^4 is on (9.3) except for such k-cells as it has in common with D_k^1. If D_{k+1} has been sufficiently finely divided D_k^4 would also lie on R, as we now suppose. Thus

$$D_k^1 + D_k^4 \tag{10.5}$$

is on R, and if reduced, mod 2, satisfies (9.3). But since D_k^4 is without boundary, we have from (10.1)

* Including the null cycle.

$$(10.6)\qquad D_{k-1} \equiv D_k^1 + D_k^4.$$

Thus D_{k-1} would bound the complex (10.5) which is on R and (9.3), contrary to the choice of D_{k-1}.

The lemma is thereby proved.

To proceed further it will be convenient to divide the complexes C_k of §9 ($k \neq 0$) into two classes according to the nature of the complex C_{k-1}^1 associated with C_k by (9.4) and (9.5). The distinctive properties of these classes are the following:

Class I. $C_{k-1}^1 \sim 0$ on R and (9.3).

Class II. $C_{k-1}^1 \sim D_{k-1}$ on R and (9.3).

That this division is exhaustive follows from Theorem 2.

LEMMA 10.2. *In Case α, each k-cycle C_k on B is homologous on B to a linear combination of the cycle D_k and cycles on A.*

C_k *in Class* I. Here there exists a complex C_k^3 such that

$$(10.7)\qquad C_{k-1}^1 \equiv C_k^3 \quad \text{on } R \text{ and } (9.3).$$

According to (9.4) and (10.7), $C_k^1 + C_k^3$ is a k-cycle. Since it is on R we have

$$(10.8)\qquad C_k^1 + C_k^3 \sim 0 \quad \text{on } B.$$

From (9.1) and (10.8) we have respectively

$$(10.9)\qquad C_k \sim C_k^1 + C_k^2 \sim C_k^3 + C_k^2 \quad \text{on } B.$$

Now $C_k^3 + C_k^2$ satisfies (9.3), and so can be J-deformed on B into a k-cycle H_k on A. We thus have

$$C_k \sim H_k \quad \text{on } B.$$

Thus the lemma is proved when C_k is in Class I.

C_k *in Class* II. Here we have

$$(10.10)\qquad C_{k-1}^1 + D_{k-1} \equiv C_k^4 \quad \text{on } R \text{ and } (9.3),$$

where C_k^4 is a complex on R and (9.3). Adding (9.4) for $k=j$, (10.1), and (10.10), we have

$$(10.11)\qquad D_k^1 + C_k^1 + C_k^4 \equiv 0 \quad \text{on } R.$$

Now the complex (10.11) is a k-cycle on R so that

$$(10.12)\qquad D_k^1 + C_k^1 + C_k^4 \sim 0 \quad \text{on } B.$$

From the definition of D_k and from (9.1) we have

$$(10.13)\qquad D_k + C_k \sim D_k^1 + D_k^2 + C_k^1 + C_k^2 \quad \text{on } B.$$

Eliminating C_k^1 from (10.13) with the aid of (10.12) we have

$$D_k + C_k \sim D_k^2 + C_k^4 + C_k^2, \quad \text{on } B. \tag{10.14}$$

But the right member of (10.14) satisfies (9.3) and is accordingly homologous on B to a cycle H_k on A. With the aid of (10.14) we have then

$$C_k \sim D_k + H_k \quad \text{on } B,$$

and the lemma is proved when C_k is in Class II.

LEMMA 10.3. *In Case β each k-cycle C_k on B is homologous on B to a k-cycle on A.*

If C_k belongs to Class I the proof in the preceding lemma applies. We shall show that in Case β, Class II is empty.

If C_k belonged to Class II, (10.10) would hold as described. Adding (10.10) and (9.5) for $j=k$, and noting that $C_{k-1}=0$, we have

$$D_{k-1} \equiv C_k^2 + C_k^4 \quad \text{on } B \text{ and } (9.3),$$

contrary to the hypothesis distinguishing Case β.

11. **Homologies among $(k-1)$-circuits, $k\neq 0$.** In this section we prove three lemmas.

LEMMA 11. 1. *In Case α any cycle C_{k-1} on A which is homologous to zero on B will be homologous to zero on A.*

If C_{k-1} is homologous to zero on B it bounds a complex C_k which we identify with the complex C_j of §9. We turn next to C_{k-1}^1 defined by (9.4) and (9.5). According to Theorem 2, C_{k-1}^1 is either homologous to zero or to D_{k-1}, on R and (9.3). If C_{k-1}^1 is homologus to D_{k-1} we note that, in Case α, D_{k-1} is homologous to zero on B and (9.3). It is always true then in Case α that there exists a k-complex C_k^5 such that

$$C_{k-1}^1 \equiv C_k^5 \quad \text{on } B \text{ and } (9.3). \tag{11.1}$$

Upon adding (9.2), (9.4) and (11.1) we have

$$C_{k-1} \equiv C_k^2 + C_k^5 \quad \text{on } B \text{ and } (9.3). \tag{11.2}$$

Since $C_k^2+C_k^5$ satisfies (9.3) it can be J-deformed into a complex K_k on A without altering C_{k-1}. Thus C_{k-1} bounds K_k on A and the lemma is proved.

Definition of D_{k-1}^1. Denote by D_{k-1}^1 a $(k-1)$-cycle obtained by deforming D_{k-1} on B and (9.3) into a $(k-1)$-circuit on A. Such a deformation is possible since D_{k-1}^1 satisfies (9.3). According to the definition of D_{k-1}^1 we have

$$D_{k-1}^1 + D_{k-1} \equiv D_k^5 \quad \text{on } B \text{ and } (9.3), \tag{11.3}$$

where D_k^5 is a k-complex on B and (9.3). We note that

$$D_{k-1}^1 \sim 0 \quad \text{on } B. \tag{11.4}$$

For upon adding (11.3) and (10.1) we have

$$D_{k-1}^1 \equiv D_k^1 + D_k^5 \quad \text{on } B.$$

LEMMA 11.2. *In Case α, D_{k-1}^1 is homologous to zero on A, but in Case β, D_{k-1}^1 is not homologous to zero on A.*

That D_{k-1}^1 is homologous to zero on A in Case α follows from (11.4) and Lemma 11.1.

If in Case β we had a congruence

$$D_{k-1}^1 \equiv D_k^6, \tag{11.5}$$

where D_k^6 is a k-complex on A, then upon adding (11.3) and (11.5) we would have

$$D_{k-1} \equiv D_k^5 + D_k^6.$$

But $D_k^5 + D_k^6$ satisfies (9.3) contrary to the nature of D_{k-1} in Case β, and the lemma is proved.

LEMMA 11.3. *In Case β, any cycle C_{k-1} on A which is homologous to zero on B is either homologous to zero or to D_{k-1}^1 on A.*

If C_{k-1} is homologous to zero on B it bounds a complex C_k which can be identified with the complex C_j of §9.

If C_k is of Class I we repeat the reasoning of Lemma 10.2 under Class I except that we here follow (10.9) with the statement that $C_k^3 + C_k^2$ can be J-deformed on B into a complex K_k on A without altering C_{k-1}. Thus C_{k-1} bounds K_k on A and the lemma is proved if C_k is in Class I.

If C_k is in Class II, (10.10) holds as before. Upon adding (9.2) and (9.4), for $j = k$, to (10.10) and (11.3), we have

$$D_{k-1}^1 + C_{k-1} \equiv D_k^5 + C_k^4 + C_k^2. \tag{11.6}$$

But the right member of (11.6) satisfies (9.3), from which fact we infer that

$$D_{k-1}^1 \sim C_{k-1} \quad \text{on } A$$

and the proof is complete.

12. The changes in the connectivities when $k \neq 0$. In this section we prove two theorems.

THEOREM 3. *Let a and b be two non-critical values of* $J(\pi)$, $a<b$, *between which J takes on a critical value corresponding to just one extremal of type k. The connectivities* R_j *of the domains* $A=(a, \rho)$ *and* $B=(b, \rho)$ *differ at most when* $j=k$ *or* $k-1$.

If $(K)_j$ be a complete j-set for A, it follows from Lemma 9.1 for $j\neq k$ that any j-cycle on B is homologous on B to a linear combination of members of $(K)_j$. From the same lemma it follows for $j\neq k-1$ that there can be no proper homologies on B among the members of $(K)_j$ without there being proper homologies among the members of $(K)_j$ on A. Thus $(K)_j$ is a complete j-set for B and the theorem is proved.

THEOREM 4. *If* $k\neq 0$ *the connectivities* R_j *of the domains A and B differ only as follows:*

$$\text{(12.1)} \qquad \textit{Case } \alpha: \qquad \Delta R_k = 1;$$

$$\text{(12.2)} \qquad \textit{Case } \beta: \qquad \Delta R_{k-1} = -1,$$

where Δ *refers to the change as we pass from A to B.*

We shall first treat Case α.

Let $(K)_k$ be a complete k-set for A. I say that $(K)_k$ together with D_k of §10 will form a complete k-set for B.

For it follows first from Lemma 10.2 that every k-cycle on B is homologous on B to a linear combination of D_k and the members of $(K)_k$.

Secondly there can be on B no proper homologies involving the members of $(K)_k$ and D_k. For if a combination of members of $(K)_k$ alone were bounding on B, with the aid of Lemma 9.1 we could infer that the same combination would be bounding on A, contrary to the nature of $(K)_k$. But no linear combination of D_k and the members of $(K)_k$ which actually involved D_k could be homologous to zero on B without contradicting Lemma 10.1.

Thus $(K)_k$ with D_k forms a complete k-set for B so that $\Delta R_k=1$.

We must now show that $\Delta R_j=0$ when $j\neq k$. This is already established in Theorem 3 provided $j\neq k-1$. When $j=k-1$, Lemmas 9.1 and 11.1 show that a complete j-set for A is a complete j-set for B. The theorem is thus established for Case α.

We can now treat Case β.

Turning to Lemma 10.3 and Theorem 3 we see readily that $\Delta R_j=0$ for $j\neq k-1$.

We shall show finally that (12.2) holds.

According to Lemma 11.2, D_{k-1}^1 is not homologous to zero on A, and so there exists a complete $(k-1)$-set for A, say $(K)_{k-1}$, of which D_{k-1}^1 is a member. Let $(H)_{k-1}$ be the set of $(k-1)$-circuits in this set after D_{k-1}^1 has been removed. I say that $(H)_{k-1}$ is a complete $(k-1)$-set for B.

For according to Lemma 9.1 and the homology (11.4) any $(k-1)$-cycle on B is homologous on B to a combination of $(k-1)$-cycles from $(H)_{k-1}$. That there are no homologies on B among members of $(H)_{k-1}$ follows readily from Lemma 11.3. Thus (12.2) holds as required and the theorem is proved.

13. **The case of minima, $k=0$.** Here the value $J(\pi)=c$ furnishes a relative minimum to $J(\pi)$ on B. We understand a relative minimum to include an absolute minimum as a special case.

Let P be a point of the critical set ω.

LEMMA 13.1. *If c is a relative minimum for $J(\pi)$ on B, then any 0-cycle C_0 on B is homologous on B to a 0-cycle on A, or a combination of 0-cycles on A and the 0-cycle P.*

According to Lemma 7.3 we have the homology

$$C_0 \sim C_0^1 + C_0^2 \quad \text{on } B, \tag{13.1}$$

where C_0^1 is a set of r points on R', and C_0^2, possibly null, is a set of points on B at which $J(\pi)<c$. Note first that C_0^2 is homologous on B to a set K_0 of points on A so that

$$C_0 \sim C_0^1 + K_0 \quad \text{on } B. \tag{13.2}$$

We distinguish between the cases r even and r odd.

If r is even, C_0^1, according to the choice of R', is homologous to zero on B, so that C_0 is homologous to K_0 on B and the lemma is proved. If r is odd, we have

$$C_0 \sim P + K_0 \quad \text{on } B,$$

and the lemma is proved for all cases.

LEMMA 13.2. *If $J(\pi)=c$ is a relative minimum for $J(\pi)$ on B, the point P on ω cannot be connected on B with any point on A.*

To prove this, note that at the boundary points of R' which are not also boundary points of B, $J(\pi)>c+e$, where e is a positive constant.

Suppose P could be joined to a point on A by a continuous curve C_1 on B. A sufficient number of iterations of the deformation T would carry C_1 into a curve C_1^1 on B on which $J(\pi)<c+e$. But it would also carry P into a point on ω in R', so that C_1^1 would still join a point on ω in R' to a point on A. This is impossible, since on C_1^1, $J(\pi)<c+e$. Thus the lemma is proved.

If $J(\pi)=c$ gives an absolute minimum to $J(\pi)$ on B, A will be devoid of points. In that case we say that the *connectivities of A are all zero*. With this understood we have the following theorem.

THEOREM 5. *If $k=0$ the connectivities R_i of the domains A and B differ only in that $\Delta R_0=1$, where ΔR_0 is the change in R_0 as we pass from A to B.*

That R_0 is the only connectivity to change as we pass from A to B follows from Theorem 3.

Now if $(K)_0$ is a complete 0-set for A, null if A is null, then $(K)_0$ with P will be a complete 0-set $(H)_0$ for B.

That all 0-cycles on B are homologous on B to members of $(H)_0$ is affirmed by Lemma 13.1.

That there are no homologies on B among members of $(H)_0$ can be seen as follows. There are no homologies on B among members of $(K)_0$. For such homologies would lead, upon applying Lemma 9.1 for $j=1$, to the conclusion that there were homologies on A among the same members of $(K)_0$. Any homology on B among the members of $(H)_0$ must then explicitly involve P which would mean that P on ω would be connected on B to some point on A contrary to Lemma 13.2. Thus there are no homologies on B among the members of $(H)_0$.

Thus $(K)_0$ and P form a complete 0-set for B. The theorem follows at once.

III. THE EXISTENCE OF EXTREMALS AND THEIR RELATIONS

14. Relations between the extremals and the connectivities. *It will be convenient to call the extremal appearing in Theorem 4 of increasing or decreasing type according as $\Delta R_k=1$ or $\Delta R_{k-1}=-1$.*

In accordance with Theorem 5 each extremal of type zero should be classified as increasing in type.

We come now to the following principal theorems. As previously we suppose a is a non-critical value of J, and that $P\neq Q$.

THEOREM 6. *If among the extremals g on which $J<a$ there are none on which P is conjugate to Q, then between the integers M_k which give the total number of these extremals of type k, and the connectivities R_i of the domain (a,ρ), the following relations hold:*

$$\begin{aligned} M_0 &\geqq R_0,\\ M_0-M_1 &\leqq R_0-R_1,\\ M_0-M_1+M_2 &\geqq R_0-R_1+R_2, \qquad (14.1)\\ &\cdots\cdots\cdots\\ M_0-M_1+\cdots+(-1)^rM_r &= R_0-R_1+\cdots+(-1)^rR_r, \end{aligned}$$

where r is the maximum of the type numbers k.

Consider first the case where the J-lengths of the different extremals g are all different.

Let the extremals of type i be further divided into p_i extremals of increasing type, and q_i extremals of decreasing type. We have at once

$$M_i = p_i + q_i, \tag{14.2}$$

$$R_i = p_i - q_{i+1} \qquad (i = 0, 1, \cdots, r), \tag{14.3}$$

where, in particular, q_0 and q_{r+1} equal zero. From the relations (14.2) and (14.3) we may eliminate p_i and find that

$$M_0 - M_1 + \cdots + (-1)^i M_i = R_0 - R_1 + \cdots + (-1)^i R_i + (-1)^i q_{i+1}. \tag{14.4}$$

Relations (14.1) follow at once from (14.4).

If now two or more of the extremals joining P to Q have the same J-length it follows from Lemma 16.2 that Q can be replaced by a point Q', arbitrarily near Q, and such that for Q' the J-lengths of the different extremals g are all different.

For the pair PQ' the relations (14.1) now hold. But as Q' approaches and takes the position Q, the numbers M_k will not change, since P is not conjugate to Q on any of the extremals g. Neither will the connectivities R_i change, as follows from Theorem 13 of §18.

Thus the relations (14.1) must hold in all cases and the theorem is proved.

We have the following corollary.

COROLLARY. *If the ith connectivity of* (a, ρ) *is* R_i *there exist at least* R_i *extremals joining* P *to* Q *of type* i.

To proceed further it will be convenient to introduce the conception of the *conjugate sequence* for the pair PQ.

Let P and Q be any pair of points on S. Let N_k be the number of closed extremal segments joining P to Q on which there are k points conjugate to P, counting conjugate points according to their orders. If there are an infinity of such extremals we replace N_k by ∞.

The sequence

$$N_0 N_1 N_2 \cdots \tag{14.5}$$

will be called the conjugate sequence for P *and* Q.

We shall investigate this sequence first for pairs P and Q which are non-specialized in the following sense.

Two distinct points P *and* Q *which are not conjugate to each other on any extremals will be termed non-specialized.*

In §16 we shall prove the following.

There exist non-specialized pairs of points PQ in the respective neighborhoods of any two points on S.

We shall begin by supposing S restricted to the most important particular case.

15. **The case where the region S is elementary.** We suppose now that S is an elementary region in the sense that it is homeomorphic with an m-sphere and its interior.

For the moment let us term a curve on s *admissible* if it has a continuously turning tangent except at most at a finite number of points, and joins P to Q. The fact that S is elementary as well as extremal-convex leads to the following lemma.

LEMMA 15.1. *On S there exists a continuous deformation δ that deforms all admissible curves for which $J<a$ through admissible curves of J-lengths less than some constant b, into a single admissible curve, thereby deforming continuous families of such curves continuously.*

Since S is elementary there exists a continuous deformation, say δ_1, of S on itself, leaving P and Q fixed, that carries S into a set of points on a curve γ joining P to Q. We can moreover suppose, for simplicity, that γ is a minimizing extremal joining P to Q.

Now let each admissible curve g for which $J<a$ be divided into $r+1$ successive segments g_i of equal J-length. Let us suppose r taken so large that the end points of g_i and their images under δ_1 can be joined by elementary extremals.

We now define the deformation δ. On each curve g let each segment g_i be replaced by the elementary extremal g_i' that joins its end points, or more particularly let a point on g_i dividing g_i in a certain ratio be replaced by that point on g_i' which divides g_i' in the same ratio. Each curve g can be readily deformed through admissible curves into the corresponding curve g' by a deformation that deforms continuous families of such curves continuously.* To deform further the resultant curves g' into a single admissible curve we require the end points of g_i' to move according to the deformation δ_1, requiring a point on g_i' which divides g_i' in a certain ratio to be deformed into that point of the moving elementary extremal g_i'' which divides g_i'' in the same ratio.

Each curve g will thereby be deformed into a broken extremal g'' with its vertices on γ. Finally let these vertices move along γ each at a constant J-rate into a set of vertices which divide γ into a set of elementary extremals

* See the deformation D'; Morse II p. 263.

of equal J-length, thereby deforming g'' into a single admissible curve γ in the desired manner.

If the elementary extremals used have J-lengths at most ρ, the curves used in the deformation will have J-lengths at most $r\rho$. Thus b can be any constant $b > \rho r$.

LEMMA 15.2. *The constant b of Lemma* 15.1 *can be chosen independently of the position of P and Q on S.*

To prove this let P' and Q' be any other pair of points on S and let g be an admissible curve joining them. Let p^+ and q^+ be minimizing extremals joining PP' and QQ' respectively in the senses indicated. Holding P' and Q' fast let g be deformed into the curve g_1 consisting of the sequence of curves $(p^-p^+gq^-q^+)$ joining P' to Q'. Note that this deformation can be made without increasing J-lengths by more than $4d$, where d is the maximum of the J-lengths of minimizing extremals joining any two points of S.

Now the curve (p^+gq^-) gives a subsegment of g_1 joining P to Q. The class of such subsegments can be deformed through admissible curves joining P to Q into a single curve γ joining P to Q, using thereby curves all of J-lengths less than some constant b_1, as is affirmed by the preceding lemma.

The class of curves g will thereby be deformed into the curve $(p^-\gamma q^+)$ through the mediation of admissible curves joining P' to Q' all of J-lengths less than b_1+4d. The latter constant can be taken as our choice of b.

Thus the statement in italics is proved.

LEMMA 15.3. *All circuits on the domain (a, ρ) are homologous to zero on a domain (b, ρ) where b is any constant sufficiently large chosen independently of P and Q.*

It will be sufficient to choose b as any constant at least as great as the constant b of Lemma 15.2. As previously we suppose J_0 greater than a and b, and n chosen as in §3.

Let C_i be any i-circuit on (a, ρ). If the broken extremals corresponding to C_i be subjected to the deformation δ of the preceding lemmas they will be carried into the curve γ. We can set up a corresponding deformation of the points (π) of C_i as follows.

We first apply the deformation D of §5 to C_i. The end points of the resulting elementary extremals will divide each original curve g defined by by C_i into $n+1$ segments of equal J-lengths. Let these end points now take those positions on the variable curve g' replacing g during δ which divide g' into segments of equal J-length. The resulting deformation will carry C_i into a point $(\pi)_0$ corresponding to γ.

The lemma follows directly.

We come now to the following theorem.

THEOREM 7. *Corresponding to the set of extremals which join two non-specialized points P and Q on which $J<a$, there always exists a complementary set of extremals on which $a<J<b$ such that for the combined sets*

$$\begin{aligned} m_0 &\geqq 1, \\ m_0 - m_1 &\leqq 1, \\ &\cdots\cdots\cdots \\ m_0 - m_1 + \cdots + (-1)^r m_r &= 1 + (-1)^r R_r, \end{aligned} \tag{15.1}$$

where m_i is the number of extremals of type i in the combined sets and r is the maximum of the type numbers of the extremals of the original set and where R_r is the rth connectivity of (a, ρ).

As previously we need prove the theorem only for the case where the J-lengths of the different extremals joining P to Q are different.

The relations (14.1) of Theorem 6 furnish our starting point.

From Lemma 15.3 we see that all circuits on (a, ρ) are homologous to zero on a domain (b, ρ) for which b is large enough.

If R_i is the ith connectivity of (a, ρ) we see that there must be at least R_i extremals for $i>0$, and R_0-1 for $i=0$, of decreasing type $i+1$ with $a<J<b$. We take these extremals $(i=0, \cdots, r)$ as the complementary set. In terms of the integers M_i and R_i of Theorem 6 we have

$$m_0 = M_0, \quad m_1 = M_1 + (R_0 - 1),$$
$$m_i = M_i + R_{i-1} \qquad (i = 1, \cdots, r).$$

Relations (15.1) follow now with the aid of (14.1).

We note that, in Theorem 7, b is a sufficiently large positive constant which may, in particular, be chosen independently of the position of P and Q on S.

We turn now to the conjugate sequences (14.5).

THEOREM 8. *If none of the integers N_i of the conjugate sequence of a non-specialized pair of points are infinite, they satisfy the infinite set of inequalities*

$$\begin{aligned} N_0 &\geqq 1, \\ N_0 - N_1 &\leqq 1, \\ N_0 - N_1 + N_2 &\geqq 1, \\ &\cdots\cdots \end{aligned} \tag{15.2}$$

If all of the integers of the conjugate sequence are finite up to N_{k+1}, the first $k+1$ relations in (15.2) *still hold.*

The first statement in the theorem is a consequence of the last. Let us turn then to the last.

Suppose all of the integers up to N_{k+1} are finite. Let a be a non-critical value of J greater than the J-lengths of all extremals with types at most k. If now we apply Theorem 7 we find that we must have

$$m_i = N_i \qquad (i = 0, 1, \cdots, k)$$

and the first $k+1$ relations in (15.2) follow from (15.1).

The theorem follows directly.

COROLLARY. *If there are no extremals joining P and Q upon which there are k conjugate points, then there are either an infinite number of extremals upon which there are fewer than k conjugate points, or else the numbers* $N_0, \cdots, N_{k-1}$ *satisfy*

$$N_0 - N_1 + \cdots + (-1)^{k-1} N_{k-1} = 1.$$

This follows from the kth and $(k+1)$st inequalities of (15.2).

We come now to a theorem in which we shall not require the numbers N_i *to be finite.*

If N_r represents ∞ we shall understand $N_i - 1$ and $N_i + 1$ as also representing ∞.

THEOREM 9. *Let P and Q be any two non-specialized points on S.*

(a) *If there are* N_k *extremals of type* $k>1$ *there are at least* N_k *extremals of the two adjacent types.*

(b) *If there are* N_1 *extremals of type one there are at least* $N_1 - 1$ *extremals of the two adjacent types.*

(c) *If there are* N_0 *extremals of type zero there are at least** $N_0 - 1$ *extremals of the adjacent type one.*

We shall first prove (a).

From the first inequality in (15.1) which involves m_{i+1}, and the third preceding inequality, we find that

$$m_i \leqq m_{i-1} + m_{i+1}, \quad i > 1.$$

If N_{i-1}, N_i, and N_{i+1} are finite, and we take the constant a in Theorem 7 large enough, these N's become equal to m_{i-1}, m_i and m_{i+1} respectively, and (a) is proved for this case.

* Part (c) is the representation here of the "minimax principle" of Birkhoff. See *Dynamical systems with two degrees of freedom*, these Transactions, vol. 18 (1917), p. 249.

If N_i is infinite there must be extremals of type i of arbitrarily great J-length. If we take a successively as the constants of a sequence of constants becoming infinite, m_i will become infinite, and hence from (15.1) either m_{i-1} or else m_{i+1}. Part (a) then follows in this case as well.

Parts (c) and (b) follow similarly from the second and third inequalities in (15.1).

IV. The density of conjugate points, and invariance of the connectivities

16. Specialized points P and Q. We wish to show that the so called non-specialized pairs of points are really general. We begin with the following lemma.

LEMMA 16.1. *The conjugate points of a point P at distances from P along the corresponding extremals not exceeding a positive constant d, form a set which is nowhere dense.*

Let (u) represent a point in an auxiliary m-space. Let each extremal issuing from P be represented in the space (u) by a ray issuing from the origin with a direction parallel to its direction at P, and with points on the extremal at distances s from P, corresponding to points on the ray at distances s from the origin. The corresponding functions

$$x_i = x_i(u) \qquad (i = 1, 2, \cdots, m)$$

will be analytic except at the origin. The jacobian $D(u)$ of these functions will vanish at the conjugate points. Its rank r at such points will be between 0 and m (Morse III §7).

Suppose $(x)_0$ is a conjugate point corresponding to a point $(u)_0$. If $D(u_0)$ is of rank r one sees that there is an r-plane X through $(x)_0$ in the space (x), such that the distance of the points $[x(u)]$ from X is an infinitesimal of at least the second order with respect to the distance ρ of (u) from $(u)_0$.

Let S_1 be the interior of an $(m-1)$-sphere of radius ρ with center at $(u)_0$. One sees that the points (x) corresponding to points (u) on S_1 can be enclosed in a volume V whose ratio to that of S_1 will approach zero as ρ approaches zero. For V one could take a generalized cylinder consisting of the points P of X at a distance $c\rho$ from $(x)_0$ together with the points on perpendiculars to X, points at a distance $h\rho^2$ from these points P, where c and h are suitably chosen positive constants independent of the choice of $(u)_0$.

Let e now be an arbitrarily small positive constant. Let us break the space (u) up into congruent m-cubes. If the diameter of each of these m-cubes be sufficiently small, then such of the corresponding sets $[x(u)]$ as contain conjugate points with $s \leqq d$ can be enclosed in elementary volumes such as V whose ratios to that of the cubes will be less than e. The sum of these volumes V will be less than e times the total volume of the corresponding cubes. The sum of these elementary volumes will then be arbitrarily small. This is possible only if the conjugate points in question are nowhere dense.

THEOREM 10. *The set of all conjugate points of a fixed point P is nowhere dense on S.*

Let $d_1, d_2, \cdots$ be an increasing set of positive constants which become infinite with their subscripts. Let Q be any point on S. It follows from the preceding lemma that there exists in any neighborhood of Q a sequence of $(m-1)$-spheres

$$S_1, S_2, \cdots$$

each within the other, and such that there are within S_i no points conjugate to P for which $s<d_i$. These $(m-1)$-spheres will have at least one common interior point, say A. The point A cannot be a conjugate point of P without violating the principle under which the $(m-1)$-spheres S_i were stated to exist.

Thus the theorem is proved.

This theorem amounts to the statement already made that there exist non-specialized pairs of points P and Q in the respective neighborhoods of any two given points of S.

The following theorem is a strong aid in proving the existence of extremals joining specialized pairs of points.

THEOREM 11. *If each pair of points in a set of non-specialized points can be joined by an extremal γ which is bounded in J-length for the set, and of type k, then any limit pair P_0Q_0, $P_0 \neq Q_0$, of pairs of the set can be joined by an extremal g on which there are at least k, and at most $k+m-1$, points* conjugate to P_0.*

Let PQ represent any pair of points in the set. As PQ approaches P_0Q_0, the initial directions of the extremals γ will have at least one limit direction. Let g be the extremal with this limit direction.

The extremal g will have at least k conjugate points on it. For otherwise there would be fewer than k conjugate points on extremals γ neighboring g. See Morse IV §9.

Suppose there were $k+r+s$ conjugate points on g, where s is the number of conjugate points to be counted at Q_0. Now $r \leqq 0$, for otherwise extremals γ neighboring g would have at least $k+r>k$ conjugate points on them. Finally

* Counting conjugate points according to their orders.

$s \leq m-1$ as was shown in Morse III §7. Thus there are at most $k+m-1$ conjugate points on g.

The following two lemmas have already been used. They can be conveniently proved here.

LEMMA 16.2. *Let PQ be a pair of distinct points, and d any positive constant. There exists in the neighborhood of the point Q at least one point which is joined to P by no two extremals whose J-lengths are equal and less than d.*

According to Lemma 16.1 there exists in the neighborhood of Q at least one point Q' which is joined to P by no extremals on which $J \leq d$ and on which Q' is conjugate to P. Suppose, however, that there are at least two extremals g and g' joining P to Q' with equal J-lengths. Suppose their direction cosines at Q' are (p) and (q).

Since Q' is not conjugate to P, a slight variation of Q' will cause a slight variation of the extremals from P to Q' neighboring the initial g and g'. The J-lengths of these extremals will be analytic functions of the coördinates of Q'.

The difference of the J-lengths of the extremals g and g' will have partial derivatives given by (4.1), and as seen in §4, not all of these partial derivatives are zero for $(p) \neq (q)$. The locus of points Q' neighboring the initial position of Q' for which two or more of the extremals have J-lengths which are equal and less than d or near d, will thus lie on a finite number of analytic $(m-1)$-dimensional manifolds without singularities. The lemma follows at once.

With the aid of Lemmas 16.1 and 16.2 one proves the following lemma. The proof is similar to that of Theorem 10.

LEMMA 16.3. *For a fixed point P there exists in the neighborhood of every point $Q \neq P$ at least one point which is not a conjugate point of P nor which is joined to P by more than one extremal of any one J-length.*

17. **Example. Geodesics on a knob.** Suppose we have an analytic surface S, without singularities, with boundary B, and homeomorphic to a circular disc. On S consider the integral of arc length. Suppose S is extremal-convex. Suppose S possesses a knob-shaped protuberance. More exactly suppose there is a portion of S homeomorphic to a circular disc, and bounded by a closed geodesic g that is shorter than nearby closed curves.

Let R be the region between g and B. Let P and Q be any pair of non-specialized points on R. The geodesics γ which give an absolute minimum to the arc length relative to all other admissible curves on R which join P to Q, and are deformable into γ on R, are infinite in number and of type zero. From Theorem 9 we have the following.

If P and Q are any two non-specialized points on R there will be an infinite set of geodesics of type one joining P to Q on S.

By the use of Theorems 7 and 11 it is not difficult to prove the following more general theorem.

If P and Q are any two points whatsoever on R, there will be an infinite set of geodesics joining P to Q, upon each of which there will be at least one point conjugate to P.

18. **The invariance of the connectivities.** We shall first concern ourselves with the dependence of the connectivities of a domain (a, ρ) upon the choice of n, the number of vertices in (π). It will be convenient to indicate the apparent dependence of (a, ρ) on n by now representing this domain by (a, ρ, n). We shall prove the following.

THEOREM 12. *The connectivity numbers are independent of n, in the sense that the connectivity numbers of (a, ρ, n) equal those of (a, ρ, n'), where n and n' are any two admissible choices of n.*

Let there be given a point $(\pi)'$ on (a, ρ, n'), determining a broken extremal g'. Let $a(\pi')$ stand for the point (π) whose n vertices divide g' into $n+1$ successive segments h_i of equal J-length. Let g be the admissible broken extremal determined by (π). We shall prove the following statement.

(a) *The curve g' can be continuously deformed on S into the curve g through the mediation of ordinary curves which join P to Q, and whose J-lengths do not exceed that of g'.*

We shall take the time t as the parameter of our deformation, and let it vary from 0 to 1.

Suppose the end points of the segment h_i of g' are joined by an elementary extremal k_i of g. For each value of t we suppose h_i divided into two successive segments the ratio of whose J-lengths is that of t to $1-t$. For each value of t from 0 to 1 we now replace the second of these segments of h_i by itself, while we replace the first by an elementary extremal that joins its end points. In this manner h_i will be deformed into k_i, and thereby g' into g. Thus the statement (a) is proved.

Each point (π) determines a point $b(\pi)=(\pi)'$ exactly as (π) determines $a(\pi')=(\pi)$. Statement (b) will now be proved.

(b) *For every point (π) on the domain (a, ρ, n) the point $(\pi)''=a(b(\pi))$ lies on the domain (a, ρ, n) also. Moreover, there exists a deformation F on (a, ρ, n), of the points (π) on (a, ρ, n), which carries these points into the corresponding points $(\pi)''$.*

Let g, g' and g'' be respectively the broken extremals determined by (π), $b(\pi)$ and $(\pi)''$. By deformations similar to those described under (a),

g can be deformed into g', and thence into g'' without increasing J. Let t be a parameter of the resultant deformation, and vary from 0 to 1. Let $Z(t)$ be the curve that thereby replaces g at the time t. A point U on $Z(t)$ may be *determined* by giving t and the J-coördinate u of the point U on $Z(t)$. See §5.

A deformation F of the point (π) into the corresponding point $(\pi)''$ will now be defined. In the space of the points (x) let each vertex of (π) move to the corresponding vertex of $(\pi)''$ in such a manner that the pair (t, u) which *determines* (see above) this variable vertex moves on a straight line in the (t, u) plane at a constant velocity equal to the distance to be traversed. The corresponding deformation F is readily seen to have the desired properties.

If C_j is any complex on the domain (a, ρ, n) we shall denote by $b(C_j)$ that complex on (a, ρ, n') which consists of the images $b(\pi)$ of points (π) on C_j. Reciprocally if C_j' is a complex on (a, ρ, n'), $a(C_j')$ will denote the set of images $a(\pi')$ of points $(\pi)'$ on C_j'. With this understood we now prove a final statement (c).

(c) *If the cycles*

$$C_j^1, \cdots, C_j^r \tag{18.1}$$

form a complete j-set for (a, ρ, n'), *the cycles*

$$a(C_j^1), \cdots, a(C_j^r) \tag{18.2}$$

will form a complete j-set for the domain (a, ρ, n).

Let C_j be any j-cycle on the domain (a, ρ, n). Then $b(C_j)$ will be a j-cycle on the domain (a, ρ, n'). Because (18.1) gives a complete j-set for (a, ρ, n') we have

$$b(C_j) + \Sigma C_j^i \equiv C_{j+1} \quad \text{on } (a, \rho, n'), \tag{18.3}$$

where Σ stands for a suitable sum of cycles (18.1) and C_{j+1} is a complex on (a, ρ, n'). From (18.3) we see that

$$a(b(C_j)) + \Sigma a(C_j^i) \equiv a(C_{j+1}) \quad \text{on } (a, \rho, n). \tag{18.4}$$

From (b) we see however that

$$a(b(C_j)) \sim C_j \quad \text{on } (a, \rho, n). \tag{18.5}$$

From (18.4) and (18.5) we see finally that

$$C_j \sim \Sigma a(C_j^i) \quad \text{on } (a, \rho, n). \tag{18.6}$$

Thus the jth connectivity number of (a, ρ, n) will be at most r. We can reverse the rôles of (a, ρ, n) and (a, ρ, n'). We infer then that the connectivities of (a, ρ, n) and (a, ρ, n') are the same. Thus (c) and the theorem are proved.

We now give another theorem on the invariance of the connectivities.

THEOREM 13. *The connectivities of the domain (a, ρ) remain unchanged during any continuous variation of the end points P and Q on S $(P \neq Q)$, provided the constant a remain a non-critical value of J.*

Let the domain (a, ρ) of admissible points (π), set up for a pair of points, PQ, now be indicated by

$$(a, \rho, P, Q). \tag{18.7}$$

If e be a sufficiently small positive constant the domain (18.7) can be J-deformed onto the domain (see § 6)

$$(a - e, \rho - e, P, Q). \tag{18.8}$$

If e_1 now be chosen as a positive constant less than e, and $P'Q'$ be a pair of points sufficiently near PQ, the domain

$$(a - e_1, \rho - e, P', Q') \tag{18.9}$$

will be included in the domain (18.7) and will include the domain (18.8).

Now the domain (18.7) can be J-deformed on itself into a set of points on (18.8), and hence into a set of points on (18.9). It follows that the connectivities of (18.7) equal those of (18.9).

This remains true for any smaller choice of e_1 and choice of $P'Q'$ sufficiently near PQ. But if e_1 be sufficiently small, and $P'Q'$ so near PQ that a remains a non-critical value of J, the domain

$$(a, \rho, P', Q') \tag{18.10}$$

can be J-deformed on itself into a set of points on (18.9) and hence has the connectivities of (18.9). Hence (18.10) and (18.7) have the same connectivities, and the theorem is proved.

19. **Extremals on closed, regular, analytic manifolds.** All of the previous developments go through for this case as in the case of the region S, except for obvious changes of which we will enumerate the most important.

All references to the boundary, or the extremal-convex hypothesis are to be omitted.

Instead of having one set of coördinates (x), it will in general be necessary to give the manifolds by a finite set of overlapping parametric representations.

All the extremals through a point P may go through a second point Q.

This case should be treated separately. It can never occur for a non-specialized pair of points.

The theory of extremals on elementary regions will apply to elementary regions on the manifold. The relations between the extremals and the connectivities of the domain (a, ρ) are given by Theorem 6.

The most important case is the case where the manifold is homeomorphic with an m-sphere. The following theorem will be proved in a later paper.

THEOREM 14. *On a manifold homeomorphic with an m-sphere there are infinitely many extremals joining any two fixed points, including extremals of arbitrarily great length, with arbitrarily many conjugate points on them.*

In this theorem it is understood that two extremals are counted as different if they have different lengths, even if they overlap. The proof of this theorem depends upon the preceding work together with a method of determining the connectivities R_i, called *the topological continuation of extremals.*

HARVARD UNIVERSITY,
CAMBRIDGE, MASS.

Reprinted from the Proceedings of the National Academy of Sciences,
Vol. 20, No. 1, pp. 46–50. January, 1934.

DOES INSTABILITY IMPLY TRANSITIVITY?

By Marston Morse

Department of Mathematics, Harvard University

Communicated December 1, 1933

Much has been written of late concerning the hypotheses of metric transitivity. One can refer for example to the historical summary[1] by Birkhoff and Koopman. Here references are given to the contributions of various writers including the recent important papers by Birkhoff, Hopf, Koopman, v. Neumann and P. Smith. However, it has not yet been proved that there is a distinction in the analytic case between metric and regional transitivity. Outside of very simple examples of integrable motion or motion on a surface of negative curvature,[4] the work prior to the present paper has not shown that metric or regional transitivity exists in general analytic cases. The present paper partially fills this gap with a general theorem on the existence of regional transitivity.

A dynamical system is said to be regionally transitive if there exists at least one motion g whose closure in phase space is identical with the phase space. In such a case we term g a *completely transitive motion.* This paper is concerned with geodesic motion on a generalized surface R^* of genus $p > 1$. We shall introduce a conception of *uniform instability* of the geodesics on R^*, and show that uniform instability of the geodesics on R^* implies regional transitivity. Surfaces on which the geodesics are uniformly unstable include all surfaces of negative curvature, as well as surfaces which contain large areas of positive curvature. Professor Birkhoff has made the very interesting conjecture that regional transitivity is implied by the condition that all closed motions be unstable and of minimum type. We are unable to confirm this opinion. Such evidence as we possess seems to be slightly against the conjecture. The present result of the author is the first in which a hypothesis on the separate geodesics leads to a general conclusion as to transitivity.

We shall say that a motion g is transitive relative to a class Ω of motions in phase space if the closure of g includes Ω as a subset. As an important example we consider a fundamental class A of minimizing geodesics on R^* and show that there always exist geodesics transitive to the motions defined by A. The class A is so chosen that it contains all motions on R^* when the system is uniformly unstable. We also state a general theorem on the deferring of transitivity in a regionally transitive system.

We suppose that we have a Fucksian group G of non-singular fractional linear transformations of the plane $z = u + iv$ carrying the interior S of the unit circle into S. We suppose all members of the group are of hyperbolic type with fixed points on the unit circle, and that the group has a finite number of generators. We regard S as a hyperbolic plane of non-euclidean geometry with the circles orthogonal to the unit circle representing the H-straight lines (H is written for hyperbolic). The H-length may be given by the integral

$$\int \frac{d\sigma}{1 - u^2 - v^2} \tag{1}$$

where $d\sigma$ is the euclidean differential of arc length. We assume that there exists a fundamental domain for the group G in the form of a region S_0 bounded by a finite set of H-straight lines and that the closure of S_0 has no points on the unit circle. Certain of the boundary points of S_0 belong to S_0. We shall not give the details here.

Instead of assigning S the H-metric (1) we can also assign S an R-metric (R is written for Riemannian) given by a positive definite form

$$ds^2 = E(u, v)du^2 + 2F(u, v)dudv + G(u, v)dv^2 \tag{2}$$

with coefficients of class C^3. We suppose the form (2) is invariant under G and defines a Riemannian manifold R. We can, if we choose, regard congruent points on the boundary of S_0 as identical. So regarded, S_0 taken with the metric (2), will define a generalized closed surface R^*. The genus of R^* will be an integer $p > 1$.

To any two points on S we assign an R-distance equal to the minimum length of R-geodesics joining the two points. A point P on S with a direction θ at P, measured in the usual way in the (u, v) plane, will define an *element* E on S. If E' and E'' are any two elements on S with directions represented by θ' and θ'', the R-angle between E' and E'' will be defined as the minimum of

$$|\theta' - \theta'' + 2n\pi|$$

for all integers n, positive, negative or zero. Two elements E' and E'' on S will be said to have an R-distance which is the sum of the R-distance between their initial points and their R-angle. The elements congruent to E' and E'' on S_0 will be said to have an R^*-distance which is the greatest lower bound of the R-distance between all pairs of elements which are respectively congruent to E' and E''.

The present paper is based on an earlier paper by the writer.[2] We take over the terminology of the earlier paper. We showed in this paper that there always exist two "boundary" geodesics on R of class A of the type of any given H-straight line.

By the type hypothesis on R we here mean the hypothesis that any two boundary geodesics of the same type are identical.

In the earlier paper we showed that the boundary geodesics were limit geodesics of the set of periodic boundary geodesics. With the aid of this fact we can readily establish the following theorem.

THEOREM 1. *If the type hypothesis holds the geodesics on R^* which are completely transitive have the power of the continuum.*

In case R^* is a surface of constant negative curvature Hedlund[3] has shown that the above geodesics define motions in phase space (the space of their elements) which have the measure of the space. No such theorem has been proved for the present general case. Theorem 1 was announced by the author in February, 1932, at the Harvard Colloquium.

Let M denote the phase space corresponding to geodesic motion on R^*, that is, the ensemble of all elements tangent to geodesics on R^* with the distance between elements in phase space defined as previously. The proof of Theorem 1 leads to Theorem 2.

THEOREM 2. *If the type hypothesis holds and h is any arc in the space M such that the motions on M which intersect h define more than one geodesic on R^*, then the completely transitive motions which intersect h have the power of the continuum.*

Uniform Instability.—The two preceding theorems show the importance of the type hypothesis. We shall give a sufficient condition for this hypothesis to hold.

To that end let g be any geodesic belonging to R. We refer R to normal geodesic coördinates (x, y) such that neighboring g

$$ds^2 = C^2\,(xy)dx^2 + dy^2 \tag{3}$$

where $y = 0$ along g and $C(x, 0) = 1$. The parametric curves, x constant, are geodesics normal to g, and y is the arc length along these geodesics measured from g. The equation of normal variation from g takes the form

$$\frac{d^2w}{dx^2} + K(x)w = 0 \tag{4}$$

where $K(x)$ is the curvature of R at the point x on g. We say that the differential equation (4) and its solutions are *based* on g. Let $w(x)$ be any solution of (4) such that

$$w^2(o) + w'^2(o) = 1. \tag{5}$$

We shall say that the geodesic system on R^* is *uniformly unstable* if the following two conditions are satisfied by every geodesic g on R.

(a) *No solution of* (4) *which is not identically zero vanishes more than once.*

(b) *There shall exist a positive function M (s) defined for all positive values of s and a positive constant s^*, both independent of g, such that M (s) becomes positively infinite with s, and such that any solution of* (4) *for which* (5) *holds shall satisfy the relation*

$$|\,w(-s_1)\,| + |\,w(s_2)\,| \geqq M(s) \tag{6}$$

provided

$$s_1 > s > s^*$$
$$s_2 > s > s^*.$$

It is understood that (6) is independent of the geodesic g upon which the solution $w(x)$ is based and of the point on g from which $x = s$ is measured. We show that uniform instability implies the type hypothesis and hence regional transitivity on R^*. We thus have the following theorem.

THEOREM 3. *If the geodesics on R^* are uniformly unstable the set of completely transitive geodesics has the power of the continuum.*

By a *geodesic ray* on R we mean a geodesic which starts from a point P of R and is continued indefinitely on R in one sense from P. In general there are two boundary geodesic rays of class A issuing from P of the "type" of any H-ray issuing from P. We show that the boundary geodesic rays of a given type are unique whenever the type hypothesis holds. The elements on a geodesic ray possess congruent elements on S_0, and these define a motion in the phase space M. If the closure of this motion on M is identical with M we term the geodesic ray completely transitive. We state the following theorem.

THEOREM 4. *If the unending geodesics on R^* are uniformly unstable the set of completely transitive geodesic rays issuing from a fixed point P on R has the power of the continuum.*

The Transitivity Function.—We now take up the question of the deferring of the degree of transitivity. Let g be a transitive geodesic ray issuing from a point P. If e is any positive number less than a suitably chosen positive constant p we show that there exists a least positive number

$$L_g(e) \qquad\qquad 0 < e < p$$

such that a segment of g of length $L_g(e)$ with initial point at P defines a curve γ in the phase space M at a distance at most e from each element of M. We term $L_g(e)$ the transitivity function defined by g.

Our theorem on the deferring of transitivity is the following:

THEOREM 5. *Let P be any arbitrary point on S and $\psi(e)$ an arbitrary positive function of e defined on the interval $0 < e < p$ and such that $\psi(e)$ becomes infinite as e tends to zero. There then exists a completely transitive geodesic ray issuing from P for which the corresponding transitivity function exceeds $\psi(e)$ at an infinite sequence of positive values of e tending to zero, provided the geodesics on R^* are uniformly unstable.*

Relative Transitivity.—We now drop the type hypothesis and the hypothesis of uniform instability. We say that a geodesic ray g on R^* is transitive relative to a set H of geodesics on R^* if the geodesics of H are limit geodesics of g. We have proved the following theorem.

THEOREM 6. *If P is an arbitrary point on S, the geodesic rays issuing from P which are transitive relative to the set of all boundary geodesics of class A have the power of the continuum.*

If the geodesics of the system possess uniform instability the boundary geodesics of class A are the only geodesics and Theorem 6 implies Theorem 4. This matter of relative transitivity is obviously related to the question of the existence of invariant sets with intermediate measures in the phase space M.

[1] Birkhoff and Koopman, "Recent Contributions to the Ergodic Theory," *Proc. Nat. Acad. Sci.*, **18**, 279 (1932). References are here given to Hopf, Koopman, v. Neumann and P. Smith.

[2] Morse, Marston, "A Fundamental Class of Geodesics on Any Closed Surface of Genus Greater Than One," *Trans. Amer. Math. Soc.*, **26**, 25–61 (1924).

[3] Hedlund, Arnold, "On the Measure of the Non-Special Geodesics on a Surface of Constant Negative Curvature," *Proc. Nat. Acad. Sci.*, **19**, 345 (1933).

[4] Birkhoff, "Dynamical Systems Colloquium Lectures." A brief proof of regional transitivity in the case of a special closed surface of genus 2 of non-constant negative curvature is given here.

[5] Birkhoff, "Probability and Physical Systems," *Bull. Amer. Math. Soc.*, **34** (1932). The author's "Transitivity Functions" and Birkhoff's "Ergodic Function" are different but related functions.

Reprinted from the Proceedings of the NATIONAL ACADEMY OF SCIENCES,
Vol. 20, No. 5, pp. 282–287. May, 1934.

ON CERTAIN INVARIANTS OF CLOSED EXTREMALS

BY MARSTON MORSE AND EVERETT PITCHER

DEPARTMENT OF MATHEMATICS, HARVARD UNIVERSITY

Communicated April 12, 1934

Morse has classified closed extremals by what we shall call indices of periodicity. See reference 3, Ch. III. We propose in this paper to present briefly some facts about the determination of the indices of periodicity of a closed extremal and some necessary conditions on these indices in terms of a generalized Poincaré rotation number.

1. *Index of Periodicity.*—We are studying a calculus of variations problem in the usual parametric form with an integrand which is positive, of class C^3, homogeneous and positively regular. The space of the dependent variables is a regular, orientable, analytic $(n + 1)$-dimensional manifold. The non-orientable case is readily reducible to the orientable case.

We suppose there are closed extremals, of which g of length ω is one.

We first map g and its neighborhood on the x axis and its neighborhood in the Euclidean space $(x, y_1, \ldots, y_n)$ by a transformation with period ω in x. When the integral of our problem is transformed in the corresponding manner, the x axis is an extremal. We denote the segment $o \leqq x \leqq \omega$ of the x axis by γ.

We write the transformed integral in non-parametric form so that x is the variable of integration. The *Jacobi equations*, in the space (x, η), are then

$$\frac{d}{dx}\Omega_{\eta'_i} - \Omega_{\eta_i} = 0 \qquad (i = 1, \ldots, n). \tag{1}$$

See reference 3, Ch. I, ¶ 4. We are interested also in the *modified Jacobi equations*

$$\frac{d}{dx}\Omega_{\eta'_i} - \Omega_{\eta_i} + \lambda\eta_i = 0 \qquad (i = 1, \ldots, n). \tag{2}$$

See reference 3, Ch. II, ¶ 4.

The *index of periodicity* of g is the number of negative values of λ for which the modified Jacobi equations have a non-null periodic solution of period ω. We count a value of λ a number of times equal to the number of such solutions which are linearly dependent. See reference 3, Ch. III.

The *order of degeneracy* of g is the number of linearly independent solutions of the Jacobi equations which have the period ω. If the order of degeneracy is zero, we call the extremal g non-degenerate. *In this paper we shall suppose the extremal* g *is non-degenerate.*

A point $x = b$ on the axis of x is a conjugate point of the distinct point $x = a$ of index r if exactly r linearly independent solutions of the Jacobi equations vanish at $x = a$ and also at $x = b$. In counting conjugate points on an interval, we shall count each conjugate point a number of times equal to its index.

2. *Order of Concavity.*—Let G be the family of solutions

$$\eta_i = y_i(x) \qquad (i = 1, \ldots, n)$$

of the Jacobi equations joining those points in the n-planes $x = 0$ and $x = \omega$ whose coördinates (η) are the same. When g is non-degenerate, the solutions of G depend linearly on n linearly independent solutions of the Jacobi equations, which may be represented as the columns of the matrix

$$||z_{ij}(x)|| \qquad (i, j = 1, \ldots, n).$$

Corresponding to a solution $\eta_i = \eta_i(x)$, $(i = 1, \ldots, n)$, of the Jacobi equations, we set

$$\Omega_{\eta'_i}(\eta, \eta') = \zeta^{\eta}_i(x) \qquad (i = 1, \ldots, n) \tag{3}$$

The quadratic form*

$$K(u) = a_{\mu\nu}u_\mu u_\nu \qquad (\mu, \nu = 1, \ldots, n) \tag{4'}$$

in which

$$a_{\mu\nu} = z_{i\nu}(0)\,[\zeta^{z}_{i\mu}(x)]_{x=0}^{x=\omega} \qquad (i, \mu, \nu = 1, \ldots\ n) \tag{4''}$$

is termed the *concavity form* of g. It is a symmetric form. Its index[5] and nullity[6] are independent of the matrix $\|z_{ij}(x)\|$ chosen as above. We term its index the *order of concavity* of g. Its nullity is the index of $x = \omega$ as a conjugate point of $x = 0$. See reference 3, Ch. III, ¶ 10, for a discussion of the concavity form.

If $x = \omega$ is not conjugate to $x = 0$, we may interpret the concavity form geometrically in the following manner. Let S be the manifold of points, in the space (x, η) of the solutions of the Jacobi equations, which lie on solutions belonging to G and passing through points (u) in the n-planes $x = 0$ and $x = \omega$ for which

$$u_1^2 + \ldots + u_n^2 = 1. \tag{5}$$

This manifold S is roughly a tube enclosing γ. The exterior normals to S at point (u) in the n-plane $x = 0$ and point (u) in the n-plane $x = \omega$ lie in a 2-plane through the x axis. Let θ be the angle between these normals, measuring θ from the normal at $x = 0$ and counting that sense of rotation positive which leads from the positive x axis to either normal. *Then for points on* (5) *the concavity form $K(u)$ and the angle θ satisfy the relation*

$$r(u) \sin \theta = K(u), \tag{6}$$

where $r(u)$ is a positive continuous function of the variables (u).

The following theorem is fundamental. See reference 3, Ch. III, ¶ 11.

THEOREM 1. *The index of periodicity of g is equal to the number of conjugate points of $x = 0$ on γ plus the order of concavity of g.*

The extremal g traced m times is a closed extremal. We shall denote it by g^m. We shall refer to the index of periodicity of g^m as the *mth index of periodicity* of g and denote it by T_m. We shall call the order of concavity of g^m the *mth order of concavity* of g.

We shall refer to the segment $0 \leq x \leq m\omega$ of the x axis as γ^m.

3. *Index of Repletion.*—If $m > 1$ and $x = \omega$ and $x = (m-1)\omega$ are not conjugate to $x = 0$, the matrices

$$\|u_{ij}(x)\| \qquad \|w_{ij}(x)\| \qquad (i, j = 1, \ldots, n)$$

exist with columns which are unique solutions of the Jacobi equations satisfying the conditions†

$$\begin{aligned} u_{ij}(0) &= 0 & w_{ij}(0) &= \delta_i^j \\ u_{ij}(\omega) &= \delta_i^j & w_{ij}([m-1]\omega) &= 0 \end{aligned} \qquad (i, j = 1, \ldots, n).$$

Then we shall call the quadratic form

$$R(z) = [\zeta_{ij}^u(\omega) - \zeta_{ij}^w(0)]z_i z_j \qquad (i, j = 1, \ldots, n) \tag{7}$$

the *mth repletion form* of g and its index the *mth index of repletion* of g. Its nullity is the index of $x = m\omega$ as a conjugate point of $x = 0$.

The following theorem brings out the meaning of the indices of repletion.

THEOREM 2. *If $x = \omega$, $x = (m - 1)\omega$, and $x = m\omega$ are not conjugate to $x = 0$, the number of conjugate points of $x = 0$ on γ^m exceeds the sum of the number on γ and the number on γ^{m-1} by the mth index of repletion of g.*

4. *Determination of Indices of Periodicity.*—We let

$$\|p_{ij}(x) \quad q_{ij}(x)\| \qquad (i, j = 1, \ldots, n),$$

whose columns are solutions of the Jacobi equations, be the base of solutions satisfying the initial conditions

$$\begin{aligned} p_{ij}(0) &= \delta_i^j & q_{ij}(0) &= 0 \\ \zeta_{ij}^p(0) &= 0 & \zeta_{ij}^q(0) &= \delta_i^j \end{aligned} \qquad (i, j = 1. \ldots, n).$$

We define the matrix A by the equation

$$A = \begin{Vmatrix} p_{ij}(\omega) & q_{ij}(\omega) \\ \zeta_{ij}^p(\omega) & \zeta_{ij}^q(\omega) \end{Vmatrix} \qquad (i, j = 1, \ldots, n). \tag{8}$$

We shall use the matrix A first in these two theorems.

THEOREM 3. *If for each positive integer m, g^m is non-degenerate and $x = m\omega$ is not conjugate to $x = 0$, then the mth order of concavity and the mth index of repletion of g are determined when A is known.*

THEOREM 4. *If for each positive integer m, g^m is non-degenerate and $x = m\omega$ is not conjugage to $x = 0$, then the mth index of periodicity of g is determined when A and the number of conjugate points of $x = 0$ on γ are known.*

We let W be a non-singular n-square matrix of constants. We set

$$\begin{aligned} P &= \|p_{ij}(\omega)\| & Q &= \|q_{ij}(\omega)\| \\ Z^p &= \|\zeta_{ij}^p(\omega)\| & Z^q &= \|\zeta_{ij}^q(\omega)\| \end{aligned} \qquad (i, j = 1, \ldots, n). \tag{9'}$$

We set‡

$$\begin{aligned} R &= WPW^{-1} & S &= WQW^* \\ Y^p &= W^{-1*}Z^pW^{-1} & Y^q &= W^{-1*}Z^qW^*. \end{aligned} \tag{9''}$$

We show under the conditions of Theorem 4 that the problem of determining the mth index of periodicity of g depends only on the number of conjugate points of $x = 0$ on γ and on invariants depending on the sets (R, S, Y^p, Y^q), where W is arbitrary subject to the conditions in the preceding paragraph.

5. *The Frequency Number μ.*—Any two solutions $\eta_i = v_i(x)$ and $\eta_i = z_i(x)$, $(i = 1, \ldots, n)$, of the Jacobi equations satisfy a relation

$$v_i(x)\zeta_i^z(x) - z_i(x)\zeta_i^v(x) \equiv \text{constant} \qquad (i = 1, \ldots, n)$$

and if this constant is zero, the two solutions will be termed *conjugate*. See reference 4, p. 626. There are at most n linearly independent solutions of the Jacobi equations in a set of mutually conjugate solutions. All the solutions which are linearly dependent on n mutually conjugate solutions will be said to form a *conjugate family*, and any set of n linearly independent solutions belonging to a conjugate family will be said to form a *conjugate base* of the conjugate family. A determinant whose columns are the solutions of a conjugate base will be termed a *focal determinant* of the conjugate family. The zeros of a focal determinant are isolated and of integral order from 1 to n. Any two focal determinants of a conjugate family vanish at the same points and to the same order. These points will be called *focal points* of the conjugate family. The order of vanishing of a focal determinant at a point is equal to its nullity at the point and will be referred to as the *order* of the focal point. In counting focal points of a conjugate family on an interval, each focal point will be counted a number of times equal to its order. This material is treated at length for a more general problem in reference 3, Ch. III.

Let F be a conjugate family with ν focal points on the interval $a < x \leqq a + m\omega$, $(m = 1, 2, \ldots)$. We define the frequency number $\mu(F,a)$ of g with respect to the conjugate family F and the initial point $x = a$ as the limit

$$\lim_{m=\infty} \frac{\nu}{m} \tag{10}$$

if this limit exists.

THEOREM 5. *The frequency number of g with respect to a conjugate family and an initial point exists and is independent of the conjugate family and the initial point.*

The number μ whose existence is affirmed in this theorem depends only on g and will be called the *frequency number* of g. The number $\frac{\omega}{\mu}$ is a generalization of the so-called Poincaré rotation number of a point transformation on a closed curve. See reference 7. Hedlund has shown the existence of this frequency number in the case $n = 1$ and has shown its relation to the problem of determining the indices of periodicity T_m in that case. See reference 8.

The following theorem gives a necessary condition on the indices of periodicity of g in terms of its frequency number.

THEOREM 6. *The limit*

$$\lim_{m=\infty} \frac{T_m}{m} \tag{11}$$

exists and is equal to the frequency number μ of g.

6. *Necessary Conditions on the Frequency Number.*—With the extremal

g we shall associate $2n$ numbers, $\rho_1, \ldots, \rho_{2n}$, which we shall call the *multipliers* of g. These numbers are the roots of the characteristic equation[9] of the matrix A.

There are no multipliers equal to 0. The non-degeneracy of g implies there are no multipliers equal to 1. We assume for convenience that there are no multipliers equal to -1.

The $2n$ numbers

$$\alpha_j = \delta_j + i\epsilon_j \qquad (j = 1, \ldots, 2n) \tag{12}$$

defined by the relations

$$\log \rho_j = \delta_j + i\epsilon_j \quad (-\pi < \epsilon_j \leqq \pi) \quad (j = 1, \ldots, 2n) \tag{13}$$

we shall term the *exponents* of g. Compare with reference 10.

Through application of the theory of lambda-matrices we have established these theorems.

Theorem 7. *If none of the exponents of g is pure imaginary, the frequency number of g is an integer.*

Theorem 8. *If the ratio of each pure imaginary exponent of g to $2\pi i$ is a rational number, the frequency number of g is rational.*

* We adopt the convention that a repeated subscript indicates summation with respect to that subscript.

† The symbol δ_i^j is 1 when $i = j$ and 0 when $i \neq j$.

‡ If B is a matrix, we denote its conjugate by B^* and its inverse by B^{-1}.

1 Morse, "Closed Extremals," these Proceedings, **15**, 856–859 (1929).

2 Morse, "Closed Extremals (first paper)," *Ann. Math.*, 2nd ser., **32**, 549–566 (1931).

3 Morse, *Calculus of Variations in the Large*, Am. Math. Soc., *Coll. Publ.* (1934).

4 Bolza, O., *Vorlesungen uber Variationsrechnung*, B. G. Teubner, Leipzig and Berlin, (1909).

5 Dickson, L. E., *Modern Algebraic Theories*, Benj. H. Sanborn and Co., Chicago, 71 (1926).

6 Reference 3, Ch. III, ¶ 2.

7 Poincaré, H., "Sur les Courbes Définies par les Équations Differentielles," *J. Math.*, (4), **1**, Ch. XV (1885).

8 Hedlund, G. A., "Poincare's Rotation Number and Morse's Type Number," *Trans. Am. Math. Soc.*, **34**, 75–97 (1932).

9 Bocher, M., *Introduction to Higher Algebra*, The Macmillan Co., New York, 282 (1912).

10 Birkhoff, G. D., "Stability and the Equations of Dynamics," *Am. J. of Math.*, **49**, 1–38 (1927).

Reprinted from the
JOURNAL DE MATHÉMATIQUES
Pures et Appliquées
Vol. xiv, pp. 49–71

Instability and transitivity;

By Marston MORSE,

Harvard University (U. S. A.).

The question as to the general existence of topological or metric transitivity has been much emphasized of late. One can refer to the historical summary by Birkhoff and Koopman, Birkhoff [II]. References are here given to the contributions of various writers including the recent important papers by Birkhoff, Hopf, Koopman, v. Neumann, and P. A. Smith. Birkhoff's Ergodic Theorem and v. Neumann's Mean Ergodic Theorem stan out at basic conclusions. The importance of these two theorems is however conditioned by the validity of the hypothesis that metric transitivity exists in general. If the hypothesis of metric transitivity fails in general, the theory will appear relatively incomplete and complex, at least until further illuminating contributions are made.

In non-analytic problems one can show by simple geodesic problems that metric transitivity fails in many cases. It would seem however that problems which include the analytic case offer a fairer test. Metric transitivity fails in all cases among geodesics on closed surfaces of revolution. Metric transitivity holds for a simple type of spiral-like motion on the torus, as is easy to prove. Birkhoff conjectured that geodesic motion on closed surfaces of constant negative curvature would offer an example of metric transitivity, and a number of mathematicians have been seeking to verify this opinion. Hedlund [I] has recently announced a proof of the desired theorem.

The theorem of Hedlund while most interesting is by no means a justification of the hypothesis of metric transitivity. The problem of whether geodesics on closed surfaces of non-constant negative curvature form a metrically transitive system has been reduced by the author to the problem of determining whether a certain topological transformation is *absolutely continuous* or not. This result can be obtained by combining Hedlund's theorem with the theorems of the author in the paper Morse [II] cited below.

But even if it turns out that geodesics on closed surfaces of negative curvature form a metrically transitive system, the hypothesis of metric transitivity will not then be justified in general. For a casual study of geodesics on surfaces of mixed curvature, partly positive and partly negative, shows that surfaces of negative curvature stand apart as an extreme case of non-general simplicity.

Let M^* be a non-singular r-dimensional manifold without boundary. On M^* suppose we have a system T of trajectories which includes on and only one trajectory through each point of M^*. We regard the time t as the parameter along these trajectories and suppose each trajectory can be continued over the time interval $-\infty < t < \infty$. So continued a trajectory will be said to be complete. The system T will be said to be *topologically transitive* if there exists a complete trajectory whose closure (the trajectory and its limit points) is the manifold M^*. Topological transitivity has been referred to as regional or geometric transitivity as well as topological transitivity. Metric transitivity implies topological transitivity as follows from the theorems of v. Neumann and Birkhoff. In examples for which topological transitivity holds there is thus a possibility that metric transitivity holds.

It is one of the purposes of the present paper to show that among the geodesics on closed surfaces R^* *of genus* $p > 1$ *uniform instability implies topological transitivity.*

Uniform instability, as it will be defined, is a property of the individual geodesics holding uniformly for all geodesics. Birkhoff has conjectured that the hypothesis of instability relative to the closed geodesics of R^* implies topological transitivity. The developments of this paper do not tend to support this conjecture.

The surfaces admitted are Riemannian manifolds which are closed in a topological sense. Those whose geodesics possess uniform instability include all topologically closed surfaces of negative curvature, as well as surfaces which possess regions of positive curvature. As previously stated our principal theorem connects instability with transitivity. Questions as to the distribution and number of transitive geodesics are answered, and a theorem on the deferring of transitivity is established.

From the point of view of the calculus of variations the hypothesis of uniform instability may be roughly regarded as the hypothesis that the first conjugate point of each finite point on a given surface *lies beyond the point at infinity*.

A proof of the existence of topological transitivity has been indicated by Birkhoff for a special surface of negative curvature of genus 2. See Birkhoff [III], p. 248. The methods of the author are based on the paper Morse [II] and are different from those of Birkhoff.

1. THE COVERING SURFACE R. — We suppose that we have a Fuchsian group G of non-singular, fractional, linear transformations of the plane $z = u + iv$ carrying the interior S of the unit circle

$$u^2 + v^2 = 1$$

into S. We suppose all members of the group are of hyperbolic type with fixed points on the unit circle, and that the group has a finite number of generators. We regard H as a hyperbolic plane of non-Euclidean geometry with the circles orthogonal to the unit circle representing the H-straight lines (H is written for hyperbolic). The H-length of a curve of S may be given by the integral

$$\int \frac{d\sigma}{1 - u^2 - v^2},$$

where $d\sigma$ is the differential of arc lengt in the (u, v)-plane.

We suppose that the fundamental domain for G consiste of a convex region S_0 on S bounded by a sequence, $p > 1$,

$$a_1 b_1 c_1 d_1, \quad a_2 b_2 c_2 d_2, \quad \ldots, \quad a_p b_p c_p d_p,$$

of segments of H-straight lines. The closure of S_0 contains no points on the unit circle. The successive H-lines in (1.2) shall form angles on S_0 equal to $\frac{\pi}{2p}$. The side a_k will be termed *conjugate* to c_k and the side b_k conjugate to d_k. There will be a transformation of the group carrying each side of S_0 into its conjugate side, carrying S_0 into an adjacent region. Points which are images of each other under G are termed *congruent*.

We make the usual convention that just one side of each pair of conjugate sides of S_0 shall be considered as belonging to S_0 and just one of S_0's vertices. The application of the tranformations of G to S_0 will yield a set of regions covering S in a one-to-one manner, as is well known.

Instead of assigning S the H-metric defined by (1.1) we can assign S an R-metric (R is written for Riemannian) defined by a positive definite form

$$(1.3) \qquad ds^2 = E(u, v)\, du^2 + 2F(u, v)\, du\, dv + G(u, v)\, dv^2$$

with coefficients of class C^3 for (u, v) on S. The form (1.3) will define a Riemannian manifold R. We suppose that the form (1.3) is invariant under G. We then regard congruent points as identical. With this understood S_0 taken with the form (1.3) defines a *topologically closed surface* R^*. The genus of R^* will be the integer p. The manifold R is the *covering surface* belonging to R^*.

The form

$$ds^2 = \frac{du^2 + dv^2}{(1 - u^2 - v^2)^2},$$

derived from (1.1) is a special instance of (1.3) and delines a topologically closed surface of constant negative curvature.

2. The phase-spase M. — Let angles θ in the (u, v)-plane be measured in the usual way from the positive u axis. A point (u, v) on S together with an angle θ will define an element E at (u, v) on S. Elements (u, v, θ) whose angles θ differ by an integral multiple of π will be regarded as identical. The set of all such elements E will define the *phase-space* M corresponding to S or R. The elements tangent to a regular curve g on R will make up a curve on M which we regard as the representative of g on M.

To any two points on R we assign an R-distance equal to the minimum length of R-geodesics joining the two points on R. If E′ and E″ are any two elements on M with directions represented by θ' and θ'', the R-angle between E′ and E″ will be taken as the minimum of

$$|\theta'' - \theta' + n\pi|,$$

for all integers n positive, negative, or zero. Two elements E′ and E″ on M will be said to possess an R-distance which is the sum of the R-angle and the R-distance between their initial points.

The set of elements with initial points on S_0 will be said to define the phase-space M* corresponding to R*. The *distance* between two elements E′ and E″ on M* will be taken as the minimum of the R-distance between all pairs of elements respectively congruent to E′ and E″ on R.

3. Previous theorems. — The present paper is based on an earlier paper Morse [II]. We recall a few définitions and theorems of the earlier paper.

The hyperbolic plane H consists of the interior S of the unit circle. Let γ be a simble open arc lying on S. Let $\bar{\gamma}$ be the closure of γ. Suppose $\bar{\gamma}$ is a simple arc with end points P and Q. If P or Q lies on the unit circle, P or Q respectively will be termed *ideal end points* of γ.

Let A and B be two point sets on R. Suppose tehre exists a number λ such that A is a subset of the points at most an R-distance λ from B, and that B is a subset of points at most an R-distance λ from A. The greatest lower bound of such numbers λ will be termed the type-distance between A and B. Two simple open arcs on R will be said to be of the *same type* if they possess a finite type-distance.

An unending geodesic g on R will be said to be of *class* A if every finite segment of g affords an absolute minimum to the R-length relative to oll rectifiable curves which join its end points on R. We state the following theorem.

Theorem 3.1. — *Every geodesic of class* A *on* R *is of the type of some* H-*straight line. Conversely there is at least one geodesic of class* A *of the type of each* H-*straight line on* S.

Moreover there exists a universal constant K *dependent only on* R *such that the type-distance between geodesics of class* A *of the same type or between geodesics of class* A *and* H-*straight lines of the same type never exceeds* K.

The surfaces admitted in Morse [II] are somewhat less general than the surfaces admitted here. Nevertheless the reader can readily see that the proofs in the earlier paper hold here practically unchanged. See Theorem 1 and Lemma 8, Morse [II].

If a curve h on S is invariant under a transformation T of the group G, h will be termed *periodic* mod T. The curve h will then define a closed curve on R*.

Corresponding to each H-straight line h there will either exist a unique geodesic g of class A of the type of h, or else two non-intersecting geodesics g' and g'' of class A of the type of h between which lie all other geodesics of class A of the type of h. The geodesics g' and g'' will be called the *boundary geodesics* of the type of h. We apply this term even in the case where $g' = g''$.

We state the following theorem. Morse [II], Theorem 11.

THEOREM 3.2. — *The boundary geodesics of the type of a periodic* H-*straight line are themselves periodic and are invariant under the same transformations of* G *as h.*

A geodesic on R which starts from a point P on R and is continued indefinitely in one sense from P will be called a *geodesic ray*. An H-straight line which emanates from P and is continued indefinitely in one sense will be called an H-*ray*. Two geodesics rays or a geodesic ray and an H-ray which emanate from a common point P on R and possess a finite type-distance will be said to be of the *same* type.

A geodesic ray g will be said to be of *class* A if every finite segment of g affords an absolute minimum to the R-length relative to rectifiable curves on R which join its end points.

With this understood we state the following theorem.

THEOREM 3.3. — *Corresponding to each* H-*ray issuing from a point* P *on* R *there exists a geodesic ray of class* A *of the same type. Conversely every geodesic ray of class* A *issuing from* P *is of the type of some* H-*ray*.

There exists a universal constant K *dependent only on* R *such that the type distance between two geodesic rays of class* A *of the same type or a geodesic ray of class* A *and an* H-*ray of the same type never exceeds* K.

Let k be an H-ray issuing from a fixed point P. There will either exist a unique geodesic ray of class A of the type of k, or there will exist two geodesic rays g' and g'' of class A of the type of k between which lie all other geodesic rays of class A of the type of k. The geodesics g' and g'' will be termed *boundary geodesic rays* of the type of k. We apply this term even in the case where $g' = g''$. The geodesic rays g' and g'' are either identical or else have at most the point P in common.

We state the following theorem.

THEOREM 3.4 — *Let b be a periodic* H-*straight line and* B *one of its ideal end points. Let g be a geodesic ray of class* A *whose initial point* P *does not lie between or on the boundary geodesics of the type of b and whose ideal end point coincides with* B. *The geodesic ray g will be asymptotic to one of the boundary geodesics of the type of b.*

That g is asymptotic to a periodic geodesic k of the type of b is stated in Lemma 10, Morse [II]. That k must be a boundary geodesic of the type of b follows from the affirmation in Lemma 11, Morse [II] that g cannot cross any periodic geodesic of class A of the type of b.

A geodesic g will be said to be a *limit geodesic* of a set of geodesics L not containing g if every element on g is a limit element of elements on geodesics of the set L.

In Morse [II] on p. 32 we stated the following lemma.

LEMMA A. — *There exists a transformation of the group* G *which has fixed points arbitrarily near the end points of any preassigned arc of the unit circle.*

The following theorem is a consequence of this lemma. Cf. Theorem 16, Morse [II].

THEOREM 3.5. — *The set of all boundary geodesics of periodic type includes all of the boundary geodesics amongits limit geodesics.*

This theorem has an immediate corollary.

COROLLARY. — *The set of all geodesic rays of class* A *through a fixed point* P *asymptotic to periodic boundary geodesics includes all of the boundary geodesics among its limit geodesics.*

4. THE HYPOTHESIS OF UNICITY. — *We shall say that* R *satisfies the hypothesis of unicity if there is but one boundary geodesic of class* A *of the type of each* H*-straight line.*

The hypothesis of unicity always holds if the surface is a surfacs of negative curvature. For on surfaces of negative curvature it is impossible to have two unending geodesics of the same type. See Hadamard [I]. More generally the hypothesis of unicity will hold if the geodesics on R possess uniform instability, as we shall presently show.

We shall prove the following theorem.

THEOREM 4.1. — *If the hypothesis of unicity holds, there is but one geodesic ray of class* A *of a given type issuing from a given point* P *of* R.

Suppose the theorem is false and that g' and g'' are two geodesic rays of class A of the same type issuing from a point P.

Exactly as in the proof of Theorem 6, Morse [II], so here it follows that g' and g'' cannot be asymptotic. Let s be the arc length on g' measured from P. Let a be any positive constant. Since the geodesics g' and g'' are not asymptotic, the minimum R-distance from the point s on g' to g'' must have a positive lower bound k for $s > a$.

Let

$$s_1, \quad s_2, \quad \ldots$$

be a sequence of points s on g' such that s_n becomes positively infinite with n and let

$$E_1, \quad E_2, \quad \ldots$$

be the corresponding elements on g'. Les

$$e_1, \quad e_2, \quad \ldots$$

be a set of elements e_n on g'' whose initial points on g'' are respectively at most the R-distance K from the corresponding points s_n on g'. The constant K is the universal constant of Theorem 3.3.

Let E_n be carried by a transformation T_n of the group G into an element E'_n with initial point ou S_0. Under T_n, e_n will be carried into an element e'_n. The pairs.

$$(E'_n, e'_n),$$

will have at least one cluster pair (E, e) on R since their initial points lie on the domain consisting of the points on R at most an R distance K from the points of S_0.

Let γ' and γ'' be the unending geodesics on R defined by the elements E and e respectively. The type-distance between γ' and γ'' exists and cannot be less than k. The geodesics γ' and γ'' are thus of the same type but not identical.

From this contradiction we infer the truth of the theorem.

When the hypothesis of unicity holds the correspondence between elements at P which respectively define H-rays and geodesic rays of the same type is readily seen to be continuous as well as one-to-one.

In such a case each geodesic ray through P is of class A.

Since P is an arbitrary point on R we can say more generally that when the hypothesis of unicity holds each unendig geodesic is of class A.

5. Topological transitivity. — The set of all elements on R* defines the phase-space M*. The elements on a geodesic g define a curve on M*, the representative of g on M*. If a geodesic or geodesic ray is represented by a curve on M* whose closure is M*, the geodesic or geodesic ray is termed *transitive*.

We shall prove the following theorem.

Theorem 5.1. — *Let* P *be an arbitrary point on* R *and* λ *an arbitrary open segment of the unit circle. If the hypothesis of unicity holds, there is at least one transitive geodesic ray issuing from* P *with an ideal end point on* λ.

Let

$$\gamma_1, \quad \gamma_2, \quad \gamma_3, \quad \ldots \tag{5.1}$$

be a set of periodic geodesics on R which includes at least one geodesic congruent to each periodic geodesic on R. From (5.1) we form the sequence

(5.2) $$\gamma_1, \quad \gamma_1\gamma_2, \quad \gamma_1\gamma_2\gamma_3, \quad \gamma_1\gamma_2\gamma_3\gamma_4, \quad \ldots,$$

in which each geodesic γ_i occurs infinitely many times. We denote the geodesics of the sequence (5.2) more simply by

(5.3) $$g_1, \quad g_2, \quad g_3, \quad \ldots$$

Let

(5.4) $$e_1, \quad e_2, \quad e_3, \quad \ldots,$$

be a sequence of positive numbers which tend to zero as the subscript becomes infinite.

We shall choose a sequence

(5.5) $$h_1, \quad h_2, \quad h_3, \quad \ldots,$$

of geodesic rays emanating from P and terminating on λ.

The choice of h_1. — Let θ denote the arc length on λ measured on λ from one end point of λ. Let a point on λ be denoted by the corresponding value of θ. Let g'_1 be an unending geodesic congruent to g_1 with at least one end point θ_1 on λ. That such a geodesic exists follows from Lemma A in § **3**. For we have merely to choose a transformation T of G with at least one fixed point on λ and apply T or its inverse a sufficient number of times to g_1 to obtain the required geodesic g'_1.

It follows from Theorem 3.4 and the hypothesis of unicity that there is a geodesic ray which issues from P, terminates at the point θ_1 on λ, and is asymptotic to g'_1. There will accordingly exist a positive constant η_1, so small that each geodesic ray which issues from P and terminates on the segment of λ for which

(5.6) $$\theta_1 - \eta_1 < \theta < \theta_1 + \eta_1,$$

will possess at least one element within an R-distance e_1 of some element on g'_1. We choose η_1 so small that the closure of the segment defined by (5.6) is interior to λ, and then choose h_1 as any geodesic ray which issues from P and terminates on the segment (5.6) of λ.

The choice of h. — Proceeding inductively we suppose that the entities

$$\theta_{n-1}, \quad \eta_{n-1}, \quad h_{n-1},$$

have been defined. We let g'_n be a periodic geodesic congruent to g_n with at least one end point θ_n on the segment of λ for which

$$(5.7) \qquad \theta_{n-1} - \eta_{n-1} < \theta < \theta_{n-1} + \eta_{n-1}.$$

There will then exist a positive constant η_n so small that the closure of the segment

$$(5.8) \qquad \theta_n - \eta_n < \theta < \theta_n + \eta_n,$$

of λ is a subsegment of (5.7) and is such that each geodesic ray which issues from P and terminates at a point θ on (5.8) will possess at least one element within an R-distance e_n of some element on g'_n. We choose h_n as any such geodesic ray.

Let E_n be the element on h_n whose initial point lies at P. Let E be a cluster element of the elements E_n, and g a geodesic ray issuing from P with the direction of E. The geodesic ray g will terminate on λ. I say moreover that g will be transitive.

For g will possess at least one element within an R-distance e_n of some element on g'_n. Let γ_m be an arbitrary geodesic of the set (5.1). By virtue of the choice of the sequence (5.2) the set of all elements on g or on geodesic rays congruent to g will include at least one element arbitrarily near some element on γ_m, and hence will include an element arbitrarily near *each* element on γ_m.

But according to Theorem 3.5 each geodesic on R is a limit geodesic of the set of all periodic geodesics on R. Thus each geodesic on R is a limit geodesic of the set of geodesics congruent to g. Hence g is a transitive ray, and the theorem is proved.

6. THE TRANSITIVITY FUNCTION $\varphi_g(e)$. — We shall prove the following lemma.

LEMMA. — *Let g be a transitive geodesic ray issuing from a point* P. *Corresponding to each positive number* α *there exists a number* $L_g(\alpha) \geq 0$ *such that the segment of g consisting of points at most an* R-*distance* $L_g(\alpha)$

from P *defines a curve in the phase-space* M^* *at most an* R*-distance* α *from each element of* M^*.

Let q_n be a segment of g with initial point at P and with an R-length n. Let q_n^* be the representative of q_n in the phase-space M^*.

Suppose the lemma fails to hold for a positive number α. Corresponding to each positive integer m there must then exist an element E_m on M^* at an R-distance from q_m^* greater than α. Let E be a cluster element of the element E_m. The distance of E from the representative of g on M^* will be at least α, contrary to the fact that g is transitive.

The lemma is according true.

Corresponding to a transitive geodesic ray g and a positive number α there will exist a greatest lower bound $\varphi_g(\alpha)$ of the number $L_g(\alpha)$ of the lemma.

We term $\varphi_g(\alpha)$ *the transitivity function defined by* g.

The transitivity function is monotonically decreasing. For large values of α it is zero. For all positive values of α less than a positive constant α_g it is positive. It become infinite as α tends to zero. Cf. Birkhoff's Ergodic Function, Birkhoff [II].

We shall prove the following theorem on the *deferring* of transitivity.

THEOREM 6.1. — *Let* $\psi(\alpha)$ *be an arbitrary positive function which becomes positively infinite as* α *tends to zero. If the hypothesis of unicity holds, there exists a transitive geodesic ray* g *issuing from an arbitrary point* P *of* R *such that the corresponding transitivity function* $\varphi_g(\alpha)$ *satisfies the relation*

$$\varphi_g(\alpha) > \psi(\alpha).$$

for an infinite sequence of positive values of α *which tend to zero as a limit.*

We shall prove this theorem by suitably altering the proof of Theorem 5.1.

We begin by choosing the sets (5.1) to (5.4) as in § 5. We also choose θ_1 and g'_1 as in § 5.

Let a_1 be a geodesic ray issuing from P with its ideal end point at the point θ_1 on λ. The geodesic ray a_1 will be asymptotic to g'_1. In the phase-space M^* the set A_1 of element determined by a_1 will have at most the element determined by g_1 as limit elements. There will accordingly exist an element E_1 on M^* and a positive constant $\alpha_1 < e_1$ such that

$$D(E_1, A_1) > \alpha_1, \tag{6.1}$$

where $D(E_1, A_1)$ denotes the R-distance on M^* between the element E_1 and the set A_1. We choose a positive constant L_1 such that

$$L_1 > \psi(\alpha_1), \tag{6.2}$$

and let H_1 denote the set of elements on M^* determined by an R-length L_1 on a geodesic ray with initial end point at P and ideal end point θ on an interval of the form (5.6). We subject the constant η_1 appearing in (5.6) to the restrictions imposed in § 5, requiring further that η_1 be so small that on M^*

$$D(E_1, H_1) > \alpha_1.$$

This is possible by virtue of (6.1). With η_1 so restricted the geodesic ray h_1 is chosen as any transitive geodesic ray which issues from P and terminates on the segment of λ defined by (5.6). We observe that

$$\varphi_{h_1}(\alpha_1) > L_1,$$

and hence that

$$\varphi_{h_1}(\alpha_1) > \psi(\alpha_1). \tag{6.3}$$

We proceed inductively, supposing that

$$\theta_{n-1}, \quad \eta_{n-1}, \quad h_{n-1}, \alpha_{n-1}, L_{n-1},$$

have already been defined. Let g'_n be a periodic geodesic congruent to g_n with at least one ideal end point ath the point θ_n on the segment (5.7) of λ. Let a_n be a geodesic ray issuing from P with its ideal end point at the point θ_n on λ. We continue exactly as in the preceding paragraph, replacing all symbols with subscript 1 by the same symbol with subscript n. For the geodesic ray h_n thereby defined we have

$$\varphi_{h_n}(\alpha_n) > \psi(\alpha_n). \tag{6.4}$$

As in § 5, so here it follows that the geodesic rays h_n possess a limit geodesic ray g issuing from P, and that g is transitive. The geodesic g is an admissible choice for each of the geodesics h_n so that

$$\varphi_g(\alpha_n) > \psi(\alpha_n).$$

The sequence α_n tends to zero as a limit as n becomes infinite, and the theorem is satisfied as stated.

We continue with the following theorem.

THEOREM 6.2. — *If the hypothesis of unicity holds, the transitive geodesic rays issuing from a given point* P *of* R *possess directions at* P *which are everywhere dense and non-denumerable.*

That the directions of transitive geodesic rays issuing from P are everywhere dense follows from the fact stated in Theorem 3.1 that the ideal end points of transitive geodesic rays can be chosen on arbitrary segments of the unit circle.

To complete the proof of the theorem let us suppose that the set of transitive geodesics issuing from P are denumerable and can accordingly be given by a sequence

$$b_1, \quad b_2, \quad \ldots. \tag{6.5}$$

Let

$$\varphi_1, \quad \varphi_2, \quad \ldots.$$

be the transitivity functions corresponding to the respective geodesic rays of the set (6.5). Let

$$\alpha_1 > \alpha_2 > \alpha_3 > \ldots$$

be a sequence of positive numbers which tend to zero as the subscript becomes infinite. Let $\psi(\alpha)$ be a function which is given by the sum

$$\psi(\alpha) = \varphi_1(\alpha) + \ldots + \varphi_n(\alpha),$$

for α on the interval

$$\alpha_n \geqq \alpha > \alpha_{n+1} \qquad (n = 1, 2, \ldots),$$

and which is 1 for values of $\alpha > \alpha_1$.

According to the preceding theorem there will exist a transitive geodesic ray g issuing from P for which the corresponding transitivity function will satisfy the relation

$$\varphi_g(\alpha) > \psi(\alpha),$$

for an infinite sequence of positive values of α tending to zero as the enumerating subscript becomes infinite. Hence for a fixed m,

$$\varphi_g(\alpha) > \varphi_m(\alpha)$$

for some value of α, so that g cannot be identical with b_m. Thus g cannot appear in the sequence (6.5), contrary to our hypothesis.

We conclude that the theorem is true.

Recall that the set of elements on R determines the phase-space M. The preceding theorem concerning the geodesic rays through a fixed point P are special cases of a more general class of theorems. The set of elements with a common initial point P determines a curve μ on M. This curve has the property that the geodesics determined by its elements are not identical with a single geodesic. In general any arc in the phase-space M whose elements define more than one geodesic on R will be termed *general.*

We state the following theorem.

THEOREM 6.3. — *If the hypothesis of unicity holds, the transitive geodesics defined by elements on a general arc* μ *in the phase-space* M *are everywhere dense on* μ *and non-denumerable.*

As the element E ranges over a general arc μ, the geodesic g defined by E will vary through a one-parameter family of geodesics. At least one of the ideal end points of these geodesics g will cover the whole of a segment of the unit circle at least once. For otherwise the geodesics g would reduce to a single geodesic.

The proof of the theorem can be obtained by making obvious modifications in the proofs of Theorems 5.1, 6.1 and 6.2, in particular by replacing the phrase « a geodesic issuing from P » by the phrase « a geodesic defined by an element on μ ». The geodesic rays h_n in the present proof will not be uniquely determined by their ideal end points on the unit circle, nor can these ideal end points in general be chosen arbitrarily subject to the earlier restrictions. Each such ideal end point must here be chosen among the totality of ideal end points which satisfy the earlier restrictions and are defined by elements on μ. Further details are unnecessary.

7. UNIFORM INSTABILITY. — If there is no conjugate point of a point P on any geodesic ray issuing from P, the geodesic rays issuing from P will form a field covering R in a one-to-one manner P alone excepted. If there are no pairs of conjugate points on any geodesic on R, it follows from the Weierstrass field theory of the calculus of variations that each geodesic on R is of class A.

But the hypothesis of unicity is not necessarily valid even when each geodesic on R is of class A as one can readily show by examples. To insure the validity of the hypothesis of unicity we must introduce more stringent restrictions on the geodesics g.

Let g be an arbitrary geodesic. Let g be referred to normal geodesic coordinates (x, y). For such coordinates

$$ds^2 = C^2(x, y)\, dx^2 + dy^2, \tag{7.0}$$

where $y = 0$ along g and $C(x, 0) \equiv 1$, while the curves $x =$ constant are geodesics normal to g and y gives the R-distance along these geodesic normals measured in an arbitrary sense from g. The equation of normal variation from g takes the form

$$\frac{d^2 w}{dx^2} + K(x)\, w = 0, \tag{7.1}$$

where x is the arc length along g measured from an arbitrary point of g and $K(x)$ is the curvature of the surface at the point x on g. Solutions $w(x)$ of (7.1) will be said to be *based* on g. A solution $w(x)$ of (7.1) such that

$$w^2(0) + w'^2(0) = 1, \tag{7.2}$$

will be termed *normal* at $x = 0$.

We shall say that the geodesics g on R are *uniformly* unstable if there are no pairs of conjugate points on g and if there exists a function $M(x)$ with the following properties :

a. The function $M(x)$ *is positive and continuous for all values of* x *exceeding some positive constant* λ, *and becomes infinite with* x.

b. Any solution of (7.1) *based on g and normal at* $x = 0$ *satisfies the relation*

$$|w(x_1)| + |w(x_2)| > M(x), \tag{7.3}$$

for $x > \lambda$ *and* $-x_1$ *and* x_2 *greater than* x.

c. The function $M(x)$ *is independent of the choice of the point* P *on g from which* x *is measured and of the choice of g on* R.

Geodesics on surfaces of negative curvature are uniformly unstable as one can readily show. But one can clearly replace considerable areas on surfaces of negative curvature by areas of positive curvature and still retain the property of uniform instability for the geodesics thereby defined.

The field F. — Let g' and g'' be two distinct boundary geodesics of the same type. Under the assumption that there are no pairs of conjugate points on any geodesic on R we shall construct a field F of geodesics related to g' and g''.

Let A_1 and A_2 be two points on g'. Let h_1 and h_2 be two sensed geodesics wich join the point A_1 and A_2 on g' to points B_1 and B_2 respectively on g''. Suppose moreover that h_1 and h_2 give paths from A_1 and A_2 to g'' as short as possible. The R-lengths of h_1 and h_2 will be at most the universal constant K of Theorem 3.1.

The points A_1, A_2, B_1, B_2 are the vertices of a geodesic quadrilateral which we shall cover with a special field of geodesics. To that end we regard points on h_1 and h_2 which divide h_1 and h_2 in the same ratio with respect fo R-length as corresponding.

Let F *be the family of geodesics which join corresponding points of* h_1 *and* h_2. *We shall investigate the representation of this family.*

We turn to an arbitrary geodesic g of the family F. Let α represent the arc length on h_2 measured along h_2 from g. Let the neighborhood of g be referred to normal geodesic coordinates (x, y) as in (7.0). Let g^α be the geodesic of F wich joins the point α on h_2 to the corresponding point on h_1. For α near zero the geodesics g^α can be represented in the form

$$y = \varphi(x, \alpha).$$

The functions $\varphi(x, \alpha)$ will be of class C^2 for α near o and x on an interval of the form

$$x(\alpha) \leqq x \leqq \bar{x}(\alpha)$$

where the functions $x(\alpha)$ and $\bar{x}(\alpha)$ give the values of x as functions of α at the respective intersections of g^α with h_1 and h_2.

A. *We shall show that*

$$\varphi_\alpha(x, o) \neq o \tag{7.4}$$

along g.

Turning to the final end point of g we have

$$\varphi_\alpha[x_2(o), o] \neq o. \tag{7.5}$$

For if h_2 is represented in terms of the arc length α, for α near o, in the form

$$y = \bar{y}(\alpha), \qquad x = \bar{x}(\alpha),$$

we obtain the identity

$$\bar{y}(\alpha) \equiv \varphi[\bar{x}(\alpha), \alpha],$$

from which it follows that

$$\frac{d\bar{y}}{da} \equiv \varphi_x \frac{d\bar{x}}{d\lambda} + \varphi_\alpha. \tag{7.6}$$

If φ_α were zero at the intersection of g and h_2 it would follow from (7.6) that h_2 and g would be tangent and hence be continuations of identical geodesics. But this is impossible, since in passing from h_2 to h_1 g would then intersect g' or g'' twice. Hence (7.5) holds as stated.

We could prove in a similar manner that $\varphi_\alpha(x, o)$ does not vanish at the intersection of g and h_1. Thus $\varphi_\alpha(x, o)$ vanishes at neither end point of g.

We can now establish (7.4). For if $\varphi_\alpha(x, o)$ vanished at an intermediate point of g the geodesics of F neighboring g would cross g near the zero of $\varphi_\alpha(x, o)$ and hence cross g twice, which is impossible. Thus (A) holds as stated.

A second representation of the field F. — It follows from (7.4) that the trajectories orthogonal to the field F are well defined at each point of F. In fact the equation

$$y - \varphi(x, \alpha) = 0,$$

can be solved for α as a function $\alpha(x, y)$ of class C^2 neighboring each point on g. The differential equation of the trajectories orthogonal to the geodesics of F can then be locally represented in the form

$$C^2(x, y)\, dx + \varphi_x[x, y)]\, dy = 0,$$

These orthogonal trajectories are accordingly without singularity.

We shall obtain a new representation of the field F. Let γ be a trajectory orthogonal to the geodesics of F issuing from a point p on g' midway between A_1 and A_2. Let $d(p, g'')$ be the R-distance from p to g''. We know that

$$0 < d(p, g'') \leqq K, \tag{7.7}$$

where K is the universal constant of Theorem 3.1. We shall *admit* only those geodesic quadrilaterals whose sides A_1 A_2 on g' have an R-length greater than 4K.

On an admissible quadrilateral γ can be continued on F over an R-length at least $d(p, g'')$ without passing off from F. For during such a continuation γ can at most reach g'', since the R-distance from p to g'' is $d(p, g'')$. Moreover during such a continuation γ cannot reach either h_1 or h_2. For if h_1, for example, were reached, A_1 and p could be joined by a combination of arcs of h_1 and γ of R-length at most 2K contrary to the choice of A_1 A_2 on g'. Thus γ can be continued on F for an R-length at least $d(p, g'')$.

Let μ be the R-length along γ measured from p. Let μ_0 be a value of μ on the interval

$$0 \leqq \mu \leqq d(p, g''), \tag{7.8}$$

and let g^0 be the geodesic of the field F which passes through the point μ_0 on γ. Let the neighborhood of g^0 be referred to normal geodesic coordinates (x, y) with g^0 as the base. The geodesics of F neighboring g^0 can be represented in the form

$$y = \psi(x, \mu), \tag{7.9}$$

where the function $\psi(x, \mu)$ is of class C^2 in its arguments. The function

$$w(x) = \psi_{\mu}(x, \mu_0) \tag{7.10}$$

will be a solution of the equation of variation

$$\frac{d^2 w}{dx^2} + K(x) w = 0. \tag{7.11}$$

based on g^0. Concerning $w(x)$ we shall prove the following statement.

B. *The solution* $w(x)$ *of* (7.11) *given by* (7.10) *satisfies the initial condition* $w(0) = 1$.

For μ near μ_0, γ can be given in the form

$$y = y^*(\mu), \qquad x = x^*(\mu), \tag{7.12}$$

and we have the identity

$$y^*(\mu) \equiv \psi[x^*(\mu), \mu].$$

From this identity we obtain a second identity

$$\frac{dy^*}{d\mu} \equiv \psi_x \frac{dx^*}{d\mu} + \psi_{\mu}. \tag{7.13}$$

But

$$\psi_x(x, \mu_0) \equiv 0.$$

Moreover at the intersection of g^0 and γ, $\frac{dy^*}{d\mu} = 1$ as follows from the choice of γ and of our coordinate system (x, y). Statement (B) follows from (7.13).

We come to a basic theorem.

THEOREM 7.1. — *Uniform instability of the geodesics on* R *implies that the hypothesis of unicity holds on* R.

We suppose that the theorem is false and that g' and g'' are two different geodesics of the same type. We shall prove that $g' = g''$ thereby arriving at a contradiction.

We identify g' and g'' with the preceding geodesics g' and g'' and construct a geodesic quadrilateral $A_1 A_2 B_1 B_2$ and field F as before.

Neighboring the geodesic g^0 of F we make use of the second representation (7.9) of the field F. Let s_1 and s_2 be the arc lengths on h_1 and h_2 respectively, measured from g'. Let g_μ be the geodesic of the field F which intersects the curve γ at the point on γ with parameter μ. Let

$$s_1(\mu), \qquad s_2(\mu)$$

be the respective values of the arc lengths on h_1 and h_2, measured from g^0, at the points on h_1 and h_2 at which the geodesic g_μ meets h_1 and h_2. In the normal geodesic coordinate system (x, y) based on g^0 suppose x is measured from the point of intersection of g^0 and γ in the sense that leads from h_1 to h_2. Let

$$x_1(\mu), \qquad y_1(\mu),$$

be the coordinates of the point of intersection of g_μ with h_1. For $\mu = \mu_0$ we have

$$\left(\frac{ds_1}{d\mu}\right)^2 = \left(\frac{dx_1}{d\mu}\right)^2 + \left(\frac{dy_1}{d\mu}\right)^2 \geqq \left(\frac{dy_1}{d\mu}\right)^2,$$

so that

$$\frac{ds_1}{d\mu} \geqq \psi_\mu[a_1, \mu_0] = w(a_1) > 0$$

where $a_1 = x_1(\mu_0)$.

Similarly let $x_2(\mu)$, $y_2(\mu)$ be the point of intersection of g_μ with h_2. We can show as above that for $\mu = \mu_0$

$$\frac{ds_2}{d\mu} \geqq w(a_2) > 0,$$

where $a_2 = x_2(\mu_0)$. We thus find that for $\mu = \mu_0$

$$(7.14) \qquad \frac{d}{d\mu}[s_1(\mu) + s_2(\mu)] \geqq w(a_1) + w(a_2).$$

Let the R-distance between A_1 and A_2 on g' be 2σ. We have denoted the intersection of g' and γ by p. The segment A_2p of g' has the R-length σ. It can be joined by a broken geodesic arc consisting of a segment of h_2 of R-length at most K, a segment of g^0 of R-length a_2, and a segment of γ of R-length at most K. From the minimizing property of the segment A_2p of g' we see that

$$\sigma < K + a_2 + K.$$

A similar relation holds upon replacing a_2 by $-a_1$. Thus

$$a_2 > \sigma - 2K, \quad -a_1 > \sigma - 2K.$$

According to (B), $w(0)=1$. Hence $w(x)$ will be « normal » at $x=0$ if multiplied by a suitable positive constant at most 1. It follows from the hypothesis of uniform instability that

$$w(a_1) + w(a_1) > M(\sigma - 2K, \tag{7.15}$$

if σ is sufficiently large. If σ is sufficiently large, the right member of (7.15) will be arbitrarily large according to the nature of $M(x)$, and in particular will be greater than $\frac{2K}{a}$ where

$$a = d(p, g'').$$

For such a choice of σ, (7.15) gives the relation

$$a[w(a_1) + w(a_2)] > 2K. \tag{7.16}$$

By virtue of (7.14) and (7.16) the sum of the R-lengths of h_1 and h_2 will be greater than 2K. From this contradiction we infer that $g' = g''$.

The proof of the theorem is complete.

Theorems 6.2 and 7.1 combine into the following theorem.

THEOREM 7.2. — *If the geodesics on* R *are uniformly unstable the transitive geodesics rays issuing from an arbitrary fixed point* P *on* R *have directions at* P *which are everywhere dense and non-denumerable.*

REFERENCES.

I. BIRKHOFF (G. D.) with SMITH (P. A.). — Structure analysis of surface transformations (*Journal de Mathématiques*, 7, 1928, p. 345-379).

II. BIRKHOFF (G. D.) with KOOPMAN (B. O.). — Recent contributions to the ergodic theory (*Proceedings of the National Academy of Sciences*, **18**, 1932, p. 279-281). References are here given to Hopf, Koopman, v. Neumann and P. A. Smith.

III. BIRKHOFF (G. D.). — Dynamical Systems (*American Mathematical Society Colloquium Publications*, New-York).

I. Hadamard (J.). — Les surfaces à courbures opposées et leurs lignes géodésiques (*Journal de Mathématiques*, **4**, 1898).

I. Hedlund (G. A.). — Metric transitivity on surface of constant negative curvature (*Proceedings of the National Academy of Sciences*, **20**, 1934).

I. Morse (M.). — Recurrent geodesics on a surface of negative curvature (*Transactions of the American Mathematical Society*, **22**, 1921, p. 84-101).

II. Morse (M.). — A fundamental class of geodesics on any closed surface of genus greater than one (*Transactions of the American Mathematical Society*, **26**, 1924, p. 25-61).

III. Morse (M.). — Does instability imply transitivity (*Proceedings of the National Academy of Sciences*, **20**, 1934).

I. Poincaré (H.). — Théorie des groupes fuchsiens (*Acta Mathematicæ*, **1**, 1882).

II. Poincaré (H.). — Sur les courbes définies par les équations différentielles (*Journal de Mathématiques*, 3ᵉ série, **7-8**, 1881-1882, et 4ᵉ série, **1-2**, 1885-1886).

Reprinted from the Proceedings of the NATIONAL ACADEMY OF SCIENCES,
Vol. 21, No. 5, pp. 258–263. May, 1935.

ABSTRACT CRITICAL SETS

BY MARSTON MORSE AND GEORGE B. VAN SCHAACK

DEPARTMENT OF MATHEMATICS, HARVARD UNIVERSITY

Communicated April 12, 1935

This paper is a summary of various results to be published in full at a later time.

PART I. THE GENERAL THEORY. We are concerned with an arbitrary metric space R and a single-valued function $f(P)$ of the point P on R. We suppose that f is continuous and admits a finite absolute minimum. If c is a value assumed by f the domain $f \leqq c$ shall be compact.

Critical points of f are defined as follows: Let P be a point at which $f(P) = c$ and let H be the set of points Q on R at which $f(Q) = c$ and which are at a distance at most $2e$ from P. Let τ be a variable which ranges on the interval

$$c - \eta \leqq \tau \leqq c + \eta, \tag{1}$$

where η is a positive constant. Let $H \times \tau$ represent the product of H and the interval (1). The point P is termed *ordinary* if for any sufficiently small e and η there exists a homeomorphism T between a neighborhood N of P and $H \times \tau$ with the following property. If the point X on N corresponds under T to the pair (Q, τ) on $H \times \tau$, $f(X) = \tau$. Any point of R which is not ordinary is termed *critical.*

The set of all critical points at which $f = c$, c a constant, is shown to be closed. A *critical set* is defined as a closed set of critical points on which f equals a constant c and which is at a positive distance from other critical points at which $f = c$. Let σ be a critical set and VW an ordered pair of neighborhoods of σ. Indicating closures by adding bars we suppose that $\overline{W} < V$ and that $\overline{V} - \sigma$ contains no critical points at which $f = c$. We shall have frequent occasion to use the phrase "corresponding to an ordered pair of neighborhoods VW" of the above sort, and shall use the expression corr VW as an abbreviation for that phrase.

The definitions of the type number of σ require the introduction of a number of preliminary terms. Let V_c be the set of points on V at which $f < c$.

By a *spannable k-cycle corr* VW we shall mean a k-cycle on W, below c, ~ 0 on W, but not ~ 0 on V_c.

By a *critical k-cycle corr* VW we shall mean a k-cycle on $W + V_c$, not $\sim$ on V to a k-cycle on V_c.

We shall deal with sums of cycles mod 2. Let A, be a class of k-cycles on R which does not include the null cycle. By a *submaximal* set M of cycles of A is meant a subset of elements of A, every proper sum of which belongs to A. The cardinal number r of the cycles in M will be called the *count* of M. The set M will be termed a *maximal* set of cycles of A if it is a proper subset of no submaximal set M' of cycles of A. If M is maximal its count r will be termed the count of cycles of A.

By a *family* F of cycles of A will be meant all proper sums of the cycles of a submaximal set M of cycles of A. We say that M defines F. We show that the count of F is the count of any submaximal set of cycles of A defining F.

We shall say that *a property B holds for any sufficiently limited ordered pair of neighborhoods* VW of σ if there exists a fixed neighborhood N of σ and corresponding to any $V < N$ a neighborhood $M(V)$ such that B holds when $V < N$, $W < M(V)$.

Let t_k be a cardinal number. The critical set σ will be said to have a k-th *subtype number* t_k if for any sufficiently limited VW there exists a family F' of spannable $(k$-$1)$-cycles corr VW and a family F'' of critical

k-cycles corr VW such that the count of F' plus the count of F'' equals t_k. The cardinal number of the class of all k-th subtype numbers of σ will be called the k-th *type number* of σ.

Excess and Absolute Type Numbers.—A constant b will be termed a *homology limit* of a cycle z if z lies on $f < b$ and is non-bounding on $f < b$. A cycle which is bounding on R but possesses a homology limit b will be termed an *excess* cycle. Let f_c denote the domain $f < c$. By a *new k-cycle corr VW* is meant a k-cycle on $W + f_c$, not homologous on $V + f_c$ to a cycle on f_c. By a *newly-bounding* (k-1)-*cycle corr VW* is meant a spannable (k-1)-cycle corr VW, not ~ 0 on f_c. If corresponding to sufficiently limited neighborhoods VW there exist families of excess new k-cycles and newly-bounding (k-1)-cycles corr VW of counts α and β, respectively, $\alpha + \beta$ will be called a k-th *excess subtype number* of σ.

Let z be a k-cycle which is non-bounding on R. The greatest lower bound of constants a such that z is homologous on R to a cycle on $f < a$ will be called the *inferior level* of z. By an *absolute* new k-cycle corr VW is meant a new k-cycle which is non-bounding on R and whose inferior level is greater than or equal to the given critical value c. The count γ of a family of absolute new k-cycles corr VW will be termed a k-th *absolute subtype number* of σ.

We show that $\alpha + \beta + \gamma$ *is a k-th subtype number of* σ.

The cardinal number of the class of all k-th excess (absolute) subtype numbers of σ will be called the k-th *excess* (*absolute*) *type number* of σ.

Let t_k, r_k, e_k be respectively the k-th type number, absolute type number and excess type number of σ. If two of these numbers are finite the third is finite and $e_k + r_k = t_k$. Moreover, t_k is infinite if either e_k or r_k is infinite. In this connection we prove the following theorem.

THEOREM 1. *Let σ be a critical set which is at a positive distance from critical points not on σ. If σ possesses a neighborhood which is coverable by a complex, all but a finite number of its type numbers are null.*

This theorem applies at once to functions which are defined in euclidean space, which are differentiable or merely continuous, and which possess an isolated critical set.

By a *complete* critical set is meant the set of all critical points at which f assumes a critical value c.

The following theorem is fundamental. It holds regardless of whether the critical values are isolated or not, or whether the respective critical sets have infinite type numbers or not.

THEOREM 2. *Let (σ) be an ensemble of complete critical sets with distinct critical values, and with finite excess subtype number sums m_0, m_1, ... There exists a set (σ') > (σ) of complete critical sets with distinct critical values and with finite k-th excess subtype number sums $M_k \geqq m_k$ such that the following infinite set of (excess) relations holds:*

$$\begin{aligned} M_0 &\geqq 0 \\ M_0 - M_1 &\leqq 0 \\ M_0 - M_1 + M_2 &\geqq 0 \\ &\cdots\cdots \end{aligned} \tag{2}$$

The following theorem is valid even if the connectivities of R are infinite.

THEOREM 3. *Let F be a finite family of k-cycles which are non-bounding on R. The complete critical sets at the respective inferior levels of the cycles of F have absolute k-th type numbers which sum to at least the count of F. If the k-th connectivity R_k of R is finite the k-th absolute type number sum of f is R_k. If R_k is infinite this sum is infinite.*

The preceding theorem is concerned with the topologically necessary critical sets. The following theorem is concerned with the critical sets in excess of those topologically necessary. The numbers E_k appearing therein may be finite even when the connectivities R_k are infinite.

THEOREM 4. *Let E_k be the k-th excess type number sum of f. If the numbers $E_0, E_1, \ldots$ are finite they satisfy the excess relations* (2). *If the numbers $E_0, \ldots, E_r$ are finite they satisfy the first $r + 1$ excess relations.*

THEOREM 5. *Let T_k be the k-th type number sum of f and R_k the k-th connectivity of R ($k = 0, 1, \ldots$). If T_k and R_k are finite the numbers $M_k = T_k - R_k$ satisfy the excess relations. If T_k and R_k are finite for $k = 0, \ldots, r$ the numbers $M_0, \ldots, M_r$ satisfy the first $r + 1$ excess relations.*

PART II. THE CASE OF REGIONS ON REGULAR MANIFOLDS. The preceding theory applies in particular to a function defined on a regular manifold M. It will also apply if the metric space R is a closed region on M. As a function on R, f will in general have critical points on the boundary of R which are not critical points of f considered as a function on M. This follows from our definition of critical points and in the general case has the great advantage that it enables us to proceed without a separate analysis of the boundary and the boundary conditions satisfied by f. If, however, the boundary of R is a set of regular manifolds it is possible to give a more intimate analysis of the critical sets of f on the boundary and their relation to the critical sets of f within R.

The theory of functions on regular regions on regular manifolds is not essentially different from the theory of functions on regular regions of a euclidean space. In Part II we shall accordingly deal with a real single-valued function $f(x_1, \ldots, x_n)$ defined on a limited, open n-dimensional region S of euclidean n-space. We shall say that the function f is of class C^1L on S if it is of class C^1 on S and if its first partial derivatives satisfy Lipschitz conditions neighboring each point of S. In Part II we shall assume that f is of class C^1L on S.

For the moment let us term critical points, sets or type numbers as defined in Part I *abstract* critical points, sets or type numbers, and those defined in the Colloquium Lectures (Morse) *ordinary*. An abstract critical point of $f(x)$ is an ordinary critical point of $f(x)$, but an ordinary critical point of $f(x)$ is not always an abstract critical point of $f(x)$. An abstract critical set σ' of $f(x)$ is a proper or an improper subset of an ordinary critical set σ. If σ is a given ordinary critical set, the set of abstract critical points of σ forms an abstract critical set σ' which can be shown to have the same type numbers as σ. If σ is an ordinary isolated critical set its type numbers in accordance with Theorem 1 are finite. We make use of this fact throughout Part II. For the remainder of Part II we shall refer to critical sets and type numbers in the ordinary sense and shall omit the word ordinary.

Admissible Functions.—Let Σ be an open subregion of S whose closure $\bar{\Sigma}$ lies on S and whose boundary B is a closed point-set consisting of a finite number of connected, regular, non-intersecting $(n\text{-}1)$-spreads of class C^3. Let f_ν denote the directional derivative of f on the normal to B in the sense that leads from points on Σ to points not on Σ. Let f° be the function defined by f on B. Let the existence and type numbers of critical sets of f° be determined in terms of f° as a function of the parameters of B.

The function f will be termed *admissible on* Σ if it satisfies the following conditions.

I. *The function f shall be of class C^1L on an open region containing $\bar{\Sigma}$ and shall have only a finite number of critical values on Σ.*

II. *The function f shall be of class C^3 neighboring B and shall have no critical points on B. The boundary function f° shall have a finite number of critical values.*

The theorem here is as follows.

THEOREM 6. *Let R_i be the i-th connectivity of $\bar{\Sigma}$. Let M_i be the sum of the i-th type numbers of the critical sets of f on Σ and of the boundary function f° defined by f at the points on the boundary of Σ at which $f_\nu < 0$. The numbers $M_i - R_i$ satisfy the first $n + 1$ excess relations* (2), *the* $(n + 1)$*-st relation becoming an equality.*

PART III. GROUP THEORY ASPECTS. Certain concepts of the theory in Part I may be developed from the point of view of group theory as follows.

Let $G = A/B$ be the quotient of groups A and B of cycles in which the operation is addition mod 2. The elements of G are classes of cycles, two cycles belonging to the same class if their sum mod 2 belongs to the group B, the zero-element of G. Any class which is not the zero-element of G will be termed a *proper* class. A cycle belonging to a class C will be ter ned a *representative* of C.

Let VW be an arbitrary pair of neighborhoods of the critical set σ of which $\bar{W} < V$. Let A be the group of k-cycles on W below c, ~ 0 on W, and B the subgroup of cycles ~ 0 on V below c. We term A/B the k-th *spannable group corr* VW. A representative of one of its proper classes is a spannable k-cycle corr VW.

Let A be the group of k-cycles on $W + V_c$, and B the subgroup of cycles homologous on V to cycles below c. We term A/B the k-th *critical group corr* VW.

Let f_c denote the domain $f < c$. Let A be the group of k-cycles on $W + f_c$ and B the subgroup of cycles homologous on $V + f_c$ to cycles below c. We term A/B the k-th *new group corr* VW.

Let A be the k-th spannable group corr VW. The elements of A are classes of k-cycles. Let B be the subgroup of A whose classes are represented by cycles which bound below c. We term A/B the k-th *newly-bounding group corr* VW.

Let A be the group of all k-cycles bounding on R, and B the subgroup of cycles without homology limits. We term A/B the k-th *excess group*.

Let A be the group of k-cycles on $W + f_c$ which are bounding on R, and B the subgroup of cycles which are homologous on $V + f_c$ to cycles below c. We term A/B the k-th *excess new group corr* VW.

Let A be the homology group of R and B the subgroup of A whose classes are represented by cycles with inferior levels less than c. We term A/B the k-th *absolute new group corr* VW.

The groups introduced in Part III correspond to the families introduced in Part I. The families seem to offer the more concise characterization of the categories of cycles needed, while the group theory representation throws a new light on their structure.

Morse, "The Calculus of Variations in the Large," *Amer. Math. Soc. Colloquium Publications*, **18**, New York (1934).

Birkhoff, G. D., and Hestenes, M. R., "Generalized Minimax Principle in the Calculus of Variations," these PROCEEDINGS, **21**, 96–99 (1935).

Reprinted from BULLETIN OF THE AMERICAN MATHEMATICAL SOCIETY
December, 1936

A SPECIAL PARAMETERIZATION OF CURVES

BY MARSTON MORSE

1. *Introduction*. Parameterization of curves by means of arc length fails when the arc length is infinite. The present paper develops the properties of a special parameterization of curves which never fails to exist and which is most useful in applications. The special parameter is called a μ-length and is an extension of a function of sets defined by H. Whitney* and applied by Whitney to families of simple non-intersecting curves. The curves employed in the present paper are general continuous images of a line segment. This necessitates a slight modification of the definition of Whitney taking order into account. The properties of μ-length developed here are directed largely towards applications in abstract variational theory. While many of Whitney's ideas go over, there are nevertheless certain sharp differences both in the proofs and in the results. In particular the variational theory of general families of closed curves requires concepts not readily suggested either by the functions of Whitney or the extension developed here. The results of the present paper will be applied in the author's paper entitled *Abstract variational theory*.

2. *The μ-Length*. Let N be a space of points p, q, r with a metric which is not in general symmetric. That is, to each ordered pair p, q of points of N there shall correspond a number denoted by pq such that $pp=0$, $pq>0$ if $p\neq q$, and

$$pq \leqq pr + rq. \tag{1}$$

We term pq the *distance* from p to q. Let $|pq|$ denote the maximum of pq and qp. We term $|pq|$ the *absolute distance* between p and q. We see that $|pq| = |qp|$ and that $|pq| \leqq |pr| + |rq|$. The points of N taken with the distance $|pq|$ form a metric space which we denote by $|N|$. We shall use the metric of N in defining μ-lengths. For other purposes, in particular in defin-

* H. Whitney, *Regular families of curves*, Annals of Mathematics, vol. 34 (1933), pp. 244–270. Also Proceedings of the National Academy of Sciences, vol. 18 (1932), pp. 275–278 and pp. 340–342.

ing continuity on N and $|N|$, we shall use the metric of $|N|$.

Let t be a number on a closed interval* $(0, a)$ Let $f(p, t)$ be a single-valued (numerical) function of p and t for p on $|N|$ and t on $(0, a)$. The function $f(p, t)$ will be termed *continuous* at (q, τ) if $f(p, t)$ tends to $f(q, \tau)$ as a limit as $|pq| + |t-\tau|$ tends to 0. For each value of t on $(0, a)$ and point p on $|N|$ let $\phi(p, t)$ be a point on $|N|$. The (point) function $\phi(p, t)$ will be termed continuous at (q, τ) if the distance $|\phi(p, t)\phi(q, \tau)|$ tends to 0 as a limit as $|pq| + |t-\tau|$. tends to 0.

Let $p(t)$ be a continuous point function of t for t on $(0, a)$. We term $p = p(t)$ a *parameterized curve* λ (written p-curve) and regard two parameterized curves as *identical* if they are defined by the same point function $p(t)$. We also say that λ is the continuous image on $|N|$ of $(0, a)$. In general curves on $|N|$ will be denoted by Greek letters $\alpha, \beta, \cdots$, while points on $|N|$ will be denoted by letters $p, q, r, \cdots$.

Let λ be a p-curve on $|N|$ given as the continuous image $p = p(t)$ of an interval $0 \leqq t \leqq a$. For $n \geqq 2$ let $t_1 \leqq t_2 \leqq \cdots \leqq t_n$ be a set of n values of t on $(0, a)$ and let $(p_1, \cdots, p_n) = S_n$ be the set of corresponding points p on λ. We term S_n an *admissible* set of n points on λ. Let the minimum of the numbers $p_i p_{i+1}$ as i ranges from 1 to $n-1$ be denoted by $d(S_n)$ and let the least upper bound of all such numbers $d(S_n)$ for a fixed n be denoted by $\mu_n(\lambda)$. Following Whitney we then set

$$\mu_\lambda = \frac{\mu_2(\lambda)}{2} + \frac{\mu_3(\lambda)}{4} + \frac{\mu_4(\lambda)}{8} + \cdots. \tag{2}$$

We term the μ_λ the *μ-length* of λ.

We enumerate certain properties of $\mu_n(\lambda)$ and μ_λ.

✓ (a) $\mu_2(\lambda) = d$, *the diameter of* λ.

✓(b) $\mu_n(\lambda) \leqq d$, *and* $\mu(\lambda) \leqq d$.

✓(c) $\mu_n(\lambda)$ *tends to* 0 *as* n *becomes infinite.*

---

* For the sake of simplicity we assume that $a > 0$ and that our p-curves do not reduce to points. One could however admit p-curves which reduce to points. The μ-lengths of such curves is zero and we would thus admit intervals $(0, a)$ for which $a = 0$. For such exceptional curves Fréchet distance is defined in the obvious manner. Theorem 2 is obvious if ζ reduces to a point. In the proof of Theorem 4, in case λ_0 reduces to a point, (17) is an easy consequence of $\lambda\lambda_0 < \delta$ provided δ is sufficiently small. Otherwise the theorems and proofs hold as written even when the p-curves reduce to points.

(d) $\mu_n(\lambda) \geqq \mu_{n+1}(\lambda)$.

(e) *If $p(t)$ is not identically constant none of the numbers $\mu_n(\lambda)$ is* 0.

Statements (a), (b), (c), and (e) are obvious. To establish (d) let S_{n+1} be any admissible set of $n+1$ points $p_1, \cdots, p_{n+1}$ on λ, $(n>1)$. There is an integer k such that $p_k p_{k+1} = d(S_{n+1})$. We shall form an admissible set S_n on λ by removing one point from S_{n+1}. If $k \neq 1$ we remove p_1. If $k=1$ we remove p_{n+1}. In both cases the pair p_k, p_{k+1} remains and $d(S_n) = d(S_{n+1})$. The set of all numbers $d(S_{n+1})$ is thus a subset of the numbers $d(S_n)$. Hence (d) is true.

Let $\mu(t)$ be the μ-length of the p-curve on λ defined by $p = p(\tau)$ for τ on the interval $(0, t)$. We shall show that $\mu(t)$ has the following properties.

(f) *$\mu(t)$ is a continuous, non-decreasing function of t.*

(g) *$\mu(t)$ is constant on each interval on which $p(t)$ is constant.*

(h) *$\mu(t)$ is constant on no interval on which $p(t)$ is not constant.*

It is clear that $\mu(t)$ is non-decreasing. To show that $\mu(t)$ is continuous let η be the p-curve defined by $p = p(t)$ for $0 \leqq t \leqq \tau < a$, and let e be an arbitrary positive number. There exists a positive number δ so small that

$$|\, p(t)p(\tau)\,| \leqq \frac{e}{2}, \qquad (\tau \leqq t \leqq \tau + \delta \leqq a). \tag{3}$$

Let ζ be the p-curve defined by $p = p(t)$ for $0 \leqq t \leqq \tau + \delta$. Let S_n^2 be an admissible set of n points on ζ. Let S_n^1 be an admissible set of n points on η obtained by replacing each point of S_n^2 for which $t > \tau$ by $p(\tau)$. No point of S_n^2 is thereby moved a distance greater than $e/2$. Hence

$$\begin{aligned} d(S_n^2) &\leqq d(S_n^1) + e, \\ \mu_n(\zeta) &\leqq \mu_n(\eta) + e, \\ \mu(\tau + \delta) &\leqq \mu(\tau) + e. \end{aligned} \tag{4}$$

Since $\mu(t)$ is non-decreasing, (4) implies continuity on the right. Continuity on the left is similarly established, and the proof of (f) is complete.

Statement (g) requires no proof. To establish (h) we assume that there are values τ and τ' of t on $(0, a)$ with $\tau' > \tau$ such that

$p(0)$, $p(\tau)$, and $p(\tau')$ are distinct. Let h and k be the p-curves defined by $p(t)$ for t on $(0, \tau)$ and $(0, \tau')$, respectively. We shall prove that $\mu_k > \mu_h$. To that end, let $2c$ be the minimum value of $p(t)p(\tau) + p(t)p(\tau')$ as t ranges on $(0, \tau)$. We observe that $c \neq 0$ since $p(\tau) \neq p(\tau')$. Let S_n be an admissible set of points $p_1, \cdots, p_n$ on h. We form S'_{n+1} on k by adding one point p_{n+1} to S_n as follows. If

$$p_n p(\tau) \geqq c, \tag{5}$$

we add $p(\tau)$. If (5) does not hold we add $p(\tau')$, and note that

$$p_n p(\tau') \geqq c \tag{6}$$

by virtue of our choice of c. We suppose N so large an integer that $d(S_n) < c$ for $n > N$. For such values of n it follows from (5) and (6) that $d(S'_{n+1}) = d(S_n)$, and hence

$$\mu_{n+1}(k) \geqq \mu_n(h). \tag{7}$$

Since $\mu_n(h)$ is not 0 and tends to 0 as n becomes infinite, for some value of $n > N$,

$$\mu_n(h) > \mu_{n+1}(h), \tag{8}$$

and for such an n it follows from (7) and (8) that $\mu_{n+1}(k) > \mu_{n+1}(h)$. From this relation and from (2) it follows that $\mu_k > \mu_h$ as stated. The proof of (h) is complete.

3. *Equivalent p-Curves.* Let η and ζ be two p-curves given by the respective equations

$$p = p(t), \qquad (0 \leqq t \leqq a), \tag{9}$$

$$q = q(u), \qquad (0 \leqq u \leqq b). \tag{10}$$

Let H be a sense-preserving homeomorphism between the closed intervals $(0, a)$ and $(0, b)$. A homeomorphism of the nature of H will be termed *admissible*. Let $u = u(t)$ be the value of u corresponding to t under H and let $D(H)$ be the maximum of the distances $|p(t)q[u(t)]|$ as t ranges over $(0, a)$. The Fréchet *distance* $\eta\zeta$ between η and ζ is the greatest lower bound of the numbers $D(H)$ as H ranges over the set of all admissible homeomorphisms H between η and ζ. We observe that $\eta\zeta = \zeta\eta \geqq 0$, and one readily proves that for any three p-curves η, ζ, λ, $\eta\lambda \leqq \eta\zeta + \zeta\lambda$.

We shall say that η is *derivable* from ζ if there exists a continuous non-decreasing function $u=u(t)$ which maps the closed interval $(0, a)$ onto the closed interval $(0, b)$ and such that $p(t)\equiv q[u(t)]$. Two p-curves which are derivable from the same p-curve η will be said to be *equivalent*.

Let $\mu(t)$ be the μ-length of η from the point 0 to the point t. Let $t(\mu)$ be the function inverse to $\mu(t)$. To each value of μ on the interval $0\leqq\mu=\mu_\eta$ there corresponds a value or interval of values $t(\mu)$. We set

$$p[t(\mu)] = r(\mu), \qquad (0 \leqq \mu \leqq \mu_\lambda), \tag{11}$$

and observe that $r(\mu)$ is single-valued by virtue of (h), §2. We shall prove the following proposition.

(A) *The function $r(\mu)$ is continuous in μ.*

Let μ_0 be a value of $\mu(t)$. Corresponding to μ_0, let $\tau_1\leqq t\leqq\tau_2$ be the interval of values taken on by $t(\mu)$ at μ_0 (τ_1 may equal τ_2). Corresponding to a positive constant e there exists a positive constant δ so small that

$$|\,p(\tau_2)p(t)\,| < e, \qquad (\tau_2 \leqq t \leqq \tau_2 + \delta). \tag{12}$$

As t ranges over the interval in (12), $\mu(t)$ increases from μ_0 to a value $\mu_1>\mu_0$. From (12) we infer that $|r(\mu_0)r(\mu)|\leqq e$ for $\mu_0\leqq\mu\leqq\mu_1$. Thus $r(\mu)$ is continuous on the right at μ_0. Continuity on the left is established similarly, and the proof of (A) is complete. We shall now prove the following statement.

(B) *The curve η is derivable from the curve $r=r(\mu)$, $(0\leqq\mu\leqq\mu_\eta)$.*

It follows from the definition of $r(\mu)$ in (11) that $p(t)\equiv r[\mu(t)]$, and the proof of (B) is complete. We term $r=r(\mu)$ a *μ-parameterization* of η and state the following theorem.

THEOREM 1. *Two p-curves are equivalent if and only if they have the same μ-parameterization.*

If η and ζ are equivalent, they are derivable from a common p-curve λ and will have μ-parameterizations identical with that of λ. Conversely, if η and ζ have a common μ-parameterization $r=r(\mu)$, η and ζ are both derivable from $r=r(\mu)$ in accordance with (B) and hence are equivalent. We shall prove the following theorem.

THEOREM 2. *The Fréchet distance between a p-curve ζ and a p-curve η derivable from ζ is null.*

Suppose η and ζ have the representations (9) and (10), respectively. Suppose η is derivable from ζ under the substitution $u=u(t)$, so that $p(t)\equiv q[u(t)]$ for t on $(0, a)$. Let c be an arbitrarily small positive constant and consider the transformation

$$u = [u(t) + ct]\left[\frac{b}{b + ca}\right], \qquad (0 \leqq t \leqq a). \tag{13}$$

This transformation establishes a homeomorphism between the closed intervals $(0, a)$ and $(0, b)$. Denote the right member of (13) by $\phi(t, c)$, and let λ_c be the p-curve

$$q = q[\phi(t, c)], \qquad (0 \leqq t \leqq a). \tag{14}$$

The Fréchet distance $\zeta\lambda_c=0$. For under the transformation (13) corresponding points of ζ and λ_c are identical. To show that $\eta\zeta=0$ we make use of the relation $\eta\zeta \leqq \eta\lambda_c+\lambda_c\zeta=\eta\lambda_c$. If c is sufficiently small, points on η and λ_c determined by the same values of t on $(0, a)$ are arbitrarily and uniformly near since $\phi(t, 0)\equiv u(t)$. Hence $\eta\lambda_c$ tends to 0 with c. But $\eta\zeta$ is independent of c and must then be 0. The proof of the theorem is complete.

We shall now prove Theorem 3.

THEOREM 3. *If η and ζ are two p-curves for which $\eta\zeta<e$, then $|\mu_\eta-\mu_\zeta| \leqq 2e$.*

Let S_n be an admissible set of n points on η. There exists an admissible set S_n' of n points on ζ with distances from the correspondingly numbered points of S_n less than e. Hence $|d(S_n')-d(S_n)| \leqq 2e$, and we infer that $|\mu_n(\zeta)-\mu_n(\eta)| \leqq 2e$. Upon referring to the definition (2) of μ-length we conclude that Theorem 3 holds as stated. We state the following corollary.

COROLLARY 1. *If $\eta\zeta=0$, $\mu_\eta=\mu_\zeta$.*

4. *Curves. A class of equivalent p-curves will be called a curve class or a curve.*

Let α and β be two curves. Let η and η' be p-curves in the class α, and ζ and ζ' p-curves in the class β. I say that

$$\eta\zeta = \eta'\zeta'. \tag{15}$$

Relation (15) follows from the relation

$$\eta\zeta \leqq \eta\eta' + \eta'\zeta' + \zeta'\zeta.$$

For $\eta\eta' = 0$, since η and η' are at a distance 0 from their common μ-parameterized curve λ in accordance with Theorem 2. Similarly $\zeta\zeta' = 0$. Upon reversing the roles of $\eta\zeta$ and $\eta'\zeta'$ we infer that (15) holds as stated. We are accordingly led to the following statement and definition.

The distance between any two p-curve classes α and β equals the distance between any other two p-curves in the classes α and β, respectively, and will be taken as the distance $\alpha\beta$ between the curves α and β.

Let λ be an arbitrary curve with μ-length μ_λ. A pair (λ, μ) will be termed *admissible* if $0 \leqq \mu \leqq \mu_\lambda$. For admissible pairs (λ, μ) let $q(\lambda, \mu)$ be the point on λ which determines the μ-length μ on λ. The following theorem is fundamental.

THEOREM 4. *The point function $q(\lambda, \mu)$ is continuous* in its arguments on the domain of admissible pairs (λ, μ).*

We shall prove (λ, μ) continuous at (λ_0, μ_0) understanding that (λ_0, μ_0) is admissible. Let e be an arbitrary positive constant. We shall show that there exists a positive constant δ such that if (λ, μ) is admissible and

$$\lambda\lambda_0 < \delta, \qquad |\mu - \mu_0| < \delta, \tag{16}$$

then

$$| q(\lambda, \mu)q(\lambda_0, \mu_0) | < e. \tag{17}$$

To that end we shall subject δ to two conditions as follows:

(i) We take $\delta < e/2$. If $\lambda\lambda_0 < \delta$, there will exist a homeomorphism T_δ between μ-parameterizations of λ and λ_0 in which corresponding points have distances less than δ. If the point μ on λ thereby corresponds to μ_1 on λ_0,

$$| q(\lambda, \mu)q(\lambda_0, \mu_1) | < \frac{e}{2}. \tag{18}$$

* We have not yet shown that for two curves α and β, $\alpha\beta = 0$ only if $\alpha = \beta$. But this is not necessary to speak of continuity. The proof of our theorem will imply that $q(\alpha, \mu) = q(\beta, \mu)$ when $\alpha\beta = 0$.

(ii) The second condition on δ is that δ be so small that when $|\mu_1-\mu_0|<3\delta$,

$$(19) \qquad |q(\lambda_0, \mu_1)q(\lambda_0, \mu_0)| < \frac{e}{2}.$$

This condition can be satisfied by virtue of the continuity of $q(\lambda_0, \mu)$ in μ.

With δ so chosen I say that (17) holds for admissible pairs (λ, μ) satisfying (16). We introduce T_δ and μ_1 on λ_0 as in (i), and recall that

$$(20) \quad |q(\mu, \lambda)q(\mu_0, \lambda_0)| \leqq |q(\lambda, \mu)q(\lambda_0, \mu_1)| + |q(\lambda_0, \mu_1)q(\lambda_0, \mu_0)|.$$

The first term on the right of (20) is less than $e/2$ by virtue of (18) and the second term is likewise, provided (19) is applicable; that is, provided $|\mu_1-\mu_0|<3\delta$. But under T_δ a point μ on λ will correspond to a point μ_1 on λ_0 such that $|\mu-\mu_1|<2\delta$ in accordance with Theorem 3. Hence $|\mu_1-\mu_0|<3\delta$, (19) is applicable, and the right member of (20) is less than e. The proof of the theorem is complete. We conclude this section with the following theorem.

THEOREM 5. *A necessary and sufficient condition that two p-curves η and ζ have the same μ-parameterization is that $\eta\zeta=0$.*

To prove the condition necessary, let λ represent a μ-parameterized curve determined by η and ζ. Then $\eta\zeta \leqq \eta\lambda+\lambda\zeta$. But $\eta\lambda=\lambda\zeta=0$ by virtue of Theorem 2. Hence the condition is necessary. To prove the condition sufficient suppose that $\eta\zeta=0$ and let $p=p(\mu)$, $(0\leqq\mu\leqq\mu_\eta)$, $q=q(\mu)$, $(0\leqq\mu\leqq\mu_\zeta)$, be μ-parameterizations of η and ζ, respectively. By virtue of Corollary 1 to Theorem 3, $\mu_\eta=\mu_\zeta$. It then follows from Theorem 4 that $p(\mu)$ and $q(\mu)$ differ by a quantity arbitrarily small in absolute value. Hence $p(\mu)=q(\mu)$, and the condition is proved sufficient. Theorem 5 taken with Theorem 1 gives the following corollary.

COROLLARY 2. *A necessary and sufficient condition that two p-curves η and ζ be equivalent is that $\eta\zeta=0$.*

Another way of stating the corollary is to say that for two curves α and β, $\alpha\beta=0$ if and only if $\alpha=\beta$.

INSTITUTE FOR ADVANCED STUDY

ANNALS OF MATHEMATICS
Vol. 38, No. 2, April, 1937

FUNCTIONAL TOPOLOGY AND ABSTRACT VARIATIONAL THEORY

BY MARSTON MORSE

(Received September 5, 1936)

INTRODUCTION

The critical points of a real single-valued differentiable function $f(x)$ of n real variables (x) are the points at which all of the partial derivatives of f vanish. Critical points present themselves in many different ways. If $V(x, y, z)$ is a Newtonian potential function the critical points of V are the points where the gravitational force is null. In such examples critical points appear as points of equilibrium. They also appear in problems involving the equilibria of floating bodies, and more generally in any problem in dynamics where there is a generalization of potential.

The following theorem in geography is an old and interesting example of critical point theory. Suppose the surface of the earth is such that the level points are finite in number. Let the points of minimum and maximum distance from the center of the earth be respectively p_0 and p_2 in number. Let the number of passes or saddle points be p_1. If these saddle points are of the "general" non-degenerate type, then

$$p_0 - p_1 + p_2 = 2. \tag{0.1}$$

This example will suggest the following questions. Is (0.1) the only relation between the numbers p_i involved? What happens in n-dimensional spaces? What becomes of the theorem when the number of critical points is infinite? And still more generally when one turns to the critical points of functionals, for example to extremals in boundary problems in the calculus of variations, what becomes of the classification into types of critical points?

As in the study of functions in general, the theory depends in its superficial form upon the class of functions presented. These functions may be analytic, finitely differentiable, continuous, or even semi-continuous. The answer to the first of the preceding questions may be given by presenting a general theorem in the so-called non-degenerate case.[1]

Suppose that f is a single-valued function of a point p on a general n-dimensional manifold R with local coordinates (x). Cf. §10. We suppose that f is of class C^2 in terms of the coordinates (x). We assume that f is *non-degenerate*, that is that the Hessian of f vanishes at no critical point of f. We assume that R is compact. It follows that the number of critical points of f is finite. Let

---

[1] Morse, M. and van Schaack, G. B., *The critical point theory under general boundary conditions*, Annuals of Mathematics, 35 (1934).

(a) be a critical point of f. By the *index* of (a) is meant the *index*[2] of the quadratic form Q whose coefficients form the Hessian of f at (a). The index is the number of minus signs in the usual canonical form for Q. We see that there are $n + 1$ possible values for the index. This is the simplest form of classification of critical points. It corresponds in dynamics to various degrees of instability at the point of equilibrium. With this understood, we state the following theorem.

THEOREM 1. *The numbers p_i of critical points of index i of the non-degenerate function f, and the connectivities R_i of R satisfy the following relations*:

$$\begin{aligned} p_0 &\geqq R_0 \\ p_1 - p_0 &\geqq R_1 - R_0 \\ p_2 - p_1 + p_0 &\geqq R_2 - R_1 + R_0 \\ &\cdots\cdots\cdots \\ p_n - p_{n-1} + \cdots + (-1)^n p_0 &= R_n - R_{n-1} + \cdots + (-1)^n R_0 . \end{aligned} \tag{0.2}$$

Upon considering the successive relations in (0.2) one finds that

$$p_i \geqq R_i \qquad (i = 1, \cdots, n). \tag{0.3}$$

Relations (0.3) imply the existence of what we may call the topologically necessary critical points. If, for example, R were a torus the relations (0.3) would imply that $p_0 \geqq 1$, $p_1 \geqq 2$, and $p_2 \geqq 1$. The relations (0.2) however contain much more. They imply relations between the numbers p_i in excess of those which are topologically necessary. If one is dealing with the problem under general boundary conditions there is a sense in which the preceding relations are complete. See M1, p. 145. We shall leave the non-degenerate problem for the present.

The analogue of the critical point in the calculus of variations is the extremal satisfying given boundary conditions, for example joining two points. The definition and evaluation of the corresponding indices of these critical points have been taken up in M1. Studies of this sort are not the main object of this memoir.

Instead we recall that the theory of functions and functionals can be unified by the study of functions f of a point on an abstract space. Sufficient generality will be obtained if the functions f are lower semi-continuous and the space is a general metric space. See §1. The curves admitted in the calculus of variations become points in this abstract space. The theory as applied to functions and to functionals would still be fairly divergent were it not for the fact that it is possible to introduce a topological definition of a critical point. In the special case of a function which is continuous in the neighborhood of a point (a) of a coordinate space (x), this definition is as follows: *The point (a) is non-critical if there exists a continuous deformation of the domain $f \leqq f(a)$ neighboring*

[2] See Morse, M., *The calculus of variations in the large*, American Mathematical Society Colloquium Publications (1934). We shall refer to these lectures as M1.

(*a*) *under which f is decreased whenever a point p is displaced, and under which the point* (*a*) *is actually displaced.*

By far the greatest difficulty and the greatest interest arise when one admits the cases where the number of critical values is not finite. The corresponding numbers m_i and R_i will then in general be infinite. They can be defined as cardinal numbers, but the real generalization lies deeper. It depends upon the fact that the entities p_i are merely dimensions of groups M_i and that these groups can be compared at most superficially by comparing their dimensions. The real comparisons are by means of isomorphisms termed "natural" because they are limited in character by the nature of f. In this part of the theory we shall find that the relations (0.2) are replaced by an infinite set of relations of the same formal character, with the numbers p_i and R_i replaced by groups, and the signs $+$ and $-$ replaced by suitably defined operations in abelian group theory of the nature of reductions with respect to a group modulus. The relations (0.2) in the non-degenerate case are a very special case. Even in the non-degenerate case the group theory relations contain new general results.

An extensive bibliography cannot be given here. The reader may however turn to M1 and to the papers[3] cited below for further references. References will be found in M1 to Birkhoff, Brown, Fréchet, Lefschetz, Hadamard, Menger, Poincaré, Schnirelman and Lusternik, Tonelli, and others.

The results here presented form a part of a set of lectures given at the Institute for Advanced Study at Princeton during the year 1935–36. Among those who were present and made helpful suggestions were Drs. Reinhold Baer and E. Čech. See §5. The author's assistant, Dr. Pitcher, rendered valuable service in the reading and preparation of the manuscript.

I. The Underlying Topology and Group Theory

1. **The space M and function F.** Let M be a space of elements $p, q, r, \cdots$ in which a number pq is assigned to each ordered pair of points such that

I. $pp = 0$,
II. $pq \neq 0$ if $p \neq q$,
III. $pr \leqq pq + rq$.

[3] Some of the more recent references are as follows:

Birkhoff, G. D. and Hestenes, M., *Generalized minimax principle in the calculus of variations*, Duke Mathematical Journal, 1 (1935), 413–432.

Lefschetz, S., *On critical sets*, Duke Mathematical Journal, 1 (1935), 392–412.

Menger, K., *Untersuchungen über allgemeine Metrik*, Mathematische Annalen, 100 (1928), 75–163; 103 (1930), 466–501; *Metrische Geometrie und Variationsrechnung*, Fundamenta Mathematicae, 25 (1935), 441–458.

Morse, M. and van Schaack, G. B., *Abstract critical sets*, Proceedings of the National Academy of Sciences, 21 (1935), 258–263.

Morse, M. and van Schaack, G. B., *Critical point theory under general boundary conditions*, Duke Mathematical Journal, 2 (1936).

Morse, M., *Functional topology and abstract variational theory*, Proceedings of the National Academy of Sciences, 22 (1936).

Lusternik, L. and Schnirelman, L., *Méthodes topologiques dans les problèmes variationnels*, Paris, Hermann & Cie.

Upon setting $r = p$ in III we see that $pq \geqq 0$ and upon setting $p = q$ that $qr = rq$. The space M is termed a metric space. It is called symmetric since $qr = rq$. The elements p, q, r, etc. are termed "points" and pq the "distance" from p to q.

Neighborhoods, limit points, sets relatively open or closed, can now be defined in the usual way.[4] In particular if e is a positive number the e-neighborhood A_e of a point set A shall consist of all points p with a distance from A less than e. A set $B \subset M$ will be said to be *compact* if every infinite sequence in A contains a Cauchy subsequence which converges to a point in A. We do not assume that the space M is compact.

Let $f(p)$ be a real single-valued function of the point p on a subset A of M. We include $\pm\infty$ among the values which f may assume. We say that f is *lower semi-continuous* at a point q of A if corresponding to each constant $c < f(q)$, there exists an e-neighborhood q_e of q such that $f(p) > c$ on the intersection $A \cdot q_e$ of A and q_e. Upper semi-continuity of f at q is similarly defined, reversing the preceding inequalities. We say that $f(p)$ is continuous at q if its value at q is finite and $f(p)$ is both upper and lower semi-continuous at q.

We shall consider a function $F(p)$ lower semi-continuous on M. We shall suppose that

$$0 \leqq F \leqq 1 \tag{1.1}$$

and that for any constant c such that $0 \leqq c < 1$ the domain $F \leqq c$ is compact.

The functions f which we shall encounter in our abstract variational theory will be bounded below and lower semi-continuous on M. They may take on the value $+\infty$. For each finite constant b the domain $f \leqq b$ will be seen to be compact. Such functions f lead to functions F as follows. Let m be the absolute minimum of f on M. Set $\varphi = f - m$ and

$$F_1 = \frac{\varphi}{1 + \varphi} \tag{1.2}$$

understanding that $F_1 = 1$ when $\varphi = +\infty$. The function F_1 defined by (1.2) is readily seen to have the properties ascribed to the function F.

The set of points at which $F < a$ or $F \leqq a$ will be denoted respectively by F_a or F_{a+}. If A is an arbitrary point set of M we shall say that A is "*definitely on*" F_a (written $\text{on}_o\ F_a$) if F is less than a constant $c < a$ at points of A. We note that A may be on F_a without being $\text{on}_o\ F_a$ even when A is closed. To illustrate this consider the following example.

Example 1.1. Let M be the curve

$$x^2 + (y - \tfrac{1}{2})^2 = \tfrac{1}{4} \tag{1.3}$$

and suppose that $F = y$ at each point (x, y) of M except at the point $(0, 1)$. At $(0, 1)$ let $F = 0$. The space M is on $F < 1$ but not $\text{on}_o\ F < 1$.

[4] See Hausdorff, F., *Mengenlehre*, Leipzig, Walter de Gruyter (1927).

2. Vietoris cycles. We shall use Vietoris[5] cycles in our topology. Singular cycles in the classical sense are inadequate in a number of ways. We shall illustrate this fact by an example. Let V be a compact subset of M and V_e the e-neighborhood of V. Let u be an arbitrary singular k-cycle not on V. Corresponding to each positive e we suppose that u is homologous to a cycle on V_e. It is not always true that u is homologous to a cycle on V as examples would show. The corresponding theorem for Vietoris cycles however is true.

We proceed with a systematic outline of the Vietoris theory generalized and modified to meet our needs.

Let (A) be a set of $k + 1$ points of M. For $k > 0$ the orderings of (A) will be divided into two classes, any ordering of one class being obtained from any other of the same class by an even permutation. The set (A) taken with one of these classes of orderings will be termed a positively oriented k-cell and taken with the other class a negatively oriented k-cell. An oriented k-cell α_k may be represented by a succession

$$\eta A_0 \cdots A_k \tag{2.0}$$

of its vertices preceded by η, where $\eta = 1$ or -1 according as the ordering $A_0 \cdots A_k$ belongs to α_k or not. In case $k = 0$ we use (2.0) to define an oriented 0-cell understanding the cell is positively or negatively oriented according as η is 1 or -1. We say that the oriented $(k - 1)$-cell

$$\eta(-1)^i A_0 \cdots A_{i-1} \cdot A_{i+1} \cdots A_k$$

is positively related to α_k. We shall say that α_k is of *norm e* if the distances between the vertices of α_k are less than e.

The cell α_k will be termed *degenerate* or *null* if at least two of its vertices are coincident. Let δ be an element in an arbitrary field[6] Δ. By a k-chain of norm e is meant a symbolic[7] sum Σ of the form $\delta^i a^i$, $i = 1, \cdots, m$, in which δ^i is an element of Δ and a^i is an oriented k-cell of norm e. The chain Σ will be termed *reduced* if none of the a^i are null, if each is positively oriented, and no two of the a^i are identical.

An arbitrary chain will be *reduced* as follows. Let a be an arbitrary positively oriented k-cell and b the corresponding negatively oriented k-cell. Any term of the form δb in Σ will be replaced by $-\delta a$. All terms involving a will be replaced by a single term $\delta' a$ in which δ' is the sum of the coefficients δ in the terms replaced. Finally all terms involving null cells or null coefficients will be dropped. The resulting reduced chain will be regarded as formally equal to the original chain Σ.

---

[5] Vietoris, L., *Über den höheren Zusammenhang kompakter Räume und eine Klasse von zusammenhangstreuen Abbildungen*, Mathematische Annalen, 97 (1927), 454–472.

See also Lefschetz, S., *Topology*, American Mathematical Society Colloquium Publications, New York (1930); and *Deformations*, Duke Mathematical Journal, 1 (1935).

[6] See van der Waerden, B. L., *Moderne Algebra*, p. 41, Berlin, Julius Springer (1930).

[7] We adopt the summation conventions of tensor analysis.

By the sum of two chains $\delta^i a^i$ and $\delta^{i'} a^{i'}$ is meant the chain $\delta^i a^i + \delta^{i'} a^{i'}$. We assume that

$$\delta[\delta^i x^i] = (\delta\delta^i)x^i.$$

It is readily seen that k-chains of norm e form an additive abelian operator group.[8]

We shall now define the boundary operator β. If a is an arbitrary oriented k-cell with $k > 0$, and e is the unit element in Δ, ea will be a k-chain and $\beta(ea)$ shall be the $(k-1)$-chain $\sum_i eb_i$ where b_i is an arbitrary $(k-1)$-cell positively related to a. More generally we set

$$\beta(\delta^i a^i) = \delta^i[\beta(ea^i)].$$

If $k = 0$ we understand that $\beta(ea) = 0$. If u is an arbitrary k-chain one sees[9] that $\beta\beta u = 0$. One terms βu the *boundary* chain defined by u.

The preceding k-chains are termed *algebraic* k-chains to distinguish them from Vietoris chains to be defined later. The term algebraic will be omitted when it is clear from the context that the chain is algebraic. In particular this will be the case whenever the norm e of the chain is mentioned.

Let B and C be compact subsets of M such that $B \subset C$. An algebraic k-chain u will be termed a cycle *modulo* B on C if βu is on B. We term u an absolute k-cycle or a relative k-cycle according as B is or is not null. The adjective absolute will ordinarily be omitted. We shall term u, e-homologous to 0 *modulo* B on C and write

$$u \sim_e 0 \qquad (\text{mod } B \text{ on } C)$$

if there exists a $(k+1)$-chain z of norm e on C such that $\beta z = u + v$, where v is a k-chain of norm e on B. If $B = 0$, then $v = 0$ and the phrase mod B is omitted.

Let $u = (u_n)$ be a sequence of algebraic k-cycles u_n, $n = 0, \cdots$, mod B on C, with norms at most e_n respectively. If the numbers e_n tend to zero as n becomes infinite, and if for each integer n there exist "connecting" homologies of the form

$$u_n \sim_{e_n} u_{n+1} \qquad (\text{mod } B \text{ on } C), \tag{2.1}$$

u is termed a (Vietoris) k-cycle mod B on C, and C a *carrier* of u. We term u homologous to zero mod B on C, and write $u \sim 0$ mod B on C if corresponding to each positive number e there exists an integer N so large that $u_n \sim_e 0$ mod B on C whenever $n > N$. The set C is termed a *carrier* of the homology $u \sim 0$ mod B on C. Vietoris k-cycles u and v are termed homologous, $u \sim v$, mod B on C, if $u - v \sim 0$ mod B on C. (See below for definition of $u - v$.)

The algebraic k-cycles u_n will be termed the *components* of $u = (u_n)$. If u and v are Vietoris k-cycles mod B on C the algebraic k-cycles $u_n \pm v_n$ are the components of Vietoris k-cycle which we denote by $u \pm v$. If δ is an ele-

[8] See van der Waerden, l.c. pp. 132–144.

[9] See Seifert-Threlfall, *Lehrbuch der Topologie*, p. 60, Leipzig, Teubner (1934).

ment in the field Δ the algebraic k-cycles δu_n are the components of Vietoris k-cycle which we denote by δu.

In the remainder of this memoir the term k-cycle mod B on C shall mean a Vietoris k-cycle mod B on C unless otherwise stated.

If (u_n) is a k-cycle mod B on C any infinite subsequence (v_n) of (u_n) defines a k-cycle $v \sim u$ mod B on C. With this understood let ζ be an algebraic k-cycle mod B on C of norm e such that $u_n \sim_e \zeta$ mod B on C for all integers n exceeding some integer N. We then write

$$u \sim_e \zeta \qquad \text{(mod } B \text{ on } C). \tag{2.2}$$

By virtue of these conventions we have the following lemma.

Lemma 2.1. *If $u = (u_n)$ and $v = (v_n)$ are k-cycles mod B on C such that for every n,*

$$u_n \sim_{e_n} v_n \qquad \text{(mod } B \text{ on } C) \tag{2.3}$$

where e_n tends to zero as n becomes infinite, then $u \sim v$ mod B on C.

In the preceding paragraphs we have spoken of algebraic and Vietoris k-cycles and homologies mod B on C where B and C were compact. If the phrase mod B on C is replaced by mod B' on C' where B' and C' are compact or not at pleasure, mod B' on C' shall mean mod B'' on C'', where B'' and C'' are respectively compact sets on B' and C'.

Letting X_e denote the e-neighborhood of X we state the following lemma.

Lemma 2.2. *If $u = (u_n)$ is a k-cycle mod B on C such that $u \sim 0$ mod B_e on C_e for every positive e, then $u \sim 0$ mod B on C.*

Corresponding to an arbitrary positive e there exists by hypothesis an integer N so large that for $n > N$, $u_n \sim_e 0$ mod B_e on C_e. For such values of n we infer the existence of an algebraic $(k + 1)$-chain w_n on C_e of norm e and an algebraic k-chain v_n on B_e of norm e such that

$$\beta w_n = u_n + v_n .$$

Let v_n' be the algebraic k-chain obtained by replacing each vertex of v_n not on B by a nearest point on B. It is easily seen that there exists an algebraic $(k + 1)$-chain z_n on B_e of norm $3e$ such that $\beta z_n = v_n' - v_n$. If we set $w_n' = w_n + z_n$ we find that

$$\beta w_n' = u_n + v_n' .$$

We now replace each vertex of w_n' on C_e but not on C by a nearest vertex on C obtaining thereby a $(k + 1)$-chain y_n on C of norm at most $5e$. We have the relation

$$\beta y_n = u_n + v_n'$$

and since v_n' is on B, $u \sim 0$ mod B on C.

3. Theorems on Vietoris k-cycles. Let g be an arbitrary additive abelian operator group with coefficients in a field Δ. With such a group there can be associated in many ways a *maximal linear set* of elements of g, that is a set S

of elements of g no proper linear combination of which with coefficients in Δ is null, and which is a proper subset of no other such set. The cardinal number m of the set S equals that of any other maximal linear set of g. This is readily proved in case m is finite. It is true even when m is infinite but we shall make no use of this fact. We term m the *dimension* of g.

Let V be a compact subset of M. We are dealing with Vietoris k-cycles. If u were an ordinary singular k-cycle homologous to zero mod V it is clear that u would be homologous to a singular k-cycle on V. If however u is a Vietoris k-cycle the corresponding theorem requires proof. Theorems 3.1, 3.2, and 3.3 are general theorems of this nature. If the field Δ consisted merely of the integers mod 2, use could be made of the compactness of the relative homology groups to prove these theorems. But with our general group Δ the proof is more difficult.

The basic ideas appear in two lemmas whose proof was communicated to the author in slightly different form by Professor E. Čech. Professor Čech had previously developed similar lemmas and theorems in his abstract theory of spaces.[10]

We begin with a definition.

Reduction sets W. Let $U \subset V$ be compact subsets of M. Corresponding to each positive δ let $W(\delta)$ be a group of algebraic k-cycles mod U on V of norm δ, with the property that for each positive $\eta < \delta$, $W(\eta)$ is a subgroup of $W(\delta)$. Such a set of groups $W(\delta)$ will be called a *reduction set W*.

The first lemma here is as follows.

LEMMA 3.1. *Corresponding to an arbitrary positive constant e there exists a positive constant $\delta < e$ such that to each cycle $w(\delta)$ in $W(\delta)$ there corresponds for every $\eta < \delta$ at least one cycle $w(\eta)$ in $W(\eta)$ such that*

$$w(\delta) \sim_e w(\eta) \qquad \text{(mod } U \text{ on } V). \tag{3.1}$$

Let $e_n = e\,3^{-n}$. Let ω denote the subgroup of cycles of $W(e_1)$ which are e-homologous to zero mod U on V. Let h_1 denote the group $W(e_1)$ mod ω. We shall begin with a proof of statement (α).

(α). The dimension of h_1 is finite.[11]

Let A be a finite net of points on U within a distance e_1 of each point of U. Let $B \supset A$ be a similar net for V. Let g denote the group of algebraic k-cycles mod A on B. It is clear that the dimension of g is finite.

Corresponding to an arbitrary algebraic k-cycle $w = w(e_1)$ of $W(e_1)$ we can obtain an algebraic k-cycle u of g of norm $e = 3e_1$ by replacing each vertex of w on U by a nearest point of A and each vertex of w not on U by a nearest point of B. We infer that $w \sim_e u$ mod U on V. But the group g admits a finite base with elements z^i so that $u = \delta^i z^i$ and hence

$$w \sim_e \delta^i z^i \qquad \text{(mod } U \text{ on } V).$$

[10] Čech, E., *Théorie générale de l'homologie dans un espace quelconque*, Fundamenta Mathematicae, 19 (1932), 149–183.

[11] Cf. Vietoris, l.c., p. 461 (3).

It follows that the dimension of h_1 is at most the number of elements z^i, and hence is finite.

The proof of (α) is complete.

Recall that h_1 is a group of classes of k-cycles. For each integer $n > 1$ let h_n be the subgroup of those classes of h_1 which contain at least one cycle of $W(e_n)$. We see that

$$h_1 \supset h_2 \supset h_3 \supset \cdots .$$

There must accordingly exist a finite integer r such that

$$\dim h_r = \dim h_{r+1} = \cdots . \tag{3.2}$$

But two operator groups with coefficients in a field and with equal finite dimensions will be identical if one group is a subgroup of the other. Hence

$$h_r = h_{r+1} = \cdots . \tag{3.3}$$

The constant e was arbitrary and $e_n = e\,3^{-n}$. We now let δ be any positive constant such that $\delta < e_r$. The cycle $w(\delta)$ of the lemma is in $W(e_r)$ and hence in some class of h_r. Corresponding to the constant $\eta < \delta$ of the lemma let p be an integer so large that $e_p < \eta$. Then $e_p < \delta < e_r$ so that $p > r$ and $h_p = h_r$. The cycle $w(\delta)$ is in a class of h_r and hence of h_p, and by virtue of its definition this class of h_p contains at least one k-cycle $w(e_p)$ of $W(e_p)$. Since $e_p < \eta$ the cycle $w(e_p)$ is a cycle $w(\eta)$. The cycles $w(\delta)$ and $w(\eta)$ are in the same class of h_p and hence of h_1. The relation (3.1) is satisfied by virtue of the definition of h_1.

The proof of the lemma is complete.

We now define ensembles Z closely related to the reduction sets W.

Ensembles Z. Corresponding to each positive δ let $Z(\delta)$ be a non-null class of algebraic k-cycles $z(\delta)$ mod U on V of norm δ. The ensemble of the classes $Z(\delta)$ as δ ranges over all positive values will be denoted by Z and assumed to have the following properties:

(1). For an arbitrary $\eta < \delta$, $Z(\eta)$ shall be a subset of $Z(\delta)$.

(2). With Z there shall be associated a reduction set W such that if $z(\delta)$ is an arbitrary element of $Z(\delta)$ any other element of $Z(\delta)$ is of the form

$$z(\delta) + w(\delta) \qquad [w(\delta) \subset W(\delta)], \tag{3.4}$$

and every sum of the form (3.4) is an element of $Z(\delta)$.

We shall illustrate the ensembles Z and reduction sets W by the following example.

Example 3.1. Let $U \subset V \subset C$ be compact subsets of M. Let u be a Vietoris k-cycle mod U on C homologous to zero mod V on C. Let $Z(\delta)$ and $W(\delta)$ consist respectively of the algebraic k-cycles $z(\delta)$ and $w(\delta)$ mod U on V of norm δ such that

$$z(\delta) \sim_\delta u, \qquad w(\delta) \sim_\delta 0 \qquad (\text{mod } U \text{ on } C). \tag{3.5}$$

That there exist algebraic k-cycles $z(\delta)$ in (3.5) may be seen as follows. Since $u \sim 0 \bmod V$ on C there exists a sequence of positive constants e_n tending to zero, and algebraic chains u_{n+1} of norm e_n on C with algebraic k-chains z_n of norm e_n on V such that

$$\beta u_{n+1} = u_n - z_n .$$

Hence $\beta u_n = \beta z_n$. It follows that z_n is an algebraic k-cycle mod U on V and that

$$z_n \sim_{e_n} u_n \qquad (\text{mod } U \text{ on } C).$$

The classes $Z(\delta)$ are accordingly not empty.

The remaining conditions on ensembles Z and reduction sets W are easily seen to be satisfied.

We shall now prove the following lemma.

LEMMA 3.2. *Corresponding to an admissible ensemble Z there exists a Vietoris k-cycle (v_n) mod U on V whose components v_n belong to $Z(e_n)$ for suitable positive constants e_n approaching zero as n becomes infinite.*

We recall that there is a reduction set W associated with Z. Beginning with an arbitrary positive number e_0 we shall give an inductive definition of a sequence of positive numbers e_n tending to zero as n becomes infinite. The constant e of Lemma 3.1 is arbitrary. In particular we can set $e = e_{n-1}$. Lemma 3.1 then affirms the existence of a constant $\delta < e$. We take $e_n < \delta$. We suppose also that e_n tends to zero as n becomes infinite.

By definition of Z there is an algebraic k-cycle z_n mod U on V in $Z(e_n)$ for each n. We shall give an inductive definition of a Vietoris k-cycle $v = (v_n)$ mod U on V. We begin by setting $v_1 = z_1$. Suppose that the components v_i, $i = 1, \cdots, m$, have been defined in such fashion that v_i is in $Z(e_i)$ and

$$v_{i+1} \sim_{e_{i-1}} v_i \qquad (\text{mod } U \text{ on } V;\ i = 1, \cdots, m-1).$$

We shall now define v_{m+1}. First recall that $z_{m+1} - v_m$ is in $W(e_m)$ in accordance with property (2) of ensembles Z. We can apply Lemma 3.1 with $e = e_{m-1}$, with $e_m < \delta < e$ and with

$$w(\delta) = z_{m+1} - v_m .$$

Setting $\eta = e_{m+1}$ Lemma 3.1 affirms the existence of a cycle $w(\eta)$ in $W(e_{m+1})$ such that

$$w(\delta) \sim_{e_{m-1}} w(\eta) \qquad (\text{mod } U \text{ on } V). \tag{3.6}$$

We set

$$v_{m+1} = z_{m+1} - w(\eta). \tag{3.7}$$

We note that v_{m+1} is in $Z(e_{m+1})$ by virtue of property (2) of ensembles Z. Upon adding the respective members of (3.6) and (3.7) we find that

$$v_{m+1} \sim_{e_{m-1}} v_m \qquad (\text{mod } U \text{ on } V).$$

The algebraic k-cycles v_n are accordingly the components of a Vietoris k-cycle v.

The proof of the lemma is complete.

We shall prove three theorems which parallel facts obvious in the theory of singular cycles. Letting $U \subset V \subset C$ be three compact subsets of M we state the first theorem as follows.

THEOREM 3.1. *If u is a k-cycle mod U on C homologous to zero mod V on C then u is homologous mod U on C to a k-cycle z mod U on V.*

This theorem follows at once from Lemma 3.2 with the choice of the ensemble Z that of Example 3.1.

If w and u are two Vietoris k-cycles we shall find it convenient at times to write $w \equiv u \bmod A$. We mean thereby that $w_n = u_n + v_n$ where v_n is a chain on A and $n = 1, 2, \cdots$.

Letting $A \subset U \subset V$ be three compact subsets of M the second theorem is as follows.

THEOREM 3.2. *If a $(k-1)$-cycle u mod A on U is homologous to zero mod A on V there exists a k-cycle v mod U on V such that $\beta v \equiv u$ mod A.*

This theorem also follows from Lemma 3.2 We take $Z(\delta)$ as the class of all k-cycles $z(\delta)$ mod U on V of norm δ such that

$$\beta z(\delta) \sim_\delta u \qquad (\text{mod } A \text{ on } U).$$

The cycles $z(\delta)$ exist by virtue of the hypothesis of the theorem. The corresponding reduction group $W(\delta)$ consists of all k-cycles $w(\delta)$ mod U on V of norm δ such that

$$\beta w(\delta) \sim_\delta 0 \qquad (\text{mod } A \text{ on } U).$$

With this choice of Z it follows from Lemma 3.2 that there exists a k-cycle z mod U on V such that $\beta z \sim u$ mod A on U. There accordingly exists a sequence of positive constants e_n tending to zero, and algebraic $(k+1)$-chains w_n of norm e_n on U with algebraic k-chains x_n of norm e_n on A such that

$$\beta w_n = u_n - \beta z_n + x_n$$

If we set $v_n = w_n + z_n$ we see that (v_n) is a Vietoris k-cycle v mod U on V and that

$$\beta v \equiv u \qquad (\text{mod } A).$$

The proof of the theorem is complete.

The third theorem is as follows.

THEOREM 3.3. *If u is a k-cycle mod A on V and $\beta u \sim 0$ on A, there exists a k-cycle v on V such that $u \sim v$ mod A on V.*

Before proving this theorem it will be convenient to establish the following.

(α). *If x is an algebraic k-cycle x of norm e mod A on V, and $\beta x \sim_e 0$ on A, there exists an algebraic k-cycle y on V such that $x \sim_e y$ mod A on V.*

It follows from the hypothesis in (α) that there exists an algebraic k-chain ω

of norm e on A such that $\beta\omega = \beta x$. We set $y = x - \omega$ and note that y is a k-cycle on V of norm e. To establish (α) we observe that

$$\beta 0 = -\omega + \omega = (y - x) + \omega,$$

so that $x \sim_e y$ mod A on V.

To prove the theorem we shall apply Lemma 3.2 setting $U = 0$. The set $Z(\delta)$ shall be the class of all k-cycles $z(\delta)$ of norm δ on V such that $z(\delta) \sim_\delta u$ mod A on V. The corresponding group $W(\delta)$ will consist of all k-cycles $w(\delta)$ of norm δ on V such that $w(\delta) \sim_\delta 0$ mod A on V. That the set $Z(\delta)$ is not empty follows from (α), and the theorem follows from Lemma 3.2.

4. **The rank conditions.** In this section we shall define an F-non-bounding k-cycle and assign to such a cycle a rank r. These ranks are fundamental. They lead to a group theoretic formulation of the structure beneath our theory. They are introduced here for the first time.

Recall that the sets of points at which $F < c$ and $F \leqq c$ are respectively denoted by F_c and F_{c+}. The phrase on$_o$ F_c of §1 will sometimes be replaced by the phrase below$_o$ c. Mod$_o$ F_c shall mean mod some compact set below$_o$ c.

If z is a k-cycle below$_o$ b, not homologous to zero below$_o$ b we term b a *homology limit* of z, and z F-non-bounding. By the *superior cycle limit* $s(z)$ of an F-non-bounding k-cycle z is meant the least upper bound of the homology limits of z. If z is non-bounding on M we set $s(z) = +\infty$. It follows from the definitions involved that $s(z)$ is a homology limit of z.

Let z be a k-cycle with superior cycle limit $s(z) = s$. By the *inferior cycle limit* $t = t(z)$ of z is meant the greatest lower bound of constants c such that $z \sim 0$ mod$_o$ F_c below$_o$ s. We do not define $t(0)$. We observe that $z \not\sim 0$ mod$_o$ F_t below$_o$ s.

If z is an F-non-bounding k-cycle we term the pair $[s(z), t(z)]$ the *rank* $r(z)$ of z. We order the pairs $[s, t]$ lexicographically. That is, we write $[s, t] = [s', t']$ if and only if $s = s'$ and $t = t'$, and write $(s, t) > (s', t')$ if $s > s'$, or if $s = s'$ and $t > t'$. In this way we are led to a simply ordered set of ranks r. These ranks are fundamental.

A bounding k-cycle which is F-non-bounding will be termed *ambiguous*. The superior cycle limit of such a cycle will be finite.

In §5 we shall give the theory of ranks an abstract formulation and we have there listed four conditions which we term *rank conditions*. These rank conditions are satisfied by the ranks $r(u)$, by the superior cycle limits $s(u)$, and by several other basic ordered sets. Our first theorem in this connection is the following.

THEOREM 4.1. *The superior cycle limits $s(u)$ of F-non-bounding k-cycles satisfy the rank conditions* I, II, III, IV *with $\rho(u)$ replaced by $s(u)$.*

I. To show that $s(\delta u) = s(u)$ for $\delta \neq 0$ we must show that the condition $u \sim 0$ below$_o$ b is equivalent to the condition $\delta u \sim 0$ below$_o$ b. The first condition clearly implies the second. But the second condition also implies the first. For any coefficient in the field Δ can be divided by a non-null δ. In particular if $\delta \neq 0$ the coefficients in any sequence of algebraic bounding relations which imply that $\delta u \sim 0$, can be divided by δ. Thus I holds as stated.

II. To establish II we suppose that $s(u)$ and $s(v)$ are both finite. The validity of II when $s(u)$ or $s(v)$ is infinite is trivial. Let e be an arbitrary positive constant. By virtue of the definition of a superior cycle limit, $u \sim 0$ below$_\circ$ $s(u) + e$ and $v \sim 0$ below$_\circ$ $s(v) + e$. Hence $u + v \sim 0$ below$_\circ$ $s^* + e$ where s^* is the maximum of $s(u)$ and $s(v)$. Property II follows as stated.

III. To establish III we suppose that $s(u) > s(v)$ and let b be any homology limit of u between $s(u)$ and $s(v)$. Observe that $v \sim 0$ and $u \not\sim 0$ below$_\circ$ b so that $u + v \not\sim 0$ below$_\circ$ b. Thus b is a homology limit of $u + v$, and III holds as stated.

IV. We set $\rho_0 = s_0$. Observe that u and v in IV are below$_\circ$ s_0. Since $s(u)$ and $s(v)$ do not exist $u \sim 0$ and $v \sim 0$ below$_\circ$ s_0. If $s(u + v)$ exists it must be less than s_0, and IV is established.

The proof of Theorem 4.1 is complete.

It might be thought that the inferior cycle limits $t(u)$ also satisfy the rank axioms. To show that this is not the case and to illustrate the preceding theorems we present several examples.

Let the metric space M be the curve

$$y = -e^{-x} \sin x \qquad (0 \leqq x \leqq 4\pi) \tag{4.1}$$

in the (x, y) plane. On M let $F = y$. The function F is continuous on M. We waive the condition $0 \leqq F \leqq 1$. We shall denote the points of M at which $x = 0, \pi, 2\pi$, and 4π by a, b, c, e respectively, and the points at which x equals

$$\frac{5\pi}{4}, \quad \frac{9\pi}{4}, \quad \frac{13\pi}{4}$$

respectively by n, p, q. The reader will do well to trace the curve M and note that n, p, q are relative extrema of F. We observe that

$$F(p) < F(q) < F(n).$$

Let A, B, C, E be 0-cycles mod 2 whose components are respectively identical with a, b, c, e.

Example 4.1. The cycle $B - C$ has the superior cycle limit $F(n)$ while $B - A$ has no homology limit.

Example 4.2. In this example we shall show that cycles u and v may have homology limits while $u + v$ has no homology limit. To that end let

$$u = C - A, \qquad v = B - C, \qquad u + v = B - A.$$

We observe that $s(u) = s(v) = F(n)$ while $s(u + v)$ does not exist.

Example 4.3. We here show that the equality is possible in rank axiom II as applied to s. We set

$$u = B - C, \qquad v = C - E,$$

and note that $u + v = B - E$. We find that

$$s(u) = s(u + v) = F(n), \qquad s(v) = F(q) < s(u).$$

Example 4.4. We shall show that the sign $<$ is possible in rank axiom II. To this end we make use of the cycles u and v in Example 4.3 and set $u' = u + v$, $v' = -u$. We observe that $u' + v' = v$. Referring to Example 4.3 we find that

$$s(u' + v') < \max [s(u'), s(v')].$$

In the next three examples we shall show that for inferior cycle limits each one of the three cases

$$t(u + v) \gtreqless \max [t(u), t(v)] \tag{4.2}$$

is possible.

Example 4.5. We shall first show the possibility of the sign $>$ in (4.2). We make use of the space M and function F of the preceding examples and set $u = B - C$, $v = E - B$. We observe that $u + v = E - C$ and that

$$s(u) = s(v) = F(n), \qquad s(u + v) = F(q),$$

$$t(u) = t(v) = F(p), \qquad t(u + v) = 0 > t(u) = t(v).$$

Example 4.6. We here illustrate the possibility of the equality in (4.2). Let the space M consist of two points a and b. Let the function F be 1 at a and 0 at b. Let u and v be two 0-cycles mod 2 with components respectively identical with a and b. We find that

$$s(u) = s(v) = s(u + v) = +\infty,$$

$$t(u) = t(u + v) = 1, \qquad t(v) = 0.$$

Example 4.7. We shall show finally that the sign $<$ is possible in (4.2). We make use of the space M and function F of the preceding example, and set $z = v - u$. We find that

$$t(z) = t(u) = 1, \qquad t(u + z) = 0.$$

We return to the general theory and state a basic theorem.

THEOREM 4.2. *The ranks $r(u)$ of F-non-bounding k-cycles satisfy the four rank conditions of* §5, *$r(u)$ replacing $\rho(u)$.*

Let b be a homology limit of a k-cycle u. Let $I[u, b]$ denote the greatest lower bound of constants c such that $u \sim 0 \bmod_o F_c \text{ on}_o F_b$.

We state the following lemma.

LEMMA 4.1. *If b is a homology limit of u, v and u + v, then*

$$I[u + v, b] \leqq \max [I(u, b), I(v, b)]. \tag{4.3}$$

If b is a homology limit of u and of v and if

$$I(u, b) > I(v, b),$$

then b is a homology limit of u + v and

$$I(u + v, b) = I(u, b).$$

The lemma follows from the definitions involved.

We shall establish Theorem 4.2.

I. That $t(\delta u) = t(u)$ if $\delta \neq 0$ follows as in the proof that $s(\delta u) = s(u)$. Hence $r(\delta u) = r(u)$.

II. To establish II we may assume without loss of generality that $r(v) \leqq r(u)$. We have then to prove that $r(u + v) \leqq r(u)$. According to Theorem 4.1 we have two cases:

(a) $s(u + v) < s(u)$.

(b) $s(u + v) = s(u)$.

In case (a), $r(u + v) < r(u)$ by virtue of the ordering of pairs (s, t). In case (b) we distinguish between the two subcases:

(b)′ $s(v) < s(u)$.

(b)″ $s(u) = s(v)$.

In case (b)′ we observe that $v \sim 0$ below$_\circ$ $s(u)$ so that $u + v \sim u$ below$_\circ$ $s(u)$. Hence $t(u + v) = t(u)$. Thus $r(u + v) = r(u)$ in case (b)′, and II is true.

In case (b)″, u, v, and $u + v$ have common superior cycle limits. It follows from the preceding lemma that

$$t(u + v) \leqq \max [t(u), t(v)]. \tag{4.4}$$

Hence $r(u + v)$ is at most the maximum of $r(u)$ and $r(v)$, and II is established.

III. We here show that $r(u + v)$ exists, provided $r(u)$ and $r(v)$ exist and are unequal. We assume without loss of generality that $r(u) > r(v)$ and consider the two cases:

(α) $s(u) > s(v)$.

(β) $s(u) = s(v)$, $\qquad t(u) > t(v)$.

In case (α) it follows from Theorem 4.1 that $s(u + v)$ exists, and in case (β), $s(u + v)$ exists by virtue of the preceding lemma. Hence $r(u + v)$ exists in both cases, and III is proved.

IV. If the cycles $u = \Sigma u_i$ and $v = \Sigma v_j$ in IV have no rank, $s(u)$ and $s(v)$ fail to exist. But $r(u + v)$ and hence $s(u + v)$ exist by hypothesis. We set $\rho_0 = r_0 = (s_0, t_0)$. It follows from the fact that superior cycle limits satisfy IV that $s(u + v) < s_0$, and hence $r(u + v) < r_0$. Hence IV holds as stated.

The proof of Theorem 4.2 is complete.

5. **The underlying group structure.** The implications of the rank conditions will be best understood if incorporated in an abstract group theory.

We suppose then that we have an additive abelian operator group G with operators which form a field Δ. We shall be concerned with various subgroups g of G. We suppose throughout that these subgroups are operator subgroups, that is if u is in g and δ is in Δ, δu is in g. Our isomorphisms shall be operator isomorphisms, that is, if u corresponds to v, δu corresponds to δv.

By a subgroup g of G with property A is meant an operator subgroup of G every element of which, with the possible exception of the null element, has the property A. A group g with property A will be termed *maximal* if it is a proper subgroup of no other subgroup g' of G with property A. The following example shows that there may be several maximal groups with a given property A.

Let G be a group generated by three elements a, b, c with coefficients in the field of integers mod 2. Suppose that a, b, $a + b$, c are the only elements with the property A. Then a and b together generate a maximal group with property A as does c.

Suppose that the respective elements u of G have the property A if and only if for each operator $\delta \neq 0$, δu has the property A. Let g be a maximal (operator) subgroup of G with property A. Let v be an arbitrary element with property A. There will then exist an operator $\delta \neq 0$ and an element v_1 in g, together with an element v_2 which fails to have the property A, or is null, such that $\delta v = v_1 + v_2$. Now division by δ is permissible, v_1/δ is in g, and v_2/δ still fails to have the property A or is null. We can accordingly infer the existence of an element z in g, and an element w which fails to have the property A or is null, such that

$$v = z + w. \tag{5.1}$$

The rank conditions. Let (ρ) be a simply ordered set of elements, symbols, numbers, points, etc. With some but not all of the elements u of G we suppose there is associated an element $\rho(u)$ in the set (ρ). We term $\rho(u)$ the rank of u. The rank $\rho(0)$ shall not be defined.

We assume that the ranks ρ satisfy the following conditions:

I. *If u has a rank and $\delta \neq 0$, $\rho(u) = \rho(\delta u)$.*

II. *If u, v, and $u + v$ have ranks, $\rho(u + v) \leqq \max\ [\rho(u), \rho(v)]$.*

III. *If u and v have unequal ranks, $\rho(u + v)$ exists.*

IV. *If $u_1, \cdots, u_m, v_1, \cdots, v_n$ have ranks at most ρ_0 while the sums $u = \Sigma u_i$ and $v = \Sigma v_j$ have no rank and $u + v$ has a rank, then $\rho(u + v) < \rho_0$.*

The elements with rank (with the null element added) will not in general form a group. For example consider the group G of 0-cycles in Example 4.2 letting the superior cycle limit $s(z)$ be the rank of the 0-cycle z in case $s(z)$ exists. In this example we have seen that $s(u)$ and $s(v)$ exist while $s(u + v)$ does not exist nor does $u + v = 0$.

We have already remarked that the primitive concept of counting is to be replaced by comparisons involving isomorphisms of groups. A comparison by means of isomorphisms in general would not be very effective. Our comparisons will always be by means of isomorphisms of a special sort the abstract counterpart of which we now define.

Two elements of G will be said to be in the same *ρ-class* if they have the same rank while their difference has no rank or a lesser rank. An isomorphism between two subgroups of G of elements with rank will be termed a *ρ-isomorphism* if corresponding non-null elements are in the same ρ-class. These ρ-isomorphisms will be exemplified by three classes of *natural isomorphisms* defined by F.

We state the following theorem.

THEOREM 5.1. *Any two maximal groups of elements of G with a fixed rank σ are ρ-isomorphic.*

Let g_σ be the group generated by the elements of G with rank at most σ. Let H_σ be the subset of elements of g_σ without rank or with rank less than σ. We continue by proving the following statement.

(α). *The elements of H_σ form a group.*

Let u and v be arbitrary elements of H_σ. Writing E for exists and $\sim E$ for does not exist we distinguish four cases:

(1) $\rho(u)\ E,\ \rho(v)\ E.$ (2) $\rho(u)\ E,\ \rho(v) \sim E.$

(3) $\rho(u) \sim E,\ \rho(v)\ E.$ (4) $\rho(u) \sim E,\ \rho(v) \sim E.$

We shall prove that $u + v$ is in H_σ. If $\rho(u + v)$ does not exist $u + v$ is in H_σ. We assume therefore that $\rho(u + v)$ exists and seek to prove that $\rho(u + v) < \sigma$. This will follow from the rank conditions II, III, III, IV respectively in the cases (1), (2), (3), (4). In case (1), $\rho(u)$ and $\rho(v)$ are less than σ since u and v are in H_σ. It follows from II that $\rho(u + v) < \sigma$. In case (2), $\rho(u) < \sigma$ and $\rho(u + v)$ must be less than σ for otherwise $\rho(v)$ would exist. Case (3) is similar. The result in case (4) follows from IV, and the proof of (α) is complete.

We return to the theorem and let m_σ be a maximal group of elements of G with rank σ. Each element of m_σ is in a coset of the residue group h_σ defined by g_σ mod H_σ, and different elements u and v of m_σ are in different cosets of h_σ, for otherwise $u - v$ would be in H_σ and hence not in m_σ. Moreover there s an element of m_σ in each coset z of h_σ for otherwise an element u of z taken with m_σ would generate a group of elements of G with rank σ containing m_σ as a proper subgroup. There is thus an isomorphism between m_σ and h_σ in which each element u of m_σ corresponds to the coset of h_σ which contains u. We term m_σ and h_σ *ρ-isomorphic* since u corresponds to a coset of elements in the same ρ-class as u.

Let m'_σ be a second maximal group of elements of G with rank σ. The group m'_σ is also ρ-isomorphic with h_σ. It follows that m_σ and m'_σ are ρ-isomorphic with each other, and the proof of the theorem is complete.

A particular consequence of the theorem is that the dimensions of any two maximal groups of elements of G with rank σ are equal, but the theorem of course implies much more than this equality of dimension.

We note another consequence of statement (α).

(β). *The property of elements with rank being in the same ρ-class is transitive.*

For if u, v, and w have the rank σ and if u is in the ρ-class of v and w, $u - v$ and $u - w$ belong to H_σ. It follows from (α) that $v - w$ belongs to H_σ so that v and w are in the same ρ-class.

We state the following lemma.

LEMMA 5.1. *If $u_1, \cdots, u_m$ are elements of G with ranks such that*

$$\rho(u_1) > \rho(u_i) \qquad (i = 2, \cdots, m) \tag{5.2}$$

then $\rho(u_1 + \cdots + u_m)$ exists and equals $\rho(u_1)$.

Suppose first that $m = 2$. By virtue of rank condition III $\rho(u_1 + u_2)$ exists. From I we see that $\rho(-u_2) = \rho(u_2)$. Since $u_1 = (u_1 + u_2 - u_2)$ we can infer from II and (5.2) that

$$\rho(u_1) \leqq \max\,[\rho(u_1 + u_2),\, \rho(u_2)] \leqq \rho(u_1 + u_2),$$

and that

$$\rho(u_1 + u_2) \leqq \max\,[\rho(u_1),\, \rho(u_2)] \leqq \rho(u_1)$$

so that $\rho(u_1) = \rho(u_1 + u_2)$. The proof of the lemma can be completed by induction with respect to m.

Let there be given a subgroup g of G and a set of subgroups $h(\alpha)$ of g, α being an enumerating index. The group g is said to be a *direct sum*

$$g = \sum_\alpha h(\alpha)$$

of the groups $h(\alpha)$ if each element u of g is a finite sum of elements from the groups $h(\alpha)$, and if there exists no relation of the form

$$u_{\alpha_1} + \cdots + u_{\alpha_m} = 0$$

in which the α_i's are distinct and u_{α_i} is a non-null element from the group $h(\alpha_i)$.

From this point on we shall frequently employ the hypothesis that the ranks ρ are well-ordered. We shall not mean thereby that the ranks *can* be well-ordered but that taken in their given orders they *are* well-ordered.[12]

We state the following theorem.

THEOREM 5.2. *If the ranks of elements of G are well-ordered a maximal subgroup g of the elements of G with rank is a direct sum*

$$\sum_\rho h(\rho) \tag{5.3}$$

of arbitrary maximal subgroups $h(\rho)$ of elements of g with rank ρ, and such groups $h(\rho)$ are maximal subgroups of elements of G with rank ρ.

It follows from the preceding lemma that the groups $h(\rho)$ admit a direct sum and that this direct sum g' is a group of elements with rank. For if $u_1, \cdots, u_m$ are non-null elements of different groups $h(\rho)$, Lemma 5.1 affirms

[12] See Hausdorff, l.c., p. 55.

that the rank of their sum v exists. Hence in particular $v \neq 0$. It remains to prove that $g' = g$.

To that end let u be an arbitrary non-null element of g. Set $\rho(u) = \sigma$. Since $h(\sigma)$ is a maximal subgroup of elements of g with rank σ there exists, in accordance with (5.1), an element v_1 in $h(\sigma)$ and an element w_1 not an element with rank σ, such that $u = v_1 + w_1$. It follows from the rank condition II that

Case I. Either $\rho(w_1)$ fails to exist,

Case II. Or $\rho(w_1) < \sigma$.

In Case I, $w_1 = 0$ since $w_1 = u - v_1$ and is in g. In Case II we treat $u - v_1$ as we originally treated u.

Proceeding inductively we infer the existence of a set of elements v_i, $i = 1, 2, \cdots, n$, respectively with decreasing ranks ρ_i, such that v_i is in $h(\rho_i)$ and

$$u = v_1 + \cdots + v_n + w_n$$

where either $\rho(w_n)$ fails to exist or $\rho(w_n) < \rho_n$. But a decreasing sequence of ranks is finite since the ranks are well-ordered, so that $\rho(w_n)$ fails to exist for some least integer n, say m. Since w_m is in g, $w_m = 0$ and $u = v_1 + \cdots + v_m$. Thus g equals the sum (5.3) as stated.

To prove the final statement of the theorem let $k(\rho)$ be any subgroup of elements of G with rank ρ, so chosen that $k(\rho) \supset h(\rho)$. It follows from Lemma 5.1 that the groups $k(\rho)$ have a direct sum J which is a group of elements with rank. Moreover $J \supset g$. Since g is maximal we conclude that $J = g$. It follows that $k(\rho) = h(\rho)$ since $h(\rho)$ is maximal in g, and we conclude that $h(\rho)$ is a maximal subgroup of elements of G with rank ρ.

The proof of the theorem is complete.

The following example shows that the omission of the hypothesis that the ranks are well-ordered in Theorem 5.2 would lead to a false conclusion.

Example 5.1. Let G be generated by a countable infinity of symbols $a_1, a_2, \cdots$ with coefficients which are integers mod 2. We suppose that a_n has a rank ρ_n and that $\rho_1 > \rho_2 > \cdots$ while the rank of a finite sum of different symbols a_i equals the maximum of the ranks of the summands. Every element of G except the null element thereby possesses a rank, and the four rank conditions are satisfied. The element $v_n = a_n + a_{n+1}$ taken with 0 forms a maximal subgroup (v_n) of elements of G with rank ρ_n. But the group G is not the direct sum $\sum_n (v_n)$ since the element a_1 is not in this direct sum.

Theorem 5.2 is complemented by the following theorem.

THEOREM 5.3. *If the ranks of elements of G are well-ordered, the direct sum of arbitrary maximal subgroups of elements of G with the respective ranks ρ is a maximal subgroup of elements of G with rank.*

Let $h(\rho)$ be a maximal subgroup of elements of G with rank ρ. As we have seen in the preceding proof the direct sum

$$H = \sum h(\rho)$$

is permissible and defines a group of elements with rank. We seek to prove that H is maximal in G.

To that end let g be a subgroup of elements of G with rank, with $g \supset H$. The group $h(\rho)$ is maximal in G and hence in g. It follows from the preceding theorem that

$$g = \sum_{\rho} h(\rho),$$

and the proof of the theorem is complete.

Note. The proofs of Theorems 5.2, 5.3 and 5.5 do not use rank condition IV.

We shall prove the following theorem.

THEOREM 5.4. *If the ranks of elements of G are well-ordered any two maximal subgroups g and g' of elements of G with rank are ρ-isomorphic.*

By virtue of Theorem 5.2 g and g' are direct sums

$$g = \sum_{\rho} g(\rho), \qquad g' = \sum_{\rho} g'(\rho)$$

of subgroups $g(\rho)$ and $g'(\rho)$ of g and g' respectively, which are maximal subgroups of elements of G with rank ρ. According to Theorem 5.1 $g(\rho)$ and $g'(\rho)$ are ρ-isomorphic. These isomorphisms between the respective groups $g(\rho)$ and $g'(\rho)$ imply a unique isomorphism T between g and g' defined as follows. An arbitrary element u of the group g is of the form

$$u = u_1 + \cdots + u_m,$$

where the elements u_i are from different groups $g(\rho_i)$. If u_i' is the element of $g'(\rho_i)$ which corresponds to u_i, then the element u shall correspond under T to the element

$$u' = u_1' + \cdots + u_m'.$$

We shall show that T is a ρ-isomorphism. Without loss of generality we can suppose that

$$\rho_1 > \rho_2 > \cdots > \rho_m.$$

It follows from Lemma 5.1 that u and u' are respectively in the ρ-classes of u_1 and u_1'. But u_1 and u_1' are in the same ρ-class. Hence u and u' are in the same ρ-class since the property of elements being in the same ρ-class is transitive.

The proof of the theorem is complete.

Extensions of the preceding theorems. If H is an arbitrary subgroup of G it is readily seen that the ranks ρ of elements of H by themselves satisfy the four rank conditions. The preceding theorems accordingly admit an immediate extension in which G is replaced by H. Thus regardless of whether the ranks of G are well-ordered or not as a whole, the preceding theorems hold upon replacing G by H, provided the ranks of H alone are well-ordered. The case where all the elements of H except the null element have rank will be met in the proof of the next theorem.

Bi-lexical ranks. Let s and t be variables which range over a simply ordered set of elements (x). We suppose that the ranks $\rho(u)$ of elements u of G are given by lexicographically ordered pairs (s, t). If an element u of G has a rank $\rho = (s, t)$ we shall write $s = s(u)$ and $t = t(u)$ and term s and t the first and second components of ρ respectively. We shall assume that $\rho(u)$ exists if and only if $s(u)$ exists, and that $s(u)$ in turn exists if and only if $t(u)$ exists. Ranks $\rho = (s, t)$ satisfying these conditions will be termed *bi-lexical.*

The ranks $r(u)$ of F-non-bounding k-cycles are bi-lexical.

We shall prove the following theorem.

THEOREM 5.5. *When the ranks $\rho = (s, t)$ of elements u of G are bi-lexical and well-ordered each maximal group of elements of G with rank is a direct sum of maximal groups of elements of G with the respective values of t, and arbitrary maximal groups of elements of G with the respective values of t sum to a maximal group of elements of G with rank.*

A maximal group g of elements of G with rank is a direct sum of maximal groups $g(\rho)$ of elements of G with the respective ranks ρ, as we have seen in Theorem 5.3. Let $h(t)$ be the direct sum of those groups $g(\rho)$ for which the second component of ρ is t. Then g is a direct sum

$$g = \sum_t h(t). \tag{5.4}$$

Moreover each group $h(t')$ must be a maximal group of elements of G for which $t(u) = t'$. For otherwise there would exist an element v such that $t(v) = t'$ and such that the direct sum of the group (v) generated by v and $h(t')$, is a group of elements u for which $t(u) = t'$. If u is an element of the group,

$$H = \sum h(t) \qquad (t \neq t') \tag{5.5}$$

then $t(u) \neq t'$, so that the ranks of elements of $(v) + h(t')$ and H are different. Hence we can form the direct sum

$$K = (v) + h(t') + H,$$

obtaining thereby a group of elements with rank. Moreover K contains g as a proper subgroup. From this contradiction we infer that the group $h(t')$ is a maximal group of elements of G for which $t(u) = t'$, and the proof of the first affirmation in the theorem is complete.

We shall now prove the second affirmation in the theorem.

To that end let $k(t)$ be an arbitrary maximal subgroup of elements u of G for which $t(u) = t$. Let $g(s, t)$ be a maximal subgroup of elements v of $k(t)$ for which $s(v) = s$. By virtue of the extension of the preceding theorems noted just before the present theorem, $k(t)$ is a direct sum

$$k(t) = \sum_s g(s, t).$$

I say that $g(s, t)$ is a maximal group of elements u of G for which $\rho = (s, t)$. Otherwise we could reason as in the proof that $h(t)$ was maximal and obtain a group $p(t)$ of elements v for which $t(v) = t$ and which would contain $k(t)$ as a proper subgroup. Hence $g(s, t)$ is a maximal group of elements of G with rank $\rho = (s, t)$. It follows from Theorem 5.3 that the group

$$\sum_{t,s} g(s, t) = \sum_{t} k(t)$$

is a maximal group of elements of G with rank, and the proof of the theorem is complete.

$*$-Telescopic group sequences. Our remarks in the remainder of this section are largely by way of orientation.

In earlier papers on critical points of functions the author has been concerned with integers $\mu_k = M_k - R_k$, positive or zero, where M_k was the kth type number sum of the critical sets and R_k was the kth connectivity of M. The numbers μ_k, when finite, were found to be of the form

$$\mu_k = \gamma_k + \gamma_{k-1} \qquad (\gamma_k \geqq 0;\ k = 0, 1, \cdots) \tag{5.6}$$

where γ_k is an integer and $\gamma_{-1} = 0$.

We continue with the following lemma.

LEMMA 5.2. *A necessary and sufficient condition that a sequence of non-negative integers $\mu_0, \mu_1, \cdots$ be representable in the form* (5.6) *is that these integers satisfy the inequalities*

$$\mu_n - \mu_{n-1} + \cdots + (-1)^n \mu_0 \geqq 0 \qquad (n = 0, 1, \cdots). \tag{5.7}$$

If the numbers μ_i are representable in the form (5.6) we find that

$$\mu_n - \mu_{n-1} + \cdots + (-1)^n \mu_0 = \gamma_n \tag{5.8}$$

so that the inequalities (5.7) are a consequence of (5.6). Conversely if the relations (5.7) are satisfied and we define γ_k by means of (5.8), we find that $\gamma_k \geqq 0$ and that (5.6) holds.

The inequalities (5.7) were fundamental in the earlier theory and were the source of numerous existence theorems for critical points.

The present more general theory is formally analogous to the earlier theory with groups m_k replacing the numbers μ_k and the group operation of reduction with respect to a modulus replacing the operation of subtraction in (5.7). In this connection it should be noted that (5.7) can be written in the form

$$\mu_n - [\mu_{n-1} - (\mu_{n-2} \cdots \mu_0) \cdots] \geqq 0, \tag{5.9}$$

employing subtractions only.

More explicitly we shall be concerned with an infinite sequence of groups $m_0, m_1, \cdots$. The function F will give rise to certain natural isomorphisms termed $*$-isomorphisms, which map a group g_k onto a group denoted by g_k^*. In terms of these $*$-isomorphisms a sequence of groups m_k will be termed *$*$-telescopic* if there exists an auxiliary sequence of groups g_k such that

$$m_k = g_k + g_{k-1}^* \qquad (g_{-1} = 0;\ k = 0, 1, \cdots). \tag{5.10}$$

The relations (5.10) are the formal analogues of the relations (5.6). In fact if μ_k and γ_k are the dimensions of m_k and g_k respectively the relations (5.6) are a consequence of the relations (5.10).

We can also give a formal analogue of the relations (5.9). To that end we read mod* A, "mod a group $*$-isomorphic with A." We then replace (5.9) by the relations

$$(5.11)\quad m_n \text{ mod}^* (m_{n-1} \text{ mod}^* (m_{n-2} \cdots \text{ mod}^* (m_0) \cdots) \geqq 0, \quad (n = 0, 1, \cdots)$$

reading $\geqq 0$ "exists." In (5.11) as in (5.9) the operations indicated are to be successively performed starting with the innermost parenthesis. The symbolic expressions

$$(5.12)\quad \begin{aligned} &m_1 \text{ mod}^* m_0 , \\ &m_2 \text{ mod}^* m_1 \text{ mod}^* m_0 , \\ &\cdots\cdots\cdots\cdots\cdots\cdots , \end{aligned}$$

appear in infinitely many of the relations (5.11), but in each of these relations they shall indicate the same groups.

It is easily seen that a necessary and sufficient condition that the groups m_k satisfy relations of the form (5.11) is that there exist auxiliary groups g_k such that the groups m_k are of the form (5.10) for suitably defined $*$-isomorphisms. These $*$-isomorphisms will presently receive a concrete exemplification.

II. Natural Isomorphisms

6. **k-caps and a-isomorphisms.** The object of chapter II is to establish certain general relations between groups of F-non-bounding i-cycles and groups of j-caps. We shall make use of three basic types of isomorphisms, r-isomorphisms between groups of F-non-bounding k-cycles, a-isomorphisms between groups of k-caps and $*$-isomorphisms between groups of F-non-bounding k-cycles and groups of (k + 1-caps. These isomorphisms depend in an essential way upon the nature of F.

We first introduce several new concepts.

k-caps. A k-cycle u on F_{a+} mod$_o$ F_a will be termed a k-cap *belonging* to a if

$$u \not\sim 0 \qquad (\text{on } F_{a+} \text{ mod}_o\, F_a).$$

The constant a will then be termed a *k-cap limit* $a(u)$ of u. The k-cap limits $a(u)$ satisfy the four rank axioms $a(u)$ replacing $\rho(u)$, provided the underlying group G of the preceding section be correctly defined.

To define G we term a sequence w of algebraic k-chains w_n with norms e_n tending to zero a *formal* k-chain. To add two formal k-chains we add their components, while δw shall be the formal k-chain with components δw_n. Formal k-chains make up an additive abelian operator group with operators δ. The components of a formal k-chain w may in particular be the components of a k-cap z. In such a case it is easily seen that $a(z)$ is uniquely determined by w.

To add two k-caps u and v consider the formal k-chain $w = u + v$. If w defines a k-cap z we regard z as the cap sum of u and v. If w defines no k-cap we understand that there is no cap sum $u + v$. In the group of formal k-chains G certain formal k-chains are thus k-caps and possess cap limits. Concerning those cap limits we have the following theorem.

THEOREM 6.1. *The k-cap limits $a(u)$ satisfy the rank conditions, $a(u)$ replacing $\rho(u)$ with the underlying group G the group of formal k-chains.*

The proof of this theorem is immediate.

We shall give certain definitions which concern k-caps. We shall term an homology

$$u \sim 0 \qquad (\text{on}_\circ\ F_{c+} \ \text{mod}_\circ\ F_c)$$

a *c-homology*. The earlier ρ-class and ρ-isomorphism are here replaced by their formal analogues *a-class* and *a-isomorphism*. It is seen that two k-caps with cap limits c belong to the same a-class if and only if their difference is c-homologous to zero.

Let p be a point on M and c a number such that $0 \leqq c \leqq 1$. We term (p, c) a *limiting pair* if c is a limit at p of values of $F(p_n)$ for some sequence p_n tending to p. We include pairs of the form $[p, F(p)]$ as limiting pairs. There are certain limiting pairs which are particularly distinguished and which will be called critical or semi-critical pairs. They will be defined in §9. Their existence depends upon a series of theorems which we shall now develop.

A k-cap which is a k-cycle will be termed a *k-cycle-cap*. The following theorem may be regarded as a first existence theorem in the theory of k-caps.

THEOREM 6.2. *Corresponding to an arbitrary F-non-bounding k-cycle u there is a k-cycle-cap z in the r-class of u such that the cap limit $a(z)$ differs arbitrarily little from the inferior cycle limit $t(u)$.*

If $t(u) = 1$, u is a cycle-cap and the theorem is true. We suppose then that $t(u) < 1$.

If the r-class of u contains a cycle-cap belonging to $t(u)$ the theorem is again true. If the r-class of u contains no such cycle-cap let β be an arbitrary constant less than 1 such that $t(u) < \beta < s(u)$.

There exists a k-cycle u' in the given r-class below$_\circ$ β with $u' \not\sim 0$ on $F \leqq \beta$. Let α be the inferior cycle limit of u' on $F \leqq \beta$. It follows from Lemma 2.2 that $u' \sim 0 \bmod F \leqq \alpha$ on $F \leqq \beta$. Upon setting $F_{\alpha+} = V$ and $F_{\beta+} = C$ with $U = 0$ it follows from Theorem 3.1 that there exists a k-cycle z on V such that $u' \sim z$ on C. But z is not α-homologous to zero for otherwise α would not be an inferior cycle limit of z on $F \leqq \beta$. Hence z is a cycle-cap belonging to α. The theorem follows directly.

An F-non-bounding k-cycle u which lies on the domain $F \leqq t(u)$ will be termed *canonical*. We have seen that there is a cycle-cap in the r-class of each F-non-bounding k-cycle u, but there need be no canonical k-cycle in the r-class of u as the following example shows.

Example 6.1. Let the space M consist of the closure of the curve

$$y = (1 - x) \sin \frac{1}{x} \qquad (0 < x \leqq 1)$$

together with a straight line segment joining the point (0, 1) to the point (−1, −1) in the (x, y) plane. On M let $F = (y + 1)/2$ so that $0 \leqq F \leqq 1$. Let u be a 0-cycle mod 2 whose components u consist of the two points (−1, −1) and (1, 0). We see that $t(u) = 0$. There is no 0-cycle on the domain $F = 0$ in the r-class of u.

Example 6.2. We can modify the preceding example to show that an inferior cycle limit may equal no cap limit. To that end we join the bottom points (−1, −1), (−1, 0) of the preceding point set M by a semi-circle on the domain $y \leqq -1$ and let the resulting point set be denoted by M'. On M' let $F = y$ waiving the conditions that $0 \leqq F \leqq 1$. The 0-cycle u of the preceding example now has an inferior cycle limit which equals no cap limit, since the semi-circle can be deformed on itself away from its end points.

7. ***-Isomorphisms.** If v is a k-cap such that $\beta v \sim 0$ below$_\circ$ its cap limit $a(v)$, v will be termed *linkable*. A non-linkable k-cap v will be said to *cap* its boundary $u = \beta v$ and u will then be said to admit the cap v. A group of F-non-bounding k-cycles will be said to be **-isomorphic* with a group g_k^* of $(k + 1)$-caps if g_k and g_k^* admit an isomorphism in which each k-cycle of g_k is capped by its correspondent in g_k^*.

We shall make use of the following lemma.

Lemma 7.1. *If u is a k-cycle mod$_\circ$ F_a on $F \leqq a$ and is a-homologous to zero, then $\beta u \sim 0$ below$_\circ$ a.*

Under the hypotheses of the lemma we can affirm the existence of a positive constant e with the following property. Corresponding to the nth component u_n of u there exists an algebraic k-chain w_n on $F \leqq a - e$ and an algebraic $(k + 1)$-chain z_n on $F \leqq a$ such that $\beta z_n = u_n + w_n$ where u_n, w_n, and z_n possess norms e_n which tend to zero as n becomes infinite. Hence $\beta u_n + \beta w_n = 0$ and $\beta u_n \sim_{e_n} 0$ on $F \leqq a - e$.

The proof of the lemma is complete.

An F-non-bounding k-cycle u which admits a k-cap v is bounding. A k-cycle which is both F-non-bounding and bounding has been termed ambiguous. With this understood we state the following theorem.

Theorem 7.1. *An ambiguous k-cycle u admits at least one $(k + 1)$-cap v, with $a(v) = s(u)$.*

Let $s(u)$ be denoted by a. If e is an arbitrary positive constant then $u \sim 0$ below$_\circ$ $a + e$ by virtue of the definition of $s(u)$. If $a < 1$ it follows from Lemma 2.2 that $u \sim 0$ on $F \leqq a$. This homology is trivial if $a = 1$.

We shall apply Theorem 3.2. To that end we set $A = 0$, and let V be a carrier of the homology $u \sim 0$ on $F \leqq a$. The cycle u has a carrier U on V below$_\circ$ a. It follows from Theorem 3.2 that there exists a $(k + 1)$-cycle

v mod U on V such that $\beta v = u$. Moreover v is not a-homologous to zero for otherwise $\beta v \sim 0$ below$_o$ a by virtue of Lemma 7.1, and we could infer that $u \sim 0$ below$_o$ a contrary to the nature of $a = s(u)$.

Thus v caps u, and the proof is complete.

Let g_k be a group of ambiguous k-cycles representable as a direct sum

$$g_k = \sum_c g_k(c) \tag{7.1}$$

of subgroups $g_k(c)$ of k-cycles u for which $s(u) = c$, where c ranges over an arbitrary set of superior cycle limits s. We term g_k *s-neuclear*, and state the following theorem.

THEOREM 7.2. *An s-neuclear group g_k of ambiguous k-cycles is $*$-isomorphic with at least one group g_k^* of $(k + 1)$-caps.*

Let $(z)_c$ be a maximal linear set in $g_k(c)$ and let (z) be a set theoretic sum of the sets $(z)_c$. The set (z) will form a maximal linear set for g_k. With each cycle z_i in (z) we now associate a $(k + 1)$-cap $K(z_i)$, as is possible by virtue of Theorem 7.1. We set $K(0) = 0$. Let $z_1, \cdots, z_n$ be a set of distinct cycles in (z). If $u = \delta_i z_i$ we set

$$K(u) = \delta_i K(z_i),$$

and prove statements (α) and (β).

(α). *The cycle u is capped by $K(u)$ if $u \neq 0$.*

In proving (α) we can assume without loss of generality that no δ_i is null. Set $a = s(u)$. Recalling that $a = \max s(z_i)$ we see that each $K(z_i)$ lies on $F \leqq a$, and hence $K(u)$ lies on this same domain. Moreover

$$\beta K(u) = \delta_i z_i = u.$$

Finally $K(u)$ is not a-homologous to zero, for otherwise $u \sim 0$ below$_o$ a by virtue of Lemma 7.1.

The proof of (α) is complete.

(β). *The $(k + 1)$-caps $K(u)$ form a group g_k^* $*$-isomorphic with the group g_k.*

Let x and y be two cycles of g_k of the form

$$x = \delta_i z_i, \qquad y = \delta_i' z_i,$$

where the z_i's form a set of distinct cycles of (z). We see that

$$K(x + y) = \Sigma(\delta_i + \delta_i') K(z_i) = K(x) + K(y).$$

Hence the caps $K(u)$ define a group homeomorphic with g_k. That this homeomorphism is simple follows from the fact that $K(u) = 0$ only if $u = 0$, in accordance with (α).

The proof of the theorem is complete.

8. **Well-ordered cap limits.** The hypothesis that the cap limits are well-ordered has important consequences which will be studied in this section.

THEOREM 8.1. *Let u be an F-non-bounding k-cycle. If the k-cap limits* *greater than* $t(u)$ *are bounded away from* $t(u)$ *there is a canonical k-cycle in the r-class of u.*

By virtue of Theorem 6.2 there exists a k-cycle-cap z in the r-class of u such that $a(z)$ differs arbitrarily little from $t(u)$. Under the hypotheses of the present theorem $a(z) = t(u)$ if $a(z) - t(u)$ is sufficiently small. But if $a(z) = t(u)$, z will be a canonical k-cycle in the r-class of u and the proof of the theorem is complete.

COROLLARY. *If the k- and* $(k + 1)$*-cap limits are well-ordered each k-cycle limit is a k- or* $(k + 1)$*-cap limit and the ranks of F-non-bounding k-cycles are well-ordered.*

That each superior cycle limit of a k-cycle is a $(k + 1)$-cap limit follows from Theorem 7.1, while inferior cycle limits of k-cycles are k-cap limits by virtue of the preceding theorem.

The ranks $r = (s, t)$ of F-non-bounding k-cycles can be grouped in classes with common s, and these classes will be well-ordered among themselves by virtue of the well-ordering of the $(k + 1)$-cap limits s. The ranks r in each such class can now be further ordered according to the values of t. Each class will then be well-ordered. The ranks thus receive their prescribed lexicographic order and it is readily seen that they are thereby well-ordered.

Let u be a k-cycle $\bmod_o H$ where H is a domain F_a or F_{a+}. A constant $b > a$ will be called a *homology limit* of u $\bmod_o H$ if u is on_o F_b and $u \not\sim 0$ below_o b $\bmod_o H$. A k-cap u with cap limit a will be termed *degenerate* if u possesses no homology limit $\bmod_o F_a$. That degenerate k-caps exist is shown by the following example.

Example 8.1. Let the space M consist of the points (x, y) on the curve $y = x \sin 1/x$ for $0 \leqq x \leqq 1$ and let $F = (y + 1)/2$ on M. Let u_n be an algebraic 0-cycle mod 2 which coincides with the point $(0, 0)$. The Vietoris 0-cycle $u = (u_n)$ is a degenerate 0-cap.

We continue with the following theorem.

THEOREM 8.2. *Let u be a k-cap with cap limit a. If the* $(k + 1)$*-cap limits greater than a are bounded away from a, u is non-degenerate.*

The theorem is trivial when $a = 1$.

Suppose $a < 1$. Let b be a lower bound between 1 and a of the $(k + 1)$-cap limits greater than a. We shall prove the following statement.

(i). *The constant b is a homology limit of u* $\bmod_o F_a$.

Suppose (i) false. Then for some constant c between a and b and some constant $\alpha < a$, $u \sim 0$ on $F \leqq c \bmod F \leqq \alpha$. Upon setting

$$A = F_{\alpha+}, \qquad U = F_{a+}, \qquad V = F_{c+}, \tag{8.1}$$

it follows from Theorem 3.2 that there exists a $(k + 1)$-cycle v mod U on V such that $\beta v \equiv u \bmod A$. Moreover

$$v \not\sim 0 \qquad (\text{mod } F_{a+} \text{ on } F_{c+}), \tag{8.2}$$

for otherwise $\beta v \sim 0$ on U, and hence $u \sim 0 \bmod A$ on U contrary to the nature of u as a k-cap.

From (8.2) it follows that there is a greatest lower bound t of numbers $e > a$ such that $v \sim 0 \bmod F \leqq e$ on $F \leqq c$. Upon referring to Lemma 2.2, setting $B = F_{t+}$ and $C = F_{c+}$ we see that

$$v \sim 0 \qquad (\text{mod } F_{t+} \text{ on } F_{c+}). \tag{8.3}$$

It follows from (8.2) and (8.3) that $t > a$. We next apply Theorem 3.1, setting

$$U = F_{a+}, \qquad V = F_{t+}, \qquad C = F_{c+},$$

and infer from (8.3) that there exists a $(k + 1)$-cycle z mod $F \leqq a$ on $F \leqq t$ homologous to v mod $F \leqq a$ on $F \leqq c$. It follows from the choice of t that v and hence $z \not\sim 0 \text{ mod}_0\, F_t$ on F_{c+}. Since z is on $F \leqq t$, t is a cap limit of z with $a < t < b$, contrary to the choice of b.

The proof of (i), and of the theorem is complete.

We shall prove the following theorem.

THEOREM 8.3. (1). *There is a k-cycle-cap v in the a-class of each linkable k-cap u.* (2). *If u is non-degenerate v is F-non-bounding.* (3). *If the k- and (k + 1)-cap limits are well-ordered v may be chosen so as to be canonical.*

We suppose that a is the cap limit of u and let B be a carrier of u on $F \leqq a$. Since u is linkable there exists a constant $\alpha < a$ such that $\beta u \sim 0$ on $F \leqq \alpha$. If we denote the latter domain by A, set $V = A + B$, and apply Theorem 3.3, statement (1) of Theorem 8.3 follows directly.

If u is non-degenerate v is likewise. The cycle v thus possesses a homology limit $\text{mod}_0\, F_a$ and a fortiori a homology limit so that (2) is true.

We now turn to the proof of (3).

If the $(k + 1)$-cap limits are well-ordered, the $(k + 1)$-cap limits greater than a are in particular bounded away from a so that u is non-degenerate in accordance with Theorem 8.2. As stated in (2), v is then F-non-bounding. If the k- and $(k + 1)$-cap limits are well-ordered, the ranks of F-non-bounding k-cycles are well-ordered, and by virtue of Theorem 8.1 there is a canonical k-cycle in each r-class of k-cycles. A maximal group $g(r)$ of k-cycles with rank r can accordingly be made up of canonical k-cycles. If $h(t)$ is the sum of the groups $g(r)$ for which the second component of r is t, $\sum_t h(t)$ will be a maximal group of k-cycles with rank. Each non-null cycle z in a group $h(t)$ will lie on $F \leqq t$ and have t as an inferior cycle limit. Hence each such cycle z will be canonical.

There accordingly exists a set $v_1, \cdots, v_n$ of canonical k-cycles from different groups $h(t)$ such that the k-cycle $w = v - (v_1 + \cdots + v_n)$ is without rank. The cycle w is not a k-cap, for otherwise w would be a non-degenerate k-cap and hence have rank. Hence v is in the a-class of $(v_1 + \cdots + v_n) = z$. Without loss of generality we can suppose that $t(v_1) > \cdots > t(v_n)$. The cycle-cap z is not necessarily canonical, but it is in the a-class of v_1, and v_1 is canonical. Thus u is in the a-class of the canonical cycle v_1, and the proof of the theorem is complete.

An a-class of k-caps will be said to be of *excess* type if it contains no cycle or contains an ambiguous k-cycle. Otherwise the a-class will contain a non-bounding k-cycle but no ambiguous k-cycle and be said to be of *space* type. A k-cap will be said to be of excess or space type according as it is in an a-class of excess or space type. A linkable k-cap may be either an excess or space cap. A non-linkable k-cap is always an excess cap. An excess k-cap and a space k-cap are never in the same a-class, and accordingly their difference is always a k-cap.

We shall now prove four theorems on the composition of maximal groups of k-caps.

THEOREM 8.4. *If the k-cap limits are well-ordered, a maximal group M_k of k-caps is a direct sum $M_k = M_k' + M_k''$ of arbitrary maximal subgroups respectively of excess and space k-caps of M_k. The groups M_k' and M_k'' are maximal among k-caps in general. Conversely arbitrary maximal groups of excess and space k-caps sum to a maximal group of k-caps.*

We shall begin by proving the following statement.

(α). *A maximal group μ_k of k-caps with a cap limit c is a direct sum*

$$\mu_k = \mu_k' + \mu_k''$$

of maximal groups μ_k' and μ_k'' of excess and space k-caps respectively with cap limit c.

The subset of linkable k-caps of μ_k together with the null element form a group a_k so that μ_k is a direct sum $\mu_k = a_k + c_k$ in which c_k is a maximal subgroup of non-linkable k-caps of μ_k. The subset of excess k-caps of a_k together with the null element form a group α_k so that a_k is a direct sum $\alpha_k + \beta_k$ in which β_k is a maximal subgroup of space k-caps of a_k. Hence μ_k is a direct sum $\alpha_k + \beta_k + c_k$. We set $\mu_k' = \alpha_k + c_k$ and $\mu_k'' = \beta_k$. That μ_k' is a group of excess k-caps with cap limit c follows from the fact that a non-linkable and a linkable k-cap with cap limits c sum to a non-linkable k-cap with cap limit c. That μ_k' and μ_k'' are respectively maximal groups of excess and space k-caps with cap limit c follows from the relation $\mu_k = \mu_k' + \mu_k''$, the fact that μ_k is maximal, and the fact that an excess and a space k-cap with cap limit c sum to a k-cap with cap limit c.

The proof of (α) is complete.

We can now establish the first two statements in the theorem.

Since the cap limits satisfy the rank conditions it follows from Theorem 5.2 that a maximal group M_k of k-caps is a direct sum of groups $\mu_k(c)$ which are maximal groups of k-caps with the respective cap limits c. Each such group $\mu_k(c)$ is a sum $\mu_k'(c) + \mu_k''(c)$ of the nature described in (α). Upon setting

$$m_k' = \sum_c \mu_k'(c), \qquad m_k'' = \sum_c \mu_k''(c)$$

we see that $M_k = m_k' + m_k''$. The groups m_k' and m_k'' are respectively groups of excess and space k-caps. That they are maximal follows from the relation $M_k = m_k' + m_k''$ and the fact that an excess and a space k-cap always sum to a k-cap.

The proof of the first two statements in the theorems is complete.

To prove the final statement in the theorem we assume that M'_k and M''_k are respectively maximal groups of excess and space k-caps. The sum

$$L_k = M'_k + M''_k \tag{8.4}$$

is composed of k-caps. Let M_k be a maximal group of k-caps such that $M_k \supset L_k$. The groups M'_k and M''_k are maximal among formal k-chains and hence in M_k. It follows from the first statement in our theorem that

$$M_k = M'_k + M''_k.$$

The proof of the theorem is complete.

THEOREM 8.5. *If the k-cap limits are well-ordered for each k,*

(i) *there exists a maximal group g of ambiguous k-cycle-caps which is a maximal group of ambiguous k-cycles,*

(ii) *such a group g_k admits a $*$-isomorphic group of $(k + 1)$-caps $\overset{*}{g_k}$,*

(iii) *while the groups g_k and $\overset{*}{g}_{k-1}$ admit a direct sum*

$$m_k = g_k + \overset{*}{g}_{k-1} \qquad (k = 0, 1, \cdots) \tag{8.5}$$

which is a maximal group of excess k-caps.

We shall prove statement (i) with the aid of Theorem 5.5.

To that end we shall introduce a new set of ranks σ, assigning ranks σ only to ambiguous k-cycles u and for such cycles setting $\sigma(u) = r(u)$. The ranks σ clearly satisfy rank conditions I, II, III.

There is a canonical k-cycle in each r-class by virtue of Theorem 8.1. It follows that there exists a maximal group $k(t)$ of ambiguous k-cycles with a prescribed inferior cycle limit t with the property that the non-null cycles of $k(t)$ are canonical. We now apply Theorem 5.5 employing ranks σ. It follows that $\Sigma\, k(t)$ is a maximal group g of ambiguous k-cycles. But each cycle of $k(t)$ is a cycle-cap with cap limit t. The group g is thus a group of cycle-caps. Since g is maximal as a group of ambiguous k-cycles it is a fortiori maximal as a group of ambiguous k-cycle-caps.

The proof of statement (i) is complete.

Statement (ii) is a consequence of Theorem 7.2.

We continue with a proof of the following statement.

(α). *The groups g_k and $\overset{*}{g}_{k-1}$ admit a direct sum m_k which is a group of excess k-caps.*

Let u and v be k-caps in g_k and $\overset{*}{g}_{k-1}$ respectively. We seek to prove that $u + v$ is an excess k-cap. If $a(u) > a(v)$, $u + v$ is a k-cap in the a-class of u, and since u is of excess type $u + v$ is of excess type. The case where $a(v) > a(u)$ is similar. If $a(u) = a(v)$, $u + v$ is non-linkable and hence of excess type. Thus (α) is true.

(β). *If u is an arbitrary non-linkable k-cap there is an element w in $\overset{*}{g}_{k-1}$ such that $u - w$ is not a non-linkable k-cap with cap limit $a(u)$.*

Let $\beta u = v$. The cycle v is ambiguous so that there exists a cycle z in g_{k-1}

such that $v - z$ has no rank. Let w be the image of z in g_{k-1}^*. Since $v - z$ has no rank $v - z \sim 0$ below$_o$ $a(u)$. Hence

$$\beta(u - w) = v - z \sim 0 \qquad [\text{below}_o\ a(u)]. \tag{8.6}$$

If $a(u - w) = a(u)$, $u - w$ is linkable by virtue of (8.6) and (β) is true. If $a(u - w)$ does not exist or is less than $a(u)$ we see that (β) is again true.

(γ). *If u is an excess linkable k-cap there is an element w in g_k such that $u - w$ is not a k-cap with cap limit $a(u)$.*

Let v be an ambiguous k-cycle in the a-class of u. There exists a k-cycle w in g_k such that $v - w$ is not an ambiguous k-cycle and hence not a k-cap. Thus v, w, and u are in the same a-class. Since u and w are in the same a-class, $a(u - w)$ does not exist or is less than $a(u)$, and (γ) is true.

(δ). *If u is an excess k-cap there is an element w in m_k such that $u - w$ is not an excess k-cap with cap limit $a(u)$.*

In case u is linkable (δ) follows from (γ).

We accordingly assume that u is non-linkable. According to (β) there exists a k-cap w in g_{k-1}^* such that $u - w$ is not a non-linkable k-cap with cap limit $a(u)$. The difference $u - w$ may satisfy (δ). Otherwise $a(u - w) = a(u)$ and $u - w$ is an excess linkable k-cap to which (γ) applies. Hence (δ) holds in all cases.

We can now prove (iii). That m_k is a group of excess k-caps has been proved in (α). It remains to show that m_k is maximal. To that end let u be an arbitrary excess k-cap and w the corresponding k-cap in (δ). If $u - w$ is not an excess k-cap (iii) is true. If $u - w$ is an excess k-cap $a(u - w) < a(u)$ in accordance with (δ). Set $w = w_1$. We now treat $u - w_1$ as we did u. Proceeding inductively and recalling that a decreasing sequence of k-cap limits is finite we conclude that there exists a set $w_1, w_2, \cdots w_m$ of k-caps in m_k with decreasing cap limits such that $u - (w_1 + \cdots + w_m)$ is not an excess k-cap.

The proof of (iii) and of the theorem is complete.

Upon referring to Lemma 5.2 we have the following corollary.

COROLLARY. *The dimension μ_k of the maximal groups m_k of excess k-caps, if finite, satisfy the relations* (5.7). *More generally the groups m_k are $*$-telescopic and satisfy relations of the form* (5.11).

We come finally to maximal groups of space k-caps. We shall prove the following theorem.

THEOREM 8.6. *If the k- and $(k + 1)$-cap limits are well-ordered each maximal group of canonical non-bounding k-cycles is*

(1) *a maximal group of non-bounding k-cycles,*

(2) *a maximal group of space k-caps.*

We shall first prove statement (1) of the theorem.

To that end we assign ranks σ to canonical non-bounding k-cycles setting $\sigma(u) = r(u)$. For such cycles $s(u) = +\infty$. The rank conditions I, II, III are satisfied by ranks σ.

Let $g(\sigma)$ be a maximal group of canonical non-bounding k-cycles with rank σ. It follows from Theorems 5.2 and 5.3 that the sum

$$g = \sum_{\sigma} g(\sigma) \tag{8.7}$$

is a maximal group of canonical non-bounding k-cycles and each maximal group of canonical non-bounding k-cycles is of the form (8.7). But there is a canonical k-cycle in each r-class of k-cycles. It follows that $g(\sigma)$ is a maximal group of non-bounding k-cycles with rank $r = \sigma$. Moreover the ranks r of non-bounding k-cycles satisfy the rank conditions I, II, III as is readily seen. Hence the preceding group g is a maximal group of non-bounding k-cycles, and statement (1) of the theorem is proved.

We continue with the following lemma.

LEMMA 8.1. *A necessary and sufficient condition that a linkable k-cap u be of space type is that on M*

$$u \not\sim 0 \qquad [\text{mod}_\circ\ F < a(u)]. \tag{8.8}$$

Any k-cycle in the a-class of a space k-cap is canonical and non-bounding, and conversely any canonical non-bounding k-cycle is a space cap.

Suppose u is a space k-cap. Let v be a k-cycle in the a-class of u. If (8.8) were false $v \sim 0$ mod$_\circ$ $F < a(u)$. There would then exist a k-cycle w below$_\circ$ $a(u)$ such that $v - w \sim 0$. But $v - w$ would be in the a-class of v and hence of u contrary to the hypothesis that u is a space k-cap. Hence the condition (8.8) is necessary.

To prove the condition (8.8) sufficient we again let v be a k-cycle in the a-class of u. It follows from (8.8) that

$$v \not\sim 0 \qquad [\text{mod}_\circ\ F < a(u)] \tag{8.9}$$

and a fortiori that v is non-bounding. Thus u is a space cap. The condition (8.8) is accordingly sufficient.

The second statement in the lemma follows at once from the first.

To establish (2) in Theorem 8.6 we assign ranks $b(u)$ to space k-caps u setting $b(u) = a(u)$. It is clear that the ranks $b(u)$ satisfy the first three rank conditions.

We shall prove that the group g in (8.7) is a maximal group of space k-caps.

Observe that $g(\sigma)$ in (8.7) is a group of space k-caps with cap limit t where $\sigma = (s, t) = (\infty, t)$. The group $g(\sigma)$ is a maximal group of space k-caps with cap limit t; for there is a canonical non-bounding k-cycle in the a-class of each space k-cap u. Moreover the cap limits of space k-caps satisfy the first three rank conditions so that the groups $g(\sigma)$ must sum to a maximal group of space k-caps as stated.

The proof of the theorem is complete.

The three preceding theorems are combined and extended in the following theorem.

THEOREM 8.7. *If the k-cap limits are well-ordered for all integers k, an arbi-*

trary maximal group M_k of k-caps is a-isomorphic with a maximal group of k-caps of the form

$$g_k + \overset{*}{g}_{k-1} + Q_k \tag{8.11}$$

where g_k is a suitably chosen maximal group of ambiguous k-cycles, $\overset{*}{g}_{k-1}$ is a group of k-caps $*$-isomorphic with g_{k-1}, and Q_k is an arbitrary maximal group of canonical non-bounding k-cycles. The group Q_k is isomorphic with the kth homology group. When the dimensions p_k and R_k of M_k and Q_k respectively are finite the following relations hold

$$\begin{aligned} p_0 &\geqq R_0, \\ p_1 - p_0 &\geqq R_1 - R_0, \\ p_2 - p_1 + p_0 &\geqq R_2 - R_1 + R_0, \\ &\cdots\cdots\cdots\cdots . \end{aligned} \tag{8.12}$$

There is a group L_k of the form (8.11) which is a maximal group of k-caps by virtue of the three preceding theorems. It follows from Theorem 5.4 that L_k is a-isomorphic with M_k. The group Q_k is a maximal group of non-bounding k-cycles in accordance with Theorem 8.6 and so is isomorphic with the kth homology group. If γ_k is the dimension of g,

$$p_k - R_k = \gamma_k + \gamma_{k-1} \qquad (\gamma_{-1} = 0,\ k = 0, 1, \cdots)$$

provided these dimensions are finite. The inequalities (8.12) follow at once.

III. Critical Points

9. Homotopic critical points. As stated in the introduction we distinguish between differential, homotopic, combinatorial, and essential critical points. Of these types of points the last three are topological. To define homotopic critical points we shall admit deformations of the following type.

Deformations. Let E be a subset of M. We shall consider deformations D of E which replace a point p initially on E at the time $t = 0$ by a point

$$q = q(p, t) \qquad (p \subset E;\ 0 \leqq t \leqq \tau)$$

on M at the time t, where t varies on the closed interval $(0, \tau)$. We shall suppose that τ is a positive constant and that $q(p, t)$ is a continuous point function of its arguments when t is on $(0, \tau)$ and p is restricted to any compact subset of E. The curve $q = q(p, t)$ obtained by holding p fast and varying t on $(0, \tau)$ will be termed the *trajectory* T defined by p. If a point q precedes a point r on a trajectory T, q will be termed an *antecedent* of r.

We shall say that a deformation D of a set S admits a *displacement* function $\delta(e)$ on S if whenever q is an antecedent of r such that $qr > e > 0$ then $F(q) - F(r) > \delta(e)$ where $\delta(e)$ is a positive single-valued function of e.

Consider the special case in which F is continuous, S compact, and D is such that $F(q) > F(r)$ whenever q is an antecedent of r distinct from r. In this case D always admits a displacement function $\delta(e)$. If however F is merely lower semi-continuous more stringent conditions are needed to establish the existence of a displacement function, as the following example shows.

Example 9.1. Let the space M consist of the points (x, y) in the square $0 \leqq x \leqq 1$, $0 \leqq y \leqq 1$. On this square let $F = -xy$ except when $y = 0$. When $y = 0$ let $F = -x$. The function F is lower semi-continuous on M. Let E be the points of M on the y axis and let D be a deformation in which each point of E moves parallel to the x axis at a unit velocity for a unit of time. The deformation D is such that $F(q) > F(r)$ whenever q is an antecedent of r distinct from r. Nevertheless the deformation D admits no displacement function on E.

F-deformations. A deformation D of E which admits a displacement function on each compact subset of E will be termed an *F-deformation* of E, and E will be said to be F-deformable under D. We shall say that E is F-deformable *definitely onto* F_c (written onto$_0$ F_c) if E is F-deformable onto a point set on $F \leqq c - e$ where e is a positive constant.

In the special case where F is continuous and M compact, a point p at which $F(p) = c$ will be said to be *homotopically ordinary* if there exists a neighborhood of p on $F \leqq c$ which is F-deformable onto$_0$ F_c. In the general case where F is lower semi-continuous and M not necessarily compact the definitions are more complex. The preceding definition is a special case of a definition which we shall now give.

A limiting pair (p, c) (see §6) will be termed *homotopically ordinary* if there exists a neighborhood of p on $F \leqq c$ which is F-deformable onto$_0$ F_c. A limiting pair (p, c) which is not homotopically ordinary will be termed *homotopically critical,* and p will then be termed *semi-critical* at the level c. A point p which is semi-critical at the level $F(p)$ will be termed *critical.*

A point p may be semi-critical at different levels c. For example the point $(0, 1)$ in Example 1.1 is semi-critical at the levels $c = 0$ and $c = 1$. When F is continuous a semi-critical point p is a critical point. We shall presently apply our theory to the calculus of variations and shall thereby be concerned with a function F which is lower semi-continuous but not in general continuous. The most important special case in this variational theory is the locally convex case and in this case it will appear that each semi-critical point p is a critical point p.

Differential critical points. Let F be a function $F(x_1, \cdots, x_n)$ of class C^2 in an open region R of the space (u). We term a point $(x) = (a)$ *differentially critical* if all the first partial derivatives of F vanish at (a). Otherwise we term (a) *differentially ordinary.*

We shall show that a point which is differentially ordinary is homotopically ordinary. To that end let

$$x_i = x_i(a_1, \cdots, a_n, t) = x_i(a, t) \qquad (i = 1, 2, \cdots, n)$$

be the trajectory defined by the differential equation

$$\frac{dx_i}{dt} = -F_{x_i}$$

with the initial conditions $x_i(a, 0) = a_i$. Upon replacing the point (a) by the point $[x(a, t)]$ we obtain an F-deformation of the points (a) neighboring any ordinary point (a^0) of R, provided t be restricted to a sufficiently small interval $(0, \tau)$. With the aid of this F-deformation it is seen that each point of R which is differentially ordinary is homotopically ordinary. Hence a homotopic critical point of $F(x)$ is a differential critical point. The converse is not true as the differential critical point $x = 0$ of the function x^3 proves.

The preceding section has affirmed the existence of cycle-caps. The existence of cycle-caps implies the existence of semi-critical points in accordance with the following fundamental theorem.

THEOREM 9.1. *Corresponding to each cycle-cap u there is at least one homotopic semi-critical point p at the level $a(u)$ on each carrier of u.*

We shall prove this theorem with the aid of two lemmas. The first lemma is as follows.

LEMMA 9.1. *Let D be an F-deformation of an e-neighborhood p_e of a point p. Let A be a compact set. There exists an F-deformation D' of A which is identical with D for points initially on $p_{e/3}$.*

The deformation D' of the lemma can be defined as follows. Suppose that the given deformation D is defined for a time interval $(0, \tau)$. For the deformation D' the time t shall vary on the same interval $(0, \tau)$. Points of A initially on $p_{e/3}$ shall be deformed in D' as in D, while points of A not on $p_{2e/3}$ shall be held fast during D'. For points q of A for which

$$\frac{e}{3} < qp \leqq \frac{2e}{3} \tag{9.1}$$

we define D' as follows. Let t_q divide the time interval $(0, \tau)$ in the ratio inverse to that in which qp divides the interval $(e/3, 2e/3)$. In the deformation D', points q of A which satisfy (9.1) initially shall be deformed as in D until the time t reaches t_q and shall be held fast thereafter. It follows that D' deforms points initially on A continuously. Further A admits a displacement function under D', namely the displacement function belonging to the intersection of A with $p_{2e/3}$ under the deformation D. The deformation D' is thus an F-deformation of A, and the proof of the lemma is complete.

Our second lemma is as follows.

LEMMA 9.2. *If A is a compact set on $F \leqq a$ which contains no semi-critical points at the level a there exists an F-deformation Δ of A onto$_o$ F_a.*

If there are no points on A in limiting pairs of the form (p, a) the set A lies on$_o$ F_a and we can take Δ as the identity.

If (p, a) is a limiting pair with p on A, (p, a) is homotopically ordinary by hypothesis. There accordingly exists an F-deformation D_p of a spherical neighborhood V_p of p on $F \leqq a$ onto $F \leqq a - e_p$ where e_p is a positive constant. Let U_p and W_p be respectively spherical neighborhoods of p with radii one-third and one-sixth that of V_p. The set π of points p on A in limiting pairs (p, a) is closed and hence compact. There accordingly exists a finite set of the neighborhoods W_p which cover π. Let W_{p_i}, $i = 1, \cdots, n$, be a set of such neighborhoods.

Let B_1 be an F-deformation of A which deforms the points of A initially on U_{p_1} as in D_{p_1}. Such a deformation exists by virtue of the preceding lemma. Let $G_1 = A$. Proceeding inductively with $i > 1$ let G_i be the final image of G_{i-1} under B_{i-1} and let B_i be an F-deformation of G_i which deforms the points of G_i initially on U_{p_i} as in D_{p_i}. Such a deformation exists by virtue of the preceding lemma.

We introduce the product deformation

$$\Delta = B_n \cdots B_1, \tag{9.2}$$

applying this deformation to points of A. We understand thereby that a point q of A is first deformed under B_1 into a final image $B_1 q$. The point $B_1 q$ is then deformed under B_2 into a final image $B_2 B_1 q$. The deformation Δ is the result of the application of the deformations $B_1, \cdots, B_n$ in this order. It is clear that Δ deforms points of A continuously.

We shall show that Δ is an F-deformation. To that end let $\delta_i(e)$ be the displacement function for G_i under B_i. Let q be an antecedent of r under Δ on a trajectory λ whose initial point is p. If $qr > e$ at least one of the deformations B_i must have displaced an image of p on λ between q and r, a distance greater than e/n. Hence

$$F(q) - F(r) > \min \delta_i\left(\frac{e}{n}\right) \qquad (i = 1, \cdots, n). \tag{9.3}$$

The deformation Δ thus admits a displacement function. We denote this displacement function by $\delta(e)$.

Let ρ be the minimum of the radii of the neighborhoods W_{p_i} and let β be the minimum of the numbers e_{p_i} and $\delta(\rho)$. We continue with a proof of the following statement.

(i). *The set A is F-deformed under Δ onto$_0$ F_a.*

It will be sufficient to show that the final images on W_{p_i} of points of A lie on $F \leqq a - \beta$. To that end we need only prove that the final images on W_{p_i} of points of G_i under $B_n \cdots B_i = \Delta_i$ lie on $F \leqq a - \beta$. Points of G_i will be divided into two classes.

Class I. Points on U_{p_i}.

Class II. Points not on U_{p_i}.

Points of G_i in Class I are deformed onto $F \leqq a - e_p \leqq a - \beta$ under B_i and hence under Δ_i. Points of G_i in Class II whose final images ζ under Δ_i are on W_{p_i} are displaced under Δ_i a distance at least ρ so that F is thereby decreased at least $\delta(\rho)$. These points ζ thus lie on $F \leqq a - \beta$, and the proof of statement (i) is complete.

Lemma 9.2 follows from statement (i) and the theorem follows from Lemma 9.2.

Spaces locally F-connected. We shall give a brief description of a special hypothesis of interest. The space M will be said to be *locally F-connected* for the order n if corresponding to n, an arbitrary point p on M, and an arbitrary positive constant e, there exists a positive constant δ with the following property. For $c \geqq F(p)$ any singular n-sphere on $F \leqq c$ (the continuous image on $F \leqq c$ of an ordinary n-sphere) on the δ-neighborhood p_δ of p is the boundary of a singular $(n + 1)$-cell on $F \leqq c + e$ and on p_e. The hypothesis of local F-connectedness is not implied by the ordinary form of local connectedness at least if F is merely lower semi-continuous.

In the locally convex case of our abstract variational theory we shall be dealing with a space Ω and function F defined on Ω such that Ω is locally F-connected. In this application the values $F = 1$ will correspond to curves of infinite length.

Let c be any constant less than 1 and d be any constant greater than c. Let $R_k(c, d)$ denote the number of k-cycles in a maximal linear set of k-cycles on $F \leqq c$ not homologous to zero on $F \leqq d$. We state the following theorem.

THEOREM 9.2. *If the space M is locally F-connected for orders at most $k + 1$, the numbers $R_k(c, d)$ are finite for $c < 1$ and the inferior cycle limits of non-bounding k-cycles cluster at most at $c = 1$.*

The proof of the first statement of the theorem while not difficult involves too great detail to be presented here. The essential ideas are those in analogous proofs in Lefschetz, ref. 3. The inferior cycle limits of non-bounding k-cycles can have no cluster value $b < 1$. Otherwise corresponding to any constant c such that $b < c < 1$ there would exist an infinite sequence u^n of non-bounding k-cycles on $F \leqq c$ with distinct inferior cycle limits less than c. The ranks of the cycles u^n would be different so that the cycles u^n would generate a group of non-bounding k-cycles of infinite dimensions. It would follow that $R_k(c, 1)$ would be infinite contrary to the first statement in the theorem.

The proof of the theorem is complete.

10. **Homotopic critical sets.**

We are concerned in this section with homotopic critical points. With this understood the word homotopic will ordinarily be omitted.

By a *complete critical set* ω at the level c is meant the set of all semi-critical points at the level c. It follows from the definition of a semi-critical point that ω is closed. The set ω is compact if $c < 1$ since the domain $F \leqq c$ is then compact and ω is a closed subset of this domain. Recall that the components of a point set A are the maximal connected subsets of A and are closed in A. The components of A are disjunct and sum to A. We shall term any sum of components of ω a *critical set* σ at the level c.

If σ is at a positive distance from $\omega - \sigma$, σ will be termed *separate*. A neighborhood of σ which is at a positive distance from $\omega - \sigma$ will also be termed separate.

Let R be a subset of M. If we regard R as a space M, waiving the condition that the subspace $F \leqq c$ of R be compact when $c < 1$, a k-cap on R has a new meaning dependent on R and will be specially designated as a *k-cap rel R*.

The following theorem is useful.

THEOREM 10.1. *Let σ be a critical set at the level c with a separate neighborhood U.*

(a). *If u is a k-cap with cap limit c rel U, u is a k-cap rel M.*

(b). *The k-cap u is c-homologous on U to a k-cap v on an arbitrarily small neighborhood of σ.*

We shall prove Theorem 10.1 with the aid of two lemmas, the first of which follows. In these lemmas the e- or η-neighborhood of a point set X will be denoted by X_e or X_η respectively.

LEMMA 10.1. (i). *Let U be a separate neighborhood of a critical set σ at the level c. Let b be a constant less than c, and η an arbitrary positive constant. A k-cap u with cap limit c and with carrier κ on $U + F_b$ is c-homologous on κ to a k-cap on U_η.*

(ii). *Let the complete critical set ω at the level c be the sum of critical sets $\sigma^i (i = 1, \cdots, n)$ with neighborhoods U^i at positive distances from one another. Let u^i be a k-cycle mod$_o$ $F < c$ on $F \leqq c$ and U^i, such that the sum $u^1 + \cdots + u^n = u$ is c-homologous to 0 on $F_b + \Sigma U^i$. Then u^i is c-homologous to 0 on U^i_η for each i.*

We shall first prove (i).

Let u_p be the pth component of u, and H_p the relative algebraic homology connecting u_p and u_{p+1} in the definition of u. Of the cells on the complement of U we drop each k-cell of u_p and each $(k + 1)$-cell in H_p. Let u'_p and H'_p be the k-chain and relative homology which thereby replace u_p and H_p respectively. The sequence u'_p together with the homologies H'_p defines a k-cap u', c-homologous to u on κ. If m is a sufficiently large integer the subsequence u'_p for which $p \geqq m$ together with the corresponding homologies H'_p will define a k-cap z on U_η. Moreover z and u' and hence z and u will be c-homologous on κ.

The proof of (i) is complete.

We come to the proof of (ii).

We set $\Sigma U^i = V$ and let e be an arbitrary positive constant less than one-half the minimum distance between pairs of neighborhoods U^i. By hypothesis v is c-homologous to 0 on $V + F_b$. From the algebraic homologies which define this c-homology we drod all $(k + 1)$-cells which are on the complement of V and all algebraic homologies whose norms are greater than $e/2$. The resulting algebraic homologies will hold on $V_{e/2}$ and we infer that u is c-homologous to 0 on V_e. It then follows from the choice of e that the cycle u^i is c-homologous to 0 on the corresponding neighborhood U^i_e.

The proof of (ii) is complete.

We come to a deformation lemma.

LEMMA 10.2. *Let κ be a compact subset of $F \leqq c$ on a separate neighborhood of a critical set σ at the level c. Corresponding to an arbitrary positive constant e there exists a positive constant $b < c$ and an F-deformation T of κ onto $\sigma_e + F_b$ in which points of κ which have been displaced a distance at least e lie on F_b.*

Set $e = 3\rho$. Let κ_1 and κ_2 be respectively the subsets of κ not on σ_ρ and $\sigma_{2\rho}$.

There are no semi-critical points on κ_1 at the level c. It follows from Lemma 9.2 that there exists an F-deformation Δ of κ_1 onto $F < c - \eta$ where $\eta > 0$. The methods used in the proof of Lemma 9.1 make it clear that the deformation Δ properly modified will yield an F-deformation T of κ in which $\kappa - \kappa_1$ is held fast while κ_2 is deformed as in Δ. Let $\delta(e)$ be the displacement function for κ under T.

Points of κ are on κ_2 or $\sigma_{2\rho}$. Points of κ on κ_2 are F-deformed under T onto $F < c - \eta$. Points of κ on $\sigma_{2\rho}$ are F-deformed on $\sigma_{3\rho} = \sigma_e$ or else displaced a distance at least ρ, and hence deformed onto $F \leqq c - \delta(\rho)$. Under T points of κ which have been displaced a distance at least e will lie on $F \leqq c - \delta(e)$. If b is less than c and greater than $c - \eta$, $c - \delta(\rho)$ and $c - \delta(e)$ the lemma holds as stated.

We shall now prove Theorem 10.1 beginning with a proof of (a).

Let θ be a carrier of u on U and $F \leqq c$. We assume (a) false. It follows that u is c-homologous to 0 with homology carrier $\kappa \supset \theta$ on $F \leqq c$. Corresponding to the complete critical set ω, and to the compact set κ Lemma 10.2 affirms the existence of a constant $b < c$ and an F-deformation T of κ onto $\omega_e + F_b$. Under T, θ is deformed on $\theta_e + F_b$. Hence u is c-homologous to 0 on $\omega_e + \theta_e + F_b$.

We distinguish between the cases $\omega = \sigma$ and $\omega \neq \sigma$. If $\omega \neq \sigma$ we apply Lemma 10.1 (ii) with $n = 2$, setting

$$\sigma^1 = \sigma, \quad \sigma^2 = \omega - \sigma, \quad u^1 = u, \quad u^2 = 0, \quad U^1 = \sigma_e^1 + \theta_e, \quad U^2 = \sigma_e^2,$$

taking e so small that $\bar{U}^1 \subset U$, and U^2 is at a positive distance from U^1. We see that u is c-homologous to 0 on $F_b + \Sigma U^i$ and infer that u is c-homologous to 0 on U. If $\omega = \sigma$ we again apply Lemma 10.1 (ii) with $n = 1$. In both cases we conclude that u is c-homologous to 0 on U.

From this contradiction we infer the truth of (a).

We turn to a proof of (b).

Let e be so small a positive constant that σ_{2e} lies on U. The k-cap u lies on U and it follows from Lemma 10.2 that u is c-homologous to a k-cap w on $\sigma_e + F_b$. Since σ_e is a separate neighborhood of σ it follows from Lemma 10.1 (i) that w is c-homologous to a k-cap v on σ_{2e} (setting $\eta = e$). Thus $u - v$ is c-homologous to 0. But from our choice of e, $u - v$ is on U. It follows from (a) that $u - v$ is c-homologous to 0 on U. But v is on σ_{2e} and e may be arbitrarily small.

Statement (b) follows directly.

A k-cap u with cap limit c will be said to be *associated* with a critical set σ if there is a k-cap in the a-class of u with a carrier on an arbitrary neighborhood of σ on $F \leqq c$. With this understood we state the following corollary.

COROLLARY 10.1 *If σ is a critical set at the level c with a separate neighborhood U, a maximal group g of k-caps rel U with cap limits c is a maximal group of k-caps associated with σ.*

It follows from (a) in Theorem 10.1 that each k-cap u of g is a k-cap rel M

and from (b) that u is "associated" with σ. Thus g is a group of k-caps associated with σ. That g is maximal among k-caps v associated with σ is seen as follows. The k-cap v is in the a-class of a k-cap u rel U by virtue of its association with σ. But g is maximal among k-caps rel U with cap limit c, so that u and hence v is in the a-class of a k-cap of g. The corollary thus holds as stated.

It follows from Theorem 5.1 that any two maximal groups g of k-caps with cap limit c are a-isomorphic. The following theorem adds an important mode of determination of a particular maximal group of k-caps.

THEOREM 10.2. *Let the complete critical set ω at the level c be the sum of critical sets σ^i $(i = 1, \cdots, n)$ with neighborhoods V^i at positive distances from one another and let g^i be a maximal group of k-caps rel V^i with cap limit c. The groups g^i admit a direct sum g and this sum is a maximal group of k-caps with cap limit c.*

We begin by proving the following statement.

(α). *If v^i is an element of g^i the sum $v = v^1 + \cdots + v^n$ is c-homologous to* 0 *only if for each i, v^i is c-homologous to* 0 *on V^i.*

If v is c-homologous to 0 it follows from (a) Theorem 10.1 that v is c-homologous to 0 on ΣV^i. Since the neighborhoods V^i are at positive distances from one another we infer that v^i is c-homologous to 0 on V^i, and the proof of (α) is complete.

It remains to prove statement (β).

(β). *A k-cap v with cap limit c is c-homologous to a sum $v^1 + \cdots + v^n$ of elements v^i in the respective groups g^i.*

It follows from Theorem 10.1 (b) that v is c-homologous to a k-cap on an arbitrarily small neighborhood of ω and hence to a sum of cycles u^i mod$_0$ $F < c$ on $F \leqq c$ and on the respective neighborhoods V^i. But each such cycle u^i is c-homologous on V^i to an element v^i of g^i in accordance with the definition of g^i, and (β) follows immediately.

Theorem 10.2 follows from (α) and (β).

The dimension of a maximal group of k-caps associated with a critical set σ will be termed the kth *type number* of σ. It follows from Corollary 10.1 that the kth type number of a separate critical set σ is the dimension of a maximal group of k-caps with cap limit c relative to a separate neighborhood of σ. We thus have the following corollary of Theorem 10.2.

COROLLARY 10.2. *The kth type number of the complete critical set ω is the sum of the kth type numbers of the critical sets $\sigma^1, \cdots, \sigma^n$.*

A k-cap u with cap limit c which is associated with a critical set σ will be said to *belong* to σ if u is associated with no proper critical subset of σ. The basic theorem here is as follows.

THEOREM 10.3. *A k-cap u with a cap limit $c < 1$ belongs to at least one critical set at the level c.*

We shall not use this theorem and accordingly omit its proof.

Non-degenerate differential critical points. We shall consider the important case in which M is compact and in which there is a neighborhood of each point

p of M homeomorphic with a region U of a euclidean n-space of rectangular coördinates (x). Neighboring any point q of U we admit any system of coördinates obtainable from the coördinates (x) by a non-singular transformation $z_i = z_i(x)$ with $z_i(x)$ of class C^2, and we admit only such coördinate systems neighboring q. A space M of this character will be said to be *locally of class* C^2. We suppose that our function F equals a function $\varphi(x)$ of class C^2 corresponding to each admissible set of coördinates (x).

A differential critical point (x_0) of $\varphi(x)$ will be termed a *differential critical point* of F. The point (x_0) will be termed degenerate if the Hessian of φ at (x_0) is zero. We *assume* here that the differential critical points of F are non-degenerate. As in the introduction the index of a differential critical point (x_0) shall be the *index* of the quadratic form whose coefficients make up the Hessian of φ at (x_0). From the hypothesis of non-degeneracy it follows that the differential critical points are isolated, and hence finite in number.

As in §9 the trajectories orthogonal to the manifolds φ constant in the local coördinate spaces (x) will serve to prove that homotopic critical points are differential critical points. A proof of the following theorem is more difficult.

THEOREM 10.4. *A non-degenerate differential critical point σ is a homotopic critical set and the jth type number of σ equals the Kronecker delta δ_k^j where k is the index of σ.*

We suppose that σ is represented by the point $(x) = (0)$ in the coördinate system (x) and that in this system $F \equiv \varphi(x)$. We also suppose that $\varphi(0) = 0$, waiving the condition that $0 \leqq F \leqq 1$. Setting

$$f(x) = -x_1^2 - \cdots - x_k^2 + x_{k+1}^2 + \cdots + x_n^2,$$

we introduce a preliminary transformation as follows:

(a). *If r is a sufficiently small positive constant the space A defined by the conditions $\varphi \leqq 0$ and $x_i x_i < r^2$, and the space B defined by the conditions $f \leqq 0$ and $x_i x_i < r^2$ admit a homeomorphism in which the subspaces $f < 0$ and $\varphi < 0$ correspond.*

It follows from methods developed in detail on pp. 169–170 of M1 that the space $\varphi \leqq 0$, $x_i x_i = r^2$ and the space $f \leqq 0$, $x_i x_i = r^2$ admit a homeomorphism in which the subspaces $f < 0$ and $\varphi < 0$ correspond. A use of the radial trajectories of pp. 156–157 of M1 then gives an easy proof of (a).

We shall make use of Corollary 10.1, determining the jth type number of σ, if such a number exists, by determining the dimension μ of a maximal group of j-caps with cap limit 0 relative to a separate neighborhood of σ on $\varphi \leqq 0$. The space A is such a neighborhood if r is sufficiently small, since each homotopic critical point is a differential critical point and the only differential critical point on A at the level 0 is σ if r is sufficiently small. It follows from (a) that μ equals the dimension of a maximal group of j-caps with cap limit 0 rel B with f replacing φ. We continue, replacing A by B and φ by f.

If $k = 0$, B consists of the point σ, and the theorem is immediate. We therefore suppose that $k > 0$. When $k > 0$ we shall prove the following:

(b). *The homology groups of the subspace* $f < 0$ *of* B *are isomorphic with those of the* $(k - 1)$*-sphere.*

(c). *The only non-null connectivity of* B *is* $R_0 = 1$.

(d). *Each* j*-cap* u *rel* B *with* $a(u) = 0$, *is non-linkable.*

(e). *The only* j*-caps* u *rel* B *with* $a(u) = 0$ *are* k*-dimensional, and a maximal group of such* k*-caps has the dimension* 1.

The space B and its subspace $f < 0$ can be respectively deformed on themselves onto the disc

$$x_1^2 + \cdots + x_k^2 \leq \frac{r^2}{2}, \qquad x_{k+1} = \cdots = x_n = 0 \tag{10.1}$$

and its spherical boundary S, holding the disc and S fast respectively. The truth of (b) and (c) follows.

Proof of (d). If there were a linkable j-cap e rel B with $a(u) = 0$ there would be a j-cycle v in the a-class of u. But any j-cycle on B is homologous on B to a cycle on S, and $f = -r^2/2$ on S. Hence v cannot be a j-cap with cap limit 0 and u cannot be linkable.

Proof of (e). If u is a j-cap rel B with $a(u) = 0$, u is non-linkable. By definition then, βu is on $f < 0$ and is there non-bounding. It follows from (*b*) that $j = 1$ or k. If $k = 1$, the first statement in (*e*) is true. If $k > 1$, I say that $j = k$. For if j were 1 it would follow from the connectedness of the subspace $f < 0$ of B when $k > 1$ that $\beta u \sim 0$ on$_o$ this subspace, contrary to the fact that u is non-linkable. Hence $j = k$ in all cases and the proof of the first statement in (e) is complete.

To prove the second statement in (e) recall that the subspace $f < 0$ of B contains a $(k - 1)$-cycle u which is non-bounding on$_o$ this subspace but bounding on B. Hence there is at least one k-cap u rel B with $a(u) = 0$.

A maximal set of k-caps rel B with $a(u) = 0$ will contain at most one k-cap. To prove this let u and v be two such k-caps. It follows from (*b*) that there exists a proper homology of the form

$$\delta_1 \beta u + \delta_2 \beta v \sim 0 \qquad (\text{on}_o\ f < 0). \tag{10.2}$$

The relative k-cycle $\delta_1 u + \delta_2 v$ cannot be a k-cap at the level 0 since it would then be linkable in accordance with (10.2). Hence $\delta_1 u + \delta_2 v$ is not a k-cap at the level 0, and the proof of (e) is complete.

We can conclude that σ is a homotopic critical point and set at the level 0. For no other point on A at the level 0 is a homotopic critical point as noted previously. But 0 is a cap limit rel A as the preceding paragraphs show. It follows from Theorem 9.1 that σ is a homotopic critical point. Since σ is an isolated critical point it must also be a critical set, and the proof of the theorem is complete.

A compact space M locally of class C^2 has finite connectivities (cf. Lefschetz, *Deformations*, l.c.), and if M is locally n-dimensional the connectivities of order greater than n are null. If F has at most non-degenerate critical points the number of critical points is finite and it follows from Theorems 10.4, 10.2 and 5.3 that a maximal group of k-caps will have a dimension m_k equal to the number of critical points of index k. In particular $m_j = 0$ if $j > n$. Referring to the relations (8.12) we have the following theorem.

THEOREM 10.5. *Let F be a function of class C^2 on an n-dimensional compact space M locally of class C^2. If the differential critical points of F are non-degenerate the numbers p_k of critical points with index k and the connectivities R_j of M satisfy the relations* (0.2) *of the introduction.*

Essential critical points. A homotopic critical set σ to which at least one k-cap belongs will be termed *homotopically essential* and in such a case a semi-critical point of σ will be termed *homotopically essential.* The class of homotopic semi-critical points B thus has a subclass C of points which are termed essential. Thus $B \supset C$. That the class C may be a proper subclass of B is shown by the following example.

Example 10.2. Let M be a curve $y = f(x)$ in the (x, y)-plane defined as follows:

$$f(x) = \sin x \qquad (-\pi \leqq x \leqq 0),$$

$$f(x) = 0 \qquad (0 \leqq x \leqq \pi),$$

$$f(x) = \sin x \qquad (\pi \leqq x \leqq 2\pi).$$

On M we set $F = (y + 1)/2$ so that $0 \leqq F \leqq 1$. The points of the curve at which $0 \leqq x \leqq \pi$ are homotopically critical and form a critical set but not an essential critical set.

We have already seen that the class A of differential critical points in problems where F is differentiable is such that $A \supset B$. We thus have the relation

$$A \supset B \supset C, \tag{10.3}$$

but B may be a proper subset of A and C a proper subset of B.

IV. ABSTRACT VARIATIONAL THEORY

11. Length. Let N be a space of points $p, q, r, \cdots$ with a metric which in general is not symmetric. That is to each ordered pair p, q of points of N there shall correspond a number denoted by pq and called the distance from p to q. These distances shall satisfy the conditions:

I. $pp = 0$,

II. $pq > 0$ if $p \neq q$,

III. $pq \leqq pr + rq$.

The relation III is not to be written $pq \leqq pr + qr$ inasmuch as this would imply that $pq = qp$ upon setting $r = p$. The spaces N satisfying I–III include those with a symmetric distance. For various applications, for instance to the calculus of variations, it is not desirable to restrict ourselves to the symmetric case. The non-symmetric case is illustrated by the integral

$$J = \int (x\dot{y} + \sqrt{\dot{x} + \dot{y}})\, dt \qquad (0 < x < 1). \tag{11.1}$$

The value of J along a curve in general is changed when the sense of integration along the curve is reversed. Curves which are extremals in one sense need not be extremals in the other sense. The absolute minimum of J from one point to another is a distance which satisfies I–III but which is not in general symmetric.

Let $| pq |$ denote the maximum of pq and qp. We shall call $| pq |$ the *absolute distance* between p and q. With $| pq |$ as a distance N becomes a metric space with a symmetric distance function. We shall denote this new space by $| N |$. We shall use the metric of N when defining length. For all other purposes, in particular in defining point functions on N or $| N |$, or general topological properties, we shall use the metric of $| N |$. One reason for this is that pq does necessarily approach zero when qp approaches zero, unless indeed we add further conditions to the conditions I–III.

We understand that a function $f(p_1, \cdots, p_n)$ of points $(p_1, \cdots, p_n)$ on $| N |$ is continuous at $(q_1, \cdots, q_n)$ if $f(p_1, \cdots, p_n)$ tends to $f(q_1, \cdots, q_n)$ as a limit as max $| p_i q_i |$ tends to zero. This is the familiar concept of continuity on $| N |$. A function is called continuous on a point set A if continuous at each point of A. We note the following:

(a). *The function pq is a uniformly continuous function of p and q on $| N |$.*

Postulate III implies that for points p, q, p', q',

$$pq \leqq pp' + p'q' + q'q. \tag{11.2}$$

If we assume that $| pp' | < \delta$ and $| qq' | < \delta$ then (11.2) implies that $pq < p'q' + 2\delta$. Upon interchanging p and p', q and q' we find that pq and $p'q'$ differ by less than 2δ so that (*a*) is true.

For each value of t on a closed interval $(0, a)$ let $q(t)$ be a point on $| N |$. The function $q(t)$ is termed *continuous* at t if the absolute distance $| q(t') \, q(t) |$ tends to zero as t' tends to t. If $q(t)$ is continuous at each point t on $(0, a)$ the set of points $q(t)$ taken in the order of the corresponding values of t will be termed a parameterized curve λ (written p-curve λ). We say that λ is the continuous image on $| N |$ of the line segment $(0, a)$. In general curves on $| N |$ will be denoted by Greek letters $\alpha, \beta, \eta, \lambda$, etc., while points on $| N |$ will be denoted by letters p, q, r, etc.

The length of λ. Let t_i be a set of values on $(0, a)$ such that $0 = t_0 < t_1 < \cdots < t_n = a$, and let $p_0, p_1, \cdots, p_n$ be the corresponding points on λ. The sets t_i and p_i will be termed *partitions* of λ of *norm* δ where δ is the maximum of the differences $t_{i+1} - t_i$. We term

$$\Sigma = \sum_{i=0}^{n-1} p_i p_{i+1},$$

a sum *approximating the length* of λ and take the length L of λ as the least upper bound of the sums Σ for all partitions of λ. The proof of the following theorem can be readily supplied by the reader familiar with the ordinary theory of length in euclidean spaces. Explicit use of course must be made of postulate III.

THEOREM 11.1. *The length of λ is the limit of any sequence Σ_n of approximating sums Σ corresponding respectively to partitions of λ whose norms tend to zero as n becomes infinite.*

Fréchet distance. Let η and ζ be two p-curves given by the respective equations

$$p = p(t) \qquad (0 \leqq t \leqq a), \tag{11.3}$$

$$q = q(u) \qquad (0 \leqq u \leqq b). \tag{11.4}$$

Let H be a sense preserving homeomorphism between the closed intervals $(0, a)$ and $(0, b)$. The homeomorphism between η and ζ defined by H will be termed *admissible*. Set

$$d(H) = \max \mid p(t)\, q(u) \mid,$$

where t and u correspond under H. Then the Fréchet distance $\eta\zeta$ between η and ζ is the greatest lower bound of the numbers $d(H)$ as H ranges over all admissible homeomorphisms between η and ζ.

We observe that $\eta\zeta = \zeta\eta \geqq 0$. If moreover η, ζ, λ are three p-curves,

$$\eta\lambda \leqq \eta\zeta + \zeta\lambda. \tag{11.5}$$

To establish (11.5) let H be an admissible homeomorphism between η and ζ under which a point p of η corresponds to a point q of ζ. Let K be an admissible homeomorphism between ζ and λ in which q of ζ corresponds to r of λ. Let KH be the resultant transformation of η into λ carrying p of η into r of λ. From the relation $\mid pr \mid \leqq \mid pq \mid + \mid qr \mid$ we find that

$$d(KH) \leqq d(H) + d(K). \tag{11.6}$$

Relation (11.5) follows from (11.6).

By the *space* of the *p-curves* will be meant a space in which the elements are p-curves and the distance between elements the Fréchet distance.

We continue with a proof of a well-known theorem.

THEOREM 11.2. *The length of a p-curve η is a lower semi-continuous function of η in the space of p-curves η.*

We are to show that if $a < L(\eta)$ there exists a positive δ such that $L(\zeta) > a$ whenever $\eta\zeta < \delta$.

Let b be a constant between a and $L(\eta)$. By virtue of the definition of L there exists a partition $p_0, p_1, \cdots, p_n$ of η such that

$$\sum p_i p_{i+1} > b \qquad (i = 0, 1, \cdots, n - 1). \tag{11.7}$$

We choose δ so that $2n\delta < b - a$. If $\eta\zeta < \delta$ there exists a homeomorphism T between η and ζ such that corresponding points have an absolute distance less than δ. In particular if q_i is the correspondent of p_i under T the points q_i determine a partition of ζ such that $| p_i q_i | < \delta$ and hence

$$q_i q_{i+1} > p_i p_{i+1} - 2\delta \qquad (i = 0, 1, \cdots, n - 1). \tag{11.8}$$

With the aid of (11.8) we see that

$$L(\zeta) \geqq \sum q_i q_{i+1} > \sum p_i p_{i+1} - 2n\delta.$$

It follows from (11.7) and our choice of δ that

$$L(\zeta) \geqq \sum q_i q_{i+1} > b - (b - a) = a,$$

and the proof is complete.

If $p_1, \cdots, p_n$ is an arbitrary sequence of points on $| N |$ it follows from postulate III that

$$p_1 p_n \leqq p_1 p_2 + p_2 p_3 + \cdots + p_{n-1} p_n .$$

This inequality together with the definition of length leads to the following theorem.

THEOREM 11.3. *The length of a p-curve joining two points p and q on* $| N |$ *is at least pq.*

μ-Length. Let η and ζ be the p-curves given by (11.3) and (11.4) respectively. We shall say that ζ is *derivable from* η if there exists a continuous non-decreasing function $u = u(t)$ which maps the closed interval $(0, a)$ onto the closed interval $(0, b)$ and under which $p(t) \equiv q[u(t)]$. Two p-curves which are derivable from the same p-curve η will be said to be *p-equivalent.* The two p-curves η and ζ will be said to be *identical,* $\eta \equiv \zeta$, if in the representations (11.3) and (11.4) $a = b$ and $p(t) \equiv q(t)$. To show that p-equivalence is transitive it will be sufficient to verify the truth of the following statement.

(a). *Corresponding to each arbitrary p-curve η there exists a p-curve φ_η from which η is derivable such that $\varphi_\alpha \equiv \varphi_\beta$ whenever the p-curves α and β are p-equivalent.*

If a p-curve of the form (11.3) has finite length, the length $s(t)$ of η from the point $t = 0$ to a general point t is a continuous non-decreasing function of t. If $t = t(s)$ is the function inverse to $s(t)$, $p[t(s)]$ is a single-valued continuous point function of s for s on the interval $0 \leqq s \leqq s(a)$. We set $q(s) \equiv p[t(s)]$. The curve $q = p(t)$ is derivable from the curve $q = q(s)$ upon setting $s = s(t)$. We term $q = q(s)$ the s-curve *determined* by η. If ζ is derivable from η it is clear that the s-curve determined by ζ is identical with the s-curve determined by η. For curves with finite length the curve φ_η corresponding to η in statement (a) could be taken as the s-curve determined by η. It follows that p-equivalence is transitive for p-curves of finite length.

But for curves with infinite length there are no s-curves. Nevertheless it is possible to replace length s by what will be termed μ-length and to associate with each p-curve η a p-curve φ_η such that statement (a) is satisfied without exception. The parameter of these new curves will be denoted by μ and the curves themselves will be termed μ-curves.[13] These μ-curves will not only serve as the p-curves φ_η of (a) but will have other important properties as follows.

(b) *The parameter μ on a p-curve φ_η varies from 0 to a value $\mu(\eta)$. The function $\mu(\eta)$ is continuous in η in the space of p-curves η and never exceeds the diameter of η.*

(c). *The condition that η and ζ be p-equivalent is equivalent to the condition that the μ-curves φ_η and φ_ζ be identical and to the condition $\eta\zeta = 0$.*

(d). *Let α be a parameter which ranges from 0 to 1 inclusive. Let the point on φ_λ at which $\mu = \alpha\mu(\lambda)$ be denoted by $q(\lambda, \alpha)$. Then $q(\lambda, \alpha)$ is continuous in its arguments.* See ref. 16. Page 441.

The following is a particular consequence of (d).

(e). *Corresponding to a preassigned (fixed) p-curve η and an arbitrary positive e there exists a positive δ such that if $\eta\delta < \delta$, points on φ_η and φ_ζ which divide φ_η and φ_ζ in the same ratio with respect to variation of μ have an absolute distance at most e.*

12. **The space Ω.** A class of p-equivalent p-curves will be termed a *curve*. The p-curves of a given class λ have a 0-distance from each other. The distance $\lambda\tau$ between two curves λ, τ shall be defined as the distance $\eta\zeta$ between any two p-curves η and ζ in the respective classes λ and τ. It is clear that the distance $\lambda\tau$ is independent of the curves η and ζ chosen to represent λ and τ respectively. It follows from property (c) of μ-curves in §11 that $\lambda\tau = 0$ if and only if the classes λ and τ are identical. Curves $\lambda, \tau, \omega, \cdots$ accordingly form a symmetric metric space M of the same character as the space M of §1.

Let a and b be two points on $|N|$. The set of curves joining a to b will form a subspace $\Omega(a, b)$ of the space M of all curves on $|N|$. We shall investigate the properties of $\Omega(a, b)$.

Recall that a subset A of a metric space is termed complete if every Cauchy sequence in A converges to a point in A, and conditionally compact if every infinite sequence of A contains a Cauchy subsequence. (See Hausdorff, l.c.) If A is conditionally compact and complete A is compact.

Apart from trivial cases where $a = b$, or the set $\Omega(a, b)$ is empty, $\Omega(a, b)$ is not conditionally compact. In fact if $\Omega(a, b)$ contains a curve λ it is possible to define an infinite sequence of curves λ_n each of which joins a to b and is defined by a point $p(t)$ which varies continuously on λ but is such that the sequence λ_n contains no Cauchy subsequence.

We have denoted the length of a p-curve by L. We note that the value of L will be the same for each p-curve of a class λ of p-curves which are p-equivalent.

[13] Morse, M., *A special parameterization of curves*, Bulletin of the American Mathematical Society, 42 (1936). This paper contains a proof of statements (a), (b), (c), (d) and a definition of μ-curves. This definition is a modification of a definition of a function of sets given by H. Whitney and applied by Whitney to families of simple non-intersecting curves. See Whitney, H., *Regular families of curves*, Annals of Mathematics, 34 (1933), pp. 244–270.

This common value of L will be termed the length of λ and again denoted by L. We understand that L may be infinite.

We shall prove the following theorem. See Fréchet ref. 17, page 441.

THEOREM 12.1. *If* $|N|$ *is compact and* c *is a finite constant, the subspace* $L \leqq c$ *of* $\Omega(a, b)$ *is compact.*

We shall begin with a proof of statement (i).

(i). *If* $|N|$ *is complete,* $\Omega(a, b)$ *is complete.*

Let λ_1, λ_2, be a Cauchy sequence Λ of curves of $\Omega(a, b)$. Given $e > 0$, there exists an integer N so large that $\lambda_i\lambda_j < e$ whenever $i > N$ and $j > N$. We choose a subsequence of Λ as follows. Let η_1 be chosen from Λ so that the distance of η_1 from every curve λ_n following η_1 in Λ is less than $\frac{1}{2}$. Proceeding inductively, for $i > 1$ choose η_i from curves λ_j following η_{i-1} in Λ so that if λ_m follows η_i in Λ,

$$\eta_i\lambda_m < 2^{-i}.$$

The sequence $\eta_1, \eta_2, \cdots$ is a subsequence of Λ with the property that

$$\eta_i \eta_{i+1} < 2^{-i}. \tag{12.1}$$

Let $\tau_1, \tau_2, \cdots$ be a sequence of homeomorphisms of which τ_i is a homeomorphism between η_i and η_{i+1} in which corresponding points p_i and p_{i+1} have an absolute distance less than 2^{-i}. Such homeomorphisms exist by virtue of (12.1). For $j > i$ the homeomorphisms $\tau_i \tau_{i+1} \cdots \tau_{j-1}$ taken in the order written yield a homeomorphism τ_{ij} between η_i and η_j by virtue of which corresponding points p_i and p_j have an absolute distance

$$|p_ip_j| < \frac{1}{2^i} + \frac{1}{2^{i+1}} + \cdots + \frac{1}{2^{j-1}} < \frac{1}{2^{i-1}}. \tag{12.2}$$

Suppose η_1 is represented by the p-curve $p = p_1(t)$, $0 \leqq t \leqq 1$. By virtue of the homeomorphism τ_{1j} a p-curve $p = p_j(t)$ representing η_j is uniquely determined, in such manner that $p_i(t)$ and $p_j(t)$ correspond under τ_{ij}. Relation (12.2) now takes the form

$$|p_i(t)\, p_j(t)| < \frac{1}{2^{i-1}}. \tag{12.3}$$

It follows from the completeness of $|N|$ that the sequence of points $p_i(t)$ converges uniformly with respect to t to a continuous point limit $p(t)$. The sequence η_i thus converges to the curve represented by $p = p(t)$, and the proof of (i) is complete.

We continue with a proof of statement (ii).

(ii). *If* $|N|$ *is conditionally compact, the subspace* $L \leqq c$ *of* $\Omega(a, b)$ *is conditionally compact.*

We shall make use of the n-fold product $|N|^n$ of $|N|$. The distance between two points

$$(p) = (p_1, \cdots, p_n); \qquad (q) = (q_1, \cdots, q_n)$$

of $| N |^n$ shall be defined as the maximum of the distances $| p_i q_i |$ and shall be denoted by $(p)(q)$.

Let Λ^0 be an infinite sequence of curves $\lambda_1^0, \lambda_2^0, \cdots$ on $L \leqq c$. Let $e_1, e_2, \cdots$ be a sequence of positive constants tending to zero. Let $n_1 > 0$ be an integer so large that the division of the curve λ_i^0 into parts of equal variation of length by the ordered set of points

$$(p^i) = (p_1^i, \cdots, p_{n_1}^i) \qquad (i = 1, 2, \cdots) \tag{12.4}$$

yields segments of λ_i^0 with absolute diameters less that $e_1/3$. An integer n_1 exists since we are concerned with curves of length at most c.

Since $| N |$ is conditionally compact, $| N |^{n_1}$ is conditionally compact. The points (p^i) in the space $| N |^{n_1}$ will thus have a Cauchy subsequence (r^j), $j = 1, 2, \cdots$, and we can assume that the sequence (r^j) has been so chosen that

$$(r^i)(r^k) < \frac{e_1}{3} \qquad (i, k = 1, 2, \cdots). \tag{12.5}$$

Let λ_i^1 be the curve of Λ^0 of which (r^i) is a partition. It follows from (12.5) and from the fact that the set of points (r^i) divide λ_i^1 into segments of absolute diameter less than $e_1/3$ that

$$\lambda_j^1 \lambda_k^1 < e_1 \qquad (j, k = 1, 2, \cdots). \tag{12.6}$$

The sequence λ_i^1 has been selected from the sequence λ_i^0. We repeat this process replacing e_1 by e_2 and λ_i^0 by λ_i^1, thereby obtaining a subsequence λ_i^2 of λ_i^1. More generally we obtain a subsequence $\lambda_1^n, \lambda_2^n, \cdots$ of the sequence $\lambda_1^{n-1}, \lambda_2^{n-1}, \cdots$ of such a nature that

$$\lambda_i^n \lambda_k^n < e_n \qquad (i, k = 1, 2, \cdots).$$

The sequence

$$\lambda_1^1, \lambda_2^2, \lambda_3^3, \cdots \tag{12.7}$$

is a subsequence of Λ^0 and has the property that

$$\lambda_n^n \lambda_m^m < e_r \qquad (\text{if } n, m > r). \tag{12.8}$$

It is accordingly a Cauchy subsequence.

The proof of (ii) is complete.

We can now prove the theorem. It follows from (i) that the space $\Omega(a, b)$ is complete. The subspace $L \leqq c$ of $\Omega(a, b)$ is closed since L is lower semi-continuous. Hence this subspace is complete. The theorem follows from (ii).

Since F is lower semi-continuous Theorem 12.1 has the following corollary.

COROLLARY. *If $\Omega(a, b)$ is not empty there exists at least one curve λ of $\Omega(a, b)$ whose length (possibly infinite) is a minimum*[14] *among lengths of curves of $\Omega(a, b)$.*

[14] See Tonelli, *Fondamenti di calcolo delle Variazioni* I and II. References will here be found to the work of Hilbert on the absolute minimum.

If H is a component (in the point set sense) of $\Omega(a, b)$ there is at least one curve ζ of H whose length (possibly infinite) is a minimum among lengths of curves of H.

A class of curves of $\Omega(a, b)$ continuously deformable on $| N |$ among curves of $\Omega(a, b)$ into a given curve of $\Omega(a, b)$ will be termed a *homotopy* class. Among curves joining two fixed points a and b on a torus for example there are infinitely many homotopy classes, and on a torus these homotopy classes are identical with the components of $\Omega(a, b)$. Under the condition of local convexity to be developed in the next chapter we shall see that the homotopy classes are identical with the components of Ω. Our corollary will then affirm the existence of at least one curve in each homotopy class K of Ω with length a minimum among lengths of curves of K.

Let $H^k(a, b)$ be the kth homology group of $\Omega(a, b)$. Let $x \ominus y$ be read "x is simply isomorphic with y." We see that $H^k(a, b) \ominus H^k(b, a)$. We shall extend this result by proving the following theorem.

THEOREM 12.2. *If $| N |$ is arcwise connected,*

$$H^k(a, b) \ominus H^k(c, d), \tag{12.9}$$

where (a, b) and (c, d) are arbitrary pairs of points on $| N |$.

We shall begin by proving statement (i).

(i). *If there is an arc on $| N |$ joining b to c then $H^k(a, b) \ominus H^k(a, c)$.*

Let η be a curve joining b to c on $| N |$. Let λ be a curve joining a to b. By the *extension* of λ by η we mean the curve obtained by tracing λ and η successively. We denote this extension by $(\lambda)_\eta$ and observe that $(\lambda)_\eta$ joins a to c.

Let a_k be an oriented k-cell on $\Omega(a, b)$. By the extension $(a_k)_\eta$ of a_k by η we mean the oriented k-cell on $\Omega(a, c)$ obtained by replacing each vertex of a_k by its extension by η, understanding of course that a vertex of a_k is a curve of $\Omega(a, b)$. Let $z_k = \delta^i a_k^i$ be an algebraic k-chain on $\Omega(a, b)$. By the *extension* $(z_k)_\eta$ of z_k is meant the k-chain

$$(z_k)_\eta = \delta^i (a_k^i)_\eta \qquad [\text{on } \Omega(a, c)]. \tag{12.10}$$

We observe that

$$\beta(z_k)_\eta = (\beta z_k)_\eta, \tag{12.11}$$

that is boundary and extension operators are commutative. If u_j is the jth component of a Vietoris cycle u, the algebraic j-cycles $(z_j)_\eta$ form a new Vietoris k-cycle $(u)_\eta$ on $\Omega(a, c)$. This follows in part from (12.11) and in part from the fact that the distance between two curves of $\Omega(a, b)$ equals the distance between their extensions on $\Omega(a, c)$, so that norms are not altered by the process of extension.

Let η' be the curve obtained by tracing η in the opposite sense. If z_k is an algebraic k-cycle of norm e on $\Omega(a, b)$, set

$$z_k^* = [(z_k)_\eta]_{\eta'}. \tag{12.12}$$

We see that z_k^* is an algebraic k-cycle of norm e on $\Omega(a, b)$.

We shall define a deformation D which continuously deforms z_k into z_k^* on $\Omega(a, b)$.

To that end let η be given in the form $p = p(\tau)$ with $0 \leqq \tau \leqq 1$ and let η_t denote the subarc of η on which $0 \leqq \tau \leqq t \leqq 1$. Under D suppose that an initial point λ (i.e. curve) of $\Omega(a, b)$ is replaced at the time t by the point (i.e. curve)

$$[(\lambda)_{\eta_t}]_{\eta_t'} \qquad (0 \leqq t \leqq 1)$$

of $\Omega(a, b)$ where η_t' represents η_t taken in the opposite sense. The deformation D carries z_k into z_k^* as stated. Hence

$$z_k^* \sim_{e+\delta} z_k \qquad [\text{on } \Omega(a, b)], \tag{12.13}$$

where e is the norm of z_k^* and z_k and δ is an arbitrary positive constant.

We return to the proof of (i) *and set up an isomorphism between the groups $H^k(a, b)$ and $H^k(a, c)$.*

To a k-cycle z of $\Omega(a, b)$ corresponds the cycle $(z)_\eta$ of $\Omega(a, c)$. It follows from (12.11) that k-cycles which are homologous on $\Omega(a, b)$ thereby correspond to k-cycles which are homologous on $\Omega(a, c)$. Each homology class of $H^k(a, b)$ is thus mapped onto a homology class of $H^k(a, c)$. We designate this mapping by T. That this mapping is a homeomorphism follows at once from the definition of the operation of extension by η. It remains to prove that T is an isomorphism.

The mapping T leads to each class of $H^k(a, c)$. If u is a k-cycle on $\Omega(a, c)$, $[(u)_{\eta'}]_\eta$ is a k-cycle on $\Omega(a, c)$, homologous to u as we see from (12.13) upon interchanging (a, b) and (a, c). But $(u)_{\eta'}$ is on $\Omega(a, b)$. Thus T leads to each class of $H^k(a, c)$.

To show that the mapping T is one-to-one we have merely to show that the null class in $H^k(a, c)$ corresponds to the null class in $H^k(a, b)$. Suppose a k-cycle on $H^k(a, c)$ is the extension $(z)_\eta$ of a k-cycle z on $H^k(a, b)$. If $(z)_\eta \sim 0$, we have

$$0 \sim (z_\eta)_{\eta'} \sim z \qquad [\text{on } \Omega(a, b)].$$

Hence T is an isomorphism as stated, and the proof of (i) is complete.

To prove the theorem we note that the operations replacing b by c in $H^k(a, b)$, or of interchanging a and b lead to groups isomorphic with $H^k(a, b)$. We can thus successively replace (a, b) by (a, c), (c, a), (c, d) and obtain a group $H^k(c, d)$ isomorphic with $H^k(a, b)$.

The proof of the theorem is complete.

We state the following corollary of the theorem.

COROLLARY. *The connectivities of $\Omega(a, b)$ are topological invariants of $| N |$.*

Whether or not the connectivities of $\Omega(a, b)$ are determined by arithmetic topological invariants of $| N |$ now known is an interesting unsolved problem. It seems probable that they are not so determined.

If $| N |$ is a space homeomorphic with an n-sphere with $n > 1$, the connectivities R_k of $\Omega(a, b)$ are 1 if $k \equiv 0 \bmod n - 1$, otherwise null. For example, for $n = 2$ and 3 respectively, the numbers R_k form the sequences 1 1 1 $\cdots$ and 1 0 1 0 $\cdots$. See M1, Theorem 15.1, p. 247.

The function F. Let η be an arbitrary curve of $\Omega(a, b)$ and let $L(\eta)$ be the length of η. We set

$$F(\eta) = \frac{L(\eta)}{1 + L(\eta)} \tag{12.14}$$

understanding thereby that $F = 1$ when L is infinite. The function F is lower semi-continuous with $0 \leqq F \leqq 1$. For $c < 1$ the subdomains $F \leqq c$ of $\Omega(a, b)$ are compact in accordance with Theorem 12.1. A curve λ of $\Omega(a, b)$ will be called a *homotopic pseudo-extremal* or *extremal* respectively if λ is a homotopic semi-critical point or a critical point of F respectively.

Our theory of functions F defined on spaces M is now applicable.

We turn to the development of conditions under which a pseudo-extremal is always an extremal, that is conditions under which a critical pair (p, c) is of the form $[p, F(p)]$.

V. LOCALLY CONVEX SPACES

13. Elementary arcs. In this chapter we shall suppose that the space of §11 is conditioned by two new postulates IV and V. In stating V we shall need the notion of an *elementary arc*. If p and q are two distinct points of $|N|$, an arc λ joining p to q will be termed *elementary* if it has the following two properties. It is simple, that is it is defined by a one-to-one continuous image $p(t)$ of a line segment $0 \leqq t \leqq a$; a point r shall lie on λ if and only if

$$pq = pr + rq. \tag{13.1}$$

Postulates IV and V are as follows.

IV. *The space $|N|$ shall be compact and connected.*

V. *There shall exist a positive constant ρ such that whenever $0 < pq \leqq \rho$ there exists at least one elementary arc which joins p to q.*

The elementary arc postulated in V will be denoted by (pq). We shall term V the postulate of *local convexity* and ρ the *norm of convexity*. We shall restrict ourselves to elementary arcs (pq) for which $pq \leqq \rho$.

The postulates I–V are independent as can be readily shown. They are satisfied on any ordinary closed manifold without singularity and with a sufficiently regular local representation. The elementary arcs are then geodesics, and geodesic arcs issuing from a point p and cut off at a sufficiently small distance ρ from p cover a neighborhood of p in a one-to-one continuous manner, p alone excepted. On $|N|$ however the elementary arcs issuing from a point p cover a neighborhood of p in general multiply as examples would show.

We regard two elementary arcs as identical if they are similarly sensed and consist of the same points. With this understood we have the following theorem.

THEOREM 13.1. *There is at most one elementary arc joining a point p to a point q.*

Let λ and ζ be elementary arcs joining p to q. Let r be a point on λ between p and q. Then $pq = pr + rq$. By virtue of the definition of an elementary arc this implies that r lies on ζ as well as λ, and the proof of Theorem 13.1 is complete.

The following theorem is of importance.

THEOREM 13.2. *Any subarc of an elementary arc* (pq) *is an elementary arc.*

We shall begin by proving statements (1) and (2) below. In these statements and elsewhere we shall make the special convention that the elementary arc joining a point p to p is the point p itself.

(1). *If* r *is a point on* (pq), *then the arc* pr *on* (pq) *is elementary.*

(2). *If* r *is a point on* (pq), *then the arc* rq *on* (pq) *is elementary.*

To establish (1) we first note that $pr \leqq \rho$ since $pr + rq = pq \leqq \rho$. It follows from V that there is an elementary arc (pr) joining p to r. Let s be any point on (pr). Then $pr = ps + sr$. Upon adding rq to both sides of this equality we find that

$$pr + rq = ps + sr + rq. \tag{13.2}$$

Since r is on (pq) we can replace the left member of (13.2) by pq. In the right member it follows from postulate III that $sr + rq \geqq sq$ so that (13.2) takes the form

$$pq \geqq ps + sq. \tag{13.3}$$

Upon comparing (13.3) with postulate III we infer that $pq = ps + sq$ so that s must lie on (pq).

We have yet to show that s is on the arc pr of (pq). To this end let (pq) and (pr) be respectively the one-to-one continuous images of the segments $0 \leqq t \leqq 1$ and $0 \leqq x \leqq 1$. Given any value of x on $0 \leqq x \leqq 1$ there is determined a unique point P on (pr) and, since P lies on (pq) as we have just seen, a unique value of t. We denote this value of t by $t(x)$. It is readily seen that the relation $t = t(x)$ is one-to-one and continuous for x on $0 \leqq x \leqq 1$. Hence (pr) must coincide with the arc pr of (pq), and (1) is proved.

The proof of (2) is similar to that of (1) interchanging the rôles of p and q in the proof of (1) and appropriately changing the order of the points involved. To establish the theorem let p, r, s, q be points on (pq) in the order written. By virtue of (1) the arc ps of (pq) is elementary. It follows from (2) that the arc rs on (ps) is elementary, and the proof is complete.

Theorem 13.2 leads to the following.

THEOREM 13.3. If $r = r(\tau)$, $0 \leqq \tau \leqq 1$, *defines an elementary arc joining* p *to* q, *the distance* $pr(\tau)$ *is a continuous increasing function of* τ.

The continuity of $pr(\tau)$ is a consequence of the continuity of $r(\tau)$ and the continuity of pr as a function of p and r. To show that $pr(\tau)$ is an increasing function of τ we note that

$$pr(\tau') = pr(\tau) + r(\tau)r(\tau') \qquad (0 \leqq \tau < \tau' \leqq 1), \tag{13.4}$$

so that $pr(\tau') > pr(\tau)$.

We are able to prove a theorem which is much stronger than Theorem 13.3.

To state this theorem let (pq) be an elementary arc with variable end points p, q. Let t be a number between 0 and pq inclusive. Let the point r on (pq) at the distance t from p be denoted by $f(p, q, t)$. Our theorem is the following.

THEOREM 13.4. *The point function $f(p, q, t)$ is continuous for $pq \leqq \rho$ and $0 \leqq t \leqq pq$.*

Let (p_n, q_n, t_n) be a sequence of admissible sets (p, q, t) which tend to a set (p^0, q^0, t^0) as a limit set. Set $r_n = f(p_n, q_n, t_n)$. Since $| N |$ is compact there is a subsequence of the points r_n with a limit point r^0. For simplicity we assume that the sequence r_n itself converges to r^0. We shall prove that

$$r^0 = f(p^0, q^0, t^0). \tag{13.5}$$

Upon letting n become infinite in the relation $p_nq_n = p_nr_n + r_nq_n$ we infer that

$$p^0q^0 = p^0r^0 + r^0q^0. \tag{13.6}$$

We also have the relations

$$p^0r^0 = \lim_{r=\infty} p_nr_n = \lim_{n=\infty} t_n = t^0. \tag{13.7}$$

From (13.6) and (13.7) we see that r^0 is the point on the elementary arc p^0r^0 which is at the distance t^0 from p^0. Hence (13.5) holds, and the theorem is true.

The following theorem depends upon the compactness of $| N |$ and postulates I–III.

THEOREM 13.5. *The absolute distance $| pq |$ approaches zero uniformly with pq, and the absolute diameter of any curve approaches zero uniformly with its length.*

Suppose the first affirmation of the theorem false. There will then exist a sequence of pairs of points (p_n, q_n) whose absolute distances are bounded from zero but whose distances p_nq_n tend to zero as n becomes infinite. There will exist at least one limit pair p^*, q^* of the pairs p_n, q_n. For such a pair $p^*q^* = 0$ so that $p^* = q^*$. But $| p^*q^* | \neq 0$ since the distances $| p_nq_n |$ are bounded from zero. From this contradiction we infer that $| pq |$ approaches zero uniformly with pq.

The concluding affirmation of the theorem follows from the definition of length and the first statement in the theorem.

Recall that a metric space M is termed *separable* if there is a subset of points of M whose closure is M and which is at most countably infinite. We state the following theorem.

THEOREM 13.6. *Under postulates* I–V *the space $\Omega(a, b)$ is separable.*

The space $| N |$ is compact and therefore separable. Let (q) be a set of points of $| N |$ at most countably infinite with closure $| N |$. We suppose that the points a and b belong to (q). By a broken extremal ζ is meant a curve composed of a finite succession of elementary arcs. The end points of these elementary arcs will be called the *vertices* of ζ. Let H be the set of all broken extremals which join a to b and whose vertices belong to (q). It is clear that the set H is at most countably infinite. Corresponding to an arbitrary curve λ of $\Omega(a, b)$ there is a broken extremal ζ within a prescribed distance e of λ. We suppose ζ so chosen that the lengths of its elementary arcs are at most $\rho/2$. Let the successive vertices of ζ be designated by $a, p_1, \cdots, p_n, b$. We replace the vertices p_i by points q_i of (q) so near the respective points p_i that the successive points $a, q_1, \cdots, q_n, b$ can still be joined by elementary arcs, and so near p_i that the distance of the resulting broken extremal η from ζ is at most e. The curve η belongs to the set H, and $\lambda\eta \leqq 2e$.

The proof of the theorem is complete.

14. **Metric extremals.** We begin this section with the following theorem.

THEOREM 14.1. *The length of an elementary arc (pq) equals pq and is a proper minimum relative to the lengths of all curves which join p to q.*

Let $p_1, \cdots, p_n$ be a set of points which appear in the order written on (pq). Any segment of an elementary arc is an elementary arc. It then follows inductively for $n = 1, 2, \cdots$ that

$$pq = pp_1 + p_2p_3 + \cdots + p_nq.$$

We infer that the length of (pq) equals pq.

Let λ be any curve joining p to q. It follows from Theorem 11.3 that $L(\lambda) \geqq pq$. It remains to show that $L(\lambda) > pq$ when λ is not the elementary arc (pq).

The proof falls into two cases.

Case I. The curve λ contains a point s not on (pq).

Case II. Each point of λ is on (pq), but $\lambda \not\equiv (pq)$.

In Case I we see that $L(\lambda) \geqq ps + sq > pq$, and the proof is complete.

In Case II there must be distinct points r and s on λ and on (pq) which appear in the order rs on (pq) but in the order sr on λ. Then

$$L(\lambda) \geqq ps + sr + rq > ps + rq > ps + sq = pq,$$

and the proof of the theorem is complete.

A *metric extremal* λ shall be the continuous image on $| N |$ of a closed or open interval T of the t-axis such that every inner point t of T is an inner point of a closed subinterval of T whose image on λ is an elementary arc. A curve which is not a metric extremal will be termed *metrically ordinary*. We continue with the following theorem.

THEOREM 14.2. *Let a and b be two points on $| N |$. Under the hypothesis of local convexity a curve λ of minimum length joining a to b is a metric extremal.*

The successive points of a sufficiently fine partition of λ can be joined by elementary arcs thereby forming a curve of finite length. Hence the length of λ must be finite. Moreover each arc η of λ with a length less than ρ must be an elementary arc. For if p and q are the end points of η, $pq \leqq L(\eta) < \rho$ in accordance with Theorem 14.1 so that p and q can be joined by an elementary arc (pq). If η is not an elementary arc $pq < L(\eta)$ by virtue of Theorem 14.1

and λ can be shortened by replacing η by (pq). It follows that λ is a metric extremal, and the proof is complete.

Homotopic extremals and pseudo-extremals were defined in §12 before the hypothesis of local convexity was introduced, while metric extremals were defined after this hypothesis was introduced. The principal theorem of this section is the following. It includes Theorem 14.2 as a special case.

THEOREM 14.3. *Under the postulates* I–V *the class of homotopic extremals on* $\Omega(a, b)$ *is identical with the class of homotopic pseudo-extremals on* $\Omega(a, b)$ *and each such extremal is a metric extremal.*

The proof of this theorem will require the introduction of a lemma and a basic deformation θ. We proceed with the necessary definitions and proofs.

Admissible sets π. Let λ be a curve on $|N|$ with end points a and b. If r is a sufficiently large integer, there is a set of $r - 1$ points on λ dividing λ into a sequence of arcs

$$\pi = (\lambda_1, \cdots, \lambda_r) \tag{14.1}$$

on each of which the distance from the initial point to the terminal point is at most ρ. A set π of curves λ_i obtained in this manner from a curve λ joining a to b on M will be termed *admissible*. We term λ_i the ith component of π. Corresponding to an admissible set π we denote the curve λ which π determines by $\lambda(\pi)$.

Let π be an admissible set (λ_i). We denote the elementary arc which joins the initial point of λ_i to the terminal point of λ_i by λ_i^*. The set of curves λ_i^* will then be denoted by π^* The set π^* is determined by π. The curve $\lambda(\pi^*)$ is a broken extremal.

The following lemma is a broad generalization of Osgood's Theorem.[15] In it we refer to the function $F(\lambda)$ defined in §13.

LEMMA 14.1. *For admissible sets* π *which possess a fixed number* r *of components and for which the distance of* $\lambda(\pi)$ *from* $\lambda(\pi^*)$ *is not less than a positive constant* e *the value of* F *on* $\lambda(\pi)$ *exceeds the value of* F *on* $\lambda(\pi^*)$ *by an amount which is bounded away from zero.*

In proving this lemma admissible sets π will function as points in a metric space. The distance between two admissible sets π,

$$\pi = (\lambda_1, \cdots, \lambda_r), \qquad \pi' = (\lambda_1', \cdots, \lambda_r')$$

with the same number of components shall be defined as the maximum distance $\lambda_i\lambda_i'$ and denoted by $\pi\pi'$.

---

[15] See Tonelli, l.c.

[16] Cf. Fréchet Sur une représentation paramétrique intrinsèque de la courbe continue la plus générale Journal de Mathématique t. IV, 1925, pp. 281–297. This very interesting representation could not be used in the present paper because it lacked the property (d) of our representations.

[17] Fréchet *Sur quelques points du calcul fonctionnel*, Rendiconti del Circolo Matematico di Palermo, vol. 22 (1906), pp. 1–74.

We consider a fixed positive integer r and observe that $L[\lambda(\pi^*)] \leqq r\rho$ for all admissible sets π with r components. Hence there is a constant m dependent on r such that

$$F[\lambda(\pi^*)] \leqq m < 1.$$

Let k be an arbitrary number between m and 1. We divide admissible sets π with r components into two classes I and II according as

$$\text{I. } F[\lambda(\pi)] > k, \qquad \text{II. } F[\lambda(\pi)] \leqq k.$$

We then introduce the function

$$H(\pi) = F[\lambda(\pi)] - F[\lambda(\pi^*)].$$

If π is in Class I,

$$H(\pi) > k - m > 0. \tag{14.2}$$

We turn to Class II. There is a constant c dependent on k such that each component λ_i of an admissible set π in Class II has a length at most c. The set of all curves on $|N|$ with lengths at most c forms a compact space G (the proof of this is similar to the proof of Theorem 12.1). Let G^r denote the product space in which G is an r-fold factor. The space G^r is compact. For r fixed, the sets π which are admissible and belong to Class II, and for which the distance $\lambda(\pi)\lambda(\pi^*) \geqq e$ form a closed subset G_0 of G^r.

The function $H(\pi)$ is lower semi-continuous. For $F(\lambda)$ is lower semi-continuous in λ, $\lambda(\pi)$ varies continuously with admissible sets π and $F[\lambda(\pi^*)]$ is a continuous function of admissible sets π. By virtue of Theorem 14.1, $H(\pi)$ is positive on G_0. At some point π_0 of G_0, $H(\pi)$ then takes on a positive absolute minimum. That is

$$H(\pi) \geqq H(\pi_0) > 0 \qquad (\pi \subset G_0).$$

Combining this result with (14.2) we find that

$$H(\pi) \geqq \min [k - m, H(\pi_0)] > 0$$

for the sets π of the lemma.

The proof of the lemma is complete.

We shall define an F-deformation $\theta(\eta, m)$ of the neighborhood of an arbitrary curve η of $\Omega(a, b)$. To that end we shall need a representation of curves λ of $\Omega(a, b)$ in terms of μ-length. See §11. Let $\mu(\lambda)$ be the μ-length of λ and let α be a parameter which ranges from 0 to 1 inclusive. Let the point q on λ at which the μ-length of λ equals $\alpha\lambda(\mu)$ be designated by

$$q = q(\lambda, \alpha). \tag{14.3}$$

The function $q(\lambda, \alpha)$ is conditioned by properties (b), (d) and (e) of μ-length as given in §11.

The deformation $\theta(\eta, m)$. Let m be a positive integer so large that η admits a partition into m successive subarcs with the following properties. The diameter of each of these arcs shall be less that $\rho/2$. If η is not an extremal at least one of these subarcs shall not be an elementary arc. Let V be a neighborhood of η on $\Omega(a, b)$ and λ a curve on V. The curve λ shall be divided into m successive subarcs on which the variation of the parameter α in $q(\lambda, \alpha)$ in (14.3) shall equal its variation on the corresponding subarcs of η. It follows from property (e) of μ-length in §11 that the neighborhood V can be chosen so small that the diameter of each of the m-subarcs of λ is less than $\rho/2$. We suppose V so chosen.

Let ω be the ith subarc of λ. We shall define θ for ω, understanding that the deformation θ of λ is obtained by simultaneously applying θ to each of the subarcs ω of λ. In the deformation θ the time t shall range from 0 to 1 inclusive. Let $t_n = 2^{-n}$, $n = 0, 1, 2, \cdots$. We shall first define θ for the times t_n.

For $t = t_n$ the deformation θ shall replace ω by a broken extremal ω_n whose end points are the end points of ω and whose successive vertices divide ω into 2^n successive segments of equal variation of the parameter α of λ. Let $(\alpha)_n$ denote the ordered set of parameter values α at the end points and points of division of ω.

Let the curve which replaces ω at the time t be denoted by ω^t. We shall now define ω^t when t is on the interval

$$t_n < t < t_{n-1}.$$

Let α' and α'' denote any pair of successive values of α in the set $(\alpha)_{n-1}$. In the set $(\alpha)_n$ there is just one value α^* of α between α' and α''. Let g denote the elementary arc of ω_n which joins the point α^* on ω to the point α'' on λ. Let t divide the interval (t_n, t_{n-1}) in an arbitrary ratio with respect to length. Let $p(t)$ denote the point on g which divides g in the same ratio with respect to length. Corresponding to each pair α', α'' chosen as above we replace the arc ξ of ω_n with end points at the points α' and α'' on ω and intermediate vertex at α^* on ω by the broken extremal with end points of ξ and one intermediate vertex $p(t)$, thereby obtaining the curve ω^t.

The definition of $\theta(\eta, m)$ is complete.

In showing that θ is an F-deformation we shall use the concept of an *adjoin* of a point set A of $|N|$. By the adjoin of A is meant the set of points on all elementary arcs whose end points lie on A. By the *2-fold adjoin* of A is meant the adjoin of the adjoin of A. We shall make use of the following lemma.

LEMMA 14.2. *Corresponding to an arbitrary positive constant e there exists a positive constant δ such that the diameter of the 2-fold adjoin of A is at most e whenever the diameter of A is at most δ.*

This lemma is a consequence of our theorems on elementary arcs.

To prove that θ is an F-deformation of the neighborhood V of η we must establish the following:

(i). *The curve λ_t replacing λ at the time t varies continuously with λ and t for λ on any compact subset S of V and for $0 \leq t \leq 1$.*

(ii). *The deformation θ admits a displacement function on S.*

Let ζ denote an arbitrary point of S and τ an arbitrary value of t on the interval $0 \leqq t \leqq 1$. Regarding λ_t as defined for the moment only for curves λ on S we shall prove that λ_t is continuous in λ and t at the pair (ζ, τ). We shall distinguish two cases: I. $\tau = 0$; II. $\tau > 0$.

We first consider Case I. The point q on λ has been given the representation $q = q(\lambda, \alpha)$. The function $q(\lambda, \alpha)$ is continuous in its arguments in accordance with (d) of §11. It follows that for λ on S the 2^n successive arcs k into which ω was divided in defining θ will be uniformly small in diameter provided $n \geqq N$, and N is a sufficiently large integer. Let e be an arbitrary positive constant. It follows from Lemma 14.2 that if N is sufficiently large the 2-fold adjoin of two successive arcs k will have a diameter at most $e/2$. In defining θ, λ_t was obtained from λ by replacing (certain) pairs k', k'' of adjacent arcs k of λ by two successive elementary arcs of the 2-fold adjoin of $k' + k''$. For $t < 2^{-N}$ each of the above arcs k involved in the construction of λ_t belongs to at least a 2^N-fold partition of an arc ω. Set $\delta = 2^{-N}$. From the choice of N we infer that

$$\lambda_t\lambda \leqq \frac{e}{2} \qquad (\lambda \subset S, t < \delta). \tag{14.4}$$

We next observe that

$$\lambda_t\zeta \leqq \lambda_t\lambda + \lambda\zeta \tag{14.5}$$

so that we see from (14.4) and (14.5) that

$$\lambda_t\zeta \leqq e \qquad \text{when } [\lambda\zeta < e/2; \lambda \subset S; t < \delta].$$

The continuity of λ_t is established at (ζ, τ) when $\tau = 0$.

We turn to Case II. For $t \geqq \tau$ and sufficiently near τ, or for $t \leqq \tau$ and sufficiently near τ, the curve λ_t is a broken extremal whose vertices vary continuously with λ and t for λ on S in accordance with (d) of §11. Hence λ_t is continuous at (λ, τ) in Case II.

The proof of (i) is complete.

We shall continue with a proof of (ii).

Let λ be a curve of S. Let ζ' be an antecedent of ζ'' on the trajectory T of λ under θ. Let e be a positive constant. Setting $F(\zeta') - F(\zeta'') = \Delta F$ we must show that whenever $\zeta'\zeta'' > e$,

$$\Delta F > \delta(e) \tag{14.6}$$

where $\delta(e)$ is a positive constant which depends only on S and e.

Let ζ_n denote the image of λ at the time t_n. We shall say that the pair ζ', ζ'' is in Class 1 if for no n is ζ_n on T between ζ' and ζ'', and if $\zeta'\zeta'' > e$. We shall first establish (14.6) for pairs ζ', ζ'' in Class 1.

As constructed in the definition of θ, ζ'' is a succession of r elementary arcs ζ_i''. The curve ζ' is a succession of r curves ζ_i' whose end points are joined by the respective elementary arcs ζ_i''. We can apply Lemma 14.1 to ζ' and ζ'' identifying ζ' with $\lambda(\pi)$ and ζ'' with $\lambda(\pi^*)$ and infer the existence of a function $\omega(r, e)$ such that

$$\Delta F > \omega(r, e) > 0 \qquad (\zeta'\zeta'' > e). \tag{14.7}$$

Relation (14.6) will follow from (14.7) provided the integer r in (14.7) is bounded for pairs ζ', ζ'' in Class 1. The integer r will fail to be bounded only if there are pairs ζ', ζ'' in Class 1 arbitrarily near λ, as one sees upon representing the curves λ of S in the form $q = q(\lambda, \alpha)$ of (14.3), and recalling that S is compact. But pairs ζ', ζ'' arbitrarily near λ would be arbitrarily near each other contrary to the hypothesis that $\zeta'\zeta'' > e$. We conclude that r in (14.7) is at most an integer R. We can accordingly take $\delta(e)$ in (14.6) as the minimum, say $\delta_1(e)$, of the numbers

$$\omega(1, e), \omega(2, e), \cdots, \omega(R, e).$$

Thus (14.6) holds for pairs ζ', ζ'' in Class 1.

We turn to pairs ζ', ζ'' not in Class 1.

As previously we admit pairs ζ', ζ'' such that $\zeta'\zeta'' > e$ and such that ζ' is an antecedent of ζ'' on a trajectory T emanating from a curve λ of S. If ζ' and ζ'' are not in Class 1 there exist integers p and s such that the images

$$\zeta_{p+1}, \quad \zeta', \quad \zeta_p, \quad \zeta_s, \quad \zeta'', \quad \zeta_{s-1} \tag{14.8}$$

of λ appear in the order written on T. We admit the possibility that consecutive curves in (14.8) may be identical. The function $F(\zeta)$ decreases monotonically as ζ moves along T in the sense of increasing t. Hence the left member of(14.6) is at least as great as each of the three numbers

$$F(\zeta') - F(\zeta_p), \qquad F(\zeta_p) - F(\zeta_s), \qquad F(\zeta_s) - F(\zeta'').$$

Admissible pairs ζ', ζ'' not in Class 1 will be assigned to Classes 2, 3, or 4 according as

$$2.\ \ \zeta'\zeta_p \geqq \frac{e}{3}, \qquad 3.\ \ \zeta_p\zeta_s \geqq \frac{e}{3}, \qquad 4.\ \ \zeta_s\zeta'' \geqq \frac{e}{3}.$$

Since $\zeta'\zeta'' > e$ each pair ζ', ζ'' will belong to at least one of these classes. Proceeding as under Class 1 we can infer the existence of a function $\delta_i(e)$, $i = 2, 3, 4$, such that $\Delta F > \delta_i(e)$ when ζ', ζ'' is in Class i.

For a fixed e let $\delta(e)$ denote the minimum of the numbers $\delta_i(e)$ for $i = 1, 2, 3, 4$. For this $\delta(e)$, (14.6) holds as stated, and the proof of (ii) is complete.

The deformation $\theta(\eta, m)$ is accordingly an F-deformation of the neighborhood V of η.

Proof of Theorem 14.3. Let η be a curve of $\Omega(a, b)$ and let $\theta(\eta, m)$ be a corresponding F-deformation of curves λ on a neighborhood of η. Let η_1 and λ_1 be respectively the final images of η and λ under θ. We shall begin by proving the following statement.

(i). *Corresponding to an arbitrary positive constant e there exists a positive constant* δ *such that for* $\lambda\eta < \delta$,

$$F(\lambda_1) \leqq F(\eta_1) + e. \tag{14.9}$$

The final image λ_1 of λ is a succession of m elementary arcs. Let $(p) = (p_0, \cdots, p_m)$ be the successive end points of these elementary arcs. Designate the set (p) determined by η_1 by (q). Let the distance $(p)(q)$ be defined as in §12 as the maximum of the distances $| p_i q_i |$. According to the construction of θ the points p_i divide λ in the same ratios with respect to μ-length as the points q_i divide η. For η fixed it follows from property (d) of μ-length in §11 that the distance $(p)(q)$ tends to zero with $\lambda\eta$. But $F(\lambda_1)$ may be regarded as a function $\varphi(p)$ of the set (p) determining (λ_1). The function $\varphi(p)$ is continuous in (p) at least if $(p)(q)$ is so small that (p) determines a sequence of elementary arcs. Hence

$$\varphi(q) \leqq \varphi(p) + e$$

provided $(p)(q)$ is sufficiently small. Statement (i) follows directly.

We continue with a proof of (ii).

(ii). *A homotopic pseudo-extremal is a homotopic extremal.*

Let η be a semi-critical point in the space $\Omega(a, b)$ at the level c. We shall show that $F(\eta) = c$. It follows from the lower semi-continuity of F that either $F(\eta) = c$ or $F(\eta) < c - e_1$ where $e_1 > 0$. If η_1 is the final image of η under θ and $F(\eta) < c - e_1$ it would follow with the aid of (14.9) that

$$F(\lambda_1) \leqq F(\eta_1) + e \leqq F(\eta) + e < c - e_1 + e \qquad (\lambda\eta < \delta). \tag{14.10}$$

If δ in (i) is sufficiently small e can be taken less than e_1 and (14.10) implies that

$$F(\lambda_1) < c - (e_1 + e) < c \qquad (\lambda\eta < \delta).$$

We can thus F-deform the neighborhood $\lambda\eta < \delta$ of η onto a domain definitely below c. The limiting pair (η, c) must then be homotopically ordinary contrary to hypothesis, and we conclude that (ii) is true.

We conclude the proof of Theorem 14.3 by establishing (iii).

(iii). *A homotopic extremal is a metric extremal.*

To prove (iii) it will be sufficient to show that if η is metrically ordinary η is homotopically ordinary. When η is metrically ordinary the m arcs into which η is initially divided in defining $\theta(\eta, m)$ are so chosen that at least one arc is not elementary. It follows that $F(\eta_1) < F(\eta)$ and upon making use of the arbitrariness of e it follows from (14.9) that θ F-deforms a sufficiently small neighborhood of η onto a domain definitely below $F(\eta)$. Thus η is homotopically ordinary.

The proof of Theorem 14.3 is complete.

15. **Properties of $\Omega(a, b)$ and of metric extremals.** In §9 we have defined locally F-connected spaces. We shall here prove the following theorem.

THEOREM 15.1. *Under postulates* I–V *the space* $\Omega(a, b)$ *is locally F-connected for all orders n.*

Let η be a curve of $\Omega(a, b)$. To establish the theorem it will be sufficient to establish the following.

(a). *Corresponding to* η *and a positive constant e there exists a positive constant* δ *and a deformation of* η_δ *on* η_e *into a curve* η_1 *on* η_e, *in which any compact subset S of* η_δ *on* $F \leqq c \geqq F(\eta)$ *is continuously deformed on* $F \leqq c + e$.

We shall apply the deformation $\theta(\eta, m)$ of the preceding section to η. Recall that the definition of θ began with the division of η into m successive arcs. If these arcs are sufficiently small in diameter and V is a sufficiently small neighborhood of η, θ will deform V on the $e/2$ neighborhood of η. We suppose θ and V so chosen. Let λ be a curve on V and let η_1 and λ_1 be final images of η and λ respectively under θ.

Each broken extremal λ_1 which is sufficiently near η_1 can be deformed into η_1 as follows. The vertex q_i of λ_1 will be joined to the corresponding vertex p_i of η_1 by an elementary arc E_i, and as t varies from 0 to 1, q_i shall be replaced by a point p_t which divides E_i in the same ratio with respect to length as that in which t divides the interval (0, 1). The broken extremal λ_1 will thereby be deformed into η_1. Denote this deformation by Λ.

Let δ be a positive constant such that $\eta_\delta < V$. Let H be the final image of η_δ under θ. The diameter of H tends to 0 with δ. We suppose δ so small that Λ deforms H on η_e and on $F \leqq c + e$.

The deformation θ applied to η_δ, followed by Λ applied to H, yields a deformation of η_δ which satisfies (*a*). Theorem 15.1 follows directly.

Referring to Theorem 9.2 we infer the following corollary of Theorem 15.1.

COROLLARY 15.1. *Under postulates* I–V *the lengths of extremals of* $\Omega(a, b)$ *which correspond to inferior cycle limits of non-bounding k-cycles, have no finite cluster value.*

Analytic functionals. We shall give a brief indication of how the preceding theory can be applied to functionals which are analytically defined. We suppose that the underlying space M is *compact, connected, locally analytic,* and *n-dimensional.* To be locally analytic M shall satisfy the conditions of a space locally of class C^2 as defined as §9, together with the condition that the transformations of coördinates $z_i = z_i(x)$ shall be defined by functions $z_i(x)$ which are analytic. In each local coördinate system we suppose that we have given an invariant function $f(x, r)$ defining an integral

$$J = \int f(x, \dot{x})\, dt$$

as in M1, p. 111. We suppose that $f(x, r)$ is analytic, positive, positive homogeneous of order 1 in the variables (r), and positive regular. See M1, p. 121. We presuppose the analyticity of M and $f(x, r)$ for the sake of simplicity, although it will be clear to the reader that less stringent conditions would suffice.

As is well known there will exist a positive constant ρ with the following properties. The extremals of J which issue from a point p of M and on which $0 \leqq J \leqq \rho$ will form a field covering a neighborhood of p in a one-to-one manner, p alone excepted, with no conjugate points of p on the extremal arcs $0 < J \leqq \rho$. We introduce a metric N in which the distance pq equals the absolute minimum of J along rectifiable curves which join p to q. With N and ρ so defined, we proceed as in the abstract variational theory noting that conditions I–V are satisfied.

As previously we are concerned with the space $\Omega(a, b)$. The integral J defines "length" and leads as at the end of §12 to the functional F on $\Omega(a, b)$. A metric extremal will here be locally regular and analytic, and will be termed a *differential extremal*. We shall prove the following theorem.

THEOREM 15.2. *Let η be a differential extremal joining a to b on which there are k conjugate points of a, but on which b is not conjugate to a. The extremal g is a homotopic critical set whose j^{th} type number is δ_k^j.*

In proving this theorem we shall make use of a subset of curves of Ω termed *canonical curves* and defined as follows. Let

$$a, a_1, \cdots, a_{m-1}, b \tag{15.1}$$

be a sequence of points on η which divide η into m successive elementary arcs with diameters less than $\rho/2$. Let M^q be a regular analytic $(n - 1)$-manifold cutting η at the point a_q without being tangent to η at a_q. Broken extremals of Ω whose intermediate vertices lie on the successive manifolds M^q will be termed *canonical*, and a set of all such curves with vertices on neighborhoods of the respective points a_q will be denoted by S. The value of F on curves λ of S will be an analytic function $\varphi(u)$ of the local coördinates of M neighboring the respective vertices a_q. As shown in M1 (cf. p. 228) the point $(u) = (u_0)$ which determines η is a non-degenerate critical point of φ of index k. Upon referring to Theorem 10.4 we infer the truth of the following.

(α). *If S consists of canonical curves determined by points (u) sufficiently near (u_0) the differential extremal η is an isolated homotopic extremal rel S, and the jth type number of η rel S equals δ_k^j.*

The curve η is isolated among differential extremals on Ω at the level $c = F(\eta)$. There are accordingly no homotopic extremals on Ω at the level c in a sufficiently small neighborhood of η, η excepted as we shall see.

We continue with a proof of statement (β).

(β). *Corresponding to the set S of (α) there exists a separate neighborhood U of η on $\Omega(a, b)$ which can be deformed into a subset of curves of S without increasing F beyond its initial value on each curve of U.*

We shall make use of the deformation $\theta(\eta, m)$ of the preceding section understanding that the m arcs into which η is initially divided in defining $\theta(\eta, m)$ are the m arcs whose vertices are given by (15.1). We take U so small that U is on the domain upon which $\theta(\eta, m)$ operates as an F-deformation. Let λ be a curve of U and μ the final image of λ under $\theta(\eta, m)$. The curve μ will be a broken extremal with intermediate vertices neighboring the respective points

a_q of η. If U is sufficiently small there will be a unique curve in S whose successive vertices lie on μ. We denote this curve by $\zeta(\mu)$.

The deformation Z. Under Z we shall deform μ into $\zeta(\mu)$. We thereby replace each intermediate vertex p of μ by a variable vertex p_t. We cause p_t to move from p along μ to the correspondingly numbered vertex r of $\zeta(\mu)$ in such fashion that the distance along μ from p to p_t changes at a rate equal to the distance along μ between p and r. The curve μ is thereby deformed into $\zeta(\mu)$. Moreover $F(\zeta) \leqq F(\mu)$. The deformation $\theta(\eta, m)$ followed by the deformation Z taken with a sufficiently small neighborhood U of η will satisfy (β).

We continue with a proof of (γ).

(γ). *A j-cap u associated with η rel S is a j-cap associated with η rel Ω.*

Since u is associated with η rel S, u is c-homologous on S to a j-cap on an arbitrarily small neighborhood of η, in particular to a j-cap v on U. If v were c-homologous to 0 on Ω, v would be c-homologous to 0 on U, by virtue of Theorem 10.1 (b). Let v^* be the final image of v under the deformation D of (β). It is clear that v^* is c-homologous to 0 on S if v is c-homologous to O on U. But v and v^* are c-homologous on S if U is sufficiently small as one sees with the aid of a deformation similar to Z operating on the formal deformation chain defined by v under D holding v and v^* fast. We infer that u is c-homologous to 0. From this contradiction we infer the truth of (γ).

We come finally to a proof of (δ).

(δ). *Any j-cap u associated with η rel Ω is c-homologous on Ω to a j-cap associated with η rel S.*

The k-cap u is c-homologous on Ω to a k-cap on the neighborhood U. It follows from (β) that u is c-homologous on Ω to a j-cap on S, and the proof of (δ) is complete.

We turn to the proof of the theorem. It follows from (γ) and (δ) that a maximal group of j-caps associated with η rel S is a maximal group of j-caps associated with η rel Ω. It then follows from (α) that the dimension of such a group is δ_k^j.

The proof of Theorem 15.2 is complete.

With the aid of this theorem and theorems of a like nature the reader can make connections with the principal results in M1.

THE INSTITUTE FOR ADVANCED STUDY.

Reprinted from DUKE MATHEMATICAL JOURNAL
Vol. 4, No. 1, March, 1938

THE INDEX THEOREM IN THE CALCULUS OF VARIATIONS

BY MARSTON MORSE

Introduction. The calculus of variations in the large is concerned with boundary problems in the large. Of these problems the simplest is that of finding extremals joining two points A and B on a regular m-manifold M. The theory[1] obtains relations between the local characteristics of the solutions of the problem and the topological characteristics (connectivities, etc.) of the space of admissible curves. The case where A and B are not conjugate on any extremal solution is termed the *non-degenerate case*. This case is the general case in the sense that for A fixed the set of points B which are conjugate to A on at least one extremal issuing from A has a null m-dimensional measure on M. The unrestricted case can be treated as a limiting case of the non-degenerate case (M, p. 239).

It appears that the most significant characteristic of an extremal solution g in the non-degenerate case is the number μ of conjugate points of A on g. This fact becomes most evident in terms of the "Index Theorem". Recall that the "index" of a critical point $(z) = (0)$ of a function $J(z)$ of a finite number of variables (z) is the number of negative characteristic roots of the Hessian of $J(z)$ at the point $(z) = (0)$. As we shall see $J(z)$ will represent the value of the integral J along a "canonical" broken extremal neighboring g with vertices determined by (z). The Index Theorem affirms that μ equals the index of the critical point (0) of this function $J(z)$. It is by means of this theorem that the topological characteristics of the neighborhood of g among admissible curves are determined.

The Index Theorem was first established in the non-parametric case in 1929 by Morse.[2] It was established in the parametric case by a reduction to the non-parametric case (M, p. 138). Recently the author has discovered a new and simpler method of proving the theorem. This method can be applied

Received December 22, 1937.

[1] M. Morse, *The Calculus of Variations in the Large*, American Mathematical Society Colloquium Publications, vol. 18, New York, 1934. A reference to these lectures will be indicated by the letter M.

Classical treatments of the conjugate point condition can be found in the following references.

G. A. Bliss, *Jacobi's condition for problems of the calculus of variations in parametric form*, Transactions of the American Mathematical Society, vol. 17 (1916), pp. 195–206.

C. Carathéodory, *Variationsrechnung und partielle Differentialgleichungen erster Ordnung*, Berlin, Teubner, 1935.

[2] M. Morse, *The foundations of the calculus of variations in the large in m-space*, Transactions of the American Mathematical Society, vol. 31 (1929), pp. 379–404.

directly to the parametric case and will be presented in this paper. The lemmas involved moreover have been useful in making advances in the theory of the "index form" under general boundary conditions.

1. **The space and integral.** We shall be concerned with a regular m-manifold M of class C^{r+1} with $r > 2$. Such a manifold is a Hausdorff topological space with the following properties. If p is a point of M, there exists a neighborhood of p which is the homeomorph of a region N in a Euclidean space E of dimension m, with coördinates (x) (termed *preferred* coördinates), such that any two sets (x) and (z) of preferred coördinates neighboring a point p on M are related by a non-singular transformation

$$z^i = z^i(x) \qquad (i = 1, \cdots, m) \tag{1.1}$$

of class C^{r+1}. Any system of coördinates obtained from preferred coördinates (x) by a non-singular transformation of the form (1.1) and of class C^n, $0 < n \leqq r + 1$, will be termed a C^n coördinate system.

In each preferred system (x) we suppose there is defined a function

$$F(x^1, \cdots, x^m, r^1, \cdots, r^m) = F(x, r)$$

of class C^r in (x, r) for (x) in the system (x) and for all sets $(r) \neq (0)$. We suppose that F is an invariant. More precisely, if $Q(z, \sigma)$ replaces F in a preferred system (z) and (x) and (z) are related as in (1.1), and if (r) in the system (x) is transformed as a contravariant tensor into (σ) in the system (z), then

$$F(x, r) = Q(z, \sigma).$$

We suppose that $F(x, r)$ is positive homogeneous of order 1 in the variables (r). Recall that

$$| F_{r^i r^j}(x, r) | = 0 \qquad (i, j = 1, \cdots, m). \tag{1.2}$$

We assume that the rank of the determinant (1.2) is $m - 1$. (Cf. M, pp. 111–112).

A subset of M locally representable in the form

$$x^i = x^i(u_1, \cdots, u_p) \qquad (i = 1, \cdots, m)$$

in terms of functions of class C^n, $n > 0$, with a functional matrix of rank p will be termed a *regular* m-manifold of class C^n on M. A regular 1-manifold (or curve) on M of class C^2 with a local parameter t will be called an *extremal* if it satisfies the Euler equations

$$\frac{d}{dt} F_{r^i} - F_{x^i} = 0 \qquad (i = 1, \cdots, m) \tag{1.3}$$

in each coördinate system (x) in which it enters. Such an extremal can be shown to be a regular curve of class C^r in terms of a suitably chosen parameter. We shall suppose our extremals so represented.

Let g be a closed arc of such an extremal. If $\gamma^i(t)$ is a regular representation of g in the system (x), then

$$F^0_{r^i r^j}\dot{\gamma}^i(t) \equiv 0, \tag{1.4}$$

where the superscript 0 indicates that the arguments are $\gamma^i(t)$ and $\dot{\gamma}^i(t)$. We assume that

$$F^0_{r^i r^j} w^i w^j > 0 \tag{1.5}$$

for every non-null set (w) independent of $[\dot{\gamma}(t)]$.

Recall that any extremal arc g lies in a C^r coördinate system (x). (Cf. M, p. 108, Theorem 1.1.) We shall define the conjugate points of a point A on g. Suppose that $t = t_0$ at A. The extremals issuing from A with directions neighboring that of g can be represented in the system (x) in the form (M, p. 117),

$$x^i = x^i(t; u) \qquad (i = 1, \cdots, m), \tag{1.6}$$

where (u) is a set of $n = m - 1$ parameters u_j and is constant on each extremal, and where t is a parameter in terms of which the extremal (u) is regularly represented. We suppose that the set $(u) = (0)$ determines g and that on g, $t_1 \leqq t \leqq t_2$. A representation of this character exists in which the functions $x^i(t, u)$ and $x^i_t(t, u)$ are of class C^{r-1} in terms of their arguments for (u) neighboring $(u) = (0)$ and t on any interval $(t_1 - e, t_2 + e)$ for which e is positive and sufficiently small, while the Jacobian

$$\Delta(t, t_0) = \frac{D(x^1, \cdots, x^m)}{D(t, u_1, \cdots, u_n)} \not\equiv 0, \qquad (u) = (0), \tag{1.7}$$

on g neighboring $t = t_0$. In terms of any representation of this character we define the conjugate points of $t = t_0$ on g as the points t on g at which $\Delta(t, t_0) = 0$. The order of vanishing of $\Delta(t, t_0)$ at a conjugate point t will be termed the *order* of that conjugate point. It is easy to show that the conjugate points and their orders are independent of the particular coördinate system (x) in which g lies and of representations of the character (1.6) in terms of which the conjugate points are defined.

2. **The second variation.** We suppose that $F(x, \dot{x}) > 0$ along g. No generality is lost in making this assumption. If this hypothesis were not satisfied by a given integrand $G(x, \dot{x})$, we could set

$$F(x, \dot{x}) = G(x, \dot{x}) + k^2\dot{x}^i\dot{\gamma}^i(t),$$

and obtain a new integrand F which would be positive along g provided k were sufficiently large. Moreover the extremals and conjugate points corresponding to the two integrands would be the same, and F and G would satisfy the condition (1.5) together.

With $F > 0$ along g we can represent any curve h on which (x) and $(\dot{x})$ are sufficiently near similar elements on g in terms of a parameter t which equals the value of J along h. Such a parameter t will be called the J-length along h. Along a curve on which t is the *J-length* F will be identically 1. We suppose g represented in terms of its J-length in the form

$$x^i = \gamma^i(t) \qquad (t_1 \leqq t \leqq t_2). \tag{2.0}$$

Recall that the second variation in the fixed end point problem corresponding to an extremal arc g has the form

$$I(\eta) = \int_{t_1}^{t_2} 2\Omega(\eta, \dot{\eta})\, dt, \tag{2.1}$$

where

$$2\,\Omega(\eta, \dot{\eta}) = F^0_{x^ix^j}\eta^i\eta^j + 2F^0_{x^ir^j}\eta^i\dot{\eta}^j + F^0_{r^ir^j}\dot{\eta}^i\dot{\eta}^j, \tag{2.2}$$

and that the Jacobi equations (written J. E.) have the form

$$L_i(\eta) = \frac{d}{dt}\Omega_{\dot{\eta}^i} - \Omega_{\eta^i} = 0 \qquad (i = 1, \cdots, m). \tag{2.3}$$

As is well known any two solutions $u^i(t)$ and $z^i(t)$ of the J. E. of class C^2 satisfy the integral

$$u^i\Omega_{\dot{\eta}^i}(z, \dot{z}) - z^i\Omega_{\dot{\eta}^i}(u, \dot{u}) \equiv \text{constant}. \tag{2.4}$$

Recall also that the equations $L_i(\eta) = 0$ are not independent, in fact satisfy the relation (cf. M, p. 123)

$$\dot{\gamma}^i(t)L_i(\eta) \equiv 0 \tag{2.5}$$

for all functions $\eta^i(t)$ of class C^2. Because of this we replace the system (2.3) by the system

$$L_i(\eta) = 0, \qquad \frac{d}{dt}M(\eta) = 0, \qquad (i = 1, \cdots, m), \tag{2.6}$$

where

$$M(\eta) = F^0_{x^i}\eta^i + F^0_{r^i}\dot{\eta}^i. \tag{2.7}$$

The operator $M(\eta)$ is suggested as follows. If $x^i(t, e)$ is a family of extremals of class C^2 yielding g for $e = 0$, we have

$$\frac{\partial}{\partial e}F[x(t, e), \dot{x}(t, e)]\bigg|^0 = M(\eta), \tag{2.8}$$

where

$$\eta^i = x^i_e(t, 0). \tag{2.9}$$

In particular if $F \equiv 1$ on members of the family, the corresponding variations η^i will satisfy the system (2.6). We term the system (2.6) the *restricted* Jacobi equations (written R. J. E.).

The R. J. E. are equivalent to the system

$$L_i(\eta) + \mu F^0_{r^i} = 0, \tag{2.10$'$}$$

$$\frac{d}{dt} M(\eta) = 0, \tag{2.10$''$}$$

as a set of conditions on (η). For if we multiply the i-th equation in (2.10)$'$ by $\dot{\gamma}^i$ and sum, it follows from (2.5) and the relation

$$F^0_{r^i}\dot{\gamma}^i \equiv F^0 \equiv 1$$

that $\mu \equiv 0$. The functional determinant of the system (2.10) with respect to the variables $\ddot{\eta}^i$ and μ is

$$\begin{vmatrix} F_{r^ir^j} & F_{r^i} \\ F_{r^j} & 0 \end{vmatrix}^0 = -F^0_1(F^0_{r^i}\dot{\gamma}^i)^2 = -F^0_1 \neq 0,$$

in accordance with M, p. 112. Hence the R. J. E. can be written in the form

$$\ddot{\eta}^i = H^i(\eta, \dot{\eta}), \tag{2.11}$$

where the functions H^i are linear and homogeneous in the variables η^i, $\dot{\eta}^i$ with coefficients which are functions of t of class C^{r-2}.

If $\rho(t)$ is any function of class C^2,

$$\eta^i = \rho(t)\dot{\gamma}^i(t) \tag{2.12}$$

is a solution of the J. E., (2.3). In fact for values of the constant e sufficiently near zero the functions

$$x^i = \gamma^i[t + e\rho(t)] \tag{2.13}$$

represent g and hence satisfy the Euler equations. According to the theorem of Jacobi the partial derivatives of the right members of (2.13) with respect to e, evaluated for $e = 0$, afford a solution of the J. E. These partial derivatives reduce to the form (2.12), and our statement is proved. We term solutions of the J. E. of the form (2.12) *tangential variations*. We continue with the following lemma.

LEMMA 2.1. *The only tangential variations which are solutions of the R. J. E. are of the form* $(a + bt)\dot{\gamma}^i$ *where* a *and* b *are constants.*

For functions $\eta^i(t)$ of the form (2.12), $M(\eta)$ takes the form

$$M(\eta) = F^0_{r^i}\dot{\rho}\dot{\gamma}^i + \rho[F^0_{r^i}\ddot{\gamma}^i + F^0_{x^i}\dot{\gamma}^i]. \tag{2.14}$$

But the factor in the bracket vanishes, as we see upon differentiating the identity

$$F[\gamma(t), \dot{\gamma}(t)] \equiv 1$$

with respect to t. Hence (2.14) takes the form

$$M(\eta) \equiv \dot{\rho},$$

from which the lemma follows.

We return to the determinant $\Delta(t, t_0)$ of (1.7). We suppose now that the extremals of the family (1.6) have been so represented that t is the J-length on each extremal. We then multiply the first column of $\Delta(t, t_0)$ by $t - t_0$ and denote the resulting determinant by $D(t, t_0)$. We term D the *conjugate point determinant.* Each of its columns is a solution of the R. J. E. and vanishes at $t = t_0$. The columns of D are independent solutions of the R. J. E. since $D \not\equiv 0$ for t near t_0. It follows from the form (2.11) of the R. J. E. that the columns of D form a base for solutions of the R. J. E. which vanish at $t = t_0$.

We continue with the following theorem.

THEOREM 2.1. *A necessary and sufficient condition that a point $t = a$ be conjugate to $t = t_0$ on g is that there exist a solution of the R. J. E. which vanishes at $t = t_0$ and $t = a$ but which does not vanish identically.*

To prove the condition necessary we suppose that $D(a, t_0) = 0$ with $a \neq t_0$. There will then exist a proper linear combination (η) of the columns of $D(t, t_0)$ which vanishes at $t = a$. We note that $(\eta) = (0)$ at t_0 but that $(\eta) \not\equiv (0)$ since $D(t, t_0) \not\equiv 0$. Hence the condition is necessary.

To prove the condition sufficient we assume that (w) is a solution of the R. J. E. which vanishes at $t = t_0$ and $t = a$ with $(w) \not\equiv (0)$. Since the columns of $D(t, t_0)$ form a base for solutions of the R. J. E. which vanish at $t = t_0$, (w) is dependent on the columns of D. But $(w) = (0)$ at a so that $D(a, t_0) = 0$, and the condition is proved sufficient.

Understanding that the nullity ν of a determinant equals its order minus its rank r we shall prove the following theorem. (Cf. M, p. 47, Theorem 3.1.)

THEOREM 2.2. *If $\theta(t)$ is a determinant whose columns are m independent solutions of the R. J. E. which vanish at $t = t_0$, the order of vanishing of $\theta(t)$ at any point $t = a$ equals the nullity ν of $\theta(a)$.*

Let r be the rank of $\theta(a)$ so that $\nu + r = m$. Without loss of generality we can suppose that the rank of the first r columns of $\theta(a)$ is r, for this would result after a suitable reordering of the columns. We lose no generality if we also suppose that the rank of the last ν columns of $\theta(a)$ is zero, since this would result upon adding suitable linear combinations of the first r columns to the remaining columns. With this understood let

$$u_h^i(t) \qquad (h = 1, \cdots, r),$$

$$z_k^i(t) \qquad (k = 1, \cdots, \nu)$$

represent the first r and last ν columns respectively of $\theta(t)$. Upon applying the integral form of the law of the mean to the elements in the last ν columns of $\theta(t)$, we find that

$$\theta(t) = (t - a)^\nu B(t), \tag{2.15}$$

where $B(t)$ is continuous in t and

$$B(a) = | u_h^i(a) \quad \dot{z}_k^i(a) | \qquad (h = 1, \cdots, r; k = 1, \cdots, \nu). \tag{2.16}$$

The theorem will follow from (2.15) if we show that $B(a) \neq 0$.

Suppose that $B(a) = 0$. I say first that $r > 0$. Otherwise there would exist a proper linear combination $\eta^i(t)$ of the columns of $\theta(t)$ such that $\eta^i(a) = \dot{\eta}^i(a) = 0$ for each i. It would follow that $(\eta) \equiv (0)$ contrary to the hypothesis that the columns of $\theta(t)$ are independent.

We suppose then that $B(a) = 0$ and $r > 0$. There will exist a proper linear combination (w) of the columns of $B(a)$ with coefficients

$$c_1, \cdots, c_r, -d_1, \cdots, -d_\nu,$$

such that $(w) = (0)$. Moreover, the constants d_k are not all null. Otherwise

$$c_h u_h^i(a) = 0 \qquad (h = 1, \cdots, r)$$

for each i, and the rank of $\theta(a)$ would be less than r.

We set

$$u^i(t) = c_h u_h^i(t), \qquad z^i(t) = d_k z_k^i(t), \qquad (i = 1, \cdots, m).$$

Then

$$u^i(a) = \dot{z}^i(a), \qquad z^i(a) = 0, \qquad (i = 1, \cdots, m). \tag{2.17}$$

We note that

$$[u(a)] \neq [0]. \tag{2.18}$$

Otherwise the relations (2.17) would imply that $[z(t)] \equiv [0]$, contrary to the hypothesis that the columns of $\theta(t)$ are independent.

The relation (2.4) is satisfied by the solutions $u^i(t)$ and $z^i(t)$ with its right member null. For $u^i(t)$ and $z^i(t)$ vanish at $t = t_0$ for each i. Recalling that

$$\Omega_{\dot{\eta}^i}(z, \dot{z}) = F^0_{r^i r^j} \dot{z}^j + F^0_{r^i x^j} z^j$$

and making use of (2.17), we see that relation (2.4) reduces to the relation

$$F^0_{r^i r^j} u^i(a) u^j(a) = 0$$

at $t = a$. It follows from (1.5) that

$$u^i(a) = c\dot{\gamma}^i(a) \qquad (c \neq 0; i = 1, \cdots, m).$$

This is impossible. For there would then exist a solution of the R. J. E. of the form

$$\eta^i(t) = z^i(t) - c(t - a)\dot{\gamma}^i \tag{2.19}$$

with $\eta^i(t) \equiv 0$ for each i, since

$$\eta^i(a) = \dot{\eta}^i(a) = 0 \qquad (i = 1, \cdots, m).$$

But upon setting $t = t_0$ in (2.19) we find that

$$0 = c(t_0 - a)\dot{\gamma}^i(t_0) \qquad (i = 1, \cdots, m). \tag{2.20}$$

Now $t_0 \neq a$ in the case $r > 0$, so that relations (2.20) are impossible, and the proof is complete.

The conjugate point determinant $D(t, t_0)$ satisfies the conditions on $\theta(t)$. In particular its columns are independent since $D(t, t_0) \not\equiv 0$ near $t = t_0$. We accordingly have the following corollary of the theorem.

COROLLARY 2.1. *The conjugate points of $t = t_0$ on g are isolated and possess orders equal to the nullity of the conjugate point determinant $D(t, t_0)$ at the respective zeros $t \neq t_0$ of this determinant.*

The columns of $D(t, t_0)$ form a base for solutions of the R. J. E. which vanish at $t = t_0$. The maximum number of independent solutions of the R. J. E. which vanish at $t = t_0$ and at a conjugate point $t = a$ of $t = t_0$ will accordingly equal the nullity of $D(a, t_0)$. We thus have a second corollary of the theorem.

COROLLARY 2.2. *The maximum number of independent solutions of the R. J. E. which vanish at $t = t_0$ and at a conjugate point $t = a$ of $t = t_0$ equals the order of the conjugate point $t = a$.*

The conjugate point determinant can be given a representation of the form

$$D(t, t_0) = (t - t_0)^m B(t, t_0),$$

where $B(t, t_0)$ is continuous in its arguments for t and t_0 on the interval (t_1, t_2). Moreover,

$$B(t_0, t_0) \neq 0,$$

as we have seen in connection with (2.15). It follows that the first conjugate point of $t = t_0$ following t_0 is bounded away from t_0 on g. We shall make use of this fact in the next section.

3. **The Index Theorem.** We continue with the extremal g represented in terms of its J-length as in (2.0). Let A and B be respectively the initial and final points of g. Let ω be a positive lower bound of J-lengths on g between a point on g and the first following conjugate point. Let

$$a_0 < a_1 < \cdots < a_{p+1} \qquad (a_0 = t_1 ; a_{p+1} = t_2)$$

be a set of values of t such that

$$a_s - a_{s-1} < \omega \qquad (s = 1, \cdots, p + 1). \tag{3.0}$$

Let $A_0, \cdots, A_{p+1}$ be the corresponding points on g. Let M_q be a regular manifold of class C^{r-1} (cf. §1) which intersects g at A_q, $q = 1, \cdots, p$, but which is not tangent to g at A_q. We suppose M_q regularly represented neighboring A_q in terms of parameters

$$u_q^j \qquad (j = 1, \cdots, n = m - 1) \tag{3.1}$$

in such a manner that the set $(u_q) = (0)$ determines A_q on g. We shall term the manifolds M_q a set of *intermediate* manifolds.

It will be convenient to write the ensemble of the sets u_q^j in the form

$$(z^1, \cdots, z^{pn}) = (u_1^1, \cdots, u_1^n, \cdots, u_p^1, \cdots, u_p^n). \tag{3.2}$$

The ensemble (z) determines a set of points

$$P_1, \cdots, P_p \tag{3.3}$$

on the respective manifolds M_q provided (z) is sufficiently near the set $(z) = (0)$.

For (z) sufficiently near (0) we can join the successive points

$$A, P_1, \cdots, P_p, B \tag{3.3$'$}$$

by extremal arcs forming a broken extremal $E(z)$. The J-length of $E(z)$ will be denoted by $J(z)$ and will be a function of class C^{r-1} of its arguments for (z) sufficiently near (0). The Index Theorem can now be stated in full as follows.

INDEX THEOREM. *The point $(z) = (0)$ is a critical point of $J(z)$ with an index equal to the number of conjugate points of $t = t_1$ on g preceding $t = t_2$, and a nullity equal to the order of $t = t_2$ as a conjugate point of $t = t_1$ on g.*

It is understood that each conjugate point is to be counted a number of times equal to its order. Recall also that the *nullity* of a critical point is defined as the nullity of the Hessian of the function at the critical point.

We shall represent the broken extremal $E(z)$ in terms of a parameter t which equals a_0 and a_{p+1} respectively at the end points of g and which equals a_q on the respective manifolds M_q. We suppose t chosen so that between any two successive vertices (3.3)′ the rate of change of t with respect to the J-length is constant. We thereby obtain a representation of $E(z)$ of the form

$$x^i = x^i(t, z) \qquad (t_1 \leqq t \leqq t_2), \tag{3.4}$$

where $x^i(t, z)$ is continuous in its arguments for (z) sufficiently near (0) and of class C^{r-1} for t on the respective intervals (a_k, a_{k+1}). For t fixed the functions $x^i(t, z)$ are of class C^{r-1} in (z) without exceptional values of t.

A computation involving the usual differentiation and integration by parts discloses the fact that for $(z) = (0)$,

$$\frac{\partial J}{\partial z^\mu} = \sum_{k=0}^{p} \left[F^0_{r^i} \frac{\partial x^i}{\partial z^\mu} \right]_{a_k}^{a_{k+1}} \qquad (\mu = 1, \cdots, pn). \tag{3.5}$$

Since the contributions to the right members from the fixed end points are null and the contributions from the two extremal arcs with end points on M_q cancel, it appears that $(z) = (0)$ is a critical point of $J(z)$, as stated.

We set

$$Q(z) = \frac{\partial^2 J^0}{\partial z^\mu \partial z^\nu} z^\mu z^\nu \qquad (\mu, \nu = 1, \cdots, pn), \tag{3.6}$$

where the superscript 0 indicates evaluation for $(z) = (0)$, and term $Q(z)$ an *index form* corresponding to g.

Let e be a parameter. For (z) fixed,

$$x^i = x^i(t, ez) \qquad (i = 1, \cdots, m) \tag{3.7}$$

is a 1-parameter family of broken extremals. We find that

$$\frac{d^2}{de^2} J(ez) \Big|^{e=0} = Q(z),$$

and infer the following.

THEOREM 3.1. *The index form $Q(z)$ admits the representation*

$$Q(z) = \int_{t_1}^{t_2} 2\,\Omega(\eta, \dot{\eta})\, dt, \tag{3.8}$$

with

$$\eta^i(t) = a^i_\mu(t) z^\mu \qquad (\mu = 1, \cdots, pn), \tag{3.9}$$

where

$$a^i_\mu(t) = \frac{\partial x^i(t, z)}{\partial z^\mu} \Big|^{(z)=(0)}. \tag{3.10}$$

In order to make full use of this theorem we shall need certain properties of quadratic forms. To that end let

$$q(x) = a_{ij} x_i x_j \qquad (i, j = 1, \cdots, m)$$

be a symmetric quadratic form. Let r be the rank of $| a_{ij} |$. The number $m - r$ is termed the *nullity* of q and will be denoted by $N(q)$. The *index* of $q(x)$ is the number of negative characteristic roots belonging to $| a_{ij} |$ and will be denoted by $I(q)$. Recall that the characteristic roots will vary continuously with the coefficients a_{ij}. The index $I(q)$ is an integer, and, although not a continuous function of the coefficients a_{ij}, is easily seen to be *lower semi-continuous.*

Let π_h be an h-plane passing through the origin in the space (x). The h-plane π_h can be represented in the form

$$x_i = b_{ij} u_j \qquad (j = 1, \cdots, h;\ i = 1, \cdots, m), \tag{3.11}$$

where the matrix $\| b_{ij} \|$ has the rank h. Upon replacing the variables (x) in $q(x)$ by the right members of (3.11), we obtain a form $H(u)$ in the variables (u). We term $H(u)$ a form defined by $q(x)$ on π_h. We say that $q(x)$ is negative definite or semi-definite on π_h if $H(u)$ is negative definite or semi-definite. With this understood we state the following lemma (M, p. 61, Lemma 7.1).

LEMMA 3.1. (a) *A necessary and sufficient condition that the index of $q(x)$ be at least h is that $q(x)$ be negative definite on some h-plane π_h through the origin.* (b) *A necessary and sufficient condition that the index plus the nullity of $q(x)$ be at least p is that $q(x)$ be negative semi-definite on some p-plane through the origin.*

Let $A(y_1, \cdots, y_n)$ be a quadratic form in the n variables (y) defined by the relation

$$A(y) = q(y_1, \cdots, y_n, 0, \cdots, 0) \qquad (0 < n \leqq m).$$

Concerning $A(y)$ we state the following lemma.

LEMMA 3.2. (a) *The index of the form* $q(x)$ *is at least that of* $A(y)$. (b) *If* $q(x)$ *is non-degenerate,*

$$I(q) \geqq I(A) + N(A). \tag{3.12}$$

The first statement in the lemma is a consequence of the first statement in Lemma 3.1 since $A(y)$ is negative definite on some h-plane through the origin with $h = I(A)$. To prove the relation (3.12) let p equal the right member of (3.12). According to Lemma 3.1 (b) there exists a p-plane through the origin on which $A(y)$ is negative semi-definite. Upon identifying the space (y) with the subspace of the first n coördinates x_i, we see that $q(x)$ is negative semi-definite on this p-plane. It follows from Lemma 3.1 (b) that

$$I(q) + N(q) \geqq p.$$

But $N(q) = 0$ by hypothesis, and relation (3.12) follows.

4. **Proof of the Index Theorem.** The second variation is evaluated in Theorem 3.1 along a family of curves given by (3.9). We regard the sets (z) as parameters in this family. On each interval

$$a_k \leqq t \leqq a_{k+1} \qquad (k = 0, \cdots, p), \tag{4.1}$$

it is seen that (3.9) defines a solution of the R. J. E. That the J. E. are satisfied follows from the fact that the functions $x^i(t, z)$ of (3.4) define extremal arcs λ for t on (4.1). But on each such extremal arc λ, F is identically constant by virtue of our choice of t. Hence the last condition in (2.6) is satisfied by the subarcs of (3.9) for which (4.1) holds. For (z) fixed, (3.9) thus defines a "broken" solution of the R. J. E.

For q fixed the variables u_q^j in the set (z) are parameters in our representation of M_q. Let the partial derivative of $x^i(t, z)$ with respect to u_q^j be indicated by adding the subscripts j and q. Note that the matrix (q fixed)

$$\| x_{jq}^i(a_q, 0) \| \qquad (i = 1, \cdots, m; j = 1, \cdots, n) \tag{4.2}$$

has the rank n since M_q is regularly represented in terms of the parameters u_q^j. From the fact that M_q is not tangent to g at its intersection with g it follows that

$$| \dot{\gamma}^i(a_q) \quad x_{jq}^i(a_q, 0) | \neq 0, \tag{4.3}$$

where the determinant in (4.3) is obtained from the matrix (4.2) by adjoining the column $\dot{\gamma}^i(a_q)$.

We continue with the following lemma.

LEMMA 4.1. *Linearly independent sets* (z) *determine linearly independent broken solutions of the R. J. E. in* (3.9).

If $(z) \neq (0)$, there will be some q for which the set u_q^j is not null. At the point $t = a_q$ on the broken solution $\eta^i(t)$ determined by (z) in (3.9) the set

$$\eta^i(a_q) = x_{jq}^i(a_q, 0)u_q^j \qquad (i = 1, \cdots, m; j = 1, \cdots, n)$$

is not null since the rank of the matrix (4.2) is n. The lemma follows directly.

We shall say that a broken solution of the family (3.9) is a *special* solution if corresponding to each corner $t = a_q$ there is a constant c such that

$$\Delta\dot{\eta}^i = \dot{\eta}^i \Big|_{a_q^-}^{a_q^+} = c\dot{\gamma}^i(a_q) \qquad (i = 1, \cdots, m). \tag{4.4}$$

Such corners will be called *tangential* corners. A broken solution of (3.9) whose subarcs are tangential solutions will have tangential corners as one sees readily. With this understood we state the following lemma.

LEMMA 4.2. *A necessary and sufficient condition that a set* $(z) \neq (0)$ *be a critical point of* $Q(z)$ *is that* (z) *determine a special solution of the family* (3.9).

Let (z) be a critical point of $Q(z)$. For q fixed we shall show that the n conditions

$$Q_{u_q^j} = 0 \qquad (j = 1, \cdots, n) \tag{4.5}$$

imply that $t = a_q$ is a tangential corner of the broken solution determined by (z). For the conditions (4.5) can be written more explicitly in the form (q not summed)

$$\Omega_{\dot{\eta}^i} \Big|_{a_q^-}^{a_q^+} x^i_{jq}(a_q, 0) = 0 \qquad (i = 1, \cdots, m; j = 1, \cdots, n), \tag{4.6$'$}$$

as follows from (3.8). Furthermore

$$\Omega_{\dot{\eta}^i} \Big|_{a_q^-}^{a_q^+} \dot{\gamma}^i(a_q) = 0. \tag{4.6$''$}$$

In fact the left member of (4.6)″ takes the form

$$(\dot{\gamma}^i F^0_{r^ir^j})\Delta\dot{\eta}^j \qquad (t = a_q; i, j = 1, \cdots, m), \tag{4.7}$$

and the parentheses in (4.7) are null. As a system of equations on the variables

$$\Omega_{\dot{\eta}^i} \Big|_{a_q^-}^{a_q^+} \qquad (q \text{ fixed}), \tag{4.8}$$

the combined equations (4.6) have a determinant (4.3) which does not vanish. Hence the variables (4.8) are null, or written more explicitly

$$F^0_{r^ir^j}\Delta\dot{\eta}^j = 0 \qquad (t = a_q; i, j = 1, \cdots, m). \tag{4.9}$$

Relations of the form (4.4) follow. The condition of the lemma is accordingly necessary.

Conversely, if (z) determines a special solution (η) in the family (3.9), then, for each q, equations (4.4), (4.9) and (4.6) are seen to hold. Hence (4.5) holds by virtue of its form (4.6)′. The set (z) defines a critical point of $Q(z)$, and the proof is complete.

If we recall that the broken solutions (η) in (3.9) are linearly independent if and only if the sets (z) which determine them in (3.9) are linearly independent, we can infer from the preceding lemma that the nullity of $Q(z)$ equals the number of linearly independent special solutions. We come then to the proof of the following theorem.

THEOREM 4.1. *The nullity of $Q(z)$ equals the order of $t = t_2$ as a conjugate point of $t = t_1$.*

Let q be any one of the integers $1, \cdots, p$. We define a broken solution $\eta_q^i(t)$ as follows (q fixed):

$$\begin{aligned} \eta_q^i &\equiv (t - a_0)\dot{\gamma}^i, & a_0 \leqq t \leqq a_q, \\ \eta_q^i &\equiv k(t - a_{q+1})\dot{\gamma}^i, & a_q \leqq t \leqq a_{q+1}, \\ \eta_q^i &\equiv 0, & a_{q+1} \leqq t \leqq a_{p+1}, \end{aligned} \tag{4.10}$$

where k is a constant so chosen that the solution η_q^i hereby defined is continuous at $t = a_q$. The choice of k is

$$k = \frac{a_q - a_0}{a_q - a_{q+1}} < 0. \tag{4.11}$$

The solution η_q^i has a tangential corner at $t = a_q$. For at $t = a_q$ we have

$$\Delta \dot{\eta}_q^i = (k - 1)\dot{\gamma}^i,$$

and $(k - 1) \neq 0$ as follows from (4.11).

To establish the theorem we begin by showing how each special broken solution $\eta^i(t)$ which is determined in (3.9) by a critical point (z) of $Q(z)$ determines an ordinary solution of the R. J. E. which vanishes at t_1 and t_2. In fact the broken solution

$$\bar{\eta}^i = \eta^i(t) - c_q \eta_q^i(t) \tag{4.12}$$

vanishes at t_1 and t_2 and becomes an ordinary solution if the constants $c_1, \cdots, c_p$ are successively chosen so that $\bar{\eta}^i$ has no corners at the points a_q. Such a choice of the constants c_q is possible since the choice of a constant c_q so as to eliminate a corner at $t = a_q$ does not introduce a new corner for which $t < a_q$. Moreover $(\bar{\eta}) \equiv (0)$ only if $(\eta) \equiv (0)$ in (4.12). For the non-vanishing of the determinants (4.3) and the condition $(\bar{\eta}) \equiv (0)$ imply that (η) is null at each of the points $t = a_q$. Hence $(\eta) \equiv (0)$ since no point $t = a_k$ is conjugate to its successor $t = a_{k+1}$. Thus linearly independent special solutions (η) determine ordinary linearly independent solutions $(\bar{\eta})$ vanishing at t_1 and t_2.

Conversely, let $(\bar{\eta})$ be an ordinary solution of the R. J. E. which vanishes at t_1 and t_2. We seek a special solution (η) which with suitable constants c_q satisfies (4.12). That is, we seek a set (z) and constants c_q such that

$$\bar{\eta}^i(t) \equiv \frac{\partial x^i(t, 0)}{\partial z^\mu} z^\mu - c_q \eta_q^i(t) \qquad (\mu = 1, \cdots, pn). \tag{4.13}$$

We start with the point $t = a_p$ and determine c_p and the last n of the variables (z) so that (4.13) holds at $t = a_p$. This is possible since the determinant (4.3) does not vanish. We next determine c_{p-1} and the n variables of the set (z) belonging to M_{p-1} so that (4.13) holds when $t = a_{p-1}$. We continue with c_{p-2}, etc., until all the constants c_q and the variables (z) have been determined. We note that the choice of c_q and the subset of (z) belonging to M_q will not affect the equations (4.13) at the points $t = a_r > a_q$. With (z) and the constants c_q so chosen, (4.13) holds at each of the points $t = a_q$, and accordingly holds identically in t.

Moreover, linearly independent solutions $(\bar{\eta})$ determine linearly independent special broken solutions (η). To establish this it is sufficient to show that a null solution (η) satisfies (4.12) with a solution $(\bar{\eta})$ and constants c_q only if $(\bar{\eta}) \equiv (0)$. But when $(\bar{\eta})$ and (η) are both without corners, the constants c_q in (4.12) must be null. Hence $(\bar{\eta}) \equiv (0)$ as stated.

The theorem follows with the aid of Corollary 2.2.

We continue with a proof of Theorem 4.2.

THEOREM 4.2. *The index of $Q(z)$ equals the sum of the orders of the conjugate points of t_1 on $t_1 < t < t_2$.*

It will be convenient to replace g by the subarc g_b on which $t_1 \leqq t \leqq b$, where $t_1 < b \leqq t_2$. On g_b we introduce the same number of intermediate vertices as previously with

$$(4.13)' \qquad a_0 < \cdots < a_{p+1} = b.$$

With g replaced by g_b we replace $J(z)$ and $Q(z)$ by $J^b(z)$ and $Q^b(z)$, respectively. The family of broken extremals $x^i(t, z)$ will here be denoted by $x^i(t, z, b)$. As b increases the functions $x^i(t, z, b)$ will vary continuously and retain the differentiability properties of $x^i(t, z)$ at least as long as the differences $a_s - a_{s-1}$ are less than the constant ω of (3.0).

We shall prove the theorem by proving the following statement.

(α) *The index of $Q^b(z)$ equals the sum of the orders of the conjugate points of t_1 on the interval $t_1 < t < b$.*

We continue with a proof of four lemmas.

LEMMA A. *Statement (α) is true when b lies between t_1 and the first conjugate point of t_1 on g.*

The Legendre condition (1.5) and the condition that b precede the first conjugate point of t_1 on g are sufficient for the segment $t_1 \leqq t \leqq b$ of g to afford a weak minimum to J. Hence

$$J^b(z) - J^b(0) \geqq 0$$

when (z) is sufficiently near (0). Whence $Q^b(z) \geqq 0$. But $Q^b(z)$ is non-degenerate by virtue of Theorem 4.1, so that $Q^b(z) > 0$ when $(z) \neq (0)$. This establishes Lemma A.

Corresponding to any admissible set (4.13)′ we now denote a_{p+1} by a_{p+2} and b, and insert a new point a_{p+1} between a_p and b. We add a new intermediate manifold M_{p+1} cutting g_b at $t = a_{p+1}$. The set of parameters replacing (z) will be denoted by (ζ). They will be $n(p+1)$ in number and will be chosen so that the first np components of (ζ) form the set (z). The set (ζ) will determine a broken extremal $\mathfrak{E}(\zeta)$ as previously. The broken extremal $\mathfrak{E}(\zeta)$ will coincide with the broken extremal $x^i(t, z, b)$ for $t_1 \leqq t \leqq a_p$. For the new construction the functions replacing $J^b(z)$ and $Q^b(z)$ will be denoted by $J^b_*(\zeta)$ and $Q^b_*(\zeta)$, respectively. We continue with the following lemma.

LEMMA B. $I(Q^b) \geqq I(Q^b_*)$.

We introduce a parameter e. The inequality

$$J^b_*(e\zeta) - J^b(ez) \geqq 0 \tag{4.14}$$

holds for e sufficiently small and for the first pn components of (ζ) given by (z). Regarded as a function $\varphi(e)$, the left member of (4.14) has a minimum when $e = 0$. Hence

$$\varphi''(0) = J^b_{*\zeta^r\zeta^s}(0)\zeta^r\zeta^s - J^b_{z^\mu z^\nu}(0)z^\mu z^\nu \geqq 0, \tag{4.15}$$

$$[\mu, \nu = 1, \cdots, np; r, s = 1, \cdots, n(p+1)].$$

The lemma follows from Lemma 3.1 (a).

Statement (α) is true for b sufficiently near t_1 as we have seen in Lemma A. Let c be a constant such that $t_1 < c \leqq t_2$. Statement (α) will follow for all values of b for which $t_1 < b \leqq t_2$ once we have established the following lemmas.

LEMMA C. *If (α) is true for $b < c$, it is true for $b = c$.*

LEMMA D. *If (α) is true for $b \leqq c$, it is true for $b > c$ and sufficiently near (c).*

As b increases $I(Q^b)$ changes at most when b passes through a conjugate point a of t_1 and will then increase by at most the order ν_a of a as a conjugate point of t_1. Hence for each value of c,

$$I(Q^c) \leqq \sum \nu_a \qquad (t_1 < a < c). \tag{4.16}$$

To apply Lemma B we take $a_{p+2} = c$ and a_{p+1} nearer c than any conjugate point of t_1. We then have

$$I(Q^c) \geqq I(Q^c_*) \geqq I(Q^{a_{p+1}}). \tag{4.17}$$

The first inequality follows from Lemma B. The second follows upon putting the last n of the variables in Q^c_* equal to zero and applying Lemma 3.2 (a). But since we are assuming that statement (α) is true for $b < c$, we have

$$I(Q^{a_{p+1}}) = \sum \nu_a \qquad (t_1 < a < c).$$

It follows from (4.17) that

$$I(Q^c) \geqq \sum \nu_a \qquad (t_1 < a < c). \tag{4.18}$$

Lemma C follows from (4.16) and (4.18).

To establish Lemma D let b be a constant $> c$ but nearer c than any conjugate point of t_1, possibly excepting c. We introduce the manifold M_{p+1} cutting g_b at $t = c$, and note that

$$I(Q^b) \geqq I(Q^b_*) \geqq I(Q^c) + N(Q^c). \tag{4.19}$$

The first inequality follows from Lemma B. The second follows upon setting the last n of the variables in Q^b_* equal to zero and applying Lemma 3.2 (b). But $N(Q^c)$ is the order of c as a conjugate point of t_1. Hence (4.19) yields the result

$$I(Q^b) \geqq \sum \nu_a \qquad (t_1 < a < b). \tag{4.20}$$

Lemma D follows upon comparing (4.20) with (4.16).

The proof of the theorem is complete. The Index Theorem follows from Theorems 4.1 and 4.2.

The index and nullity of $Q(z)$ depend only on the conjugate points of t_1 on g and their orders, and are accordingly independent of the number, position and representation of the manifolds M_q, provided these manifolds are admissibly distributed and represented.

INSTITUTE FOR ADVANCED STUDY.

Reprinted from the
AMERICAN JOURNAL OF MATHEMATICS
Vol. LX, No. 4, pp. 815–866

SYMBOLIC DYNAMICS.

By Marston Morse and Gustav A. Hedlund.

1. Introduction. In both its qualitative and quantitative aspects, modern theoretical dynamics is largely concerned with recurrence and transitivity. The qualitative study of dynamics owes its inception to Hill and Poincaré. Its development in the past three decades has been dominated by the work of Birkhoff, to whom belongs the principal credit for the trend of modern research in dynamics. The quantitative (metrical) study of dynamical systems is of more recent origin, but with the establishment of the ergodic theorems of Birkhoff and von Neumann, its development has been rapid.

The methods used in the study of recurrence and transitivity frequently combine classical differential analysis with a more abstract symbolic analysis. This involves a characterization of the ordinary dynamical trajectory by an unending sequence of symbols termed a *symbolic trajectory* such that the properties of recurrence and transitivity of the dynamical trajectory are reflected in analogous properties of its symbolic trajectory. For general research of this character one may refer to papers by Hadamard, Birkhoff and the authors. Papers which are special in that they refer to spaces of constant curvature, but which are significant by virtue of the symbolisms involved have been written by Artin, Koebe, Martin, Myrberg and Nielsen. Martin and Robbins have contributed papers on purely symbolic aspects.

In the general theory as developed up to the present, the dividing line between symbolic and differential methods has not been sharply defined. It is a first aim of the present paper to separate the symbolic aspects from the differential aspects and to develop the symbolic theory on its own account. In this way many interesting and important dynamical questions can be given an algebraic formulation quite apart from the classical theories of differential equations.

A symbolic trajectory T is formed from symbols taken from a finite set of generating symbols subject to rules of admissibility which are dependent on the underlying Poincaré fundamental group. The trajectory T taken with one of its symbols is called a *symbolic element.* Symbolic elements are analogues of line elements on an ordinary trajectory. A metric is assigned to the space of symbolic elements and this space turns out to be perfect, compact, and totally disconnected. The necessary first theorems on symbolic recurrence and transitivity are then rapidly developed in a way suggestive of the differential treatment of Birkhoff.

We come to the more novel developments of the paper. Recall that the recurrency index $R(n)$ of T is a minimum number m such that each block of m successive symbols of T contains every block of n successive symbols of T. If $R(n)$ exists for each n, T is termed recurrent (cf. Morse [2], p. 94). In § 7 we are concerned with conditions on $R(n)$. A typical condition is that $R(n) > 2n$ for every non-periodic recurrent T. In § 8 we make a special study of the Morse recurrent trajectory (Morse [2], p. 95) obtaining $R(n)$ explicitly. This is the first and only known evaluation of $R(n)$ for a non-periodic recurrent trajectory. We give a definition of almost periodicity of symbolic trajectories somewhat similar to the ordinary definition of Besicovitch ([1], p. 77) and show that the Morse trajectory is not almost periodic. Most of the known recurrent trajectories are almost periodic. In § 9 we give general methods for deriving new recurrent trajectories from a given recurrent trajectory. For the Morse trajectory, $R(n)/n$ is shown to have an inferior limit of 5.5 and a superior limit of 10. In § 10 we define a recurrent trajectory for which the inferior limit of $R(n)/n$ is finite while the superior limit is $+\infty$.

We continue with a study of transitivity. An admissible symbolic sequence of the form $a_1a_2a_3 \cdots$ is termed a *ray*. We assign the Baire metric (cf. Hausdorff [1], p. 101) to the space of rays. Let $\theta(n)$ be the minimum length of a block which contains a copy of each admissible n-block. A ray R which contains a copy of each admissible block is termed transitive. If R is transitive, let $\phi(n)$ be the minimum length of an initial block of R which contains a copy of each admissible n-block. We generalize a theorem of Martin [1] and include a hitherto unpublished theorem of R. Oldenburger on the impossibility that $\phi(n) = \theta(n)$ for two successive values of n $(n > 1)$. A typical general theorem states that the set of transitive rays for which

$$\lim_{n\to\infty} \inf. \frac{\phi(n)}{\theta(n)} = 1$$

is a residual set of the second category.

Symbolic dynamics as the authors conceive it forms one of the three divisions

(1) representation theory,
(2) symbolic dynamics,
(3) existence of space forms,

of the whole theory. The representation theory is concerned with the conditions on space forms under which trajectories admit a one-to-one symbolic representation in terms of which the recurrence or transitivity of the trajectory can be determined. These conditions will involve the Poincaré fundamental group of the space and differential conditions such as that of uniform instability (cf. Morse [4], p. 64). In (3) one is concerned with the existence of space forms satisfying the conditions discovered in (1). The questions involved are rather deep extensions of the Hilbert, Koebe theory of spaces of negative curvature (cf. Hilbert [1], and Koebe [1]). A simple typical theorem is that there exists no two-dimensional Riemannian manifold of the topological type of the torus satisfying the condition of uniform geodesic instability. The bearing of such studies on questions of topological and metric transitivity will be made clear in later papers.

2. Symbolic sequences. Having seen the significance of symbolic sequences we shall now develop their theory on an independent basis. We start with a finite set of μ distinct symbols termed *generating symbols.* We suppose that $\mu > 1$. If a_n represents a generating symbol, sequences of the form

$$\cdots a_{-1}\, a_0\, a_1 \cdots, \tag{2.1}$$

$$a_r\, a_{r+1}\, a_{r+2} \cdots, \tag{2.2'}$$

$$\cdots a_{m-2}\, a_{m-1}\, a_m, \tag{2.2''}$$

$$a_{r+1}\, a_{r+2} \cdots a_{r+n}, \tag{2.3}$$

will be termed indexed *sequences* (written I-sequences). More particularly I-sequences of the forms (2.1), (2.2) and (2.3) will be termed *I-trajectories, I-rays,* and *I-blocks* respectively. It will be convenient to term the generating symbol represented by a_n the *value* of a_n. The I-blocks (2.3) will be said to have the *length* n and will frequently be termed n-blocks.

Suppose A and B are I-sequences of symbols a_n and b_m respectively. We shall regard A and B as identical and write $A \equiv B$ if and only if the ranges of the indices n and m are the same and a_n and b_n have the same value. Let r be an integer, positive, negative or zero. Suppose that the ranges of n and m are such that symbols a_n and b_m correspond in a one-to-one manner with a_n corresponding to b_{n+r}. If then a_n and b_{n+r} have the same value we shall say that A and B are *similar* and write $A \sim B$. We shall also write

$$B \equiv D_r A, \qquad A \equiv D_{-r} B.$$

If $r = 0$, $D_r A \equiv A$, but in general $D_r A$ is different from A when $r \neq 0$.

The class of I-sequences similar respectively to an I-trajectory, I-ray, or I-block will be termed a *trajectory, ray,* or *block,* and will be represented by the sequence of values involved without indices. For example if 1 and 2 are generating symbols, the unending sequence

$$\cdots 121212 \cdots$$

is a trajectory but not an I-trajectory.

Corresponding to an I-sequence A we shall refer to *I-subblocks* or *I-subrays.* We shall thereby mean subsets of consecutive symbols of A 'indexed as' in A with successive indices differing by unity. Thus if A is of the form

$$a_1\, a_2\, a_3\, a_4,$$

a_2a_3 is an I-subblock, but a_1a_3 is not an I-subblock.

In classical dynamics an element based on a motion T is a point P of T together with the direction tangent to T at P. The symbolic analogue of such an element is defined as follows. Let (a) be an I-trajectory and a_r the r-th symbol of (a). The I-trajectory (a) taken with a_r will be termed an I-element based on (a) at a_r, and will be denoted by $E(r,a)$. We term r the *index* of $E(r,a)$ and a_r the *preferred* symbol. Let $E(s,b)$ be an I-element of index s based on an I-trajectory (b). We shall regard $E(r,a)$ and $E(s,b)$ as identical, and write

$$E(r,a) \equiv E(s,b)$$

if and only if $(a) \equiv (b)$ and $r = s$. We shall say that $E(r,a)$ is similar to $E(s,b)$ and write $E(r,a) \sim E(s,b)$ if

$$D_{-r}(a) \equiv D_{-s}(b).$$

The class of I-elements similar to a given I-element $E(r,a)$ will be termed a *symbolic element* E. Two symbolic elements are regarded as identical if the corresponding classes of I-elements are identical. We say that E is based on (a) and represented by $E(r,a)$.

We shall give an illustrative example. Let (a) be an I-trajectory in which a_n equals 1 or 2 according as n is even or odd, and let (b) be an I-trajectory in which $b_{n+1} = a_n$. Evidently

$$E(0,a) \not\equiv E(0,b), \qquad E(0,a) \sim E(1,b).$$

More generally if n is even and m is odd,

$$E(n,a) \sim E(m,b)$$

and these similar elements serve to define a symbolic element E. There are just two different symbolic elements based on (a) or (b).

3. The space of symbolic elements. We shall assign a distance E_1E_2 to any two symbolic elements E_1, E_2. To that end we say that an I-block of the form

$$c_{r-m} \cdots c_r \cdots c_{r+m}$$

is *centered* at c_r. Let E_1 and E_2 be represented by indexed elements $E(r,a)$ and $E(s,b)$ respectively. If $E_1 \not\equiv E_2$, let m be the length of the longest I-blocks of (a) and (b) which are similar and centered at a_r and b_s respectively. We then set

$$E_1E_2 = E(r,a)E(s,b) = 1/m \qquad (m \neq 0). \tag{3.1}$$

In the special case where $m = 0$ we regard the distance (3.1) as infinite. If $E_1 \equiv E_2$, the distance (3.1) is taken as zero. It is clear that the distance E_1E_2 is independent of the particular I-elements $E(r,a)$ and $E(s,b)$ chosen to represent E_1 and E_2 respectively.

The distances so defined satisfy the usual metric axioms:

$$E_1E_2 = 0 \text{ if and only if } E_1 \equiv E_2, \tag{3.2}$$

$$E_1E_2 = E_2E_1, \tag{3.3}$$

$$E_1E_3 \leqq E_1E_2 + E_2E_3. \tag{3.4}$$

Relation (3.4) follows from the stronger relation

$$E_1E_3 \leqq \max(E_1E_2, E_2E_3). \tag{3.5}$$

In case any one of the distances involved in (3.5) is 0 or ∞, (3.5) holds trivially as is readily seen. Excepting these cases, let h, k, and m be the reciprocals of E_1E_3, E_1E_2 and E_2E_3 respectively. Consider the case $k \geqq m$. Evidently then $h \geqq m$ so that $h \geqq \min(k, m)$. The latter relation yields the condition

$$1/h \leqq \max(1/k, 1/m)$$

from which (3.5) follows. The case $m \geqq k$ leads to the same result.

Recall that a metric space M is compact if each infinite sequence of points of M contains at least one subsequence which converges to a point of M. With this understood we shall prove the following.

THEOREM 3.1. *The space M of all symbolic elements is compact.*

Let E_n be the n-th element in an arbitrary sequence of symbolic elements. Suppose E_n is based on the I-trajectory (a^n). Without loss of generality we can suppose that the preferred symbol in (a^n) has the index 0. We shall define a doubly infinite matrix of I-trajectories

$$T_{ij} \qquad (i, j = 1, 2, \cdots) \tag{3.6}$$

as follows. Since there is at most a finite number of generating symbols there exists an infinite subset of the symbols

$$a_0^n \qquad (n = 1, 2, \cdots)$$

which have the same "value" c_0. Let the I-trajectories T_{1j} be a subsequence of the I-trajectories (a^n) in which $a_0{}^n$ has the value c_0. Proceeding inductively we suppose that the first $k-1$ rows of the matrix (3. 6) have been determined. The k-th row shall then be a subsequence T_{ki} of the $(k-1)$-st row such that the symbols of T_{ki} of index $k-1$ have a common value c_{k-1}, while the symbols of index $1-k$ have a common value c_{1-k}.

Let (c) be the I-trajectory

$$\cdots c_{-1}\, c_0\, c_1 \cdots.$$

Then (c) has an I-block of length $2k-1$, with center at c_0, in common with T_{kk}. Let $\mathcal{E}_k$ be the element of index 0 based on T_{kk}, and let $\mathcal{E}$ be the element of index 0 based on (c). Then

$$\mathcal{E}\mathcal{E}_k \leqq \frac{1}{2k-1},$$

so that the elements $\mathcal{E}_k$ converge to $\mathcal{E}$. Moreover the elements $\mathcal{E}_k$ are a subsequence of the given elements $\mathcal{E}_n$, and the proof of the theorem is complete.

A metric space is termed *totally disconnected* if its connected subsets all reduce to points. Cf. Hausdorff [1], p. 152. Recall also that if C is a connected set and d is a distance assumed by a pair of points of C, there exists a pair of points of C which assumes an arbitrary distance between 0 and d. In the case of the space M of symbolic elements, the distances take on only the discrete values $1/n$, so that the space M is totally disconnected.

Finally, the space M is perfect. We already know that M is compact. To prove that M is perfect it is necessary to show that in every neighborhood of an indexed element $E(0,a)$ there is an I-element $E(0,b)$ not similar to $E(0,a)$. To that end we take (b) so that it has a prescribed block A of (a) centered at a_0, in common with (a), but that all longer blocks of (a) with centers at a_0 differ from corresponding blocks of (b). The element $E(0,b)$ will not be similar to $E(0,a)$ but will have an arbitrarily small distance from $E(0,a)$ if the block A is sufficiently long. We accordingly have the following theorem.

THEOREM 3. 2. *The space of all symbolic elements is compact, perfect and totally disconnected.*

By virtue of a well known theorem in point set theory (cf. Hausdorff [1], p. 160), Theorem 3. 2 has the following corollary.

COROLLARY. *The space of all symbolic elements is the homeomorph of the Cantor, perfect, nowhere dense, linear set.*

4. The space of rays and the space of *I*-trajectories. Let R_1 and R_2 be rays with indexed representations of the form

$$a_r\, a_{r+1} \cdots .$$

If $R_1 \not\equiv R_2$, the rays R_1 and R_2 have common initial blocks of finite maximum length m. With Baire (cf. Hausdorff, p. 101) we then set

$$R_1 R_2 = \frac{1}{m+1}.$$

If $R_1 \equiv R_2$, we set $R_1R_2 = 0$. The distance so defined satisfies the usual metric axioms including a stronger triangle relation similar to (3. 5). One readily sees that the space of rays is compact, perfect and totally disconnected.

This distance function in the space of rays is similar to that previously defined in the space of elements. We define a distance between I-trajectories which is based on an entirely different concept and is analogous to the Besicovitch metric of the theory of almost periodic functions (cf. Besicovitch [1], p. 73).

Let (a) and (b) be I-trajectories. Let $| a_i - b_i |$ be 0 if a_i and b_i have the same symbolic value, and 1 if this is not the case. We then set

$$[a, b] = \operatorname*{lim.\,sup.}_{n\to\infty} \frac{1}{2n+1} \sum_{-n}^{n} | a_i - b_i |.$$

It is easily shown that this distance function satisfies the usual metric axioms with the exception that $[a, b] = 0$ does not imply $(a) \equiv (b)$. A simple example proving this last statement is obtained by letting all the symbols of (a) be 1, and all those of (b) be 1 with the exception of the symbol of index 0 which shall be 2.

If $(a') \sim (a)$ and $(b') \sim (b)$, it will not in general be true that $[a', b'] = [a, b]$, but corresponding to (a') there always exists a $(b') \sim (b)$ such that the equality does hold. Recalling the notation of § 2, let $D_r(a)$ denote the I-trajectory (a') in which $a'_{i+r} = a_i$. Thus $D_r(a) \sim (a)$. Furthermore, if $(a') \sim (a)$, there exists an r such that $D_r(a) \equiv (a')$. We shall establish the following lemma.

LEMMA 4. 1. *If (a) and (b) are two I-trajectories and r is integral,*

$$(4.1) \qquad [D_r(a), D_r(b)] = [a, b].$$

Since $D_r(D_{-r}(a)) = (a)$, it is sufficient to prove (4. 1) under the assumption that $r > 0$. Denoting $D_r(b)$ by (b') we have

$$\frac{1}{2n+1}\sum_{-n}^{n}|a'_i - b'_i| = \frac{1}{2n+1}\sum_{-n-r}^{n-r}|a_i - b_i|$$
$$= \frac{1}{2n+1}\sum_{-n}^{+n}|a_i - b_i| + \frac{1}{2n+1}\sum_{-n-r}^{-n-1}|a_i - b_i| - \frac{1}{2n+1}\sum_{n-r+1}^{n}|a_i - b_i|.$$

Since $0 \leqq |a_i - b_i| \leqq 1$, the last two terms approach 0 as n becomes infinite. Thus

$$\lim_{n\to\infty}\sup.\frac{1}{2n+1}\sum_{-n}^{n}|a'_i - b'_i| = \lim_{n\to\infty}\sup.\frac{1}{2n+1}\sum_{-n}^{n}|a_i - b_i|,$$

which is the stated result.

Let T be a trajectory and let (a) be an I-trajectory representing T. Given $\epsilon > 0$, T will be said to admit the *ϵ-translation* r if $[D_r(a), a] < \epsilon$. From (4.1), this definition is independent of the I-trajectory chosen to represent T. Any trajectory admits the ϵ-translation 0, and this will be termed the *trivial* ϵ-translation.

As in the theory of almost periodic functions, a set of integers $\cdots < r_{-1} < r_0 < r_1 < r_2 < \cdots$ is termed *relatively dense* if the numbers $|r_i - r_{i-1}|$, $i = 0, \pm 1, \cdots$, form a bounded set. We define the trajectory T as *almost periodic* if given $\epsilon > 0$, T admits a relatively dense set of ϵ-translations. A periodic trajectory is evidently almost periodic. We shall presently define recurrent trajectories and shall use the notion of almost periodicity in the classification of such trajectories. Most of the known examples of recurrent trajectories turn out to be almost periodic, but the Morse recurrent trajectory of § 8 is not almost periodic, as we shall see.

5. Restricted sequences. In the applications to dynamical theory the sequences admitted are not formed freely from the generating symbols, but are subject to certain limitations. The simplest of these restrictions may be called the *restriction of inverses.* In this case the generating symbols form a set of the form

$$a_1, \cdots, a_p, a_1^{-1}, \cdots, a_p^{-1}, \qquad (p > 1)$$

in which a_i^{-1} is termed the inverse of a_i. In this case a sequence is inadmissible if a symbol and its successor are inverses. In other cases a circular order C of the generating symbols may appear and an integer r, with the restriction that p-blocks in the order C with $p > r$ are to be excluded. Another type of restriction prohibits successive blocks of certain definite forms.

In general problems restrictions of all of these types will appear. What these restrictions are in any problem depends upon the underlying group theory and topology. A detailed study of these restrictions will appear in a later paper. In the present paper we are fortunately able to abstract the essentials of these conditions as follows.

Certain blocks of the generating symbols will be distinguished and termed admissible. A ray or trajectory all of whose subblocks are admissible will be termed admissible as well as any element based on an admissible trajectory. The conditions of admissibility on blocks are as follows.

(α). *A block is admissible if and only if all its subblocks are admissible.*

(β). *Corresponding to any admissible block A there exist at least two generating symbols e, dependent at most upon the last two symbols of A and such that Ae is admissible. A similar condition holds replacing last by first and Ae by eA.*

(γ). *Corresponding to any two admissible blocks A and B there exists at least one generating symbol e dependent at most upon the last two symbols of A and the first two symbols of B and such that AeB is admissible.*

(δ). *There exist at least two generating symbols α and β such that all blocks formed from α and β alone are admissible.*

These conditions are satisfied when there are at least two generating symbols and their inverses and the only restriction is the restriction of inverses. They are also satisfied by the well known Nielsen symbolism (cf. Nielsen [2], pp. 211-217) for surfaces of genus $p > 1$. Various symbolisms developed by the authors satisfy these conditions, including one which replaces the Nielsen symbolism to advantage.

We state the following theorem.

THEOREM 5.1. *The set of admissible elements form a compact, perfect, totally disconnected subset Ω of the space M of all elements.*

That Ω is compact may be seen as follows. Let E_n be a sequence of admissible elements based respectively on trajectories t_n. Let E be a limit element of the sequence E_n. Suppose that E is based on the trajectory t. Each block in t must appear in some trajectory t_n and accordingly be admissible. Hence t and E are admissible, and Ω is compact.

To show that Ω is perfect let E be an arbitrary admissible element. Represent E by an I-element $E(0, a)$. Let A be a $(2n+1)$-block of (a) centered at a_0. By virtue of admissibility condition (β), the block A can be extended to form an admissible I-trajectory of the form

$$(5.1) \qquad \cdots e_{-n-2}e_{-n-1}Ae_{n+1}e_{n+2}\cdots,$$

where the symbols e_i and e_{-i}, $i > n$, can be successively chosen in two different ways. Let E' be an element of index 0 based on (5.1). With the freedom of choice of the symbols e_i and e_{-i} it appears that the elements E' include elements different from E. But these elements also include elements arbitrarily near E since A is arbitrarily long.

The set Ω is accordingly perfect. It is totally disconnected since the space M is totally disconnected, and the proof of the theorem is complete.

A theorem similar to the above is true of the set of admissible rays, taking the metric of rays into account. Unless otherwise stated, the only trajectories and rays which will be used will be those which are admissible in the sense of this section.

6. Limit trajectories and minimal trajectories. Let t be a trajectory represented by an I-trajectory (a). If there exists a positive integer r such that

$$a_{n+r} = a_n \qquad (n = 0, \pm 1, \pm 2, \cdots),$$

t will be said to have the *period* r. A trajectory of period r is determined by each of its r-blocks. There are accordingly at most countably infinite many distinct periodic trajectories.

Let Σ be an arbitrary set of admissible trajectories. A trajectory t will be termed a *limit* trajectory of Σ if the elements based on trajectories of Σ include among their limit elements at least one element based on t. The set Σ can consist of a single trajectory, for example of the trajectory t given by

$$(6.1) \qquad \cdots 2220220202220222 \cdots.$$

The trajectory $\cdots 222 \cdots$ is a limit trajectory of the trajectory (6.1).

A periodic trajectory has no limit trajectory because it bears at most a finite number of elements. Let t be a non-periodic trajectory, and (a) an I-trajectory representing t. Any two I-elements with indices n and r respectively based on (a), with $n > r$, are dissimilar. For a similarity relation between these two elements would imply that (a) had the period $n - r$, contrary to hypothesis. The elements based on t with positive indices must then have at least one limit element, and we term such elements *positive limit elements* of t. *Negative limit elements* are similarly defined.

A trajectory can have more than one distinct limit trajectory. Letting a^n stand for a block $a \cdots a$ of length n we see that a trajectory of the form

$$(6.2) \qquad \cdots \beta^4\alpha^3\beta^2\alpha\beta^2\alpha^3\beta^4 \cdots$$

has the trajectories $\cdots \alpha\alpha\alpha \cdots$ and $\cdots \beta\beta\beta \cdots$ as limit trajectories. More generally let the set of admissible blocks be enumerated in the form

$$A_0, A_1, \cdots.$$

If generating symbols e_n and b_n are successively chosen in accordance with block condition (γ) the trajectory

$$\cdots A_2 b_2 A_1 b_1 A_0 e_1 A_1 e_2 A_2 \cdots \tag{6.3}$$

will be admissible. It will have every admissible trajectory as a limit trajectory. The elements based on the trajectory (6. 3) are everywhere dense in the set Ω of admissible elements. Such a trajectory is termed *transitive.*

In the remainder of this section we shall be concerned with a fundamental type of limit trajectories termed *minimal* trajectories. The dynamical counterpart of such trajectories was first introduced by Birkhoff [1]. Birkhoff also developed their general theory. In this connection the symbolic method was first used by Morse ([1], [2]) to establish the existence of non-periodic recurrent motions. Theorems 6. 1 and 6. 3 of this section and Theorem 7. 2 of the following section are the abstract symbolic analogues of dynamical theorems of Birkhoff. The proofs presented here are independent of the theory of differential equations.

Let s be an admissible trajectory and n an arbitrary positive integer. The number of different n-blocks of s is finite. It will be denoted by $P_s(n)$ and termed the n-th *permutation number* of s. If t is a limit trajectory of s,

$$P_s(n) \geqq P_t(n). \tag{6.4}$$

A trajectory s such that the equality holds in (6. 4) for every n and every limit trajectory t of s will be termed *minimal.*

We understand that periodic trajectories are minimal. They satisfy the above definition vacuously since they have no limit trajectories. We continue with a proof of the following theorem.

THEOREM 6. 1. *Among the positive (or negative) limit trajectories of an admissible non-periodic trajectory T there is at least one admissible minimal trajectory.*

We shall give the proof for the case of positive limit trajectories. To that end we shall define certain symbols inductively as follows:

H_0 = the set of positive limit trajectories of T.
p_1 = the minimum of $P_t(1)$ for t in H_0.
H_1 = the subset of H_0 for which $P_t(1) = p_1$.
. . . .
p_n = the minimum of $P_t(n)$ for t in H_{n-1}.
H_n = the subset of H_{n-1} for which $P_t(n) = p_n$.

The set H_0 is not empty since T is non-periodic. Hence H_n is not empty. Moreover

$$H_0 \supset H_1 \supset H_2 \cdots.$$

We continue with a proof of the following statement.

(α). *The set H_n is closed in the sense that it includes its limit trajectories.*

That H_0 is closed follows from the definition of limit trajectories. Proceeding inductively we shall assume that H_{n-1} is closed and prove that H_n is closed. In case the set H_n has no limit trajectories, H_n is certainly closed. Suppose then that H_n has a limit trajectory t. To show that t is in H_n, recall that $H_{n-1} \supset H_n$ and that t is accordingly a limit trajectory of H_{n-1}. Since H_{n-1} is assumed to be closed t is in H_{n-1}. By virtue of the definition of p_n,

$$P_t(n) \geqq p_n. \tag{6.5}$$

On the other hand, t is a limit trajectory of trajectories of H_n, and the n-th permutation numbers of trajectories of H_n equal p_n. Hence

$$P_t(n) \leqq p_n. \tag{6.6}$$

Hence the equality prevails in (6.5) and (6.6), and t is in H_n. Thus H_n is closed.

The sets H_n form a decreasing sequence of closed sets. It follows from the compactness of the space of admissible elements that the intersection H of the sets H_n is closed and non-void. Let s be a trajectory of H. If s is periodic, s is minimal and the proof is complete. If s is not periodic, let t be a limit trajectory of s. Since t and s are both in H_n for every n, t and s have the same n-th permutation number for every n. Hence s is minimal, and the proof is complete.

THEOREM 6.2. *If s is a minimal trajectory and t a limit trajectory of s, then t is a minimal trajectory and s a limit trajectory of t.*

Every n-block of s is in t since s is minimal, so that s is a limit trajectory of t. To show that t is minimal, let r be one of its limit trajectories. Then r is also a limit trajectory of s. Since s is minimal, the n-th permutation numbers of t and r equal that of s, and so equal each other. Hence t is minimal, and the proof is complete.

The set of limit trajectories of a trajectory t will be termed the *derived set* of t. If s is a non-periodic minimal trajectory, each member of its derived set S has s as a limit trajectory, and hence has the same derived set. That this property is characteristic of non-periodic minimal trajectories is implied by the following theorem.

THEOREM 6.3. *A set S of trajectories which is the derived set of each of its members consists of non-periodic minimal trajectories.*

Since the derived set of a periodic trajectory is empty, no trajectory of the set S can be periodic.

If s is a trajectory of S and t is a limit trajectory of s,

$$P_s(n) \geqq P_t(n). \tag{6.7}$$

But t is in S and thus s is a limit trajectory of t, so that the equality sign must hold in (6.7). Hence s is minimal and the proof is complete.

The elements based on trajectories of S form a compact, perfect, totally disconnected set H. This is behind an earlier result of Morse that there exist discontinuous recurrent motions. Cf. Morse [2], p. 99. The set H is the homeomorph of the Cantor, perfect, nowhere dense, linear set. In particular H has the power $\aleph$ of the continuum. Since there is at most a countable infinity $\aleph_0$ of elements based on each trajectory of S we can conclude that the trajectories of S have a power a such that $\aleph_0 a = \aleph$. But $\aleph_0 a = a$ (cf. Hausdorff, pp. 30-31) so that $a = \aleph$. The set S thus has the power of the continuum.

We shall refer to the following definition. A set of trajectories which is the derived set of each of its members or which reduces to a single periodic trajectory is termed a *minimal set* of trajectories.

7. Recurrent trajectories. Let h be an admissible sequence, finite or infinite, and n a positive integer. A least integer m (if any exists) such that each m-block of h contains every n-block of h will be called the n-th *recurrency index* $R_h(n)$ of h. If h is a ray or a trajectory and $R_h(n)$ exists for each n, h will be termed *recurrent* and $R_h(n)$ the *recurrency function* of h.

We begin with the following theorem.

THEOREM 7.1. *If s is a recurrent trajectory and t is a limit trajectory of s, then t is recurrent and s is a limit trajectory of t. Moreover*

$$R_s(n) = R_t(n).$$

Each block of t appears in s since t is a limit trajectory of s. Let A be an arbitrary n-block of s. A copy of A is found in each block of s of length $R_s(n)$, and hence in each block of t of length $R_s(n)$. Hence $R_t(n)$ exists with

$$R_t(n) \leqq R_s(n). \tag{7.1}$$

Since each block A of s appears in t, s is a limit trajectory of t. The rôles of s and t can then be interchanged so that the equality prevails in (7.1). The proof of the theorem is complete.

The following theorem is the symbolic analogue of a theorem of Birkhoff. Cf. Birkhoff [2], p. 199.

THEOREM 7.2. *A necessary and sufficient condition that a trajectory s be recurrent is that it be minimal.*

We begin by assuming that s is minimal but not recurrent, and seek a contradiction. If s is not recurrent, there exists a block U of s with the following property. There exists a sequence A_n of blocks of s whose lengths become infinite with n and which do not contain U. Without loss of generality we can suppose that the length of A_n is odd. Let E_n then be the element based on s with preferred symbol at the center of A_n. Since the space of admissible elements is compact there exists a subsequence of the elements E_n which converges to an admissible element E. Let t be the trajectory on which E is based. It is clear that t does not contain the block U. Hence $t \neq s$. It follows that t is a limit trajectory of s. Since s is minimal, t and s must then contain the same blocks. From this contradiction we infer that s is recurrent.

It remains to show that s is minimal if recurrent. If s is periodic, s is minimal. Assume then that s is not periodic and let S be the derived set of s. It follows from Theorem 7.1 that S is the derived set of each of its members. We conclude from Theorem 6.3 that s is minimal, and the proof is complete.

It is natural to ask what limitations on recurrency indices $R(n)$ and permutation indices $P(n)$ are imposed by conditions of periodicity or recurrency. We give a partial answer in the following lemmas and theorems.

LEMMA 7.1. *If $P(n)$ is the number of different n-blocks in a trajectory t, $P(n+1) \geqq P(n)$.*

Let the different n-blocks in t be enumerated in the form

$$B_i \qquad [i = 1, 2, \cdots, P(n)], \tag{7.2}$$

and let e_i be the successor of B_i in some position in t. The $(n+1)$-blocks $B_i e_i$ are all different, and the lemma follows immediately.

If (a) is an indexed trajectory, a block B of (a) whose first and last symbols are a_r and a_m will be denoted by $[r, m]$. We shall term r the *first index* of B and m the *final index* of B.

THEOREM 7.3. *A trajectory t for which $P(m+1) = P(m)$ for some m has a period $\omega \leqq P(m)$.*

Let (a) be an indexed representation of t. There are $P(m)$ dissimilar m-blocks B_i in (a). The elements immediately following copies of B_i in (a) must always be the same. Otherwise there would be at least $P(m)+1$ dissimilar $(m+1)$-blocks. Similarly the elements immediately preceding copies of B_i must always be the same.

Consider the blocks $B=[i+1, i+\mathrm{m}]$. Since there are just $P(m)$ dissimilar m-blocks, for some i there must exist a similarity relation of the form

$$(7.3) \qquad [i+1, i+m] \sim [i+1+\omega, i+m+\omega],$$

in which $0<\omega \leqq P(m)$. We shall show that (a) has a period ω. We begin with an inductive proof of the relations,

$$(7.4) \qquad a_{i+r}=a_{i+r+\omega} \qquad (r=1,2,\cdots).$$

That (7.4) holds for $r=1,2,\cdots, m$ follows from (7.3). Assuming that (7.4) holds for $r=1,2,\cdots, n \geqq m$, we shall prove that (7.4) holds for $r=n+1$. The m-blocks with final indices $i+n$ and $i+n+\omega$ respectively are similar by virtue of our inductive hypothesis. It then follows from the principle stated in the first paragraph of this proof that (7.4) holds for $r=n+1$.

It follows similarly that $a_{i+r}=a_{i+r+\omega}$ for $r \leqq 0$. We thereby make use of the fact that two similar m-blocks must be preceded by symbols with equal values. One starts with the similarity relation (7.3) and proceeds inductively. We conclude that (a) has the period ω, and the proof is complete.

COROLLARY. *For a non-periodic trajectory* $P(m+1)>P(m)$ *for each* $m>0$.

There exist trajectories for which $P(n+1)=P(n)+1$, so that the relation $P(n+1)>P(n)$ of the corollary is in a sense the best possible. An example of a recurrent trajectory for which $P(n+1)=P(n)+1$ for each $n>0$ will be given in a later paper.

LEMMA 7.2. *If t is a trajectory which involves μ generating symbols and has no period less than ω, then*

$$(7.5) \qquad P(n) \geqq n+\mu-1$$

for all values n for which $P(n)<\omega$.

Relation (7.5) is clearly true when $n=1$. From the preceding theorem $P(m+1)>P(m)$ for each m for which $P(m)<\omega$. Relation (7.5) follows inductively.

4

THEOREM 7.4. *If t is a non-periodic trajectory involving μ generating symbols,*

$$P(n) \geqq n + \mu - 1.$$

LEMMA 7.3. *If t is a recurrent trajectory with no period less than ω, then*

$$R(n) \geqq P(n) + n \geqq 2n + \mu - 1 \tag{7.6}$$

for each n such that $P(n) < \omega$.

Suppose that $P(n) < \omega$ and set $R(n) = m$. Since t is recurrent, each m-block B contains each of the $P(n)$ n-blocks of t. Hence

$$m \geqq P(n) + n - 1. \tag{7.7}$$

We shall exclude the equality in (7.7).

Suppose B has the form $[r + 1, r + m]$ in an indexed representation (a) of t. Suppose $m = P(n) + n - 1$. Then no two n-blocks of B are similar. In particular the initial n-block A of B does not appear twice in B. But $R(n) = m$ so that A appears in the I-block $[r + 2, r + m + 1]$, and since A appears but once in B, A appears with final index $r + m + 1$. Since A appears with final index $r + n$ as well, $a_{r+n} = a_{r+m+1}$ for arbitrary r. Hence t has a period

$$m - n + 1 = [P(n) + n - 1] - n + 1 = P(n) < \omega.$$

This is contrary to the hypothesis that t has no period less than ω. Hence $m > P(n) + n - 1$, and (7.6) follows with the aid of (7.5).

Lemma 7.3 and Theorem 7.4 yield the following theorem.

THEOREM 7.5. *If t is a non-periodic recurrent trajectory involving μ generating symbols,*

$$R(n) \geqq P(n) + n \geqq 2n + \mu - 1, \tag{7.8}$$

where $R(n)$ and $P(n)$ are respectively the recurrency and permutation indices of t.

It will be convenient to set $P(0) = 0$ for each trajectory t.

THEOREM 7.6. *If a trajectory t involves μ generating symbols and has a minimum period ω,*

$$R(n) \leqq n + \omega - 1 \qquad (n = 1, 2, \cdots). \tag{7.9}$$

Moreover there exists a largest integer m such that $P(m) < \omega$. For such an m,

$$R(n) \geqq 2n + \mu - 1 \qquad (n = 1, 2, \cdots, m \leqq \omega - \mu), \tag{7.10}$$

$$R(n) = n + \omega - 1 \qquad (n = m + 1, m + 2, \cdots). \tag{7.11}$$

Relation (7.9) is immediate. To establish the existence of a largest integer m such that $P(m) < \omega$, observe that each integer $n > 0$ for which $P(n) < \omega$ satisfies the relations

$$2n + \mu - 1 \leqq R(n) \leqq n + \omega - 1,$$

by virtue of (7.6) and (7.9). It follows that m exists and is at most $\omega - \mu$. Relation (7.10) follows from (7.6).

To establish (7.11) let n be any integer $> m$. Then $P(n) \geqq \omega$ and

$$R(n) \geqq P(n) + n - 1 \geqq \omega + n - 1. \tag{7.12}$$

Relation (7.11) follows from (7.12) and (7.9).

Let (a) be an arbitrary I-trajectory. Let n be a positive integer, and let m be an integer, positive, negative or zero. For n and m fixed, let $H(m)$ be the minimum length of those blocks of (a) with initial index m which contain copies of all n-blocks of (a). If no such block with initial index m exists, let $H(m) = +\infty$. If, as m ranges over all integers, the set $H(m)$ is bounded, it is clear that the n-th recurrency index $R(n)$ of (a) exists and

$$R(n) = \max. H(m). \tag{7.13}$$

Let r be any integer such that $H(r+1)$ assumes the maximum of $H(m)$. The $R(n)$-block $[r+1, r+R(n)]$ will then be termed a *minimax n-covering* in (a) or in the trajectory t represented by (a). The length of a minimax n-covering B in t is $R(n)$ and the final n-block in B does not otherwise appear in B. These two properties are characteristic provided $R(n)$ exists.

We are led to the following theorem.

THEOREM 7.7. *If t is a trajectory for which $R(n+1)$ and $R(n)$ exist, $R(n+1) > R(n)$.*

To prove the theorem let B be a minimax n-covering in t. Let A be the final n-block in B and e the symbol following A. The $(n+1)$-block Ae does not appear in B since A has no second appearance in B. Hence $R(n+1) > R(n)$, as stated.

It may happen that $R(n+1) = R(n) + 1$ for infinitely many values of n as the recurrent trajectory of the following section shows.

8. The Morse recurrent trajectory. In this section we shall recall the definition of the non-periodic I-trajectory T first defined by Morse [2], p. 95, and shall give the explicit values of the corresponding recurrency function, denoting this function by $\rho(n)$.

Recall that the symbols α and β can be combined in any way to form admissible blocks. We here take α and β as 1 and 2 respectively. Set

$$(8.0)\qquad \begin{array}{ll} a_0 = 1, & b_0 = 2, \\ a_1 = a_0 b_0, & b_1 = b_0 a_0, \\ \cdot\ \cdot\ \cdot\ \cdot & \cdot\ \cdot\ \cdot\ \cdot \\ a_{n+1} = a_n b_n, & b_{n+1} = b_n a_n, \end{array}$$

understanding that the blocks $a_n b_n$ and $b_n a_n$ are to be expanded as blocks of 1's and 2's of length 2^{n+1}. The I-trajectory T shall have the representation

$$(8.1)\qquad \cdots c_{-2} c_{-1} c_0 c_1 c_2 \cdots.$$

In T we set

$$(8.2)\qquad c_0 c_1 \cdots c_{2^n-1} \equiv a_n.$$

Note that (8. 2) is consistent for all n, since a_m is an initial subblock of a_n for $m < n$. The symbols c_i are thereby defined for $i > 0$. To complete the definition of (8. 1) we set

$$(8.3)\qquad c_{-r} = c_{r-1} \qquad (r > 0).$$

The subray

$$(8.4)\qquad R = c_0 c_1 c_2 \cdots$$

of T thus starts as follows:

(8. 5) 1221 2112 2112 1221 2112 1221 1221 2112 $\cdots$.

Corresponding to an arbitrary block A, the block which consists of the symbols of A in inverse order will be denoted by A^{-1}. We note that

$$(8.6')\qquad a_n = a_n^{-1}, \quad b_n = b_n^{-1}, \quad (n \text{ even})$$
$$(8.6'')\qquad a_n = b_n^{-1}, \quad b_n = a_n^{-1}, \quad (n \text{ odd}).$$

Relations (8. 6) are clearly true when $n = 0$, and their truth in general follows inductively from the relations

$$a^{-1}_{m+1} = (a_m b_m)^{-1} = b_m^{-1} a_m^{-1},$$
$$b^{-1}_{m+1} = (b_m a_m)^{-1} = a_m^{-1} b_m^{-1}.$$

From (8. 6) we infer that the 2^n-block of (8. 1) whose final index is -1 is $a_n^{-1} = a_n$ when n is even and $a_n^{-1} = b_n$ when n is odd. Thus the 2^{n+1}-block whose initial index is -2^n is $a_n a_n$ when n is even and $b_n a_n$ when n is odd. Each block of T appears in such a 2^{n+1}-block for n sufficiently large. Since

$$a_{n+3} = a_n b_n b_n a_n b_n a_n a_n b_n,$$

each block of T appears in the subray R of (8.4). The recurrency indices of T (if they exist) are thus completely determined by this subray.

Let A be an arbitrary block of 1's and 2's. The block obtained from A by replacing 1 by 2 and 2 by 1 will be denoted by A'. An easy inductive proof shows that

$$a'_n = b_n, \qquad b'_n = a_n. \tag{8.7}$$

The trajectories T' and T_n. Let T' be a trajectory with an indexed representation (e) in which $e_i = c_i$ for $i \geqq 0$, and for $i < 0$ equals 1 or 2 according as c_i is 2 or 1 respectively. The 2^n-block of T whose final index is -1 is a_n or b_n according as n is even or odd. The corresponding 2^n-block of T' is then $a'_n = b_n$ or $b'_n = a_n$ according as n is even or odd. We infer that each block of T' has a copy in the subray R of (8.4). The recurrency indices of T' (if they exist) are then completely determined by R. Since this is also true of T we have proved the following lemma.

LEMMA 8.1. *The recurrency indices of T and T', if they exist, are equal and are uniquely determined by the ray R in* (8.4).

We introduce a trajectory T_n, $n \geqq 0$, whose generating symbols are a_n and b_n, and which is obtained from T or T' according as n is even or odd, by replacing 1 by a_n and 2 by b_n. It follows from the relations (8.0), (8.3) and (8.6) that an expansion of the symbols a_n and b_n in T_n into blocks of 1's and 2's will yield T. Unless otherwise stated T_n shall be regarded as a trajectory in the symbols a_n and b_n unexpanded as blocks of 1's and 2's. For example $a_n b_n$ will be regarded as a 2-block of T_n, or, if we choose, as a 2^{n+1}-block of T.

We shall make use of the following properties of the ray R of (8.4).

(a). *There are no more than two successive* 1's *or* 2's *in R.*

(b). *Each* 5-*block of R contains the blocks* 12 *and* 21.

(c). *Each* 5-*block of R contains either the block* 11 *or else the block* 22.

To establish (a) we note that each 3-block of R will be found in one of the expanded 2-blocks of T_1; that is, in one of the blocks

$$a_1 b_1 = 1221, \qquad a_1 a_1 = 1212,$$
$$b_1 a_1 = 2112, \qquad b_1 b_1 = 2121,$$

and no more than two successive 1's or 2's appear in these blocks.

To verify (b) observe that a 5-block B which did not contain 12 would contain 21 in its first 3-block by virtue of (a). Its final 3-block would then be 111 contrary to (a). We infer that B contains 12, and similarly 21.

To prove (c) we infer from (a) that there can be no more than two consecutive a_1's or b_1's in T_1. In particular R cannot contain the block 12121 since R would then contain the block $a_1a_1a_1$ or $b_1b_1b_1$ according as 12121 possessed an even or odd first index. Similarly R contains no block 21212, and (c) follows directly.

Statements (a), (b) and (c) are true if R is replaced by T or T' since each block in T or T' appears in the subray R.

We continue with the following lemma.

LEMMA 8.2. *The recurrency index $\rho(n)$ of the trajectory T has the values*

$$\rho(1) = 3, \qquad \rho(2) = 9, \qquad \rho(3) = 11.$$

That $\rho(1) = 3$ follows from (a). We continue by showing that $\rho(2) \leqq 9$. To that end consider a 5-block B of T_1. The block B will contain the blocks

$$a_1b_1 = 1221, \qquad b_1a_1 = 2112, \tag{8.8}$$

by virtue of (b) and either

$$a_1a_1 = 1212 \quad \text{or} \quad b_1b_1 = 2121, \tag{8.9}$$

by virtue of (c). The only 2-blocks in T are 12, 21, 11 and 22. If B_0 denotes the block B expanded in terms of 1's and 2's, we see that each 2-block of T appears in B_0 *interior* to B_0. Thus the end symbols of B_0 can be deleted and the resulting block B'_0 contains all 2-blocks of T. But any 9-block of T will contain some block of the nature of B'_0 so that $\rho(2) \leqq 9$.

That $\rho(2)$ is not less than 9 follows upon considering the 9-block of T with initial index 10. This block has the form

$$12\ 1221\ 211,$$

and will contain no block 11, if the final 1 is removed. Thus $\rho(2) = 9$.

To show that $\rho(3) = 11$, recall that there are no more than two successive 1's or 2's in T so that the only 3-blocks are

$$112, \quad 122, \quad 121, \quad 211, \quad 221, \quad 212. \tag{8.10}$$

Consider the block B_0 of the preceding proof. It contains the blocks (8.8) and at least one of the blocks (8.9). It follows that B_0 contains each 3-block (8.10). But any 11-block of T will contain a B_0, so that $\rho(3) \leqq 11$.

That $\rho(3) \geqq 11$ follows upon noting that the 11-block of T with initial index 3 has the form

$$1\ 2112\ 2112\ 12,$$

and that removal of the final symbol from this block leaves no subblock of the form 212. Hence $\rho(3) = 11$, and the proof of the lemma is complete.

Let T'_1 be the trajectory obtained from T' by grouping the symbols of T' into 2-blocks with even first indices, regarding T'_1 as generated by the blocks a_1, b_1. It is readily seen that T'_1 is formed from the blocks a_1, b_1 exactly as T is formed from the symbols 1 and 2 respectively. We shall make use of this fact in proving the following lemma.

LEMMA 8. 3. *No two r-blocks of $T(T')$ for which $r \geqq 4$ are equal, unless their first indices are equal mod 2.*

It is sufficient to prove the lemma for the case $r = 4$. To that end observe that each 4-block of $T(T')$ as it stands is in some 6-block A of $T(T')$ with even first index. Let A_1 be the 3-block of $T_1(T'_1)$ determined by A. Since $\rho(3) = 11$, each 3-block of $T_1(T'_1)$ appears in the 11-block of $T_1(T'_1)$ with first index 0. The lemma follows upon verifying its truth for the 22-block of T with first index 0. For each 4-block of T appears in this 22-block with a first index which is unchanged mod 2.

The following lemma is useful.

LEMMA 8. 4. *If $\rho(m+1)$ exists and $m \geqq 2$,*

$$\rho(2m) \leqq 2\rho(m+1) - 1.$$

We need only to show that each block B of T of length $2\rho(m+1) - 1$ contains a copy of each $2m$-block x of T. We distinguish two cases.

Case 1. *The first index of x is even.* Here x determines an m-block of T_1 and so appears in each expanded $\rho(m)$-block of T_1. Hence x appears in each $[2\rho(m) + 1]$-block of T. Thus $\rho(2m)$ exists and

$$2\rho(m) + 1 \geqq \rho(2m).$$

But

$$\rho(m+1) \geqq \rho(m) + 1,$$

in accordance with Theorem 7. 7. Hence

$$2\rho(m+1) - 1 \geqq 2\rho(m) + 1 \geqq \rho(2m).$$

This completes the proof in Case 1.

Case 2. *The first index of x is odd.* Let c' and c'' be respectively the elements of T which precede and follow the above block B. If B has an odd first index, $c'B$ has an even first index and a length $2\rho(m+1)$ and so contains a copy of each $(m+1)$-block of T_1. Hence $c'B$ must contain a copy y of x. But the index of y is odd in accordance with Lemma 8.3, so that y does not include c'. Hence x appears in B. If B has an even first index, Bc'' contains a copy y of x. But the index of c'' is odd, while the final index of x is even. Hence y does not include c'' and lies in B.

The proof of the lemma is complete.

Since $\rho(3)$ exists it follows from the lemma that $\rho(4)$ exists, then $\rho(6)$, $\rho(10)$, etc. But if $\rho(n)$ exists, $\rho(m)$ exists for $m < n$. We conclude that $\rho(n)$ exists for all n.

We come to a principal theorem.

THEOREM 8.1. *For $m \geqq 2$ the recurrency function $\rho(m)$ of T satisfies the relations*

$$\rho(2m) = 2\rho(m+1) - 1, \tag{8.11}$$

$$\rho(2m+1) = 2\rho(m+1). \tag{8.12}$$

Proof of (8.11). It will be sufficient to show that the equality prevails in Lemma 8.4. To that end let g be a minimax $(m+1)$-covering in T_1 as defined in § 7. The length of g is $\rho(m+1)$, and the final $(m+1)$-block of g appears only once in g. Let x be an arbitrary $2m$-block of T of odd first index, and let y be an $(m+1)$-block of T_1 which when expanded contains x. When $m \geqq 2$, x appears only as an interior block of the expanded y in accordance with Lemma 8.3. Since a 2-block of T of even index is uniquely determined by either of its symbols we see that y is uniquely determined by x. It thus requires copies of all $(m+1)$-blocks of T_1 to cover all $2m$-blocks of T of odd index. In particular it will require a copy of the final $(m+1)$-block of g. As a block of T, g can then be shortened on the right by at most one symbol and, when expanded, still cover all $2m$-blocks of T. The equality accordingly holds in Lemma 8.4, and the proof of (8.11) is complete.

Proof of (8.12). Recall that $\rho(2m+1)$ exceeds $\rho(2m)$. It follows then from (8.11) that

$$\rho(2m+1) \geqq 2\rho(m+1). \tag{8.13}$$

It remains to show that the equality prevails in (8.13). To that end let B be a $2\rho(m+1)$-block of T, and let c' and c'' be respectively the symbols preceding and following B. Let x be an arbitrary $(2m+1)$-block of T. We distinguish between several cases:

Case 1. The index of c' is even, that of x odd.
Case 2. The index of c' is even, that of x even.
Case 3. The index of c' is odd.

Case 1. The block x appears in some expanded $(m+1)$-block of T_1. Hence x has a copy y in $c'B$, for the first $2\rho(m+1)$-block of $c'B$ determines a $\rho(m+1)$-block of T_1. But the index of y is odd so that y does not include c' and thus lies in B.

Case 2. Here c'' has an odd index and the last $2\rho(m+1)$-block of Bc'' determines a $\rho(m+1)$-block of T_1. Hence x has a copy y in Bc''. But the final index of y is even so that y lies in B.

Case 3. In this case B determines a $\rho(m+1)$-block of T_1, hence contains a copy of each $(m+1)$-block of T_1, and accordingly a copy of x.

Thus a copy of x appears in B in all cases. The equality accordingly prevails in (8.13), and the proof of (8.12) is complete.

With the aid of the preceding theorem we can derive a general formula for $\rho(n)$ when $n \geqq 3$. To that end we first show that integers $n \geqq 3$ can be represented in the form

$$n = 2^r + p, \qquad (p = 2, 3, \cdots, 2^r + 1), \tag{8.14}$$

where $r \geqq 0$ and r and p are uniquely determined by n. The smallest and largest values of n given by (8.14) for a fixed r are respectively

$$2^r + 2, \qquad 2^{r+1} + 1,$$

and when r is replaced by $r+1$ the smallest value of n is

$$2^{r+1} + 2.$$

Thus as r ranges over the values $0, 1, 2, \cdots$ and p over the corresponding values listed in (8.14), n takes on each of the integers $3, 4, \cdots$ once and only once. We continue with the proof of the following theorem.

THEOREM 8.2. *For values of n of the form* (8.14) *the recurrency function $\rho(n)$ of T has the value*

$$\rho(n) = 10 \cdot 2^r + p - 1. \tag{8.15}$$

When $r = 0$ in (8.14) $p = 2$ and $n = 3$. We have seen that $\rho(3) = 11$, so that (8.15) holds when $n = 3$. When $r = 1$, we see that $p = 2$ and 3 while $n = 4$ and 5 respectively in (8.14). Relation (8.15) affirms that $\rho(4) = 21$ and $\rho(5) = 22$, and these values are correct as follows from (8.11) and (8.12) respectively upon setting $m = 2$. We proceed inductively assuming that (8.15) holds for a fixed r and the related values of p, and seek to prove that

(8.16) $$\rho(\mu) = 10 \cdot 2^{r+1} + q - 1$$

where

(8.17) $$\mu = 2^{r+1} + q \qquad (q = 2, \cdots, 2^{r+1} + 1).$$

We distinguish between the cases where q is even and q is odd.

Case 1. *q even.* Here q has the form $2s$ and

$$\mu = 2(2^r + s).$$

Upon setting $m = 2^r + s$ in (8.11) we find that

(8.18) $$\rho(\mu) = 2\rho(2^r + s + 1) - 1.$$

Upon referring to (8.17) we see that $s + 1$ ranges over the values

(8.19) $$2, 3, \cdots, 2^r + 1.$$

When $n = 2^r + s + 1$, $p = s + 1$ in (8.14) and the range (8.19) is also the range prescribed in (8.14). We can accordingly use our inductive hypothesis to infer that

(8.20) $$\rho(2^r + s + 1) = 10 \cdot 2^r + p - 1 = 10 \cdot 2^r + s.$$

Upon returning to (8.18) we conclude that

$$\rho(\mu) = 2(10 \cdot 2^r + s) - 1 = 10 \cdot 2^{r+1} + q - 1.$$

Relation (8.16) thus holds when q is even.

Case 2. *q odd.* Let q and μ be values of q and μ given by (8.17) with q odd. Corresponding to q and μ there exists a pair of even integers q_1, μ_1 again satisfying (8.17) with $\mu_1 = 2m$ and

$$q = q_1 + 1, \qquad \mu = 2m + 1.$$

Upon using the result established under Case 1,

$$\rho(\mu_1) = \rho(2m) = 10 \cdot 2^{r+1} + q_1 - 1.$$

But

$$\rho(2m + 1) = \rho(2m) + 1$$

by virtue of (8.11) and (8.12), so that in Case 2

$$\rho(\mu) = 10 \cdot 2^{r+1} + q_1 = 10 \cdot 2^{r+1} + q - 1.$$

Thus (8.16) holds in Case 2, and the proof of the theorem is complete.

This explicit determination of $\rho(n)$ is the first such determination for a non-periodic, recurrent trajectory.

The manner in which $\rho(n)$ varies with n is illustrated by the following table:

$n = 1, 2, 3, 4, 5, 6, 7, 8, 9, 10, 11, 12, 13, 14, 15, 16, 17, 18,$
$\rho(n) = 3, 9, 11, \underline{21, 22,} \underline{41, 42, 43, 44,} \underline{81, 82, 83, 84, 85, 86, 87, 88,} 161,$

where the groups underlined are related in an obvious manner. For a given r in (8.14) the smallest and largest values of $\rho(n)$ are respectively

$$10 \cdot 2^r + 1, \qquad 11 \cdot 2^r.$$

We see that for a fixed r,

$$10 > \frac{10 \cdot 2^r + 1}{2^r + 2} \geqq \frac{\rho(n)}{n} = \frac{10 \cdot 2^r + p - 1}{2^r + p} \geqq \frac{11 \cdot 2^r}{2^{r+1} + 1}.$$

We are thus led to the following theorem.

THEOREM 8.3. *If $\rho(n)$ is the recurrency function of T,*

$$\text{(8.21)} \qquad \lim_{n\to\infty}\text{inf.} \frac{\rho(n)}{n} = \frac{11}{2}, \qquad \lim_{n\to\infty}\text{sup.} \frac{\rho(n)}{n} = 10,$$

and 10 *and* 11/2 *are the limits respectively of $\rho(n)/n$ for sequences of values of n given by*

$$n = 2^r + 2, \qquad n = 2^{r+1} + 1, \qquad (r = 0, 1, \cdots).$$

If e is a prescribed positive constant and r is a sufficiently large positive integer, the ratios

$$\frac{\rho(n)}{n} = \frac{10 \cdot 2^r + p - 1}{2^r + p} \qquad (p = 2, 3, \cdots, 2^r + 1)$$

assume values within e of each value k between the limits (8.21). There accordingly exists a subsequence of values of n for which

$$\lim_{n\to\infty} \frac{\rho(n)}{n} = k.$$

For all values of n, $\rho(n) < 10n$, while for each positive e and sufficiently large values of n, $2\rho(n) > (11 - e)n$.

That the trajectory T is not periodic follows from Theorem 7.6.

In § 4 a trajectory has been defined as almost periodic if it admits a relatively dense set of ϵ-translations for any positive ϵ. We show that a recurrent trajectory is not necessarily almost periodic by proving that the Morse recurrent trajectory admits only the trivial ϵ-translation if $\epsilon < 1/4$. To that end we first prove two lemmas.

LEMMA 8.5. *If* $\cdots \delta_1\delta_0\delta_1 \cdots$ *is a sequence of the numbers* 0 *or* 1, *and* p *and* κ *are integers,* $\kappa > 0$, *then*

$$\limsup_{n\to\infty} \frac{1}{2n+1}\sum_{i=-n}^{n} \delta_i = \limsup_{n\to\infty} \frac{1}{2n+1}\sum_{i=-n}^{n+p} \delta_i = \limsup_{m\to\infty} \frac{1}{2\kappa m+1}\sum_{i=-\kappa m}^{\kappa m} \delta_i.$$

The first equality follows at once from the relationship

$$\lim_{n\to\infty} \left| \frac{1}{2n+1}\sum_{i=-n}^{n} \delta_i - \frac{1}{2n+1}\sum_{i=-n}^{n+p} \delta_i \right| \leqq \lim_{n\to\infty} \frac{p}{2n+1} = 0.$$

As to the second, let $n = \kappa m + r$, $0 \leqq r < \kappa$. Then

$$\frac{1}{2n+1}\sum_{i=-n}^{n} \delta_i = \frac{1}{2n+1}\sum_{i=-\kappa m}^{\kappa m} \delta_i + \frac{1}{2n+1}\sum_{i=-\kappa m-r}^{-\kappa m-1} \delta_i + \frac{1}{2n+1}\sum_{i=\kappa m+1}^{\kappa m+r} \delta_i,$$

where the last two sums are not present if $r = 0$. Each of these last two sums approaches 0 as n becomes infinite. It follows that

$$\begin{aligned}\limsup_{n\to\infty} \frac{1}{2n+1}\sum_{i=-n}^{n} \delta_i &= \limsup_{n\to\infty} \frac{1}{2n+1}\sum_{i=-\kappa m}^{\kappa m} \delta_i \\ &= \limsup_{n\to\infty} \frac{2\kappa m+1}{2n+1}\cdot\frac{1}{2\kappa m+1}\sum_{i=-\kappa m}^{\kappa m} \delta_i = \limsup_{m\to\infty} \frac{1}{2\kappa m+1}\sum_{i=-\kappa m}^{\kappa m} \delta_i.\end{aligned}$$

The proof of the lemma is complete.

We recall that T_n is the trajectory with generating symbols a_n and b_n obtained from T or T', according as n is even or odd, by replacing 1 by a_n and 2 by b_n and that an expansion of the symbols a_n and b_n in T_n into blocks of 1's and 2's will yield T.

LEMMA 8.6. *If* $r = 2^p q$, $p \geqq 0$, *and* (c) *and* (d) *are I-trajectories representing* T *and* T_p, *respectively,*

$$[D_r(c), (c)] = [D_q(d), (d)].$$

It follows from Lemma 4.1 that the value of $[D_r(c), (c)]$ is independent of the particular indexed representation (c) of T which we use. We shall find it convenient to take (c) as the I-trajectory (8.1). Then $D_r(c)$ is an I-trajectory (c') in which $c'_i = c_{i-r}$. Similarly we can suppose that in (d), d_i in expanded form is given by

$$d_i = c_{i2^p} \cdots c_{(i+1)2^p-1},$$

for (d) is then an indexed representation of T_p. Since $r = 2^p q$, $D_q(d)$ has the form (d') where $d'_i = d_{i-q} = c_{i2^p-r} \cdots c_{(i+1)2^p-r-1}$, in expanded form.

Since d_i and d'_i are either a_p or b_p, it follows from (8.7) that in their expanded form, d_i and d'_i are either identical, or one is obtained from the other by replacing 1 by 2 and 2 by 1. If we set

$$\{d'_i - d_i\} = \sum_{j=i2^p}^{(i+1)2^p-1} | c'_j - c_j |,$$

it follows that

$$\{d'_i - d_i\} = 2^p \, | d'_i - d_i |.$$

With the aid of Lemma 8.5, we then see that

$$\begin{aligned}
[D_r(c), (c)] &= \limsup_{n\to\infty} \frac{1}{2n+1} \sum_{i=-n}^{n} | c'_i - c_i | \\
&= \limsup_{m\to\infty} \frac{1}{2 \cdot 2^p m + 1} \sum_{i=-m2^p}^{m2^p} | c'_i - c_i | \\
&= \limsup_{m\to\infty} \frac{1}{2^{p+1} m + 1} \sum_{i=-m2^p}^{m2^p+2^p-1} | c'_i - c_i | \\
&= \limsup_{m\to\infty} \frac{1}{2^{p+1} m + 1} \sum_{i=-m}^{m} \{d'_i - d_i\} \\
&= \limsup_{m\to\infty} \frac{2^p}{2^{p+1} m + 1} \sum_{i=-m}^{m} | d'_i - d_i | \\
&= \limsup_{m\to\infty} \frac{1}{2m+1} \sum_{i=-m}^{m} | d'_i - d_i | = [D_r(d), (d)].
\end{aligned}$$

The proof of the lemma is complete.

We come to the proof of the desired theorem.

THEOREM 8.4. *The Morse recurrent trajectory admits only the trivial* ϵ*-translation* 0 *if* $\epsilon < 1/4$.

It is sufficient to show that

$$[D_r(c), (c)] \geqq \tfrac{1}{4}, \qquad r \neq 0, \tag{8.22}$$

where (c) is again the indexed representation (8.1) of the Morse recurrent trajectory.

Case 1. *r odd.* Let

$$\sigma_n = \sum_{i=n}^{n+3} | c'_i - c_i | = \sum_{i=n}^{n+3} | c_{i-r} - c_i |.$$

Since r is odd, the first indices of $c_{n-r}c_{n-r+1}c_{n-r+2}c_{r-r+3}$ and $c_n c_{n+1} c_{n+2} c_{n+3}$ are not congruent mod 2. It follows from Lemma 8. 3 that these two 4-blocks cannot be similar, so that $\sigma_n \geqq 1$. Applying Lemma 8. 5, we find that

$$\begin{aligned}[D_r(c), (c)] &= \limsup_{n\to\infty} \frac{1}{2n+1} \sum_{i=-n}^{n} | c'_i - c_i | \\ &= \limsup_{n\to\infty} \frac{1}{8n+1} \sum_{i=-4n}^{4n+3} | c'_i - c_i | \\ &= \limsup_{n\to\infty} \frac{1}{8n+1} \sum_{i=-n}^{n} \sigma_{4i}.\end{aligned}$$

Since $\sigma_n \geqq 1$, it follows that $\sum_{i=-n}^{n} \sigma_{4i} \geqq 2n+1$, and hence

$$(8.23) \quad [D_r(c), (c)] = \limsup_{n\to\infty} \frac{1}{8n+1} \sum_{i=-n}^{n} \sigma_{4i} \geqq \lim_{n\to\infty} \frac{2n+1}{8n+1} = \tfrac{1}{4}.$$

We have proved (8. 22) when r is odd.

Let $(\bar{c})$ be the I-trajectory obtained from (c) by replacing c_n, $n < 0$, by 1 if c_n is 2, and by 2 if c_n is 1. Then $(\bar{c})$ is an indexed representation of the trajectory T'. Since Lemma 8. 3 holds for T', as well as T, a proof similar to that given for (c) yields the relation

$$(8.24) \qquad [D_r(\bar{c}), (\bar{c})] \geqq \tfrac{1}{4}, \qquad r \text{ odd}.$$

Case 2. *r even.* Since $r \neq 0$, $r = 2^p q$, where q is odd. From Lemma 8. 6,

$$(8.25) \qquad [D_r(c), (c)] = [D_q(d), (d)].$$

The symbols of (d) are a_p and b_p, and if in (d) we set $a_p = 1$, $b_p = 2$, we obtain (c) or $(\bar{c})$ according as p is even or odd. Since q is odd, (8. 23) or (8. 24) applied to the right side of (8. 25) yields

$$[D_q(d), (d)] \geqq \tfrac{1}{4}.$$

Combining this inequality with (8. 25) we conclude that

$$[D_r(c), (c)] \geqq \tfrac{1}{4}, \qquad r \text{ even}.$$

The proof of the theorem is complete.

The following corollary is an immediate consequence of the preceding theorem.

COROLLARY. *The Morse recurrent trajectory is not almost periodic.*

Upon distinguishing recurrent trajectories according to whether they are almost periodic or not, we can state that recurrent trajectories of the type of

Morse are the only ones so far defined which are not almost periodic. In particular, the recurrent trajectories defined by Birkhoff ([4], p. 22) are almost periodic. The proof of this and the further development of the theory of almost periodic trajectories will be given in a later paper by the authors.

9. The derivation of recurrent trajectories. Having exhibited a non-periodic recurrent trajectory T, we shall describe four methods of deriving a recurrent trajectory from T, termed respectively derivation by projection, reduction, association and substitution. We also introduce a general method of derivation which includes each of the preceding. If the given trajectory consists exclusively of the free symbols α and β, these derived trajectories will all be admissible. In general they may or may not be admissible. If the given trajectory T is recurrent, the derived trajectories will be recurrent although usually differing in high degree from T. We suppose that T has an indexed representation (c) in terms of generating symbols $a_1, \cdots, a_\mu$.

Projection. The trajectory obtained from (c) by omitting a_k whenever it occurs in (c) will be said to be the k-th projection of (c).

Reduction. If the generating symbols consist of a finite set of integers, the trajectory obtained by reducing the elements c_j mod p will be said to be derived by reduction mod p.

Association. Let m be a positive integer at least 2, and let B_i be the m-block of (c) whose first index is i. There is at most a finite set of different blocks B_i. We introduce the I-trajectory

$$\text{(B)} \qquad \cdots B_{-1}B_0B_1 \cdots,$$

regarding the blocks B_i as the generating symbols.

Substitution. Let $A_1, \cdots, A_\mu$ be s-blocks of the generating symbols, $s > 0$. The trajectory (c^*) of a_i's obtained by replacing a_i in (c) by A_i and expanding will be said to be derived from (c) by block substitution.

Generalized substitution. Let $r_1, \cdots, r_m$ be a fixed set of integers, positive, negative, or zero. Let $\phi(x_1, \cdots, x_m)$ be a single-valued function (either numerical or symbolic) of its arguments x_i as each x_i ranges over the set $(a_1, \cdots, a_\mu)$ of generators of (c). We include the convention that a null set may be one of the symbolic values of ϕ. We introduce the symbols

$$E_i = \phi[c_{i+r_1}, c_{i+r_2}, \cdots, c_{i+r_m}],$$

and thereby define an I-trajectory (E). The number of different symbols E_i is finite. We add the convention that a null set E_i is to be dropped from the trajectory (E). Derivation by generalized substitution includes all of the preceding types of derivation. We shall illustrate this fact for the case of derivation by association. Derivation by association is obtained if we set

$$\phi(x_1, \cdots, x_m) = (x_1, \cdots, x_m),$$

and understand that

$$(r_1, \cdots, r_m) = (0, 1, \cdots, m-1).$$

The symbol E_i reduces to the block

$$B_i = c_i c_{i+1} \cdots c_{i+m-1},$$

as in the case of derivation by association.

We shall illustrate these processes further by deriving two recurrent trajectories from the trajectory T of the preceding section. We shall use the method of association letting $m = 2$. The generating symbols are the 2-blocks of T, namely 11, 12, 22, 21. We shall denote these blocks respectively by 1, 2, 3, 4 and write a few terms of the resulting trajectory. The symbols with indices $i \geqq 0$ written above the corresponding symbols of T begin as follows:

2342412341242342412
12212112211212212112.

Upon reducing the first of these trajectories mod 2 we obtain a trajectory which contains

0100010101000100010.

This trajectory is distinct from T since it has sequences of three equal symbols. These new trajectories are recurrent as we shall see.

A trajectory T' derived from a recurrent trajectory T by projection or reduction is clearly recurrent. The trajectory T' may however be periodic when T is non-periodic. The facts for the case of derivation by association are otherwise, as the following theorem shows.

THEOREM 9.1. *If (c') is a recurrent I-trajectory derived from a recurrent I-trajectory (c) by association in r-blocks, and $R'(n)$ and $R(n)$ are the recurrency functions of (c') and (c) respectively, then*

$$R'(n) = R(n + r - 1) - r + 1. \tag{9.1}$$

Moreover (c') *is periodic if and only if* (c) *is periodic.*

A block $[k+1, k+n]$ of (c') is determined by the $(n+r-1)$-block $[k+1, k+n+r-1]$ of (c) and conversely. We term these blocks *corresponding.* Each $(n+r-1)$-block of (c) appears in each $R(n+r-1)$-block H of (c). But H corresponds in the above sense to a block (c') of length

$$R(n+r-1) - (r-1), \tag{9.2}$$

while an $(n+r-1)$-block of (c) corresponds to an n-block of (c'). Hence $R'(n)$ is at most the integer (9.2).

To show that $R'(n)$ is at least the integer (9.1), let H be a minimax $(n+r-1)$-covering in (c), that is an $R(n+r-1)$-block of (c) whose final $(n+r-1)$-block appears just once in H. Let K be the block of (c') corresponding to H. The final n-block of K corresponds to the final $(n+r-1)$-block of H and appears just once in K, so that $R'(n)$ is at least the integer (9.1), and the proof of (9.1) is complete.

If (c) is non-periodic, $R(n) > 2n$ in accordance with Theorem 7.5. It follows then from (9.1) that

$$R'(n) > 2n + r - 1.$$

The I-trajectory (c') is accordingly non-periodic, and the proof of the theorem is complete.

The preceding theorem leads to the following extension of Theorem 7.5.

THEOREM 9.2. *If* $P(r)$ *is the number of different* r*-blocks in a recurrent non-periodic* I*-trajectory* (c), *then for* $m \geqq r$,

$$R(m) \geqq 2m + P(r) - r. \tag{9.3}$$

Let (c') be derived from (c) by association in r-blocks. If $R'(n)$ is the recurrency function of (c'), it follows from Theorem 7.5 that

$$R'(n) \geqq 2n + P(r) - 1.$$

But from the preceding theorem we then see that

$$R(n+r-1) - r + 1 = R'(n) \geqq 2n + P(r) - 1,$$

from which (9.3) follows upon setting $m = n + r - 1$.

Before coming to the next theorem we shall need the following definition.

5

A definition. Let there be given a finite set S of q-blocks $B_1, \cdots, B_p$ of a set of generating symbols. Let n be a positive integer less than q. A least integer $r(n) \leqq q$ (if such exists) such that each n-block appearing in a block of the set S appears in each $r(n)$-subblock of each of the blocks B_j will be called the n-th *recurrency index* of the set S. For example, let the blocks B_i be the blocks

$$B_1 = 1221, \quad B_2 = 2112, \quad B_3 = 1212, \quad B_4 = 2121. \tag{9.4}$$

These blocks have a recurrency index $r(1) = 3$. The recurrency index $r(2)$ does not exist.

The following theorem concerns the I-trajectory (c^*) derived from the I-trajectory (c) by substitution. The I-trajectory (c^*) is formed as described above by substituting the respective s-blocks A_i for the generating symbols a_i appearing in (c). These blocks A_i are then expanded to form the I-trajectory (c^*) of the symbols a_i. We shall apply the preceding definition of the recurrency index of a set of blocks to the set S of $2s$-blocks A_iA_j. Each m-block of (c^*) for which $m \leqq s+1$ lies in one of these $2s$-blocks of (c^*) so that in the case where the set S has a recurrency index $r(m) \leqq s+1$ it is clear that the recurrency index $R^*(m)$ of (c^*) exists and is at most $r(m)$. This is affirmed in part (a) of the theorem. Part (b) conditions $R^*(n)$ on the lower side instead of on the upper side. This theorem is useful in the following section.

THEOREM 9.3. *Let (c^*) be an I-trajectory derived from an I-trajectory (c) by substitution as described above.*

(a) *If the set of $2s$-blocks A_iA_j of (c^*) has an m-th recurrency index $r(m) \leqq s+1$, then*

$$R^*(m) \leqq r(m). \tag{9.5}$$

(b) *If some block A_i appears r times consecutively in (c^*) with $r > 1$, and if (c^*) does not have the period s, then*

$$R^*(s+1) > sr, \tag{9.6}$$

or else $R^*(s+1)$ fails to exist.

Statement (a) has already been proved. We therefore turn to the proof of (b).

It is assumed that the block $B = A_i \cdots A_i$, of length sr, appears in (c^*). Since the length of A_i is s, any $(s+1)$-block of B has equal terminal elements.

If $R^*(s+1)$ exists and $R^*(s+1) \leqq sr$, every $(s+1)$-block of (c^*) has a copy in B. Thus every $(s+1)$-block of (c^*) has equal terminal elements and (c^*) has the period s, contrary to hypothesis.

The proof of the theorem is complete.

The following theorem has a special interest in view of the mode of formation of the Morse recurrent trajectory. The trajectory H thereby affirmed to exist is not necessarily admissible. It will be admissible if the only symbols employed are α and β.

THEOREM 9.4. *Corresponding to an arbitrary recurrent ray*

$$R = a_0 a_1 a_2 \cdot \cdot \cdot ,$$

there exists a recurrent trajectory H with R as subray and with the recurrency indices of R. The trajectory H is in general not uniquely determined by R.

Let B_r be the initial r-block of R, $r = 1, 2, \cdot \cdot \cdot$. The block B_r appears later in R immediately preceded by some r-block A_r. Of the blocks $A_1, A_2, \cdot \cdot \cdot$ there exists an infinite subset S_1 with last symbol in common. We take a_{-1} as this last symbol. Of the blocks of S_1 there exists an infinite subset S_2 with the second from the last symbol in common. We take a_{-2} as this common symbol. Proceeding in this way we define sets $S_3, S_4, \cdot \cdot \cdot$ and symbols $a_{-3}, a_{-4}, \cdot \cdot \cdot$ respectively. That the resulting trajectory H is recurrent and has the same recurrency indices as R follows from the fact that each block of H appears in R.

That H is not uniquely determined is shown by the trajectories T and T' of § 8 which have the recurrent ray R of (8.4) as a common subray.

10. Deferred recurrence. In this section we shall exhibit a recurrent trajectory E for which $R(n)$ exceeds a prescribed function $f(n)$ for infinitely many values of n but which is such that the limit inferior of $R(n)/n$ is finite. That a recurrent trajectory exists for which $R(n)$ exceeds a prescribed function $f(n)$ for all values of n has already been shown by Robbins ([1], Theorem 2). Our example shows the possibility of extreme variability in the indices $R(n)$. The trajectories E here defined form an extended class of which the trajectory T of § 8 is a member.

The generating symbols shall be the free symbols α and β. Let $r_0, r_1, \cdot \cdot \cdot$ be an arbitrary set of positive integers. We shall make use of powers of blocks, understanding that these powers are to be formally expanded. Set

$$\begin{array}{ll} A_0 = \alpha, & B_0 = \beta, \\ A_1 = A_0^{r_0} B_0^{r_0}, & B_1 = B_0^{r_0} A_0^{r_0}, \\ \cdot\ \cdot\ \cdot\ \cdot\ \cdot\ \cdot & \cdot\ \cdot\ \cdot\ \cdot\ \cdot\ \cdot \\ A_{n+1} = A_n^{r_n} B_n^{r_n}, & B_{n+1} = B_n^{r_n} A_n^{r_n}. \end{array}$$

Let s_n be the length of A_n and B_n. Observe that

$$s_{n+1} = 2s_n r_n. \tag{10.1}$$

Let (c) be an I-trajectory such that the initial s_n-block of the ray $c_0c_1c_2\cdots$ is A_n for each n. This definition is self consistent since A_{n-1} is an initial block of A_n. We further set

$$c_{-m-1} = c_m \qquad (m = 0, 1, \cdots). \tag{10.2}$$

Proceeding inductively we see that

$$\begin{aligned} A_n^{-1} &= A_n, & B_n^{-1} &= B_n, & &(n \text{ even}) \\ A_n^{-1} &= B_n, & B_n^{-1} &= A_n, & &(n \text{ odd}). \end{aligned}$$

Hence the s_n-block with final index -1 is A_n when n is even and B_n when n is odd. If the symbols of (c) are grouped into s_r-blocks with first indices $\equiv 0 \bmod s_r$, one obtains an I-trajectory E_r with generating symbols A_r and B_r.

If the integers r_i are all 1 and $\alpha = 1$, $\beta = 2$, (c) yields the recurrent trajectory T of § 8. We shall make use of the recurrency function $\rho(n)$ of T. The trajectory defined by (c) will be denoted by E. It is uniquely determined by the symbols α and β and the integers r_i.

THEOREM 10. 1. *The trajectory E is recurrent and non-periodic with a recurrency function $R^*(n)$ such that*

$$s_{n+1} \leqq R^*(s_n + 1) \leqq 2s_{n+3}. \tag{10.3}$$

Let B be an arbitrary $(s_n + 1)$-block of E. The block B appears in some expanded 2-block of E_n. But each such 2-block of E_n appears in both A_{n+3} and B_{n+3} as is easily seen. The latter two blocks are s_{n+3}-blocks of E and one of them at least is in each $2s_{n+3}$-block of E. Hence the right-hand condition in (10. 3) is valid. It follows that E is recurrent.

The left condition in (10. 3) follows from (b) in Theorem 9. 3. For the block A_n appears $2r_n$ consecutive times in E. Moreover E does not have the period s_n since E contains the block A_nB_n and B_n is different from A_n. From (9. 6) we then find that

$$R^*(s_n + 1) \geqq 2r_n s_n = s_{n+1},$$

and the proof of (10. 3) is complete. Moreover we see that

$$R^*(s_n + 1) \geqq 2s_n$$

and it follows from Theorem 7. 6 that E cannot be periodic.

COROLLARY. *If $f(n)$ is an arbitrary positive integral function, it is possible to choose the r_n's or any subset of the r_n's so that for corresponding values of n,*

$$R^*(s_n+1) > f(s_n+1). \tag{10.4}$$

Recall that $s_{n+1} = 2s_n r_n$. In order that (10.4) may hold we have merely to take r_n so large that

$$s_{n+1} > f(s_n+1)$$

and (10.4) will follow from (10.3).

We continue with the following theorem.

THEOREM 10.2. *If the sequence of integers r_i used in defining E contains arbitrarily long blocks of 1's,*

$$\lim_{n\to\infty}\inf \frac{R^*(n)}{n} \leqq \frac{11}{2}. \tag{10.5}$$

If $\rho(m)$ is the recurrency function of the trajectory T of § 8, there exists a sequence of integers m_i becoming infinite with i and such that

$$\lim_{i\to\infty}\inf \frac{\rho(m_i)}{m_i} = \frac{11}{2}. \tag{10.6}$$

We suppose integers m_i so chosen. The sequence of integers r_i contains arbitrarily long blocks of 1's. It follows that there exists a sequence of positive integers

$$h_1 < k_1 < h_2 < k_2 < \cdots, \tag{10.7}$$

such that for each i,

$$r_{h_i} = r_{h_i+1} = \cdots = r_{k_i} = 1. \tag{10.8}$$

Moreover these integers can be chosen so that

$$\mu_i = k_i - h_i > \rho(m_i).$$

We shall prove the following statement.

(α). *The m_i-th recurrency index of E_{h_i} is at most the recurrency index $\rho(m_i)$.*

The trajectory E_{h_i} is generated from the blocks A_{h_i}, B_{h_i}. These blocks combine to form the blocks A_{k_i}, B_{k_i} in exactly the way in which the symbols 1 and 2 combine to form the blocks a_{μ_i} and b_{μ_i} of the Morse trajectory T. In E_{h_i} the blocks

(10.9) $$A_{k_i}B_{k_i}, \quad A_{k_i}A_{k_i}, \quad B_{k_i}A_{k_i}, \quad B_{k_i}B_{k_i},$$

regarded as blocks of the symbols A_{h_i} and B_{h_i}, accordingly have an m_i-th recurrency index $\rho(m_i)$. For

$$\rho(m_i) < k_i - h_i < 2^{k_i - h_i} = 2^{\mu_i},$$

and 2^{μ_i} is the length of the blocks A_{k_i} and B_{k_i} regarded as blocks of the symbols A_{h_i}, B_{h_i}. It follows from (a) of Theorem 9.3 that E_{h_i} has an m_i-th recurrency index at most $\rho(m_i)$, and the proof of (α) is complete.

We now regard E as derived from E_{h_i} by expansion of the blocks A_{h_i} and B_{h_i} of E_{h_i} into blocks of α's and β's. Set

(10.10) $$n_i = s_{h_i}(m_i - 1).$$

An arbitrary n_i-block of E is found in some expanded m_i-block of E_{h_i}, and hence in each expanded $\rho(m_i)$-block of E_{h_i}, and finally in each $[s_{h_i}\rho(m_i) + s_{h_i}]$-block of E. Hence

(10.11) $$R(n_i) \leqq s_{h_i}[\rho(m_i) + 1].$$

It follows from (10.10), (10.11) and (10.6) that

$$\lim_{i\to\infty}\text{inf.}\, \frac{R(n_i)}{n_i} \leqq \lim_{i\to\infty}\text{inf.}\, \frac{\rho(m_i)+1}{m_i - 1} = \frac{11}{2},$$

and the proof of the theorem is complete.

The preceding theorems and corollary combine to give use the following.

THEOREM 10.3. *If $f(n)$ is an arbitrary positive integral function of n, then for suitable choices of the integers r_i defining E, the recurrency function $R^*(n)$ of E exceeds $f(n)$ for infinitely many values of n while*

$$\lim_{n\to\infty}\text{inf.}\, \frac{R^*(n)}{n} \leqq \frac{11}{2}.$$

11. Transitive rays and their ergodic functions. We shall concern ourselves with rays of the form

$$c_0c_1c_2 \cdots$$

Such a ray will be termed transitive if it contains a copy of each admissible subblock. Transitive rays exist. If $A_1, A_2, \cdots$ is an enumeration of admissible blocks, the ray

$$X = A_1e_1A_2e_2 \cdots$$

is admissible provided the symbols e_i are successively chosen in a suitable way. The ray X is clearly transitive. Let R be an arbitrary ray and let n be a positive integer. Suppose an initial r-block B of R contains each admissible n-block while no initial subblock of B has this property. The integer r will then be termed the n-th *transitivity index* $\phi(n)$ of R. If R is transitive, the n-th transitivity index of R exists for each value of n and will be termed the *ergodic function* $\phi(n)$ of R.

A block $H(n)$ of minimum length containing each admissible n-block will be termed a *minimum n-covering*. The length $\theta(n)$ of $H(n)$ will be termed the n-th *covering index*. In the case where the generating symbols are all free, Martin [1] has shown that for each n there exists a minimum n-covering in which no n-block is repeated. We shall presently extend Martin's result to the case where successive inverses are prohibited. We are concerned here with the existence of transitive rays whose ergodic function is in some sense relatively small. Noting that each ergodic function $\phi(n)$ satisfies the condition

$$\phi(n) \geqq \theta(n),$$

we are led to the question, does there exist a transitive ray such that $\phi(n) = \theta(n)$ for each n? The answer is no, in general, as we shall see.

We begin with the case where there are just two generating symbols, namely the free symbols α and β. In this case it follows from Martin's theorem that

$$\theta(n) = 2^n + n - 1.$$

The only minimum 1-coverings are $\alpha\beta$ and $\beta\alpha$. The only minimum 2-covering which starts with $\alpha\beta$ is $\alpha\beta\beta\alpha\alpha$. If we wish a minimum 3-covering which starts with this 5-block we may continue with β or α, but we must then continue with $\alpha\beta$ or $\beta\alpha\beta$ respectively, thereby obtaining the blocks

$$(\alpha\beta\beta\alpha\alpha)\beta\alpha\beta, \qquad (\alpha\beta\beta\alpha\alpha)\alpha\beta\alpha\beta$$

Further continuation subject to the condition that no 3-block repeats is impossible. Since $\theta(3) = 10$, there exists no transitive ray for which

$$\phi(n) = \theta(n) \qquad (n = 1, 2, 3).$$

It is conceivable that there might exist a transitive ray for which $\phi(n) = \theta(n)$ for all values of n exceeding some fixed integer m. That this is impossible is shown by the following previously unpublished theorem of R. Oldenburger.

THEOREM 11.1. *If the generating symbols reduce to two free symbols* α *and* β *there exists no transitive ray whose ergodic function* $\phi(n)$ *equals the covering index* $\theta(n)$ *for two successive values of* $n > 1$.

The proof as given by Oldenburger is essentially as follows.

We assume the theorem false. There then exists a minimum n-covering $H(n)$, $n > 2$, which starts with a minimum $(n-1)$-covering $H(n-1)$. No n-block repeats in $H(n)$, nor $(n-1)$-block in $H(n-1)$. We shall arrive at a contradiction.

Since α and β can be interchanged throughout it is no essential restriction to assume that in $H(n-1)$ the $(n-1)$-block β^{n-1} occurs before α^{n-1}. We shall then show that $H(n-1)$ has the form

$$H(n-1) = [\beta^{n-1}\alpha \cdots \beta\alpha^{n-1}], \tag{11.1}$$

and that $H(n)$ has the form

$$H(n) = H(n-1)[\alpha\beta \cdots \alpha\beta^{n}]. \tag{11.2}$$

The block of $H(n)$ which remains when $H(n-1)$ is removed will be denoted by J. To establish the correctness of the representations (11.1) and (11.2) one reasons as follows.

The block α^n appears in $H(n)$. It cannot appear wholly in $H(n-1)$, for this would imply repetition of α^{n-1} in $H(n-1)$, and thus $\beta\alpha^n$ appears in $H(n)$. Since β^{n-1} precedes α^{n-1} in $H(n-1)$, $\beta\alpha^{n-1}$ appears in $H(n-1)$. Since $\beta\alpha^{n-1}$ appears just once in $H(n)$, it must appear as the initial n-block of $\beta\alpha^n$, and since this latter block is not wholly in $H(n-1)$, $H(n-1)$ ends with $\beta\alpha^{n-1}$ and J begins with α. This α is of course followed by β.

The block β^n is in $H(n)$ and not wholly in $H(n-1)$. Since J begins with α, β^n must appear wholly in J, and hence $\alpha\beta^{n-1}$ and $\alpha\beta^n$ appear in J. It follows that β^{n-1}, which appears in $H(n-1)$, must appear at the left of $H(n-1)$; otherwise it would be preceded by α, and $\alpha\beta^{n-1}$ would appear twice in $H(n)$. Hence $H(n-1)$ begins with $\beta^{n-1}\alpha$. If J does not end with β^n, $\beta^n\alpha$ appears in J, and $\beta^{n-1}\alpha$ appears both in J and $H(n-1)$. Hence J ends with $\alpha\beta^n$ and the representations (11.1) and (11.2) are correct.

We now make use of the assumption that $n > 2$. Then $n-2 > 0$ and $\alpha\beta^{n-2}$ occurs in $H(n-1)$. This block cannot be followed by β without repetition of β^{n-1} in $H(n-1)$. Hence $\alpha\beta^{n-2}\alpha$ appears in $H(n-1)$. But $\beta^{n-2}\alpha$ already appears at the beginning of $H(n-1)$. Hence $H(n)$ cannot start with a minimum $(n-1)$-covering, and the proof of the theorem is complete.

We turn to the case where the only condition on the generating symbols is the restriction of inverses. This is the simplest case of strict dynamical significance. It arises in the case of surfaces of negative curvature with boundaries. Cf. Morse [1]. In this case one has a set of generating symbols of the form

$$a_1, \cdots, a_p; a_1^{-1}, \cdots, a_p^{-1} \qquad (p > 1), \tag{11.3}$$

with the restriction that successive symbols shall not be inverses. We shall prove the following theorem by suitably modifying Martin's proof.

THEOREM 11.2. *For generating symbols of the form* (11.3), *subject to no condition beyond the restriction of inverses, there exists a minimum r-covering in which no r-block repeats.*

The following rule for the formation of a minimum r-covering $H(r)$ was first given by Martin for the case of free generating symbols $c_1, \cdots, c_n$. Start with c_1^{r-1} and continue successively adjoining the generating symbol of highest index consistent with the condition that no r-block be repeated. This will lead to a minimum r-covering in the case of free symbols c_i. For the case of inverses we adjoin the condition that successive symbols shall not be inverses. With this condition added the process will not always lead to a minimum r-covering. We are able however to prove the following lemma.

LEMMA 11.1. *A necessary and sufficient condition that the application of the Martin rule, subject to the restriction of inverses, shall lead to a minimum r-covering is that c_1 and c_2 shall not be inverses in the set $c_1, \cdots, c_n$ of generating symbols.*

Let M_r be the block of maximum length obtained by applying the Martin rule subject to the restriction of inverses. If A is any admissible $(r-1)$-block, the r-blocks

$$Ac_n, \cdots, Ac_1 \tag{11.4}$$

appear in M_r, if at all, in the order written. One of the blocks (11.4) is inadmissible since the last symbol in A cannot be followed by its inverse. If Ac_1 is in M_r, all other admissible blocks in (11.4) are in M_r. If Ac_1 is not admissible, Ac_2 is admissible and is in M_r only if all of its predecessors in (11.4) are in M_r. We set $E = c_1^{r-1}$. We continue with a proof of (a) and (b).

(a). *The block M_r ends with E.* If M_r ended with an $(r-1)$-block $A \neq E$, M_r could not begin with A, since M_r begins with E. A would then appear in M_r at most $n-1$ times; otherwise some one of the $n-1$ admissible blocks $c_i A$ would appear twice. Following A's last appearance we can continue M_r with c_2 (or c_1 if Ac_2 is inadmissible), contrary to the hypothesis that M_r ends with A. Hence M_r ends with E.

(b). *The block M_r ends with Ec_1.* Suppose M_r ends with an r-block $B \not\equiv Ec_1$. Set $B = bE$ in accordance with (a). We are supposing that $b \neq c_1$. The block E appears n times in M_r; otherwise M_r could be continued. Hence Ec_1 appears in M_r but not at the end. But Ec_1 cannot be continued in accordance with the Martin rule. From this contradiction we infer the truth of (b).

We shall now make use of the assumption that c_1 and c_2 are not inverses and show that M_r contains all r-blocks.

We first observe that each admissible block Ec_j appears in M_r since M_r ends with Ec_1. Hence M_r contains n copies of E. It follows that M_r contains each admissible block $c_j E$. Let $B = b_1 \cdots b_r$ be an arbitrary admissible r-block. We shall show that M_r contains B.

Suppose that M_r does not contain B. Set $b_2 \cdots b_r = D$. Then $D \not\equiv E$ by virtue of (b). Hence D appears at most $n-2$ times in M_r; otherwise each admissible block $c_j D$ would appear in M_r, including B. Hence Dc_1 does not appear in M_r (or Dc_2, if Dc_1 is inadmissible). We now apply the same reasoning to Dc_1 (or Dc_2) that we have applied to B and infer that $b_3 \cdots b_r c_1 c_1$ (or $b_3 \cdots b_r c_2 c_1$) does not appear in M_r. Continuing we arrive at the conclusion that $c_1{}^r$ does not appear in M_r, contrary to (b).

The condition of the lemma is accordingly sufficient.

To prove the condition necessary we suppose that $c_1 = c_2{}^{-1}$. If M_r were an r-covering $c_2{}^r$ would appear in M_r, but not in the last position, by virtue of (a). But $c_2{}^r$ cannot be continued without violating the Martin rule, or else the condition on inverses. Hence the condition is necessary, and the proof of the lemma is complete.

Theorem 11. 2 is an immediate consequence of the lemma.

If there are n generating symbols subject merely to the restriction of inverses, there are $n(n-1)^{r-1}$ different admissible r-blocks so that the r-th covering index $\theta(n)$ has the value

$$\theta(r) = n(n-1)^{r-1} + r - 1.$$

With this understood we shall prove the following theorem.

THEOREM 11. 3. *When there are n generating symbols subject merely to* *the restriction of inverses, each ergodic function* $\psi(r)$ *is at least* $\theta(r)$ *and there exists an ergodic function* $\phi(r)$ *such that*

$$\phi(r) < \left(\frac{n-1}{n-2}\right) n(n-1)^{r-1} < \frac{n-1}{n-2}\theta(r). \tag{11.5}$$

Let H_r be a minimum r-covering and let H^*_r be the block obtained from H_r by omitting the first $r-1$ cymbols c_1. The ray

$$R = H_1 H^*_2 H^*_3 \cdots$$

is admissible and transitive. We note in particular that H^*_r is preceded by c_1^{r-1} in R. For this ray $\phi(r)$ is at most the length of the block $H_1 H^*_2 \cdots H^*_r$, that is

$$\phi(r) \leqq \sum_{m=1}^{r} n(n-1)^{m-1}.$$

Hence

$$\phi(r) \leqq n\left[\frac{1-(n-1)^r}{1-(n-1)}\right] = \left(\frac{n-1}{n-2}\right) n\left[(n-1)^{r-1} - \frac{1}{n-1}\right],$$

and (11.5) follows at once.

We conclude this section by proving the following theorem.

THEOREM 11.4. *If* $N(r)$ *is the number of admissible r-blocks under our general conditions of admissibility, there exists an ergodic function* $\phi(r)$ *such that*

$$\phi(r) < (r+1)N(r), \tag{11.6}$$

while each ergodic function is at least $N(r) + r - 1$.

We begin by proving the following.

(a). *There exists an r-covering of length less than* $(r+1)N(r)$.

Statement (a) is true if $r = 1$. For if $a_1, \cdots, a_n$ *is* the set of generating symbols, there exists a 1-covering of the form $a_1 e_1 a_2 e_2 \cdots e_{n-1} a_n$ provided the symbols e_i are suitably chosen. This 1-covering has the length $2n-1 < 2N(1)$. We assume then that A_r is an r-covering of length less than $(r+1)N(r)$. We shall show how to form an $(r+1)$-covering of length less than $(r+2)N(r+1)$.

There are $N(r)$ distinct r-blocks in A_r and hence at least $N(r)-1$ distinct $(r+1)$-blocks. There remain at most

$$\mu = N(r+1) - N(r) + 1$$

$(r+1)$-blocks, say $B_1, \cdots, B_\mu$ not in A. The block

$$A_{r+1} = A_r e_1 B_1 e_2 B_2 e_3 \cdots e_\mu B_\mu,$$

in which the symbols e_i are chosen so as to make A_{r+1} admissible, is an $(r+1)$-covering. Let s_i be the length of A_i. We see that

$$s_{r+1} = s_r + (r+2)\mu.$$

By virtue of the hypothesis that $s_r < (r+1)N(r)$ we infer that

$$\begin{aligned} (11.7) \qquad s_{r+1} &< (r+1)N(r) + (r+2)[N(r+1) - N(r) + 1] \\ &= (r+2)N(r+1) - N(r) + r + 2. \end{aligned}$$

But for $r > 1$,

$$N(r) \geqq 2^r \geqq r + 2,$$

so that (11.7) yields the relation

$$s_{r+1} < (r+2)N(r+1),$$

and the proof of (a) is complete.

In proving (a) we have defined r-coverings A_r, $r = 1, 2, \cdots$, such that A_r is an initial block of A_{r+1}. A ray whose first s_r symbols are those of A_r for each r will satisfy (11.6). The final statement of the theorem is immediate.

12. Category and sets of transitive rays. In this section we shall determine the category of the set of transitive rays or of sets of relatively transitive rays. More particularly, we shall determine the category of various subsets of transitive rays defined by limitations on the ergodic functions. In this way we obtain an idea of the distribution of transitive rays. A similar use can be made of measure theory (cf. § 13), but the development of this phase of the theory is deferred to a later paper.

In the category theory, the space used will be the space S of admissible rays with the Baire metric. Cf. § 4. Recall that the space S is compact, perfect, and totally disconnected. Let H be a perfect subset of S. A set A is of the *first category* relative to H if it is the sum of an at most enumerable set of nowhere dense (relative to H) subsets of H. These subsets will be called the components of A. The components of A may vary from the extreme case where they are all null to the case where they are perfect. The complement $H - A$ of a set A of the first category is termed a *residual set* of the second category. Such a set is everywhere dense in H and has the power of the continuum. Cf. Hobson [1], p. 132.

Let R be a given ray. The rays obtained by deleting a finite initial block of R will be said to be *based* on R, those obtained by the inverse process of adding a finite block to R will be termed *extensions* of R. The rays which are limit rays in the space S of the set of rays based on R will be called *limit* rays of R. A ray R which is a limit ray of itself will be said to be *transitive relative to its limit rays.* It is thereby necessary and sufficient that R be non-periodic and that every block of R appear infinitely many times in R. The limit rays of a relatively transitive ray R form a closed set H. This set is identical with its derived set since each ray based on R is a limit ray of the set. The set H is accordingly perfect. Concerning the set H we have the following theorem.

THEOREM 12.1. *In the set H of limit rays of a relatively transitive ray R, the rays which are transitive relative to H form a residual set in H.*

Let $A_1, A_2, \cdots$ be an enumeration of the blocks of R. Let E_n be the subset of rays of H which contain no copy of A_n. The set E_n is closed and contains no rays based on R. It is nowhere dense in H since the rays based on R are everywhere dense in H. Finally the set S of rays which are transitive relative to H is of the form

$$S = H - \Sigma E_n,$$

and the theorem follows from the definition of a residual set in H.

The above sets E_n may all be null, in fact are all null if R is a subray of a non-periodic recurrent trajectory. In the case where R is transitive, that is transitive relative to the set of all admissible rays, the sets E_n are never null as the following lemma shows.

LEMMA 12.1. *The set K of rays which exclude an admissible r-block B for which $r > 2$ is closed, nowhere dense, and of the power of the continuum.*

It is clear that K is closed. That K is nowhere dense in S follows from the existence of a transitive ray R. For the rays based on R are everywhere dense in S and transitive. Since K is closed, it must then be nowhere dense.

It remains to prove that K is of the power of the continuum. To that end we consider the admissible rays

$$\alpha^{r_1}\beta\alpha^{r_2}\beta\alpha^{r_3}\beta\cdots, \qquad r_i = 2 \text{ or } 3, \tag{12.1}$$

formed from the free symbols α and β. This set is in one-to-one correspondence with the dyadic decimals

$$.r_1r_2r_3\cdots, \qquad r_i = 2 \text{ or } 3,$$

and hence is of the power of the continuum.

Similarly, the set

$$\beta^{r_1}\alpha\beta^{r_2}\alpha\beta^{r_3}\cdots, \qquad r_i = 2 \text{ or } 3, \tag{12.2}$$

consists of admissible rays and is of the power of the continuum.

A ray of the form (12. 1) has no 3-blocks in common with a ray of the form (12. 2). Since the length of B is at least 3, it cannot appear in a ray of (12. 1) and a ray of (12. 2). Hence at least one of the sets (12. 1) and (12. 2) is contained in K. Thus K must be of the power of the continuum.

The set of non-transitive rays is of the first category. But much more than this can be said, as the following theorem shows.

THEOREM 12. 2. *The set Σ of non-transitive rays is everywhere dense in S and is the sum of countably many nowhere dense, closed sets, each of which is of the power of the continuum.*

That the set Σ is everywhere dense follows from the fact that an arbitrary block B and a sequence A of the symbols α can be combined in the form BeA to form a non-transitive ray. Rays of this form are everywhere dense in S. Let $A_1, A_2, \cdots$ be an enumeration of admissible r-blocks with $r > 2$. The set E_n of rays which exclude A_n is closed, nowhere dense and of the power of the continuum, in accordance with Lemma 12. 1. The set Σ is the sum of the sets E_n, and the proof is complete.

THEOREM 12. 3. *If $f(n)$ is an arbitrary positive function of the positive integers n, the set F of transitive rays with ergodic function $\phi(n) > f(n)$, for all values of n, has the power of the continuum.*

Let $A_1, A_2, \cdots$ be an enumeration of admissible blocks consistent with their lengths. Let $r_1, r_2, \cdots$ be an arbitrary sequence of positive integers and let $c_1c_2\cdots$ be an arbitrary sequence of the symbols α and β. We introduce the block

$$B_i = \alpha^{r_i}c_i\alpha^{r_i}e_iA_i \qquad (i = 1, 2, \cdots),$$

choosing e_i so that B_i is admissible. We then introduce the ray

$$E = B_1 b_1 B_2 b_2 \cdots,$$

choosing the symbols b_i so that the ray E is admissible. It is clear that E is transitive.

Let $\phi(n)$ be the ergodic function of E. We choose r_1 so that $\phi(n) > f(n)$ for $n = 1, 2, 3$. Let B be the initial $f(4)$-block of E. If r_2 is chosen sufficiently large, blocks $A_1, A_2, \cdots, A_p$ entering into the definition of B will all be 1-blocks. In fact, if r_2 is sufficiently large only A_1 will enter into the definition of B. If A_i is a 1-block, $e_i A_i b_i$ is a 3-block so that B cannot contain β^4. With r_2 so chosen, $\phi(4) > f(4)$.

Proceeding inductively with $m > 2$ we choose r_m so large that in the initial $f(m+2)$-block C of E, at most $(m-1)$-blocks A_i are employed. The corresponding blocks $e_i A_i b_i$ are at most $(m+1)$-blocks so that β^{m+2} does not appear in C. Hence

$$\phi(m+2) > f(m+2) \qquad (m = 3, 4, \cdots). \tag{12.3}$$

Thus $\phi(n) > f(n)$ for all values of n.

The above symbols c_i can be chosen arbitrarily from the set α, β, the r_i's remaining fixed. For fixed r_i's there are then as many different rays E as there are different sequences $c_1 c_2 \cdots$, that is a number equal to the power of the continuum.

The proof of the theorem is complete.

The set F of the theorem is not closed. The set of transitive rays not in F is open in the set of all transitive rays so that F is everywhere dense only if it includes all transitive rays.

THEOREM 12. 4. *Given an arbitrary function $f(n)$, the set ω of transitive rays R with ergodic functions $\phi(n) \leqq f(n)$ for n exceeding some integer dependent on R is a set of the first category.*

Let D_m be the set of transitive rays for which $\phi(n) \leqq f(n)$ for n exceeding m. The set D_m is closed. But we have seen in Theorem 12. 2 that the set of intransitive rays is everywhere dense. Hence D_m is nowhere dense. But $\omega = \Sigma D_m$, and the proof is complete.

Let $H(n)$ be a block of minimum length $\theta(n)$ covering all admissible n-blocks. We have termed $\theta(n)$ the n-th covering index. As we have seen in § 11, there is in general no transitive ray with ergodic function $\phi(n) = \theta(n)$. We have however the following theorem.

THEOREM 12. 5. *The set G of transitive rays with ergodic function $\phi(n)$ such that*

$$\lim_{n\to\infty} \inf. \frac{\phi(n)}{\theta(n)} = 1, \tag{12.4}$$

where $\theta(n)$ is the n-th covering index, is a residual set of the second category.

We begin with a proof of the following.

(a). *The set G is not empty.*

To prove (a) let δ_n be a sequence of positive constants converging to zero. We shall consider a ray of the form

$$R = H(r_0)e_1H(r_1)e_2H(r_2)\cdots,$$

where $H(n)$ is defined as above and the symbols e_i are chosen so that R is admissible. The integers r_i are chosen as follows. Let $r_0 = 1$. Proceeding inductively we suppose that r_{n-1} is defined and that the block

$$H(r_0)e_1H(r_1)e_2\cdots e_{n-1}H(r_{n-1})e_n$$

has the length m_n. We then choose r_n so large that

$$\frac{\theta(r_n) + m_n}{\theta(r_n)} < 1 + \delta_n. \tag{12.5}$$

If $\phi(n)$ is the ergodic function of the resulting ray R, it is clear that

$$\phi(r_n) \leqq \theta(r_n) + m_n,$$

since the initial block of length $\theta(r_n) + m_n$ in R contains $H(r_n)$ and hence all r_n-blocks. The function $\phi(n)$ will satisfy (12.4) by virtue of (12.5), and the proof of (a) is complete.

If A is an arbitrary admissible r-block, the ray $X = AeR$ is admissible if e is suitably chosen. Moreover the ray R and the ray X have ergodic functions ϕ and ψ respectively such that

$$\theta(n) \leqq \psi(n) \leqq \phi(n) + r + 1.$$

Hence X satisfies the condition (12.4) and is in G. We conclude that the rays of G are everywhere dense.

To come to a proof of the theorem let m be a positive integer and p a rational number > 1. The rays for which the transitivity index $\psi(m)$ satisfies the condition

$$\psi(m) > \theta(m)p$$

form a closed set $D_m{}^p$. The intersection

$$E_q{}^p = D_q{}^p D^p{}_{q+1} D^p{}_{q+2} \cdots$$

is closed. The complement of G is the sum of the sets $E_q{}^p$ as q ranges over all positive integers and p ranges over all rational numbers > 1. But G is everywhere dense so that $E_q{}^p$ is nowhere dense, and the proof of the theorem is complete.

THEOREM 12.6. *The set Z of transitive rays with ergodic functions $\phi(n)$ such that*

$$\lim_{n\to\infty}\inf. \frac{\phi(n)}{\theta(n)} > 1$$

is everywhere dense, has the power of the continuum, and is of the first category.

The set Z includes the set of transitive rays for which $\phi(n) > 2\theta(n)$, and so has the power of the continuum in accordance with Theorem 12.3. The set Z is a subset of the complement of the residual set G of Theorem 12.5 and so is of the first category. It remains to show that Z is everywhere dense.

If R is a ray of Z, any ray R' based on R is again transitive. If the initial symbol of R' is the $(r+1)$-st symbol of R, the ergodic functions $\psi(n)$ and $\phi(n)$ of R' and R, respectively, satisfy the conditions

$$\psi(n) \geqq \phi(n) - r, \qquad \lim_{n\to\infty}\inf. \frac{\psi(n)}{\theta(n)} > 1.$$

Hence R' belongs to Z. But the rays R' based on the transitive ray R are everywhere dense so that Z is everywhere dense as stated.

The proof of the theorem is complete.

13. Trajectories with the separation property. The results of this section will form the basis of a more extensive treatment to be presented later. No proofs will be given in this place.

Let t be a trajectory involving two generating symbols α and β. By the α-length (β-length) of a block B of t we shall mean the number of α's (β's) in B. The trajectory t will be said to have the *separation property* and be of *type S* if the following holds.

The α-lengths of any two n-blocks of t differ by at most 1.

6

The preceding condition implies the corresponding condition for β-lengths.

Trajectories of type S arise in many ways. In particular let

$$\frac{d^2y}{dx^2}+f(x)y=0 \tag{13.1}$$

be a differential equation in which $f(x)$ is continuous and has the period 1. Let $u(x)$ be a solution of (13.1) which vanishes at least once, but not identically. We are concerned with the distribution of the zeros of $u(x)$. Without loss of generality we can suppose that none of these zeros occur at the points $x=n$, where n is an integer, positive, negative or zero. For the zeros of $u(x)$ are countable, and for our purposes any transformation of the form $\bar{x}=x+c$ is admissible. With this understood let a trajectory t be defined as follows. Let each point on the x-axis at which $u(x)=0$ be replaced by the symbol α, and each point $x=n$ be replaced by the symbol β. The resulting trajectory t will be of type S. With trivial exceptions every trajectory of type S can be obtained in this way. The study of trajectories of type S is thus a study of the zeros of solutions of equations of the form (13.1).

Let $\alpha_n(\beta_n)$ denote the α-length (β-length) of some n-block of a trajectory t of type S. It can be shown that there exists a number θ, $0\leqq\theta\leqq 1$, such that

$$\lim_{n\to\infty}\frac{\alpha_n}{n}=\theta,\qquad \lim_{n\to\infty}\frac{\beta_n}{n}=1-\theta.$$

The number θ will be termed the *rotation number* of t. Cf. Poincaré [1].

The trajectory is not uniquely determined by the rotation number. If θ is irrational, there are infinitely many trajectories of type S having θ as rotation number and these trajectories are all recurrent. Thus trajectories of type S are in general recurrent. A simple exceptional example is the trajectory in which all the symbols are α except for one β, in which case the rotation number is 1.

Given θ, $0\leqq\theta\leqq 1$, the recurrent trajectories of type S which have θ as rotation number form a minimal set which consists of a single periodic trajectory if and only if θ is rational. Since all of the trajectories of one of these minimal sets have the same recurrency function, it can be denoted by $R_\theta(n)$. The behavior of $R_\theta(n)$ varies widely with the choice of θ, but in all cases

$$\liminf_{n\to\infty}\frac{R_\theta(n)}{n}\leqq 4.$$

There exist irrational values of θ, and thus non-periodic recurrent trajectories such that

$$\lim_{n\to\infty}\text{sup.}\ \frac{R_\theta(n)}{n} \leqq 6.5.$$

On the other hand, given an arbitrary function $f(n)$, there exist irrational values of θ such that $R_\theta(n) > f(n)$ for infinitely many values of n.

If certain exceptional θ-sets of (Lebesgue) measure zero are excluded, $R_\theta(n)$ displays great regularity in its asymptotic behavior. We can state that, except for a θ-set of measure zero,

$$\lim_{n\to\infty}\text{sup.}\ \frac{R_\theta(n)}{n} = +\infty,$$

while if $c > 1$, except for a θ-set of measure zero,

$$\lim_{n\to\infty}\text{sup.}\ \frac{R_\theta(n)}{n^c} \leqq 1.$$

14. Typical open questions. Let R be a ray of the form $c_0c_1c_2\cdots$ and let t be a trajectory containing R as a subray. Among the positive limit trajectories of t are found one or more distinct minimal sets of trajectories. Cf. Theorem 6.1. Suppose v_1 of these sets are periodic trajectories, and v_2 of these sets general minimal sets. The number v_1 is at most aleph null, and v_2 is at most aleph. That v_2 may equal aleph will be shown by an example in a later paper. The numbers v_1 and v_2 are uniquely determined by R and will be called the *limit indices* of R of the first and second types respectively. Our first question is as follows.

(1). What are the limit indices of algebraic numbers, regarding decimal representations as rays? Are these rays ever transitive or recurrent? What are the answers to similar questions concerning π and e?

(2). Let t be a non-periodic recurrent trajectory. In all known examples $R(n)/n$ fails to converge. Is this true in general?

(3). If t is a non-periodic recurrent trajectory, $R(n) \geqq kn$ where $k \geqq 2$. Cf. Theorem 7.5. What is the least value of k such that $R(n) \geqq kn$ for each non-periodic recurrent trajectory?

(4). In the case of general admissibility certain blocks and finite sequences of blocks are prohibited. Knowing these prohibitions, is it possible to give a constructive rule for the formation of a minimum block covering all admissible r-blocks, and what is the most general such r-covering?

(5). Does there exist an ergodic function $\phi(n)$ such that the ratio of $\phi(n)$ to $\theta(n)$ tends to the limit 1? Cf. Theorem 12.5.

(6). Is a theorem similar to that of Oldenburger true in case the generating symbols suffer the restriction of inverses, or in the general case?

(7). What is the most general minimal set with a given recurrency function $R(n)$? With permutation index $P(n)$? With $P(n)$ and $R(n)$ both given? These questions should be answered first when there are just two generating symbols.

These questions seem to point to an algebraic analysis. Questions more intimately connected with measure theory, or the analysis of space forms will be presented in a later paper.

THE INSTITUTE FOR ADVANCED STUDY,
AND
BRYN MAWR COLLEGE.

BIBLIOGRAPHY.

ARTIN, E.

1. "Ein mechanisches System mit quasiergodischen Bahnen," *Abhandlungen aus dem Mathematischen Seminar der Hamburgischen Universität*, Bd. 3 (1924), pp. 170-175.

BESICOVITCH, A. S.

1. *Almost periodic functions*, Cambridge, University Press (1932).

BIRKHOFF, G. D.

1. "Quelques théorèmes sur le mouvement des systèmes dynamiques," *Bulletin de la Société Mathématique de France*, t. 40 (1912), pp. 303-323.
2. "Dynamical systems," *American Mathematical Society Colloquium Publications*, vol. 9 (New York, 1927).
3. "Nouvelles recherches sur les systèmes dynamiques," *Memoriae Pont. Acad. Scient. Novi Lyncaei* (Série 3), t. 1, pp. 85-216.
4. "Sur le problème restreint des trois corps" (Second Mémoire), *Annali della R. Scuola Normale Superiore di Pisa* (Serie II), t. 5 (1936), pp. 1-42.

HADAMARD, J.

1. "Les surfaces à courbure opposées et leur lignes géodésiques," *Journal de Mathématiques pures et appliquées* (5th series), t. 4 (1898), pp. 27-73.

HAUSDORFF, F.

1. *Mengenlehre,* Leipzig, Walter de Gruyter (1927).

HEDLUND, G. A.

1. "On the metrical transitivity of the geodesics on closed surfaces of constant negative curvature," *Annals of Mathematics,* vol. 35 (1934), pp. 787-808.
2. "A metrically transitive group defined by the modular group," *American Journal of Mathematics,* vol. 57 (1935), pp. 668-678.
3. "Two-dimensional manifolds and transitivity," *Annals of Mathematics,* vol. 37 (1936), pp. 534-542.

HILBERT, D.

1. "Über Flächen von konstanter Gaussscher Krümmung," *Transactions of the American Mathematical Society,* vol. 2 (1901), pp. 87-99.

HOBSON, E. W.

1. *The Theory of Functions of a Real Variable,* Cambridge, University Press (1921).

KOEBE, P.

1. "Riemannsche Mannigfaltigkeiten und nicht euklidische Raumformen," *Sitzungsberichte der Preussischen Akademie der Wissenschaften* (1927), pp. 164-196; (1928), pp. 345-442; (1929), pp. 414-457; (1930), pp. 304-364, 505-541; (1931), pp. 506-534.

MARTIN, M. H.

1. "A problem in arrangements," *Bulletin of the American Mathematical Society,* vol. 40 (1934), pp. 859-864.
2. "The ergodic function of Birkhoff," *Duke Mathematical Journal,* vol. 3 (1937), pp. 248-278.

MORSE, M.

1. "A one-to-one representation of geodesics on a surface of negative curvature," *American Journal of Mathematics,* vol. 43 (1921), pp. 33-51.
2. "Recurrent geodesics on a surface of negative curvature," *Transactions of the American Mathematical Society,* vol. 22 (1921), pp. 84-100.
3. "A fundamental class of geodesics on any closed surface of genus greater than one," *Transactions of the American Mathematical Society,* vol. 26 (1924), pp. 25-61.
4. "Instability and transitivity," *Journal de Mathématiques pures et appliquées* (9th series), t. 14 (1935), pp. 49-71.

MYRBERG, P. J.

1. "Einige Anwendungen der Kettenbrüche in der Theorie der binären quadratischen Formen und der elliptischen Modulfunktionen," *Annales Acad. Sc. Fennicae,* t. 23 (1924).
2. "Ein Approximationssatz für die Fuchsschen Gruppen," *Acta Mathematica,* t. 57 (1931), pp. 389-409.

NIELSEN, J.

1. "Om geodaetiske Linier i lukkede Mangfoldigheder med konstant negativ Krumning," *Matematisk Tidskrift* B, (1925), pp. 37-44.
2. "Untersuchungen zur Topologie der geschlossenen zweiseitigen Flächen," *Acta Mathematica*, t. 50 (1927), pp. 189-358.

POINCARÉ, H.

1. "Sur les courbes définies par les équations différentielles," *Journal de Mathématiques pures et appliquées* (Série 4), t. 1 (1885), pp. 167-244.

ROBBINS, H. E.

1. "On a class of recurrent sequences," *Bulletin of the American Mathematical Society*, vol. 43 (1937), pp. 413-417.

CONFERENCE

FAITE AU COURS DE LA SEANCE DU 11 MAI 1938.

LA DYNAMIQUE SYMBOLIQUE.

PAR MARSTON MORSE.

1. *Introduction.* — La dynamique moderne commence avec Poincaré, Birkhoff, Hadamard et leurs successeurs ont poursuivi avec succès son développement. Un des procédés techniques les plus importants dans ce domaine a été l'emploi de suites infinies (c) de symboles, de la forme

$$(1.1) \qquad \ldots, \quad c_{-1}, \quad c_0, \quad c_1, \quad \ldots.$$

Ces suites représentent des trajectoires et seront appelées des *trajectoires symboliques*. Les symboles c_i sont pris dans un ensemble fini de symboles $a_1, \ldots, a_r$, qu'on appelle les symboles *générateurs*. En général, ces symboles générateurs sont déterminés par la topologie de l'espace dans lequel se trouvent les trajectoires.

Nous allons examiner un exemple d'une portée générale. Considérons les géodésiques sur une surface S de courbure négative. Nous supposons que S est limitée par une frontière formée de $\nu + 1$ géodésiques fermées C_i, et que S est topologiquement équivalente à un domaine plan limité par $\nu + 1$ cercles. Soit $(b_1, \ldots, b_\nu)$ une suite de ν arcs de géodésiques, simples et distincts, l'arc b_i joignant C_i à C_{i+1} pour $i = 1, \ldots, \nu$. Des coupures suivant les arcs b_i rendraient S simplement connexe. Maintenant, nous ne nous occuperons que des géodésiques qui demeurent toujours dans S, qu'on les prolonge dans l'un ou dans l'autre sens. De telles géodésiques peuvent être caractérisées par la manière dont elles traversent les arcs b_i. Nous dénoterons la traversée de b_i dans un sens par b_i, dans l'autre sens par b_i^{-1}. La géodésique est alors représentée par une suite (c) de symboles de la forme (1.1), les symboles générateurs étant les éléments de l'ensemble

$$(1.2) \qquad b_1, \quad \ldots, \quad b_\nu, \quad b_1^{-1}, \quad \ldots, \quad b_\nu^{-1}.$$

Dans une suite (c), b_i n'est jamais suivi ni précédé par b_i^{-1}. En effet, dans le cas contraire, il y aurait deux points A et B sur b_i qui seraient joints par un arc h de géodésique distinct de l'arc AB sur b_i, et ces deux arcs de géodésique formeraient la frontière d'une région simplement connexe de S. C'est impossible sur une surface à courbure négative, comme il résulte de la formule Gauss-Bonnet.

Des suites (c) de la forme (1.1), engendrées par les symboles générateurs (1.2) et soumises à la condition que b_i n'est jamais suivi ou précédé par b_i^{-1}, seront appelées *suites permises*. Après Hadamard, nous pouvons énoncer le théorème suivant :

Théorème (1.I). — *Sur la surface à courbure négative* S, *la correspondance entre géodésiques permises et trajectoires symboliques permises, qui fait correspondre à une géodésique g la trajectoire symbolique définie par les intersections de g avec les coupures qui rendent* S *simplement connexe, est biunivoque.*

En outre, non seulement g est représentée par sa trajectoire symbolique (c) dans le sens du théorème (1.I), mais les propriétés dynamiques essentielles de g peuvent aussi être caractérisées au moyen de (c). (En particulier, on peut caractériser la transitivité topologique au moyen de la transitivité symbolique, comme nous verrons plus loin.) On trouvera dans l'*American Journ. of Math.* (*Hall. Memorial number*, automne, 1938) un mémoire de A. Hedlund et Morse, qui développe ce point de vue. Ce mémoire éclaire la division de la dynamique en ses composantes symboliques et différentielles souvent confondues dans des travaux antérieurs, où une partie de la dynamique purement symbolique a été développée en termes de la théorie des équations différentielles.

On peut commencer par étudier les trajectoires symboliques (c) de la forme (1.1) et développer une théorie de la dynamique symbolique complètement indépendante de la théorie des équations différentielles. Par exemple, (c) sera dite transitive si chaque n-bloc permis formé avec les symboles générateurs (1.2) apparaît dans (c). (Un n-bloc est un bloc de n symboles). Cette théorie formelle est seulement une des deux parties qui composent la théorie générale.

1° *Existence de systèmes dynamiques admettant une représentation symbolique.*

2° *Dynamique symbolique.*

Nous ne ferons ici que quelques remarques sur la partie 1. Les conditions suffisantes qu'on connaît pour le problème 1 se rapportent à des géodésiques sur des variétés. Par exemple, il est suffisant que la surface soit orientable, que son premier nombre de Betti soit au moins égal à 3 et que la condition d'instabilité uniforme soit satisfaite. Ces conditions peuvent être aussi étendues à des variétés de dimensions n. Les recherches dans cette direction sont récentes et encore imcomplètes. Les conditions cherchées doivent être d'une nature telle que :

1° *Les trajectoires permises* g *aient une correspondance biunivoque avec leur représentation par des trajectoires symboliques* (c);
2° *Les propriétés dynamiques essentielles, comme la transitivité des trajectoires, puissent être caractérisées entierement symboliquement.*

La condition 2° sera plus claire si nous définissons la transitivité ordinaire. Une trajectoire g sera appelée transitive relativement à un système de trajectoires H si la fermeture de g contient H. Quand nous nous occupons de géodésiques, la définition doit être interprétée comme suit :

Soit Z une surface, u, v des coordonnées locales sur Z. Soit $\left(\frac{du}{ds}, \frac{dv}{ds}\right)$ une direction sur Z au point (u, v). L'élément $\left(u, v, \frac{du}{ds}, \frac{dv}{ds}\right)$ constitue un point de ce que nous appellerons l'*espace des phases* correspondant à Z. Les éléments de contact le long d'une géodésique g définissent une courbe g' de l'espace des phases; un ensemble H de géodésiques définit un ensemble H′ de courbes dans l'espace des phases. La géodésique g est dite *transitive* relativement à H si la courbe g' est transitive relativement à H′. Quand H est l'ensemble de toutes les géodésiques sur Z, ce type de transitivité est appelé *transitivité topologique* (notation T. T.), pour le distinguer de la *transitivité métrique* (notation T. M.). Nous ne définirons pas T. M. complètement. Nous nous contentons de dire que c'est une propriété statistique de toutes les géodésiques, ce qui la distingue de T. T., où une certaine géodésique joue un rôle à part. Naturellement l'étude de T. T. précède celle de T. M., car on peut montrer que T. M. entraîne toujours T. T. La réciproque est douteuse, même dans le cas analytique. Dans l'état actuel de la théorie, les cas connus de T. M. pour les géodésiques, sont relatifs uniquement à des surfaces de courbure négative constante, et de tels exemples ne nous éclairent pas sur la généralité de T. M. Au contraire, nous verrons que T. T. existe dans des cas très généraux. La question de l'existence de T. T. pour une géodésique donnée a un intérêt physique immédiat. Considérons par exemple le système des planètes et du Soleil regardés comme des masses ponctuelles soumises à la loi de Newton. Est-ce que l'orbite de la Terre est transitive par rapport à une certaine région de l'espace des phases ? On n'a encore aucune réponse à ce sujet.

2. *Instabilité uniforme.* — Les conditions (1) et (2) sont satisfaites par les géodésiques de surfaces compactes, de courbure négative, sans frontières, ou ayant pour frontières des géodésiques fermées, quand le premier nombre de Betti de ces surfaces est au moins égal à 3. A notre point de vue, l'importance de la condition que la courbure soit négative, réside en cette propriété : Si g est une géodésique orientée, et g' une autre géodésique quelconque, g' tend à s'éloigner de g pour au moins l'un des deux sens sur g. Mais cette propriété subsiste sous des conditions beaucoup moins restrictives que celle de la courbure négative. *L'instabilité uniforme* est une telle condition, elle est définie en se servant de la première variation des géodésiques, plutôt que des géodésiques elles-mêmes, et l'on peut ainsi plus facilement vérifier si elle est satisfaite.

Soient g une géodésique de la surface Z, s la longueur d'arc sur g, $K(s)$ la courbure de Gauss de Z au point S de g. L'équation

$$\frac{d^2 w}{ds^2} + K(s) w = 0 \tag{2.1}$$

est complètement définie quand g est donnée. C'est l'équation différentielle de Jacobi de g. Soit $w(s)$ une solution de 2.1, non identiquement nulle, et k une constante arbitraire. La géodésique g sera dite *instable* si aucune des solutions $w(s)$ ne s'annule plus d'une fois et si pour toute solution $w(s)$ et pour tout point $s = a$ il existe une constante positive H ne dépendant que de k et de g, telle que

$$|w(a-x) + w(a+y)| > k\, w(a) \tag{2.2}$$

quand x et y sont supérieurs à H. Si H ne dépend que de k et non de g, nous dirons que la condition *d'instabilité uniforme* (I.U) est satisfaite.

Par exemple, on vérifie immédiatement que la condition I.U. est satisfaite pour $K(s) \equiv -1$. On peut, sans restriction de généralité, supposer $w(a) = 1$ dans (2.2). Plus généralement la condition I.U. est satisfaite $k(s) < 0$, mais elle l'est aussi dans des cas plus étendus, quand la courbure n'est positive que dans des régions qui n'ont pas une distribution trop dense sur la surface, ou qui n'ont pas des diamètres trop larges. Elle n'est jamais satisfaite si la courbure est partout positive. On peut aller plus loin et montrer que la condition I.U. n'est jamais satisfaite sur des surfaces de genre 0 ou 1. La formule de Gauss-Bonnet montre facilement qu'une telle surface ne peut avoir partout une courbure négative. Il est plus difficile de montrer que la condition I.U. ne peut pas être satisfaite.

L'étude des types topologiques de variétés qui satisfont à I. U. conduit à d'importantes généralisations de problèmes classiques de Hilbert, Kœbe, etc. sur l'existence et la nature de surfaces à courbure négative. On démontre que, pour tout entier $p > 1$, il existe une surface de genre p satisfaisant à la condition I. U.

3. *Rayons symboliques.* — Nous allons maintenant donner ici quelques résultats typiques parmi ceux obtenus dans le mémoire cité sur la Dynamique symbolique. Nous considérons des suites R de symboles

$$(3.1) \qquad c_1, \quad c_2, \quad c_3, \quad \ldots$$

pris parmi un ensemble fini de symboles générateurs et soumis, à l'occasion, à un nombre fini de conditions, par exemple celles imposées aux symboles générateurs (1.2). Nous appellerons une telle suite un *rayon symbolique*.

Un tel rayon R est dit *transitif* s'il contient tout n bloc permis. Si R est transitif, nous définissons une *fonction ergodique* $\varphi(n)$ appartenant à R, de la façon suivante : $\varphi(n)$ est l'entier minimum m tel que le m-bloc initial de R contienne chaque n-bloc permis. Tout bloc fini qui contient chaque n-bloc permis est appelé un *n-système complet*.

Si les symboles générateurs sont les symboles (1.2) assujettis à la seule condition que b_i n'est jamais précédé ou suivi de b_i^{-1}, il existe un n-système complet dans lequel chaque n-bloc apparaît juste une fois. On peut construire un tel n-système complet (n fixe) de la façon suivante : Nous écrivons d'abord b_i $n-1$ fois. Nous ajoutons alors successivement le dernier symbole de (1.2) qui permet de satisfaire aux deux conditions suivantes : le nouveau bloc est permis, et aucun n-bloc n'est reproduit. De cette façon, on arrive à un n-système complet dans lequel aucun n-bloc n'est répété. Cette règle de construction est due à Martin, qui en établit la validité dans le cas où il n'y a pas d'inverses.

Comme illustration de cette règle prenons le cas de deux symboles générateurs 1, 2, n'ayant pas d'inverses, et cherchons un 3-système complet dans lequel aucun 3-bloc n'est répété. La règle nous donne le 3-système complet :

$$1\quad 1\quad 2\quad 2\quad 2\quad 1\quad 2\quad 1\quad 1\quad 1$$

qui satisfait bien à la condition qu'aucun 3-bloc n'y apparaît deux fois.

Soit $\theta(n)$ la longueur minimum d'un n-système complet. Alors, pour chaque fonction ergodique, on a $\varphi(n) \geqq \theta(n)$. Dans le cas du système de générateurs (1.2), il existe un n-système complet dans lequel aucun n-bloc ne se répète, de sorte que

$$\theta(n) = p(p-1)^{n-1} + n - 1,$$

où p est le nombre de symboles générateurs.

Nous posons maintenant la question : Existe-t-il toujours un rayon transitif tel que

$$\varphi(n) = \theta(n), \tag{3.2}$$

quel que soit n ?

Un contre-exemple, formé par Oldenberger, montre qu'il n'en est rien. Il considère le cas où il y a $p > 1$ générateurs libres, n'ayant pas d'inverses, et il montre que pour de tels systèmes il n'y a pas de rayon et de fonctions ergodiques telles que (3.2) soit vrai pour deux valeurs successives de n plus grandes que un. Dans ce cas, il n'y a donc pas de moindre fonction ergodique, c'est-à-dire que pour toute fonction ergodique on peut en trouver une autre qui est plus petite que la première pour une infinité de valeurs de n.

4. *La catégorie des rayons transitifs.* — La question de l'existence d'une meilleure fonction ergodique peut être posée d'une façon bien différente. Assignons la métrique de Baire à l'espace Ω de tous les rayons permis : B et C étant deux rayons permis,

$$B = b_1,\ b_2,\ \ldots,$$
$$C = c_1,\ c_2,\ \ldots,$$
$$\operatorname{dist}(B, C) = \frac{1}{n+1},$$

n étant le plus grand entier tel que

$$(c_1,\ \ldots,\ c_n) \equiv (b_1,\ \ldots,\ b_n).$$

On peut montrer que l'espace Ω est parfait, compact et totalement discontinu.

Rappelons qu'un sous-ensemble A de points d'un espace Ω est appelé un ensemble de *première catégorie* s'il est réunion dénombrable de sous-ensembles nulle part denses de Ω. Un ensemble de la forme $\Omega - A$ est dit *ensemble résiduel* ou de *deuxième catégorie*. Il est bien connu que tout ensemble résiduel est partout dense dans Ω et qu'il a la puissance du continu. Si un sous-ensemble B de Ω est résiduel, cela veut dire, dans un sens topologique, que B contient la plupart des points de Ω. Ceci fait comprendre l'intérêt du théorème suivant :

Théorème 4.II. — *Les rayons transitifs qui possèdent des fonctions ergodiques* $\varphi(n)$ *telles que*

$$\liminf_{n \to \infty} \frac{\varphi(n)}{\theta(n)} = 1 \tag{3.3}$$

forment un ensemble résiduel dans l'espace de Baire Ω.

Rappelons-nous que $\varphi(n) \geqq \theta(n)$ pour tout n, puisque $\theta(n)$ est le minimum de $\varphi(n)$, pour n fixe et des rayons transitifs variables. Quand la condition (3.3) est satisfaite pour un rayon transitif, la fonction ergodique $\varphi(n)$ est équivalente à $\theta(n)$ sur au moins une suite infinie de valeurs entières de n. Les rayons transitifs pour lesquels la relation (3.3) est satisfaite comprennent « la plupart » des rayons dans un sens topologique. A l'opposé, on peut aussi montrer que l'ensemble des rayons non transitifs est partout dense dans Ω et qu'il a la puissance du continu.

Revenons à nos géodésiques. Soit O un point arbitraire sur une surface close de genre $p > 1$. Si la condition I. U. est satisfaite, la correspondance entre les géodésiques issues de O et leurs rayons symboliques représentatifs est biunivoque; les rayons symboliques transitifs correspondent à des géodésiques transitives et réciproquement. De plus, la fonction ergodique $\varphi(n)$, exprimant la rapidité de la transitivité symbolique, a une interprétation immédiate.

Les théorèmes et les remarques donnés ci-dessus sont typiques de la théorie, mais ne prétendent pas en donner une idée exhaustive. Pour un exposé plus complet, nous renvoyons à l'article sur la Dynamique symbolique, cité plus haut. On y trouvera les références aux travaux de Poincaré, Birkhoff, Hadamard, Hedlund, Morse, Martin, Artin, Hilbert et Kœbe dont il a été question.

ANNALS OF MATHEMATICS
Vol. 41, No. 2, April, 1940

RANK AND SPAN IN FUNCTIONAL TOPOLOGY

BY MARSTON MORSE

(Received August 9, 1939)

1. Introduction.

The analysis of functions F on metric spaces M of the type which appear in variational theories is made difficult by the fact that the critical limits, such as absolute minima, relative minima, minimax values etc., are in general infinite in number. These limits are associated with relative k-cycles of various dimensions and are classified as 0-limits, 1-limits etc. The number of k-limits, suitably counted is called the kth type number m_k of F. The theory seeks to establish relations between the numbers m_k and the connectivities p_k of M. The numbers p_k are finite in the most important applications. It is otherwise with the numbers m_k.

The theory has been able to proceed provided one of the following hypotheses is satisfied. The critical limits cluster at most at $+\infty$; the critical points are isolated;[1] the problem is defined by analytic functions; the critical limits taken in their natural order are well-ordered. These conditions are not generally fulfilled. The generality of the theory rested upon the fact that the cases treated approximate in a certain sense the most general problems which it is desirable to treat. From this point of view it became necessary to study the way in which the critical points of the approximating function approximate[2] the critical points of the given function in number and in type. This has been followed with success in certain cases. It should be pursued much further but it becomes increasingly difficult.

As an example we refer to the study of minimal surfaces[3] of the topological type of the circular disc bounded by a fixed simple closed curve g. Even when g is analytic and regular the number of minimal surfaces bounded by g may be infinite as far as is known at the moment. No useful or simple conditions are known under which a minimal surface bounded by g is isolated from other such minimal surfaces. Moreover a study of geodesics shows that the difficulties described here are inherent in all but the analytic case.

The present paper introduces a completely new way of meeting these difficulties. It is not by way of restrictive hypotheses. The hypotheses are those

[1] See for example the excellent book by Seifert and Threlfall, *Variationsrechnung im Grossen*, Berlin, Teubner (1939), or earlier papers of the author.

[2] See Morse, *The calculus of variations in the large*, American Mathematical Society Colloquium Publications 18, New York (1934). In particular Chapter VI.

[3] See Morse and Tompkins, *Minimal surfaces of general critical type*, Annals of Mathematics 40 (1939), pp. 443–472.

which are fulfilled in all of the general problems so far treated. The notion of the *span* of a relative cycle u is introduced. It is the difference of two values of F intrinsically associated with u. (See §8.) It measures the minimum length of an interval of values of F essential to a description of the bounding relations of u. Only certain relative k-cycles termed k-caps are counted and among these k-caps only those whose spans exceed a fixed positive constant e. The number of k-caps of span $> e$ suitably counted is now denoted by m_k^e. The numbers m_k^e are finite if the connectivities are finite, and the whole theory proceeds on a finite basis, obtaining relations between the type numbers m_k^e and the connectivities p_k similar to those which occur in the earlier analytic and non-degenerate cases. Except for being positive the number e is arbitrary. Upon taking different values of e one gets relations between different sets of k-limits.

A very special example may be helpful. Let the space M be a simple closed curve in the xy-plane. On M let $F = y$. If F has a finite number of relative extrema on M the number m_0 of relative minima equals the number m_1 of relative maxima. If F has an infinite number of relative extrema this relation is apparently lost. It appears again however in the new theory in a more general form. The number m_1^e of 1-caps with spans $> e$ equals the number m_0^e of 0-caps with spans $> e$. Even when the relative extrema of F are finite in number the relations $m_1^e = m_0^e$ contain more than the ordinary relation $m_1 = m_0$.

In general each k-cap u possesses a "cap height" $a(u)$ such that u is on the subset $F \leqq a$ of M and on no subset $F \leqq b$ for which $b < a$. The function $a(u)$ is called a *rank function*.[4] It is one of a number of rank functions defined for subsets of chains or cycles. These rank functions satisfy four rank conditions whose investigation is a matter of abstract group theory. The two fundamental questions involved are the following. I. Let A be an arbitrary set of k-caps u over which a rank function $\rho(u)$ is defined. Let $g(\rho)$ be a maximal group of k-caps of A with a fixed rank ρ. When is it true that a maximal group g of k-caps of A is a direct sum of the form

$$g = \sum_{\rho} g(\rho),$$

where ρ ranges over all ranks of k-caps of A? II. When is it true that the dimension of the maximal group g is independent of the choice of g as maximal in A? Once these questions are answered the finiteness of the numbers m_k^e is proven, and the relations between the numbers m_k^e and p_k are established.

The present paper is a theory of critical limits, not of critical points. A critical limit is always assumed by the function F at some critical point provided certain conditions such as the compactness of the sets $F \leqq c$ and the upper-reducibility of F are satisfied. This part of the theory is developed[5] in M1

---

[4] Rank functions of this type were introduced for the first time in Morse, *Functional topology and abstract variational theory*, Annals of Mathematics 38 (1937), pp. 386–449. We shall refer to this paper as M2.

[5] See Morse, *Functional topology and abstract variational theory*, Mémorial des Sciences Mathématiques XCII, Paris, Gauthier-Villars (1938). We shall refer to this paper as M1.

and will be further extended in a later paper. The difficulties are not those of group theory but of analysis.

2. The rank conditions R

We shall be concerned with an additive abelian operator group G with coefficients δ in a field Δ. Let $[\rho]$ be a simply ordered set of elements. To certain elements u of G there shall be assigned an element $\rho(u)$ of $[\rho]$. The elements of G for which $\rho(u)$ is not defined shall include the null element of G. We term $\rho(u)$ the *rank* of u.

Let $G(\rho, \omega)$ denote the subgroup of G generated by the elements of G with ranks at most ω using arbitrary coefficients δ in Δ. The group $G(\rho, \omega)$ will in general contain non-null elements without rank. We shall assume that the rank function $\rho(u)$ satisfies rank conditions R as follows:

I. *If u has a rank and $\delta \neq 0$, $\rho(u) = \rho(\delta u)$.*

II. *If u, v and $u + v$ have ranks, $\rho(u + v) \leqq \max [\rho(u), \rho(v)]$.*

III. *If u and v have different ranks, $u + v$ has a rank.*

IV. *If u and v are elements in $G(\rho, \omega)$ without rank, $\rho(u + v)$ cannot exist and equal ω.*

Conditions I, II, III are independent as is readily seen. That condition IV is independent of conditions I, II, III is shown by the following example.

EXAMPLE 2.1. Let G be generated by elements a, b, c with Δ the field of integers mod 2. Suppose that a, b, $a + b$, c have the rank 1, but that no other elements of G have a rank. Rank conditions I, II, III are satisfied. The group $G(\rho, 1)$ is identical with G. Condition IV is not satisfied. The elements $a + c$ and $b + c$ have no rank while their sum $a + b$ has the rank 1.

EXAMPLE 2.2. Let M be a compact metric space and F a function which is continuous on M. Let F_s be the subset of M on which $F \leqq s$. Let G be the group of all singular k-cycles on M taken mod 2. Each non-bounding cycle u of G will be assigned a rank $t(u)$. Let H be the homology class of u, and v an arbitrary member of H. Let $b(v)$ be the maximum of F on v. The greatest lower bound of the numbers $b(v)$ as v ranges over H will be called the *inferior cycle limit* of u or of H and will be termed the rank $t(u)$ of u. These ranks $t(u)$ satisfy the rank conditions I to IV.

The groups which we shall study are *operator* subgroups, that is subgroups which contain δu whenever they contain u. Our isomorphisms will be operator isomorphisms, that is isomorphisms in which δu corresponds to δv whenever u corresponds to v.

We shall be concerned with various classes of elements of G, for example, the class of elements having rank. A class A of elements of G will be termed an *operator* class if whenever u is in A and $\delta \neq 0$, δu is in A. It will be convenient to exclude the null element from operator classes. A group g in A shall mean a subgroup of G every element of which except the null element is in A. Such a group g will be termed *maximal* in A if g is a proper subgroup of no subgroup of G in A.

A *generating* set for a subgroup g of G is a set of elements of g linear combinations of which, using coefficients in Δ, include all elements of g. A generating set of g no subset of which is a generating set of g will be called a *base* for g. We shall be concerned with operator subgroups whose bases contain at most aleph null elements. It is easy to show that the cardinal number of elements in a base of such groups g is the same for all bases of g. This number is called the *dimension* of g.

Let A be an operator class of G. Maximal groups of elements in A may have different dimensions. As an example let A consist of the elements

$$a, \quad b, \quad a + b, \quad c,$$

in Example 2.1. The group generated by a and b and the group generated by c are maximal in A, but these groups have different dimensions.

Any operator subgroup of G all of whose elements except the null element have ranks ρ will be termed a *ρ-subgroup* of G. We shall be concerned with the conditions under which any two maximal ρ-subgroups of G have the same dimension. In this connection the concepts of ρ-class and ρ-isomorphism are important.

Two elements u and v of G with the same rank ρ will be said to be in the same *ρ-class* or rank class if $u - v$ has no rank or has a rank less than ρ. An isomorphism between two ρ-subgroups of G will be termed a *ρ-isomorphism* if corresponding non-null elements are in the same ρ-class.

LEMMA 2.1. *In a ρ-subgroup of G with a finite dimension r the number of different ranks is at most r.*

The proof of this lemma is given in M1, p. 17. It depends merely on rank conditions I, II, III. The following lemma is a consequence of Lemma 4.1 of M1.

LEMMA 2.2. *Let g be a subgroup of G in an operator class A. A necessary and sufficient condition that g be maximal in A is that each element in A have the form*

$$u = z + w,$$

where z is in g and w not in A.

Two groups of elements with rank are not in general ρ-isomorphic if their dimensions are equal. The condition of ρ-isomorphism is thus much stronger than that of isomorphism. The significance of the following theorem will now be apparent.

THEOREM 2.1. *When rank conditions R are satisfied, any two maximal groups of elements of G with the same fixed rank ρ are ρ-isomorphic.*

This theorem is proved on p. 15 of M1, and depends essentially upon all four rank conditions R. In this connection the following two theorems were also proved.

THEOREM 2.2. *Under rank conditions R the property of elements with rank being in the same ρ-class is transitive.*

That is, if u is in the ρ-class of v, and v is in the ρ-class of z, u is in the ρ-class of z.

THEOREM 2.3. *The subset of elements of the group $G(\rho, \omega)$ with no rank or with a rank $\rho < \omega$ form a group H_ω.*

Let g be a subgroup of G. Let α be an enumerating index in a simply ordered set. For each α let $h(\alpha)$ be a subgroup of g. The group g is said to be a *direct sum*

$$g = \sum_\alpha h(\alpha),$$

if the groups $h(\alpha)$ generate g and if there exists no relation of the form

$$u(\alpha_1) + \cdots + u(\alpha_m) = 0,$$

in which the α_i's are distinct and $u(\alpha_i)$ is a non-null element from the group $h(\alpha_i)$.

The following three theorems depend only on rank conditions I, II, III.

THEOREM 2.4. *If g is a ρ-subgroup of G whose ranks form a discrete ascending sequence, g is a direct sum,*

$$g = \sum_\rho g(\rho), \tag{2.1}$$

of arbitrary maximal subgroups $g(\rho)$ of g with the respective ranks ρ of elements of g.

Let $\rho_1 < \cdots < \rho_n$ be a subset of ranks of elements of g, and let $u_1, \cdots, u_n$ be non-null elements of the respective groups $g(\rho_1), \cdots, g(\rho_n)$. It follows from II and III that the sum

$$v = u_1 + \cdots + u_n$$

has the rank ρ_n. Hence $v \neq 0$. The groups $g(\rho)$ thus have a ρ-subgroup g' of g as a direct sum. It remains to show that $g' = g$.

To that end let u be an arbitrary non-null element of g. The theorem will follow if assumption (α) leads to a contradiction.

(α). *Whatever the element v in g', $u - v \neq 0$.*

Set $\rho(u) = r$. Since $g(r)$ is a maximal group of elements of g with rank r it follows from Lemma 2.2 that there exists an element v_1 in $g(r)$ such that $u - v_1$ is an element u_1 not in $g(r)$. The element u_1 has a rank by virtue of (α), and it follows from II that $\rho(u_1) < r$. We now treat u_1 as we treated u. It follows from (α) that (α) is satisfied if u_1 replaces u. Proceeding inductively we infer the existence of an infinite sequence of elements $u_1, u_2, \cdots$ with decreasing ranks. This is contrary to the assumption that the ranks of elements in g form a discrete ascending sequence.

Hence (α) is false, $g = g'$, and the theorem is true.

THEOREM 2.5. *If the ranks of elements of G form a discrete ascending sequence, arbitrary maximal groups $g(\rho)$ of elements with the respective ranks ρ have as a direct sum a maximal ρ-subgroup of G.*

The groups $g(\rho)$ clearly have a ρ-subgroup g of G as a direct sum. It remains to show that g is maximal. Let g^* be any maximal ρ-subgroup of G containing g. Then $g(\rho)$ is a maximal subgroup of elements of g^* with rank ρ. It follows from the preceding theorem that g^* is the direct sum of the groups $g(\rho)$, and the proof is complete.

THEOREM 2.6. *Suppose the ranks ρ of G form a discrete ascending sequence, and that g is a maximal ρ-subgroup of G. If $g(\rho)$ is a subgroup of elements of g with rank ρ, chosen so as to be maximal in g, then $g(\rho)$ is also maximal in G.*

The group g is the sum

$$g = \sum_{\rho} g(\rho) \tag{2.2}$$

of the respective groups $g(\rho)$. If one of these groups, say $g(r)$, is not maximal in G there will exist a subgroup $g'(r)$ of elements of G with rank r such that $g'(r)$ contains $g(r)$ and is maximal in G. In (2.2) let $g(r)$ be replaced by $g'(r)$. Let the resulting sum (2.2) then be denoted by g^*. It is clear that g^* contains g and is a ρ-subgroup of G. But g^* is not identical with g because $g'(r)$ is a maximal subgroup of elements of g^* with rank r. Hence g is not a maximal ρ-subgroup of G. From this contradiction we infer that $g(r)$ is maximal in G.

Let g be a ρ-subgroup of G. A base B for g will be termed *ρ-proper* if each finite linear combination,[6]

$$v = \delta_i u_i, \qquad \delta_i \neq 0, \qquad (i = 1, \cdots, r)$$

of elements of B satisfies the condition

$$\rho(v) = \max [\rho(u_1), \cdots, \rho(u_r)].$$

In general a minimal base for g will not be ρ-proper, and g may even admit no such base.

LEMMA 2.3. *A ρ-subgroup g of G whose ranks form a discrete ascending sequence admits a ρ-proper base.*

For g is a direct sum,

$$g = \sum_{\rho} g(\rho),$$

of the maximal subgroups of elements of g with the respective ranks ρ. If $B(\rho)$ is a base for $g(\rho)$, it is clear that the ensemble of the bases $B(\rho)$ is a ρ-proper base for g. The proof of the lemma is complete.

Let $(u_1, u_2, \cdots)$ be a finite or infinite set of independent elements of G. Let $(v_1, v_2, \cdots)$ be a second such set with u_i corresponding to v_i. Let $g(u)$ and $g(v)$ be the operator groups generated respectively by the elements u_i and v_i. If the finite proper linear combination $\delta_i u_i$ of $g(u)$ be made to correspond to $\delta_i v_i$ in $g(v)$, $g(u)$ will thereby be isomorphically mapped on $g(v)$. We shall say that the mapping of the set (u) onto the set (v) has been *isomorphically extended* to the groups $g(u)$ and $g(v)$.

THEOREM 2.7. *Let (u) be the set of elements in a ρ-proper base of a group $g(u)$. Let each element u_i of (u) be mapped on an element v_i in the same rank class. The elements v_i are independent and generate a group $g(v)$. The isomorphic extension of the mapping of the base (u) on the base (v) defines a ρ-isomorphism between $g(u)$ and $g(v)$.*

[6] We adopt the summation convention of tensor analysis.

The mapping of $g(u)$ onto $g(v)$ is a homeomorphism. Under this homeomorphism a linear combination $x = \delta_i x_i$ $(i = 1, \cdots, r)$ of the elements of (u) corresponds to the element $y = \delta_i y_i$ of $g(v)$ where x_i and y_i are in the same rank class. If $\delta_i = 0$ for no i, and if we set

$$\omega = \max [\rho(x_1), \cdots, \rho(x_r)],$$

then $\rho(x) = \omega$ since (u) is a ρ-proper base. The element $x - y$ is in the subgroup H_ω of Theorem 2.3 so that $x - y$ has no rank or a rank $< \omega$. Hence x and y are in the same rank class and y is not null. The homeomorphic mapping of $g(u)$ onto $g(v)$ is accordingly an isomorphism, and since corresponding elements x and y are in the same rank class this isomorphism is a ρ-isomorphism.

THEOREM 2.8. *If the ranks ρ of elements of G form a discrete ascending sequence, any two maximal ρ-subgroups of G are ρ-isomorphic.*

Let g and g' be the two maximal groups. If $g(\rho)$ and $g'(\rho)$ are respectively maximal subgroups of g and g' with the rank ρ, then

$$g = \sum_\rho g(\rho), \qquad g' = \sum_\rho g'(\rho),$$

as we have seen in Theorem 2.4. It follows from Theorem 2.6 that $g(\rho)$ and $g'(\rho)$ are maximal groups of elements of G with rank ρ, and we can conclude from Theorem 2.1 that $g(\rho)$ and $g'(\rho)$ admit a ρ-isomorphism $I(\rho)$. Let $B(\rho)$ be a base for $g(\rho)$ and let $B'(\rho)$ be the base for $g'(\rho)$ which arises by virtue of the preceding ρ-isomorphism $I(\rho)$. The logical sums

$$B = \sum_\rho B(\rho), \qquad B' = \sum_\rho B'(\rho)$$

form ρ-proper bases for g and g'. The isomorphisms $I(\rho)$ map B on B'. If this mapping be isomorphically extended g will be mapped ρ-isomorphically on g' in accordance with the preceding theorem.

The proof of the theorem is complete.

The following lemma will be useful.

LEMMA 2.4. *If g is an operator subgroup of G and g_1 an operator subgroup of g, then g is represented by the direct sum*

$$g = g_1 + g_2,$$

where g_2 is a maximal group of elements of g, not in g_1.

The groups g_1 and g_2 have a direct sum. For if u_1 and u_2 are respectively non-null elements in g_1 and g_2 and $u_1 + u_2 = 0$, u_2 would be in g_1.

Moreover $g = g_1 + g_2$. For if u is an arbitrary element in g but not in g_1 or g_2, there exists an element v in g_2 such that $u - v$ is in g_1. Thus u is the sum of an element $u - v$ in g_1 and an element v in g_2.

The proof of the lemma is complete.

THEOREM 2.9. *If g is an operator subgroup of G, the rank function $\theta(u)$ defined by $\rho(u)$ as u ranges over g satisfies the rank conditions R.*

That $\theta(u)$ satisfies rank conditions I, II, and III is an immediate consequence of the fact that $\rho(u)$ satisfies R, and that g is an operator subgroup of g. Condition IV stated for $\theta(u)$ will be denoted by IVg.

Condition IVg concerns the group $g(\theta, \omega)$ generated by elements of g for which $\theta(u) \leqq \omega$. Condition IV concerns the group $G(\rho, \omega)$ generated by elements of G for which $\rho(u) \leqq \omega$. Observe that $g(\theta, \omega)$ is a subgroup of $G(\rho, \omega)$. If u and v are elements in $g(\theta, \omega)$ without rank θ, condition IVg affirms that $\theta(u + v) \neq \omega$. We shall see that this true. For u and v are elements in $G(\rho, \omega)$ without rank ρ, and it follows from IV that $\rho(u + v) \neq \omega$. Hence $\theta(u + v) \neq \omega$, and the proof of IV$g$ is complete.

3. Bi-lexical ranks

Let s and t be real variables which range on the intervals $0 \leqq s \leqq \infty$, $0 \leqq t \leqq \infty$, with $s > t$. Let the pairs (s, t) be ordered lexicographically. That is let $(s, t) = (s', t')$ if and only if $s = s'$, $t = t'$, and

$$(s, t) < (s', t'), \qquad (s', t') > (s, t),$$

if $s < s'$ or if $s = s'$ and $t < t'$. In this section we suppose that for certain non-null elements u of G rank functions $s(u)$ and $t(u)$ are defined. We understand that $t(u)$ exists if and only if $s(u)$ exists. We set

$$r(u) = [s(u), t(u)],$$

and prove the following theorem.

THEOREM 3.1. *Necessary and sufficient conditions that the bi-lexical ranks* $r(u) = [s(u), t(u)]$ *satisfy rank conditions* I, II, III *are as follows*:

(a). *The ranks* $s(u)$ *satisfy conditions* I, II, III.

(b). *For each fixed* s *the rank function* $t^s(u)$ *defined by* $t(u)$ *when* $s(u) = s$ *shall satisfy conditions* I, II, III.

(c). *When* $s(v) < s(u)$, $t(u + v) = t(u)$.

We shall first assume that the ranks $r(u)$ satisfy I, II, III, and show that (a), (b), and (c) are satisfied.

That (a) is satisfied follows directly from the fact that $r(u)$ satisfies I, II, III, and that the pairs (s, t) are ordered lexicographically. The verification of (b) is similar. Finally under the hypothesis of (c),

$$r(u + v) = r(u),$$

since $r(u)$ satisfies II and III. Hence $t(u + v) = t(u)$.

Conversely we shall show that conditions (a), (b), and (c) imply that the ranks $r(u)$ satisfy conditions I to III.

I. It follows from (a) that for $\delta \neq 0$, $s(\delta u) = s(u)$. Setting $s = s(u)$ in (b) we infer from (b) that $t(\delta u) = t(u)$. Hence

$$r(\delta u) = r(u),$$

and I is satisfied by ranks $r(u)$.

II. To establish II for ranks $r(u)$ we may assume without loss of generality that $r(v) \leqq r(u)$. We must then prove that $r(u + v) \leqq r(u)$. By virtue of (a) we have two cases:

(α). $s(u + v) < s(u)$,

(β). $s(u + v) = s(u)$.

In case (α), $r(u + v) < r(u)$ by virtue of the ordering of the pairs (s, t). In case (β) we distinguish two subcases:

(β'). $s(v) < s(u)$,

(β''). $s(u) = s(v)$.

In case (β'), $t(u + v) = t(u)$ in accordance with (c) so that $r(u + v) = r(u)$, and II holds. In case (β''), $t(u + v) \leqq t(u)$ by virtue of (b). Hence

$$r(u + v) \leqq r(u).$$

Thus II holds in all cases.

III. To establish III we must show that $r(u + v)$ exists, provided $r(u)$ and $r(v)$ exist and are unequal. Without loss of generality we can assume that $r(u) > r(v)$. We consider two cases:

(m). $s(u) > s(v)$,

(n). $s(u) = s(v)$, $t(u) > t(v)$.

In case (m), it follows from (a) that $s(u + v)$ exists. Hence $r(u + v)$ exists. In case (n), $t(u + v)$ exists by virtue of (b). Hence $r(u + v)$ exists in both cases, and the proof of III is complete.

The ranks $t(u)$ need not satisfy conditions I to III even when the ranks $r(u)$ satisfy these conditions. This is shown by the following example.

EXAMPLE 3.1. Let G consist of the elements

$$0, \quad u, \quad v, \quad u + v,$$

the field Δ being the integers mod 2. Suppose that

$$r(u) = (3, 0), \qquad r(v) = (3, 0), \qquad r(u + v) = (2, 1).$$

The ranks r satisfy I, II, and III. However

$$t(u) = 0, \qquad t(v) = 0, \qquad t(u + v) = 1$$

so that

$$t(u + v) > \max\,[t(u), t(v)].$$

Thus the ranks t do not satisfy II.

Spans $s - t$. An element u with rank $r = (s, t)$ will be said to have the span $s - t$, understanding that $s - t$ is infinite if s is infinite. It is clear that $s - t$ is constant as u varies over any r-class.

4. Invariant classes

Let A be the class of all elements of G with ranks in the set $[\rho]$. A subset E of A will be termed *ρ-invariant* if whenever u is in E every element in the ρ-class of u is in E. Let $E(\rho, \omega)$ denote the group generated by elements of E with

$\rho \leqq \omega$. We shall be concerned with ρ-invariant subsets E which satisfy the following condition:

V. *If u and v are in $E(\rho, \omega)$, if*

$$\rho(u) = \rho(v) = \rho(u + v) = \omega,$$

and u and v are not in E, then $u + v$ is not in E.

Example 4.1. Let A be a set of elements of G with bi-lexical ranks (s, t) satisfying rank conditions R. The subset of elements of A with spans $s - t > e$ forms an invariant subset E of A satisfying V.

Theorem 4.1. *Let E be an invariant subset of A which satisfies* V. *The rank function $\eta(u)$ defined by $\rho(u)$ when u is in E satisfies the rank conditions R.*

That $\eta(u)$ satisfies conditions I and II of R follows immediately from the fact that the rank function $\rho(u)$ satisfies I and II, and that E is an operator subset of A. In III, $\eta(u) \neq \eta(v)$ by hypothesis. Hence $\rho(u) \neq \rho(v)$. Suppose $\rho(u) > \rho(v)$. Then $\rho(u + v)$ exists, and $u + v$ and u are in the same ρ-class. Since E is a ρ-invariant subclass of A, $u + v$ must be in E because u is in E. Hence $\eta(u + v)$ exists, and III is satisfied.

Let IVρ and IVη denote the conditions IV stated respectively for ranks η and ranks ρ. Note that $G(\eta, \omega)$ in IVη is a subgroup of $G(\rho, \omega)$ in IVρ. We seek to establish IVη.

Condition IVη may be put in the following form:

(α). *If u and v are in $G(\eta, \omega)$ and $\eta(u + v) = \omega$, at least one of the elements u and v is in E.*

The elements u and v are in $G(\rho, \omega)$ and $\rho(u + v) = \omega$. It follows from IVρ that not both $\rho(u)$ and $\rho(v)$ can fail to exist. We consider two cases:

$$\text{Case a.} \quad \rho(u) < \omega, \qquad \text{Case b.} \quad \rho(u) = \omega,$$

the remaining cases arising upon replacing u by v.

Case a. Here $\rho(u) < \omega$, $\rho(u + v) = \omega$ so that v and $u + v$ are in the same ρ-class. But $u + v$ is in E since $\eta(u + v)$ exists in (α), and E is ρ-invariant. Hence v is in E.

Case b. Here $\rho(u) = \omega$. If $\rho(v)$ fails to exist or $\rho(v) < \omega$, then $u + v$ and u are in the same ρ-class and u is in E as in Case a. There remains the case

$$\rho(u) = \rho(v) = \eta(u + v) = \omega.$$

It follows from V that at least one of the elements u and v is in E.

Thus (α) is true in all cases and the proof of IVη is complete. This completes the proof of the theorem.

Let b be an operator subgroup of G and c be a maximal group of elements of G not in g. Then $b + c = G$ as is well-known. Theorem 4.2 which follows is a theorem of this type, but of a less regular form. The reader may find it simpler to postpone the study of Theorem 4.2 until its application in the proof of Theorem 11.3. The appropriateness of the form of Theorem 4.2 will then be clear. It has been included in this section to bring out the fact that it is independent of the topological aspects of this paper.

Let A be the class of all elements of G with rank, and suppose that A is the sum of two disjoint invariant subclasses B and C with the following properties:

(i). The sum of two elements in B in different ρ-classes is in B.

(ii). The sum of two elements in C with different ranks ρ is in C.

(iii). The sum of an element in B and an element in C is in A.

THEOREM 4.2. *Let B and C be invariant subclasses of A satisfying the above conditions, and let b and c be respectively maximal groups of elements in B and C. If the ranks of C form a discrete ascending sequence, $b + c$ is a maximal group of elements of A.*

That $b + c$ is a group of elements in A follows from (iii). It remains to show that $b + c$ is maximal in A. To that end let u be an arbitrary element in A, and assume that the following statement is true.

(α). *If v is an arbitrary element in $b + c$, $u - v$ is in A.*

We shall show that (α) leads to a contradiction.

If u is in C, there exists an element x in c such that $u - x$ is not in C, since c is maximal. Setting $v_1 = u - x$ we see that v_1 is in B by virtue of (α). The elements u and x are both in C, but v_1 is not in C. It follows from (ii) that $\rho(x) = \rho(u)$. But $v_1 = u - x$, and it follows from rank condition II that

$$\rho(v_1) \leqq \rho(u) \qquad (v_1 \subset B). \tag{4.1}$$

In case u is in B we set $v_1 = u$ and $x = 0$, so that (4.1) holds for any element u in A.

Since v_1 is in B and b is maximal there exists an element y in b such that $v_1 - y$ is not in B. Set

$$u_1 = v_1 - y = u - x - y.$$

It follows from (α) that u_1 is in C. Since v_1 and y are in B but $v_1 - y$ is not in B, it follows from (i) that v_1 and y are in the same ρ-class. But $u_1 = v_1 - y$ and it follows from the definition of a ρ-class that $\rho(u_1) < \rho(v_1)$. Hence

$$\rho(u_1) < \rho(u).$$

Observe that (α) holds if u is replaced by u_1. Proceeding inductively we infer the existence of an infinite sequence of elements u_n in C with ranks $\rho(u_n)$ which decrease with n. This is impossible since the ranks of C form a discrete ascending sequence.

We conclude that (α) is false and that $b + c$ is maximal in A. The proof of the theorem is complete.

5. Vietoris chains and cycles[7]

The space M shall be a metric space of elements $p, q, r, \cdots$ with a distance function pq satisfying the usual conditions:

$$\text{I. } pp = 0, \qquad \text{II. } pq \neq 0 \text{ if } p \neq q, \qquad \text{III. } pr \leqq pq + rq.$$

[7] Vietoris, L., *Über den höheren Zusammenhang kompakter Räume und eine Klasse von Zusammenhangstreuen Abbildungen*, Mathematische Annalen 97 (1927), pp. 454–472. Further references can be found in M1.

We shall be concerned with formal k-chains and k-cycles (Vietoris) defined as in M1, §1. We take over the definitions of pages 4–7 including definitions of the terms algebraic k-chains, algebraic k-cycles mod B on C, (Vietoris) k-cycles mod B on C, homologies $u \sim 0$ and $u \sim v$ mod B on C, carriers of cycles and homologies, homology classes, the boundary βu of an algebraic k-chain u or a formal k-chain u. We combine k-chains and k-cycles linearly with coefficients δ in an arbitrary field Δ.

Let B and C be compact sets on M with $B \subset C$. Let B_e and C_e be respectively e-neighborhoods of B and C. Let $u = (u_n)$ be a k-cycle mod B on C with algebraic component k-cycles u_n mod B on C. We shall make use of the following well-known lemma.

LEMMA 5.1. *If $u = (u_n)$ is a k-cycle* mod B *on* C *such that* $u \sim 0$ mod B_e *on* C_e *for each positive e, then* $u \sim 0$ mod B *on* C.

Corresponding to an arbitrary positive constant e there exists by hypothesis an integer N so large that for $n > N$,

$$u_n \sim_e 0 \qquad \text{(mod } B_e \text{ on } C_e).$$

For such values of n we infer the existence of an algebraic $(k + 1)$-chain w_n on C_e of norm e and an algebraic k-chain v_n on B_e of norm e such that

$$\beta w_n = u_n + v_n .$$

From this relation we see that $\beta u_n + \beta v_n = 0$, so that βv_n is on B. Let fv_n be the algebraic chain obtained from v_n by replacing each vertex x of v_n not on B by a nearest point fx on B. Let Dv_n be the deformation chain of M1, page 9, corresponding to this mapping fx. As in M1, (1.3),

$$\beta Dv_n = v_n - fv_n - D\beta v_n .$$

But βv_n is on B so that $D\beta v_n = 0$. If we set $w_n' = w_n - Dv_n$ we find that

$$\beta w_n' = u_n + fv_n ,$$

where w_n' has the norm $3e$, and fv_n is on B.

We now replace each vertex of w_n' on C_e but not on C by a nearest vertex on C, obtaining thereby a $(k + 1)$-chain y_n on C of norm at most $5e$. The chain $u_n + fv_n$ is on C and is unchanged. Hence

$$\beta y_n = u_n + fv_n ,$$

and since fv_n is on B, $u \sim 0$ mod B on C.

Let A, B, and C be three compact sets of M with $A \subset B \subset C$. We shall make use of the three following theorems. Proofs can be found in M2, page 396, making use of two fundamental lemmas due to E. Čech.

THEOREM 5.1. *If u is a k-cycle* mod A *on* C, *homologous to* 0 mod B *on* C, *then the homology class of* u mod A *on* C *contains a k-cycle* mod A *on* B.

THEOREM 5.2. *If a k-cycle u* mod A *on* B *is homologous to* 0 mod A *on* C, *there exists a* $(k + 1)$*-cycle* mod B *on* C *such that* $\beta v - u$ *is on* A.

THEOREM 5.3. *If u is a k-cycle* mod A *on* C *and* $\beta u \sim 0$ *on* A, *there exists an absolute k-cycle v on C such that* $u \sim v$ mod A *on* C.

The dimension of a maximal group of n-cycles on M, non-bounding on M will be termed the nth *connectivity* of M. If $A \subset B \subset M$, the dimension of a maximal group of n-cycles on A non-bounding on B will be termed the nth *connectivity of A on B*. We shall be concerned with cases where this dimension is finite and obviously uniquely defined.

6. The function F

The function $F(p)$ shall be single-valued on M and such that $0 \leqq F < 1$. A function J on M such that $0 \leqq J < \infty$ can be reduced to a function F by a transformation

$$F = \frac{J}{1 + J}.$$

Let c be a finite constant. The subsets of M on which $F \leqq c$ will be denoted by F_c. A set A will be said to be *definitely on* F_c (written d-on F_c) if A is on F_{c-e} for some positive e. The phrase d-mod F_c shall mean mod F_{c-e} for some positive e. A set d-on F_c will be said to be *d-below* c.

The space M and the function F shall satisfy three principal hypotheses of which the first is as follows.

Bounded compactness. For each constant $c < 1$, the sets F_c shall be compact.

This postulate implies that the function F is lower semi-continuous. It is always fulfilled when M is compact and F is lower semi-continuous.

Let p be a point of M at which $F(p) = c$. The set M will be said to be *locally F-connected* of order r at p if corresponding to each positive constant e there exists a positive constant δ such that each singular r-sphere on the δ-neighborhood of p and on $F_{c+\delta}$ bounds an $(r + 1)$-cell of norm e on F_{c+e}. It might be thought that local F-connectedness of order r is the same as local connectedness of order r in the space of the pairs $[p, F(p)]$. That this is not the case is shown by the following example.

EXAMPLE 6.1. Let the space M be the semi-circular disc $x^2 + y^2 \leqq 1$, $y \geqq 0$. Suppose that $F(x, y) = y$ except when $x = 0$. Let $F(0, y) \equiv 0$. The space M is locally F-connected of all orders, but the set of points $[x, y, F(x, y)]$ regarded as a set in euclidean three-space $[x, y, z]$ is not locally 0-connected at the points $[0, y, 0]$ at which $y > 0$.

Local F-connectedness. We assume that M is locally F-connected of all orders at each point p of M.

If M is locally connected of order r and if F is continuous, M is locally F-connected of order r.

The whole space M is not in general compact. That each homology class contains a cycle on a compact subset F_c of M is part of the implication of the following hypotheses.

F-regularity. Let a and b be constants with $a < b < 1$. In each non-trivial *homology class on M* mod F_b *there shall exist a cycle* mod F_b *on some compact subset F_c of M where $c < 1$. Moreover each relative k-cycle v* mod F_a *on F_b which satisfies an homology*

$$v \sim 0 \qquad (\text{mod } F_b) \tag{6.1}$$

on M shall satisfy a similar homology on some compact subset F_c of M where $c < 1$.

A simple example in which this postulate is not satisfied follows.

EXAMPLE 6.2. Let M be the circle

$$x^2 + (y - 1)^2 = 1. \tag{6.2}$$

We see that $0 \leqq y \leqq 2$. Let $F = y/2$ except at the point $(x, y) = (0, 2)$. Set $F(0, 2) = 0$. The circle is a carrier of a non-bounding 1-cycle which is homologous to no cycle with carrier on which F is bounded from 1.

These three postulates are satisfied for all ordinary variational problems. See M1, Part III. They are also satisfied by a suitably chosen space and function F in minimal surface theory. See Morse and Tompkins, loc. cit.

Let a, b, c be three constants such that

$$a < b < c < 1.$$

We shall prove two theorems which would be false if stated for singular cycles, and which embody the principle of the "accessibility" of the sets F_c with respect to homologies.

THEOREM 6.1. *If a k-cycle u* mod F_a *on F_c is such that*

$$u \sim 0 \qquad (\text{on } F_c \text{ mod } F_{b+e}) \tag{6.3}$$

for every positive e, the homology class of u mod F_a *on F_c contains a k-cycle* mod F_a *on F_b.*

It follows from Lemma 5.1 and (6.3) that

$$u \sim 0 \qquad (\text{on } F_c \text{ mod } F_b).$$

We now turn to Theorem 5.1 and set

$$A = F_a, \qquad B = F_b, \qquad C = F_c, \tag{6.4}$$

noting that these sets are compact. Theorem 6.1 follows from Theorem 5.1.

If $a < 0$, u becomes an absolute cycle and there is a cycle in its homology class on F_b.

THEOREM 6.2. *If a k-cycle u* mod F_a *on F_b is homologous to* 0 mod F_a *on F_{c+e} for every positive e, there exists a $(k + 1)$-cycle v* mod F_b *on F_c such that $\beta v - u$ is on F_a.*

It follows from Lemma 5.1 that $u \sim 0$ mod F_a on F_c. We now apply Theorem 5.2, setting A, B, C equal to F_a, F_b, F_c respectively. Theorem 6.2 results.

If $a < 0$, u is an absolute cycle and there exists a $(k + 1)$-cycle v mod F_b on F_c such that $\beta v = u$.

THEOREM 6.3. *Let a and c be positive constants such that $a < c < 1$. The kth connectivity $R^k(a, c)$ of F_a on F_c is finite.*

This is a consequence of the hypotheses of bounded compactness and of local F-connectedness, as follows from Theorem 6.1 of M1, page 28.

7. F-homology classes and ranks r

An absolute k-cycle u which lies definitely on some subset F_c of M for which $c < 1$ and satisfies the condition

$$u \not\sim 0 \qquad \text{(d-on } F_c) \tag{7.1}$$

will be termed an *F-cycle*, and will be said to possess the *homology bound* c. The least upper bound of the homology bounds of u will be called the *superior cycle limit*[8] $s(u)$ *of* u. In case u is non-bounding we understand that $s(u)$ is infinite. In case u is bounding it follows from the condition of F-regularity that $s(u) < 1$. Thus $s(u)$ never equals 1. It follows from the definitions involved that $s(u)$ is an homology bound of u in case u is bounding. Given an F-cycle u, the greatest lower bound of numbers $b < s(u)$ such that

$$u \sim 0 \qquad \text{(d-on } F_s \text{ mod } F_b) \tag{7.2}$$

will be called the *inferior cycle limit* $t(u)$ of u.

If $s(u)$ is regarded as a rank, two F-k-cycles u and v are in the same s-class if and only if for $s = s(u)$

$$u \sim v \qquad \text{(d-on } F_s). \tag{7.3}$$

An s-class becomes an homology class when $s = \infty$. An s-class will accordingly be termed an *F-homology* class.

If u is an F-cycle the pair $[s(u), t(u)]$ will be regarded as a rank $r(u)$ of u. We order these ranks lexicographically. If two elements u and v are in the same F-homology class they are in the same r-class but not conversely. A necessary and sufficient condition that elements u and v with the same rank $r = (s, t)$ be in the same r-class is that

$$u \sim v \qquad \text{(d-on } F_s \text{ d-mod } F_t). \tag{7.4}$$

It is seen that (7.3) implies (7.4), but not conversely.

Observe that null cycles have no s-rank or t-rank.

We continue with the following lemma.

LEMMA 7.1. *The ranks $s(u)$ of k-cycles satisfy rank conditions R.*

The underlying group G is here the group of all absolute k-cycles.

I. To show that $s(\delta u) = s(u)$ for $\delta \neq 0$ we must show that the condition $u \sim 0$ d-on F_c is equivalent to the condition $\delta u \sim 0$ d-on F_c. The first condition clearly implies the second. But the second condition implies the first. For any coefficient in the field Δ can be divided by a non-null δ. Thus I holds.

II. In II, it is assumed that u, v, and $u + v$ have ranks s. Let c be the maximum of $s(u)$ and $s(v)$. By virtue of the definition of superior cycle limits

$$u \sim 0, \qquad v \sim 0, \qquad \text{(on } F_{c+e})$$

[8] In M1, $s(u)$ was inadvertently used in another sense. We revert to the usage of M2.

for each positive constant e, so that

$$u + v \sim 0 \qquad \text{(on } F_{c+e}\text{)}.$$

Hence $s(u + v) \leqq c$, and II is satisfied.

III. We suppose that $s(u) > s(v)$ and let c be any homology bound of u between $s(u)$ and $s(v)$. Then $v \sim 0$ and $u \not\sim 0$ d-on F_c so that $u + v \not\sim 0$ d-on F_c. Thus c is an homology bound of $u + v$. Hence $s(u + v)$ exists.

IV. In the notation of IV let u and v be elements in $G(s, \omega)$. The cycles u and v are d-on F_b where b is some constant $< \omega$. In IV, b is an homology bound of neither u nor v, and so it cannot be an homology bound of $u + v$. Hence $s(u + v) < b < \omega$ if $s(u + v)$ exists. We infer the truth of IV.

LEMMA 7.2. *Let c be a value assumed by the rank function s. The rank function $t^c(u)$ defined by $t(u)$ for those k-cycles for which $s = c$ satisfies rank conditions I, II, III.*

I. Condition I follows from the fact that for $\delta \neq 0$ relation (7.2) holds if and only if it holds when u is replaced by δu.

II. Let b be any constant such that

$$c > b > \max [t(u), t(v)] \qquad (s(u) = s(v) = c). \tag{7.5}$$

Then

$$u \sim 0, \qquad v \sim 0, \qquad \text{(d-on } F_c \bmod F_b\text{)}$$

in accordance with (7.2). Hence $u + v$ satisfies a similar homology. We conclude that $t(u + v) \leqq b$. But b is any constant for which (7.5) holds so that

$$t(u + v) \leqq \max [t(u), t(v)],$$

and the proof of II is complete.

III. We here assume that $s(u) = s(v) = c$, and that $t(u) \neq t(v)$. We seek to prove that $s(u + v) = c$ so that $t^c(u + v)$ exists.

We suppose that $t(u) < t(v)$. Then for each constant b between $t(u)$ and $t(v)$,

$$u \sim 0, \qquad u + v \sim v \not\sim 0, \qquad \text{(d-on } F_c \bmod F_b\text{)}$$

in accordance with (7.2). In particular

$$u + v \not\sim 0 \qquad \text{(d-on } F_c\text{)}$$

so that $s(u + v)$ exists and equals c. Hence $t^c(u + v)$ exists, and the proof of III is complete.

THEOREM 7.1. *The ranks $r = (s, t)$ of F-k-cycles satisfy the rank conditions R.*

That ranks r satisfy conditions I, II, III will follow from Theorem 3.1. For conditions (a) and (b) are satisfied in Theorem 3.1 by virtue of the two preceding lemmas. Condition (c) of Theorem 3.1 is that when $s(v) < s(u)$, $t(u) = t(u + v)$. But when $s(v) < s(u)$, $s(u + v) = s(u)$. Setting $a = s(u)$,

$$v \sim 0, \qquad u + v \sim u, \qquad \text{(d-on } F_a\text{)}.$$

Hence $t(u + v) = t(u)$, and condition (c) is satisfied. The ranks r thus satisfy conditions I, II, III.

To establish IV let $\omega = (\omega_1, \omega_2)$ be a particular rank r. The group $G(r, \omega)$ is a subgroup of the group $G(s, \omega_1)$. If u and v are two elements of $G(r, \omega)$, u and v are in $G(s, \omega_1)$. Since ranks s satisfy IV, the hypothesis that $s(u + v) = \omega_1$ implies that $s(u)$ or $s(v)$ exists. Hence $r(u)$ or $r(v)$ exists, and the proof of IV is complete.

An F-cycle u with rank $r = (s, t)$ will be termed *canonical* if u is on F_t. Let c be an arbitrary constant $c < 1$. A k-cycle on F_c will be termed *canonical relative* to F_c if, when F_c is regarded as the whole space M, u is canonical. If τ is the inferior cycle limit of u relative to F_c it is clear that $\tau \geqq t(u)$.

LEMMA 7.3. *Let H be an F-homology class of k-cycles with rank $r = (s, t)$ and let c be a constant between s and t and < 1. Let η be an arbitrary positive constant. The class H contains at least one k-cycle which is canonical relative to F_c, and which possesses an inferior cycle limit τ relative to F_c such that $\tau < t + \eta$.*

By virtue of the definition of t there exists a constant $b < t + \eta$ and $< c$ such that

$$u \sim 0 \qquad \text{(d-on } F_s \text{ mod } F_b).$$

It follows from Theorem 5.1 that the homology class of u d-on F_s contains a k-cycle v on F_b. Let τ be the inferior cycle limit of v relative to F_c. It is clear that

$$t \leqq \tau \leqq t + \eta.$$

By virtue of the definition of τ,

$$v \sim 0 \qquad \text{(on } F_c \text{ mod } F_{\tau+e})$$

for every $e > 0$. It follows from Theorem 6.1 that the homology class of v on F_c contains a k-cycle z on F_τ. The cycle z is canonical relative to F_c, and the proof of the lemma is complete.

THEOREM 7.2. *Each F-homology class of k-cycles contains at least one canonical k-cycle.*

Let (s, t) be the cycle limits of the given F-homology class H, and let $c < 1$ be a constant such that $s > c > t$. By virtue of the preceding lemma H contains a sequence of k-cycles z_n, $n = 1, 2, \cdots$, with the following properties. The cycle z_n is canonical relative to F_c and the inferior cycle limit τ_n of z_n relative to F_c tends to t as n becomes infinite.

If the theorem is false $\tau_n > t$ for each n, and we can suppose without loss of generality that τ_n decreases with n. Let a be a constant such that $t < a < c$. We can suppose that $\tau_n < a$ for each n. Hence z_n is on F_a. Relative to F_c the bi-lexical rank of z_n is (∞, τ_n). Every proper linear combination of the cycles z_n has a rank relative to F_c of the form (∞, τ). Thus the cycles z_n are on F_a and independent with respect to bounding on F_c. This is impossible by virtue of Theorem 6.3.

We infer the truth of the theorem.

We shall give an example showing that the preceding theorem would not be true were M not locally F-connected of the different orders r.

EXAMPLE 7.1. Let M be the closure in the space (x, y) of the curve

$$3y = (1 - x) \sin \frac{1}{x} + 1 \qquad (0 < x \leqq 1).$$

On M, $0 \leqq y \leqq 2/3$, and M is compact. Let $F = y$ on M. The space M is not locally connected of order 0. Let u and v be respectively Vietoris 0-cycles with the points (0, 0), (1, 1/3) as carriers. Let z be the cycle $u - v$. It appears that $s(z) = 2/3$, $t(z) = 0$. But there is no 0-cycle on $F = 0$ homologous to z d-on $F \leqq 2/3$.

As in the general theory of bi-lexical ranks $r = (s, t)$ we term the difference $s - t$ the *span* of an F-cycle. Theorem 4.1 thus gives the following. (See Example 4.1).

THEOREM 7.3. *Let E be the subset of F-cycles with spans $s - t > e$. The rank function $r_e(u)$ defined by $r(u)$ for u in E satifies the rank conditions R.*

We shall continue with the following lemma.

LEMMA 7.4. *The ranks r of bounding F-k-cycles with spans $> e > 0$ are at most finite in number.*

Since the F-cycles u of the lemma are bounding $s(u) < 1$, and since their spans exceed e,

$$t(u) < s(u) - e < 1 - e.$$

Without loss of generality we can suppose that the F-cycles of the lemma are canonical, since there is a canonical F-cycle in each F-homology class, and hence a canonical F-cycle with each different rank. Let b and a be constants such that

$$0 < b - a < e, \qquad b \leqq 1 - e.$$

Each rank $t(u)$ of a cycle of the lemma lies on an interval of the form

$$a < t < b. \tag{7.6}$$

The set T of such ranks can be covered by a finite set of intervals of this form.

For a and b fixed the number of different ranks t in T and on (7.6) must be finite. For

$$t < b, \qquad s > t + e > a + e,$$

so that the corresponding canonical F-cycles lie on F_b and are independent on F_{a+e}. The number of such cycles is at most $R(b, a + e)$. Since $a + e > b$ this number is finite in accordance with Theorem 6.3, and the proof of the lemma is complete.

THEOREM 7.4. *The dimension of a maximal group g of bounding F-cycles with a given rank r is finite.*

Suppose r has the form (s, t). Let B be a base for g. Let each element v_i

in B be replaced by a canonical F-cycle u_i in the F-homology class of v_i. We have

$$u_i \sim v_i \qquad \text{(d-below } s).$$

The cycles v_i are independent d-below s. Hence the cycles u_i are independent d-below s. Let c be any constant between s and t and < 1. The cycles u_i are on F_t and independent on F_c. Their number is at most $R(t, c)$, and the theorem follows from the finiteness of $R(t, c)$.

THEOREM 7.5. *The dimension of a maximal group g of bounding F-cycles with spans $> e > 0$ is finite.*

By virtue of Lemma 7.4 the ranks r of elements of g are finite in number. The group g is accordingly a direct sum,

$$g = \sum_r g(r),$$

of maximal subgroups of g with the respective ranks of elements of g. The dimension of each such group $g(r)$ is finite in accordance with the preceding lemma. Hence the dimension of g is finite, and the proof of the theorem is complete.

THEOREM 7.6. *In any infinite sequence (s_n, t_n) of distinct ranks of F-cycles with spans $> e > 0$, t_n tends to 1 as n becomes infinite.*

It follows from Theorem 7.2 that there exists a canonical k-cycle u_n with rank (s_n, t_n). The cycle u_n is on $F \leqq t_n$.

If the theorem were false, the values of t_n would have a cluster value $a < 1$. Let b and c be constants such that

$$a < b < c < a + e \qquad (c < 1).$$

There will exist an infinite subset of the values of n for which $t_n < b$ and $s_n > c$. The corresponding cycles u_n will be on F_b and be independent on F_c. This is impossible since the connectivities of F_a on F_c are finite.

The theorem follows.

8. k-Caps and their cap heights

Let a be a constant at most 1. If u is a k-cycle d-mod F_a on F_a, an homology of the form

$$u \sim 0 \qquad (\text{on } F_a \text{ d-mod } F_a) \tag{8.1}$$

will be called an *a-homology*. A k-cycle on F_a d-mod F_a not a-homologous to 0 will be called a k-cap with *cap height a*. We write $a = a(u)$. We note that $a(u)$ is uniquely determined by the formal k-chain w whose components are those of u. In fact $a(u)$ is the greatest lower bound c of numbers b such that w is on F_b. For it is impossible that $a(u) > c$ since u would then satisfy an a-homology. It is equally impossible that $a(u) < c$ since this would imply that w is on F_a, contrary to the definition of c. That no cap height equals 1 follows at once from the condition of F-regularity.

We regard $a(u)$ as a new rank function defined for a subset of elements of the group G of formal chains. It is important that we show that $a(u)$ satisfies the rank conditions R, for from this fact it follows that the maximal groups which appear later have dimensions which are independent of their mode of formation.

THEOREM 8.1. *The cap heights $a(u)$ satisfy the rank conditions R.*

That I is satisfied follows as for the ranks $s(u)$. That II is satisfied follows from the fact that $a(u)$ is the greatest lower bound of numbers b such that the components of u are on F_b. Turning to III, suppose that $a(u) < a(v)$. Then u is $a(v)$-homologous to zero and v is not $a(v)$-homologous to zero. Hence $u + v$ is not $a(v)$-homologous to zero. Since $u + v$ is on F_a for $a = a(v)$, $a(v)$ is a cap height of $u + v$, and III is proved. In IV, u and v are elements of $G(a, \omega)$, not k-caps. Hence u and v are ω-homologous to zero. It follows that $u + v$ is ω-homologous to zero, and IV is true.

The proof of the theorem is complete.

LEMMA 8.1. *Let u be a k-cycle* d-mod F_a *on F_a. If u is a-homologous to zero βu is homologous to zero d-below a.*

Under the hypothesis of the lemma, there exists a formal $(k + 1)$-chain w on F_a such that

$$\beta w = u - x, \tag{8.2}$$

where x is a formal k-chain d-below a. Applying the operator β to both sides of (8.2) we find that $\beta u = \beta x$, and the lemma follows directly.

Cap class. In accordance with the general definitions of a rank class, two k-caps with the same cap height a are said to be in the same a-class or *cap class* if $u - v$ satisfies an a-homology.

A k-cap u with cap height a will be said to be *linkable* if

$$\beta u \sim 0 \qquad \text{(d-below } a\text{)}. \tag{8.3}$$

Two k-caps u and v in the same cap class will both be linkable or both non-linkable. For if u and v define the same cap class, $u - v$ is a-homologous to zero, and

$$\beta u - \beta v \sim 0 \qquad \text{(d-below } a\text{)}$$

in accordance with the preceding lemma. It follows that (8.3) holds for u if and only if (8.3) holds for v.

Superior cap limit $\sigma(u)$. Let u be a k-cap with cap height a. The least upper bound σ of numbers $b \geqq a$ such that

$$u \not\sim 0 \qquad (\text{on } F_b \text{ d-mod } F_a) \tag{8.4}$$

will be called the *superior cap limit* $\sigma(u)$ of u. It is clear that

$$\sigma(u) \geqq a(u).$$

If the numbers b satisfying (8.4) are unbounded, we set $\sigma(u) = \infty$. It appears that two k-caps in the same cap class have the same superior cap limit. There exist k-caps for which $\sigma(u) = a(u)$, as the following example shows.

EXAMPLE 8.1. Let M be the interval $0 \leqq x \leqq 1$. Let $F(0) = 1/2$ and

$$F(x) = \frac{x}{2} \sin \frac{1}{x} + \frac{1}{2} \qquad (0 < x \leqq 1).$$

The Vietoris 0-cycle u with carrier $x = 0$ is such that $\sigma(u) = a(u) = F(0)$.

No k-cap u has a superior cap limit $\sigma(u) = 1$. For that would imply that for each constant $b > 1$, and for $a = a(u)$,

$$u \sim 0 \qquad (\text{on } F_b \text{ d-mod } F_a).$$

But $F_b = M$, and $a(u) < 1$, and it would follow from the condition of regularity that for some constant c between a and 1,

$$u \sim 0 \qquad (\text{on } F_c \text{ d-mod } F_a)$$

so that $\sigma(u) < 1$.

Inferior cap limit $\tau(u)$. A k-cap u which is non-linkable satisfies (8.4) with each constant $b \geqq a$, as follows from Lemma 8.1. For such a k-cap, $\sigma(u) = \infty$. But when u is non-linkable βu is an F-cycle v with $s(v) = a(u)$. In such a case the inferior cycle limit of βu will be called the *inferior cap limit* of u, and denoted by $\tau(u)$. Thus $\tau(u) = t(v)$. The function $\tau(u)$ is not defined when u is linkable.

The boundaries of non-linkable k-caps u in the same cap class have the same superior cycle limit $a(u)$, and are in the same F-homology class by virtue of Lemma 8.1. They accordingly have a common inferior cap limit τ. Thus inferior and superior cap limits depend only on the cap class in which a k-cap is found.

The span of a k-cap. If u is linkable k-cap, its cap span shall be

$$S(u) = \sigma(u) - a(u).$$

If u is non-linkable, its cap span shall be

$$S(u) = a(u) - \tau(u).$$

As we have seen in Example 8.1 the cap span of a k-cap may be null. The span of a non-linkable k-cap u is the span of βu as an F-cycle. Hence $S(u) \neq 0$.

Let u and v be k-caps. If $a(u) < a(v)$, $u + v$ is in the cap class of v so that

$$S(u + v) = S(v).$$

If $a(u) = a(v)$, and u is linkable but v non-linkable, $u + v$ is non-linkable and

$$S(u + v) = S(v).$$

The span of an F-cycle u with bi-lexical ranks (s, t) is $s - t$. This will now be termed the *cycle span* of u to distinguish it from the *cap span* $S(u)$ in case u is a k-cap.

Let e be a positive constant.

THEOREM 8.2. *The rank function $a_e(u)$ defined by the cap heights $a(u)$ of k-caps with cap spans $> e$ satisfies the rank conditions R.*

We rely on Theorem 4.1, taking A as the set of all k-caps and E as the subset of

k-caps with cap spans $> e$. The cap heights $a(u)$ shall replace the ranks $\rho(u)$. The set E is an a-invariant subset of A since caps in the same cap class have the same cap span. It remains to verify condition V of §4.

Condition V is concerned with k-caps u and v generated by k-caps x of E for which $a(x) \leqq \omega$. Condition V affirms that if

$$a(u) = a(v) = a(u + v) = \omega, \tag{8.5}$$

and if $S(u) \leqq e$ and $S(v) \leqq e$, then $S(u + v) \leqq e$.

To establish V we shall first show that u and v in V are linkable. Observe that the generating k-caps x in V satisfy the homology

$$\beta x \sim 0 \qquad (\text{d-on } F_\omega \text{ d-mod } F_{\omega-e}). \tag{8.6}$$

This is trivial if x is linkable. If x is non-linkable, it is a consequence of the fact that $S(x) > e$. It follows that u and v satisfy (8.6) as well as x. If u were non-linkable, $\tau(u) < \omega - e$ since u satisfies (8.6), so that

$$e < \omega - \tau(u) = S(u),$$

contrary to hypotheses in V. Hence u is linkable. Similarly v is linkable.

It follows from (8.5) that $u + v$ is linkable. To establish V we observe that when (8.5) holds,

$$\sigma(u + v) \leqq \max [\sigma(u), \sigma(v)]. \tag{8.7}$$

Without loss of generality we can suppose that $\sigma(u) \geqq \sigma(v)$. Then

$$S(u + v) = \sigma(u + v) - \omega \leqq \sigma(u) - \omega = S(u) \leqq e. \tag{8.8}$$

Thus V holds. Theorem 8.2 follows from Theorem 4.1.

COROLLARY 8.2. *The rank function $a_1(u)$ defined by the cap heights of k-caps with infinite cap spans satisfies the rank conditions R.*

This follows from the theorem upon setting $e = 1$. For the only spans which exceed 1 are infinite. In this connection recall that the only k-caps with infinite cap spans are linkable.

In the following theorem $a_e'(u)$ should be read with "linkable" and $a_e''(u)$ with "non-linkable."

THEOREM 8.3. *The rank function $a_e'(u)$ [$a_e''(u)$] defined by the cap heights $a(u)$ of linkable [non-linkable] k-caps with finite spans $> e$ satisfies the rank conditions R.*

We again turn to Theorem 4.1, taking A as the set of all k-caps with spans $>$ e, and E as the subset of linkable [non-linkable] k-caps with finite spans $> e$. The ranks $a_e(u)$ of the preceding theorem will serve as the ranks $\rho(u)$. The set E is an a_e-invariant subset of A since two k-caps in the same cap class will belong to E, if either belongs to E. It remains to verify condition V of §4.

Condition V affirms that if u and v are generated by k-caps x of E for which $a_e(x) \leqq \omega$, if

$$a_e(\omega) = a_e(v) = a_e(u + v) = \omega,$$

and if u and v are not in E, then $u + v$ is not in E. In proving V we distinguish between the two theorems which may be read from Theorem 8.3.

CASE I. Linkable k-caps. In this case the generating caps x are linkable, and it is clear that u, v, and $u + v$ are linkable and have finite spans. Condition V will be true if whenever u and v are generated as in V, and $S(u) \leqq e$, $S(v) \leqq e$, then $S(u + v) \leqq e$. But this follows from (8.7) and (8.8) as in the preceding proof.

CASE II. Non-linkable k-caps. In this case the generating caps x in V are non-linkable, and it follows from (8.6) as in the proof of Theorem 8.2 that u, v, and $u + v$ are linkable. Hence $u + v$ is not in E, and V is true.

Theorem 8.3 follows from Theorem 4.1.

9. Relation between k-caps and F-cycles

A k-cycle u which is also a k-cap will be called a *k-cycle cap*. Each k-cycle cap u is an F-cycle. Otherwise

$$u \sim 0 \qquad \text{(d-on } F_b) \tag{9.0}$$

for each positive constant $b > a(u)$. It would follow from Lemma 5.1 that u would be a-homologous to zero, contrary to the fact that u is a k-cap with cap height a. Since u is both a k-cap and an F-cycle the following terms are well defined:

$a(u)$,	the cap height of u,
$\sigma(u)$,	the superior cap limit of u,
$t(u)$,	the inferior cycle limit of u,
$s(u)$,	the superior cycle limit of u,
$S(u)$,	the cap span $\sigma(u) - a(u)$ of u,
$s - t$,	the cycle span of u.

EXAMPLE 9.1. Let M be the interval $-2\pi \leqq x \leqq 2\pi$. Let $F(x)$ be a function such that

$$7F(x) = 2 \cos x + 3 \qquad (-\pi/2 \leqq x \leqq \pi/2)$$

with $7\,F(x) = -\cos x + 3$ on the remainder of M. The function F is positive and continuous on M. Let x and y be Vietoris 0-cycles with the points $x = \pm \pi/2$ respectively as carriers. Let v be the cycle cap $x - y$. One sees that

$$a(v) = \tfrac{3}{7}, \qquad \sigma(v) = \tfrac{4}{7}, \qquad s(v) = \tfrac{5}{7}, \qquad t(v) = \tfrac{2}{7}.$$

LEMMA 9.1. *Each k-cycle cap u is an F-cycle with*

(a). $$a(u) \geqq t(u), \qquad s(u) \geqq \sigma(u).$$

(b). $$\text{If } s(u) = \sigma(u), \qquad a(u) = t(u).$$

(c). $$\text{If } a(u) = t(u), \qquad s(u) = \sigma(u).$$

That u is an F-cycle has just been shown. We turn to the proof of (a).

(a). The cycle u is on F_a so that $a(u) \geqq t(u)$. If $\sigma(u) = a(u)$, $s(u) > \sigma(u)$ since u is an F-cycle. We therefore suppose that $\sigma(u) > a(u)$. Let b be any constant between $\sigma(u)$ and $a(u)$. It follows from the definition of $\sigma(u)$ that

$$u \not\sim 0 \qquad \text{(on } F_b \text{ d-mod } F_a).$$

These constants b are homology bounds of u so that $s(u) \geqq \sigma(u)$ as stated in (a).

(b). If $s(u) = \sigma(u)$, $t(u) = a(u)$. For if $t(v) < a(u)$, it would follow from the definition of $t(u)$ that

$$u \sim 0 \qquad \text{(d-on } F_s \text{ d-mod } F_a).$$

But $s = \sigma$ so that this is contrary to the definition of σ. Hence $t(u) = a(u)$.

(c). If $t(u) = a(u)$, $\sigma(u) = s(u)$. For if $s(u) > \sigma(u)$,

$$u \sim 0 \qquad \text{(d-on } F_s \text{ d-mod } F_a)$$

by virtue of the definition of $\sigma(u)$. It would follow from the definition of $t(u)$ that $t(u) < a(u)$. Hence $\sigma(u) = s(u)$.

The proof of the lemma is complete.

Lemma 9.2. *Let η be an arbitrary positive constant. The cap class of each linkable k-cap u contains at least one k-cycle cap z such that*

$$s(z) \leqq \sigma(z) + \eta. \tag{9.1}$$

There are two cases according as $\sigma(u)$ is finite or infinite.

Case I. $\sigma(u) < \infty$. Here $\sigma < 1$. By virtue of the definition of $\sigma(u)$ and of $a(u)$,

$$u \sim 0 \qquad \text{(on } F_{\sigma+\eta} \text{ d-mod } F_a).$$

Without loss of generality we can suppose that $\sigma + \eta < 1$ so that $F_{\sigma+\eta}$ is compact. It follows from Theorem 5.2 that there exists a $(k + 1)$-cycle v mod F_a on $F_{\sigma+\eta}$ such that $\beta v - u$ is d-below a. Set $\beta v = z$. Then z is a k-cycle on F_a. It is also a k-cap in the cap class of u since $z - u$ is d-below a. Moreover z is an F-cycle by virtue of the preceding lemma. Hence $s(z)$ exists. But $z = \beta v$ so that $z \sim 0$ on $F_{\sigma+\eta}$. Hence (9.1) holds, and the proof of the lemma is complete.

Case II. $\sigma(u) = \infty$. Since u is linkable there exists a constant $c < a$ such that

$$\beta u \sim 0 \qquad \text{(on } F_c).$$

We apply Theorem 5.3, setting $A = F_c$ and $C = F_a$. We thereby infer the existence of an absolute k-cycle z on F_a such that $u \sim z$ mod F_c on F_a. Hence z is a k-cycle cap in the cap class of u. Since $\sigma(u) = \infty$, it follows from Lemma 9.1 that $s(u) = \infty$, and (9.1) is satisfied.

Theorem 9.1. *The cap class of each linkable k-cap u of positive cap span contains at least one canonical k-cycle cap v. For any such cycle,*

$$t(v) = a(v), \qquad \sigma(v) = s(v). \tag{9.2}$$

CASE I. $\sigma(u) = \infty$. There exists a k-cycle cap v in the cap class of u by virtue of the preceding lemma. In this case $\sigma(v) = \infty$ so that $s(v) = \sigma(v)$. It follows from (b) in Lemma 9.1 that $t(v) = a(v)$.

CASE II. $\sigma(u) < \infty$. Let η_n be a sequence of positive constants converging to zero. It follows from the preceding lemma that there exists a k-cycle cap z_n in the cap class of u such that

$$s(z_n) \leqq \sigma(u) + \eta_n .$$

It follows from Lemma 9.1 (a) that

$$s(z_n) - t(z_n) \geqq \sigma(u) - a(u). \tag{9.3}$$

But the left member of (9.3) is the cycle span of z_n and the right member the cap span $S(u)$ of u. Recall that $S(u) > 0$ by hypothesis.

At most a finite number of the numbers $s(z_n)$ are different from $\sigma(u)$. Otherwise the ranks of the cycles z_n would form an infinite set of distinct ranks with cycle spans at least $S(u)$. Since $t(z_n)$ does not tend to 1 as n becomes infinite this is contrary to Theorem 7.6. We conclude that $s(v) = \sigma(u)$ for some cycle v of the set z_n.

That v is canonical and that (9.2) holds now follows from Lemma 9.1.

The capping of F-cycles. Let u be an F-k-cycle with a finite superior cycle limit s. A $(k + 1)$-cycle v d-mod F_s on F_s such that

$$\beta v = u$$

will be said to *cap u*. This term is justified by the fact that when v caps u, v is a non-linkable $(k + 1)$-cap with $a(v) = a(s)$. For v is on F_s and satisfies no s-homology. Otherwise it would follow from Lemma 8.1 that

$$\beta v \sim 0 \qquad \text{(d-below } s\text{)}$$

contrary to the nature of $s(u)$. Concerning the existence of a relative $(k + 1)$-cycle v capping u we have the following theorem.

THEOREM 9.2. *Each F-k-cycle u with a finite superior limit s is capped by at least one relative $(k + 1)$-cycle v. Moreover v is non-linkable and*

$$a(v) = s(u).$$

The limit s is finite so that $s < 1$. Thus F_s is compact. By virtue of the definition of s,

$$u \sim 0 \qquad (\text{on } F_{s+e})$$

for each positive constant e. It follows from Lemma 5.1 that

$$u \sim 0 \qquad (\text{on } F_s).$$

Moreover u lies on some compact set F_b for which $b < s$. We now apply Theorem 5.2, setting

$$A = 0, \qquad B = F_b, \qquad C = F_s.$$

We thereby infer the existence of a $(k + 1)$-cycle v mod F_b on F_s such that $\beta v = u$.

The first statement of the theorem follows. The concluding statement of the theorem has already been proved.

THEOREM 9.3. *The cap heights of k-caps u with cap spans $> e > 0$ cluster at most at the value* 1.

If u is linkable, it follows from Theorem 9.1 that there exists a k-cycle cap v with rank $[\sigma(u), a(u)]$ and accordingly with cycle span

$$\sigma(u) - a(u) > e.$$

It follows from Theorem 7.6 that the cap heights $a(u)$ of this type cluster at most at the value 1.

If u is non-linkable, βu is a $(k - 1)$-cycle v with rank $[a(u), \tau(u)]$ and accordingly with cycle span

$$\tau(u) - a(u) > e \qquad [\tau(u) < a(u) < 1].$$

It follows again from Theorem 7.6 that the values $\tau(u)$ and hence the values $a(u)$ cluster at most at the value 1.

THEOREM 9.4. *The cap heights of k-caps u with finite cap spans $> e > 0$ are finite in number.*

The proof of this theorem is similar to the proof of Theorem 9.3 making use of Lemma 7.4.

10. Cap-isomorphisms and F-isomorphisms

Recall that a canonical k-cycle u is an F-cycle cap such that $t(u) = a(u)$. We continue with the following lemma.

LEMMA 10.1 *The cap span of a canonical k-cycle u equals its cycle span. The cap span of the sum,*

$$v = v_1 + \cdots + v_n,$$

of a finite set of canonical k-cycles v_i with increasing ranks $t(v_i)$ is at least the cycle span of v_n.

If u is canonical $t(u) = a(u)$, and then $s(u) = \sigma(u)$ by virtue of Lemma 9.1 (a). Thus

$$S(u) = \sigma(u) - a(u) = s(u) - t(u),$$

and the first statement of the lemma is true.

To prove the second statement of the lemma set $t_i = t(v_i)$, $i = 1, \cdots, n$, and recall that v_i is on F_{t_i} since v_i is canonical. Observe that v is a k-cap with $a(v) = t_n$. Let b be any constant between t_n and $\sigma(v_n)$. For $i = 1, 2, \cdots, n - 1$,

$$v_i \sim 0 \qquad \text{(on } F_b \text{ d-mod } F_{t_n}\text{)}$$

since v_i is on F_{t_i} and $t_i < t_n$. Moreover

$$v_n \not\sim 0 \qquad \text{(on } F_b \text{ d-mod } F_{t_n}\text{)} \tag{10.1}$$

by virtue of the choice of $b < \sigma(v_n)$ and the definition of $\sigma(v_n)$. Hence (10.1) holds for v as well as for v_n. Taking account of the arbitrariness of b between t_n and $\sigma(v_n)$ we see that $\sigma(v) \geqq \sigma(v_n)$. Hence

$$\sigma(v) - a(v) \geqq \sigma(v_n) - a(v_n) = s(v_n) - t(v_n).$$

Thus $S(v)$ is at least the cycle span of v_n, and the proof of the lemma is complete.

Recall that two F-cycles are in the same F-homology class if in the same s-class. Here $s(u)$ is the rank function defined by the superior cycle limit of u. An s-isomorphism will be termed an *F-isomorphism.* Recall that two F-cycles which are in the same F-homology class are in the same r-class, but not conversely. Here $r(u)$ is the rank function defined by $[s(u), t(u)]$.

THEOREM 10.1. *Each group g of (bounding) k-cycles with cycle spans $> e > 0$ if F-isomorphic with a group γ of k-cycle caps with (finite) cap spans $> e > 0$.*

We shall make use of Theorem 2.7. To that end let (u) be an r-proper base for g. That g possesses an r-proper base follows from the fact (see Theorem 7.6) that its ranks r form a discrete ascending sequence. But an r-proper base is at once seen to be an s-proper base since the ranks $r = (s, t)$ are ordered lexicographically. To apply Theorem 2.7 we map each element u_i of the base (u) onto a canonical k-cycle v_i in the F-homology class of u_i. This is possible by virtue of Theorem 7.2. This mapping will be isomorphically extended to the group γ defined by the base (v). So extended the mapping defines an F-isomorphism of g with γ.

It remains to show that the non-null cycles z of γ are k-caps with (finite) cap spans $> e$. Let $x_1, \cdots, x_m$ be distinct elements in the base (v), and let $\delta_1, \cdots, \delta_m$ be arbitrary non-null elements of Δ. We distinguish between two cases.

CASE I. $z = \delta_i x_i$ with $t(x_1) = \cdots = t(x_m) = \text{const.} = a$.

CASE II. Not Case I.

In Case I let

$$s = \max [s(x_1), \cdots, s(x_m)].$$

Then $r(z) = (s, a)$ since (u) is an r-proper base. In particular $t(z) = a$. But z is on F_a. Thus z is canonical so that its cycle span $s - a$ equals its cap span $S(z)$. In the special case in which the F-cycles of g are bounding, $s - a$ is finite. Hence $S(z)$ is finite in this special case.

In Case II, z has the form

$$z = y_1 + \cdots + y_n, \qquad (n > 1)$$

where y_i is a canonical k-cycle of γ of the type of z in Case I and the ranks $t_i = t(v_i)$ are different. Suppose that t_i increases with i. It follows that z is a k-cap with cap height t_n, and we infer from Lemma 10.1 that the cap span of z is at least the cycle span of v_n, that is exceeds e.

In the special case in which the F-cycles of g are bounding, we must show that

$S(z)$ is finite. But in this case the cycle span of y_i is finite. Thus y_i is bounding for each i. Hence z is bounding so that $s(z) - t(z)$ is finite. But

$$S(z) = \sigma(z) - a(z) \leqq s(z) - t(z).$$

We conclude that $S(z)$ is finite.

The proof of the theorem is complete.

We refer to the ranks $a(u)$ defined by the cap heights of k-caps u. An a-isomorphism will be called a *cap-isomorphism.* Under an a-isomorphism elements correspond which are in the same cap class.

THEOREM 10.2. *Each group g of linkable k-caps u with (finite) cap spans $> e > 0$ is cap-isomorphic with a group of (bounding) F-cycle caps with cycle spans $> e > 0$.*

The different cap heights of k-caps with spans $> e > 0$ form a discrete ascending sequence by virtue of Theorem 9.3. It follows from Lemma 2.3 that g admits an a-proper base (u). We map each k-cap u_i of (u) on a k-cycle cap v_i in the same cap class with a cycle span equal to $S(u_i)$. This is possible by virtue of Theorem 9.1. By virtue of Theorem 2.7 this mapping can be extended isomorphically so as to give a cap-isomorphism of g with the group γ generated by the set (v).

The cycle span of a k-cycle cap is at least its cap span in accordance with Lemma 9.1 (a), so that the cycle spans of γ exceed e. Moreover if the cap spans of g are finite, $S(u_i)$ in particular is finite. But the cycle span of v_i equals $S(u_i)$ so that v_i is bounding. Hence γ is a group of bounding k-cycles if the cap spans of g are finite.

The proof of the theorem is complete.

COROLLARY 10.2. *The dimension of a maximal group of linkable k-caps with finite cap spans $> e > 0$ is finite.*

This follows from Theorem 10.2 and Theorem 7.5 which states that the dimension of a maximal group of bounding F-cycles with spans $> e > 0$ is finite.

β-isomorphisms. Let g be a group of formal k-chains u. The group of boundary chains βu will be denoted by βg. The groups g and βg will be said to be β-isomorphic if g and βg are isomorphic with u in g corresponding to βu in βg.

THEOREM 10.3. *A group g of non-linkable k-caps u is β-isomorphic with its boundary group βg. The cap span of a k-cap u in g equals the cycle span of its image βu in βg.*

If u in g corresponds to βu in βg, g is mapped homeomorphically on βg. This homeomorphism is an isomorphism if whenever $\beta u = 0$, $u = 0$. If $\beta u = 0$ with $u \neq 0$, u would be linkable contrary to the nature of g. Hence $u = 0$ when $\beta u = 0$, and g is isomorphic with βg.

If $u \neq 0$,

$$s(\beta u) = a(u), \qquad t(\beta u) = \tau(u),$$

so that the cap span $a(u) - \tau(u)$ equals the cycle span

$$s(\beta u) - t(\beta u)$$

of βu. This completes the proof of the theorem.

COROLLARY 10.3. *The dimension of a maximal group g of non-linkable k-caps with spans $> e > 0$ is finite.*

For the isomorphic group βg is a group of bounding F-cycles with cycle spans $> e$, and has a finite dimension in accordance with Theorem 7.5.

THEOREM 10.4. *Each group γ of bounding F-k-cycles with cycle spans $> e > 0$ is the boundary group βg of a group g of non-linkable k-caps with spans $> e$.*

The elements u of γ have ranks $s(u)$. The dimension of γ is finite by virtue of Theorem 7.5 so that the ranks s of γ are finite in number. It follows from Lemma 2.3 that γ has an s-proper base (x). Let x be an element in (x). The F-cycle x is capped by a relative $(k + 1)$-cycle v on $F \leqq s(x)$ so that $\beta v = x$ in accordance with Theorem 9.2. Let (v) be the set of relative $(k + 1)$-cycles thereby capping the cycles x of (x). Let g be the group generated by the elements of (v). It is clear that $\beta g = \gamma$.

Let $(x_1, \cdots, x_r)$ be distinct cycles in (x) capped respectively by relative $(k + 1)$-cycles $(v_1, \cdots, v_r)$ of (v). Set

$$u = \delta_k x_k, \qquad v = \delta_k v_k \qquad (\delta_k \neq 0;\ k = 1, \cdots, r).$$

We understand that u in γ is mapped on v in g. I say that v is a non-linkable k-cap capping u. For

$$s(u) = \max\,[s(x_1), \cdots, s(x_r)]$$

since (x) is an s-proper base. Moreover v is on $F \leqq s(u)$ and $\beta v = u$. Thus v caps u by definition.

The cap span of v is

$$a(v) - \tau(v) = s(u) - t(v).$$

But $s(u) - t(u) > e$ by hypothesis, and the proof of the theorem is complete.

11. Maximal groups

We shall be concerned with the following groups ($k \geqq 0$, $e > 0$):

F_k^e, a maximal group of bounding F-cycles with cycle spans $> e$,

Λ_k^e, a maximal group of linkable k-caps with finite cap spans $> e$,

N_k^e, a maximal group of non-linkable k-caps with cap spans $> e$,

C_k^e, a maximal group of k-caps with finite cap spans $> e$.

The dimensions of the first three of these groups are finite in accordance with Theorem 7.5, and Corollaries 10.2 and 10.3. And as we have seen, rank functions $r_e(u)$, $a_e'(u)$, $a_e''(u)$ can be respectively defined for elements u of the type appearing in these three groups, and so defined satisfy the rank conditions R. Theorem 2.8 accordingly yields the following.

THEOREM 11.1. *Any group F_k^e, Λ_k^e, or N_k^e is ρ-isomorphic with any other* *group* F_k^e, Λ_k^e, *or* N_k^e *respectively, provided* $\rho(u)$ *represents the rank function appropriate to elements in the respective groups.*

COROLLARY 11.1. *The dimensions of maximal groups of the above types are independent of the choice of the maximal groups among groups of the same types.*

LEMMA 11.1. *The groups* F_k^e *and* Λ_k^e *have the same dimension.*

By virtue of Theorem 10.1, F_k^e is F-isomorphic with a group of k-caps of the type appearing in Λ_k^e so that

$$\dim F_k^e \leqq \dim \Lambda_k^e .$$

On the other hand it follows from Theorem 10.2 that Λ_k^e is cap-isomorphic with a group of F-cycles of the type appearing in F_k^e so that

$$\dim F_k^e \geqq \dim \Lambda_k^e .$$

The lemma follows directly.

LEMMA 11.2. *The groups* N_{k+1}^e *and* F_k^e *have the same dimension.*

The proof of this lemma is similar to that of the preceding. It depends upon Theorems 10.3 and 10.4.

THEOREM 11.2. *There exists a maximal group* N_{k+1}^e *of non-linkable* $(k + 1)$-*caps with spans* $> e > 0$, *such that* βN_{k+1}^e *is a group* F_k^e *and a group* Λ_k^e.

Starting with an arbitrary group F_k^e, Theorem 10.1 affirms the existence of an F-isomorphic group γ_k^e of bounding k-cycle caps with cap spans $> e$. The group γ_k^e will be a group Λ_k^e if it has the dimension of Λ_k^e. That this is the case now follows from Lemma 11.1. The group γ_k^e is thus a group F_k^e and a group Λ_k^e.

It follows from Theorem 10.4 that γ_k^e is the boundary group βM_{k+1}^e of a group of non-linkable $(k + 1)$-caps with spans $> e$. The group M_{k+1}^e will be a group N_{k+1}^e if it has the dimension of N_{k+1}^e. That this is the case follows from Lemma 11.2. Hence M_{k+1}^e is a group N_{k+1}^e. For such a choice of N_{k+1}^e, the group βN_{k+1}^e is a group F_k^e and a group Λ_k^e as stated.

The proof of the theorem is complete.

THEOREM 11.3. *Each group* C_k^e *is a direct sum of the form*

$$C_k^e = \Lambda_k^e + N_k^e , \tag{11.1}$$

where Λ_k^e *and* N_k^e *are suitably chosen groups of their respective types.*

Let $a^*(u)$ be the rank function defined by the cap heights of elements of C_k^e. The function $a^*(u)$ satisfies conditions R by virtue of Theorem 2.9. We shall now apply Theorem 4.2, taking $a^*(u)$ as the rank function ρ. Let A be the set of k-caps in C_k^e. Let B and C be respectively the a^*-invariant subsets of linkable and non-linkable k-caps of A.

Conditions (i), (ii), and (iii) preceding Theorem 4.2 here take the form:

(i). The sum of two linkable k-caps in A in different cap classes is a linkable k-cap in A.

(ii). The sum of two non-linkable k-caps in A with different cap heights is a non-linkable k-cap in A.

(iii). The sum of a linkable k-cap and a non-linkable k-cap in A is in A.

These conditions are obviously fulfilled. It follows from Theorem 4.2 that C_k^e has the form

$$C_k^e = b + c,$$

where b and c are respectively maximal groups of elements in B and C. But b and c must be groups Λ_k^e and N_k^e respectively. Otherwise one of the groups b and c, say b, will be a proper subgroup of a group b' of the type Λ_k^e. The group $b' + c$ would then be of the type of C_k^e, and contain a group of the type C_k^e as a proper subgroup. This is impossible since each group C_k^e is maximal. Thus b is a group Λ_k^e. Similarly c is a group N_k^e.

The proof of the theorem is complete.

COROLLARY 11.3. *For a fixed k and $e > 0$, the dimensions of all groups of the type C_k^e are the same.*

This is true of groups C_k^e because each group C_k^e has the form (11.1) and groups of the types Λ_k^e and N_k^e have dimensions independent of their mode of formation.

THEOREM 11.4. *If N_k^e is the maximal group of non-linkable k-caps of span $> e$ affirmed to exist in Theorem* 11.2, *the group*

$$N_k^e + \beta N_{k+1}^e \qquad (k = 0, 1, \cdots) \tag{11.2}$$

is a maximal group C_k^e of k-caps with finite spans $> e$.

By virtue of the choice of the group N_{k+1}^e, the group (11.2) is of the form

$$N_k^e + \Lambda_k^e.$$

As such it is clearly a group of k-caps with finite spans $> e$. It will be a group C_k^e if its dimension is that of a group C_k^e. That this is the case follows from Theorem 11.3 and its corollary.

Consider the special case where $0 = R_1 = R_2 = \cdots$. For $k > 0$ there will then be no non-bounding k-cycles. There can then be no linkable k-cap u with an infinite cap span. For there is an F-cycle cap v in the cap class of u, and $s(v) \geqq \sigma(u)$. If $\sigma(u) = \infty$, $s(v) = \infty$ and v is non-bounding, contrary to hypothesis. If then $0 = R_1 = R_2 = \cdots$, the word finite can be deleted from the preceding theorem for $k > 0$, giving the following corollary.

COROLLARY 11.4. *In the case in which $R_1 = R_2 = \cdots = 0$, the groups*

$$C_k^e = N_k^e + \beta N_{k+1}^e \qquad (k = 1, 2, \cdots) \tag{11.3}$$

of the theorem are maximal groups of k-caps with spans $> e$.

The case $0 = R_1 = R_2 = \cdots$ is particularly important. It occurs in the theory of minimal surfaces spanning a simple closed curve.

Let c_k^e and n_k^e be respectively the dimensions of the groups C_k^e and N_k^e of the theorem. It follows from the theorem that except for $k = 0$,

$$c_k^e = n_k^e + n_{k+1}^e \qquad (k = 0, 1, \cdots). \tag{11.4}$$

The dimension $n_0^e = 0$ since there are no non-linkable 0-caps. A necessary and sufficient condition that a sequence of non-negative integers $c_0^e, c_1^e, \cdots$ be representable in the form (11.4) with each $n_k^e \geqq 0$ is that

$$c_m^e - c_{m-1}^e + \cdots + (-1)^m c_0^e \geqq 0 \qquad (m = 0, 1, \cdots). \tag{11.5}$$

For when the numbers c_k^e are representable in the form (11.4) we find that the left member of (11.5) equals n_{m+1}^e so that (11.5) holds. Conversely if (11.5) holds for each m and we define n_{m+1}^e as the left member of (11.5) with $n_0^e = 0$, we find that (11.4) holds. Thus (11.4) and (11.5) may be regarded as equivalent sets of conditions on the numbers c_k^e.

Many other interesting relations follow from (11.4), in particular the relations

$$c_{m+1}^e + c_{m-1}^e \geqq c_m^e \qquad (m = 1, 2, \cdots). \tag{11.6}$$

In the theory of critical points of functions, (11.6) would lead to the following theorem. Among critical points "of span" $e > 0$, the "number" of critical points of types $m + 1$ and $m - 1$ is at least the number of critical points of type m. This is for the special case in which $0 = R_1 = R_2 = \cdots$. We now return to the more general case.

12. Type number relations

We were concerned in the preceding section with k-caps with finite cap spans. We now bring in k-caps of infinite cap span. Such k-caps u are linkable with $\sigma(u) = \infty$.

THEOREM 12.1. *Each maximal group P_k of k-caps with infinite cap spans is cap-isomorphic with a group γ of canonical non-bounding k-cycles containing one and only one k-cycle in each homology class.*

Any cap span which is at least 1 is infinite. Hence P_k is a maximal group of k-caps with cap spans > 1. It follows from Theorem 10.2 that P_k is cap-isomorphic with a group of k-cycle caps with cycle spans > 1. The cycles v of γ accordingly have infinite cycle spans and so are non-bounding. Thus $s(v) = \sigma(v) = \infty$. It follows from Lemma 9.1 that $t(v) = a(v)$ so that v is canonical.

Since the cycles of γ are non-bounding γ contains at most one cycle in each homology class. Let H be the set of homology classes which contain no cycle of γ. It remains to show that the set H is empty.

Let r_0 be the minimum rank of cycles in classes of H. Such a minimum exists by virtue of Theorem 7.6. Let z be a canonical k-cycle with rank r_0 and in some class of H.

Since P_k is a maximal group of k-caps with infinite cap spans there exists a k-cap y in P_k such that $z - y$ is not a k-cap with infinite cap span. Setting $a = a(z)$ we infer that

$$z - y \sim 0 \qquad (\text{d-mod } F_a).$$

Let x be the image of y in γ. Since x and y are a-homologous

$$z - x \sim 0 \qquad \text{(d-mod } F_a). \tag{12.1}$$

The cycle x is in an homology class of γ. The cycle $z - x$ is in no such class since z is in no such class. Hence $z - x$ is non-bounding. Thus $s(z - x) = \infty$, and it follows from (12.1) that $t(z - x) < a$. Hence

$$r(z - x) < r(z).$$

This is contrary to the choice of $r_0 = r(z)$ as the minimum of the ranks of cycles in classes of H. We infer that H is empty.

The proof of the theorem is complete.

COROLLARY 12.1. *If the kth connectivity p_k is infinite, there exist infinitely many distinct cap heights belonging to k-caps with infinite cap spans. These cap heights cluster only at the value $a = 1$.*

Let (∞, t_n) be the ranks r of a base of the group γ of the theorem. If the cap heights t_n assume at most a finite set of values let b be the maximum of such values. Then the cycles of γ lie on F_b. If c is a constant such that $b < c < 1$ the connectivities $R(b, c)$ would then be infinite contrary to fact.

We infer that t_n assumes infinitely many different values. It follows from Theorem 7.6 that $\text{Lim } t_n = 1$, and the proof of the corollary is complete.

THEOREM 12.2. *Let C_k^e and P_k be respectively maximal groups of k-caps with finite cap spans $> e > 0$ and with infinite cap spans. The sum*

$$g = C_k^e + P_k \tag{12.2}$$

is a maximal group of k-caps with spans $> e$.

The group g is composed of k-caps with spans $> e$. Let g^* be a maximal group of k-caps with spans $> e$, containing g. We shall prove that $g^* = g$.

To that end we apply Theorem 4.2, taking our ranks as the cap heights of caps of g^*, and identifying the set A with g^*. The sets B and C shall consist respectively of k-caps of g^* with finite and with infinite spans. Conditions (i), (ii), and (iii) of Theorem 4.2 are immediately verified. Moreover the different ranks of elements of C form an ascending sequence. It follows from Theorem 4.2 that

$$g^* = C_k^e + P_k,$$

and we conclude that $g = g^*$.

The proof of the theorem is complete.

Theorems 11.4 and 12.2 lead to the following corollary.

COROLLARY 12.2. *If N_k^e is a suitably chosen maximal group of non-linkable k-caps with cap spans $> e > 0$, then*

$$M_k^e = N_k^e + \beta N_{k+1}^e + P_k \qquad (k = 0, 1, \cdots)$$

is a maximal group of k-caps with cap spans $> e$.

Corresponding to each cap height a, the dimension of a maximal group of

k-caps with cap heights a and with spans $> e$ will be called the kth *e-type number* $m_k^e(a)$ of a. The rank function $a_e(u)$ defined by cap heights of k-caps with cap spans $> e$ satisfies conditions R. It follows from Theorem 2.1 that the number $m_k^e(a)$ is independent of the particular maximal group chosen. We set

$$m_k^e = \sum_a m_k^e(a),$$

where a ranges over all cap heights and term m_k^e the kth *e-type number sum.*

The cap heights $a_e(u)$ form a discrete ascending sequence and satisfy the rank conditions R. Hence M_k^e is the direct sum of maximal groups of k-caps with the respective cap heights $a_e(u)$ and with cap spans $> e$. It follows that

$$\dim M_k^e = m_k^e .$$

If we set

$$\dim N_k^e = n_k^e , \qquad \dim P_k = p_k ,$$

we have the following theorem.

THEOREM 12.3. *For $e > 0$, the kth e-type number sums m_k^e are given by relations of the form*

$$m_k^e = n_k^e + n_{k+1}^e + p_k \qquad (k = 0, 1, \cdots). \tag{12.3}$$

The numbers n_k^e are finite and $n_0^e = 0$. The numbers p_k are the connectivities of M in accordance with Theorem 12.1 and may be infinite. If p_k is infinite, the right member of (12.3) is regarded as infinite.

COROLLARY 12.3. *If the connectivities p_k are finite and if we set $\mu_k = m_k^e - p_k$, then*

$$\mu_n - \mu_{n-1} + \cdots + (-1)^n \mu_0 \geqq 0 \qquad (n = 0, 1, \cdots).$$

COROLLARY 12.4. *If for a particular m, the connectivities p_{m-1}, p_m, and p_{m+1} are finite,*

$$\mu_{m+1} + \mu_{m-1} \geqq \mu_m \qquad (m > 0).$$

COROLLARY 12.5. *The numbers m_k^1 give the type number sums corresponding to k-caps with infinite cap spans and are such that*

$$m_k^1 \geqq p_k \qquad (k = 0, 1, \cdots).$$

COROLLARY 12.6. *If the connectivities $p_0, \cdots, p_{m-1}$ are finite and $\mu_{m+1} = 0$,*

$$\mu_0 \geqq 0,$$

$$\mu_1 - \mu_0 \geqq 0,$$

$$\cdots\cdots\cdots\cdots$$

$$\mu_m - \mu_{m-1} + \cdots + (-1)^m \mu_0 = 0.$$

The conditions of this corollary are satisfied if the underlying space is a finite m-dimensional complex as is readily seen.

We call attention to the following important corollaries.

COROLLARY 12.7. *If for some positive e every cap of span > e is linkable, then* m_k^e *is the* k^{th} *connectivity of M.*

COROLLARY 12.8. *If every cap is linkable, then* m_k^e *is independent of e and equals the* k^{th} *connectivity of M.*

If every cap is linkable $m_k^e = m_k^\infty$, so that every cap has an infinite cap span. The connectivities then take the form

$$p_k = \sum_c m_k(c),$$

where $m_k(c)$ is the dimension of a maximal group of linkable k-caps with cap heights c. This mode of determination of p_k has sufficed in certain applications where all other methods failed.

13. k-Cap heights finite in number

The dimension $m_k(c)$ of a maximal group of k-caps with cap heights c will be called the kth *type number* of c.

THEOREM 13.1. *If a k-cap height c is bounded from* $(k - 1)$*- and* $(k + 1)$*-cap heights other than c, the* k^{th} *type number of c is finite, and no k-cap u with cap height c has a cap span* 0.

Let λ_k be a group of linkable k-caps with cap heights c. If dim λ_k is finite, one sees that λ_k is cap-isomorphic with a group of k-cycle caps u. In accordance with Lemma 9.1 each such k-cycle u is an F-cycle with

$$t(u) \leqq c, \qquad s(u) \geqq c. \tag{13.0}$$

Moreover $s(u) > c$. Otherwise $t(u) = c = s(u)$ in accordance with Lemma 9.1 (b) and u would not be an F-cycle.

Suppose there are no $(k + 1)$- or $(k - 1)$-cap heights on the closed interval $(c - e, c + e)$, possibly excepting c. Then $s(u) > c + e$, since $s(u)$ is a $(k + 1)$-cap height. The cycles u are thus on F_c and non-bounding on F_{c+e}. Hence

$$\dim \lambda_k \leqq R\,(c, c + e). \tag{13.1}$$

Let ω_k be a group of non-linkable k-caps with cap heights c. Then $\beta\omega_k$ is an isomorphic group of $(k - 1)$-cycles v with $s(v) = c$. Moreover $t(v) < c - e$ since the interval $c - e \leqq t < c$ contains no $(k - 1)$-cap heights. Let b be a constant between $c - e$ and c. Then v is on F_{c-e} and non-bounding on F_b. Hence

$$\dim \omega_k \leqq R(c - e, b). \tag{13.2}$$

If λ_k and ω_k are maximal, one sees that $\lambda_k + \omega_k$ is a maximal group of k-caps with cap heights c. But the dimension of $\lambda_k + \omega_k$ will be finite in accordance with (13.1) and (13.2). The first affirmation of the theorem is accordingly proved.

No k-cap u with $a(u) = c$ has a cap span 0. For if $S(u) = 0$, u would be linkable. It would follow from Lemma 9.2 that the cap class of u would contain at least one k-cycle cap z such that

$$s(z) < c + e,$$

with $c < s(z)$ as we have seen in connection with (13.0). But this is contrary to the nature of e since there are no $(k + 1)$-cap heights on the open interval $(c, c + e)$. Hence no k-cap u with cap height c has a cap span 0.

The proof of the theorem is complete.

LEMMA 13.1. *If the $(k - 1)$-, (k)- and $(k + 1)$-cap heights are finite in number, the cap spans of k-caps are bounded from zero.*

There will be no k-caps with cap spans 0 by virtue of the preceding theorem. Let e be a positive constant less than the absolute difference between any $(j + 1)$-cap height and distinct j-cap height, for $j = k$ and $k - 1$. Then the cap span of each k-cap is readily seen to exceed e.

The sum of the kth type numbers $m_k(c)$ of the various k-cap heights c will be called the kth *type number sum*.

THEOREM 13.2. *If the k-cap heights are finite in number for each k, the kth type number sums m_k and the kth connectivities p_k are finite. Moreover the differences $\mu_k = m_k - p_k$ satisfy the relations*

$$\mu_n - \mu_{n-1} + \cdots + (-1)^n \mu_0 \geqq 0 \qquad (n = 0, 1, \cdots). \tag{13.3}$$

For a given k and c, $m_k(c)$ is finite by virtue of Theorem 13.1. But the number of k-cap heights is finite by hypothesis so that the sum

$$m_k = \sum_c m_k(c)$$

is finite. That the connectivity p_k is finite follows at once from Corollary 12.1.

To establish the relations (13.3) let N be an arbitrary positive integer. Let e be a positive constant less than the cap spans of k-caps for which k is at most N. That such a constant e exists follows from the preceding lemma. Recalling that $m_k^e(c)$ is the kth e-type number of c we see that for $k = 0, 1, \cdots, N$,

$$m_k^e(c) = m_k(c),$$

so that

$$m_k^e = m_k .$$

But if we set $\mu_k = m_k^e - p_k$ the relations (13.3) are satisfied in accordance with Corollary 12.3. Hence (13.3) holds for $\mu_k = m_k - p_k$ and for n at most N. But N is an arbitrary positive integer. Relations (13.3) thus hold as stated.

COROLLARY 13.2. *Under the conditions of the theorem*

$$m_k \geqq p_k , \qquad (k = 0, 1, \cdots)$$

and for $k \geqq 0$,

$$\mu_{k+1} + \mu_{k-1} \geqq \mu_k \qquad (\mu_{-1} = 0).$$

THE INSTITUTE FOR ADVANCED STUDY.

Reprinted from Duke Mathematical Journal
Vol. 6, No. 2, June, 1940

THE FIRST VARIATION IN MINIMAL SURFACE THEORY

By Marston Morse

Introduction. We are concerned with harmonic surfaces S bounded by $n + 1$ non-intersecting simple closed curves[1] γ_k in a Euclidean m-space (x). The coördinates x^j on such surfaces, which are necessarily orientable and of genus 0, shall be harmonic functions of parameters (u, v) ranging over a connected region[2] B in the (u, v)-plane bounded by $n + 1$ circles $C_0, \cdots, C_n$. Let (r_k, θ_k) be polar coördinates in the (u, v)-plane with pole at the center (u_k, v_k) of C_k. Let $g_k(\theta_k)$ be a vector[3] defined on C_k giving an admissible representation of γ_k $(k = 0, 1, \cdots, n)$. Let $h(u, v)$ be the vector defining S. We suppose that $h(u, v)$ is continuous on the closure $\bar{B}$ of B and that its boundary values, represented in terms of the respective angular coördinates θ_k, determine admissible representations of the curves (γ) of the form

$$[g_0(\theta_0), \cdots, g_n(\theta_n)] = (g). \tag{0.1}$$

Conversely, each admissible representation (0.1) of the set (γ) determines a harmonic surface S defined over B. We admit no other surfaces.

Let $h(u, v)$ represent the harmonic surface defined by an admissible representation (g) of the curves (γ) and let $D(g)$ be one-half the sum (finite or infinite) of the classical[4,5] Dirichlet integrals of the components $h^i(u, v)$ of $h(u, v)$. We term $D(g)$ the *Dirichlet sum.* The radii σ_k of the circles C_k and the coördinates (u_k, v_k) of their centers will be called the *circle parameters* of B. Set[6]

$$(\sigma_1, u_1, v_1, \cdots, \sigma_n, u_n, v_n) = (\eta).$$

Received April 10, 1940.

[1] We start with a 1-1 representation of γ_k on a circle C. We shall admit any other representation of γ_k which is obtained from the given representation by a monotone transformation (not necessarily 1-1) of C into itself.

[2] Although we restrict ourselves to surfaces of the topological type of B, the methods introduced are of such general character as to admit extension to the most general type of representation. For such extensions it is sufficient that the boundaries γ_k of the type regions B be circles. It is even sufficient that there exist a conformal map of a neighborhood of γ_k on B into an annular region, γ_k going into a circle under this conformal map. The introduction of subregions of multiply-sheeted Riemann surfaces naturally makes no difficulty.

[3] Curves and surfaces in our Euclidean m-space will invariably be represented as vectors as will the various Fourier coefficients which will presently enter.

[4] Jesse Douglas I: *Solution of the problem of Plateau,* Transactions of the American Mathematical Society, vol. 33(1931), pp. 263–321. Douglas II: *The problem of Plateau for two contours,* Journal of Mathematics and Physics, vol. 10(1931), pp. 315–359. Douglas III: *Minimal surfaces of higher topological structure,* Transactions of the American Mathematical Society, vol. 39(1939), pp. 205–298. Various additional references are here given.

[5] T. Radó, *On the problem of Plateau,* Ergebnisse der Mathematik, vol. 2(1933).

[6] We do not include the parameters (σ_0, u_0, v_0) in the set (η) because without changing the value of $D(g)$, C_0 can be taken as the unit circle with center at the origin.

We shall regard D not only as a function of (g), but also as a function of (η). The representation (g) will be a representation on the respective circles defined by (η). The fundamental theorem is as follows.

A necessary and sufficient condition that a harmonic surface, for which D is finite, be minimal is that the first variation of D be identically zero.

We understand thereby that the first variation vanishes for a class of variations now to be defined.

Let (g) be an admissible representation of the circles (γ) in terms of the respective angles $(\theta_0, \cdots, \theta_n) = (\theta)$. We shall admit transformations of θ_k of the form

$$\mu_k = \theta_k + e_k\lambda_k(\theta_k), \tag{0.2}$$

where $\lambda_k(\theta_k)$ is analytic and has a period 2π, and e_k is a constant restricted by a condition of the form

$$| e_k | \leqq e_k(\lambda_k) < 1 \tag{0.3}$$

to be so small that for each value of θ_k

$$1 + e_k\lambda_k'(\theta_k) \neq 0. \tag{0.4}$$

Here $e_k(\lambda_k)$ is a constant dependent on λ_k. Subject to (0.2) we obtain a new[7] representation (p) of (γ) such that

$$p_k(\mu_k, e_k) \equiv g_k(\theta_k). \tag{0.5}$$

The boundary vectors $[p(\theta, e)]$ define a new harmonic surface $H(u, v, e)$. The corresponding Dirichlet sum $D(p)$ will be written as a function

$$D(g, \eta, \lambda, e), \qquad | e_k | \leqq e_k(\lambda_k), \tag{0.6}$$

where

$$(e_0, \cdots, e_n) = (e), \qquad (\lambda_0, \cdots, \lambda_n) = (\lambda).$$

The differential

$$V = \sum_i D_{\eta_i} d\eta_i + \sum_k D_{e_k} de_k \qquad (i = 1, \cdots, 3n; k = 0, \cdots, n), \tag{0.7}$$

evaluated[8] at $e = 0$, will be called the *first variation* of D at (g, η), and the coefficients in this differential will be called the *variational coefficients.* We shall say that the first variation of D at (g, η) is identically zero if these coefficients vanish for all admissible sets (λ) and for the given set (g, η). It is in this sense that the fundamental theorem is to be understood.

This theorem has been completely proved only for the case of one boundary

[7] We term (p) the (λ, e) "transform" of (g).

[8] Subscripts used to indicate partial differentiation shall always refer to the *original* arguments of the function. Thus if $f(x, y)$ is given, $f_x(x, x)$ means the x-derivative of $f(x, y)$ evaluated for $y = x$.

curve. The only proof known to the author is that[9] in Douglas I. When the class of admissible surfaces is enlarged to include surfaces which are no longer harmonic, Radó (loc. cit.), Courant[10] and others[11] have given proofs of the corresponding theorem. But the theorem here stated does not follow in any immediate way from the proofs by Radó and Courant. If one adds the hypothesis that the given surface is a minimizing surface, then connections can be easily made between the two types of proof, but the requirements of the theory in the large preclude such a hypothesis. This is not to say that it is impossible to proceed in the general theory except in the class of harmonic surfaces, but rather to say that in the present state of development, the class of harmonic surfaces forms the simplest and most elegant medium for the general theory.

Our proof of the theorem may be briefly described as follows. The harmonic vector $h(u, v)$ is the real part of a vector $F(w)$, whose components are analytic functions of the complex variable $w = u + iv$, the imaginary part in general being multiple-valued. With Weierstrass we introduce the scalar product[12]

$$f(w) = F'(w) \cdot F'(w), \tag{0.8}$$

whose vanishing is the condition that $h(u, v)$ be minimal. In case the boundary representation (g) is analytic, classical methods yield formulas for the variational coefficients involving $f(w)$ and (g, η, λ), and the theorem follows with ease. Our contribution is to show that when $D(g)$ is finite the variational coefficients are *continuous* in (g, η, λ) provided the metric for the sets (g, η, λ) is suitably defined. An obvious limiting process then enables us to infer the theorem in the general case as a consequence of the formulas in the special analytic case.

1. **Continuity of the variational coefficient when** $n = 0$. When $n = 0$, there is just one boundary curve $\gamma_0 = \gamma$ represented on the circle $C_0 = C$. Without loss of generality we can suppose that C is the unit circle with center at the origin. For the Dirichlet sum $D(g)$ is independent of the position of the circle C since any circle is obtainable from any other by conformal transformations of the plane. The circle parameters (η) do not enter in the case $n = 0$.

Let (r, θ) be polar coördinates with pole at the origin, and let $g(\theta)$ be an admis-

[9] In the case of two contours (Douglas II) there is a serious gap on page 328 in inferring the fundamental formula (2.18) from the preceding formula. A similar difficulty arises on page 248 of Douglas III. That this gap can be bridged by an extended use of the appropriate theta functions seems most likely. Douglas is of like opinion. Our proofs make no use of theta functions.

[10] R. Courant, *Plateau's problem and Dirichlet's principle*, Annals of Mathematics, vol. 38(1937), pp. 679–724.

[11] M. Shiffman, *The Plateau problem for non-relative minima*, Annals of Mathematics, vol. 40(1939), pp. 834–854.

[12] We term $f(w)$ the *Weierstrass modul* belonging to the region B and the boundary values (g).

sible representation of γ as an image of C. As shown in Douglas I the Dirichlet sum $D(g)$ has the form

$$(1.0) \qquad A(g) = \frac{1}{16\pi} \iint_T \frac{[g(\alpha) - g(\beta)]^2}{\sin^2 [\frac{1}{2}(\alpha - \beta)]} d\alpha\, d\beta,$$

where T is the domain[13]

$$(1.1) \qquad \begin{aligned} 0 \leqq \alpha \leqq 2\pi, \\ \beta - \pi \leqq \alpha \leqq \beta + \pi. \end{aligned}$$

The integral $A(g)$ has a singularity on T when $\alpha = \beta$. Let T_m be the subdomain of T on which

$$| \alpha - \beta | > \frac{1}{m},$$

and let $A^{(m)}(g)$ be the value of the integral (1.0) when T is replaced by T_m. By definition

$$A(g) = \lim_{m=\infty} A^{(m)}(g).$$

In the case $n = 0$ there is but one variational coefficient, namely D_e. We shall obtain a formula for this coefficient. Let $\lambda(\theta)$ be real and analytic in θ with a period 2π. Set

$$(1.2) \qquad \mu = \theta + e\lambda(\theta), \qquad | e | \leqq e(\lambda),$$

where $e(\lambda)$ is a positive constant so small that the θ-derivative of the right member of (1.2) never vanishes. Under (1.2) let

$$p(\mu, e) \equiv g(\theta),$$

and let $H(u, v, e)$ be the harmonic surface determined by (p) for each fixed e. We have

$$H(\cos \mu, \sin \mu, e) \equiv p(\mu, e).$$

The corresponding integral $A^{(m)}(p)$ has the form

$$(1.3) \qquad A^{(m)}(p) = \frac{1}{16\pi} \iint_{T_m} \frac{[p(\mu, e) - p(\nu, e)]^2}{\sin^2 [\frac{1}{2}(\mu - \nu)]} d\mu\, d\nu.$$

Upon setting

$$(1.4) \qquad \mu = \alpha + e\lambda(\alpha), \qquad \nu = \beta + e\lambda(\beta),$$

[13] Douglas takes T as the domain $0 \leqq \alpha \leqq 2\pi$, $0 \leqq \beta \leqq 2\pi$. This yields the same value of $D(g)$ because of the periodicity of the integrand. From the point of view of a treatment of the singularities of the integrand our choice of T seems simpler.

(1.3) takes the form

$$A^{(m)}(p) = \frac{1}{16\pi} \iint_{T_m} \frac{[g(\alpha) - g(\beta)]^2 [1 + e\lambda'(\alpha)][1 + e\lambda'(\beta)]}{\sin^2 [\frac{1}{2}(\alpha - \beta) + \frac{1}{2}e(\lambda(\alpha) - \lambda(\beta))]} \, d\alpha \, d\beta. \tag{1.5}$$

We shall assume that $A(g)$ is finite and obtain a formula for $A(p) - A(g)$. To that end we shall analyze the denominator of the integrand in (1.5).

Set

$$\omega = \tfrac{1}{2}[\lambda(\alpha) - \lambda(\beta)],$$

and note that on T, for $\alpha \neq \beta$ and $e \neq 0$,

$$\begin{aligned} &\sin [\tfrac{1}{2}(\alpha - \beta) + e\omega] \\ &= \sin \tfrac{1}{2}(\alpha - \beta) \left[\cos e\omega + e \cos \tfrac{1}{2}(\alpha - \beta) \frac{\sin e\omega}{e \sin \frac{1}{2}(\alpha - \beta)} \right]. \end{aligned} \tag{1.6}$$

Observe that

$$\frac{\sin e\omega}{e \sin \frac{1}{2}(\alpha - \beta)} = \frac{\sin e\omega}{e\omega} \frac{\omega}{\alpha - \beta} \frac{\alpha - \beta}{\sin \frac{1}{2}(\alpha - \beta)}. \tag{1.7}$$

Recall that $\lambda(\alpha)$ is analytic. The form of the right member of (1.7) shows that the function thereby defined is analytic in (α, β, e) for (α, β) on T, except for removable singularities when $\alpha = \beta$ or $e = 0$. Hence (1.6) yields the relation

$$\sin [\tfrac{1}{2}(\alpha - \beta) + e\omega] = \sin \tfrac{1}{2}(\alpha - \beta) [1 + ek(\alpha, \beta, e)], \tag{1.8}$$

where $k(\alpha, \beta, e)$ is analytic in its arguments. Upon differentiating the respective members of (1.8) with respect to e and setting $e = 0$, we find that

$$\omega \cos \tfrac{1}{2}(\alpha - \beta) = k(\alpha, \beta, 0) \sin \tfrac{1}{2}(\alpha - \beta), \tag{1.9}$$

a fact of use later.

To continue, we need to know that on T,

$$1 + ek(\alpha, \beta, e) \neq 0, \qquad |e| \leqq e(\lambda). \tag{1.10}$$

We can infer this from (1.8). For

$$\sin [\tfrac{1}{2}(\alpha - \beta) + e\omega] = 0$$

only if

$$\tfrac{1}{2}(\alpha - \beta) + e\omega \equiv 0 \qquad (\text{mod } \pi),$$

or only if

$$\alpha + e\lambda(\alpha) \equiv \beta + e\lambda(\beta) \qquad (\text{mod } 2\pi). \tag{1.11}$$

Since $\alpha + e\lambda(\alpha)$ is an increasing function of α for $|e| \leqq e(\lambda)$, (1.11) holds only if $\alpha \equiv \beta$ (mod 2π). But (1.10) holds even when $\alpha \equiv \beta$, as one infers from the first order vanishing of the left member of (1.8) when $\alpha \equiv \beta$.

By virtue of the binomial theorem we can write

$$
\begin{aligned}
(1.12)\qquad [1 + ek(\alpha, \beta, e)]^{-2} &= 1 - 2ek(\alpha, \beta, e) + 3e^2k^2(\alpha, \beta, e) + \cdots \qquad (|e| \leqq e(\lambda)) \\
&= 1 - 2ek(\alpha, \beta, 0) + e^2B(\alpha, \beta, e),
\end{aligned}
$$

where $B(\alpha, \beta, e)$ is analytic in its arguments. For brevity we set

$$\frac{[g(\alpha) - g(\beta)]^2}{\sin^2[\frac{1}{2}(\alpha - \beta)]} = K(\alpha, \beta).$$

Referring to (1.5), (1.8) and (1.12), we see that

$$
\begin{aligned}
(1.13)\qquad A^{(m)}(p) &= \frac{1}{16\pi}\iint_{T_m} K(\alpha, \beta)[1 - 2ek(\alpha, \beta, 0) + e^2B(\alpha, \beta, e)] \\
&\qquad\qquad \cdot[1 + e\lambda'(\alpha)][1 + e\lambda'(\beta)]\, d\alpha\, d\beta \\
&= A^{(m)}(g) + eS^{(m)}(g, \lambda) + e^2r^{(m)}(g, \lambda, e),
\end{aligned}
$$

where

$$(1.14)\qquad S^{(m)}(g, \lambda) = \frac{1}{16\pi}\iint_{T_m} K(\alpha, \beta)[\lambda'(\alpha) + \lambda'(\beta) - 2k(\alpha, \beta, 0)]\, d\alpha\, d\beta,$$

$$(1.15)\qquad r^{(m)}(g, \lambda, e) = \frac{1}{16\pi}\iint_{T_m} K(\alpha, \beta)W(\alpha, \beta, e)\, d\alpha\, d\beta,$$

and where $W(\alpha, \beta, e)$ is analytic for $|e| \leqq e(\lambda)$.

The integral

$$(1.16)\qquad A^{(m)}(g) = \frac{1}{16\pi}\iint_{T_m} K(\alpha, \beta)\, d\alpha\, d\beta$$

converges by hypothesis as m becomes infinite. The integrals $S^{(m)}(g, \lambda)$ and $r^{(m)}(g, \lambda, e)$ also converge, as one sees by comparison with the integral $A^{(m)}(g)$, noting that the factors

$$|\lambda'(\alpha) + \lambda'(\beta) - 2k(\alpha, \beta, 0)|, \qquad |W(\alpha, \beta, e)|$$

are bounded. Upon letting m become infinite in (1.13), we infer that $A(p)$ is finite and that

$$(1.17)\qquad A(p) = A(g) + eS(g, \lambda) + e^2r(g, \lambda, e),$$

where $S(g, \lambda)$ and $r(g, \lambda, e)$ are of the form (1.14) and (1.15) respectively with T_m replaced by T.

We note that

$$r(g, \lambda, e) \leqq LA(g),$$

where L is a bound for $|W(a, b, e)|$. Recalling that $A(g)$ is finite, we see that

$$A_e(p) = S(g, \lambda) \qquad (e = 0).$$

Upon referring to (1.9) and (1.14), we obtain the following theorem.

THEOREM 1.1. *When $A(g)$ is finite, the variational coefficient $A_e(p)$ has the form*

$$S(g, \lambda) = \frac{1}{16\pi} \iint_T \frac{[g(\alpha) - g(\beta)]^2}{\sin^2 [\frac{1}{2}(\alpha - \beta)]} E(\alpha, \beta)\, d\alpha\, d\beta,$$

where

$$E(\alpha, \beta) = \lambda'(\alpha) + \lambda'(\beta) - [\lambda(\alpha) - \lambda(\beta)] \cot [\tfrac{1}{2}(\alpha - \beta)] \qquad (\alpha \not\equiv \beta, \bmod 2\pi).$$

We shall show that the integrand of $S(g, \lambda)$ has a removable singularity when $\alpha = \beta$. On T set

$$A(\alpha, \beta) = \tfrac{1}{2}(\alpha - \beta) \cot [\tfrac{1}{2}(\alpha - \beta)] \qquad (\alpha \neq \beta) \tag{1.18}$$

with $A(\alpha, \alpha) \equiv 1$. So defined $A(\alpha, \beta)$ is analytic. Moreover, $A(\alpha, \beta)$ is an even function of $(\alpha - \beta)$ so that $A_\beta(\alpha, \alpha) \equiv 0$.

We have

$$\lambda(\alpha) - \lambda(\beta) = (\alpha - \beta) \int_0^1 \lambda'[\alpha + t(\beta - \alpha)]\, dt,$$

and making use of (1.9) and (1.18) we see that

$$E(\alpha, \beta) = \lambda'(\alpha) + \lambda'(\beta) - 2A(\alpha, \beta) \int_0^1 \lambda'[\alpha + t(\beta - \alpha)]\, dt. \tag{1.19}$$

Using the relations $A(\alpha, \alpha) = 1$, $A_\beta(\alpha, \alpha) = 0$, we find that

$$E(\alpha, \alpha) = 0, \qquad E_\beta(\alpha, \alpha) = 0.$$

We represent $E(\alpha, \beta)$ by Taylor's formula, regarding $E(\alpha, \beta)$ as a function of β, and expanding $E(\alpha, \beta)$ about the point $\beta = \alpha$ with a remainder as a term of the second order. Thus

$$E(\alpha, \beta) = \tfrac{1}{2}(\alpha - \beta)^2 \int_0^1 E_{\beta\beta}[\alpha, \alpha + t(\beta - \alpha)]\, dt.$$

Upon setting

$$\tfrac{1}{2}(\alpha - \beta) \{\sin [\tfrac{1}{2}(\alpha - \beta)]\}^{-1} = M(\alpha, \beta) \qquad (\alpha \neq \beta)$$

with $M(\alpha, \alpha) = 1$, we see that $M(\alpha, \beta)$ is analytic on T, and that

$$S(g, \lambda) = \frac{1}{8\pi} \iint_T \left\{[g(\alpha) - g(\beta)]^2 M^2(\alpha, \beta) \int_0^1 E_{\beta\beta}[\alpha, \alpha + t(\beta - \alpha)]\, dt\right\} d\alpha\, d\beta. \tag{1.20}$$

It appears from (1.19) that $E_{\beta\beta}$ depends on λ and its first three derivatives. To two admissible functions λ, say λ and λ^*, we shall assign a distance $\lambda\lambda^*$ equal to the maximum value of

$$(1.21)\qquad \begin{aligned} &| \lambda(\theta) - \lambda^*(\theta) | + | \lambda'(\theta) - \lambda^{*\prime}(\theta) | + | \lambda''(\theta) - \lambda^{*\prime\prime}(\theta) | \\ &\qquad + | \lambda'''(\theta) - \lambda^{*\prime\prime\prime}(\theta) |. \end{aligned}$$

To two boundary vectors $g(\theta)$ and $g^*(\theta)$ we shall assign a distance gg^* equal to the maximum of the vector length

$$(1.22)\qquad | g(\theta) - g^*(\theta) |.$$

In the space of the sets (g, λ), distance will be defined by the sum

$$gg^* + \lambda\lambda^*.$$

A function of the sets (g, λ) will be regarded as continuous if continuous with respect to the metric of the sets (g, λ). Upon referring to (1.19), we see that $E_{\beta\beta}$ is continuous in (λ). Making use of (1.20), we obtain the following basic theorem.[14]

THEOREM 1.2. *The functional $S(g, \lambda)$ is continuous in its arguments, and when $A(g)$ is finite, equals the derivative $A_e(p)$ evaluated when $e = 0$.*

The functional $S(g, \lambda)$ is thus well defined and continuous even when $A(g)$ is infinite, and in many ways is more amenable to analysis than $A(g)$. The implications of this fact are very significant, and do not seem to have been sufficiently exploited.

2. The fundamental theorem when $n = 0$. We continue with the one-parameter family $H(u, v, e)$ of harmonic surfaces of §1, defined as in §1 by the (λ, e) "transform" $p(\theta, e)$ of $g(\theta)$. It will be convenient, however, to suppose that C is a circle with arbitrary radius a and center at the origin. Recall that $p(\theta, 0) \equiv g(\theta)$. We shall begin with the case where the components of $g(\theta)$ are analytic in θ. The vector $H(u, v, e)$ will be analytic on $\bar{B}$, and in particular will possess continuous partial derivatives on $\bar{B}$ of all orders. We have

$$A(p) = A(g, \lambda, e) = \frac{1}{2}\iint_B \left\{\left(\frac{\partial H}{\partial u}\right)^2 + \left(\frac{\partial H}{\partial v}\right)^2\right\} du\, dv.$$

If we differentiate under the integral sign with respect to e, integrate by parts in the classical way, and finally make use of the fact that H is harmonic, we find that

$$(2.1)\qquad A_e = \int_C \frac{\partial H}{\partial e}\left(\frac{\partial H}{\partial u}\, dv - \frac{\partial H}{\partial v}\, du\right).$$

[14] Cf. M. Morse and C. B. Tompkins, *The existence of minimal surfaces of general critical types*, Annals of Mathematics, vol. 40(1939), pp. 443–472.

In terms of the polar coördinates (r, θ), (2.1) yields the formula

$$A_e = \int_0^{2\pi} \frac{\partial H}{\partial e} \frac{\partial h}{\partial r} r\, d\theta, \qquad (r = a, e = 0), \tag{2.2}$$

where $h(u, v) = H(u, v, 0)$.

Recall that

$$p(\mu, e) \equiv g(\theta), \tag{2.3}$$

subject to the relation

$$\mu = \theta + e\lambda(\theta), \qquad |e| \leqq e(\lambda). \tag{2.4}$$

In (2.3) and (2.4) we regard e and μ as independent and θ as dependent, and differentiate with respect to e, holding μ fast. We then set $e = 0$. Thus

$$p_e(\mu, 0) = g'(\theta) \frac{\partial \theta}{\partial e},$$

$$0 = \frac{\partial \theta}{\partial e} + \lambda(\theta) \qquad (e = 0).$$

But

$$H(a \cos \mu, a \sin \mu, e) \equiv p(\mu, e),$$

so that

$$\frac{\partial H}{\partial e}\Big|^{e=0} = p_e(\mu, 0) = g'(\theta) \frac{\partial \theta}{\partial e} = -g'(\theta)\lambda(\theta) \qquad (r = a).$$

Finally

$$h(a \cos \theta, a \sin \theta) \equiv g(\theta),$$

so that when $r = a$

$$\frac{\partial h}{\partial \theta} = g'(\theta), \qquad \frac{\partial H}{\partial e} = -\frac{\partial h}{\partial \theta} \lambda(\theta) \qquad (e = 0).$$

Hence (2.2) takes the form

$$A_e = -\int_0^{2\pi} \lambda(\theta) \frac{\partial h}{\partial \theta} \frac{\partial h}{\partial r} r\, d\theta \qquad (r = a, e = 0). \tag{2.5}$$

In terms of the function $F(w)$ of the introduction, $h(u, v) = RF(w)$. Recall that

$$F'(w) = e^{-i\theta}\left(\frac{\partial h}{\partial r} - \frac{i}{r} \frac{\partial h}{\partial \theta}\right) \qquad (r \neq 0). \tag{2.6}$$

We introduce the "Weierstrass modul" $f(w) = F'(w) \cdot F'(w)$. Multiplying both sides of (2.6) by $w = re^{i\theta}$ and taking the scalar product of the respective members by themselves, we find that

$$w^2 f(w) = \left(r\frac{\partial h}{\partial r} - i\frac{\partial h}{\partial \theta}\right)^2. \tag{2.6)'$$

The imaginary parts of the respective members take the form

$$I[w^2 f(w)] = -2r\frac{\partial h}{\partial r}\frac{\partial h}{\partial \theta}.$$

Referring to (2.5), we have the following lemma.

LEMMA 2.1. *When the boundary vector $g(\theta)$ is analytic, the variational coefficient $A_e(g, \lambda, 0)$ takes the form*

$$A_e(g, \lambda, 0) = \frac{1}{2}\int_0^{2\pi} \lambda(\theta) I[w^2 f(w)]\, d\theta \qquad (r = a), \tag{2.7}$$

where $f(w)$ is the Weierstrass modul (0.8) belonging to the circular region bounded by C and to the boundary vector $g(\theta)$.

We turn to the case where $g(\theta)$ is merely continuous. Let $h^*(u, v)$ be the vector which is harmonic for $r < 1$, continuous for $r \leqq 1$, and such that

$$h^*(\cos\theta, \sin\theta) \equiv g(\theta).$$

Set

$$g_a(\theta) = h^*(a\cos\theta, a\sin\theta) \qquad (0 < a \leqq 1). \tag{2.8}$$

The vector $g_a(\theta)$ is analytic for $a < 1$. Let $z = \rho e^{\psi i}$ be an arbitrary complex point with $\rho < 1$. Hold z fast and set

$$\lambda_a(\theta) = \frac{1}{\pi}\frac{a^2 - \rho^2}{a^2 - 2a\rho\cos(\theta - \psi) + \rho^2} \qquad (\rho < a). \tag{2.9}$$

We have in $\lambda_a(\theta)$ the Poisson kernel for the circle $r = a$. In Lemma 2.1 take C as a circle $r = a < 1$. From (2.7) and the properties of the Poisson integral we find that

$$A_e(g_a, \lambda_a, 0) = I[z^2 f_a(z)],$$

where $f_a(z)$ is the Weierstrass modul belonging to the region $|z| < a$ and to the boundary vector $g_a(\theta)$. But $f_a(z) \equiv f_1(z)$, so that

$$A_e(g_a, \lambda_a, 0) = I[z^2 f_1(z)]. \tag{2.10}$$

For z fixed, the right member of (2.10) is independent of a, while the left member of (2.8) is continuous in its arguments by virtue of Theorem 1.2. Upon letting a tend to 1 as a limit, we obtain the following theorem.

THEOREM 2.1. *Corresponding to an arbitrary choice of the complex point z with $|z| < 1$ and choice of λ as the Poisson kernel (2.9) for the unit circle with pole $\rho e^{\psi i} = z$,*

$$A_e(g, \lambda, 0) = I[z^2 f(z)],$$

where $f(z)$ is the Weierstrass modul belonging to the region $|z| < 1$ and to the boundary vector $g(\theta)$.

The fundamental theorem of the introduction can now be established.

Suppose first that $A_e(g, \lambda, 0) = 0$ for every admissible λ. In particular λ can be chosen as in the preceding theorem. Hence $I[z^2 f(z)] = 0$ at each point $|z| < 1$. It follows that $z^2 f(z)$ is a constant. Upon setting $z = 0$, we see that this constant is null. Hence $f(z) \equiv 0$, and the harmonic surface $h^*(u, v)$ defined by g for $r \leqq 1$ is minimal.

Conversely, suppose that $h^*(u, v)$ is minimal. Then $f(z) \equiv 0$. We shall make use of (2.7), taking g as the function g_a of (2.8), $a < 1$, and recalling that $f_a(w) \equiv f(w)$. We infer that

$$A_e(g_a, \lambda, 0) = 0$$

for each admissible λ. Upon letting a tend to 1 and using Theorem 1.2, we infer that

$$A_e(g, \lambda, 0) = 0,$$

as required.

3. A series for $D(g)$. We return to the notation of the introduction. We suppose that C_0 is the unit circle and contains B in its interior. Let

$$(g_0, g_1, \cdots, g_n) = (g)$$

be the given boundary vectors, and $h(u, v)$ the harmonic surface thereby defined over B.

Let Γ_k be a circle on B concentric with C_k, with a radius ρ_k. Let R_k be the annular region bounded by C_k and Γ_k $(k = 0, \cdots, n)$. Set $w = u + iv$. The harmonic vector $h(u, v)$ is the real part of an analytic vector function $F(w)$. The function $F'(w)$ is single-valued and so admits a Laurent expansion on R_k in powers of $w - w_k$, where w_k is the center of C_k. From this expansion for $F'(w)$ we find that $F(w)$ has the form

$$F(w) = c_k \log (w - w_k) + \sum_m \gamma_k(m)(w - w_k)^m \qquad (m = 0, \pm 1, \pm 2, \cdots)$$

on R_k. Since $RF(w)$ is single-valued, we infer that c_k is real. We set[15]

$$\gamma_k(m) = A_k(m) - iB_k(m) \qquad (k = 0, \cdots, n),$$

[15] Cf. Douglas II, p. 338 for a similar procedure.

where $A_k(m)$ and $B_k(m)$ are real. Then on R_k,

$$\text{(3.1)}\quad \begin{aligned} h(u, v) &= c_k \log r_k + A_k(0) + \sum_{m=1}^{\infty} r_k^m [A_k(m) \cos m\theta_k + B_k(m) \sin m\theta_k] \\ &\quad + \sum_{m=1}^{\infty} r_k^{-m} [A_k(-m) \cos m\theta_k - B_k(-m) \sin m\theta_k]. \end{aligned}$$

Let τ_k be taken between ρ_k and σ_k, the radius of C_k, and set

$$\text{(3.2)}\quad (\tau_0, \tau_1, \cdots, \tau_n) = (\tau).$$

Let $B(\tau)$ be a subregion of B in which for each k the boundary circle C_k of B is replaced by a concentric circle $C_k(\tau)$ of radius τ_k. Let

$$\text{(3.3)}\quad d(\tau) = \frac{1}{2} \iint_{B(\tau)} \left\{ \left(\frac{\partial h}{\partial u}\right)^2 + \left(\frac{\partial h}{\partial v}\right)^2 \right\} du\, dv,$$

so that

$$\text{(3.4)}\quad D(g) = \lim_{(\tau) = (\sigma)} d(\tau).$$

Upon integrating the right member of (3.3) by parts in the usual way and making use of the fact that $h(u, v)$ is harmonic, we find that

$$d(\tau) = \frac{1}{2} \sum_k \int_{C_k(\tau)} h \left(\frac{\partial h}{\partial u} dv - \frac{\partial h}{\partial v} du \right),$$

where the sense of integration on C_0 is counterclockwise, and along $C_1, \cdots, C_n$, clockwise. We introduce the multipliers

$$\epsilon_0 = 1, \qquad \epsilon_1 = \cdots = \epsilon_n = -1,$$

and write $d(\tau)$ in the form

$$d(\tau) = \frac{1}{2} \sum_k \int_0^{2\pi} \epsilon_k h \frac{\partial h}{\partial r_k} r_k\, d\theta_k \Big|^{r_k = \tau_k} \qquad (k = 0, \cdots, n),$$

$$\text{(3.5)}\quad d(\tau) = \frac{1}{4} \sum_k \epsilon_k \tau_k \frac{\partial}{\partial r_k} \int_0^{2\pi} h^2\, d\theta_k \Big|^{r_k = \tau_k}.$$

To evaluate[15] the integrals in (3.5) we make use of Parseval's theorem, regarding (3.1) as a Fourier development with fixed r_k. For r_k between ρ_k and σ_k,

$$\text{(3.6)}\quad \begin{aligned} \int_0^{2\pi} h^2\, d\theta_k &= 2\pi [c_k \log r_k + A_k(0)]^2 \\ &\quad + \pi \sum_{m=1}^{\infty} \{ r_k^{2m} [A_k^2(m) + B_k^2(m)] + 2A_k(m) A_k(-m) \\ &\quad - 2B_k(m) B_k(-m) + r_k^{-2m} [A_k^2(-m) + B_k^2(-m)] \}. \end{aligned}$$

The right member of (3.6) will be shown to converge uniformly for r_k on any closed subinterval I_k of the values between ρ_k and σ_k exclusive. For the subseries of terms in (3.6) which involve *positive* powers of r_k converges for r_k between ρ_k and σ_k since one can apply Parseval's theorem to the sum of the corresponding terms in (3.1). Hence this subseries converges uniformly on I_k. Similarly, the subseries of (3.6) involving negative powers of r_k converges uniformly on I_k, and so (3.6) can be differentiated termwise with respect to r_k.

It follows from (3.5) and (3.6) that

$$d(\tau) = \pi \sum_k \epsilon_k c_k[c_k \log \tau_k + A_k(0)]$$

$$(3.7) \qquad + \pi \sum_m \tfrac{1}{2}m \sum_k \epsilon_k\{[A_k^2(m) + B_k^2(m)]\tau_k^{2m} - [A_k^2(-m) + B_k^2(-m)]\tau_k^{-2m}\}$$

$$(k = 0, 1, \cdots, n;\ m = 1, 2, \cdots).$$

Consider the subseries $T(\tau)$ of (3.7) whose m-th term is

$$T(m, \tau) = \tfrac{1}{2}\pi m\left\{[A_0^2(m) + B_0^2(m)]\tau_0^{2m} + \sum_{k=1}^{n} [A_k^2(-m) + B_k^2(-m)]\tau_k^{-2m}\right\},$$

and the residual subseries,

$$S(\tau) = \pi \sum_{k=0}^{n} \epsilon_k c_k[c_k \log \tau_k + A_k(0)]$$

$$- \pi \sum_m \tfrac{1}{2}m\left\{\sum_{k=1}^{n} [A_k^2(m) + B_k^2(m)]\tau_k^{2m} + [A_0^2(-m) + B_0^2(-m)]\tau_0^{-2m}\right\}.$$

Let J_k be a closed interval of values of τ_k between σ_k and ρ_k which includes σ_k as an end point but which excludes ρ_k. We wish to show that $T(\tau)$ converges uniformly for τ_k on J_k. Now $S(\tau)$ converges uniformly on J_k, as one sees upon comparison with the convergent series Σ of corresponding terms in (3.1), or better, with the series obtained by applying Parseval's theorem to Σ. Hence

$$(3.8) \qquad |S(\tau)| \leqq M \qquad (\tau_k \text{ on } J_k,\ k = 0, 1, \cdots, n),$$

where M is a constant independent of (τ). We note that $T(m, \tau) \geqq 0$, and conclude from (3.7) that

$$(3.9) \qquad \sum_{m=1}^{\infty} T(m, \tau) \leqq d(\tau) + M \leqq D(g) + M.$$

From the continuity of $T(m, \tau)$ we infer that

$$(3.10) \qquad \sum_{m=1}^{\infty} T(m, \sigma) \leqq D(g) + M.$$

But for τ_k on J_k, $T(m, \tau) \leqq T(m, \sigma)$. Hence the series (3.9) is termwise dominated by the convergent series (3.10) and so converges uniformly. The right member of (3.7) is accordingly continuous at $(\tau) = (\sigma)$, and we conclude that

$$D(g) = \pi \sum_k \epsilon_k c_k [c_k \log \sigma_k + A_k(0)]$$

$$(3.11) \qquad + \pi \sum_m \tfrac{1}{2} m \sum_k \epsilon_k \{[A_k^2(m) + B_k^2(m)]\sigma_k^{2m} - [A_k^2(-m) + B_k^2(-m)]\sigma_k^{-2m}\}$$

$$(k = 0, 1, \cdots, n;\ m = 1, 2, \cdots;\ \epsilon_0 = 1,\ \epsilon_1 = \cdots = \epsilon_n = -1).$$

This formula is fundamental.

The formula when $n = 0$. In this case C_0 is the only circle. We take C_0 as the unit circle and let (r, θ) be polar coördinates with pole at the center of C_0. As is well known, the harmonic function $h(u, v)$ defined by the boundary vector $g(\theta)$ on C_0 has the form

$$h(u, v) = \tfrac{1}{2}a_0 + \sum_{m=1}^{\infty} r^m(a_m \cos m\theta + b_m \sin m\theta) \qquad (r < 1),$$

where (a_i, b_i) are Fourier vector coefficients of $g(\theta)$. Upon comparing this representation of $h(u, v)$ with the representation (3.1), we see that $c_0 = 0$, and that for positive values of m,

$$A_0(m) = a_m, \qquad B_0(m) = b_m, \qquad A_0(-m) = B_0(-m) = 0.$$

Hence (3.11) gives the formula

$$(3.12) \qquad A(g) = \tfrac{1}{2}\pi \sum_m m(a_m^2 + b_m^2),$$

as is well known.

4. **A second representation of** $D(g)$. We shall evaluate the coefficients in (3.1) in terms of the Fourier vector coefficients $a_k(m)$, $b_k(m)$ of $h(u, v)$ on C_k, and of the Fourier vector coefficients $\alpha_k(m)$, $\beta_k(m)$ of $h(u, v)$ on Γ_k. We multiply the respective members of (3.1) by $\cos m\theta_k$ and $\sin m\theta_k$, and integrate between 0 and 2π with r_k fixed between σ_k and ρ_k. By virtue of the continuity of the resulting integrals as functions of r_k, r_k can take on the values σ_k and ρ_k so that

$$(4.1) \qquad \begin{cases} 2[A_k(0) + c_k \log \sigma_k] = a_k(0), \\ 2[A_k(0) + c_k \log \rho_k] = \alpha_k(0), \end{cases}$$

$$(4.2) \qquad \begin{cases} A_k(m)\sigma_k^m + A_k(-m)\sigma_k^{-m} = a_k(m), \\ A_k(m)\rho_k^m + A_k(-m)\rho_k^{-m} = \alpha_k(m), \end{cases}$$

$$(4.3) \qquad \begin{cases} B_k(m)\sigma_k^m - B_k(-m)\sigma_k^{-m} = b_k(m), \\ B_k(m)\rho_k^m - B_k(-m)\rho_k^{-m} = \beta_k(m), \end{cases}$$

$$(k = 0, \cdots, n).$$

Set

$$t_0 = \frac{\rho_0}{\sigma_0}, \qquad t_h = \frac{\sigma_h}{\rho_h} \qquad (h = 1, \cdots, n).$$

From the preceding relations we see that

$$(4.4)\qquad \begin{cases} 2c_h \log t_h = a_h(0) - \alpha_h(0), \\ 2c_0 \log t_0 = \alpha_0(0) - a_0(0). \end{cases} \qquad (h = 1, \cdots, n),$$

In (3.10) we set

$$(4.5)\qquad \begin{aligned} A_h(-m)\sigma_h^{-m} &= a_h(m) - A_h(m)\sigma_h^m, \\ A_0(m)\sigma_0^m &= a_0(m) - A_0(-m)\sigma_0^{-m}, \\ B_h(-m)\sigma_h^{-m} &= -b_h(m) + B_h(m)\sigma_h^m, \\ B_0(m)\sigma_0^m &= b_0(m) + B_0(-m)\sigma_0^{-m}, \end{aligned} \qquad (h = 1, \cdots, n),$$

in accordance with (4.2) and (4.3). Making use of (4.1) and (4.4), we then reduce (3.11) to the form

$$(4.6)\qquad \begin{aligned} D(g) = \frac{\pi}{4} \sum_k a_k(0) \frac{[\alpha_k(0) - a_k(0)]}{\log t_k} &+ \frac{\pi}{2} \sum_m m \sum_k [a_k^2(m) + b_k^2(m)] \\ &- \pi \sum_m m \sum_h \{a_h(m)A_h(m) + b_h(m)B_h(m)\}\sigma_h^m \\ &- \pi \sum_m m\{a_0(m)A_0(-m) - b_0(m)B_0(-m)\}\sigma_0^{-m} \end{aligned}$$

$$(h = 1, \cdots, n;\ k = 0, \cdots, n;\ m = 1, 2, \cdots).$$

From (4.2) and (4.3) we see that, for $h = 1, \cdots, n$,

$$(4.7)\qquad \begin{cases} A_h(m)\sigma_h^m = t_h^m \dfrac{[\alpha_h(m) - t_h^m a_h(m)]}{1 - t_h^{2m}}, \\[2ex] A_0(-m)\sigma_0^{-m} = t_0^m \dfrac{[\alpha_0(m) - t_0^m a_0(m)]}{1 - t_0^{2m}}, \end{cases}$$

$$(4.8)\qquad \begin{cases} B_h(m)\sigma_h^m = t_h^m \dfrac{[\beta_h(m) - t_h^m b_h(m)]}{1 - t_h^{2m}}, \\[2ex] -B_0(-m)\sigma_0^{-m} = t_0^m \dfrac{[\beta_0(m) - t_0^m b_0(m)]}{1 - t_0^{2m}}. \end{cases}$$

Upon using (3.12), (4.6), (4.7), and (4.8), we find that

$$(4.9)\qquad D(g) = A(g_0) + \cdots + A(g_n) + R(g),$$

where

$$(4.10)'\qquad R(g) = T(g) + R_0(g) + \cdots + R_n(g),$$

$$(4.10)''\qquad T(g) = \frac{\pi}{4} \sum_k a_k(0) \frac{[\alpha_k(0) - a_k(0)]}{\log t_k} \qquad (k = 0, \cdots, n),$$

$$R_k(g) = \pi \sum_m \left[m t_k^m a_k(m) \frac{[t_k^m a_k(m) - \alpha_k(m)]}{1 - t_k^{2m}} \right.$$

(4.10)'''

$$\left. + m t_k^m b_k(m) \frac{[t_k^m b_k(m) - \beta_k(m)]}{1 - t_k^{2m}} \right]$$

$$(k = 0, \cdots, n; m = 1, 2, \cdots).$$

Let M and μ be constants such that $M > 1$, $\mu < 1$, and suppose that

(4.11) $$| g_k | < \tfrac{1}{2}M, \qquad t_k < \mu < 1 \qquad (k = 0, 1, \cdots, n).$$

Then

(4.12) $$1 - t_k^{2m} > 1 - \mu,$$

and the Fourier coefficients $a_k(m)$, $\alpha_k(m)$, etc. have magnitudes at most M. Moreover,

(4.13) $$| t_k^m a_k(m) - \alpha_k(m) | < 2M, \qquad | t_k^m b_k(m) - \beta_k(m) | < 2M,$$

so that the m-th term in R_k is in magnitude at most

(4.14) $$\frac{4\pi M^2 m \mu^m}{1 - \mu} \qquad (\mu < 1).$$

The series of terms (4.14) is convergent so that the series R_k converges absolutely. This fact justifies the regrouping of the terms of (3.11) in the form (4.9).

We shall examine the dependence of R_k on the boundary vectors (g) and the circles (C). The distance $d(gg^*)$ between two admissible sets (g) and (g^*), both given in terms of the angles (θ), will be defined as the sum

$$g_0 g_0^* + g_1 g_1^* + \cdots + g_n g_n^*,$$

where $g_k g_k^*$ has been defined in (1.22). The circles (C) depend on the circle parameters (η) of the introduction. The distance $d(\eta\eta^*)$ between two admissible sets of circle parameters (η) and (η^*) will be taken as the Euclidean distance between (η) and (η^*), regarded as points in the space of their coördinates. Finally the distance between two admissible sets (g, η) and (g^*, η^*) will be taken as the sum

$$d(gg^*) + d(\eta\eta^*).$$

The sets (g, η) may then be regarded as points on an abstract metric space. The concept of continuity of a function $\varphi(g, \eta)$ will be referred to the general notion of continuity of a function on an abstract metric space.

The coefficients $a_k(m)$, $b_k(m)$ are continuous in (g), but independent of (η). The coefficients $\alpha_k(m)$ and $\beta_k(m)$ are continuous in (g, η) at least as long as the circles (Γ) remain fixed and interior to B. For under such circumstances $h(u, v)$, evaluated on (Γ), depends continuously on (g, η). If (4.11) holds for a given set (g, η), it will hold for a sufficiently small neighborhood Σ of (g, η) so that the m-th term in R_k will still be dominated on Σ by the term (4.14). Thus R_k will converge uniformly on Σ.

Hence $R_k(g)$ varies continuously with (g, η), (g, η) remaining admissible. The same is clearly true of $T(g)$, and hence of $R(g)$.

Under circumstances to be described in the next section, the vectors (g), the circles (C), and the ratios t_k will vary with a parameter e in such a way that

$$\left|\frac{\partial a_k(m)}{\partial e}\right| < mN, \quad \left|\frac{\partial b_k(m)}{\partial e}\right| < mN, \quad \left|\frac{\partial \alpha_k(m)}{\partial e}\right| < mN, \quad \left|\frac{\partial \beta_k(m)}{\partial e}\right| < mN, \tag{4.15}$$

$$\left|\frac{\partial t_k}{\partial e}\right| < N \qquad (m = 1, 2, \cdots; k = 0, 1, \cdots, n), \tag{4.16}$$

where N is a constant > 1, independent of e, k, and m. When (4.11), (4.15), and (4.16) hold, the e-derivative of the m-th term of R_k is in magnitude at most

$$32\pi N M^2 \frac{m^2 \mu^{m-1}}{(1 - \mu)^2},$$

as one sees upon first verifying that

$$\left|\frac{\partial}{\partial e}\left(\frac{t_k^m}{1 - t_k^{2m}}\right)\right| < \frac{2mN\mu^{m-1}}{(1 - \mu)^2}.$$

Hence the series of e-derivatives of the terms of R_k converges uniformly, and we have the following fundamental lemma.

LEMMA 4.1. *When (4.11), (4.15), and (4.16) hold, the remainder R in (4.9) has an e-derivative which is continuous in e and in any other arguments upon which the Fourier coefficients $a_k(m)$, $\alpha_k(m)$, etc., and the ratios t_k depend continuously with their e-derivatives.*[16]

The remainder $R(g)$ is continuous in (g, η). If we define $D(g)$ as in (3.4) and use (4.9) to represent the approximation $d(\tau)$, we arrive at the following theorem.

THEOREM 4.1. *A necessary and sufficient condition that $D(g)$ be finite is that $A(g_0), \cdots, A(g_n)$ be finite.*

The case of two contours, $n = 1$. Formulas (4.10) can be considerably simplified in case $n = 1$. In fact (4.10) remains valid if Γ_0 and Γ_1 are replaced by C_0 and C_1 respectively. This amounts to letting ρ_1 tend to σ_0 and ρ_0 tend to σ_1 as a limit. The coefficients $\alpha_1(m)$ and $\alpha_0(m)$ will then tend to $a_0(m)$ and $a_1(m)$ respectively. Similarly, $\beta_1(m)$ and $\beta_0(m)$ tend to $b_0(m)$ and $b_1(m)$ respectively. The ratios t_0 and t_1 will have a common limit t given by σ_1/σ_0. The series $R_k(g)$ given by (4.10)''' is dominated termwise at all stages of this limiting process

[16] The parameter e is not to be confused with the set of constants (e) appearing as arguments of the function $D(g, \eta, \lambda, e)$. However, we shall have occasion to set $e = e_k$.

by a series of constant terms of the form (4.14). We are accordingly justified in setting

$$t_0 = t_1 = t,$$
$$\alpha_1(m) = a_0(m), \qquad \alpha_0(m) = a_1(m),$$
$$\beta_1(m) = b_0(m), \qquad \beta_0(m) = b_1(m)$$

in (4.10). We thereby find (cf. Douglas II, loc. cit., p. 339) that

(4.17) $$D(g) = A(g_0) + A(g_1) + T(g) + R^*(g),$$

where

$$T(g) = -\frac{\pi}{4}\frac{[a_1(0) - a_0(0)]^2}{\log t},$$

$$R^*(g) = \pi \sum_{m=1}^{\infty} mt^{2m} \frac{[a_0^2(m) + b_0^2(m) + a_1^2(m) + b_1^2(m)]}{1 - t^{2m}}$$
$$- 2\pi \sum_{m=1}^{\infty} mt^m \frac{[a_0(m)a_1(m) + b_0(m)b_1(m)]}{1 - t^{2m}}.$$

5. **The first variation of D.** Recall that $\lambda_k(\theta_k)$ is an analytic function of θ_k with a period 2π. The set $(\lambda_0, \cdots, \lambda_n)$ has been denoted by (λ). The distance between two sets (λ) and (λ^*) will be taken as the sum

(5.1) $$d(\lambda_0\lambda_0^*) + \cdots + d(\lambda_n\lambda_n^*),$$

where $d(\lambda_k\lambda_k^*)$ is defined in (1.21). The distance between sets (g, η, λ, e) and sets $(g^*, \eta^*, \lambda^*, e^*)$ will be taken as

(5.2) $$d(gg^*) + d(\eta\eta^*) + d(\lambda\lambda^*) + d(ee^*),$$

where $d(ee^*) = | e - e^* |$. The concept of continuity of functions of sets (g, η, λ, e) will then be referred to the general concept of continuity of a point on an abstract metric space. Continuity of functions of sets (g, η, λ) or (g, η, e) will be similarly defined.

We shall make a study of the function $D(g, \eta, \lambda, e)$ of (0.6) and of its derivatives

$$D_{e_k}(g, \eta, \lambda, e), \qquad D_{\eta_i}(g, \eta, \lambda, e),$$

showing that these derivatives are continuous in their arguments at admissible sets (g, η, λ, e) at which D is finite.

We shall begin with a study of the Fourier coefficients $a_k(m)$, $b_k(m)$ of the (λ_k, e_k) transform $p(\theta_k, e_k)$ of g_k. These coefficients will be functions of the sets (g, λ, e) as well as of k and m, but will be independent of (η). In particular for $m > 0$,

(5.3) $$a_k(m) = \frac{1}{\pi}\int_0^{2\pi} p_k(\mu, e_k) \cos m\mu \, d\mu.$$

Upon setting $\mu = \theta + e_k\lambda_k(\theta)$ subject to the condition (0.3) on e_k, we see that

$$(5.4) \qquad a_k(m) = \frac{1}{\pi}\int_0^{2\pi} g_k(\theta)\cos[m\theta + e_k m\lambda_k(\theta)][1 + e_k\lambda_k'(\theta)]\,d\theta.$$

We see from (5.4) that $a_k(m)$ and its e_j-derivatives are continuous in (g, λ, e). A similar conclusion holds for $b_k(m)$.

We turn to the Fourier coefficients $\alpha_k(m)$ and $\beta_k(m)$. Let $G(x, y, u, v, \eta)$ be the Green's function with pole at the point (u, v), set up for the region B^η bounded by the respective circles C_k^η with circle parameters (η). The point (x, y) shall vary near (C^η) with $(x, y) \neq (u, v)$. With respect to the variables (x, y), $\partial G/\partial N$ shall represent the inner normal derivative of G, evaluated near (C^η). The partial derivatives

$$(5.5) \qquad \frac{\partial^2 G}{\partial x\,\partial N}, \qquad \frac{\partial^2 G}{\partial y\,\partial N}, \qquad \frac{\partial^2 G}{\partial \eta_i\,\partial N} \qquad (i = 1, \cdots, 3n)$$

exist and are continuous in their arguments for (x, y) near (C^η), and (u, v) interior to B^η, $(x, y) \neq (u, v)$. This can be established by an elementary analysis of the properties of G. The vector $H(u, v, e)$ is given by the formula[17]

$$(5.6) \quad H(u, v, e) = \frac{1}{2\pi}\sum_k \int_{C_k^\eta} p_k(\mu_k, e_k)\frac{\partial}{\partial N}G(x, y, u, v, \eta)\rho_k\,d\mu_k \quad (k = 0, 1, \cdots, n),$$

where on C_k^η

$$(5.7) \qquad x = u_k + \sigma_k\cos\mu_k, \qquad y = v_k + \sigma_k\sin\mu_k.$$

We set

$$(5.8) \qquad \mu_k = \theta_k + e_k\lambda_k(\theta_k)$$

in (5.6), subject to (0.3), so that $H(u, v, e)$ takes the form

$$(5.9) \qquad H(u, v, e) = \frac{1}{2\pi}\sum_k \int_{C_k^\eta} g_k(\theta)[1 + e_k\lambda_k'(\theta)]L_k\,d\theta_k,$$

where, subject to (5.7) and to (5.8),

$$\rho_k\frac{\partial G}{\partial N} = L_k(e_k, \theta_k, u, v, \eta) \qquad [(x, y) \text{ on } C_k^\eta;\ (u, v) \text{ on } B^\eta].$$

By virtue of the existence and continuity of the partial derivatives (5.5), L_k is continuously differentiable in the arguments e_k and η_i. We can accordingly obtain the e_k-derivative or η_i-derivative of $H(u, v, e)$ by differentiating under the integral sign in (5.9). Recalling that $\alpha_k(m)$ and $\beta_k(m)$ are Fourier coefficients of H taken on the respective circles Γ_k, we conclude that the functions $\alpha_k(m)$ and $\beta_k(m)$ as well as their e_j- and η_i-derivatives are continuous in (g, η, λ, e).

We shall use Lemma 4.1 to prove a lemma concerning $R_k(p)$, where (p) is the (λ, e) transform of (g). We write $R_k(p)$ as a function $R_k(g, \eta, \lambda, e)$.

[17] The sense of integration is such that μ_0 varies from 0 to 2π, and $\mu_1, \cdots, \mu_n$ vary from 2π to 0.

LEMMA 5.1. *The e_j-derivative of $R_k(g, \eta, \lambda, e)$ exists and is continuous at each admissible set* $(g^0, \eta^0, \lambda^0, e^0)$.

Corresponding to the given set $(g^0, \eta^0, \lambda^0, e^0)$, there exists a neighborhood Σ of $(g^0, \eta^0, \lambda^0, e^0)$ so small that on Σ, for each k,

$$|g_k(\theta)| < \tfrac{1}{2}M, \qquad |\lambda_k(\theta)| < \Lambda, \qquad |\lambda_k'(\theta)| < \Lambda, \tag{5.10}$$

where M and Λ are constants independent of k and θ. It will be convenient to suppose M and Λ at least 1. We see that $|a_k(m)|$ and $|b_k(m)|$ are less than M. From (5.4) and (5.10) we infer that

$$\left|\frac{\partial a_k(m)}{\partial e_k}\right| < 4mM\Lambda^2, \qquad \left|\frac{\partial b_k(m)}{\partial e_k}\right| < 4mM\Lambda^2, \tag{5.11}$$

making use of the fact that $|e_k| < 1$.

We seek limitations similar to (5.11) on the e_j-derivatives of $\alpha_k(m)$ and $\beta_k(m)$. To that end we shall differentiate under the integral sign in (5.9). If the circles (Γ) are fixed and the neighborhood Σ of $(g^0, \eta^0, \lambda^0, e^0)$ is sufficiently small, the circles (C^η) will be bounded from the circles (Γ), so that the point (u, v) on (Γ) will be bounded from the point (x, y) on (C^η). Hence the e_j-derivative of L_k will have an absolute value which is bounded on Σ. Hence $|H_{e_j}|$ will admit a bound A for (u, v) on (Γ) independent of (g, η, λ, e) on Σ. Upon differentiating H under the integral sign in the Fourier integrals giving $\alpha_k(m)$ and $\beta_k(m)$, we infer that

$$\left|\frac{\partial \alpha_k(m)}{\partial e_j}\right| \leq 2A, \qquad \left|\frac{\partial \beta_k(m)}{\partial e_j}\right| \leq 2A \qquad (k, j = 0, 1, \cdots, n). \tag{5.12}$$

We can apply Lemma 4.1. For we have seen that the coefficients

$$a_k(m), \qquad b_k(m), \qquad \alpha_k(m), \qquad \beta_k(m)$$

and their e_j-derivatives are continuous in (g, η, λ, e), the circles (Γ) remaining fixed, and (g, η, λ, e) admissible. In Lemma 4.1 set $e = e_j$. Conditions (4.11) and (4.15) of Lemma 4.1 are satisfied for (g, η, λ, e) on Σ, by virtue of (5.10), (5.11) and (5.12). Condition (4.16) of Lemma 4.1 holds since t_k is independent of e_j. Hence $R_k(p)$ has an e_j-derivative which is continuous on Σ, and the proof of the lemma is complete.

We continue with a proof of Lemma 5.2 again using Lemma 4.1.

LEMMA 5.2. *The η_i-derivative of $R_k(g, \eta, \lambda, e)$ exists and is continuous at each admissible set* $(g^0, \eta^0, \lambda^0, e^0)$.

We choose a neighborhood Σ of $(g^0, \eta^0, \lambda^0, e^0)$ as in the preceding lemma. Relations (5.10) then hold. The coefficients $a_k(m)$, $b_k(m)$ are independent of (η). The coefficients $\alpha_k(m)$, $\beta_k(m)$ and their η_i-derivatives are the Fourier coefficients respectively of $H(u, v, e)$ and $H_{\eta_i}(u, v, e)$ evaluated on Γ_k. The function $H(u, v, e)$ is given by the integral (5.9), and $H_{\eta_i}(u, v, e)$ can be obtained by differentiating (5.9) under the integral sign. For sets (g, η, λ, e) on Σ the circles (C^η) are bounded from the circles (Γ), so that the point (u, v) on (Γ) will be bounded from the point (x, y) on (C^η). With the variables so restricted, the η_i-derivative of L_k in (5.9) will be bounded in absolute value so that $| H_{\eta_i}(u, v, e) |$ will be bounded by a constant B. The derivatives

$$\left| \frac{\partial \alpha_k(m)}{\partial \eta_i} \right|, \qquad \left| \frac{\partial \beta_k(m)}{\partial \eta_i} \right| \qquad [(g, \eta, \lambda, e) \text{ on } \Sigma]$$

will be bounded by $2B$ independently of k and i. Finally if Σ is sufficiently small, the ratios

$$t_0 = \frac{\rho_0}{\sigma_0}, \qquad t_h = \frac{\sigma_h}{\rho_h} \qquad (h = 1, \cdots, n)$$

will be bounded and will have bounded η_i-derivatives, since the denominators σ_0 and ρ_h will be bounded from zero.

We can thus apply Lemma 4.1, setting e in Lemma 4.1 equal to η_i, and infer that the η_i-derivative of $R_k(g, \eta, \lambda, e)$ exists and is continuous at each admissible set $(g^0, \eta^0, \lambda^0, e^0)$.

We complete these results in the fundamental theorem.

THEOREM 5.1. *The variational coefficients D_{e_k} and D_{η_i} exist and are continuous in the arguments (g, η, λ, e) at each admissible set of these arguments for which D is finite.*

Let (p) be the (λ, e) transform of (g), defined on the circles (C^η). It follows from (4.9) and (4.10) that

$$D(p) = A(p_0) + \cdots + A(p_n) + T(p) + R_0(p) + \cdots + R_n(p). \tag{5.13}$$

We have seen in Theorem 1.2 that $A_{e_k}(p_k)$ is continuous in its arguments (g_k, λ_k, e_k), and that $A(p_k)$ does not depend on (η). According to Lemmas 5.1 and 5.2, $R_k(p)$ has continuous e_k- and η_i-derivatives. The same is true of $T(p)$, as follows at once from the continuity and differentiability of the coefficients a_k, α_k, etc., and the ratios t_k.

The theorem follows from (5.13).

6. The fundamental theorem.

We begin this section with a lemma.

Let $\psi(r, \theta)$ be a scalar function which is harmonic on an annular region of the (u, v)-plane of the form $r_0 \leqq r < 1$. Let (ρ, φ) be polar coördinates of a point for which $r_0 \leqq \rho < 1$, and suppose that $\rho < r$. Let $P(r, \theta, \rho, \varphi)$ be the Poisson kernel for the disc $u^2 + v^2 < r^2$ with pole at the point (ρ, φ). Set

$$\int_0^{2\pi} P(r, \theta, \rho, \varphi)\psi(r, \theta)\, d\theta = \zeta^{(r)}(\rho, \varphi). \tag{6.1}$$

Regarded as a function of (ρ, φ), $\zeta^{(r)}(\rho, \varphi)$ is harmonic for $\rho < r$ and, if extended by the definition

$$\zeta^{(r)}(r, \varphi) = \psi(r, \varphi),$$

is continuous for $\rho \leqq r$.

LEMMA 6.1.[18] *If for each fixed* $\rho < 1$

$$\lim_{r=1} \zeta^{(r)}(\rho, \varphi) = 0 \tag{6.2}$$

uniformly in φ, $\psi(r, \theta)$ *tends to* 0 *uniformly with respect to* θ *as* r *tends to* 1.

The difference

$$\psi(\rho, \varphi) - \zeta^{(r)}(\rho, \varphi) = M^{(r)}(\rho, \varphi) \qquad (r_0 < r < 1) \tag{6.3}$$

is harmonic in the variables (ρ, φ) for $r_0 \leqq \rho < r$ and vanishes for $\rho = r$. Hence $M^{(r)}(\rho, \varphi)$ can be continued by reflection so as to be harmonic over the ring $\Sigma(r)$,

$$r_0 \leqq \rho \leqq \frac{r^2}{r_0}.$$

The maximum m_r of $| M^{(r)}(\rho, \varphi) |$ on $\Sigma(r)$ will be assumed when $r = r_0$. It is the maximum of

$$| \psi(r_0, \varphi) - \zeta^{(r)}(r_0, \varphi) |$$

with respect to φ and tends to the maximum of $\psi(r_0, \varphi)$ as r tends to 1, by virtue of (6.2).

Let r_1 be a constant such that $r_0 < r_1 < 1$. For $r_1 \leqq r < 1$, m_r will admit a bound N independent of r. Suppose r_1 so near 1 that $r_1^2 > r_0$. Then $\Sigma(r_1)$ will contain the unit circle in its interior. Let S be a ring interior to $\Sigma(r_1)$, concentric with $\Sigma(r_1)$, containing the unit circle in its interior. Let d be the distance of the boundary of S from that of $\Sigma(r_1)$. Then on S

$$| M_\rho^{(r)}(\rho, \varphi) | < \frac{2N}{d}, \tag{6.4}$$

in accordance with the theory of harmonic functions.

It follows from (6.4) and from the fact that $M^{(r)}(r, \varphi) = 0$ that on S

$$| \psi(\rho, \varphi) - \zeta^{(r)}(\rho, \varphi) | \leqq \frac{2N}{d} (r - \rho).$$

We hold (ρ, φ) fast and let r tend to 1. By virtue of (6.2), $\zeta^{(r)}(\rho, \varphi)$ tends to 0 so that

$$| \psi(\rho, \varphi) | \leqq \frac{2N}{d} (1 - \rho).$$

The proof of the lemma is complete.

The lemma admits various extensions. It is clear that the unit circle can be replaced by an arbitrary circle C, and that the annular region adjoining C may be either interior or exterior to C. Further, under the hypothesis of the lemma, $\psi(\rho, \varphi)$ can be harmonically continued over C so as to be zero on C.

[18] Cf. Courant, loc. cit., p. 712.

We return to a study of the Dirichlet sum $D(g, \eta, \lambda, e)$ of (0.6). Corresponding to the relation (2.5) we here have the relation

$$D_{e_k}(g, \eta, \lambda, 0) \equiv -\int_{C_k^\eta} \lambda_k(\theta_k) \frac{\partial h}{\partial \theta_k} \frac{\partial h}{\partial r_k} \sigma_k \, d\theta_k .$$

When the boundary vectors (g) are analytic, $D_{e_k}(g, \eta, \lambda, 0)$ takes the form

$$D_{e_k}(g, \eta, \lambda, 0) = \frac{1}{2} \int_{C_k^\eta} \lambda_k(\theta_k) I[(w - w_k)^2 f(w)] \, d\theta_k , \tag{6.5}$$

where C_0^η is to be taken counterclockwise, and $C_1^\eta, \cdots, C_n^\eta$ clockwise, essentially as in Lemma 2.1.

We can now prove the following theorem.

THEOREM 6.1. *Corresponding to the boundary vectors (g) and circle parameters (η), let $f(w)$ be the Weierstrass modul of the harmonic function $h(u, v)$ determined by (g). A necessary and sufficient condition that $D_{e_k}(g, \eta, \lambda, 0) = 0$ for every admissible set (λ) and for the given set (g, η) is that the function*

$$I[(w - w_k)^2 f(w)] \qquad (k = 0, 1, \cdots, n) \tag{6.6}$$

admit a harmonic continuation over C_k^η, and in particular be null on C_k^η.

Let $[C(\tau)]$ be a set of circles interior to B^η of radii $(\tau_0, \cdots, \tau_n) = (\tau)$, concentric with the respective circles (C^η), and so near (C^η) that they bound a connected region $B(\tau)$. Let (g^τ) be the boundary vectors defined by $h(u, v)$ on the respective circles $[C(\tau)]$ in terms of the respective angles θ_k.

Let (η^τ) be the circle parameters determined by the circles $[C(\tau)]$. Referring to (6.1), let

$$\lambda_k^\tau(\theta_k) = P(\tau_k, \theta_k, \rho_k, \varphi_k) \qquad (\rho_k < \tau_k \leqq \sigma_k) \tag{6.7}$$

holding (ρ_k, φ_k) fast. For $\tau_k < \sigma_k$, the vectors (g^τ) are analytic in the parameters (θ) so that (6.5) is applicable. Hence for $w - w_k = \tau_k e^{\theta_k i}$

$$D_{e_k}(g^\tau, \eta^\tau, \lambda^\tau, 0) = \frac{1}{2} \int_{C_k(\tau)} \lambda_k^\tau(\theta_k) I[(w - w_k)^2 f(w)] \, d\theta_k \qquad (\tau_k < \sigma_k). \tag{6.8}$$

Set $(\lambda^\sigma) = (\lambda)$. If (τ) tends to (σ), $(g^\tau, \eta^\tau, \lambda^\tau)$ tends to (g, η, λ). By virtue of Theorem 5.1 the left member of (6.8) then tends to $D_{e_k}(g, \eta, \lambda, 0)$. Moreover, this approach is uniform with respect to the parameter φ_k of (6.7). For if ρ_k is fixed, if (τ) ranges on the closure of a sufficiently small neighborhood of (σ) and φ_k is arbitrary, the sets $(g^\tau, \eta^\tau, \lambda^\tau)$ form a compact ensemble on which the left member of (6.8) is continuous.

Under the hypothesis that $D_{e_k}(g, \eta, \lambda, 0) = 0$, we conclude that, for each k,

$$\lim_{(\tau) = (\sigma)} \int_{C_k(\tau)} \lambda_k^\tau(\theta_k) I[(w - w_k)^2 f(w)] \, d\theta_k = 0$$

uniformly with respect to the parameter φ_k of (6.7). It follows from Lemma 6.1 and its extensions that the function (6.6), harmonically continued, is null on C_k^η as stated, and we conclude that the condition of the theorem is necessary.

To prove the condition of the theorem sufficient we start with relation (6.8), assuming that the function (6.6) on $C_k(\tau)$ tends to zero uniformly as (τ) tends to (σ). By virtue of the continuity of the left member of (6.8), we can conclude that $D_{e_k}(g, \eta, \lambda, 0) = 0$, and the proof of the theorem is complete.

Formulas[19] *for* $D_{\eta_i}(g, \eta, \lambda, 0)$. We shall hold (g) fast and begin by varying the radii τ_k of the circles C_k, denoting these circles as previously by $C_k(\tau)$ and holding the centers $w_k = u_k + iv_k$ fast. The initial value of (τ) is denoted by (σ). As previously let (r_k, θ_k) be polar coördinates with pole at the point w_k. Let[20] $M(r_k, \theta_k, \tau)$ represent the harmonic surface determined by (g) on the region $B(\tau)$ bounded by the circles $[C(\tau)]$. Let $d(g, \tau)$ be the Dirichlet sum over $B(\tau)$ determined by $M(r_k, \theta_k, \tau)$. We shall begin by supposing the boundary vectors (g) analytic, and shall obtain a formula for $d_{\tau_k}(g, \sigma)$.

The integral $d(g, \tau)$ depends on τ_k both in its limits and in its integrand, so that d_{τ_k} consists of two principal terms. Indicating evaluation when $(\tau) = (\sigma)$ by a superscript 0, we have

$$d^0_{\tau_k} = -\frac{1}{2}\int_{C_k(\sigma)}\left[M^{02}_{r_k} + \frac{1}{\sigma_k^2}M^{02}_{\theta_k}\right]ds - \int_{C_k(\sigma)} M^0_{r_k}M^0_{\tau_k}\,ds \quad (k = 1, 2, \cdots, n). \tag{6.9}$$

Of these integrals the first arises from the fact that the limits depend on τ_k, while the second arises upon differentiating $d(g, \tau)$ under the integral sign and integrating by parts in the usual way.

Upon differentiating the relation

$$M(\tau_k, \theta_k, \tau) \equiv g_k(\theta_k)$$

with respect to τ_k, we find that

$$M_{r_k} + M_{\tau_k} = 0 \qquad (r_k = \tau_k),$$

so that (6.9) takes the form

$$d^0_{\tau_k} = \frac{1}{2}\int_{C_k(\sigma)}\left[M^{02}_{r_k} - \frac{M^{02}_{\theta_k}}{\sigma_k^2}\right]ds.$$

The relation (2.6)′ is here replaced by the relation

$$(w - w_k)^2 f(w) = (r_k M^0_{r_k} - iM^0_{\theta_k})^2,$$

from which it follows that

$$d^0_{\tau_k} = \frac{1}{2}\int_{C_k(\sigma)} R[(w - w_k)^2 f(w)]\,\frac{ds}{\sigma_k^2}. \tag{6.10}$$

[19] Cf. Courant, loc. cit., p. 714.

[20] $M(r_k, \theta_k, \tau)$ is a vector in the space $(x^1, \cdots, x^m)$. This function depends upon k, but we have omitted a subscript k for the sake of simplicity of notation.

We shall show that (6.10) holds even when the boundary vectors are merely continuous, whenever the Weierstrass modul $f(w)$ is analytically continuable over the circles $C_k(\sigma)$. To establish this we use the boundary vectors (g^τ) and circles $[C(\tau)]$ of the proof of Theorem 6.1. For $(\tau) \neq (\sigma)$, the vectors (g^τ) are analytic in the angles (θ) and (6.10) yields the relation

$$d_{\tau_k}(g^\tau, \tau) = \frac{1}{2}\int_{C_k(\tau)} R[(w - w_k)^2 f(w)] \frac{ds}{\tau_k^2} \qquad (k = 1, \cdots, n). \tag{6.11}$$

When $f(w)$ is analytically continuable over $C_k(\sigma)$, the right member of (6.11) is continuous in (τ), even when $(\tau) = (\sigma)$. The left member of (6.11) is likewise continuous in (τ) by virtue of Theorem 6.1. It follows that (6.11) holds when $(\tau) = (\sigma)$, that is, (6.10) holds when the vectors (g) are merely continuous in (θ).

We shall now vary the coördinates (u, v) of the centers of the circles C_k^η ($k = 1, \cdots, n$), denoting the variable coördinates by (a_k, b_k) and the initial coördinates by (u_k, v_k) as previously. We hold the radii of the circles C_k^η fast as well as the boundary vectors (g). Let $N(u, v, a, b)$ be the harmonic vector determined by (g) on the region $B(a, b)$ bounded by the circles with centers (a_k, b_k). Let $\delta(g, a, b)$ be the Dirichlet sum over $B(a, b)$ determined by $N(u, v, a, b)$. Evaluation when $(a_k, b_k) = (u_k, v_k)$ ($k = 1, \cdots, n$) will be indicated by adding the superscript 0. In the case where the vectors $g_k(\theta_k)$ are analytic we find that

$$\delta^0_{a_k} = -\int_{C_k^\eta} N^0_{a_k} \frac{\partial N^0}{\partial r_k}\, ds + \frac{1}{2}\int_{C_k^\eta} [N_u^{0^2} + N_v^{0^2}]\, dv \qquad (k = 1, 2, \cdots, n), \tag{6.12}$$

the first integral arising from differentiating $\delta(g, a, b)$ under the integral sign and integrating by parts, and the second integral arising from the way in which a_k enters into the limits. The sense of integration on C_k^η is clockwise.

From the identity

$$N(a_k + \sigma_k \cos\theta_k, b_k + \sigma_k \sin\theta_k, a, b) = g_k(\theta_k),$$

we infer, by differentiating with respect to a_k, that on C_k^η,

$$N^0_u + N^0_{a_k} = 0,$$

so that (dropping η from C_k^η for simplicity of notation)

$$\delta^0_{a_k} = \int_{C_k} N^0_u \frac{\partial N^0}{\partial r_k}\, ds - \frac{1}{2}\int_{C_k} [N_u^{0^2} + N_v^{0^2}] \cos\theta_k\, ds. \tag{6.13}$$

Upon recalling that

$$\frac{\partial N}{\partial r_k} = N_u \cos\theta_k + N_v \sin\theta_k,$$

we reduce (6.13) to the form

$$\delta^0_{a_k} = \int_{C_k} N^0_u N^0_v \sin \theta_k \, ds + \frac{1}{2} \int_{C_k} [N^{02}_u - N^{02}_v] \cos \theta_k \, ds,$$

$$\delta^0_{a_k} = \frac{1}{2\sigma_k} \int_{C_k} R[(w - w_k) f(w)] \, ds \qquad (k = 1, 2, \cdots, n). \tag{6.14}$$

Similarly one shows that when the vectors (g) are analytic in the variables (θ),

$$\delta^0_{b_k} = \frac{1}{2\sigma_k} \int_{C_k} I[(w_k - w) f(w)] \, ds \qquad (k = 1, \cdots, n). \tag{6.15}$$

When the Weierstrass modul $f(w)$ is analytically continuable over the circles (C^η), it follows, as in the proof of (6.10), that (6.14) and (6.15) hold even if (g) is merely continuous in the variables (θ). We thus have the following lemma.

Lemma 6.2. *When the Weierstrass modul $f(w)$ is analytically continuable over the circles (C^η), the partial derivatives D_{η_i} are given by* (6.10), (6.14) *and* (6.15) *according as η_i is respectively the radius r_k of C^η_k, the coördinate a_k, or b_k, of the center of C^η_k.*

Theorem 6.2. *When D is finite at (g, η), a set of necessary and sufficient conditions that the first variation V of D at (g, η) be "identically null" is as follows:*

(i) *For each k, $I[(w - w_k)^2 f(w)]$ tend uniformly to 0 as the point on B tends to C^η_k $(k = 0, 1, \cdots, n)$.*

$$\text{(ii)} \qquad \int_{C_k} f(w) \, dw = 0 \qquad (k = 1, \cdots, n),$$

$$\text{(iii)} \qquad \int_{C_k} w f(w) \, dw = 0 \qquad (k = 1, \cdots, n).$$

We assume that $V \equiv 0$. Then condition (i) holds as stated in Theorem 6.1. It follows that $f(w)$ is analytically continuable over C_k so that (6.10), (6.14) and (6.15) hold in accordance with Lemma 6.2. We are assuming that these partial derivatives are all null, so that from (6.14) and (6.15) we infer that

$$\int_{C_k} (w - w_k) f(w) \, ds = 0 \qquad (k = 1, \cdots, n). \tag{6.16}$$

Condition (ii) follows. From the vanishing of the partial derivatives (6.10) and from condition (i) we infer that

$$\int_{C_k} (w - w_k)^2 f(w) \, ds = 0 \qquad (k = 1, \cdots, n). \tag{6.17}$$

Relation (iii) follows and the conditions of the theorem are necessary.

Conversely, we assume that conditions (i), (ii) and (iii) hold. Then $D_{e_k}(g, \eta, \lambda, 0) \equiv 0$ in (λ) by virtue of Theorem 6.1. From (i) it also follows that $f(w)$

is analytically continuable over C_k so that (6.10), (6.14) and (6.15) hold in accordance with Lemma 6.2. Relation (6.16) follows from (ii), and (6.17) from (ii) and (iii). Relations (6.10), (6.14) and (6.15) then show that the derivatives D_{η_i} vanish as stated. Thus the conditions of the theorem are proved sufficient.

The proof of the following lemma is due to Hans Lewy.[21]

LEMMA 6.3. *Under conditions* (i), (ii) *and* (iii) *of Theorem* 6.2, $f(w) \equiv 0$.

Lemma 6.3 and Theorem 6.2 combine in the fundamental theorem.

THEOREM 6.3. *A necessary and sufficient condition that the first variation of a finite Dirichlet sum D be identically zero at* (g, η) *is that the harmonic surface determined by* (g) *on the region with circle parameters* (η) *be minimal.*

The case of two contours, $n = 1$. The fact that two non-intersecting circles can be carried by a conformal map of the plane into two concentric circles suggests that in case $n = 1$, Theorem 6.3 can be replaced by a simpler theorem with a simpler proof. One begins by establishing Theorem 6.1 and equation (6.10) as before, (6.10) holding when the Weierstrass modul $f(w)$ is analytically continuable over the boundary of B. No further calculations are needed.

Take C_0 as the unit circle with center at the origin. Take C_1 as a circle of radius $\sigma < 1$, concentric with C_0. The region B is a ring bounded by C_0 and C_1. Of the circle parameters $(\eta) = (\sigma, u_1, v_1)$, σ alone shall vary. With this understood, set

$$D(g, \eta, \lambda, e) = d(g, \sigma, \lambda, e).$$

Theorem 6.3 is replaced by the following theorem.

THEOREM 6.4. *The harmonic surface determined by* (g) *on* B *is minimal if, for each admissible* (λ),

$$d_{e_0}(g, \sigma, \lambda, 0) = d_{e_1}(g, \sigma, \lambda, 0) = 0, \tag{6.18}$$

$$d_{\sigma}(g, \sigma, \lambda, 0) = 0. \tag{6.19}$$

As previously, let $f(w)$ be the Weierstrass modul of the harmonic function $h(u, v)$ determined by (g) on B. If (6.18) holds, $I[w^2f(w)]$ admits a harmonic continuation over C_0 and C_1, and is null on C_0 and C_1 in accordance with Theorem 6.1. Hence $w^2f(w)$ equals a real constant K on B. If (6.19) also holds, it follows from (6.10) that $K = 0$. Hence $f(w) \equiv 0$, and S is minimal.

THE INSTITUTE FOR ADVANCED STUDY.

[21] See Courant, loc. cit., p. 715. One notes that the conditions of Theorem 6.2 imply that (ii) and (iii) hold even when $k = 0$. This follows from Cauchy's theorem upon integrating $f(w)$ or $wf(w)$ around the boundary of B.

Reprinted from the Proceedings of the NATIONAL ACADEMY OF SCIENCES,
Vol. 26, No. 12, pp. 713–716. December, 1940.

UNSTABLE MINIMAL SURFACES OF HIGHER TOPOLOGICAL TYPES

BY MARSTON MORSE AND C. B. TOMPKINS

INSTITUTE FOR ADVANCED STUDY

Communicated October 24, 1940

1. We are ultimately concerned with extending the calculus of variations in the large to multiple integrals. The problem of the existence of minimal surfaces of unstable type contains many of the typical difficulties, especially those of a topological nature. In the case of m contours ($m > 1$) new difficulties appear not found either in the general theory when $m = 1$ or in the extensive minimum theory when $m > 1$. The case $m = 2$, however, appeared to contain the essentially new difficulties, and for simplicity we present this case, leaving the extensions for later papers.

2. The Curves g_0 and g_1.—We shall be concerned with two simple closed rectifiable curves g_0 and g_1 in n-space separated by an $(n - 1)$-plane and satisfying the chord arc condition. Under the chord arc condition the ratio of an arbitrary chord length to the lesser of the two corresponding arc lengths shall be bounded from 0. In certain aspects of our work not all these hypotheses are needed. For example, in the homotopy theorem the curves g_0 and g_1 need only be simple and closed.

3. Canonical Pairs (φ, h).—The curves g_k, $k = 0, 1$, shall each be of the form $x = g(t)$, in vector notation, where $0 \leqq t \leqq 2\pi$ with t proportional to the arc length and $g(t)$ with a period 2π in t. We shall use other representations of g of the form $g(h(\alpha))$ where $h(\alpha)$ is a non-decreasing transformation of α with the property that $h(\alpha + 2\pi) \equiv h(\alpha) + 2\pi$. The case where the discontinuities of $h(\alpha)$ occur exclusively at a set of points α which are congruent mod 2π to a constant c is called *degenerate*. Let $h(\alpha)$ and $k(\alpha)$ be two transformations and r any rational integer. The minimum of the Lebesgue integral

$$\frac{1}{2\pi}\int_0^{2\pi} (h(\alpha) - k(\alpha) + 2\pi r)^2 d\alpha$$

with respect to r will be denoted by $[h, k]$. We regard $[h, k]$ as a distance in a metric space I, proving that $[h, k]$ satisfies the triangle axiom and identifying transformations h and k if $[h, k] = 0$. The space I is compact.

We choose three constants $\alpha_1 < \alpha_2 < \alpha_3$ on the interval $0 \leqq \alpha < 2\pi$. Continuous transformations $\varphi(\alpha)$ which satisfy the condition $\varphi(\alpha_i) = \alpha_i$ are termed *restricted* and form a subspace I^* of I. Such transformations will be denoted by Greek letters. A directly conformal 1 to 1 transformation of the disc $u^2 + v^2 \leqq 1$ into itself induces a transformation $T(\theta)$ of the point $(1, \theta)$ on the circle $r = 1$ into a point $(1, T(\theta))$. We term $T(\theta)$ a *Mobius* *transformation.* A restricted transformation $\varphi(\alpha)$ and a general transformation $h(\alpha)$ such that $\varphi(\alpha) \equiv h(T(\alpha))$ are said to be *connected* by T. A Mobius transformation T connecting a restricted transformation φ with a continuous transformation h is uniquely determined on I by φ and h, and maps a subspace of $I^* \times I$ continuously onto a subspace of I. The space of Mobius transformations completed with the degenerate transformations is a compact subspace of I. Pairs (φ, h) are termed *canonical* if φ is restricted, and if h is either connected with φ or is degenerate. The limit (φ, h) on I^2 of a sequence of canonical pairs is canonical whenever φ is continuous.

4. *The Dirichlet Sum D.*—Corresponding to a vector $p(\theta)$ defining a simple closed curve, we introduce the Douglas functional

$$A(p) = \frac{1}{16\pi} \int\int_G \frac{[p(\alpha) - p(\beta)]^2}{\sin^2 \frac{(\alpha - \beta)}{2}} d\alpha\, d\beta \tag{1}$$

with G the parallelogram $(0 \leqq \alpha \leqq 2\pi)$, $(-\pi \leqq \alpha - \beta \leqq \pi)$.

Let B be a ring region bounded by a circle C_0: $r = 1$ and a circle $C_1 : r = \rho$, $0 < \rho < 1$. Let $H(u, v)$ be the harmonic surface in vector form defined over B with the respective boundary values

$$p_k(\theta) = g_k[h_k(\theta)] \quad (k = 0, 1) \tag{2}$$

where $h_k(\theta)$ is a continuous transformation. The usual Dirichlet integral sum for $H(u, v)$ will be denoted by $D(p_0, p_1, \rho)$. Its value [6] is

$$D(p_0, p_1, \rho) = A(p_0) + A(p_1) + R(p_0, p_1, \rho) \qquad (0 < \rho < 1) \tag{3}$$

where R is given in [2], p. 280. We define $D(p_0, p_1, \rho)$ when $\rho = 0$ or when h_0 or h_1 is degenerate, by (3), noting that R is defined in these cases in [2].

The function D and space of sets (p_0, p_1, ρ) is inadequate because the degenerate points offer trivial minima, the space is not simply connected, D is not weakly upper reducible (defined in [1]) at degenerate points, and because the representation of minimal surfaces by critical sets is needlessly multiple, thereby complicating the topology.

5. *The Space* Π *and function* $W(P)$.—A product of metric spaces will here be metricized by a distance function which is the sum of the component distance functions. Let J denote the interval $0 \leqq \rho < 1$. The space Π shall consist of sets or points $P = (\varphi_0, \varphi_1, h_0, h_1, \rho)$ in which ρ varies on $J, h(0) = 0$, and (φ_k, h_k), $k = 0, 1$, is a canonical pair of transformations. The pair $\{\varphi_0, \varphi_1\}$ will be termed the *restricted projection* of P and will be regarded as a point on I^2. When $\rho = 0$, two points P with the same restricted projection are identified. When $\rho > 0$ points P are identified as on I^4J. Let $d(P, Q)$ be the distance between P and G as points of I^4J. With $Q = (\psi_0, \psi_1, k_0, k_1, \sigma)$ let

$$\delta(P, Q) = [\varphi_0\psi_0] + [\varphi_1\psi_1] + \rho + \sigma.$$

The points P form a space Π on which our final choice of distance function PQ shall be the minimum of $d(P, Q)$ and $\delta(P, Q)$. The usual axioms are satisfied by PQ.

Set

$$q_k(\theta) = g_k[\varphi_k(\theta)] \qquad A(g_k, \varphi_k) = A(q_k).$$

To define $W(P)$ we use (2) and set

$$W(P) = A(g_0, \varphi_0) + A(g_1, \varphi_1) + R(p_0, p_1, \rho).$$

As one sees $W(P) > D$ when h_0 or h_1 is degenerate. We show that $W(P)$ is regular at infinity, and boundedly compact in the sense of [1]. Moreover, the connectivity R_0 of the subspace on which W is finite is 1, and the other connectivities are 0. *A* point P at which $\rho > 0$ is termed *differentially critical* when the boundary vectors (2) define a ring minimal surface over B; or, if $\rho = 0$, when these vectors define disc minimal surfaces. The homotopy theorem states that a homotopic critical point is differentially critical.

We show that $W(P)$ is weakly upper reducible at each ordinary or degenerate point, and if $g'_k(t)$ satisfies a Lipschitz condition at critical points. *The general theory thus applies with the consequent relations between the critical sets classified as to type members.* The following special theorem is provable by a limiting process [11] under the original weak hypothesis. These hypotheses are weaker than those of Shiffman [4] in the case of one contour.

THEOREM. *If g_1 and g_1 bound a ring minimal surface belonging to a minimizing set of critical points of W, there either exists a ring minimal surface of non-minimizing type bounded by g_0 and g_1, or else a disc minimal surface of non-minimizing type bounded by g_0 or by g_1.*

If g_0 and g_1 possess convex plane projections, the first alternative of the theorem alone holds.

When $\rho = 0$, W is the sum of the functions $A(g_k, \varphi_k)$ defined on the product space $I^* \times I^*$. Critical points in this subspace correspond to two disc minimal surfaces. Let a function $f(x) + f(y) = F$ be defined on a product of two metric spaces of points x and y, respectively. Are the type numbers of critical points of F obtainable by summation from those of $f(x)$ and $f(y)$ in the natural way, that is, as in the case where $f(x)$ and $f(y)$ are non-degenerate forms? This question has been answered affirmatively by A. E. Pitcher under conditions which admit a wide range of applications. Pitcher's result will presently be published.

[1] Morse and Tompkins, "The Existence of Minimal Surfaces of General Critical Types," *Ann. Math.*, **40** (1939). Corrections for this paper will appear in *Ann. Math.*, Jan., 1941.

[2] Morse, "The First Variation in Minimal Surface Theory," *Duke Math. Jour.*, **6** (1940).

[3] Morse and Tompkins, "Minimal Surfaces of Unstable Type," *Bull. Amer. Math. Soc.*, **46,** 222 (1940). This is an abstract of the present paper presented at the Columbus meeting in Dec., 1939. See also *Science*, **91,** 457 (1940) for a fuller abstract.

[4] Shiffman, "The Plateau Problem for Non-relative Minima," *Ann. Math.*, **40** (1939).

[5] Courant, "The Existence of Minimal Surfaces of Given Topological Structure under Prescribed Boundary Conditions," *Acta Mathematica*, **72** (1940).

[6] Douglas, "The Problem of Plateau for Two Contours," *Jour. Math. and Physics*, **10** (1931).

[7] Douglas, "Minimal Surfaces of Higher Topological Structure," *Ann. Math.*, **40,** 205–298 (1939).

[8] Radò, "On the Problem of Plateau," *Ergebnisse der Mathematik*, **2,** 75 (1933).

[9] Morse, ***Functional Topology and Abstract Varational Theory***, Gauthier-Villars, Paris.

[10] Morse, "Rank and Span in Functional Topology," *Ann. Math.*, **41,** 419–454 (1940)

[11] Morse and Tompkins, "Minimal Surface of Unstable Type by a New Mode of Approximation." To be published in *Ann. Math.*, **42** (1941).

Annals of Mathematics
Vol. 42, No. 1, January, 1941

MINIMAL SURFACES NOT OF MINIMUM TYPE BY A NEW MODE OF APPROXIMATION

Marston Morse and C. Tompkins

(Received May 31, 1940)

1. Introduction. We begin with a study of a non-negative function $f^0(p)$ defined at each point of a metric space M. We suppose that $f^0(p)$ is the limit of a sequence of functions $f^n(p)$ also defined over M. Let p_n be a homotopic critical point p_n of $f^n(p)$. We study the limit points q of sequences p_n, regarding q as a new type of critical point of the limit function $f^0(p)$.

We give a simple application.

Let g^0 be a simple, closed, rectifiable curve in a euclidean space with the property that the ratio of an arbitrary chord length of g^0 to the smaller of the corresponding two arc lengths of g^0 is bounded from zero. This chord arc condition on g^0 is less restrictive than the condition imposed by Shiffman[1] in his study of unstable minimal surfaces. We make use of Douglas's Dirichlet function $f^0(\varphi)$. The point φ is a monotone transformation of the arc length along g^0. The basic result is as follows.

If g^0 bounds two minimal surfaces of disc type belonging to disjoint minimizing sets of such surfaces (cf. §2), *then g^0 also bounds another minimal surface of disc type not of minimum type.*

Results of this nature but under more restrictive hypotheses have been obtained previously by Morse and Tompkins[2] and by Shiffman, loc. cit. We have found that less restrictive hypotheses on g^0 are adequate, because they can be compensated by more restrictive hypotheses on the approximating functions $f^n(p)$, $n > 0$, making the analysis of the functions $f^n(p)$ much simpler. This applies particularly to the study of upper-reducibility.[3] For it is easy to show that the restricted functions $f^n(p)$ are upper-reducible, while we do not need to show that $f^0(p)$ is upper-reducible.

2. The mode of approximation. Let M be a metric space with points p, q, r and distances pq, pr, etc. satisfying the usual metric axioms. On M there shall be given a sequence of functions $f^n(p)$ converging to a function $f^0(p)$. Thus

$$\lim_{n=\infty} f^n(p) = f^0(p). \tag{2.1}$$

[1] Shiffman. *The Plateau problem for non-relative minima.* Annals of Mathematics, vol. 40 (1939), pp. 834–854. See (2) §3.

[2] Morse and Tompkins. *The existence of minimal surfaces of general critical types.* Annals of Mathematics, vol. 40 (1939), pp. 443–472. The letters MT will be used in the text to refer to this paper. See corrections elsewhere in this number of the Annals.

[3] Morse. *Functional topology and abstract variational theory.* Mémorial des Sciences Mathématiques, Fascicule 92 (1939). See §8. The letter M will refer to this pamphlet.

This convergence is in general not uniform. The functions $f^m(p)$, $m = 0, 1, \cdots$, shall satisfy conditions I to VI to be described. If c is an arbitrary constant and f is a function defined on M, f_c shall denote the subset of points of M at which $f \leqq c$. Our first condition is as follows.

HYPOTHESIS I. *Bounded compactness. The sets* f_c^m, $m = 0, 1, \cdots$, *shall be compact for each constant c.*

It follows from this hypothesis that each function f^m is lower semi-continuous.

HYPOTHESIS II. *If* $\lim p_n = p$,

$$\text{inf. lim. } f^n(p_n) \geqq f^0(p). \tag{2.2}$$

HYPOTHESIS III. *Corresponding to any subsequence* F^r *of the functions* f^n *and points* p_r *such that* $F^r(p_r)$ *is bounded independently of r, there exists a subsequence of the points* p_r *which converges to a point of M.*

We shall prove the following lemma.

LEMMA 2.1. *Let e and c be positive constants. If n is sufficiently large, and b is a constant sufficiently near c,* f_b^n *is on the e-neighborhood* N_e *of* f_c^0.

Suppose the lemma false. There will then exist a constant $e > 0$ and a subsequence F^r of the functions f^n together with points p_r, $r = 1, 2, \cdots$, such that p_r is not on N_e, while

$$c \geqq \sup_{r=\infty} \lim F^r(p_r). \tag{2.3}$$

By virtue of Hypothesis III a subsequence of the points p_r converges to a point q of M.

Without loss of generality we may suppose that $p_1, p_2, \cdots$ is this subsequence. By hypothesis q is not on N_e so that $f^0(q) > c$. But it follows from II that

$$\inf_{r=\infty} \lim F^r(p_r) \geqq f^0(q). \tag{2.4}$$

From (2.3) and (2.4) it follows that $c \geqq f^0(q)$. From this contradiction we infer the truth of the lemma.

HYPOTHESIS IV. *Corresponding to each compact subset A of M, there exists a constant* κ^A *such that for each integer n,* $f^n \leqq \kappa^A f^0$ *on A.*

We shall make use of chains and cycles taken in the sense of Vietoris with coefficients in the field of integers mod 2. Cf. M, §1. If p is a point of M, p^* shall represent a 0-cycle each of whose component 0-cycles is on p. This convention will be permanent. The following lemma is an immediate consequence of our definitions.

LEMMA 2.2. *A necessary and sufficient condition that two points p and q belong to the same component of a compact subset A of M is that* $p^* + q^* \sim 0$ *on A.*

Let f be a function defined on M. A set A will be termed *minimizing* relative to f if f equals a constant c on A, and if there exists an e-neighborhood A_e of A such that $f > C$ on $A_e - A$. If f satisfies Hypothesis I, each minimizing set ω of f is closed. For at a limit point q of ω, $f(q) \leqq c$ by virtue of the lower semi-continuity of f. It follows from the definition of a minimizing set q that q belongs to ω. Hence ω is closed.

A subset of M on which f is bounded will be termed *f-bounded*. We shall prove the following theorem.

THEOREM 2.1. *Let y and z be points respectively of two disjoint minimizing sets ω_1 and ω_2 of f^0. If y and z are connected by an f^0-bounded subset of M and Hypotheses I to IV are satisfied, there exists a subsequence F^r of the functions f^n with the following properties. There exists a least value μ_r of c for which*

$$p^* + q^* \sim 0 \qquad (\text{on } F^r_c),$$

with

$$\mu_r > \max [F^r(y), F^r(z)],$$

while the numbers μ_r converge to a limit μ such that

$$\mu > \max [f^0(y), f^0(z)].$$

We begin the proof of this theorem with certain lemmas. The first lemma concerns a compact subset C of M. A proof may be found in Morse,[4] page 430.

LEMMA 2.3. *If u is a k-cycle on C such that $u \sim 0$ on C_e for each positive e, then $u \sim 0$ on C.*

We shall prove the following lemma.

LEMMA 2.4. *Under the hypotheses of the theorem there exists a least value ν_0 of c such that $y^* + z^* \sim 0$ on f^0_c.*

It follows from the first hypothesis of the theorem and from Lemma 2.2 that there is at least one value of c such that $y^* + z^* \sim 0$ on f^0_c. Let ν_0 be the greatest lower bound of such values of c. There then exist values of c arbitrarily near ν_0 such that $y^* + z^* \sim 0$ on f^0_c. But it follows from Hypothesis I that if c is sufficiently near ν_0, f^0_c lies on an arbitrarily small neighborhood N_e of $f^0_{\nu_0}$. Hence $y^* + z^* \sim 0$ on N_e. We set $C = f^0_{\nu_0}$. We see that $y^* + z^*$ is on C. It follows from Lemma 2.3 that $y^* + z^* \sim 0$ on C, and the proof of Lemma 2.4 is complete.

The points y and z are connected on $f^0_{\nu_0}$. But y and z belong to disjoint minimizing sets. These sets are compact, and hence at a positive distance from each other. It follows that

$$\nu_0 > f^0(y), \qquad \nu_0 > f^0(z). \tag{2.5}$$

The set $f^0_{\nu_0}$ is compact. Hence on this set in accordance with IV, $f^n \leqq \kappa\nu_0$, where κ is a constant. Hence $y^* + z^* \sim 0$ on $f^n_{\kappa\nu_0}$. Let ν_n be the least number c such that $y^* + z^* \sim 0$ on f^n_c. That ν_n exists follows as in the proof of Lemma 2.4. We see that

$$\nu_n \leqq \kappa\nu_0 \qquad (n = 1, 2, \cdots). \tag{2.6}$$

[4] Morse. *Rank and span in functional topology.* Annals of Mathematics, vol. 41 (1940), pp. 419–454.

LEMMA 2.5. *If ν_n is the least number c such that $y^* + z^* \sim 0$ on f_c^n, $n = 0, 1, \cdots$, and if $\mu = \inf \lim_{n=\infty} \nu_n$, then*

$$\mu \geqq \nu_0 . \tag{2.7}$$

Let μ_r be a subsequence of the numbers ν_n such that

$$\lim_{r=\infty} \mu_r = \mu. \tag{2.8}$$

Let F^r be the corresponding subsequence of the functions f^n so that $y^* + z^* \sim 0$ on $F^r_{\mu_r}$. Let e be an arbitrary positive constant, and set $f^0_\mu = C$. It follows from Lemma 2.1 that $F^r_{\mu_r} \subset C_e$, provided r is sufficiently large. Hence $y^* + z^* \sim 0$ on C_e. It follows from Lemma 2.3 that $y^* + z^* \sim 0$ on C. From the definition of ν_0 we infer that $\mu \geqq \nu_0$, and the lemma is proved.

We shall now prove Theorem 2.1, making use of the sequence F^r and the numbers μ_r of the proof of Lemma 2.5. Observe that

$$\mu > \max [f^0(y), f^0(z)]$$

in accordance with (2.5) and (2.7). Recall that F^r and μ_r converge respectively to f^0 and μ. Hence if ρ is a sufficiently large integer

$$\mu_r > \max [F^r(y), F^r(z)] \qquad (r \geqq \rho).$$

Theorem 2.1 is accordingly satisfied by sequences F^r and μ_r, starting with integers $r \geqq \rho$.

3. *h*- and *fh*-critical points. Let F be a function defined on M. We refer to the definition of an F-deformation and a homotopic critical point of F (written h-critical point) as given on page 30 of M. Let $F^1, F^2, \cdots$ be a subsequence of the sequence $f^1, f^2, \cdots$ converging to f^0, and let q_r be an h-critical point of F^r. Any limit point q of the point set q_r will be termed an *fh-critical point* of f^0. This term is relative to the sequence f^n, to which the letter f refers in the term fh-critical point.

LEMMA 3.1. *The set of fh-critical points of f^0 is closed, and the set of h-critical points of f^0 is closed among points at any given level μ.*

That the set of h-critical points of f^0 is closed at the level μ follows at once from the definition of an h-critical point.

To prove that the set of fh-critical points of f^0 is closed, let p be a limit point of a sequence p_r of fh-critical points of f^0. For each fixed r, p_r is by definition the limit of a sequence $p(r, s)$, $s = 1, 2, \cdots$, of h-critical points of functions $f^{n(r,s)}$ where the integer $n(r, s)$ increases with s. Hence if integers $s_1, s_2, \cdots$ are successively chosen with s_r sufficiently large, the sequence $p(r, s_r)$ will converge to p as r becomes infinite, while the integers $n(r, s_r)$ will increase with r. Hence p is an fh-critical point of f^0, and the proof of the lemma is complete.

To continue we need two additional hypotheses. We refer to MT, page 445, for the definition of weak upper-reducibility.

HYPOTHESIS V. *The functions $f^1, f^2, \cdots$ shall be weakly upper-reducible.*

Lemma 8.1 of M clearly holds if weak upper-reducibility replaces upper-reducibility. We shall use Lemma 8.1 of M in proving the following lemma. We refer to Theorem 2.1 and the constants μ, μ_r and functions F^r described therein.

LEMMA 3.2. *The function F^r of Theorem 2.1 assumes the value μ_r in at least one h-critical point w_r such that y, z, w_r are in the same component of $F^r_{\mu_r}$.*

Let κ be the component of $F^r_{\mu_r}$ which contains y and z. The set κ is compact. Suppose there is no h-critical point of $F^r_{\mu_r}$ on κ at the level μ_r. By virtue of Lemma 8.1 of M there will then exist an F^r-deformation of κ into a set definitely below μ_r. Hence $y^* + z^* \sim 0$ on F^r_c for some $c < \mu_r$, contrary to the definition of μ_r.

We infer the truth of the lemma.

HYPOTHESIS VI. (a). *Let p_r be an h-critical point of f^{n_r}, $n_1 < n_2 < \cdots$, such that p_r converges to p as r becomes infinite. Then $f^{n_r}(p_r)$ shall converge to $f^0(p)$.* (b) *Let p_r be an h-critical point of f^0 such that p_r converges to p. Then $f^0(p_r)$ shall converge to $f^0(p)$.*

A point q will be said to be of *non-minimizing* type relative to $f(p)$ if there is a point p in every neighborhood of q such that $f(p) < f(q)$. With this understood we come to a major theorem.

THEOREM 3.1. *If f^0 admits two disjoint minimizing sets ω_1 and ω_2 and if there are points y and z in ω_1 and ω_2 respectively which are connected by an f^0-bounded set, if moreover Hypotheses I to VI are satisfied, then f^0 possesses at least one fh- or h-critical point of non-minimizing type.*

We refer to the numbers μ_r and μ of Theorem 2.1, and prove the following:

(i). *The set σ of h- and fh-critical points of f^0 at the level μ is closed.*

This follows from Lemma 3.1.

(ii). *The set σ is not empty.*

The h-critical point w_r of Lemma 3.2 is a point at which $F^r(w_r) = \mu_r$. These points w_r have at least one cluster point w by virtue of Hypothesis III. It follows from Hypothesis VI (a) that $f^0(w) = \mu$. Thus w is an fh-critical point belonging to σ, so that σ is not empty.

(iii). *There is at least one point of σ of non-minimizing type.*

The set σ, on which $f^0 = \mu$, fails to contain the points y and z of Theorem 2.1, since μ exceeds $f^0(y)$ and $f^0(z)$ as stated in Theorem 2.1. Observe that y, z, and w lie on the same component of f^0_μ since y, z, and w_r lie on the same component of $F^r_{\mu_r}$, and the maximum distance of points of $F^r_{\mu_r}$ from f^0_μ tends to zero as μ_r tends to μ. The set σ cannot be a minimizing set, otherwise σ would be at a positive distance from its complement on f^0_μ, contrary to the fact that y, z, and w are connected on f^0_μ, while y and z are not in σ and w is in σ. Since σ is compact and not a minimizing set of f^0, there exists at least one point of σ of non-minimizing type.

The proof of the theorem is complete.

4. The application to minimal surface theory. Let g^0 be a simple, closed, rectifiable curve in a space of coordinates $(x_1, \cdots, x_n)$. We suppose g^0 has the vector form

$$\mathbf{x} = \mathbf{g}^0(s),$$

where s is the arc length along g^0. The curve g^0 is given with a sense and an origin $s = 0$. So given, g^0 will be termed *coordinated*. It will be convenient to suppose that the total length of g^0 is 2π. We shall assume that g^0 satisfies the following condition.

Chord arc hypothesis. Under this hypothesis the ratio of the length of an arbitrary chord of g^0 to the length of the minimum subtended arc of g^0 shall be bounded from zero for all chords of g^0.

Any simple, closed, regular curve of class C^1 satisfies this hypothesis, as does a simple closed curve composed of a finite set of regular arcs of class C^1, provided at each corner p the two tangent rays directed from p never make a null angle. In particular any simple closed polygon is admissible. This condition is less restrictive than the corresponding condition of Shiffman (loc. cit.). For under Shiffman's condition two tangent rays at a corner cannot make angles less than a right angle, while it is easy to show that under Shiffman's condition the chord arc hypothesis is always satisfied.

A sequence h^n of closed, rectifiable, coordinated curves $\mathbf{x} = \mathbf{h}^n(s)$ will be said to *converge in length*[5] to h^0: $\mathbf{x} = \mathbf{h}^0(s)$ if

$$\lim_{n=\infty} \mathbf{h}^n(s) = \mathbf{h}^0(s)$$

uniformly on each bounded interval for s. It is clear that there exists a sequence of simple, closed, coordinated polygons which converge in length to g^0. The corners of these polygons can be "rounded off" to obtain a sequence of simple, regular, coordinated, closed curves g^n of class C^2 converging in length to g^0. Without loss of generality we can suppose that the length of each curve g^n is 2π, since if this is not already the case, a magnification from the origin in the ratio ρ_n to 1 will bring this about, where ρ_n must be suitably chosen and so chosen converges to 1. Let $\mathbf{x} = \mathbf{g}^n(s)$ be the vector representation of g^n.

Let $\varphi(\alpha)$ be a continuous non-decreasing function of α such that

$$\varphi(\alpha + 2\pi) \equiv \varphi(\alpha) + 2\pi. \tag{4.1}$$

---

[5] The coördinated curves h^n converge in length to h^0 if the following three conditions are fulfilled. The sensed curve h^n converges to the sensed curve h^0 according to Fréchet, the length of h^n converges to the length of h^0, and the point $s = 0$ on h^n converges to the point $s = 0$ on h^0. For related, but variant, definitions of convergence in length, see the following three sources. Adams and Lewy. *On convergence in length.* Duke Mathematical Journal, vol. 1 (1935), pp. 19–26. McShane. *Curve space topologies associated with variational problems.* Ann. Scuola Norm. Super. Pisa, (2) vol. 9 (1940), pp. 45–60. Morse. *The calculus of variations in the large.* American Mathematical Society Colloquium Publications (1934), p. 209.

We suppose that $\varphi(\alpha)$ satisfies a *three point condition* defined as follows. The relation $\alpha = \varphi(\alpha)$ shall hold for three given distinct values $\alpha_1, \alpha_2, \alpha_3$, of α on the interval $0 \leqq \alpha < 2\pi$, where $\alpha_1, \alpha_2, \alpha_3$ are independent of φ. A function $\varphi(\alpha)$ of this nature will be termed *admissible*. We shall impose another condition on the functions φ when we define the space M.

We admit representations of g^n of the form

$$\mathbf{x} \equiv \mathbf{g}^n[\varphi(\alpha)] \equiv \mathbf{p}^n(\alpha) \qquad (n = 0, 1, \cdots). \tag{4.2}$$

We shall consider the function $\varphi(\alpha)$ as a point in an abstract metric space in which the distance between two points $\varphi(\alpha)$ and $\psi(\alpha)$ shall be the number

$$\psi\varphi = \max |\varphi(\alpha) - \psi(\alpha)| \qquad (0 \leqq \alpha \leqq 2\pi). \tag{4.3}$$

The function $f^n(\varphi)$ shall be defined as the Douglas function[6]

$$f^n(\varphi) = \frac{1}{16\pi}\iint_\omega \frac{[\mathbf{p}^n(\alpha) - \mathbf{p}^n(\beta)]^2}{\sin^2\dfrac{(\alpha-\beta)}{2}}\,d\alpha\,d\beta, \quad (n = 0, 1, \cdots), \tag{4.4}$$

where ω denotes the parallelogram

$$0 \leqq \beta \leqq 2\pi,$$

$$\beta - \pi \leqq \alpha \leqq \beta + \pi.$$

The Douglas function is an improper integral with an integrand which is singular when $\alpha = \beta$.

As in MT we introduce the improper integral

$$H(\varphi) = \frac{1}{16\pi}\iint_\omega \frac{[\varphi(\alpha) - \varphi(\beta)]^2}{\sin^2\dfrac{(\alpha-\beta)}{2}}\,d\alpha\,d\beta. \tag{4.5}$$

It follows from the chord arc hypothesis on g^0 and the ordinary properties of length that there exists a constant κ such that

$$[\mathbf{p}^0(\alpha) - \mathbf{p}^0(\beta)]^2 \leqq [\varphi(\alpha) - \varphi(\beta)]^2 \leqq \kappa[\mathbf{p}^0(\alpha) - \mathbf{p}^0(\beta)]^2, \tag{4.6}$$

where κ is independent of the choice of φ. From (4.6) we see that

$$f^0(\varphi) \leqq H(\varphi) \leqq \kappa f^0(\varphi). \tag{4.7}$$

Upon comparing $f^n(\varphi)$ with $H(\varphi)$ we find that

$$f^n(\varphi) \leqq H(\varphi) \leqq \kappa f^0(\varphi). \tag{4.8}$$

Moreover g^n is regular and of class C^2 from which we infer that the ratio of an arbitrary chord length of g^n to either of the corresponding arc lengths exceeds

[6] Douglas. I. *Solution of the problem of Plateau*. Transactions of the American Mathematical Society, vol. 33 (1931), pp. 263–321. II. *The mapping theorem of Koebe and the problem of Plateau*. Journal of Mathematics and Physics, vol. 10 (1931), pp. 106–130.

some positive constant κ_n. It follows as in the proof of (4.8) that $f^0 \leqq \kappa_n f^n$. From this result and from (4.8) we conclude that $f^n(\varphi)$ is finite if and only if $f^0(\varphi)$ is finite.

We shall restrict the space M of points φ to points φ for which $f^0(\varphi)$ is finite.

We shall prove the following theorem.

THEOREM 4.1. *At each point of M,*

$$\lim_{n=\infty} f^n(\varphi) = f^0(\varphi). \tag{4.9}$$

Let e be a positive constant less than π and let ω_e be the subset of the parallelogram ω on which $| \alpha - \beta | < e$. We set

$$\omega = \omega_e + \omega_e^*, \qquad (\omega_e \cdot \omega_e^*) = 0,$$

and

$$f^n(\varphi) = F(\varphi, n, e) + F^*(\varphi, n, e), \qquad (n = 0, 1, \cdots), \tag{4.10}$$

where $F(\varphi, n, e)$ and $F^*(\varphi, n, e)$ are defined as is $f^n(\varphi)$ except that ω_e and ω_e^* shall replace ω as the respective domains of integration. From (4.6) we see that

$$F(\varphi, n, e) \leqq \kappa F(\varphi, 0, e). \tag{4.11}$$

By hypothesis $f^0(\varphi)$ is convergent so that $F(\varphi, 0, e)$ is convergent. Noting that the integral $F^*(\varphi, n, e)$ is proper, we see that

$$\lim_{n=\infty} F^*(\varphi, n, e) = F^*(\varphi, 0, e). \tag{4.12}$$

To establish the theorem, observe that

$$| f^n(\varphi) - f^0(\varphi) | \leqq | F(\varphi, n, e) - F(\varphi, 0, e) | + | F^*(\varphi, n, e) - F^*(\varphi, 0, e) |. \tag{4.13}$$

Corresponding to an arbitrary positive constant η let $e > 0$ be chosen so small that $| F(\varphi, 0, e) | < \eta$. This is possible since $F(\varphi, 0, e)$ is a convergent integral. Making use of (4.11) we see that

$$| F(\varphi, n, e) - F(\varphi, 0, e) | \leqq \kappa\eta + \eta.$$

With e so chosen, let n be so large that

$$| F^*(\varphi, n, e) - F^*(\varphi, 0, e) | < \eta.$$

Making use of (4.13), we then find that

$$| f^n(\varphi) - f^0(\varphi) | \leqq \kappa\eta + \eta + \eta,$$

and the theorem follows at once.

5. Verification of hypotheses. We shall show that the Douglas functions $f^n(\varphi)$ defined in the preceding section satisfy Hypotheses I to VI of §§1 to 3.

HYPOTHESIS I. That M is boundedly compact relative to $f^n(\varphi)$ follows from the work of Douglas II. See also Rado[7] V, 17, and Courant.[8]

HYPOTHESIS II. To verify this hypothesis we recall the origin of the Douglas function. Let g be a simple, closed, rectifiable, coordinated curve given with a vector representation $\mathbf{x} = \mathbf{g}(s)$ in terms of arc length s. Let $\varphi(\alpha)$ be an admissible function φ. Corresponding to the representation $\mathbf{x} = \mathbf{p}(\alpha) = \mathbf{g}[\varphi(\alpha)]$ there exists a harmonic surface $\mathbf{x} = \mathbf{x}(u, v)$ defined and continuous for $u^2 + v^2 \leqq 1$ and such that

$$\mathbf{x}\,(\cos\theta, \sin\theta) \equiv \mathbf{p}(\theta),$$

where r and θ are polar coordinates in the (u, v)-plane. We shall say that the harmonic surface $\mathbf{x} = \mathbf{x}(u, v)$ is *defined* by the pair (g, φ).

The Douglas function $f(\varphi)$ corresponding to g equals the Dirichlet sum

$$D(g, \varphi) = \frac{1}{2}\iint\limits_{R}\left[\left(\frac{\partial \mathbf{x}}{\partial u}\right)^2 + \left(\frac{\partial \mathbf{x}}{\partial v}\right)^2\right] du\, dv,$$

where R represents the region $u^2 + v^2 < 1$.

Let φ^n be a sequence of points on M converging to φ^0. Recall that g^n converges in length to g^0. Hypothesis II is satisfied if

$$\text{(5.1)} \qquad \inf \lim_{n=\infty} D(g^n, \varphi^n) \geqq D(g^0, \varphi^0).$$

Relation (5.1) is a consequence of the lower semi-continuity of $D(g, \varphi)$ of which a conventional proof may be indicated as follows. Let $D(g, \varphi, r)$ denote the integral obtained from $D(g, \varphi)$ upon replacing R by the region $u^2 + v^2 < r < 1$. By definition

$$D(g, \varphi) = \lim_{r=1} D(g, \varphi, r).$$

The lower semi-continuity of $D(g, \varphi)$ follows from the fact that $D(g, \varphi, r)$ is continuous in (g, φ), positive and non-decreasing in r.

Relation (5.1) holds, and Hypothesis II follows.

HYPOTHESIS III. Let $\varphi^r(\alpha)$ be an infinite sequence of points φ on M such that

$$f^{n_r}(\varphi^r) \leqq c \qquad (n_1 < n_2 < \cdots; r = 1, 2, \cdots)$$

for some constant c and suitable choices of the integers n_r. To establish III we must prove that some subsequence of the sequence φ^r converges to a point on M.

According to the theory of bounded monotone functions there exists a sub-

---

[7] Radó. *On the problem of Plateau.* Ergebnisse der Mathematik und ihrer Grenzgebiete, Berlin (1933).

[8] Courant. *Plateau's problem and Dirichlet's principle.* Annals of Mathematics, vol. 38 (1937), pp. 679–724.

sequence of the functions φ^r which converges for each value of α to a non-decreasing function $\theta(\alpha)$. The function $\theta(\alpha)$ satisfies the three point condition and the condition

$$\theta(\alpha + 2\pi) \equiv \theta(\alpha) + 2\pi.$$

It remains to show that $\theta(\alpha)$ is continuous.

Whether continuous or not $\theta(\alpha)$ possesses right and left limits a_1 and a_2 at each point α_0. Moreover $| a_1 - a_2 | \neq 2\pi$, since $\theta(\alpha)$ satisfies the three point condition. If $| a_1 - a_2 | \neq 0$, it follows as in the proof of Theorem III of Douglas II, page 113, that

$$\sup \lim_{r=\infty} D(g^{n_r}, \varphi^r) = \infty.$$

But

$$f^{n_r}(\varphi^r) = D(g^{n_r}, \varphi^r) \leqq c.$$

From this contradiction we infer that $a_1 = a_2$ so that $\theta(\alpha)$ is continuous. Thus $\theta(\alpha)$ defines a point of M, and Hypothesis III is verified.

Hypothesis IV. This hypothesis is an immediate consequence of (4.8).

Hypothesis V. The functions $f^n(\varphi)$ are weakly upper-reducible for $n > 0$ in accordance with Theorem 5.1 of MT.

The proof of Hypothesis VI is more involved.

Hypothesis VI. Let g be a simple, closed, coordinated, rectifiable curve, and $\varphi(\alpha)$ a transformation which is admissible in the sense of §4. Let $A(g, \varphi)$ be the area of the harmonic surface defined by the pair (g, φ). In a forthcoming paper we shall prove the following theorem of use in establishing VI.

(i) *The area $A(g, \varphi)$ is continuous in its arguments.*

The notion of continuity of $A(g, \varphi)$ is relative to the concept of convergence of g and φ. Convergence of curves g shall be convergence in length, and convergence of φ, convergence defined by the distance function $\varphi\psi$ of (4.3).

A pair (g, φ) which defines a minimal surface will be called *differentially critical.* When (g, φ) is differentially critical,

$$D(g, \varphi) = A(g, \varphi), \tag{5.2}$$

as is well-known. Cf. Douglas II, Theorem I. When φ is an h-critical point of the function $f(\varphi) = D(g, \varphi)$, (g, φ) will be termed *h-critical.* That an h-critical pair is a differentially critical pair follows from Theorem 6.2 of MT. Hence (5.2) holds for h-critical pairs. From (i) we draw the following conclusion.

(ii). *The function defined by $D(g, \varphi)$ on the subset H of h-critical pairs (g, φ) is continuous on H.*

Hypothesis VI (b) is satisfied by virtue of (ii). To verify Hypothesis VI (a) we return to the sequence $f^n(\varphi)$, and prove the following lemma.

Lemma 5.1. *If ψ^0 is an fh-critical point of f^0, the harmonic surface defined by (g^0, ψ^0) is minimal.*

By definition of an fh-critical point of f^0, ψ^0 is the limit of a sequence of h-critical points ψ^r belonging respectively to a subsequence f^{n_r} of the function f^n. Let

$$S^r: \mathbf{x} = \mathbf{x}^r(u, v) \qquad (r = 1, 2, \cdots)$$

be the minimal surface defined by (g^{n_r}, ψ^r), and let

$$S^0: \mathbf{x} = \mathbf{x}^0(u, v)$$

be the harmonic surface defined by (g^0, ψ^0). The harmonic surface S^0 will be minimal if its differential coefficients satisfy the conditions

$$E(u, v) \equiv G(u, v), \qquad F(u, v) \equiv 0 \qquad (u^2 + v^2 < 1). \tag{5.3}$$

Let E^r, F^r, G^r denote the corresponding differential coefficients of S^r. Recall that

$$\lim_{r\to\infty} (g^{n_r}, \psi^r) = (g^0, \psi^0).$$

Upon representing $\mathbf{x}^r(u, v)$ by means of the Poisson integral, one finds that E^r, F^r, G^r converge to E, F, G as r becomes infinite. But S^r is minimal, so that conditions of the form (5.3) hold for S^r. It follows that (5.3) holds for S^0, so that S^0 is minimal.

The proof of the lemma is complete.

By virtue of the lemma, (5.2) holds on the set of fh-critical points of f^0. Hypothesis VI (a) is satisfied by virtue of (i).

Our principal theorem is as follows.

THEOREM 5.1. *Let g^0 be a simple, closed, rectifiable curve g^0 which satisfies the chord arc hypothesis. If g^0 bounds two minimal surfaces of disc type defined respectively by points on disjoint minimizing sets of the Douglas function $f^0(\varphi)$, then g^0 also bounds a minimal surface of disc type but not of minimum type.*

It follows from the hypotheses of the theorem that $f^0(\varphi)$ possesses two disjoint minimizing subsets ω_1 and ω_2. Let φ and ψ be points on ω_1 and ω_2 respectively. The 1-parameter family of points

$$t\psi(\alpha) + (1 - t)\varphi(\alpha), \qquad (0 \leqq t \leqq 1),$$

is an f^0-bounded arc connecting φ with ψ as has been shown in MT, page 451. The first hypothesis of Theorem 3.1 is accordingly satisfied.

Hypotheses I to VI hold for the sequence $f^n(\varphi)$ as we have seen. Theorem 3.1 thus applies, and we infer the existence of at least one fh- or h-critical point ψ^0 of f^0 not of minimum type. But we have just seen that such a point ψ^0 defines a minimal surface of disc type bounded by g^0; this surface is not of minimum type relative to the Douglas function f^0.

The proof of the theorem is complete.

THE INSTITUTE FOR ADVANCED STUDY AND
PRINCETON UNIVERSITY

Reprinted from DUKE MATHEMATICAL JOURNAL
Vol. 11, No. 1, March 1944

UNENDING CHESS, SYMBOLIC DYNAMICS AND A PROBLEM IN SEMIGROUPS

BY MARSTON MORSE AND GUSTAV A. HEDLUND

1. **Introduction.** The question of the possibility of an unending game of chess (with modified rules concerning a draw) leads to the problem of constructing an unending sequence of symbols from a given finite set so that the infinite sequence does not contain a block of the form EEe, where E is itself a block and e is the first symbol in E. Each block is supposed composed of a subsequence of *successive* symbols of the unending sequence. It was discovered [3] that an unending sequence generated from two symbols as previously defined and used to establish the existence of non-periodic recurrent motions in dynamics (cf. [1], [2]) has the desired property. The proof of this is published here for the first time.

The conventional rules governing chess contain rules concerning a draw by virtue of which every game contains a finite number of moves. Continental chess players have speculated upon various modifications of these rules and the possibility of an unending game of chess resulting therefrom. One such set of rules governing a draw was brought to the attention of Morse a number of years ago by members of the mathematics faculty of the University of Muenster. The set of pawns and pieces on the board prior to a move constitute what may be called a configuration. The *identity of two configurations* is understood to be complete identity both as to position and character of pieces and pawns. With this understood we see that there is at most a finite number of different configurations. These configurations are put together in a game in a sequence. In the terminology of the first paragraph a draw results if the sequence defined by the game contains a block of the type EEe. The conventional rules requiring that a pawn be moved or a piece taken after a certain number of moves are dropped in this modified game. The problem of the unending game under these modified rules had apparently been studied for some time in Germany without a solution being obtained.

It has also been pointed out to us by R. P. Dilworth that the existence of unending sequences of the type which yield a solution of the chess problem makes possible the construction of useful examples of semigroups. A semigroup is a set S of elements $a, b, c, \cdots$ in which an associative operation ab is defined. The element z is a zero element if $za = az = z$ for all a in S. If A and B are subsets of S, let AB denote the set of all products ab, where a and b are arbitrary elements of A and B, respectively. The semigroup S is said to be nilpotent if there exists an integer k such that $S^k = z$. It is possible to construct a non-nilpotent semigroup S generated by three elements, such that the square of

Received November 20, 1943.

every element in S is zero, provided it is possible to construct an unending sequence of three symbols which contains no block of the form EE, where E is a block. We prove that such an example can be constructed with the aid of the sequence constructed by Morse.

In order that the paper be self-contained we repeat the definition of the Morse recurrent symbolic trajectory and develop its properties ab initio.

2. **Definition and properties of the Morse symbolic trajectory.** The Morse recurrent symbolic trajectory is made up of two generating symbols and is defined as follows [4; 832]. Let

$$\begin{aligned} a_0 &= 1, & b_0 &= 2, \\ a_1 &= a_0 b_0, & b_1 &= b_0 a_0, \\ &\cdots, & &\cdots, \\ a_{n+1} &= a_n b_n, & b_{n+1} &= b_n a_n, \end{aligned} \tag{2.0}$$

with the understanding that the blocks $a_n b_n$ and $b_n a_n$ are to be expanded as blocks of 1's and 2's of length 2^{n+1}. The indexed trajectory T will then have the representation

$$\cdots c_{-2} c_{-1} c_0 c_1 c_2 \cdots, \tag{2.1}$$

where

$$c_0 c_1 \cdots c_{2^n - 1} = a_n \qquad (n = 0, 1, 2, \cdots), \tag{2.2}$$

and

$$c_{-i} = c_{i-1} \qquad (i = 1, 2, 3, \cdots). \tag{2.3}$$

The subray

$$R = c_0 c_1 c_2 \cdots \tag{2.4}$$

of T thus starts as follows:

(2.5) 1221 2112 2112 1221 2112 1221 1221 2112 2112 1221.

Corresponding to an arbitrary block A, the block which consists of the symbols of A in inverse order will be denoted by A^{-1}. We note that

$$a_n = a_n^{-1}, \quad b_n = b_n^{-1} \qquad (n \text{ even}), \tag{2.6'}$$

$$a_n = b_n^{-1}, \quad b_n = a_n^{-1} \qquad (n \text{ odd}). \tag{2.6''}$$

Relations (2.6) are clearly true when $n = 0$ and their truth in general follows inductively from the relations

$$a_{m+1}^{-1} = (a_m b_m)^{-1} = b_m^{-1} a_m^{-1},$$

$$b_{m+1}^{-1} = (b_m a_m)^{-1} = a_m^{-1} b_m^{-1}.$$

We infer from (2.6) that the 2^n-block of (2.1) the index of whose final symbol is -1 is $a_n^{-1} = a_n$ when n is even and $a_n^{-1} = b_n$ when n is odd. Thus the 2^{n+1}-block the index of whose initial symbol is -2^n is $a_n a_n$ when n is even and $b_n a_n$ when n is odd. Each block of T appears in such a 2^{n+1}-block for n sufficiently large. Since

$$a_{n+3} = a_n b_n b_n a_n b_n a_n a_n b_n ,$$

each block of T appears in the subray R of (2.4).

Let A be an arbitrary block of 1's and 2's. The block obtained from A by replacing 1 by 2 and 2 by 1 will be denoted by A'. An easy inductive proof shows that

$$a_n' = b_n , \qquad b_n' = a_n .$$

Let T' be a trajectory with indexed representation (e) in which $e_i = c_i$, $i \geq 0$, while, for $i < 0$, e_i equals 1 or 2 according as c_i is 2 or 1 respectively. The 2^n-block of T the index of whose final symbol is -1 is a_n or b_n according as n is even or odd. The corresponding 2^n-block of T' is then $a_n' = b_n$ or $b_n' = a_n$ according as n is even or odd. We infer that each block of T' has a copy in the subray R of (2.4).

We introduce the trajectory T_n, whose generating symbols are a_n and b_n, and which is obtained from T or T' according as n is even or odd, by replacing 1 by a_n and 2 by b_n. It follows from relations (2.0), (2.3) and (2.6) that an expansion of the symbols of a_n and b_n in T_n into blocks of 1's and 2's will yield T. Unless otherwise stated, T_n will be regarded as a trajectory in the symbols a_n and b_n unexpanded as blocks of 1's and 2's. For example, $a_n b_n$ will be regarded as a 2-block of T_n, or, if we choose, as a 2^{n+1}-block of T.

We shall make use of the following properties of the trajectories T and T'.

(2.7) There are no more than two successive 1's or 2's in T or T'.

(2.8) Each 5-block of T or T' contains either the block 11 or else the block 22.

To establish (2.7) we note that each 3-block of R will be found in one of the expanded 2-blocks of T_1; that is, in one of the blocks

$$a_1 b_1 = 1221, \qquad a_1 a_1 = 1212,$$

$$b_1 a_1 = 2112, \qquad b_1 b_1 = 2121,$$

and no more than two successive 1's or 2's appear in these blocks. Since each block of T or T' occurs in R, statement (2.7) is proved.

To prove (2.8) we infer from (2.7) that there can be no more than two consecutive a_1's or b_1's in T_1. In particular, R cannot contain the block 12121, since, if this were the case, T_1 would contain the block $a_1 a_1 a_1$ or $b_1 b_1 b_1$ according as 12121 possessed an even or odd first index. Similarly R contains no block 21212. Since each block of T or T' occurs in R, statement (2.8) is proved.

3. Solution of the "chess problem".

THEOREM 3.1. *The Morse symbolic trajectory T does not contain a block B of the form $D\overline{D}d$, where $D = \overline{D}$ and d is the initial symbol of D.*

Suppose that the theorem is false and that T contains a block $D\overline{D}d$ fulfilling the stated conditions. Let ω be the length of D.

Case I. ω is odd.

If $\omega = 1$, B is either 111 or 222. According to (2.7), neither of these blocks can occur in T and we arrive at a contradiction.

If $\omega \geq 3$, the length of the block B is at least 7 and from (2.8) we infer that B contains at least one of the blocks 11 or 22. We consider only the case where B contains 11, since the case where B contains 22 can be treated similarly.

Since $B = D\overline{D}d$ with $D = \overline{D}$, and d the initial symbol of D, any 2-block appearing in B must appear in the initial $(\omega + 1)$-block of B. It follows that if $c_i c_{i+1}$ is the first occurrence of 11 in B, when B is considered as occurring in the indexed representation (2.1) of T, then $c_{i+\omega} c_{i+1+\omega}$ is also in B and is the block 11. Since ω is odd, either i is even or else $i + \omega$ is even. But in view of the fact that T can be obtained from T_1 by expanding the symbols a_1 and b_1 of T_1 into 12 and 21 respectively, and the fact that the initial indices of the 2-blocks obtained by expanding the symbols of T_1 are all even, it follows that in the representation (2.1) of T, $c_j c_{j+1}$ can be 11 or 22 only if j is odd. Thus ω cannot be odd.

Case II. ω is even.

Let us suppose that in the indexed representation (2.1) B occurs with the initial index $i + 1$ and hence

$$D = c_{i+1} c_{i+2} \cdots c_{i+\omega},$$

$$\overline{D} = c_{i+\omega+1} c_{i+\omega+2} \cdots c_{i+2\omega},$$

$$d = c_{i+2\omega+1}.$$

Subcase IIa. $i + 1$ is odd.

Since $c_j c_{j+1}$ can be 11 or 22 only if j is odd, it follows that, if $i + 1$ is odd, $c_i c_{i+1}$ must be either 12 or 21, and since ω is even we infer that $i + \omega$ is even and hence $c_{i+\omega} c_{i+\omega+1}$ must be either 12 or 21. Since $c_{i+1} = c_{i+\omega+1}$, it follows that $c_i = c_{i+\omega}$. Thus the block

$$c_i c_{i+1} \cdots c_{i+\omega-1} c_{i+\omega} \cdots c_{i+2\omega-1} c_{i+2\omega} c_{i+2\omega+1} \tag{3.1}$$

satisfies the conditions

$$c_j = c_{j+\omega}, \qquad j = i, i + 1, \cdots, i + \omega + 1,$$

and i is even. But then there is a block of T_1 which when expanded yields (3.1) and hence is of the form

$$D^* \overline{D}^* d^*, \tag{3.2}$$

where $D^* = \overline{D}^*$ and d^* is the initial symbol of D^*. The length of the block D^* is $\frac{1}{2}\omega$.

But the symbolic trajectory T_1 is obtained from the symbolic trajectory T' by replacing 1 by a_1 and 2 by b_1. We have seen that each block of T' has a copy in the subray R of T and we conclude that in the case under consideration, T must contain a block of the form $B^* = E\overline{E}e$, where $E = \overline{E}$, e is identical with the initial symbol of E, and E is of length $\frac{1}{2}\omega$. If $\frac{1}{2}\omega$ is odd, we are led to the impossible Case I; if $\frac{1}{2}\omega$ is even and B^* occurs with odd initial index we are led to a repetition of Case IIa; if B^* occurs only with even initial index we are led to the final Case IIb which we now consider.

Subcase IIb. $i + 1$ is even.

Since $c_j c_{j+1}$ can be 11 or 22 only if j is odd, it follows that, if $i + 1$ is even, $c_{i+2\omega+1}c_{i+2\omega+2}$ must be either 12 or 21, and since ω is even we infer that $i + \omega + 1$ is even and $c_{i+\omega+1}c_{i+\omega+2}$ must be either 12 or 21. Since $c_{i+\omega+1} = c_{i+2\omega+1}$, it follows that $c_{i+\omega+2} = c_{i+2\omega+2}$. Thus the block

$$c_{i+1} \cdots c_{i+\omega}c_{i+\omega+1} \cdots c_{i+2\omega}c_{i+2\omega+1}c_{i+2\omega+2} \tag{3.3}$$

satisfies the conditions

$$c_j = c_{j+\omega}, \qquad j = i + 1, i + 2, \cdots, i + \omega + 2$$

and, moreover, $i + 1$ is even. But then there is a block of T_1 which when expanded yields (3.3) and hence is of the form (3.2) with D^* of length $\frac{1}{2}\omega$. Again we have reduced the situation either to the impossible Case I or to one of the Subcases IIa or IIb with ω only one-half of its original value. Since a positive integer contains the factor 2 at most a finite number of times, we must eventually be led to the impossible Case I.

The proof of the theorem is complete.

4. Solution of the "group problem". We now make use of a process called association [4; 843] by means of which a given symbolic trajectory determines another symbolic trajectory. Let (2.1) be an indexed representation of the Morse symbolic trajectory T and let B_i be the indexed 2-block of (2.1) with first index i. Since T has just two generating symbols, 1 and 2, there are at most four non-similar blocks B_i and from (2.5) it is clear that all the possible 2-blocks 11, 12, 21, and 22 occur in T. The indexed trajectory

$$\cdots B_{-1}B_0B_1B_2 \cdots \tag{4.1}$$

thus has four generating symbols. Let S be the symbolic trajectory of which (4.1) is an indexed representation. If we denote the 2-blocks 11, 12, 21, 22 by 1, 2, 3, 4, respectively, a simple computation shows that $B_0B_1B_2 \cdots$ begins as follows:

2432 3124 3123 2432 3123 2431.

THEOREM 4.1. *The symbolic trajectory* S *contains no block of the form* $A\bar{A}$ *with* $A = \bar{A}$.

If the theorem were not true, it would follow from the construction of S that T would contain a block of the form $D\bar{D}d$, where $D = \bar{D}$ and d is the initial symbol of D. But according to Theorem 3.1, this is not possible, and Theorem 4.1 is proved.

Let U be the symbolic trajectory which is obtained from S by changing 4 in S to 1. Thus, in particular, U contains the block

$$2132 \quad 3121 \quad 3123 \quad 2132 \quad 3123 \quad 2131.$$

THEOREM 4.2. *The symbolic trajectory* U *has three generators and contains no block of the form* $E\bar{E}$ *with* $E = \bar{E}$.

That U has just three generators is obvious from the construction of U from S.

Suppose that U contains a block $E\bar{E}$ with $E = \bar{E}$. This would imply the existence of a block $F\bar{F}$ in S such that F and $\bar{F}$ become identical when 4 is replaced by 1. Suppose that, in the indexed representation (4.1) of S, F has the representation $B_{i+1}B_{i+2} \cdots B_{i+\omega}$ and $\bar{F}$ has the representation $B_{i+\omega+1}B_{i+\omega+2} \cdots B_{i+2\omega}$. Since 1 in S corresponds to 11 in T and 4 in S corresponds to 22 in T, while T contains neither of the blocks 111 or 222, it follows that 1 in S must be preceded by 3 and followed by 2, while 4 in S must be preceded by 2 and followed by 3.

If $\omega = 1$, it follows from the hypothesis $E = \bar{E}$ that either $F = \bar{F}$ or $F\bar{F} = 14$ or $F\bar{F} = 41$. But Theorem 4.1 implies that the first case is impossible and since 1 is followed by 2 and preceded by 3 in S, the second and third possibilities are excluded. Thus we cannot have $\omega = 1$.

Assuming that $\omega > 1$, we let j be an integer such that $i + 1 \leq j < i + \omega$. From the hypothesis $E = \bar{E}$ we infer that $B_j = B_{j+\omega}$ unless $B_j = 4$ and $B_{j+\omega} = 1$, or vice versa. But $B_j = 4$ implies $B_{j+1} = 3$; therefore $B_{j+1+\omega} = 3$; and consequently $B_{j+\omega} = 4 = B_j$. If $B_j = 1$, then $B_{j+1} = 2$, $B_{j+1+\omega} = 2$ and $B_{j+\omega} = 1 = B_j$. Thus we have

$$B_j = B_{j+\omega}, \qquad i + 1 \leq j < i + \omega. \tag{4.2}$$

The hypothesis $E = \bar{E}$ implies that $B_{i+\omega} = B_{i+2\omega}$ unless $B_{i+\omega} = 4$ and $B_{i+2\omega} = 1$ or vice versa. But if $B_{i+\omega} = 4$, then $B_{i+\omega-1} = 2$; therefore $B_{i+2\omega-1} = 2$; and consequently $B_{i+2\omega} = 4 = B_{i+\omega}$. If $B_{i+\omega} = 1$, then $B_{i+\omega-1} = 3 = B_{i+2\omega-1}$, and consequently $B_{i+2\omega} = 1 = B_{i+\omega}$. Combining this with (4.2) we obtain

$$B_j = B_{j+\omega}, \qquad i + 1 \leq j \leq i + \omega.$$

But this implies that $F = \bar{F}$, which is impossible according to Theorem 4.1.

The assumption that U contains a block of the form EE with $E = E$ leads to a contradiction and the proof of Theorem 4.2 is complete.

Bibliography

1. Marston Morse, *A one-to-one representation of geodesics on a surface of negative curvature*, American Journal of Mathematics, vol. 43(1921), pp. 33–51.
2. Marston Morse, *Recurrent geodesics on a surface of negative curvature*, Transactions of the American Mathematical Society, vol. 22(1921), pp. 84–100.
3. Marston Morse, *A solution of the problem of infinite play in chess*, Abstract 360, Bulletin of the American Mathematical Society, vol. 44(1938), p. 632.
4. Marston Morse and G. A. Hedlund, *Symbolic dynamics*, American Journal of Mathematics, vol. 60(1938), pp. 815–866.

The Institute for Advanced Study and The University of Virginia.

Reprinted from
ACTA MATHEMATICA
Vol. 79, pp. 51–103

DEFORMATION CLASSES OF MEROMORPHIC FUNCTIONS AND THEIR EXTENSIONS TO INTERIOR TRANSFORMATIONS.

By

MARSTON MORSE and MAURICE HEINS
Institute for Advanced Study PRINCETON, N. J.
Brown University PROVIDENCE, R. I.

§ 1. *Introduction.* The first objective of this paper is to use topological methods and concepts to enlarge the store of knowledge of meromorphic functions. Deformation classes of meromorphic functions are defined. The extension to interior transformations results in new homotopy theorems and contrasts between interior and conformal maps. Earlier papers have shown that there is a considerable body of theorems which can be formulated so as to retain meaning and validity after arbitrary homeomorphisms of the z- or w-spheres. Many theorems involving the relations between zeros, poles, and branch point antecedents and the images under f of boundaries are of this character. See Morse and Heins, (1) and Morse (2).

The second objective is to distinguish between the properties of meromorphic functions which are shared by interior transformations and those which are not. For the transformations from $\{|z| < 1\}$ to the w-sphere which are considered we find no difference[1] between meromorphic functions and interior transformations with respect to the invariants necessary to characterize a deformation class of functions with prescribed zeros, poles, and branch point antecedents. However, sequences $[f_k(z)]$ of meromorphic functions properly taken from different deformation classes cover the w-plane in a manner suggestive of the Picard theorem on essential singularities but with no counterpart for sequences of interior transformations. The discovery of such properties points to the problem of finding the non-topological assumptions which must be imposed upon interior transformations in order that they may share the non-topological properties discovered.

[1] For domains for z other than the disc differences may arise.

Topologically we are concerned with the homotopy classes of open simply-connected Riemann surfaces with prescribed properties, conformally with the existence of deformations of prescribed type through meromorphic functions when it is known that such deformations are possible through interior transformations. We seek to construct models of all deformation classes by »composing» homeomorphisms of $\{|z| < 1\}$ with an interior transformation with the prescribed zeros, poles, and branch point antecedents. Conformally this is possible only in trivial cases.

Interior transformations are used essentially in the sense of Stoilow. (4) With Whyburn (3) they are »interior» and »light». Whyburn has studied the underlying point set characteristics of these transformations in a very general setting. The uniformization theorems of Stoilow are found useful. See also, v. Kerékjártó, (5) pp. 173—184.

The following section gives a summary of the principal results without details or proofs. A detailed exposition follows.

§ 2. *The problem and the principal results.* In the sense in which we shall use the term an *interior transformation* $w = f(z)$ will be a sense-preserving continuous map of the open disc $S\{|z| < 1\}$ into the complex w-sphere with the following characteristic property. If z_0 is any point on S there exists a sense-preserving homeomorphism $\varphi(z)$ from a neighborhood N of z_0 to another neighborhood N_1 of z_0 with z_0 fixed, such that the function $f[\varphi(z)] = F(z)$ is analytic on N except at most for a pole at z_0, and is not identically constant. The trans, formation f is said to have a zero or pole at z_0 if F has a zero or pole at z_0- more precisely f is said to have a *zero* or *pole* of the *order* of the zero or pole of F at z_0.

If z_0 is the antecedent of a branch point of the r-th order of the inverse of F, z_0 is said to be an antecedent of a *branch point* of the r-th order of the inverse of f. In any case it is seen that $r + 1$ is the number of times a neighborhood of $w_0 = f(z_0)$ is covered by $f(z)$ for $|z - z_0|$ sufficiently small.

We shall restrict ourselves to the case in which f has a finite set of zeros

$$a_0, a_1, \ldots, a_r \qquad (r > 0), \tag{2.1}$$

and poles

$$a_{r+1}, \ldots, a_n \qquad (n > 1)^1, \tag{2.2}$$

---

[1] In all cases $m = n + 1$ shall denote the total number of zeros and poles. The cases $m = 1$ and $m = 0$ are treated in § 14.

with branch point antecedents

$$b_1, \ldots, b_\mu \qquad (\mu \geqq 0), \tag{2.3}$$

with $n > 1$ until § 14 is reached. In the principal study we shall assume that the zeros, poles, and branch points are *simple*, that is, have the order 1. The zeros, poles, and branch point antecedents will form a set of points

$$(\alpha) = (a_0, \ldots, a_n, \; b_1, \ldots, b_\mu)$$

termed *characteristic.* Two points in the set (α) both of which are zeros, or poles, or branch point antecedents are said to be of *like character.* Any reordering of the points of (α) in which points of like character are reordered among themselves will be termed *admissible.* Except in the case $\mu = 0$, $m = 2$ we shall use no reordering in which a_0 is changed in relative order.

Admissible f-deformations. We shall admit deformations D of f of the form

$$w = F(z, t) \qquad (|z| < 1) \qquad (0 \leqq t \leqq 1)$$

in which t is the deformation parameter and

$$F(z, 0) \equiv f(z). \qquad (|z| < 1)$$

We require that F map (z, t) continuously into the w-sphere and reduce to an interior transformation for each fixed t. Let (α^t) be the characteristic set of F at the time t. We require that the points of (α^t) vary continuously with t, remain simple, distinct and constant in number and character as t varies from 0 to 1, return respectively, when $t = 1$, to some one but not necessarily the same one of the characteristic points of $f(z)$ of like character. The terminal transformation $F(z, 1)$ has the same characteristic set as f with a possible reordering.

If (α^t) is independent of t the deformation is termed *restricted.* If $(\alpha^0) = (\alpha^1)$, the deformation is termed *terminally restricted.* If (α^1) is an admissible reordering of (α^0), D is termed *semi-restricted.* We let X denote any one of these three types of admissible deformations. Interior transformations which admit a deformation into each other of type X will be said to belong to the *same X-deformation class* and to be *X-equivalent.*

The invariants J_i. We shall define a set (J) of n numbers $J_i(f, \alpha)$ $(i = 1, \ldots, n)$ associated respectively with the respective pairs (a_0, a_i). The J_i's are invariant under any restricted deformation of f. The set (J) is thus associated with an ordered set (α) in which the first zero a_0 plays a special role. If F is a transformation with the ordered set (α), then all sets (J) belonging to transformations f with the same ordered set (α) have the form

$$J_i(f, \alpha) = J_i(F, \alpha) + r_i \qquad (i = 1, \ldots, n) \tag{2.4}$$

where r_i is an arbitrary integer. All such sets are realizable.

Topological models with prescribed sets (α) *and* (J). One begins by exhibiting an interior transformation F with the given ordered set (α). Then (2.4) defines the ensemble of sets (J) which are »associated» with (α). Any one of these sets (J) associated with (α) belongs to a transformation f obtainable from F in a simple way.

To obtain these models use is made of sense-preserving homeomorphisms $\eta(z)$ of the disc $S = \{|z| < 1\}$ onto itself. We term η *restricted* if (α) is point-wise invariant under η, *semi-restricted* if points of (α) are transformed into points of (α) of like character. Set $n + 1 = m$. Except[1] when $m = 2$ and $\mu = 0$, one can prescribe the ordered set (α) and any one of the associated sets (J) and affirm the existence of a semi-restricted homeomorphism η, dependent on (α) and (J), such that the function[2] $F\eta$ has (α) as an ordered characteristic set and (J) as its set of invariants. The case $m = 2$, $\mu = 0$ is exceptional.

Meromorphic models. The preceding models f are not in general meromorphic. Nevertheless there exists an explicit formula for a meromorphic function f with a prescribed ordered characteristic set (α) and associated invariants (J).

Restricted deformation classes C. A necessary and sufficient condition that two transformations f_1 and f_2 with the same ordered set (α) be in the same class C is that f_1 and f_2 possess the same invariants (J) with respect to (α). If f_1 and f_2 are meromorphic, the restricted deformation of f_1 into f_2 which is affirmed to exist can be made through meromorphic functions. This is equally true of the deformation classes C' and C'' to which we now refer.

Terminally restricted deformation classes C'. A new invariant is needed here. Two transformations f with the same ordered set (α) will be said to belong to the same *category* if the sums

$$J_1 + \cdots + J_n$$

for the two functions are equal mod. 2. There are three principal cases (assuming $m > 1$ until § 14).

[1] We are excluding the case $m < 2$ until § 14 in order to avoid complexity of statement. When $m < 2$ there are no invariants (J).

[2] We write $F(\eta(z)) = F\eta$.

I $\mu > 0$ or m odd.

II $\mu = 0$, $m = 4, 6, 8, \ldots$

III $\mu = 0$, $m = 2$.

In Case I any two functions f_1 and f_2 with the same set (α) are in the same class C'. In Case II, f_1 and f_2 are in the same class C' if and only if they are of the same category. In Case III f_1 and f_2 are in the same class C' if and only if they have the same invariant J_1 with respect to (α).

Semi-restricted deformation classes C''. If $m > 2$ or $\mu > 0$ there is but one class C'' with the given set (α). The case $\mu = 0$ and $m = 2$ is again exceptional.

Other exceptional cases. The cases in which $m < 2$, are excluded until § 14. We have required that $r > 0$ in (2. 1), thus excluding the possibility that f have poles but no zeros. This exceptional case is readily brought under the preceding by replacing f by its reciprocal. The case where there are no poles is admitted.

Cases in which the characteristic points are not simple can be treated essentially as in the simple case provided the deformations which are admitted are required to preserve the multiplicity of the respective characteristic points. If one permits the multiplicities of the characteristic points to change by virtue of various types of coalescence, an interesting theory of degeneracy arises analogous to the theory which describes the degeneracy of elliptic functions into trigonometric functions. It is planned to return to this problem in a later paper.

Topological and conformal differences. A first difference has already been indicated. Except when $m = 2$ and $\mu = 0$ models of all restricted deformation classes with a given set (α) can be obtained as functions F by composing a particular model F_0 with characteristic set (α) with suitably chosen semi-restricted homeomorphisms η of S. This is impossible if one operates only with meromorphic functions.

A second difference appears in the extent to which the w-sphere is covered by an infinite sequence $[f_k]$ of transformations with a prescribed set (α) and at most one representative from each restricted deformation class. We term $[f_k]$ a *model sequence.* Let R be any connected region of the w-sphere which contains $w = 0$ and $w = \infty$ and whose closure does not cover the w-sphere. If the models $[f_k]$ are not required to be meromorphic a model sequence $[f_k]$ can be defined so as to cover no points on the complement of R and with no subsequence converging continuously on S to 0 or ∞. This is impossible if the functions f_k are meromorphic.

To give a more revealing statement in the meromorphic case let a point z_0 of S be termed a *covering point* of $[f_k]$ if corresponding to any arbitrary neighborhood N of z_0 the set of images of N under $[f_k]$ covers the finite w-plane ($w = 0$ excepted) infinitely many times. Let H be any connected neighborhood on S of the points

$$(a_0, a_1, \ldots, a_n) = (a)$$

with the points (a) excluded. Any model sequence $[f_k]$ of meromorphic functions no subsequence of which »converges continuously» to 0 or to ∞ on H possesses at least one covering point on H.

This theorem follows readily from a theorem of Julia on normal families (see Montel (6), p. 37) once the appropriate properties of meromorphic models have been derived. Other theorems concerning the covering points z_0 of $[f_k]$ are obtained.

We begin the detailed study with the development of homotopy properties of locally simple arcs k and introduce invariants $d(k)$ designed to make possible a topological definition of the invariants J_i.

Part I. The Topological Theory.

§ 3. *The difference order $d(k)$ of a locally simple arc k.* The object of this section is to attach a number $d(k)$ to a locally simple, sensed arc k with prescribed end points in such a manner that $d(k)$ remains invariant under a class of deformations of k (to be defined) and characterizes k as representative of its deformation class.

Local simplicity. See Morse and Heins (1) I. The arcs k which are admitted are »represented» by continuous and locally 1—1 images

$$w(t) = u(t) + iv(t) \qquad (0 \leqq t \leqq t_0)$$

of a line interval $(0, t_0)$ and shall intersect their end points $w(0)$ and $w(t_0)$ only when $t = 0$ and t_0 respectively. The condition of local simplicity implies that there exists a constant $e > 0$ such that any subarc of k whose diameter is less than e is simple; such a constant e is called a *norm of local simplicity of k.* Any set of locally simple curves which admit the same norm of local simplicity will be termed *uniformly* locally simple.

Strictly speaking, $w(t)$ is a *representation* of k and not identical with k. We admit any other representation of k obtained by mapping the interval $(0 \leqq t \leqq t_0)$ homeomorphically onto another interval $(0, t_1)$. The arc k can be identified with a class of such representations. The property of local simplicity of an arc k is independent of the representation of k which is used, because the existence of a norm of local simplicity is independent of the representation used.

Let the end points of k be $w = a$ and $w = b$ respectively.

Admissible deformations of k. We shall admit deformations D of k with the following properties. The arcs of D shall be represented in a locally $1-1$ way in the form

$$w = H(t, \lambda) \qquad (0 \leqq t \leqq t_0) \qquad (0 \leqq \lambda \leqq \lambda_0) \tag{3.1}$$

where H maps the (t, λ) rectangle continuously into the finite w-plane with

$$H(0, \lambda) \equiv a \qquad H(t_0, \lambda) \equiv b$$

The arc k shall be represented by $H(t, 0)$. The arc k is »deformed» at the »time» λ into the arc k^λ represented by $H(t, \lambda)$. *The arcs k^λ shall be uniformly locally simple and shall intersect a and b only as end points.*

If the requirement of uniform local simplicity were replaced by the requirement that the arcs k^λ be separately locally simple, there would be but one deformation class of arcs with the end points a and b, instead of the countably infinite set of deformation classes which actually exists.[1]

Any two arcs k^λ appearing in the above deformation are said to be in the *same deformation class.* Actually the deformation is one of »representations», but it is immediately obvious that any two representations of the same arc can be admissibly deformed into each other. See (1) I p. 603. Accordingly the property of two arcs of being in the same deformation class is independent of their representations.

To avoid misunderstanding, the following should be pointed out. The possibility that $a = b$ is admitted. Let k_a and k_b be proper simple subarcs of k with a the initial point of k_a and b the terminal point of k_b. The arcs k_a and k_b can intersect in infinitely many points, even coincide. Fig. 1 shows four examples of arcs k in the same deformation class in the case $a = b$. The figures must of course be superimposed so that the point $w = a$ coincides with $w = b$ in all four cases. The lower right curve is designed to indicate the possibility of a curve

[1] Cf. Lemma 3.3.

8—46545. *Acta mathematica.* 79. Imprimé le 1 mars 1947

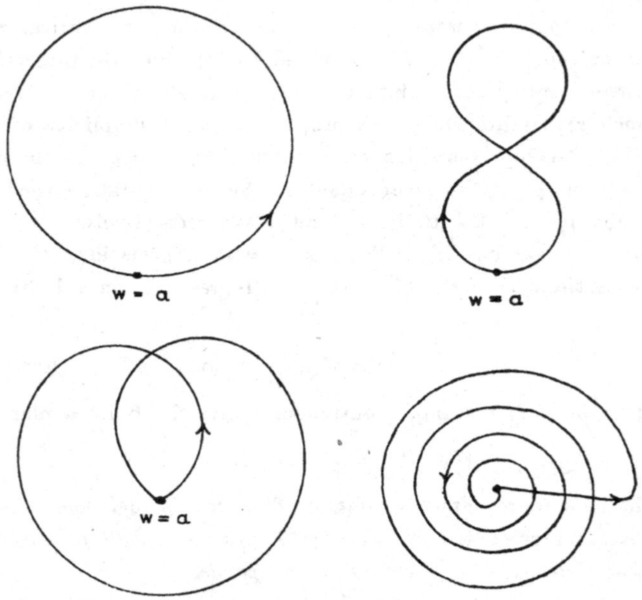

Figure I.

in the given deformation class with a simple spiral terminal arc. Another figure could be drawn indicating an arc in the same deformation class with spiral like simple subarcs at both ends. Reversed in sense the four arcs in Fig. I belong to a second and different deformation class.

The regular case $a \neq b$. The arc k is termed *regular* if it admits a representation $w(t)$ in which $w'(t)$ exists, is continuous and is never zero. For many purposes it will be convenient to measure angles in *rotational units*, that is, in units which equal 2π radians. The algebraic increment in

$$\frac{1}{2\pi} \arg w'(t) \qquad (0 \leqq t \leqq t_0) \tag{3.2}$$

as t increases from 0 to t_0 and $\arg w'(t)$ varies continuously will be denoted by $P(k)$. The angle P is in rotational units and represents the total angular variation of the tangent to k as the point of tangency $w(t)$ traverses k. Let c be either end point of k. The limiting algebraic increment of

$$\frac{1}{2\pi} \arg [w(t) - c] \qquad [0 < t < t_0]$$

as t increases from 0 to t_0 and the argument varies continuously will be denoted by $Q_c(k)$.

In the regular case with $a \neq b$ we tentatively define $d(k)$ by the equation

$$d(k) = P(k) - Q_a(k) - Q_b(k) \qquad (a \neq b) \tag{3.3}$$

and show that $d(k)$ is an integer.

The number $d(k)$ equals the rotational units in the total angular variation of a unit vector X which varies continuously as follows from

$$\frac{b-a}{|b-a|} \tag{3.4}$$

back to the same vector. Let X start with the vector (3.4) and coincide with

$$\frac{w(t)-a}{|w(t)-a|} \qquad (t_0 \geqq t > 0) \tag{3.5}$$

as t decreases from t_0 to 0, excluding 0. Let X then coincide with

$$\frac{w'(t)}{|w'(t)|} \qquad (0 \leqq t \leqq t_0) \tag{3.6}$$

as t increases from 0 to t_0. Finally let X coincide with

$$\frac{b-w(t)}{|b-w(t)|} \qquad (t_0 > t \geqq 0) \tag{3.7}$$

as t decreases from t_0 to 0. The initial and final vectors coincide with (3.4). The algebraic increments of $\arg\left(\frac{X}{2\pi}\right)$ corresponding to the variations associated with (3.5), (3.6) and (3.7) respectively are

$$-Q_a, \quad P, \quad -Q_b,$$

so that the resultant algebraic increment in angle is given by (3.3). The variation in X is continuous so that the statement in italics follows.

The general case $a \neq b$. The arc k is no longer assumed regular. A subarc of k will be defined by an interval (σ, τ) for t. If this subarc is simple the vector

$$w(\tau) - w(\sigma) \tag{3.8}$$

has a well defined direction. This direction will vary continuously with σ and τ so long as the arc (σ, τ) remains simple and $\sigma < \tau$. We term such a variation an *admissible chord variation*. In such a variation we suppose that the angle

$$\frac{1}{2\pi} \arg [w(\tau) - w(\sigma)] \tag{3.9}$$

has been chosen so as to vary continuously with σ and τ. For an admissible chord variation the algebraic increment of (3.9) depends only on the initial and final simple subarcs.

Let k_a and k_b be respectively proper simple subarcs of k of which the initial point of k_a is a and the terminal point of k_b is b. Let the chord subtending k_a vary admissibly into the chord subtending k_b. Let

$$P(k, k_a, k_b)$$

represent the resulting algebraic increment of (3.9). Let the algebraic increment of

$$\frac{1}{2\pi} \arg [w(t) - a] \tag{3.10}$$

as t increases from its terminal value t_a on k_a to t_0 be denoted by $Q_a(k, k_a)$ and let the algebraic increment of

$$\frac{1}{2\pi} \arg [w(t) - b] \tag{3.11}$$

as t increases from 0 to its initial value t_b on k_b be denoted by $Q_b(k, k_b)$.

The difference order $d(k)$ when $a \neq b$ is defined by the equation

$$d(k) = P(k, k_a, k_b) - Q_a(k, k_a) - Q_b(k, k_b) \qquad (a \neq b) \tag{3.12}$$

We state the lemma.

Lemma 3.1. *The difference order $d(k)$ as defined by* (3.12), *where $a \neq b$, is an integer which is independent of the choice of k_a and k_b among proper simple subarcs of k with end points as prescribed.*

The proof that $d(k)$ is an integer is essentially the same as in the regular case. As before, $d(k)$ measures the angular variation of a unit vector which initially and terminally coincides with (3.4) and may be broken up into the variations which define $-Q_a$, P, and $-Q_b$ respectively in (3.12).

It should be noted that any one of these three component angular variations may be arbitrarily large in numerical value and increase without limit as k_a or k_b tend to a or b respectively. This would certainly happen if k had spiral like terminal simple subarcs. We have the following lemma.

Lemma 3. 2. *If k is regular*

$$d(k) = P - Q_a - Q_b \qquad (a \neq b) \tag{3.13}$$

consistent with the earlier definition (3. 3).

The following lemma is also immediately obvious.

Lemma 3. 3. *The difference order $d(k)$ $(a \neq b)$ is independent of any admissible deformation of k. There are accordingly at least as many deformation classes as there are different values of $d(k)$ for admissible arcs joining a to b.*

For any integer m a »model» arc K_m joining a to b and with $d(k) = m$ can be defined as follows. For $n = 0$ take K_0 as the straight line segment from a to b. Let c be the mid point of K_0 and C be a unit circle tangent to K_0 at c and to the left of K_0. To define K_n for $n > 0$ trace K_0 from a to c, then trace C n times in the positive sense, and finally trace K_0 from c to b. Take K_{-n} as the reflection of K_n in K_0. That $d(K_m) = m$ follows from (3. 13). Since $d(k)$ is invariant under admissible deformations no two of the models K_m are in the same deformation class. It can be shown[1] that each admissible curve k which joins a to b and for which $d(k) = m$ is admissibly deformable into K_m.

The case $b = \infty$. We suppose that a is finite as previously, and make the obvious extensions of previous definitions as follows.

The arc k may be given in the w-plane. By convention it joins $w = a$ to $w = \infty$ if its closure $\bar{k}$ on the w-sphere joins $w = a$ to $w = \infty$. It is termed locally simple if $\bar{k}$ is locally simple on the w-sphere. Admissible deformations are defined as previously but on the w-sphere. If k is locally simple there exists a value τ with $0 < \tau < t_0$ such that the subarc $\tau \leqq t < t_0$ is simple in the finite w-plane. On this arc $|w(t)|$ becomes infinite as t tends to t_0.

To define $d(k)$ let k^s represent the subarc $(0, s)$ of k, with $0 < s < t_0$. Then $d(k^s)$ is well defined. It is clearly independent of s provided $s > \tau$. We accordingly set

$$d(k) = d(k^s) \qquad (\tau < s < t_0).$$

That $d(k)$ is invariant under admissible deformations follows as previously.

The arc k is termed *regular* if its closure $\bar{k}$ on the w-sphere is regular; one then has the lemma.

[1] We make no use of this theorem and accordingly omit the proof.

Lemma 3.4. *In case $\bar{k}$ is regular on the w-sphere, a is finite and $b = \infty$, then*

$$d(k) = P(k) - Q_a(k). \tag{3.14}$$

In case $\bar{k}$ is regular on the w-sphere k has a definite *asymptote* in the finite w-plane as t tends to t_0. The subarc k^s of k is regular, and if $w(t)$ represents k,

$$d(k^s) = P(k^s) - Q_a(k^s) - Q_{w(s)}(k^s).$$

On letting s tend to t_0 the last term tends to zero and the first two terms tend to the corresponding terms in (3.14). Thus Lemma 3.4 holds as stated.

The case $a = b$. We here suppose that a is finite. The values

$$Q_a(k, k_a) \qquad Q_b(k, k_b)$$

are undefined since they involve the null vector $b - a$. Moreover, the value of $d(k)$ which would be obtained by taking an appropriate limit as a tends to b is not the invariant which is useful. Instead we proceed as follows.

In the regular case one sets

$$d(k) = P(k) - Q_a(k). \qquad (a = b) \tag{3.15}$$

In the general case let

$$Q_a(k, k_a, k_b)$$

equal the algebraic increment of

$$\frac{1}{2\pi} \arg [w(t) - a] \tag{3.16}$$

as t varies monotonically from its terminal value t_a on k_a to its initial value t_b on k_b. In the general case one sets

$$d(k) = P(k, k_a, k_b) - Q_a(k, k_a, k_b). \qquad (a = b) \tag{3.17}$$

The first relevant facts are as follows:

Lemma 3.5. *The value of $d(k)$ when $a = b$ equals $\frac{1}{2}$ mod 1. It is independent of the choice of k_a and k_b among proper simple subarcs of k with end points as prescribed, and is invariant under admissible deformations of k.*

The angular variation which defines $d(k)$ is that of a vector Y which starts with $w(t_a) - a$ and ends with this vector reversed in sense. In fact it is sufficient if Y first varies admissibly as a chord from its initial position to the vector $b - w(t_b)$ subtending k_b; this variation gives P in (3.17). One continues with a variation of

$$b - w(t)$$

in which t decreases from t_b to t_a and in which the angular variation is measured by Q_a in (3.17). But the terminal vector Y coincides with the initial vector Y reversed in sense so that $d(k)$, measured in rotational units, equals $\frac{1}{2}$ mod 1.

The independence of $d(k)$ of the choice of k_a and k_b is clear as is its invariance under admissible deformations.

We note the following.

Lemma 3.6. *If k is regular and if $a = b$ (finite), or if a is finite and $b = \infty$, then*

$$d(k) = P(k) - Q_a(k). \tag{3.18}$$

Models for the different deformation classes when $b = \infty$ are obtainable from those for which $b = b_1$ (b_1 finite) by making a directly conformal transformation of the w-sphere which leaves a fixed and carries b_1 to ∞.

In case $a = b$ (finite) the values of $d(k)$ are found to be

$$\ldots, \frac{-3}{2}, -\frac{1}{2}, \frac{1}{2}, \frac{3}{2}, \ldots$$

A positively sensed circle C through a has a difference order $\frac{1}{2}$. Reversing sense always changes the sign of the difference order. To obtain a curve K_r with a difference order $n = \dfrac{2r+1}{2}$, and with $r > 0$, one attaches a small positively sensed circle C_1 to C within C, and tangent to C at some point c other than $a = b$; one then traces C until c is reached, then traces C_1 r times in the positive sense, and continues to b on C. To obtain a curve with difference order $-n$ one can reverse the sense of K_n, or more symmetrically reflect K_n in the tangent to C at a.

It can be shown that these models represent all possible deformation classes of admissible curves when $a = b$. We shall not use this fact.

§ 4. *Three deformation lemmas.* Let h be a simple arc joining two different points z_1 and z_2 in the finite z-plane. »Admissible» deformations of such an arc keep z_1 and z_2 fixed. If in addition only simple arcs are employed, the deformation will be termed an *isotopic* deformation of h. A first lemma is as follows.

Lemma 4.1. *Any simple arc h, joining z_1 and z_2 ($z_1 \neq z_2$) in the finite z-plane, can be isotopically deformed on an arbitrary neighborhood N of h into a simple, regular, analytic arc joining z_1 to z_2.*

We begin by proving the following:

(a) *The arc h can be isotopically deformed on N into a simple arc h^* on which sufficiently restricted terminal subarcs are straight.*

For simplicity take $z_1 = 0$. Choose e so that $0 < 2e < |z_2|$. Let C be the circle $|z| = e$ and E the open disc $\{|z| < e\}$. Let h_1 be a maximal connected subarc of h on E with $z = 0$ as an initial point. Let h_2 be a simple open arc of which h_1 is a subarc, which lies on E and whose closure is an arc joining diametrically opposite points of C. There exists a sense-preserving homeomorphism T of $\bar{E}$ which leaves $z = 0$ and C pointwise fixed, and maps $\bar{h}_2$ onto a diameter of $\bar{E}$. It follows from a theorem of Tietze (7) that T may be generated by an isotopic deformation Δ of $\bar{E}$ from the identity leaving C and $z = 0$ pointwise fixed.

If e is sufficiently small, Δ will deform $h \vee \bar{E}$ only on N and yield an image of h with a straight initial subarc. Such a deformation of h will be isotopic. Neighboring z_2, h can be similarly deformed, so that (a) follows.

To complete the proof of the lemma, let g be a closed Jordan curve of which the arc h^* in (a) is a subarc, and which is analytic and regular neighboring z_1 and z_2. Let R be the Jordan region bounded by g. Let $\bar{R}$ be mapped homeomorphically onto a circular disc, conformally at points of R. On the disc the pencil of circles through the images of z_1 and z_2 will have antecedents on $\bar{R}$ which will suffice to deform h^* isotopically on N into a simple, regular, analytic arc joining z_1 to z_2.

This completes the proof of the lemma.

In deriving deformation theorems for h when h joins z_1 to z_2 on S and $z_1 \neq z_2$, no generality is lost if it is supposed that a is real and positive and that, with $a < 1$,

$$z_1 = -a \qquad z_2 = a \qquad (a \neq 0).$$

For S can be mapped conformally onto itself so that z_1 and z_2 go into $-a$ and a respectively for a suitable value of a.

A non-singular transformation of coordinates. Let the z-plane be referred to polar coordinates (r, θ). For the above a and for each constant b on the interval $0 \leqq b < 1$ a transformation from polar coordinates (ϱ, φ) to (r, θ) will be defined by the equations

$$r = \varrho\left(\frac{1 - a^2 b^2}{1 - \varrho^2 b^2}\right), \quad \theta = \varphi \qquad (0 \leqq \varrho^2 b^2 < 1). \tag{4.1}$$

For $b = 0$, (4. 1) reduces to the identity. For $b \neq 0$ the interval $0 \leqq r < \infty$ and the interval

$$(4.2) \qquad 0 \leqq \varrho < \frac{1}{b}$$

correspond in a $1-1$ manner. The circle $\varrho = a$ corresponds to the circle $r = a$. Subject to (4. 2) the rectangular transformation of coordinates defined by (4. 1) is $1-1$, analytic and non-singular. The circle $\varrho = 1$ corresponds to the circle for which

$$(4.3) \qquad r = \frac{1 - a^2 b^2}{1 - b^2}.$$

We shall make use of the transformation (4. 1) and prove the following lemma.

Lemma 4. 2. *A simple, regular, analytic arc h joining $-a$ to a on S can be isotopically deformed on S through regular analytic arcs h^t into a circular arc joining $-a$ to a on S under a deformation in which the point with length parameter s on h corresponds to a point*

$$(4.4) \qquad z = z(s, t) \qquad \begin{Bmatrix} 0 \leqq t \leqq 1 \\ 0 \leqq s \leqq s_0 \end{Bmatrix}$$

at the time t, where $z(s, t)$ is analytic in the real variables (s, t), and $z_s \neq 0$.

To define the deformation (4. 4) let g be a Jordan curve on S which includes h as a subarc and which is analytic and regular neighboring a and $-a$. Let R be the Jordan region on S bounded by g. Let C be a circle on R and let R_1 be the subregion of R exterior to C. The region R_1 can be mapped $1-1$ and conformally on an annulus A in such a manner that the mapping can be extended $1-1$ and conformally over C and h. Suppose that the image k of h is on the outer circular boundary of A. We shall deform k through concentric circular arcs k_t, $0 \leqq t \leqq 1$, on A into a prescribed arc k_1 on the inner circular boundary of A. Such a deformation is readily defined in terms of the polar coordinates of the annulus so as to have a regular analytic representation in terms of t and the arc length on k. The antecedent h_t on R_1 of k_t on A will deform $h = h_0$ into a subarc h_1 of C.

During the deformation h_t of h the end points of h_t move. To remedy this defect let h_t be carried by a linear transformation $az + d$ (unique) into an arc H_t joining $-a$ to a. Note that $H_0 = h_0 = h$, and that H_1 will be *on* S provided the above arc k_1 has been chosen sufficiently short, and this we suppose done.

9—46545. *Acta mathematica.* 79. Imprimé le 3 mars 1947.

The deformation H_t satisfies the lemma except that H may not remain on S, although it begins and ends on S. We shall accordingly modify H_t as follows. Let $r(t)$ be the maximum value of r on H_t. Note that $r(0)$ and $r(1)$ are < 1. Let $[t]$ be the set of values of t on which $r(t) \geqq 1$. Let $b(t)$ be a real analytic function of t for $0 \leqq t \leqq 1$ such that

$$0 \leqq b(t) < 1 \qquad b(0) = b(1) = 0, \tag{4.5}$$

and so near 1 on $[t]$ that

$$r(t) < \frac{1 - a^2 b^2(t)}{1 - b^2(t)} \qquad (0 \leqq t \leqq 1). \tag{4.6}$$

With $b(t)$ so chosen let T_t be the transformation from the z-plane of (r, θ) defined by the inverse of (4.1) when $b = b(t)$, and let h^t be the image $T_t H_t$ of H_t at the time t. It follows from the choice of $b(t)$ and from (4.6) that $\varrho < 1$ on h^t, so that h^t is on S. Moreover, $h^0 = h$ and for every t the end points of h^t are a and $-a$ respectively. The arc h^1 is the circular arc H_1.

The deformation h^t, suitably represented, satisfies the lemma. The representation of h^t is completely determined by the requirement that $z(s, t)$ represent the point on h^t at the time t into which the point s on h has been deformed. The condition $z_s \neq 0$ is obviously satisfied and the proof of the lemma is complete.

(b) *In the preceding lemma at most a finite number of arcs h^t pass through any point of S not on h.*

To verify (b) let z_0 be a point on S not on h. The set of pairs (s, t) on the (s, t) rectangle which satisfy the condition $z(s, t) = z_0$ is empty, or consists of a finite number of pairs, or includes at least one analytic arc $s = s(t)$. In the last case the fact that $z_s(s, t) \neq 0$ insures that the arc $s(t)$ can be continued analytically, in either sense, and in particular in the sense of decreasing t until a boundary point (s^*, t_0) of the (s, t) rectangle is reached. But $z(s^*, t_0)$ is then a point of h, so that z_0 is a point of h, contrary to hypothesis. Hence (b) holds as stated.

The following lemma is a consequence of Lemma 4.2 and (b).

Lemma 4.3. *Any two simple, regular, analytic arcs h_1 and h_2 which join $-a$ and a on S $(a \neq 0)$ can be isotopically deformed into each other on S through simple, regular, analytic arcs no more than a finite number of which pass through any point of S not on h_1 or h_2.*

The arc h_i, $i = 1, 2$, can be deformed on S in the manner stated in Lemma 4.1 and in (b), into a circular arc k_i joining $-a$ to a. But k_1 can be deformed into k_2 through a pencil of circles joining $-a$ to a on S. Lemma 4.3 follows.

§ 5. *The invariants J_i.* Before coming to the definition of the invariants J_i a theorem on interior transformations will be recalled. See ref. (1) II p. 653. Let g be a locally simple, sensed, closed plane curve: $w = w(t)$, with $w(t + 2\pi) \equiv w(t)$. If e_1 is a sufficiently small positive constant, and $0 < e < e_1$

$$\frac{1}{2\pi} \arg [w(t + e) - w(t)] \qquad (0 \leqq t \leqq 2\pi)$$

will be well defined,[1] and as t increases from 0 to 2π, will change by an integer $p(g)$ independent of $e < e_1$. We term p the *angular order* of g. If g does not pass through $w = 0$, its ordinary order with respect to $w = 0$ will be denoted by $q(g)$. The theorem which we shall use is as follows:

Theorem 5.1. *Let B be a closed Jordan curve in the z-plane with interior G and let $F(z)$ be an interior transformation of some region R which includes $\bar{G}$ in its interior, and which maps R into the w-sphere. If B is free from branch point antecedents and has an image g which does not intersect $w = 0$ or ∞, then*

$$n(0) + n(\infty) - \mu = 1 + q(g) - p(g) \tag{5.1}$$

where $n(0)$, $n(\infty)$ and μ are respectively the numbers of zeros, poles, and branch point antecedents of F on G counting these points with their multiplicities.

To come to the definition of the invariants J_i, let f be an interior transformation from $S = \{|z| < 1\}$ to the w-sphere with the characteristic set

$$(a) = (a_0, a_1, \ldots, a_n, b_1, \ldots, b_\mu) \qquad (m = n + 1).$$

Let h_i be a simple curve joining a_0 to a_i on S with $i > 0$. Two curves h_i will be said to be of the same *topological type* if they can be isotopically deformed into each other on S without intersecting the set (a) other than in h_i's end points a_0 and a_i. Let h_i^f denote the image of h_i under f. The curve h_i^f will join $f(a_0)$ to $f(a_i)$ in the w-plane and be locally simple. If h_i is isotopically deformed for $0 \leqq t \leqq 1$ through curves of the same topological type, h_i^f will be admissibly deformed in the sense of § 3 through a uniformly locally simple family of arcs joining $f(a_0)$ to $f(a_i)$ and intersecting $f(a_0)$ and $f(a_i)$ only as end points. The difference order $d(h_i^f)$ is accordingly invariant under such deformations.

However $d(h_i^f)$ will change in general with the topological type of h_i. It is possible to define another function $V(h_i)$ of h_i which changes with the type of h_i exactly as does $d(h_i^f)$. To that end set

[1] Here and elsewhere the argument of any continuously varying non-null function F will be taken as a branch which varies continuously with F.

$$A(z,\alpha) = (z-a_0)(z-a_1)\ldots(z-a_n)$$

$$(5.2)\qquad B(z,\alpha) = (z-b_1)(z-b_2)\ldots(z-b_\mu) \qquad (\mu > 0)$$

$$B(z,\alpha) \equiv 1 \qquad (\mu = 0)$$

$$(5.3)\qquad C_i(z,\alpha) = \frac{(z-a_0)(z-a_i)B(z,\alpha)}{A(z,\alpha)} \qquad (i > 0).$$

The right member of (5.3) has a removable singularity at $z = a_0$ and at $z = a_i$. We suppose that $C_i(z,\alpha)$ takes on its limiting values at a_0 and a_i so that

$$(5.4)\qquad C_i(a_0,\alpha) = \frac{(a_0-a_i)B(a_0,\alpha)}{A'(a_0,\alpha)}$$

$$(i > 0)$$

$$(5.5)\qquad C_i(a_i,\alpha) = \frac{(a_i-a_0)B(a_i,\alpha)}{A'(a_i,\alpha)}.$$

Corresponding to a variation of z along h_i set

$$(5.6)\qquad V(h_i) = \frac{1}{2\pi}[\arg C_i(z,\alpha)]_{z=a_0}^{z=a_i}.$$

Regardless of the arguments used

$$(5.7)'\qquad V(h_i) \equiv \frac{1}{2\pi}[\arg C_i(a_i,\alpha) - \arg C_i(a_0,\alpha)] \qquad (\text{mod } 1).$$

We shall prove the following theorem:

Theorem 5.2. *The value of the difference*

$$(5.7)\qquad d(h_i^t) - V(h_i) \qquad (i = 1,\ldots,n)$$

is independent of h_i among simple curves which join a_0 to a_i without intersecting the other points of the characteristic set (α).

The value of the difference (5.7) is clearly independent of isotopic deformations of h_i through curves of the same topological type, since this is true of both terms in (5.7). By virtue of Lemma 4.1 we can accordingly restrict attention to arcs h_i which are admissible in the lemma and in addition are regular and analytic. If h_i^0 and h_i^1 are two such curves we seek to prove that (5.7) has the same value for h_i^0 as for h_i^1.

In accordance with Lemma 4.3 there exists an isotopic deformation of h_i^0 into h_i^1 through simple, regular, analytic arcs h_i^t $(0 \leqq t \leqq 1)$ such that h_i^t intersects the set $(\alpha) - (a_0, a_i)$ for at most a finite set of values t_0 of t. It is only as t passes through such a value t_0 that

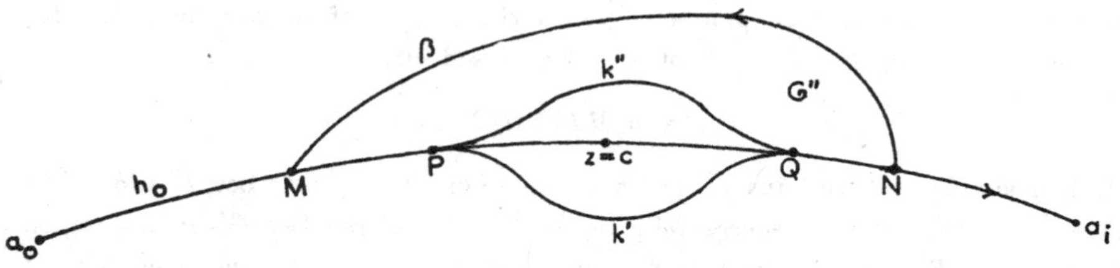

Figure 2.

(5.8) $$d(h_i^{t\,f}) - V(h_i^t)$$

could possibly change. Suppose that $h_i^{t_0}$ passes through just one point $z = c$ of the set $(\alpha) - (a_0, a_i)$. The proof in the case in which there are several points $z = c$ on $h_i^{t_0}$ will be seen to be similar. We shall show that there is no net change in (5.8) as $t - t_0$ changes sign.

Let h', h'', and h^0 be respectively the arcs h_i^t for which $t = t_0 - e$, $t_0 + e$ and t_0. For e_1 sufficiently small and $0 < e < e_1$, the curves h' and h'' will intersect (α) in a_0 and a_i only. We suppose that $e < e_1$. Without loss of generality we can suppose that h' passes[1] $z = c$ to the right of h^0. If h'' likewise passes $z = c$ to the right of h^0, it is possible to deform h' into h'' without intersecting $z = c$. (One moves each point of h' along a normal[2] to h^0 until h'' is met.) In this case (5.7) has the same value on h' as on h''.

Suppose then that h'' passes $z = c$ to the left of h^0. Without changing the topological type of h' one can deform the points of h' along normals to h^0 so that h' comes to coincide with h_0 except on a short open arc k' which lies to the right of h^0 neighboring $z = c$. Similarly one can deform h'' so that it comes to coincide with h_0 except on a short open arc k'' which lies to the left of h^0 neighboring $z = c$. We can also suppose that the end points of k' and k'' on h_0 coincide in points P and Q. See Fig. 2.

Let M and N be points of h^0 such that the arc (MN) of h^0 contains the arc (PQ) of h^0 on its interior. Let β be a simple open arc which joins M to N to the left of h'' so near h'' that

$$B'' = \beta(MP)k''(QN)$$

[1] More definitely, we suppose that h' intersects the normal to h^0 at $z = c$ to the right of h_0.
[2] Provided e_1 is sufficiently small.

is a Jordan curve with no points of (a) on the closure of its interior. We refer to the subarcs (MP) and (QN) of h^0. The closed curve

$$B' = \beta(MP)k'(QN)$$

is then simple and contains no point of (a) other than $z = c$. Let B' and B'' be taken in their positive senses relative to their interiors. Let G' be the region bounded by B', and G'' the region bounded by B''. We shall apply Theorem 5.1 to f as defined on $\bar{G}'$, and to f as defined on $\bar{G}''$.

Case I. f on G'. Let g' be the image of B' under f. The point $z = c$ is the only zero, pole, or branch point antecedent on G'. The numbers $n(0)$, $n(\infty)$ and μ refer to $z = c$. One only of these numbers differs from zero. In accordance with Theorem 5.1,

$$n(0) + n(\infty) - \mu = 1 + q(g') - p(g'). \tag{5.9}$$

Case II. f on G''. Let g'' be the image of B' under f. There are no zeros, poles or branch point antecedents on G'' so that

$$0 = 1 + q(g'') - p(g''). \tag{5.10}$$

From (5.9) and (5.10) one obtains the relation

$$n(0) + n(\infty) - \mu = [p(g'') - q(g'')] - [p(g') - q(g')]. \tag{5.11}$$

Equation (5.11) will give us our basic equality.

On taking account of the fact that B' coincides with B'' except along the arcs k' and k'' respectively, and that h' (as altered) similarly coincides with h'' (as altered) except along k' and k'' respectively, it appears that the right member of (5.11) has the value

$$d(h''^f) - d(h'^f) \tag{5.12}$$

in accordance with the definition of the difference order d.

On the other hand, $V(h') - V(h'')$ reduces to the variation of arg C_i along the closed curve $k'k''$, and so equals the number of zeros minus the number of poles of C_i within this curve. Thus

$$V(h') - V(h'') = \mu - n(0) - n(\infty) \tag{5.13}$$

in accordance with the definition of C_i. From (5.11) then

$$d(h''^f) - d(h'^f) = V(h'') - V(h'). \tag{5.14}$$

There is thus no change in (5.7) as t passes through t_0, and the theorem follows.

The invariants J_i. As suggested by Theorem 5.2 we set

$$J_i(f, \alpha) = d(h_i^f) - V(h_i) \qquad (i = 1, \ldots, n) \tag{5.15}$$

for any simple arc h_i which joins a_0 to a_i without intersecting $(\alpha) - (a_0, a_i)$. The numbers J_i are independent of the choice of h_i among admissible arcs h_i and of restricted deformations of f.

A necessary condition that two interior transformations with the same characteristic set (α) be in the same restricted deformation class is accordingly that their invariants J_i be respectively equal. It will be shown in § 12 that this condition is sufficient.

If (β) is an admissible reordering of (α), then

$$J_i(f, \alpha) \neq J_i(f, \beta)$$

in general. In case of ambiguity we shall refer to $J_i(f, \alpha)$ as the ith invariant J_i *with respect to* (α).

The value of $V(h_i)$ is independent of f and depends on (α). The values of $d(h_i^f)$ obtainable by changing f, differ by integers. One can accordingly include all values of J_i with respect to (α) in the set

$$J_i = J_i(f^0, \alpha) + m_i \tag{5.16}$$

where m_i is an integer and f^0 an interior transformation with the characteristic set (α). We thus have the theorem,

Theorem 5.3. *The invariants $J_i(f, \alpha)$ for a given i belonging to two different transformations f with the same characteristic set (α) differ by an integer m_i.*

It will be seen that there are functions with the prescribed set (α) for which the integers m_i are prescribed.

We term the sets (J) given by (5.16) with m_i an arbitrary integer the sets (J) associated with (α).

§ 6. *The existence of at least one interior transformation f with a prescribed characteristic set (α).* In this section we shall establish the existence of at least one f with the characteristic set (α) by exhibiting the Riemann image of S with respect to f. In the next section we shall show that except when $\mu = 0$ and

$m = 2$, f can be composed with suitably chosen semi-restricted homeomorphisms of S to obtain composite functions $f\eta$ with invariants $J_i(f, \alpha) + r_i$ where r_i is an arbitrary integer. These models are meromorphic only in special cases. Meromorphic models will be described by formula in § 10.

The following lemma will be used.

Lemma 6.1. *There exists a sense-preserving homeomorphism T of S onto itself in which the image of an arbitrary point set*

$$z_1, \ldots, z_s \tag{6.0}$$

of s distinct points on S is a prescribed set

$$w_1, \ldots, w_s \tag{6.1}$$

of s distinct points on S.

Let g be a simple arc which joins two points on the boundary of S but which otherwise lies on S and passes through the points (6.0) in the order written. Let k be a similar arc passing through the points (6.1). The closed domains into which g divides $\bar{S}$ can be carried respectively by sense-preserving homeomorphisms T_1 and T_2 into the closed domains into which k divides $\bar{S}$ and these homeomorphisms can then be modified so as to transform g into k and in particular to carry the respective points z_i into the corresponding points w_i. The lemma follows.

Theorem 6.1. *There exists at least one interior transformation f of S with a prescribed characteristic set (α).*

The set (α) is proposed as an ordered set of r zeros, s poles and μ branch point antecedents.

The function f affirmed to exist will be defined by describing the Riemann image H of S with respect to f over the w-sphere Σ. As defined H will be the homeomorph of S, will cover the point $w = 0$ of Σ r times, the point $w = \infty$ s times, and possess μ simple branch points covering points of Σ distinct from $w = 0$ and $w = \infty$. One starts with an arbitrary open disc-like piece k of Σ which does not cover $w = 0$ or $w = \infty$. One then extends $r + s$ narrow open tongues from k over Σ with tips covering $w = 0$ r times and $w = \infty$ s times, keeping the extended surface free from branch points and simply connected. To the boundary of k so extended one joins μ two-sheeted branch elements making each junction along a short simple boundary arc of the branch element so that these elements do not cover

$w=0$ or $w=\infty$. We can suppose that the boundary of the resulting open Riemann surface does not cover $w=0$ or $w=\infty$.

The Riemann surface H so obtained can be mapped homeomorphically in a sense-preserving fashion onto S and, by virtue of the preceding lemma, in such a manner that the r points of H covering $w=0$, the s points covering $w=\infty$, and the μ branch points of H go respectively into the proposed zeros, poles, and branch point antecedents of the given set (α).

The proof of the theorem is complete.

The case $\mu=0$ $m=2$. In all cases except this one the composition of a particular transformation f possessing a prescribed characteristic set, with a suitably chosen homeomorphism η of S will yield (cf. § 8) a transformation $f\eta$ with the given set (α) and invariants J_i differing from those of f by arbitrary integers.

This method of composition fails when $\mu=0$ and $m=2$. In this case there is but one function $C_i(z,\alpha)$, namely $C_1\equiv 1$, so that $V(h_i)$ as given by (5.6) reduces to 0. There is but one invariant J_i, namely J_1, and

$$J_1 = d(h_1^f). \tag{6.2}$$

The set (α) reduces to (a_0, a_1) and there are two cases according as a_1 is a zero or pole. In the case of two zeros a_0 and a_1, the only possible values of J_1 are

$$\cdots -\frac{3}{2},\ -\frac{1}{2},\ \frac{1}{2},\ \frac{3}{2},\ \cdots \tag{6.3}$$

and in the case of a zero a_0 and a pole a_1

$$\cdots -2,\ -1,\ 0,\ 1,\ 2, \cdots. \tag{6.4}$$

We state the following theorem:

Theorem 6.2. *In case* $\mu=0$ *and* $m=2$ *there exists an interior transformation* f *of* S *with a prescribed characteristic set* (a_0, a_1) *with* J_1 *arbitrarily chosen from among the values* (6.3) *when* a_1 *is a zero, and from among the values* (6.4) *when* a_1 *is a pole.*

A proof of this theorem may be given by constructing a Riemann surface for the inverse of a function of the required type. However, the theorem is also established by the formulas of § 10 and the more topological proof will be omitted.

10—46545. *Acta mathematica.* 79. Imprimé le 3 mars 1947.

§ 7. *The variation in $J_i(f, \alpha)$ caused by variation in (α).* Let f be an interior transformation of S with the characteristic set (α). Let f^t, $0 \leqq t \leqq 1$, be an admissible deformation of f in which (α^t) is the characteristic set of f^t at the time t. Recall that

$$J_i(f, \alpha) = d(h_i^f) - V(h_i), \tag{7.0}$$

by definition. As f is deformed, $d(h_i^f)$ remains invariant while $V(h_i)$ depends on (α) but not on f. We have the important result:

The algebraic increment

$$\Delta J_i = J_i(f^1, \alpha^1) - J_i(f^0, \alpha^0) \tag{7.1}$$

in the invariants J_i in an admissible deformation f^t depends only on the path (α^t) and not on f.

More explicit formulas for J_i and ΔJ_i are needed. To that end set

$$d(h_i^f) \equiv u_i \qquad (\text{mod } 1), \tag{7.2}$$

and recall that u_i can be taken as $\frac{1}{2}$ when a_i is a zero, and 0 when a_i is a pole. Let e_i be 1 or -1 according as a_i is a zero or a pole. Then

$$u_i - \frac{1}{2\pi} \arg e_i \equiv \frac{1}{2} \qquad (\text{mod } 1). \tag{7.3}$$

To make the formula for J_i more definite we shall use a branch $\overline{\arg} X$ of the argument for which

$$0 \leqq \overline{\arg} X < 2\pi$$

signalling this branch by the addition of the bar. In accordance with the definition of C_i in (5.3)

$$C_i(a_i, \alpha) = \frac{B(a_i, \alpha)}{A'(a_i, \alpha)}(a_i - a_0) \qquad (i \neq 0)$$

$$C_i(a_0, \alpha) = \frac{B(a_0, \alpha)}{A'(a_0, \alpha)}(a_0 - a_i).$$

On referring to the formula (5.7)′ for $V(h_i)$ and making use of (7.0), (7.2) and (7.3), one finds that

$$J_i(f, \alpha) = \frac{\overline{\arg}}{2\pi}\left[\frac{e_i A'(a_i, \alpha)}{B(a_i, \alpha)}\right] - \frac{\overline{\arg}}{2\pi}\left[\frac{A'(a_0, \alpha)}{B(a_0, \alpha)}\right] + I_i \qquad (i = 1, \ldots, n) \tag{7.4}$$

where I_i is an integer, and $e_i = 1$ when a_i is a zero, and -1 when a_i is a pole.

The integers $I_i(f, \alpha)$ are invariant under restricted deformations of f and are uniquely determined by f and (α). They are fundamental in the meromorphic theory.

From (7.4) and the continuity of $J(f^t, \alpha^t)$ one obtains the following result:

Lemma 7.0. *If f^t, $0 \leqq t \leqq 1$, is an admissible deformation of a transformation f with characteristic set (α) and with (α^t) the characteristic set of f^t, then the difference*

$$\Delta J_i = J_i(f^1, \alpha^1) - J_i(f^0, \alpha^0) \qquad (i = 1, \ldots, n) \tag{7.5}$$

is given by

$$\theta_i(\lambda) = \left[\sum_k \frac{\arg}{2\pi} \left(\frac{a_i^t - a_k^t}{a_0^t - a_k^t} \right) - \sum_j \frac{\arg}{2\pi} \left(\frac{a_i^t - b_j^t}{a_0^t - b_j^t} \right) \right]_{t=0}^{t=1} \tag{7.6}$$

where

$$k = 1, \ldots, i-1, i+1, \ldots, n; \; j = 1, \ldots, \mu;$$

where λ represents the path (α^t); and where the argument in (7.6) is immaterial as long as a branch which varies continuously with t is used.

A path (α^t), $0 \leqq t \leqq 1$, which leads from a set (β) back to (β) will be called an *α-circuit*. We shall also use paths λ in which (α^1) is an admissible reordering of $(\alpha^0) = (\beta)$. We term such a path *an admissible α-circuit* mod β. As previously, admissibility of a path λ requires that $a_0^1 = a_0^0$.

The difference ΔJ_i in (7.5) will concern us not so much as the difference

$$\delta J_i = J_i(f^1, \beta) - J_i(f^0, \beta) \qquad (i = 1, \ldots, n)$$

because it is necessary to compare the J_i's with reference to the same set (β). When λ is an α-circuit,

$$\delta J_i = \Delta J_i,$$

and in this case δJ_i depends only on λ and not on the initial and final sets (J). If λ is an admissible α-circuit mod β which is not an α-circuit, this is never the case, as we shall see.

The group Ω of J-displacement vectors. Let $\{J, \beta\}$ denote the set of invariants (J) realizable as invariants of an interior transformation with the characteristic set (β). It will presently be seen that the set $\{J, \beta\}$ is identical with the complete set of (J)'s »associated» with (β) in (5.16).

When (α^t), $0 \leqq t \leqq 1$, represents an α-circuit λ leading from (β) to (β), the differences ΔJ_i given by (7.6) are integers r_i. If one sets

$$J_i = J_i(f^0, \beta) \qquad T_\lambda(J_i) = J_i(f^1, \beta)$$

it is seen that the α-circuit λ *induces* a transformation

$$T_\lambda(J) = (J) + (r)$$

of $\{J, \beta\}$ onto itself, in fact a translation. We term (r) the J-displacement vector *induced* by λ. The J-displacement vectors induced by α-circuits from (β) to (β) form an *additive abelian group* Ω. The group Ω is a subgroup of the additive group G of all integral vectors (r);. Ω may coincide with G, be a proper subgroup of G and even reduce to the null element.

We shall seek a set of generators of Ω.

To that end let z_p and z_q be any two distinct points of (β). An α-circuit $G(z_p, z_q)$ leading from (β) to (β) will be defined in which all points of (β) except the pair (z_p, z_q) remain fixed while the paths z_p^t, z_q^t $(0 \leqq t \leqq 1)$ are such that $z_p^t - z_q^t$ rotates through an angle -2π. We suppose, moreover, that these paths lie on a topological disc on S which does not intersect the set $(\beta) - (z_p, z_q)$. Such paths clearly exist. The disc can then be isotopically deformed on itself, into a disc arbitrarily close to a point. If this α-circuit is used in (7.6), the only terms which will make a non-null contribution are those which involve $z_p^t - z_q^t$ or $z_q^t - z_p^t$.

The case $\mu > 0$. In this case we introduce the α-circuit

$$\lambda_k = G(a_k, b_1) \qquad (k = 1, \ldots, n)$$

and obtain the following lemma.

Lemma 7.1. *When* $\mu > 0$, *there exist* α-*circuits* λ_k, $k = 1, \ldots, n$, *for which the components of the corresponding* J-*displacement vectors are* δ_i^k,[1] $i = 1, \ldots, n$. *These vectors generate* Ω *as the complete group* G *of integral vectors* (r).

The case $\mu = 0$ *and* $m > 2$. In this case the α-circuits

$$\lambda_{rs} = G(a_r, a_s) \qquad (r < s)$$

are introduced. The following lemma results.

Lemma 7.2. *When* $\mu = 0$ *and* $m > 2$ *there exist* α-*circuits* $\lambda_{rs}(r, s = 0, 1, \ldots, n;$ $r < s)$, *for which the corresponding* J-*displacement vector* D_{rs} *has the components*

$$-\delta_i^r - \delta_i^s \qquad (i = 1, 2, \ldots, n) \tag{7.7}$$

when $rs \neq 0$, *and when* $r = 0$ *has all components* 1 *except the* s-*th, which is zero.*

The vectors D_{rs} *generate the group* Ω. *When* m *is odd,* Ω *is the group* G *of all integral vectors* (r). *When* $m = 4, 6, 8, \ldots$, Ω *is the subgroup of* G *of vectors* (r) *for which* Σr_i *is even.*

---

[1] The Kronecker delta gives the i-th component.

The components of D_{rs} are obvious from (7.6).

When m is odd, the matrix whose columns are the components of the vectors

$$D_{12}, D_{23}, \ldots \quad D_{n-1\,n}\, D_{01} \qquad (m = n + 1) \tag{7.8}$$

has the determinant $\omega = -1$. For example, when $m = 5$:

$$\omega = \begin{vmatrix} -1 & 0 & 0 & 0 \\ -1 & -1 & 0 & 1 \\ 0 & -1 & -1 & 1 \\ 0 & 0 & -1 & 1 \end{vmatrix} = \begin{vmatrix} -1 & 0 & 0 & 0 \\ -1 & -1 & 0 & 0 \\ 0 & -1 & -1 & 0 \\ 0 & 0 & -1 & 1 \end{vmatrix} = -1. \tag{7.9}$$

When m is odd, the vectors accordingly generate G and hence Ω.

To treat the case $m = 4, 6, 8, \ldots$, let us term a vector (r) for which Σr_i is even, of *even* category, otherwise of *odd* category. When m is even, each vector D_{rs} is of even category, and hence the vectors generated by the vectors D_{rs} are of even category.

The vectors D_{rs} generate the group Ω.

To see this let λ be an arbitrary admissible α-circuit. The vector $a_r^t - a_s^t$ $(r < s)$ rotates through 2π an integral number m_{rs} of times (possibly zero) as t increases from 0 to 1. It follows from (7.6) that the J-displacement vector induced by λ has the form

$$-\Sigma m_{rs} D_{rs}$$

where the summation extends over the pairs (r, s) with $r < s$. Thus the vectors D_{rs} generate Ω.

It remains to show that every vector (r) of even category is in Ω. To that end we introduce a vector E whose components are δ_i^1. The matrix whose columns are the components of the vectors

$$D_{12}, D_{23}, \ldots, D_{n-1\,n}, E \tag{7.10}$$

is of odd order n and has a determinant 1 so that the vectors (7.10) generate G. Thus (r) is of the form

$$(r) = sE + D \tag{7.11}$$

where s is an integer and D is in Ω. Observe that

$$D_{12} - D_{23} + D_{34} - + \cdots - D_{n-1\,n} + D_{1\,n} = -2E$$

so that s in (7.11) can be taken as 1 or 0. The vector D is of even category, and if (r) is of even category, s in (7.11) cannot be 1. Thus (r) is in Ω, if of even category. This completes the proof when $m = 2, 4, 6, 8, \ldots$.

The case $\mu = 0$, $m = 2$. In this case there is but one value of i in (7.6), and $\Delta J_1 = 0$. Hence Ω reduces to the vector $(r) = 0$.

§ 8. *The generation of interior transformations by composition $f\eta$ with restricted homeomorphisms η.* We have seen in § 6 that there is at least one interior transformation f with a prescribed characteristic set (β). We shall see to what extent one can choose restricted homeomorphisms η of S onto itself so that $f\eta$ has invariants (J) arbitrarily prescribed from those associated with (β).

To that end we first connect restricted homeomorphisms η leaving (β) fixed with α-circuits from (β) to (β).

Any sense-preserving homeomorphism η of S may be generated as the terminal homeomorphism of an isotopic deformation η^t of S from the identity. More explicitly there exists a 1-parameter family η^t of homeomorphisms S onto S of the form

$$\eta^t \equiv \varphi(z, t) \qquad (0 \leqq t \leqq 1), \tag{8.1}$$

where φ is continuous in z and t,

$$z \equiv \varphi(z, 0)$$

and

$$\eta(z) \equiv \varphi(z, 1).$$

Let (α^t) be the antecedent of (β) under η^t. If η is a restricted homeomorphism leaving (β) fixed, (α^t) determines an α-circuit λ from (β) to (β). We shall say that η *induces* this α-circuit. If f is an interior transformation with the characteristic set (β), the composite function of z, $f\eta^t$, affords a terminally restricted deformation f^t of f in which the characteristic set of f^t at the time t is (α^t).

Formula (7.6) is applicable to the f-deformation $f^t = f\eta^t$ with its associated α-circuit (α^t), $0 \leqq t \leqq 1$, and yields the result

$$J_i(f\eta, \beta) - J_i(f, \beta) = r_i \qquad (i = 1, 2, \ldots, n) \tag{8.2}$$

where (r) is the J-displacement vector determined by (α^t). This displacement vector is independent of the choice of α-circuits λ induced by η since for the same f, (β), and η in (8.2), a second choice of a λ induced by η cannot change (r). The vector (r) is a J-displacement vector $D_1(\eta)$ *determined* by η in the group Ω. If $D_2(\lambda)$ is the vector in Ω *determined* by the α-circuit λ, then

$$D_1(\eta) = D_2(\lambda)$$

whenever λ is induced by η.

It can be shown that any α-circuit (α^t), $0 \leqq t \leqq 1$, from (β) to (β) is induced by *some* restricted homeomorphism of S leaving (β) fixed; to establish this one must show that there exists an isotopic deformation η^t of S from the identity in which (α^t) is the antecedent of (β) at the time t and in which η^1 is a restricted homeomorphism leaving (β) fixed. The details of a proof of this need not be given. It is sufficient to suggest to the reader that η^t can be defined by a sequence of deformations in each of which just one point of (α^t) is moved from an initial point z_0 to a nearby point z_1. One can make use of a deformation Δ from the identity of a small circular neighborhood N of z_0, defining the deformation Δ as the identity outside of N.

We summarize as follows:

Lemma 8.1. *Each restricted homeomorphism η of S leaving (β) fixed induces a class of α-circuits λ leading from (β) to (β), and every α-circuit leading from (β) to (β) is induced by a class of restricted homeomorphisms η leaving (β) fixed. If η induces λ and (r) is the J-displacement vector determined in* (7.6) *by λ, then* (8.2) *holds for every interior transformation with (β) as a characteristic set. The vector (r) depends only on η and not on the choice of an α-circuit λ induced by η.*

The reciprocal relations between restricted homeomorphisms η and their induced α-circuits and the theorems on the nature of the group Ω of J-displacement vectors induced by α-circuits yield the following theorem:

Theorem 8.1. *If f^0 is an interior transformation of S with the characteristic set (β) and invariants (J^0), suitably chosen restricted homeomorphisms η of S leaving (β) fixed will yield interior transformations $f^0\eta$ with invariants $(J^0) + (r)$ where*

(1) *(r) is an arbitrary integral vector when $\mu > 0$, or when $\mu = 0$ and m is odd,*

(2) *(r) is an arbitrary integral vector of even category when $\mu = 0$ and $m = 4, 6, 8, \ldots$,*

(3) *$(r) = (0)$ only, when $\mu = 0$ and $m = 2$.*

No other values of (J) can be obtained by composition $f^0\eta$ of f^0 with restricted homeomorphisms η.

§ 9. *α-circuits mod (β) and semi-restricted homeomorphisms*[1] *η of S.* We resume the theory of admissible α-circuits mod (β) initiated in § 7. As in § 7 we are concerned with an admissible deformation f^t, $0 \leqq t \leqq 1$, for which (α^t) is the characteristic set of f^t at the time t. We suppose that (α^1) is an admissible reordering of (α^0). In particular $a_0^1 = a_0^0$.

[1] The present section could be omitted by a reader who wishes to comprehend first the main theory.

Such a reordering defines a permutation π of $(1, \ldots, n)$ in which i is replaced by $\pi(i)$ and

$$a_i^1 = a_{\pi(i)}^0 \qquad (i = 1, \ldots, n). \tag{9.1}$$

For any interior transformation f with characteristic set (a^0),

$$J_i(f, a^1) = J_{\pi(i)}(f, a^0) \qquad (i = 1, \ldots, n) \tag{9.2}$$

in accordance with the definition of (J). If $(x_1, \ldots, x_n)$ is an arbitrary set of n symbols, we shall write

$$(x_{\pi(1)}, \ldots, x_{\pi(n)}) = \pi(x).$$

Thus (9.2) takes the form

$$[J(f, a^1)] = \pi[J(f, a^0)].$$

It follows from (7.6) that

$$J_i(f^1, a^1) = J_i(f^0, a^0) + \theta_i(\lambda) \qquad [\lambda = (a^t)]. \tag{9.3}$$

From (9.2) and (9.3) one sees that

$$J_{\pi(i)}(f^1, \beta) = J_i(f^0, \beta) + \theta_i(\lambda) \qquad (\beta) = (a^0). \tag{9.4}$$

Equations (9.4) may be written in the vector forms

$$\pi[J(f^1, \beta)] = [J(f^0, \beta)] + [\theta]. \tag{9.5}$$

$$[J(f^1, \beta)] = \pi^{-1}\{J(f^0, \beta) + [\theta]\}. \tag{9.6}$$

We thus have the following lemma:

Lemma 9.1. *Any admissible a-circuit mod (β) of the form $\lambda = \{(a^t), 0 \leq t \leq 1\}$ in which*

$$(a_1^1, \ldots, a_n^1) = \pi(a_1^0, \ldots, a_n^0),$$

induces a transformation

$$T_\lambda(J) = \pi^{-1}\{(J) + (\theta(\lambda))\} \tag{9.7}$$

of $\{J, \beta\}$ such that for any admissible deformation f^t of an interior transformation in which (a^t) is the characteristic set of f^t the invariants (J^1) of f^1 with respect to (β) are the transforms $T_\lambda(J^0)$ of the invariants (J^0) of f^0 with respect to (β).

The transformation T_λ of $\{J, \beta\}$ is compounded of a translation $(J) + (\theta)$ and a permutation π^{-1} of the components of the translated vector. It is a translation if and only if π is the identity. If π is not the identity, the numbers $\theta_i(\lambda)$ are not integers in general, so that $(J) + (\theta)$ is not in $\{J, \beta\}$ in general.

It follows from (9.4) that

$$\sum_i [J_i(f^1, \beta) - J_i(f^0, \beta)] = \Sigma \theta_i = q(\lambda), \tag{9.8}$$

introducing q. Here q is an odd or even integer depending only on λ. The transformation T_λ is termed of *odd* or *even category* according as q is odd or even. When $\mu = 0$ and $m = 4, 6, 8, \ldots$, all translations of $\{J, \beta\}$ induced by α-circuits have been seen to be of even category. Transformations of odd category are sought in this case. We shall prove the following lemma:

Lemma 9.2. *When $\mu = 0$ and m is even, a necessary and sufficient condition that an admissible α-circuit λ mod (β) induce a transformation U_λ of $\{J, \beta\}$ of odd category is that (α^1) be an admissible odd permutation of (α^0).*

To prove the lemma we evaluate $q(\lambda)$ in (9.8). When $\mu = 0$, a summation of the right members of (7.6) yields the result

$$q(\lambda) = \left[\frac{\arg}{\pi} \prod_{i,j} (a_i^t - a_j^t) - \frac{(n-1)}{2\pi} \arg \prod_k (a_0^t - a_k^t)\right]_{t=0}^{t=1} \tag{9.9}$$

where $i, j, k = 1, 2, \ldots, n$ with $i < j$. Since $a_0^1 = a_0^0$ and $n - 1$ is even, the contribution of the last term in (9.9) is an even integer. Observe that

$$\prod_{i,j} (a_i^1 - a_j^1) = \pm \prod_{i,j} (a_i^0 - a_j^0) \tag{9.10}$$

according as (α^1) is an even or odd permutation of (α^0) so that $q(\lambda)$ is correspondingly odd or even. This completes the proof of the lemma.

It is a consequence of this lemma that, when $\mu = 0$ and $m = 4, 6, 8, \ldots$, there exists a transformation U_λ of $\{J, \beta\}$ of odd category. For two at least of the points a_1, a_2, a_3, are of like type (zeros or poles) and there accordingly exists an admissible α-circuit mod (β) which interchanges these two points but which otherwise leaves (β) fixed.

The following lemma can now be proved:

Lemma 9.3. *Let J^0 be an arbitrary set in $\{J, \beta\}$ and let (r) be an arbitrary set of n integers. When $\mu = 0$ and $m = 4, 6, 8, \ldots$, there exists a transformation T_λ of $\{J, \beta\}$ induced by an admissible α-circuit λ mod. (β) such that the relation*

$$T_\lambda(J) = (J) + (r) \tag{9.11}$$

holds for $(J) = (J^0)$ and the given set (r).

The lemma is a consequence of Lemma 7.2 when (r) is of even category. There then exists a T such that (9.11) holds for *every* (J) in $\{J, \beta\}$.

Suppose then that (r) is of odd category. It follows from the preceding lemma that, when $\mu = 0$ and $m = 4, 6, 8, \ldots$, there exists a transformation U of $\{J, \beta\}$ of odd category induced by an admissible α-circuit mod (β) which replaces (β) by an admissible reordering (β'). Set

$$U(J^0) - (J^0) = (s). \tag{9.12}$$

The set (s) is of odd category and hence $(r) - (s)$ is of even category. In accordance with Lemma 7.2 there exists a translation V of $\{J, \beta'\}$ of even category induced by an α-circuit from (β') to (β') such that

$$V[J^0 + (s)] = [J^0 + (s)] + (r) - (s).$$

On making use of (9.12) it is seen that

$$VU(J^0) = (J^0) + (r).$$

Hence the transformation $T = VU$ satisfies the lemma.

The case $\mu = 0$, $m = 2$. In this case there is but one invariant J, and this invariant has the form

$$J(f, a) = J_1(f, a_0, a_1) = d(h_1^f)$$

where h_1 is a simple curve joining a_0 to a_1 on S. The term $V(h_1) = 0$ since $C_1(z) \equiv 1$ in this case. When a_0 is a zero and a_1 is a pole, we admit no relative α-circuits which interchange a_0 and a_1.

When both a_0 and a_1 are zeros, we shall admit relative α-circuits which interchange a_0 and a_1. When $m = 2$ and $\mu = 0$, the right member of (7.6) is devoid of terms so that

$$J_1(f^1, a_1, a_0) = J_1(f^0, a_0, a_1) \tag{9.13}$$

if (a^t) interchanges a_1 and a_0 during the deformation f^t.

A simple but striking illustration may be given. Let f be an interior transformation of S with $\mu = 0$, $m = 2$, and with zeros at real points a and $-a$. $(0 < a < 1)$. Let $F(z) = f(-z)$. There exists no admissible deformation of f into F which returns each zero into its initial position. This is a consequence of the fact that

$$J_1(F, a_0, a_1) = -J_1(f, a_0, a_1) \neq 0 \tag{9.13$'$}$$

while the existence of the deformation would require that

$$J_1(F, a_0, a_1) = J_1(f, a_0, a_1)$$

contrary to (9.13)′.

Let k_1 be h_1 reversed in sense. Then in accordance with the definition of J_1

(9. 14) $$J_1(f, a_0, a_1) = d(h_1^f) = -d(k_1^f) = -J_1(f, a_1, a_0).$$

It follows from (9. 13) and (9. 14) that when (a^t) interchanges the zeros in a deformation f^t

(9. 15). $$J_1(f^1, a_0, a_1) = -J_1(f^0, a_0, a_1).$$

We summarize in the lemma:

Lemma 9. 4. *When $\mu = 0$ and $m = 2$, admissible deformations f^t with characteristic sets (a^t), $0 \leqq t \leqq 1$, may interchange two zeros a_0 and a_1 but must return a zero a_0^t and pole a_1^t to their initial positions. In any case*

(9. 16) $$J_1(f^1, a_0, a_1) = \pm J_1(f^0, a_0, a_1)$$

where the minus sign prevails if and only if (a^t) interchanges two zeros.

Contrast this result with the fact that, when f is an even admissible function with just two zeros on S and no poles, f can be restrictedly deformed into $f(-z)$; in fact by the identity. The reason is found in the fact that an even function has a branch point antecedent at the origin so that $\mu > 0$.

Semi-restricted homeomorphisms η. The developments of § 8 showing the reciprocal relationship between a-circuits, (a^t) and restricted homeomorphisms η which induce them is paralleled here by the relationship between admissible a-circuits λ mod β and semi-restricted homeomorphisms η which admissibly reorder (β). Let η^t be an isotopic deformation of S generating η and let (a^t) be the antecedent of (β) under η^t. We say that η *induces* the relative a-circuit $\lambda = \{(a^t)\ 0 \leqq t \leqq 1\}$. The f-deformation $f^t = f\eta^t$ has the characteristic set (a^t) and with λ comes under Lemma 9. 1, so that for the transformation T_λ of $\{J, \beta\}$ given by (9. 7) and induced by λ

(9. 17) $$[J(f\eta, (\beta)] = T_\lambda [J(f, \beta)].$$

As in § 8 we infer that each such relative a-circuit λ is »induced» by some semi-restricted homeomorphism η. Hence if λ is an arbitrary admissible a-circuit mod β, there exists a semi-restricted homeomorphism η admissibly reordering (β) such that (9. 17) holds for every f with the characteristic set (β).

The results of the present section on transformations T_λ of $\{J, \beta\}$ induced by relative a-circuits λ together with the results of the preceding section on translations in the group Ω lead to the following theorem:

Theorem 9.1. *Let f be an interior transformation of S with the characteristic set (β) and let (r) be an arbitrary set of n integers. Except in the case in which $\mu = 0$ and $m = 2$ there exists a semi-restricted homeomorphism η of S, admissibly reordering (β) but leaving a_0 fixed, such that*

$$J_i(f\eta, \beta) = J_i(f, \beta) + r_i \qquad (i = 1, \ldots, n). \tag{9.18}$$

When $\mu = 0$, $m = 2$, and η is an arbitrary homeomorphism leaving a_0 and a_1 fixed, or interchanging a_0 and a_1 in case a_0 and a_1 are zeros, then

$$J_1(f\eta, a_0, a_1) = \pm J_1(f, a_0, a_1)$$

where the minus sign holds if and only if η interchanges a_0 and a_1.

It is of interest to add that when η in (9.18) is restricted, and (r) is fixed, (9.18) holds for all transformations f with the characteristic set (β) if it holds for one such f. This is not the case in general if η is not restricted.

Part II. Meromorphic Functions.

§ 10.1. *The residual function $\varphi(z)$ and the canonical functions $F(z, \alpha, r)$.* Suppose that f is meromorphic on S and possesses the characteristic[1] set (α). The function $\varphi(z)$ defined by the equation

$$\frac{f'(z)}{f(z)} = \varphi(z)\frac{B(z, \alpha)}{A(z, \alpha)} \tag{10.1}$$

is analytic on S except for removable singularities, and never zero. We term $\varphi(z)$ the *residual function* of f.

The algebraic increment of arg φ along any simple regular arc h_i joining a_0 to a_i on S equals $2\pi J_i(f, \alpha)$, as we shall see. To establish this fact a lemma is needed.

Let h_i be referred to its arc length s measured from $z = a_0$. Suppose that the total length of h_i is σ. Let e be a constant with $0 < e < \sigma$. Suppose that $z(s)$ and $z_1(s)$ are functions of s of which $z(s)$ represents the point s on h_i and $z_1(s)$ the point $s + \Delta s$ on h_i, where $\Delta s = e$. The parameter s shall vary on the interval

$$0 \leqq s \leqq \sigma - e. \tag{10.2}$$

[1] In the preceding we have supposed that $m > 1$. The results of the present section hold for $m = 1$ and with obvious interpretations for $m = 0$.

With $z(s)$ and $z_1(s)$ so determined, set

$$\Delta z = z_1(s) - z(s). \tag{10.3}$$

We shall consider the increment of angle given by

$$U(h_i, e) = \left[\arg \frac{\Delta z}{(z_1(s) - a_0)(z(s) - a_i)}\right]_{s=0}^{s=\sigma-e} \tag{10.4}$$

Lemma 10.1. *The value of $U(h_i, e)$ is zero.*

Reference to the definition (3.12) of the difference order shows that

$$2\pi d(h_i) = U(h_i, e)$$

regardless of the choice of $e < \sigma$. On letting e tend to σ it appears that $U(h_i, e) = 0$.

The following theorem is one of the bridges between the theory of interior transformations and meromorphic functions.

Theorem 10.1. *The algebraic increment $E(h_i)$ of the argument of the residual function φ of f as z traverses a simple, regular arc h_i leading from a_0 to a_i on S equals $2\pi J_i(f, \alpha)$.*

The value of $E(h_i)$ is independent of the choice of h_i among regular arcs leading from a_0 to a_i since $\varphi \neq 0$ on S. Hence no generality will be lost if h_i does not intersect the set $(\alpha) - (a_0, a_i)$. We suppose h_i so chosen.

We make use of the terminology preceding Lemma 10.1. Set

$$\Delta f = f[z_1(s)] - f[z(s)] \qquad (0 \leq s \leq \sigma - e).$$

Recall that

$$\varphi = \frac{f'}{f}\frac{A}{B}.$$

It is clear that $E(h_i)$ is the limit as e tends to 0 of

$$\left[\arg \frac{\Delta f}{\Delta z} - \arg f - \arg \frac{B}{A}\right]_{s=0}^{s=\sigma-e}$$

where $z = z(s)$ in f, B and A.[1] Recall that

$$\frac{B}{A} = \frac{C_i(z, \alpha)}{(z - a_0)(z - a_i)}.$$

It thus appears that $E(h_i)$ is equally the limit of

$$[\arg \Delta f - \arg f - \arg C_i - U(h_i, e)]_{s=0}^{s=\sigma-e}. \tag{10.5}$$

[1] Strictly $\arg f$ and $\arg \frac{B}{A}$ are not defined for $s = 0$, but limiting values exist. The appropriate conventions are understood.

But $U(h_i, e) = 0$ according to Lemma 10.1, and

$$V(h_i) = \frac{1}{2\pi} \arg C_i \Big|_{s=0}^{s=\sigma}.$$

In the terminology of (3.3), it follows that

$$E(h_i) = 2\pi \{P(h_i^f) - Q_0(h_i^f) - V(h_i)\}.$$

By virtue of Lemma 3.6,

$$d(h_i^f) = P(h_i^f) - Q_0(h_i^f)$$

so that

$$E(h_i) = 2\pi [d(h_i^f) - V(h_i)] = 2\pi J_i(z, \alpha).$$

This completes the proof of the theorem.

On multiplying the members of (10.1) by $z - a_j$ and letting z tend to a_j as a limit, one finds that

$$(10.6) \qquad \varphi(a_j) = e_j \frac{A'(a_j, \alpha)}{B(a_j, \alpha)} \qquad (j = 0, 1, \ldots, n)$$

where $e_j = 1$ if a_j is a zero and -1 if a_j is a pole.

The following lemma is basic.

Lemma 10.2. *Corresponding to an arbitrary admissible characteristic set* (α) *and to a function* $\psi(z)$ *which is non-null and analytic on* S *and satisfies* (10.6) *in terms of* (α), *there exists a function* $F(z)$ *which is meromorphic on* S, *possesses the characteristic set* (α), *and for which the residual function is* $\psi(z)$.

Such a function F is obtained as an integral of (10.1) in the form

$$(10.7) \qquad F = Ce^{\int \frac{\psi B}{A} dz} \qquad (C = \text{const} \neq 0)$$

upon removing singularities at the proposed zeros and poles of (α), and any function F with the residual function ψ and characteristic set (α) is of this form. Because $\psi(z)$ satisfies (10.6) the residue of $\frac{\psi B}{A}$ at a_j is e_j, so that a_j is a zero or pole of F as required. Moreover

$$\frac{F'}{F} = \psi \frac{B}{A},$$

and since $\psi \neq 0$, $F' = 0$ only at the zeros of B. Thus F possesses the characteristic set (α) and has the residual function $\psi(z)$.

It will simplify the notation if one sets

$$e_j \frac{A'(a_j, \alpha)}{B(a_j, \alpha)} = g_j(\alpha) \qquad (j = 0, \ldots, n) \tag{10.8}$$

where $e_j = 1$ if a_j is a zero and -1 if a_j is a pole. The g_j's are functions of (α) which one can suppose given before one has knowledge of the existence of a function f from which they are derived. In taking arguments of these g_j's we have referred to a choice of the argument for which

$$0 \leqq \overline{\arg} X < 2\pi \tag{10.9}$$

signalling this choice by adding the bar. Similarly we shall write

$$\overline{\log} X = \log |X| + \sqrt{-1}\, \overline{\arg} X. \tag{10.10}$$

With this terminology we write (7.4) *in the basic form*

$$J_i(f, \alpha) = \frac{\overline{\arg}\, g_i(\alpha)}{2\pi} - \frac{\overline{\arg}\, g_0(\alpha)}{2\pi} + I_i(f, \alpha) \tag{10.11}$$

recalling that I_i is an integer. Given (α), (I) uniquely determines (J) and conversely.

The fundamental existence theorem for meromorphic functions follows.

Theorem 10.2. *Corresponding to any admissible characteristic set (α) and arbitrary set (r) of n integers there exists a function $F(z, \alpha, r)$ which is meromorphic in z on S, whose characteristic set is (α) and whose invariants $[I(F, \alpha)] = (r)$.*

In terms of the given integers (r) with $r_0 = 0$ adjoined, set

$$c_j(\alpha, r) = \log g_i(\alpha) + 2\pi r_j \sqrt{-1} \qquad (j = 0, \ldots, n). \tag{10.12}$$

The Lagrange interpolation formula suffices to yield a polynomial[1] $P(z)$ such that

$$P(a_j) = c_j \qquad (j = 0, \ldots, n).$$

For fixed (α) and (r) we shall show that the function[1]

$$\psi(z) = e^{P(z)} \tag{10.13}$$

is admissible as a residual function and yields a solution F of our problem. In fact,

$$\psi(a_j) = e^{P(a_j)} = e^{c_j} = g_j = e_j \frac{A'(a_j, \alpha)}{B(a_j, \alpha)} \qquad (j = 0, 1, \ldots, n)$$

[1] More explicitly we could write

$$P = P(z, \alpha, r) \qquad \psi = \psi(z, \alpha, r).$$

so that conditions (10.6) on a residual function are satisfied. To apply Lemma 10.2 we note also that $\psi(z)$ is non-vanishing and analytic on S. Hence ψ is the residual function of a function F with the given characteristic set (α).

It remains to prove that $I_i(F,\alpha)=r_i$. Since $\psi\neq 0$ on S there exists a single-valued continuous branch of $\arg\psi$ over S, and for any such continuous branch

$$2\pi J_i(F,\alpha)=\arg\psi(a_i)-\arg\psi(a_0) \tag{10.14}$$

in accordance with Theorem 10.1. For the moment let $\hat{X}$ denote the coefficient of $\sqrt{-1}$ in X. Then $\hat{P}(z)$ is a continuous branch of $\arg\psi$ by virtue of (10.13).

On setting $\arg\psi=\hat{P}(z)$ in (10.14) we find that

$$2\pi J_i(F,\alpha)=\hat{P}(a_i)-\hat{P}(a_0)=\hat{c}_i-\hat{c}_0.$$

From the definition of c_i it then follows that

$$J_i(F,\alpha)=\left(\frac{\arg g_i}{2\pi}+r_i\right)-\frac{\arg g_0}{2\pi}.$$

Reference to (10.11) shows that

$$I_i(F,\alpha)=r_i \qquad (i=1,\ldots,n).$$

Thus the characteristic set of F is (α), and its invariants (I) with respect to (α) equal (r). This completes the proof of the theorem.

The preceding polynomials P will be taken in the explicit Lagrangian form

$$P(z,c)=A\left[\frac{c_0}{(z-a_0)A_0'}+\cdots+\frac{c_n}{(z-a_n)A_n'}\right] \tag{10.15}$$

where

$$A=(z-a_0)(a-a_1)\ldots(z-a_n)$$

and A_j' is the value of A' when $z=a_j$. With ψ given by (10.13) we shall refer to the *canonical functions*

$$F(z,\alpha,r)=Ce^{\int\frac{\psi B}{A}dz} \tag{10.16}$$

and will determine C by the condition that

$$F_z(a_0,\alpha,r)=1.$$

In (10.16)

$$A=A(z,\alpha)\quad B=B(z,\alpha)\quad \psi=\psi(z,\alpha,r).$$

§ 11. *The monogenic continuation with respect to* (α) *of the functions* $F(z, \alpha, r)$. Starting with a particular value (β) of (α), a point z and set (r), one can continue $F(z, \alpha, r)$ as a function of (α), into a monogenic function $M(z, \alpha)$ of (α). As we shall see, $M(z, \alpha)$ may be single-valued in (α), as happens in the case $\mu = 0$, $m = 2$, or it may be infinitely multiple-valued. The branches of $M(z, \alpha)$ form a subset of the functions $F(z, \alpha, r)$ including the extreme possibilities that this subset consists of all of the functions $F(z, \alpha, r)$ or just one. For different sets (r) no two functions $F(z, \alpha, r)$ are identical in z; identity of the functions would imply identity of their invariants (J) and through (10. 11), identity of their invariants $(I) = (r)$.

The mechanism of the continuation is best understood by noting that $M(z, \alpha)$ is a function of (α) defined by (10. 16) through the mediation of the c_j's. The c_j's enter through the polynomial $P(z, c)$ of (10. 15). Finally, the c_j's are given as functions

$$c_j(\alpha, r) = \log g_j + 2\pi i r_j \qquad (r_0 = 0). \tag{11. 1}$$

For each set of integers $(r) = (r_1, \ldots, r_n)$, (11. 1) defines $[c(\alpha, r)]$ as an analytic vector function, more precisely as a branch of an infinitely multiple-valued function of (α). It is not implied that all elements $[c(\alpha, r)]$ are branches of the same monogenic function of (α). As (α) varies continuously through a point (β) the functions $c_j(\alpha, r)$ may suffer jumps of the form

$$\Delta c_j = 2\pi i \sigma_j \qquad (j = 0, \ldots, n) \tag{11. 2}$$

where $\sigma_j = \pm 1$. We need the following lemma.

Lemma 11. 1. *A jump of the* $c_j(\alpha, r)$*'s in which all* c_j*'s change by* $2\pi i$ *or by* $-2\pi i$ *corresponds to no singularity in the right member of* (10. 16).

Corresponding to any change in the c_j's in which Δc_j is a constant k independent of j set $c_j' = c_j + k$. Then

$$P(z, c) + k = P(z, c').$$

When in particular $k = \pm 2\pi i$ the new residual function is

$$\psi(z, c') = e^{P(z, c) \pm 2\pi i} = \psi(z, c)$$

so that there is no change in the right member of (10. 16).

Any continuation of an element $[c(\alpha, r)]$ by itself alone or through other elements $[c(\alpha, s)]$ in which the only singularities are a finite number of jumps of the type in Lemma 11.1 will be termed *effectively analytic;* such a continuation causes no singularity in the right member of (10. 16). The set of elements $[c(\alpha, r)]$ defined by (11.1) will not in general include all of the analytic continuations with respect to (α) of a given element $[c(\alpha, s)]$, because we have taken $r_0 = 0$. However, the set of elements $[c(\alpha, r)]$ does permit »effective» analytic continuation of any given element $[c(\alpha, s)]$. A continuous variation of (α) in which the functions $c_j(\alpha, s)$ suffer jumps (11.2) at a point (β) will correspond to no singularity in a continuation of $M(z, \alpha)$ in which one changes at (β) from $[c(\alpha, s)]$ to $[c(\alpha, r)]$ with

$$r_i = s_i - \sigma_i + \sigma_0 \qquad (i = 1, \ldots, n)$$

in (10. 12).

The following theorem is a consequence of this result.

Theorem 11. 1. *Any one of the monogenic functions of (α) obtained by analytic α-continuation of a particular element $F(z, \alpha, s)$ is composed of single-valued branches forming a subset of the canonical functions $F(z, \alpha, r)$.*

There remains the problem of determining how the functions $F(z, \alpha, r)$ combine to make up the monogenic functions $M(z, \alpha)$ of (α). A necessary and sufficient condition that $F(z, \alpha, r)$ be continuable with respect to (α) into $F(z, \alpha, s)$ is that there exist an α circuit $\lambda = \{(\alpha^t), 0 \leqq t \leqq 1,\}$ from (β) to (β) and a function $M(z, \alpha)$ monogenic in (α) such that the family of functions

$$f^t = M(z, \alpha^t) \tag{11.3}$$

obtained by analytic α-continuation from the branch

$$M(z, \beta) = F(z, \beta, r)$$

terminate with the branch

$$M(z, \beta) = F(z, \beta, s).$$

This is possible if and only if there exists an α-circuit λ from (β) to (β) for which the induced J-displacement vector in Ω is

$$\Delta J = \Delta I = (s) - (r). \tag{11.4}$$

The necessity of this condition is obvious since f^t is merely a special case of the more general restricted f-deformations considered in § 7. The existence of such an α-circuit $\lambda = (\alpha^t)$ is also sufficient. For one can continue $F(z, \alpha, r)$ along the path (α^t) through the family (11.3). Relation (11.4) will then hold so that the invariants (I), with respect to (β), of the terminal canonical function must be (s).

The lemmas of § 7 on the group Ω accordingly yield the following theorem.

Theorem 11. 2. *The canonical functions $F(z, \alpha, r)$ combine as branches of functions $M(z, \alpha)$ monogenic in (α) as follows:*

(1) *When $\mu > 0$, or $\mu = 0$ and m is odd, all canonical functions F are branches of a single monogenic function $M(z, \alpha)$;*

(2) *When $\mu = 0$ and $m = 4, 6, 8, \ldots$, all canonical functions $F(z, \alpha, r)$ for which (r) is of even category are branches of one monogenic function of (α), while all for which (r) is of odd category are branches of another;*

(3) *When $\mu = 0$ and $m = 2$, each canonical function $F(z, \alpha, r)$ is of itself monogenic in (α).*

§ 12. *Equivalence of meromorphic functions.* Restricted, terminally restricted, and semi-restricted equivalence of two interior transformations f_1 and f_2 of S have been defined in § 2, in terms of admissible f-deformations. If the deformations employ meromorphic functions only, the equivalence is said to be of *meromorphic type.* The first theorem on equivalence follows.

Theorem 12. 1. *Necessary and sufficient conditions that meromorphic functions f_1 and f_2 with the same characteristic set (α) be restrictedly and meromorphically equivalent are that*

$$J_i(f_1, \alpha) = J_i(f_2, \alpha) \qquad (i = 1, \ldots, n). \tag{12.1}$$

That the conditions are necessary has already been established. We shall accordingly prove the conditions sufficient assuming that (12. 1) holds.

Let φ_1 and φ_2 be residual functions of f_1 and f_2 respectively. Set

$$\psi(z, t) = e^{(1-t)\log\varphi_1 + t\log\varphi_2} \qquad (0 \leqq t \leqq 1) \tag{12.2}$$

using continuous branches of $\log \varphi_1$ and $\log \varphi_2$ with

$$\log \varphi_1(a_0) = \log \varphi_2(a_0). \tag{12.3}$$

The condition (12. 3) can be fulfilled since

$$\varphi_1(a_0) = \varphi_2(a_0)$$

in accordance with (10. 6). The resulting function $\psi(z, t)$ satisfies the conditions on a residual function in Lemma 10. 2 regardless of the value of t, as will now be shown.

First, for $0 \leqq t \leqq 1$ and for z on S, $\psi(z, t)$ is non-null and analytic, since φ_1 and φ_2 are non-null and analytic on S. The proof that conditions (10.6) are satisfied by $\psi(z, t)$ is as follows.

The relation

$$\log \varphi_1 \Big|_{a_0}^{a_i} = \log \varphi_2 \Big|_{a_0}^{a_i} \qquad (i = 1, \ldots, n) \tag{12.4}$$

will first be verified. The real parts of the logarithms are equal at a_0 as well as at a_i, since φ_1 and φ_2 satisfy (10.6).

The hypothesis (12.1) in the form

$$\arg \varphi_1 \Big|_{a_0}^{a_i} = \arg \varphi_2 \Big|_{a_0}^{a_i}$$

then insures the truth of (12.4). It follows from (12.4) that for branches of the logarithm for which (12.3) holds

$$\log \varphi_1(a_i) = \log \varphi_2(a_i) \qquad (i = 1, \ldots, n). \tag{12.5}$$

Using (12.5) in (12.2) we find that

$$\psi(a_i, t) = e^{\log \varphi_1(a_i)} = \varphi_1(a_i) \qquad (i = 1, \ldots, n).$$

Thus $\psi(z, t)$ satisfies (10.6).

In accordance with Lemma 10.2 $\psi(z, t)$ is the residual function of a meromorphic function of z

$$f(z, t) = e^{\int \frac{\psi B}{A} dz} \qquad (0 \leqq t \leqq 1) \tag{12.6}$$

with characteristic set (α).

It follows from (12.6) that

$$f(z, 0) = C_1 f_1(z) \qquad C_1 \neq 0$$

$$f(z, 1) = C_2 f_2(z) \qquad C_2 \neq 0$$

where C_1 and C_2 are constants. The function

$$\frac{f(z, t)}{C_1^{1-t} C_2^t} \qquad (0 \leqq t \leqq 1) \tag{12.7}$$

gives the required meromorphic deformation of f_1 into f_2.

The canonical (α)*-projection of a transformation* f. If an interior transformation f has a characteristic set (α) and integral invariants (I) with respect to (α), $F(z, \alpha, I)$ will be called the (α)*-projection* of f. If f is meromorphic it has just been seen that f admits a restricted deformation of meromorphic type into its (α)-projection.

If (α^1) is any admissible reordering of (α) it will be shown that the (α)- and (α^1)-projections of f are identical as functions of z. If (I^1) is the set of integral invariants of f with respect to (α^1) this statement takes the form of an identity in z.

$$(12.8) \qquad F(z, \alpha, I) \equiv F(z, \alpha^1, I^1).$$

To establish (12.8) recall that the definition of a canonical function involves the functions g_i and c_i of (α) as well as I_i. If

$$(a_1^1, \ldots, a_n^1) = \pi(a_1, \ldots, a_n),$$

one verifies that in passing from the (α)- to the (α^1)-projection of f a similar permutation π of g_i, c_i and I_i, $(i = 1, \ldots, n)$ is made with g_0 and c_0 unchanged. The Lagrange polynomial $P(z, c)$ is invariant under this change and (12.8) follows.

If f is admissibly deformed through a family f^t, $0 \leqq t \leqq 1$, of interior transformations with (α^t) the characteristic set of f^t, the (α^t)-projection F^t of f^t will admissibly deform the (α)-projection F^0 of f through a family of canonical functions. If $M(z, \alpha)$ is the monogenic function of (α) of which F^0 is a branch, then

$$F^t = M(z, \alpha^t) \qquad 0 \leqq t \leqq 1$$

provided the appropriate branches of $M(z, \alpha^t)$ are selected for each t. This deformation F^t is termed *canonical.* In these terms one can state that a necessary and sufficient condition that two meromorphic functions f^1 and f^2 with the same characteristic set (β) be terminally restrictedly and meromorphically equivalent is that their (β)-projections admit a canonical deformation into each other. The latter condition is merely that the two (β)-projections be branches of the same function $M(z, \alpha)$ monogenic in (α). Hence Theorem 11.2 yields the following.

Theorem 12. 2. *Necessary and sufficient conditions that two admissible meromorphic functions f^1 and f^2 with the same characteristic set (β) be terminally restrictedly and meromorphically equivalent are that either*

(i) $\mu > 0$ *or* $\mu = 0$ *and* m *be odd, or*

(ii) $\mu = 0$, $m = 4, 6, 8, \ldots$, *and the invariants* (I^1) *and* (I^2) *of* f^1 *and* f^2 *with respect to* (α) *be in the same category, or*

(iii) $\mu = 0$, $m = 2$ *and* $(I^1) = (I^2)$.

We complete this section with the following theorem.

Theorem 12. 3. (a) *Any two admissible meromorphic functions f^1 and f^2 with the same characteristic set (β) are semi-restrictedly and meromorphically equivalent except in the case $\mu = 0$ and $m = 2$.*

(b) *In case* $\mu = 0$ *and* $m = 2$, *and* a_1 *is a pole*, f^1 *and* f^2 *are equivalent in the sense of* (a) *if and only if their invariants* J_1 *with respect to* (a_0, a_1) *are equal. If* a_1 *is a zero*, f^1 *and* f^2 *are equivalent if and only if their invariants* J_1 *with respect to* (a_0, a_1) *are equal in absolute value.*

Statement (a) follows from Theorem 12.2 except when $\mu = 0$, $m = 4, 6, 8, \ldots$, and (I^1) and (I^2) are of opposite category. In this case one first deforms f^1 into its (β)-projection F^1. One then continues F^1 with respect to (α) along any admissible path (α^t) that leads to a set (α^1) which is an admissible odd permutation of (β). In accordance with Lemma 9.2, the canonical function F^* into which F^1 is thereby continued will have an invariant set (I^*) with respect to (β) of a category opposite to that of (I^1), and hence of a category which is the same as that of (I^2). It follows from Theorem 12.2 (ii) that F^* can be terminally restrictedly and meromorphically deformed into f^2. This completes the proof of (a).

The necessary conditions for equivalence of f^1 and f^2 in (b) have already been affirmed in Lemma 9.4.

We come to sufficient conditions. When the invariants J_1 with respect to (β) are equal, a restricted deformation carries f^1 into f^2. There remains the case of two zeros in which the invariants J_1 with respect to (β) are equal but opposite in sign. In this case we regard S as a hyperbolic plane and make a hyperbolic rotation about the hyperbolic mid-point of a_0 and a_1, carrying a_0 into a_1. Let η^t be the analytic, isotopic deformation of S thereby defined, $0 \leqq t \leqq 1$. The composite function $f^1 \eta^t$ deforms f^1 into a function f^* which in accordance with (9.15) has the same invariant J_1 with respect to (β) as does f^2. Hence f^* can be restrictedly deformed into f^2.

Statement (b) follows.

§ 13. *Equivalence of interior transformations.* Let f be an interior transformation of S with a characteristic set (β). By virtue of a theorem of Stoilow (4) there exists a homeomorphism ζ from a simply-connected region R of the z-plane to S such that the composite function $f\zeta$ is meromorphic on R. When the boundary of R consists of more than a point, R can be mapped directly conformally onto S. In such a case we can suppose that ζ is a homeomorphism from S to S, and we say that f then comes under *Case A*.

When the boundary of R consists of a point, one can first restrictedly deform f on S into a function f_1 that comes under Case A. To this end let e be so small a positive constant that all points of (β) lie on the disc $\{|z| < 1 - 2e\}$.

Let ζ^t be an isotopic deformation of S onto $S' = \{|z| < 1 - e\}$ which leaves points of S at which $|z| \leqq 1 - 2e$ fixed. Thus ζ^1 is a homeomorphism from S to S'. The composite function $f\zeta^t$ deforms f into a function f_1 again defined over S. The function f_1 comes under Case A. For the Riemann image of S under $w = f_1(z)$ simply covers the Riemann image under $w = f(z)$ of S'. This Riemann image is then conformally equivalent to a subregion of R whose boundary consists of more than one point.

We have the following lemma concerning admissible interior transformations of S.

Lemma 13.1. *Any such transformation f^* of S admits a restricted deformation into a function f for which a homeomorphism ζ of S exists such that $f\zeta$ is meromorphic on S.*

Use will be made of the canonical functions $M(z, \alpha)$ monogenic in (α) of the preceding sections. Let η^t be an isotopic deformation of S from the identity and let (α^t) be the image of (β) under η^t. The characteristic set of

$$M(\eta^t, \alpha^t) \qquad (0 \leqq t \leqq 1) \tag{13.1}$$

remains constantly (β). For example, if a_i represents a zero of (β), then $\eta^t(a_i) = a_i^t$ and

$$M(a_i^t, \alpha^t) = 0$$

since (α^t) is the characteristic set of $M(z, \alpha^t)$.

The basic theorem here is as follows.

Theorem 13.1. *Any interior transformation f with a characteristic set (β) admits a restricted deformation into its canonical β-projection.*

In accordance with Lemma 13.1 one can suppose that a homeomorphism ζ of S exists such that $f\zeta$ is meromorphic on S.

The proof of the theorem is relatively simple once the appropriate auxiliary functions are defined. These functions are as follows:

$\zeta(z)$ = A homeomorphism of S such that $f\zeta$ is meromorphic on S.

$\eta(z)$ = The inverse of ζ on S.

$\lambda(z)$ = The meromorphic function $f\zeta$. The characteristic set of λ is the image of (β) under η.

η^t = An isotopic deformation of S from the identity, generating η. $(0 \leqq t \leqq 1)$.

(α^t) = The image of (β) under η^t at the time t. The characteristic set of λ is (α^1).

F_1 = The (α^1)-projection $F(z, \alpha^1, r)$ of λ.

λ^t = A restricted deformation of meromorphic type of λ into F_1. $(0 \leqq t \leqq 1)$.

$M(z, \alpha)$ = The function $M(z, \alpha)$ monogenic in (α) of which $F(z, \alpha, r)$ is a branch.

The required deformation of f will be given as a sequence of two deformations of t in both of which (β) remains the characteristic set.

The first deformation is defined by the composite function,

$$\lambda^t \eta \qquad (0 \leqq t \leqq 1). \tag{13.2}$$

When $t = 0$, this function reduces to f and when $t = 1$ it becomes $F_1 \eta$. The characteristic set of $\lambda^t \eta$ remains constantly that of $\lambda \eta$, namely (β).

The second deformation is defined by the functions

$$M(\eta^t(z), \alpha^t) \qquad (0 \leqq t \leqq 1) \tag{13.3}$$

with t however decreasing from 1 to 0. As noted in connection with (13.1), the characteristic set remains constantly (β). When $t = 1$ one starts with the branch

$$M(z, \alpha) = F(z, \alpha, r)$$

at the set (α^1). When $t = 1$, the function (13.3) reduces to

$$M(\eta, \alpha^1) = F_1 \eta.$$

When $t = 0$, the function (13.3) becomes $M(z, \beta)$. This is the (β)-projection of f. It has the characteristic set (β); it also has the invariants (I) of f since it is obtained from f by a restricted deformation.

The deformation (13.2) followed by the deformation (13.3), with decreasing t, thus deforms f into its (β)-projection, as required.

The deformation classes into which meromorphic functions may be divided are, in the sense of the following theorem, not affected by the inclusion or exclusion of interior transformations which are not meromorphic.

Theorem 13.2. *Two meromorphic functions f_1 and f_2 with the same characteristic set (β) are restrictedly, terminally restrictedly, or semi-restrictedly equivalent only if they are meromorphically equivalent in the same sense.*

Let F_1 and F_2 be the (β)-projections of f_1 and f_2 respectively. The functions f_1 and f_2 are restrictedly and meromorphically deformable into F_1 and F_2 respectively. Any admissible deformation f^t of f_1 into f_2 in which (α^t) is the characteristic set of f^t implies a deformation F^t of F_1 into F_2 through the (α^t)-projection of f^t. The deformation F^t will be restricted, terminally restricted, or semi-restricted in the same sense as f^t. In addition, F^t is of meromorphic type.

The theorem follows.

Corollary 13.1. *Conditions on the invariants* (J) *of two interior transformations* f_1 *and* f_2 *with the same characteristic set* (β) *together with conditions on* μ *and* m *which are necessary and sufficient for the equivalence of* f_1 *and* f_2 *in any one of the senses of the theorem are precisely those stated in § 12 for equivalence in the same sense of meromorphic functions.*

The functions f_1 and f_2 have the same characteristic set (β), and characteristic integers μ and m, and invariants (J^1) and (J^2) with respect to (β) as their respective (β)-projections F_1 and F_2. In accordance with Theorem 13.2 the equivalence of f_1 and f_2 in a given sense implies and is implied by the meromorphic equivalence in the same sense of F_1 and F_2. The conditions for the latter equivalence in terms of μ, m, (J^1) and (J^2) are thus the conditions for the equivalence of f_1 and f_2 in the given sense.

The theorem follows.

§ 14. *The special cases* $m = 0$ *and* $m = 1$. We admit μ branch point antecedents as previously. In case $m = 0$ or 1 there are no invariants J_i. The construction of a function with the given characteristic set as given in § 6 is still valid here. When $m = 1$ we assume that a_0 is a zero. The case of a pole is treated by considering f^{-1}.

In the meromorphic case the residual function φ is defined by the equation

$$\frac{f'}{f} = \frac{\varphi B}{(z - a_0)^m}$$

both when $m = 0$ and $m = 1$. A canonical function with a given characteristic set may be given in the form

$$F_m(z, a) = e^{\int \frac{B(z)}{B(a_0)} \frac{dz}{(z-a_0)^m}} \qquad (m = 0 \text{ or } 1).$$

The residual function must satisfy the condition

$$\varphi(a_0) = \frac{1}{B(a_0)} \qquad (m = 1);$$

it must be analytic and non-null on S.

If f_1 and f_2 are two meromorphic functions with the same characteristic set (a) and residual functions φ_1 and φ_2 the function

$$e^{\int \frac{\varphi_1^{1-t} \varphi_2^t B \, dz}{(z-a_0)^m}} \qquad (0 \leqq t \leqq 1)$$

suffices to deform restrictedly and meromorphically a constant multiple of f_1 into a constant multiple of f_2. Hence f_1 and f_2 are restrictedly and meromorphically equivalent. Any interior transformation f with the characteristic set (α) can be restrictedly deformed into its (α)-projection $F_m(z, \alpha)$ following the proof of Theorem 13.1.

We summarize what is essential as follows:

Theorem 14.1. *When $m = 0$ or 1, any interior transformation f with the characteristic set (α) can be restrictedly deformed into its (α)-projection $F_m(z, \alpha)$, meromorphically if f is meromorphic.*

In the very special case in which $m = 0$ and $\mu = 0$ the characteristic set (α) is empty, $B(z)$ may be taken as 1 and e^z as the canonical function into which each interior transformation with an empty characteristic set can be admissibly deformed.

§ 15. *Covering properties of sequences of meromorphic transformations of S.* We shall be concerned with infinite sequences $[f_k]$ of meromorphic transformations of S with the same characteristic set (α). The number m of zeros and poles in the set

$$(a_0, a_1, \ldots, a_n) = (a) \tag{15.1}$$

shall be at least two. A sequence $[f_k]$ in which each of the functions f_k is meromorphic and no two functions f_k are in the same restricted deformation class will be termed a *model sequence*. We shall be concerned with the set W of points $w = f_k(z)$ on the w-sphere given by a model sequence for z on S. When the set (15.1) includes both zeros and poles, we shall see that the set W covers each point of the w-sphere infinitely many times. When the set (15.1) consists of zeros alone, the set W will cover each point of the w-sphere infinitely many times ($w = \infty$ excepted) provided f_k does not converge uniformly to zero on every compact subset of S.

A term due to Carathéodory is convenient. A sequence $[F_n]$ of functions of z meromorphic on a region R is said to *converge continuously* to F on R if, with respect to the metric on the w-sphere, $[F_n]$ converges uniformly to F on every closed subset of R. The function F is necessarily meromorphic on R, or reduces to the function $F = \infty$.

Suppose that each function F_n is meromorphic on S and has the characteristic set (α). If $[F_n]$ converges continuously on S to F, then F is identically o or ∞ or else has the characteristic set (a). If $[F_n]$ converges continuously on $S-(a)$ to a function F not identically o or ∞, then $[F_n]$ converges continuously on S without exception. If F_n has no poles, the preceding statement remains valid with the condition $F \not\equiv 0$ omitted. These results follow from classical theorems.

The principal results of this section are an immediate consequence of theorems on normal families of functions. A family M of functions meromorphic on a region R of the z-plane is termed *normal* if corresponding to an arbitrary sequence $[F_n]$ of functions of the family there exists a subsequence which converges continuously on R to a meromorphic function F or to ∞. According to a theorem of Julia a necessary and sufficient condition that M be normal on R is that M be normal on some neighborhood of each point of R. Cf. (6), p. 37. The property which connects the theory of normal families with Picard's theorem is the following. If the family M is not normal on R, then every point w on the w-sphere, with the possible exception of two points, is covered infinitely many times by points $f(z)$ defined by members of the family for z on R. In particular if M is not normal on R, there exists at least one point z_0 on R in no neighborhood of which M is normal. Such a point z_0 is called *a point* J, and the images under the functions of M of an arbitrary neighborhood of a point J cover the w-sphere infinitely many times, two points at most excepted.

Let R be a subregion of S. A model sequence $[f_k]$ no subsequence of which converges continuously to o or ∞ on R will be termed *proper on* R. We shall prove the following lemma.

Lemma 15.1. *A model sequence which is proper on* $S-(a)$ *is not a normal family on* $S-(a)$.

Suppose the lemma false and that a subsequence $[F_n]$ of the given model sequence converges continuously on $S-(a)$ to a function F. The function F is analytic on $S-(a)$ and never zero. It follows from the definition of a residual function that the sequence $[\varphi_n]$ of residual functions of the functions F_n converges continuously on $S-(a)$ to a function φ which satisfies the relation

$$\frac{F'}{F} = \varphi \frac{B}{A} \tag{15.2}$$

on $S-(a)$, and is accordingly analytic and never zero on $S-(a)$. Hence the functions φ_n converge continuously on S without exception to a function φ that is analytic and never zero on S. The invariants J_i^n of F_n must then converge in accordance with Theorem 10.1. Since the invariants J_i^n differ for different values of n by integers, it follows that for n exceeding some integer n_0, J_i^n is independent of n.

This is contrary to the hypothesis that the members of the sequence $[f_k]$ belong to different restricted deformation classes. We infer the truth of the lemma.

We state a lemma concerning quasi-normal sequences. Cf. (6) p. 66.

Lemma 15.2. *Let* $[F_n]$ *be a sequence of functions which are meromorphic on a neighborhood* N *of a point* z_0 *and analytic and not zero on* $N - z_0$. *If* $[F_n]$ *converges continuously to* ∞ (0) *on* $N - z_0$ *but fails to so converge on* N, *then for* n *sufficiently large*[1] *and for some* z *on* N, F_n *assumes any given value* w *except* ∞ (0).

The two cases involved in this lemma are reducible to each other on replacing F_n by its reciprocal. The case in which $F \equiv \infty$ is treated in the above reference.

A point z_0 will be called a *covering point* relative to a model sequence $[f_k]$ if the totality of w-images of the functions f_k, $k = 1, 2, \ldots$, for z on an arbitrary neighborhood of z_0 cover the points of the w-sphere infinitely many times, $w = 0$ and $w = \infty$ at most excepted. The first covering theorem follows.

Theorem 15.1. *Let* $[f_k]$ *be a model sequence of meromorphic functions.*

I. *If* $[f_k]$ *is proper on* $S - (a)$, *there exists at least one covering point on* $S - (a)$ *relative to* $[f_k]$.

II. *If a subsequence of* $[f_k]$ *converges continuously to* ∞ [0] *on* $S - (a)$, *then any zero* [*pole*[2]] a_i *is a covering point relative to* $[f_k]$.

Case I *or* II *always occurs.*

In Case I $[f_k]$ is not normal on $S - (a)$ in accordance with Lemma 15.1. There accordingly exists a point z_0 of type J on $S - (a)$. For any neighborhood of z_0 relative to $S - (a)$, $w = 0$ and $w = \infty$ are values not taken on by $[f_k]$. Hence every other value is taken on infinitely often. Thus z_0 is a covering point relative to $[f_k]$. In Case II the theorem follows from Lemma 15.2.

Corollary. *If* $[f_k]$ *is a model sequence there exists at least one covering point on* S, *excepting only the case in which there are no poles in* (a) *and* $[f_k]$ *converges continuously to zero on* S.

Recall that the case in which there are no zeros in (a) has been excluded. If admitted, this case would parallel the case in which there are no poles.

---

[1] That is for n exceeding an integer n_0 depending on N and w.

[2] If there are any poles.

Given a model sequence $[f_k]$ which is not proper, a proper model sequence can be readily obtained by replacing each f_k by a constant multiple $c_k f_k$ with c_k suitably chosen. In particular it is sufficient to choose the constants c_k so that $|c_k f_k|$ is bounded from 0 and ∞ at some point not in (a).

The following theorem is a consequence of Bloch's theorem. Cf. (9) p. 230.

Theorem 15. 2. *If* $[f_k]$ *is a model sequence no subsequence of which converges to* 0 *on* S, *and if the characteristic set* (a) *includes no poles, there then corresponds to any positive constant* r, *no matter how large, a member* $f_{k(r)}$ *of the sequence and a circular disc* D_r *of radius* r *in the* w*-plane such that* D_r *is the one-to-one image under* $f_{k(r)}$ *of some subdomain of* S.

This theorem follows from Bloch's theorem provided the values

$$|f_k'| \qquad (k = 1, \ldots,) \tag{15.3}$$

are unbounded on some compact subset of S.

Let S_c be the subdisc of S concentric with S and of radius c. If the values (15.3) admitted a bound M_c on S_c for each c, $0 < c < 1$, $|f_k|$ would be bounded independently of k on each S_c and $[f_k]$ would be normal on S. Since no subsequence of $[f_k]$ converges continuously to zero by hypothesis, the sequence $[f_k]$ would then be proper. This is contrary to Lemma 15. 1. Thus for some $c < 1$ the values (15. 3) are unbounded on S_c. Hence there are points z on this S_c and values of k for which $|f_k'|$ is arbitrarily large.

The theorem follows.

The preceding theorems can be extended as follows. Let J_0 be defined as 0. Let a sequence $[f_k]$ of meromorphic functions with the given characteristic set be termed *model* with respect to two points (a_r, a_s) in (a) if no two pairs (J_r, J_s) and (J_r^*, J_s^*) belonging to different functions in the sequence $[f_k]$ have the property that

$$J_s^* - J_r^* = J_s - J_r. \tag{15.4}$$

If a_r and a_s are not both poles, we could suppose that a_r is a zero and change the notation so that $a_r = a_0$. The condition (15. 5) would then take the form

$$J_s^* = J_s.$$

If a_r and a_s are both poles one could replace each function f_k by its reciprocal, noting that the residual function of f_k and its reciprocal are negatives of each other.

Lemma 15.1 can now be replaced by the following.

Lemma 15.3. *A sequence* $[f_k]$ *which is model with respect to* a_r *and* a_s *and which is proper on* $N-(a_r, a_s)$, *where* N *is any connected neighborhood of* a_r *and* a_s, *is not a normal family on* $N-(a_r, a_s)$.

With a_r taken as a_0 the proof is essentially the same as that of Lemma 15.1. Lemma 15.2 is unchanged for present purposes. Theorem 15.1 takes the following form.

Theorem 15.3. *Let* $[f_k]$ *be a sequence which is model with respect to* a_r *and* a_s *and let* N *be any connected neighborhood of* a_r *and* a_s.

I. *If* $[f_k]$ *is proper on* $N-(a_r, a_s)$, *there exists at least one covering point on* $N-(a_r, a_s)$ *relative to* $[f_k]$.

II. *If a subsequence of* $[f_k]$ *converges continuously to* ∞ $[0]$ *on* $N-(a_r, a_s)$, *then any zero* [*pole*] *in the pair* (a_r, a_s) *is a covering point relative to* $[f_k]$.

Case I *or* II *always occurs.*

The set of points J taken relative to a model sequence $[f_k]$ is *closed* on S by virtue of the definition of a point J. One can then establish the following.

Theorem 15.4. *If a model sequence* $[f_k]$ *is proper on every subregion of a region* R *then the set* E *of points* J *on* R *is perfect (possibly empty) relative to* R. *Each non-empty component* E_1 *of* E *contains at least one zero and one pole of* (a), *or else has a limit point on the boundary of* R.

No point z_0 of E can be isolated in E. To see this let N be a neighborhood of z_0 such that $N-z_0$ contains no point of (a). Continuous convergence of a subsequence $[F_n]$ of $[f_k]$ on $N-z_0$ to a function F implies that F is analytic and not null on $N-z_0$ since $[f_k]$ is proper on $N-z_0$ and no f_k has a zero or pole on $N-z_0$. It follows that $[F_n]$ converges continuously on N so that z_0 cannot be a point J. We infer that z_0 is not isolated in E and that E is accordingly perfect relative to R.

Suppose that E_1 is on R. We shall show that E_1 must contain at least one zero and one pole in (a). Suppose in particular that E_1 contained no pole in (a). Then E_1 could be separated on R from the poles in (a) and from the boundary of R by a finite set (g) of regular Jordan curves on R which do not intersect E or (a) and bound a subregion R_1 of R on which E_1 lies. The sequence $[f_k]$ would be normal on a neighborhood N of (g). On N a subsequence $[F_n]$ of $[f_k]$ would converge continuously to a function F which would be analytic and never zero on $N-(a)$. It follows that $[F_n]$ would converge continuously on R_1 so that E_1 would be empty.

The supposition that E_1 is on R and contains no zero in (a) would similarly lead to the conclusion that E_1 is empty. The theorem follows.

References.

(1) MORSE, M. and HEINS, M. Topological methods in the theory of functions of a single complex variable.

I. Deformation types of locally simple curves, Annals of Math. 46 (1945) pp. 600—624.

II. Boundary values and integral characteristics of interior transformations and pseudo-harmonic functions, Ibid., pp. 625—666.

III. Causal isomorphisms in the theory of pseudo-harmonic functions, Ibid., 47 (1946) pp. 233—274.

(2) MORSE, M. The topology of pseudo-harmonic functions, Duke Math. Jour. 13 (1946) pp. 21—42.

(3) WHYBURN, G. T. Analytic topology, Amer. Math. Soc. Coll. Lect. New York 1942.

(4) STOILOW, S. Leçons sur les principes topologiques de la théorie des fonctions analytiques, Paris 1938.

(5) v. KERÉKJÁRTÓ, B. Vorlesungen über Topologie, Springer, Berlin 1923.

(6) MONTEL, PAUL. Leçons sur les familles normales de fonctions analytiques, Gauthier-Villars, Paris 1927.

(7) TIETZE, H. Sur les représentations continues des surfaces sur elles-mêmes, Comptes Rendus 157 (1913) pp. 509—512.

(8) CARATHÉODORY, C. Conformal representation, Cambridge University Press, London.

(9) BIEBERBACH, L. Lehrbuch der Funktionentheorie, Vol. II, Leipzig, 1923.

(10) MORSE, M. Topological methodes in the theory of meromorphic functions. Annals of Mathematics Studies. Princeton N. J.

ANNALS OF MATHEMATICS
Vol. 50, No. 4, October, 1949

FUNCTIONALS F BILINEAR OVER THE PRODUCT $A \times B$ OF TWO P-NORMED VECTOR SPACES

I. The representation of F*

BY MARSTON MORSE AND WILLIAM TRANSUE

(Received October 7, 1948)

I. THE REPRESENTATION OF F

1. Introduction

Paper I is the first of several papers on functionals F bilinear[1] on the product $A \times B$ of two p-normed (pseudo-normed) vector spaces A and B. References are given at the end of this paper. Paper I is concerned with the representation of F by means of a repeated Lebesgue-Stieltjes integral (written L-S-integral) when the spaces A and B satisfy Conditions I–V here given. Paper II will show that these hypotheses are satisfied by p-normed spaces extending the well known spaces c, C, l_p, L_p $(p \geqq 1)$, certain subspaces of M, the spaces defined with the aid of the complementary convex functions Φ and Ψ of Zygmund and Orlicz, and other more general spaces. The pair A, B can be any pair chosen from the spaces satisfying I–V. The choice of A imposes no restriction on the choice of B. Later papers will show how these results specialize, and will develop a variational and spectral theory for F.

To indicate the principal objectives in the representation theory the results for the space $C \times C$ will first be reviewed. Fréchet has shown the following. For $(x, y) \in C \times C$, and F bilinear over $C \times C$,

$$F(x, y) = \int_0^1 x(s)\, d_s \int_0^1 y(t)\, d_t\, k(s, t), \tag{1.1}$$

where the Fréchet variation $h(k)$ of k is finite and where the integral is a repeated S-integral. Here x and y map the interval $[0, 1]$ continuously into R_1, the real axis, and k is defined in the product $[0, 1] \times [0, 1]$. If further, $k(s, t) = 0$ on the s and t axes (not a real restriction),

$$F(x, y) = \int_0^1 y(s)\, d_s \int x(t)\, d_t\, k(t, s). \tag{1.2}$$

---

* This paper was written in partial fulfillment of a contract with the Office of Naval Research under which Dr. Transue received financial support.

[1] To be bilinear F shall be linear in the sense of Banach over A and B separately, for $y \in B$ and $x \in A$ held fast respectively, while for some constant N

$$| F(x, y) | \leqq | x |_A \, | y |_B \, N.$$

Cf. Theorem 6.1. Definitions of a p-norm and of p-normed antecedents appear near the end of this introduction.

Moreover if $| x |_C$ and $| y |_C$ are norms in C,

$$| F(x, y) | \leqq | x |_C | y |_C h(k). \tag{1.3}$$

The *functional modulus* $\| F \|$ of F is by definition

$$\| F \| = \sup_{(x,y)} F(x, y) \qquad [| x |_C = | y |_C = 1]. \tag{1.4}$$

If $h(k) = \| F \|$, k will be said to be of *minimum type*.

The results of a preliminary paper by the authors (see ref. 5) permit a completion of Fréchet's theorem on $C \times C$ by characterizing the subclass of functionals k for which $h(k)$ is finite and k is of minimum type. Given a k for which $h(k)$ is finite, the authors show how to derive a k_1 of minimum type from k. Given a bilinear F, an explicit process is indicated for finding a k of minimum type from F such that (1.1) and (1.2) hold. This process is in terms of a linear extension of $A \times B$ and a corresponding extension of F.

The general theory of functionals F, bilinear on $A \times B$, will proceed as follows. The elements of A (or B) will be functionals x mapping an axis of reals R_1 into R_1, and will be determined in a special way by the values $x(s)$ on a closed subset $E_A(E_B)$ of R_1. Representations of F will be of the form

$$F(x, y) = \int_{R_1} x(s)\, d_s \int_{R_1} y(t)\, d_t\, k(s, t), \qquad [(x, y) \in A \times B] \tag{1.5}$$

employing *L-S*-integrals. The variation $h(A, B; k)$ of k will be taken in a newly defined sense, and assumed finite. Here k maps R_2, the space of real pairs (s, t) into R_1. If k belongs to a well defined class $H(A, B)$, the order of integration in (1.5) can be interchanged so that

$$F(x, y) = \int_{R_1} y(s)\, d_s \int_{R_1} x(t)\, d_t\, k(t, s). \tag{1.6}$$

If $| x |_A$ and $| y |_B$ are the norms of x and y in A and B respectively,

$$| F(x, y) | \leqq | x |_A | y |_B h(A, B; k), \tag{1.7}$$

and $h(A, B; k)$ is the modulus of F.

Moreover $F(x, y)$ is proved representable as the limit of a finite double sum partially analogous to a double *R-S*-integral, but without assuming that k has a bounded variation in the classical sense of Vitali. The definition and value of $h(A, B; k)$ will vary with A and B. The condition that $h(A, B; k)$ be finite is frequently less restrictive than that the Vitali variation of k be finite. The values $F(x, y)$ are the analogues of conditionally convergent double series which can be summed to $F(x, y)$ in the sense of Pringsheim.

The choice of functionals k to be admitted and definition of $h(A, B; k)$ shall be such that

(a). *any F which is bilinear over $A \times B$ admits a representation* (1.5),

(b). *and a representation* (1.6),

(c). *for any admissible k,* (1.5) *and* (1.6) *should define an F bilinear over $A \times B$,*
(d). *k is uniquely determined by F,*
(e). $\| F \| = h(A, B; k)$.

It is found that admissible k can be characterized without any reference to measurability, although an admissible k can be proved to be measurable.

Various theorems of Fréchet, of Littlewood, and of Hilbert theory follow from the results of the present paper. The representation will permit an association of the Jacobi equations of variational theory with the integro-differential equations of very general form in a setting more general than either of its classical components. On the one hand a characterization of the Jacobi equations in this setting can be given, and on the other hand many of the geometric concepts and results of variational theory can be carried over to integral equation theory.

We conclude the introduction with certain general remarks and definitions necessary for the understanding of our representation theory and its applications.

Pseudo-norms. Let $\dot{Q}$ be a linear vector space satisfying the conditions of Banach, Chapter II. In such a space the null element θ is well defined. For each $x \in \dot{Q}$, let $N(x)$ be a non-negative number such that

$$N(cx) = | c | N(x) \tag{1.8}$$

for every constant c, and such that for arbitrary x_1 and x_2 in $\dot{Q}$

$$N(x_1 + x_2) \leqq N(x_1) + N(x_2). \tag{1.9}$$

We then term $N(x)$ a *p-norm* (*pseudo-norm*) over $\dot{Q}$. We do not require that the condition $N(x) = 0$ implies $x = \theta$.

Let H be the subgroup of $\dot{Q}$ of elements x for which $N(x) = 0$, and introduce the group

$$Q = \dot{Q}, \qquad \text{mod } H.$$

It is clear that Q is a linear vector space of "cosets" $\bar{x}$ of $\dot{Q}$ such that x_1 and x_2 in $\dot{Q}$ are in the same coset $\bar{x}$ if and only if $N(x_1 - x_2) = 0$. One sees that $N(x_1) = N(x_2)$ when x_1 and x_2 are in the same coset $\bar{x} \in Q$. If to each $\bar{x} \in Q$, a norm

$$| \bar{x} |_Q = N(x) \tag{1.10}$$

is assigned whenever $x \in \dot{Q}$ is in the coset $\bar{x}$, Q becomes a linear space which is normed in the sense of Banach. In particular the condition $| \bar{x} |_Q = 0$ implies that $\bar{x}$ is the null element in Q representing the null coset $H \subset \dot{Q}$.

A linear p-normed vector space $\dot{Q}$ which leads to a normed vector space Q in the above way will be called a *p-normed antecedent* of Q, while Q will be called the *normed descendant* of $\dot{Q}$.

Let $\dot{Q}$ be a p-normed antecedent of a normed vector space Q, and let f be a functional in $\bar{Q}$, that is a functional which is linear over Q in the sense of Banach. Given $\bar{x} \in Q$, let $x \in \dot{Q}$ be an arbitrary element in the coset of $\dot{Q}$ represented by $\bar{x}$. If one sets

$$f(\bar{x}) = \dot{f}(x), \tag{1.11}$$

a functional $\dot{f}$ is defined over $\dot{Q}$ that is linear over $\dot{Q}$. In particular $\dot{f}$ is additive and homogeneous over $\dot{Q}$ and satisfies the relation

$$(1.12) \qquad |\dot{f}(x)| \leqq N(x)\|f\|.$$

It follows from (1.12) that $\dot{f}$ is continuous over $\dot{Q}$ and in particular that $\dot{f}(x)$ is constant over any coset $\bar{x}$.

Suppose conversely that $\dot{f}$ is independently given as linear over $\dot{Q}$ in the sense that $\dot{f}$ is additive and homogeneous over $\dot{Q}$ and that for x_0 and x_n in $\dot{Q}$ the condition $N(x_n - x_0) \to 0$ as $n \to \infty$ implies that $\dot{f}(x_n) \to \dot{f}(x_0)$. It follows that $\dot{f}(x)$ is constant over any coset $\bar{x}$. The equation (1.11) can now be used to define f over Q. One sees that f is linear over Q when $\dot{f}$ is linear over $\dot{Q}$. According to a well known theorem f then possesses a modulus $\|f\|$ such that

$$f(\bar{x}) \leqq |\bar{x}|_Q \|f\|,$$

so that $\dot{f}$ satisfies (1.12) whenever it is given as linear over $\dot{Q}$.

Let $\bar{x}_0$ be any element in Q. According to a theorem of Hahn (Cf. Banach p. 55), there exists a functional $F \in \bar{Q}$ such that

$$F(\bar{x}_0) = |\bar{x}_0|_Q, \qquad \|F\| = 1.$$

Let $x_0 \in \dot{Q}$ be any element in the coset $\bar{x}_0$. The functional $\dot{F}$ defined as above over $\dot{Q}$ in terms of F is linear over $\dot{Q}$ and is such that

$$\dot{F}(x_0) = N(x_0), \qquad \|\dot{F}\| = 1.$$

Since *any* functional $\dot{F}$ independently given as linear over $\dot{Q}$ defines as above a functional F linear over the normed descendant Q of $\dot{Q}$ we have shown that the theorem of Hahn generalizes to linear functionals defined over a linear vector space with a p-norm.

Example. Let $\dot{Q}$ be the vector space of functionals x defined over the interval $[0, 1]$ and L-integrable. Such functionals are to be added by adding their values at each point $s \in [0, 1]$. In $\dot{Q}$ the null element is the functional with null values. The L-integral $N(x)$ over $[0, 1]$ of $|x(s)|$ is a p-norm. Elements x_1 and x_2 in the same coset $\bar{x}$ have values which differ at most on a set of L-measure zero. Examples can easily be given in which elements in the same coset have values which differ on sets of very general character.

Part I. Admissible Spaces A

2. Preliminary definitions

We shall be concerned with real variables s, t, etc., and points (s, t). The Cartesian space of points s will be denoted by R_1, that of points (s, t) by R_2. Let E be a closed subset of R_1. We suppose that $s \geqq 0$ in E, and that E contains at least two points, including the point $s = 0$. We do not suppose that E is bounded. In particular E may be the set on which $s \geqq 0$.

Definition. *Functions g, canonical* rel E. Such functions shall map R_1 into R_1 and satisfy two conditions.

(1). They shall be *based* on E in the following sense. Let $[a, b]$ be any closed interval whose end points are in E while the open interval (a, b) is in $R_1 - E$. On $[a, b]$ g shall be linear. For $s \leqq 0$, $g(s)$ shall equal 0. Set

$$\omega_E = \sup s \in E. \tag{2.1}$$

For $s \geqq \omega_E$, $g(s)$ shall equal $g(\omega_E)$. This condition is empty if ω_E is not finite.

(2). Except at most when $s = \omega_E$, g shall be left continuous.

Distribution functions (written *d*-functions) based on E will now be defined.

DEFINITION. *d-Functions g based on E.* Such functions shall be canonical rel E and have finite total variations over each finite interval $[0, b]$.

The total variation of g over R_1 will be denoted by $T(g)$.

DEFINITION. *Functions x belonging to E.* A function x with values $x(s)$ mapping R_1 into R_1 will be said to *belong* to E if, whenever (a, b) is a maximal finite subinterval of $R_1 - E$, $x(s)$ is constant on $[a, b)$ if $b \neq \omega_E$, and on $[a, b]$ if $b = \omega_E$. Moreover $x(s) = 0$ for $s < 0$ and $s > \omega_E$.

Functions x belonging to E and functions g based on E are uniquely determined by their values on E.

The above concepts will be made clearer by the following theorem generalizing a theorem of F. Riesz. Let $C(E)$ be a p-normed vector space of functions x mapping R_1 into R_1 and restricted as follows. Each x shall 'belong' to E in the above sense. The function defined by $x(s)$ when s is confined to E shall be continuous over E, and as $s \to \infty$, $x(s) \to 0$. The p-norm $| x |$ of x shall be $\sup_s | x(s) |$.

THEOREM 2.1. *A functional f which is linear over $C(E)$ has the form*

$$f(x) = \int_{R_1} x(s)\, dg(s), \qquad [x \in C] \tag{2.2}$$

where g is a d-function based on E and is uniquely determined by f with $\| f \| = T(g)$.

The classical theorem of F. Riesz can be stated as follows: a functional f which is linear over $C\,[E = [0, 1]]$ has the form (2.2) where g is of bounded variation over $[0, 1]$; if g is a d-function based on $[0, 1]$, g is uniquely determined by f and $\| f \| = T(g)$.

Because of its special character the proof of Theorem 2.1 will be deferred to Paper II. This theorem with a proper interpretation and choice of E not only includes the Riesz theorem but also the Steinhaus theorem (Cf. Banach, p. 65) on the representation of a function which is linear over the space c of convergent sequences.

DEFINITION. *Partitions π belonging to E.* Let a finite set of points s_r of E be given in the order

$$0 = s_0 < s_1 < \cdots s_{p-1} < s_p. \tag{2.3}$$

In case ω_E in (2.1) is finite s_p shall equal ω_E. Let R_1 be divided into *disjoint* intervals by the points s_r, including each $s_r \neq \omega_E$ only as a left end point, and including ω_E, if finite, only as a right end point. The resulting partition π of R_1 will be said to *belong* to E.

If ω_E is infinite π consists of the successive intervals

$$(2.4) \qquad (-\infty, s_0)[s_0, s_1)[s_1, s_2) \cdots [s_{p-1}, s_p)[s_p, \infty).$$

If ω_E is finite the last two intervals are of the form

$$(2.5) \qquad [s_{p-1}, \omega_E](\omega_E, \infty).$$

A first motivation for the inclusion or exclusion of a point s_r in an interval of π and for the definition of a d-function based on E is found in the desire that (2.6) and (2.7) hold as stated. A d-function g based on E defines a regular measure function m_g. (Cf. McShane p. 227.) If I_r is the interval of π limited by s_r and s_{r-1}, then

$$(2.6) \qquad m_g(I_r) = g(s_r) - g(s_{r-1}) = \Delta_r(g),$$

introducing $\Delta_r(g)$. (Cf. McShane, pp. 253–254.) If the above conventions as to the inclusion of end points were altered at a point s_r of discontinuity of g, then (2.6) would fail to hold. The definition of a d-function based on E is such that whenever s_i is a point of E (whether ω_E or not) the exclusion of s_i as an end point of I_r is always accompanied by the continuity of g at s_i relative to an approach to s_i from within I_r.

DEFINITION. *Step-functions* (s-functions) *belonging to* E. An s-function η *belonging* to E will be *associated* with any partition π belonging to E by assigning to η constant values η_r on the respective intervals I_r of π ($r = 1, \cdots, p$), and the value 0 on the extreme unbounded intervals of π.

If g is a d-function based on E and η an s-function belonging to E then, employing an L-S-integral and the summation convention of tensor algebra,

$$(2.7) \qquad \int_{R_1} \eta(s)\, dg(s) = \eta_r \Delta_r(g) \qquad [r = 1, \cdots, p].$$

This relation follows from (2.6).

Observe that the conditions that an s-function η belong to E are consistent with the definition that a more general function x *belong* to E. These definitions will be illustrated in §5 in the case in which A is equivalent to l_p ($p > 0$).

3. Conditions I and II on p-normed vector spaces A

The class of admissible spaces A will be defined by listing their characteristic properties I–V. The elements $x \epsilon A$ are functionals belonging (in the sense of §2) to a closed set E_A (restricted as in §2). In considering these limitations on the representation of elements in A it should be understood that the bilinear theory which will be developed applies not only to the admissible spaces A, but also to spaces *equivalent* to spaces A. (Cf. Banach, p. 180.)

An admissible space A shall then be a p-normed vector space of functionals belonging as above to a closed set E_A. The p-norm of $x \epsilon A$ shall be denoted by $|x|_A$. We suppose that A contains elements for which $|x|_A \neq 0$. With this supposed, A contains *unit* elements, that is elements x for which $|x|_A = 1$.

Let f be linear over A. The *modulus* of f by definition is

$$\| f \| = \sup_x | f(x) | \qquad [| x |_A = 1].$$

DEFINITION. *A proper norm for a vector space A_d of d-functions.* Let $N(g)$ be a non-negative number (possibly ∞) defined for every g which is canonical rel E_A with

$$N(g_1 + g_2) \leqq N(g_1) + N(g_2), \qquad N(cg) = | c | N(g), \qquad [\text{for } N(g) < \infty]$$

for c an arbitrary constant. Suppose moreover that $g(s) \equiv 0$ whenever $N(g) = 0$. Then N will be termed a *proper* norm for a vector space A_d of d-functions based on E if g is in A_d when and only when $N(g)$ is finite.

The norms $N(g)$ which enter in practice are defined with the aid of generalized Hellinger integrals. No á priori assumption as to the measurability of g is required. The finiteness of $N(g)$ assures this in every case considered.

The following are examples of norms:

(3.1) The total variation $T(g)$ of g,

The qth root of the Hellinger integral,

$$\left[\int_{R_1} \frac{[dg(s)]^q}{[ds]^{q-1}} \right]^{1/q} = H_q(g), \qquad [q > 1] \tag{3.2}$$

$$\text{The } \sup \frac{| g(s'') - g(s') |}{| s'' - s' |} = H(g), \tag{3.3}$$

taking (3.3) over all pairs (s'', s') for which $s'' \neq s'$. Given an admissible closed set E, a possible normed vector space A_d is the space of all functions g canonical rel E, for which one of the above norms is finite.

We shall use the Banach notation in which $\bar{A}$ is the space of functionals linear over A. The norm of an element $f \in \bar{A}$ is taken as the previously defined modulus $\| f \|$. The first two conditions on admissible spaces A can now be stated.

CONDITION I. *A shall be a p-normed vector space of functionals x 'belonging' to a closed subset E_A of R_1. $\bar{A}$ shall be equivalent to a space A_d of d-functions based on E_A with a proper norm, with $g \in A_d$ corresponding to $f_g \in \bar{A}$ if and only if for each $x \in A$, $f_g(x)$ is given by the L-S-integral*

$$f_g(x) = \int_{R_1} x(s)\, dg(s). \tag{3.4}$$

CONDITION II. *There exists a monotone increasing function M_A with values $M_A(\alpha) \geqq 0$ defined for $\alpha \geqq 0$ such that whenever η is an s-function belonging to E_A and $\eta(s) = 0$ for $s > \alpha$, then for every $g \in A_d$,*

$$\left| \int_{R_1} \eta(s)\, dg(s) \right| \leqq M_A(\alpha)\, | g |_{A_d} \max | \eta(s) |. \tag{3.5}$$

Condition I implies that $f \in \bar{A}$ has a representation (3.4) with $g_f \in A_d$ uniquely determined by f among d-functions in A_d, and that

$$\|f\| = |g_f|_{A_d}, \tag{3.6}$$

while conversely each $g \in A_d$ determines an $f_g \in \bar{A}$ such that

$$f_g(x) = \int_{R_1} x(s)\, dg(s) \qquad [x \in A].$$

Condition II is referred to in (4.6), (5.4), (12.7)′, and (12.13).

The simplest examples which illustrate I and II are obtained by taking E_A as the interval [0, 1] and A as one of the spaces C, $L_p(p > 1)$ and L, conceived first as an unnormed linear vector space of functionals and converted into a p-normed space by defining a p-norm so as to have a *value* equal to that of the norm as classically defined in this space. We make the trivial modification[2] noted below. The norms of an element $g \in A_d$ will then be given by (3.1), (3.2), and (3.3) respectively. This will be established in the next paper in a more general setting. Admissible spaces A equivalent to the spaces $l_p(p \geqq 1)$ will be defined in §5.

4. An extension A_e of A which includes s-functions

Linear extensions of a given vector space A and of the linear functionals defined over A can in general be made in several ways. In the case at hand A is to be extended so that it includes all s-functions η belonging to E_A.

The s-functions η belonging to E_A and the elements $x \in A$ combine linearly to form a unique vector space A_e. It remains to define a p-norm for elements $u \in A_e$. Given $f \in \bar{A}$, a functional f^e will be defined for $u \in A_e$ by setting

$$f^e(u) = \int_{R_1} u(s)\, dg_f(s). \tag{4.1}$$

Since u is of the form $\eta + x$, where η is an s-function belonging to E_A and x is in A, the L-S-integral (4.1) exists. In fact, from (2.7),

$$f^e(u) = \eta_r \Delta_r(g_f) + f(x), \qquad [r = 1, \cdots, p]$$

and in particular $f^e(x) = f(x)$ for $x \in A$. Our definition of the p-norm of u in A_e is motivated by the following lemma.

LEMMA 4.1. *For fixed x_0 in a p-normed vector space B*

$$|x_0|_B = \sup_F \frac{|F(x_0)|}{\|F\|}, \tag{4.2}$$

taking the sup *over all $F \in \bar{B}$ for which $\|F\| \neq 0$.*

---

[2] The functionals x are ordinarily defined only over [0, 1] in the spaces C and L_p. In this paper they will be understood as defined over R_1, setting $x(s) = 0$ on the complement of [0, 1].

Let m be the sup in (4.2). That $| x_0 |_B \geqq m$ follows at once from the relation

$$| F(x_0) | \leqq | x_0 |_B \, || F || . \tag{4.3}$$

It remains to show that $| x_0 |_B = m$. According to an extension of a theorem of Hahn (Cf. §1) there exists an $F \in \bar{B}$ such that

$$F(x_0) = | x_0 |_B , \qquad || F || = 1.$$

The existence of such an F shows that it is impossible that $| x_0 |_B > m$, and (4.2) follows.

DEFINITION. *For fixed* $u \in A_e$, *set*

$$| u |_{A_e} = \sup_f \frac{| f^e(u) |}{|| f ||}, \tag{4.4}$$

taking the sup *over all* $f \in \bar{A}$ *for which* $|| f || \neq 0$.

LEMMA 4.2. *With* $| u |_{A_e}$ *defined by* (4.4), A_e *is a p-normed vector space extending* A.

Since A_e is a vector space it remains to show that $| u |_{A_e}$, as defined by (4.4), is finite, and has the other properties required of a p-norm. That this norm has the usual homogeneity follows from the homogeneity of the L-S-integral $f^e(u)$. If y and z are in A_e

$$| f^e(y + z) | = | f^e(y) + f^e(z) | \leqq | f^e(y) | + | f^e(z) |,$$

and from this and (4.4) it follows that

$$| y + z |_{A_e} \leqq | y |_{A_e} + | z |_{A_e} . \tag{4.5}$$

The finiteness of $| u |_{A_e}$. The proof of (4.5) does not depend upon the finiteness of the A_e-norm of y or z. With the aid of (4.5) one can prove that the A_e-norm of u is finite for every $u \in A_e$. To that end observe that the definition (4.4) gives

$$| x |_{A_e} = | x |_A , \qquad \text{[for } x \in A]$$

by virtue of Lemma 4.1 (with $B = A$). Hence $| x |_{A_e}$ is finite. But every $u \in A_e$ is of the form $\eta + x$ where η is an s-function belonging to E_A. By virtue of (4.5)

$$| u |_{A_e} \leqq | \eta |_{A_e} + | x |_{A_e} ,$$

so that one has merely to show that $| \eta |_{A_e}$ is finite. If s_p is the last vertex of the partition π associated with η, then $\eta(s) = 0$ for $s > s_p$. It follows from Condition II (3.5) and equation (3.6) that

$$| f^e(\eta) | \leqq M_A(s_p) \, || f || \max | \eta(s) |,$$

so that

$$| \eta |_{A_e} \leqq M_A(s_p) \max | \eta(s) | , \tag{4.6}$$

in accordance with the definition (4.4). Thus $| u |_{A_e}$ is finite.

This completes the proof of the lemma.

From (4.4) one infers the fundamental relation

$$|f^e(u)| \leqq |u|_{A_e} \|f\| \qquad [u \in A_e] \tag{4.7}$$

We can also prove that

$$\|f^e\| = \|f\|. \tag{4.8}$$

For (4.7) implies that $\|f^e\| \leqq \|f\|$ while the relation

$$|f^e(u)| \leqq |u|_{A_e} \|f^e\| \qquad [u \in . \,.] \tag{4.9}$$

shows that

$$|f(x)| \leqq |x|_A \|f^e\|,$$

so that $\|f\| \leqq \|f^e\|$. We conclude that (4.8) holds.

Definition. *The space* $\bar{A}_{LS}$. The integral f^e given by (4.1) will be termed the *L-S-extension* of $f \in \bar{A}$. The modulus $\|f^e\|$ will be assigned to f^e as norm. For $f \in \bar{A}$ the elements f^e then form a normed vector space $\bar{A}_{LS}$.

We summarize and complete the preceding results as follows.

Theorem 4.1. *Under Conditions I and II on A there exists a unique minimal linear extension A_e of A which includes the s-functions belonging to A and is such that the p-norm of $u \in A_e$ is given by* (4.4). *With A_e so defined the spaces*

$$A_d, \qquad \bar{A}, \qquad \bar{A}_{LS},$$

are equivalent with $g \in A_d$ corresponding to $f_g \in \bar{A}$ and to $f_g^e \in \bar{A}_{LS}$, where

$$f_g(x) = \int_{R_1} x(s)\, dg(s), \qquad f_g^e(u) = \int_{R_1} u(s)\, dg(s), \tag{4.10}$$

for $x \in A$ and $u \in A_e$.

As a consequence

$$\|f_g\| = \|f_g^e\| = |g|_{A_d}, \tag{4.11}$$

$$|f_g^e(u)| \leqq |u|_{A_e} \|f_g\| = |u|_{A_e} |g|_{A_d}. \tag{4.12}$$

5. Conditions III, IV, V on A

Definition. *Sub-unit elements* η. An s-function η belonging to E_A for which $|\eta|_{A_e} \leqq 1$ will be termed *sub-unit* in A_e.

The following lemma prepares the way for Condition III. This lemma together with Condition III characterizes elements $g \in A_d$.

Lemma 5.1. *If A satisfies I, II, a necessary condition that a functional g which is canonical* rel E_A *be in A_d is that for some finite constant K,*

$$\sup_{\eta} |\eta_r \Delta_r(g)| \leqq K, \tag{5.1}$$

taking the sup *over all sub-unit s-functions $\eta \in A_e$.*

In fact for $g \in A_d$,

$$|\eta_r \Delta_r(g)| = \left| \int_{R_1} \eta(s)\, dg(s) \right| \leqq |\eta|_{A_e} |g|_{A_d}, \qquad \text{[by (4.12)]} \tag{5.2}$$

so that in (5.1) K can be taken as $|g|_{A_d}$.

Condition[3] III. *If A satisfies I, II, and g is canonical* rel E_A *the condition* (5.1) *is sufficient that g be in A_d.*

Condition IV. *The s-functions η belonging to E_A shall be everywhere dense in A_e.*

When $A = A_e$ the Condition III can be established with the aid of I, II, IV. This fact simplifies the verification of III in practice. The theorem follows.

Theorem 5.1. *If A satisfies I, II, IV and $A = A_e$, then a necessary and sufficient condition that a functional g which is canonical* rel E_A *be in A_d is that* (5.1) *hold.*

That (5.1) is necessary has already been proved under Lemma 5.1. To establish the sufficiency of the condition (5.1), statements (a) to (d) will be proved. We assume all conditions of the theorem except that g be in A_d.

(a). *The total variation $T_0^b(g)$ of g over any finite interval of the form* $[0, b]$ *is finite.*

Since g is assumed canonical rel E_A the truth of (a) will follow from the truth for any point $b \in E_A$. Suppose then that b is in E_A. Specialize η by supposing that each $|\eta_r| = 1$ $(r = 1, \cdots, p)$ and that $\eta(s) \equiv 0$ for $s \in$ comp $[0, b]$. With η so restricted in A_e, (5.1) implies that

$$T_0^b(g) = \sup_{\eta} |\eta_r \Delta_r(g)| \leqq K |\eta|_{A_e}. \tag{5.3}$$

For these special η it follows from (4.6) that

$$|\eta|_{A_e} \leqq M_A(b) \max |\eta(s)| = M_A(b),$$

so that (5.3) takes the form

$$T_0^b(g) \leqq K M_A(b). \tag{5.4}$$

Statement (a) follows.

With g known to be of bounded variation over any finite interval, one infers for any s-function $\eta \in A_e$ that

$$\eta_r \Delta_r(g) = \int_{R_1} \eta(s)\, dg(s). \tag{5.5}$$

This integral will be denoted by $\phi(\eta)$ and extended in definition over A_e.

(b). *If a sequence of s-functions $\eta^{(n)} \in A_e$ $(n = 1, 2, \cdots)$ converges to $u \in A_e$, then $\phi[\eta^{(n)}]$ converges to a limit independent of the sequence $[\eta^{(n)}]$ converging to u.*

According to (5.1)

$$|\phi(\eta)| \leqq K |\eta|_{A_e}, \tag{5.6}$$

[3] It can be shown by example that Condition III is independent of Conditions I and II.

and in particular

$$(5.7) \qquad |\phi[\eta^{(n)}] - \phi[\eta^{(m)}]| = |\phi[\eta^{(n)} - \eta^{(m)}]| \leqq K\,|\eta^{(n)} - \eta^{(m)}|_{A_e}.$$

Hence $\phi[\eta^{(n)}]$ converges. The uniqueness of the limit is similarly proved taking account of the convergence of $\eta^{(n)}$ to u.

Corresponding to an arbitrary $u \in A_e$ there exists by IV a sequence of s-functions $\eta^{(n)} \in A_e$ which converges in A_e to u. Using (b) set

$$(5.8) \qquad \operatorname*{limit}_{n\to\infty} \phi[\eta^{(n)}] = \phi(u).$$

(c). *The functional ϕ is linear over A_e.*

For s-functions η and $\bar\eta \in A_e$ it is clear from (5.5) that

$$(5.9) \qquad \phi(\eta + \bar\eta) = \phi(\eta) + \phi(\bar\eta).$$

If $\eta^{(n)} \to u \in A_e$ and $\bar\eta^{(m)} \to \bar u \in A_e$, we infer from (5.8) and (5.9) that

$$\phi(u + \bar u) = \phi(u) + \phi(\bar u).$$

Thus ϕ is additive over A_e. The homogeneity follows similarly from the relation $\phi(c\eta) = c\phi(\eta)$, c constant. Finally if $\eta^{(n)} \to u$ in A_e, the norm of $\eta^{(n)}$ in A_e tends to the norm of u in A_e and it follows from (5.6) that

$$|\phi(u)| \leqq K\,|u|_{A_e}.$$

This completes the proof of (c).

(d) *The element g is in A_d.*

Since ϕ is linear over A, and A satisfies Condition I, there exists a $g_\phi \in A_d$ such that for $x \in A$,

$$\phi(x) = \int_{R_1} x(s)\, dg_\phi(s).$$

In particular when $x = \eta$ (since $A = A_e$),

$$(5.10) \qquad \phi(\eta) = \eta_r \Delta_r(g) = \int_{R_1} \eta(s)\, dg_\phi(s) = \eta_r \Delta_r(g_\phi).$$

On taking $\eta_1 = 1$ and $\eta_r = 0$ for $r > 1$, (5.10) implies that

$$g(s_1) = g_\phi(s_1)$$

for every $s_1 > 0$ in E_A. Since g_ϕ and g are based on E_A, $g_\phi = g$.

This completes the proof of the theorem.

Once g is known to be in A_d, either by virtue of the preceding theorem when $A = A_e$, or by Condition III when $A \neq A_e$, the following theorem gives an evaluation of $|g|_{A_d}$.

THEOREM 5.2. *When A satisfies I, II, IV,*

$$(5.11) \qquad \sup_\eta |\eta_r \Delta_r(g)| = |g|_{A_d} \qquad [r = 1, \cdots, r_\eta]$$

for $g \in A_d$, taking the sup *over all sub-unit $\eta \in A_e$.*

In the equivalence between A_d, $\bar{A}$ and $\bar{A}_{LS}$ affirmed in Theorem 4.1, suppose that $g \epsilon A_d$ has images $f_g \epsilon \bar{A}$ and $f_g^e \epsilon \bar{A}_{LS}$. By virtue of (4.10) and (2.7)

$$f_g^e(\eta) = \eta_r \Delta_r(g), \tag{5.12}$$

so that from (4.12), and the sub-unit character of η

$$\sup_\eta | \eta_r \Delta_r(g) | \leqq \sup_\eta | \eta |_{A_e} | g |_{A_d} \leqq | g |_{A_d}. \tag{5.13}$$

It remains to exclude the inequality in (5.13).

From (4.11) and the definition of $\| f_g \|$ respectively,

$$| g |_{A_d} = \| f_g \| = \sup_{x \epsilon A} | f_g(x) | \qquad [| x |_A = 1]. \tag{5.14}$$

By IV, s-functions $\eta \epsilon A_e$ are everywhere dense in A_e, and consequently sub-unit s-functions exist arbitrarily near each unit $x \epsilon A$. From the continuity of f_g^e over A_e and from (5.14)

$$\sup_\eta | f_g^e(\eta) | \geqq | g |_{A_d} \qquad [| \eta |_{A_e} \leqq 1].$$

Hence the equality must prevail in (5.13), and the proof of the theorem is complete.

Lemma 5.2. *When A satisfies Condition I, the condition*

$$f_g(x) = \int_{R_1} x(s)\, dg(s) = 0$$

for a fixed $g \epsilon A_d$ and for every $x \epsilon A$ implies that $g(s) \equiv 0$.

It follows from Condition I that $| g |_{A_d} = \| f_g \|$. For the given g, $\| f_g \| = 0$ so that $| g |_{A_d} = 0$. Since A_d-norms are "proper" by hypothesis in Condition I, $g(s) \equiv 0$.

Definition. *The* l.s.c. (*lower semi-continuity*) *of the norm of g*. Let G_A be the linear vector space of all functionals g canonical rel E. Pointwise convergence of a sequence of functions $g_n \epsilon G_A$ to a function $g \epsilon G_A$ shall mean that for each s

$$\operatorname*{limit}_{n \to \infty} g_n(s) = g(s).$$

Let R_1^+ denote the space R_1 extended by adding the ideal value $+\infty$. Let ϕ map G_A into R_1^+. Then ϕ is termed l.s.c. over G_A if whenever g_n converges pointwise to g in G_A

$$\liminf_{n \to \infty} \phi(g_n) \geqq \phi(g). \tag{5.15}$$

It is understood (Cf. §3) that the A_d-norm $| g |_{A_d}$ is defined for each element g which is canonical rel E_A, and that this norm is finite if and only if $g \epsilon A_d$. It follows from Theorem 5.2 that for $g \epsilon G_A$, regardless of the finiteness of $| g |_{A_d}$, but with I, II, III, IV holding,

$$\sup_\eta | \eta_r \Delta_r(g) | = | g |_{A_d}, \tag{5.16}$$

taking the sup over all sub-unit s-functions η in A_e. With l.s.c. defined and (5.16) understood, the following theorem can be proved.

THEOREM 5.3. *The A_d-norm of elements g in the space G_A of all functions g which are canonical relative to E_A is lower semi-continuous when* I, II, III, IV *hold.*

For a fixed partition π belonging to E_A and associated s-functions η it is clear that

$$\eta_r \Delta_r(g) \qquad [r = 1, \cdots, p_\pi]$$

is continuous with respect to pointwise convergence of g_n to g in G_A. It follows from a well known theorem on l.s.c. that the A_d-norm (5.16) is l.s.c. (Cf. McShane, l.c. p. 41.)

EXAMPLE 5.0. With the spaces C, $L_p(p > 1)$, L understood as in footnote (2), elements g in the corresponding spaces C_d, $(L_p)_d$, and L_d have norms given respectively by (3.1), (3.2), and (3.3). That these norms are l.s.c. is directly verifiable.

The notation involving A may be summarized as follows:

A = A p-normed vector space of functionals belonging to E_A eventually required to satisfy I–V.

E_A = A closed subset of R_1 restricted as in §2.

A_d = A subspace of d-functions based on E_A. Characterized in I.

A_e = A minimal linear extension of A including all s-functions belonging to E_A. Characterized in Theorem 4.1.

$\bar{A}$ = The normed vector space of functionals linear over A.

$\bar{A}_{LS}$ = The space of L-S-extensions f^e of elements $f \in \bar{A}$.

ω_A = Superior limit of $s \in E_A$.

To state Condition V on A two definitions are required.

DEFINITION. *Convergence in norm.* Let B be a p-normed vector space. A sequence of elements $x^{(n)} \in B$ will be said to converge in norm to an element $u \in B$ if

$$\lim_{n \to \infty} | x^{(n)} |_B = | u |_B.$$

DEFINITION. *Relative convergence.* Let B be a p-normed vector space. A sequence of elements $x^{(n)} \in B$ will be said to converge to an element $u \in B$ *relative* to $f \in \bar{B}$ if

$$\lim_{n \to \infty} f[x^{(n)}] = f(u).$$

Convergence relative to *every* element $f \in \bar{B}$ is the so called *weak* convergence. (Cf. Banach, p. 133.)

Condition V will now be stated, preceded, for convenience, by a restatement of Conditions I–IV.

CONDITION I. *A shall be a p-normed vector space of functionals x 'belonging' to a closed subset E_A of R_1. $\bar{A}$ shall be equivalent to a space A_d of d-functions based on E_A with a proper norm, with $g \in A_d$ corresponding to $f_g \in \bar{A}$ if and only if for each $x \in A$, $f_g(x)$ is given by the L-S-integral*

$$f_g(x) = \int_{R_1} x(s)\, dg(s).$$

CONDITION II. *There exists a monotone increasing function M_A with values $M_A(\alpha) \geqq 0$ defined for $\alpha \geqq 0$ such that whenever η is an s-function belonging to E_A and $\eta(s) = 0$ for $s > \alpha$, then for every $g \in A_d$*

$$\left| \int_{R_1} \eta(s)\, dg(s) \right| \leqq M_A(\alpha) \mid g \mid_{A_d} \max \mid \eta(s) \mid.$$

CONDITION III. *If A satisfies I, II, and g is canonical* rel E_A *the condition* (5.1) *is sufficient that g be in A_d.*

CONDITION IV. *The s-functions η belonging to E_A shall be everywhere dense in A.*

CONDITION V. *Corresponding to an arbitrary element $u \in A_e$ there shall exist a sequence of elements $x^{(n)} \in A$ which converges to $u \in A_e$, both in norm and relative to every f^e for each $f \in \bar{A}$.*

The proof of Lemma 9.4 will make essential use of Condition V. If $A = A_e$, Condition V is trivially satisfied on taking $x^{(n)} = u$ for every n. Taking account of Theorem 5.2 one has the following theorem.

THEOREM 5.4. *If $A = A_e$, Conditions I, II, and IV imply III, and V is trivially satisfied.*

The preceding will be illustrated by two very general examples.

EXAMPLE 5.1. *An admissible space $A = T[l_p]$ equivalent to $l_p (p \geqq 1)$.*

Case $p > 1$. Recall that l_p is the space of vectors $u = (u_1, u_2, \cdots)$ with norm

$$(5.17) \qquad \mid u \mid_{l_p} = [\sum_n \mid u_n \mid^p]^{1/p} \qquad [n = 1, 2, \cdots].$$

The conjugate space is the space l_q, where $1/p + 1/q = 1$. Any functional f which is linear over l_p has the form

$$(5.18) \qquad f(u) = u_n b_n, \qquad [n = 1, 2, \cdots]$$

where b is an element $b_f \in l_q$ uniquely determined by f with

$$\| f \| = \mid b \mid_{l_q}.$$

The mapping of l_p onto l_q determined by b_f is an equivalence.

Let E^0 be the set $s = 0, 1, 2, \cdots$. Given $u \in l_p$, let x be a functional mapping R_1 into R_1, belonging to E^0, and denoted by $T(u)$, with values $x(s)$ for which

$$(5.19) \qquad x(n) = u_{n+1} \qquad [n = 0, 1, 2, \cdots].$$

For such an x, $x(s)$ is constant on each interval $[n, n + 1)$. The operation T is additive, homogeneous and biunique, and maps l_p onto a vector space $A = T[l_p]$ which is equivalent to l_p provided one introduces the norm

$$(5.20) \qquad \mid x \mid_A = \mid u \mid_{l_p}.$$

If $f \in \bar{l}_p$ is made to correspond to $F \in \bar{A}$ provided $f(u) = F(x)$ when $x = T(u)$, an equivalence between $\bar{l}_p$ and $\bar{A}$ is implied.

To show that A is admissible Lemma 5.3 is required.

LEMMA 5.3. *The image $A = T[l_p]$ of $l_p (p > 1)$ is a space equivalent to l_p, and if one sets $E_A = E^0$, satisfies Conditions I–V.*

The equivalence of A and l_p is already clear.

To proceed it is necessary to establish the existence of a space A_d satisfying I with A. To that end, given $b \in l_q$, let g be a functional mapping R_1 into R_1, based on $E_A = E^0$, denoted by $S(b)$, with values $g(s)$ such that

$$g(n) = g(n-1) + b_n \qquad [n = 1, 2, \cdots]. \tag{5.21}$$

With $g(0) = 0$, $g(s)$ is thereby determined by every $s \in R_1$. The graph of g is a continuous broken line with vertices at the points of E^0. Under S, l_q is mapped in an additive, homogeneous, biunique manner onto a vector space $A' = S[l_q]$, equivalent to l_q, provided one sets

$$|g|_{A'} = |b|_{l_q},$$

when $g = S(b)$. Under T the functional $f \in \bar{l}_p$ represented in (5.18) becomes a functional $F \in \bar{A}$ such that

$$F(x) = f(u) = u_n b_n = \int_{R_1} x(s)\, dg(s), \tag{5.22}$$

provided of course, $x = T(u)$ and $g = S(b)$. Moreover

$$\| F \| = \| f \| = |b|_{l_q} = |g|_{A'}.$$

The above equivalences $\bar{A}$ with $\bar{l}_p$, $\bar{l}_p$ with l_q, l_q with A' imply an equivalence of $\bar{A}$ with A' in which $F \in \bar{A}$ corresponds to $g \in A'$ if and only if (5.22) holds. Thus A' can serve as the required space A_d, and Condition I is satisfied by A.

To show that A satisfies Condition II set

$$u^{(n)} = (u_1, \cdots, u_n, 0, 0, \cdots)$$

for arbitrary $u \in l_p$. Then $u^{(n)}$ is in l_p and $T[u^{(n)}]$ is an arbitrary s-function η belonging to $E^0 = E_A$. We have

$$F(\eta) = \int_{R_1} \eta(s)\, dg(s) = \sum_{r=1}^{n} u_r b_r,$$

$$|F(\eta)| \leqq |u^{(n)}|_{l_p} |b|_{l_q} = |u^{(n)}|_{l_p} \| F \|,$$

$$|F(\eta)| \leqq n^{1/p} \| F \| \underset{0<r<n+1}{\text{maximum}} |u_r|,$$

$$|F(\eta)| \leqq n^{1/p} \| F \| \text{ maximum } |\eta(s)| \qquad [\eta(s) = 0 \text{ for } s > n]. \tag{5.23}$$

From (5.23) it follows that (3.5) is satisfied by a function M_A with values

$$M_A(\alpha) = \alpha^{1/p}, \qquad [\alpha \geqq 0]$$

so that Condition II is satisfied.

It is clear that $A = A_e$. To show that Condition IV is satisfied, let $x = T(u)$ be an arbitrary element $x \in A$. Then

$$\lim_{n\to\infty} u^{(n)} = u, \qquad \lim_{n\to\infty} T[u^{(n)}] = T(u) = x.$$

But $T[u^{(n)}]$ is an admissible s-function $\eta \in A_e$ so that IV is satisfied.

According to Theorem 5.4, Conditions III and V are also satisfied and the proof of the lemma is complete.

EXAMPLE 5.2. *An admissible space $A = T[l]$ equivalent to l.* Recall that l is the space of vectors $u = (u_1, u_2, \cdots)$ with norm $|u|_l = \sum |u_n|$. The conjugate space (m) is the space of vectors $b = (b_1, b_2, \cdots)$ with norm $|b|_{(m)} = \sup_n |b_n|$. A functional $f \in \bar{l}$ has the form (5.18) with b an element $b_f \in (m)$ uniquely determined by f and with $\|f\| = |b|_{(m)}$. The mapping of $\bar{l}$ onto (m) defined by b_f is an equivalence.

For $u \in l$ let $T(u)$ be defined as in (5.19). Then T maps l onto a vector space $A = T[l]$ which is equivalent to l if one sets $|x|_A = |u|_l$ when $x = T(u)$. An equivalence of $\bar{l}$ with $\bar{A}$ is implied.

LEMMA 5.4. *The image $T[l]$ of l is a space A equivalent to l, and if one sets $E_A = E^0$ satisfies Conditions I–V.*

A space A_d satisfying Condition I with A is defined as follows. For every $b \in (m)$ set $g = S(b)$. The image $S[m]$ of (m) is a space A' equivalent to (m) provided one sets $|g|_{A'} = |b|_{(m)}$ when $g = S(b)$. Under T a functional f represented as in (5.18) becomes a functional $F \in \bar{A}$ such that (5.22) holds, and $\|F\| = \|f\| = |b|_{(m)} = |g|_{A'}$. The space A′ is thus the required space A_d and Condition I holds.

Condition II is satisfied since (5.23) holds for $p = 1$.

Condition IV is satisfied as in the case $p > 1$.

Conditions III and V are satisfied by virtue of Theorem 5.4.

The proof of the lemma is complete.

A statement which will serve by analogy as a guide in the representation of bilinear functionals is as follows. The functions $g \in A_d$ which have proved adequate in the L-S-representation of elements in $\bar{A}$ have two characteristic properties, (given the essential properties of A)

(a). *They are canonical* rel E_A,

(b). *Their A_d-norms* (5.16) *are finite.*

Property (a) depends upon A only through E_A. Property (b) depends upon the nature of s-functions in A_e and their norms.

In the bilinear theory functions g mapping R_1 into R_1 will be replaced by functions k mapping R_2 into R_1. In terms which will presently be defined, the functions k which will prove adequate in the L-S-representation

$$\int_{R_1} x(s)\, d_s \int_{R_1} y(t)\, d_t k(s, t) = F(x, y) \qquad [x \in A,\ y \in B] \tag{5.24}$$

of functionals F bilinear on $A \times B$, have two characteristic properties:

(α). *They are canonical* rel $E_A \times E_B$,

(β). *Their variations $h(A, B; k)$ are finite.*

PART II. FUNCTIONALS BILINEAR OVER $A \times B$

6. Functions canonical rel $E_A \times E_B$

We are concerned with the product of two p-normed vector spaces A and B satisfying Conditions I — V. The elements $x \in A$ and $y \in B$ are then functionals 'belonging' to closed subsets E_A and E_B of R_1 restricted as in §2. When A or B is complete the definition of an F bilinear over $A \times B$ can be weakened as the following theorem shows.

THEOREM 6.1. *If $F(\cdot, y)$ and*[4] *$F(x, \cdot)$ are linear over A and B respectively and A (or B) is complete, then F is bilinear over $A \times B$.*

It is sufficient to show that

$$(6.1) \qquad | F(x, y) | \leqq | x |_A | y |_B M \qquad [x \in A, y \in B]$$

for some constant M.

Since $F(\cdot, y)$ is linear over A for fixed $y \in B$, there exists a constant $N(y)$ such that (with θ the null element in B)

$$(6.2) \qquad \sup_x | F(x, y) | = \| F(\cdot, y) \| = N(y) | y |_B, \qquad [N(\theta) = 0]$$

taking the sup over all unit $x \in A$. If $N(y)$ has a finite bound K independent of $y \in B$, the desired constant M may be taken as K. Otherwise there exists a sequence of non-null elements $y^{(n)} \in B$ for which $N[y^{(n)}] \to +\infty$. For each n introduce the functional $V_n \in \bar{A}$ for which

$$V_n(x) = \frac{F[x, y^{(n)}]}{| y^{(n)} |_B} \qquad [x \in A].$$

Since $F(x, \cdot)$ is linear over B with

$$| F(x, y) | \leqq | y |_B \| F(x, \cdot) \|,$$

$| V_n(x) |$ is bounded by $\| F(x, \cdot) \|$ for fixed $x \in A$ and variable $n = 1, 2, \cdots$. It follows from Theorem V of Banach, p. 80, that when A is complete there exists a constant K independent of n such that $\| V_n \| \leqq K$. Since $\| V_n \| = N[y^{(n)}]$ we infer that $N(y)$ is bounded independent of $y \in B$, contrary to hypothesis. Hence (6.1) holds as stated. One arrives at the same conclusion when B is complete.

In seeking to represent the functionals F bilinear over $A \times B$ as L-S-integrals (5.24) we shall define the conditions (α) and (β) indicated by analogy at the end of §5. That our conditions on the distribution function k are neither too restrictive nor insufficiently restrictive will be clear only when the major objectives outlined in the introduction have been obtained. We first define the conditions (α).

DEFINITION. *Functions k based on $E_A \times E_B$.* The function k shall map R_2 into R_1 in such a manner that for each fixed s, $k(s, \cdot)$ is based on E_B, and for each fixed t, $k(\cdot, t)$ is based on E_A.

The following lemma will be needed.

LEMMA 6.1. (a). *Corresponding to an arbitrary definition of $k(s, t)$ over $E_A \times E_B$ with $k(s, t) = 0$ on the s and t axes, $k(s, t)$ can be further defined on $C(E_A \times E_B)$ in one and only one way so that k is based on $E_A \times E_B$.* (b). *If k maps R_2 into R_1*

[4] $F(\cdot, y)$ represents the functional in $\bar{A}$ whose value for arbitrary $x \in A$ and fixed y is $F(x, y)$. One similarly defines $F(x, \cdot)$.

in such a manner that $k(\cdot, t)$ is based on E_A for each $t \in E_B$ and $k(s, \cdot)$ is based on E_B for each $s \in R_1$, then k is based on $E_A \times E_B$.

The values prescribed in (a) for $k(s, t)$ over $E_A \times E_B$ and the condition that $k(\cdot, t)$ be based on E_A for each $t \in E_B$ determine the values of $k(s, t)$ over $R_1 \times E_B$. The condition that $k(s, \cdot)$ be based on E_B for each $s \in R_1$ can then be satisfied and determines $k(s, t)$ where previously undefined. The unused condition that $k(\cdot, t)$ be based on E_A for each $t \in CE_B$ gives a second condition on each point $(s, t) \in CE_A \times CE_B$ but no other independent condition. That this final condition is consistent with the preceding conditions is readily verified. Thus (a) holds.

Statement (b) follows from (a). For the conditions in (b) together with the values of $k(s, t)$ over $E_A \times E_B$ determine $k(s, t)$ uniquely over R_2 without contradiction. Starting again with the values of $k(s, t)$ as given over $E_A \times E_B$, the condition that k be based on $E_A \times E_B$ likewise leads to a unique function k' by virtue of (a). Since k' satisfies all conditons (b) on k, $k = k'$.
Since k' is based on $E_A \times E_B$, k must likewise be based on $E_A \times E_B$.

DEFINITION. *The functions k_1^A and k_2^B.* Let k_1^A and k_2^B be functions mapping R_2 into R_1 and associated with k, A and B by setting

$$\text{(6.3)}\qquad \begin{aligned} k_1^A(s, t) &= \lim_{u \to 0+} k(s - u, t), && [s \neq \omega_A = \sup s \in E_A] \\ k_2^B(s, t) &= \lim_{u \to 0+} k(s, t - u), && [t \neq \omega_B = \sup t \in E_B] \end{aligned}$$

provided the indicated limits exist. In case ω_A or ω_B is finite set

$$k_1^A(\omega_A, t) = k(\omega_A, t)$$

$$k_2^B(s, \omega_B) = k(s, \omega_B).$$

It should be clear that the subscript 1, or 2 refers to the first or second argument respectively.

An extension of the above notation will be useful. Set $k_1^A = p$ and $k_2^B = q$ if the limits implied exist. If p_2^B and q_1^A exist set

$$p_2^B = k_{12}^{AB}, \qquad q_1^A = k_{21}^{BA}.$$

The condition

$$\text{(6.4)}\qquad k_{12}^{AB} = k_{21}^{BA}$$

will play a crucial role in §§9 and 12. If

$$\text{(6.5)}\qquad k = k_1^A = k_2^B,$$

relation (6.4) follows. On the other hand if $k = k_{12}^{AB}$ and (6.4) holds, then (6.5) follows, since

$$k_2^B = k_{122}^{ABB} = k_{12}^{AB} = k,$$

$$k_1^A = k_{211}^{BAA} = k_{21}^{BA} = k.$$

DEFINITION. *Functions k canonical* rel $E_A \times E_B$ shall be functions based on $E_A \times E_B$ for which

$$k = k_1^A = k_2^B. \tag{6.6}$$

Functions k canonical rel to $E_A \times E_B$ could be defined as functions k mapping R_2 into R_1 for which $k(\cdot, t)$ is canonical rel E_A for each t, and for which $k(s, \cdot)$ is canonical rel E_B for each s. (For the existence of canonical k see Morse and Transue, ref. 5.)

7. The variation $h(A, B; k)$ of k

At the end of §5 it has been indicated by broad analogy that the conditions on the distribution function k adequate for the L-S-representations (5.24) should be as follows:

(α). *The functions k should be canonical* rel $E_A \times E_B$,

(β). *The variations $h(A, B; k)$ should be finite.*

Let k then be canonical rel $E_A \times E_B$. We shall define the variation $h(A, B; k)$ of k.

DEFINITION. *Mixed differences.* Let π_A be a partition of R_1 belonging to E_A in the sense of §2, with vertices

$$0 \leqq s_0 < s_1 < \cdots < s_p \tag{7.1$'$}$$

in E_A, and π_B a partition of R_1 belonging to E_B, with vertices

$$0 \leqq t_0 < t_1 < \cdots < t_q \tag{7.1$''$}$$

in E_B. Let I_r^A $(r = 1, \cdots, p)$ be the rth of the finite intervals of π_A, and I_n^B $(n = 1, \cdots, q)$ the nth of the finite intervals of π_B. We shall associate the mixed difference

$$\begin{aligned}\Delta_{rn}(k) &= k(s_r, t_n) - k(s_{r-1}, t_n)\\ &\quad - k(s_r, t_{n-1}) + k(s_{r-1}, t_{n-1})\end{aligned}$$

with the product interval $I_r^A \times I_n^B$.

For a functional g mapping R_1 into R_1 we have defined $\Delta_r(g)$ by setting

$$g(s_r) - g(s_{r-1}) = \Delta_r(g) \qquad [r = 1, \cdots, p].$$

In the case of k set

$$k(s_r, t) - k(s_{r-1}, t) = \Delta_r^{(1)}[k(\cdot, t)],$$

$$k(s, t_n) - k(s, t_{n-1}) = \Delta_n^{(2)}[k(s, \cdot)],$$

the superscript (1) or (2) indicating that the first or second argument in k is differenced. Let the functional mapping R_1 into R_1 whose value at t is $\Delta_r^{(1)}[k(\cdot, t)]$ be denoted by $\Delta_r^{(1)}(k)$ and the functional whose value at s is $\Delta_n^{(2)}[k(s, \cdot)]$ be denoted by $\Delta_n^{(2)}(k)$. The following relation is basic:

$$\Delta_{rn}(k) = \Delta_r[\Delta_n^{(2)}(k)] = \Delta_n[\Delta_r^{(1)}(k)]. \tag{7.2}$$

DEFINITION. *The variations* h, h', h''. Let k be canonical rel $E_A \times E_B$. Taking the "sups" over all sub-unit s-function pairs $(\eta, \zeta) \in A_e \times B_e$, that is with

$$|\eta|_{A_e} \leqq 1, \qquad |\zeta|_{B_e} \leqq 1,$$

set

$$(7.3)' \qquad \sup_{\eta,\zeta} |\eta_r \zeta_n \Delta_{rn}(k)| = h(A, B; k), \qquad [r = 1, \cdots, p_\eta; n = 1, \cdots, q_\zeta]$$

$$(7.3)'' \qquad \sup_{\eta} |\eta_r \Delta_r^{(1)}(k)|_{B_d} = h'(A, B; k),$$

$$(7.3)''' \qquad \sup_{\zeta} |\zeta_n \Delta_n^{(2)}(k)|_{A_d} = h''(A, B; k),$$

admitting the possibility that the A_d- or B_d-norms may have infinite values and that accordingly any or all of the three sups may be $+\infty$.

It is useful to note that if one sets $k(s, t) = k^*(t, s)$, then

$$(7.4) \qquad h'(A, B; k) = h''(B, A; k^*).$$

DEFINITION. *The subclasses of functions* k, *canonical* rel $E_A \times E_B$, *for which the variations* (7.3) *are finite will be denoted respectively by*

$$H(A, B), \qquad H'(A, B), \qquad H''(A, B).$$

LEMMA 7.1. (a). *If* $k \in H'(A, B)$, *then* $k(s, \cdot) \in B_d$ *for each fixed* s, *and for* $s \in E_A$

$$(7.5)' \qquad |k(s, \cdot)|_{B_d} \leqq h'(A, B; k) M_A(s).$$

(b). *If* $k \in H''(A, B)$, *then* $k(\cdot, t) \in A_d$ *for each fixed* t, *and for* $t \in E_B$

$$(7.5)'' \qquad |k(\cdot, t)|_{A_d} \leqq h''(A, B; k) M_B(t).$$

Let η be an s-function belonging to E_A and associated with a partition in which the first interval has s as a right end point. Suppose that $\eta_1 = 1$ and that $\eta_r = 0$ for $r > 1$. Then $(7.3)''$ and (4.6) imply that

$$(7.6) \qquad |k(s, \cdot) - k(0, \cdot)|_{B_d} \leqq h'(A, B; k)\, M_A(s).$$

Since k is canonical rel E_A, $k(0, t) \equiv 0$. Thus

$$(7.7) \qquad |k(s, \cdot)|_{B_d} \leqq h'(A, B; k)\, M_A(s).$$

Hence $k(s, \cdot) \in B_d$, since $k(s, \cdot)$ is canonical rel E_B and has a finite B_d-norm. This proves (a).

The proof of (b) is similar.

The following theorem is fundamental.

THEOREM 7.1. *The subclasses* $H(A, B)$, $H'(A, B)$ *and* $H''(A, B)$ *of the class* $K(A, B)$ *of all distribution functions* k *which are canonical* rel $E_A \times E_B$, *are identical. For any* $k \in H(A, B)$,

$$(7.8) \qquad h(A, B; k) = h'(A, B; k) = h''(A, B; k).$$

Let η, ζ be s-functions belonging to E_A, E_B respectively. Set

$$\eta_r \zeta_n \Delta_{rn}(k) = S_k(\eta, \zeta). \tag{7.9}$$

From the nature of sups in general,

$$\sup_{\eta,\zeta} S_k(\eta, \zeta) = \sup_{\eta} [\sup_{\zeta} S_k(\eta, \zeta)] = \sup_{\zeta} [\sup_{\eta} S_k(\eta, \zeta)], \tag{7.10}$$

taking the sups as in the definition (7.3). From (7.2)

$$S_k(\eta, \zeta) = \eta_r \Delta_r[\zeta_n \Delta_n^{(2)}(k)], \tag{7.11}$$

so that from (7.10),

$$h(A, B; k) = \sup_{\zeta} \{\sup_{\eta} \eta_r \Delta_r [\zeta_n \Delta_n^{(2)}(k)]\}. \tag{7.12}$$

For fixed ζ the inner sup in (7.12) may be evaluated by (5.16) with

$$\sup_{\eta} \eta_r \Delta_r [\zeta_n \Delta_n^{(2)}(k)] = | \zeta_n \Delta_n^{(2)}(k) |_{A_d}.$$

Hence

$$h(A, B; k) = \sup_{\zeta} | \zeta_n \Delta_n^{(2)}(k) |_{A_d} = h''(A, B; k).$$

The proof of the remaining equality in (7.8) is similar, and the theorem follows.

8. The bilinear functionals $M_k(x, y)$

Corresponding to any distribution function $k \in H(A, B)$, we shall define a special functional M_k, bilinear over $A \times B$ and such that for $x \in A$ and $y \in B$,

$$| M_k(x, y) | \leqq | x |_A | y |_B \, h(A, B; k). \tag{8.1}$$

In (12.15) it will appear that

$$h(A, B; k) = \| M_k \|,$$

that is that $h(A, B; k)$ is the smallest constant such that (8.1) holds for every $x \in A$ and $y \in B$.

Corresponding to an arbitrary $k \in H(A, B)$ and s-functions $\eta \in A_e$ and $\zeta \in B_e$, set

$$\eta_r \zeta_n \Delta_{rn}(k) = S_k(\eta, \zeta) \qquad [r = 1, \cdots, r_\eta; n = 1, \cdots, n_\zeta]. \tag{8.2}$$

We shall prove the following theorem.

THEOREM 8.1. *For any $k \in H(A, B)$ and s-functions $\eta \in A_e$ and $\zeta \in B_e$,*

$$| S_k(\eta, \zeta) | \leqq | \eta |_{A_e} | \zeta |_{B_e} \, h(A, B; k). \tag{8.3}$$

It follows from (7.2) that

$$| S_k(\eta, \zeta) | = | \eta_r \Delta_r [\zeta_n \Delta_n^{(2)}(k)] | .$$

Set $g = \zeta_n \Delta_n^{(2)}(k)$. Since g is canonical rel E_A, (7.3)$'''$ implies that

$$|g|_{A_d} \leqq |\zeta|_{B_e} h''(A, B; k). \tag{8.4}$$

Since $h(A, B; k) = h''(A, B; k)$ by Theorem 7.1, one can replace h'' by h in (8.4). Thus the A_d-norm of g is finite and $g \in A_d$. Hence

$$\begin{aligned} |\eta_r \Delta_r(g)| &\leqq |\eta|_{A_e} |g|_{A_d} \qquad \text{[by (4.12)]} \\ &\leqq |\eta|_{A_e} |\zeta|_{B_e} h(A, B; k), \end{aligned}$$

using (8.4). Hence (8.3) holds as stated.

LEMMA 8.1. *For $k \in H(A, B)$ and s-function pairs (η, ζ) and (η', ζ') in $A_e \times B_e$,*

$$\begin{aligned} &|S_k(\eta, \zeta) - S_k(\eta', \zeta')| \\ &\quad\leqq [|\eta|_{A_e} |\zeta - \zeta'|_{B_e} + |\zeta|_{B_e} |\eta - \eta'|_{A_e}] h(A, B; k). \end{aligned} \tag{8.5}$$

In the special case in which η and η' are associated with a common partition π_A of R_1 belonging to E_A, and ζ and ζ' are associated with a common partition π_B belonging to E_B, it is obvious that

$$S_k(\eta, \zeta) - S_k(\eta', \zeta') = S_k(\eta, \zeta - \zeta') + S_k(\eta - \eta', \zeta'),$$

and (8.5) follows with the aid of (8.3).

In the general case let π_A^* be a partition of R_1 belonging to E_A whose vertices include the vertices of the partitions associated with η and η', and let π_B^* similarly combine the vertices of the partitions associated with ζ and ζ'. Let η^* and $\eta^{*\prime}$ be s-functions associated with π_A^* such that

$$\eta(s) = \eta^*(s), \qquad \eta'(s) = \eta^{*\prime}(s),$$

and let ζ^* and $\zeta^{*\prime}$ be similarly associated with π_B^*, ζ and ζ'. It is clear that (8.5) will hold when a $*$ is added to each of the arguments. But

$$\begin{aligned} S_k(\eta, \zeta) &= S_k(\eta^*, \zeta^*), & S_k(\eta', \zeta') &= S_k(\eta^{*\prime}, \zeta^{*\prime}), \\ |\eta|_{A_e} &= |\eta^*|_{A_e}, & |\zeta|_{B_e} &= |\zeta^*|_{B_e}, \\ |\eta - \eta'|_{A_e} &= |\eta^* - \eta^{*\prime}|_{A_e}, & |\zeta - \zeta'|_{B_e} &= |\zeta^* - \zeta^{*\prime}|_{B_e}. \end{aligned}$$

Hence the validity of (8.5) with the arguments starred implies its validity as written.

THEOREM 8.2. *For $k \in H(A, B)$ and sequences of s-functions $\eta^{(n)} \in A_e$, $\zeta^{(m)} \in B_e$ which converge respectively in A_e and B_e to $x \in A$* and *$y \in B$,*

$$\lim_{\substack{n \to \infty \\ m \to \infty}} S_k[\eta^{(n)}, \zeta^{(m)}] \tag{8.6}$$

exists and equals a number

$$M_k(x, y) \tag{8.7}$$

independent of the choice of the sequences of s-functions converging to x and y respectively.

Given $e > 0$, it follows from Lemma 8.1 that

$$| S_k[\eta^{(n)}, \zeta^{(m)}] - S_k[\eta^{(\mu)}, \zeta^{(\nu)}] | < e,$$

provided n, m, μ, ν exceed a sufficiently large integer $N(e)$. Hence the limit (8.6) exists. It follows from a second use of Lemma 8.1 that the limit $M_k(x, y)$ is independent of the choice of the sequences of s-functions converging to x and y respectively.

THEOREM 8.3. *The functional M_k of Theorem 8.2 is bilinear over $A \times B$, and for $x \in A$ and $y \in B$,*

$$| M_k(x, y) | \leqq | x |_A | y |_B \, h(A, B; k). \tag{8.8}$$

By virtue of Condition IV, §5 on A (B), s-functions η (ζ) are everywhere dense in A_e (B_e). Relation (8.8) accordingly follows from (8.3), and the definition of M_k in Theorem 8.2.

The additivity of $M_k(\cdot, y)$ for fixed y will follow from the relation

$$S_k(\eta + \eta', \zeta) = S_k(\eta, \zeta) + S_k(\eta'. \zeta), \tag{8.9}$$

obviously valid when η and η' are associated with a common partition of E_A. Corresponding to x and x' in A and y in B, there exist sequences of elements $\eta^{(n)} \in A_e$, $\eta'^{(n)} \in A_e$, $\zeta^{(n)} \in B_e$, converging respectively in A_e, A_e, and B_e, to x, x', and y, while $\eta^{(n)}$ and $\eta'^{(n)}$ are assocated with the *same* partition $\pi_A^{(n)}$ of E_A. Relation (8.9) will thus remain valid if the superscript (n) is added to each argument, and on letting $n \to \infty$, it follows that

$$M_k(x + x', y) = M_k(x, y) + M_k(x', y).$$

The additivity of $M_k(x, \cdot)$ for fixed x is similarly proved.

For any real constant a,

$$S_k(a\eta, \zeta) = a\, S_k(\eta, \zeta),$$

from which it follows from a limiting process similar to that used above that

$$M_k(ax, y) = a\, M_k(x, y) \qquad [(x, y) \in A \times B].$$

Similarly $M_k(x, \cdot)$ is homogeneous for fixed $x \in A$. Thus M_k is additive and homogeneous while (8.8) holds. Thus M_k is bilinear over $A \times B$.

9. The k-B-transform Φ_y of $y \in B$

In order to obtain representation of functionals F bilinear over $A \times B$ in the form of repeated integrals (1.1) and (1.2) it is necessary to study the L-S-integral

$$\Phi_y(s) = \int_{R_1} y(t)\, d_t k(s, t), \tag{9.0}$$

transforming $y \in B$ into a functional Φ_y mapping R_1 into R_1, with values $\Phi_y(s)$ defined for each $s \in R_1$. This transform exists under the conditions laid down in the following definition.

Definition. *The k-B-transform Φ_y. Subject to the conditions*

$$k(s, \cdot) \in B_d \text{ for each fixed } s \in R_1, \tag{9.1)'$$

$$| k(s, \cdot) |_{B_d} \leqq N(s), \quad [s \in R_1] \tag{9.1)''$$

for some $N(s)$ monotonically increasing with s, we shall term Φ_y the k-B-transform of $y \in B$.

The fact that the L-S-integral (9.0) exists for each s follows from Condition I on B, since y is in B by hypothesis and $k(s, \cdot)$ is in B_d by (9.1)′. Condition (9.1)″ will be used in the proof of Lemmas 9.2 and 12.1. Properties of Φ_y such as being based on E_A, being canonical rel E_A, or in A_d, are reflected (under suitable hypotheses) in similar properties of $k(\cdot, t)$ and conversely.

Lemma 9.1. *A necessary and sufficient condition that a k-B-transform Φ_y of $y \in B$ be based on E_A for each $y \in B$ is that $k(\cdot, t)$ be based on E_A for each $t \in R_1$.*

(α). *The condition is necessary.* By definition of the k-B-transform Φ_y, $k(s, \cdot)$ is in B_d for each $s \in R_1$. It follows from Condition I that for fixed s, $k(s, \cdot)$ is uniquely determined by $\Phi_.(s)$ as an element in B_d, and therefore by Lemma 5.2 has uniquely determined values $k(s, t)$. In particular, if for each y, $\Phi_y(s)$ is linear, zero, or constant on the closure of a maximal interval J of CE_A, then for fixed t, $k(s, t)$ can be, and therefore must be linear, zero, or constant respectively on J. Hence $k(\cdot, t)$ is based on E_A for each $t \in R_1$ if Φ_y is based on E_A for each $y \in B$.

The condition of the lemma is accordingly necessary. That the condition is sufficient requires no detailed proof. Note that the condition (9.1)″ has not been used in the proof of (α).

Lemma 9.2. *If Φ_y is the k-B-transform of $y \in B$ and if the limits*

$$k_{12}^{AB}, \qquad k_{21}^{BA}, \tag{9.2}$$

exist and are equal, then the limits $\Phi_y(s-)$ and $k(s-, t)$ exist and

$$\Phi_y(s-) = \int_{R_1} y(t)\, d_t\, k(s-, t) \qquad [s \neq \omega_A]. \tag{9.3}$$

We proceed by proving (i), (ii), (iii).

(i). *For each $s \neq \omega_A$, $k_1^A(s, \cdot)$* equals *$k_{21}^{BA}(s, \cdot)$.*

The definition of Φ_y as the k-B-transform of $y \in B$ has been made contingent in (9.1)′ on $k(s, \cdot)$ being in B_d. With $k(s, \cdot)$ in B_d, $k_2^B = k$. From this relation and the assumed existence of k_{21}^{BA} we have

$$k_{21}^{BA} = k_1^A, \tag{9.4}$$

(ii). *For each $s \neq \omega_A$, $k_1^A(s, \cdot)$ is canonical* rel E_B.

By definition of k_1^A in §6,

$$k_1^A(s, t) = \lim_{u \to 0+} k(s - u, t) \qquad [s \neq \omega_A].$$

As the pointwise limit of $k(s - u, \cdot)$ based on E_B, $k_1^A(s, \cdot)$ is itself based on E_B. Moreover $k_1^A(s, \cdot)$ is then canonical rel E_B, by definition, provided

$$k_1^A(s, \cdot) = k_{12}^{AB}(s, \cdot). \tag{9.5}$$

But (9.5) follows from (9.4) and the hypothesis that the limits (9.2) exist and are equal. Thus (ii) holds.

(iii). *For each* $s \neq \omega_A$, $k_1^A(s, \cdot)$ *is in* B_d.

Recall that the B_d-norm of $k(s, \cdot)$ satisfies (9.1)″, and that the B_d-norm is l.s.c. in the sense of Theorem 5.3. Hence the limit $k_1^A(s, \cdot)$ has a finite B_d-norm at most $N(s)$ of (9.1)″, and (iii) holds as stated.

PROOF OF (9.3). For fixed $s \neq \omega_A$, set

$$k(s - u, t) - k_1^A(s, t) = f_s(u, t),$$

and observe that $f_s(u, \cdot)$ is in B_d since $k(s - u, \cdot)$ and $k_1^A(s, \cdot)$ are in B_d. To prove (9.3) observe that

$$\Phi_y(s-) = \operatorname*{limit}_{u \to 0+} \int_{R_1} y(t)\, d_t k(s - u, t)$$

so that (9.3) will follow if

$$\operatorname*{limit}_{u \to 0+} \int_{R_1} y(t)\, d_t f_s(u, t) = 0. \tag{9.6}$$

To that end observe that the definition of k_1^A implies that

$$\operatorname*{limit}_{u \to 0+} f_s(u, t) = 0. \tag{9.7}$$

We shall also need the inequality

$$| f_s(u, \cdot) |_{B_d} \leqq | k(s - u, \cdot) |_{B_d} + | k_1^A(s, \cdot) |_{B_d} \leqq 2N(s),$$

which follows from (9.1)″.

Let ζ be any s-function in B_e. Then for fixed $s \neq \omega_A$,

$$\begin{aligned} \left| \int_{R_1} y(t)\, d_t f_s(u, t) \right| &\leqq \left| \int_{R_1} [y(t) - \zeta(t)]\, d_t f_s(u, t) \right| + \left| \int_{R_1} \zeta(t)\, d_t f_s(u, t) \right| \\ &\leqq | y - \zeta |_{B_e} | f_s(u, \cdot) |_{B_d} + | \zeta_n \Delta_n [f_s(u, \cdot)] | \\ &\leqq | y - \zeta |_{B_e}\, 2N(s) + | \zeta_n \Delta_n [f_s(u, \cdot)] |. \end{aligned} \tag{9.8}$$

Let $e > 0$ be given. Let ζ be chosen so that for fixed s

$$| y - \zeta |_{B_e}\, 2N(s) < e,$$

as is possible, since the s-functions ζ are everywhere dense in B_e. With ζ fixed, (9.7) implies that for $u > 0$ and sufficiently small the last term in (9.8) is less than e in absolute value. Hence (9.6) holds, and (9.3) follows.

The following lemma is of basic importance.

LEMMA 9.3. *If the limits* k_{12}^{AB} *and* k_{21}^{BA} *exist and are equal, then* $k(\cdot, t)$ *is canonical* rel E_A *for each* $t \in R_1$, *if and only if the k-B-transform* Φ_y *is canonical* rel E_A *for each* $y \in B$.

(i). Suppose first that Φ_y is canonical rel E_A for each $y \in B$. Then by definition of "canonical rel E_A", Φ_y is based on E_A and

$$\Phi_y(s-) = \Phi_y(s) \qquad [s \neq \omega_A]. \tag{9.9}$$

It follows from Lemma 9.1 that $k(\cdot, t)$ is based on E_A for each $t \in R_1$, and from (9.3) and (9.9) that

$$\int_{R_1} y(t)\, d_t k(s-, t) = \int_{R_1} y(t)\, d_t k(s, t) \tag{9.10}$$

for fixed $s \neq \omega_A$ and every $y \in B$. It has been seen in the preceding proof that $k(s-, \cdot)$ like $k(s, \cdot)$ is in B_d. From (9.10),

$$\int_{R_1} y(t)\, d_t [k(s, t) - k(s-, t)] = 0 \qquad [s \neq \omega_A]$$

for every $y \in B$. It follows from Lemma 5.2 that

$$k(s, t) = k(s-, t) \qquad [s \neq \omega_A]. \tag{9.11}$$

This completes the proof that $k(\cdot, t)$ is canonical rel E_A.

(ii) Conversely suppose that $k(\cdot, t)$ is canonical rel E_A. Then Φ_y is based on E_A by virtue of Lemma 9.1. Moreover (9.11) holds by definition of $k(\cdot, t)$ as canonical rel E_A, and with the aid of (9.3), (9.9) is implied. Hence Φ_y is canonical rel E_A.

This completes the proof of the lemma.

For Φ_y to be in A_d it is sufficient that Φ_y be canonical rel E_A and have a finite A_d-norm. This leads to the next lemma.

LEMMA 9.4. *The condition $k \in H(A, B)$ implies that the k-B-transform Φ_y is in A_d for fixed $y \in B$, and that*

$$\sup_y |\Phi_y|_{A_d} = h(A, B; k), \tag{9.12}$$

taking the sup *over sub-unit $y \in B$.*

(α). *When k is in $H(A, B)$, the k-B-transform Φ_y exists and is canonical* rel E_A.

By virtue of Theorem 7.1 $h(A, B; k) = h'(A, B; k)$ so that the condition $k \in H(A, B)$ implies (7.5)′, that is,

$$|k(s, \cdot)|_{B_d} \leq h(A, B; k)\, M_A(s). \tag{9.13}$$

for $s \in E_A$. Condition (9.1)″ is then satisfied by a function N based on E_A which equals $h(A, B; k)\, M_A(s)$ $[s \in E_A,\ s > 0]$. Thus conditions (9.1) upon which the existence of Φ_y are contingent are satisfied. By Lemma 9.3 Φ_y is canonical rel E_A; for $k \in H(A, B)$ implies that k_{12}^{AB} and k_{21}^{BA} exist and are equal and $k(\cdot, t)$ is canonical rel E_A. This completes the proof of (α).

To establish (9.12) set

$$\sup_y |\Phi_y|_{A_d} = \lambda(k) \qquad [\text{for } |y|_B \leq 1]. \tag{9.14}$$

Statements (β) and (γ) will be proved.

(β). *The condition* $\lambda(k) \leqq G$, *a finite constant, implies that* $h(A, B; k) \leqq G$.

Let ζ be a sub-unit s-function in B_e. By Condition V on B there exists a sequence of elements $y^{(m)} \in B$ which converges to ζ in norm and relative to f^e for each $f \in \bar{B}$. Since $\zeta \in B_e$ is sub-unit, and since $|y^{(m)}|_B \to |\zeta|_{B_e}$, $y^{(m)}$ can be supposed sub-unit in B. For convenience in this proof write

$$\Phi_y(s) = \phi(y, s).$$

As defined $\phi(\cdot, s)$ is in $\bar{B}$ for fixed s. Its L-S-extension over B_e [Cf. (4.10)] for fixed s has the representation

$$\phi^e(v, s) = \int_{R_1} v(t)\, d_t k(s, t) \qquad [v \in B_e]. \tag{9.15}$$

For fixed s the above relative convergence of $y^{(m)}$ to ζ implies that

$$\operatorname*{limit}_{m \to \infty} \phi[y^{(m)}, s] = \phi^e(\zeta, s).$$

Since the elements $y^{(m)} \in B$ are sub-unit, the hypothesis in (β) implies that

$$|\phi[y^{(m)}, \cdot]|_{A_d} \leqq G, \qquad [m = 1, 2, \cdots]$$

and it follows from the l.s.c. of A_d-norms that

$$|\phi^e(\zeta, \cdot)|_{A_d} \leqq G. \tag{9.16}$$

But from (9.15)

$$\phi^e(\zeta, \cdot) = \zeta_n \Delta_n^{(2)}(k), \qquad [n = 1, \cdots, n_\zeta]$$

so that from (9.16)

$$\sup_{\zeta} |\zeta_n \Delta_n^{(2)}(k)|_{A_d} \leqq G, \tag{9.17}$$

taking the sup over sub-unit $\zeta \in B_e$. From (9.17) we infer that $h''(A, B; k) \leqq G$, and (β) follows.

(γ). *The condition* $h(A, B; k) \leqq G$, *a finite constant, implies* $\lambda(k) \leqq G$.

From the hypothesis in (γ)

$$|\zeta_n \Delta_n^{(2)}(k)|_{A_d} \leqq G \tag{9.18}$$

for every sub-unit s-function $\zeta \in A_e$. Given a sub-unit $y \in B$ there exists, by Condition IV on B, a sequence of sub-unit s-functions $\zeta^{(m)} \in B_e$ which converges to $y \in B$. Hence for each fixed s

$$\operatorname*{limit}_{m \to \infty} \phi^e(\zeta^{(m)}, s) = \phi(y, s).$$

In accordance with (9.18)

$$|\phi^e(\zeta^{(m)}, \cdot)|_{A_d} \leqq G.$$

The l.s.c. of A_d-norms then implies that for fixed y

$$|\phi(y, \cdot)|_{A_d} \leqq G,$$

and (γ) follows.

The relation (9.12) follows from (β) and (γ). Since Φ_y is canonical rel E_A by (α), and has a finite A_d-norm by (9.12), it is in A_d. This completes the proof.

10. Definition of the L-S-integrals $A_k(x, y)$ and $A_k^*(x, y)$

For $k \in H(A, B)$ the k-B-transform Φ_y of $y \in B$ exists and is in A_d, as has just been seen. Hence the L-S-integral

$$A_k(x, y) = \int_{R_1} x(s)\, d_s \Phi_y(s) \tag{10.1}$$

exists for $(x, y) \in A \times B$ and may be written

$$A_k(x, y) = \int_{R_1} x(s)\, d_s \int_{R_1} y(t)\, d_t k(s, t). \tag{10.2}$$

Set $k(t, s) = k^*(s, t)$. As observed in (7.4),

$$h'(A, B; k) = h''(B, A; k^*).$$

It follows from Theorem 7.1 that

$$h(A, B; k) = h(B, A; k^*). \tag{10.3}$$

Thus the condition $k \in H(A, B)$ implies the condition $k^* \in H(B, A)$. We infer the following from Lemma 9.4. For $k \in H(A, B)$ the k^*-A-transform Φ_x^* of $x \in A$ exists, and is in B_d. Hence the L-S-integral

$$A_k^*(x, y) = \int_{R_1} y(s)\, d_s \Phi_x^*(s) \tag{10.4}$$

exists for $(x, y) \in A \times B$ and may be written

$$A_k^*(x, y) = \int_{R_1} y(s)\, d_s \int_{R_1} x(t)\, d_t k(t, s). \tag{10.5}$$

LEMMA 10.1 *If k is in $H(A, B)$ and if the s-function pair (η, ζ) converges in $A_e \times B_e$ to (x, y) in $A \times B$, then*

$$A_k(x, y) = \operatorname*{limit}_{\eta \to x} \operatorname*{limit}_{\zeta \to y} S_k(\eta, \zeta), \tag{10.6}$$

$$A_k^*(x, y) = \operatorname*{limit}_{\zeta \to y} \operatorname*{limit}_{\eta \to x} S_k(\eta, \zeta). \tag{10.7}$$

By virtue of its definition (10.1), $A_k(x, y)$ equals

$$\int_{R_1} x(s)\, d_s \Phi_y(s) = \operatorname*{limit}_{\eta \to x} \int_{R_1} \eta(s)\, d_s \Phi_y(s) \qquad [\text{since } \Phi_y \in A_d]$$

$$= \underset{\eta \to x}{\text{limit}}\ \eta_r \Delta_r(\Phi_y) \qquad \text{[by (2.7)]}$$

$$= \underset{\eta \to x}{\text{limit}}\ \eta_r \Delta_r \int_{R_1} y(t)\, d_t k(\cdot\,, t) \qquad \text{[by definition of } \Phi_y]$$

$$= \underset{\eta \to x}{\text{limit}}\ \eta_r \Delta_r \underset{\zeta \to y}{\text{limit}} \int_{R_1} \zeta(t)\, d_t k(\cdot\,, t) \qquad [\text{since } k(s, \cdot) \in B_d]$$

$$= \underset{\eta \to x}{\text{limit}}\ \underset{\zeta \to y}{\text{limit}}\ \eta_r \Delta_r \int_{R_1} \zeta(t)\, d_t k(\cdot\,, t)$$

$$= \underset{\eta \to x}{\text{limit}}\ \underset{\zeta \to y}{\text{limit}}\ \eta_r \Delta_r[\zeta_n \Delta_n^{(2)}(k)] \qquad \text{[by (2.7)]}$$

$$= \underset{\eta \to x}{\text{limit}}\ \underset{\zeta \to y}{\text{limit}}\ \eta_r \zeta_n \Delta_{rn}(k) \qquad \text{[by (7.2)]}$$

$$= \underset{\eta \to x}{\text{limit}}\ \underset{\zeta \to y}{\text{limit}}\ S_k(\eta, \zeta) \qquad [\text{by definition of } S_k].$$

The proof of (10.7) is similar.

LEMMA 10.2. *For $k \in H(A, B)$, A_k and A_k^* are bilinear over $A \times B$.*

Suppose $x \in A$ and y_1 and $y_2 \in B$. By definition of A_k,

$$A_k(x, y_1 + y_2) = \int_{R_1} x(s)\, d_s \Phi_{y_1+y_2}(s)$$

$$= \int_{R_1} x(s)\, d_s[\Phi_{y_1}(s) + \Phi_{y_2}(s)] \qquad \text{[Cf. (9.0)]}$$

$$= A_k(x, y_1) + A_k(x, y_2).$$

That $A_k(\cdot, y)$ is additive for fixed y is trivial. The homogeneity of $A_k(\cdot, y)$ is trivial, while the homogeneity of $A_k(x, \cdot)$ follows from the homogeneity of $\Phi_y(s)$ in y for fixed s.

It remains to establish the inequality which proves A_k continuous. It follows from (10.1) that

$$\begin{aligned} A_k(x, y) &\leqq |\, x\,|_A\, |\, \Phi_y\,|_{A_d} \\ &\leqq |\, x\,|_A\, |\, y\,|_B\, h(A, B; k) \end{aligned} \tag{10.8}$$

using (9.12). Thus A_k is continuous, and hence bilinear.

The proof that A_k^* is bilinear is similar.

THEOREM 10.1. *For $k \in H(A, B)$ and $(x, y) \in A \times B$,*

$$M_k(x, y) = A_k(x, y) = A_k^*(x, y). \tag{10.9}$$

By its definition in Theorem 8.2,

$$M_k(x, y) = \underset{\substack{\eta \to x \\ \zeta \to y}}{\text{limit}}\ S_k(\eta, \zeta), \tag{10.10}$$

where (η, ζ) is an s-function pair in $A_e \times B_e$. As in the Pringsheim (See ref. 7) theory of double series the existence of the limit in (10.10) implies that the limit (10.10) equals the limits

$$\lim_{\eta \to x} \lim_{\zeta \to y} S_k(\eta, \zeta),$$

$$\lim_{\zeta \to y} \lim_{\eta \to x} S_k(\eta, \zeta),$$

so that (10.9) follows from Lemma 10.1.

11. The extensions F_{12}^e and F_{21}^e over $A_e \times B_e$ of F bilinear over $A \times B$

These extensions will be defined and characterized in this section without using Condition V. In §12 we shall prove that $F_{12}^e = F_{21}^e$, making use of theorems which depend upon Condition V.

For fixed $y \in B$ and $x \in A$, $F(\cdot, y) \in \bar{A}$ admits a representation (Cf. I, §2),

$$F(x, y) = \int_{R_1} x(s)\, d_s X(s, y)$$

with $X(\cdot, y) \in A_d$. The L-S-extension (Cf. §4) of $F(\cdot, y)$ over A_e is given for $u \in A_e$ by

$$F_1^e(u, y) = \int_{R_1} u(s)\, d_s X(s, y). \tag{11.1}$$

The subscript 1 is added to indicate the preference for the first argument of F in this extension.

LEMMA 11.1. *The functional F_1^e is bilinear over $A_e \times B$ with*

$$F_1^e(u, y) \leqq |u|_{A_e} |y|_B \| F \| \qquad [(u, y) \in A_e \times B] \tag{11.2}$$

For fixed $y \in B$, (11.1) shows that $F_1^e(\cdot, y)$ is additive and homogeneous over A_e.

To show that $F_1^e(u, \cdot)$ is additive over B for fixed $u \in A_e$, we shall first show that $X(s, \cdot)$ is additive over B for fixed s. To that end take $y_1, y_2 \in B$. For arbitrary $x \in A$,

$$\begin{aligned}\int_{R_1} x(s)\, d_s X(s, y_1 + y_2) &= F(x, y_1 + y_2) = F(x, y_1) + F(x, y_2) \\ &= \int_{R_1} x(s)\, d_s X(s, y_1) + \int_{R_1} x(s)\, d_s X(s, y_2) \\ &= \int_{R_1} x(s)\, d_s[X(s, y_1) + X(s, y_2)].\end{aligned}$$

It follows from Lemma 5.2 that

$$X(s, y_1 + y_2) = X(s, y_1) + X(s, y_2). \tag{11.3}$$

With the aid of (11.1) we can conclude that $F_1^e(u, \cdot)$ is additive over B for fixed $u \in A_e$. To show that $F_1^e(u, \cdot)$ is homogeneous over B one proceeds similarly, first proving $X(s, \cdot)$ homogeneous over B. Thus F_1^e is additive and homogeneous over $A_e \times B$.

Proof of (11.2). From Theorem 4.1, for fixed $y \in B$,

$$|F_1^e(u, y)| \leqq |u|_{A_e} \| F(\cdot, y) \|, \tag{11.4}$$

where

$$\| F(\cdot, y) \| = \sup_x |F(x, y)| \qquad [\text{for } |x|_A = 1].$$

Since F is bilinear, $\| F \|$ is finite and

$$|F(x, y)| \leqq |y|_B \| F \|, \qquad [\text{for } |x|_A = 1]$$

so that

$$\| F(\cdot, y) \| \leqq |y|_B \| F \|. \tag{11.5}$$

Relation (11.2) follows from (11.5) and (11.4).

The proof of Lemma 11.1 is complete.

Definition. *The extension F_{12}^e of F over $A_e \times B_e$.* For fixed $u \in A_e$, $F_1^e(u, \cdot)$ is in $\bar{B}$ by the preceding lemma, and so has an L-S-extension over B_e which will be denoted by $F_{12}^e(u, \cdot)$.

If for fixed $u \in A_e$, one sets

$$F_1^e(u, y) = \int_{R_1} y(t)\, d_t Y(u, t) \qquad [y \in B]$$

with $Y(u, \cdot) \in B_d$, our L-S-extension has the form

$$F_{12}^e(u, v) = \int_{R_1} v(t)\, d_t Y(u, t) \qquad [(u, v) \in A_e \times B_e]. \tag{11.6}$$

The subscript 2 is added to indicate the preference in this extension for the second argument of F_1^e.

Lemma 11.2. *The functional F_{12}^e is bilinear over $A_e \times B_e$ and for $(u, v) \in A_e \times B_e$,*

$$|F_{12}^e(u, v)| \leqq |u|_{A_e} |v|_{B_e} \| F \|. \tag{11.7}$$

For fixed $u \in A_e$, $F_{12}^e(u, \cdot)$ is additive and homogeneous over B_e by virtue of (11.6).

To prove $F_{12}^e(\cdot, v)$ additive over A_e for fixed $v \in B_e$ we shall show that $Y(\cdot, t)$ is additive over A_e for fixed t. To that end take $u_1, u_2 \in A_e$. For arbitrary $y \in B$,

$$\int_{R_1} y(t)\, d_t Y(u_1 + u_2, t) = F_1^e(u_1 + u_2, y) = F_1^e(u_1, y) + F_1^e(u_2, y)$$
$$= \int_{R_1} y(t)\, d_t[Y(u_1, t) + Y(u_2, t)].$$

It follows from Lemma 5.2 that

$$Y(u_1 + u_2, t) = Y(u_1, t) + Y(u_2, t).$$

One now sees from (11.6) that $F_{12}^e(\cdot, v)$ is additive over A_e for fixed $v \in B_e$, and the proof of homogeneity is similar. Thus F_{12}^e is additive and homogeneous over $A_e \times B_e$.

PROOF OF (11.7). From Theorem 4.1 for fixed $u \in A_e$,

$$| F_{12}^e(u, v) | \leq | v |_{B_e} \| F_1^e(u, \cdot) \|, \tag{11.8}$$

where

$$\| F_1^e(u, \cdot) \| = \sup_y | F_1^e(u, y) | \qquad [| y |_B = 1].$$

From (11.2),

$$| F_1^e(u, y) | \leq | u |_{A_e} \| F \|, \qquad [| y |_B = 1]$$

so that

$$\| F_1^e(u, \cdot) \| \leq | u |_{A_e} \| F \|. \tag{11.9}$$

Relation (11.7) follows from (11.8) and (11.9).

This completes the proof of Lemma 11.2.

DEFINITION. *The extension F_{21}^e of F over $A_e \times B_e$.* One first introduces the L-S-extension $F_2^e(x, \cdot)$ over B_e of $F(x, \cdot)$ for each $x \in A$. Then having proved that F_2^e is bilinear over $A \times B_e$ one introduces the L-S-extension $F_{21}^e(\cdot, v)$ over A_e of $F_2^e(\cdot, v)$ for each $v \in B_e$.

The proof of Lemma 11.3 is similar to that of Lemma 11.2.

LEMMA 11.3. *The functional F_{21}^e is bilinear over $A_e \times B_e$ and for $(u, v) \in A_e \times B_e$,*

$$| F_{21}^e(u, v) | \leq | u |_{A_e} | v |_{B_e} \| F \|. \tag{11.10}$$

DEFINITION. *The elementary s-functions z_A^u.* For each fixed $u \in E_A$ the s-function z_A^u is defined as follows:

(1). For $0 < u < \omega_A$, z_A^u shall be the characteristic function of $[0, u)$,

(2). For $u = \omega_A$, z_A^u shall be the characteristic function of $[0, \omega_A]$,

(3). For $u = 0$, $z_A^u(s)$ shall vanish for every s.

12. The L-S-representation of functionals F bilinear over $A \times B$

We shall make use of the elementary s-functions $z_A^s \in A_e$ and $z_B^t \in B_e$.

LEMMA 12.1. *A functional F which is bilinear over $A \times B$ admits a representation*

$$F(x, y) = \int_{R_1} x(s)\, d\Phi_y(s), \qquad [(x, y) \in A \times B] \tag{12.1}$$

in which Φ_y is the k-B-transform of $y \in B$ with

$$k(s, t) = F_{12}^e(z_A^s, z_B^t) \qquad [(s, t) \in E_A \times E_B]. \tag{12.2}$$

By virtue of Theorem 4.1, $F(\cdot, y)$ admits a representation

$$F(x, y) = \int_{R_1} x(s)\, d_s X(s, y), \qquad [x \in A] \tag{12.3}$$

for fixed $y \in B$ with $X(\cdot, y) \in A_d$ and

$$| X(\cdot, y) |_{A_d} = \| F(\cdot, y) \| .$$

We continue with a proof of (a), (b), (c).

(a). *For each s, $X(s, \cdot)$ is in $\bar{B}$.*

It follows from (12.3) that

$$F_1^e(z_A^s, y) = X(s, y) \qquad [s \in E_A]. \tag{12.4}$$

From (12.4) and Lemma 11.1 we conclude that $X(s, \cdot)$ is in $\bar{B}$ for $s \in E_A$. Suppose then $s \in CE_A$. In particular suppose that

$$s = ms_1 + (1 - m)s_2, \qquad [0 < m < 1]$$

where s_1 and s_2 are in E_A, but the interval (s_1, s_2) is in CE_A. Since $X(\cdot, y)$ is based on E_A,

$$X(s, y) = mX(s_1, y) + (1 - m) X(s_2, y). \tag{12.5}$$

But $X(s_1, \cdot)$ and $X(s_2, \cdot)$ are in $\bar{B}$ as (12.4) shows, so that $X(s, \cdot)$ is in $\bar{B}$ by virtue of (12.5). In case $s \leqq 0$ and $s \geqq \omega_A$ respectively,

$$X(s, y) = 0, \qquad X(s, y) = X(\omega_A, y), \tag{12.6}$$

since $X(\cdot, y)$ is based on E_A. Hence (a) holds for these s as well, and thus holds in general.

(b). *$X(\cdot, y)$ is a k-B-transform of $y \in B$.*

Since $X(s, \cdot)$ is in $\bar{B}$ for fixed s, by (3.4)

$$X(s, y) = \int_{R_1} y(t)\, d_t k(s, t) \qquad [y \in B] \tag{12.7}$$

for a suitable k with $k(s, \cdot) \in B_d$. The functional $X(\cdot, y)$ is the k-B-transform Φ_y of y provided k satisfies (9.1)′ and (9.1)″. But (9.1)′ requires that $k(s, \cdot)$ be in B_d, already verified. To verify (9.1)″ observe that for $s \in E_A$,

$$\begin{aligned} | k(s, \cdot) |_{B_d} &= \| X(s, \cdot) \| && \text{[by (4.11)]} \\ &= \sup_y | X(s, y) | && [\text{for } | y |_B = 1] \\ &\leqq | z_A^s | \, \| F \| && \text{[from (12.4) and (11.2)]} \\ &\leqq M_A(s) \, \| F \| && \text{[from (4.6)].} \end{aligned} \tag{12.7$'$}$$

Thus (9.1)″ holds for $s \in E_A$, if one sets $N(s) = M_A(s) \| F \|$ for $s \in E_A$. But $X(\cdot, y)$ is in A_d as defined in (12.3), and it follows from the proof of (α) under Lemma 9.1 that $k(\cdot, t)$ is based on E_A for each $t \in R_1$. If then $N(s)$, as defined for $s \in E_A$, $s > 0$ is extended over R_1 so as to be based on E_A, (9.1)″ will hold for $s \in R_1$. The proof of (b) is complete.

(c) *Relation* (12.2) *holds.*

If $X^*(s, \cdot)$ is the L-S-extension of $X(s, \cdot)$, (12.7) implies that

$$X^*(s, z_B^t) = k(s, t) \qquad [t \in E_B]. \tag{12.8}$$

Relation (12.4) leads to the equation

$$F_{12}^{e}(z_A^{s}, v) = X^{e}(s, v), \qquad [s \in E_A, v \in B_e] \tag{12.9}$$

and (12.2) follows from (12.8) and (12.9). Thus (12.2) holds.
With Φ_y taken as $X(\cdot, y)$ the lemma follows.

The following lemma is crucial. It makes essential use of the principal theorem of Morse and Transue l.c. (hereinafter referred to as MT).

LEMMA 12.2. *The functional k of Lemma 12.1 is in $H(A, B)$ with*

$$h(A, B; k) \leqq \| F \|. \tag{12.10}$$

To show that k is in $H(A, B)$ it is sufficient to verify the following:
(α). *For fixed s, $k(s, \cdot)$ is canonical* rel E_B.
(β). $k_{12}^{AB} = k_{21}^{BA}$ (Cf. §6).
(γ). *For fixed t, $k(\cdot, t)$ is canonical* rel E_A.
(δ). *$h(A, B; k)$ is finite.*

(α). That (α) is satisfied is clear since $k(s, \cdot)$ is in B_d as stated in (b) of the preceding proof.

(β). To establish (β) use will be made of (12.2). Let (η, ζ) be an s-function pair in $A_e \times B_e$ associated with partitions of R_1,

$$0 \leqq s_1 < \cdots < s_{r_\eta},$$
$$0 \leqq t_1 < \cdots < t_{n_\zeta},$$

belonging to E_A and E_B respectively. On account of (12.2)

$$F_{12}^{e}[z_A^{s_r} - z_A^{s_{r-1}}, z_B^{t_n} - z_B^{t_{n-1}}] = \Delta_{rn}(k). \tag{12.11}$$

One can represent η and ζ as sums

$$\eta = \eta_r[z_A^{s_r} - z_A^{s_{r-1}}], \qquad [r = 1, \cdots, r_\eta]$$
$$\zeta = \zeta_n[z_B^{t_n} - z_B^{t_{n-1}}], \qquad [n = 1, \cdots, n_\zeta]$$

so that (12.11) leads to the result

$$F_{12}^{e}(\eta, \zeta) = \eta_r \zeta_n \, \Delta_{rn}(k).$$

From (11.7) and the preceding equality

$$| \eta_r \zeta_n \Delta_{rn}(k) | \leqq | \eta |_{A_e} | \zeta |_{B_e} \| F \|. \tag{12.12}$$

In accordance with (4.6)

$$| \eta |_{A_e} \leqq M_A(s_{r_\eta}) \max | \eta(s) |,$$
$$| \zeta |_{B_e} \leqq M_B(t_{n_\zeta}) \max | \zeta(s) |,$$

so that (12.12) implies that

$$| \eta_r \zeta_n \Delta_{rn}(k) | \leqq \max | \eta(s) | \max | \zeta(s) | K, \tag{12.13}$$

where

$$K = M_A(s_{r_\eta}) \, M_B(t_{n_\zeta}).$$

From Theorem 6.2 *of* MT *and from* (12.13) *we can infer that* (β) *holds.*

To that end recall that Theorem 6.2 of MT applies to functionals k defined on a rectangle $[a, a'] \times [b, b']$. To make the present application we shall take $a = b = 0$, and a' and b' as any points in E_A and E_B, respectively, for which $a' > 0$ and $b' > 0$. In (12.13) let (η, ζ) be associated with partitions for which

$$s_{r_\eta} = a', \qquad t_{n_\zeta} = b',$$

in other words be s-functions $\eta \in A_e$ and $\zeta \in B_e$ for which $\eta(s) = 0$ for $s > a'$, and $\zeta(t) = 0$ for $t > b'$. In particular (12.13) implies that

$$e_r e'_n \Delta_{rn}(k) \leqq M_A(a')\, M_B(b'), \tag{12.13$'$}$$

$$[r = 1, \cdots, r_\pi; n = 1, \cdots, n_\pi]$$

where $|e_r| = |e'_n| = 1$ and any partition π of $R_1 \times R_1$ is *admitted* whose vertices are on the intersection of $E_A \times E_B$ with $[0, a'] \times [0, b']$. We shall verify the following:

(i). *The* sup *of the left member of* (12.13)$'$ *over all admissible* π *is not increased if taken over all finite partitions* π' *of* $[0, a'] \times [0, b']$.

Let J be an open interval of $CE_A \cap [0, a']$ whose end points s_{p-1}, s_p are in E_A and suppose that π is an admissible partition in which s_{p-1} and s_p are vertices. Let s' be a point of J dividing J in the ratio λ to μ, with $\lambda > 0, \mu > 0, \lambda + \mu = 1$. Let a partition π' be formed from π by adding the vertex s'. A new sum $S(\pi')$ in (12.13)$'$ corresponding to π' is obtained from the given sum for π on replacing e_p by

$$\bar{e}_p = \lambda\, e_p^{(1)} + \mu e_p^{(2)}, \qquad [|e_p^{(1)}| = |e_p^{(2)}| = 1\,]$$

with no change in $\Delta_{pn}(k)$, on account of the linearity of $k(\cdot, t)$ over J. To maximize $S(\pi')$, holding π fast and varying $e_p^{(1)}$ and $e_p^{(2)}$ only [among values ± 1], requires that $|\bar{e}_p| \leqq 1$. Hence the addition of s' does not increase the sup in (12.13)$'$. Cf. Lemma 3.4 MT. Moreover J can be further subdivided without increasing the sup in (12.13)$'$. If π' were given with $s' \in J$ a vertex, but s_{p-1} or s_p not vertices, one could add s_p or s_{p-1} to π' to obtain a partition π'' with

$$S(\pi') \leqq S(\pi'') \leqq M_A(a')\, M_B(b')$$

as before, provided the constants e_r and e'_n are suitably chosen in the sum $S(\pi'')$. Similar conclusions are obtained when E_A is replaced by E_B. Statement (i) follows.

The conclusions of MT apply to the distribution function k' defined by k on $[0, a'] \times [0, b']$. Observe that a' and b' are arbitrary in E_A and E_B respectively, and in particular one can take $a' = \omega_A$ or $b' = \omega_B$ if ω_A or ω_B is finite. Relation (β) follows.

(γ) With (β) established, (γ) follows on applying Lemma 9.3 to the k-B-transform $\Phi_y = X(\cdot, y)$ of the preceding lemma. Recall that $X(\cdot, y)$ is in A_d as noted in connection with (12.3), so that in particular Φ_y is canonical rel E_A. According to Lemma 9.3 $k(\cdot, t)$ is likewise canonical rel E_A.

(δ) With k known to be canonical rel $E_A \times E_B$, (12.12) implies that $h(A, B; k)$, as defined, is finite and that (12.10) is true.

A principal theorem will now be proved.

THEOREM 12.1. *A functional F bilinear over $A \times B$ admits a representation*

$$F(x, y) = A_k(x, y) \tag{12.14}$$

for a $k \in H(A, B)$ uniquely determined, as in (12.2), *by F. For this k,*

$$\| F \| = h(A, B; k). \tag{12.15}$$

The representation (12.1) takes the form

$$F(x, y) = A_k(x, y)$$

where, according to Lemma 12.2, k is in $H(A, B)$ and

$$h(A, B; k) \leqq \| F \|. \tag{12.16}$$

By virtue of Theorem 10.1, $A_k(x, y) = M_k(x, y) = F(x, y)$ and from (8.8)

$$h(A, B; k) \geqq \| M_k \|. \tag{12.17}$$

Relations (12.16) and (12.17), with $\| F \| = \| M_k \|$, imply (12.15). The uniqueness of k, when F has the form (12.14) is equivalent to the following:

(i). *The condition $\| F \| = h(A, B; k) = 0$ implies that $k(s, t) = 0$ for every $(s, t) \in R_2$.*

To establish (i), let (η, ζ) be an s-function pair in $A_e \times B_e$ for which $\eta_1 = 1$, $\eta_r = 0$ for $r > 1$, and $\zeta_1 = 1$, $\zeta_n = 0$ for $n > 1$. Then $\eta_r \zeta_n \Delta_{rn}(k) = k(s_1, t_1)$, so that the condition

$$\frac{1}{| \eta |_{A_e} | \zeta |_{B_e}} | \eta_r \zeta_n \Delta_{rn}(k) | \leqq h(A, B; k) = 0$$

implies that $k(s_1, t_1) = 0$ for every point $(s_1, t_1) \in E_A \times E_B$. Hence $k(s, t) = 0$ for every $(s, t) \in R_2$.

The proof of the theorem is complete.

Lemmas 12.1 and 12.2 combined, are paralleled by the following lemma with a parallel proof. Recall that κ^* is the functional with values $\kappa^*(s, t) = \kappa(t, s)$.

LEMMA 12.3. *A functional F which is bilinear over $A \times B$ admits a representation*

$$F(x, y) = \int_{R_1} y(s)\, d\Phi_x^*(s) = A_\kappa^*(x, y) \qquad [(x, y) \in A \times B] \tag{12.18}$$

in which Φ_x^* is the κ^*-A-transform of $x \in A$ with

$$\kappa(s, t) = F_{21}^s(z_A^s \,(\, z_B^t) \qquad [(s, t) \in E_A \times E_B] \tag{12.19}$$

and κ is in $H(A, B)$ with $h(A, B; \kappa) = \| F \|$.

The final equality in the lemma depends upon an argument similar to the one just used in proving that $h(A, B; k) = \| F \|$ in Theorem 12.1.

COROLLARY 12.1. *The functions* κ *and* k *in Theorem* 12.1 *and Lemma* 12.3 *are equal so that*

$$k(s, t) = F_{21}^e(z_A^s, z_B^t) = F_{12}^e(z_A^s, z_B^t) \qquad [(s, t) \in E_A \times E_B]. \tag{12.20}$$

It follows from (12.14), (12.18) and Theorem 10.1 that $F = A_k = A_\kappa^* = A_\kappa$. From the uniqueness of k in Theorem 12.1, $k = \kappa$, and (12.20) follows from (12.2) and (12.19).

THEOREM 12.2. *The extensions* F_{12}^e *and* F_{21}^e *over* $A_e \times B_e$ *of a functional* F *bilinear over* $A \times B$ *are equal.*

Let (η, ζ) be an s-function pair in $A_e \times B_e$. It follows from (12.20) and the additive and homogeneous character of F_{12}^e and F_{21}^e that

$$F_{12}^e(\eta, \zeta) = F_{21}^e(\eta, \zeta). \tag{12.21}$$

Since the s-function pairs (η, ζ) are everywhere dense in $A_e \times B_e$, (12.21) implies that

$$F_{12}^e(u, v) = F_{21}^e(u, v)$$

for arbitrary $(u, v) \in A_e \times B_e$.

DEFINITION. *The D-function k of F.* Corresponding to any functional F which is bilinear over $A \times B$ the unique $k \in H(A, B)$ for which

$$F(x, y) = M_k(x, y) = A_k(x, y) = A_k^*(x, y)$$

will be called the D-function (distribution function) k of F.

DEFINITION. The *L-S-extension of F over* $A_e \times B_e$. For $(u, v) \in A_e \times B_e$ the functional F^e for which

$$F^e(u, v) = F_{12}^e(u, v) = F_{21}^e(u, v)$$

will be termed the L-S-extension of F over $A_e \times B_e$.

THEOREM 12.3. *If* k *is the D-function of a functional* F *which is bilinear over* $A \times B$, *the L-S-extension* F^e *of* F *over* $A_e \times B_e$ *is the only bilinear extension* F' *of* F *over* $A_e \times B_e$ *for which*

$$k(s, t) = F'[z_A^s, z_B^t] \qquad [(s, t) \in E_A \times E_B].$$

One has

$$F'[z_A^s, z_B^t] = F^e[z_A^s, z_B^t]$$

for every (s, t) in $E_A \times E_B$ and it follows as in the proof of the equality $F_{12}^e = F_{21}^e$ in Theorem 12.2 that $F^e = F'$ over $A_e \times B_e$.

The following theorem is added for completeness.

THEOREM 12.4. *If* k *is the D-function of a functional* F *which is bilinear over* $A \times B$, *the L-S-extension* F^e *of* F *over* $A_e \times B_e$ *has the representations*

$$F^e(u, v) = \int_{R_1} u(s)\, d_s \int_{R_1} v(t)\, d_t\, k(s, t), \tag{12.22}$$

$$F^e(u, v) = \int_{R_1} v(s)\, d_s \int_{R_1} u(t)\, d_t\, k(t, s). \tag{12.23}$$

It will be sufficient to prove (12.22), as the proof of (12.23) is similar. Let $J(u, v)$ denote the right member of (12.22). It has not yet been shown that J is continuous over $A_e \times B_e$, but it is already clear that $J(u, v)$ exists[5] as a repeated L-S-integral and is additive and homogeneous over $A_e \times B_e$. To establish (12.22) it will accordingly be sufficient to verify that

$$F^e(x, y) = J(x, y), \qquad [(x, y) \in A \times B] \tag{12.24}$$

$$F^e(z_A^s, y) = J(z_A^s, y), \qquad [s \in E_A] \tag{12.25}$$

$$F^e(x, z_B^t) = J(x, z_B^t), \qquad [t \in E_B] \tag{12.26}$$

$$F^e(z_A^s, z_B^t) = J(z_A^s, z_B^t). \tag{12.27}$$

Equation (12.24) follows from (12.14). To establish (12.25) note that

$$F^e(z_A^s, y) = F_1^e(z_A^s, y) = \int_{B_1} y(t)\, d_t k(s, t), \tag{12.28}$$

using the relation $F = A_k$ and (10.2). Equation (12.25) follows on evaluating $J(z_A^s, y)$ in accordance with the definition of J. To establish (12.26) note that

$$F^e(x, z_B^t) = F_2^e(x, z_B^t) = \int_{B_1} x(t)\, d_t k(t, s)$$

from the relation $F = A_k^*$ and (10.5). Equation (12.26) follows on evaluating $J(x, z_B^t)$ in accordance with the definition of J. Equation (12.27) holds since its left member equals $k(s, t)$ by (12.20), and its right member also equals $k(s, t)$, as one sees from the definition of J.

This establishes (12.22), and the theorem follows.

THE INSTITUTE FOR ADVANCED STUDY.
KENYON COLLEGE.

BIBLIOGRAPHY

1. BANACH, *Théorie des opérations linéaires*, Warsaw, 1932.
2. FRÉCHET, *Sur les fonctionnelles bilinéaires*, Trans. Amer. Math. Soc., 16 (1915), 215–234.
3. HOBSON, *The theory of functions of a real variable*, vol. I, Cambridge University Press, 1927.
4. MCSHANE, *Integration*, Princeton University Press, 1944.
5. MORSE AND TRANSUE, *Functionals of bounded Fréchet variaticn*, Canadian J. Math., 1 (1949), 153–165.
6. ORLICZ, *Über eine gewisse Klasse von Räumenn vom Typus* B, Bull. Internat. Acad. Polonaise 1932, 207–220.
7. PRINGSHEIM, *Zur Theorie der zweifach unendlichen Zahlenfolgen*, Math. Ann., 53 (1900), 289–321.
8. ZYGMUND, *Trigonometrical series*, Warsaw, 1935, Chapter 4.

[5] The existence of $J(u, v)$ follows on setting $u = x + \eta$, $v = y + \zeta$ with $(x, y) \in A\ B$ and (η, ζ) an s-function pair.

ANNALS OF MATHEMATICS
Vol. 51, No. 3, May, 1950

FUNCTIONALS F BILINEAR OVER THE PRODUCT $A \times B$ OF TWO PSEUDO-NORMED VECTOR SPACES

II. Admissible spaces* A

BY MARSTON MORSE AND WILLIAM TRANSUE

(Received March 7, 1949)

1. Introduction

This paper is the second of several papers on functionals F bilinear[1] on the product $A \times B$ of two pseudo-normed vector spaces A and B. (See end of §1.) Paper I is concerned with the representation of F by means of a repeated Lebesgue-Stieltjes integral (written L-S-integral) of the form

$$(1.1)\quad F(x, y) = \int_{R_1} x(s)\, d_s \int_{R_1} y(s)\, d_t\, k(s, t) = \int_{R_1} y(s)\, d_s \int_{R_1} x(t)\, d_t\, k(t, s),$$

where (x, y) is in $A \times B$, R_1 is the s or t axis, and k maps the (s, t)-plane R_2 into R_1 subject to certain additional conditions (given in §2) determined by A and B. To serve our purposes a distribution function k has been assigned a variation $h(A, B; k)$ and has been further restricted so that

(a) an F which is bilinear over $A \times B$ admits both representations (1.1),
(b) conversely for any admissible k, the integrals (1.1) define an F which is bilinear over $A \times B$,
(c) k is uniquely determined by F,
(d) the functional modulus $\| F \|$ equals $h(A, B; k)$.

Recall that $\| F \|$ is the least number M such that

$$| F(x, y) | \leqq | x |_A | y |_B M,$$

where $| x |_A$ and $| y |_B$ are pseudo-norms of x and y in A and B respectively.

In Paper I the problem of finding distribution functions k for which (a), (b), (c), (d) hold has been solved under the assumption that the spaces A and

* The contribution of Dr. Transue to this paper is in partial fulfillment of a contract between the Office of Naval Research and the Institute for Advanced Study.

[1] The obvious connection between linear transformations from the space A (or B) to the space of distribution functions $B_d(A_d)$ and bilinear theory motivates Axiom I in Paper I and §9 of that paper. Transformations from one concrete linear space of functions into another have been extensively studied. See, for example, L. Kantorovitch et B. Vulich, *Sur la représentation des opérations linéaires*, Compositio Mathematica 5 (1937), p. 117–165. References are here given to Bochner, Dunford, Gelfand and others to whose later papers the reader may wish to turn. Our methods and intended applications are rather different. The representation theorem in Paper I can be used to modify and extend a few of the earlier concrete representations of operations. A brief note indicating these relations will be published.

B satisfy Conditions I to V to be stated in §2. The spaces A admitted are associated with a *closed subset E of R_1 which includes $s = 0$ and at least one other point, and on which $s \geq 0$.* The elements $x \in A$ are functions mapping R_1 into R_1 and determined (with certain restrictions) by the values of $x(s)$ when s is in E. Broadly speaking the Conditions I to V on A are concerned with the nature of the representation of elements $f \in \bar{A}$ (the space of functionals linear over A).

It is the obiect of the present paper to show that certain general spaces $C(E)$, $L_p(E)$, $\Phi(E)$ are admissible, that is satisfy Conditions I *to* V.

In the special cases in which E is the interval [0, 1], or the set of points $s = 0, 1, 2, \cdots$, or the set of points $s = 1$ and $s = 0, \frac{1}{2}, \frac{3}{4}, \frac{7}{8}, \cdots$ the spaces studied lead to those conventionally denoted by C, c, L_p, l_p, L_Φ. The use of a generalized Hellinger integral to specify the norm of the space A_d conjugate to A unifies the theory and permits, for example, the theory of C and c to be subsumed under the theory of $C(E)$, and the theory of L_p and l_p under that of $L_p(E)$. New methods involving generalized Hellinger integrals in the Orlicz theory lead to representations of linear functionals f on a broader and more explicit basis than that on which the previous theory has been built. In §10 a more technical statement is made as to the sense in which our representation theory extends the Orlicz theory.

The general representation theory of Paper I will lead[2] to a comparison of $h(A, B; k)$ for different products $A \times B$, and a comparison of $h(A, B; k)$ with the Vitali variation of k ordinarily used in L-S-integral theory. In general we shall show that the finiteness of $h(A, B; k)$ does not imply the finiteness of $V(k)$, nor conversely. The product $C \times C$ is exceptional in that $h(C, C; k) \leq V(k)$. The memoir[3] depends upon the results in the present paper.

Pseudo-norms. Let A be a linear vector space with a unit element θ. Let $N(x) \geq 0$ be a number assigned to each $x \in A$ and such that $N(cx) = |c| N(x)$ for each constant c, and

$$N(x_1 + x_2) \leq N(x_1) + N(x_2)$$

for each pair of elements x_1 and x_2 in A. We do not assume that the condition $N(x) = 0$ implies that $x = \theta$. Under these conditions $N(x)$ is termed a *pseudo-norm* for A, and A is said to be *pseudo-normed*. The spaces A and B shall be pseudo-normed.

2. Conditions I – V on A

In this section the definitions used in stating Conditions I–V will be reviewed and these conditions restated. The closed set E is restricted as in §1. We abbre-

[2] *M. Morse* and *W. Transue*, *Norms of Distribution Functions Associated with Bilinear Functionals*, Contributions to Fourier Analysis, Annals of Mathematics Study No. 25. To be published in 1950.

[3] *M. Morse* and *W. Transue*, *A Characterization of the Bilinear Sums Associated with the Classical Second Variation*, Ann. Matematica, 29 (1950).

viate complement by comp rather than by C. We shall consider functions g mapping R_1 into R_1.

Definition. *Functions g based on E.* If (a, b) is a maximal open interval of comp E, g shall be linear on the closure $[a, b]$ of (a, b). We require that $g(s) = 0$ for $s \leqq 0$, and that $g(s) = g(\omega_E)$ for

$$s \geqq \omega_E = \sup s \in E. \tag{2.1}$$

The constant ω_E may be infinite and condition (2.1) empty.

Definition. *Functions g canonical* rel E are based on E and left continuous except at most when $s = \omega_E$.

Definition. *d-Functions g based on E* are canonical rel E and of bounded variation on each finite subinterval of R_1.

Definition. *Functions x belonging to E* map R_1 into R_1 and whenever (a, b) is a maximal finite open interval in comp E, are constant on $[a, b)$ in case $b \neq \omega_E$, and on $[a, b]$ in case $b = \omega_E$. Moreover $x(s) = 0$ when $s < 0$ or $s > \omega_E$.

Definition. *Partitions π of R_1 belonging to E.* Let a finite set of points $s_r \in E$ be given in the order

$$0 = s_0 \leqq s_1 < \cdots < s_{r(\pi)}.$$

In case ω_E is finite $s_{r(\pi)}$ shall equal ω_E. Let R_1 be divided into successive *disjoint* intervals by the points s_r, including s_r as a left end point if $s_r \neq \omega_E$, and including ω_E, if finite, as a right end point. The resulting partition π of R_1 will be said to *belong* to E. A step function η will be *associated* with π by assigning values η_r to η on the respective finite intervals I_r of π $[r = 1, \cdots, r(\pi)]$, and the value 0 to the two extreme unbounded intervals of π. We term η an *s-function belonging to E*.

Definition. *A proper norm for a vector space A_d.* Let $N(g)$ be a non-negative number (possibly $+\infty$) defined for every function g which is canonical rel E_A with

$$N(g_1 + g_2) \leqq N(g_1) + N(g_2),$$
$$N(cg) = |c| N(g), \qquad [\text{for } N(g) < \infty]$$

for canonical g_1, g_2, and g and arbitrary constant c. Then N will be termed a norm which is proper relative to the vector space A_d of d-functions based on E if g is in A_d when and only when $N(g)$ is finite, and if $N(g) = 0$ implies that $g(s) \equiv 0$.

The Conditions I–V on admissible spaces A are as follows:

Condition I. *A shall be a pseudo-normed vector space of functionals x belonging to an admissible closed subset E of R_1. The space $\bar{A}$ conjugate to A shall be equivalent* (Cf. Banach, p. 180) *to a space A_d of d-functions g based on E with a proper norm, where $g \in A_d$ corresponds to $f_g \in A_d$ if and only if for each $x \in \bar{A}$,*

$$f_g(x) = \int_{R_1} x(s)\, dg(s). \tag{2.2}$$

CONDITION II. *There exists a monotone increasing function M_A with values $M_A(\alpha) \geqq 0$ defined for $\alpha \geqq 0$ such that when η is an s-function belonging to E and $\eta(s) = 0$ for $s > \alpha$, then for every $g \in A_d$,*

$$\left| \int_{R_1} \eta(s)\, dg(s) \right| \leqq M_A(\alpha) \mid g \mid_{A_d} \max \mid \eta(s) \mid. \tag{2.3}$$

DEFINITION. Let A_e be the vector space obtained on extending A linearly by the s-functions belonging to E. With Conditions I and II holding set

$$f_g^e = \int_{R_1} u(s)\, dg(s), \qquad [u \in A_e]$$

and term f_g^e the *L-S-extension* of $f_g \in \bar{A}$. The space A_e will be converted into a pseudo-normed vector space by setting (Paper I, §4),

$$\mid u \mid_{A_e} = \sup_f \frac{f^e(u)}{\| f \|}, \qquad [u \in A_e] \tag{2.4}$$

taking the sup over all $f \in \bar{A}$ for which $\| f \| \neq 0$.

Corresponding to the partition π of R_1 with vertices (2.2) and associated s-function η, we shall set

$$g(s_r) - g(s_{r-1}) = \Delta_r(g) \qquad [r = 1, \cdots, r(\pi)].$$

If g is a d-function based on E,

$$\eta_r \Delta_r(g) = \int_{R_1} \eta(s)\, dg(s), \tag{2.5}$$

using the summation convention of tensor algebra. If $\mid \eta \mid_{A_e} \leqq 1$, we shall term η *subunit*.

CONDITION III. *If A satisfies Conditions* I *and* II, *a sufficient condition that a function g which is canonical* rel *E be in A_d, is that for some finite constant K*

$$\sup_\eta \mid \eta_r \Delta_r(g) \mid = K, \tag{2.6}$$

taking the sup over all subunit s-functions $\eta \in A_e$.

DEFINITION. *Convergence in pseudo-norm and* rel *convergence.* Let B be a pseudo-normed vector space. A sequence of elements $u^{(n)} \in B$ will be said to converge to an element in pseudo-norm if $\mid u^{(n)} \mid_B \to \mid u \mid_B$ as $n \to \infty$, and to converge to $u \in B$ rel to $F \in \bar{B}$ if $F(u^{(n)}) \to F(u)$.

CONDITION IV. *The s-functions η belonging to E shall be everywhere dense in A_e.*

CONDITION V. *Corresponding to an arbitrary element $u \in A_e$ there shall exist a sequence of elements $x^{(n)} \in A$ which converges to u in A_e both in pseudo-norm and relative to the L-S-extension f^e of every $f \in \bar{A}$.*

Condition (2.6) is necessary if A satisfies Conditions I and II (Cf. Paper I, Lemma 5.1). The following theorem is useful in verifying the admissibility of a space A. (Theorem 5.4 of Paper I.)

THEOREM 2.1. *If A satisfies Conditions* I, II, IV *and if* $A = A_e$, *Conditions* III *and* V *are satisfied by* A.

We shall verify Conditions I–V for spaces $C(E)$, $L_p(E)$ and $\Phi(E)$ generalizing spaces conventionally denoted by C, L_p and L_Φ. The spaces $C(E)$ and $L_p(E)$ reduce to c and l_p for proper choice of E. As previously E is an arbitrary closed subset of R_1 on which $s \geqq 0$. The set E shall include the point $s = 0$ and at least one other point $s > 0$.

3. The space $C(E)$

Let $C(E)$ denote the vector space of functionals x belonging to E such that $x(s) \to 0$ as $s \to \infty$, and such that the function defined by x over E is continuous. Introduce the pseudo-norm

$$|x|_{C(E)} = \max_s |x(s)| \qquad [s \in R_1]. \tag{3.1}$$

This is an exceptional case in which a pseudo-norm is also a norm in the sense of Banach. Observe that the max $|x(s)|$ here obtained would be equally well obtained if s varied over E rather than over R_1. The space $C(E)$ is a normed vector space.

Let $C^*(E)$ be the space of d-functions based on E with finite total variation $T(g)$. The norm of g in $C^*(E)$ shall be $T(g)$. Since $T(g)$ is defined for each g which is canonical rel E and infinite if g is not in $C^*(E)$, it is clear that $T(g)$ is a proper norm for $C^*(E)$. We shall show that $C(E)$ satisfies Conditions I to V and that if one sets $A = C(E)$, the space A_d of Conditions I–V can be taken as $C^*(E)$.

The approximation $x^{(n)}$ *of* η. Let π be a partition of R_1 belonging to E, and η an associated s-function. Let n be an arbitrary positive integer. Let (z^n) be any sequence of functions mapping R_1 continuously into R_1 with values at the vertices of π equal to those of η, with absolute bound equal to max $|\eta(s)|$ and with $z^{(n)}(s)$ converging to $\eta(s)$ for each $s \in R_1$ as $n \to \infty$. Many such sequences of functions exist. There then exists a function $y^{(n)}$ belonging to E such that $x^{(n)}(s) = z^{(n)}(s)$ for $s \in E$. The only difficulty arises when ω_E is an isolated point of E. In this case let (b, ω_E) be the interval of comp E preceding ω_E. To belong to E, $x^{(n)}(s)$ must be constant on $[b, \omega_E]$. This is consistent with the requirement that $x^{(n)}(s) = z^{(n)}(s)$ for $s \in E$, if and only if $\eta(b) = \eta(\omega_E)$. Since $\eta(s)$ is constant on the last finite interval $[a, \omega_E]$ of π and $a \leqq b$, $\eta(b) = \eta(\omega_E)$ and the definition of $x^{(n)}$ is possible. We have established the following lemma.

LEMMA 3.1. *Given an s-function* η *belonging to* E, *these exists a sequence of functions* $x^{(n)}$ *in* $C(E)$ *with norms equal to* max $|\eta(s)|$ *and such that* $x^{(n)}(s)$ *converges pointwise to* $\eta(s)$ *as* $n \to \infty$ *with* $x^{(n)}(s) = \eta(s)$ *at the vertices of the partition* π.

Lemma 3.2 is complementary to Lemma 3.1.

LEMMA 3.2. *Given* $x \in C(E)$, *these exists a sequence of s-functions* $\eta^{(n)}$ *belonging to* E *and converging uniformly to* x *as* $n \to \infty$.

Such a sequence can be constructed as follows. Corresponding to each positive integer n, let $0 \leqq s_0 < s_1 < \cdots < s_{r(n)}$ be a partition π_n belonging to E, and $\eta^{(n)}$ an associated s-function restricted as follows. Setting

$$\eta_r^{(n)} = x(s_r), \qquad [r = 1, \cdots, r(n)] \tag{3.2}$$

the vertices s_r of π_n shall be so numerous and so dense in E that

$$\sup_{s \in E} | x(s) - \eta^{(n)}(s) | < \frac{1}{n}. \tag{3.3}$$

This is possible because of the continuity of the function defined by x over E and because $x(s) \to 0$ as $s \to +\infty$. It then follows that (3.3) will hold for arbitrary $s \in R_1$ and not merely for $s \in E$. Thus $\eta^{(n)}$ converges uniformly to x as $n \to \infty$.

We shall make use of the Banach notation whereby $\bar{A}$ denotes the space of functionals linear over A.

The proof that $C(E)$ satisfies Condition I is contained in the following two theorems.

THEOREM 3.1. (i). *For each $g \in C^*(E)$ the L-S-integral*

$$f(x) = \int_{R_1} x(s)\, dg(s) \qquad [x \in C(E)] \tag{3.4}$$

defines an element $f \in \overline{C(E)}$ for which $\| f \| = T(g)$. (ii). *If $f(x) = 0$ for every $x \in C(E)$, then $g(s) = 0$ for every $s \in R_1$.*

PROOF OF (i). From the theory of L-S-integrals,

$$| f(x) | \leqq | x |_{C(E)} T(g),$$

so that f is in $\overline{C(E)}$ and $\| f \| \leqq T(g)$. It remains to show that $\| f \| \geqq T(g)$.

Recall that

$$T(g) = \sup_{\eta} \eta_r \Delta_r(g), \qquad [r = 1, \cdots, r(\pi)] \tag{3.5}$$

taking the sup over all partitions π belonging to E and associated s-functions η for which each $| \eta_r | = 1$. To obtain $T(g)$ in (3.5) it is not necessary to take the sup over *all* partitions of R_1. One can omit vertices which are not in E on account of the definition of g over the intervals in comp E. Using (2.5) and the elements $x^{(n)}$ in Lemma 3.1 converging pointwise and boundedly to η,

$$| \eta_r \Delta_r(g) | = \left| \int_{R_1} \eta(s)\, dg(s) \right| = \lim_{n \to \infty} \left| \int_{R_1} x^{(n)}(s)\, dg(s) \right|$$
$$= \lim_{n \to \infty} | f[x^{(n)}] | \leqq \| f \|,$$

since the norm of $x^{(n)}$ in $C(E)$ is max $| \eta(s) | = 1$. Reference to (3.5) then shows that $T(g) \leqq \| f \|$. We conclude that $\| f \| = T(g)$.

PROOF OF (ii). If $f(x) = 0$ for every $x \in C(E)$, $\| f \| = 0$, and hence by (i) $T(g) = 0$. Since $g(0) = 0$ and g is canonical, it follows that $g(s) \equiv 0$.

The proof of the next theorem resembles in part the proof of the Riesz theorem given by Banach, p. 59, excepting what seems to be an error of Banach in the definition of the distribution function g, with a consequent error in the proof.

THEOREM 3.2. *Corresponding to each $f \in \overline{C(E)}$, there exists a unique element $g \in C^*(E)$ such that for each $x \in C(E)$,*

$$f(x) = \int_{R_1} x(s)\, dg(s). \tag{3.6}$$

DEFINITION *of* γ. Let $C'(E)$ be the space obtained by linearly extending $C(E)$ to include the s-functions η belonging to E, assigning a norm,

$$| u |_{C'(E)} = \sup_s | u(s) |,$$

to each element $u \in C'(E)$. By virtue of Theorem 2 of Banach, there exists a functional $F \in \overline{C'(E)}$ such that $F(x) = f(x)$ for each $x \in C(E)$ and $\| f \| = \| F \|$. Let γ be a function based on E such that for $t \in E$,

$$F(z_E^t) = \gamma(t), \tag{3.7}$$

where z_E^t is the characteristic function of $[0, t)$ if $t < \omega_E$, and of $[0, t]$ if $t = \omega_E$. We understand that $\gamma(0) = 0$. We shall prove statements (a), (b), and (c).

(a) $T(\gamma) \leqq \| f \|$.

For an arbitrary partition π of R_1 belonging to E set

$$\eta = \eta_r(z_E^{s_r} - z_E^{s_{r-1}}) \qquad [r = 1, \cdots, r(\pi)].$$

For this s-function,

$$| F(\eta) | = | \eta_r \Delta_r(\gamma) | \leqq | \eta |_{C'(E)} \| F \| = | \eta |_{C'(E)} \| f \|. \tag{3.8}$$

Note that

$$\sup_\eta \eta_r \Delta_r(\gamma) = T(\gamma), \tag{3.9}$$

taking the sup over all partitions of R_1 belonging to E and all associated s-functions η with each $| \eta_r | = 1$. In (3.9) it is legitimate to restrict the partitions to those with vertices in E on account of the definition of γ on the intervals of comp E. One can thus use (3.8) to limit the sup in (3.9), obtaining the result $T(\gamma) \leqq \| f \|$. This completes the proof of (a).

To proceed let g be the canonical distribution function defined by γ, that is let

$$g(s) = \gamma(s-) \text{ for } s \neq \omega_E; \qquad g(\omega_E) = \gamma(\omega_E).$$

(b) *If* $\eta^{(n)}$ *is the sequence of s-functions of Lemma* 3.2 *converging uniformly to* $x \in C(E)$, *then*

$$\lim_{n \to \infty} \eta_r^{(n)} \Delta_r(g - \gamma) = 0 \qquad [r = 1, \cdots, r(n)]. \tag{3.10}$$

The function $x \in C(E)$ is discontinuous at most at right end points, $s = \sigma$ of maximal intervals of comp E, and at the point $s = \omega_E$ (if finite). At a point σ, $g(\sigma) = \gamma(\sigma)$ since γ is based on E and so left continuous at σ. Moreover $g(\omega_E) = \gamma(\omega_E)$ by definition of g. For brevity set $h = g - \gamma$, $h(s_r) = h_r$, $[r = 1, \cdots, r(n)]$.

Then

$$\begin{aligned} \eta_r^{(n)} \Delta_r(h) &= \eta_1^{(n)} h_1 + \eta_2^{(n)}[h_2 - h_1] + \cdots + \eta_r^{(n)}(n) \, [h_{r(n)} - h_{r(n)-1}] \\ &= h_1[\eta_1^{(n)} - \eta_2^{(n)}] + \cdots + h_{r(n)}[\eta_{r(n)}^{(n)} - \eta_{r(n)+1}^{(n)}], \end{aligned} \tag{3.10$'$}$$

where $\eta_{r(n)+1}^{(n)} = 0$. When $s_r = \sigma$ or ω_E, $h_r = 0$.

Let e be any positive constant. If n is sufficiently large, say $n > N_e$,

$$| x(s) - \eta^{(n)}(s) | < e, \qquad [s \in R_1]$$

in accordance with Lemma 3.2. Among the points s_r $[r = 1, \cdots, r(n)]$ let s_i be any point not a point σ or ω_E. Since x is continuous at s_i it follows from the preceding inequality that

$$| x(s_i) - \eta_{i+1}^{(n)} | \leqq e, \qquad | x(s_i) - \eta_i^{(n)} | \leqq e,$$

so that for the subscript i of s_i, $| \eta_i^{(n)} - \eta_{i+1}^{(n)} | \leqq 2e$.

We infer from (3.10)′ that for $n > N_e$,

$$| \eta_r^{(n)} \Delta_r(h) | \leqq 2e \sum h_r \leqq 2e\, T(h) \leqq 4e\, T(\gamma),$$

recalling that $T(g) \leqq T(\gamma)$. Relation (3.10) follows, thus establishing (b).

(c) *The function g is a d-function belonging to E such that* (3.6) *holds.*

As defined g is canonical rel E, while g is a d-function rel E by virtue of (a). To show that (3.6) holds we shall show that both members of (3.6) are the limit of $F(\eta^{(n)})$ as $n \to \infty$. Recall that $f(x) = F(x)$ for $x \in C(E)$ and that F is continuous over $C'(E)$. Since $\eta^{(n)} \to x$ in $C'(E)$, $F(\eta^{(n)}) \to F(x) = f(x)$. On the other hand

$$F(\eta^{(n)}) = \eta_r^{(n)} \Delta_r(\gamma) \qquad \text{[by definition of } \gamma\text{]}.$$

Hence

$$\lim_{n\to\infty} F(\eta^{(n)}) = \lim_{n\to\infty} \eta_r^{(n)} \Delta_r(g) \qquad \text{[by (b)]}$$

$$= \lim_{n\to\infty} \int_{R_1} \eta^{(n)}(s)\, dg(s) \qquad \text{[by (2.5)]}$$

$$= \int_{R_1} x(s)\, dg(s),$$

since $\eta^{(n)}$ converges uniformly to x and $T(g) < \infty$. Thus (3.6) holds, and (c) is established.

That g is unique follows from (ii) of Theorem 3.1. This completes the proof of Theorem 3.2.

It follows from Theorems 3.1 and 3.2 that $A = C(E)$ satisfies Condition I with $A_d = C^*(E)$.

Corresponding to any $g \in C^*(E)$, an arbitrary $x \in C(E)$ and $u \in C'(E)$, set

$$f_g(x) = \int_{R_1} x(s)\, dg(s), \qquad f_g^e(u) = \int_{R_1} u(s)\, dg(s)$$

as in §2, terming $f_g^e(u)$ the *L-S-extension* of f_g. To establish Conditions IV and V it is necessary to prove the following:

LEMMA 3.3. *The space $C'(E) = C_e(E)$ where $C_e(E)$ is defined as in* §2.

The vectors u in $C'(E)$ and $C_e(E)$ are the same. To establish the lemma one has merely to show that the norm of u is the same in $C'(E)$ as in $C_s(E)$, that is that

$$\sup_s |u(s)| = \sup \frac{|f_g^e(u)|}{\|f_g\|}, \tag{3.11}$$

where f_g ranges over all functionals in $\overline{C(E)}$ for which $\|f_g\| > 0$.

It follows from the relation $\|f_g\| = T(g)$ of Theorem 3.1 and the definition of f_g^e that when $\|f_g\| > 0$,

$$\sup_s |u(s)| \geqq \frac{|f_g^e(u)|}{\|f_g\|}. \tag{3.12}$$

To establish the lemma it is sufficient to show that the equality holds in (3.12).

Let b be an arbitrary point of E. We shall define a $g \in C^*(E)$ by giving the values of $g(s)$ for $s \in E$ as follows:

$$[g(s) = 0 \mid s < b], \qquad [g(s) = 1 \mid s > b],$$
$$[g(b) = 0 \mid b \neq \omega_E], \qquad [g(b) = 1 \mid b = \omega_E].$$

With g so defined it is clear that $\|f_g\| = T(g) = 1$ and it will be shown that

$$|u(b)| = |f_g^e(u)|. \tag{3.13}$$

CASE I. $\omega_E \neq b$. Consider the subcase I′ of I in which b is the left end point of a maximal interval (b, d) of comp E. In case I′ $g(d) = 1$, $g(s)$ increases linearly on $[b, d]$ from 0 to 1 and is otherwise constant over R_1. Since $u(s) = u(b)$ on $[b, d)$, one sees that

$$\int_{R_1} u(s)\, dg(s) = u(b)$$

in Case I′, so that (3.13) holds. In Case I $g(b) = g(b-) = 0$ without exception, and when I′ does not occur, $g(b+) = 1$ so that (3.13) again holds.

CASE II. $\omega_E = b$. Consider the subcase II′ of II in which b is the right end point of a maximal subinterval (a, b) of comp E. In case II′, g increases linearly from 0 to 1 on $[a, b]$ while $u(s) = u(b)$ on $[a, b]$ so that (3.13) holds. In Case II $g(b+) = g(b) = 1$ without exception, and when II′ does not occur, $g(b-) = 0$ so that (3.13) again holds.

Hence (3.13) holds for arbitrary choice of $b \in E$ and the above g. The lemma follows.

The fundamental theorem can now be proved.

THEOREM 3.3. *The space $C(E)$ is a space A satisfying Conditions* I–V *with $C^*(E)$ the associated space A_d.*

Condition I is satisfied as a consequence of Theorems 3.1 and 3.2.

Condition II is satisfied in the form

$$\left| \int_{R_1} \eta(s)\, dg(s) \right| \leqq T(g) \max |\eta(s)|$$

for every $g \in C^*(E)$.

Condition III, when $A = C(E)$, can be stated as follows. When g is canonical rel E a sufficient condition that $T(g)$ be finite is that for some finite constant K,

$$\sup_{\eta} | \eta_r \Delta_r(g) | = K, \tag{3.14}$$

where the sup is taken over all partitions π belonging to E and associated s-functions η for which $\max | \eta(s) | \leqq 1$. When (3.14) holds for partitions restricted to those belonging to E, it also holds for unrestricted finite partitions of R_1 because of the definition of g on the intervals of comp E. Hence (3.14) implies that $T(g) < \infty$, and Condition III holds.

Condition IV is satisfied by virtue of Lemmas 3.2 and 3.3, since the uniform convergence of $\eta^{(n)}$ to x in Lemma 3.2 implies the convergence of $\eta^{(n)}$ to x in $C_e(E)$.

Condition V is satisfied. To see this, first note that an arbitrary $u \in C_e(E)$ is of the form $x + \eta$ where x is in $C(E)$ and η an s-function belonging to E. It is sufficient that the sequence of elements $x^{(n)}$ of Lemma 3.1 converges in norm to η in $C_e(E)$ and that $x^{(n)}$ converges to η relative to the L-S-extension f^e of each $f \in \overline{C(E)}$; that is, that

$$\operatorname*{limit}_{n \to \infty} \int_{R_1} x^{(n)}(s)\, dg(s) = \int_{R_1} \eta(s)\, dg(s) \tag{3.15}$$

for arbitrary $g \in C^*(E)$. The relation (3.15) follows readily from the bounded convergence of $x^{(n)}$ to η. Hence Condition V is satisfied.

This completes the proof of Theorem 3.3.

The space $C(E)$ for which E is the interval [0, 1] will be denoted by C.

Corollary 3.1. *The space C is an admissible space A with A_d the space of d-functions based on* [0, 1].

This corollary implies in particular the classical theorem of F. Riesz on the representation of functions $f \in \bar{C}$.

The space c. Let E_0 be the set of points in R_1 with coordinates

$$t_0 = 1, \qquad t_n = 1 - \frac{1}{n} \qquad [n = 1, 2, \cdots].$$

An element $u \in c$ is a vector $u = (u_1, u_2, \cdots)$ whose components converge to a number u_0 and whose norm is

$$| u |_c = \sup_r | u_r | \qquad [r = 1, 2, \cdots]. \tag{3.16}$$

If one makes $x \in C(E_0)$ correspond to $u \in c$ when and only when

$$x(t_n) = u_n, \qquad [n = 0, 1, 2, \cdots] \tag{3.17}$$

an equivalence between $C(E_0)$ and c is defined. To a functional $F \in \bar{c}$ corresponds an $f \in \overline{C(E_0)}$ for which $F(u) = f(x)$ subject to (3.17). One has the theorem.

Theorem 3.4. *The space c is equivalent to the space $C(E_0)$ under the mapping* (3.17).

The representation theorem for an $F \in \bar{c}$ here finds a new setting and proof.

THEOREM 3.5. *An element $F \in \bar{c}$ determines an absolutely convergent sequence of real numbers $C_1, C_0, \cdots$ such that*

$$F(u) = u_n C_n, \qquad \| F \| = \sum | C_n | \qquad [n = 0, 1, \cdots] \tag{3.18}$$

for every $u \in c$. Conversely (3.18) *defines an $F \in \bar{c}$ whenever $\sum | C_n |$ converges.*

Given an $F \in \bar{c}$, Theorem 3.2 implies that the $f \in \overline{C(E_0)}$ for which $F(u) = f(x)$ under (3.17) has a representation

$$f(x) = \int_{R_1} x(s)\, dg(s), \tag{3.19}$$

where g is based on E_0. The graph of g is continuous except possibly at $s = 1$. Otherwise this graph is a continuous broken line with corners where $s = t_r$ ($r = 1, 2, \cdots$). Hence (3.19) reduces to the equation

$$f(x) = \sum_{r=1}^{\infty} u_r[g(t_{r+1}) - g(t_r)] + u_0[g(1) - g(1-)]. \tag{3.19$'$}$$

Recall that $t_1 = 0$ so that $g(t_1) = 0$. If one sets

$$C_0 = g(1) - g(1-), \qquad C_r = g(t_{r+1}) - g(t_r), \quad [r = 1, 2, \cdots] \tag{3.20}$$

the relation $F(u) = u_n C_n$ results from (3.19)$'$. Moreover with f given by (3.19), $\| f \| = T(g)$ in accordance with Theorem 3.1, so that on using (3.20),

$$\| F \| = T(g) = \sum_{n=0}^{\infty} | C_n |.$$

Thus (3.18) holds as stated.

Conversely, when an absolutely convergent sequence $[C_n]$ is given, (3.20) is satisfied by a function g based on E_0. With $g(t_1) = g(0) = 0$ one finds from (3.20) that

$$g(t_{r+1}) = C_1 + C_2 + \cdots + C_r, \qquad [r = 1, 2, \cdots]$$
$$g(1-) = C_1 + C_2 + \cdots,$$
$$g(1) = C_0 + C_1 + C_2 + \cdots.$$

For such a g, $T(g) = \sum | C_n |$ ($n = 0, 1, \cdots$), so that g is in $C^*(E_0)$, and defines an $f_g \in \overline{C(E_0)}$ given by (3.19). Under the equivalence between c and $C(E_0)$ defined by (3.17), an $F \in \bar{c}$ given by (3.18) is derived from (3.19)$'$.

4. The space $L_p(E)$ $(p > 1)$

Let E be a closed set restricted as in §3. Let $L_p(E)$ be a vector space of measurable functions x belonging to E and such that the L-integral

$$\int_{R_1} | x(s) |^p\, ds \tag{4.1}$$

is finite. The pth root of the integral (4.1) is taken as the pseudo-norm of x in $L_p(E)$. Thus $L_p(E)$ is a pseudo-normed vector space. To determine the appropriate conjugate space Hellinger integrals will be used. Let q be chosen so that $1/p + 1/q = 1$.

The norm $H_q(g)$ $(q > 1)$. Let g be an arbitrary function mapping R_1 into R_1. We shall refer to *general partitions* of R_1 as distinguished from partitions belonging to E. Such general partitions are defined by vertices $s_0 < s_1 < \cdots < s_{r(\pi)}$, and determine $r(\pi)$ successive finite intervals. As previously we write

$$\Delta_r(g) = g(s_r) - g(s_{r-1}) \qquad [r = 1, 2, \cdots, r(\pi)].$$

Set

$$\sup_{\pi} \frac{|\Delta_r(g)|^q}{|\Delta_r(s)|^{q-1}} = I_q(g), \tag{4.2}$$

summing with respect to r and taking the sup over all general partitions of R_1. One terms $I_q(g)$ the *Hellinger integral* of g, of index q, taken over R_1 with respect to s. The qth root of $I_q(g)$ will be denoted by $H_q(g)$, understanding that $H_q(g) = \infty$ when $I_q(g) = \infty$. Certain properties (a), $\cdots$, (e) of $H_q(g)$ will be recalled.

(a) *Summing over the intervals of a partition* π,

$$a \leqq s_0 < s_1 < \cdots < s_m = b, \tag{4.3}$$

of $[a, b]$,

$$\frac{|\Delta_r(g)|^q}{|\Delta_r(s)|^{q-1}} \geqq \frac{|g(b) - g(a)|^q}{|b - a|^{q-1}} \qquad [r = 1, 2, \cdots, m;\ r \text{ summed}]. \tag{4.4}$$

Let $y(s)$, with $a \leqq s \leqq b$, represent the ordinate of a point $[s, y(s)]$ on a broken line in the (s, y)-plane with vertices $[s_i, g(s_i)]$ $(i = 0, 1, \cdots, m)$. Then

$$|y(b) - y(a)|^q = \left| \int_a^b \dot{y}(s)\, ds \right|^q \leqq |b - a|^{q/p} \int_a^b |\dot{y}(s)|^q\, ds \tag{4.5}$$

by the Hölder lemma. Since y is linear between vertices s_i of π, and since $q = p(q - 1)$, this implies that

$$\frac{|g(b) - g(a)|^q}{|b - a|^{q-1}} \leqq \left| \frac{\Delta_r(g)}{\Delta_r(s)} \right|^q \Delta_r(s), \tag{4.6}$$

from which (4.4) follows.

(b) *If* $g(s)$ *is linear over an interval* $[a, b]$ *of a partition* π' *of* R_1, *and if the sum*

$$\frac{|\Delta_r(g)|^q}{|\Delta_r(s)|^{q-1}} = S_{\pi'}$$

is taken over the intervals of π', *further subdivision of* $[a, b]$ *will not increase this sum* $S_{\pi'}$.

We refer to (a) and note that when $y(s)$ is linear over $[a, b]$ the equality holds in (4.5), and hence in (4.6) and (4.4), so that (b) follows.

The analysis of the sup (4.4) as given in Titchmarsh, pp. 384–386, leads to statement (c) (also a consequence of the more general Theorem 6.1 of this paper).

(c) *When* $q > 1$ *and* $I_q(g)$ *is finite,* g *is* a.c. (*absolutely continuous*), $\dot{g}$ *is in* L_q, *and*

$$H_q(g) = \left[\int_{R_1} |\dot{g}(s)|^q \, ds\right]^{1/q}. \tag{4.7}$$

Conversely if g *is* a.c. *and* $\dot{g}$ *is in* L_q, (4.7) *holds.*

(d) *Whether* $H_q(g_1)$ *and* $H_q(g_2)$ *are finite or not*

$$H_q(g_1 + g_2) \leqq H_q(g_1) + H_q(g_2). \tag{4.8}$$

If $H_q(g)$ is infinite for either g_1 or g_2, (4.8) holds trivially. If both of these norms are finite, (4.8) follows from Minkowski's theorem.

(e) *The function* g *is constant if and only if* $H_q(g) = 0$.

DEFINITION. *The space* $C_q^*(E)$. The vector space of functions which are based on E and for which $H_q(g)$ is finite will be denoted by $L_q^*(E)$. The norm of an element g in $L_q^*(E)$ shall be $H_q(g)$.

It appears from the preceding that $H_q(g)$ is a norm which is proper relative to $L_q^*(E)$. It is defined (although possibly infinite) for arbitrary g mapping R_1 into R_1, and for functions based on E is finite if and only if g is in $L_q^*(E)$. The space $L_q^*(E)$ is a normed vector space.

We shall show that $L_p(E)$ is an admissible space A and that $L_q^*(E)$ is the associated space A_d. The following lemma is useful.

LEMMA 4.1. *The* sup (4.2) *which defines the Hellinger integral* $I_q(g)$ *is unaltered if the partitions* π *are restricted to those which belong to* E.

Let π_1 be a "general" finite partition of R_1, and let s' be an arbitrary partition point in π_1. If s' is in a maximal interval (a, b) of comp E, let a (if finite) and b (if finite) be added to π_1 (if not already in π_1) to form a partition π_2. By virtue of (a),

$$\sum_r \frac{|\Delta_r(g)|^q}{|\Delta_r(s)|^{q-1}} \tag{4.9}$$

will not be smaller over π_2 than over π_1. From π_2 let the point s' (if in comp E) be dropped. If a and b are finite, no change will be induced in the sum (4.9) by virtue of (b). If $a = -\infty$, $b = 0$ and the dropping of s' makes no change since $g(s) = 0$ for $s \leqq 0$. If $b = \infty$, $a = \omega_E$ and the dropping of s' again makes no change since $g(s) = g(\omega_E)$ for $s \geqq \omega_E$. After a finite number of changes of the above sort a partition π belonging to E will be obtained for which the sum (4.9) is not smaller than for π_1. The lemma follows.

THEOREM 4.1 (i). *For fixed $g \in L_q^*(E)$ and arbitrary $x \in L_p(E)$ the L-S-integral*

$$f(x) = \int_{R_1} x(s)\, dg(s) \tag{4.10}$$

exists and defines an element $f \in \overline{L_p(E)}$ for which $\|f\| = H_q(g)$.

(ii) *If $f(x) = 0$ for every $x \in L_p(E)$, $g(s) = 0$ for every $s \in R_1$.*

PROOF OF (i). When $H_q(g)$ is finite, g is a.c. by (c) so that the integral (4.10) exists provided the integral

$$\int_{R_1} |x(s)|\,|\dot{g}(s)|\, ds \qquad \text{(Cf. McShane, p. 311]} \tag{4.11}$$

exists. Since $\dot{g}$ is in $L_q(E)$ and x is in $L_p(E)$ the integral (4.11) exists, and

$$H_q(g) \geqq \|f\| \tag{4.12}$$

by Hölder's Lemma. We exclude the inequality in (4.12) as follows.

Let π be an arbitrary partition belonging to E and η an associated s-function $\eta = (\eta_1, \cdots, \eta_m)$. If η is a unit s-function in $L_p(E)$, by definition of $\|f\|$,

$$|f(\eta)| = |\eta_r \Delta_r(g)| \leqq \|f\| \qquad [r = 1, \cdots, m]. \tag{4.13}$$

In the space R_m of vectors with m components introduce vectors a and b for which

$$a_r = \eta_r[\Delta_r(s)]^{1/p}, \qquad b_r = \Delta_r(g)[\Delta_r(s)]^{-1/p} \qquad [r \text{ not summed}].$$

Then from (4.13),

$$\|f\| \geqq |\eta_r \Delta_r(g)| = |a_r b_r| \qquad [r \text{ summed}]. \tag{4.14}$$

Since η is a unit s-function in $L_p(E)$, we may define

$$|a|_p = [\sum_r |a_r|^p]^{1/p} = 1. \tag{4.14$'$}$$

With $|b|_q$ similarly defined it follows from the relation $\|f\| \geqq |a_r b_r|$ and the arbitrariness of a as a unit vector in R_m that

$$\|f\| \geqq |b|_q = \left[\sum \frac{|\Delta_r(g)|^q}{|\Delta_r(s)|^{q-1}}\right]^{1/q}. \tag{4.15}$$

Although the partitions admitted in (4.15) are restricted to those which belong to E, it follows from (4.15) and Lemma 4.1 that

$$\|f\| \geqq H_q(g). \tag{4.16}$$

The equality is thereby excluded in (4.12).

This establishes (i).

PROOF OF (ii). If $f(x) = 0$, for every $x \in L_p(E)$, $\|f\| = 0$. Hence $H_q(g) = 0$ by (i), and so $g(s) \equiv 0$ by property (e) of $H_q(g)$, since $g(0) = 0$.

Before coming to Theorem 4.2 the following lemma is needed.

LEMMA 4.2. *Corresponding to an arbitrary* $x \in L_p(E)$ $(p \geqq 1)$ *there exists a sequence of s-functions* $\eta^{(n)}$ *belonging to* E *such that* $\eta^{(n)} \to x$ *in* $L_p(E)$ *and* $\eta^{(n)}(s) \to x(s)$ p.p. (presque partout) *in* R_1.

By the general theory of L-integrals there exists a sequence $[\zeta^{(n)}]$ of general step-functions such that $\zeta^{n)} \to x$ in $L_p(R_1)$ and $\zeta^{(n)}(s) \to x(s)$ p.p. in R_1. The step-function $\zeta^{(n)}$ admits points of discontinuity at an arbitrary finite set of points π_r of R_1. We shall modify $\zeta^{(n)}$ so as to obtain an s-function $\eta^{(n)}$ which belongs to E and for which

$$\int_{R_1} | x(s) - \eta^{(n)}(s) |^p \, ds \leqq \int_{R_1} | x(s) - \zeta^{(n)}(s) |^p \, ds. \tag{4.17}$$

In the first place we can suppose that $\zeta^{(n)}(s)$ like $x(s)$ vanishes for $s < 0$ and $s > \omega_E$ and that $s = 0$ and $s = \omega_E$ (if finite) are the terminal vertices of π_n.

Let J be any one of the intervals defined by π_n on which $\zeta^{(n)}$ is constant. If a vertex a of π_n is a left end point of J and a is in E, we modify J, if necessary, so that a is always included in J. If a vertex b of π_n is a right end point of J and b is in E, we modify J, if necessary, so that b is included in J only if $b = \omega_E$. On the modified J, $\zeta^{(n)}$ shall equal its original constant value.

Let H be any maximal finite interval whose open subinterval belongs to comp E and which would be admissible in a partition belonging to E. If any such interval H contains a vertex or vertices of π_n in its open subinterval, we remove such vertices of π_n, adding H to the intervals of the partition defined by π_n and modifying $\zeta^{(n)}$ so that $\zeta^{(n)}$ will equal the constant value of x on H.

After a finite number of modifications of this character at most equal to the number of vertices in π_n, π_n will be replaced by a partition π_n' belonging to E, and $\zeta^{(n)}$ will be replaced by an s-function $\eta^{(n)}$ associated with π_n' and belonging to E. The modifications have been of such a character that (4.17) holds so that $\eta^{(n)} \to x$ in $L_p(E)$, while $\eta^{(n)}(s) \to x(s)$ p.p. in R_1.

THEOREM 4.2. *Corresponding to each* $f \in \overline{L_p(E)}$ $(p > 1)$ *there exists a unique* $g \in L_q^*(E)$ *such that*

$$f(x) = \int_{R_1} x(s) \, dg(s). \tag{4.18}$$

Let g be a functional based on E such that at each point $t \in E$,

$$f(z_E^t) = g(t) \qquad \text{[Cf. (3.7)].} \tag{4.19}$$

(λ). *g is in* $L_q^*(E)$. A continuous variation of $t \in E$ causes z_E^t to vary continuously in $L_p(E)$ and hence $g(t)$ to vary continuously. Hence g maps R_1 into R_1 continuously. Thus g is canonical rel E. With g defined by (4.19), equation (4.13) holds for a unit s-function $\eta \in L_p(E)$. As in the proof of (i), (4.16) is implied, so that $H_q(g) < \infty$, and g is in $L_q^*(E)$.

PROOF OF (4.18). Corresponding to the given $x \in L_p(E)$ let $\eta^{(n)}$ be a sequence of

s-functions belonging to E and converging in $L_p(E)$ to $x \in L_p(E)$ as in Lemma 4.2. Set

$$\eta^{(n)} = \eta_r^{(n)}(z_E^{s_r} - z_E^{s_{r-1}}) \qquad [r = 1, \cdots, r(\pi)].$$

With g defined by (4.19),

$$f(\eta^{(n)}) = \eta_r^{(n)}\Delta_r(g) = \int_{R_1} \eta^{(n)}(s)\, dg(s). \tag{4.20}$$

As $n \to \infty$, $f(\eta^{(n)}) \to f(x)$ since f is continuous over $L_p(E)$. Recall that g is a.c. since in $L_q^*(E)$. By Hölder's lemma,

$$\left| \int_{R_1} [x(s) - \eta^{(n)}(s)]\, \dot{g}(s)\, ds \right| \leqq | x - \eta^{(n)} |_{L_p(E)} H_q(g),$$

so that

$$f(x) = \operatorname*{limit}_{n \to \infty} \int_{R_1} \eta^{(n)}(s)\, dg(s). = \int_{R_1} x(s)\, dg(s).$$

The uniqueness of g in (4.18) is a consequence of (ii) in Theorem 4.1.

This completes the proof of the theorem.

The principal theorem of this section follows.

THEOREM 4.3. *The space $L_p(E)$ $(p > 1)$ satisfies Conditions* I *to* V *on A with $L_q^*(E)$ as the space A_d and $1/p + 1/q = 1$.*

Condition I is satisfied by $A = L_p(E)$ with $A_d = L_q^*(E)$ as a consequence of the two preceding theorems.

Condition II is satisfied in the form

$$\left| \int_{R_1} \eta(s)\, dg(s) \right| \leqq \alpha^{1/p} H_q(g) \max | \eta(s) |, \qquad [\alpha \geqq 0]$$

where η is an s-function for which $\eta(s) = 0$ when $s > \alpha$.

Condition IV is satisfied in accordance with Lemma 4.2.

Condition V holds trivially since $A = A_e$.

Condition III is satisfied as a consequence of Theorem 2.1, since $A = A_e$.

This completes the proof of the theorem.

When $E = [0, 1]$, $L_p(E)$ will be denoted by L_p and $L_q^*(E)$ by L_q^*. With this understood the following corollary is obtained.

COROLLARY 4.1. *The space L_p $(p > 1)$ satisfies Conditions* I *to* V, *with L_q^* the associated space A_d* $(1/p + 1/q = 1)$.

The Riesz-Fréchet representation of a functional $f \in L_p$ is a special consequence of Theorems 4.1 and 4.2. If H is the vector subgroup of functionals $x \in L_p$ which vanish p.p., the classical space L_p is equivalent to L_p mod H.

The space l_p $(p > 1)$. Let E° be the set of points $s = 0, 1, \cdots$ of R_1. A function $x \in L_p(E^\circ)$ is determined by its values $x(n)$ $(n = 0, 1, \cdots)$ and has a norm,

$$\left[\int_{R_1} | x(s) |^p\, ds\right]^{1/p} = [\sum | x(n) |^p]^{1/p} \qquad [n = 0, 1, \cdots]. \tag{4.21}$$

A $g \in L_q^*(E^\circ)$ is determined by its values $g(n)$ $(n = 1, 2, \cdots)$ and has a norm

$$(4.22) \qquad \left[\sup_\pi \frac{|\Delta_r(g)|^q}{|\Delta_r(s)|^{q-1}}\right]^{1/q} = [\sum |g(n+1) - g(n)|^q]^{1/q}, \quad [n = 0, 1, \cdots]$$

where the sup has been taken over all partitions π of R_1 belonging to E°. According to Theorem 4.2 any $f \in \overline{L_p(E^\circ)}$ determines a unique $g \in L_q^*(E^\circ)$ such that for any $x \in \overline{L_p(E^\circ)}$, with n summed

$$(4.23) \quad f(x) = \int_{R_1} x(s)\, dg(s) = x(n)[g(n+1) - g(n)] \quad [n = 0, 1, \cdots].$$

Conversely Theorem 4.1 implies that any $g \in L_q^*(E^\circ)$ determines an $f \in \overline{L_p(E^\circ)}$ of the form (4.23).

The space $L_p(E^\circ)$ is equivalent to l_p under a mapping in which $x \in L_p(E^\circ)$ corresponds to $u = (u_1, u_2, \cdots) \in l_p$ if and only if

$$(4.24) \qquad x(n) = u_{n+1} \qquad [n = 0, 1, \cdots].$$

The space $L_q^*(E^\circ)$ is equivalent to l_q under a mapping in which $g \in L_q^*(E^\circ)$ corresponds to a vector $b \in l_q$ if and only if

$$(4.25) \qquad g(n+1) - g(n) = b_{n+1} \qquad [n = 0, 1, 2, \cdots].$$

Given $F \in l_p$, there exists an $f \in L_p(E^\circ)$ such that $F(u) = f(x)$ under (4.24). According to (4.23) then, $F \in l_p$ is of the form

$$F(u) = u_n b_n, \qquad [n = 1, 2, \cdots]$$

where b is in l_q, confirming a known result. According to Theorem 4.1,

$$\|f\| = H_q(g) = [\sum |g(n-1) - g(n)|^q]^{1/q}, \qquad [n = 0, 1, \cdots]$$

so that one obtains the classical result,

$$\|F\| = [\sum |b_n|^q]^{1/q} \qquad [n = 1, 2, \cdots].$$

Given $b \in l_q$ the values $g(n)$ which satisfy (4.25) with $g(0) = 0$ have the form

$$g(n) = b_1 + b_2 + \cdots + b_n \qquad [n = 1, 2, \cdots].$$

It is in this way that the theory of $F \in l_p$ is derivable from the theory of $f \in \overline{L_p(E^\circ)}$.

5. The space $L(E)$

Let $L(E)$ be a vector space of measurable functions x belonging to E such that the L-integral

$$(5.1) \qquad \int_{R_1} |x(s)|\, ds$$

is finite. This integral is taken as the pseudo-norm of $x \in L(E)$. To determine the associated space A_d the norm $H(g)$ is introduced.

DEFINITION. *The norm* $H(g)$. Corresponding to an arbitrary function g mapping R_1 into R_1 set

$$H(g) = \sup_{(s,t)} \frac{|g(s) - g(t)|}{|s - t|}, \tag{5.2}$$

taking the sup over all pairs (s, t) for which $s > t$.

The following properties of $H(g)$ are readily verified.

(α) *If* $H(g)$ *is finite*, g *is* a.c..

(β) *If* g *is linear over an interval* $[a, b]$, *the* sup $H(g)$ *in* (5.2) *may be obtained on restricting* s *and* t *to the range* $R_1 - (a, b)$ *with* $s > t$.

(γ) *For arbitrary* g_1 *and* g_2 *mapping* R_1 *into* R_1,

$$H(g_1 + g_2) \leqq H(g_1) + H(g_2).$$

(δ) *If* $H(g) = 0$ *and* $g(0) = 0$, $g(s) \equiv 0$.

DEFINITION. *The space* $L^*(E)$. The space of functions g based on E for which $H(g)$ is finite, with $H(g)$ the norm of g, is termed the space $L^*(E)$. It is clear that $H(g)$ is a proper norm relative to $L^*(E)$ and that $L^*(E)$ is a normed vector space.

The theorems which show that $L(E)$ satisfies Condition I follow.

THEOREM 5.1 (i). *For any fixed* $g \in L^*(E)$ *and* $x \in L(E)$ *the L-S-integral*

$$f(x) = \int_{R_1} x(s)\, d\, g(s) \tag{5.3}$$

gives the values of an element $f \in \overline{L(E)}$ *for which* $\|f\| = H(g)$.

(ii) *If* $f(x) = 0$ *for every* $x \in L(E)$, $g(s) = 0$ *for every* $s \in R_1$.

PROOF OF (i). When $H(g)$ is finite, g is a.c. and $|\dot{g}(s)| \leqq H(g)$ p.p. so that

$$f(x) = \int_{R_1} x(s)\, \dot{g}(s)\, ds. \tag{5.4}$$

Hence

$$\|f\| \leqq H(g). \tag{5.5}$$

It remains to establish the equality in (5.5).

If $H(g) = 0$, the equality holds trivially in (5.5). If $H(g) > 0$, the inequality in (5.5) can be excluded as follows. Let e be a positive constant. It follows from property (β) of $H(g)$ that for some pair of points, s_1, s_2 in E with $s_1 < s_2$,

$$\frac{|g(s_2) - g(s_1)|}{|s_2 - s_1|} > H(g) - e. \tag{5.6}$$

Let a partition π belonging to E be defined by the points $0 < s_1 < s_2$ in case $0 < s_1$, and by the points $0 = s_1 < s_2$ in case $s_1 = 0$. Let η be an s-function associated with π such that η is the characteristic function of the interval of π determined by s_1 and s_2. Then

$$(5.7) \qquad |f(\eta)| = |g(s_2) - g(s_1)| \leqq |\eta|_{L(E)} \|f\| = |s_2 - s_1| \|f\|.$$

It follows from (5.6) and (5.7) that

$$H(g) - e < \|f\|.$$

Since e is arbitrary > 0, $\|f\| \geqq H(g)$, so that the equality must hold in (5.5).

PROOF OF (ii). If $f(x) = 0$ for every $x \in L(E)$, $\|f\| = 0$. Hence $H(g) = 0$ by (i), so that $g(s) \equiv 0$ by property (δ) of $H(g)$.

This completes the proof of Theorem 5.1.

THEOREM 5.2. *Corresponding to any* $f \in \overline{L(E)}$ *there exists a unique* $g \in L^*(E)$, *such that for every* $x \in L(E)$,

$$f(x) = \int_{R_1} x(s)\, dg(s).$$

Let g be a function based on E such that at each point t of E, $g(t) = f(z_E^t)$. [Cf. (3.7)]. Then for any s-function η belonging to E, $\eta_r \Delta_r(g) = f(\eta)$, and since f is linear by hypothesis,

$$|\eta_r \Delta_r(g)| \leqq |\eta|_{L(E)} \|f\|.$$

As a consequence (5.7) holds for s_1 and s_2 in E with $s_1 < s_2$ and η chosen as in (5.7). Hence g is a.c. and in $L^*(E)$. The remainder of the proof is practically identical with the proof of (4.18), setting $p = 1$ and replacing $H_q(g)$ by $H(g)$. Note that Lemma 4.2 covers the case $p = 1$.

The principal theorem on $L(E)$ follows.

THEOREM 5.3. *The space* $A = L(E)$ *satisfies Conditions* I *to* V *with* $A_d = L^*(E)$.

Condition I is satisfied by $A = L(E)$ with $A_d = L^*(E)$ by virtue of the two preceding theorems.

Condition II is satisfied in the form

$$\int_{R_1} \eta(s)\, dg(s) \quad \leqq \alpha\, H(g) \max |\eta(s)|, \qquad [\alpha \geqq 0]$$

where $\eta(s) = 0$ for $s > \alpha$.

Condition IV is satisfied by virtue of Lemma 4.2.

Condition V holds trivially since $A = A_e$.

Condition III holds in accordance with Theorem 2.1, since $A = A_e$.

See H. Steinhaus for the representation of $F \in \overline{L}$ in the case in which $E = [0, 1]$.

The space l. Let E° be the set of points $s = 0, 1, 2, \cdots$ of R_1. A function $x \in L(E^\circ)$ is determined by its values $x(n)$ $(n = 0, 1, \cdots)$ and has a norm

$$(5.8) \qquad \int_{R_1} |x(s)|\, ds = \sum |x(n)| \qquad [n = 0, 1, \cdots].$$

A function $g \in L^*(E^\circ)$ is determined by its values $g(n)$ $(n = 1, 2, \cdots)$ and has a norm (r not summed),

$$\sup_{\pi} \left| \frac{\Delta_r(g)}{\Delta_r(s)} \right| = \sup_{n} | g(n+1) - g(n) |, \qquad [n = 0, 1 \cdots] \tag{5.9}$$

where π ranges over all partitions of R_1 belonging to E°. According to Theorem 5.2 any element $f \in \overline{L(E^\circ)}$ determines a unique element $g \in L^*(E)$ such that for each $x \in L(E^\circ)$,

$$f(x) = \int_{R_1} x(s)\, d\, g(s) = x(n)[g(n+1) - g(n)], \qquad [n = 0, 1, \cdots] \tag{5.10}$$

while

$$\| f \| = \sup_{n} | g(n+1) - g(n) | \qquad [n = 0, 1, \cdots].$$

Conversely any $g \in L^*(E^\circ)$ determines a unique $f \in \overline{L(E^\circ)}$ given by (5.10).

The space $L(E^\circ)$ is equivalent to l under a mapping in which $x \in L(E^\circ)$ corresponds to $u = (u_1, u_2, \cdots)$ in l, if and only if $x(n) = u_{n+1}$ $(n = 0, 1, \cdots)$. The representation of $F \in l$ obtained from (5.10) agrees with classical results otherwise obtained.

6. Complementary Φ and Ψ and associated generalized Hellinger integrals

The spaces which we shall study in the next section are given in terms of complementary functions Φ and Ψ introduced by W. H. Young and used by Orlicz to define spaces of which our spaces $A = \Phi(E)$ are an extension. The Orlicz spaces were further studied by Zygmund, Chapter IV. Our introduction of a generalized Hellinger integral $I_\Psi(g)$ to aid in the definition of a norm of g in A_d is very useful. The principal objective is the verification for $\Phi(E)$ of Conditions I to V. In this section we review certain properties of Φ and Ψ, and introduce a generalized Hellinger integral $I_\Psi(g)$.

As with Young let φ and ψ be two functionals with values $\varphi(u)$ and $\psi(v)$ defined for $u \geqq 0$, $v \geqq 0$, continuous, vanishing when $u = v = 0$, strictly increasing, tending to ∞, and inverse to each other. (Cf. Zygmund, p. 64). Then for $a, b \geqq 0$, Young's inequality gives

$$ab \leqq \Phi(a) + \Psi(b), \tag{6.1}$$

where

$$\Phi(x) = \int_0^x \phi(u)\, du, \quad \Psi(y) = \int_0^y \psi(v)\, dv \qquad [x, y \geqq 0]. \tag{6.2}$$

The geometrical proof of (6.1) in terms of $\Phi(a)$ and $\Psi(b)$ as areas in the (a, b)-plane is obvious. It will simplify the notation if the definitions of Φ and Ψ are extended by setting

$$\Phi(-x) = \Phi(x), \qquad \Psi(-y) = \Psi(y) \qquad [x \geqq 0, y \geqq 0].$$

In terms of its geometrical interpretation one sees that the equality holds in (6.1) if and only if $b = \varphi(a)$, or equivalently $a = \psi(b)$. The functions Φ and Ψ as given by (6.2) for $x, y \geqq 0$ are continuous, strictly increasing, tending to ∞, and convex.

The generalized Hellinger integral $I_\Psi(g)$. Let g be a function mapping R_1 into R_1. Let a "general" partition π of R_1 be defined by the points

$$s_0 < s_1 < \cdots < s_{r(\pi)} . \tag{6.3}$$

With this partition given we refer to quotients (s not summed)

$$\zeta_r^g = \frac{\Delta_r(g)}{\Delta_r(s)} = \frac{g(s_r) - g(s_{r-1})}{s_r - s_{r-1}}, \qquad [r = 1, \cdots, r(\pi)] \tag{6.4}$$

corresponding to the respective intervals $[s_{r-1}, s_r)$. In the special case in which $\Psi(v) = |v|^q$ (treated in §4) we have introduced the Hellinger integral (r summed)

$$\sup_\pi \frac{|\Delta_r(g)|^q}{|\Delta_r(s)|^{q-1}} = \sup_\pi \left| \frac{\Delta_r(g)}{\Delta_r(s)} \right|^q \Delta_r(s), \qquad [r = 1, \cdots, r(\pi)] \tag{6.5}$$

taking the sup over all partitions π of the above type. The generalization is to introduce (see W. H. Young Theorem 5)

$$\sup_\pi \Psi\left(\frac{\Delta_r(g)}{\Delta_r(s)}\right) \Delta_r(s) = I_\Psi(g), \qquad [r = 1, \cdots, r(\pi)] \tag{6.6}$$

summing with respect to s, and taking the sup as in (6.5). It is clear that $I_\Psi(g)$ exists either as a finite number or as $+\infty$.

We term $I_\Psi(g)$ *the generalized Hellinger integral of* g *of index* Ψ, *taken over* R_1 *with respect to* s.

For any measurable function z mapping R_1 into R_1 set

$$\int_{R_1} \Psi[z(s)]\, ds = J_\Psi(z), \tag{6.7}$$

understanding that $J_\Psi(z) = +\infty$ if $\Psi(z)$ is not summable. Recall that $\Psi[z(s)] = \Psi[\,|z(s)|\,]$, so that the integrand in (6.7) is never negative. If $|z(s)|$ is bounded and vanishes everywhere except for a set of finite L-measure, $J_\Psi(z)$ is finite.

The following theorem generalizes the properties of the Hellinger integral $I_q(z)$ of §4.

THEOREM 6.1. *If* g *is any function mapping* R_1 *into* R_1, *a necessary and sufficient condition that* $I_\Psi(g)$ *be finite is that* g *be* a.c. *and* $J_\Psi(\dot g)$ *be finite. When* g *is* a.c.,

$$I_\Psi(g) = J_\Psi(\dot g). \tag{6.8}$$

It is not assumed that g is measurable in the first sentence of the theorem, nor that either member of (6.8) is finite.

LEMMA 6.1. *If* g *is finitely*[4] a.c. *and* $J_\Psi(\dot g)$ *is finite, then* $I_\Psi(g)$ *is finite and*

$$I_\Psi(g) \leqq J_\Psi(\dot g). \tag{6.9}$$

[4] We say that g is finitely a.c. if a.c. over every finite interval. Theorem 6.1 holds with "a.c." replaced by "finitely a.c." as the proof shows.

We make use of the preceding partition π. For $r = 1, 2, \cdots, r(\pi)$, r not summed,

$$\Psi\left[\frac{\int_{s_{r-1}}^{s_r} \dot{g}(s)\, ds}{\Delta_r(s)}\right] \leqq \Psi\left[\frac{\int_{s_{r-1}}^{s_r} |\dot{g}(s)|\, ds}{\Delta_r(s)}\right] \leqq \frac{\int_{s_{r-1}}^{s_r} \Psi\, [\dot{g}(s)]\, ds}{\Delta_r(s)},$$

using the Jensen inequality for convex functions Ψ (Cf. Zygmund, p. 68, $p(t) \equiv 1$). Multiplying the extreme members by $\Delta_r(s)$ and summing with respect to r,

$$\Psi\left[\frac{\Delta_r(g)}{\Delta_r(s)}\right] \Delta_r(s) \leqq J_\Psi(\dot{g}). \tag{6.10}$$

Relation (6.9) follows from (6.10) on taking the sup of the left member of (6.10) over all admissible partitions π of R_1.

LEMMA 6.2. *If g is an arbitrary function mapping R_1 into R_1, and if the Hellinger integral $I_\Psi(g)$ is finite, then g is* a.c.

Corresponding to an arbitrary partition π of R_1 of the above type, let $\sum$ refer to a sum over an arbitrarily selected subset of intervals of π whose union is a point set B of R_1. Let ζ^g be a step-function associated with π, with a value ζ_r^g given by (6.4) on the rth interval of π. Then

$$\begin{aligned}
\sum |\Delta_r(g)| &= \sum |\zeta_r^g|\, \Delta_r(s) = \int_B |\zeta^g(s)|\, ds \\
&= \int_B \left|\frac{\zeta^g(s)}{c}\right| c\, ds \qquad [c = \text{constant} > 1] \\
&\leqq \int_B \Psi\left(\frac{\zeta^g(s)}{c}\right) ds + \int_B \Phi\,(c)\, ds \qquad [\text{using } (6.1)] \\
&= \sum \Psi\left(\frac{\zeta_r^g}{c}\right) \Delta_r(s) + \Phi\,(c)\, m(B) \\
&\leqq \frac{1}{c} \sum \Psi\,(\zeta_r^g)\, \Delta_r(s) + \Phi\,(c)\, m(B) \qquad [\text{by convexity of } \Psi] \\
&\leqq \frac{1}{c} I_\Psi(g) + \Phi\,(c)\, m(B).
\end{aligned} \tag{6.11}$$

Given an $e > 0$ one can choose c so large that the first term on the right of (6.11) is $< e$, and with c fixed take $m(B)$ so small that $\Phi(c)\, m(B) < e$. Then $\sum |\Delta_r(g)| \leqq 2\, e$ for this choice of B. The lemma follows.

LEMMA 6.3. *If the Hellinger integral $I_\Psi(g)$ is finite, then g is* a.c. *and*

$$J_\Psi(\dot{g}) \leqq I_\Psi(g). \tag{6.12}$$

If $I_\Psi(g)$ is finite it follows from the preceding lemma that g is a.c. As shown in Titchmarsh, pp. 385–386, the absolute continuity of g implies the existence of a sequence of step-functions $\eta^{(m)}$ ($m = 1, 2, 3, \cdots$) defined for partitions $\pi^{(m)}$ with r^{th} interval $[r_{r-1}^{(m)}, r_r^{(m)})$ $[r = 1, \cdots, r^{(m)}]$ and value thereon

$$\eta_r^{(m)} = \frac{g(s_r^{(m)}) - g(s_{r-1}^{(m)})}{s_r^{(m)} - s_{r-1}^{(m)}} \qquad [m = 1, 2, \cdots].$$

with $\eta^{(m)}(s) = 0$ on the extreme unbounded intervals of $\pi^{(m)}$ and such that $\eta^{(m)}(s)$ converges pointwise p.p. to $\dot{g}(s)$. Setting $\Delta_r^{(m)}(s) = s_r^{(m)} - s_{r-1}^{(m)}$, $[r = 1, \cdots, r^{(m)}]$

$$J_\Psi(\eta^{(m)}) = \Psi(\eta_r^{(m)})\, \Delta_r^{(m)}(s) \leqq I_\Psi(g) \qquad \text{(r summed).} \tag{6.13}$$

Since $\eta^{(m)}(s)$ converges pointwise p.p. to $\dot{g}(s)$, Fatou's theorem implies that

$$\liminf_{m\to\infty} J_\Psi(\eta^{(m)}) \geqq J_\Psi(\dot{g}), \tag{6.14}$$

so that (6.12) follows from (6.13) and (6.14).

Proof of Theorem 6.1. If $I_\Psi(g)$ is finite, g is a.c. and $J_\Psi(\dot{g})$ finite by Lemma 6.3. Both (6.9) and (6.12) then hold implying (6.8). Conversely if g is a.c. and $J_\Psi(\dot{g})$ finite, $I_\Psi(g)$ is finite by Lemma 6.1.

To establish (6.8) when g is a.c. observe that when $I_\Psi(g)$ is finite, (6.8) holds as stated above. When g is a.c. and $I_\Psi(g)$ is infinite, $J_\Psi(\dot{g})$ is likewise infinite by Lemma 6.1. Hence (6.8) holds whenever g is a.c.

Let E be a closed set of the type admitted in §1. We generalize Lemma 4.1 as follows.

Lemma 6.4. *For any g based on E the Hellinger integral $I_\Psi(g)$ may be obtained as a*

$$\sup_\pi \Psi\left[\frac{\Delta_r(g)}{\Delta_r(s)}\right]\Delta_r(s), \qquad [r = 1, \cdots, r(\pi)] \tag{6.15}$$

in which r is restricted to partitions belonging to E.

That the range of partitions in (6.15) can be restricted to those belonging to E may be seen on extending (a) and (b) of §4 as follows.

(a) *Summing over the intervals of a partition π,*

$$a \leqq s_0 < s_1 < \cdots < s_m = b, \tag{6.16}$$

of $[a, b]$,

$$\Psi\left[\frac{\Delta_r(g)}{\Delta_r(s)}\right]\Delta_r(s) \geqq \Psi\left[\frac{g(b) - g(a)}{b - a}\right](b - a). \tag{6.17}$$

Let $y(s)$, defined for $a \leqq s \leqq b$ give the ordinate on a broken line in the (s, y)-plane joining successive points $[s_i, g(s_i)]$ $(i = 0, 1, \cdots, m)$. It follows from the Jensen integral inequality (loc. cit.) that

$$\frac{\int_a^b \Psi[\dot{y}(s)]\, ds}{b - a} \geqq \Psi\left[\frac{\int_a^b |\dot{y}(s)|\, ds}{b - a}\right] \geqq \Psi\left[\frac{\int_a^b \dot{y}(s)\, ds}{b - a}\right]. \tag{6.18}$$

Relation (6.17) is obtained on taking account of the linear character of y between vertices and multiplying the extreme members of (6.18) by $b - a$.

(b) *If $g(s)$ is linear over an interval $[a, b]$ belonging to a finite partition π' of R_1, further subdivision of π' by adding vertices in (a, b) will not increase the sum*

$$\Psi\left[\frac{\Delta_r(g)}{\Delta_r(s)}\right]\Delta_r(s), \qquad [r = 1, 2, \cdots, r(\pi')] \tag{6.19}$$

taken over the intervals of π'.

Since $\dot{g}(s)$ is constant over (a, b) on setting $y(s) = g(s)$ in (6.18) the equality holds. The same must then be true of (6.17) and (b) follows.

The lemma follows as in the proof of Lemma 4.1 on modifying a general partition π_1 of R_1 so that it becomes a partition belonging to E. This can be done without decreasing the sum (6.13) by virtue of (a) and (b), so that the restricted sup in (6.15) is not less than $I_\Psi(g)$. Since the restricted partitions are included among the general partitions, the restricted sup equals the unrestricted sup as stated in the lemma.

THEOREM 6.2. *The integral $I_\Psi(g) = 0$ if and only if $g(s) \equiv 0$.*

By Theorem 6.1, $I_\Psi(g) = 0$ if and only if $J_\Psi(\dot{g}) = 0$, that is, if and only if $\dot{g} = 0$ p.p., or finally if and only if $g(s) \equiv 0$.

If g_1 and g_2 are arbitrary functions mapping R_1 into R_1, then for $m_1 \geqq 0$, $m_2 \geqq 0, m_1 + m_2 = 1$,

$$I_\Psi(m_1 g_1 + m_2 g_2) \leqq m_1 I_\Psi(g_1) + m_2 I_\Psi(g_2). \tag{6.20}$$

If either term on the right of (6.20) is infinite, (6.20) is trivial. If both terms on the right of (6.20) are finite, (6.20) follows from the relation

$$I_\Psi(g) = J_\Psi(\dot{g}) = \int_{R_1} \Psi[\dot{g}(t)]\, dt,$$

and the convexity of Ψ on setting $g = m_1 g_1 + m_2 g_2$.

The following theorem affirms that I_Ψ is lower semi-continuous.

THEOREM 6.3. *If a sequence of functions g_n mapping R_1 into R_1 converges pointwise to a function g mapping R_1 into R_1, then*

$$\liminf_{n\to\infty} I_\Psi(g_n) \geqq I_\Psi(g). \tag{6.21}$$

For a fixed partition π of R_1 the sum

$$\Psi\left[\frac{\Delta_r(g)}{\Delta_r(s)}\right]\Delta_r(s) = S_\pi(g)$$

is continuous with respect to pointwise convergence of g in the sense that

$$\operatorname*{limit}_{n\to\infty} S_\pi(g_n) = S_\pi(g).$$

It follows from a general theorem on lower semi-continuity that $I_\Psi(g)$, as defined by $\sup_\pi S_\pi(g)$, is lower semi-continuous (See McShane, p. 41).

7. The space $\Psi^*(E)$ and norm $H_\Psi(g)$

We shall see that the space $\Psi^*(E)$ is a space of d-functions which includes the preceding spaces $L_p^*(E)$ as special cases.

DEFINITION. Let E be a closed subset of R_1 as in §1. Let $\Psi^*(E)$ be the space of all functions g based on E for which

$$I_\Psi(kg) < \infty \tag{7.1}$$

for some positive k.

That the space $\Psi^*(E)$ is *additive* is seen as follows. If g_1 and g_2 are based on E, if for some positive k_1 and k_2, $I_\Psi(k_1g_1)$ and $I_\Psi(k_2g_2)$ are finite, and if one sets $k = \min(k_1, k_2)$, then from (6.20) with $m_1 = m_2 = 1/2$ therein,

$$I_\Psi\left[\frac{k}{2}(g_1 + g_2)\right] \leqq \tfrac{1}{2}[I_\Psi(kg_1) + I_\Psi(kg_2)] \leqq \tfrac{1}{2}[I_\Psi(k_1 g_1) + I_\Psi(k_2 g_2)] < \infty.$$

Hence $g_1 + g_2$ is in $\Psi^*(E)$ with g_1 and g_2. It is trivial that cg is in $\Psi^*(E)$ with g (c = constant). Thus $\Psi^*(E)$ is a linear vector space.

DEFINITION. *The limit* k_g. For $g \in \Psi^*(E)$ let k_g be the sup (possibly ∞) of all numbers k such that (7.1) holds.

Given Ψ and g, set $I_\Psi(kg) = f(k)$. Examples at the end of this section will show that there exist admissible Ψ and $g \in \Psi^*(E)$ for which k_g and $f(k_g)$ are independently prescribed on the interval $[0, \infty)$. In these examples the set E will be a closed interval $[0, h]$.

In order to define a norm for g the following lemma is needed.

LEMMA 7.1. *For $g \in \Psi^*(E)$ and $g \not\equiv 0$, $f(k)$ increases strictly and continuously from 0 to $f(k_g)$ as k increases from 0 to $k_g < \infty$; as $k \to k_g = \infty$ from below, $f(k) \to +\infty$.*

When g is in $\Psi^*(E)$, g is a.c. according to Theorem 6.1 and

$$f(k) = J_\Psi(k\dot{g}) \qquad \text{[possibly } +\infty] \tag{7.2}$$

for every $k > 0$. It follows from the monotonicity of Ψ that $f(k)$ is monotone increasing. That $f(k)$ varies continuously with k for $0 \leqq k \leqq k_g$ is seen as follows.

Suppose that k_n is on the interval $[0, k_g)$ and converges to $k_0 < \infty$ as $n \to \infty$. Then

$$\Psi[k_n \dot{g}(s)] \to \Psi[k_0 \dot{g}(s)]. \tag{7.3}$$

In case $k_0 < k_g$ it follows immediately that $f(k_n) \to f(k_0)$. In the case that $k_0 = k_g$ it follows from Fatou's theorem that $\liminf f(k_n) \geqq f(k_0)$ while the monotonicity of Ψ implies that $\limsup f(k_n) \leqq f(k_0)$. Hence $f(k_n) \to f(k_g)$ (possibly ∞) as $k_n \to k_0$.

That $f(k)$ is strictly increasing as k increases from 0 to k_g (assuming always that $g \not\equiv 0$) follows from the fact that $\Psi[k\dot{g}(s)]$ is strictly increasing with k for each s for which $0 < |\dot{g}(s)| < \infty$, that is for a set of values of s of positive L-measure, while for other values of s, $\dot{g}(s) = 0$, excepting at most a set of L-measure zero. That

$f(k) \to \infty$ in case $k_g = \infty$ and $k \to \infty$ follows from the relation

$$\Psi(ku) > k\Psi(u) \qquad [u > 0]\ [k > 1].$$

The preceding lemma shows that $f(k)$ defines a function which is $+\infty$ when $k > k_g$, but which is left continuous at k_g when k_g is finite. When k_g and $f(k_g)$ are finite, $f(k)$ jumps from $f(k_g)$ to $+\infty$ at k_g. This lemma validates the following definition.

DEFINITION. *The norm* $H_\Psi(g)$. For $g \in \Psi^*(E)$ and $g \not\equiv 0$, the values of $k > 0$ for which

$$I_\Psi(kg) \leqq 1 \tag{7.4}$$

have a finite positive maximum $M(g)$. Whee $g \not\equiv 0$ set $H_\Psi(g) = [M(g)]^{-1}$. When $g \equiv 0$ set $H_\Psi(g) = 0$.

It follows that

$$I_\Psi\left[\frac{g}{H_\Psi(g)}\right] \leqq 1 \tag{7.5}$$

when $g \not\equiv 0$, where the inequality occurs if and only if $k_g < \infty$ and

$$I_\Psi(k_g g) < 1, \tag{7.6}$$

and in this case $k_g H_\Psi(g) = 1$. For g based on E but not in $\Psi^*(E)$, $I_\Psi(kg) = +\infty$ for every $k > 0$. For such g set $H_\Psi(g) = \infty$. With this understood the following lemma can be proved.

LEMMA 7.2. *For functionals g_1 and g_2 based on E,*

$$H_\Psi(g_1 + g_2) \leqq H_\Psi(g_1) + H_\Psi(g_2), \tag{7.7}$$

while for every constant $|c| \neq 0$,

$$H_\Psi(cg_1) = |c|\, H_\Psi(g_1). \tag{7.8}$$

Relation (7.8) is immediate even when $H_\Psi(g_1) = \infty$. Relation (7.7) is trivial if either term on the right is $+\infty$, or 0. If for example $H_\Psi(g_1) = 0$, $g_1 \equiv 0$ and (7.7) holds. Introducing constants ρ, a, b, suppose then that

$$0 < \rho = H_\Psi(g_1) + H_\Psi(g_2) < \infty,$$

$$a\rho = H_\Psi(g_1) > 0, \qquad b\rho = H_\Psi(g_2) > 0.$$

Then $a + b = 1$, and

$$I_\Psi\left(\frac{g_1 + g_2}{\rho}\right) = I_\Psi\left[a\left(\frac{g_1}{a\rho}\right) + b\left(\frac{g_2}{b\rho}\right)\right]$$

$$\leqq aI_\Psi\left(\frac{g_1}{a\rho}\right) + bI_\Psi\left(\frac{g_2}{b\rho}\right) \leqq a + b = 1$$

by virtue of the convexity of Ψ and the relations

$$I_\Psi\left(\frac{g_1}{a\rho}\right) \leqq 1, \qquad I_\Psi\left(\frac{g_2}{b\rho}\right) \leqq 1.$$

Thus

$$I_\Psi\left(\frac{g_1 + g_2}{\rho}\right) \leqq 1,$$

so that $H_\Psi(g_1 + g_2) \leqq \rho$. Thus (7.7) holds.

We summarize and complete the results as follows.

THEOREM 7.1. *The vector space $\Psi^*(E)$ of functions g based on E for which the Hellinger integral $I_\Psi(kg)$ is finite for some $k > 0$ and for which $H_\Psi(g)$ is the norm satisfies the conditions of Banach on a normed vector space. Moreover $H_\Psi(g)$ is a proper norm for $\Psi^*(E)$, finite for g based on E, if and only if g is in $\Psi^*(E)$. For g based on E, a necessary and sufficient condition that $H_\Psi(g) \leqq 1$ is that g be* a.c. *and $J_\Psi(\dot{g}) \leqq 1$.*

It is only the final statement which is as yet unproved.

For brevity set $H_\Psi(g) = c$. By definition $c = 0$ if and only if $g \equiv 0$, and in this case $J_\Psi(\dot{g}) = 0$. Suppose then that $c > 0$. We have

$$I_\Psi\left(\frac{g}{c}\right) \leqq 1$$

by definition of $H_\Psi(g) = c$. By Theorem 6.1 g is then a.c. and

$$J_\Psi\left(\frac{\dot{g}}{c}\right) = I_\Psi\left(\frac{g}{c}\right) \leqq 1.$$

If $c \leqq 1$,

$$J_\Psi(\dot{g}) \leqq J_\Psi\left(\frac{\dot{g}}{c}\right) \leqq 1$$

because of the monotonicity of Ψ. Thus the condition $H_\Psi(g) \leqq 1$ implies the condition $J_\Psi(\dot{g}) \leqq 1$.

Conversely, if g is a.c. and $J_\Psi(\dot{g}) \leqq 1$,

$$I_\Psi(g) = J_\Psi(\dot{g}) \leqq 1,$$

so that $H_\Psi(g) \leqq 1$ by its definition.

THEOREM 7.2. *The norm H_Ψ is lower semi-continuous over $\Psi^*(E)$. That is*

$$\liminf_{n\to\infty} H_\Psi(g_n) \geqq H_\Psi(g), \tag{7.9}$$

whenever a sequence of functions $g_n \in \Psi^*(E)$ converges pointwise to $g \in \Psi^*(E)$.

If $g_n \equiv 0$ for infinitely many integers n, $g \equiv 0$, $H_\Psi(g) = 0$ and (7.9) is trivially true. Suppose then that $g_n \not\equiv 0$ $(n = 1, 2, \cdots)$. Let $M(g_n)$ be the reciprocal of $H_\Psi(g_n)$. Set

$$M_0 = \limsup_{n\to\infty} M(g_n).$$

Case I. $M_0 = 0$. In this case, the lim inf in (7.9) is ∞ so that (7.9) holds trivially.

Case II. $M_0 = \infty$. For all values of n for which $M(g_n) > 1$,

$$M(g_n)\, I_\Psi(g_n) \leqq I_\Psi[M(g_n)\, g_n] \leqq 1, \qquad [n \text{ not summed}]$$

so that

$$\liminf_{n\to\infty} I_\Psi(g_n) = 0.$$

By virtue of the lower semi-continuity of I_Ψ (Theorem 6.3), $I_\Psi(g) = 0$ so that $g \equiv 0$ by Theorem 6.2. In this case $H_\Psi(g) = 0$, and (7.9) again holds trivially.

Case III. $0 < M_0 < \infty$. For any constant k for which $0 < k < M_0$, $I_\Psi(kg_n) \leqq 1$ for infinitely many integers n, so that $I_\Psi(kg) \leqq 1$ by the lower semi-continuity of I_Ψ. Hence $M(g) \geqq k$, and since k is any constant between 0 and M_0, $M(g) \geqq M_0$. Hence

$$H_\Psi(g) \leqq \frac{1}{M_0} = \liminf_{n\to\infty} H_\Psi(g_n).$$

EXAMPLE 7.1. If $\Psi(y) = |y|^q$ $(q > 1)$, then for finite $I_\Psi(g)$,

$$I_\Psi(g) = J_\Psi(\dot g) = \int_{R_1} |\dot g(s)|^q\, ds.$$

For $g \in \Psi^*(E)$ and $g \not\equiv 0$, the limit $k_g = \infty$. Hence the norm $H_\Psi(g)$ equals the constant c such that

$$I_\Psi\left(\frac{g}{c}\right) = \int_{R_1} \left|\frac{\dot g(s)}{c}\right|^q ds = 1.$$

Thus

$$H_\Psi(g) = \left[\int_{R_1} |\dot g(s)|^q\, ds\right]^{1/q}.$$

EXAMPLE 7.2. We shall define an admissible Ψ, set E, and $g \in \Psi^*(E)$ for which k_g and $f(k_g)$ are prescribed positive numbers a and b.

Set $\Psi(u) = e^u$ for $u \geqq \log 4$. Observe that the tangent to the graph $y = e^u$ at the point u_0 with abscissa $u_0 = \log 4$ intersects the u-axis at a point u_1 at which $u_1 > 0$. It follows that $\Psi(u)$ can be extended in definition over the interval $[0, u_0)$ so as to be admissible. In particular the extension can be made so that $\dot\Psi(u)$ is continuous and increasing for $0 \leqq u < \infty$ and

$$\dot\Psi(0) = \Psi(0) = 0.$$

An a.c. functional g will be defined as an integral with $g(0) = 0$. To that end let $0 = s_1 < s_2 < s_3 < \cdots$ be an infinite sequence of points on the s-axis such that

$$s_{n+1} - s_n = \frac{b}{n(n+1)2^{n+1}} \qquad [n = 1, 2, \cdots]. \tag{7.10}$$

It is clear that s_n converges to a finite value $h(b)$. The functional g shall be the integral of $\lambda(s)/a$ for which $g(0) = 0$ and for which

$$\lambda(s) = \log 2^{n+1}, \qquad [s_n \leqq s < s_{n+1}, n = 1, 2, \cdots] \tag{7.11}$$

while $\lambda(s) = 0$ at all other points of R_1. On the nth interval

$$\Psi[\lambda(s)] = \Psi(\log 2^{n+1}) = 2^{n+1}, \qquad [n = 1, 2, \cdots] \tag{7.12}$$

so that

$$A_\Psi(a\dot{g}) = \int_{R_1} \Psi[\lambda(s)]\, ds = \frac{b}{1\cdot 2} + \frac{b}{2\cdot 3} + \frac{b}{3\cdot 4} + \cdots = b.$$

If e is a positive constant one sees that

$$J_\Psi[a(1 + e)\dot{g}] = b\left[\frac{2^{2e}}{1\cdot 2} + \frac{2^{3e}}{2\cdot 3} + \frac{2^{4e}}{3\cdot 4} + \cdots\right] = \infty.$$

The set E shall be any interval $[0, h]$ which contains each point s_n. It appears that g is in $\Psi(E)$ and that $k_g = a$ and $f(k_g) = b$.

EXAMLPE 7.3. We shall define an admissible Ψ, set E and $g \in \Psi^*(E)$, for which k_g is a prescribed number $a > 0$ and $f(k_g) = \infty$.

Take Ψ as in the preceding example, but replace (7.10) by the equation

$$s_{n+1} - s_n = \frac{1}{n2^{n+1}} \qquad [n = 1, 2, \cdots]. \tag{7.13}$$

Define λ as in (7.11). Then (7.12) holds so that

$$J_\Psi(a\dot{g}) = \tfrac{1}{1} + \tfrac{1}{2} + \tfrac{1}{3} + \cdots = +\infty.$$

If e is a positive constant,

$$J_\Psi[a(1 - e)\dot{g}] = \frac{2^{-2e}}{1} + \frac{2^{-3e}}{2} + \frac{2^{-4e}}{3} + \cdots < \infty.$$

Let E be any interval $[0, h]$ which contains each point s_n. It appears that $k_g = a$ and $f(k_g) = +\infty$.

8. The space $\Phi(E)$

This space will generalize the spaces $L_p(E)$ $(p > 1)$ of §4. We refer to the complementary functions Φ and Ψ of §6 and to the closed set E of §1.

The space $\Phi(E)$ shall be the linear vector space of all functions x belonging to E for which $J_\Phi(kx) < \infty$ for every $k > 0$.

This definition should be contrasted with that of $\Psi^*(E)$ which implies that $\Psi^*(E)$ is the linear vector space of all functions g based on E which are a.c. and for which $J_\Psi(k\dot{g}) < \infty$ for *some* $k > 0$.

The space $\Phi(E)$ is not empty. In particular it includes all s-functions belonging to E. It is additive; with x_1, $x_2 \in \Phi(E)$,

$$J_\Phi\left[\frac{k}{2}(x_1 + x_2)\right] \leqq \tfrac{1}{2}[J_\Phi(kx_1) + J_\Phi(kx_2)] < \infty \tag{8.1}$$

for every k, by virtue of the convexity of Φ. Clearly for any constant c, cx is in $\Phi(E)$ with x. Thus $\Phi(E)$ is a linear vector space.

The definition of a pseudo-norm of $x \in \Phi(E)$ may be motivated as follows. The norm of an element x in a pseudo-normed vector space A is given by $\sup_f |f(x)|$ taking the sup over all $f \in \bar{A}$ for which $\|f\| = 1$. (Cf. Lemma 4.2, Paper I). When A satisfies Condition I of §2 this becomes

$$|x|_A = \sup_g \left| \int_{R_1} x(s)\, dg(s) \right|, \tag{8.2}$$

taking the sup over all $g \in A_d$ for which $|g|_{A_d} = 1$.

In the case at hand x is in $\Phi(E)$ and g in $\Psi^*(E)$. Then g is a.c., and if g has a unit norm,

$$\left| \int_{R_1} x(s)\dot{g}(s)\, ds \right| \leqq \int_{R_1} |x(s)|\,|\dot{g}(s)|\, ds \tag{8.3}$$

$$\leqq J_\Phi(x) + J_\Psi(\dot{g}) \leqq J_\Phi(x) + 1, \qquad \text{[Cf. Theorem 7.1]} \tag{8.4}$$

using the Young inequality. The following definition thus seems appropriate.

The norm of $x \in \Phi(E)$ shall be (Cf. Zygmund p. 96)

$$|x|_{\Phi(E)} = \sup_g \left| \int_{R_1} x(s)\, dg(s) \right|, \tag{8.5}$$

taking the sup over all $g \in \Psi^*(E)$ with subunit (or unit) norms, or equivalently over all g in $\Psi^*(E)$ for which $J_\Psi(\dot{g}) \leqq 1$. (Cf. Theorem 7.1).

Another form of this definition can be derived as follows. Observe that for $x \in \Phi(E)$ and $g \in \Psi^*(E)$,

$$\left| \int_{R_1} x(s)\, dg(s) \right| \leqq \int_{R_1} |x(s)|\,|\dot{g}(s)|\, ds = \int_{R_1} x(s)\, dh(s), \tag{8.6}$$

where

$$h(s) = \int_0^s |\dot{g}(s)| \operatorname{sign} x(s)\, ds \qquad [\operatorname{sign} 0 = 1].$$

Since $|\dot{h}(s)| = |\dot{g}(s)|$ p.p., $J_\Psi(\dot{h}) = J_\Psi(\dot{g})$. Since h is continuous and linear on the closure of each maximal interval of comp E, while $h(s) = 0$ for $s \leqq 0$ and $h(s) = h(\omega_E)$ for $s \geqq \omega_E$, h is in $\Psi^*(E)$ with g. Hence the sup of the extreme members of (8.6) as g ranges over all elements in $\Psi^*(E)$ with unit norm is the same, and equals $|x|_{\Phi(E)}$. We have established the following. (For a special case see Orlicz, p. 210.)

The pseudo-norm of an $x \in \Phi(E)$ is given by the equation

$$\sup_g \int_{R_1} |x(s)|\,|\dot{g}(s)|\, ds = |x|_{\Phi(E)}, \tag{8.7}$$

taking the sup over all $g \in \Psi^*(E)$ with unit norms.

From this evaluation of $| x |_{\Phi(E)}$ it follows that

$$| x + y |_{\Phi(E)} \leqq | x |_{\Phi(E)} + | y |_{\Phi(E)} \tag{8.8}$$

for x and y in $\Phi(E)$, while for each constant c,

$$| cx |_{\Phi(E)} = | c | \, | x |_{\Phi(E)} . \tag{8.9}$$

The space $\Phi(E)$ thus satisfies the conditions on a pseudo-normed vector space.

The following lemma serves to establish convergence in pseudo-norms in $\Phi(E)$ in important cases.

LEMMA 8.1. *If $[u^{(n)}]$ is a sequence of elements in $\Phi(E)$ for which*

$$\operatorname*{limit}_{n \to \infty} J_\Phi[r\, u^{(n)}] = 0 \tag{8.10}$$

for every positive integer r, then $| u^{(n)} |_{\Phi(E)} \to 0$ as $n \to \infty$.

Let g be an element of unit norm in $\Psi^*(E)$. Then $J_\Psi(\dot g) \leqq 1$ by Theorem 7.1, and hence

$$J_\Psi\left(\frac{\dot g}{r}\right) \leqq \frac{1}{r} J_\Psi(\dot g) \leqq \frac{1}{r} \qquad [r = 1, 2, \cdots]. \tag{8.11}$$

Hence

$$\begin{aligned} \int_{R_1} | u^{(n)}(s) | \, | \dot g(s) | \, ds &= \int_{R_1} | r\, u^{(n)}(s) | \left| \frac{\dot g(s)}{r} \right| ds \\ &\leqq J_\Phi(r\, u^{(n)}) + J_\Psi\left(\frac{\dot g}{r}\right) \\ &\leqq J_\Phi(r\, u^{(n)}) + \frac{1}{r} \leqq \frac{2}{r}, \end{aligned} \tag{8.12}$$

provided n exceeds a sufficiently large integer $n(r)$ in accordance with hypothesis (8.10). Taking the sup of the left member of (8.12) over all g of unit norm in $\Psi^*(E)$,

$$| u^{(n)} |_{\Phi(E)} \leqq \frac{2}{r} \qquad [n > n(r)] . \tag{8.13}$$

Since r is an arbitrary positive integer the lemma follows.

The nth truncated image x_n of $x \in \Psi(E)$. Given $x \in \Psi(E)$ and a positive integer n, $x_n(s)$ will be defined as follows. Among values $s \in E$ for which $s \leqq n$, let ω_n be the maximum. If $\omega_n < \omega_E$, set $x_n(s) = 0$ on $[\omega_n, \infty)$. If $\omega_n = \omega_E$, set $x_n(s) = 0$ on (ω_E, ∞). At other points of R_1 set $x_n(s) = x(s)$ when $| x(s) | < n$, and set $x_n(s) = n \operatorname{sign} x(s)$ when $| x(s) | \geqq n$. One sees that $x_n(s) \to x(s)$ as $n \to \infty$, that $| x_n(s) |$ increases monotonically with n, and that

$$| x_n(s) | \leqq | x(s) |, \qquad | x(s) - x_n(s) | \leqq | x(s) | . \tag{8.14}$$

Finally x_n is in $\Phi(E)$ with x. In particular when $\omega_n < \omega_E$, $x_n(s)$ is constant with $x(s)$ on any subinterval of $(-\infty, \omega_n)$, and when $\omega_n = \omega_E$, $x_n(s)$ is constant with $x(s)$ on any subinterval of $(-\infty, \omega_E]$.

LEMMA 8.2. *If x is in $\Phi(E)$ and x_n is the nth truncated image of x, $|x - x_n|_{\Phi(E)} \to 0$ as $n \to \infty$.*

Lemma 8.2 will follow from Lemma 8.1 if for every positive integer r,

$$\lim_{n\to\infty} J_\Phi[r(x - x_n)] = 0. \tag{8.15}$$

Recall that the integrand $\Phi[rx]$ of $J_\Phi(rx)$ is summable and that

$$\Phi[r\{x(s) - x_n(s)\}] \leqq \Phi[rx(s)].$$

Since $x_n(s) \to x(s)$ p.p., a conventional theorem on L-integration implies that (8.15) holds. The lemma follows.

The following theorem shows that $\Phi(E)$ is a space satisfying Condition IV of §2.

THEOREM 8.1. *Given an $x \in \Phi(E)$, there exists a sequence of s-functions $\eta^{(n)}$ belonging to E such that $\eta^{(n)} \to x$ in $\Phi(E)$ and $\eta^{(n)}(s) \to x(s)$ p.p. in R_1.*

By virtue of Lemma 8.2 it is sufficient to prove the theorem when x is replaced by its mth truncated image x_m. Denote x_m by y. Recall that $y(s) = 0$ for $s > m$ and $s < 0$. (As in the proof of Lemma 4.2 the reasoning of the next paragraph is valid either when y is used or y replaced by x.)

As in the proof of Lemma 4.2, there exists a sequence of step-functions $\zeta^{(r)}$ $(r = 1, 2, \cdots)$ not belonging to E in general, such that $\zeta^{(r)}(s) \to y(s)$ p.p. as $r \to \infty$, and such that for each r,

$$J_\Phi[r(y - \zeta^{(r)})] < \frac{1}{r}. \tag{8.16}$$

As in the proof of Lemma 4.2, one modifies $\zeta^{(r)}$ so as to obtain a step-function $\eta^{(r)}$ with the above properties of $\zeta^{(r)}$ and in addition belonging to E. With $\eta^{(r)}$ so chosen,

$$J_\Phi[r(y - \eta^{(n)})] \leqq J_\Phi[n(y - \eta^{(n)})] < \frac{1}{n}, \qquad [\text{for } r < n]$$

so that for fixed r,

$$\lim_{n\to\infty} J_\Phi[r(y - \eta^{(n)})] = 0.$$

Theorem 8.1 now follows from Lemma 8.2 when x is replaced by x_m, and so follows for arbitrary $x \in \Phi(E)$.

NOTE. *If in the theorem $|x(s)| \leqq M$ for every s, it is clear that $\eta^{(n)}$ can be so chosen that $|\eta^{(n)}(s)| \leqq M$ for every s. If $x(s) = 0$ for $s > m > 0$, $\eta^{(n)}$ can be so chosen that $\eta^{(n)}(s) = 0$ for $s > m$.*

Theorem 8.2 is a consequence of Lemma 8.1. It will be used to establish the absolute continuity of a d-function g in §9.

THEOREM 8.2. *Let* $[u^{(n)}]$ *be a sequence of functions in* $\Phi(E)$ *which are respectively characteristic functions of subsets* $S^{(n)}$ *of* R_1 *for which* $m[S^{(n)}] \to 0$ *as* $n \to \infty$. *Then* $| u^{(n)} |_{\Phi(E)} \to 0$ *as* $n \to \infty$.

For each integer r,

$$J_\Phi(ru^{(n)}) = \int_{R_1} \Phi(ru^{(r)})\, ds = \Phi(r)m[S^{(n)}] \to 0$$

as $n \to \infty$. Lemma 8.1 implies the conclusion of the theorem.

The mixed Dini derivative $\hat{g}$ *of* g. The derivative $\dot{g}$ of a function $g \in \Psi^*(E)$ does not in general belong to E because $\dot{g}(s)$ does not in general exist at the end points of a maximal interval J of comp E, although constant on J itself. If, however, one introduces a mixed Dini derivative $\hat{g}$ of g such that

$$\hat{g}(s) = D_+[g(s)], \qquad [s \neq \omega_E]$$

$$\hat{g}(s) = D_-[g(s)], \qquad [s = \omega_E]$$

then $\hat{g}$ belongs to E, at least if it be assigned an arbitrary value, say 0, on the point set of null L-measure where it is still undefined. If g is in $\Psi^*(E)$, g is a.c. and $\hat{g}(s) = \dot{g}(s)$ p.p. Hence the integral of $\hat{g}$ which vanishes when $s = 0$ reduces to g; moreover

$$J_\Psi(\hat{g}) = J_\Psi(\dot{g}) = J_\Psi(| \dot{g} |).$$

9. Verification of Conditions I – V for $\Phi(E)$

We continue with the space $\Phi(E)$ of §8, recalling that Φ and Ψ are complementary functions. We shall consider the L-S-integral

$$f_g(x) = \int_{R_1} x(s)\, d\, g(s) = \int_{R_1} x(s)\dot{g}(s)\, ds \tag{9.1}$$

for $g \in \Psi^*(E)$ and $x \in \Phi(E)$. Recall that g is a.c., $J_\Psi(k\dot{g}) < \infty$ for *some* $k > 0$, and $J_\Phi(kx) < \infty$ for *every* $k > 0$. Young's inequality in the form,

$$\int_{R_1} | x(s) | \, | \dot{g}(s) | \, ds \leqq J_\Phi\left(\frac{x}{k}\right) + J_\Psi(k\dot{g}), \qquad [k > 0]$$

shows that the final integral in (9.1) exists. Hence the first integral in (9.1) exists. In studying the final integral in (9.1) it will be simpler if one can assume that $\dot{g}(s) \geqq 0$. To this end the following lemma is useful.

LEMMA 9.1. *For fixed* $g \in \Psi^*(E)$ *there is an equivalence of* $\Phi(E)$ *with* $\Phi(E)$ *in which* $x \in \Phi(E)$ *corresponds to* $u \in \Phi(E)$ *under a transformation*

$$u(s) = x(s) \operatorname{sign} \hat{g}(s), \qquad [\operatorname{sign} 0 = 1] \tag{9.2}$$

for which

$$\int_{R_1} x(s)\dot{g}(s)\, ds = \int_{R_1} u(s)\, \dot{\lambda}(s)\, ds, \tag{9.3}$$

where

$$(9.4)\qquad \lambda(s) = \int_0^s |\hat{g}(s)|\, ds,$$

λ is in $\Psi^*(E)$ with g, and $H_\Psi(\lambda) = H_\Psi(g)$.

That the correspondence (9.2) is one-to-one is seen on exhibiting the single-valued inverse

$$(9.5)\qquad x(s) = u(s) \operatorname{sign} \hat{g}(s).$$

One notes that the image u of $x \in \Phi(E)$ is in $\Phi(E)$, and conversely. This is a consequence of the properties of the mixed Dini derivative $\hat{g}$ of g. Moreover $|u(s)| = |x(s)|$, so that

$$(9.6)'\qquad |x|_{\Phi(E)} = |u|_{\Phi(E)},$$

and the transformation is an equivalence. The verification of (9.3) is immediate. One sees moreover that λ is in $\Psi^*(E)$ with g, and since $|\dot{\lambda}(s)| = |\dot{g}(s)|$ p.p. that

$$(9.6)''\qquad H_\Psi(\lambda) = H_\Psi(g).$$

Theorems 9.1 and 9.2 which follow are the *basic* representation theorems for elements in $\overline{\Phi(E)}$. Together they show that $\Phi(E)$ satisfies Condition I of §2.

THEOREM 9.1. (i). *For any $g \in \Psi^*(E)$ the functional f_g defined by* (9.1) *is in* $\overline{\Phi(E)}$ *with*

$$(9.7)\qquad \| f_g \| = H_\Psi(g).$$

(ii). *If $f_g(x) = 0$ for every $x \in \Phi(E)$, $g(s) \equiv 0$ for every $s \in R_1$.*

Without loss of generality we can suppose that $\dot{g}(s) \geqq 0$ p.p. For the truth of the theorem with g replaced by λ and f_g by f_λ as in Lemma 9.1 implies its truth for g. For simplicity of notation set $c = H_\Psi(g)$. That f_g is additive and homogeneous over $\Phi(E)$ follows immediately from the form of the integral (9.1).

Proof that $\| f_g \| \leqq c$. If $c = 0$, $g(s) \equiv 0$ and hence f_g is linear in the sense of Banach, with $\| f_g \| = 0$. In this case (9.7) is trivial.

Suppose that $c > 0$. It then follows from the homogeneity of $H_\Psi(g)$ that

$$(9.8)\qquad H_\Psi\left(\frac{g}{c}\right) = \frac{1}{c} H_\Psi(g) = 1.$$

With (9.8) holding it follows from the definition of the pseudo-norm of $x \in \Phi(E)$ that

$$|x|_{\Phi(E)} \geqq \left| \int_{R_1} x(s)\, d\left(\frac{g(s)}{c}\right) \right|,$$

so that

$$\left| \int_{R_1} x(s)\, dg(s) \right| \leqq c\, |x|_{\Phi(E)}.$$

Hence f_g is linear over $\Phi(E)$ and $\| f_g \| \leqq c$. It remains to show that $\| f_g \| \geqq c$, assuming that $c > 0$ and $\dot{g}(s) \geqq 0$ p.p.

Proof that $\| f_g \| \geqq c(c > 0)$. We begin with the special case in which $\dot{g}(s)$ is bounded over R_1, and vanishes for $s > n$ an integer. To show that $\| f_g \| \geqq c$ in this special case, use will be made of Young's equality

$$ab = \Phi(a) + \Psi(b) \tag{9.9}$$

holding when $a = \psi(b)$. Suppose $\hat{g}$ defined as at the end of §8. Set $x(s) = \psi[\hat{g}(s)/c]$. Then $x(s)$ is bounded over R_1, vanishes for $s > n$ and $s < 0$, and x is in $\Phi(E)$. In this special case $k_g = \infty$ (Cf. §7) so that

$$J_\Psi\left[\frac{\hat{g}}{c}\right] = J_\Psi\left[\frac{\dot{g}}{c}\right] = I_\Psi\left[\frac{g}{c}\right] = 1 \qquad \text{[Cf. (7.5)].} \tag{9.10}$$

We have

$$\begin{aligned}
\int_{R_1} x(s)\dot{g}(s)\,ds &= c\int_{R_1} x(s)\,\frac{\dot{g}(s)}{c}\,ds \\
&= c\left[J_\Phi(x) + J_\Psi\left(\frac{\dot{g}}{c}\right)\right] && \text{[by (9.9)]} \\
&= c\,[J_\Phi(x) + 1] && \text{[by (9.10)]} \\
&\geqq c \mid x \mid_{\Phi(E)} && \text{[by (8.4)].}
\end{aligned}$$

Hence for our special g, $\| f_g \| \geqq c$.

To treat the general case let γ_n be the nth truncated image of $\hat{g}$, and set

$$g_n(s) = \int_0^s \gamma_n(t)\,dt.$$

Recall that $\gamma_n(s)$ converges pointwise to $\hat{g}$ and that $0 \leqq \gamma_n(s) \leqq \hat{g}(s)$. It follows that g_n converges pointwise to g. Set $H_\Psi(g_n) = c_n$.

By the lower semi-continuity of H_Ψ we see that $\lim\inf c_n \geqq c > 0$. Without loss of generality we can then suppose that each $c_n > 0$. According to the result of the preceding paragraph, $\| f_{g_n} \| \geqq c_n$.

Proof that $\| f_g \| = c$. Note that

$$\| f_g \| = \sup_x \int_{R_1} x(s)\dot{g}(s)\,ds = \sup_x \int_{R_1} \mid x(s) \mid \dot{g}(s)\,ds,$$

taken over all unit $x \in \Phi(E)$. This relation is valid since $\dot{g}(s) \geqq 0$ p.p., and since $\mid x(s) \mid$ defines a unit element in $\Phi(E)$. Moreover

$$\| f_{g_n} \| = \sup_x \int_{R_1} \mid x(s) \mid \dot{g}_n(s)\,ds \geqq 0 \qquad [\mid x \mid_{\Phi(E)} = 1].$$

Since

$$\dot{g}(s) = \hat{g}(s) \geqq \gamma_n(s) = \dot{g}_n(s) \geqq 0, \qquad \text{[p.p.]} \tag{9.11}$$

we can conclude that $\| f_g \| \geqq \| f_{g_n} \|$ so that

$$\| f_g \| \geqq \| f_{g_n} \| = c_n = H_\Psi(g_n).$$

From the lower semi-continuity of H_Ψ we infer that

$$\| f_0 \| \geqq H_\Psi(g) = c.$$

It has already been shown that $\| f_0 \| \leqq c$. We conclude that $\| f_0 \| = c$.

PROOF OF (ii). If $f_0(x) = 0$ for every $x \in \Phi(E)$, $\| f_0 \| = 0$. Hence $H_\Psi(g) = 0$ by (9.7) and $g \equiv 0$ by definition of $H_\Psi(g)$. This completes the proof of Theorem 9.1.

In the proof of Theorem 9.2 a modification of Lemma 9.1 will be useful in the same way in which Lemma 9.1 was useful in proving Theorem 9.1. In this modification $\Phi(E)$ will be replaced by a vector subspace $\Phi_0(E)$ and $\Psi^*(E)$ by a vector space $A(E)$ containing $\Psi^*(E)$. Let $\Phi_0(E)$ be the linear subspace of those functionals $x \in \Phi(E)$ for which $| x(s) | < M$ for some constant M and every $s \in R_1$, and for which $x(s) = 0$ for $s > n$ for some positive integer n. The constant M and integer n are permitted to vary with x. Let $A(E)$ be the vector space of functionals g finitely a.c. and based on E. The proof of the following lemma is immediate.

LEMMA 9.2. *For fixed $g \in A(E)$ there is an equivalence of $\Phi_0(E)$ with $\Phi_0(E)$ in which $x \in \Phi_0(E)$ corresponds to $u \in \Phi_0(E)$ under the transformation* (9.2) *where* (9.3) *holds subject to* (9.2) *and* (9.4), *while g is in $\Psi^*(E)$ with λ and $H_\Psi(\lambda) = H_\Psi(g)$.*

THEOREM 9.2. *Corresponding to any $f \in \overline{\Phi(E)}$ there exists a unique $g \in \Psi^*(E)$ such that for each $x \in \Phi(E)$,*

$$f(x) = \int_{R_1} x(s)\, d\, g(s). \tag{9.12}$$

Let g be the function based on E such that for each point $t \in E$,

$$g(t) = f(z_E^t) \qquad \text{[Cf. (3.7)].} \tag{9.13}$$

If η is any s-function belonging to E it follows from (9.13) that

$$f(\eta) = \eta_r \Delta_r(g) \qquad [r = 1, \cdots, r(\pi)].$$

Since f is linear by hypothesis,

$$| \eta_r \Delta_r(g) | \leqq | \eta |_{\Phi(E)} \| f \|. \tag{9.14}$$

We continue by proving (a), (b), (c).

(a) *The function g is finitely* a.c.

In (9.14) set $\eta_r = 1$ or 0, and let B be the subset of R_1 on which $\eta(s) = 1$. Then (9.14) takes the form

$$\left| \sum{}' \Delta_r(g) \right| \leqq | \eta |_{\Phi(E)} \| f \|, \tag{9.15}$$

where the summation $\sum'$ is over those values of r for which $\eta_r = 1$. Since η is the characteristic function of B, it follows from Theorem 8.2 that $| \eta |_{\Phi(E)}$ in (9.15) tends to zero as $m(B) \to 0$. Thus $\sum' | \Delta_r g | \to 0$ as $m(B) \to 0$ when the end points of the intervals are in E. A *general* finite set of subintervals is (after finite partition) the sum of a set of intervals with end points in E and a set of intervals in comp E. These intervals are assumed in a fixed finite interval I. The reader can supply details from this point on.

(b) *If y is in $\Phi(E)$, $|y(s)| \leqq M$ for $s \in R_1$ and $y(s) = 0$ for $s > m > 0$, then*

$$f(y) = \int_{R_1} y(s)\, dg(s). \tag{9.16}$$

In accordance with Theorem 8.1 and Note there exists a sequence $\eta^{(n)}$ of s-functions in $\Phi(E)$ such that $\eta^{(n)} \to y$ in $\Phi(E)$, $\eta^{(n)}(s) \to y(s)$ p.p. while $|\eta^{(n)}(s)| \leqq M$ for $s \in R_1$, and $\eta^{(n)}(s) = 0$ for $s > m$. The linearity of f over $\Phi(E)$ implies that $f(\eta^{(n)}) \to f(y)$. But

$$f(\eta^{(n)}) = \eta_r^{(n)} \Delta_r(g) = \int_{R_1} \eta^{(n)}(s)\, dg(s) \to \int_{R_1} y(s)\, dg(s),$$

using the L-theorem on bounded convergence on a bounded interval. Hence the two members of (9.16) are limits of $f(\eta^{(n)})$ as $n \to \infty$, and so are equal.

(c) *The function g is in $\Psi^*(E)$.* It is known that g is in $A(E)$, and that for any $y \in \Phi_0(E)$, (9.16) holds. That g is in $\Psi^*(E)$ will be established using no other conditions on g. It follows from Lemma 9.2 that one can suppose that $\dot{g}(s) \geqq 0$ p.p. $s \in R_1$. Otherwise one could make the transformations (9.2) and (9.4) of $y \in \Phi_0(E)$ and $g \in A(E)$ respectively, and establish (c) for the transform λ of g. From Lemma 9.2 it would follow that g is in $\Psi^*(E)$ with λ.

Suppose then that $\dot{g}(s) \geqq 0$. To show that g is in $\Psi^*(E)$ it is sufficient to show that $J_\Psi(k\dot{g}) < \infty$ for some $k > 0$. (Cf. Theorem 6.1.) For that purpose Young's equality will be useful. We pass over the trivial case $\|f\| = 0$.

Let α_n be the nth truncated image of $\hat{g}/\|f\|$, and set $u_n = \psi(\alpha_n)$. Then α_n and u_n are bounded and measurable, with $\alpha_n(s) = u_n(s) = 0$ when $s < 0$ and $s > n$. Since $\hat{g}$ belongs to E, α_n and hence u_n belong to E, and hence u_n is in $\Phi_0(E)$. By Young's equality,

$$\begin{aligned} J_\Phi(u_n) + J_\Psi(\alpha_n) &= \int_{R_1} u_n(s)\, \alpha_n(s)\, ds \\ &\leqq \int_{R_1} u_n(s) \frac{\hat{g}(s)}{\|f\|}\, ds = \frac{f(u_n)}{\|f\|} && \text{[by (b)]} \\ &\leqq |u_n|_{\Phi(E)} \leqq J_\Phi(u_n) + 1 && \text{[by (8.4)].} \end{aligned}$$

Hence $J_\Psi(\alpha_n) \leqq 1$. But

$$J_\Psi\left[\frac{\dot{g}}{\|f\|}\right] = J_\Psi\left[\frac{\hat{g}}{\|f\|}\right] = \operatorname*{limit}_{n\to\infty} J_\Psi(\alpha_n) \leqq 1.$$

Hence $I_\Psi(g/\|f\|) \leqq 1$ so that g is in $\Psi^*(E)$ by definition of $\Psi^*(E)$. Statement (c) follows.

FINAL PROOF OF THE THEOREM. Let x_n be the nth truncated image of $x \in \Phi(E)$. Then

$$f(x_n) = \int_{R_1} x_n(s)\, dg(s)$$

by (b). Moreover $x_n(s) \to x(s)$ p.p. while $0 \leqq |x_n(s)| \leqq |x(s)|$, while $x\dot{g}$ is summable, since x is in $\Phi(E)$ by hypothesis, and g in $\Psi^*(E)$ by (c.) By a conventional theorem on L-integration, as $n \to \infty$,

$$\int_{R_1} x_n(s)\dot{g}(s)\,ds \to \int_{R_1} x(s)\dot{g}(s)\,ds = \int_{R_1} x(s)\,dg(s).$$

On the other hand $x_n \to x$ in $\Phi(E)$ by Lemma 8.2, so that $f(x_n) \to f(x)$ as $n \to \infty$, since f is linear over $\Phi(E)$. Hence

$$f(x) = \int_{R_1} x(s)\,dg(s). \tag{9.17}$$

In fact both members of (9.17) are the limit of $f(x_n)$ as $n \to \infty$.

This completes the proof of the theorem, apart from a proof of the uniqueness of g. The uniqueness of a $g \in \Psi^*(E)$ such that (9.17) holds follows from Theorem 9.1 (ii).

The principal theorem concerning $\Phi(E)$ follows.

THEOREM 9.3. *The space* $A = \Phi(E)$ *satisfies Conditions* I *to* V *with* $A_d = \Psi^*(E)$.

Condition I is satisfied by virtue of the two preceding theorems.

Condition II is satisfied in the form

$$\left| \int_{R_1} \eta(s)\,dg(s) \right| \leqq [\Phi(1)\alpha + 1] H_\Psi(g) \max_s |\eta(s)|, \qquad [\alpha \geqq 0] \tag{9.18}$$

where $\eta(s) = 0$ for $s > \alpha$ and g is in $\Psi^*(E)$. To see this recall that

$$\left| \int_{R_1} \eta(s)\,dg(s) \right| \leqq |\eta|_{\Phi(E)} H_\Psi(g). \tag{9.19}$$

Let ζ be the characteristic function of the point set on which $|\eta(s)| > 0$. Set $k = |\max \eta(s)|$. From (8.7) and the homogeneity of norms in $\Phi(E)$,

$$\begin{aligned} |\eta|_{\Phi(E)} &\leqq |k\zeta|_{\Phi(E)} = k|\zeta|_{\Phi(E)} \\ &\leqq k[J_\Phi(\zeta) + 1] \leqq k[\alpha\Phi(1) + 1], \end{aligned} \tag{9.20}$$

using (8.4). Relation (9.18) follows from (9.19) with the aid of (9.20).

Condition IV is satisfied in accordance with Theorem 8.1.

Condition V is trivial since $A = A_e$.

Condition III is a consequence of the other conditions when $A = A_e$, as implied by Theorem 2.1.

10. Relation to Orlicz representation theory.

The space L_Φ^* considered by Orlicz is a space of measurable functions x belonging (in our sense) to a finite interval $[0, a]$ with $J_\Phi(kx)$ finite for some $k > 0$. To obtain a representation of linear functionals over L_Φ^* Orlicz adds the condition that for large u,

$$\Phi(2u) < C\Phi(u) \tag{10.1}$$

for C a constant independent of u. With the condition (10.1) added it is easily shown that if x is in L_Φ^*, $J_\Phi(kx) < \infty$ for every k. Thus the space finally considered by Orlicz is our space $\Phi(E)$ specialized by the condition (10.1) and setting $E = [0, a]$. Orlicz has also noted the possibility of dropping the requirement that φ be strictly monotone.

Our representation theory goes beyond the Orlicz theory in the following senses: cf. Zygmund p. 97

(1) No requirement (10.1) is imposed. Our spaces $\Phi(E)$ admit all measurable x belonging to E for which $J_\Phi(kx) < \infty$ for every k. With Φ chosen as in Example 7.2 it is easy to show that the resulting space $\Phi(E)$ is not found among the spaces admitted by Orlicz when $E = [0, a]$.

(2) The admission of arbitrary closed sets E restricted as in (1) goes beyond the Orlicz case of an interval $[0, a]$.

(3) The use of the generalized Hellinger integral to define the space $\Psi^*(E)$ and our definition of d-functions based on E permit us to show that $A = \Phi(E)$ with $A_d = \Psi^*(E)$ satisfies Conditions I to V.

THE INSTITUTE FOR ADVANCED STUDY

BIBLIOGRAPHY

S. BANACH, Théorie des opérations linéaires, Warsaw, 1932.

Z. W. BIRNBAUM and W. ORLICZ, *Über die Verallgemeinerung des Begriffes der zueinander konjugierten Potenzen*, Studia Math. 3 (1931), 1–67.

M. FRÉCHET, *Sur les fonctionnelles bilinéaires*, Trans. Amer. Math. Soc. 16 (1915), 215–234.

E. J. MCSHANE, Integration, Princeton Universit Press, 1944.

M. MORSE and W. TRANSUE, *Functionals of bounded Fréchet variation*, Canadian Jour. Math. 1 (1949), 153–165.

M. MORSE and W. TRANSUE, *Functionals F bilinear over the product A × B of two pseudo-normed vector spaces*, I. *The representation of F*, Ann. of Math. 50 (1949), 777–815.

M. MORSE and W. TRANSUE, *Integral representations of bilinear functionals*, Proc. Nat. Acad. Sci. 35 (1949), 136–143.

W. ORLICZ, *Über eine gewisse Klasse von Räumen vom Typus B*, Bull. Internat. Acad. Polonaise, 1932, 207–220.

F. RIESZ, *Sur les opérations fonctionnelles linéaires*, C. R. Acad. Sci. 149 (1909), 974–977.

H. STEINHAUS, *Additive und stetige Funktionaloperationen*, Math. Zeit. 5 (1913), 186–221.

W. H. YOUNG, *On successions with subsequences converging to an integral*. Proc. London Math. Soc. 24 (1926), pp. 1–20.

A. ZYGMUND, *Trigonometrical series*, Warsaw, 1935, Chapter IV.

Reprinted for private circulation from
BULLETIN OF THE ATOMIC SCIENTISTS
Vol. XV, No. 2, February 1959

Mathematics and the Arts

MARSTON MORSE

TO TALK about art other than in the impersonal sense of history, is to talk about the moments when one has been confronted with beauty. Every essay on art that lights a hidden niche has its source in the life of the writer. You will then perhaps understand why I start with the mood of my childhood.

One hundred miles northeast of Derry, New Hampshire, lie the Belgrade Lakes, and out of the last and longest of these lakes flows the Messalonskee. I was born in its valley, "north of Boston" in the land of Robert Frost. The "Thawing Wind" was there, the "Snow," the "Birches," and the "Wall" that had to be mended: I was born on a sprawling farm cut by a pattern of brooks that went nowhere—and then somewhere. A hundred acres of triangles of timothy and clover, and twisted quadrilaterals of golden wire grass, good to look at, and good riddance. At ten I combed it all with horse and rake, while watching the traffic of mice beneath the horse's feet.

All that Frost has described was there—the meanness and generosity of men and women. A neighbor's house burned down in a wind, and everyone knew who held the grudge. The woman who must have killed her lover (so everyone thought) stood up in prayer meeting and testified, and there was no more judgment against her than was proper. The autumn winds were the prelude to the loon's strange song. There was time to think in the winter, to like some things better than others.

My mother's world was the world of music, and her world became mine. At thirteen I was playing the organ in church and wished the time in summer to study and practice. Somewhat reluctantly my father conceded me the mornings. He said that the grass was too wet to rake in the morning, and that I could walk the three miles from church to farm. And so I learned some of the Bach Fugues for the organ, and the moving Sonatas which Mendelssohn had written to honor the memory of Bach.

Marston Morse is professor of mathematics, Institute for Advanced Study, Princeton University.

Grecian art first became real to me in the shop of an old cabinet maker. I began to learn from him about cabinet making, and the history of his art. Sheraton chairs and tables were scattered about his shop, with their fluted columns and acanthus leaves. It came to me, all of a sudden, that *these* were fragments of Grecian temples.

Mathematics the Sister of the Arts

There was a copy of the *Cabinet Maker of Sheraton* in a remote library. It started with descriptive geometry and continued with a theory of ornaments. Cornices were constructed with ruler and compass; symmetry and perfection reigned throughout. Here was a meeting of mathematics and art, something final and universal, as it seemed to me then. It was very alive, because it was so new. But it was not mathematics as I know it today, and as it should be known; it was matter without the spirit. I made the same mistake that artists have made since the time of the Greeks, and placed mathematics alongside of the arts as their *handmaiden*. It is a humble and honorable position and very necessary; for one must begin with exactness in all the arts. But mathematics is the *sister*, as well as the *servant* of the arts and is touched with the same madness and genius. This must be known.

There was a German painter and engraver born in the fifteenth century with the name of Albrecht Dürer who wanted mathematics to be more than a handmaiden of art. His discontent on this account was unique among artists of all time. More completely than any other artist he formulated the rules of symmetry, perspective, and proportion, and used them in his art. But any one who thinks Dürer's spirit is bound by rules is mistaken. There is almost a shock in passing from his rugged, first engravings to the radiant classical beauty and slender proportions of his *Adam and Eve* of 1507.

Dürer was a creative mathematician as well as an artist. He wanted his geometric theories to measure up to his art. His great engraving *Melencolia I* is a psychological self-portrait. The perplexed and thoughtful heroine is the figure of geometry. Everything I have to say today is hidden in this engraving or may be derived from it by projection into the future. Let me quote from my colleague Erwin Panofsky.

The engraving *Melencolia I*, he says, "... typifies the artist of the Renaissance who respects practical skill, but longs all the more fervently for mathematical theory—who feels 'inspired' by celestial influences and eternal ideas, but suffers all the more deeply from his human frailty and intellectual finiteness ... Dürer was an artist-geometer, and one who suffered from the very limitations of the discipline he loved. In his younger days, when he prepared the engraving *Adam and Eve*, he had hoped to capture absolute beauty by means of a ruler and a compass. Shortly before he composed the *Melencolia I* he was forced to admit: 'But what absolute beauty is, I know not. Nobody knows it except God.'"

In his dependence upon geometry Dürer was inspired by Leonardo but repudiated by Michelangelo. Later artists followed Dürer only half way or not at all; it is indeed hard to follow an inspiration. Leonardo himself had little of Dürer's divine discontent.

Back of Dürer and Leonardo in the distant past stands the Roman architect and geometer Vitruvius. The Mesopotamian artists also looked on geometry as an aid to art, and this was well known to the prophet

Courtesy of The Art Institute of Chicago

Melencolia I (1514) ALBRECHT DÜRER

Isaiah. Chapter 44 of Isaiah is written against idolatry; it is also an essay on aesthetics. The thirteenth verse reads: "The carpenter stretcheth out his rule; he marketh it out with a line; he fitteth it with planes, and he marketh it out with the compass, and maketh it after the figure of a man, according to the beauty of a man; that it may remain in the house."

Isaiah would minimize geometry in the arts, Dürer would maximize it. Neither Isaiah nor Dürer was content.

Let us turn to the relation between mathematics and music. The evolution of the scales, from the archaic sequences of tones of Euripedes to the whole tone scale of Debussy, shows that mathematics and music have much in common. And there is also the arithmetical basis for harmony. It is not too difficult to compose in the technical scheme of Debussy and thereby to get some of his naturalistic effects, but no one can explain the profound difference between the opera, *Péléas et Mélisande*, on the one hand, and *Tristan and Isolde*, on the other, by reference to whole tone scales or any other part of musical theory.

Geometric form imposed on music can have a null effect. As an example, I shall compare the First Prelude of Bach, as found in the *Well-Tempered Clavichord*, with the First Prelude of Chopin. The First Prelude of Bach is without melody, and consists of repeating ascending arpeggios with similar form and length. It is intended that the effect shall be harp-like. The musical text as a whole exhibits a design that appears in no one of the other forty-eight preludes. Looked at geometrically, the First Prelude of Chopin has a very similar geometric design, and if the Chopin prelude is played an octave higher than written, with perfect evenness of tone and tempo, the actual musical similarity of the two preludes is most striking. If, however, the Chopin prelude is played with the color and pulsating rhythm which it demands, all similarity to the Bach disappears.

Most convincing to me of the spiritual relations between mathematics and music, is my own very personal experience. Composing a little in an amateurish way, I get exactly the same elevation from a prelude that has come to me at the piano, as I do from a new idea that has come to me in mathematics.

Nature of Affinity

My thesis is prepared. It is that *the basic affinity between mathematics and the arts is psychological and spiritual and not metrical or geometrical.*

The first essential bond between mathematics and the arts is found in the fact that discovery in mathematics is not a matter of logic. It is rather the result of mysterious powers which no one understands, and in which the unconscious recognition of beauty must play an important part. Out of an infinity of designs a mathematician chooses one pattern for beauty's sake, and pulls it down to earth, no one knows how. Afterwards the logic of words and of forms sets the pattern right. Only then can one tell someone else. The first pattern remains in the shadows of the mind.

All this is like Robert Frost's "figure a poem makes." The poet writes: "I tell how there may be a better wildness of logic, than of inconsequence. But the logic is backward, in retrospect after the act. It must be more felt, than seen ahead like prophecy." Or again, "For me the initial delight is in the surprise of remembering something I didn't know I knew. I am in a place, in a situation, as if I had materialized from cloud, or risen out of the ground."

Compare this with the account of how the French mathematician Henri Poincaré came to make one of his greatest discoveries. While on a geologic excursion a mathematical idea came to him. As he says it came "without anything in my former thoughts seeming to have paved the way." He did not then have the time to follow up this idea. On returning from his geologic excursion he sought to verify the idea. He had no immediate success, and turned to certain other questions which interested him, and which seemed at the time to have no connection with the idea which he wished to verify. Here again he was unsuccessful. Disgusted with his failure he spent a few days at the seaside and thought of something else. One morning while walking on the bluff the final solution came to him with the same characteristics of brevity and suddenness as he had experienced on sensing the initial idea, and quite remarkably he had a sense of complete certainty. He made his great discovery.

An account of Gauss is similar. He tells how he came to establish a theorem which had baffled him for two years. Gauss writes: "Finally, two days ago, I succeeded, not on account of my painful efforts, but by the grace of God. Like a sudden flash of lightning, the riddle happened to be solved. I myself cannot say, what was the conducting thread, which connected what I previously knew, with what made my success possible."

These words of Frost, Poincaré, and Gauss show how much artists are in agreement as to the psychology of creation.

A second affinity between mathematicians and other artists lies in a psychological necessity under which both labor. Artists are distinguished from their fellows who are not artists by their overriding instinct of self-preservation as creators of art. This is not an economic urge as everyone knows who has a variety of artist friends. I shall illustrate this by the case of Johann Sebastian Bach and his son Philipp Emanuel.

Johann Sebastian's work culminates and closes a religious and musical epoch. It is inconceivable that Philipp Emanuel could have continued as a composer in the same sense as his father and have *lived* as an artist. He did in fact reject his father's musical canons. There is considerable evidence that his environment called for a new musical spirit. History justifies Philipp Emanuel; Mozart said of him, "He is the father, we are the children"; Haydn was inspired by him and Beethoven admired him. With all this to his credit, posterity can perhaps forgive him for calling his father an old wig.

Quite analogous to the son's turning away from his father is the story of the relation of mathematician Henri Poincaré to his younger colleague, Lebesgue. Poincaré had used the materials of the nineteenth-century mathematics to revolutionize much of mathematics. He had gone so far in mathematics that it is doubtful whether his younger colleagues in France could go on in the same sense without introducing essentially new techniques. This was in fact what several of them did. One of the new fields was what is called "set theory," and one of the innovators Lebesgue.

Poincaré criticized the members of the new school rather severely. It is on record that at a Congress in Rome he made this prediction. "Later generations will regard set theory as a malady from which one has recovered." (One may remark parenthetically that the history of art records many maladies from which art has recovered.)

The response of Lebesgue to Poincaré was given on his elevation to a Professorship at the Collège de France. An older eminent colleague had praised the school of Lebesgue. Lebesgue made public reference to the "precious encouragement which had largely compensated for the reproaches" which his school had had to suffer.

I regard the reactions of both Poincaré and Lebesgue as dictated by instincts of self-preservation, typical of the artist. Such self-preservation was clearly to the advantage of mathematics as well. I am also one of the few mathematicians who think that Poincaré as well as Lebesgue was right, in that mathematics will return more completely to the great ideas of Poincaré with full appreciation of the innovations of Lebesgue, but with a truer understanding of the relation of mathematical technique to mathematical art.

Before coming to the third type of evidence of the affinity of mathematics with the rest of the arts it might be well to ask what is it that a mathematician wants as an artist. I believe that he wishes merely to *understand* and to *create*. He wishes to understand, simply, if possible—but in any case to understand; and to create, beautifully, if possible—but in any case to create. The urge to understand is the urge to embrace the world as a unit, to be a man of integrity in the Latin meaning of the word. A world which values great works of art, music, poetry, or mathematics, can only approve and honor the urge of any man capable of such activities, to create.

The third type of evidence of the affinity of mathematics with the arts is found in the comparative history of the arts. The history of the arts is the history of recurring cycles and sharp antitheses. These antitheses set pure art against mixed art, restraint against lack of restraint, the transient against the permanent, the abstract against the nonabstract. These antitheses are found in all of the arts, including mathematics.

In particular the antithesis of pure art and mixed art is very much in evidence in the relations between poetry and music. There have been those who wished to keep poetry and music separate at all times. Plato took sides when he said, "Poetry is the Lord of the Lyre," and music had to fight a long battle to obtain complete autonomy.

Quite analogously in mathematics there are those who would like to keep algebra and geometry apart, or would like to subordinate one to the other. The battle became acute when the discovery of analytic geometry by Descartes made it finally possible to represent all geometry by algebra. The battle between algebra and geometry has been waged from antiquity to the present.

Grecian art was of course restrained and a departure from restraint has always brought a reaction. Berlioz gave an example of extreme lack of restraint. To get the maximum effect of Doomsday trumpets in *The Last Day of the World*, Berlioz devised four full-fledged brass bands to play high in the four corners of St. Peters. One American composer even wanted to fire cannon on the beat.

Mathematicians too, are often unrestrained. In this direction are the grandiose cosmologies with more generality than reality. These fantasies are sometimes based neither on nature or logic. Mathematicians of today are perhaps too exuberant in their desire to build new logical foundations for everything. Forever the foundation and never the cathedral. Logic is now so well understood that the laying of foundations is not very difficult. The thing has gone so far that one of my Polish colleagues recently suggested that the right to lay foundations should be rationed, or put on the basis of the right to build one foundation for every genuine classical effort.

The antithesis between logic and intuition manifested itself in the days of the Greeks. Pythagoras had a mystical preference for whole numbers. The irrational numbers were not understood by the Greeks and hence avoided as much as possible. History has made a full turn and the nineteenth century saw the meteoric rise of a more sophisticated Pythagoras by the name of Kronecker. Kronecker laid down the rule "all results of mathematical analysis must ultimately be expressible in properties of integers."

This proclamation cut deeply into the life and work of Kronecker's colleague Weierstrass. Here are a few lines from Weierstrass's reproach.

> But the worst of it is that Kronecker uses his authority to proclaim, that all those who up to now have labored to establish the theory of functions, are sinners before the Lord—truly it is sad, and it fills me with a bitter grief, to see a man, whose glory is without flaw, let himself be driven by the well-justified feeling of his own worth, to utterances whose injurious effect upon others he seems not to perceive.

The human documents which I have put before you are not concerned with processes which a machine can duplicate. One cannot decide between Kronecker and Weierstrass by a calculation. Were that the case, many of us would turn to another and truer art. As Dürer knew full well, there is a center and final substance in mathematics whose perfect beauty is rational, but rational "in retrospect." The discovery which comes before, those rare moments which elevate man, and the searchings of the heart which come after are not rational. They are gropings filled with wonder and sometimes sorrow.

Often, as I listen to students as they discuss art and science, I am startled to see that the "science" they speak of and the world of science in which I live are different things. The science that they speak of is the science of cold newsprint, the crater-marked logical core, the page that dares not be wrong, the monstrosity of machines, grotesque deifications of men who have dropped God, the small pieces of temples whose plans have been lost and are not desired, bids for power by the bribe of power secretly held and not understood. It is science without its penumbra or its radiance, science after birth, without intimation of immortality.

The creative scientist lives in "the wildness of logic" where reason is the handmaiden and not the master. I shun all monuments that are coldly legible. I prefer the world where the images turn their faces in every direction, like the masks of Picasso. It is the hour before the break of day when science turns in the womb, and, waiting, I am sorry that there is between us no sign and no language except by mirrors of necessity. I am grateful for the poets who suspect the twilight zone.

The more I study the interrelations of the arts the more I am convinced that every man is in part an artist. Certainly as an artist he shapes his own life, and moves and touches other lives. I believe that it is only as an artist that man knows reality. *Reality is what he loves, and if his love is lost it is his sorrow.*

References

Erwin Panofsky. *Albrecht Dürer*. Vol. I, pp. 157–71. Princeton University Press.

Erwin Panofsky. *The Codex Huygens and Leonardo da Vinci's Art Theory*. See page 107. London: The Warburg Institute.

Curt Sachs. *The Commonwealth of Art*. See page 244 for Berlioz. New York: W. W. Norton and Co., Inc.

Jacques Hadamard. *The Psychology of Invention in the Mathematical Field*. Princeton University Press.

E. T. Bell. *Men of Mathematics*. See chapter on Kronecker. New York: Simon and Schuster.

Author's Note

This essay was read at a conference in honor of Robert Frost in 1950 at Kenyon College. The subject of the Conference was "The Poet and Reality" and there were four speakers besides myself. Among these was L. A. Strong, a distinguished Irish poet and novelist. I single out this poet because it was certain sentences of his which caused me to rewrite the concluding paragraphs of my essay. Strong read the lines of one of his poems with great feeling and beauty of diction. When he had finished, he had occasion to refer to science, and I have no doubt he had in mind the opening of the atomic era. Placing his hand on his forehead he said, "Science is here," and then placing his hand upon his heart, he continued, "And poetry is here." The poet's words and gestures were pregnant with misunderstanding. Each scientist will have his own interpretation of what the speaker meant. I felt moved to rewrite much of what I had written, in the interest of science, and of poetry.

It is with some misgivings that I conceive of an essay originally intended for an audience of poets as now presented to an audience primarily of physicists. My doubts are somewhat lessened by a belief that many physicists, like mathematicians, are guided in the discovery and shaping of their theories by their sense of harmony and of beauty. I know, however, that there are some mathematicians and presumably some physicists, who do not feel any such guidance or at least do not regard the influence of the aesthetic in their groping for scientific law as important.

I do not understand such reluctance to admit the extra-empirical mental processes, particularly when one approaches such problems as those of particle physics, or of a reformulation of quantum mechanics. One cannot believe that the mathematical forms chosen to represent experimental fact are always uniquely determined by the empirical data. I shall illustrate this by referring to the bases of the general relativity theory. It is clear that this theory is much more flexible and more likely than the Newtonian theory, and that it is consistent with important experimental evidence. But admitting this, the form of the general relativity is not thereby uniquely determined. There is in the background of general relativity theory a tacit assumption that the paths of light which it is desired to represent correspond to a solution of the inverse problem of the calculus of variations, a problem which in general is known to have no solution, and which, with even less mathematical generality, has the geodesic solution which Einstein presupposes. The solution which Einstein has accepted may be a very good approximation to the observed physical universe for bounded space-time but at the same time may be infinitely in error when applied to a region of space-time which is unbounded; that is when applied to a cosmology which is not known to correspond to a closed and bounded universe.

There is here a tension between the aesthetically simple, and the mathematically general (sometimes less simple). One has to learn how properly to resolve this tension. But first of all one must admit the uncertainty and seek to understand it. To the extent to which there is a multiplicity of mathematical forms *a priori* available to express an empirically anchored physical law, to that extent one must call on further experimental evidence, or logic, or aesthetic judgment.

At stake is not only truth, but freedom. Such freedom of choice as exists must be acknowledged and comprehended, or else it is lost. In a mileu in which freedom of hypothesis is well understood the likelihood of intuitive discovery of high order will certainly be increased. The satisfactions of the physicist and the artist may be combined.—M. M.

November 20, 1958

"The roads by which men arrive at their insights into celestial matters seem to me almost as worthy of wonder as those matters themselves."

—Johannes Kepler, tr. Arthur Koestler
(*Encounter*, December 1958)

EXTRAIT DE »FUNDAMENTA MATHEMATICAE« T. XXXIX (1952).

The Existence of Pseudoconjugates on Riemann Surfaces.

By

M. Morse and J. Jenkins.

§ 1. Introduction. Among the characteristics of a function U which is harmonic on a Riemann surface G^* are the topological interrelations of the level lines of U. One has merely to look at the level lines of $\mathcal{R}z$, $\mathcal{R}e^z$, $\mathcal{R}\log z$, etc. to sense both complexity and order. The dual level lines of a conjugate V of U add to this order and complexity. It seems likely that outstanding problems in Riemann surface theory, such as the type problem, the nature of essential singularities, the existence of functions on the Riemann surface with restricted properties cannot be thoroughly understood in the absence of a complete analysis of the topological characteristics of these level lines.

Such a topological study properly belongs to a somewhat larger study namely that of PH (pseudoharmonic) functions and their pseudoconjugates (defined in § 2). A first problem is that of the existence of PH functions U on G^* with prescribed level sets locally topologically like families of parallel straight lines except in the neighbourhood of points of a discrete set ω of points z_0. At a point z_0 the level curves of U may cross after the manner of the level curves of a harmonic function with a critical point at z_0.

Let E be the finite z-plane. In case $G^* = E$ and $\omega = 0$, and excluding all recurrent level curves other than periodic curves, Kaplan has solved the above problem in [5]. More recently Boothby has extended Kaplan's result to the case of a general ω. As explained in [4] the *a priori* exclusion of recurrent level curves other than periodic curves does not seem justified. The writers of this paper have accordingly established the existence of PH functions without this hypothesis of non-recurrence [4].

The present paper gives a solution of a second major problem in the topological theory, namely the existence of a function V, PC (pseudoconjugate) to a given PH function U. When $\omega=0$ **Kaplan** has affirmed that the set of level curves of U are topologically equivalent to the trajectories of a differential system of the form

$$\frac{dx}{dt}=p(x,y), \quad \frac{dy}{dt}=q(x,y) \qquad (p^2+q^2\neq 0),$$

where p and q are of class C' over E. The method of proof of this theorem as outlined by **Kaplan** in a few paragraphs in [6] does not seem to form an adequate basis for proving the existence of V.

We here give an explicit topological proof of the existence of a V, PC to U. The key to this proof lies in the proper use of the theory of μ-length as developed for arbitrary curves in a general metric space [8, § 27].

The results and methods of this paper will lead in a subsequent paper to theorems on the existence of PH functions and their pseudoconjugates on an arbitrary open Riemann surface.

The reader familiar with the geometric theory of dynamical systems as initiated by **Poincaré** [9] with its attention to singularities, periodicity, recurrence, etc., will recognize that the underlying topological theory of the level lines of a PH function is a form of 2-dimensional topological dynamics.

§ 2. Review of earlier theorems. Let S be the complex z-sphere with Z the point $z=\infty$. Let ω be a set of isolated points in $G^*=S-Z$ with no point other than Z as limit point on S.

The family F, the sets G^* and G. An arc, open arc, or topological circle in G^* is for us the $1-1$ continuous image in G^* of a closed interval, open interval, or circle respectively. These elements are to be distinguished from parameterized curves (p-curves) which are mappings, and not sets. Let F be a family of elements α, β, γ etc., which are open arcs or topological circles in $G=G^*-\omega$ and which include one and only one α meeting each point $p\in G$.

The family F shall have the following local property. Let D be the open disc ($|w|<1$) in the complex w-plane. With each point $p\in S-Z$ there shall be associated an "F-neighbourhood" X_p of p with $\bar{X}_p\subset G\cup p$, and a homeomorphic mapping T_p of $\bar{X}_p$ onto $\bar{D}$ under which p goes into $w=0$ and the maximal open arcs of $F|X_p$ go into the maximal open level arcs of $\mathcal{R}w^n$ in D when $n=1$, and in D with the origin deleted when $n>1$. We term n the *exponent* of p. The exponent of p shall be 1 for $p \in G$, and exceed 1 for $p \in \omega$. Points $p \in \omega$ are termed *F-singular points.* A value $w \in D$ is termed the *canonical parameter* of its antecedent in X_p.

The open arcs of $F|X_p$ which have limiting end points at $p \in \omega$ are termed *F-rays* of X_p incident with p. They are $2n$ in number, thereby showing that n is independent of the choice of X_p. These F-rays of X_p divide $X_p - p$ into $2n$ open regions termed *F-sectors* incident with p.

Right N. With each $p \in G$ one can also associate a neighbourhood N_p of p with $\bar{N}_p \subset G$, and a *homeomorphic sense-preserving mapping* of $\bar{N}_p$ onto a square K: $(-1 \leqslant u \leqslant 1)$ $(-1 \leqslant v \leqslant 1)$ such that p goes into the origin in K and the maximal subarcs of $F|\bar{N}_p$ go into arcs $u=c$; $-1 \leqslant v \leqslant 1$, where the constant c ranges over the interval $[-1, 1]$. We refer to N_p as a right N of p and term u and v *canonical coordinates* of the antecedent in $\bar{N}_p$ of (u,v) in K.

Transversals. By a *transversal* λ is meant an open arc in G whose intersection with any right N with canonical coordinates (u,v) has locally the form $v=\varphi(u)$ where φ is single-valued and continuous. By the *principal transversal* of a right N is meant the open arc in N on which $v=0$, $-1<u<1$.

F-vectors. Any sensed subarc A of an $\alpha \in F$ will be called an *F-vector*. By definition a F-vector is simple, closed, and never a topological circle. The following lemma is established in [4, § 3].

Lemma 2.1. *Each F-vector is in some right N.*

Coherent sensing. Let each $\alpha \in F$ be given a sense. The resulting family F^s of sensed α will be called a *sensed image* of F. We shall refer to a continuous deformation Δ of an F-vector A in the space of F-vectors metricized by means of the Fréchet distance between any two sensed arcs. We shall understand that each image of A under Δ is sensed by Δ, that is that the sense of the image arc shall be determined by the images under Δ of the initial and final points of A. We say that F^s is *coherently sensed* if any continuous deformation Δ of an F-vector A, initially sensed by F^s, through F-vectors sensed by Δ, is necessarily a deformation through F-vectors sensed by F^s. In the case at hand, where $G^* = S - Z$ is simply connected, there are two distinct coherently sensed images F^s of F [4, § 5].

The following theorem came next in our development:

Theorem 2.1. *When $G^* = S - \omega$ each $\alpha \in F$ is the homeomorphic image in G of an open interval with limiting end points on S, distinct unless both are coincident with Z. Each finite end point is a singular point of F.*

The family F^*. It is shown in [4, § 7] that when $G^* = S - Z$ there are no topological circles in G^* formed from the union of the closures of a finite set of $\alpha \in F$. In this case we define F^* as the set of all open arcs h, k, m, etc. in G^* with the following properties. If p is an F-non-singular point of h, h shall contain the $\alpha_p \in F$ meeting p, and any finite limiting end point or end points of α_p; if an F-singular point q is in h, h shall contain just two of the $\alpha \in F$ with q as a limiting end point. An $h \in F^*$ may be identical with an $\alpha \in F$, or it may be formed from a sequence of $\alpha \in F$ of one of the forms

$$\ldots \alpha_{-2}\, \alpha_{-1}\, \alpha_0\, \alpha_1\, \alpha_2 \ldots$$

$$\ldots \alpha_{-2}\, \alpha_{-1}\, \alpha_0; \quad \alpha_0\, \alpha_1\, \alpha_2 \ldots$$

$$\alpha_0\, \alpha_1 \ldots \alpha_n.$$

The following theorem bears on the behaviour of subarcs of $h \in F^*$ [4, Cor. 7.5].

Theorem 2.2. *When $G^* = S - Z$ an arc g in $S - Z$ which is the closure of a finite sequence of $\alpha \in F$ intersects the closure of an F-neighbourhood X_p or of a right N, if at all, in a single arc.*

Corollary 2.2. *When $G^* = S - Z$ each $h \in F^*$ has Z as a limiting end point in both senses.*

Bands $R(N)$. Given a right N the union of the sets $\alpha \in F$ which intersect N will be called the *band* $R(N)$. In accordance with Lemma 8.1 [4] each band $R(N)$ is simply connected when $G^* = S - Z$, and in accordance with Theorem 8.1 [4] the boundary $\beta R(N)$ of $R(N)$ in S is the union of Z and at most a countable set of non-intersecting open arcs $h \in F^*$ of which at most a finite number have diameters on S which exceed a finite constant d. An $h \in F^*$ in $\beta R(N)$ is either *concave towards* $R(N)$ in that $R(N)$ contains just one F-sector incident with each singular point of F in h, or *semi-concave towards* $R(N)$ in that $R(N)$ includes just one F-sector incident with each singular point of F in h except for one singular point q of F in h; corresponding to q, $R(N)$ contains just two F-sectors incident with q and the F-ray incident with q between the two F-sectors incident with q. An $h \in F^* | \beta R(N)$ which meets βN is always concave towards $R(N)$ (Cf. Theorem 8.1 [4]).

A singular point p of F in $\beta R(N)$ is termed of *type* 1 or *type* 2 relative to $R(N)$ if $R(N)$ contains one or two F-sectors respectively incident with p.

An entrance to $R(N)$. An open arc η in $\beta R(N)$ will be called an *entrance to* $R(N)$ if η is the union of a finite or countably infinite set of $\alpha \in F$ each joined to its successor or predecessor at a singular point in F, and if η intersects βN in an open arc. An entrance η to $R(N)$ is an open subarc of an $h \in F^*$ in $\beta R(N)$ such that h intersects βN and is accordingly concave towards $R(N)$ [4, Th. 8.1].

We shall make use of the following decomposition of G^* [4, Th. 9.1].

Theorem 2.3. *When $G^* = S - Z$ there exists a sequence of non-intersecting bands $R(N_r)$, $r = 1,2,\ldots$, and for each $R(N_r)$ with $r > 1$ an open entrance η_r to $R(N_r)$, such that the η_r do not intersect each other or any of the bands, and such that $G^* = \text{Union } \Sigma_n$, $n = 1,2,\ldots$, where $\Sigma_1 = R(N_1)$ and for $n = 2,3,\ldots$*

$$\Sigma_n = \Sigma_{n-1} \cup R(N_n) \cup \eta_n; \quad \bar{\Sigma}_{n-1} \cap \bar{R}(N_n) = \bar{\eta}_n \cup Z \tag{2.1}$$

and where the N_r are so chosen that any compact subset of G^* is included in Σ_n for n sufficiently large.

We note that each Σ_n is open and simply connected. If Σ_n contains a point $p \in G$, it contains the $\alpha \in F$ which meets p.

Let U_0 be harmonic in a neighbourhood H of $q \in G^*$ and not identically constant. Let Ψ be a topological mapping of a neighbourhood H_0 of q onto H. Then $U_0\Psi$ is termed *PH (pseudoharmonic)* at q. A function U is termed *PH over a region* $R \subset G^*$ if U is PH at each point of R.

Let φ be analytic in a neighbourhood H of $q \in G^*$, and not identically constant, and let Ψ be a sense-preserving topological mapping of a neighbourhood H_0 of q onto H. Then $\varphi\Psi$ is termed *interior at* q. A function f is termed *interior over* a region $R \subset G^*$ if f is interior at each point of R.

If U is PH over a region R and V real, and if $U + iV$ is interior over R then V is said to be *PC (pseudoconjugate) to U over R*.

§ 3. Bands $R(N)$ conformally represented. Let U be given as PH over $R(N)$ with the $\alpha \in F$ as level lines. It is relatively easy to use signed μ-length along the $\alpha \in F | R(N)$ measured algebraically from the principal transversal λ of N to obtain a function V PC to U. To extend the definition of V to $\beta R(N)$ is more difficult because $\beta R(N)$ is not readily given as a p-curve. The arcs of $\beta R(N)$ meeting Z require a definite parameterization which is related to the parameterization of the $\alpha \in F | R(N)$. To this end we find it useful to map a unit disc $D | (|z|<1)$ directly conformally onto $R(N)$, and then by continuously extending the mapping of D over $\bar{D}$ to obtain the desired parameterization of arcs in $\beta R(N)$.

Canonical coordinates (u,v) in a right N. Let (u,v) be canonical coordinates in a right N. Set $u+iv=w$. To each point $z \in \bar{N}$ corresponds a complex canonical parameter $w=\Psi(z)$ given with N. The mapping Ψ is given as 1—1 continuous and sense-preserving in N. The value of U at the point $w \in \bar{N}$ reduces to a strictly monotone function of u.

Convention. *In this paper we shall choose the canonical coordinate u in each right N so that $U(u)$ is strictly increasing. Let F^s be a coherently sensed image of F so chosen that in each right N the canonical coordinate v increases in the positive sense of $F^s | \bar{N}$ along each level arc of U.*

This relation of F^s and U to the canonical coordinates (u,v) in one right N implies a similar relation to the canonical coordinates in an arbitrary right N. This follows from the orientability of S and the fact that F^s is a coherently sensed image of F. In this connection recall that canonical coordinates (u,v) in a right N are to be chosen so that the mapping $u+iv=\varphi(z)$ which carries $z \in N$ into the canonical point (u,v) is an interior transformation.

The open arc β_u. Let λ be the principal transversal of N. Let the point on $\bar{\lambda}$ with canonical coordinates $(u,0)$ in $\bar{N}$ be denoted by $\lambda(u)$. The coordinate u ranges on an interval $I=[-1,1]$. Let $\beta_u \in F^s | (u \in I)$ meet $\bar{\lambda}$ in the point $\lambda(u)$. In particular β_1 and β_{-1} are in F^s and on $\beta R(N)$.

The conformal mapping f. Let f be a direct conformal mapping of the open unit disc D onto $R(N)$. An arc $h \in F^*$ in $\beta R(N)$ is a *free boundary arc* of $R(N)$ in the sense that it is an open arc which does not meet the closure of $\beta R(N)-h$. The mapping f can be extended to a homeomorphism which maps an open arc h^D on

βD onto h [2, p. 86]. If f is further extended so as to map each point in the set

$$Q = \beta D - \mathit{Union}\ h^D \qquad (h \in F^* | \beta R(N)) \tag{3.0}$$

into Z, then the extended f maps $\bar{D}$ continuously into S.

Indeed if $\{z_n\}$ is a sequence of points in $\bar{D}$ tending to a point in Q, the image sequence $\{f(z_n)\}$ can have no accumulation point in $R(N)$ or on an arc $h \in F^*$ in $\beta R(N)$. Thus $f(z_n) \to Z$ as $n \uparrow \infty$ and f maps $\bar{D}$ continuously into S.

Without loss of generality we may assume that f carries $z = +1$ and $z = -1$ in βD into $\lambda(1)$ and $\lambda(-1)$ in $\beta R(N)$ respectively.

If Y is any set in $\bar{R}(N)$ we will denote $f^{-1}(Y)$ by Y^D. The sensed antecedents β_1^D and β_{-1}^D in βD of β_1 and β_{-1} respectively in $\beta R(N)$ are open arcs in βD which appear in βD in counter-clockwise and clockwise sense respectively, with $z = 1$ an inner point of β_1^D and $z = -1$ an inner point of β_{-1}^D.

Let (α) be the set of $\alpha \in F | \beta R(N)$, and (α^D) the set of the respective antecedents of $\alpha \in (\alpha)$.

Let $d(\alpha)$ and $d(\alpha^D)$ be respectively the diameter in S of $\alpha \in (\alpha)$ and of α^D in (α^D). As stated in § 2 there are at most a finite number of the $\alpha \in (\alpha)$ with $d(\alpha) > 1/n$. The same is true of the $\alpha^D \in (\alpha^D)$ since $\Sigma d(\alpha^D) \leqslant 2\pi$, the length of βD. It follows that $d(\alpha^D) \to 0$ if $d(\alpha) \to 0$, and conversely that $d(\alpha) \to 0$ if $d(\alpha^D) \to 0$.

The antecedent Z^D. This antecedent of $Z \in S$ is closed. It is nowhere dense in βD since it can contain no subarc of βD.

The antecedent β_u^D. Since β_{-1} and β_1 are in $\beta R(N)$, β_{-1}^D and β_1^D are in βD. For $-1 < u < 1$, β_u^D is in D since β_u is in $R(N)$. The point $(r, \Theta) = (1, 0)$ in βD is in β_1^D. We shall restrict $\Theta = \operatorname{arc} z | (z \in \bar{D})$ to values on the interval $0 \leqslant \Theta < 2\pi$.

The end points $\operatorname{Exp}(i\Theta_1(u))$ and $\operatorname{Exp}(i\Theta_2(u))$ of β_u^D. For $z \in \beta_u^D | (u \in I)$, $\Theta = \operatorname{arc} z$ has limiting initial and final values which we denote by $\Theta_1(u)$ and $\Theta_2(u)$ respectively. That $\Theta_1(u)$ and $\Theta_2(u)$ exist is immediate when $u = \pm 1$. If β_u^D $(-1 < u < 1)$ did not have unique limiting end points in βD, $\bar{\beta}_u^D - \beta_u^D$ would include some subarc of an $\alpha^D \in (\alpha^D)$ since Z^D is nowhere dense in βD. Since f is continuous over $\bar{D}$, $\bar{\beta}_u - \beta_u$ would then include some subarc of an $\alpha \in (\alpha)$ contrary to the fact that $\bar{\beta}_u - \beta_u$ contains at most two points of S. Hence the limits $\Theta_1(u)$ and $\Theta_2(u)$ exist and are unique. The limiting end points of β_u^D are then $\operatorname{Exp}(i\Theta_1(u))$ and $\operatorname{Exp}(i\Theta_2(u))$.

Lemma 3.1. *The functions Θ_1 and Θ_2 are monotone on the interval I: $-1 \leqslant u \leqslant 1$, with Θ_1 increasing, Θ_2 decreasing, and*

$$0 < \Theta_2(u) < \pi, \quad \pi < \Theta_1(u) < 2\pi. \tag{3.1}$$

The principal transversal λ of N is in $R(N)$ and has an antecedent λ^D in D which is a cross cut of D with initial and final end points $z=-1$ and $z=1$ in βD. Each $\beta_u^D|(u \in I)$ is divided by λ^D

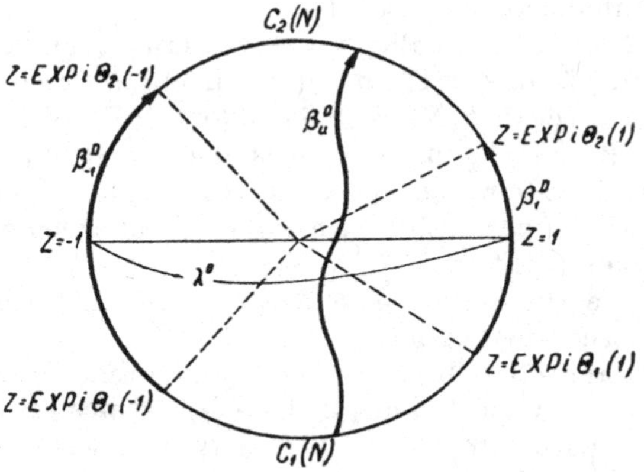

into a half open arc with initial point $\lambda^D(u)$ in λ^D and final point $\operatorname{Exp}(i\Theta_2(u))$ in βD, and a half open arc with final point $\lambda^D(u)$ in λ^D and initial point $\operatorname{Exp}(i\Theta_1(u))$ in βD. The relations (3.1) clearly hold for $u = \pm 1$ and hence hold for each $u \in I$.

For $-1 < u < 1$, β_u is in $R(N)$ and has limiting end points on $\beta R(N)$. Since $R(N)$ is simply connected β_u separates $R(N)$. In particular if $u' < u'' < u'''$ in I, $\beta_{u''}$ separates $\beta_{u'}$ and $\beta_{u'''}$ in $R(N)$, for otherwise $\lambda(u')$ and $\lambda(u''')$ would be connected on $R(N) - \beta_{u''}$ and $\beta_{u''}$ would not separate $R(N)$. The monotonicity of Θ_1 and Θ_2 follows. It is clear moreover that

$$0 < \Theta_2(1) \leqslant \Theta_2(-1) < \pi < \Theta_1(-1) \leqslant \Theta(1) < 2\pi$$

from which we infer that Θ_1 is increasing and Θ_2 decreasing.

Let $C_1(N)$ and $C_2(N)$ be subarcs of βD conditioned as follows:

$$C_2(N): \ \Theta_2(-1) \geqslant \Theta \geqslant \Theta_2(1), \tag{3.2$'$}$$

$$C_1(N): \ \Theta_1(1) \geqslant \Theta \geqslant \Theta_1(-1). \tag{3.2$''$}$$

The arc $C_2(N)$ bears all the final end points of the $\beta_u^D | (u \in I)$, and $C_1(N)$ all the initial end points. Then $C_2(N)$ and $C_1(N)$ clearly contain each $a^D \in (a^D)$ except β_1^D and β_{-1}^D. We shall be more explicit in the case of $C_2(N)$.

Lemma 3.2. *Corresponding to each $a \in F$ with $a^D \in C_2(N)$ there is a unique value $u \in I$ such that a^D lies in an arc on βD of one of the forms*

$$\Theta_2(u-) > \Theta > \Theta_2(u), \tag{3.3}$$

$$\Theta_2(u) > \Theta > \Theta_2(u+). \tag{3.4}$$

Let J be the interval of values of Θ such that $\operatorname{Exp}(i\Theta)$ is in a^D. Then J is open and contains no value $\Theta_2(s) | (s \in I)$; for the final end point $\operatorname{Exp}(i\Theta_2(s))$ of β_s^D has an f-image which is either an F-singular point or Z, and accordingly not in a. Hence J must be in an interval of the type (3.3) or (3.4). Since no two intervals of type (3.3) or (3.4) intersect, the interval (3.3) or (3.4) containing J must be unique.

The p-curve B_u in S. Let the half open arc of β_u with initial point $\lambda(u)$ be closed by its final end point in S, and be parameterized by μ-length in S to define a p-curve B_u.

If g is a p-curve it is convenient to introduce the carrier $|g|$ of g defined as the union of the points in the range of g.

The extensions B_{u+} and B_{u-} of B_u in S. Let B_u^D be extended as a p-curve by the arc of points $\operatorname{Exp}(i\Theta)$ of βD on which

$$\Theta_2(u) \geqslant \Theta \geqslant \Theta_2(u+) \qquad (-1 \leqslant u < 1) \tag{3.5}$$

taken in the sense of decreasing Θ and admitting the possibility that (3.5) may reduce to a point and B_u^D coincide with its extension. Let B_{u+}^D be the resultant extension of B_u^D as a p-curve, and $B_{u+} = f(B_{u+}^D)$ its p-curve image in $\bar{R}(N)$. Recall that a p-curve is not a set, and that $|B_{u+}^D|$ is in general not the full antecedent of the set $|B_{u+}|$, since Z is in $|B_{u+}|$ (see Lemma 3.3) and $|B_{u+}^D|$ will not in general contain the whole set Z^D. We shall suppose that B_{u+} is parameterized by μ-length measured along B_{u+} from $\lambda(u)$, and note that B_{u+} is a p-curve in $\bar{R}(N)$ extending B_u.

Let B_u^D be similarly extended in $\bar{D}$ as a p-curve by adding the arc of βD (if any exists) on which

$$\Theta_2(u-) \geqslant \Theta \geqslant \Theta_2(u) \qquad (-1 < u \leqslant 1) \tag{3.6}$$

taken in the sense of increasing Θ. Let B^D_{u-} be this extension of B^D_u. Set $B_{u-}=f(B^D_{u-})$, parameterizing B_{u-} by means of μ-length in S to form a p-curve B_{u-} extending B_u.

We understand that $\Theta_2(u+)$ and B_{u+} are defined for $-1\leqslant u<1$, and that $\Theta_2(u-)$ and B_{u-} are defined for $-1<u\leqslant 1$.

We shall need the following lemma.

Lemma 3.3. *When $G^*=S-Z$ the terminal point of B_{u+} or B_{u-} is Z.*

The f-image of the point $\operatorname{Exp}[i\Theta_2(u)]$ in βD is either Z or a singular point of F. Since the singular points of F have Z as their sole limit point in S the f-image of $\operatorname{Exp}(i\Theta_2(u+))$ or $\operatorname{Exp}(i\Theta_2(u-))$ must be Z. But these points are respectively the terminal points of B_{u+} and B_{u-} so that the lemma follows.

Note that the relation $\Theta_2(u)=\Theta_2(u+)$ implies $B_u=B_{u+}$ and the relation $\Theta_2(u)=\Theta_2(u-)$ implies that $B_u=B_{u-}$. In either case the terminal point of B_u is Z by virtue of Lemma 3.3.

§ 4. B_{u+} and B_{u-} as Fréchet limits. We treat the case of B_{u+}. The case of B_{u-} is similar.

Lemma 4.1. *Given $e>0$ and u in the interval $-1\leqslant u<1$, an e-neighbourhood of $|B^D_{u+}|$ contains $|B^D_t|$ provided $t>u$, and $t-u$ is sufficiently small.*

(a) If the lemma were false there would exist a sequence of values t_n tending to u from above as $n\uparrow\infty$, and points $z_n\in|B^D_{t_n}|$ such that $z_n\to z_0$ not in $|B^D_{u+}|$ as $n\uparrow\infty$. Two cases are distinguished.

Case (i). $z_0\in D$. Set $p_n=f(z_n)$, $n=0,1,\ldots$, where f is the conformal mapping defined in § 3. For a suitable $s\in(-1,1)$, $\beta_s\in F\,|\,R(N)$ meets p_0. The point $p_0\in\beta_s$ is a limit point of points p_n only if $t_n\to s$. Hence $s=u$. But p_0 is not in $|B_u|\subset|B_{u+}|$ by hypothesis (a). Hence p_0 is in $\beta_u-|B_u|$. Hence p_0 is not a limit point of points in $|B_{t_n}|$ as $t_n\to u$, so that z_0 is not a limit point of the sequence $z_1,z_2,\ldots$ We infer that case (i) is impossible.

Case (ii). $z_0\in\beta D$. We shall show here that z_0 must lie in $|B^D_{u+}|$ contrary to hypothesis (a).

For $u<t<1$ let λ_t be the subarc of λ on which $z=\lambda(s)$ $(u\leqslant s\leqslant t)$. Let η_t be the arc or point of βD on which $\Theta_2(u+)\geqslant\Theta\geqslant\Theta_2(t)$. Let C_t be the topological circle on $\bar{D}$ formed by the circular sequence of the four arcs (with η_t possibly a point)

$$|B^D_{u+}|,\quad \eta_t,\quad |B^D_t|,\quad \lambda_t.$$

Let X_t be the closure of the Jordan region bounded by C_t. For $u<s<t$, B_s^D is in X_t. As $t \downarrow u$, X_t contracts on itself as a point set, and $C_t \cap \beta D$ is an arc of $X_t \cap \beta D$ which contracts on itself to $|B_{u+}^D| \cap \beta D$ as a limiting arc or point. The point z_0 is in X_t for each $t>u$. In case (ii) z_0 must then be in $C_t \cap \beta D$ for each $t>u$ and hence in $|B_{u+}^D| \cap \beta D$. From this contradiction to (a) we infer the truth of the lemma.

A right neighbourhood relative to $\bar{R}(N)$. Let g be an arc in $\beta R(N)-Z$ whose end points are F-non-singular and whose F-singular points are finite in number and of type 1 relative to $R(N)$.

Lemma 4.2. *There exists a neighbourhood M of g relative to $\bar{R}(N)$ such that $\bar{M}$ is the homeomorph of a square* $(0 \leqslant u \leqslant 1)$ $(0 \leqslant v \leqslant 1)$ *in which the open edge* $u=0$, $0<v<1$, *contains the image of g and each arc* $u = \text{const} \neq 0$, $0 \leqslant v \leqslant 1$, *corresponds to an F-vector in $\bar{R}(N)$.*

The lemma is true if there are no singular points of F on g, by virtue of Lemma 2.1. It is also clearly true if g is the closure of two F-rays on the boundary of an F-sector. The lemma is readily established in the general case following the pattern of the proof of Lemma 3.1 of [4].

The neighbourhood of g in Lemma 4.2 is termed a *right neighbourhood of g relative to $\bar{R}(N)$.*

In the proof of Theorem 4.1 use will be made of various distance functions. If A and B are non-empty subsets of S let $d(A,B)$ denote the inferior limit of the distances $d(p,q)$ between $p \in A$ and $q \in B$. Let

$$\delta(A,B) = \sup_{p \in A} d(p,B).$$

If A, B, C are subsets of S,

$$\delta(A,C) \leqslant \delta(A,B) + \delta(B,C). \tag{4.1}$$

If $B = B_1 \cup B_2$,

$$\delta(A,B) \geqslant \min\,[\delta(A,B_1),\ d(A,B_2)]. \tag{4.2}$$

If g and h are p-curves in S, $\varrho(g,h)$ shall denote the ***Fréchet distance*** between g and h.

Theorem 4.1 is basic in the proof of the existence of a V pseudoconjugate to U. We prepare for its proof.

An e-decomposition of B_{u+}. We suppose $u \in [-1,1)$. Given $e>0$ the number of open arcs $\gamma \in F^*$ in B_{u+} such that $\delta(\gamma, Z) \geqslant e/4$ is finite (see § 3). Let $\gamma_1, \ldots, \gamma_m$ be these open arcs ordered as they appear in the p-curve B_{u+}. No one of the γ_i, $i=1,\ldots,m$ intersects B_u. Let γ_0 be the maximal initial half open arc of B_{u+} which does not meet Z. The initial point of γ_0 is $\lambda(u)$ and $\bar{\gamma}_0 \supset |B_u|$. Let

$$g_0, g_1, \ldots, g_m \tag{4.3}$$

be subarcs of $\gamma_0, \gamma_1, \ldots, \gamma_m$ respectively, whose end points are F-nonsingular and are such that the initial point of g_0 is $\lambda(u)$ and

$$\delta(\gamma_r - g_r, Z) < e/4 \qquad (r=0,\ldots,m).$$

Parameterizing g_r as in B_{u+} it appears that B_{u+} is a sequence of sub-p-curves

$$g_0, k_0, g_1, k_1, \ldots, g_m, k_m. \tag{4.4}$$

No p-curve g_r intersects any other p-curve g_i or any p-curve k_i.

A right neighbourhood M_r of g_r. Recall that $-1 \leqslant u < 1$. The set $N - \beta_u$ is the union of two open sets M^u and N^u of which $\overline{N^u}$ contains $\lambda(1)$. Then $R(N^u) \subset R(N)$. Moreover $\gamma_1, \ldots, \gamma_m$ are in $\beta R(N) \cap \beta R(N^u)$ while γ_0 is in $\beta R(N^u)$. Each F-singular point of γ_i, $i=1,\ldots,m$, is of type 1 relative to $R(N)$ and $R(N^u)$, by virtue of the definition of the extension B_{u+} of B_u and nature of Θ_2. Each F-singular point of γ_0 is of type 1 relative to $R(N^u)$ by virtue of Theorem 8.1 (b) [4]. By Lemma 4.2 then, the respective g_r in (4.4) have right neighbourhoods

$$M_0, M_1, \ldots, M_m, \tag{4.5}$$

relative to $\bar{R}(N^u)$. Each M_r in (4.5) will be so restricted that each maximal sub-p-curve of B_t in $\bar{M}_r$ (when $B_t \cap M_r \neq 0$) is at a Fréchet distance less than e from g_r (sensed as in F^*), and that $\bar{M}_r \cap \bar{M}_i = 0$ for $r \neq i$.

Lemma 4.3. *Corresponding to the given $u \in [-1,1)$ and $e>0$ a constant t_1 $(1>t_1>u)$ for which t_1-u is sufficiently small has the following property. Each simple p-curve B_t for which $u<t<t_1$ admits a decomposition*

$$g_0^t, k_0^t, g_1^t, k_1^t, \ldots, g_m^t, k_m^t \tag{4.6}$$

into successive non-intersecting p-curves such that $\varrho(g_r^t, g_r) < e$, and, setting

$$K^t = \text{Union } |k_r^t|, \quad K = \text{Union } |k_r| \qquad (r=0,\ldots,m);$$
$$\delta(K^t, K) < e/4. \tag{4.7}$$

We shall refer to $\bar{D}$ and make use of the fact that B_{u+}^D is simple. Set

$$\operatorname{Exp}[i\Theta_2(u+)]=q.$$

The point q is the final end point of B_{u+}^D. As $t\downarrow u$ the final end point $\operatorname{Exp}[i\Theta_2(t)]$ of B_t^D tends to q while the initial end point $\lambda^D(t)$ of B_t^D tends to the initial end point $\lambda^D(u)$ of B_{u+}^D; at the same time

$$\delta(|B_t^D|,|B_{u+}^D|)\to 0$$

in accordance with Lemma 4.2. The simple arc B_{u+}^D meets the disjoint sets

(4.8) $$M_0^D, M_1^D, \ldots, M_m^D, q$$

in the order written. It follows from Lemma 4.2 that if t_1-u is sufficiently small and if $u<t<t_1<1$, B_t^D will meet each of the sets $\bar{M}_r^D$ in (4.8) in at least one maximal p-curve $(g_r^t)^D$ and by virtue of Theorem 2.2 in at most one p-curve, there will be a terminal p-curve $(k_m^t)^D$ of B_t^D which intersects no $\bar{M}_r^D$, and the initial point $\lambda^D(t)$ of B_t^D will be in M_0^D. We suppose t_1 and t so conditioned.

It follows from this choice of g_r^t and M_r that $\varrho(g_r^t, g_r)<e$. The residual p-curves k_r^t in (4.6) are determined. Set $\operatorname{Union}_r |g_r|=\Gamma$, $r=0,1,\ldots,m$. In the relation

(4.9) $$\delta(K^t, K\cup\Gamma)\geqslant \min[\delta(K^t,K), d(K^t,\Gamma)] \qquad \text{[Cf. (4.2)]}$$

the left member tends to 0 as $t\downarrow u$ by virtue of Lemma 4.2, while $d(K^t,\Gamma)$ remains bounded from 0. Hence for t_1-u sufficiently small and $u<t<t_1$

(4.10) $$\delta(K^t,K)\leqslant\delta(K^t,K\cup\Gamma).$$

Relation (4.7) then follows for suitable choice of t_1.

Theorem 4.1. *If $t\downarrow u\in[-1,1)$ then B_t tends to B_{u+} in the sense of Fréchet.*

Corresponding to $e>0$ let B_{u+} be given the decomposition (4.4), where $\delta(K,Z)<e/4$. If t_1 is chosen as in Lemma 4.3 and if $u<t<t_1$, then $\delta(K^t,K)<e/4$. Moreover by (4.1)

$$\delta(k_r^t,Z)\leqslant\delta(K^t,Z)\leqslant\delta(K^t,K)+\delta(K,Z)<e/2 \quad (r=0,\ldots,m).$$

Hence both k_r^t and k_r are within a distance $e/2$ of Z so that $\varrho(k_r^t,k_r)<e$. By virtue of Lemma 4.3 $\varrho(g_r^t,g_r)<e$. We infer that

$$\varrho(B_t,B_{u+})<e \qquad (u<t<t_1)$$

and the theorem follows.

§ 5. Definition of $V|(\bar{R}(N)-Z)$**.** We are supposing that U is PH over $S-Z$, each $a \in F$ a level set of U and that each F-singular point in ω is a topological critical point of U.

Lemma 5.1. *Let N_0 be a right neighbourhood with canonical coordinates (u,v). If V is continuous over N_0 and strictly increasing with v for u constant, then V is* PC *to U over N_0.*

Since U has each $a \in F$ as level set the value of U at the point (u,v) in N_0 depends only on u. Let $U_1(u)$ be this value and let $V_1(u,v)$ be the value of V at (u,v). By hypotheses $V_1(u,v)$ increases in a strict sense with v for u constant and $u \in (-1,1)$, so that $U_1(u)+iV_1(u,v)$ defines an interior transformation from the $(u+iv)$-plane to the (U_1+iV_1)-plane. The mapping from the $(u+iv)$-plane to the z-plane in which $(u+iv)$ corresponds to the point $z \in N$ represented by (u,v) is interior by convention as to admissible coordinates (u,v). The mapping from N_0 in the z-plane into the $(U+iV)$-plane is accordingly interior, and hence V is PC to U.

Lemma 5.2. *Given a band $R(N)$ there exists a function V,* PH *and bounded over $R(N)$,* PC *to U with boundary values on $\beta R(N)-Z$ which are strictly monotone on each $a \in F$ in $\beta R(N)$, increasing in the positive sense of a as derived from F^s, and such that V is continuous over $\bar{R}(N)-Z$.*

Definition of $V|R(N)$**.** We refer to the principal transversal λ of N with its 1—1 continuous parameterization $\lambda(u)$, $-1<u<1$, and to the open arc β_u of F^s meeting λ in the point $\lambda(u)$. For $z \in \beta_u$ let $V(z)$ denote the signed μ-distance measured from $\lambda(u)$ along β_u to z; taking $V(z)>0$ when z follows $\lambda(u)$ on β_u, and negative when z precedes $\lambda(u)$ on β_u. The parameter μ is bounded.

The mapping

$$U(z)+iV(z)=\varphi(z)=w' \qquad (z \in R(N)) \tag{5.1}$$

of $R(N)$ into the $w'=U+iV$ plane is continuous over $R(N)$, since U and V are continuous over $R(N)$. It is 1—1 since the value of $U(z)=u$ uniquely determines the open arc β_u on which z lies, and the signed μ-length $V(z)$ then uniquely determines the point $z \in \beta_u$.

The mapping φ from the z-plane to the w'-plane is also sense preserving, as we shall now prove. It is sufficient to prove that $\varphi|N$ is sense preserving. Let (u,v) be canonical coordinates in N as previously. By convention as to canonical coordinates v increases in the positive sense of F^s along each arc $u=$ constant in N. But V as defined above similarly increases in the positive sense of F^s along each arc $u=$ constant in N. It follows from Lemma 5.1 that $U+iV$ is interior over N. Hence $U+iV$ is interior over $R(N)$ and V is PC to U.

Definition of $V|(\beta R(N)-Z)$. For $z \in \beta_1$ or β_{-1} $V(z)$ is defined as the algebraic μ-value at z on β_1 or β_{-1} as measured from $\lambda(1)$ or $\lambda(-1)$ respectively. The continuity of $V|\bar{R}(N)$ at such a point z is clear. There remain points z on $f[C_2(N)]-Z$ or $f[C_1(N)]-Z$. Suppose $z \in f[C_2(N)]-Z$. Then z is either in an $\alpha \in F^s|\beta R(N)$ or is an F-singular point $z \in \omega \cap \beta R(N)$.

I. Consider $\alpha \in F^s|\beta R(N)$. Such an α is in a unique p-curve B_{u+} or B_{u-}, say B_{u+}. Parameterize B_{u+} by μ-length measured from $\lambda(u)$. At each point $z \in \alpha$ whose parameter in B_{u+} is μ set $V(z)=\mu$. Since $B_t|(t>u)$ converges in the sense of Fréchet to B_{u+} as $t \downarrow u$, it follows from the theory of μ-length that V, as just defined over $R(N)$, takes on continuous boundary values on α. The boundary values of V on α as given by μ-length on B_{u+} are strictly increasing in the positive sense of α as derived from F^s. The case of an $\alpha \in F^s|B_{u-}$ is similar.

II. Consider $z \in \omega \cap \beta R(N)$ with z of type 1 relative to $R(N)$. In this case z is the common end point of two $\alpha \in F^s|\beta R(N)$, say α' and α''. Then any sufficiently restricted open subarc of $\alpha' \cup z \cup \alpha''$ containing z is in a unique p-curve B_{u+} or B_{u-}. One defines $V(z)$ as in I and proves that $V|\bar{R}(N)$, as defined on a neighbourhood of z relative to $\bar{R}(N)$ is continuous at z.

III. Consider $z \in \omega \cap \beta R(N)$ with z of type 2 relative to $R(N)$. In this case there exists a unique $u \in (-1,1)$ such that z is the terminal point of β_u. With N^u defined as in § 4 let M^u be $N-(N^u \cup \beta_u)$. Then z is in

$$\beta R(N^u) \cap \beta R(M^u) = \bar{R}(N^u) \cap \bar{R}(M^u) = \bar{\beta}_u \cup Z.$$

But $V|\beta_u$ is already defined, and extends both $V|R(N^u)$ and $V|R(M^u)$ continuously. As in I and II unique extensions of $V|R(N^u)$ and $V|R(M^u)$ over neighbourhoods of z relative to $\bar{R}(N^u)$ and $\bar{R}(M^u)$ respectively exist and make V continuous over such neighbourhoods. It follows that these two definitions of $V|\beta_u$ and hence of $V(z)$ agree, and make $V|\bar{R}(N)$ continuous over a neighbourhood of z relative to $\bar{R}(N)$.

The case of a $z \in f[C_1(N)]-Z$ is similar and this completes the proof of Lemma 5.2.

§ 6. The definition of V over $S-Z$. An $h \in F^*$ will be said to have a *sense derived* from F^s if each $a \in F^s$ in h has the sense in h which is proper to it in F^s. Clearly an $h \in F^*$ can derive a sense from F^s only if at each F-singular point p in h the two F-rays in h incident with p are separated among F-rays incident with p by an even number of F-rays incident with p, including a null set of such F-rays. In this connection we shall use the following lemma.

Lemma 6.1. *Each $h \in F^*$ in $\beta R(N)$ which is concave towards $R(N)$ admits a sense derived from F^s. On h so sensed, the function V, defined in § 5 over $\bar{R}(N)-Z$, is strictly increasing.*

The function V, as defined in $R(N)$ in § 5, increases in the positive sense of each $a \in F^s|R(N)$ and so by virtue of Lemma 5.2 increases in the positive sense of each $a \in F^s|\beta R(N)$. Since the h in Lemma 6.1 clearly admits a sense derived from F^s, Lemma 6.1 follows.

Once V has been formally defined over $S-Z$ it is necessary to show that V is PH, and PC to U. For this purpose Lemma 6.2 is needed.

The following lemma has been established in [3 p. 26].

Lemma 6.2. *If a mapping φ from S to a w-plane is continuous in a neighbourhood H of a point $p \in S$ and interior at each point of $H-p$, then φ is interior at p.*

Definition of $V|\Sigma_n$. Use will be made of the notation of Theorem 2.3 beginning with N_1. From § 5 a definition of $V|(\bar{R}(N_1)-Z)$ is obtained on setting $N=N_1$. Proceeding inductively we shall assume that this definition of V has been extended over $\bar{\Sigma}_{n-1}-Z$ for $n>1$, so that V is PH over Σ_{n-1} and PC to U over Σ_{n-1}, continuous and bounded over $\bar{\Sigma}_{n-1}-Z$ with values on each $a \in F$ on which V has been defined which are strictly increasing in the positive sense of F^s. This induction will be completed at the end of § 6.

Since $\bar{\Sigma}_{n-1} \supset \bar{\eta}_n$ by virtue of Theorem 2.3, $V|(\bar{\eta}_n-Z)$ is defined. Let V^* be defined over $\bar{R}(N_n)-Z$ as in § 5, so as to be PH and bounded over $R(N_n)$ and PC to U. The open arc η_n serving as an "entrance" to $R(N_n)$ is a subarc of an open arc $h \in F^*$ in $\beta R(N_n)$ and by definition meets βN_n. According to Theorem 8.1 (b) of [4], h is then concave toward $R(N_n)$ and hence admits a sense derived from F^s. In accordance with Lemma 6.1 the entrance η_n may be parameterized by the values $t(z)=V(z)|(z \in \eta_n)$. These are values already assigned to η_n by our inductive hypothesis.

The open arc η_n, sensed by F^s, can be equally well parameterized by the values

$$\mu(z)=V^*(z)\,|\,(z\in\eta_n),$$

where $\mu(z)$ is the signed μ-length, measured along η_n from the intersection of η_n with the closure of the principal transversal of N_n. Values $\mu(z)$ and $t(z)$ which parameterize the same point $z\in\eta_n$ stand in a relation

$$t=T(\mu) \qquad (\mu_1<\mu<\mu_2),$$

where (μ_1,μ_2) is the finite range of $V^*(z)$ over η_n, and T is a continuous strictly increasing function mapping (μ_1,μ_2) onto the finite range (t_1,t_2) of $t(z)$ over η_n. Let T as defined over (μ_1,μ_2) be extended over the μ-axis so as to map the μ-axis homeomorphically onto the t-axis. Set

$$V(z)=TV^*(z) \qquad (z\in\bar{R}(N_n)-Z).$$

In particular this definition of V over $\bar{\eta}_n-Z$ agrees with the definition of V over $\bar{\eta}_n-Z$ as derived from the earlier definition of $V|(\bar{\Sigma}_{n-1}-Z)$. This completes the definition of $V|(\bar{\Sigma}_n-Z)$. It is clear that V is bounded and continuous over $\bar{\Sigma}_n-Z$ with values on each $a\in F$ on which V has been defined which are strictly increasing in the sense of F^s. A definition of $V|(S-Z)$ is implied.

Theorem 6.1. *Corresponding to any function U given as* PH *over $S-Z$ there exists a function V,* PH *over $S-Z$ and* PC *to U.*

The function V as just defined over $S-Z$ is clearly PH over each $R(N_n)$ and there PC to U. It remains to consider V in a neighbourhood of a point $p\in\eta_n$. If p is F-non-singular we introduce a right N_0 of p with canonical coordinates (u,v). We note that $\eta_n\cap N_0$ satisfies the condition $u=0$. As defined V is increasing over each $a\in F^s|(\bar{\Sigma}_{n-1}-Z)$ in the positive sense of a, by our inductive hypothesis. By construction V is increasing over each $a\in F^s|(\bar{R}(N_n)-Z)$ in the positive sense of a. In N_0 this is the sense of increasing v by virtue of our choice of F^s. It follows from Lemma 5.1 that V is PC to U on N_0.

If $p\in\eta_n$ is F-singular the preceding analysis shows that $U+iV$ is continuous in a neighbourhood X_p of p and interior at each point of X_p-p. It follows from Lemma 6.2 that $U+iV$ is interior at p.

This completes the induction and the proof of the theorem.

§ 7. Uniformization of PH functions. A PH function is obtained locally from a harmonic function by composition with a topological transformation. The question naturally arises whether a similar result holds in the large. This is the case for a PH function when defined in a simply connected domain as stated in Theorem 7.1.

Theorem 7.1. *Let U be* PH *in a simply connected domain J_z in the z-plane. There then exists a homeomorphism φ from J_z onto a domain J_w in the w-plane and a function U^* harmonic on J_w such that $U = U^* \varphi$.*

Indeed J_z is homemorphic to $S-Z$ under a transformation ψ from J_z to $S-Z$. Hence $U' = U\psi^{-1}$ is PH over $S-Z$. By Theorem 6.1 there exists V', PH over $S-Z$ and PC to U'. Then $V = V'\psi$ is PH over J_z and PC to $U = U'\psi$ [3, p. 39].

The function $U + iV = g$ is interior on J_z and thus $\zeta = g(z)$ maps J_z on a simply connected Riemann surface R_ζ spread over the $\zeta = (U + iV)$-plane. By the general uniformization theorem there exists a domain J_w in the w-plane and a function f analytic on J_w such that $\zeta = f(w)$ maps J_w onto R_ζ. The mapping $f^{-1}g$ is a homeomorphism φ of J_z^1 onto J_w and if one sets $f = U^* + iV^*$ with U^* and V^* real, we have $g = f\varphi$ and $U = U^*\varphi$. This proves Theorem 7.1.

It is clear that the theorem holds for a U which is PH in any simply connected Riemann domain.

Recall that V' and thus V may be constructed so as to be bounded. If this is done the Riemann surface R_ζ is of hyperbolic type. From this we deduce the following corollary.

Corollary 7.1. *If J_z is of hyperbolic type the mapping φ of Theorem* 7.1 *may be chosen so as to map J_z onto itself.*

Bibliography.

[1] W. M. Boothby, *The topology of the level curves of harmonic functions with critical points*, American Journal of Mathematics **73** (1951), pp. 512-538.

[2] C. Carathéodory, *Conformal representation*, Cambridge 1932.

[3] J. A. Jenkins and M. Morse, *Contour equivalent pseudoharmonic functions and pseudoconjugates*, American Journal of Mathematics **74** (1952), pp. 23—51.

[4] — *Topological methods on Riemann surfaces. Pseudoharmonic functions*, Annals of Mathematics Studies, Princeton 1952.

[5] W. Kaplan, *Regular curve families filling the plane*, Duke Mathematical Journal I **7** (1940), pp. 154-185, II **8** (1941), pp. 11-45.

[6] — *Differentiability of regular curve families on the sphere. Lectures in Topology*, Ann Arbor 1941, pp. 299-301.

[7] — *Topology of level curves of harmonic functions*, Transactions of the American Mathematical Society **63** (1948), pp. 514-522.

[8] M. Morse, *Topological methods in the theory of functions of a complex variable*, Annals of Mathematics Studies, Princeton 1947.

[9] H. Poincaré, *Mémoire sur les courbes définies par une équation différentielle*, Journal de Mathématique **7**, 3-rd series, (1881), pp. 375-422, and **8** (1882), pp. 251-296.

[10] S. Stoïlow, *Leçons sur les principes topologiques de la théorie des fonctions analytiques*, Paris 1938.

[11] G. T. Whyburn, *Analytic topology*, American Mathematical Society Colloquium Lectures, New York 1942.

Institute for Advanced Study and Johns Hopkins University.

ANNALS OF MATHEMATICS
Vol. 64, No. 3, November, 1956
Printed in U.S.A.

C-BIMEASURES Λ AND THEIR INTEGRAL EXTENSIONS

BY MARSTON MORSE AND WILLIAM TRANSUE

(Received December 26, 1955)

§1. Introduction

Let **C** be the field of complex numbers. **C**-measures on a locally compact topological space E are defined in Ref. 5. [Cf. Ref. 3, Chap. III.] Let E' and E'' be two locally compact topological spaces. Let $\mathcal{K}'_{\mathbf{C}}$ $[\mathcal{K}''_{\mathbf{C}}]$ be the vector space over **C** of the continuous mappings of E' $[E'']$ into **C** with compact support. Let $\mathcal{K}'$ and $\mathcal{K}''$ be similarly defined with **R** replacing **C**. For each pair $(u, v) \in \mathcal{K}'_{\mathbf{C}} \times \mathcal{K}''_{\mathbf{C}}$ let $\Lambda\ (u, v)$ be a complex number. For fixed u the mapping $v \to \Lambda(u, v)$ is denoted by $\Lambda\ (u, \cdot)$. The definition of $\Lambda(\cdot, v)$ is similar. With this understood the definition of a **C**-bimeasure is as follows.

The mapping Λ *of* $\mathcal{K}'_{\mathbf{C}} \times \mathcal{K}''_{\mathbf{C}}$ *into* **C** *is termed a* **C**-*bimeasure on* $E' \times E''$ *if* $\Lambda(\cdot, v)$ *and* $\Lambda(u, \cdot)$ *are* **C**-*measures on* E' *and* E'', *respectively, for fixed* $v \in \mathcal{K}''_{\mathbf{C}}$ *and* $u \in \mathcal{K}'_{\mathbf{C}}$.

Let K' $[K'']$ be an arbitrary compact subset of E' $[E'']$, and let $\mathcal{K}'_{K'}$ $[\mathcal{K}''_{K''}]$ be the vector subspace of $\mathcal{K}'_{\mathbf{C}}$ $[\mathcal{K}''_{\mathbf{C}}]$ of mappings $u[v]$ with support in K' $[K'']$. Let

$$(1.1) \qquad U(u) = \max_{s \in E'} |\, u(s) \,| \qquad U(v) = \max_{t \in E''} |\, v(t) \,| .$$

A necessary and sufficient condition that a bilinear mapping Λ of $\mathcal{K}'_{\mathbf{C}} \times \mathcal{K}''_{\mathbf{C}}$ into **C** be a **C**-bimeasure is that for each choice of the compact subsets $K' \subset E'$ and $K'' \subset E''$ there exists a constant $M(K', K'')$ such that

$$(1.2) \qquad |\, \Lambda(u, v) \,| \leqq M(K', K'')\ U(u)\ U(v)$$

for each $(u, v) \in \mathcal{K}'_{K'} \times \mathcal{K}''_{K''}$. [See Ref. 6, §2.]

Bimeasures. If the field **C** is replaced by the field **R** of real numbers, the definition of a **C**-measure reduces to the definition of an **R**-measure or simply a "measure" in the sense of Bourbaki, Ref. 3. In our terminology the definition of a **C**-bimeasure $\Lambda_{\mathbf{R}}$ similarly reduces to the definition of a bimeasure when **R** replaces **C**.

Absolute measures. It should be pointed out that the theory of **C** or **R**-bimeasures is not a theory of **C** or **R**-measures on $E' \times E''$. In this respect bimeasure theory differs from product measure theory. [Cf. Ref. 3, p. 89.] The most striking indication of this is found [Ref. 6, §10] in the fact that a bimeasure $\Lambda_{\mathbf{R}}$ on $E' \times E''$ does not in general admit a decomposition of the form

$$(1.3) \qquad \Lambda_{\mathbf{R}} = F_1 - F_2$$

where F_1 and F_2 are two positive bimeasures on $E' \times E''$. Recall that with each measure μ on E there is associated a unique decomposition $\mu = \mu^+ - \mu^-$ and an absolute measure $|\, \mu \,| = \mu^+ + \mu^-$. [See Ref. 3, p. 54.] Since (1.3) is not in general valid one cannot use a relation $|\, \Lambda_{\mathbf{R}} \,| = F_2 + F_2$ to define an absolute bimeasure.

It might seem that one could extend the formula

$$|\mu|(f) = \sup_{|u| \leqq f} |\mu(u)| \tag{1.4}$$

where (in the notation of Bourbaki) u and f are in $\mathcal{K}(E)$ and $f \geqq 0$. If $\mathcal{K}'_+$ and $\mathcal{K}''_+$ denote the subspaces of $\mathcal{K}'$ and $\mathcal{K}''$ of real positive mappings, one can in fact introduce the number

$$|\Lambda_{\mathbf{R}}|(f, g) = \sup_{|u| \epsilon f, |v| \epsilon g} |\Lambda_{\mathbf{R}}(u, v)| \tag{1.5}$$

for

$$(u, v) \in \mathcal{K}' \times \mathcal{K}'' \qquad (f, g) \in \mathcal{K}'_+ \times \mathcal{K}''_+. \tag{1.6}$$

With $|\Lambda_{\mathbf{R}}|$ so defined $|\Lambda_{\mathbf{R}}|(\cdot, g)$ and $|\Lambda_{\mathbf{R}}|(f, \cdot)$, unlike $|\mu|$, are not in general additive (cf. Ref. 6, §11), so that $|\Lambda_{\mathbf{R}}|$ does not in general admit an extension over $\mathcal{K}' \times \mathcal{K}''$ which is a bimeasure. This means that the theory of the integral extensions of a bimeasure or **C**-bimeasure must depart considerably from the methods used in defining and characterizing integral extensions of a measure.

Let $\mathbf{C}^{E'}$ $[\mathbf{C}^{E''}]$ denote the vector space of mappings of E' $[E'']$ into **C**. Given a pair of mappings

$$(x, y) \in \mathbf{C}^{E'} \times \mathbf{C}^{E''} \tag{1.7}$$

and a **C**-bimeasure Λ, our problem is to extend the definition of Λ so as to obtain a unique and natural definition of $\Lambda(x, y)$ as a limit (when possible) of values of Λ on $\mathcal{K}'_{\mathbf{C}} \times \mathcal{K}''_{\mathbf{C}}$. Unlike the corresponding problem for measures this problem involves at least four steps, and these steps may be ordered in different ways.

The Λ-integral of (x, y). Set the given **C**-measure $\Lambda(u, \cdot) = \alpha'_u$. Similarly set $\Lambda(\cdot, v) = \alpha''_v$. The integral $\alpha'_u(y)$ may exist for each $u \in \mathcal{K}'_{\mathbf{C}}$ and the mapping $u \to \alpha'_u(y)$ define a **C**-measure β'_y on E', or this may fail to happen. If the **C**-measure β'_y exists, the integral $\beta'_y(x)$ may or may not exist. The existence of the integral $\alpha''_v(x)$, the **C**-measure $\beta''_x : v \to \alpha''_v(x)$, and the integral $\beta''_x(y)$, are similarly in question. Finally if $\beta'_y(x)$ and $\beta''_x(y)$ exist one may or may not have the equality

$$\beta'_y(x) = \beta''_x(y). \tag{1.8}$$

In case (1.8) is defined and is valid, we set $\Lambda(x, y)$ equal to the common value of the two members of (1.8), call $\Lambda(x, y)$ the Λ-integral of (x, y), and term (x, y) Λ-integrable.

The superior integral Λ^*. In Ref. 6 we have associated a "superior integral" $\Lambda^*(h, k)$ with Λ for each pair (h, k) of positive numerical functions (finite or not), defined on E' and E'' respectively. The functions $\Lambda^*(\cdot, k)$ and $\Lambda^*(h, \cdot)$ are monotone and convex for fixed k and h, respectively. Moreover Λ^* satisfies two fundamental limit theorems established in Ref. 6.

A typical theorem using Λ^* is as follows. If the integrals $\beta'_y(x)$ and $\beta''_x(y)$ exist in (1.8), then (1.8) holds whenever $\Lambda^*(|x|, |y|) < \infty$. This is an analogue of the Fubini theorem for double integrals.

We shall obtain conditions sufficient, and in certain cases necessary, that

β'_y and β''_x exist as **C**-measures and that (x, y) be Λ-integrable. The most satisfactory conditions are in terms of Λ^*.

In this paper a **C**-bimeasure Λ is given and conditions are sought that a pair of mappings (x, y) be Λ-integrable. In a sequel to this paper on "Bilinear functionals on products $A \times B$ of function spaces with duals of integral type" it is the spaces A and B which are given. **C**-bimeasures Λ are sought such that for *each* $(x, y) \in A \times B$, (x, y) is Λ-integrable. The latter study leads to the representation of bounded operators from A to B' the dual of A, or from B to A'.

§2. Λ-integrability

We shall refer to

$$(x, y) \in \mathbf{C}^{E'} \times \mathbf{C}^{E''}, \qquad (u, v) \in \mathcal{K}'_{\mathbf{C}} \times \mathcal{K}''_{\mathbf{C}}. \tag{2.1}$$

Let Λ be a **C**-bimeasure. By hypothesis $\Lambda(u, \cdot)$ is a **C**-measure on E'' and $\Lambda(\cdot, v)$ a **C**-measure on E'.

Definition (i). *If* (a), *the integral* $\Lambda(\cdot, v)(x)$ *exists for each* $v \in \mathcal{K}''_{\mathbf{C}}$ *and if*

(b) *the mapping* $v \to \Lambda(\cdot, v)(x)$ *is a* **C**-*measure on* E'' *this measure will be denoted by* $\Lambda(x, \cdot)$ *so that*

$$\Lambda(\cdot, v)(x) = \Lambda(x, \cdot)(v). \tag{2.2}$$

The **C**-*measure* $\Lambda(\cdot, y)$ *is similarly defined.*

Definition (ii). *We shall say that* (x, y) *is* Λ-*integrable if the following three conditions are fulfilled.*

I. *The measures* $\Lambda(x, \cdot)$ *and* $\Lambda(\cdot, y)$ *exist.*

II. *The integrals* $\Lambda(x, \cdot)(y)$ *and* $\Lambda(\cdot, y)(x)$ *exist and*

$$\Lambda(x, \cdot)(y) = \Lambda(\cdot, y)(x). \tag{2.3}$$

Definition (iii). *In case* (x, y) *is* Λ-*integrable we set*

$$\Lambda(x, y) = \Lambda(x, \cdot)(y) = \Lambda(\cdot, y)(x). \tag{2.4}$$

The algebra of Λ-integrals depends upon the following lemma and theorems.

Lemma 2.1. (α) *If* $\Lambda(\cdot, y_1)$ *and* $\Lambda(\cdot, y_2)$ *exist as* **C**-*measures on* E', *then* $\Lambda(\cdot, y_1 + y_2)$ *is a* **C**-*measure on* E' *and*

$$\Lambda(\cdot, y_1 + y_2) = \Lambda(\cdot, y_1) + \Lambda(\cdot, y_2). \tag{2.5}$$

(β) *If* $\Lambda(x_1, \cdot)$ *and* $\Lambda(x_2, \cdot)$ *exist as* **C**-*measures on* E'', *then* $\Lambda(x_1 + x_2, \cdot)$ *is a* **C**-*measure on* E'' *and*

$$\Lambda(x_1 + x_2, \cdot) = \Lambda(x_1, \cdot) + \Lambda(x_2, \cdot). \tag{2.6}$$

We shall establish (β). For $v \in \mathcal{K}''_{\mathbf{C}}$, $\Lambda(x_1, \cdot)$ and $\Lambda(x_2, \cdot)$ exist as **C**-measures on E'' by hypothesis, so that

$$\Lambda(x_i, \cdot)(v) = \Lambda(\cdot, v)(x_i) \qquad (i = 1, 2) \tag{2.7}$$

in accordance with (2.2). For the **C**-measure $\Lambda(\cdot, v)$

$$\Lambda(\cdot, v)(x_1 + x_2) = \Lambda(\cdot, v)(x_1) + \Lambda(\cdot, v)(x_2). \tag{2.8}$$

The **C**-measures $\Lambda(x_1, \cdot)$ and $\Lambda(x_2, \cdot)$ have a **C**-measure α as a sum and

$$\alpha(v) = \Lambda(x_1, \cdot)(v) + \Lambda(x_2, \cdot)(v).$$

By (2.7) and (2.8) $\alpha(v) = \Lambda(\cdot, v)(x_1 + x_2)$. With $v \to \alpha(v)$, $v \to \Lambda(\cdot, v)(x_1 + x_2)$ is a **C**-measure. In accordance with Definition (i) its value at v is denoted by $\Lambda(x_1 + x_2, \cdot)(v)$. Reference to the preceding equations shows that

$$\Lambda(x_1 + x_2, \cdot)(v) = \Lambda(x_1, \cdot)(v) + \Lambda(x_2, \cdot)(v)$$

so that (2.6) holds.

The proof of (α) is similar.

THEOREM 2.1. (α) *If* (x, y_1) *and* (x, y_2) *are* Λ*-integrable, then* $(x, y_1 + y_2)$ *is* Λ*-integrable and*

$$\Lambda(x, y_1 + y_2) = \Lambda(x, y_1) + \Lambda(x, y_2). \tag{2.9}$$

(β) *Similarly if* (x_1, y) *and* (x_2, y) *are* Λ*-integrable then* $(x_1 + x_2, y)$ *is* Λ*-integrable and*

$$\Lambda(x_1 + x_2, y) = \Lambda(x_1, y) + \Lambda(x_2, y). \tag{2.10}$$

We shall prove (β). To this end we verify the conditions of Definitions (i) and (ii) as follows.

(i) (a) The integral $\Lambda(\cdot, v)(x_1 + x_2)$ exists since $\Lambda(\cdot, v)(x_1)$ and $\Lambda(\cdot, v)(x_2)$ exist by hypothesis. The integral $\Lambda(u, \cdot)(y)$ exists by hypothesis.

(i) (b) The mapping $v \to \Lambda(\cdot, v)(x_1 + x_2)$ is a **C**-measure on E'' by virtue of Lemma 2.1(β). The mapping $u \to \Lambda(u, \cdot)(y)$ is a **C**-measure on E' by hypothesis.

(ii) I. The measures $\Lambda(x_1 + x_2, \cdot)$ and $\Lambda(\cdot, y)$ exist as stated above.

(ii) II. The integral $\Lambda(x_1 + x_2, \cdot)(y)$ exists, since (2.6) holds and the integrals $\Lambda(x_1, \cdot)(y)$ and $\Lambda(x_2, \cdot)(y)$ exist. The integral $\Lambda(\cdot, y)(x_1 + x_2)$ exists since the integrals $\Lambda(\cdot, y)(x_1)$ and $\Lambda(\cdot, y)(x_2)$ exist by hypothesis.

(ii) III. $$\Lambda(x_1 + x_2, \cdot)(y) = \Lambda(\cdot, y)(x_1 + x_2),$$

since

$$\Lambda(x_1, \cdot)(y) = \Lambda(\cdot, y)(x_1) \qquad \Lambda(x_2, \cdot)(y) = \Lambda(\cdot, y)(x_2) \tag{2.11}$$

by hypothesis, while

$$\begin{aligned} \Lambda(x_1 + x_2, \cdot)(y) &= \Lambda(x_1, \cdot)(y) + \Lambda(x_2, \cdot)(y) && \text{(by 2.6)} \\ &= \Lambda(\cdot, y)(x_1) + \Lambda(\cdot, y)(x_2) && \text{by (2.11)} \\ &= \Lambda(\cdot, y)(x_1 + x_2). \end{aligned} \tag{2.12}$$

Thus $(x_1 + x_2, y)$ is Λ-integrable. That (2.10) holds follows from (2.12) and Definition (iii).

The proof of (α) is similar.

For completeness we add the following.

THEOREM 2.2. *If* (x, y) *is* Λ*-integrable and if* a *and* b *are complex constants then* (ax, by) *is* Λ*-integrable and* $\Lambda(ax, by) = ab\Lambda(x, y)$.

Partial maps into **C**. Let $E'_1[E''_1]$ be a subset of $E'[E'']$. Let $z[w]$ map $E'_1[E''_1]$ into **C**. For $s \in E' - E'_1[t \in E'' - E''_1]$ suppose $z[w]$ either undefined or such that $\mathcal{R}z(s)$ and $\mathcal{I}z(s)[\mathcal{R}w(t)$ and $\mathcal{I}w(t)]$ are numerical values in $\bar{\mathbf{R}}$, possibly infinite. We term $z[w]$ a *partial map* of $E'[E'']$ into **C**.

The definitions (i), (ii), (iii) will be understood as extended to the case in which (x, y) is replaced by the pair (z, w). A **C**-measure $\Lambda(z, \cdot)$, $[\Lambda(\cdot, w)]$ may then exist in the sense of this extended definition; the integrals $\Lambda(z, \cdot)(w)$ and $\Lambda(\cdot, w)(z)$ may also exist. Finally (z, w) is termed Λ-*integrable* if these integrals exist and if

$$\Lambda(z, \cdot)(w) = \Lambda(\cdot, w)(z). \tag{2.13}$$

If (2.13) holds one introduces the Λ-*integral* $\Lambda(z, w) = \Lambda(z, \cdot)(w) = \Lambda(\cdot, w)(z)$.

Lemma 2.1 and Theorems 2.1 and 2.2 then hold with (x, y) replaced by (z, w).

Λ-*negligible* sets. A subset E'_0 of $E'[E''_0$ of $E'']$ will be said to be Λ-negligible if $\Lambda(\cdot, v)$-negligible $[\Lambda(u, \cdot)$-negligible$]$ in the sense of Ref. 5, §5 for each $v \in K''_{\mathbf{C}}[u \in K'_{\mathbf{C}}]$. If $E'_0[E''_0]$ is Λ-negligible and if the above mapping $z[w]$ restricted to $E' - E'_0[E'' - E''_0]$ maps $E' - E'_0[E'' - E''_0]$ into **C**, then $z[w]$ will be said to be a mapping $z \in \mathbf{C}^{E'}[w \in \mathbf{C}^{E''}](p, p, \Lambda)$ and finite (p, p, Λ).

Two mappings z' and z'' in $\mathbf{C}^{E'}(p, p, \Lambda)$ will be said to be *equal* (p, p, Λ) if $z'(s) = z''(s)$ except on a Λ-negligible subset of E'. The equality $w'(t) = w''(t)(p, p, \Lambda)$ of two mappings in $\mathbf{C}^{E''}(p, p, \Lambda)$ is similarly defined. Given a partial map z of $E'[w$ of $E'']$ into **C** let $\tilde{z} \in \mathbf{C}^{E'}[\tilde{w} \in \mathbf{C}^{E''}]$ be mappings such that $\tilde{z}(s) = z(s)[\tilde{w}(t) = w(t)]$ whenever $z(s) \in \mathbf{C}[w(t) \in \mathbf{C}]$ and such that $\tilde{z}(s) = 0$ $[\tilde{w}(t) = 0]$ for other points $s \in E'[t \in E'']$. We term $\tilde{z}[\tilde{w}]$ the *finite projection* into $\mathbf{C}^{E'}[\mathbf{C}^{E''}]$ of $z[w]$.

If z is a partial map of E' into **C** such that $\Lambda(z, \cdot)$ is a **C**-measure on E' then z is in $\mathbf{C}^{E'}(p, p, \Lambda)$. If z' and z'' are two partial maps of E' into **C** which are equal $[p, p, \Lambda]$ then $\Lambda(z', \cdot)$ is a **C**-measure on E'' if and only if $\Lambda(z'', \cdot)$ is a **C**-measure on E'' and

$$\Lambda(z', \cdot) = \Lambda(z'', \cdot) = \Lambda(\tilde{z}', \cdot) = \Lambda(\tilde{z}'', \cdot).$$

§3. Approximating and majorizing the integrals $\Lambda(x, \cdot)(y)$ and $\Lambda(\cdot, y)(x)$

Let (x, y) be given as in (2.1). With the aid of Theorem 8.1 of Ref. 5, the basic approximation Theorem 3.1 can be established.

Notation. We shall refer to the positive real axis $\mathbf{R}_+$ and to its completion $\bar{\mathbf{R}}_+$ by the point $+\infty$. Let $\mathcal{I}'_+[\mathcal{I}''_+]$ be the space of lower semicontinuous mappings of $E'[E'']$ into $\bar{\mathbf{R}}_+$. In addition to the pairs of mappings (x, y) and (u, v) limited as in (2.1) we shall make permanent use of

$$(h, k) \in \bar{\mathbf{R}}_+^{E'} \times \bar{\mathbf{R}}_+^{E''}, \qquad (p, q) \in \mathcal{I}'_+ \times \mathcal{I}''_+, \qquad (f, g) \in \mathcal{K}'_+ \times \mathcal{K}''_+. \tag{3.0}$$

The filtering sets ${}_-\phi'_p$, ${}_-\phi''_q$, ${}_+\phi'_h$, ${}_+\phi''_k$. In Ref. 6 we have introduced filtering sets of mappings as follows. [Cf. Ref. 2, p. 35.] For $p \in \mathcal{I}'_+$ let ${}_-\phi'_p$ be the ensemble of mappings $f \in \mathcal{K}'_+$ with $f \leq p$, filtering for the relation $\geqq$. For $q \in \mathcal{I}''_+$ let ${}_-\phi''_q$ be similarly composed of $g \in \mathcal{K}''_+$. For an arbitrary mapping h of E' into $\bar{\mathbf{R}}_+$ let ${}_+\phi'_h$ be the ensemble of mappings $p \in \mathcal{I}'_+$ with $p \geqq h$, filtering for the relation $\leqq$. For an arbitrary mapping k of E'' into $\bar{\mathbf{R}}_+$ let ${}_+\phi''_k$ be similarly composed of $q \in \mathcal{I}''_+$.

We assume throughout this paper that Λ is a C-bimeasure.

THEOREM 3.1. *If the integral* $\Lambda(\cdot, y)(x)[\Lambda(x, \cdot)(y)]$ *exists, then for any* $(p, q) \in {}_+\phi'_{|x|} \times {}_+\phi''_{|y|}$ *and for arbitrary* $\delta > 0$ *there exists* $(u, v) \in \mathcal{K}'_C \times \mathcal{K}''_C$ *with* $|u| \leqq p$, $|v| \leqq q$ *such that*

$$(3.1) \qquad | \Lambda(u, v) - \Lambda(\cdot, y)(x) | < \delta \; [| \Lambda(u, v) - \Lambda(x, \cdot)(y) | < \delta].$$

Set $\Lambda(\cdot, y) = \alpha$. In our definition in §2 the existence of the integral $\Lambda(\cdot, y)(x)$ presupposes that α is a C-measure, and that x is α-integrable. By Theorem 8.1 of Ref. 5 there accordingly exists a $u \in \mathcal{K}'_C$ with $|u| \leqq p$ such that $|\alpha(u) - \alpha(x)| < \frac{1}{2}\delta$, or in terms of Λ

$$(3.2) \qquad | \Lambda(\cdot, y)(u) - \Lambda(\cdot, y)(x) | < \tfrac{1}{2}\delta.$$

By definition of $\Lambda(\cdot, y)$, $\Lambda(u, \cdot)(y) = \Lambda(\cdot, y)(u)$. We now set $\Lambda(u, \cdot) = \beta$ and apply Theorem 8.1 of Ref. 5 to $\beta(y)$, inferring thereby that there exists $v \in \mathcal{K}''_C$ such that $|v| \leqq q$ and

$$(3.3) \qquad | \Lambda(u, \cdot)(v) - \Lambda(u, \cdot)(y) | < \tfrac{1}{2}\delta.$$

Since $\Lambda(u, \cdot)(v) = \Lambda(u, v)$ we conclude from (3.3) and (3.2) that the first inequality in (3.1) is valid.

The proof of the second inequality is similar.

DEFINITION OF Λ^*. We have defined Λ^* in Ref. 6 by setting

$$(3.4) \qquad \Lambda^*(p, q) = \sup_{|u| \leqq p,\, |v| \leqq q} | \Lambda(u, v) |$$

$$(3.5) \qquad \Lambda^*(h, k) = \inf_{p \geqq h,\, q \geqq k} \Lambda^*(p, q)$$

where the notation of (3.0) is employed.

In Ref. 6 we have shown that $\Lambda^*(h, \cdot)$ is monotone and convex on the space of positive numerical functions $k \in \bar{\mathbf{R}}''_+$. Similarly $\Lambda^*(\cdot, k)$ is monotone and convex.

The approximation Theorem 3.1 leads to the following "majorizing" theorem.

THEOREM 3.2. *If the integral* $\Lambda(x, \cdot)(y)$ *exists*

$$(3.6) \qquad | \Lambda(x, \cdot)(y) | \leqq \Lambda^*(|x|, |y|).$$

If the integral $\Lambda(\cdot, y)(x)$ *exists*

$$(3.7) \qquad | \Lambda(\cdot, y)(x) | \leqq \Lambda^*(|x|, |y|).$$

It will be sufficient to establish (3.6) when $\Lambda^*(|x|, |y|) < \infty$. Let (u, v) and (p, q) be chosen as in Theorem 3.1 with $|u| \leqq p$, $|v| \leqq q$. Then $|\Lambda(u, v)| \leqq \Lambda^*(p, q)$ in accordance with the definition of $\Lambda^*(p, q)$. Because of (3.1)

$$| \Lambda(x, \cdot)(y) | \leqq | \Lambda(u, v) | + \delta \leqq \Lambda^*(p, q) + \delta.$$

Since δ is arbitrary $| \Lambda(x, \cdot)(y) | \leqq \Lambda^*(p, q)$. Hence

$$| \Lambda(x, \cdot)(y)| \leqq \inf_{p \geqq |x|,\, q \geqq |y|} \Lambda^*(p, q) = \Lambda^*(| x |, | y |)$$

establishing (3.6). The proof of (3.7) is similar.

Theorems 3.1 and 3.2 can be extended as follows.

THEOREM 3.3. *Let z and w be partial maps of E' and E'' respectively into* **C** *such that the integral $\Lambda(\cdot, w)(z)[\Lambda(z, \cdot)(w)]$ exists. Then for $| \bar{z} | = h$ and $| \bar{w} | = | k |$, for any $(p, q) \in {}_+\phi_h' \times {}_+\phi_k''$, and for arbitrary $\delta > 0$, there exists $(u, v) \in \mathcal{K}_C' \times \mathcal{K}_C''$ with $| u | \leqq p$, $| v | \leqq q$ such that*

$$(3.8) \qquad | \Lambda(u, v) - \Lambda(\cdot, w)(z) | < \delta \qquad [| \Lambda(u, v) - \Lambda(z, \cdot)(w) | < \delta].$$

If the integral $\Lambda(\cdot, w)(z)$ exists the integral $\Lambda(\cdot, \bar{w})(\bar{z})$ exists and equals $\Lambda(\cdot, w)(z)$. Similarly the existence of $\Lambda(z, \cdot)(w)$ implies that $\Lambda(\bar{z}, \cdot)(\bar{w})$ exists and equals $\Lambda(z, \cdot)(w)$. Theorem 3.1 applies to the pair $(\bar{z}, \bar{w})$ and Theorem 3.3 is inferred.

THEOREM 3.4. *If the integral $\Lambda(z, \cdot)(w)$ exists*

$$(3.9) \qquad | \Lambda(z, \cdot)(w) | \leqq \Lambda^*(| \bar{z} |, | \bar{w} |).$$

If the integral $\Lambda(\cdot, w)(z)$ exists

$$(3.10) \qquad | \Lambda(\cdot, w)(z) | \leqq \Lambda^*(| \bar{z} |, | \bar{w} |).$$

The proof of Theorem 3.4 differs from that of Theorem 3.2 only in that the choice of (u, v) and (p, q) in the present proof are in accordance with Theorem 3.3 rather than Theorem 3.1.

§4. Λ-integrability by the method of decomposition

Given a mapping $x \in \mathbf{C}^E$, the mappings $\mathcal{R}x$ and $\mathcal{I}x$ are well defined, as well as their positive and negative components,

$$(4.0) \qquad [\mathcal{R}x]^+, \qquad [\mathcal{R}x]^-, \qquad [\mathcal{I}x]^+, \qquad [\mathcal{I}x]^-.$$

The respective mappings (x_1, x_2, x_3, x_4) appearing in (4.0) will be termed the *Riesz components* of x. The Riesz components (y_1, y_2, y_3, y_4) of a mapping $y \in \mathbf{C}^{E''}$ are similarly defined.

We shall seek conditions for the Λ-integrability of a pair of mappings $(x, y) \in \mathbf{C}^{E'} \times \mathbf{C}^{E''}$. Let

$$(4.1) \qquad (x_1, x_2, x_3, x_4) \qquad (y_1, y_2, y_3, y_4)$$

be respectively the Riesz components of x and y. It would seem a priori likely that conditions for the Λ-integrability of (x, y) could be obtained by the *method of decomposition*, that is by a study of the Λ-integrability of the pairs (x_i, y_j) $i, j = 1, \cdots, 4$. This method is useful for some purposes. It is limited by the following fact. The hypothesis that the integral $\Lambda(x, \cdot)(y)[\Lambda(\cdot, y)(x)]$ exists does not imply that the respective integrals $\Lambda(x_i, \cdot)(y_j)[\Lambda(\cdot, y_j)(x_i)]$ exist, as we shall show by counter example. Thus the fundamental theorem that the existence of $\Lambda(x, \cdot)(y)$

and $\Lambda(\cdot, y)(x)$ and the finiteness of $\Lambda^*(|x|, |y|)$ implies the Λ-integrability of (x, y) does not appear capable of proof by the direct method of decomposition.

In this section we shall establish a lemma and theorem on Λ-integrability, using the method of decomposition. In the next section we present the basic limit theorems on Λ^* in vector form using methods which are apparently more powerful than the method of decomposition.

The first "decomposition" lemma follows.

LEMMA 4.1. *Let the Riesz components of the mappings x and y be given as in* (4.1).

(i) *If each $\Lambda(x_i, \cdot)$ is a* **C**-*measure on E'', then $\Lambda(x, \cdot)$ is a* **C**-*measure on E'' and*

$$\Lambda(x, \cdot) = \Lambda(x_1, \cdot) - \Lambda(x_2, \cdot) + i\Lambda(x_3, \cdot) - i\Lambda(x_4, \cdot). \tag{4.2}$$

(ii) *If each $\Lambda(\cdot, y_i)$ is a* **C**-*measure on E', $\Lambda(\cdot, y)$ is a* **C**-*measure on E' and*

$$\Lambda(\cdot, y) = \Lambda(\cdot, y_1) - \Lambda(\cdot, y_2) + i\Lambda(\cdot, y_3) - i\Lambda(\cdot, y_4). \tag{4.3}$$

This lemma follows at once from Lemma 2.1.

We are led to the following theorem.

THEOREM 4.1. *If the Riesz components of mappings $(x, y) \in \mathbf{C}^{E'} \times \mathbf{C}^{E''}$ are given in* (4.1) *and if each pair (x_i, y_j), $i, j = 1, \cdots, 4$, is Λ-integrable, then (x, y) is Λ-integrable.*

If each pair (x_i, y_j) is Λ-integrable it follows from Lemma 4.1 that the measures $\Lambda(x, \cdot)$ and $\Lambda(\cdot, y)$ exist and are given by (4.2) and (4.3) respectively. By hypothesis the integrals

$$\Lambda(x_i, \cdot)(y_j) = \Lambda(\cdot, y_j)(x_i) \qquad (i, j = 1, \cdots, 4) \tag{4.4}$$

exist and are equal as indicated. With the aid of (4.2) one sees that $\Lambda(x, \cdot)(y)$ exists and is representable in the form

$$\Lambda(x, \cdot)(y) = \textstyle\sum_{ij} a_{ij}\Lambda(x_i, \cdot)(y_j) \qquad (i, j = 1, \cdots, 4) \tag{4.5}$$

where a_{ij} is in **C** and $a_{ij} = a_{ji}$. With the aid of (4.3) one similarly sees that the integral $\Lambda(\cdot, y)(x)$ exists and that

$$\Lambda(\cdot, y)(x) = \textstyle\sum_{ij} a_{ij}\Lambda(\cdot, y_i)(x_j) \qquad (i, j = 1, \cdots, 4). \tag{4.6}$$

It follows then from (4.4), (4.5) and (4.6) and the relation $a_{ij} = a_{ji}$ that $\Lambda(x, \cdot)(y) = \Lambda(\cdot, y)(x)$.

Counter example. This example will show that the existence of the integral $\Lambda(x, \cdot)(y)$ does not in general imply the existence of each integral $\Lambda(x_i, \cdot)(y_j)$ even when $\Lambda^*(|x|, |y|) < \infty$.

Let E' be the pair of points $\{1, 2\}$ with discrete topology and E'' the interval $[0, 1]$. Let μ denote Lebesgue measure on $[0, 1]$, and let k be a positive function on $[0, 1]$ which is not Lebesgue integrable. Suppose further that $k(s)$ is bounded by a constant B for $s \in E''$. For $(u, v) \in \mathcal{K}'_{\mathbf{C}} \times \mathcal{K}''_{\mathbf{C}}$ let a **C**-bimeasure Λ on $E' \times E''$ be defined by its values

$$\Lambda(u, v) = [u(1) + u(2)]\mu_e(v), \tag{4.7}$$

where $\mu_e(v) = \mu(\mathcal{R}v) + i\mu(\mathcal{I}v)$, and defines a **C**-measure μ_e on E'' which extends the measure μ. Let $u_0 \in \mathcal{K}_{\mathbf{C}}'$ have the values $u_0(1) = 1$, $u_0(2) = -1$. Then $u_0^+(1) = 1$, $u_0^+(2) = 0$, $u_0^-(1) = 0$, $u_0^-(2) = 1$ so that $|u_0|(1) = 1$ and $|u_0|(2) = 1$. Using the definition of Λ^* we see that with $i = 1, 2$

$$(4.8)' \qquad \Lambda^*(|u_0|, q) = \sup_{|u(i)| \leqq 1} |u(1) + u(2)| \, |\mu|^*(q) = 2|\mu|^*(q)$$

$$(4.8)'' \qquad \Lambda^*(|u_0|, k) = 2 \inf_{q \geqq k} \mu^*(q) \leqq 2B$$

noting that an admissible choice of q in (4.8)″ is obtained by setting $q(t) = B$ for $t \in E''$.

Since $u_0(1) + u_0(2) = 0$, it follows from (4.7) that the **C**-measure $\Lambda(u_0, \cdot) \equiv 0$. The integral $\Lambda(u_0, \cdot)(k)$ accordingly exists. Moreover (4.7) implies that the **C**-measure $\Lambda(u_0^+, \cdot) = \mu_e$. Since k is not Lebesgue integrable the integral $\mu(k)$ does not exist. Thus $\Lambda(u_0^+, \cdot)(k)$ does not exist even though $\Lambda^*(|u_0|, k) < \infty$ in accordance with (4.8) and the integral $\Lambda(u_0, \cdot)(k)$ exists.

Because of the limitations of the method of decomposition in the study of Λ-integrability we turn in the next section to new vectorial methods.

§5. The superior integral Λ^*

With the aid of the first and second limit theorems for Λ^* as established in Ref. 6, and as here extended in vector form, and with the aid of the majorizing Theorem 3.2, conditions sufficient for Λ-integrability can be established. The extent to which these conditions are necessary will also be studied.

Use will be made of the pairs of positive mappings (h, k), (p, q) and (f, g), in (3.0).

The vectors $\mathbf{z}$ *and finite images* $\dot{\mathbf{z}}$. Functions h, k, p, q, f, g are numerical valued and positive. They are respectively limited to six well-defined function spaces. Let Z be any one of these six spaces. Let the vector $\mathbf{z}$ represent a point (z_1, z_2, z_3, z_4) in the product space

$$(5.1) \qquad Z \times Z \times Z \times Z = [Z]^4.$$

Referring to the definition of $\dot{z}_i$ in §2 set

$$(5.2) \qquad \dot{z}_1 - \dot{z}_2 + i(\dot{z}_3 - \dot{z}_4) = \dot{\mathbf{z}}.$$

Note that this *finite image* $\dot{\mathbf{z}}$ of $\mathbf{z}$ is well-defined while $z_1 - z_2$ is not defined if $z_1(s) = z_2(s) = +\infty$ for some s. Similarly $z_3 - z_4$ might not be defined.

Vector pairs

$$(5.3) \quad (\mathbf{h}, \mathbf{k}) \in [\bar{\mathbf{R}}_+^{E'}]^4 \times [\bar{\mathbf{R}}_+^{E''}]^4; \qquad (\mathbf{p}, \mathbf{q}) \in [\mathcal{I}_+']^4 \times [\mathcal{I}_+'']^4; \qquad (\mathbf{f}, \mathbf{g}) \in [\mathcal{K}_+']^4 \times [\mathcal{K}_+'']^4$$

have well-defined finite images $(\dot{\mathbf{h}}, \dot{\mathbf{k}})$, $(\dot{\mathbf{p}}, \dot{\mathbf{q}})$, $(\dot{\mathbf{f}}, \dot{\mathbf{g}})$ in $\mathbf{C}^{E'} \times \mathbf{C}^{E''}$. Given $x \in \mathbf{C}^{E'}$ $[y \in \mathbf{C}^{E''}]$, we say that a vector $\mathbf{x}[\mathbf{y}]$ *canonically represents* $x[y]$ if the components of $\mathbf{x}[\mathbf{y}]$ are the Riesz components of x [of y].

We shall refer to *product filtering sets*

$$(5.4) \qquad \Phi'(\mathbf{p}), \ {}_{-}\Phi''(\mathbf{q}), \ {}_{+}\Phi'(\mathbf{h}), \ {}_{+}\Phi''(\mathbf{k}),$$

defined as follows. The product of a finite number of filtering sets has been defined in §7 of Ref. 5. Given $(\mathbf{p}, \mathbf{q})$ and $(\mathbf{h}, \mathbf{k})$ as in (5.3) we introduce the product filtering sets,

$$\prod_{i=1}^{4} {}_{-}\phi'_{p_i} = {}_{-}\Phi'(\mathbf{p}), \qquad \prod_{i=1}^{4} {}_{-}\phi''_{q_i} = {}_{-}\Phi''(\mathbf{q}) \tag{5.5}$$

in the product spaces $[\mathcal{K}'_+]^4$ and $[\mathcal{K}''_+]^4$ respectively, as well as the product filtering sets,

$$\prod_{i=1}^{4} {}_{+}\phi'_{h_i} = {}_{+}\Phi'(\mathbf{h}), \qquad \prod_{i=1}^{4} {}_{+}\phi''_{k_i} = {}_{+}\Phi''(\mathbf{k}). \tag{5.6}$$

Following notational conventions introduced in §7 of Ref. 5, the limits with respect to the filters of sections of the respective filtering sets (5.4) will be indicated by symbols

$$\lim_{\mathbf{f}\uparrow\mathbf{p}}, \qquad \lim_{\mathbf{g}\uparrow\mathbf{q}}, \qquad \lim_{\mathbf{p}\downarrow\mathbf{h}}, \qquad \lim_{\mathbf{q}\downarrow\mathbf{k}} \tag{5.7}$$

placed before the function whose limit is desired. Given one of the above vectors $\mathbf{z}$ let

$$|\,\mathbf{z}\,| = z_1 + z_2 + z_3 + z_4\,.$$

With this understood the first limit theorem on Λ^* in Ref. 6, §6 has the following extension.

THEOREM 5.1. *If* $\Lambda^*(|\,\mathbf{p}\,|, |\,\mathbf{q}\,|) < \infty$ *then*

$$\lim_{\mathbf{f}\uparrow\mathbf{p}} \Lambda^*(|\,\dot{\mathbf{p}} - \dot{\mathbf{f}}\,|, |\,\mathbf{q}\,|) = 0; \qquad \lim_{\mathbf{g}\uparrow\mathbf{q}} \Lambda^*(|\,\mathbf{p}\,|, |\,\dot{\mathbf{q}} - \dot{\mathbf{g}}\,| = 0. \tag{5.8}$$

Note first that the hypothesis $\Lambda^*(|\,\mathbf{p}\,|, |\,\mathbf{q}\,|) < \infty$ implies that

$$\Lambda^*(p_i, |\,\mathbf{q}\,|) < \infty \qquad (i = 1, 2, 3, 4). \tag{5.9}$$

To establish the first relation in (5.8) observe further that the convexity of $\Lambda^*(\cdot, |\,\mathbf{q}\,|)$ and the relation

$$|\,\dot{\mathbf{p}} - \dot{\mathbf{f}}\,| \leqq \sum_{i=1}^{4} |\,p_i - f_i\,| \tag{5.10}$$

together imply that

$$\Lambda^*(|\,\dot{\mathbf{p}} - \dot{\mathbf{f}}\,|, |\,\mathbf{q}\,|) \leqq \sum_{i=1}^{4} \Lambda^*(p_i - f_i, |\,\mathbf{q}\,|).$$

Moreover $|\,\mathbf{q}\,|$ is clearly in $\mathcal{J}''_+$ with the mappings q_i, $i = 1, 2, 3, 4$. It follows from Theorem 6.1 of Ref. 6. that

$$\lim_{f_i\uparrow p_i} \Lambda^*(p_i - f_i, |\,\mathbf{q}\,|) = 0 \qquad (i = 1, \cdots 4).$$

The first relation in (5.8) is a consequence.

The proof of the second relation in (5.8) is similar.

The second limit theorem on Λ^* in §7 of Ref. 6 similarly leads to a theorem on $(x, y) \in \mathbf{C}^{E'} \times \mathbf{C}^{E''}$.

THEOREM 5.2. *Suppose that* $\Lambda^*(|\,x\,|, |\,y\,|) < \infty$ *and that vectors* $\mathbf{x}$ *and* $\mathbf{y}$ *canoni-*

cally represent x and y, then

$$\lim_{\mathbf{p}\downarrow\mathbf{x}} \Lambda^*(|\dot{\mathbf{p}} - \dot{\mathbf{p}}'|, |y|) = 0, \qquad \lim_{\mathbf{q}\downarrow\mathbf{y}} \Lambda^*(|x|, |\dot{\mathbf{q}} - \dot{\mathbf{q}}'|) = 0 \tag{5.11}$$

where $\mathbf{p}'$ is in the section $S'(\mathbf{p})$ of $_+\Phi'(\mathbf{x})$ and $\mathbf{q}'$ is in the section $S''(\mathbf{q})$ of $_+\Phi''(\mathbf{y})$.

We shall make use of the following corollary of Theorem 5.1.

COROLLARY 5.1. (i). *If $\Lambda^*(|\mathbf{p}|, |y|) < \infty$ then*

$$\lim_{\mathbf{f}\uparrow\mathbf{p}} \Lambda^*(|\dot{\mathbf{p}} - \dot{\mathbf{f}}|, |y|) = 0. \tag{5.12}$$

(ii). *If $\Lambda^*(|x|, |\mathbf{q}|) < \infty$ then*

$$\lim_{\mathbf{g}\uparrow\mathbf{q}} \Lambda^*(|x|, |\dot{\mathbf{q}} - \dot{\mathbf{g}}|) = 0. \tag{5.13}$$

Let (y_1, y_2, y_3, y_4) be the Riesz components of y. To establish (i) one first notes that the hypothesis $\Lambda^*(|\mathbf{p}|, |y|) < \infty$ implies that for some $\mathbf{q} \in [\mathcal{J}''_+]^4$ with $q_i \geqq y_i$, $\Lambda^*(|\mathbf{p}|, |\mathbf{q}|) < \infty$. Relation (5.12) now follows from the first relation in (5.8). The relation (5.13) is similarly established.

Theorem 5.3 extends Cor. 7.1 of Ref. 6. In stating this theorem we adopt the convention that when vectors $\mathbf{a}$ and $\mathbf{b}$ are in $[\bar{\mathbf{R}}^E]^4$ and $a_i \leqq b_i$, $i = 1, \cdots, 4$, we write $\mathbf{a} \leqq \mathbf{b}$.

THEOREM 5.3. *Suppose that $\Lambda^*(|x|, |y|) < \infty$ and that $\mathbf{y}$ canonically represents y. If $\mathbf{q}'$ varies in the section $S(\mathbf{q})$ of $_+\Phi''(\mathbf{y})$ and if $\mathbf{k}'$ and $\mathbf{k}''$ are in $[\bar{\mathbf{R}}_+^{E''}]^4$ and vary so that*

$$\mathbf{q}' \leqq \mathbf{k}' \leqq \mathbf{q}, \qquad \mathbf{q}' \leqq \mathbf{k}'' \leqq \mathbf{q} \tag{5.14}$$

then

$$\lim_{\mathbf{q}\downarrow\mathbf{y}} \Lambda^*(|x|, |\dot{\mathbf{k}}' - \dot{\mathbf{k}}''|) = 0. \tag{5.15}$$

Relation (5.15) follows from Cor. 7.1 of Ref. 6 on noting that

$$\Lambda^*(|x|, |\dot{\mathbf{k}}' - \dot{\mathbf{k}}''|) \leqq \sum_{i=1}^4 \Lambda^*(|x|, |k_i' - k_i''|).$$

Theorem 5.3 has a *dual* in which the roles of the first and second arguments are interchanged.

We make use of the following in the next sections. If p is in $\mathcal{J}'_+$ and if $v \to \Lambda(\cdot, v)(p)$ defines a measure on E'', p is finite (p, p, Λ). For the integral $\Lambda(\cdot, v)(p)$ then exists for each $v \in \mathcal{K}''_{\mathbf{C}}$, implying that p is finite $(p, p, \Lambda(\cdot, v))$. If $\Lambda(p, \cdot)$ is a $\mathbf{C}$-measure on E'', $\Lambda(\tilde{p}, \cdot)$ is a $\mathbf{C}$-measure on E'' and

$$\Lambda(p, \cdot) = \Lambda(\tilde{p}, \cdot). \tag{5.16}$$

§6. C-measures $\Lambda(\dot{\mathbf{p}}, \cdot)$ and $\Lambda(\cdot, \dot{\mathbf{q}})$

We are here concerned with pairs of vectors $(\mathbf{p}, \mathbf{q})$ limited as in (5.3). The results obtained in this section are essential in the study of the Λ-integrability of a general pair of mappings $(x, y) \in \mathbf{C}^{E'} \times \mathbf{C}^{E''}$. A major lemma follows.

LEMMA 6.1. *If p is in $\mathcal{G}'_+$, a necessary and sufficient condition that $\Lambda(p, \cdot)$ exist as a* **C***-measure on E'' is that*

$$\Lambda^*(p, g) < \infty \tag{6.1}$$

for each $g \in \mathcal{K}''_+$.

We first prove the condition sufficient.

To show that $\Lambda(p, \cdot)$ is a **C**-measure one must verify (a) and (b) of Definition (i) of §2.

(a) The integral $\Lambda(\cdot, v)(p)$ exists for $v \in \mathcal{K}''_{\mathbf{C}}$ if the superior integral $|\Lambda(\cdot, v)|^*(p) < \infty$. [Lemma 5.2, Ref. 5.] Now $|v|$ is in $\mathcal{K}''_+$, and by Lemma 3.1 of Ref. 6, and (6.1),

$$|\Lambda(\cdot, v)|^*(p) \leqq \Lambda^*(p, |v|) < \infty. \tag{6.2}$$

Hence $\Lambda(\cdot, v)(p)$ exists.

(b) That the mapping $v \to \Lambda(\cdot, v)(p)$ of $\mathcal{K}''_{\mathbf{C}}$ into **C** is a **C**-measure on E'' is seen as follows. Let K be a compact subset of E'', and $g \in \mathcal{K}''_+$ such that $g \geqq \phi_K$, where ϕ_K is the characteristic function of K. Since $\Lambda(\cdot, v)(p)$ exists

$$|\Lambda(\cdot, v)(p)| \leqq \Lambda^*(p, |v|) \qquad \text{[by Th. 3.4]}.$$

In case K contains the support of v, $U(v)g \geqq |v|$ so that

$$U(v)\Lambda^*(p, g) \geqq \Lambda^*(p, |v|) \geqq |\Lambda(\cdot, v)(p)|.$$

Thus $\Lambda(p, \cdot)$ is a **C**-measure on E''.

We now prove the condition (6.1) necessary.

Given $g \in \mathcal{K}''_+$ let K be a compact subset of E'' which includes the support of g. Let H_K be the subset of $v \in \mathcal{K}''_{\mathbf{C}}$ with support K. With the norm $v \to U(v)$, H_K is a Banach space. Note that $g \in H_K$. For each $u \in \mathcal{K}'_{\mathbf{C}}$ there exists a positive constant $M(u)$ such that

$$\sup_{v \in H_K} \frac{|\Lambda(u, v)|}{U(v)} = M(u) < \infty \qquad [v \neq 0] \tag{6.3}$$

since $\Lambda(u, \cdot)$ is a **C**-measure. Concerning $M(u)$ we shall prove the following.

(β). *For fixed K and variable $u \in \mathcal{K}'_{\mathbf{C}}$ with $|u| \leqq p \in \mathcal{G}''_+$, $M(u)$ admits a bound B.*

If (β) were false there would exist a sequence of mappings $u_n \in \mathcal{K}'_{\mathbf{C}}$ with $|u_n| \leqq p$ such that $M(u_n) \uparrow \infty$ as $n \uparrow \infty$. Consider then the sequence of linear forms V_n on H_K with values $V_n(v) = \Lambda(u_n, v)$. For fixed n, V_n is continuous on H_K.

For $v \in H_K$ set $\Lambda(\cdot, v) = \alpha$. By hypothesis on Λ, α is a **C**-measure on E''. Since $\alpha(p)$ exists by hypothesis $|\alpha|(p)$ exists by Lemma 4.1 of Ref. 5. Thus for fixed $v \in H_K$ and for $n = 1, 2, \cdots$,

$$|V_n(v)| = |\Lambda(\cdot, v)(u_n)| \leqq |\Lambda(\cdot, v)|(|u_n|) \leqq |\Lambda(\cdot, v)|(p).$$

Now $\Lambda(\cdot, v)(p)$ exists by hypothesis so that $|\Lambda(\cdot, v)|(p) < \infty$. Thus $|V_n(v)|$ is bounded for $n = 1, 2, \cdots$, and fixed $v \in H_K$. It follows from Theorem 5 of

Banach, p. 80, that for some constant B_0, $\| V_n \| \leqq B_0$ for $n = 1, 2, \cdots$. But $\| V_n \| = M(u_n)$ by (6.3). Thus $M(u_n) \leqq B_0$ for $n = 1, 2, \cdots$. From this contradiction we infer the truth of (β).

To prove that $\Lambda^*(p, g) < \infty$ we start with the definition

$$\Lambda^*(p, g) = \sup_{|u| \leq p,\ |v| \leq g} | \Lambda(u, v) | \qquad [(u, v) \in \mathcal{K}'_C \times \mathcal{K}''_C]. \tag{6.4}$$

Since v is in H_K this becomes

$$\leqq M(u)U(g) \leqq BU(g) < \infty$$

in accordance, respectively, with (6.3) and (β). This establishes the necessity of the condition (6.1).

To Lemma 6.1 corresponds a dual lemma giving a necessary and sufficient condition that $\Lambda(\cdot, q)$ be a **C**-measure on E'' for $q \in \mathcal{G}''_+$.

Let pairs $(\mathbf{p}, \mathbf{q})$ be conditioned as in (5.3).

THEOREM 6.1. (i) *A necessary and sufficient condition that* $\Lambda(p_j, \cdot)$ *be a* **C**-*measure on* E'' *for* $j = 1, 2, 3, 4$ *is that for each* $g \in \mathcal{K}''_+$

$$\Lambda^*(| \mathbf{p} |, g) < \infty. \tag{6.5}$$

(ii) *A necessary and sufficient condition that* $\Lambda(\cdot, q_j)$ *be a* **C**-*measure on* E' *for* $j = 1, 2, 3, 4$ *is that for each* $f \in \mathcal{K}''_+$

$$\Lambda^*(f, | \mathbf{q} |) < \infty. \tag{6.6}$$

We consider (i) and prove the condition (6.5) sufficient. Since $\Lambda^*(| \mathbf{p} |, g) < \infty$ we infer that $\Lambda^*(p_j, g) < \infty$, $j = 1, \cdots, 4$ for each $g \in \mathcal{K}''_+$, so that by Lemma 6.1 $\Lambda(p_j, \cdot)$ is a **C**-measure on E''.

We now prove the condition (6.5) of (i) necessary. It follows from Lemma 6.1 that $\Lambda^*(p_j, g) < \infty$ for $j = 1, 2, 3, 4$. Since $| \mathbf{p} | = p_1 + p_2 + p_3 + p_4$ the convexity of Λ^* in its first argument implies that (6.5) holds as stated.

The proof of (ii) is similar.

COROLLARY 6.1. *If* (6.5) *holds* $\Lambda(\dot{\mathbf{p}}, \cdot)$ *is a* **C**-*measure on* E''. *If* (6.6) *holds* $\Lambda(\cdot, \dot{\mathbf{q}})$ *is a* **C**-*measure on* E'.

If (6.5) holds $\Lambda(p_j, \cdot)$, $j = 1, 2, 3, 4$, is a **C**-measure on E'', $p_j < \infty (p, p, \Lambda)$ and $\Lambda(p_j, \cdot) = \Lambda(\bar{p}_j, \cdot)$ in accordance with (5.16). It follows from Lemma 2.1 that $\Lambda(\dot{\mathbf{p}}, \cdot)$ is a **C**-measure on E''. The case of $\Lambda(\cdot, \dot{\mathbf{q}})$ is similar.

COROLLARY 6.2. *If* $E'[E'']$ *is compact* $\Lambda(p, \cdot)[\Lambda(\cdot, q)]$ *is a* **C**-*measure on* $E''[E']$ *whenever* $p[q]$ *is positive, lower semi-continuous and bounded.*

Recall that $\Lambda^*(f, g) < \infty$ for each pair $(f, g) \in \mathcal{K}'_+ \times \mathcal{K}''_+$ by virtue of Theorem 2.1 of Ref. 6. This applies in particular to the case in which f is a constant mapping f_0 with the value $\sup_s p(s) \mid (s \in E')$. With this choice of f_0

$$\Lambda^*(p, g) \leqq \Lambda^*(f_0, g) < \infty$$

so that $\Lambda(p, \cdot)$ is a **C**-measure on E'' by Lemma 6.1. The case of q is similar.

§7. The relation $\Lambda(x, \cdot)(y) = \Lambda(\cdot, y)(x)$

In this section we shall be concerned with mappings $(x, y) \in \mathbf{C}^{E'} \times \mathbf{C}^{E''}$ such that the integrals $\Lambda(x, \cdot)(y)$ and $\Lambda(\cdot, y)(x)$ exist in the sense of §2, and shall set

$$\Lambda(x, \cdot)(y) - \Lambda(\cdot, y)(x) = D(x, y). \tag{7.1}$$

When these integrals exist a necessary and sufficient condition that (x, y) be Λ-integrable is that $D(x, y) = 0$.

Let $\mathbf{y}$ be a vector canonically representing y. [See §5.] We shall make repeated use of the fact that when the integral $\Lambda(x, \cdot)(y)$ exists,

$$\lim_{\mathbf{q} \downarrow \mathbf{y}} \Lambda(x, \cdot)(\dot{\mathbf{q}}) = \Lambda(x, \cdot)(y) \qquad [\mathbf{q} \in {}_{+}\Phi(\mathbf{y})] \tag{7.2}$$

in accordance with Lemma 7.1 of Ref. 5.

We shall also refer to vectors $(\mathbf{p}, \mathbf{q})$ conditioned as in (5.3). When the integral $\Lambda(x, \cdot)(\dot{\mathbf{q}})$ exists and $\Lambda(|x|, |\mathbf{q}|) < \infty$

$$\lim_{\mathbf{g} \uparrow \mathbf{q}} \Lambda(x, \cdot)(\dot{\mathbf{g}}) = \Lambda(x, \cdot)(\dot{\mathbf{q}}) \tag{7.3}$$

where $\mathbf{g}$ is in ${}_{-}\Phi(\mathbf{q})$. The formula (7.3) will result from Lemma 7.2 of Ref. 5 once the implication

$$\Lambda^*(|x|, |\mathbf{q}|) < \infty \Rightarrow |\Lambda(x, \cdot)|^*(q_i) < \infty \qquad (i = 1, \cdots, 4) \tag{7.4}$$

has been established. Since $q_i \leqq |\mathbf{q}|$

$$|\Lambda(x, \cdot)|^*(q_i) \leqq |\Lambda(x, \cdot)|^*(|\mathbf{q}|) = \sup_{|v| \leqq |\mathbf{q}|} |\Lambda(x, \cdot)(v)|$$

where v is in $\mathcal{K}_{\mathbf{C}}''$. Taken with Theorem 3.2 this becomes

$$\leqq \sup_{|v| \leqq |\mathbf{q}|} \Lambda^*(|x|, |v|) = \Lambda^*(|x|, |\mathbf{q}|) < \infty \tag{7.5}$$

thus establishing (7.4). Set $\Lambda(x, \cdot) = \alpha$ and note that

$$N_1(q_i, |\alpha|) = |\Lambda(x, \cdot)|^*(q_i) < \infty \qquad (i = 1, \cdots, 4)$$

by virtue of (7.4). Relation (7.3) follows from Lemma 7.2 of Ref. 5.

Formulas for $\Lambda(x, \cdot)(y)$ *and* $\Lambda(\cdot, y)(x)$. If the integrals $\Lambda(\cdot, y)(x)$ and $\Lambda(x, \cdot)(y)$ exist, if $\Lambda^*(|x|, |y|) < \infty$, and if $\mathbf{x}$ and $\mathbf{y}$ canonically represent x and y, then

$$\Lambda(x, \cdot)(y) = \lim_{\mathbf{q} \downarrow \mathbf{y}} \lim_{\mathbf{g} \uparrow \mathbf{q}} \lim_{\mathbf{p} \downarrow \mathbf{x}} \lim_{\mathbf{f} \uparrow \mathbf{p}} \Lambda(\dot{\mathbf{f}}, \dot{\mathbf{g}}) \tag{7.6}$$

$$\Lambda(\cdot, y)(x) = \lim_{\mathbf{p} \downarrow \mathbf{x}} \lim_{\mathbf{f} \uparrow \mathbf{p}} \lim_{\mathbf{r} \downarrow \mathbf{y}} \lim_{\mathbf{e} \uparrow \mathbf{r}} \Lambda(\dot{\mathbf{f}}, \dot{\mathbf{e}}) \tag{7.7}$$

where $\mathbf{p} \in {}_{+}\Phi'(\mathbf{x})$; $\mathbf{r}, \mathbf{q} \in {}_{+}\Phi''(\mathbf{y})$; $\mathbf{f} \in {}_{-}\Phi'(\mathbf{p})$; $\mathbf{g} \in {}_{-}\Phi''(\mathbf{q})$; $\mathbf{e} \in {}_{-}\Phi''(\mathbf{r})$.

To establish these formulas let $(\mathbf{p}_0, \mathbf{q}_0)$ be chosen (as is possible since $\Lambda^*(|x|, |y|) < \infty$) so that

$$(\mathbf{p}_0, \mathbf{q}_0) \in {}_{+}\Phi'(\mathbf{x}) \times {}_{+}\Phi''(\mathbf{y}), \qquad \Lambda^*(|\mathbf{p}_0|, |\mathbf{q}_0|) < \infty. \tag{7.8}$$

Let $S'(\mathbf{p}_0)$ and $S''(\mathbf{q}_0)$ be the sections of ${}_{+}\Phi'(\mathbf{x})$ and ${}_{+}\Phi''(\mathbf{y})$ respectively in which $\mathbf{p} \leqq \mathbf{p}_0$, $\mathbf{q} \leqq \mathbf{q}_0$. Restrict $\mathbf{p}$ to $S'(\mathbf{p}_0)$ and $\mathbf{r}, \mathbf{q}$ to $S''(\mathbf{q}_0)$.

Proof of (7.6). We start with (7.2). To $\Lambda(x, \cdot)(\dot{\mathbf{q}})$ in (7.2) one can apply (7.3) provided $\mathbf{q}$ is in $S''(\mathbf{q}_0)$, since this condition on $\mathbf{q}$ implies that

$$\Lambda^*(|x|, |\mathbf{q}|) \leqq \Lambda^*(|\mathbf{p}_0|, |\mathbf{q}_0|) < \infty.$$

Thus (7.2) and (7.3) yield the formula

$$\Lambda(x, \cdot)(y) = \lim_{\mathbf{q}\downarrow y} \lim_{\mathbf{g}\uparrow \mathbf{q}} \Lambda(x, \cdot)(\dot{\mathbf{g}}). \tag{7.9}$$

The existence of $\Lambda(x, \cdot)$ as a $\mathbf{C}$-measure implies that $\Lambda(x, \cdot)(\dot{\mathbf{g}}) = \Lambda(\cdot, \dot{\mathbf{g}})(x)$ in accordance with (2.2). To (7.2) there corresponds a dual formula with the roles of x and y interchanged. Using this dual formula, the procedure used in establishing (7.9) applied to $\Lambda(\cdot, \dot{\mathbf{g}})(x)$ shows that

$$\Lambda(x, \cdot)(\dot{\mathbf{g}}) = \lim_{\mathbf{p}\downarrow x} \lim_{\mathbf{f}\uparrow \mathbf{p}} \Lambda(\dot{\mathbf{f}}, \dot{\mathbf{g}}). \tag{7.10}$$

Together (7.9) and (7.10) establish (7.6). The proof of (7.7) is similar.

A formula for $D(x, y)$. If the integrals $\Lambda(x, \cdot)(y)$ and $\Lambda(\cdot, y)(x)$ exist and if $\Lambda^*(|x|, |y|) < \infty$, then (7.6) and (7.7) hold and with trivial modifications yield the formulas

$$\Lambda(x, \cdot)(y) = \lim_{\mathbf{q}\downarrow y} \lim_{\mathbf{g}\uparrow \mathbf{q}} \lim_{\mathbf{p}\downarrow x} \lim_{\mathbf{f}\uparrow \mathbf{p}} \lim_{\mathbf{r}\downarrow y} \lim_{\mathbf{e}\uparrow \mathbf{r}} \Lambda(\dot{\mathbf{f}}, \dot{\mathbf{g}}) \tag{7.11}$$

$$\Lambda(\cdot, y)(x) = \lim_{\mathbf{q}\downarrow y} \lim_{\mathbf{g}\uparrow \mathbf{q}} \lim_{\mathbf{p}\downarrow x} \lim_{\mathbf{f}\uparrow \mathbf{p}} \lim_{\mathbf{r}\downarrow y} \lim_{\mathbf{e}\uparrow \mathbf{r}} \Lambda(\dot{\mathbf{f}}, \dot{\mathbf{e}}) \tag{7.12}$$

$$D(x, y) = \lim_{\mathbf{q}\downarrow y} \lim_{\mathbf{g}\uparrow \mathbf{q}} \lim_{\mathbf{p}\downarrow x} \lim_{\mathbf{f}\uparrow \mathbf{p}} \lim_{\mathbf{r}\downarrow y} \lim_{\mathbf{e}\uparrow \mathbf{r}} \Lambda(\dot{\mathbf{f}}, \dot{\mathbf{g}} - \dot{\mathbf{e}}). \tag{7.13}$$

We come to a fundamental theorem.

Theorem 7.1. *If the integrals $\Lambda(\cdot, y)(x)$ and $\Lambda(x, \cdot)(y)$ exist and if $\Lambda^*(|x|, |y|) < \infty$ then*

$$\Lambda(\cdot, y)(x) = \Lambda(x, \cdot)(y). \tag{7.14}$$

Use will be made of (7.13). Let the ordered sequence of the six limits in (7.13) be denoted by L so that (7.13) takes the form $D(x, y) = L\Lambda(\dot{\mathbf{f}}, \dot{\mathbf{g}} - \dot{\mathbf{e}})$. For $\mathbf{r}$, $\mathbf{q}$ in $S''(\mathbf{q}_0)$ write

$$\Lambda(\dot{\mathbf{f}}, \dot{\mathbf{g}} - \dot{\mathbf{e}}) = \Lambda(\dot{\mathbf{f}}, \dot{\mathbf{g}} - \dot{\mathbf{q}}) + \Lambda(\dot{\mathbf{f}}, \dot{\mathbf{r}} - \dot{\mathbf{e}}) + \Lambda(\dot{\mathbf{f}}, \dot{\mathbf{q}} - \dot{\mathbf{r}}). \tag{7.15}$$

The derivation of (7.9) and (7.10) shows that the integrals $\Lambda(\dot{\mathbf{f}}, \dot{\mathbf{q}})$ and $\Lambda(\dot{\mathbf{f}}, \dot{\mathbf{r}})$ exist. Making use of $(\mathbf{p}_0, \mathbf{q}_0)$ as chosen in (7.8) and $\mathbf{p} \leqq \mathbf{p}_0$, $\mathbf{q} \leqq \mathbf{q}_0$, Theorem 3.2 implies that for $\mathbf{f} \in \underline{\Phi}'(\mathbf{p})$

$$|\Lambda(\dot{\mathbf{f}}, \dot{\mathbf{q}} - \dot{\mathbf{g}})| \leqq \Lambda^*(|\mathbf{p}_0|, |\dot{\mathbf{q}} - \dot{\mathbf{g}}|) \tag{7.16}$$

$$|\Lambda(\dot{\mathbf{f}}, \dot{\mathbf{r}} - \dot{\mathbf{e}})| \leqq \Lambda^*(|\mathbf{p}_0|, |\dot{\mathbf{r}} - \dot{\mathbf{e}}|) \tag{7.17}$$

$$|\Lambda(\dot{\mathbf{f}}, \dot{\mathbf{q}} - \dot{\mathbf{r}})| \leqq \Lambda^*(|\mathbf{p}_0|, |\dot{\mathbf{q}} - \dot{\mathbf{r}}|). \tag{7.18}$$

Taking account of the constancy of $\mathbf{p}_0$ relative to the limit operation L we find that

$$LA^*(|\mathbf{p}_0|, |\dot{\mathbf{q}} - \dot{\mathbf{g}}|) = \lim_{\mathbf{q}\downarrow \mathbf{y}} \lim_{\mathbf{g}\uparrow \mathbf{q}} \Lambda^*(|\mathbf{p}_0|, |\dot{\mathbf{q}} - \dot{\mathbf{g}}|) = 0 \tag{7.19}$$

where the inner limit is zero by virtue of Theorem 5.1. Similarly

$$LA^*(|\mathbf{p}_0|, |\dot{\mathbf{r}} - \dot{\mathbf{e}}|) = \lim_{\mathbf{r}\downarrow \mathbf{y}} \lim_{\mathbf{e}\uparrow \mathbf{r}} \Lambda^*(|p_0|, |\dot{\mathbf{r}} - \dot{\mathbf{e}}|) = 0. \tag{7.20}$$

It follows from (7.13), (7.15), (7.16), (7.17), (7.19) and (7.20) that

$$D(x, y) = L\Lambda(\dot{\mathbf{f}}, \dot{\mathbf{q}} - \dot{\mathbf{r}}). \tag{7.21}$$

From (7.21), (7.18) and the nature of L, for $\mathbf{q}$ and $\mathbf{r}$ in $S''(\mathbf{q}_0)$

$$|D(x, y)| \leqq \sup_{\mathbf{q},\mathbf{r}} \Lambda^*(|\mathbf{p}_0|, |\dot{\mathbf{q}} - \dot{\mathbf{r}}|). \tag{7.22}$$

Set $q_i' = \inf(q_i, r_i)$ for $i = 1, 2, 3, 4$. Then $\mathbf{q}'$ is in $S''(q_0)$ and $\mathbf{q}' \leqq \mathbf{q} \leqq \mathbf{q}_0$, $\mathbf{q}' \leqq \mathbf{r} \leqq \mathbf{q}_0$. By Theorem 5.3 the right member of (7.22) is at most a prescribed $e > 0$ if the section $S''(\mathbf{q}_0)$ is sufficiently advanced in ${}_+\Phi''(\mathbf{y})$. It follows that $D(x, y) = 0$, and the proof of the theorem is complete.

COROLLARY 7.1. *If z and w are partial maps of E' and E'' respectively into* **C** *such that the integrals $\Lambda(z, \cdot)(w)$ and $\Lambda(\cdot, w)(z)$ exist and $\Lambda^*(|z|, |w|) < \infty$, then*

$$\Lambda(\cdot, w)(z) = \Lambda(z, \cdot)(w). \tag{7.23}$$

If the integral $\Lambda(z, \cdot)(w)$ exists the integral $\Lambda(\bar{z}, \cdot)(\tilde{w})$ exists and equals $\Lambda(z, \cdot)(w)$. Similarly with $\Lambda(\cdot, w)(z)$. Moreover

$$\Lambda^*(|\bar{z}|, |\tilde{w}|) \leqq \Lambda^*(|z|, |w|) < \infty.$$

Corollary 7.1 follows then from Theorem 7.1.

§8. Counter examples

Each example is associated with Theorem 7.1.

EXAMPLE 1. *This example shows that the integrals $\Lambda(p, \cdot)(q)$ and $\Lambda(\cdot, q)(p)$ may exist while $\Lambda(p, \cdot)(q) \neq \Lambda(\cdot, q)(p)$.*

Sequential spaces $\mathcal{E}'$ and $\mathcal{E}''$. Let $\mathcal{E}$ be a space of the points 1, 2, $\cdots$ with discrete topology. The subset K_n of points 1, 2, $\cdots$, n of $\mathcal{E}$ is compact, and each compact subset of $\mathcal{E}$ is contained in some K_n for proper choice of n. Since any mapping of $\mathcal{E}$ into **C** is continuous, the space $K_{\mathbf{C}}(\mathcal{E})$ is the space of functions u with values $u(i)$, 1, 2, $\cdots$, such that $u(i) = 0$ for i exceeding some integer. The characteristic function of $\mathcal{E}$ is continuous but not in $\mathcal{K}_{\mathbf{C}}$. It is however the supremum of the functions $\omega_n \in \mathcal{K}_{\mathbf{C}}$ such that

$$\omega_n(i) = 1, i = 1, \cdots, n; \qquad \omega_n(i) = 0, i > n. \tag{8.1}$$

Let the spaces $\mathcal{E}'$ and $\mathcal{E}''$ be homeomorphic images of $\mathcal{E}$ with points again denoted by 1, 2, $\cdots$.

The **C**-*bimeasures* Λ_A. Let A be a matrix with complex elements a_{ij}, $(i, j = 1, 2, \cdots)$. For each $(u, v) \in \mathcal{K}_C' \times \mathcal{K}_C''$ set

$$\Lambda_A(u, v) = \sum_{i=1}^{\infty} \sum_{j=1}^{\infty} u(i)v(j)a_{ij}. \tag{8.2}$$

There exist integers m_u and n_v such that $u(i) = 0$ for $i > m_u$, and $v(j) = 0$ for $j > n_v$. Thus at most a finite number of terms in (8.2) differ from zero. Moreover

$$\Lambda_A(u, v) \leq \sum_{i=1}^{m_u} \sum_{j=1}^{n_v} | a_{ij} | U(u)U(v). \tag{8.3}$$

One infers that Λ_A is a **C**-bimeasure on $\mathcal{E}' \times \mathcal{E}''$.

Essentially finite matrices A. If at most a finite number of the constants a_{ij} in any row or column of A differ from zero, the matrix A will be termed essentially finite. We shall prove the following.

(α) *If A is essentially finite then corresponding to mappings* $(h, k) \in \mathbf{R}_+^{\mathcal{E}'} \times \mathbf{R}_+^{\mathcal{E}''}$ $\Lambda_A(\cdot, k)$ *is a* **C**-*measure on* $\mathcal{E}'$, *and* $\Lambda_A(h, \cdot)$ *a* **C**-*measure on* $\mathcal{E}''$.

The mappings (h, k) are continuous so that in accordance with the definition of Λ^*

$$\Lambda_A^*(h, k) = \sup_{|u| \leq h, |v| \leq k} | \Lambda(u, v) | \qquad [(u, v) \in \mathcal{K}_C' \times \mathcal{K}_C'']. \tag{8.4}$$

In particular let k be replaced by $g \in \mathcal{K}_+''$. For some integer n, $g(j) = 0$ for $j > n$. Since A is essentially finite, for fixed n there exists an integer $m(n)$ such that $a_{ij} = 0$ if $j \leq n$ and $i \geq m(n)$. Hence

$$| \Lambda_A^*(h, g) | \leq \sum_{i=1}^{m(n)} \sum_{j=1}^{n} | h(i)\, g(j)\, a_{ij} | < \infty.$$

It follows from Lemma 6.1 that $\Lambda_A(h, \cdot)$ is a **C**-measure on $\mathcal{E}''$. Similarly $\Lambda_A(\cdot, k)$ is a **C**-measure on $\mathcal{E}'$.

EXAMPLE 1. *The choice of A*. In this example we shall set $a_{ij} = 1$ for $i = j - 1$ and $j = 2, 3, \cdots$ and set $a_{ij} = -1$ for $i = j + 1$ and $j = 1, 2, \cdots$. Otherwise set $a_{ij} = 0$. The constants a_{ij} are the elements of the matrix

$$A = \left\| \begin{array}{ccccc} 0 & 1 & 0 & 0 & \cdots \\ -1 & 0 & 1 & 0 & \cdots \\ 0 & -1 & 0 & 1 & \cdots \\ 0 & 0 & -1 & 0 & \cdots \\ \cdot & \cdot & \cdot & \cdot & \cdots \\ \cdot & \cdot & \cdot & \cdot & \cdots \end{array} \right\| \tag{8.5}$$

Let Λ_A now be defined by (8.2) using the constants of the matrix (8.5).

The **C**-*measures* $\Lambda_A(\cdot, J)$ *and* $\Lambda_A(I, \cdot)$ *in Ex.* 1. Let I and J be the characteristic functions of $\mathcal{E}'$ and $\mathcal{E}''$ respectively. Since the matrix (8.5) is essentially finite $\Lambda_A(\cdot, J)$ and $\Lambda_A(I, \cdot)$ are **C**-measures on $\mathcal{E}'$ and $\mathcal{E}''$, respectively, in accordance with (α) above. We continue by proving (β).

(β) *In Ex.* 1 *the* **C**-*measure* $\Lambda_A(\cdot, J)$ *is a discrete* **C**-*measure of mass* 1 *at the point* $t = 1$ *of* $\mathcal{E}'$.

Recall that for $u \in \mathcal{K}_C'$ and $g \in {}_{-}\phi_J'$

$$\Lambda_A(u, \cdot)(J) = \lim_{g \uparrow J} \Lambda_A(u, \cdot)(g) \tag{8.6}$$

by Lemma 6.2 of Ref. 5. The mappings ω_n defined in (8.1), ordered according to increasing n form a filtering set in $\mathcal{K}_+''$ and define a Fréchet filter of sections [Ref. 2, p. 33] equivalent to the filter of sections of $_-\phi_J'$. Hence (8.6) implies that

$$\Lambda_A(u, \cdot)(J) = \lim_{m \uparrow \infty} \Lambda_A(u, \cdot)(\omega_m) = \lim_{m \uparrow \infty} \Lambda_A(u, \omega_m). \tag{8.7}$$

If $u(i) = v(i) = 0$ for $i \geqq n$, it is readily seen that when A is the matrix (8.5)

$$\Lambda_A(u, \omega_{n+1}) = \sum_{i=1}^{n} \sum_{j=1}^{n+1} u(i) a_{ij} = u(1)$$

$$\Lambda_A(\omega_{n+1}, v) = \sum_{i=1}^{n+1} \sum_{j=1}^{n} v(j) a_{ij} = -v(1).$$

Hence by (8.7) $u(1) = \Lambda_A(u, \cdot)(J)$. But $\Lambda_A(\cdot, J)$ is a **C**-measure by (α), so that by (2.2)

$$\Lambda_A(u, \cdot)(J) = \Lambda_A(\cdot, J)(u) = u(1). \tag{8.8}$$

Thus $\Lambda_A(\cdot, J)$ is a mapping $u \rightarrow u(1)$ of $\mathcal{K}_{\mathbf{C}}'$ into **C**, and by definition [Ref. 3, p. 51] is a **C**-measure of mass 1 at the point $s = 1$ of $\mathcal{E}'$.

Example 1 *as counter example*. Set $\Lambda_A(\cdot, J) = \alpha$. Using ($\beta$) we shall see that I, the characteristic function of $\mathcal{E}'$ is α-integrable and that $\alpha(I) = 1$. In fact the relation $\alpha(u) = u(1)$ implies that $|\alpha|^* (|u|) = |u(1)|$ and $|\alpha|^* (I) = 1$. Let $e \in \mathcal{K}_{\mathbf{C}}''$ be the characteristic function of the point $t = 1$. Since $|\alpha|^* (I) < \infty$, I is α-integrable. Hence $|\alpha|^* (I - e) = |\alpha| (I) - |\alpha| (e) = |\alpha|^* (I) - |\alpha|^* (e) = 0$. We infer that $\alpha(I) = \alpha(e)$ so that $\Lambda_A(\cdot, J)(I) = \alpha(e) = 1$.

One similarly shows that $\Lambda_A(I, \cdot)$ is a discrete **C**-measure of mass -1 at the point $t = 1$ of $\mathcal{E}''$, and that $\Lambda_A(I, \cdot)(J) = -1$. Hence

$$\Lambda_A(I, \cdot)(J) \neq \Lambda_A(\cdot, J)(I). \tag{8.9}$$

Thus the integrals $\Lambda_A(I, \cdot)(J)$ *and* $\Lambda_A(\cdot, J)(I)$ *exist and are unequal.*

EXAMPLE 2. *This example shows that it is not necessary that* $\Lambda^*(p, q) < \infty$ *in order that* (p, q) *be* Λ*-integrable.*

Let a_{ij} be defined as in Ex. 1. We introduce $b_{ij} = (-1)^{(i+1)} a_{ij}$, defining a matrix B of the form

$$B = \begin{Vmatrix} 0 & 1 & 0 & 0 & 0 & \cdots \\ 1 & 0 & -1 & 0 & 0 & \cdots \\ 0 & -1 & 0 & 1 & 0 & \cdots \\ 0 & 0 & 1 & 0 & -1 & \cdots \\ 0 & 0 & 0 & -1 & 0 & \cdots \\ \cdot & \cdot & \cdot & \cdot & \cdot & \cdots \\ \cdot & \cdot & \cdot & \cdot & \cdot & \cdots \end{Vmatrix}$$

Let Λ_B be defined in terms of B by (8.2), b_{ij} replacing a_{ij}. The analysis of Ex. 1 gives $\Lambda_B(I, \cdot)(J) = \Lambda_B(\cdot, J)(I) = 1$. In Ex. 1 $\Lambda_A^*(I, J) = \infty$ in accordance with Theorem 7.1. Directly from the definition of Λ^* one sees that $\Lambda_B^* = \Lambda_A^*$ so that $\Lambda_B^*(I, J) = \infty$. Thus (I, J) is Λ_B-integrable while $\Lambda_B^*(I, J) = \infty$.

§9. Necessary and sufficient conditions that $\Lambda(\cdot, q)(p) = \Lambda(p, \cdot)(q)$

We shall refer to pairs of mappings (p, q) and (f, g) conditioned as in (3.0). We shall assume that the integrals $\Lambda(\cdot, q)(p)$ and $\Lambda(p, \cdot)(q)$ exist in the sense of §2, and seek necessary and sufficient conditions that

$$\Lambda(\cdot, q)(p) = \Lambda(p, \cdot)(q) \tag{9.1}$$

or equivalently, necessary and sufficient conditions that (p, q) be Λ-integrable when the integrals $\Lambda(\cdot, q)(p)$ and $\Lambda(p, \cdot)(q)$ exist. We have already seen in Theorem 7.1 that a sufficient condition that (9.1) hold is that $\Lambda^*(p, q) < \infty$. But this condition is not necessary as Ex. 2 in §8 shows.

Filtering sets H'_p, H''_q, H_{pq}. Let H'_p, $[H''_q]$ be an arbitrary ensemble of mappings $f \in \mathcal{K}'_+$ with $\sup f = p$, $[g \in \mathcal{K}''_+$ with $\sup g = q]$ filtering for the relation $\leqq$. Let H_{pq} be the product on $\mathcal{K}'_+ \times \mathcal{K}''_+$ of the filtering sets H'_p and H''_q.

If the integral $\Lambda(\cdot, q)(p)$ exists then

$$\Lambda(\cdot, q)(p) = \lim_{f \in H'_p} \lim_{g \in H''_q} \Lambda(f, g) \tag{9.2}$$

as we shall see. Similarly, if the integral $\Lambda(p, \cdot)(q)$ exists,

$$\Lambda(p, \cdot)(q) = \lim_{g \in H''_q} \lim_{f \in H'_p} \Lambda(f, g). \tag{9.3}$$

To establish (9.2) we infer from Lemma 6.3 of Ref. 5 that

$$\Lambda(\cdot, q)(p) = \lim_{f \in H'_p} \Lambda(\cdot, q)(f). \tag{9.4}$$

The hypothesis that $\Lambda(\cdot, q)(p)$ exists implies, by definition of this integral in §2, that the integral $\Lambda(f, \cdot)(q)$ exists and equals $\Lambda(\cdot, q)(f)$. By a second application of Lemma 6.3 of Ref. 5.

$$\Lambda(f, \cdot)(q) = \lim_{g \in H''_q} \Lambda(f, \cdot)(g). \tag{9.5}$$

Now $\Lambda(f, \cdot)(g)$ is by definition $\Lambda(f, g)$ so that (9.2) follows from (9.4) and (9.5). The proof of (9.3) is similar.

We infer the following.

THEOREM 9.1. *If the integrals* $\Lambda(p, \cdot)(q)$ *and* $\Lambda(\cdot, q)(p)$ *exist a necessary and sufficient condition that they have a common value is that*

$$\lim_{f \in H'_p} \lim_{g \in H''_q} \Lambda(f, g) = \lim_{g \in H''_q} \lim_{f \in H'_p} \Lambda(f, g). \tag{9.6}$$

When (9.6) *holds these limits give the value*

$$\Lambda(p, q) = \Lambda(p, \cdot)(q) = \Lambda(\cdot, q)(p).$$

To obtain conditions of a different sort that (9.1) hold set

$$D(p, q) = \Lambda(p, \cdot)(q) - \Lambda(\cdot, q)(p) \tag{9.7}$$

as in (7.1).

LEMMA 9.1. *If the integrals* $\Lambda(p, \cdot)(q)$ *and* $\Lambda(\cdot, q)(p)$ *exist*

$$D(p, q) = \lim_{f \in H'_p} \Lambda(p - f, \cdot)(q) = -\lim_{g \in H''_q} \Lambda(\cdot, q - g)(p). \tag{9.8}$$

We shall establish the first relation in (9.8). The relation (9.4) can be written in the form

$$\Lambda(\cdot, q)(p) = \lim_{f \in H'_p} \Lambda(f, \cdot)(q) \tag{9.9}$$

since $\Lambda(\cdot, q)(f) = \Lambda(f, \cdot)(q)$ when the integral $\Lambda(\cdot, q)(p)$ exists. Moreover when $\Lambda(p, \cdot)(q)$ exists $\Lambda(p, \cdot)$ is a **C**-measure on E'', so that $\Lambda(p, \cdot) - \Lambda(f, \cdot) = \Lambda(p - f, \cdot)$ in accordance with Lemma 2.1 (β). Using this relation and (9.9), one obtains the first relation in (9.8).

The second relation is similarly proved.

We infer the following.

THEOREM 9.2. *If the integrals* $\Lambda(p, \cdot)(q)$ *and* $\Lambda(\cdot, q)(p)$ *exist two equivalent necessary and sufficient conditions that these integrals be equal are that*

$$\lim_{f \in H'_p} \Lambda(p - f, \cdot)(q) = 0 \qquad \lim_{g \in H''_q} \Lambda(\cdot, q - g)(p) = 0.$$

As a corollary one obtains a simple confirmation of Theorem 7.1 when $(x, y) = (p, q)$.

COROLLARY 9.1. *If the integrals* $\Lambda(p, \cdot)(q)$ *and* $\Lambda(\cdot, q)(p)$ *exist, a sufficient condition that they be equal is that* $\Lambda^*(p, q) < \infty$.

In fact it follows from Theorem 3.2 that

$$|\Lambda(p - f, \cdot)(q)| \leqq \Lambda^*(p - f, q)$$

so that in accordance with the limit Theorem 6.1 of Ref. 6

$$\lim_{f \in H'_p} |\Lambda(p - f, \cdot)(q)| \leqq \lim_{f \in H'_p} \Lambda^*(p - f, q) = 0.$$

Thus the first condition of Theorem 9.2 is satisfied, and hence $\Lambda(p, \cdot)(q) = \Lambda(\cdot, q)(p)$.

Further sufficient conditions. The preceding analysis suggests other sufficient conditions that (9.1) hold.

Uniform H''_q *convergence.* Granting that the integral $\Lambda(\cdot, q)(p)$ exists, $\Lambda(f, g)$ converges to $\Lambda(f, \cdot)(q)$ according to $g \in H''_q$, as (9.5) shows. We say that this convergence is uniform with respect to f in some section Σ' of H'_p, if for arbitrary $e > 0$ there exists a section S''_e of H''_q such that

$$|\Lambda(f, g) - \Lambda(f, \cdot)(q)| < e \qquad [\text{for } g \in S''_e, f \in \Sigma']. \tag{9.10}$$

We term such convergence uniform H''_q-convergence.

Uniform H'_p-convergence. Uniform H'_p convergence of $\Lambda(f, g)$ to $\Lambda(\cdot, g)(p)$ for g in some section Σ'' of H''_q is similarly defined.

Convergence according to H_{pq}. If F maps $\mathcal{K}'_+ \times \mathcal{K}''_+$ into **C** one says that $F(f, g)$ converges to A according to H_{pq} if the limit of F with respect to $(f, g) \in H_{pq}$ is A.

We shall establish Theorem 9.2.

THEOREM 9.2. *If the integrals $\Lambda(p, \cdot)(q)$ and $\Lambda(\cdot, q)(p)$ exist, each of the following conditions are sufficient that*

$$\Lambda(p, \cdot)(q) = \Lambda(\cdot, q)(p). \tag{9.11}$$

(a) *The convergence of $\Lambda(f, g)$ to $\Lambda(f, \cdot)(q)$ according to $g \in H''_q$ shall be uniform for f in some section Σ' of H'_p.*

(b) *The convergence of $\Lambda(f, g)$ to $\Lambda(\cdot, g)(p)$ according to $f \in H'_p$ shall be uniform for g in some section Σ'' of H''_q.*

(c) *$\Lambda(f, g)$ shall converge according to H_{pq}.*

In case (a), (b), *or* (c) *is satisfied, $\Lambda(p, q)$ equals*

$$\lim_{(f,g)\in H_{pq}} A(f, g) = \lim_{f\in H'_p} \lim_{g\in H''_q} \Lambda(f, g) = \lim_{g\in H''_p} \lim_{f\in H'_q} \Lambda(f, g). \tag{9.12}$$

We establish the following.

(1) *The condition* (c) *is sufficient that* (9.1) *hold.* This follows from the fact that the H_{pq}-convergence of Λ to a limit Λ_0 implies that the two limits appearing in (9.6) exist and are equal. Moreover, the limit Λ_0 equals the limit $\Lambda(p, q)$ of the two members of (9.6), so that (9.12) holds.

(2) *The condition* (a) *implies* (c). Let $e > 0$ be given. Since the integral $\Lambda(\cdot, q)(p)$ exists, $\Lambda(\cdot, q)(f)$ converges to $\Lambda(\cdot, q)(p)$ according to $f \in H'_p$. There accordingly exists a section S'_e of H'_p such that

$$| \Lambda(\cdot, q)(p) - \Lambda(\cdot, q)(f) | < e \qquad [f \in S'_e]. \tag{9.13}$$

Condition (a) presupposes that

$$| \Lambda(f, g) - \Lambda(f, \cdot)(q) | < e \tag{9.14}$$

for g in some section S''_e of H''_q and for f in some section Σ' of H'_p independent of e. From (9.13) and (9.14) and the relation $\Lambda(f, \cdot)(q) = \Lambda(\cdot, q)(f)$ we infer that

$$| \Lambda(f, g) - \Lambda(\cdot, q)(p) | < 2e \tag{9.15}$$

for $g \in S''_e$, and $f \in S'_e \cap \Sigma'$. Hence $\Lambda(f, g)$ converges according to H_{pq}. Thus (a) implies (c), so that (a) is sufficient that (9.1) hold.

(3) *The condition* (b) *similarly implies* (c). Hence (b) is sufficient that (9.1) hold.

§10. The Λ-integrability of mappings ($\dot{\mathbf{p}}, \dot{\mathbf{q}}$)

We return to pairs of vectors ($\mathbf{p}, \mathbf{q}$) limited as in (5.3). We are in a position to give sufficient conditions not only that $\Lambda(\dot{\mathbf{p}}, \cdot)$ and $\Lambda(\cdot, \dot{\mathbf{q}})$ exist as **C**-measures, but also that the integrals $\Lambda(\dot{p}, \cdot)(\dot{q})$ and $\Lambda(\cdot, \dot{q})(\dot{p})$ exist and are equal. The result here obtained will serve as a model for the case of Borel mappings (x, y) in §11.

THEOREM 10.1. *Sufficient conditions that a pair of mappings* $(\dot{p}, \dot{q})$, *conditioned as in* (5.3), *be* Λ*-integrable, are that* $\Lambda^*(|\mathbf{p}|, |\mathbf{q}|) < \infty$, *and that*

(1) $\Lambda^*(|\mathbf{p}|, g) < \infty$ *for each* $g \in \mathcal{K}''_+$

(2) $\Lambda^*(f, |\mathbf{q}|) < \infty$ *for each* $f \in \mathcal{K}'_+$.

We have seen in Corollary 6.1 that condition (1) is sufficient that $\Lambda(\dot{p}, \cdot)$ be a **C**-measure on E'', and (2) sufficient that $\Lambda(\cdot, \dot{q})$ be a **C**-measure on E'. Let $\bar{p}_i$ $[\bar{q}_i]$ be the finite projection of p_i $[q_i]$, $(i = 1, 2, 3, 4)$. Recall that $\dot{p} = \bar{p}_1 - \bar{p}_2 + i(\bar{p}_3 - \bar{p}_4)$, $\dot{q} = \bar{q}_1 - \bar{q}_2 + i(\bar{q}_3 - \bar{q}_4)$.

PROOF THAT $\Lambda(\bar{p}_i, \cdot)(\bar{q}_j)$ EXISTS. It follows from condition (1) and Lemma 6.1 that for $i = 1, 2, 3, 4$, that $p_i < \infty$ (p, p, Λ) and that $\Lambda(\bar{p}_i, \cdot) = \Lambda(p_i, \cdot)$ is a **C**-measure on E''. Set $\Lambda(p_i, \cdot) = \alpha_i$. If $|\alpha_i|^*(q_j) < \infty$, then the integral $\alpha_i(q_j)$ exists by Lemma 5.2 of Ref. 5, and hence the integral $\alpha_i(\bar{q}_j)$ exists. By virtue of the Bourbaki definition of the superior integral

$$|\alpha_i|^*(\bar{q}_j) \leqq |\alpha_i|^*(q_j) = \sup_{g \leqq q_j} |\alpha_i|(g) \qquad [g \in \mathcal{K}''_+].$$

Hence by our definition of $|\alpha_i|(g)$ in §3 of Ref. 5

$$|\alpha_i|^*(q_j) = \sup_{|u| \leqq q_j} |\Lambda(p_i, \cdot)(u)| \leqq \Lambda^*(p_i, q_j) \leqq \Lambda^*(|\mathbf{p}|, |\mathbf{q}|)$$

where u is in $\mathcal{K}'_{\mathbf{C}}$ and where the middle inequality follows from the relation $|\Lambda(p_i, \cdot)(u)| \leqq \Lambda^*(p_i, |u|)$ of Theorem 3.4. Hence $|\alpha_i|^*(q_j) < \infty$ so that the integral $\alpha_i(\bar{q}_j) = \Lambda(\bar{p}_i, \cdot)(\bar{q}_j)$ exists. Similarly $\Lambda(\cdot, \bar{q}_j)(\bar{p}_i)$ exists.

Thus the integrals $\Lambda(\bar{p}_i, \cdot)(\bar{q}_j)$ exist. Similarly the integrals $\Lambda(\cdot, \bar{q}_i)(\bar{p}_j)$ exist. The integrals $\Lambda(\dot{p}, \cdot)(\dot{q})$ and $\Lambda(\cdot, \dot{q})(\dot{p})$ accordingly exist. These integrals are equal by virtue of Theorem 7.1, since

$$\Lambda^*(|\dot{p}|, |\dot{q}|) \leqq \Lambda^*(|\mathbf{p}|, |\mathbf{q}|) < \infty.$$

Hence $(\dot{p}, \dot{q})$ is Λ-integrable in accordance with Definition (iii) of §2.

§11. The Λ-integrability of Borel mappings (x, y)

Theorems 10.1 and 7.1 give conditions that a pair of mappings $(x, y) \in \mathbf{C}^{E'} \times \mathbf{C}^{E''}$ be Λ-integrable. In Theorem 7.1 it is assumed that the integrals $\Lambda(x, \cdot)(y)$ and $\Lambda(\cdot, y)(x)$ exist. A general approach to the problem of the existence of these integrals can be made when x and y are Borel measurable mappings (written B-measurable mappings).

Borel measurable sets $\mathcal{B}$. A class $\mathcal{B}$ of subsets of E will be termed the class of B-measurable sets of E [cf. Ref. 4] if $\mathcal{B}$ is the smallest class containing all compact sets of E, and such that

(1) if A_1 and A_2 are in $\mathcal{B}$, then $A_1 \cap CA_2 \in \mathcal{B}$;

(2) if $A_i \in \mathcal{B}$ $(i = 1, 2, \cdots)$ then $\text{Union}_i\, A_i \in \mathcal{B}$.

B-measurable mappings. Given $f \in \mathbf{R}^E$ let E_f denote the set $E \mid (f(s) \neq 0)$. The mapping f is termed *B-measurable* if and only if for each $a \in \mathbf{R}$ the set $f_a =$

$E_f \mid (f(s) \geqq a)$ is B-measurable. A mapping $x \in \mathbf{C}^E$ is termed B-measurable if $\mathbf{R}x$ and $\mathfrak{s}x$ are B-measurable. If f is B-measurable E_f is the union of the set f_n for $n = -1, -2, \cdots$ and so is B-measurable. By the usual argument one shows that $-f$ is B-measurable with f, and $-x$ with x.

If α is a $\mathbf{C}$-measure on E a subset A of E [mapping $x \in \mathbf{C}^E$] is termed *α-measurable* if the set A [mapping x] is $|\alpha|$-measurable on E in the sense of Bourbaki [Ref. 3, p. 180]. The set E is always α-measurable.

LEMMA 11.1. *If a subset A of E [a mapping $x \in \mathbf{C}^E$] is B-measurable, it is α-measurable for each $\mathbf{C}$-measure α on E.*

We first consider A. The class H of $|\alpha|$-measurable sets on E is a clan [Ref. 3, p. 185, Cor. 4] which contains each compact subset of E [Ref. 3, p. 156, Cor. to Prop. 10 and p. 195, Cor. 1] and each denumerable union of sets in H [Ref. 3, p. 189, Cor. 2]. The class H must then contain $\mathfrak{B}$ by definition of $\mathfrak{B}$. In particular $A \in H$.

Consider a B-measurable $x \in \mathbf{C}^E$. Set $\mathfrak{R}x = f$. By hypothesis f is B-measurable so that for each $a \in \mathbf{R}$ the set f_a is B-measurable. As seen in the preceding paragraph the set f_a is α-measurable. So also then is E_f. Since E is α-measurable we infer that $E - E_f$ is α-measurable. To show that f is α-measurable note that the set $S_a = E \mid (f(s) \geqq a)$ is the set f_a when $a > 0$, and the set $f_a \cup (E - E_f)$ when $a \leqq 0$. Thus S_a is α-measurable for each $a \in \mathbf{R}$. It follows [Prop. 9, Ref. 3, p. 192] that f is α-measurable.

The mapping $\mathfrak{s}x$ is similarly proved α-measurable. This completes the proof of Lemma 11.1.

The Riesz components of $x \in \mathbf{C}^E$. If $x \in \mathbf{C}^E$ is B-measurable the Riesz components of x are B-measurable as we now recall. The definition of the B-measurability of x implies that $\mathfrak{R}x$ and $\mathfrak{s}x$ are B-measurable. Set $\mathfrak{R}x = f$. If f is B-measurable f^+ is B-measurable. For the set $f_a^+ = f_a$ when $a > 0$, and $f_a^+ = f_0$ when $a \leqq 0$. Hence f^+ is B-measurable. To show that f^- is B-measurable recall that $-f$ is B-measurable with f. Since $f^- = (-f)^+$ it is clear that f^- is B-measurable. Thus the Riesz components of x are B-measurable.

We continue with the following lemma.

LEMMA 11.2. *If h is positive and B-measurable, and if $\Lambda^*(h, g) < \infty$ for each $g \in \mathcal{K}_+''$, then $\Lambda(h, \cdot)$ is a $\mathbf{C}$-measure on E''.*

Given $g \in \mathcal{K}_+''$, we show first that the integral $\Lambda(\cdot, g)(h)$ exists. Set $\Lambda(\cdot, g) = \alpha$. Then α is a $\mathbf{C}$-measure on E'. Now h is $|\alpha|$-measurable by Lemma 11.1. For the integral $\alpha(h)$, or equivalently $|\alpha|(h)$, to exist, it is sufficient [Ref. 3, p. 194, Th. 5] that $|\alpha|^*(h)$ be finite. Since $\Lambda^*(h, g) < \infty$ there exists $p_1 \in {}_+\phi_h$ such that $\Lambda^*(p_1, g) < \infty$. With p in the section $S'(p_1)$ of ${}_+\phi_h'$ on which $p \leqq p_1$,

$$(11.1) \qquad |\alpha|^*(h) = \inf_p |\alpha|^*(p) = \inf_p |\Lambda(\cdot, g)|^*(p) \leqq \Lambda^*(p_1, g)$$

in accordance with L. 3.1 of Ref. 6. Thus $|\alpha|^*(h)$ is finite so that $\Lambda(\cdot, g)(h)$ exists.

That the integral $\Lambda(\cdot, v)(h)$ exists for each $v \in \mathcal{K}_{\mathbf{C}}''$ follows now from the fact that if g is any one of the Riesz components of v, $\Lambda(\cdot, g)(h)$ exists. Knowing that

$\Lambda(\cdot, v)(h)$ exists, it follows as in (b) of the proof of Lemma 6.1 that the mapping $v \to \Lambda(\cdot, v)(h)$ is a **C**-measure on E''.

Granting that x and y are B-measurable mappings of E' and E'' respectively into **C** we shall verify the following.

(a) *The condition* $\Lambda^*(|x|, g) < \infty$ *for each* $g \in \mathcal{K}''_+$ *implies that* $\Lambda(x, \cdot)$ *is a* **C**-*measure on* E''.

(b) *The condition* $\Lambda^*(f, |y|) < \infty$ *for each* $f \in \mathcal{K}'_+$ *implies that* $\Lambda(\cdot, y)$ *is a* **C**-*measure on* E'.

(c) *If the conditions of* (a) [(b)] *are satisfied, the condition* $\Lambda^*(|x|, |y|) < \infty$ *implies that the integral* $\Lambda(x, \cdot)(y)$ $[\Lambda\cdot, y)(x)]$ *exists.*

PROOF OF (a). Let (x_1, x_2, x_3, x_4) be the Riesz components of x. The condition $\Lambda^*(|x|, g) < \infty$ implies that $\Lambda^*(x_i, g) < \infty$ $(i = 1, \cdots, 4)$ and hence by Lemma 11.2 that $\Lambda(x_i, \cdot)$ is a **C**-measure on E''. It follows from Lemma 4.1 that $\Lambda(x, \cdot)$ is the **C**-measure,

$$\Lambda(x, \cdot) = \Lambda(x_1, \cdot) - \Lambda(x_2, \cdot) + i\Lambda(x_3, \cdot) - i\Lambda(x_4, \cdot).$$

PROOF OF (b). The proof of (b) is similar to that of (a).

PROOF OF (c). Let h and k respectively be Riesz components of x and y. To show that the integral $\Lambda(x, \cdot)(y)$ exists if the conditions of (a) are satisfied, and if $\Lambda^*(|x|, |y|) < \infty$, it will be sufficient to show that under these hypotheses each integral $\Lambda(h, \cdot)(k)$ exists. The condition $\Lambda^*(|x|, |y|) < \infty$ implies that $\Lambda^*(h, k) < \infty$, and this in turn implies that for some $q_1 \in {}_+\phi''_k$, $\Lambda^*(h, q_1) < \infty$.

Set $\Lambda(h, \cdot) = \alpha$. Then α is a **C**-measure on E'' under the conditions of (a). Moreover $\alpha(k)$, or equivalently $|\alpha|(k)$, exists if $|\alpha|^*(k) < \infty$ [Ref. 3, p. 194, Th. 5]. To show that $|\alpha|^*(k) < \infty$ recall that

$$|\alpha|^*(k) = \inf_{q \in {}_+\phi''_k} |\alpha|^*(q).$$

It will be sufficient to show that $|\alpha|^*(q_1) < \infty$. Now with $v \in \mathcal{K}''_C$ and $|v| \leqq q_1$

$$|\alpha|^*(q_1) = \sup_v |\alpha(v)| = \sup_v |\Lambda(h, \cdot)(v)| \leqq \Lambda^*(h, q_1)$$

by virtue of Theorem 3.2. Thus $|\alpha|^*(k)$ is finite so that $\alpha(k)$ exists. Thus the integral $\Lambda(x, \cdot)(y)$ exists.

The integral $\Lambda(\cdot, y)(x)$ similarly exists if the conditions of (b) are satisfied and if $\Lambda^*(|x|, |y|) < \infty$.

The results (a), (b), and (c) taken with Theorem 7.1 give a fundamental theorem.

THEOREM 11.1. *If the mappings* $(x, y) \in \mathbf{C}^{E'} \times \mathbf{C}^{E''}$ *are Borel measurable, then* $\Lambda(x, \cdot)$ *and* $\Lambda(\cdot, y)$ *are* **C**-*measures on* E' *and* E'' *respectively whenever*

(α) $\Lambda^*(|x|, g) < \infty$ *for each* $g \in \mathcal{K}''_+$,

(β) $\Lambda^*(f, |y|) < \infty$ *for each* $f \in \mathcal{K}'_+$.

If in addition $\Lambda^*(|x|, |y|) < \infty$, *the integrals* $\Lambda(x, \cdot)(y)$ *and* $\Lambda(\cdot, y)(x)$ *exist and are equal.*

INSTITUTE FOR ADVANCED STUDY
KENYON COLLEGE

References

1. Banach, S., Théorie des Opérations Linéaires, Warsaw, 1932.
2. Bourbaki, N., Éléments de Mathématique, Fasc. II, Topologie Générale, Chap. I, II, Paris, 1951.
3. Bourbaki, N., Éléments de Mathématique, Fasc. XIII, Intégration, Chap. I-IV, Paris, 1952.
4. Halmos, Paul, Measure Theory, New York, 1950.
5. Morse, M., and Transue, W., *Semi-normed vector spaces with duals of integral type*, J. Analyse Math., vol. IV, (1955), pp. 149–186.
6. Morse, M., and Transue, W., C-*bimeasures* Λ *and their superior integrals*, Rend. Circ. Mat. Palermo, Tome IV (1955), pp. 270–300.

The Existence of Polar Non-Degenerate Functions on Differentiable Manifolds

by Marston Morse

Annals of Mathematics,
Vol. 71 (1960), pp. 352–383.

Errata

Page	Line	Error	Should read
352	5	function	functions
	6	functions	function
353	−2	distance	the distance
354	8	Refer	We refer
	9	(cf., [2§1].	(cf., [2§1]).
355	−14	with	*with*
359	−14	definied	defined
	−7	equantions	equations
363	−8	from	form
364	10	from	form
366	−9	$(\frac{a}{2} < x < \frac{a}{4})$	$(\frac{a}{2} > x > \frac{a}{4})$
368	17	that	than
370	14	finial	final
371	−8	J	J_ν
	−4	κ	κ_ν
372	3	oridinary	ordinary
	7	$\nu = -1, 1)$	$(\nu = -1, 1)$
	14	$s \in I$	$s \in I_\nu$
	−11	g	g_ν
373	−11	(1.07)	(10.7)
374	−1	$\lvert {}^1x \rVert$	$\lVert {}^1x \rVert$
376	16	are	arc
381	−14	$\wedge_\epsilon(g')$	$\wedge_{\epsilon'}(g')$
383	−6	50 (1960)	46 (1950)

ANNALS OF MATHEMATICS
Vol. 71, No. 2, March, 1960
Printed in Japan

THE EXISTENCE OF POLAR NON-DEGENERATE FUNCTIONS ON DIFFERENTIABLE MANIFOLDS

BY MARSTON MORSE

(Received November 2, 1959)

1. Introduction

Let M be a compact, connected differentiable n-manifold of class C^∞, with $n > 1$. A non-degenerate function f of class C^∞ on M will be termed *polar* if f has just one critical point of index 0 and just one critical point of index n. It is the first object of this paper to establish the existence of polar function on M. From a more general point of view the object is to show how any non-degenerate functions on M can, by simple well-defined processes, be reduced to a polar function. Theorem 12.1 gives the required reduction.

The author has shown in [1] that there exist an infinity of non-degenerate functions on M. Given such a function f, a simple modification of f is defined in § 2 and yields a non-degenerate function whose critical values are distinct. Let F be such a function.

Corresponding to an arbitrary real number c let F^c be the subset of points $P \in M$ at which $F(P) \leqq c$. (In § 9, F_t will denote the subset of *ordinary* points of F at which $F(P) = t$.) Let $R_0(c)$ be the number of maximal connected subsets of F^c. A critical point A of F of index 1 is said to be of *decreasing* type, if as c increases through $F(A)$, $R_0(c)$ decreases. If F has two or more critical points of index 0 there will exist at least one critical point A_0 of index 1 and of decreasing type. The fundamental lemma is as follows:

LEMMA 1.1. *With a critical point A_0 of F of index 1 and of decreasing type there can be associated a critical point A_1 of index 0 and an open region R on M, with the following properties.*

(1) *The region R contains A_0 and A_1 and has a closure which contains no other critical point of F.*

(2) *There exists a non-degenerate function $\mathcal{F}$ of class C^∞ on M such that $\mathcal{F}$ is identical with F on $M - R$ and has no critical point on R.*

Thus $\mathcal{F}$ has one fewer critical point of index 0 than F. If $\mathcal{F}$ still has two or more critical points of index 0, $\mathcal{F}$ can be replaced by a non-degenerate function with one fewer critical point of index 0. This process can be continued until a non-degenerate function F_1 is obtained with just one critical point of index 0 and the same number of critical points of index n as F.

If F has more than one critical point of index n, $-F_1$ has more than one critical point of index 0. Operating on $-F_1$ as on F one obtains finally a polar non-degenerate function, F^*.

Conclusions can be drawn concerning the properties in the large of the critical points of F^* of indices $1, 2, n-1, n-2$. These conclusions have a bearing on the topological structure of M, particularly when $n = 2$, 3 and 4. A later paper will develop these conclusions.

The notational requirements of this paper are such that it will clarify matters if at the end of each section there is presented a summary of the symbols which form a part of the permanent notation of the paper. For this section we present the following:

Permanent notation. M, F, $\mathfrak{F}$, F^c.

2. Distinct critical values

In this paper a non-degenerate function of class C^∞ on M will be referred to simply as a non-degenerate function on M. We shall establish the following lemma.

LEMMA 2.1. *Corresponding to an arbitrary non-degenerate function f on M and to a fixed set of arbitrarily small disjoint neighborhoods N_A of the respective critical points A of f there exists a non-degenerate function $\mathfrak{f}$ on M with the following properties:*

(i) *The functions f and $\mathfrak{f}$ have the same values except at most on the neighborhoods N_A.*

(ii) *The function $\mathfrak{f}$ has distinct critical values, and the same critical points as f.*

(iii) *For each A and in a suitably chosen open subset of N_A containing A, $f - \mathfrak{f}$ is a constant.*

We shall refer at various points in this paper to diffeomorphisms (cf., [2]) of one differentiable manifold of class C^∞ into another such differentiable manifold of class C^∞. In every case we shall understand, without further mention of the fact, that *such a diffeomorphism is of class C^∞.*

The n-plane E_n. Let E_n be the euclidean n-plane with rectangular coordinates $(x^1, \cdots, x^n)$. It will be convenient to set

$$(x^1, \cdots, x^n) = \mathbf{x} \tag{2.1}$$

and to denote distance of the point $\mathbf{x}$ from the origin by $||\mathbf{x}||$. Corresponding to an arbitrary positive constant a we introduce the open n-disc,

$$\Delta_a^x = \{\mathbf{x} \mid \|\mathbf{x}\| < a\} , \tag{2.2}$$

that is the union of points $\mathbf{x}$ such that $\|\mathbf{x}\| < a$.

If A is an arbitrary critical point of f there exist arbitrarily small open neighborhoods W of A on M such that for any positive constant a some diffeomorphism

$$\Phi : \Delta_a^x \to E_n ; \qquad \mathbf{x} \to p(\mathbf{x}) \tag{2.3}$$

maps Δ_a^x onto W, where $p(\mathbf{x}) \in W$ is the image of $\mathbf{x} \in E_n$ and A is the image of the origin. Refer to W as a *coordinate domain* on M and to Δ_a^x as the corresponding *coordinate range*, (cf., [2, § 1]. Setting

$$f(p(\mathbf{x})) = \psi(\mathbf{x}) \qquad (\mathbf{x} \in \Delta_a^x) \tag{2.4}$$

we term ψ the function *representing* f in Δ_a^x *under* Φ. We suppose that f is ordinary on W except for the critical point A. The function ψ is then ordinary on Δ_a^x except for a critical point at the origin. The lemma will follow readily once statement (α) has been established.

(α) *Set* $a = 3a'$. *Corresponding to any sufficiently small positive constant* e *there exists a function* φ_e *of class* C^∞ *over* Δ_a^x *such that*

$$\varphi_e(\mathbf{x}) - \psi(\mathbf{x}) = e \qquad (\mathbf{x} \in \Delta_{a'}^x) \tag{2.5}$$

$$\varphi_e(\mathbf{x}) = \psi(\mathbf{x}) \qquad (\mathbf{x} \in \Delta_a^x - \Delta_{2a'}^x) \tag{2.6}$$

while φ_e *is ordinary except for a critical point at the origin.*

Let μ be a mapping of class C^∞ of the axis of reals into the interval $[0, 1]$, defined as an even function such that

$$\mu(t^2) = 1 \qquad |t| < a'$$

$$\mu(t^2) = 0 \qquad |t| > 2a' .$$

Such a function exists. Set

$$\varphi_e(\mathbf{x}) = \psi(\mathbf{x}) + e\mu(\|x\|^2) \qquad (\mathbf{x} \in \Delta_a^x) . \tag{2.7}$$

Observe that the partial derivatives of φ_e with respect to the coordinates x^i converge uniformly to the corresponding partial derivatives of ψ as e tends to zero. For e sufficiently small φ_e has the required properties.

Let (A_i), $i = 1, \cdots, n$ be the set of critical points of f. Let (W_i) be a set of disjoint open neighborhoods W_i of the respective critical points A_i of f, given, as above, as images under diffeomorphisms Φ_i of Δ_a^x onto W_i. Let ψ_i "represent" f on Δ_a^x under Φ_i. For arbitrarily small positive constants e_i let φ_{e_i} be related to ψ_i as φ_e was related to ψ in (α). Let f be equal to f at each point of f not in Union W_i. On W_i, $i = 1, \cdots, n$, let f be modified by replacement by the function "represented" by φ_{e_i} under Φ_i, rather than by ψ_i. For proper choices of the constants e_i, f satisfies the lemma.

Permanent notation. E_n, $\mathbf{x}$, $\|\mathbf{x}\|$, Δ_a^x.

3. Fields of $(n-1)$-manifolds on Σ_n

Let Σ_n be a Riemannian n-manifold of class C^∞. By an *admissible r-manifold in Σ_n, $0 < r \leq n$*, we mean a differentiable r-manifold of class C^∞, that is the image under a diffeomorphism into Σ_n of a differentiable r-manifold of class C^∞.

We shall refer to a parameter α on a bounded open interval I and to points

$$(u^1, \cdots, u^{n-1}) = \mathbf{u}$$

in a euclidean $(n-1)$-space E_{n-1}. For $e > 0$ set

$$\Delta_e^u = \{\mathbf{u} \mid \|\mathbf{u}\| < e\} . \tag{3.0}$$

For each $\alpha \in I$ let H_α be an admissible $(n-1)$-manifold in Σ_n. Let Λ be the *α-parameterized* aggregate of $(n-1)$-manifolds

$$\Lambda = \{H_\alpha \mid \alpha \in I\} . \tag{3.1}$$

DEFINITION. *We say that Λ is an admissible family of $(n-1)$-manifolds if, corresponding to an arbitrary $\beta \in I$ and point $q \in H_\beta$, there exists an open neighborhood $I(\beta)$ of β in I, a constant $e > 0$ and a 1-parameter family of diffeomorphisms*

$$\Phi_\alpha : \Delta_e^u \longrightarrow H_\alpha ; \quad \mathbf{u} \longrightarrow p(\mathbf{u}, \alpha) \qquad [\alpha \in I(\beta)] \tag{3.2}$$

which generate a diffeomorphism,

$$\Phi : \Delta_e^u \times I(\beta) \longrightarrow \Sigma_n ; \quad \mathbf{u} \times \alpha \longrightarrow p(\mathbf{u}, \alpha) , \tag{3.3}$$

with $q \in \Phi_\beta(\Delta_e^u)$.

We note that Φ_α is merely a restriction of Φ in which α is fixed.

By the *carrier* $|\Lambda|$ of the aggregate (3.1) is meant the union on Σ_n of the points of the respective manifolds H_α.

A field Λ. An admissible family Λ of $(n-1)$-manifolds H_α in Σ_n will be termed a *field* if there is at most one manifold H_α of Λ meeting each point of the carrier $|\Lambda|$ of Λ. We term this condition on Λ the *biuniqueness* condition on Λ.

An α-parameterized aggregate Λ of $(n-1)$-manifolds is a field if it is an admissible family of $(n-1)$-manifolds and satisfies the biuniqueness condition.

The carrier $|\Lambda|$ of a field Λ is an *open subset* of Σ_n, and as such is a *submanifold* of Σ_n which admits a differential structure derived from that of Σ_n. The carrier $|\Lambda|$ thereby admits an ensemble of local representations. But Λ *qualifies* as an admissible family of $(n-1)$-manifolds only if $|\Lambda|$ admits the above special local representations Φ generated by families of admissible local representations Φ_α of the manifolds H_α.

The following lemma will be useful in § 9. It is a direct consequence of the definition of a field.

LEMMA 3.0. *Let an aggregate of admissible $(n-1)$-manifolds be presented with two parameterizations*

$$\Lambda' = \{H_\alpha \mid \alpha \in I\} \quad \Lambda'' = \{\mathcal{H}_t \mid t \in J\} .$$

If there exists a diffeomorphism θ of I onto J and if $H_\alpha = \mathcal{H}_t$ when and only when α and t correspond under θ, then Λ' is a field if and only if Λ'' is a field.

An arc κ traversing a field Λ. Let κ be a simple, regular, open arc of class C^∞ in the carrier $|\Lambda|$, meeting each manifold H_α of a field Λ in a single point p_α, and not tangent to H_α at p_α. We say then that κ traverses Λ.

LEMMA 3.1. *If κ traverses a field of $(n-1)$-manifolds Λ of the form* (3.1), *and if one sets $p_\alpha = H_\alpha \cap \kappa$, the mapping*

$$\alpha \longrightarrow p_\alpha \qquad (\alpha \in I) \tag{3.4}$$

is a diffeomorphism of I onto κ of class C^∞.

Suppose that $q \in H_\beta$ is a point on κ. We make use of the representation (3.3) of Σ_n near q. The equation $p_\alpha = p(\mathbf{u}, \alpha)$ admits a unique solution

$$u^i = \varphi^i(\alpha) \qquad (i = 1, \cdots, n-1)$$

for α sufficiently near β. It remains to show that each φ^i is of class C^∞ for α sufficiently near β. Since κ is a regular arc of class C^∞, it can be admissibly represented near q in terms of the local coordinates $(u^1, \cdots, u^{n-1}, \alpha)$ in the form

$$u^1 = u^1(t), \cdots, u^{n-1} = u^{n-1}(t), \quad \alpha = \alpha(t) , \qquad [\alpha(0) = \beta] .$$

Now $\alpha'(0) \neq 0$ since κ is not tangent to H_β. Hence κ may be admissibly represented in the form $u^i = \psi^i(\alpha)$, $i = 1, \cdots, n-1$, for α sufficiently near β. But $\varphi^i(\alpha) = \psi^i(\alpha) : i = 1, \cdots, n-1$, for α so near β that both the functions φ^i and ψ^i are defined. It follows that φ^i is of class C^∞, $i = 1, \cdots, n-1$ for α sufficiently near β.

This establishes the lemma.

If the field Λ is traversed by κ we shall denote Λ by $\Lambda(\kappa)$.

Subfield of a field $\Lambda(\kappa)$. If γ is an open subarc of κ given by a diffeomorphism

$$\alpha \longrightarrow p_\alpha = H_\alpha \cap \kappa \qquad (\alpha \in J \subset I) \tag{3.5}$$

of J into κ then a subfield

$$\Lambda(\gamma) = \{H_\alpha \mid \alpha \in J\} \tag{3.6}$$

of $\Lambda(\kappa)$ is well-defined and is traversed by γ. If γ is a relatively compact open subarc of κ, or equivalently J is a relatively compact open subinterval of I, we shall presently define a subfield $\Lambda_\varepsilon(\gamma)$ of $\Lambda(\kappa)$ corresponding to any sufficiently small positive constant ε. We shall need results established in [3].

The n-space $\mathcal{E}_{n-1}$. Let $\mathcal{E}_{n-1}$ be the euclidean $(n-1)$-space of rectangular coordinates $(z^1, \cdots, z^{n-1})$. With $(z^1, \cdots, z^{n-1}) = \mathbf{z}$, and $e > 0$ set

$$\Delta_e^z = \{\mathbf{z} \mid \|\mathbf{z}\| < e\} \tag{3.7}$$

Lemma 3.1 of [3] gives the following :

LEMMA 3.2. *Corresponding to an arbitrary point p on a Riemannian $(n-1)$-manifold H of class C^∞, and to any sufficiently small positive constant e, there exists a diffieomorphism ψ of Δ_e^z into H such that each ray λ in Δ_e^z issuing from the origin in Δ_e^z is mapped isometrically onto a geodesic arc on H issuing from p.*

DEFINITION. *Under the conditions of Lemma 3.2, ψ will be termed a radial-geodesic isometry of Δ_e^z into H with pole $p \in H$, and the image $\psi(\Delta_e^z)$ in H will be termed a geodesic $(n-1)$-disc of radius e on H with pole p.*

It is not implied that ψ is an isometry, but merely that its restriction to the rays λ is isometric.

Corresponding to $\beta \in I$ and a positive constant e we introduce the e-neighborhood

$$I_e(\beta) = (\beta - e, \beta + e) \cap I \tag{3.8}$$

of β relative to I and state Theorem 4.1 of [3] as follows:

LEMMA 3.3. *Let $\Lambda(\kappa)$ be a field of $(n-1)$-manifolds in Σ_n, traversed by κ and of the form (3.1). Corresponding to an arbitrary $\beta \in I$, there exist a positive constant e and a diffeomorphism*

$$\Phi : \Delta_e^z \times I_e(\beta) \longrightarrow \Sigma_n\,, \tag{3.9}$$

such that for each fixed $\alpha \in I_e(\beta)$, Φ reduces to a radial-geodesic isometry Φ_α of Δ_e^z into H_α with pole $H_\alpha \cap \kappa$.

The constant e in Lemma 3.3 will be termed a *span* associated with β and $\Lambda(\kappa)$. It is clear that $e/2$ is a span associated with values of α in any sufficiently restricted neighborhood of β on I. Hence if J is a relatively compact open subinterval of I there exists a constant $\varepsilon > 0$ which is a span associated with each $\alpha \in J$. Hence Lemma 3.3 implies the corollary.

COROLLARY 3.1. *Let $\Lambda(\kappa)$ be a field traversed by κ of the form* (3.1). *Let γ be an open, relatively compact subarc of κ, and let J be an open, relatively compact subinterval of I such that γ is given by the diffeomorphism* (3.5).

Corresponding to a sufficiently small positive constant ε there exists, for an arbitrary $\beta \in J$, a diffeomorphism,

$$\Phi : \Delta_\varepsilon^x \times J_\varepsilon(\beta) \longrightarrow \Sigma_n \tag{3.10}$$

such that for fixed $\alpha \in J_\varepsilon(\beta)$, Φ reduces to a radial-geodesic isometry Φ_α of Δ_ε^x into H_α with pole $H_\alpha \cap \gamma$.

If then $\mathcal{D}_\alpha$ is the geodesic $(n-1)$-disc on H_α with pole $H_\alpha \cap \gamma$ and radius ε, the α-parameterized aggregate

$$\Lambda_\varepsilon(\gamma) = \{\mathcal{D}_\alpha \mid \alpha \in J\} \tag{3.11}$$

of $(n-1)$-manifolds $\mathcal{D}_\alpha$ is a subfield of $\Lambda(\gamma)$ traversed by γ.

Permanent notation. Φ, Φ_α, $\Lambda(\gamma)$, $\Lambda_\varepsilon(\gamma)$, (3.11).

4. f-arcs and ortho-f-arcs

We suppose again that Σ_n is a Riemannian n-manifold of class C^∞ and that f is a non-degenerate function on Σ_n. Let Ω_0 be the set of critical points of f on Σ_n. Let κ be a simple, regular open arc of class C^∞ on $\Sigma_n - \Omega_0$.

f-arcs. If κ meets no level manifold of f in more than one point and is never tangent to such a manifold, we term κ an f-arc on $\Sigma_n - \Omega_0$.

Let $\tau \to \varphi(\tau)$ be a diffeomorphism of an open interval J onto κ. If $f(\varphi(\tau)) = \tau$; $\tau \in J$, the representation $\tau \to \varphi(\tau)$ of κ will be termed *f-canonical.*

We shall translate this definition of an f-canonical representation of κ into terms of local coordinates on Σ_n. To that end let X be an open subset of the n-plane of points $(x^1, \cdots, x^n) = \mathbf{x}$ which is mapped by a diffeomorphism

$$X \longrightarrow \Sigma_n \colon \mathbf{x} \longrightarrow p(\mathbf{x}) \tag{4.1}$$

onto a coordinate domain W in $\Sigma_n - \Omega_0$. Set

$$\psi(\mathbf{x}) = f(p(\mathbf{x})) \qquad (\mathbf{x} \in X). \tag{4.2}$$

Suppose that κ is in W with an image γ on X. Given a diffeomorphism $\tau \to \varphi(\tau)$ of J onto κ, let $\tau = \mathbf{x}(\tau)$ be the diffeomorphism of J onto γ such that $p(\mathbf{x}(\tau)) = \varphi(\tau)$.

Then the representation $\tau \to \varphi(\tau)$ *of* κ *is f-canonical if and only if*

$$\psi(\mathbf{x}(\tau)) = \tau \qquad (\tau \in J) \tag{4.3}$$

that is, if and only if the representation $\tau \to \mathbf{x}(\tau)$ *of* γ *is* ψ*-canonical.*

LEMMA 4.1. *Every f-arc* κ *admits an f-canonical representation.*

By virtue of the preceding it is sufficient to prove that every ψ-arc γ in X admits a ψ-canonical representation.

Let $t \to \mathbf{y}(t)$ be a diffeomorphism of I onto γ in E_n. Set

$$\psi(\mathbf{y}(t)) = \theta(t) \qquad (t \in I). \tag{4.4}$$

We shall make a change of parameter from t to τ by setting $\tau = \theta(t)$ for $t \in I$. This change is admissible. For

$$\frac{d\theta}{dt} = \frac{\partial\psi}{\partial x^i}\frac{\partial \mathbf{y}^i}{\partial t} \neq 0 \quad (\text{summing for } i = 1, \cdots, n) \tag{4.5}$$

because γ is tangent to no level manifold of ψ. Let $\tau \to \mathbf{x}(\tau)$ be the diffeomorphism of class C^∞ of the interval $\theta(I)$ onto γ in which

$$\mathbf{x}(\tau) = \mathbf{y}(\theta^{-1}(\tau)) \qquad (\tau \in \theta(I)).$$

Then (4.4) yields the identity

$$\psi[\mathbf{x}(\tau)] = \tau .$$

The lemma follows.

Ortho-f-arcs. The trajectories orthogonal to the level manifolds of f are well-definied in $\Sigma_n - \Omega_0$. We term these trajectories *ortho-f-arcs* on Σ_n, and shall recall the differential equations defining them.

On the coordinate domain $W \subset \Sigma_n - \Omega_0$ with its local coordinates $(x^1, \cdots, x^n)$

$$ds^2 = g_{ij}(\mathbf{x})dx^i dx^j \qquad (i, j = 1, \cdots, n) \tag{4.6}$$

where the form is positive definite and the coefficients are functions of class C^∞ for $\mathbf{x} \in X$. With $\psi(x)$ defined as in (4.2), the ortho-f-arcs are represented in X by solutions of the differential equantions

$$\frac{dx^i}{dt} = \frac{g^{ij}(\mathbf{x})\psi_{x^j}(\mathbf{x})}{g^{ij}(\mathbf{x})\psi_{x^i}(\mathbf{x})\psi_{x^j}(\mathbf{x})} \qquad (i, j = 1, \cdots, n) \tag{4.7}$$

where the denominator of the quotient on the right never vanishes for $\mathbf{x} \in X$, since ψ has no critical point in X. Along a solution $\mathbf{x} = \mathbf{x}(t)$, $t \in J$, of the equations (4.7),

$$\frac{d\psi}{dt} = \psi_{x^i}\frac{dx^i}{dt} = 1 \qquad (t \in J) ,$$

so that

$$t = \psi(\mathbf{x}(t)) + c \qquad (c \text{ a constant}) . \tag{4.8}$$

If one takes $c = 0$ one limits the parameterization of the solution but not the ortho-f-arc γ represented. When $c = 0$, $t \to \mathbf{x}(t)$ is a ψ-canonical representation of γ.

The preceding study of ψ and of ortho-ψ-arcs permits us to establish a lemma needed in § 9.

LEMMA 4.2. *Let I be an open interval of values of f. For each fixed $t \in I$ let f_t be that level manifold of f in $\Sigma_n - \Omega_0$ on which $f = t$. The t-parameterized aggregate,*

$$\Lambda = \{f_t \mid t \in I\} , \tag{4.9}$$

is a field of $(n-1)$-manifolds on Σ_n, in the sense of § 3.

In § 3 there are two conditions on Λ that Λ be a field. The biuniqueness condition, that there be at most one manifold f through a given point of $|\Lambda|$, is satisfied, since f is single-valued. It remains to establish the existence of local diffeomorphisms Φ, characterized in (3.2) and (3.3), by virtue of which Λ will qualify as an admissible family of $(n-1)$-manifolds on Σ_n. To say that these local diffeomorphisms Φ exist is equivalent to affirming the following.

(i) *Let J be an open interval of values of ψ assumed by ψ in the coordinate range X in which the differential equations* (4.7) *are defined. The aggregate*

$$\Lambda^* = \{\psi_t \mid t \in J\} \tag{4.10}$$

of level manifolds ψ_t of ψ is a family of $(n-1)$-manifolds in X, admissible in the sense of § 3.

We shall establish (i).

Let $\mathbf{c}$ be a point in X and set $\psi(\mathbf{c}) = b$. We shall set up a family of solutions of (4.7) with initial points on ψ_b near $\mathbf{c}$. We begin by setting up a family of solutions of (4.7) with initial points on a relatively compact open neighborhood $N(\mathbf{c})$ of $\mathbf{c}$ in X, as follows. If $J(b)$ is a sufficiently small open subinterval of J containing b, then for each point $\mathbf{a} \in N(\mathbf{c})$ there exists a solution of (4.7) of the form

$$\mathbf{x} = \mathbf{w}(\mathbf{a}, t) \qquad (t \in J(b)) , \tag{4.11}$$

satisfying the initial condition $\mathbf{a} = \mathbf{w}(\mathbf{a}, b)$. To restrict the initial points $\mathbf{a}$ to points on ψ_b near $\mathbf{c}$ we must use an admissible representation of ψ_b near $\mathbf{c}$.

To that end we refer to the $(n-1)$-space E_{n-1} of coordinates $(u^1, \cdots, u^{n-1})$. If e is a sufficiently small positive constant there exists a diffeomorphism

$$\text{(4.12)} \qquad \mathbf{A} : \Delta_e^u \longrightarrow \psi_b\,, \quad \mathbf{u} \longrightarrow \mathbf{A}(\mathbf{u})\,, \qquad [\text{with } \mathbf{A}(\mathbf{0}) = \mathbf{c}]$$

which maps Δ_e^u onto a coordinate domain $Z(\mathbf{c})$ in ψ_b, with $Z(\mathbf{c}) \subset N(\mathbf{c})$. We obtain a family of solutions of (4.7) with initial points $\mathbf{A}(\mathbf{u}) \in Z(\mathbf{c})$ by substituting $\mathbf{A}(\mathbf{u})$ for $\mathbf{a}$ in (4.11). The required family thus has the form

$$\text{(4.13)} \qquad \mathbf{x} = \mathbf{w}(\mathbf{A}(\mathbf{u}), t) = \mathbf{B}(\mathbf{u}, t) \qquad [\mathbf{u} \in \Delta_e^u,\ t \in J(b)]$$

introducing $\mathbf{B}(\mathbf{u}, t)$. To use this family of solutions in proving (i) we need the two properties of these solutions affirmed in (ii).

(ii) *The representation $t \to \mathbf{B}(\mathbf{u}, t)$ of these solutions is ψ-canonical, and the jacobian*

$$\text{(4.14)} \qquad \frac{D(B^1, \cdots, \cdots, B^n)}{D(u^1, \cdots, u^{n-1}, t)} \neq 0 \qquad (\mathbf{u} \in \Delta_e^u,\ t = b)\,.$$

To verify the first affirmation of (ii) recall that

$$\text{(4.15)} \qquad t = \psi(\mathbf{B}(\mathbf{u}, t)) + c(\mathbf{u})\,, \qquad (\mathbf{u} \in \Lambda_e^u,\ t \in J(b))\,,$$

where $c(\mathbf{u})$ depends *a priori* on $\mathbf{u}$. If one sets $t = b$ in (4.15) and makes use of the relations

$$\mathbf{B}(\mathbf{u}, b) = \mathbf{w}(\mathbf{A}(\mathbf{u}), b) = \mathbf{A}(\mathbf{u}); \quad \psi(\mathbf{A}(\mathbf{u})) = b\,,$$

one finds that $c(\mathbf{u}) = 0$, as desired.

To verify (4.14) observe that the columns of this jacobian are the vectors

$$\frac{\partial \mathbf{B}}{\partial u^1}, \cdots, \frac{\partial \mathbf{B}}{\partial u^{n-1}}, \frac{\partial \mathbf{B}}{\partial t}\,, \qquad (t = b)$$

of which the first $n-1$ are independent vectors tangent to ψ_b at the point $\mathbf{A}(\mathbf{u})$, and the last vector is orthogonal to ψ_b at this point. Relation (4.14) follows. This establishes (ii).

The preceding results enable us to affirm statement (iii).

(iii) *If ε is a sufficiently small positive constant less than the constant e in* (4.12), *and $I(b)$ is a sufficiently restricted open subinterval of $J(b)$ containing b, the function* $\mathbf{B}$ *defines a diffeomorphism*

$$\text{(4.16)} \qquad \Phi : \Delta_\varepsilon^u \times I(b) \longrightarrow X : \mathbf{u} \times t \longrightarrow \mathbf{B}(\mathbf{u}, t)$$

which, for fixed $t \in I(b)$, reduces to a diffeomorphism

$$\text{(4.17)} \qquad \Phi_t\colon \Delta_\varepsilon^u \longrightarrow \psi_t\colon \mathbf{u} \longrightarrow B(\mathbf{u}, t)$$

with

$$\text{(4.18)} \qquad \mathbf{c} \in \Phi_b(\Delta_\varepsilon^u) \subset Z(\mathbf{c})\,.$$

Recalling the characteristics of the diffeomorphism Φ, demanded in

(3.2) and (3.3), we see that the aggregate Λ^*, given in (4.10), qualifies as an admissible family of $(n-1)$-manifolds by virtue of (iii). Lemma 4.2 then follows from (i) as previously indicated.

Permanent notation. f-arcs, ortho-f-arcs.

5. The constant $\rho = 5\sigma$ and mappings T_A

We return to the compact n-manifold M of § 1 and to the non-degenerate function F on M. We shall employ a coordinate range D_ρ in the space E_n of coordinates $(x^1, \cdots, x^n)$ setting

$$D_\rho = \{\mathbf{x} \mid \|\mathbf{x}\| < \rho\} . \tag{5.0}$$

An alternative notation for D_ρ would be Δ_ρ^x. We use D_ρ because ρ and D_ρ are to be *fixed* in (α) below, and held fast throughout this paper.

Let Ω be the set of critical points of F on M. For $A \in \Omega$ let k be the index of the critical point A. The author has established the following: cf., [4, Lemma 10.1, p. 44].

(α) *Corresponding to a sufficiently small positive constant ρ there exists a diffeomorphism*

$$T_A: D_\rho \longrightarrow M : \mathbf{x} \longrightarrow T_A(\mathbf{x}) \qquad [\text{with } T_A(0) = A] \tag{5.1}$$

which maps D_ρ onto a coordinate domain W_A in M and is such that

$$F(T_A(\mathbf{x})) - F(A) = e_i^A x^i x^i \qquad (\mathbf{x} \in D_\rho,\ i = 1, \cdots, n) \tag{5.2}$$

where

$$e_i^A = \cdots = e_k^A = -1;\ e_{k+1}^A = \cdots = e_n^A = 1 , \tag{5.3}$$

with k the index of A.

The constant ρ and diffeomorphism T_A: $A \in \Omega$ will be chosen so that the following is true.

(β) *Statement (α) holds for each $A \in \Omega$ and for a choice of ρ independent of A.*

(γ) *The domains W_A in M are disjoint.*

(δ) *The ranges of F on the respective domains W_A are disjoint*

The constant σ. Our procedure will make essential use of a constant σ such that $5\sigma = \rho$. We shall refer to the respective open n-discs $D_{j\sigma}$: $j = 1, 2, \cdots, 5$, of radii $j\sigma$. In § 6 we shall define a Riemannian metric on M in such a fashion that T_A reduces to an isometry on $D_{3\sigma}$. Once this metric is defined no further use is made of $D_{4\sigma}$ and $D_{5\sigma}$. In § 8 and thereafter we make repeated use of D_σ, $D_{2\sigma}$, and $D_{3\sigma}$ in defining the field $\Lambda_e(g)$, and finally in defining $\mathcal{F}$.

It will be convenient in § 7 to compose the mapping T_A with a suitably chosen orthogonal transformation of the coordinates $x^{k+1}, \cdots, x^n$. The resulting mapping of D_ρ into M can replace T_A in statements (α) to (δ) without alteration of these statements in any other respect. In particular the choice of ρ can remain fixed.

After this final choice of T_A the mappings T_A: $A \in \Omega$, and constant ρ will remain unchanged throughout this paper.

A detail as to future notation can be explained here. Let ζ be the subarc of the x^1-axis on which $-\rho < x^1 < \rho$. Let

$$(5.4) \qquad (a, b),\ (a, b],\ [a, b),\ [a, b]$$

be subintervals of the open interval $(-\rho, \rho)$, understood in the usual senses. If η is a subarc of ζ represented by one of the intervals (5.4), then $T_A(\eta)$ is a well-defined arc in M. By a slight but useful departure from conventional notation we shall denote the respective arcs $T_A(\eta)$ by

$$(5.5) \qquad T_A(a, b)\ ,\ T_A(a, b]\ ,\ T_A[a, b)\ ,\ T_A[a, b]\ .$$

Such arcs as

$$(5.6) \qquad T(0, 2\sigma)\ ,\ T(-2\sigma, 0]\ ,\ T[-2\sigma, 2\sigma]\ ,$$

will appear in § 8.

Permanent notation. $\rho = 5\sigma$; $D_{j\rho}$: $j = 1, \cdots, 5$; T_A: $A \in \Omega$; (5.5).

6. A Riemannian metric on M

In accord with a theorem of Whitney [6] we can suppose M regularly embedded in a euclidean space E_m of dimension $m = 2n + 1$. The metric induced on M by the euclidean metric of E_m will be termed the *E_m-metric* of M.

The E_m-metric of M is not adequate for our purpose. In particular with this choice of metric on M, the respective mappings T_A of the euclidean n-discs D_ρ are not in general isometric so that some of the advantages of the representation (5.2) of F on D_ρ are lost. To be more specific let

$$(6.1) \qquad K_A(\mathbf{x}) = e_i^A x^i x^i \qquad (i = 1, \cdots, n)$$

be the quadratic from associated with A in the right member of (5.2). The segments of the x^i-axes on which $0 < x^i < \rho$ are ortho-K_A-arcs, if orthogonality is defined by the euclidean metric on E_n, but the images of these segments on M under T_A are not in general ortho-F-arcs, if orthogonality on M is defined by its E_m-metric.

We shall modify the E_m-metric of M in the disjoint neighborhoods W_A of the respective critical points A of F, leaving the E_m-metric of M elsewhere unchanged on M.

To that end we shall use an even function η mapping the t-axis into the interval [0, 1], defining η as a function of class C^∞ in such a manner that

$$\begin{aligned} &\eta(t^2) = 0 && (0 \leqq t \leqq 3) \\ &\eta(t^2) = 1 && (t \geqq 4) \\ 1 > &\eta(t^2) > 0 && (3 < t < 4) . \end{aligned} \tag{6.2}$$

On D_ρ in E_n the euclidean metric is defined by the Riemannian form

$$ds^2 = \delta_{ij} dx^i dx^j \qquad (i, j = \cdots, n) . \tag{6.3}$$

On W_A we suppose that the E_m-metric of M at the point $T_A(\mathbf{x})$ is given for $\mathbf{x} \in D_\rho$ by the from

$$ds^2 = a_{ij}(\mathbf{x}) dx^i dx^j \qquad (i, j = 1, \cdots, m) . \tag{6.4}$$

We shall define a positive definite Riemannian form

$$g_{ij}(\mathbf{x}) dx^i dx^j \qquad (i, j = 1, \cdots, n)(\mathbf{x} \in D_\rho) , \tag{6.5}$$

for $\mathbf{x} \in D_\rho$ by setting

$$g_{ij}(\mathbf{x}) = \eta\Big(\frac{\|\mathbf{x}\|^2}{\sigma^2}\Big) a_{ij}(\mathbf{x}) + \Big(1 - \eta \frac{\|\mathbf{x}\|^2}{\sigma^2}\Big) \delta_{ij} . \tag{6.6}$$

We shall assign this form to the point $T_A(\mathbf{x}) \in W_A$. On the closure of $D_{3\sigma}$ the form (6.5) reduces to (6.3), while on $D_\rho - D_{4\sigma}$ it reduces to (6.4). The coefficients g_{ij} are clearly of class C^∞ for $\mathbf{x} \in D_\rho$. It remains to prove (i).

(i) *The form* (6.5) *is positive definite on the set of points* $\mathbf{x}$ *such that*

$$3\sigma < \|\mathbf{x}\| < 4\sigma . \tag{6.7}$$

Note that

$$0 < \eta\Big(\frac{\|\mathbf{x}\|^2}{\sigma^2}\Big) < 1 \tag{6.8}$$

when $\mathbf{x}$ satisfies (6.7). If the form (6.5) were not positive definite when $\mathbf{x}$ satisfies (6.7), it would be singular for some set $\mathbf{x}$ satisfying (6.7). For that set $\mathbf{x}$ the quotient

$$\Big(\eta\Big(\frac{\|\mathbf{x}\|^2}{\sigma^2}\Big) - 1\Big) \Big/ \eta\Big(\frac{\|\mathbf{x}\|^2}{\sigma^2}\Big) \tag{6.9}$$

would be a characteristic root of the form (6.4). But the quotient (6.9) is negative on account of (6.8). Since the form (6.4) has no negative characteristic root we infer that the form (6.5) is positive definite when $\mathbf{x}$ satisfies (6.7).

This completes the proof of (i). We modify the E_m-metric of M in this way over each of the coordinate domains W_A. The resulting form defines our metric over M.

We record the following:

LEMMA 6.1. *With M metricized as above and with $D_{3\sigma}$ assigned its ordinary euclidean metric, the mapping of $D_{3\sigma}$ into M defined by the restriction of T_A to $D_{3\sigma}$ is an isometry.*

For $\mathbf{x} \in D_{3\sigma}$ the sets $(x^1, \cdots, x^n)$ serve both as rectangular coordinates on $D_{3\sigma}$, and, under the mapping $\mathbf{x} \to T_A(\mathbf{x})$, as local coordinates on the coordinate domain $T_A(D_{3\sigma})$ of M. In terms of these coordinates both the euclidean metric on $D_{3\sigma}$ and our Riemannian metric on $T_A(D_{3\sigma})$ are defined by the form (6.3). This establishes the lemma.

Permanent notation. $K_A(\mathbf{x}) = e_i^A x^i x^i$.

7. Continuation of F-arcs

We suppose that the representation of each F-arc κ in M is F-canonical in the sense of § 4. In such a representation

$$t \longrightarrow P(t) \in \kappa\ , \qquad F(P(t)) = t\ . \tag{7.0}$$

Such a representation exists by Lemma 4.1.

Suppose now that κ is an ortho-F-arc. Let J be the interval for t on κ. The arc κ admits continuation as a canonically represented ortho-F-arc over a maximum open interval $a < t < b$ which includes J. We shall be concerned with continuation of ortho-F-arcs *in the sense of decreasing t* and shall prove that

$$\lim_{t \downarrow a} P(t) = A\ , \tag{7.1}$$

where A is a critical point of F.

PROOF OF (7.1). First consider ortho-K_A-arcs in $D_{3\sigma}$, where $K_A(\mathbf{x})$ is given by (6.1). Any ortho-K_A-arc in $D_{3\sigma}$ can be continued in $D_{3\sigma}$ to a point Q at a K_A-level below zero, or tends radially to the origin.

Returning to the ortho-F-arc κ, let t_n be a decreasing sequence of values of t which tend to a as $n \uparrow \infty$ and is such that $P(t_n)$ tends to a point Q in M. Such a sequence exists. If Q were an ordinary point of M the arc κ would meet Q and continue through Q in the sense of decreasing t. Hence Q must be a critical point A of F. It remains to show that (7.1) holds. On the neighborhood $T_A(D_{3\sigma})$ of A the ortho-F-arcs are images under T_A of ortho-K_A-arcs, by virtue of Lemma 6.1 and the representation (5.2) of F on $D_{3\sigma}$. Our remarks concerning continuation of ortho-K_A-arcs on $D_{3\sigma}$ now make it clear that (7.1) must hold.

To establish Lemma 7.2 we shall need Lemma 7.1. Lemma 7.1 concerns a function h with values

$$h(x, y) = x^2 - y^2 \tag{7.2}$$

in the (x, y)-plane. Lemma 7.1 shows how an ortho-h-arc whose continuation, in the sense of decreasing h, would terminate at the origin, can be extended as an h-arc so as to "by-pass" the origin and reach an h-level below zero.

LEMMA 7.1. *Corresponding to an arbitrary positive constant a there exists an h-arc ζ_a in the (x, y)-plane with the following properties.*

The arc ζ_a is symmetric with respect to the straight line L on which $x = y$. It is a sequence of three sensed h-arcs of which the first arc is the segment of the x-axis on which $x \geqq a$, and is taken in the sense of decreasing x. The third and final arc is the segment of the y-axis on which $y \geqq a$ and is taken in the sense of increasing y. On the second and middle arc $0 < x < a$ and $0 < y < a$.

The first and third subarcs of ζ_a are already well-defined. To define the second arc we shall make use of an even function ν, of class C^∞, mapping the t-axis into the interval $[0, 1]$ in such a fashion that

$$\begin{aligned} &\nu(t) = 1 && \left(0 \leqq t \leqq \tfrac{1}{2}\right) \\ &\nu(t) = 0 && (t \geqq 1) \\ &\nu'(t) < 0 && \left(\tfrac{1}{2} < t < 1\right). \end{aligned} \tag{7.3}$$

The second arc η will be defined as a sequence of three subarcs η_1, η_2, η_3, each taken in the sense of decreasing x.

The arc η_1 shall have the representation

$$y = \nu\left(\frac{x}{a}\right)\frac{a^2}{8x} \qquad \left(a > x > \frac{a}{2}\right).$$

The arc η_2 is the locus

$$8xy = a^2 \qquad \left(\frac{a}{2} < x < \frac{a}{4}\right).$$

The arc η_3 is the reflection of η_1 in the line L. The successive end points of the arcs η_1, η_2 and η_3, are

$$(a, 0)\,, \quad \left(\frac{a}{2}, \frac{a}{4}\right), \quad \left(\frac{a}{4}, \frac{a}{2}\right), \quad (0, a)\,.$$

One sees that ζ_a is a regular arc of class C^∞. To show that ζ_a is an h-arc it is sufficient to verify that the composite arc $\eta = \eta_1\eta_2\eta_3$ is an h-arc. This is a consequence of the fact that the slope of η is always negative while the slopes of the level curves of h at the points of η are always positive.

The arc ζ_a so defined satisfies the lemma.

LEMMA 7.2. *Any ortho-F-arc κ on M whose limiting end point Q (in the sense of decreasing F) is not a critical point of F, can be continued as an F-arc, in the sense of decreasing F, so as to form an F-arc whose limiting end point is a critical point A of index zero and whose intersection with $T_A(D_{3\sigma})$ is an ortho-F-arc.*

In accord with (7.1) κ can be continued in the sense of decreasing F so as to form an ortho-F-arc γ with a limiting end point which is a critical point A. Let k be the index of A. It is clear that $0 \leq k < n$. The case in which $k = 0$ requires no further consideration. Suppose then that $0 < k < n$. In this case κ can be continued as an F-arc so as to by-pass A in the following way.

Recall that F has the canonical representation (5.2) over the coordinate range $D_{3\sigma}$ of the coordinate domain $T_A(D_{3\sigma}) = Y_A$. In the euclidean n-plane of $D_{3\sigma}$ let π_{n-k} be the $(n-k)$-plane on which $x^1 = \cdots = x^k = 0$. The arc γ has an intersection with Y_A represented in $D_{3\sigma}$ by a ray tending to the origin in π_{n-k}. After a suitable orthogonal transformation of the coordinates $x^{k+1}, \cdots, x^n$ the terminal arc $\gamma_1 = Y_A \cap \gamma$ will be represented by an open segment ξ in $D_{3\sigma}$ of the positive x^n-axis terminating at the origin. With the coordinates $(x^1, \cdots, x^n)$ so chosen, and T_A correspondingly re-chosen, let $K_A(\mathrm{x})$ be defined over $D_{3\sigma}$ as the right member of (5.2). Let Π_2 be the 2-plane of coordinates (x^1, x^n). The restriction of K_A to $\Pi_2 \cap D_{3\sigma}$ is a function h with values

$$h(x^1, x^n) = -(x^1)^2 + (x^n)^2 \qquad [(x^1)^2 + (x^n)^2] < 9\sigma^2 . \tag{7.4}$$

Since the limiting end point of κ is a point Q which by hypothesis is not a critical point, we can assume, without loss of generality in proving the lemma, that Q is represented by a point $(x^1, x^n) = (0, a)$ in ξ with $a > 0$. Let ξ_1 be the subarc of ξ on which $a < x^n$. By virtue of Lemma 7.1, ξ_1 can be continued as an h-arc in the sense of decreasing h so as to form an h-arc ξ_a with a terminal subarc ξ_2 in $\Pi_2 \cap D_{3\sigma}$ on which $a < x^1$ and $x^n = 0$. The arc

$$\kappa \cup T_A(\xi_a)$$

is a continuation of κ as an F-arc on M. It has a terminal subarc $T_A(\xi_2)$ which is an ortho-F-arc on which $F < F(A)$.

This process of extension of κ can be continued until a critical point of index zero appears as a limiting end point. It is understood that the continuation is along ortho-F-arcs except at most in the neighborhoods Y_A of critical points A of positive index. If such a critical point A is approached as a limiting point along an ortho-F-arc it will be by-passed on an F-arc as above.

8. An elementary saddle arc and its extension g

Let B be an arbitrary critical point of F with index k such that $0 < k < n$. Set $F(B) = b$ and let b' and b'' be critical values of F such that $b' < b''$ and the interval (b', b'') contains b and no other critical values of F. Corresponding to an arbitrary real number c let F^c be the subset of points P on M such that $F(P) \leqq c$. For arbitrary choices of c' on the open interval (b', b) the sets $F^{c'}$ are homeomorphic. Similarly, for arbitrary choices of c'' on the interval (b, b'') the sets $F^{c''}$ are homeomorphic. Let $R_0(c)$ be the number of disjoint components of F^c.

If the index k of A is 1 then either

$$R_0(c'') - R_0(c') = -1\,, \qquad \text{(cf., [5, Cor. 7.4])} \tag{8.1}$$

or else $R_0(c'') = R_0(c')$. If (8.1) holds, B will be called a saddle point of decreasing type. If $1 < k < n$, then

$$R_0(c'') = R_0(c')\,.$$

LEMMA 8.1. *If F has more that one critical point of index zero, then F has at least one saddle point B of decreasing type.*

Let A' and A'' be two critical points of F of index zero. If c is a properly chosen real number, A' and A'' will be in disjoint components of F^c. If c is sufficiently large, $M = F^c$. Since M is a connected differentiable manifold, A'' and A' are arc-wise connected on M. Hence there is a smallest value b of F such that A' and A'' are arc-wise connected on F^b. It is readily seen that b is a critical value of F and that the corresponding critical point B of F is a saddle point of decreasing type.

We shall make use of the notation introduced in (5.5) and present the following definition.

If B is a saddle point of F, the arc $T_B(-2\sigma, 2\sigma)$ in M will be called an elementary saddle arc meeting B.

When A is a critical point of index zero the choice of the canonical mapping T_A of D_ρ into M in § 5 is not unique in that T_A may be replaced by a canonical mapping $T_A Z^{-1}$ in which Z is an arbitrary orthogonal transformation of the coordinates (x) in D_ρ. In the following lemma we introduce two canonical mappings corresponding to which the choice of Z is yet to be made. For convenience of notation we shall denote a preliminary canonical mapping by T_A^*, reserving T_A as a designation of a final canonical mapping $T_A^* Z^{-1}$ in which Z has been appropriately chosen.

In the next lemma we shall introduce three critical points A_r: $r = -1, 0, 1$. For brevity of notation set

$$T_{A_r} = T_r \qquad T^*_{A_r} = T^*_r \qquad (r = -1, 0, 1) \tag{8.2}$$

We shall use T^*_r instead of T_r when the choice of T^*_r is in accord with (5.2) but preliminary.

LEMMA 8.2. *If A_0 is a saddle point of decreasing type the elementary saddle arc $T_0(-2\sigma, 2\sigma)$ meeting A_0 is separated by A_0 into the two open ortho-F-arcs*

$$T_0(-2\sigma, 0)\,, \quad T_0(0, 2\sigma) \qquad \text{[cf., (5.5)]}\,. \tag{8.3}$$

These two arcs can be continued in the sense of decreasing F so as to form disjoint F-arcs κ_{-1} and κ_1, respectively, with distinct limiting end points A_{-1} and A_1 which are critical points of index zero. The intersection of κ_{-1} and κ_1 with the subsets

$$T^*_{-1}(D_{2\sigma})\,, \quad T^*_1(D_{2\sigma}) \tag{8.4}$$

are ortho-F-subarcs of κ_{-1} and κ_1, respectively, with the limiting end points A_{-1} and A_1.

Lemma 8.2 follows from Lemma 7.2 except for the assertion that κ_{-1} and κ_1 are disjoint and $A_{-1} \neq A_1$. To verify this assertion note that one starts with the disjoint ortho-F-arcs (8.3). The continuations γ_{-1} and γ_1 of these arcs, as maximal open ortho-F-arcs, will necessarily be disjoint. We must, however, exclude the possibility that γ_{-1} and γ_1 have the same critical point A as a limiting end point. Were that the case, then for any constant c such that

$$F(A_0) - 2\sigma < c < F(A_0)\,, \tag{8.5}$$

the two disjoint subsets of

$$F^c \cap T_0(D_{2\sigma})$$

would be arc-wise connected to A on F^c. It would follow that A_0 is not of decreasing type as a saddle point. Reasoning in this way one sees that the critical points of M which are approached and by-passed in the process of defining κ_{-1} and κ_1, respectively, are different, as are the intermediate ortho-F-arcs. In particular the limiting critical end points A_{-1} and A_1 are distinct.

As a matter of permanent notation there are *two cases* to be considered under the hypotheses of Lemma 8.1. They are the cases in which

$$F(A_{-1}) < F(A_1)\,, \quad F(A_{-1}) > F(A_1)\,, \tag{8.6}$$

respectively. We shall suppose that *the first of these cases occurs*. The treatment in this case will carry over to the other case with trival notational modifications.

We turn to the F-arcs κ_{-1} and κ_1 of Lemma 8.2. Their final arcs in the domains (8.4) on M are the images under T^*_{-1} and T^*_1, respectively, of open radial arcs b_{-1} and b_1 in $D_{2\sigma}$. The radial arcs b_ν: $\nu = -1, 1$, have the lengths 2σ and limiting end points at the origin in $D_{2\sigma}$. We now choose orthogonal transformations Z_ν: $\nu = -1, 1$, of the coordinates (x) in $D_{2\sigma}$ so that $Z_{-1}(b_{-1})$ and $Z_1(b_1)$ are the intervals $(0, 2\sigma)$ and $(-2\sigma, 0)$ respectively, of the x^1-axis. On setting

$$T_1 = T^*_1 Z_1^{-1} \qquad T_{-1} = T^*_{-1} Z_{-1}^{-1} ,$$

we can affirm the following.

Lemma 8.3. *The F-arcs κ_{-1} and κ_1 of Lemma* 8.2 *intersect the respective sets* (8.4) *in finial open arcs*

$$T_{-1}(0, 2\sigma) \qquad T_1(-2\sigma, 0) \qquad [\text{cf., } (5.5)] , \tag{8.7}$$

with the limiting end points A_{-1} and A_1.

An extension g of the saddle arc $T_0(-2\sigma, 2\sigma)$. Let g be the sensed, open, regular arc of class C^∞ given by the disjoint union,

$$g = T_{-1}(-2\sigma, 0] \cup \kappa_{-1} \cup A_0 \cup \kappa_1 \cup T_1[0, 2\sigma) \qquad [\text{cf., } (5.5)] \tag{8.8}$$

taking these arcs and point A_0 in the order written. Let s be the arc length measured along g from its initial limiting end point. The range of s on g is an open interval $(0, s^*)$. Let the point on g with parameter s be denoted by $g(s)$. Let s_r: $r = -1, 0, 1$, be the s-coordinate of A_r. Note that

$$s_{-1} = 2\sigma = s^* - s_1 . \tag{8.9}$$

The arc ω in $D_{2\sigma}$. Let ω be the subarc of the x^1-axis in $D_{2\sigma}$ on which $-2\sigma < x^1 < 2\sigma$ and let $\omega(c)$ be the point of ω at which $x^1 = c$. In accord with the condition (δ) of § 5 on the mappings T_A, the ranges of F on the three open subsets $T_{-1}(\omega)$, $T_1(\omega)$, $T_0(\omega)$ of g have closures with no values in common. Moreover arbitrary values on these closures have the order of the values,

$$F(A_{-1}) < F(A_1) < F(A_0) , \tag{8.10}$$

respectively.

Let $G(s)$ be the value of F at the point $g(s)$ of g. We see that G has an absolute minimum on g at A_{-1}, an absolute maximum at A_0, a relative minimum at A_1 and no other critical points on g. It follows from the results of the preceding paragraph that

$$(8.11)\quad (\sup G(s) \mid g(s) \in T_{-1}(\omega)) < (\inf G(s) \mid g(s) \in T_1(\omega)) < (\sup G(s) \mid g(s) \in T_1(\omega)) < (\inf G(s) \mid g(s) \in T_0(\omega)) .$$

The new function $\mathcal{F}$. In accord with the program outlined in § 1, F will be replaced by a non-degenerate function $\mathcal{F}$ on M. We commence here by defining $\mathcal{F}$ on g. At the point $g(s)$ on g let $\mathcal{G}(s)$ denote the value of $\mathcal{F}$. We can choose $\mathcal{G}$ of class C^∞ on $(0, s^*)$ and such that

$$(8.12)\quad \begin{aligned} \mathcal{G}(s) &= G(s) && (0 < s \leqq s_{-1} + \sigma) \\ \mathcal{G}(s) &< G(s) && (s_{-1} + \sigma < s < s_1 + \sigma) \\ \mathcal{G}(s) &= G(s) && (s_1 + \sigma \leqq s < s^*) \\ \mathcal{G}'(s) &> 0 && (s_{-1} < s < s^*) . \end{aligned}$$

Note that the first interval includes the s-coordinate s_{-1} of A_{-1} while the the third interval does not include the s-coordinate s_1 of A_1. That a function $\mathcal{G}$ exists satisfying these conditions follows at once from the properties of G as listed above. The function $\mathcal{G}$ has an absolute minimum at A_{-1} on g and no other critical point on g.

In § 12 we shall complete the definition of $\mathcal{F}$ on M. In § 9 and § 10 we shall prepare for the definition in § 11 of a field $\Lambda_\varepsilon(g)$ of $(n-1)$-manifolds traversed by g. We shall use $\Lambda_\varepsilon(g)$ in § 12 in defining $\mathcal{F}$ near g.

Permanent notation. $A_r, s_r, T_r : r = -1, 0, 1;$ $\kappa_{-1}, \kappa_1, g, g(s), \omega, \omega(c);$ $G(s), \mathcal{G}(s)$.

9. The fields $\Lambda_\varepsilon(g_\nu) : \nu = -1, 1$

Recalling that s_r: $r = -1, 0, 1$, is the s-coordinate of A_r on g, we introduce the subintervals

$$(9.0)\quad \begin{cases} I_{-1} = (s_{-1}, s_0) \\ I_1 \;\;= (s_0, s_1) \\ J_{-1} = (s_{-1} + \sigma, s_0 - \sigma) \\ J_1 \;\;= (s_0 + \sigma, s_1 - \sigma) \end{cases}$$

of the intervals $(0, s^*)$, noting that J_ν: $\nu = -1, 1$, is a relatively compact subinterval of I_ν. The interval I_ν: $\nu = -1, 1$, is the range of s on the F-arc κ_ν defined in § 8. Set

$$(9.1)\quad g_{-1} = \{g(s) \mid s \in J_{-1}\} , \quad g_1 = \{g(s) \mid s \in J_1\} .$$

Note that g_ν: $\nu = -1, 1$, is a relatively compact subarc of κ . Concerning κ_ν we shall establish the following lemma.

LEMMA 9.1. *If for each* $s \in I_\nu$: $\nu = -1, 1$, H_s *is that level manifold of* F *in* $M - \Omega$ *which meets* $g(s)$, *the* s-*parameterized aggregate*

(9.2)
$$\Lambda(\kappa_\nu) = \{H_s \mid s \in I_\nu\} \qquad (\nu = -1, 1)$$

is a field of $(n-1)$*-manifolds traversed by* κ_ν.

For each value t of F assumed by F at some oridinary point of F let F_t denote the level manifold of F in $M - \Omega$ on which $F = t$. Let $\mathcal{I}_\nu$: $\nu = -1, 1$, be the range of F on κ_ν. By virtue of Lemma 4.2 the t-parameterized aggregate

(9.3)
$$\Lambda_\nu = \{F_t \mid t \in \mathcal{I}_\nu\} \qquad \nu = -1, 1)$$

of $(n-1)$-manifolds F_t is a field. The aggregates of $(n-1)$-manifolds Λ_ν and $\Lambda(\kappa_\nu)$ are the same, but parameterized differently. We shall show that the fact that Λ_ν is a field, implies that $\Lambda(\kappa_\nu)$ is a field.

Since the arc κ_ν traverses the field Λ_ν it follows from Lemma 3.1 that the mapping

$$t \longrightarrow p_t = F_t \cap \kappa_\nu \qquad (t \in \mathcal{I}_\nu)$$

is a diffeomorphism of $\mathcal{I}_\nu$ onto κ_ν. But the mapping $s \rightarrow g(s)$ for $s \in I$, is also a diffeomorphism of I_ν onto κ_ν. It follows that there exists a diffeomorphism θ of I_ν onto $\mathcal{I}_\nu$ such that $g(s) = p_t$ if and only if s and t correspond under θ. Since H_s and F_t are level manifolds of F meeting $g(s)$ and p_t, respectively, $H_s = F_t$ if and only if s and t correspond under θ. Since Λ_ν is a field it follows from Lemma 3.0 that $\Lambda(\kappa_\nu)$ is a field.

The fields $\Lambda_\varepsilon(g_\nu) : \nu = -1, 1$. Since g_ν: $\nu = -1, 1$, is a subarc of κ_ν the subfield $\Lambda(g_\nu)$ of $\Lambda(\kappa_\nu)$ is well-defined, cf., § 3. Since g_ν is a relatively compact subarc of κ_ν a subfield $\Lambda_\varepsilon(g_\nu)$ of $\Lambda(g_\nu)$ is defined for any sufficiently small positive constant ε, as the subfield $\Lambda_\varepsilon(\gamma)$ of $\Lambda(\gamma)$ was defined in Corollary 3.1. We suppose ε taken so small that the carriers of the fields $\Lambda_\varepsilon(g\ ) : \nu = -1, 1$, do not intersect. This is possible since the closures of the arcs g_{-1} and g_1 in M do not intersect. The field $\Lambda_\varepsilon(g_\nu)$ has the form

(9.4)
$$\Lambda_\varepsilon(g_\nu) = \{\mathcal{D}_s \mid s \in J_\nu\} \qquad (\nu = -1, 1)$$

where $\mathcal{D}_s$ is a geodesic $(n-1)$-disc on H_s of radius ε with pole at $g(s)$.

Permanent notation. $\Lambda_\varepsilon(g_\nu) : \nu = -1, 1$.

10. Two fields $\lambda_\varepsilon^k(\omega) : k = 0, 1$, in E_n traversed by ω

We continue the preparation for the definition of $\Lambda_\varepsilon(g)$ in § 11. by defining the fields $\lambda_\varepsilon^k(\omega)$ in E_n. We begin with notational innovations.

Tubes E_n^a. For each point $\mathbf{x} = (x^1, \cdots, x^n)$ in E_n set

(10.0)
$$(x^2, \cdots, x^n) = {}^1\mathbf{x}\,; \quad \|{}^1\mathbf{x}\|^2 = (x^2)^2 + \cdots + (x^n)^2\,.$$

For each positive constant a let E_n^a be the "tube" in E_n given as the set

$$E_n^a = \{\mathbf{x} \mid \| {}^1\mathbf{x} \| < a\} \tag{10.1}$$

The $(n-1)$-manifolds in fields $\lambda^k(\omega)$ will be defined as sections of a tube E_n^a, symmetric with respect to the x^1-axis. That is each such $(n-1)$-manifold H shall have a representation

$$x^1 = \psi(x^2, \cdots, x^n) \qquad \{\text{for } \| {}^1\mathbf{x} \| < a\} \tag{10.2}$$

where ψ is of class C^∞ and the value of ψ depends only on the value of $\| {}^1\mathbf{x} \|$. This symmetry of H with respect to the x^1-axis implies that H is orthogonal to the x^1-axis. Because of its representation in the form (10.2), H meets the x^1-axis in just one point.

In Lemma 10.1 we shall refer to the quadratic forms

$$K^k(\mathbf{x}) = (-1)^k (x^1)^2 + \| {}^1\mathbf{x} \|^2 \qquad (\mathbf{x} \in D_\rho) \tag{10.3}$$

associated in (5.2) with the critical points of F of indices $k = 0$ and 1.

LEMMA 10.1. *There exist two fields in* E_n

$$\lambda^k(\omega) = \{H_c^k \mid -2\sigma < c < 2\sigma\} \qquad (k = 0, 1) \tag{10.4}$$

of $(n-1)$-manifolds H_c^k with the following properties:

(α.1) *Each manifold H_c^k is a section of the tube E_n^a for which $a = \sigma/2$ and is symmetric with respect to the x^1-axis.*

(α.2) *For fixed c the manifold H_c^k has the representation*

$$H_c^k = \{\mathbf{x} \mid x^1 = \psi^k(c, x^2, \cdots, x^n)\} \tag{10.5}$$

where ψ^k is of class C^∞ on the domain

$$W = \{(c, {}^1\mathbf{x}) \mid \| {}^1\mathbf{x} \| < \sigma/2;\ -2\sigma < c < 2\sigma\} \tag{10.6}$$

and

$$\psi^k(c, x^2, \cdots, x^n) = -\psi^k(-c, x^2, \cdots, x^n) \tag{1.07}$$

$$\psi_c^k(c, x^2, \cdots, x^n) > 0 \qquad \left[\psi_c^k = \frac{\partial \psi^k}{\partial c}\right] \tag{10.8}$$

$$\psi^k(c, 0, \cdots, 0) = c\,. \tag{10.9}$$

(α.3) *For $\sigma \leqq |c| < 2\sigma$ the manifold H_c^k is a submanifold of that level manifold of K^k which meets the point $\omega(c)$ on the x^1-axis.*

(α.4) *The carrier of $\lambda^k(\omega)$ is in $D_{3\sigma}$.*

Not all of the manifolds in the field $\lambda^k(\omega)$ are submanifolds of level manifolds of K^k, nor could they be such, since the level manifold of K^k meeting the origin is singular at the origin. The problem of defining these fields $\lambda^k(\omega)$ is one of starting with the level manifolds of K^k, to which reference is made in (α.3), and interpolating between these manifolds other manifolds H^k_c, for $|c| < \sigma$, which are symmetric sections of the tube E^a_n and such that $\lambda^k(\omega)$ is a field.

Definition of ψ^k. Recall first that the level manifold of K^k which meets $\omega(c)$ and is restricted to $E^{\sigma/2}_n$, has the representation

$$x^1 = \text{sign}\, c(c^2 - (-1)^k \|{}^1\mathbf{x}\|^2)^{1/2} \qquad (\|{}^1\mathbf{x}\| < \sigma/2) \tag{10.10}$$

for $|c| > \sigma$. These manifolds include the level manifolds of K^k to which reference is made in (α.3). To define ψ^k in (α.2) we introduce a mapping μ of the axis of reals into the interval [0, 1], defining μ as an even function $t \to \mu(t)$ of class C^∞ such that

$$\begin{aligned} \mu(t) &= 1 && (t \geqq 1) \\ \mu(t) &= 0 && (0 \leqq t \leqq 1/2) \\ 0 < \mu'(t) &< 4 && \left(\tfrac{1}{2} < t < 1\right). \end{aligned}$$

such a function μ clearly exists. Understanding that sign $0 = 0$, let ψ^k be defined by setting

$$\psi^k(c, x^2, \cdots, x^n) = \text{sign}\, c\left(c^2 - (-1)^k \|{}^1\mathbf{x}\|^2 \mu\left(\frac{c}{\sigma}\right)\right)^{1/2} \tag{10.11}$$

for $(c, {}^1\mathbf{x}) \in W$, where W is given by (10.6).

We verify the conditions (α.1) to (α.4).

Condition (α.1). Granting for the moment that ψ^k is well-defined, it is apparent that for fixed c its value depends only on $\|{}^1\mathbf{x}\|$, so that condition (α.1) is satisfied.

Conditions (α.2). For $|c| \leqq \sigma/2$

$$\psi^k(c, x^2, \cdots, x^n) = c \tag{10.12}$$

since $\mu(c/\sigma) = 0$ for $|c| \leqq \sigma/2$. For $|c| > \sigma/2$ the parenthesis in (10.11) of which the square root is taken, has a positive value, since within the parenthesis

$$c^2 > \frac{\sigma^2}{4}, \quad \left|\mu\left(\frac{c}{\sigma}\right)\right| \leqq 1, \quad \|{}^1\mathbf{x}\|^2 < \sigma^2/4 .$$

Thus ψ^k is well-defined and of class C^∞ on W. That (10.7) holds is immediate on recalling that μ is an even function.

Taking account of the relation (10.12) and the oddness of ψ^k for fixed ${}^1\mathbf{x}$, it is sufficient to establish (10.8) for $c > \sigma/2$. For such c, (10.8) holds if and only if

$$2c - (-1)^k \frac{\|{}^1\mathbf{x}\|^2}{\sigma}\mu'\left(\frac{c}{\sigma}\right) = Q > 0 ,$$

introducing Q. But for $|{}^1\mathbf{x}\| < \sigma/2$

$$Q > 2c - \frac{\sigma}{4}\mu'\left(\frac{c}{\sigma}\right).$$

Hence $Q > 0$, since $|\mu'(c/\sigma)| < 4$, while $2c > \sigma$ in the case at hand. The relation (10.9) is immediate.

Condition (α.3). For $\sigma \leqq |c| < 2\sigma$, $\mu(c/\sigma) = 1$ and $\psi^k(c, {}^1\mathbf{x})$ reduces to the right member of (10.10). Condition (α.3) follows.

Condition (α.4). On the carrier of $\lambda^k(\omega)$

$$\|\mathbf{x}\|^2 = [\psi^k(c, {}^1\mathbf{x})]^2 + \|{}^1\mathbf{x}\|^2 \leqq c^2 + 2\|{}^1\mathbf{x}\|^2 < 5\sigma^2$$

so that this carrier is in $D_{3\sigma}$.

Proof that $\lambda^k(\omega)$ *is a field.* Note first that there is at most one manifold H_c^k through each point of $|\lambda^k(\omega)|$, by virtue of condition (10.8), since (10.8) implies that for fixed ${}^1\mathbf{x}$, x^1 increases with c. It remains to show that the aggregate of $(n-1)$-manifolds H_c^k in $\lambda^k(\omega)$, parameterized by c, qualifies in the sense of § 3 as an admissible family of $(n-1)$-manifolds. We represent these manifolds in the parametric form

$$x^1 = \psi^k(c, u^1, \cdots, u^{n-1}); \quad x^2 = u^1, \cdots, x^n = u^{n-1}, \tag{10.13}$$

for $\mathbf{u} \in \Delta^u_{\sigma/2}$ and $c \in J = (-2\sigma, 2\sigma)$. In the notation of (3.3) we have here a diffeomorphism

$$\Phi : \Delta^u_{\sigma/2} \times J \longrightarrow E_n; \quad \mathbf{u} \times c \longrightarrow p(\mathbf{u}, c)$$

in which $p(\mathbf{u}, c)$ is the point $\mathbf{x}$ given by (10.13). It is seen that such a Φ has the properties demanded in § 3.

This completes the proof of Lemma 10.1.

The subfield $\lambda^k_\varepsilon(\omega)$ *of* $\lambda^k(\omega)$. The manifolds H_c^k of $\lambda^k(\omega)$ are defined for $c \in J$. But for $\sigma \leqq |c| < 2\sigma$ they are representable as in (10.10). It will be convenient in this paragraph to suppose these manifolds defined for $\sigma \leqq |c| < 3\sigma$ by (10.10). With this understood J becomes a relatively compact subinterval of the interval of definition of the extended field $\{H_c^k \mid -3\sigma < c < 3\sigma\}$. In accord with the definition of $\Lambda_\varepsilon(\gamma)$ in Corollary 3.1, $\lambda^k_\varepsilon(\omega)$ is a well-defined subfield of $\lambda^k(\omega)$, provided ε is a sufficiently small positive constant. This subfield has the form

$$\lambda^k_\varepsilon(\omega) = \{\mathscr{D}_c^k \mid -2\sigma < c < 2\sigma\} \qquad (k = 0, 1) \tag{10.14}$$

where $\mathscr{D}_c^k$ is a geodesic $(n-1)$-disc on H_c^k with radius ε and pole at $\omega(c)$.

The radial geodesics on the geodesic discs $\mathscr{D}_c^k$. The manifold H_c^k in $\lambda^k(\omega)$ if a "manifold of revolution" with the x^1-axis as its axis of revolution, cf., (α.1). If $(a^2, \cdots, a^n) = {}^1\mathbf{a}$ is such that $\|{}^1\mathbf{a}\| = 1$, then any arc of H_c^k with the representation

$$x^1 = \psi^k(c, ta^2, \cdots, ta^n) \qquad \left(0 \leqq t < \frac{\sigma}{2}\right) \tag{10.15}$$

on which ${}^1\mathbf{a}$ is constant, *is a generator* of H_c^k. It is also *a geodesic on* H_c^k. This is a simple property of the Euler equations of geodesics on a manifold of revolution, familiar for surfaces of revolution. It follows that $\mathcal{D}_c^k$ is not only a *submanifold* of a manifold of revolution with the x^1-axis as axis of revolution, but is itself a manifold of revolution generated by any one of its radial-geodesic arcs issuing from $\omega(c)$.

The following lemma is fundamental in defining $\mathcal{F}$ in § 12.

LEMMA 10.2. *On a radial-geodesic arc of* $\mathcal{D}_c^k$ *issuing from* $\omega(c)$, *with arc length* r *measured from* $\omega(c)$, K^k *defines a function*

$$r \longrightarrow \Theta_c^k(r) \qquad (0 \leqq r < \varepsilon) \tag{10.16}$$

with a derivative as to r *which vanishes on the interval* $(0, \varepsilon)$ *when* $|c| \geqq \sigma$, *and is positive on this interval when* $|c| < \sigma$.

If $\mathbf{x}$ is a point on $\mathcal{D}_c^k$

$$K^k(\mathbf{x}) = (-1)^k(\psi^k(c, \mathbf{x}))^2 + \|{}^1\mathbf{x}\|^2 = (-1)^k c^2 + \|{}^1\mathbf{x}\|^2\left(1 - \mu\left(\frac{c}{\sigma}\right)\right).$$

Along the geodesic are (10.15), $\|{}^1\mathbf{x}\|^2 = t^2$, so that on this arc

$$\frac{dK^k}{dr} = 2t\left(1 - \mu\left(\frac{c}{\sigma}\right)\right)\frac{dt}{dr} \qquad \left(\frac{dt}{dr} > 0\right) \tag{10.17}$$

Lemma 10.2 follows on making use of the definition of μ.

Permanent notation. $\lambda_g^k(\omega) : k = 0, 1$; (10.14).

11. The field $\Lambda_e(g)$

In the notation of § 8 the critical point A_r: $r = -1, 0, 1$, on g has the s-coordinate s_r on g. We here consider the subarcs of g on which s ranges on the following successively overlapping intervals:

$$\begin{cases} (s_{-1} - 2\sigma, s_{-1} + 2\sigma) = W_1 \\ (s_{-1} + \sigma, s_0 - \sigma) = W_2 \\ (s_0 - 2\sigma, s_0 + 2\sigma) = W_3 \\ (s_0 + \sigma, s_1 - \sigma) = W_4 \\ (s_1 - 2\sigma, s_1 + 2\sigma) = W_5 \end{cases} \tag{11.1}$$

introducing $W_1, \cdots, W_5$. These intervals are the ranges of s on the respective subarcs

$$T_{-1}(\omega), g_{-1}, T_0(\omega), g_1, T_1(\omega) \qquad \text{[cf., (8.8), (9.1)]} \tag{11.2}$$

of g. Successive arcs in this sequence overlap in arcs represented by the disjoint intervals

$$(11.3) \qquad \begin{cases} (s_{-1} + \sigma, s_{-1} + 2\sigma) = W_1 \cap W_2 \\ (s_0 - 2\sigma, s_0 - \sigma) \quad = W_2 \cap W_3 \\ (s_0 + \sigma, s_0 + 2\sigma) \quad = W_3 \cap W_4 \\ (s_1 - 2\sigma, s_1 - \sigma) \quad = W_4 \cap W_5 . \end{cases}$$

In §§ 9 and 10 we have defined disjoint fields of manifolds

$$(11.4) \qquad \Lambda_\varepsilon(g_{-1}) , \quad \Lambda_\varepsilon(g_1)$$

in M traversed by g_{-1} and g_1 respectively, and fields of manifolds

$$(11.5) \qquad \lambda^0_\varepsilon(\omega) , \quad \lambda^1_\varepsilon(\omega)$$

in E_n traversed by ω. The manifolds in these four fields have been respectively denoted by

$$(11.6) \qquad \mathcal{D}_s , \quad \mathcal{D}_s , \quad \mathcal{D}^0_c , \quad \mathcal{D}^1_c .$$

Since the mappings T_r: $r = -1, 0, 1$, are isometries on $D_{2\sigma}$,

$$(11.7) \qquad \begin{cases} g(s) = T_{-1}(\omega(c)) \mid (c = s - s_{-1}) & (s \in W_1) \\ g(s) = T_0(\omega(c)) \mid (c = s - s_0) & (s \in W_3) \\ g(s) = T_1(\omega(c)) \mid (c = s - s_1) & (s \in W_5) . \end{cases}$$

Definition of $\Lambda_\varepsilon(g)$. Subject to further restriction on the magnitude of ε we shall (*a priori*) overdefine the $(n - 1)$-manifolds $\mathcal{H}_s$: $0 < s < s^*$ in $\Lambda_\varepsilon(g)$ by setting

$$(11.8) \qquad \begin{cases} \mathcal{H}_s = T_{-1}(\mathcal{D}^0_c) \mid (c = s - s_{-1}) & (s \in W_1) \\ \mathcal{H}_s = \mathcal{D}_s & (s \in W_2) \\ \mathcal{H}_s = T_0(\mathcal{D}^1_c) \mid (c = s - s_0) & (s \in W_3) \\ \mathcal{H}_s = \mathcal{D}_s & (s \in W_4) \\ \mathcal{H}_s = T_1(\mathcal{D}^0_c) \mid (c = s - s_1) & (s \in W_5) . \end{cases}$$

We state a fundamental lemma.

LEMMA. 11.1. *If ε is a sufficiently small positive constant, $\mathcal{H}_s$ is uniquely defined in* (11.8) *for each $s \in (0, s^*)$, and the s-parameterized aggregate*

$$(11.9) \qquad \Lambda_\varepsilon(g) = \{\mathcal{H}_s \mid 0 < s < s^*\}$$

of these geodesic $(n - 1)$-discs is a field traversed by g and locally representable as is $\Lambda_\varepsilon(\gamma)$ in Corollary 3.1.

We begin by proving (i).

(i) *The manifold $\mathcal{H}_s$: $0 < s < s^*$ is uniquely defined in* (11.8).

Note that $\mathcal{H}_s$ is separately defined in (11.8) for s in each of the intervals $W_1, \cdots, W_5$, and hence has (*a priori*) two definitions for each of the intervals (11.3). Let us begin with $W_1 \cap W_2$, the first of the intervals (11.3). To show that $\mathcal{H}_s$ is uniquely defined for $s \in W_1 \cap W_2$ we must show that

$$\mathcal{D}_s = T_{-1}(\mathcal{D}_c^0) \mid (c = s - s_{-1}) \qquad (s \in W_1 \cap W_2). \tag{11.10}$$

For the two members of (11.10) are the two manifolds to which $\mathcal{H}_s$ is set equal by definition in (11.8).

PROOF OF (11.10). For $s \in W_1 \cap W_2$, $\mathcal{D}_s$ is a geodesic $(n-1)$-disc with pole $g(s)$ and radius ε on that level manifold H' of F which meets $g(s)$. Subject to the condition $c = s - s_{-1}$, $\mathcal{D}_c^0$ is a geodesic $(n-1)$-disc of radius ε and pole $\omega(c)$ on that level manifold of K^0 which meets $\omega(s)$. Since T_{-1} is an isometric mapping of $D_{3\sigma}$ into M, and since $\mathcal{D}_c^0 \subset D_{3\sigma}$ by virtue of $(\alpha.4)$, Lemma 10.1, we conclude that the manifold on the right of (11.10) is a geodesic $(n-1)$-disc of radius ε and pole $T_{-1}(\omega(c))$ on that level manifold, H'' of F, which meets $T_{-1}(\omega(c))$. According to (11.7)

$$g(s) = T_{-1}(\omega(c)) \mid (c = s - s_{-1}) \qquad (s \in W_1). \tag{11.11}$$

Hence $H' = H''$, since these manifolds both meet the point (11.11). The right member of (11.10) is thus a geodesic $(n-1)$-disc on the same level manifold of F as $\mathcal{D}_s$, with the same pole and the same radius. Hence (11.10) holds.

The proof that the two definitions of $\mathcal{H}_s$ agree for s in any other of the intervals (11.3) is similar and (i) follows.

We continue with a proof of the following.

(ii) *If ε is a sufficiently small positive constant, no two manifolds $\mathcal{H}_{s'}$ and $\mathcal{H}_{s''}$ in $\Lambda_\varepsilon(g)$ for which $s' \neq s''$ can intersect.*

Let W be the union of the intervals (11.3) and set $W_j^* = W_j - W : j = 1, \cdots, 5$. The intervals W_j^* have disjoint closures. If ε is sufficiently small the carriers of the five aggregates

$$\{\mathcal{H}_s \mid s \in W_j^*\} \qquad (j = 1, \cdots, 5)$$

will have disjoint closures. For such an ε, taken in the range of values for which the fields (11.4) and (11.5) have been defined, statement (ii) will hold. For $(0, s^*)$ is the union of the nine disjoint subintervals

$$W_1^*,\ W_1 \cap W_2,\ W_2^*,\ W_2 \cap W_3,\ W_3^*,\ W_3 \cap W_4,\ W_4^*,\ W_4 \cap W_5,\ W_5^*.$$

The corresponding nine aggregates of manifolds $\mathcal{H}_s$ are disjoint for the above choice of ε. In any one of these aggregates no two different manifolds $\mathcal{H}_{s'}$ and $\mathcal{H}_{s''}$ intersect. Statement (ii) follows.

The s-parameterized aggregate $\Lambda_\varepsilon(g)$ is a family of geodesic discs, representable as is $\Lambda_\varepsilon(\gamma)$ in Corollary (3.1), since each of the fields (11.4) and (11.5) have this property. This fact, together with (ii), implies that $\Lambda_\varepsilon(g)$ is a field if ε is sufficiently small.

Lemma 11.1 is thereby established.

Permanent notation. $\Lambda_\varepsilon(g)$, (11.9).

12. The definition and properties of $\mathcal{F}$

Under the hypothesis that the non-degenerate function F on M has at least two critical points on M of index zero, we have established the existence in § 8 of a saddle point A_0 of F of decreasing type and two critical points A_{-1} and A_1 of index zero, lying with A_0, on an extended saddle arc g with the following properties. The arc g meets the points A_{-1}, A_0, A_1 in the order written. On g, F defines a function with values which we have denoted by $G(s)$, where s is the arc length measured along g from the initial limiting end point of g. The function G has an absolute minimum, an absolute maximum, and a relative minimum at A_{-1}, A_0, A_1, respectively, and has no other critical points.

We began the definition of $\mathcal{F}$ in § 8. Along g, $\mathcal{F}$ is to reduce to a function $\mathcal{G}$ having an absolute minimum at A_{-1}, but no other critical points. $\mathcal{G}(s)$ was so defined on the interval $(0, s^*)$ for s, that $G(s) = \mathcal{G}(s)$ on initial and final subintervals of $(0, s^*)$

$$(s_{-1} - 2\sigma, s_{-1} + \sigma] , \quad [s_1 + \sigma, s_1 + 2\sigma) , \tag{12.0}$$

where s_{-1} and s_1 are the s-coordinates of A_{-1} and A_1. For other values of s, $\mathcal{G}(s) < G(s)$. Set

$$d(s) = G(s) - \mathcal{G}(s) \qquad (0 < s < s^*) \tag{12.1}$$

With ε as in $|\Lambda_\varepsilon(g)|$, let ε' be a constant such that $0 < \varepsilon' < \varepsilon$. We introduce a mapping θ of class C^∞ of the axis of reals into the interval $[0, 1]$ such that

$$\begin{cases} \theta(t) = 1 & (t \leqq 0) \\ \theta(t^2) = 0 & (t \geqq \varepsilon') \\ \theta'(t^2) < 0 & (0 < t < \varepsilon') . \end{cases} \tag{12.2}$$

Such a mapping clearly exists.

Definition of $\mathcal{F}$. At each point $p \in M$ not in $|\Lambda_\varepsilon(g)|$ set $\mathcal{F}(p) = F(p)$. If p is a point on the manifold $\mathcal{H}_s$ of the field $\Lambda_\varepsilon(g)$ at a geodesic distance r on $\mathcal{H}_s$ from the point $g(s)$, denote s and r by $s(p)$ and $r(p)$, respectively, and set

(12.3) $$\mathcal{F}(p) = F(p) - \theta(r^2(p))d(s(p)) .$$

At a point $g(s)$ on g this definition gives

(12.4) $$\mathcal{F}(g(s)) = F(g(s)) - (G(s) - \mathcal{G}(s)) = \mathcal{G}(s)$$

consistent with our preliminary definition of $\mathcal{F}$ along g in § 8.

We continue with a proof of Lemma 12.1.

LEMMA 12.1. *The function $\mathcal{F}$ so defined on M is of class C^∞ on M.*

Let $\mathcal{A}$ be the carrier of $\Lambda_\varepsilon(g)$, and let $C\mathcal{A}$ be the complement of $\mathcal{A}$ relative to M. It is clear that $\mathcal{F}$ is of class C^∞ on the interior of $C\mathcal{A}$, since $F = \mathcal{F}$ on this interior. To prove $\mathcal{F}$ of class C^∞ on M it is not sufficient (*a priori*) to prove, in addition, that $\mathcal{F}$ is of class C^∞ on $\mathcal{A}$, because the nature of $\mathcal{F}$ at points of $\beta\mathcal{A}$, the boundary of $\mathcal{A}$, will still be in question. We surmount this difficulty as follows.

The arc g' and set $\mathcal{A}'$. Let g' be the open subarc of g on which $d(s) > 0$. The arc g' is a relatively compact subarc of g. The subfield $\Lambda_\varepsilon(g')$ of $\Lambda_\varepsilon(g)$ is well-defined, as is the subfield $\Lambda_{\varepsilon'}(g')$. Set

(12.5) $$\mathcal{A}' = | \Lambda_{\varepsilon'}(g') | .$$

The set $\mathcal{A}'$ is a relatively compact subset of $\mathcal{A}$. On taking account of the fact that $\theta(t^2) = 0$ for $t \geqq \varepsilon'$ it appears that

(12.6) $$\mathcal{F}(p) = F(p) \qquad (p \in C\mathcal{A}').$$

We conclude that $\mathcal{F}$ is of class C^∞ on the interior $\mathcal{B}$ of $C\mathcal{A}'$. But $\mathcal{B}$ is an open subset of M which includes $\beta\mathcal{A}$. To prove Lemma 12.1 it is accordingly sufficient to prove statement (i).

(i) *$\mathcal{F}$ is of class C^∞ on $\mathcal{A}$.*

PROOF OF (i). A proof of (i) requires an appropriate set of local coordinates in $\mathcal{A}$. To that end let b be an arbitrary value on $(0, s^*)$ and let J be an open subinterval of $(0, s^*)$ that contains b. The subfamily of manifolds $\mathcal{H}_s$: $s \in J$ of $\Lambda_\varepsilon(g)$ has a carrier $W(b, J)$ which is an open subset of $\mathcal{A}$. Each point of $\mathcal{A}$ is in an open set of this character. We refer now to the $(n - 1)$-plane $\mathcal{E}_{n-1}$ of points $(z^1, \cdots, z^{n-1}) = \mathrm{z}$ introduced preceding Lemma 3.2. If J is sufficiently small it follows from Lemma 11.1 and Corollary 3.1 that there exists a "radial-geodesic" isometry Φ of $\Delta_\varepsilon^z \times J$ onto $W(b, J)$. The sets

(12.7) $$(\mathrm{z}, s) = (z^1, \cdots, z^{n-1}, s) \in \Delta_\varepsilon^{(z)} \times J$$

are thus admissible local coordinates on the coordinate domain $W(b, J)$ in $\mathcal{A}$.

Let p then be a point in $W(b, J)$ with the local coordinates (12.7). The point p is on $\mathcal{H}_s$ at a geodesic distance $r < \varepsilon$ from $g(s)$. Since Φ is a radial-geodesic isometry,

$$r^2 = (z^1)^2 + \cdots + (z^{n-1})^2 .$$

The last term in the right member of (12.3) accordingly has the representation

$$\theta(r^2)d(s) = \theta((z^1)^2 + \cdots (z^{n-1})^2)d(s) , \tag{12.8}$$

defining a function of class C^∞ on the coordinate range $\Delta_\varepsilon^z \times J$.

Statement (i) is thereby proved and Lemma 12.1 follows.

We continue with a proof of Lemma 12.2.

LEMMA 12.2. *The function $\mathcal{F}$ reduces to F on $C\mathcal{A}'$ and has no critical point on the closure of $\mathcal{A}'$.*

The first statement in the lemma has already been established, cf., (12.6). We continue by proving statements (a), (b), (c).

(a) *$\mathcal{F}$ has no critical point on $\beta\mathcal{A}'$.*

$\mathcal{F} - F$ is of class C^∞ over M, vanishes on $C\mathcal{A}'$, and accordingly vanishes with all of its partial derivatives with respect to local coordinates at points of $\beta\mathcal{A}'$. Thus $\mathcal{F}$ has no critical point on $\beta\mathcal{A}'$, since F has no critical point on $\beta\mathcal{A}'$.

(b) *$\mathcal{F}$ has no critical point on g'.*

This is a consequence of the definition of $\mathcal{G}(s)$ along g and in particular along g', cf., (8.12).

(c) *$\mathcal{F}$ has no critical point on $\mathcal{A}' - g'$.*

Let p be a point of $\mathcal{A}' - g'$. It is on a manifold $\mathcal{H}_s$ of the field, $\Lambda_\varepsilon(g')$, at a distance r from $g(s)$ along a geodesic γ on $\mathcal{H}_s$, issuing from $g(s)$. We note that $0 < r < \varepsilon'$. We turn to the definition of $\mathcal{F}$ in (12.3).

Differentiating both members of (12.3) as to r along γ we have

$$\frac{d\mathcal{F}}{dr} = \frac{dF}{dr} - 2r\theta'(r^2)d(s) \qquad (0 < r < \varepsilon'). \tag{12.9}$$

The last term in (12.9) is positive since for $(0 < r < \varepsilon')$

$$r > 0 , \quad -\theta'(r^2) > 0 , \quad d(s) > 0 .$$

In particular $\theta'(r^2) < 0$ for $0 < r < \varepsilon'$ in accord with the definition of θ, while $d(s) > 0$ in accord with the definition of g'. To establish (c) it is accordingly sufficient to establish the following.

(d) *Along the geodesic γ, $dF/dr \geqq 0$ for $0 < r < \varepsilon'$.*

To establish (d) we distinguish two cases.

Case I. $\mathcal{H}_s$ is a submanifold of a level manifold of F. In this case $dF/dr = 0$ along γ.

Case II. (not Case I). Recall that on g', s ranges on the interval $(s_{-1} + \sigma, s_1 + \sigma) = L$, introducing L. By virtue of its definition in (11.8), $\mathcal{H}_s$, for $s \in L$, is a submanifold of a level manifold of F, at most excepting values of s in one of the two subintervals,

$$(12.10) \qquad \begin{aligned} (s_0 - \sigma, s_0 + \sigma) &= L_0 \\ (s_1 - \sigma, s_1 + \sigma) &= L_1 \end{aligned}$$

of L. In Case II, therefore, s is in $L_0 \cup L_1$. Consider then an $s \in L_0$.

For $s \in L_0$, $\mathcal{H}_s$ is defined in (11.8) by setting

$$(12.11) \qquad \mathcal{H}_s = T_0(\mathcal{D}_c^1) \mid (c = s - s_0) .$$

Under the inverse of T_0 the radial-geodesic γ on $\mathcal{H}_s$, issuing from $g(s)$, is mapped isometrically onto a radial-geodesic κ on $\mathcal{D}_c^1$, issuing from $\omega(c)$ where $c = s - s_0$. Let the quadratic form $K^1(\mathbf{x})$ be defined as in (10.3). If $p(r)$ is the point on γ at the distance r along γ from $g(s)$, and if $\mathbf{x}(r)$ is the point in E_n on κ, at the distance r along κ from $\omega(c)$, then, in accord with (5.2),

$$(12.12) \qquad F(p(r)) - F(A_0) = K^1(\mathbf{x}(r)) = \Theta_c^1(r) ,$$

introducing Θ_c^1 as in Lemma 10.2. It follows from (12.12) and Lemma 10.2 that

$$\frac{dF(p(r))}{dr} = \frac{d\Theta_c^1(r)}{dr} > 0 \qquad (0 < r < \varepsilon') .$$

The case of an s in L_1 is treated similarly. This completes the proof of (d) and hence of Lemma 12.2.

Since A_0 and A_1 are in $\mathcal{A}'$ we infer the following.

COROLLARY 12.1. *Suppose that F has m critical points. If F has more than one critical point of index zero there exists a non-degenerate function $\mathcal{F}$ with just $m - 2$ critical points and such that $\mathcal{F}$ is identical with F in neighborhoods of each of these critical points. The critical points of F which do not thus appear as critical points of $\mathcal{F}$ are a critical point of index zero and a saddle point of decreasing type.*

If F has $\mu+1$ points of index zero one can repeat this process μ times, arriving finally at a non-degenerate function F_1 with $m - 2\mu$ critical points such that F_1 is identical with F in neighborhoods of each of F_1's critical points, and such that F_1 has just one critical point of index 0. If F has $\nu + 1$ critical points of index n, F_1 will likewise. The function $-F_1$ will have $\nu + 1$ critical points of index zero and just one critical point of index n. Our process will replace $-F_1$ by a polar non-degenerate function F_2 with $m - 2\mu - 2\nu$ critical points. We finally replace F_2 by $F^* = -F_2$ and infer the following theorem:

THEOREM 12.1. *If F is a non-degenerate function on M with m critical values, m critical points, $\mu + 1$ critical points of index zero, and $\nu + 1$ critical points of index n, there exists a polar non-degenerate function F^* with $m - 2\mu - 2\nu$ critical points such that F^* is identical with F in some neighborhood of each of F^*'s critical points. If $n > 2$ the function F^*, as defined, will have μ fewer critical points of index 1 than F, and ν fewer critical points of index $n - 1$.*

Theorem 12.1 is perhaps the first definite advance in solving the difficult problem of modifying a given non-degenerate function so as to eliminate the topologically unnecessary critical points.

INSTITUTE FOR ADVANCED STUDY

REFERENCES

1. MARSTON MORSE, *The existence of non-degenerate functions on a differentiable manifold.* Annali di Mathematica, 44 (1960)
2. ———, *Differentiable mappings in the Schoenflies theorem.* Compositio Math. (1959)
3. ———, *Fields of geodesics through a point.* Proc. Nat. Acad. Sci. U.S.A. 50 (1960).
4. ———, Functional topology and abstract variational theory, in Mémorial des Sciences mathématiques, Gauthier-Villars, Paris, 1938.
5. ———, The calculus of variations in the large. Colloquium, Amer. Math. Soc., 190 Hope St., Providence 6, R. I., 1934.
6. HASSLER WHITNEY, *Differentiable manifolds.* Ann. of Math. 37(1936), 645-680

Reprinted from the
BULLETIN OF THE AMERICAN MATHEMATICAL SOCIETY
March 1960, Vol. 66, No. 2
Pp. 113-115

A REDUCTION OF THE SCHOENFLIES EXTENSION PROBLEM

BY MARSTON MORSE

Communicated by Edwin Moise, December 28, 1959

1. In his study of the Schoenflies extension problem Mazur, [1], includes the hypothesis that the given mapping ϕ be "semilinear" in a neighborhood of some point in the shell domain of ϕ. In Mazur's paper the problem of removing this hypothesis is unsolved. We shall show how to remove this hypothesis, leaving the "shell hypothesis." In the process of doing this we establish two fundamental lemmas concerning the equivalence of extension problems.

Let E and $\mathcal{E}$ be euclidean n-spaces, $n>1$, with rectangular coordinates (x) and (y) respectively. Let (x) and (y) be represented by vectors $\mathbf{x}$ and $\mathbf{y}$ with components $(x_1, \cdots, x_n)$ and $(y_1, \cdots, y_n)$ respectively. Let $\|\mathbf{x}\|$ be the distance of the point $\mathbf{x}$ from the origin. Let I be the mapping of E onto $\mathcal{E}$ under which $\mathbf{x}=\mathbf{y}$. Let S_{n-1} be an $(n-1)$-sphere of unit radius in E with center at the origin. Let $\mathfrak{J}S_{n-1}$ be the interior of S_{n-1}. With $0<a<1$, introduce the n-shell

$$\sigma_a = \{\mathbf{x} \mid 1 - a < \|\mathbf{x}\| < 1 + a\}. \tag{1}$$

As applied to a mapping of an open subset of a euclidean space or a differentiable manifold into a similar space, a *C^m-diffeomorphism* shall have its usual meaning when $m>0$, but when $m=0$ shall be merely a *homeomorphism*.

In this sense let

$$\phi: \sigma_a \to \mathcal{E} \tag{2}$$

be a C^m-diffeomorphism of σ_a onto a subset of $\mathcal{E}$. The set $\phi(\sigma_a)$ is open in $\mathcal{E}$ and affords a neighborhood of the manifold $\phi(S_{n-1}) = \mathfrak{M}_{n-1}$. Let $\mathfrak{J}\mathfrak{M}_{n-1}$ be the interior of $\mathfrak{M}_{n-1}$. We assume that ϕ carries points of σ_a which are interior to S_{n-1} into points which are interior to $\mathfrak{M}_{n-1}$. We term these conditions on ϕ the *shell hypothesis*.

THEOREM 1. (i) *If N is a sufficiently small neighborhood of S_{n-1} there exists an extension Λ_ϕ of $\phi \mid N$ which is a homeomorphism of $N \cup \mathfrak{J}S_{n-1}$ onto $\phi(N) \cup \mathfrak{J}\mathfrak{M}_{n-1}$.*

(ii) *If $m>0$ and if Z is an arbitrary point interior to S_{n-1}, then Λ_ϕ and N may be chosen so that Λ_ϕ is, in addition, a C^m-diffeomorphism of $N \cup \mathfrak{J}(S_{n-1}) - Z$ into $\mathcal{E}$.*

In the differentiable case ($m>0$), the theorem has been proved by the author in [2]. In the topological case ($m=0$) we shall give a proof of (i) by transforming the problem into an equivalent problem solvable by the methods of Mazur in [1], or by the subsequent methods of the author in [2].

Given the C^m-diffeomorphism ϕ of Theorem 1, we term a mapping Λ_ϕ which satisfies Theorem 1 a solution of the Schoenflies extension problem $[\phi, \sigma_a]$. Let f be a C^m-diffeomorphism of the set

$$\phi(\sigma_a) \cup \mathring{J}\phi(S_{n-1}) = \Omega \tag{3}$$

into $\mathcal{E}$. It must be remembered that when $m=0$, a C^m-diffeomorphism is merely a homeomorphism.

LEMMA 1. (i) *If ϕ is a C^m-diffeomorphism of σ_a into $\mathcal{E}$ satisfying the shell hypothesis, $m \geqq 0$, then $f\phi$ is a C^m-diffeomorphism of σ_a into $\mathcal{E}$ satisfying the shell hypothesis.*

(ii) *A necessary and sufficient condition that there exist a solution Λ_ϕ of the problem $[\phi, \sigma_a]$ is that there exist a solution $\Lambda_{f\phi}$ of the problem $[f\phi, \sigma_a]$.*

PROOF OF (i). It is clear that $f\phi$ is a C^m-diffeomorphism of σ_a into $\mathcal{E}$. Set $f(\phi(S_{n-1})) = \mathcal{L}_{n-1}$. We must show that $f\phi$ carries a point $\mathbf{x}$ of σ_a interior to S_{n-1} into a point interior to $\mathcal{L}_{n-1}$.

This will follow once we have proved the relation

$$f(\mathring{J}\mathfrak{M}_{n-1}) = \mathring{J}\mathcal{L}_{n-1}. \tag{4}$$

Now $\mathring{J}\mathfrak{M}_{n-1}$ is the unique, bounded, maximal connected subset of $\mathcal{E}$ which does not meet $\mathfrak{M}_{n-1}$. Set $f(\mathring{J}\mathfrak{M}_{n-1}) = W$. The image set W is bounded and connected. Making use of the fact that f is locally a C^m-diffeomorphism of open subsets of $\mathcal{E}$ onto open subsets of $\mathcal{E}$, one can verify the definition of f^{-1} along all paths in $\mathcal{E}$ that start with points of W and do not meet $\mathcal{L}_{n-1}$. Since the image under f^{-1} remains in $\mathring{J}\mathfrak{M}_{n-1}$ these paths are in W. Hence W is a maximal, connected subset of $\mathcal{E}$ that does not meet $\mathcal{L}_{n-1}$. Since W is bounded $W = \mathring{J}\mathcal{L}_{n-1}$.

Thus (4) holds. A point $\mathbf{x}$ of σ_a interior to S_{n-1} has by hypothesis an image $\phi(\mathbf{x})$ interior to $\mathfrak{M}_{n-1}$, and by (4) $f(\phi(\mathbf{x}))$ is interior to $\mathcal{L}_{n-1}$. This establishes (i).

PROOF OF (ii). Suppose that Λ_ϕ is a solution of problem $[\phi, \sigma_a]$. I say that $f\Lambda_\phi$ is a solution of problem $[f\phi, \sigma_a]$. For $f\Lambda_\phi$ is an extension of $(f\phi)|N$ since $f(\Lambda_\phi(\mathbf{x})) = f(\phi(x)) = (f\phi)(\mathbf{x})$, $[\mathbf{x} \in N]$. Moreover $f\Lambda_\phi$ is a homeomorphism of $N \cup \mathring{J}(S_{n-1})$ onto

$$(f\Lambda_\phi)(N \cup \mathring{J}S_{n-1}) = f(\phi(N)) \cup f(\mathring{J}\mathfrak{M}_{n-1}) = (f\phi)(N) \cup \mathring{J}\mathcal{L}_{n-1} \tag{5}$$

in accord with (4). Moreover, if $m>0$, $f\Lambda_\phi$ is a C^m-diffeomorphism of $N \cup \mathring{J}S_{n-1} - Z$ into $\mathcal{E}$.

Conversely if $\Lambda_{f\phi}$ is a solution of problem $[f\phi, \sigma_a]$, defined over $N\cup\mathring{J}(S_{n-1})$, then $f^{-1}\Lambda_{f\phi}$ is a solution of problem $[\phi, \sigma_a]$, defined over $N\cup\mathring{J}S_{n-1}$. One makes use of the relation, derived from (4),

$$\mathring{J}\mathcal{M}_{n-1} = f^{-1}(\mathring{J}\mathcal{L}_{n-1}).$$

THE HYPOTHESIS A. *We say that the given mapping ϕ satisfies hypothesis* A *if ϕ reduces to I in some neighborhood X, relative to σ_a, of some point Q of S_{n-1}.*

LEMMA 2. *Corresponding to a given problem $[\phi, \sigma_a]$ of index $m \geq 0$, for suitable definition of a C^m-diffeomorphism f of the set Ω into $\mathcal{E}$ the C^m-diffeomorphism $f\phi$ of σ_a into $\mathcal{E}$ satisfies Hypothesis* A.

PROOF OF LEMMA 2. Let Q be an arbitrary fixed point of S_{n-1}. For the purposes of this proof suppose that $\phi(Q)$ is the origin in $\mathcal{E}$. Let $\mathcal{H}_\rho$ be an open n-ball in $\mathcal{E}$ with center at the origin and radius ρ. In particular let $\mathcal{H}_{2b}$ have a radius so small that $\mathcal{H}_{2b}\subset\phi(\sigma_a)$. Set

$$\phi^{-1}(\mathcal{H}_b) = X_b, \qquad \phi^{-1}(\mathcal{H}_{2b}) = X_{2b}$$

and note that $X_{2b}\subset\sigma_a$. Let $t\rightarrow\lambda(t)$ be a C^∞-diffeomorphism of the open interval $(0, 2b)$ onto the positive t-axis such that $\lambda(t)=t$ for $t\in(0, b)$. Such a mapping exists.

Let μ be the C^∞-diffeomorphism of $\mathcal{H}_{2b}$ onto $\mathcal{E}$ such that the point $y\in\mathcal{H}_{2b}$ has the image $\lambda(\|y\|)y/\|y\|$ when $\|y\|\neq 0$ and the image O when $y=O$. Set

$$\psi(x) = \mu(\phi(x)) \qquad (x \in X_{2b}), \tag{7}$$

and note that $\psi(x)=\phi(x)$ for $x\in X_b$ and that $\psi(X_{2b})=\mathcal{E}$. We can accordingly define a C^m-diffeomorphism f over $\mathcal{E}$ by setting

$$f(y) = I(\psi^{-1}(y)) \qquad (y \in \mathcal{E}). \tag{8}$$

We have

$$f(\phi(x)) = I(x) \qquad (x \in X_b). \tag{9}$$

Thus $f\phi$ satisfies Hypothesis A.

Theorem 1 (i) follows from the two lemmas taking account of the known fact that the problem $[f\phi, \sigma_a]$ admits a solution even when $m=0$, since $f\phi$ satisfies, not only the shell hypothesis, but also hypothesis A.

REFERENCES

1. Barry Mazur, *On embeddings of spheres*, Bull. Amer. Math. Soc. vol. 65 (1959) pp. 59–65.

2. Marston Morse, *Differentiable mappings in the Schoenflies theorem*. Compositio Math. vol. 14 (1959) pp. 83–151.

INSTITUTE FOR ADVANCED STUDY

FUNDAMENTA
MATHEMATICAE
L (1962)

Schoenflies problems

by

M. Morse (Princeton, N.Y.)

Dedicated to the Fundamenta Mathematicae on the occasion of the publication of its 50th volume, with grateful appreciation of what this journal of mathematics has meant to the world of mathematics during the last fifty years.

§ 1. Introduction. The theorem that the union of a Jordan curve and its interior in a 2-plane is a closed 2-cell is commonly called a Schoenflies theorem. The problems which arise in attempting to generalize this theorem in euclidean spaces of higher dimension are called *Schoenflies problems*.

The generalization which suggests itself first is false. Let M be a topological $(n-1)$-sphere in an euclidean n-space E with $n > 1$. Let $\mathring{J}M$ be the open interior of M and JM the closure of $\mathring{J}M$. It is not always true that JM is a closed n-cell for $n > 2$. See Ref. [0].

A major advance in formulating a valid Schoenflies extension theorem when $n > 2$ was made by Barry Mazur in Ref. [2]. Mazur concerned himself with a topological $(n-1)$-sphere on an euclidean n-sphere. We shall present a theorem which is essentially that of Mazur, but in which Mazur's n-sphere is replaced by the euclidean n-space E. This use of an euclidean n-space E in place of an eucliden n-sphere accords with subsequent developments which we shall present.

Mazur's theorem. Mazur made two assumptions, the second of which, as we shall see, is unnecessary. Let S be an $(n-1)$-sphere in E with center at the origin and radius 1. In our formulation of Mazur's theorem these hypotheses are as follows.

I. Let φ be a homeomorphism of an open neighborhood N of S into E under which points interior (exterior) to S go into points interior (exterior) to the $(n-1)$-manifold $\varphi(S) = M$.

II. Suppose that there exists in N a neighborhood of a point P of S in the form of a star of euclidean cells incident with P such that on each cell of the star φ is linear.

Conclusion (α). *Then $\varphi|S$ admits an extension as a homeomorphism λ_φ of JS onto JM such that λ_φ agrees with φ at all points of some neighborhood $W \subset N$ of S interior to S.*

It is trivial that λ_φ can be extended as a homeomorphism to agree with φ at points of N exterior to S. In general W cannot include all points of N interior to S.

The extension problem (φ, N). *Topological case.* Given φ and N as in Hypothesis I, and omitting Hypothesis II, the problem of finding λ_φ so as to satisfy (α) will be called the *extension problem* (φ, N) in the *topological case.*

The topological case is to be distinguished from the *differentiable case* in which φ is a C^m-diffeomorphism of S into E, $m > 0$, and the *analytic case* in which φ is an analytic diffeomorphism of S into E.

In Ref. [3] we have established the following theorem, reducing Mazur's hypotheses to I:

THEOREM 1.1 (i). *Given the topological extension problem* (φ, N), *there exists a homeomorphic mapping f of E into E such that $f\varphi$ reduces to the identity on a neighborhood of some point of S.*

(ii). *A necessary and sufficient condition that there exist a solution of the extension problem (φ, N) is that there exist a solution of problem $(f\varphi, N)$.*

Since the homeomorphism $f\varphi$ satisfies Mazur's Hypothesis II, Mazur's theorem and the above theorem have the following corollary.

COROLLARY 1.1. *Each Schoenflies extension problem (φ, N) in the topological case admits a solution λ_φ satisfying Conclusion* (α).

Theorem 1.1 was first discovered by the author in a form appropriate to the differentiable case. (Cf. Lemma 3.1 of Ref. [4].) In Ref. [4] the first solution of the extension problem in the differentiable case was announced. This solution was a diffeomorphism with one differential singularity; it was a homeomorphism. The complete exposition in the differentiable case was given in Ref. [5]. With the discovery of Theorem 1.1 and its analogue in the differentiable case it was recognized that Part II of Ref. [5], with appropriate verbal changes, gave a solution of the extension problem (φ, N) *both* in the topological and the differentiable case. A simplified exposition with explicit formulas that made this clear has been presented by Huebsch and Morse in Ref. [6].

Three proofs of Corollary 1.1 appeared at approximately the same time.

Morton Brown in Ref. [7] gave a direct solution of the extension problem (φ, N) in the topological case. Mazur's result and Theorem 1.1, as combined by the author in Ref. [3], gave another proof. These two proofs apply only to the topological case. The methods of Part II of Ref. [5]

(later simplified in Ref. [6]) afford a proof both in the topological and the differentiable case.

The explicit solution of the problem (φ, N), in Ref. [6], is the basis of what follows concerning *cell problems*. Cell problems are distinguished in § 3 from extension problems. The results of Ref. [6] lead to the theorem that the interior of M (in the differentiable case) is an open analytic n-ball, that is the image of an open euclidean n-ball under a real analytic diffeomorphism. The methods of Ref. [6] also serve as a model for a theory of families of Schoenflies extensions. See Ref. [10] and Ref. [11].

Γ-problems and simple extension problems. The data in the problem of obtaining continuous families of Schoenflies extensions includes a topological space Γ on which a parametric point p rests. Such a problem will be called a Γ-problem to distinguish it from an extension problem (φ, N). We term the latter problem a *simple* extension problem. Both in Γ-problems and in simple extension problems one distinguishes between the *topological*, the *differentiable* and the *analytic case*, first in the data, and, subsequently, in the conclusions. See §§ 4 and 5.

Extension problems and cell problems.

(i) Theorem 1.1 and its corollary concern an extension problem.

(ii) Given a C^1-diffeomorphism f of S onto $f(S) = M$, the corresponding *open cell problem* is to find an analytic diffeomorphism F of an open n-ball onto $\mathring{J}M$. It is solved by Huebsch and Morse in Ref. [12]. In general F is in no sense an extension of f.

(iii) In § 3 we define a *closed cell problem* (with differentiability index $m = \infty$). We show how this closed cell problem is related to the general extension problem. In this connection we introduce "differentiable isotopies". These isotopies differ from those introduced by Milnor in Ref. [13]. For $n = 2$, or 3, a solution of this closed cell problem exists, but the existence of a solution in the general case is an open question.

The reader familiar with the theory of extensions of continuous mappings, as developed by Borsuk (Ref. [17]), or Hurewicz, will note a formal similarity between certain of the results presented here and earlier extension theorems. In particular, there exist in the present theory strong relations between differentiable isotopies and the existence of differentiable extensions (§ 2). However the similarity is no more than formal. In fact, the Borsuk theory aims at continuous extensions over sets of great generality, whereas the theory of Schoenflies extensions starts with domains of extreme simplicity, such as S, and finds its principal difficulty in showing that extensions exist as homeomorphisms or diffeomorphisms.

In the next section we shall present some of the principal theorems concerning simple Schoenflies problems in the differentiable case.

§ 2. The extension problem (φ, N). Differentiable case. We begin with notation and conventions.

Notation. As previously E is an euclidean n-space, S an $(n-1)$-sphere in E with center at the origin 0, and M a topological $(n-1)$-sphere in E. If 0 is in $\mathring{J}M$ we set

$$JM - 0 = J_0 M .$$

Conventions. A real analytic or C^m-mapping F into E of an open subset X of E, is defined in the usual way, as is an analytic or C^m-diffeomorphism of X into E. We suppose that $m > 0$. We shall extend these definitions to any subset X of E such that $X \subset \mathrm{Cl}\,\mathring{X}$. For such a set X we say that F is an *analytic* or *C^m-mapping* into E, or an *analytic* or *C^m-diffeomorphism* of X into E, if F admits an extension F^e over an open subset $Y \supset X$ of E with the respective properties of F, when Y replaces X.

A_0 or C_0^m-diffeomorphisms, $m > 0$. Let X be a subset of E such that $\mathring{X}$ contains the origin and $X \subset \mathrm{Cl}\,\mathring{X}$. Among such sets are JS and $\mathring{J}S$. Set $X - 0 = X_0$. We understand that an *A_0 or C_0^m-mapping* F of X into E is a continuous mapping of X into E whose restriction to X_0 is an analytic or C^m-mapping, respectively, of X_0 into E. Such a mapping will be called an A_0 or C_0^m-diffeomorphism of X into E, if it is a homeomorphism of X into E, and if $F|X_0$ is an analytic or C^m-diffeomorphism, respectively, of X_0 into E.

The following theorem was established by Morse in Ref. [5]. A C^m-diffeomorphism of S into E, $m > 0$, is understood in the usual sense.

THEOREM 2.1. *A C^m-diffeomorphism f of S into E, $m > 0$, can be extended by a C_0^m-diffeomorphism F of JS onto $Jf(S)$.*

Extensions over JS without differential singularity. The question arises, when can the extension F of f in Theorem 2.1 be chosen so as to be a C^m-diffeomorphism of JS into E.

In Theorem 2.2 and its corollaries we shall suppose that all diffeomorphisms are of class C^∞. This is to avoid the detail which is necessarily associated with a treatment under weaker hypotheses.

A definition is needed.

Differentiable isotopies. Let R be the axis of reals. Two C^∞-diffeomorphisms $f, g\colon S \to E$ will be said to be *differentiably isotopic* if there exists a C^∞-diffeomorphism

$$h\colon\ S \times R \to E \times R; \qquad (x, t) \to \big(h(x, t), t\big)$$

such that $h(x, t) = f(x)$ for $t \leqslant 0$, and $h(x, t) = g(x)$ for $t \geqslant 1$. We term h a *differentiable isotopy of f into g* or simply a differentiable isotopy of f. A differentiable isotopy of two C^∞-diffeomorphisms $F, G\colon JS \to E$ is similarly defined.

This definition should be compared with that of Milnor, Ref. [13]. Under Milnor's definition a differentiable isotopy of a diffeomorphism f_0 of S onto S is a deformation of f_0 through a family f_t, $-\infty < t < \infty$, of diffeomorphisms each in S^S, but under our definition is through a family of diffeomorphisms each in E^S.

Theorem 2.2 is the fundamental theorem of this section. The necessity of the condition is relatively easy to establish. The sufficiency of the condition may be established by means of methods introduced in §§ 1-6 of Ref. [14].

THEOREM 2.2. *A necessary and sufficient condition that a C^∞-diffeomorphism f_0 of S into E be extendable as an orientation-preserving C^∞-diffeomorphism of JS into E, is that there exist a differentiable isotopy of $f_0(f_t \in E^S)$ into the identity map of S onto S.*

According to Theorem 5 of Milnor, *loc. cit.*, there exists a C^∞-diffeomorphism f_0 of degree 1 of a 6-sphere S onto S such that f_0 is not differentiably isotopic to the identity in Milnor's sense, that is with $f_t \in S^S$. This result of Milnor has an extension in the form of a corollary of Theorem 2.2. See Ref. [21].

COROLLARY 2.1. *When $n = 7$ there exists a C^∞-diffeomorphism f_0 of a 6-sphere S onto S of degree $+1$, which is not differentiably isotopic to the identity with $f_t \in E^S$.*

In establishing this corollary use is made of Milnor's manifold M_3^7 as well as of Theorem 2.2.

The following theorem gives an "extension" of a differentiable isotopy.

THEOREM 2.3. *Corresponding to a differentiable isotopy f_t, $-\infty < t < \infty$, of a C^∞-diffeomorphism f_0 into the identity f_1, with $f_t \in E^S$, there exist C^∞-diffeomorphisms F_t, $-\infty < t < \infty$, of JS into E extending the respective mappings f_t, and defining a differentiable isotopy of F_0 into the identity map of JS onto JS.*

The sufficiency of the condition in Theorem 2.2 is a corollary of Theorem 2.3.

Theorem 2.3 would be false if one added that F_0 could be taken arbitrarily among the C^∞-diffeomorphisms of JS into E which extend f_0. An example will show this.

We distinguish three differentiable extension problems.

(a) A C_0^m-extension problem, given $f \in E^S$.

(b) A C^m-extension problem, given $f \in E^S$.

(c) A C^m-extension problem, given $f \in S^S$.

Problem (a) is to find a C_0^m-diffeomorphism F of JS into E extending the given C^m-diffeomorphism of f into E. It is solvable. Problems (b) and (c) are similar, except that F is required to be a C^m-diffeomorphism of JS into E and in Problem (c), $f(S) = S$. As Milnor has shown, Problem (c) may fail to have a solution for certain values of n. Since a Problem (c) is a Problem (b), a Problem (b) may likewise fail to have a solution.

Schoenflies integers. We shall call an integer $n > 1$, a Schoenflies integer, if every C^∞-extension Problem (b) has a solution when $E = E_n$.

Milnor integers. We shall call an integer $n > 1$, a Milnor integer, if when $S = S_{n-1}$ every C^∞-extension Problem (c) has a solution.

It is obvious that each Schoenflies integer is a Milnor integer. We are ignorant as to the converse. According to Munkres in 6.4 of Ref. [15], $n = 2$ and $n = 3$ are Milnor integers (our notation). See also Smale, Ref. [19]. The author has proved that $n = 2$ and $n = 3$ are also Schoenflies integers, confirming and extending the above results of Munkres and Smale.

The integer 7 is not a Milnor integer, as one sees with the aid of Milnor's manifold M_3^7. Hence 7 is not a Schoenflies integer.

§ 3. Schoenflies cell problems. The image in E of an open n-ball in E under an analytic diffeomorphism will be called an *open analytic n-ball.* The fundamental theorem of this section is the following.

THEOREM 3.1. *If M is the image of S in E under a C^1-diffeomorphism f, the interior of M is an open analytic n-ball.*

The open-cell problem. Given M the problem of finding the analytic diffeomorphism F of an open n-ball onto $\mathring{J}M$ will be called the Schoenflies *open cell problem.* This problem is not an extension problem. The manifold M may be given as the C^1-diffeomorphic image in E of S under diffeomorphisms other than f, but the open-cell problem remains the same. The diffeomorphism F may not extend any diffeomorphism of S into E which represents M. The proof of Theorem 3.1 is given in Ref. [12], by showing that there exists a C^1-diffeomorphism G of $\mathring{J}S$ onto $\mathring{J}M$, and making use of a general theorem in Ref. [12] that any C^1-diffeomorphism of an open subset X of E onto an open subset Y of E can be approximated by a real analytic diffeomorphism of X onto Y.

The purely topological version of Theorem 3.1 involves topological $(n-1)$-spheres Σ in E, termed *elementary.* Cf. Ref. [6], § 4. By definition, Σ is elementary if it is the image of S in E under a homeomorphism F of S onto Σ which is extendable as a homeomorphism into E over an open neighborhood of S. Not every topological $(n-1)$-sphere in E is elementary. See Ref. [1]. If Σ is a topological $(n-1)$-sphere which is not elementary the interior of Σ is not always an open topological n-ball. See Ref. [0]. However if Σ is elementary, not only is $\mathring{J}\Sigma$ an open topological n-ball but $J\Sigma$ is a closed topological n-ball. This is a consequence of Corollary 1.1.

To Theorem 3.1 we add the following conjecture.

CONJECTURE. *The interior of an elementary topological $(n-1)$-sphere Σ in E is an open analytic n-ball.*

This conjecture is correct when $n = 2$ or 3. It follows from the conformal mapping theorem when $n = 2$. When $n = 3$ the proof runs as follows. By Corollary 1.1, $\mathring{J}S$ is the homeomorph of $\mathring{J}\Sigma$. Both $\mathring{J}S$ and $\mathring{J}\Sigma$ are differentiable 3-manifolds. Hence they are C^1-diffeomorphic. See Theorem 6.3, Munkres, Ref. [15]. See also Thom, Ref. [20]. By a general theorem of Huebsch and Morse in Ref. [12], $\mathring{J}S$ and $\mathring{J}\Sigma$ are then analytically diffeomorphic.

If the Hauptvermutung (cf. Munkres, loc. cit., p. 544) is here valid our conjecture for general n is provable by a similar argument. However a more likely mode of proof would be by a limiting process, applying Theorem 3.1 to C^1-topological $(n-1)$-spheres in $\mathring{J}\Sigma$ which approximate Σ.

The closed cell problem. Let M be the image of S in E under a C^∞-diffeomorphism f of S into E. A C^∞-diffeomorphism of S onto M which is extendable as an orientation-preserving C^∞-diffeomorphism of JS onto JM will be called a *preferred representation* of M. Given a C^∞-representation of M, the problem of finding a preferred representation of M will be called the *closed cell problem* (of index $m = \infty$).

Let an orientation-preserving C^∞-diffeomorphism of JS into E be termed a $(+)$ C^∞-diffeomorphism of JS into E.

The group Λ_{n-1}. Given $n > 1$ let Λ_{n-1} be the group of C^∞-diffeomorphisms f_0 of S onto S which are preferred representations of S, or equivalently (by Theorem 2.2) the group of all C^∞-diffeomorphisms f_0 of S onto S which are differentiably isotopic ($f_t \in E^S$) to the identity. We understand that the product of two diffeomorphisms in Λ_{n-1} is their composition.

The closed cell problem differs radically from the C^∞-extension problem. Let f be a C^∞-diffeomorphism of S onto M. When $M = S$ the corresponding closed cell problem always has a trivial solution obtained by representing M by the identity map g of S onto $M = S$, and extending g over JS by the identity.

We shall prove a lemma giving a relation between three Schoenflies problems.

(α) *The closed cell problem (of index $m = \infty$).*

(β) *The C^∞-extension problem, given $f \in E^S$.*

(γ) *The C^∞-extension problem, given $f \in S^S$.*

LEMMA 3.1. *Let M be the image in $E = E_n$ of $S = S_{n-1}$ under a C^∞-diffeomorphism f. If M admits a preferred representation g, then f is a preferred representation of M if and only if $g^{-1}f$ is in Λ_{n-1}.*

To verify the lemma note that $f = g(g^{-1}f)$. By hypotheses there exists a $(+)$ C^∞-diffeomorphism G of JS onto JM extending g.

If $g^{-1}f$ is in Λ_{n-1}, there exists a $(+)$ C^∞-diffeomorphism K of JS onto JS extending $g^{-1}f$. The mapping

$$F = GK \tag{3.1}$$

is a $(+)$ C^∞-diffeomorphism of JS onto JM extending f.

Conversely, the existence of a $(+)$ C^∞-diffeomorphism F of JS onto JM extending f, and the existence of G as above, implies the existence of the $(+)$ C^∞-diffeomorphism

$$K = G^{-1}F$$

of JS onto JS extending $g^{-1}f$, so that $g^{-1}f$ is in Λ_{n-1}.

We give below a proof that the closed cell problem has a solution when $n = 2$. The author has had for several years a proof that the closed cell problem for $n = 3$ has a solution. A solution for this case will presently be published. We know of no evidence that the closed cell problem for arbitrary n ever fails to have a solution. It differs in this respect from the C^m-extension problem.

We shall further clarify the relations between the problems (α), (β), (γ) by introducing the following sets of integers.

The set τ. Let τ denote the set of all integers $n > 1$ such that the closed cell problem in $E = E_n$ always has a solution.

The set σ_1. Let σ_1 be the set of Schoenflies integers.

The set σ_2. Let σ_2 be the set of Milnor integers.

The definitions of τ, σ_1, and σ_2, taken with Lemma 3.1, imply the following.

LEMMA 3.2. *Between the sets* τ, σ_1, *and* σ_2 *there exist the relations*

$$\sigma_1 \subset \sigma_2, \qquad \tau \cap \sigma_2 = \sigma_1 . \tag{3.2}$$

It is conceivable that τ includes all integers $n > 1$, and in such a case $\sigma_1 = \sigma_2$. Since it is known that $n = 2$ and $n = 3$ are contained in σ_2 and τ, it follows from the relation $\tau \cap \sigma_2 = \sigma_1$ that 2 and 3 are contained in σ_1.

Proof that $2 \in \tau$. There is given a Jordan curve M in E_2 such that M is regular and of class C^∞. There exists a Green's function G on JM with logarithmic pole at a point of $\mathring{J}M$. By definition G reduces to zero on M. Let $z = x+iy$. If e is a positive constant the level curve M_e on which $G(z) = -e$, will be a non-singular, analytic, closed curve in $\mathring{J}M$ and, if e is sufficiently small, will be so near M that M is included in an open "field" of short normals to M_e with one and only one normal through each point of the "field". It is then easy to set up an orientation-preserving C^∞-diffeomorphism T of E_2 onto E_2 such that $T(M_e) = M$.

There exists a directly conformal map K of $\mathring{J}S_1$ onto $\mathring{J}M_e$. Since M_e is regular and analytic, K is conformally extendable over an open neighborhood of JS_1. The C^∞-diffeomorphism TK maps JS_1 onto JM preserving orientation, and is extendable as a C^∞-diffeomorphism over an open neighborhood of JS_1. Thus TK is a solution of the Schoenflies closed cell problem posed by the giving of M.

Proof that $2 \in \sigma_2$. There is given a C^∞-diffeomorphism f of S_1 onto S_1 which we seek to extend over JS_1. Without loss of generality we can suppose that f maps S_1 onto S_1 preserving sense on S_1. Were this not the case a reflection of S_1 in a diameter, composed with f, would give a mapping g of S_1 onto S_1 preserving sense, and f would be extendable over JS_1 if g were so extendable.

Let (r, θ) be polar coordinates in E_2. Under f a point $(r, \theta) = (1, \theta)$ on S_1 corresponds to a point with polar coordinates $(1, k(\theta))$ on S_1, where $\theta \to k(\theta)$ can be taken as a C^∞-diffeomorphism of the θ-axis onto itself such that $k(\theta + 2\pi) = k(\theta) + 2\pi$ and $k'(\theta) > 0$. Let $t \to \mu(t)$ be a C^∞-mapping of the t-axis onto the interval $[0, 1]$ such that $\mu(t) = 0$ for $t \leqslant 0$, and $\mu(t) = 1$ for $t \geqslant 1$. A C^∞-differentiable isotopy of f into the identity is defined by the mappings

$$\theta \to (1 - \mu(t))\, k(\theta) + \mu(t)\, \theta \qquad (r = 1)\,,$$

applied to S_1 for each value of t. That 2 is in σ_2 now follows from Theorem 2.2.

The case $n = 3$. A proof that $n = 3$ is in σ_2 is relatively simple. Our proof that $n = 3$ is in τ is nore movel and involves the theory of critical points of non-degenerate functions.

The removal of the restriction to C^∞-diffeomorphisms, insofar as it occurs in this paper, is facilitated by the following theorem on "elevating manifold differentiability". See Ref. [16].

THEOREM 3.2. *Corresponding to a compact, regular, differentiable* $(n-1)$*-manifold of class* C^m *in* E, $m > 0$, *there exists an orientation-preserving* C^m*-diffeomorphism* T *of* E *onto* E *such that* $T(M)$ *is a regular differentiable* $(n-1)$*-manifold of class* C^∞ *in* E.

§ 4. Extensions of families of diffeomorphisms. Ref. [10] Theorem 2.3 concerns the extension over JS of the family of mappings f_t, $-\infty < t < \infty$, of a differentiable isotopy. In this section we consider extensions of families of mappings of much more general character. We begin with notation.

Product spaces. Let U and V be two spaces, and $U \times V$ their product. If X is a subset of $U \times V$ let $\mathrm{pr}_1 X$ and $\mathrm{pr}_2 X$ be defined as in Ref. [18]. For $v \in \mathrm{pr}_2 X$ set

$$X^v = \{u \mid (u, v) \in X\}$$

and term X^v the *v-section* of X. We have no need for u-sections of X.

Let $(u, v) \to F(u, v)$ be a mapping of X into E. For fixed $v \in \mathrm{pr}_2 X$ the partial mapping

$$u \to F(u, v) = F^v(u) \qquad (u \in X^v)$$

will be called the *v-section* F^v of F.

Γ-problems. The topological case. Let Γ be an arbitrary paracompact space with points p, q, etc. Let E be an euclidean n-space, with $n > 1$, with points x, y, etc. Let S be the unit sphere in E with center at the origin. We consider the product $E \times \Gamma$ and its subsets, such as $S \times \Gamma$.

The data required to define a Schoenflies Γ-problem is a neighborhood L of $S \times \Gamma$, open relative to $E \times \Gamma$, and a continuous mapping

$$\Phi: L \to E: (x, p) \to \Phi(x, p),$$

whose p-sections

$$\Phi^p: L^p \to E: x \to \Phi^p(x) \qquad (p \in \Gamma)$$

are homeomorphisms of L^p into E such that Φ^p maps points of L^p, interior (exterior) to S, into points which are interior (exterior) to the topological $(n-1)$-sphere $\Phi^p(S)$ in E. The fundamental extension theorem in the *topological* case is the following. Ref. [10].

THEOREM 4.1. *Corresponding to* (Φ, L, Γ) *conditioned as above, and to any sufficiently small neighborhood* N *of* $S \times \Gamma$, *open relative to* $E \times \Gamma$, *there exists a continuous mapping into* E

$$(4.1) \qquad \Lambda_\Phi: N \cup (JS \times \Gamma); \qquad (x, p) \to \Lambda_\Phi(x, p)$$

whose p-section for $p \in \Gamma$, *is a homeomorphism of* $N^p \cup JS$ *into* E *which extends* $\Phi^p | N^p$.

Γ-problems. The differentiable case. In the differentiable case there is given an integer $m > 0$ and an arbitrary C^m-manifold Γ with a countable base. The data is an ensemble (Φ, L, Γ), with Φ and L conditioned as in the topological case, and in addition such that Φ is of class C^m on L and for each $p \in \Gamma$, Φ^p a C^m-diffeomorphism of L^p into E

Conventions. Let X be a neighborhood of $JS \times \Gamma$, open relative to $E \times \Gamma$. By a C^m_Γ-mapping F, $m > 0$, of X into E we mean a continuous mapping of X into E whose restriction to $X - (0 \times \Gamma)$ is a C^m-mapping. By a C^m_Γ-mapping of $JS \times \Gamma$ into E we mean a mapping extendable as a C^m_Γ-mapping F into E of some neighborhood X of $JS \times \Gamma$ open relative to $E \times \Gamma$.

The first theorem in the differentiable case is as follows. See Ref. [10].

THEOREM 4.2. *Corresponding to* (Φ, L, Γ) *conditioned as above, and to any sufficiently small neighborhood* $N \subset L$ *of* $S \times \Gamma$, *open relative to* $E \times \Gamma$, *there exists a* C^m_Γ-*mapping* Λ_Φ *into* E, *of the form* (4.1) *whose p-section, for each* $p \in \Gamma$, *is a* C^m_0-*diffeomorphism of* $N^p \cup JS$ *into* E *which extends* $\Phi^p | N^p$.

In the differentiable case, $m > 0$, there is an equivalent theorem. Cf. Lemmas 6.1 and 6.2, Ref. [11].

THEOREM 4.3. *Given a C^m-mapping Φ of $S \times \Gamma$ into E, each of whose p-sections, for $p \in \Gamma$, is a C^m-diffeomorphism of S into E, there exists a C^m_Γ-mapping Λ_Φ of $JS \times \Gamma$ into E, whose p-section, for each $p \in \Gamma$, is a C^m_0-diffeomorphism of JS into E which extends Φ^p.*

Note that Theorem 4.2 is the analogue of Theorem 4.1 stated for the topological case. Theorem 4.3 has an analogue, Theorem 5.2, in the analytic case. Theorem 4.2 has no analogue in the analytic case, nor Theorem 4.3 in the topological case. For a more precise statement of these differences see § 6.

§ 5. The analytic case. We begin with the simple Schoenflies problem. One should recall the definition of an A_0-diffeomorphism as given in § 2. The first theorem is as follows.

THEOREM 5.1. *An analytic diffeomorphism f of S into E can be extended by an A_0-diffeomorphism F of JS onto $JF(S)$.*

This theorem was first stated and proved by Royden in Ref. [8], making use of Theorem 2.1 of Morse. Because of the many difficulties in extending this theorem to the case of analytic families Θ of diffeomorphisms f in Ref. [11], Huebsch and Morse found it necessary to give a different proof of Theorem 5.1, one that could serve as a model when Θ was given. This model is presented in Ref. [9]. Recently Theorem 5.1 has been greatly extended. Cor 7.1 [12].

Γ-problems. The analytic case. Theorem 4.3 has its analogue in the analytic case. Here Γ is a real analytic, regular, proper r-manifold in some euclidean space.

Convention. An A_Γ-mapping of $JS \times \Gamma$ into E is defined as was a C^m_Γ-mapping of $JS \times \Gamma$ into E, and*a C^m-mapping by an analytic mapping.

The principal theorem is as follows.

THEOREM 5.2. *Corresponding to a real analytic mapping Φ of $S \times \Gamma$ into E each of whose p-sections is an analytic diffeomorphism of S into E, there exists an A_Γ-mapping Λ_Φ of $JS \times \Gamma$ into E whose p-section for each $p \in \Gamma$ is an A_0-diffeomorphism of JS into E which extends Φ^p.*

§ 6. The data in simple extension problems. We shall bring out fundamental differences between the topological, differential and analytic cases of the simple extension problem.

In posing a simple extension problem the domain of definition of the given mapping f into E has been one of two types.

I. *The $(n-1)$-sphere S.* In this case an extension F of f over JS is tentatively admitted if $f = F|S$.

*Note: instead of "and" read "replacing". (Corrected 1981)

II. *An open neighborhood N of S.* In this case an extension F of f is tentatively admitted if $F|W = f$, where $W \subset N$ is some sufficiently small open neighborhood of S.

The question arises, is it possible in all three cases, the topological, the differentiable, and the analytic, to take the domain of f either of type I or II at pleasure, and be assured of an extension F of f over JS which is *admissible*, that is, one that is tentatively admissible in the sense of I or II and in addition such that F is a homeomorphism, a C_0^m-diffeomorphism, or an A_0-diffeomorphism of JS into E, respectively, according to the case at hand. The answer is no. Corresponding to the case the following table gives the type of domain of f for which f always has an extension F which is *admissible* in the above sense. We term such a domain admissible.

Case	Type of domain
Topological	II
Differentiable	I or II
Analytic	I

That S is not admissible as a domain in the topological case follows from Ref. [1]. We have seen in Theorem 1.2 that S is admissible as a domain in the differentiable case. A domain of type II is likewise admissible in the differentiable case, as is shown in Ref. [6].

In the analytic case a domain of type II is not admissible. That is, if the given mapping f is an analytic diffeomorphism into E of an open neighborhood of S, there will exist, in general, no A_0-extension of f which agrees with f in a neighborhood $W \subset N$ of S. The following example shows this.

EXAMPLE. Let (r, θ) be polar coordinates in a 2-plane E_2 of coordinates x, y. Set $z = x + iy$. Let a point $z = e^{i\theta}$ on the circle $|z| = 1$ be represented by θ and let this circle C be mapped diffeomorphically onto an ellipse G by setting

$$x = a\cos\theta\,, \quad y = b\sin\theta \quad (0 < a < b)\,,$$

the point $z = e^{i\theta} \in C$ corresponding to $(x, y) \in G$. Let this map of C onto G be denoted by g. Let g be extended over a small open neighborhood of C in such a manner that short normals to C and G at corresponding points, themselves correspond. Cf. Ref. [5], § 5. In particular let points in $\mathring{J}C$ and $\mathring{J}G$ on corresponding normals correspond if at the same distance from C and G. Let this correspondence be similarly defined for points exterior to C and G and near C and G. We understand that no normal to C extends to the origin and no normal to G extends to a focal point of G. Let the resulting analytic diffeomorphism of an open neighborhood of C onto an open neighborhood of G be denoted by f.

This mapping f does not admit an extension as an A_0-diffeomorphism of JC onto JG. In fact the analytic extension of f will fail to be a diffeomorphism at those points of JC which are antecedents of focal points of G.

Nevertheless the original analytic mapping g of C onto G admits an extension as an analytic diffeomorphism into E_2 of an open neighborhood of JC. This statement is a consequence of the following facts. The integer $n = 2$ is in σ_1 so that g can be extended as a C^∞-diffeomorphism over JC. The mapping g thus admits an extension as an analytic diffeomorphism into E_2 of an open neighborhood of JC, in accord with the following theorem. Cf. Huebsch and Morse, Ref. [9].

THEOREM 5.3. *If an analytic diffeomorphism f of S into E admits an extension as a C^m-diffeomorphism of JS into E, $m > 1$, then f admits an extension as an analytic diffeomorphism of JS into E.*

The hypothesis of this theorem may fail to be satisfied when $n > 3$.

References

[0] J. Alexander, *An example of a simply connected surface bounding a region which is not simply connected,* Proc. Nat. Acad. Sci. 10 (1924), pp. 8-10.

[1] M. L. Antoine, *Sur la possibilité d'étendre l'homéomorphie de deux figures à leur voisinage,* Comptes Rendus 171 (1920), pp. 661-663.

[2] B. Mazur, *On embeddings of spheres,* Bull. Amer. Math. Soc. 65 (1959), pp. 59-65.

[3] Marston Morse, *A reduction of the Schoenflies extension problem,* Bull. Amer. Math. Soc. 66 (1960), pp. 113-115.

[4] — *Differentiable mappings in the Schoenflies problem,* Proc. Nat. Acad. Sci., U.S.A., 44 (1958), pp. 1068-1072.

[5] — *Differentiable mappings in the Schoenflies theorem,* Compositio Mathematica 14 (1959), pp. 83-151.

[6] W. Huebsch and M. Morse, *An explicit solution of the Schoenflies extension problem,* Journ. Math. Soc. of Japan 12 (1960), pp. 271-289.

[7] Morton Brown, *A proof of the generalized Schoenflies theorem,* Bull. Amer. Math. Soc. 66 (1960), pp. 74-76.

[8] H. L. Royden, *The analytic approximation of differentiable mappings,* Math. Annalen 139 (1960), pp. 171-179.

[9] W. Huebsch and M. Morse, *The Schoenflies extension in the analytic case,* Annali di Matematica 54 (1961), pp. 359-378.

[10] — *The dependence of the Schoenflies extension on an accessory parameter,* Journ. d'Analyse Mathématique 8 (1961), pp. 209-271.

[11] — *Schoenflies extensions of analytic families of diffeomorphisms,* Math. Annalen 144 (1961), pp. 162-174.

[12] — *Schoenflies extensions without interior differential singularity,* Annales of Math. (1961), to appear.

[13] John Milnor, *On manifolds homeomorphic to the 7-sphere,* Annals of Math. 64 (1956), pp. 399-405.

[14] M. Morse and E. Baiada, *Homotopy and homology related to the Schoenflies problem,* Annals of Math. 58 (1953), pp. 142-165.

[15] James Munkres, *Obstructions to the smoothing of piecewise-differentiable homeomorphisms,* Annals of Math. 72 (1960), pp. 521-554.

[16] Marston Morse, *On elevating manifold differentiability*, Jubilee of the Indian Math. Soc. (1961), to appear.

[17] K. Borsuk, *Sur les retracts*, Fund. Math. 17 (1931), pp. 152-170.

[18] N. Bourbaki, *Théorie des ensembles*, Fascicule de résultats, Paris 1958.

[19] S. Smale. *Diffeomorphisms of the 2-sphere*, Proc. Amer. Math. Soc., 10 (1959), pp. 621-626.

[20] R. Thom, *Des variétés triangulées aux variétés différentiables*, Proc. of the Int. Cong. of Math. 1958, Cambridge Univ. Press (1960), pp. 248-255.

[21] Marston Morse, *Schoenflies extensions and differentiable isotopies*, Journ. de Math. Pures et Appliquées 41 (1962), to appear.

[22] William Huebsch and Marston Morse, *Analytic diffeomorphism approximating C^m-diffeomorphisms*, to be published.

Reçu par la Rédaction le 20. 1. 1961

Reprinted from
Differential and Combinatorial Topology,
edited by Stewart S. Cairns, Princeton University Press, 1965.

Bowls of a Non-Degenerate Function on a Compact Differentiable Manifold[1]

MARSTON MORSE

1. Introduction

Let **M** be a connected, compact, oriented, differentiable n-manifold of class C^∞, with $n > 2$. There exist non-degenerate functions f on **M** of class C^∞, as the author showed in his Colloquium Lectures in 1934 (see p. 243). The relevant theorem is that if **M** is in a euclidean m-space E_m the set of points in E_m which are not focal points of **M** in the distance problem are everywhere dense in E_m. If P is not on **M** nor a focal point the distance from P to a point on **M** defines a non-degenerate function on **M**.

In [1] it is shown that by suitable elimination of pairs of critical points with indices n and $n - 1$ or indices 1 and 0, f is replaced by a non-degenerate function of *polar* type, that is a function having only one point of relative minimum and one point of relative maximum. It is one of the objects of this paper to present conditions under which critical points w, $\dot{w}$ with respective indices $k - 1$ and k, where $1 < k < n$, can be eliminated by a modification of f that does not introduce other critical points nor alter f in sufficiently small neighborhoods of the critical points of f other than w and $\dot{w}$. We suppose that f is of polar type and that the values of f at different critical points are different.

We have found that sufficient conditions for such an elimination of critical points can be given in terms of differentiable k and $(n - k)$-submanifolds of **M**, associated with the respective critical points of f of index k, and termed *bowls* of f.

We use methods which combine the study of the critical points of f and of diffeomorphisms of level manifolds of f, with the study of bowls of f and of families of f-arcs, analogous to the orthogonal trajectories of the level manifolds of f. The first objective is to reveal in differential terms fundamental topological characteristics of **M**. The ultimate, but presently

[1] This research was supported by the Air Force Office of Scientific Research.

more remote objective, is the determination of the classes of mutually diffeomorphic n-manifolds $\mathbf{M}$.

To define the bowls of f it is necessary to define *canonical* representations of neighborhood of the critical points of f together with *canonical* representations of f thereon.

All mappings and manifolds to which reference is made in this paper will be of class C^∞.

Notation. Let E_n be a euclidean n-space of points x with rectangular coordinates $(x^1, \ldots, x^n)$. We shall regard x as a vector with components $x^1, \ldots, x^n$ and length $\|x\|$. For each positive constant a set

$$D_a^x = (x \mid \|x\| < a) \tag{1.1}$$

We shall make use of the following projections of E_n into E_n:

$$x \to pr_k x = (x^1, \ldots, x^k, 0, \ldots, 0) \tag{1.2}$$

$$x \to pr^k x = (0, \ldots, 0, x^{k+1}, \ldots, x^n) \tag{1.3}$$

$$x \to pr_k^1 x = (0, x^2, \ldots, x^k, 0, \ldots, 0) \quad (k > 0) \tag{1.4}$$

Definition 1.1. *Coordinate presentations* on $\mathbf{M}$. Let U be a non-empty open subset of E_n and

$$x \to F(x)\colon U \to X \tag{1.5}$$

a homeomorphism F of U onto an open subset X of $\mathbf{M}$. We say that F maps the coordinate *range* U onto the coordinate *domain* X of $\mathbf{M}$, and term $(F\colon U, X)$ a *coordinate presentation*, or more explicitly a *presentation of* X *on* $\mathbf{M}$ *with coordinates* $(x^1, \ldots, x^n)$.

Definition 1.2. *An oriented differentiable structure* ST *on* $\mathbf{M}$. Such a structure is defined by a set W of presentations $(F\colon U, X)$ of open subsets X of $\mathbf{M}$ whose *union* is $\mathbf{M}$, and which are *compatible* in the following sense. If

$$(F_1\colon U_1, X_1) \qquad (F_2\colon U_2, X_2)$$

are two coordinate presentations of $\mathbf{M}$ such that $X_1 \cap X_2 = X \neq \emptyset$, the homeomorphism

$$x \to F_2^{-1}(F_1(x))\colon F_1^{-1}(X) \to F_2^{-1}(X) \tag{1.6}$$

shall be of class C^∞ with a positive jacobian. An arbitrary presentation $(F'\colon U', X')$ of an open subset X' of $\mathbf{M}$ will be termed *admissible* in the given differentiable structure ST, if it is compatible, in the above sense, with each presentation in W whose domain X intersects X'.

Let Ω be the set of critical points of f. Let $\mathbf{0}$ be the origin in E_n. Let R be the axis of reals.

The following lemma is a consequence of Lemma 6.1 of [1].

LEMMA 1.1. *There exists a Riemannian metric* $\mathscr{R}^0$ *on* **M** *and a positive constant* σ *so small that the following is true.*

Corresponding to each critical point $z \in \Omega$ *there exists an admissible coordinate presentation*

$$(T_z\colon D_\sigma^x, X_z) \qquad (T_z(\mathbf{0}) = z) \tag{1.7}$$

of a domain X_z *of* **M** *which contains* z *and has the following properties:*

(i) *If* z *has the index* k *then*

$$f(T_z(x)) - fT_z(\mathbf{0}) = -\|pr_k x\|^2 + \|pr^k x\|^2 \tag{1.8}$$

(ii) T_z *maps* D_σ^x *isometrically into* **M**.

(iii) *The sets* X_z, $z \in \Omega$ *have disjoint closures in* **M**, *and the sets* $f(X_z)$ *have disjoint closures in* R.

Reference to §6 of [1] discloses the following freedom in choosing the metric $\mathscr{R}^0$ in Lemma 1.1. Let $\mathscr{R}$ be an arbitrary Riemannian metric on **M**, and Y an arbitrary open subset of **M** whose closure does not meet Ω. Then in Lemma 1.1, the metric $\mathscr{R}^0$, the constant σ and the coordinate presentations (1.7) can be chosen so that $\mathscr{R}|Y = \mathscr{R}^0|Y$.

DEFINITION 1.3. *A family* **H** *of* f*-arcs.* In terms of the Riemannian metric $\mathscr{R}^0$ of Lemma 1.1 the trajectories orthogonal to the level manifolds of f are well-defined on $\mathbf{M} - \Omega$. We term these trajectories *ortho-f-arcs*. The family of ortho-f-arcs on $\mathbf{M} - \Omega$ is a special case of a family **H** of f-arcs on $\mathbf{M} - \Omega$ satisfying the following three conditions:

I. *In each coordinate domain* X *on* $\mathbf{M} - \Omega$ *with admissible coordinates* $(x^1, \ldots, x^n)$, *the* f*-arcs of* **H** *shall be represented by solutions of differential equations*

$$\frac{dx^i}{dt} = Z^i(x) \qquad (i = 1, \ldots, n) \tag{1.9}$$

of class C^∞, *where* $Z(x)$ *is a contravariant vector on* **M** *whose orthogonal projection into the gradient* g *of* f *at the point* $p \in \mathbf{M}$ *represented by* x, *is a non-null vector with the sense of* g.

II. *The vectors* Z *are such that* $df/dt = 1$ *along each* f*-arc of* **H**. *Each* f*-arc* k *of* **H** *is so parameterized as a solution of* (1.9) *that a point* $p \in k$ *is represented by the point* $t = f(p)$ *in* R.

III. *The subarcs of an* f*-arc in the canonical domains* X_z *of Lemma* 1.1 *are ortho-f-arcs.*

H-arcs. The maximal connected arcs of the family **H** will be called H-arcs. In terms of these H-arcs we now define the bowls of f.

With each critical point z with index $k > 0$, we shall associate a descending k-bowl $B_-(z, k)$, and with each critical point z of index $k < n$, an ascending $(n - k)$-bowl $B_+(z, n - k)$.

DEFINITION 1.4. *If $k > 0$, $B_-(z, k)$ shall be the differentiable k-submanifold of* **M** *formed by the union of z and the H-arcs which have z as "upper" limiting end point.*

If $k < n$, $B_+(z, n - k)$ shall be the $(n - k)$-submanifold of **M** *formed by the union of z and the H-arcs which have z as a "lower" limiting end point.*

We term z the *pole* of the bowls just defined, and observe that in case $0 < k < n$

$$B_+(z, n - k) \cap B_-(z, k) = z \tag{1.10}$$

By definition a bowl contains no critical points other than its pole. The differential properties of a bowl in a neighborhood of its pole follow from Lemma 1.1.

We state the following theorem.

THEOREM 1.1 (α). *A descending k-bowl $B_-(z, k)$ is the C^∞-diffeomorph of an open euclidean k-ball.*

(β). *An ascending $(n - k)$-bowl $B_+(z, n - k)$ is the C^∞-diffeomorph of an open euclidean $(n - k)$-ball.*

We shall sketch the proof of (α). For full details of the proof see [2].

Without loss of generality in proving (α) we can suppose that $f(z) = 0$.

We introduce a $(k - 1)$-sphere S_{k-1} in $pr_k E_n$ by setting

$$S_{k-1} = (x \in pr_k E_n \,|\, \|x\| = \sigma) \tag{1.11}$$

A point $u \in S_{k-1}$ determines an open ray ρ_u in $pr_k E_n$, with one end at the origin and the other at the point $u \in S_{k-1}$. Moreover $T_z(\rho_u)$ is a subarc of a unique H-arc λ_u in $B_-(z, k)$. Let $(0, a(u))$ be the open interval of values of $|f(p)|^{\frac{1}{2}}$ on λ_u. Let ρ_u' be an extension of ρ_u in $pr_k E_n$ as an open ray which has one end point at the origin and the other at a distance $a(u)$ from the origin. We introduce the open subset

$$Z_z = (\rho_u' \,|\, u \in S_{k-1}) \cup z \tag{1.12}$$

of $pr_k E_n$ and affirm the following.

LEMMA 1.2 (a). *The k-bowl $B_-(z, k)$ is the C^∞-diffeomorph of Z_z under a mapping in which a point in ρ_u' at a distance $t < a(u)$ from the origin corresponds to the point p on the H-arc λ_u at which $|f(p)|^{\frac{1}{2}} = t$, while the origin in Z_z corresponds to z.*

(b) *The set Z_z is the real analytic diffeomorph of an open k-ball.*

Statement (a) follows from the definition of $B_-(z, k)$, and from the classical properties of trajectories of the system (1.9). The proof of (b) can be readily given with the aid of Corollary 9.1 of [3], as will be shown in detail in [2].

Statement (α) of Theorem 1.1 follows from Lemma 1.2. The proof of (β) in Theorem 1.1 is similar.

2. The choice of the family H of f-arcs

As indicated in §1 one can start with an arbitrary Riemannian metric $\mathscr{R}$ on **M**, and by suitable modifications of $\mathscr{R}$ near the respective critical points of f, arrive at the Riemannian metric $\mathscr{R}^0$ affirmed to exist in Lemma 1.1. Let the family of ortho-f-arcs on $\mathbf{M} - \Omega$, determined by $\mathscr{R}^0$, be denoted by $\mathbf{H}^0$. The family $\mathbf{H}^0$ satisfies the conditions I, II, III of § 1 on an admissible family of f-arcs.

Our purposes will be served if the ultimate family **H** of f-arcs on $\mathbf{M} - \Omega$ is obtained by a finite number of special modifications of $\mathbf{H}^0$, the character of which we shall presently define. We begin with notation.

If α is an arbitrary value of f set

$$f^\alpha = (p \in \mathbf{M} | f(p) = \alpha) \tag{2.1}$$

If $[\alpha, \beta]$ is a closed interval of ordinary values of f set

$$f_{\alpha\beta} = (p \in \mathbf{M} | \alpha \leq f(p) \leq \beta) \tag{2.2}$$

Definition 2.1. *An isotopy on f^a*. Cf. [8]. Let a be an ordinary value of f, and h and k two diffeomorphisms of f^a onto f^a. A C^∞-mapping,

$$(x, t) \to F(x, t): f^a \times R \to f^a \tag{2.3}$$

is called an isotopy F of h into k, if each partial mapping

$$x \to F(x, t) = F^t(x) \qquad (x \in f^a) \tag{2.4}$$

is a diffeomorphism of f^a onto f^a, if $h = F^t$ for $t \leq 0$ and if $k = F^t$ for $t \geq 1$. We say that such an isotopy is *determined* by the diffeomorphisms F^t, $0 \leq t \leq 1$.

Definition 2.2. Let Y be a subset of $\mathbf{M} - \Omega$. Two admissible families $\mathbf{H}'$ and $\mathbf{H}''$ of f-arcs will be said to *differ at most on* Y if the ensemble of subarcs on $\mathbf{M} - \Omega - \bar{Y}$ of H'-arcs is identical with the ensemble of subarcs on $\mathbf{M} - \Omega - \bar{Y}$ of H''-arcs.

Definition 2.3. *H-related points*. Corresponding to an admissible family **H** of f-arcs, two points of $\mathbf{M} - \Omega$ will be said to be *H-related* if they lie on the same H-arc.

The diffeomorphisms $H_{c,e}$. Let c and e be ordinary values of f between which there are no critical values of f. We admit the possibility that $c < e$, $c = e$, or $c > e$. Let $H_{c,e}$ denote the diffeomorphism of f^e onto f^c in which a point $p \in f^e$ is mapped into the H-related point of f^c. Note that $H_{e,c}$ is the inverse of $H_{c,e}$.

DEFINITION 2.4. *An* $[a, b]$*-replacement of* **H**. Let $[a, b]$ be a closed interval of ordinary values of f assumed at no point of the subsets $X_z|(z \in \Omega)$ given in Lemma 1.1. By an $[a, b]$*-replacement of* **H** is meant an admissible family $\mathbf{H}^*$ of f-arcs whose subarcs differ from those of **H** at most on f_{ab}. We say that this replacement *belongs* to the pair of consecutive critical values between which a and b lie.

By virtue of the following lemma such a replacement of H by H^* may properly be called *isotopic*. In Lemma 2.1 we refer to values

$$c(t) = ta + (1 - t)b \qquad (0 \le t \le 1) \tag{2.5}$$

interpolating between a and b.

LEMMA 2.1. *If* $\mathbf{H}^*$ *is an* $[a, b]$*-replacement of* **H**, *the composite diffeomorphisms*

$$\Gamma^t = H_{a,c(t)} H^*_{c(t),a} \qquad (0 \le t \le 1) \tag{2.6}$$

of f^a *onto* f^a *"determine" an isotopy* Γ *on* f^a *of the diffeomorphism* $\Gamma^0 = H_{ab}H^*_{ba}$ *into the identity* Γ^1.

Conversely if γ *is an arbitrary diffeomorphism of* f^a *onto* f^a, *admitting an isotopy* Γ *on* f^a *of* γ *into the identity, there exists a unique* $[a, b]$*-replacement* $\mathbf{H}^*$ *of* **H** *such that* (2.6) *holds*. Cf. [2] for proof.

DEFINITION 2.5. *General isotopic replacements of* $\mathbf{H}^0$. Given the family $\mathbf{H}^0$ of ortho-f-arcs determined by the Riemannian metric $\mathscr{R}^0$ of Lemma 1.1, let **H** be an admissible family of f-arcs obtained from $\mathbf{H}^0$ by a finite sequence of $[a, b]$-replacements, including at most one $[a, b]$-replacement belonging to each pair of successive critical values of f. We term such a family **H** an *isotopic replacement* of $\mathbf{H}^0$.

We shall presently condition **H** in terms of the bowls of f defined with the aid of **H**.

The intersections of bowls of f. Note first that no descending (ascending) bowl meets another descending (ascending) bowl. Apart from the special case of an ascending and descending bowl which have the same pole, two bowls intersect, if at all, in an ensemble of H-arcs. That is, if two bowls intersect in a point p, not a common pole, they intersect in the maximal H-arc g containing p. Such an H-arc has as upper limiting end point the pole of the descending bowl, and as lower limiting end point the pole of the ascending bowl.

DEFINITION 2.6. *The trace of a bowl on level manifold*. If α is a value of f, not the minimum or maximum value of f, the intersection with f^α of a k-bowl B will be called the *trace* B^α *of* B *on* f^α. If this trace is not empty, or the pole, it is a differentiable $(k - 1)$-submanifold of f^α.

The special bowls B_- *and* B_+. Let B_- be a descending k-bowl with the pole $\dot{w}$, and B_+ an ascending $(n - k + 1)$-bowl with the pole w. The point $\dot{w}$ has the index k, the point w the index $k - 1$. These are the bowls with which the principal theorem is concerned.

DEFINITION 2.7. *Regular intersections of B_- and B_+.* If the intersection of B_- and B_+ contains a point p, set $f(p) = \alpha$. The intersection of B_- and B_+ will be termed *regular* at p if the $(k - 1)$- and $(n - k)$-dimensional traces of B_- and B_+ on f^α intersect in p on f^α without tangent in common at p. It is shown in [2] that if B_- and B_+ interesect regularly at p, they intersect regularly at each point p' of the f-arc ω meeting p. In such a case we say that B_- and B_+ *intersect regularly along* ω.

Lemma 2.1 prepares the way for the main theorem, Theorem 2.1.

LEMMA 2.1. *There exists an isotopic replacement* $\mathbf{H}$ *of the family* $\mathbf{H}^0$ *of ortho-f-arcs such that the bowls of f, defined in terms of H-arcs, have the following properties.*

(i) *A descending k-bowl and an ascending r-bowl for which* $r + k \leq n$ *intersect at most in a common pole.*

(ii) *A descending k-bowl, with* $1 < k < n$, *and an ascending* $(n - k + 1)$-*bowl do not intersect, or intersect regularly in a finite ensemble of maximal f-arcs of* $\mathbf{H}$. See [4] and [11].

To state Theorem 2.1 a definition is needed.

DEFINITION 2.8. *The k-dome* $\mathscr{B}_-$ (z, k) *of* $B_-(z, k)$. Each H-arc of $B_-(z, k)$ has a critical point $z' \in \Omega$ as lower limiting end point. The corresponding values $f(z')$ are finite in number, and so have a maximum at some critical point z^0. The restriction of $B_-(z, k)$ on which $f(z) \geq f(p) > f(z^0)$ will be called the *k-dome* $\mathscr{B}_-(z, k)$ of $B_-(z, k)$. We note the following.

The k-dome of a descending k-bowl is the diffeomorph of an open k-ball.

Inverted domes of ascending bowls can be similarly defined and treated.

The principal theorem follows. See [2].

THEOREM 2.1. *Let B_- be a descending k-bowl with pole $\dot{w}$ and B_+ an ascending $(n - k + 1)$-bowl with pole w, such that B_- and B_+ intersect regularly in a single H-arc ω, and such that the closure of B_- meets no critical point between the f-levels of w and $\dot{w}$.*

Let N be a prescribed open neighborhood of the closure of the dome of B_- such that N contains w and $\dot{w}$ and $\bar{N}$ no other critical points of f.

There then exists a non-degenerate function $\hat{f}$ on $\mathbf{M}$ *without critical points on $\bar{N}$ and such that $\hat{f}(p) = f(p)$ for $p \in \mathbf{M} - N$.*

DEFINITION 2.9. *Admissible changes of coordinates* in Lemma 1.1. We shall admit changes in rectangular coordinates $(x^1, \ldots, x^n)$ in Lemma 1.1 of the following types.

The coordinates can be arbitrarily permuted, the coordinates whose squares are summed can be orthogonally transformed, the remaining coordinates orthogonally transformed, and such changes can be made any finite number of times in any order, provided the resultant transformation has the determinant 1.

Choice of $T_{\dot{w}}$. After a suitable admissible change of coordinates an isometric presentation of $X_{\dot{w}}$ in Lemma 1.1 will become an isometric presentation

$$(T_{\dot{w}}\colon D^x_\sigma, X_{\dot{w}})$$

of $X_{\dot{w}}$ such that the interval $-\sigma < x^1 < 0$ of the x^1-axis has the terminal arc of ω of length σ as its image under $T_{\dot{w}}$, while

$$f(T_{\dot{w}}(x)) - f(\dot{w}) = -\|pr_k x\|^2 + \|pr^k x\|^2 \qquad (x \in D^x_\sigma) \tag{2.7}$$

Choice of T_w. After a suitable admissible change of coordinates an isometric presentation on X_w in Lemma 1.1 will become an isometric presentation (T_w, D^y_σ, X_w) of X_w such that conditions (a) and (b) are satisfied.

(a) *The interval* $0 < y^1 < \sigma$ *of the* y^1*-axis has the initial arc of* ω *of length* σ *as its image under* T_w.

(b) $$f(T_w(y)) - f(w) = (y^1)^2 - \|pr^1_k y\|^2 + \|pr^k y\|^2 \qquad (y \in D^y_\sigma)$$

An isometric presentation T_w of X_w satisfying conditions (a) and (b) is not necessarily an adequate final choice of T_w. In particular suppose that T^*_w is such a presentation of X_w in terms of coordinates $(u^1, \ldots, u^n)$. To satisfy (i) below it may be necessary to change from coordinates $(u^1, \ldots, u^n)$ to coordinates $(y^1, \ldots, y^n)$ identical with $(u^1, \ldots, u^n)$, except that $y^2 = -u^2$ and $y^n = -u^n$. We are thus led to *two* final choices of an isometric presentation T_w of X_w satisfying (a) and (b).

Definition 2.10. Let $\mathbf{H}^*$ be an admissible family of f-arcs on $M - \Omega$ such that the corresponding bowls B^*_- and B^*_+ intersect regularly in $\omega = B_- \cap B_+$. We say that such a family $\mathbf{H}^*$ is ω*-canonical* if for one of the two choices of T_w indicated in the preceding paragraph the following is true.

(i) *Each point* $T_{\dot{w}}(x)$ *for which* $x^1 = -\sigma/2$ *and for which* x *is sufficiently near the* x^1*-axis, is* $\mathbf{H}^*$*-related to the point* $T_w(y)$ *for which* $y = \sigma/2$ *and* $(y^2, \ldots, y^n) = (x^2, \ldots, x^n)$.

With ω-canonical families $\mathbf{H}$ so defined we state the following lemma.

Lemma 2.2. *Under the conditions of Theorem* 2.1 *on* $\mathbf{H}$, B_- *and* B_+, *there exists an isotopic replacement* $\mathbf{H}^*$ *of* $\mathbf{H}$ *satisfying the following three conditions.*

(α) *The family* $\mathbf{H}^*$ *contains* $\omega = B_- \cap B_+$ *as an* f*-arc and differs from* $\mathbf{H}$ *at most on a prescribed open neighborhood* N^* *of* ω.

(β) *The new bowls* B^*_- *and* B^*_+ *satisfy the conditions on* B_- *and* B_+ *of Theorem* 2.1.

(γ) *The family* $\mathbf{H}^*$ *is* ω*-canonical.*

In §3 we make use of a family $\mathbf{H}^*$ satisfying Lemma 2.2. The presentation T_w in §3 will mean one for which (i) in Definition 2.10 is satisfied.

3. The replacement $\hat{f}$ of f

Without loss of generality in proving Theorem 2.1 we can suppose that $f(w) = 0$ and that $f(\dot{w}) = 1$. Let the dome of the k-bowl B^*_- of Lemma 2.2 be denoted by $\mathscr{B}^*_-$. The subset

$$T_w(pr_k D^y_\sigma) \tag{3.1}$$

of $\mathbf{M}$ is a k-dimensional, differentiable submanifold of $\mathbf{M}$ whose intersection with $\mathscr{B}^*_-$, includes an open subset of $\mathscr{B}^*_-$, in accord with Lemma 2.2.

A k-manifold β_e *extending the dome* $\mathscr{B}^*_-$. If $e < \sigma/2$, the open subset,

$$\beta'_e = (p = T_w(pr_k y) \,|\, |y^1| < e, \;\; f(p) > -e^2) \tag{3.2}$$

of the k-manifold (3.1) is well-defined. Moreover

$$\beta''_e = (p \in B^*_- \,|\, 1 \geq f(p) > -e^2) \tag{3.3}$$

is an open subset of $B,^*_-$ and so is a k-dimensional differentiable submanifold of $\mathbf{M}$. If e is sufficiently small

$$\beta'_e \cap \beta''_e = (p = T_w(pr_k y) \,|\, 0 < y^1 < e, f(p) > -e^2) \tag{3.4}$$

as a consequence of Lemma 2.2.

Thus if e is sufficiently small, β'_e *and* β''_e *intersect in a k-dimensional, differential manifold, namely the right member of* (3.4), *so that the union*

$$\beta_e = \beta'_e \cup \beta''_e \supset Cl\, \mathscr{B}^*_-$$

is a k-dimensional, differentiable submanifold of $\mathbf{M}$. *We suppose e so conditioned.*

In Theorem 2.1 there is given a neighborhood N of the closure of the dome of B_-. If the neighborhood N^* of ω, prescribed in Lemma 2.2, is taken sufficiently small, and if the constant e conditioning β_e, is sufficiently small, then

$$Cl\, \mathscr{B}^*_- \subset N \qquad Cl\, \beta_e \subset N \tag{3.5}$$

We suppose that N^* *and e are chosen so that* (3.5) *holds.*

Let Δ_r be an open $(n-k)$-ball in a euclidean space E_{n-k}, where Δ_r has the radius r and a center at the origin $\mathbf{0}$ in E_{n-k}.

The domain of redefinition of f. This domain of redefinition will be taken as the carrier $|W|$ of a differentiable f-fibre-bundle W with the following properties.

I. The *base* of W shall be β_e and $|W|$ a neighborhood of β_e, open relative to $\mathbf{M}$ and included in the neighborhood N of $Cl\,\mathscr{B}_-$ given in Theorem 2.1. Cf. (3.5).

II. The *fibre* of W shall be the $(n-k)$-ball Δ_2 of radius 2.

III. For $p \in \beta_e$ the fibre Y_p of W *over* p shall be a differentiable $(n-k)$-manifold in $\mathbf{M}$, orthogonal to β_e at p, and meeting β_e only in p.

IV. The *fibre-bundle* W shall be defined by a diffeomorphism

$$(p, v) \to \Lambda(p, v) \colon \beta_e \times \Delta_2 \to \mathbf{M} \tag{3.6}$$

onto $|W|$, and such that $\Lambda(p, \mathbf{0}) = p$ for $p \in \beta_e$.

V. For each $p \in \beta_e$ the partial map

$$v \to \Lambda(p, v) = \Lambda(v) \qquad [v = (v^1, \ldots, v^{n-k}) \in \Delta_2] \tag{3.7}$$

shall be a diffeomorphism of Δ_2 onto the fibre Y_p over p.

VI. W shall be an f-fibre-bundle in the following sense. For each $p \in \beta_e$ and for v restricted to an arbitrary ray λ emanating from the origin and of length 2, the partial mapping

$$v \to f(\Lambda(p, v)) \colon \Delta_2 \to R \tag{3.8}$$

defines a mapping of λ into R in which f is non-decreasing as the arc length s, measured along λ from the origin, increases.

The principal difficulty is in defining the fibre-bundle W so that property VI is realized. For the terms involved in I to V, see [6]. The term *f-fibre-bundle* is introduced here for the first time. That such an f-fibre-bundle exists follows from [5].

Redefinition of f on β_e. We state the following essential lemma.

LEMMA 3.1. *There exists a relatively compact open subset V of β_e such that w and $\dot{w}$ are in V and $\bar{V} \subset \beta_e$, and a real-valued differentiable function f^0 of class C^∞ on β_e, without critical points on β_e, and such that*

$$f^0(p) = f(p) \qquad (p \in \beta_e - V) \tag{3.9$'$}$$

$$f^0(p) < f(p) \qquad (p \in V) \tag{3.9$''$}$$

Method of proof of Lemma 3.1. We turn to the euclidean k-space E_k of points $(u_1, \ldots, u_k)$ and introduce a model function,

$$u \to \zeta(u) = 3u_1^2 - 2u_1^3 - (u_2^2 + \cdots + u_k^2) \tag{3.10}$$

The function ζ has a critical point at the origin and at the point on the u_1-axis at which $u_1 = 1$. Other points of E_k are ordinary points of ζ. The critical values of ζ are 0 and 1, respectively. These are the critical values of f on β_e, that is, the values of f at w and $\dot{w}$.

One shows in [2] that if e_0 is a sufficiently small positive constant, and if $0 < e < e_0$, the open subset

$$b_e = (u \in E_k | \zeta(u) > -e^2, u_1(3 - 2u_1)^{\frac{1}{2}} > -e) \tag{3.11}$$

of E_k is the diffeomorph of β_e under a mapping $p \to \theta(p) = u$ in which

$$\zeta(\theta(p)) = f(p) \tag{3.12}$$

The problem of redefining f on β_e so that Lemma 3.1 is satisfied, is thereby reduced to an equivalent problem of redefining ζ on b_e. The latter problem is solved in relatively explicit manner in [7].

Definition of $\hat{f}$ on **M**. We shall need an auxiliary C^∞-mapping $t \to \mu(t)$ of R onto [0, 1] such that

$$\mu(t) = 1 \qquad (t \leq 0) \tag{3.13)$'$}$$

$$\mu(t) = 0 \qquad (t \geq 1) \tag{3.13)$''$}$$

$$\mu'(t) < 0 \qquad (0 < t < 1) \tag{3.13)$'''$}$$

We shall refer to the $n - k$-ball Δ_1 of radius 1 and to the set

$$\Pi = (q = \Lambda(p, v) | p \in V, v \in \Delta_1) \tag{3.14}$$

observing that $\bar{\Pi} \subset |W|$.

The function $\hat{f}$ will be defined on **M** as an extension of the function f^0 defined on β_e in Lemma 3.1. For this purpose recall the representation

$$q = \Lambda(p, v) \qquad (p \in \beta_e, v \in \Delta_2) \tag{3.15}$$

of points $q \in |W|$. Let $\hat{f}$ be defined on **M** by setting

$$\hat{f}(q) = f(q) \qquad (q \in \mathbf{M} - \bar{\Pi}) \tag{3.16}$$

and, subject to (3.15), by setting

$$\hat{f}(q) = f(q) - \mu(\|v\|^2)[f(p) - f^0(p)] \qquad (p \in \beta_e, v \in \Delta_2) \tag{3.17}$$

The function $\hat{f}$ is thereby over defined on $|W| - \bar{\Pi}$, but consistently, as one readily verifies. The mapping $\hat{f}$ is of class C^∞ on **M**, since the domains of definition of $\hat{f}$ in (3.16) and (3.17) are open subsets of **M** and the restrictions of $\hat{f}$ to these sets are of class C^∞. Finally one shows that $\hat{f}$ has no critical points on $|W|$. For details see [2].

Since $\hat{f}(q)$ differs from $f(q)$ at most on $|W|$ and since $|W| \subset N$, the non-degenerate function $\hat{f}$ satisfies Theorem 2.1.

4. Composite manifolds

The theory of bowls of a non-degenerate function f on $\mathbf{M}$ is naturally associated with the theory of composite manifolds. Cf. [9].

Let a be an ordinary value of f. We are assuming that $\mathbf{M}$ is orientable. It follows that the level manifold f^a is orientable. Let λ then be an orientation-preserving diffeomorphism of f^a onto f^a. We could cut $\mathbf{M}$ along f^a and join a point p on the lower side of the cut to the point $\lambda(p)$ on the upper side, thereby forming a composite manifold. However, this intuitive approach is not adequate for our purposes.

We wish to form a composite manifold which, like $\mathbf{M}$, possesses an oriented differentiable structure, a Riemannian metric $\mathscr{R}^*$, a non-degenerate function f^*, and a family $\mathbf{H}^*$ of f^*-arcs, not arbitrarily defined on the new manifold, but related respectively to the oriented differentiable structure, Riemannian metric, non-degenerate function f and family $\mathbf{H}$ of f-arcs on $\mathbf{M}$. For this reason we review a general formal approach to the composition of manifolds as presented in [9].

General compositions. Let M and $\mathscr{M}$ be two abstractly given, oriented, differentiable n-manifolds, without points in common. Let W and $\mathscr{W}$ be fixed, open, non-empty subsets of M and $\mathscr{M}$ respectively. We presuppose the existence of an orientation-preserving diffeomorphism

$$\mu: W \to \mathscr{W} \qquad \text{(onto } \mathscr{W}\text{)} \tag{4.0}$$

Let each point $p \in W$ be identified with its image $\mu(p) \in \mathscr{W}$ to form a "point" $[p: \mu(p)]$. Subject to this identification let Σ be the ensemble of all points of M and $\mathscr{M}$.

The mappings Π, Π_1, Π_2. To each point $p \in M$ corresponds a "point" $\Pi(p) \in \Sigma$, represented by p if $p \notin W$ and by $[p: \mu(p)]$ if $p \in W$. To each point $q \in \mathscr{M}$ corresponds a point $\Pi(q) \in \Sigma$ represented by q if $q \notin \mathscr{W}$, and by $[\mu^{-1}(q): q]$ if $q \in \mathscr{W}$. Thus Π maps $M \cup \mathscr{M}$ onto Σ. Set

$$\Pi_1 = \Pi|M, \quad \Pi_2 = \Pi|\mathscr{M} \tag{4.1}$$

The mappings $\Pi_1: M \to \Sigma$ and $\Pi_2: \mathscr{M} \to \Sigma$ are biunique, but not in general onto Σ. The mapping $\Pi: M \cup \mathscr{M} \to \Sigma$ is onto Σ but not biunique. For $p \in W$ or $q \in \mathscr{W}$

$$\Pi_1(p) = \Pi_2(\mu(p)), \quad \Pi_2(q) = \Pi_1(\mu^{-1}(q)) \tag{4.2}$$

Observe that

$$\Pi_1(M) \cap \Pi_2(\mathscr{M}) = \Pi_1(W) = \Pi_2(\mathscr{W})$$
$$\Pi(M \cup \mathscr{M}) = \Pi_1(M) \cup \Pi_2(\mathscr{M}) = \Sigma$$

Σ *topologized.* Let X and $\mathscr{X}$ be arbitrary open subsets of M and $\mathscr{M}$ respectively. The ensemble of subsets $\Pi_1(X)$ and $\Pi_2(\mathscr{X})$ of Σ will be taken as a base for the open sets of Σ. So topologized Σ is not necessarily a Hausdorff space. Cf. [9]. In the application which we shall make Σ will be a Hausdorff space, so that we here assume that Σ is a Hausdorff space.

We note that Π_1 and Π_2 are homeomorphisms of M and $\mathscr{M}$ respectively onto open subsets $\Pi_1(M)$ and $\Pi_2(\mathscr{M})$ of Σ.

Let maximal bases for the oriented C^∞-structures of M and $\mathscr{M}$ respectively be given by coordinate presentations of the form

$$(4.3) \qquad (F_i\colon U_i, X_i)_{i\in\alpha} \qquad\qquad (F_j\colon U_j)_{j\in\beta}$$

It follows from Lemma 2.1 of [9] that the ensemble of coordinate presentations

$$(4.4) \qquad (\Pi_1 F_i\colon U_i, \Pi_1(X_i))_{i\in\alpha}, \quad (\Pi_2 F_j\colon U_j, \Pi_2(X_j))_{j\in\beta}$$

is a base for an oriented C^∞-structure on Σ. Given this structure, Π_1 and Π_2 are orientation-preserving diffeomorphisms of M and $\mathscr{M}$ respectively onto Σ.

Let the composite manifold Σ be represented by setting

$$(4.5) \qquad \Sigma = [M, \mathscr{M}, \mu, W, \mathscr{W}]$$

To properly apply the preceding general theory to composite manifolds derived from submanifolds of $\mathbf{M}$ we need to prove the following.

An oriented differentiable structure on each non-singular level manifold f^a may be determined as follows by the oriented differentiable structure of $\mathbf{M}$.

Let V be an open subset of E_{n-1} with coordinates $v^1, \ldots, v^{n-1}$ and $(G; V, Y)$ an arbitrary admissible, coordinate presentation of an open domain of f^a. If $e > 0$ is sufficiently small, the interval $J = (a - e, a + e)$ is an interval of ordinary values of f. Let g_v be a subarc of the H-arc meeting the point $G(v)$ of f^a, restricted so that for p on g_v, $f(p) \in J$. Let the point q on g_v at which $f(p)$ has the value u^n be denoted by $h(u) = (u^1, \ldots, u^n)$, where we have set $(v^1, \ldots, v^{n-1}) = (u^1, \ldots, u^{n-1})$. We identify the product $V \times J$ with an open subset U of a euclidean space of coordinates $u^1, \ldots, u^n$. Then $(h\colon U, h(U))$ is an admissible presentation of $h(U)$ in the unoriented, differentiable structure of $\mathbf{M}$. If this coordinate presentation is admissible in the oriented structure of $\mathbf{M}$ we *prefer* $(G\colon V, Y)$ as a presentation of f^a. One shows readily that the coordinate presentations $(G\colon V, Y)$ thereby preferred are orientation consistent when they "overlap," and form an oriented differentiable structure on f^a.

We suppose f^a given the oriented differentiable structure determined as above from that of $\mathbf{M}$.

5. Composite manifolds formed from M

Let $[a, b]$ be a closed interval of ordinary values of f that includes no value taken on by f on the sets $X_z|(z \in \Omega)$ of Lemma 1.1. Let $c = (a + b)/2$. Set

$$M = p \in \mathbf{M}|f(p) < b)$$
$$M' = (p \in \mathbf{M}|f(p) > a)$$
$$W = (p \in \mathbf{M}|a < f(p) < b)$$

Corresponding to each point $p \in M'$ let $\theta(p)$ be a copy of p, regarded as distinct from p. So defined θ maps M' onto a set $\theta(M') = \mathscr{M}$, disjoint from M' and M. It is trivial that $\mathscr{M}$ can be so topologized and structured that θ is an orientation-preserving diffeomorphism of M' onto $\mathscr{M}$. The set $\theta(W)$ is an open subset of $\mathscr{M}$ to be denoted by $\mathscr{W}$.

In the notation of §4 a composite manifold Σ is defined by five elements

$$[M, \mathscr{M}, \mu, W, \mathscr{W}] \tag{5.0}$$

of which $M, \mathscr{M}, W, \mathscr{W}$ have already been specially defined in this section. It remains to define the diffeomorphism μ of W onto $\mathscr{W}$.

Parameters (p, t) *on* W. Let p be an arbitrary point of f^c. Let $\mathbf{H}$ be the given family of f-arcs on $\mathbf{M} - \Omega$. Let g_p be the intersection with W of the H-arc meeting f^c in p. On g_p, $f(p)$ ranges over the interval (a, b). For $p \in f^c$ and $t \in (a, b)$ let $k(p, t)$ be the point $q \in W$ on g_p at which $f(q) = t$. Then the mapping

$$(p, t) \to k(p, t): f^c \times (a, b) \to W \tag{5.1}$$

is a diffeomorphism onto W.

Definition of μ. Let $p \to \lambda(p)$ be an orientation-preserving diffeomorphism of f^c onto f^c. Then λ can be extended as an orientation-preserving diffeomorphism $q \to \lambda^*(q)$ of W onto W, by making the point $q = k(p, t)$ in W correspond to the point $\lambda^*(q) = k(\lambda(p), t)$ in W. An orientation-preserving diffeomorphism

$$q \to \mu(q) = \theta(\lambda^*(q)): W \to \mathscr{W} \tag{5.2}$$

of W onto $\mathscr{W}$ is thereby defined.

The composite manifold $\mathbf{M}^\lambda$. The elements in (5.0) have been specially defined in this section. Let

$$\mathbf{M}^\lambda = [M, \mathscr{M}, \mu, W, \mathscr{W}] \tag{5.3}$$

be the composite manifold thereby defined, with $\mathbf{M}^\lambda$ topologized and differentiably structured in terms of these elements and their oriented structures as in §4. The structure thereby defined on $\mathbf{M}^\lambda$ is an oriented differentiable structure. The diffeomorphisms

$$\Pi_1: M \to \mathbf{M}^\lambda \qquad \Pi_2: \mathscr{M} \to \mathbf{M}^\lambda \tag{5.4}$$

are well-defined and are orientation-preserving.

A non-degenerate function f^* on $\mathbf{M}^\lambda$ derived from f on $\mathbf{M}$. Given $p \in M$ and $q \in \mathscr{M}$ set

$$f^*(\Pi_1(p)) = f(p), \quad f^*(\Pi_2(q)) = f(\theta^{-1}(q)) \tag{5.5}$$

The values $f^*(u)$ are defined for each point $u \in \mathbf{M}^\lambda$, twice at points u of $\Pi_1(W) = \Pi_2(\mathscr{W})$ but consistently, as one readily sees.

Recalling that the mappings $\Pi_1: M \to \mathbf{M}^\lambda$ and $\Pi_2: \mathscr{M} \to \mathbf{M}^\lambda$ are diffeomorphisms, one sees from (5.5) that f^* is non-degenerate. Set

$$M^a_- = (p \in \mathbf{M} | f(p) < a)$$
$$M^b_+ = (p \in \mathbf{M} | f(p) > b).$$

The critical points of f are in $M^a_- \cup M^b_+$. If z is a critical point of f in M^a_-, $\Pi_1(z)$ is a critical point of f^*. If z is a critical point of f in M^b_+, $\Pi_2(\theta(z))$ is a critical point of f^*.

Let Ω^* be the set of critical points of f^* in $\mathbf{M}^\lambda$.

A Riemannian metric $\mathscr{R}$ on $\mathbf{M}^\lambda$. In order that the basic Lemma 1.1 be extended to critical points z^* of f^*, it is necessary that a Riemannian metric $\mathscr{R}$ be defined on $\mathbf{M}^\lambda$, preferably in a manner closely related to the metric $\mathscr{R}^0$ on $\mathbf{M}$.

It is possible to define a Riemannian metric $\mathscr{R}$ on $\mathbf{M}^\lambda$ in such a manner that the restricted diffeomorphisms

$$\Pi_1 | M^a_-, \quad \Pi_2 | \theta(M^b_+)$$

into $\mathbf{M}^\lambda$ are isometries. We suppose $\mathscr{R}$ so defined. Each canonical neighborhood $X_z | (z \in \Omega)$ of Lemma 1.1 is included in one of the sets M^a_- and M^b_+. We draw the following conclusion.

If z is a critical point of f in M^a_- then $\Pi_1(z) = z^*$ is a critical point of f^* and Lemma 1.1 is satisfied if the elements

$$\mathbf{M}, f, z, T_z, X_z, \mathscr{R}^0 \tag{5.6}$$

are replaced respectively by

$$\mathbf{M}^\lambda, f^*, z^*, \Pi_1(T_z), \Pi_1(X_z), \mathscr{R} \tag{5.7}$$

If z is a critical point of M^b_+ then $\Pi_2(\theta(z)) = z^*$ is a critical point of f^*, and Lemma 1.1 is satisfied if the elements (5.6) are replaced respectively by

$$\mathbf{M}^\lambda, f^*, z^*, \Pi_2(\theta T_z)\Pi_1(\theta(X_z)), \mathscr{R} \tag{5.8}$$

An admissible family $\mathbf{H}^*$ *of* f^**-arcs on* $\mathbf{M}^\lambda - \Omega^*$. A family $\mathbf{H}$ of f-arcs is given on $\mathbf{M} - \Omega$. By an H-arc on $\mathscr{M}$ we mean the θ-image of a maximal H-arc on M'. H^*-arcs on $\Pi_1(M)$ and $\Pi_2(\mathscr{M})$ shall be images under Π_1 and Π_2 of maximal H-arcs on M and $\mathscr{M}$ respectively. There exists a unique admissible family $\mathbf{H}^*$ of maximal f^*-arcs on $\mathbf{M}^\lambda - \Omega^*$ whose subarcs on $\Pi_1(M)$ and $\Pi_2(\mathscr{M})$ are respectively the H-arcs which we have just defined.

6. Diffeomorphisms of M onto the composite manifold $\mathbf{M}^\lambda$

The diffeomorphisms of $\mathbf{M}$ onto $\mathbf{M}^\lambda$ with which we are concerned are the special diffeomorphisms which arise naturally from the way in which the composite manifold $\mathbf{M}^\lambda$ is formed from $\mathbf{M}$.

Recall that $\Pi_1|f^c$ is a diffeomorphism of the level manifold f^c of $\mathbf{M}$ onto the level manifold f^{*c} of $\mathbf{M}^\lambda$.

Diffeomorphisms of type λ. *An orientation-preserving diffeomorphism* φ *of* $\mathbf{M}$ *onto* $\mathbf{M}^\lambda$ *which induces an orientation-preserving diffeomorphism of* f^c *onto* $\Pi_1(f^c)$ *will be said to be of type* λ.

In this definition we do not require that $\varphi|f^c = \Pi_1|f^c$.

Diffeomorphisms λ *of upper type*. If the diffeomorphism Π_1 of M into $\mathbf{M}^\lambda$ is extendable as a diffeomorphism φ of $\mathbf{M}$ onto $\mathbf{M}^\lambda$, φ will be termed an *upper diffeomorphism* of $\mathbf{M}$ onto $\mathbf{M}^\lambda$ of type λ, and λ of *upper type*.

Diffeomorphism λ *of lower type*. If the diffeomorphism $\Pi_2\theta$ of M' into $\mathbf{M}^\lambda$ is extendable as a diffeomorphism Ψ of $\mathbf{M}$ onto $\mathbf{M}^\lambda$, Ψ will be termed a *lower diffeomorphism of* $\mathbf{M}$ *onto* $\mathbf{M}^\lambda$ of type λ and λ of *lower type*.

The following theorems will be proved in [10].

THEOREM 6.1 (i). *A necessary and sufficient condition that there exists an upper diffeomorphism of* $\mathbf{M}$ *onto* $\mathbf{M}^\lambda$ *of type* λ *is that the diffeomorphism* λ^* *of* W *onto* W *be extendable as a diffeomorphism of* M' *onto* M'.

(ii) *A necessary and sufficient condition that there exist a lower diffeomorphism of* $\mathbf{M}$ *onto* $\mathbf{M}^\lambda$ *of type* λ *is that the diffeomorphism* λ^* *of* W *onto* W *be extendable as a diffeomorphism of* M *onto* M.

There may exist a diffeomorphism of $\mathbf{M}$ onto $\mathbf{M}^\lambda$ of type λ even when there exist no upper or lower diffeomorphisms of $\mathbf{M}$ onto $\mathbf{M}^\lambda$ of type λ. The basis for the positive part of this affirmation is Theorem 6.2.

THEOREM 6.2. *Let* λ_1 *and* λ_2 *be diffeomorphisms of* f^c *onto* f^c *of lower and upper types respectively. If* $\lambda = \lambda_2\lambda_1$ *there exists a diffeomorphism of* $\mathbf{M}$ *onto* $\mathbf{M}^\lambda$ *of type* λ.

COROLLARY 6.1. *Let λ_1 and λ_2 be given as in Theorem 6.2. If $\lambda = \lambda_1\lambda_2$ there exists a diffeomorphism of* $\mathbf{M}$ *onto* $\mathbf{M}^{\lambda^{-1}}$ *of type* A^{-1}.

Simple examples will show the following.

If λ is an orientation-preserving diffeomorphism of f^c onto f^c, the composite manifolds $\mathbf{M}^\lambda$ *and* $\mathbf{M}^{\lambda^{-1}}$ *are not in general diffeomorphic.*

We distinguish between the four following mutually exclusive cases. These cases presuppose the existence of an orientation-preserving diffeomorphism λ of f^c onto f^c.

Case I. There exist both an upper and lower diffeomorphism of $\mathbf{M}$ *onto* $\mathbf{M}^\lambda$ *of type λ.*

Case II. There exists an upper (lower) diffeomorphism of $\mathbf{M}$ *onto* $\mathbf{M}^\lambda$ *of type λ, but no lower (upper) diffeomorphism of* $\mathbf{M}$ *onto* $\mathbf{M}^\lambda$ *of type λ.*

Case III. There exists no diffeomorphism of $\mathbf{M}$ *onto* $\mathbf{M}^\lambda$ *of type λ.*

Case IV. There exists a diffeomorphism of $\mathbf{M}$ *onto* $\mathbf{M}^\lambda$ *of type λ, but no upper or lower diffeomorphism of* $\mathbf{M}$ *onto* $\mathbf{M}^\lambda$ *of type λ.*

Case I can be realized by taking λ as a diffeomorphism of f^c onto f^c differentiably isotopic on f^c to the identity.

Case III can be realized on making use of the classic Milnor example, taking $\mathbf{M}$ as a euclidean 7-sphere and f as a non-degenerate function on $\mathbf{M}$ with just two critical points.

To show that Cases II and IV are realizable in a proper setting we shall begin the following section with a study of a 3-sphere on which a function f with just four critical points has been specially defined. The composite manifolds thereby obtained include diffeomorphs of $\mathbf{M}$ in all cases in which $n = 3$ and f on $\mathbf{M}$ has fewer than six critical points.

7. The case of at most four critical points, $n = 3$

In this section we shall take $\mathbf{M}$ as a 3-sphere S_3 on which a non-degenerate function f has four critical points with indices 0, 1, 2, 3 and critical values 0, 1, 3, 4. The section f^c of S_3, will be taken as a surface of torus type at an f-level $c = 2$. As in the geometry of inversion, S_3 will be represented by a 3-plane E_3 closed at the point at infinity.

The manifold E_3^+. Let E_3 be a euclidean space of points x with coordinates (x^1, x^2, x^3). Let E_3^+ be obtained from E_3 by adding "a point Q at infinity" to E_3. We suppose E_3^+ structured so that it is the C^∞-diffeomorph of S_3. We shall define a non-degenerate function f on E_3^+.

The function f. The function f shall have its maximum at Q on E_3^+. We suppose f so defined on E_3^+ that the level manifold f^c, with $c = 2$, is a conventional torus τ. The critical points of f of index 0 and 1 shall be on the interior of τ and the critical point of f of index 2 on the exterior of τ at the point of the axis of the torus at the center of the torus.

The torus τ. Let $z = (u, v)$ be rectangular coordinates in a 2-plane Z. For arbitrary $z \in Z$ the ensemble of points (x^1, x^2, x^3) of the form

$$P(z) = [(2\text{-}\cos u) \cos v, \sin u, (2\text{-}\cos u) \sin v] \tag{7.1}$$

is a torus τ generated by revolving the circle

$$x^1 = 2\text{-}\cos u,\ x^2 = \sin u,\ x^3 = 0 \qquad (-\infty < u < \infty)$$

about the x^2-axis. The mapping $P \to P(z)$ of Z into E_3 is real analytic, is onto τ and doubly periodic, with period 2π in u and v.

Mappings Λ of Z onto Z. Corresponding to any set $[p, q, r, s]$ of rational integers such that $ps - rq = 1$ we introduce the linear mapping $z \to \Lambda(z)$ of the form

$$\begin{aligned} u' &= pu + qv \\ v' &= ru + sv \end{aligned} \tag{7.2}$$

of Z onto Z, and the orientation-preserving diffeomorphism

$$p \to \lambda(p) = (P\ \Lambda\ P^{-1}((p) \qquad (p \in \tau) \tag{7.3}$$

of τ onto τ.

The composite manifold E_3^λ. Let Λ be an arbitrary diffeomorphism of Z onto Z of the form (7.2), and λ the corresponding diffeomorphism of τ onto τ given by (7.3). Regarding τ as the manifold f^c at the f-level 2 on the manifold E_3^+, let E_3^λ denote the corresponding composite manifold, constructed from E_3^+ and f^c as $\mathbf{M}^\lambda$ was more generally constructed from $\mathbf{M}$ and f^c in §4.

The composite differentiable manifolds E_3^λ include the topological types of the so-called *lens* spaces. Extending results in the topological case it is provable in the differentiable case that E_3^λ is the diffeomorph of a 3-sphere if and only if $|p| = 1$ in (7.2). The author has established the following.

CASE IV. Of the four Cases enumerated at the end of §6, Case IV can be realized by virtue of (a_1).

(a_1). *Necessary and sufficient condition that there exists a diffeomorphism of type λ of E_3^+ onto E_3^λ which is neither an upper nor a lower diffeomorphism of E_3^λ of type λ is that in* (7.2)

$$|p| = 1, \quad q \neq 0, \quad r \neq 0$$

CASE II. This case can be realized by virtue of (a_2).

(a_2). *Necessary and sufficient condition that E_3^+ admit a lower (upper) diffeomorphism of E_3^+ onto E_3^λ of type λ is that $|p| = 1$ and $r = 0$ $(q = 0)$ in* (7.2).

8. Homology 3-spheres

Let $c_1 < c_2 < \cdots < c_r$ be a set of ordinary values of f on $\mathbf{M}$, not in the subsets $f(X_z)$ of R (Cf. Lemma 1.1). Let λ_i be an orientation preserving diffeomorphism of f^{c_i} onto f^{c_i} $i = 1, \ldots, r$. Let the composite manifold, previously denoted by $\mathbf{M}^\lambda$, here be more explicitly denoted by $M(f, \lambda, c)$, recalling that λ was a diffeomorphism of f^c onto f^c. Generalizing $M(f, \lambda, c)$ in the obvious manner, we introduce the *composite manifold*

$$M(f;\, \lambda_1, \ldots, \lambda_r;\, c_1, \ldots, c_r) \tag{8.1}$$

We are admitting only those polar non-degenerate functions f on $\mathbf{M}$ whose values at distinct critical points are distinct. It will be convenient to term a value of such an f at a critical point of index k a *critical value of index* k.

Diffeomorphic models for homology 3-*spheres*. Let Σ_3 be a differentiable manifold which is a homology 3-sphere. On Σ_3 let Φ be an admissible polar-non-degenerate function with r-critical values

$$a_1 < a_2 < \cdots < a_r \tag{8.2}$$

of index 2. Let $c_1 < c_2 < \cdots < c_r$ be r ordinary values of Φ such that

$$c_1 < a_1 < c_2 < a_2 < \cdots < c_r < a_r \tag{8.3}$$

so chosen that for $i = 1, \ldots, r$, (c_i, a) is an interval of ordinary values of Φ. There then exists a polar-non-degenerate function on a 3-sphere S_3 with the same critical values as Φ, with the same indices. We state the following theorem.

THEOREM 8.1. *For suitable choices of the diffeomorphisms* λ_i, $i = 1, \ldots, r$ *of the level manifolds* F^{c_i} *onto themselves the composite manifold*

$$S_3(F;\, \lambda_1, \ldots, \lambda_r;\, c_1, \ldots, c_r) = \mathscr{S}_3 \tag{8.4}$$

is a diffeomorph of Σ_3.

If F^* *is the polar-non-degenerate function on the manifold* $\mathscr{S}_3$ *"derived"* (Cf. §5) *from the function* F *on* S_3, *there exists a diffeomorphism* θ *of* Σ_3 *onto* $\mathscr{S}_3$ *such that*

$$\Phi(p) = F^*(\theta(p)) \qquad (p \in \Sigma_3)$$

Comment on Theorem 8.1. If Σ_3 is *not* a homology sphere, there may exist no polar-non-degenerate function F on S_3 with the same indexed critical values as those of Φ.

The numerical order of the critical values of Φ on the homology 3-sphere Σ_3 is far from arbitrary. If c is an arbitrary number, then

$$N_1(c) \geq N_2(c) \tag{8.5}$$

where $N_k(c)$, $k = 1, 2$, is the number of critical values of Φ of index k less than c. This statement is not in general true if Σ_3 is not a homology sphere.

Consistent with (8.5) it is always possible to continuously modify Φ as a non-degenerate function on the homology 3-sphere Σ_3 so that the maximum critical value of index 1 is less than the minimum critical value of index 2. Such an order of the critical values is the least illuminating from the point of view of the next paragraph.

If the critical values of Φ on Σ_3 with indices 1 and 2 alternate as the critical values increase, and if Σ_3 is simply connected, it can be shown that Σ_3 is the diffeomorph of a 3-sphere in accord with the Poincaré conjecture. It follows that if Σ_3 is a simply-connected differentiable 3-manifold which is not the diffeomorph of a 3-sphere then any polar-non-degenerate function Φ on Σ_3 must have at least six critical points.

In justification of the principal hypothesis of Theorem 2.1 one can prove the following special theorem.

THEOREM 8.2. *In case* $\mathbf{M}$ *is 3-dimensional suppose that* f *has just four critical points. These critical points must include a critical point* $\dot{w}$ *of index 2, and a critical point* w *of index* 1.

Let $\mathbf{H}^0$ *be the family of ortho-f-arcs on* $\mathbf{M}$-Ω. *Then a necessary and sufficient condition that* $\mathbf{M}$ *be the diffeomorph of a 3-sphere is that* $f(\dot{w}) > f(w)$ *and that there exist an isotopic replacement* $\mathbf{H}$ *of* $\mathbf{H}^0$ *such that the descending bowl* B_- *with pole* $\dot{w}$, *and the ascending bowl* B_+ *with pole* w *intersect regularly in an* H-*arc.*

9. An extension theorem

The problem of giving conditions sufficient that our n-manifold $\mathbf{M}$ be the diffeomorph of another such manifold $\dot{\mathbf{M}}$, can be approached in terms of critical point theory. Such an approach leads to an extension problem, to be defined below, and an extension theorem, Theorem 9.1.

The data. We presuppose the existence on $\mathbf{M}$ and $\dot{\mathbf{M}}$ of polar-non-degenerate functions f and $\dot{f}$ with the same indexed critical values. As previously we suppose that each value is assumed at a unique critical point of $\mathbf{M}$ and $\dot{\mathbf{M}}$, respectively, with a common index k.

DEFINITION 9.1. $(f, \dot{f})$-diffeomorphisms. Let α be an ordinary value of f and $\dot{f}$. Set

$$f_\alpha = (p \in \mathbf{M} | f(p) \leq \alpha)$$

$$\dot{f}_\alpha = (p \in \dot{\mathbf{M}} | \dot{f}(p) \leq \alpha)$$

An orientation preserving diffeomorphism $p \to \theta(p)$ of f_a onto $\dot{f}_a$ such that

$$f(p) = \dot{f}(\theta(p)) \qquad (p \in f_a) \tag{9.1}$$

will be called *an* $(f, \dot{f})$*-diffeomorphism* of f_a onto $\dot{f}_a$.

DEFINITION 9.2. Let c be a critical value of f and $\dot{f}$ of index k, with c interior to an interval $[a, b]$ of ordinary values of f and $\dot{f}$. Suppose that there exists an $(f, \dot{f})$-diffeomorphism

$$\theta; f_a \to \dot{f}_a : p \to \theta(p) \tag{9.2}$$

of f_a onto $\dot{f}_a$. We suppose that $1 < k < n$.

PROBLEM P_c. *We seek sufficient conditions that* θ *admit an extension which is an* $(f, \dot{f})$*-diffeomorphism of* f_b *onto* $\dot{f}_b$.

In this definition we have referred to a diffeomorphism θ of a *closed set* f_a. We understand that such a diffeomorphism is well-defined only if it admits an *extension* as a diffeomorphism into $\dot{\mathbf{M}}$ of an open neighborhood of the given closed set.

Let z and $\dot{z}$ be the critical points of f and $\dot{f}$, respectively, such that $f(z) = \dot{f}(\dot{z}) = c$. We refer to isometric presentations,

$$x \to T_z(x) : D^x_\sigma \to X_z \tag{9.3}$$

$$y \to T_{\dot{z}}(y) : D^y_\sigma \to X_{\dot{z}} \tag{9.4}$$

of neighborhoods X_z and $X_{\dot{z}}$ of z and $\dot{z}$ on $\mathbf{M}$ and $\dot{\mathbf{M}}$, as characterized in Definition 1.3. Let B_- and $\dot{B}_-$ be the k-bowls on $\mathbf{M}$ and $\dot{\mathbf{M}}$ respectively descending from the critical points z and $\dot{z}$. Let B^a_- and $\dot{B}^a_-$ be respectively the traces of these bowls on the level manifolds f^a and $\dot{f}^a$. Let η be a constant such that $c - \sigma < \eta < c$. The trace B^η_- of the bowl B_- is included in X_z and the trace $\dot{B}^\eta_-$ of the bowl B_- is included in $X_{\dot{z}}$.

DEFINITION 9.3. *A canonical diffeomorphism* π *of* B^a_- *onto* $\dot{B}^a_-$. The traces B^a_- and B^η_- of the bowl B_- on $\mathbf{M}$ are diffeomorphic, $\mathbf{H}$-related points corresponding. Similarly the traces $\dot{B}^a_-$ and $\dot{B}^\eta_-$ of the bowl $\dot{B}^\eta_-$ on $\dot{\mathbf{M}}$ are diffeomorphic, $\dot{\mathbf{H}}$-related points corresponding. The traces B^η_- and $\dot{B}^\eta_-$ are diffeomorphic under a mapping in which points

$$T_z(x) \in B^\eta_- \qquad T_{\dot{z}}(y) \in \dot{B}^\eta_- \tag{9.5}$$

correspond if $x = y$. There is thereby induced a *canonical diffeomorphism* π

$$B^a_- \leftrightarrow B^\eta_- \leftrightarrow \dot{B}^\eta_- \leftrightarrow \dot{B}^a_-$$

of B^a_- onto $\dot{B}^a_-$.

The canonical diffeomorphism π *of* B^a_- *onto* $\dot{B}^a_-$ *can be extended by special diffeomorphism* Π *into* f^a *of any sufficiently small open neighborhood* N *of* B^a_-*, relative to* f^a.

Definition of Π. The definition of Π is similar to that of π. A sufficiently small open neighborhood N of B^a_-, relative to f^a, is the diffeomorph of an open neighborhood $\mathscr{N} \subset X_z$ of B^η_- relative to f^η, $\mathbf{H}$-related points corresponding. This neighborhood $\mathscr{N}$ of B^η_- is the diffeomorph of a neighborhood $\dot{\mathscr{N}}$ of $\dot{B}^\eta_-$, relative to $\dot{f}^\eta$, under a mapping in which points

$$T_z(x) \in \mathscr{N} \qquad T_{\dot{z}}(y) \in \dot{\mathscr{N}} \tag{9.6}$$

correspond if $x = y$. The neighborhood $\dot{\mathscr{N}}$, if sufficiently small, is the diffeomorph of a neighborhood $\dot{N}$ of $\dot{B}^a_-$ relative to $\dot{f}^a$, $\dot{\mathbf{H}}$-related points corresponding. Thus if N is a sufficiently small open neighborhood of B^a_-, relative to f^a, there results a diffeomorphism Π of N onto $\dot{N}$ extending π.

In order to satisfy condition (i) in Definition 2.10 we presented two admissible choices of T_w in the paragraph preceding Definition 2.10.

So here, in precisely the same manner we present two admissible choices of T_z.

The definition of π and Π will be affected by this choice of T_z. For example, if $k = 2$, the sense of the diffeomorphism π of the curve B^a_- onto the curve $\dot{B}^a_-$ will be reversed by a change from one of these presentations T_z to the other.

We state an extension theorem.

THEOREM 9.1. *In order that an* $(f, \dot{f})$*-diffeomorphism* θ *of* f *onto* $\dot{f}_a$ *admit an extension which is an* $(f, \dot{f})$*-diffeomorphism of* f_b *onto* $\dot{f}_b$, *it is sufficient that for one of the above two admissible choices of* T_z, *and for a sufficiently small neighborhood* N *of* B^a_-, *relative to* f^a, *the following be true.*

The diffeomorphism $\theta | f^a$ *of* f^a *onto* $\dot{f}^a$ *is differentiably isotopic among diffeomorphisms of* f^a *onto* $\dot{f}^a$ *to a diffeomorphism* φ *of* f^a *onto* $\dot{f}^a$ *such that*

$$\varphi(p) = \Pi(p) \qquad (p \in N) \tag{9.7}$$

In case the index k of the critical value c is 1, Problem P_c always has a solution, as one sees readily.

When $k = n$ Problem P_c is undefined. If we understand that c is the maximum value of f and $\dot{f}$, and a an ordinary value exceeding each critical value other than c, one has the following problem. If θ is a $(f, \dot{f})$-diffeomorphism of f_a onto $\dot{f}_a$, what are sufficient conditions that θ admit an extension which is an $(f, \dot{f})$-diffeomorphism of $\mathbf{M}$ onto $\dot{\mathbf{M}}$? This problem is equivalent to the differentiable form of the Schoenflies extension problem for an $(n - 1)$-sphere.

When $n = 3$ the only extension of the above types through a critical value c which is not always possible, occurs when the index $k = 2$, and in this case Theorem 9.1 can be reduced to the following.

Suppose that **M** *is 3-dimensional, and that the index $k = 2$. In order that an $(f, \acute{f})$-diffeomorphism θ of f_a onto $\acute{f}_a$ admit an extension which is an $(f, \acute{f})$-diffeomorphism of f_b onto $\acute{f}_b$ the following is sufficient.*

The diffeomorphism $\theta|f^a$ of f^a onto $\acute{f}^a$ is differentiably isotopic among diffeomorphisms of f^a onto $\acute{f}^a$ to a diffeomorphism φ of f^a onto $\acute{f}^a$ such that

$$\varphi(p) = \pi(p) \qquad (p \in B^a_-) \tag{9.8}$$

References

[1] M. Morse, The existence of polar non-degenerate functions on differentiable manifolds, *Annals of Math.*, *71* (1960), p. 352–383.

[2] M. Morse, The elimination of critical points of a non-degenerate function on a manifold, *Jour. D'analyse Math.* (to appear).

[3] W. Huebsch and M. Morse, Schoenflies extensions without interior differential singularities, *Annals of Math.*, *76* (1962), p. 18–54.

[4] E. Baiada and M. Morse, Homotopy and homology related to the Schoenflies problem, *Annals of Math.*, *58* (1953), p. 142–165.

[5] M. Morse, Quadratic forms Q and Q-fibre-bundles, *Annals. of Math.* To appear.

[6] N. Steenrod, *The Topology of Fibre Bundles*, Princeton University Press, 1951.

[7] M. Morse, A model non-degenerate function. To be published.

[8] J. Milnor, On manifolds homeomorphic to the 7-sphere, *Annals of Math.*, *64* (1956), p. 399–405.

[9] M. Morse, Differentiable mappings in the Schoenflies theorem, *Compositio Mathematica*, *14* (1959), p. 83–151.

[10] M. Morse, Diffeomorphisms of composite manifolds. To be published.

[11] S. Smale, A survey of some recent developments in differential topology, *Bull. Amer. Math. Soc.*, *69* (1963), p. 131–145.

BULLETIN OF THE
INSTITUTE OF MATHEMATICS
ACADEMIA SINICA
Volume 4, Number 1, June 1976

FRÉCHET NUMBERS AND GEODESICS ON SURFACES

BY

STEWART S. CAIRNS AND MARSTON MORSE

1. **Abstract.** This paper is concerned with the global relations between the Fréchet numbers R_i of M_n (Def. 2.4) and the geodesics joining two prescribed points $A_1 \neq A_2$ on M_n and $A_1 A_2$-homotopic to a prescribed curve h. The point pair $A_1 \neq A_2$ is assumed ND (non-degenerate) in that on no geodesic which joins A_1 to A_2 is A_2 conjugate to A_1. As shown in [4, p. 233], the set of points A_2 on M_n which are conjugate to A_1 on some geodesic have a Lebesgue measure 0 on M_n. See also §11 of [5].

The global relations here presented are specializations of relations recently presented in [8] on an n-dimensional compact connected Riemannian manifold M_n. We turn to the case $n=2$ because in that case the Fréchet numbers can be shown to be finite and uniquely determined by the Euler characteristic of the manifold M_2. No such finality has been achieved when $n>2$.

Each geodesic γ joining A_1 to A_2 is assigned an index equal to the "count" of conjugate points of A_1 on γ. This type of classification of geodesics leads to relations (Theorem 2.1) which go far beyond the relations classically obtained by specializing Riemannian sectional curvature of M_n. Theorem 2.1 is a consequence of Theorem 27.1 of [5].

Curves on M_n are required to be *proper*. That is, in the representation of a curve γ as a continuous mapping of an interval $[a, b]$, no point of γ is to be the image of a subinterval of $[a, b]$. Lemma 5.1 of [10] is a basic lemma on an $A_1 A_2$-homotopy in which the curves joining A_1 to A_2 are required to be proper.

The paper is presented in two parts. In Part I, that part of the general theory is presented which is needed in the study of geodesics on surfaces. Part I comprises §§1, 2 and 3. Part II presents the applications to surfaces, with particular emphasis on results not yet obtained in the general theory.

PART I. REVIEW OF GENERAL THEORY

2. **Underlying analogies.** Theorem 2.1 will present the geodesic-homology relations involving the Fréchet numbers R_i of M_n. §2 clarifies the analogies which govern the passage from the theory of critical *points* to a theory of critical *extremals*. The relations pre-

Received by the editors December 24, 1975.

sented in Theorem 30.1 of [2] are typical of (critical point)-homology relations.

In the earlier (critical point)-homology relations the most essential elements are the following:

(I) The real-valued function f whose critical points are to be studied;

(II) The nondegeneracy of these critical points;

(III) The count m_i of critical points of index i;

(IV) The connectivities of M_n over a field, here taken as the field Q of rational numbers.

When geodesics joining A_1 to A_2 on M_n replace critical points of f, the elements which replace the respective elements of (I), (II), (III), (IV) are as follows:

(I) ***The length integral L, replacing f.*** Two points $A_1 \neq A_2$ are prescribed on M_n and a curve h joining A_1 to A_2. The function f is replaced by the integral L of length, evaluated on any piecewise regular curve γ joining A_1 to A_2 and A_1A_2-homotopic to h. For A_1A_2-homotopies see Def. 3.3 of [9]. The following lemma defines a basic geodesic g in the A_1A_2-homotopy class of h.

LEMMA 2.1. *Corresponding to the point pair $A_1 \neq A_2$ and the curve h joining A_1 to A_2 on M_n, there exists at least one geodesic g which joins A_1 to A_2, is A_1A_2-homotopic to h and affords an absolute minimum to the length of piecewise regular curves which join A_1 to A_2 and are A_1A_2-homotopic to h.* (See Theorem 21.2 of [5].)

A ***prime geodesic*** *g.* The geodesic g is termed ***homotopically minimizing*** when Lemma 2.1 holds. If in addition, g's endpoints $A_1 \neq A_2$ are a ND point pair, g will be termed a ***prime geodesic*** on M_n.

(II) ***Nondegenerate point pairs*** $A_1 \neq A_2$. Let γ be a geodesic joining A_1 to A_2. Conjugate points of A_1 on γ and their multiplicities are defined in §10 of Morse [5]. The ***index*** of a geodesic γ joining A_1 to A_2 is the count of the conjugate points of A_1 on γ definitely preceding A_2. The ***nullity*** of γ is, by definition, the multiplicity of A_2 as a conjugate point of A_1. The geodesic γ is termed ND if its nullity is zero. We are assuming that the point pair (A_1, A_2) is

ND in that each geodesic which joins A_1 to A_2 is ND. It is of interest that r, reversed in sense, is a geodesic which has the same index and nullity as r. This follows from the Extended Separation Theorem 10.4 of Morse [5].

Let a point P be prescribed and fixed on M_n. The set of points Q which are conjugate to P on some geodesic joining P to Q has a measure zero on M_n. This was first proved on pages 233 and 234 of Morse [4].

According to Corollary 24.2 of Morse [5], if (A_1, A_2) is a ND point pair, the count of geodesics r joining A_1 to A_2 with lengths less than a prescribed constant c is less than some finite integer N_c. It follows that the total number of geodesics joining the ND point pair (A_1, A_2) is finite or countably infinite.

(III) *The count m_i^g replacing the count m_i.* Any geodesic which joins A_1 to A_2 and is A_1A_2-homotopic to a "prime" geodesic g will be called *g-admissible.*

The analogue of the "count" m_i of critical points of index i is the count m_i^g (possibly infinite) of g-admissible geodesics with index i. Examples show that m_i^g can be infinite for a given i. However, the following condition on M_n is sufficient that m_i^g be finite for each i. (Theorem 27.2 of Morse [5].)

DEFINITION 2.1. *A condition* $)B($ *on* M_n. A condition $)B($ is satisfied by M_n, if each geodesic on M_n of length at least a fixed constant $B > 0$ bears a first conjugate point of its initial point.

We shall prove the following.

(A) *If M_n satisfies a condition $)B($, then the type numbers m_j^g are finite for each prime geodesic g on M_n and each integer $j \geq 0$.*

Fix g and j. The condition $)B($ and Separation Theorem 10.4 of [5] imply that each g-admissible geodesic on M_n with length exceeding $(j+1)B$ has an index at least $j+1$. Each g-admissible geodesic with index j has a length less than $(j+1)B$. It follows from Corollary 24.1 of [5] that the number of g-admissible geodesics with lengths less than $(j+1)B$ is finite. Thus (A) is true.

The class of surfaces which satisfy a condition $)B($ not only includes all surfaces of positive Gaussian curvature but also a large class of surfaces whose curvature is less restricted.

(IV) ***Fréchet numbers $\mathbf{R}_i$ replacing connectivities R_i of M_n.*** If h and k are curves joining A_1 to A_2, on M_n, let hk denote the Fréchet distance(1) between h and k. It is immediate that hk satisfies the classical axioms on a distance in a metric space except that the condition $hk = 0$ does not imply that $h = k$, as examples show.

With Fréchet [1] we introduce the class, say $\mathbf{h}$, of all curves obtained from h by a 1–1 continuous sense-preserving change of parameter. Fréchet defines a distance $\mathbf{hk}$ by setting $\mathbf{hk} = hk$. It is shown in [6] that $\mathbf{hk} = 0$ if and only if $\mathbf{h} = \mathbf{k}$. That $\mathbf{hk} = \mathbf{kh}$ is immediate and that the usual triangle axiom is satisfied.

DEFINITION 2.2. Let $\mathcal{F}_{A_1}^{A_2}$ denote the metric space of ***Fréchet classes*** $\mathbf{h}$ of curves h joining A_1 to A_2 on M_n.

The metric space $\mathcal{F}_{A_1}^{A_2}$ is pathwise connected when M_n is an n-sphere. In general, $\mathcal{F}_{A_1}^{A_2}$ is not pathwise connected. The following definition and lemma prepare for Morse's definition of Fréchet numbers of M_n.

DEFINITION 2.3. *The class* $((N_n))$ *of differentiable manifolds.* Let N_n be a compact connected differentiable manifold of class C^∞. Then $((N_n))$ shall denote the class of all differentiable manifolds M_n of class C^∞ homeomorphic to N_n. M_n may not be diffeomorphic to N_n, as Milnor's exotic 7-spheres show.

The following lemma has been established in Landis-Morse [10].

LEMMA 2.2. *Let M_n be any manifold in the class $((N_n))$. Let $A_1 \neq A_2$ be an arbitrary point pair on M_n and $\mathcal{F}_{A_1}^{A_2}$ the corresponding Fréchet space. Pathwise components of $\mathcal{F}_{A_1}^{A_2}$ are then homeomorphic for every choice of $M_n \in ((N_n))$ and of a disjoint point pair (A_1, A_2) on M_n.*

DEFINITION 2.4. ***The Fréchet numbers $\mathbf{R}_i$ of M_n or $((N_n))$.*** The ith connectivity $\mathbf{R}_i$, over Q, common (by Lemma 2.2) to the pathwise components of a Fréchet space $\mathcal{F}_{A_1}^{A_2}$ of a manifold $M_n \in ((N_n))$ is independent of the choice of $M_n \in ((N_n))$ and of $A_1 \neq A_2$ on M_n. $\mathbf{R}_i$ will be called the ith ***Fréchet number*** of M_n or $((N_n))$.

(1) See Morse [6] and Landis-Morse [10].

A major theorem in the geodesic-homology theory follows. This theorem is special case of Theorem 27.1 of Morse [5]. It is obtained by taking the Weierstrass integral J of Morse [5] as the integral of length on M_n. For proof see Morse [7] and [8].

GLOBAL THEOREM 2.1. *Let m_r^g be the number (possibly infinite) of g-admissible geodesics of index r joining a* ND *point pair $A_1 \neq A_2$ on M_n.*

If the numbers m_r^g are finite for each r, then $m_r^g \geq R_r$ for each r and

$$\begin{aligned} m_0^g &\geq R_0 \\ m_1^g - m_0^g &\geq R_1 - R_0 \\ m_2^g - m_1^g + m_0^g &\geq R_2 - R_1 + R_0 \\ \cdots \quad \cdots &\geq \cdots \quad \cdots \end{aligned}$$

The principal hypothesis of Theorem 2.1 is that M_n be g-finite in the following sense.

DEFINITION 2.5. *g-Finite manifolds M_n.* A compact Riemannian manifold of class C^∞ bearing a "prime" geodesic g will be called *g-finite* if the type numbers m_i^g are finite for each $i \geq 0$.

For M_n to be g-finite for every choice of a prime geodesic g it is sufficient that M_n satisfy a condition $)B($ in the sense of Definition 2.1, as has been shown.

The computation of Fréchet numbers of M_n. It follows from Lemma 2.2 that the Fréchet numbers of M_n are pseudotopological invariants in that they are the same for any two homeomorphic differentiable manifolds M_n. A major problem is to compute these numbers. *This problem is completely solved in the present paper when $n = 2$.*

It will be shown that the Fréchet numbers of compact differentiable surfaces are finite and completely determined by their Euler characteristics. We find that the Fréchet numbers are 1 without exception when the surface is homeomorphic to a sphere or projective plane. In all other cases $R_0 = 1$ and other Fréchet numbers vanish.

The fact that $R_0 = 1$ in all cases reflects the fact that Fréchet numbers are the connectivities of pathwise components of the Fréchet space $\mathcal{G}_{A_1}^{A_2}$.

The pathwise components of $\mathcal{G}_{A_1}^{A_2}$. If a ND point pair $A_1 \neq A_2$ is prescribed on M_n there clearly exists a maximal set $(h_i) = H$ of curves h_i joining A_1 to A_2 on M_n no two of which are $A_1 A_2$-homotopic. The cardinal number of the set (h_i) is a finite integer or aleph null. Corresponding to each $h_i \in H$ there exists a homotopically minimizing geodesic g_i, $A_1 A_2$-homotopic to h_i. Theorem 2.1 is valid with $g = g_i$ and with Fréchet numbers $\boldsymbol{R}_j$ independent of the choice of $h_i \in H$. The cardinal number of (h_i) is independent of the choice of $M_n \in ((N_n))$ and of the choice of the ND point pair (A_1, A_2) on M_n. The pathwise components $\mathcal{G}_i$ which contain the respective Fréchet curve classes $\boldsymbol{h}_i$ are disjoint and have $\mathcal{G}_{A_1}^{A_2}$ as their union.

3. **The g-connectivities L_i^g of M_n.** According to Theorem 2.1, the Fréchet numbers $\boldsymbol{R}_i$ of M_n are finite if M_n is g-finite in the sense of Definition 2.5. When M_n is g-finite one can make use of g-admissible geodesics joining A_1 to A_2 to define a finite integer L_i^g termed the *i*th *g-connectivity* of M_n. The computation of the numbers $\boldsymbol{R}_i$ then reduces to the computation of the g-connectivities L_i^g in accord with the following theorem.

THEOREM 3.1. *If the manifold M_n is g-finite for a prime geodesic g, then the jth g-connectivity L_j^g equals $\boldsymbol{R}_j$ for each integer $j \geq 0$.*

Proved in [8].

It is the task of this section to recall the definition of the numbers L_i^g. Two lemmas are required. The first of these lemmas has a form well known to differential geometers. For Weierstrass extremals see [5, §19].

LEMMA 3.1. *Corresponding to the integral L of Riemannian length there exists a positive number m (termed a preferred length) such that the following is true.*

The geodesics with lengths m issuing from a point p arbitrarily prescribed in M_n intersect in no point other than p, bear no conjugate point of p, cover a closed topological n-disc on M_n with p an interior *point and have lengths which afford a proper absolute minimum to L relative to piecewise regular curves which join their endpoints on M_n.*

The preferred length m introduced in Lemma 3.1 leads to the following definition.

DEFINITION 3.1. *Elementary geodesics.* A geodesic on M_n whose length is at most the preferred length m is called an elementary geodesic.

With the aid of elementary geodesics, basic subspaces of product spaces $(M_n)^\nu$ will now be characterized.

Notation for Definition 3.2. Let M_n be a compact Riemannian manifold of class C^∞ which is g-finite for some prime geodesic g. Let β be any value in R such that $L(g) < \beta$ and β is *g-ordinary* in that it is not the length of any g-admissible geodesic. If ν is a positive integer such that

$$L(g) < \beta < m(\nu+1)\,, \qquad (m \text{ from Lemma 3.1}) \tag{3.1}$$

g can be partitioned into $\nu + 1$ successive elementary geodesics of equal length less than m. The successive endpoints of these subarcs of g form a sequence

$$A_1,\, p_1, \cdots,\, p_\nu,\, A_2$$

of points of M_n. The points $p_1, \cdots, p_\nu$ define a ν-tuple p on the ν-fold product $(M_n)^\nu$ of M_n by itself. The ν-tuple p is a "point" in a subspace $[g]^\nu_\beta$ of $(M_n)^\nu$ which is now defined.

DEFINITION 3.2. *A vertex space* $[g]^\nu_\beta$. If (3.1) holds with β g-ordinary, a maximal, pathwise connected, subspace of the product space $(M_n)^\nu$ satisfying the following three conditions is called a *vertex space* $[g]^\nu_\beta$.

Condition I. Each ν-tuple $z = (z_1, \cdots, z_\nu)$ in $[g]^\nu_\beta$ shall be such that successive points in the sequence

$$A_1,\, z_1, \cdots,\, z_\nu,\, A_2 \qquad (\nu > 0) \tag{3.2}$$

of points of M_n which are distinct can be joined by elementary geodesics on M_n.

Condition II. The *broken geodesic,* say $\zeta^\nu(z)$, joining A_1 to A_2

and defined by the successive elementary geodesics joining successive distinct points in (3.2) has a length $\leq \beta$.

Condition III. $[g]^{\nu}_{\beta}$ contains the ***normal*** ν-tuple of g, that is the ν-tuple which partitions g into $\nu+1$ elementary geodesics of equal length.

Two lemmas on vertex spaces will complete the preparation for Lemma 3.4.

The connectivities R_j of a vertex space $[g]^{\nu}_{\beta}$. Theorem 26.1 of [5] implies that these connectivities are finite. Lemma 3.2 below is a consequence of this theorem. We shall recall the terms involved. Let

$$S_{\beta} = (\gamma_0, \cdots, \gamma_r) \tag{3.3}$$

be a maximal set of g-admissible geodesics $A_1 A_2$-homotopic to g under the length level β. S_{β} contains g. A geodesic γ in the set S_{β} may be regarded as a broken geodesic $\zeta^{\nu}(z)$ (without corners) defined by the "normal" ν-tuple z of γ.

For each integer $j \geq 0$ let m^{β}_j be the number of geodesics in the set S_{β} with index j. The number m^{β}_j is termed the jth ***type number*** of the set S_{β} of geodesics. If κ is the maximum of the indices of the geodesics in S_{β}, $m^{\beta}_j = 0$ for $j > \kappa$. From Theorem 26.1 of Morse [5] we infer the following.

LEMMA 3.2. *The jth connectivity of the vertex space $[g]^{\nu}_{\beta}$, over Q, is at most m^{β}_j, that is, at most the number of g-admissible geodesics which are $A_1 A_2$-homotopic to g under the length level β.*

We interpolate a lemma on how the connectivities of a vertex space $[g]^{\nu}_{\beta}$ vary with variation of the integer ν and the g-ordinary value β. Theorem 5.1 of Morse [7] reads as follows.

LEMMA 3.3. *Vertex spaces $[g]^{\nu}_{\beta}$ and $[g]^{\mu}_{\beta}$ with the same g-ordinary bound β have isomorphic homology groups over Q of each dimension.*

Notation for Lemma 3.4. If there are no g-admissible geodesics of index i set $\pi_i = J(g)$. If, however, m^{Ω}_i is finite and positive, let π_i denote the maximum length of g-admissible geodesics of index i.

In Lemma 3.4 we shall refer to the value

$$a_i = \max(\pi_i, \pi_{i+1}) . \tag{3.4}$$

The value a_i is defined for each integer $i \geq 0$. Its definition is relative to M_n and to a prime geodesic g on M_n.

Definition 3.3 and Lemma 3.4 are taken from §6 of [7].

DEFINITION 3.3. *i-Mature values β.* Let an integer $i \geq 0$ be prescribed. When M_n is g-finite, a g-ordinary value β will be called ***i-mature*** if $\beta > a_i$ of (3.4), and if each g-admissible geodesic of index i is $A_1 A_2$-homotopic to g under the length level β.

LEMMA 3.4. *If β is i-mature, then the ith connectivity of a vertex space $[g]^\nu_\beta$ is an integer L^g_i independent of such values β and of integers ν such that $m(\nu + 1) > \beta$.*

Lemma 3.4 leads to a definition.

DEFINITION 3.4. *The g-connectivities of M_n.* A vertex space $[g]^\nu_\beta$ with b j-mature will be termed j-mature. The jth connectivity L^g_j over Q common to j-mature vertex spaces $[g]^\nu_b$ is called the jth ***g-connectivity*** of M_n.

A program for the computation of Fréchet numbers. This program is based on Theorem 3.1. There is given a class $((N_n))$ of compact connected differentiable manifolds of class C^∞, homeomorphic to one such manifold M_n. We outline a sequence of steps to be carried out to obtain the g-connectivities L^g_i when M_n is g-finite.

Step 1. To choose a manifold $M_n \in ((N_n))$ on which the computation of geodesics is relatively simple.

Step 2. To define a ND point pair (A_1, A_2) on M_n, and a homotopically minimizing geodesic g joining A_1 to A_2.

Step 3. To characterize the g-admissible geodesics joining A_1 to A_2 with their indices. When $n = 2$ this is always possible if M_2 has been suitably chosen in Step 1.

Step 4. If M_n is g-finite, to apply the Traction Theory of [9] to suitably chosen vertex spaces $[g]^\nu_\beta$ of $(M_n)^\nu$ to determine the g-connectivities L^g_j of M_n.

According to Theorem 3.1 these steps will yield the Fréchet numbers R_j common to all manifolds in $((N_n))$, provided M_n can be chosen in $((M_n))$ so as to be g-finite. If $n=2$ we show that such a choice of M_2 can always be made. If §§4, 5, 6 we carry out these steps in the case in which N_2 is a surface with Euler characteristic 2.

NOTE. The broken geodesic $\zeta^\nu(z)$ introduced in Condition II on $[g]_\beta^\nu$ will play an important role in later sections and will be termed the ***broken geodesic defined by the ν-tuple z***.

PART II. GEODESICS ON SURFACES

4. **Surfaces with Euler characteristic 2.** Our first objective is to show that each Fréchet number $R_i=1$ for such surfaces. This conclusion will be reached in §6. In this section we begin carrying out the program of §3 when M_2 is a 2-sphere S. Proposition 4.1 will presently be formulated; it will be proved in §5. Granting this proposition we prove in this section that the Fréchet numbers for a 2-sphere are 1 without exception. Steps 1, 2, 3 of our program will now be carried out.

Step 1. It is clear that a 2-sphere is the simplest of the compact surfaces with Euler characteristic 2. Suppose that a 2-sphere S is defined in a space R^3 of rectangular coordinates u, v, w, by the condition

$$u^2+v^2+w^2=1\,. \tag{4.1}$$

Step 2. Let Σ be the great circle of S on which

$$u=\cos t\,,\quad v=\sin t\,,\quad w\equiv 0\,,\qquad (t\in R) \tag{4.2}$$

and give Σ the sense of increasing t. The ND point pair (A_1, A_2) called for in Step 2, will be taken as the point pair,

$$A_1=(1,0,0)\,,\quad A_2=(0,1,0)\,, \tag{4.3}$$

on Σ. A ND homotopically minimizing geodesic g joining A_1 to A_2 is afforded by the arc of Σ of length $\pi/2$ which joins A_1 to A_2 in the (u, v)-plane.

Step 3. On S each geodesic joining A_1 to A_2 is g-admissible. Enumerated in the order of their lengths let these geodesics be denoted by

$$\gamma_0, \gamma_1, \gamma_2, \cdots . \tag{4.4}$$

Of these geodesics $\gamma_0 = g$. Each of these geodesics is carried by the great circle Σ of (4.2). For each integer $i \geq 0$, γ_{2i} traces Σ in the positive sense and passes $2i$ conjugate points of A_1 before ending at A_2. For i odd, γ_i is the geodesic which starts at A_1, traces Σ in the negative sense and passes i conjugate points of A_1 before ending at A_2.

Lengths $b_\nu = L(\gamma_\nu)$. Set $\pi/2 = \delta$. The length, say b_ν of γ_ν, has the value

$$b_\nu = \nu\pi + \delta = (\nu + 1)\pi - \delta . \qquad (\delta = \pi/2) \tag{4.5}$$

The index of γ_ν is ν. There is but one geodesic on S which is g-admissible and has a given index. Hence the sphere S is ***g-finite*** in the sense of Definition 2.5.

Step 4. As has been seen, the Fréchet number $\boldsymbol{R}_0 = 1$ for M_n and in particular for S. To prove that the Fréchet number $\boldsymbol{R}_\nu = 1$ for S when $\nu > 0$ we shall study the ordinary connectivities, over Q, of a finite set of vertex subspaces $[g]^\nu_\beta$ of S^ν. See Definition 3.2 of $[g]^\nu_\beta$.

Let $r \geq 1$ be a prescribed integer. The preferred length m of Lemma 3.1 relative to the sphere S can be taken as any positive number less than π. In particular, it can be taken so that

$$\frac{b_\nu}{\nu + 1} < m < \pi \qquad (\text{for} \quad \nu = 0, 1, \cdots, r) \tag{4.6}$$

as (4.5) shows. When (4.6) holds, values $\beta_\nu \in \boldsymbol{R}$ exist such that for the prescribed r

$$b_\nu < \beta_\nu < m(\nu + 1) . \qquad (\nu = 0, 1, \cdots, r) \tag{4.7}$$

Moreover, $b_\nu < \beta_\nu < b_{\nu+1}$ for $\nu = 0, 1, \cdots, r-1$, since (4.5) and (4.6) imply that

$$\beta_\nu < m(\nu + 1) < m(\nu + 1) + \frac{\pi}{2} < b_{\nu+1} . \tag{4.8}$$

Hence each β_ν of (4.7) is a g-ordinary value on S. The fact that $b_0 = \pi/2$ and that

$$b_0 < \beta_0 < b_1 < \beta_1 < \cdots < b_r < \beta_r \tag{4.9}$$

is noted and the following lemma inferred.

LEMMA 4.1. *For* $\nu = 1, \cdots, r$, $[g]^\nu_{\beta_\nu}$ *is a well-defined vertex subspace of the product* S^ν.

A condition of the form (3.1) is satisfied with β_ν replacing β, since $L(g) < \beta_\nu < m(\nu+1)$ by (4.7). Lemma 4.1 follows.

By making use of the Traction Theorems of [9] the following basic proposition will be proved in §5.

PROPOSITION 4.1. *For the integer* $r \geq 1$ *prescribed in Step* 4 *the nonnull connectivities, over* Q, *of the vertex space* $[g]^r_{\beta_r}$ *of the product* S^r *are*

$$R_0 = R_1 = \cdots = R_r = 1\,. \tag{4.10}$$

Lemma 3.4 and Proposition 4.1 have the following corollary:

COROLLARY 4.1. (i) *The vertex space* $[g]^r_{\beta_r}$ *of Proposition* 4.1 *is* $(r-1)$*-mature in the sense of Definition* 3.4.

(ii) *The Fréchet number* R_{r-1} *of* S *is* 1.

Proof of (i). β_r is $(r-1)$-mature in the sense of Definition 3.4, since a_{r-1}, as defined in (3.4), has the value $b_r < \beta_r$ while the geodesic γ_{r-1} is A_1A_2-homotopic to g under the length level $b_r < \beta_r$.

Proof of (ii). Three facts imply (ii):

(1) By Proposition 4.1, $R_{r-1}[g]^r_{\beta_r} = 1$.

(2) By Lemma 3.4 and Corollary 4.1 (i), $R_{r-1}[g]^r_{\beta_r} = L^g_{r-1}$.

(3) By Theorem 3.1, $R_{r-1} = L^g_{r-1}$.

Proposition 4.1, Corollary 4.1 (ii), and Lemma 2.2 imply a terminal theorem:

THEOREM 4.1. *The Fréchet numbers of any compact surface with Euler characteristic* 2 *equal* 1.

In inferring the truth of Theorem 4.1 from (ii) of Corollary 4.1 we have used the classical theorem that any compact surface with Euler characteristic 2 is homeomorphic to a 2-sphere. See [12, p. 33]. Lemma 2.2 implies that compact homeomorphic surfaces have the same Fréchet numbers. Theorem 4.1 follows.

It remains to prove Proposition 4.1. An inductive proof will be given in §5. To that end the traction theorems of [9] will be applied to mappings f^ν which we now define.

DEFINITION 4.0. *The mappings* f^ν, $\nu = 1, 2, \cdots, r$. According to Lemma 4.1 for $\nu = 1, \cdots, r$ vertex subspaces $[g]^\nu_{\beta_\nu}$ of the product S^ν are well-defined. For each such ν let f^ν be a real-valued function whose domain is $[g]^\nu_{\beta_\nu}$ and whose value at a ν-tuple $z \in [g]^\nu_{\beta_\nu}$ is the length of the broken geodesic $\zeta^\nu(z)$. The mappings

$$z \to f^\nu(z) : [g]^\nu_{\beta_\nu} \to R \qquad (\nu = 1, 2, \cdots, r) \tag{4.11}$$

are continuous and of class C^∞ on the interiors of their domains. Cf. §20 of [5].

Use will be made of sublevel subsets of $[g]^\nu_{\beta_\nu}$ of the form

$$f^\nu_c = \{z \in [g]^\nu_{\beta_\nu} \mid f^\nu(z) \leq c\}. \qquad (c \in R) \tag{4.12}$$

The set f^ν_c is empty if c is less than $L(g)$. Note that $f^\nu_{\beta_\nu}$ is the domain of f^ν. Proposition 4.1 affirms that the non null connectivities of $f^r_{\beta_r}$ are

$$R_0 = R_1 = \cdots = R_r = 1. \tag{4.13}$$

In verifying this affirmation we shall find the following special ν-tuples on S^ν particularly useful.

DEFINITION 4.1. *The normal* ν*-tuple* $\sigma^{(\nu)}$ *of* γ_ν. The g-admissible geodesics on S whose lengths are less than β_r are $\gamma_0, \gamma_1, \cdots, \gamma_r$ of which $\gamma_0 = g$. According to (4.7)

$$L(\gamma_\nu) < \beta_\nu < m(\nu + 1). \qquad (\nu = 1, \cdots, r) \tag{4.14}$$

Hence a subdivision of γ_ν into $\nu + 1$ successive elementary geodesics of equal length is possible. The resultant points of division of γ_ν are the vertices of a ν-tuple, say

$$\sigma^{(\nu)} = (\sigma^{(\nu)}_1, \cdots, \sigma^{(\nu)}_\nu) \in [g]^\nu_{\beta_\nu}, \qquad (\nu = 1, 2, \cdots, r) \tag{4.15}$$

termed the ***normal*** ν-tuple of γ_ν.

The ν vertices of the normal ν-tuples $\sigma^{(\nu)}$ are uniquely determined in position on the great circle Σ : (4.2) of S. If Σ is traced in a positive or negative sense according as ν is even or odd, starting at $A_1 = (1, 0, 0)$, the vertices of the ν-tuple $\sigma^{(\nu)}$ on Σ, in their order in (4.15) are the endpoints of geodesics of the respective lengths

$$(4.16) \qquad \pi - \frac{\delta}{\nu+1}, \ 2\pi - \frac{2\delta}{\nu+1}, \cdots, \ \nu\pi - \frac{\nu\delta}{\nu+1} \qquad \left(\delta = \frac{\pi}{2}\right)$$

in accord with (4.5). Successive vertices of $\sigma^{(\nu)}$ thus lie on Σ in the (u, v)-plane on opposite sides of the v-axis. No one of these vertices is $(1, 0, 0)$ or $(-1, 0, 0)$. No two of these vertices have the same coordinate u.

This description of the vertices of $\sigma^{(\nu)}$ is needed in completing the proof of Proposition 4.1. The role of a normal ν-tuple $\sigma^{(\nu)}$ as a traction induced critical point of f^ν will now be clarified. We shall apply [9].

5. **Traction induced critical points.** This section recalls the definition of special deformations introduced in [9] and termed ***tractions.*** Tractions are extensions of the retracting deformations of Borsuk [11]. They exist in general situations where retracting deformations fail to exist. We give the relevant definitions.

Deformations. Let t be a real variable termed the time. Let $I = [0, 1]$ denote an interval for t. A deformation D on X of a subspace A of a topological space X is a continuous mapping

$$(5.1) \qquad (p, t) \to D(p, t) : A \times I \to X$$

such that $D(p, 0) \equiv p$ for $p \in A$. For fixed p, the partial mapping

$$(5.2) \qquad t \to D(p, t) : I \to X$$

is called the ***trajectory*** of p under D. $D(p, 1)$ is termed the ***final*** image of p under D.

DEFINITION 5.1. *Tractions.* A deformation D of a subspace A of X will be termed a ***traction of*** A ***into a subspace*** B ***of*** A, if D deforms A on A ***into*** B and deforms B ***on*** B.

Each "deformation retracting" A onto B is a traction of A into B but a traction of A into B is not in general a deformation retracting A onto B. However, a traction of A into B shares with a deformation retracting A onto B a fundamental property. There exists an isomorphic mapping of the qth homology group $H_q(A)$ of A onto $H_q(B)$. For proof see §2 of [9].

Each mapping f^ν defined in §4 can be identified with a mapping F of the type analyzed in [9]. The relevant analysis in [9] of these model mappings F will now be recalled.

A mapping F. As in [9] let X be a metric space topologized in the usual way by its metric. X is the domain of a continuous real-valued function,

$$p \to F(p) : X \to R. \tag{5.3}$$

For each $c \in R$ set

$$F_c = \{p \in X \mid F(p) \leq c\}. \tag{5.4}$$

We understand that F_c may be empty.

A traction D of a subset A of the domain of F will be called an ***F-traction*** if

$$F(D(p, 0)) \geq F(D(p, t)) \qquad ((p, t) \in A \times I). \tag{5.5}$$

A special critical point σ of F will now be defined. Let an integer $\mu \geq 1$ be prescribed and fixed. We refer to μ-tuples

$$x = (x_1, x_2, \cdots, x_\mu) \in R^\mu \tag{5.6}$$

and introduce an origin-centered open μ-ball D_e of radius e in R^μ.

DEFINITION 5.2. *A traction induced critical point σ of F.* A point $\sigma \in X$ will be called a ***traction*** induced critical point of index k if the following "local" and "global" conditions are satisfied.

Condition I, *local. For some constant $b > F(\sigma)$ and some sufficiently small $e > 0$ there exists an injective homeomorphism*

$$x \to \varphi_\sigma(x) : D_e \to F_b \tag{5.7}$$

onto a topological μ-ball $\Lambda_\sigma = \varphi_\sigma(D_e)$ containing the point σ as $\varphi_\sigma(0)$ and such that

$$F(\varphi_\sigma(x)) - F(\sigma) \equiv -x_1^2 - \cdots - x_k^2 + x_{k+1}^2 + \cdots + x_\mu^2 \qquad (x \in D_e). \tag{5.8}$$

Condition II, *global. For a sufficiently small radius* e *of* D_e *and for some value* $c \in R$ *such that*

$$(5.9) \qquad F(\sigma) > c > F(\sigma) - e^2$$

there exists an F-traction of F_b *into* $\Lambda_\sigma \cup F_c$.

When the index $k > 0$ the following two definitions are added.

DEFINITION 5.3. *A critical* $(k-1)$*-cycle* w_c^{k-1} *of* σ *at the F-level* c. Set $F(\sigma) - c = \rho^2$ and introduce a $(k-1)$-sphere

$$(5.10) \qquad s_{k-1} = \{x \in D_e \mid x_1^2 + \cdots + x_k^2 = \rho^2 : x_{k+1} = \cdots = x_\mu = 0\}$$

of radius ρ and denote $\varphi_\sigma(s_{k-1})$ (2) by π_{k-1}. The F-level of π_{k-1} is c. Let w_c^{k-1} be any singular $(k-1)$-cycle on π_{k-1} whose homology class on π_{k-1} is a base for the homology group $H_{k-1}(\pi_{k-1})$ over Q.

A comparison of (5.8) and (5.10) indicates the truth of the following special lemma.

LEMMA 5.1. *In case the index* k *of* σ *equals* μ *in* (5.8), *then if* $F(\sigma)-c$ *is sufficiently small and positive, the subset of* $\varphi_\sigma(D_e)$ *on which* $F \equiv c$ *is a topological* $(k-1)$*-sphere which is the carrier of a critical* $(k-1)$*-cycle* w_c^{k-1} *of* σ.

When $k = 1$, this topological $(k-1)$-sphere at the F-level c is understood to be a pair of points $p \neq q$.

DEFINITION 5.4. *A linking cycle* w_c^{k-1}. A traction induced critical point σ of F and an associated critical $(k-1)$-cycle w_c^{k-1} at the F-level c are said to be of *linking* type if $w_c^{k-1} \sim 0$ on F_c.

From Theorem 1.1 of [9] the following can be inferred.

TRACTION THEOREM 5.1. *Let* σ *be a critical point of* F *of index* $k > 0$, *induced by an F-traction of* F_b *into* $\Lambda_\sigma \cup F_c$, *as in Definition* 5.2. *If* σ *is of linking type then the connectivities of* F_b, *over* Q, *equal those of* F_c *except that*

$$(5.11) \qquad R_k F_b = R_k F_c + 1.$$

Under the hypotheses of Theorem 5.1, σ will be said to be of ***increasing*** type. The term increasing is associated with the increase of R_k in (5.11).

(2) $\varphi_\sigma(s_{k-1})$ is a topological $(k-1)$-sphere.

Theorem 5.1 is the principal aid in proving the following.

LEMMA 5.2. *For* ν *on the range* $1, 2, \cdots, r$, *the normal* ν*-tuple* $\sigma^{(\nu)}$ *of the geodesic* γ_ν *of* §4 *is a traction induced critical point of increasing type of the mapping*

$$z \to f^\nu(z) : [g]^\nu_{\beta_\nu} \to R \qquad (\text{of } (4.11)) . \tag{5.12}$$

Lemma 5.2 implies Proposition 4.1. This will now be shown, leaving the proof of Lemma 5.2 for §6.

The proof of Proposition 4.1 is inductive. We begin by proving Proposition 4.1 when $r=1$. That is, we verify the following affirmation:

(A_1) *The nonnull connectivities of the vertex space* $[g]^1_{\beta_1}$ *of* S *are* $R_0 = R_1 = 1$.

Proof of A_1. In the notation of (4.9), $\beta_0 < b_1 < \beta_1$, where $b_1 = L(\gamma_1)$. Equivalently

$$\beta_0 < f^1(\sigma^{(1)}) < \beta_1 . \tag{5.13}$$

β_0 is an ordinary value of L between b_0 and b_1. It can be taken so that $b_1 - \beta_0$ is arbitrarily small. Granting the truth of Lemma 5.2 one can identify F, F_b, σ, F_c of Theorem 5.1, respectively, with

$$f^1, \ f^1_{\beta_1}, \ \sigma^{(1)}, \ f^1_{\beta_0} \tag{5.14}$$

and infer that the connectivities of $f^1_{\beta_1}$ are those of $f^1_{\beta_0}$ except that

$$R_1 f^1_{\beta_1} = R_1 f^1_{\beta_0} + 1 . \tag{5.15}$$

According to the final corollary of Appendix IV of [5] the connectivities of $f^1_{\beta_0}$ all vanish except that $R_0 f^1_{\beta_0} = 1$. Statement (A_1) follows.

Completion of proof of Proposition 4.1. Let Proposition 4.1m denote Proposition 4.1 when r is replaced by an integer m on the range $1, 2, \cdots, r-1$. Proceeding inductively we shall assume that Proposition 4.1m is true and prove that Proposition 4.1 is true when $r = m+1$.

In the notation of §4

$$\beta_m < f^{m+1}(\sigma^{(m+1)}) < \beta_{m+1} \qquad (b_{m+1} = f^{m+1}(\sigma^{(m+1)})) . \tag{5.16}$$

Moreover β_m can be taken so that $b_{m+1}-\beta_m$ is arbitrarily small and positive. Granting the truth of Lemma 5.2 one can identify F, b, σ, c of Theorem 5.1, respectively, with

$$f^{m+1},\ \beta_{m+1},\ \sigma^{(m+1)},\ \beta_m. \tag{5.17}$$

We infer then from Theorem 5.1 that the connectivities over Q, of the vertex space $[g]^{m+1}_{\beta_{m+1}}$ are those of the vertex space $[g]^{m+1}_{\beta_m}$ except that

$$R_{m+1}[g]^{m+1}_{\beta_{m+1}} = R_{m+1}[g]^{m+1}_{\beta_m} + 1. \tag{5.18}$$

Now by virtue of Lemma 3.3

$$R_{m+1}[g]^{m+1}_{\beta_m} = R_{m+1}[g]^{m}_{\beta_m} = 0, \tag{5.19}$$

where the final equality is an inductive hypothesis. From (5.18) and (5.19) we infer that $R_{m+1}[g]^{m+1}_{\beta_{m+1}} = 1$.

This completes the proof of Proposition 4.1 under the assumption that Lemma 5.2 is true.

A summary. It remains to prove Lemma 5.2. This will be done in §6. One should recall that the main objective of §§4, 5, 6 is to prove Theorem 4.1. If $\Rightarrow$ is read "implies," the logical relations which lead to a proof of Theorem 4.1 are as follows:

Proposition 4.1 $\Rightarrow$ Theorem 4.1 (in §4).
Lemma 5.2 $\Rightarrow$ Proposition 4.1 (in §5).

6. **Proof of Lemma** 5.2. A lemma will now be formulated that will lead to a proof of Lemma 5.2. Recall that $b_\nu = L(r_\nu)$.

LEMMA 6.1. *For* $\nu = 1, \cdots, r$, *there exists a product*

$$\Pi_\nu = C_1 \times C_2 \times \cdots \times C_\nu \subset f^\nu_{b_\nu} \tag{6.1}$$

of ν *disjoint circles on* S *whose planes are orthogonal to the* u*-axis and such that the following is true*:

(a₁) *For* ν*-tuples* $z \in \Pi_\nu$, $f^\nu(z) \leq b_\nu$,
(a₂) $f^\nu(z) = b_\nu$ *if and only if* $z = \sigma^{(\nu)}$,
(a₃) $z = \sigma^{(\nu)}$ *is a nondegenerate critical point of* $f^\nu \mid \Pi_\nu$.

The circles $C_1, \cdots, C_\nu$ *defined*. Let the u-coordinates of the successive vertices on the great circle Σ of $\sigma^{(\nu)}$ be respectively denoted by $u_1, u_2, \cdots, u_\nu$. For i on the range $1, \cdots, \nu$ let C_i be the circle on S on which $u = u_i$. As noted in §4 successive vertices of $\sigma^{(\nu)}$ on Σ are on opposite sides of the v-axis in the (u, v)-plane. No one of these vertices is $(1, 0, 0)$ or $(-1, 0, 0)$. Two circles C_i have different coordinates u_i and so are disjoint.

Proof of Lemma 6.1 (a_1). Let $d(p, q)$ denote the minimum distance on S between points p and q on S. If $(z_1, \cdots, z_\nu) \in \Pi_\nu$ we affirm that the distances on S between successive points of S in the sequence

$$A_1,\ z_1, \cdots, z_\nu,\ A_2 \tag{6.2}$$

are at most

$$\frac{b_\nu}{\nu} + 1 = \omega_\nu\,, \qquad (b_\nu \text{ from } (4.5)) \tag{6.3}$$

introducing ω_ν. To verify this, note first that

$$d(A_1, z_1) = \omega_\nu\,, \qquad (\text{for } z_1 \in C_1) \tag{6.4}$$

since A_1 is on the u-axis. Moreover, for j on the range $1, \cdots, \nu - 1$,

$$\underset{z_j \in C_j;\ z_{j+1} \in C_{j+1}}{\text{maximum}}\ d(z_j, z_{j+1}) = \omega_\nu\,, \tag{6.5}$$

as one readily verifies. Finally

$$\underset{z_\nu \in C_\nu}{\text{maximum}}\, d(z_\nu, A_2) = \omega_\nu\,. \tag{6.6}$$

According to (4.6), $\omega_\nu < m$, so that successive points in the sequence (6.2) can be joined by elementary geodesics. A ν-tuple $z \in \Pi_\nu$ is thus in the domain of f^ν and by virtue of (6.4), (6.5) and (6.6),

$$f^\nu(z) \le \omega_\nu(\nu + 1) = b_\nu \tag{6.7}$$

establishing (a_1).

The inclusion (6.1) follows.

Proof of Lemma 6.1 (a_2). For $p \in C_j$ and $q \in C_{j+1}$, $d(p, q) = \omega_\nu$ if and only if the elementary geodesic which joins p to q is orthogonal to C_j and C_{j+1} and has the length ω_ν. Moreover, for $z_\nu \in C_\nu$ $d(z_\nu, A_2) = \omega_\nu$ if and only if the elementary geodesic which joins z_ν to A_2 is orthogonal to C_ν and has the length ω_ν. Five conditions necessary that the equality hold in (a_1) for a ν-tuple $z \in \Pi_\nu$ can now be formulated.

Condition (1). The $\nu + 1$ elementary geodesics joining the successive points in (6.2) must each have the length ω_ν, since none can have a length exceeding ω_ν and since $b_\nu = (\nu + 1)\,\omega_\nu$.

Condition (2). The vertex $z_\nu = \sigma_\nu^{(\nu)}$, while the elementary geodesic joining z_ν to A_2 must have the length ω_ν and be orthogonal to C_ν.

Condition (3). The elementary geodesics of $\zeta^\nu(z)$ are arcs of a great circle, say Σ_1, that meets A_1 and is orthogonal to each of the circles C_j.

Condition (4). Since the great circle Σ_1 meets A_1 and the vertex $z_\nu = \sigma_\nu^{(\nu)}$, Σ_1 must be the great circle Σ.

Condition (5). Since no two of the circles C_j are identical, $\zeta^\nu(z)$ must trace Σ in ***one*** sense if the equality is to hold in (a_2).

We infer the truth of (a_2) of Lemma 6.1.

Proof of Lemma 6.1 (a_3). To prove (a_3) use will be made of a manifold frame $((M))_\gamma^\nu$, as defined in §25 of [5]. We now construct a special manifold frame $((M))_{\gamma_\nu}^\nu$ "based" on the geodesic γ_ν.

The manifold frame $((M))_{\gamma_\nu}^\nu$. Let ϵ be an arbitrarily small positive constant. Let $M^{(i)}$ be an open arc of C_i of length 2ϵ with the vertex $\sigma_i^{(\nu)}$ on γ_i as the mid-point of $M^{(i)}$. If ϵ is sufficiently small the product

$$M^{(1)} \times M^{(2)} \times \cdots \times M^{(\nu)} \tag{6.8}$$

is a manifold frame $((M))_{\gamma_\nu}^\nu$ defined by ν 1-dimensional manifolds $M^{(i)}$ cutting across γ_ν at the respective vertices

$$\sigma_1^{(\nu)}, \cdots, \sigma_\nu^{(\nu)}$$

of $\sigma^{(\nu)}$. The dimension of $((M))_{\gamma_\nu}^\nu$ is ν.

Corollary 25.1 of [5] will now be used to prove (a_3). In any application of [5] in this paper the integral J of [5] is taken as the integral of length on S. If z is a ν-tuple of $((M))_{\gamma_\nu}^\nu$, $\mathcal{J}^\nu(z)$, as defined in §21 of [5], accordingly means the length of the broken geodesic $\zeta^\nu(z)$. If z is a ν-tuple in S^ν in a sufficiently small neighborhood of $\sigma^{(\nu)}$, $\mathcal{J}^\nu(z) = f^\nu(z)$, in accord with Definition 4.0 of f^ν. Since $((M))^\nu_{\gamma_\nu}$ is a neighborhood of $\sigma^{(\nu)}$, relative to Π_ν, Lemma 6.1 (a_3) is equivalent to the following statement.

(B) *The* ν*-tuple* $z = \sigma^{(\nu)}$ *is a nondegenerate critical point of* $f^\nu | ((M))^\nu_{\gamma_\nu}$.

In our application of Corollary 25.1 of [5] to prove (B), the dimension n of S is 2. Hence $m = n - 1 = 1$ and μ, defined as $m\nu$, reduces to ν. The coordinates of a ν-tuple $x \in R^\nu$ are denoted by $x_1, \cdots, x_\nu$. Recall that the index of the geodesic γ_ν is ν. With this understood, Corollary 25.1 of [5] implies the following.

(B_1) *If* $D_\nu \subset R^\nu$ *is an origin-centered closed* ν*-ball of sufficiently small radius* e_ν*, there exists an injective diffeomorphism*

$$x \to \Phi_\nu(x) : D_\nu \to ((M))^\nu_{\gamma_\nu} \tag{6.9}$$

of class C^∞ *such that* $\Phi_\nu(0) = \sigma^{(\nu)}$ *and*

$$f^\nu(\Phi_\nu(x)) - b_\nu \equiv -x_1^2 - \cdots - x_\nu^2 \qquad (x \in D_\nu)\,. \tag{6.10}$$

The representation (6.10) of f^ν near $\sigma^{(\nu)}$ implies (B) and Lemma 6.1 (a_3). This completes the proof of Lemma 6.1.

The following lemma is needed in proving Lemma 5.2.

LEMMA 6.2. *For* ν *on the range* $1, \cdots, r$ *the normal* ν*-tuple* $\sigma^{(\nu)}$ *of* γ_ν *is a traction induced critical point of* f^ν *of index* ν.

The relation (6.10) shows that $\sigma^{(\nu)}$ satisfies the Local Condition of Definition 5.2 that $\sigma^{(\nu)}$ be a traction induced critical point of f^ν of index ν. That $\sigma^{(\nu)}$ satisfies the corresponding Global Condition of Definition 5.2 follows from Traction Theorem Ω_ν of Appendix IV of [5].

The following lemma is also needed in proving Lemma 5.2.

LEMMA 6.3. *The* ν*-tuple* $\sigma^{(\nu)}$ *of Lemma* 6.2 *is of linking type in the sense of Definition* 5.4.

It follows from (6.1) that $\Pi_\nu \subset f^\nu_{b_\nu}$. By virtue of (6.10) the index of $\sigma^{(\nu)}$ is ν. It follows from Lemma 5.1 that if $b_\nu - c$ is positive and sufficiently small the section of Π_ν at the f^ν-level c is the carrier of a critical $(\nu-1)$-cycle $w_e^{\nu-1}$ of $\sigma^{(\nu)}$ and hence, by Lemma 6.1, that $w_e^{\nu-1} \sim 0$ on f_e^ν.

Thus Lemma 6.3 is true.

Proof of Lemma 5.2. By virtue of Lemmas 6.2 and 6.3 the hypotheses of Theorem 5.1 are satisfied if σ, F, k of Theorem 5.1 are replaced by $\sigma^{(\nu)}$, f^ν, ν of Lemma 5.2. Theorem 5.1 then implies the truth of Lemma 5.2.

Thus Theorem 4.1 is true as indicated at end of §5.

7. **The projective plane.** We shall show that the Fréchet numbers of the projective plane are 1 without exception. Any surface homeomorphic to the projective plane will have the same Fréchet numbers in accord with Lemma 2.2. Such surfaces are the compact surfaces with Euler characteristic 1. See [12, p. 33] for Euler characteristics.

As is well known, a topological model of the projective plane is given by a sphere whose antipodal points have been identified. Explicitly, let S be the 2-sphere introduced in §4. Let p be an arbitrary point on S and $\hat{p}$ its antipode on S. Let P be the set of unordered pairs

$$\xi(p) = (p, \hat{p}) = (\hat{p}, p) = \xi(\hat{p}) \tag{7.0}$$

as p and its antipode $\hat{p}$ range over S. Then

$$p \to \xi(p) : S \to P \tag{7.1}$$

is a mapping of S onto the set P, a priori untopologized. A subset $U \subset P$ will be termed *open* if $\xi^{-1}(U)$ is open on S. See [12, p. 243]. If P is so topologized the mapping ξ is continuous and termed a *projection* of S onto P.

P is given a differentiable and Riemannian structure by requiring that the projection ξ of S onto P be regarded locally as an isometric diffeomorphism of S onto P. So topologized and structured P will be called a model *projective plane*.

Geodesics on P. Geodesics on S and P are restricted to those of finite length with compact carriers. If r is a geodesic on S, its projection on P, say r', is a geodesic on P of the same length.

Conversely, let λ be a geodesic on P with initial point an identified pair $(p, \hat{p}) = (\hat{p}, p)$. There then exists a unique geodesic on S with initial point p on S and projection λ on P, as well as a unique geodesic on S with initial point $\hat{p}$ on S and projection λ on P.

The point pair A_1' and A_2' on P. The points $A_1 = (1, 0, 0)$ and $A_2 = (0, 1, 0)$ have projections, say A_1' and A_2', on P. The geodesic g joins A_1 to A_2 on S and has the length $\pi/2$. Set $A_3 = (0, -1, 0)$ on S. The points A_2 and A_3 are antipodal so that their projections A_2' and A_3' on P coincide. On S the points A_1 and A_2 have been preferred and fixed. On P let A_1' and A_2' be similarly preferred and fixed.

The geodesics g' and k' on P. Let k be the geodesic on S which joins A_1 to A_3 and has the length $\pi/2$. Note that the geodesic k is the reflection in the (u, w)-plane of the geodesic g. Let k' be the projection of k on P. We state a lemma whose simple proof is left to the reader.

LEMMA 7.1. (i) *The geodesics g' and k' join A_1' and A_2' on P but are not $A_1'A_2'$-homotopic on P.*

(ii) *A curve which joins A_1' to A_2' on P is $A_1'A_2'$-homotopic on P to g' or else to k'.*

The following lemma will enable us to proceed with our task of finding the Fréchet numbers of the projective plane P.

LEMMA 7.2. *The geodesics on P joining A_1' to A_2' which are $A_1'A_2'$-homotopic on P to the geodesic g' are the projections*

$$r_0', r_1', r_2', \cdots \tag{7.2}$$

on P of the geodesics $r_0, r_1, r_2, \cdots$ which join A_1 to A_2 on S. In particular $r_0' = g'$.

The proof of Lemma 7.2 will make use of the following statement.

(α) *If r is a curve on S joining A_1 to A_2, r's projection on P is $A_1'A_2'$-homotopic on P to g'.*

Proof of (α). It is known that r is A_1A_2-homotopic on S to g. The projection on P of the curves on S joining A_1 to A_2 which define this homotopy define an $A_1' A_2'$-homotopy of r' to g'. Thus (α) is true.

Proof of Lemma 7.2. It follows from (α) that the geodesics (7.2) are ***among*** the geodesics on P which satisfy the conditions of Lemma 7.2. It remains to prove that there are no other geodesics on P which satisfy the conditions of Lemma 7.2. To that end let

$$\eta_0, \eta_1, \eta_2, \cdots \tag{7.3}$$

be the geodesics on S which are the reflections in the (u, w)-plane of the respective geodesics $r_0, r_1, r_2, \cdots$. Let

$$\eta_0', \eta_1', \eta_2', \cdots \tag{7.4}$$

be the projections on P of the respective geodesics (7.3).

Statement (α) has an obvious extension.

(β) *If r is a curve on S joining A_1 to the antipode A_3 of A_2, then r's projection on P is $A_1' A_2'$-homotopic to k'.*

It follows that the geodesics (7.4) are $A_1' A_2'$-homotopic to k' and hence, by Lemma 7.1, not to g'. Moreover, the geodesics listed in (7.2) and (7.4) comprise ***all*** the geodesics which join A_1' to A_2'. However, only the geodesics listed in (7.2) are $A_1' A_2'$-homotopic to g'.

Thus Lemma 7.2 is true.

A lemma on conjugate points on P is needed.

LEMMA 7.3. *Let r be a geodesic on S with initial point A_1 and let r' be the projection of r on P. The conjugate points of A_1 on r have the same distances from A_1 measured along r as the distances from A_1' measured along r' of the conjugate points of A_1' on r'. Moreover, multiplicities of corresponding conjugate points on r and r' are identical.*

If conjugate points are defined as in §§9 and 10 of [5], Lemma 7.3 is a trivial consequence of the fact that the mapping ξ of S onto P is locally an isometric diffeomorphism.

We state a corollary of Lemmas 7.2 and 7.3.

COROLLARY 7.1. *The geodesic r_ν' on P in the list* (7.2) *is nondegenerate and has the index ν.*

We note that the geodesic $r'_0 = g'$ of Lemma 7.2 is homotopically minimizing on P, since it is the shortest of the geodesics in the list (7.2). The geodesic g' is accordingly a ***prime geodesic*** on P in the sense of §2. Another choice of a prime geodesic on P is k'. Note that in the list (7.4) $\eta'_0 = k'$. The geodesics (7.4) are the geodesics on P which join A'_1 to A'_2 and are $A'_1 A'_2$-homotopic to k'.

It remains to prove our principal theorem on the projective plane.

THEOREM 7.1. *The Fréchet numbers of the projective plane are* 1.

Let $\mathcal{F}_{A_1}^{A_2}$ be the Fréchet metric space associated with the sphere S and the points A_1 and A_2 on S. Let $\mathcal{F}'^{A'_2}_{A'_1}$ be the Fréchet space similarly associated with P and the projections A'_1 and A'_2 on P of A_1 and A_2. Let $\mathcal{F}_g$ and $\mathcal{F}'_{g'}$ be the pathwise components of $\mathcal{F}_{A_1}^{A_2}$ and $\mathcal{F}'^{A'_2}_{A'_1}$ that contain g and g', respectively. The following lemma implies Theorem 7.1.

LEMMA 7.4. *$\mathcal{F}_g$ is homeomorphic to $\mathcal{F}'_{g'}$.*

To establish Lemma 7.4 we shall define a natural homeomorphism T of $\mathcal{F}_g$ onto $\mathcal{F}'_{g'}$. To that end let h be a curve on S joining A_1 to A_2, $A_1 A_2$-homotopic to g and let h' be the projection of h on P. h' is $A'_1 A'_2$-homotopic to g' on P. As in [6] let $\boldsymbol{h}$ and $\boldsymbol{h}'$ be the Fréchet curve classes respectively of h and h'. It is clear that $\boldsymbol{h}'$ has a value, say $T(\boldsymbol{h})$, determined by $\boldsymbol{h}$. It remains to show that the mapping

$$\boldsymbol{h} \to T(\boldsymbol{h}) : \mathcal{F}_g \to \mathcal{F}'_{g'} \tag{7.5}$$

is bijective and bicontinuous.

Proof that T is bijective. Recall that the mapping ξ of S onto P is locally an isometric diffeomorphism. One sees then that T is surjective.

Let k be an arbitrary curve joining A_1 to A_2 and k' the projection of k on P. To show that T is bijective suppose that $\boldsymbol{h}' = \boldsymbol{k}'$ and hence $\boldsymbol{h}'\boldsymbol{k}' = 0$. Because h and k have a common initial point A_1 we infer that $hk = 0$ and hence $\boldsymbol{hk} = 0$. It follows from Corollary 2.1 of [6] that $\boldsymbol{h} = \boldsymbol{k}$, so that T is bijective.

Proof that T is bicontinuous. Let $\boldsymbol{h}$ and $\boldsymbol{k}$ be given in $\mathcal{G}_g$ and $\boldsymbol{h}' = T(\boldsymbol{h})$ in $\mathcal{G}'_{g'}$. Corresponding to a fixed $\boldsymbol{h}$ a sufficiently small positive constant e will have the following property. For $\boldsymbol{k} \in \mathcal{G}_g$, $\boldsymbol{hk} < e$ if and only if $\boldsymbol{h}'\boldsymbol{k}' < e$, as the reader is asked to prove. It will follow that T is continuous at $\boldsymbol{h}$ and T^{-1} continuous at $\boldsymbol{h}'$.

T is accordingly a homeomorphism so that Lemma 7.4 is true.

Any compact surface with Euler characteristic 1 is homeomorphic to the projective plane P. It follows from this fact and Theorem 7.1 that the following is true.

THEOREM 7.2. *The Fréchet numbers $\boldsymbol{R}_i$ are* 1 *for each compact surface with Euler characteristic* 1.

8. **Surfaces with nonpositive Euler characteristic.** Our theorem in this case follows.

THEOREM 8.1. *The Fréchet numbers of compact surfaces M with Euler characteristics $\chi(M) \leq 0$ are zero except that $\boldsymbol{R}_0 = 1$.*

This theorem is a corollary of a theorem and lemma in Part I and two other lemmas which we now present.

Reference [13] is to one of a series of papers on "Curvature functions on 2-manifolds." On page 387 of [13] a "Table" of known results implies the following.

LEMMA 8.1. *In the topological equivalence class of an arbitrary compact surface there exists an abstract surface M of class C^∞ with a Riemannian structure of constant Gaussian curvature K.*

From the Gauss-Bonnet formula it follows that the constant K is positive, negative, or zero according as $\chi(M)$ is positive, negative or zero.

The proof of Lemma 8.1 by the differential geometers listed in [13] is by methods relevant to conformal mapping and general curvature functions on 2-manifolds. See also the survey by Gluck in [14]. It is the intention of the authors to supplement these references by a more elementary proof of Lemma 8.1.

The second lemma which aids in proving Theorem 8.1 concerns a result on conjugate points known to Jacobi and Bonnet. It can be stated as follows.

LEMMA 8.2. *On a differentiable surface N_2 of class C^∞ on which the Gaussian curvature is nonpositive there are no conjugate points of the initial point of a geodesic.*

Let r be a geodesic on the surface and s the arc length measured along r from r's initial point p. Let $K(s)$ be the Gaussian curvature of N_2 at the point s on r. One finds that the conjugate points on r of the initial point of r are the zeros $s>0$ of a solution $s \to r(s)$ of the differential equation

$$\frac{d^2 y}{ds^2} + K(s)\, y = 0 \tag{8.1}$$

which vanishes when $s=0$ but is not identically zero. See [15, pp. 228–231] for a derivation of the Jacobi differential equation (8.1).

Granting the validity of this representation of the conjugate points on r, Lemma 8.2 follows at once.

Completion of proof of Theorem 8.1. Let M_2 be the differentiable surface for which $\chi(M_2) \leq 0$. Let N_2 be a differentiable surface homeomorphic to M_2 with a constant nonpositive Gaussian curvature. If a 'prime geodesic' g has been chosen on N_2, Theorem 2.1 can be applied to N_2 and g. Since the type numbers $m_i^g = 0$ when $i>0$ in this application, the relations

$$m_i^g \geq \boldsymbol{R}_i \qquad (i = 0, 1, \cdots) \tag{8.2}$$

of Theorem 2.1 imply that $\boldsymbol{R}_i = 0$ when $i > 0$. By Lemma 2.2, M_2 has the same Fréchet numbers as N_2.

Theorem 8.1 follows.

REFERENCES

1. M. Fréchet, *Sur quelques points du calcul fonctionnel,* Rend. Circ. Mat. Palermo **22** (1906), 1-74.

2. M. Morse and S. S. Cairns, *Critical Point Theory in Global Analysis and Differential Topology,* Academic Press, New York, 1969.

3. M. Morse, *Connectivities $\boldsymbol{R}_i$ of Fréchet spaces in variational topology,* Proc. Nat. Acad. Sci. U. S. A. **72** (1975), 2069-2070.

4. ———, *The Calculus of Variations in the Large,* Colloquium Publication 18, 4th printing, Amer. Math. Soc., Providence, R. I., 1965.

5. ———, *Global Variational Analysis: Weierstrass Integrals on a Riemannian Manifold,* Mathematical Notes, Princeton University Press, Princeton, N. J., 1976.

6. ———, *Fréchet curve classes,* J. Math. Pures Appl. **53** (1974), 291-298.

7. ———, *Nondegenerate point pairs in global variational analysis*, J. Differential Geometry, to appear.

8. ———, *Frêchet Numbers in Global Variational Analysis*, Houston J. Math., to appear.

9. D. Landis and M. Morse, *Tractions in critical point theory*, Rocky Mountain J Math. **5** (1975), 379-399.

10. ———, *Geodesic joins and Frêchet curve classes*, Rend. di Matematica **8** (1975), 161-185.

11. K. Borsuk, *Sur les rêtractes*, Fund. Math. **17** (1931), 152-170.

12. W. S. Massey, *Algebraic Topology: An Introduction*, Harcourt Brace & World, New York, 1967.

13. J. L. Kazdan and F. W. Warner, *Curvature functions for compact 2-manifolds*, Ann. of Math. **99** (1974), 14-47.

14. H. Gluck, *Manifolds with preassigned curvature. A survey*, Bull. Amer. Math. Soc. **81** (1975), 313-329.

15. O. Bolza, *Vorlesungen über Variationsrechnung*, B. G. Teubner, Leipzig and Berlin, 1909.

INSTITUTE FOR ADVANCED STUDY, SCHOOL OF MATHEMATICS, PRINCETON, NEW JERSEY 08540, U. S. A.

BIBLIOGRAPHY OF THE PUBLICATIONS OF MARSTON MORSE

PAPERS

1. *Proof of a general theorem on the linear dependence of P analytic functions of a single variable*, Bull. Amer. Math. Soc. **23** (1916), 114–117.

2. *A one-to-one representation of geodesics on a surface of negative curvature*, Amer. J. Math. **43** (1921), 33–51.

3. *Recurrent geodesics on a surface of negative curvature*, Trans. Amer. Math. Soc. **22** (1921), 84–110.

4. *A fundamental class of geodesics on any closed surface of genus p greater than one*, Trans. Amer. Math. Soc. **26** (1924), 25–61.

5. *Relations between the critical points of a real function of n independent variables*, Trans. Amer. Math. Soc. **27** (1925), 345–396.

6. *The analysis and analysis situs of regular n-spreads in (n + s)-space*, Proc. Nat. Acad. Sci. **13** (1927), 813–817.

7. *The foundations of a theory of the calculus of variations in the large*, Trans. Amer. Math. Soc. **30** (1928), 213–274.

8. *Singular points of vector fields under general boundary conditions*, Amer. J. Math. **51** (1929), 165–178.

9. *The critical points of functions and the calculus of variations in the large*, Bull. Amer. Math. Soc. **35** (1929), 38–54.

10. *The foundations of the calculus of variations in the large in m-space* (1st paper), Trans. Amer. Math. Soc. **31** (1929), 379–404.

11. *Closed extremals*, Proc. Nat. Acad. Sci. **15** (1929), 856–959.

12. *The problems of Lagrange and Mayer under general end conditions*, Proc. Nat. Acad. Sci. **16** (1930), 229–233.

13. *A generalization of the Sturm-separation and comparison theorems in n-space*, Math. Ann. **103** (1930), 52–69.

14. *The critical points of a function of n variables*, Proc. Nat. Acad. Sci. **16** (1930), 777–779.

15. *The critical points of a function of n variables*, Trans. Amer. Math. Soc. **33** (1931), 72–91.

16. *The order of vanishing of the determinant of a conjugate base*, Proc. Nat. Acad. Sci. **17** (1931), 319–320.

17. *The problems of Lagrange and Mayer with variable endpoints*, by M. Morse and Sumner Byron Myers, Proc. Amer. Acad. Arts and Sci. **66** (1931), 235–253.

18. *Closed extremals* (1st paper), Ann. of Math. (2) **32** (1931), 549–566.

19. *Sufficient conditions in the problem of Lagrange with fixed endpoints*, Ann. of Math. (2) **32** (1931), 567–577.

20. *Sufficient conditions in the problem of Lagrange with variable end conditions*, Amer. J. Math. **53** (1931), 517–546.

21. *The foundations of a theory of the calculus of variations in the large in m-space* (2nd paper), Trans. Amer. Math. Soc. **32** (1930), 599–631.

22. *A characterization of fields in the calculus of variations*, by M. Morse and S. B. Littauer, Proc. Nat. Acad. Sci. **18** (1932), 724–730.

23. *The calculus of variations in the large*, Verhandlungen des Internationalen Mathematiker-Kongresses Zürich, 1932, Vol. 1, pp. 173–188.

24. *Does instability imply transitivity?*, Proc. Nat. Acad. Sci. **20** (1934), 46–50.

25. *On certain invariants of closed extremals*, by M. Morse and Everett Pitcher, Proc. Nat. Acad. Sci. **20** (1934), 282–287.

26. *The critical point theory under general boundary conditions*, by M. Morse and George Booth Van Schaack, Ann. of Math. (2) **35** (1934), 545–571.

27. *Sufficient conditions in the problem of Lagrange without assumptions of normalcy*, Trans. Amer. Math. Soc. **37** (1935), 147–160.

28. *Instability and transitivity*, Jour. de Mathématiques, Paris **14** (1935), 49–71.

29. *Abstract critical sets*, by M. Morse and George B. Van Schaack, Proc. Nat. Acad. Sci. **21** (1935), 258–263.

30. *Generalized concavity theorems*, Proc. Nat. Acad. Sci. **21** (1935), 359–362.

31. *Three theorems on the envelope of extremals*, Proc. Nat. Acad. Sci. **21** (1935), 619–621; Bull. Amer. Math. Soc. **42** (1936), 136–144.

32. *Functional topology and abstract variational theory*, Proc. Nat. Acad. Sci. **22** (1936), 313–319.

33. *Critical point theory under general boundary conditions*, by M. Morse and George B. Van Schaack, Duke Math. J. **2** (1936), 220–242.

34. *Singular quadratic functionals*, by M. Morse and Walter Leighton, Trans. Amer. Math. Soc. **40** (1936), 252–286.

35. *A special parametrization of curves*, Bull. Amer. Math. Soc. **42** (1936), 915–922.

36. *Functional topology and abstract variational theory*, Ann. of Math. (2) **38** (1937), 386–449.

37. *The index theorem in the calculus of variations*, Duke Math. J. **4** (1938), 231–246.

38. *Functional topology and abstract variational theory*, Proc. Nat. Acad. Sci. **24** (1938), 326–330.

39. *Symbolic dynamics*, by M. Morse and G. A. Hedlund, Amer. J. Math. **60** (1938), 815–866.

40. *On the existence of minimal surfaces of general critical types*, by M. Morse and C. B. Tompkins, Proc. Nat. Acad. Sci. **25** (1939), 153–158; Ann. of Math. (2) **40** (1939),443–472.

41. *Sur le calcul des variations*, Ann. Inst. H. Poincaré **9** (1939), 1–11.

42. *La dynamique symbolique*, Bull. Soc. Math. France **67** (1939), 1–7.

43. *Symbolic dynamics*. II. *Sturmian sequences*, by M. Morse and G. A. Hedlund, Amer. J. Math. **61** (1940), 1–42.

44. *Rank and span in functional topology*, Ann. of Math. (2) **41** (1940), 419–454.

45. *The first variation in minimal surface theory*, Duke Math. J. **6** (1940), 263–289.

46. *Twentieth century mathematics*, Amer. Scholar **9** (1940), 499–504.

47. *Unstable minimal surfaces of higher topological types*, by M. Morse and C. B. Tompkins, Proc. Nat. Acad. Sci. **26** (1940), 713–716.

48. *A mathematical theory of equilibrium with applications to minimal surface theory*, Science **93** (1941), 69–71.

49. *Minimal surfaces not of minimum type by a new mode of approximation*, by M. Morse and C. B. Tompkins, Ann. of Math. (2) **42** (1941), 62–72.

50. *Corrections to our paper on the existence of minimal surfaces of general critical types*, by M. Morse and C. B. Tompkins, Ann. of Math. (2) **42** (1941), 331.

51. *Mathematics in the defense program*, by M. Morse and W. L. Hart, The Math. Teacher, May 1941, pp. 195–202.

52. *Unstable minimal surfaces of higher topological structure*, by M. Morse and C. B. Tompkins, Duke Math. J. **8** (1941), 350–375.

53. *The continuity of the area of harmonic surfaces as a function of the boundary representations*, by M. Morse and C. B. Tompkins, Amer. J. Math. **63** (1941), 825–838.

54. *Report on the War Preparedness Committee of the AMS and MAA at the Chicago meeting*, Bull. Amer. Math. Soc. **47** (1941), 829–831.

55. *What is analysis in the large?*, Amer. Math. Monthly **49** (1942), 358–364.

56. *Manifolds without conjugate points*, by M. Morse and G. A. Hedlund, Trans. Amer. Math. Soc. **51** (1942), 362–386.

57. *Mathematics and the maximum scientific effort in total war*, Scientific Monthly **56** (1943), 1–6.

58. *Variational theory in the large including the non-regular case*, by M. Morse and George Ewing–First paper, Ann. of Math. (2) **44** (1943), 339–353; Second paper, Ann. of Math. (2) **44** (1943), 354–374.

59. *Functional topology*, Bull. Amer. Math. Soc. **49** (1943), 144–149.

60. *Unending chess, symbolic dynamics and a problem in semigroups*, by M. Morse and G. A. Hedlund, Duke Math. J. **11** (1944), 1–7.

61. *Topological methods in the theory of functions of a single complex variable*: I. *Deformation types of locally simple plane curves*; II. *Boundary values and integral characteristics of interior transformations and pseudoharmonic functions*, by M. Morse and M. Heins, Ann. of Math. (2) **46** (1945), 600–624; 625–666; III. *Causal isomorphisms in the theory of pseudoharmonic functions*, Ann. of Math. (2) **47** (1946), 233–273.

62. *The topology of pseudoharmonic functions*, Duke Math. J. **13** (1946), 21–42.

63. *George David Birkhoff and his mathematical work*, Bull. Amer. Math. Soc. **52** (1946), 357–391.

64. *Topological methods in the theory of functions of a complex variable*, by M. Morse and M. Heins, Bull. Amer. Math. Soc. **53** (1947), 1–15.

65. *Deformation classes of meromorphic functions and their extensions to interior transformations*, by M. Morse and M. Heins, Acta Math. **79** (1947), 51–103.

66. *Functions on a metric space and a setting for isoperimetric problems*, Studies and Essays, presented to R. Courant on his 60th birthday, January 8, 1948, pp. 253–263. Interscience Publishers Inc., New York, 1948.

67. *A positive, lower semicontinuous nondegenerate function on a metric space*, Fund. Math. **35** (1948), 47–78.

68. *L-S-homotopy classes of locally simple curves*, Ann. Soc. Polonaise Math. **21** (1948), 236–256.

69. *Equilibria in nature, stable and unstable*, Proc. Amer. Philos. Soc. **93** (1949), 222–225.

70. *The Fréchet variation and the convergence of multiple Fourier series*, by M. Morse and W. Transue, Proc. Nat. Acad. Sci. **35** (1949), 395–399.

71. *Integral representations of bilinear functionals*, by M. Morse and W. Transue, Proc. Nat. Acad. Sci. **35** (1949), 136–143.

72. *Functionals of bounded Fréchet variation*, by M. Morse and W. Transue, Canad. J. Math. **1** (1949), 153–165.

73. *Functionals F bilinear over the product $A \times B$ of two p-normed vector spaces*: I. *The representation of F*, by M. Morse and W. Transue, Ann. of Math. (2) **50** (1949), 777–815; II. *Admissible spaces A*, Ann. of Math. (2) **51** (1950), 576–614.

74. *Les progrès de l'analyse variationnelle globale et son programme*, Rendiconti di Matematica e delle sue applicazioni, Serie V, **10** (3) (4) (1948), 1–11.

75. *Topological methods in the theory of functions of a complex variable*, Ann. Matematica **28** (1949), 21–25.

76. *A characterization of the bilinear sum associated with the classical second variation*, by M. Morse and W. Transue, Ann. Matematica **29** (1949), 25–68.

77. *L-S homotopy classes on the topological image of the projective plane*, Bull. Amer. Math. Soc. **55** (1949), 981–1003.

78. *The Fréchet variation and a generalization for multiple Fourier series of the Jordan test*, by M. Morse and W. Transue, Rivista di Matematica della Universita di Parma **1** (1950), 1–16.

79. *The Fréchet variation sector limits, and left decompositions*, by M. Morse with W. Transue, Canad. J. Math. (3) **2** (1950), 344–374.

80. *A calculus of Fréchet variations*, by M. Morse with W. Transue, J. Indian Math. Soc. **14** (2) (3) (1950), 65–117.

81. *The Fréchet variation and Pringsheim convergence of double Fourier series*, Contributions to Fourier Analysis, by M. Morse and W. Transue, Ann. Math. Studies, no. 25, Princeton Univ. Press, Princeton, N. J., 1950, pp. 46–103.

82. *Norms of distribution functions associated with bilinear functionals*, by M. Morse and W. Transue, ibidem, pp. 104–144.

83. *Bilinear functionals over CXC*, by M. Morse and W. Transue, Acta Sci. Math. **12** (1950), 41–48.

84. *Recent advances in variational theory in the large*, Proc. Internat. Cong. Math. **2** (1950), 143–156.

85. *A new implication of the Young-Pollard convergence criteria for a Fourier series*, by M. Morse and W. Transue, Duke Math. J. **18** (1951), 563–571.

86. *Trends in analysis*, J. Franklin Inst. **251** (1951), 33–43.

87. *Mathematics and the arts*, Yale Review **40** (1951), 604–612. 75th birthday of Robert Frost. Essay given at Kenyon College.

88. *Homology relations on regular orientable manifolds*, Proc. Nat. Acad. Sci. **38** (1952), 247–258.

89. *Contour equivalent pseudoharmonic functions and pseudoconjugates*, by M. Morse with J. Jenkins, Amer. J. Math. **74** (1952), 23–51.

90. *Homotopy and homology related to the Schoenflies problem*, by M. Morse with E. Baiada, Ann. of Math. (2) **58** (1953), 142–165.

91. *Topological methods on Riemann surfaces. Pseudoharmonic functions*, by M. Morse with J. Jenkins, Ann. of Math. Studies, no. 30, Princeton Univ. Press, Princeton, N. J., 1953, pp. 111–139.

92. *The existence of pseudoconjugates on Riemann surfaces*, by M. Morse with J. Jenkins, Fund. Math. **39** (1953), 269–287.

93. *The generalized Fréchet variation and Riesz-Young-Hausdorff type theorems*, by M. Morse with W. Transue, Rend. Circ. Math. Palermo, Serie II, Tome II (1953), 35 pp.

94. *Conjugate nets, conformal structure, and interior transformations on open Riemann surfaces*, by M. Morse with J. Jenkins, Proc. Nat. Acad. Sci. **39** (1953), 1261–1268.

95. *Conjugate nets on an open Riemann surface*, by M. Morse with J. Jenkins, Proc. Univ. Michigan Conf., June 1953.

96. *Curve families F^* locally the level curves of a pseudoharmonic function*, by M. Morse with J. Jenkins, Acta Math. **91** (1954), 42 pp.

97. (With W. Transue) *Semi-normed vector spaces with duals of integral type*, J. Analyse Math. **4** (1954–1955), 149–186.

98. *Bimeasures and their integral extensions*, Ann. Mat. Pura Appl. **39** (1955), 345–356.

99. (With W. Transue) *C-bimeasures Λ and their superior integrals Λ^**, Rend. Circ. Math. Palermo **4** (1955), 270–300.

100. (With W. Transue) *The representation of a C-bimeasure on a general rectangle*, Proc. Nat. Acad. Sci. **42** (1956), 89–95.

101. *La construction topologique d'un réseau isotherme sur une surface ouverte*, J. Math. Pures Appl. **35** (1956), 67–75.

102. (With W. Transue) *C-bimeasures Λ and their integral extensions*, Ann. of Math. (2) **64** (1956), 480–504.

103. (With W. Transue) *Products of a C-measure and a locally integrable mapping*, Canad. J. Math. **9** (1957), 475–486.

104. *Differentiable mappings in the Schoenflies problem*, Proc. Nat. Acad. Sci. **44** (1958), 1068–1072.

105. (With W. Transue) *Vector subspaces A of C^E with duals of integral type*, J. Math. **37** (1958), 44–363.

106. (With W. Transue) *The existence of vector function spaces with duals of integral type*, Colloq. Math. **6** (1958), 95–117.

107. (With W. Transue) *The local characterization of vector function spaces with duals of integral type*, J. Analyse Math. **6** (1958), 225–260.

108. *Mathematics and the arts*, Bull. of the Atomic Scientists **15** (1959), 55–59 (Yale Rev. 1951).

109. *Mathematics, the arts and freedom*, Thought (Fordham Univ. Quarterly) **34** (1959), 16–24.

110. *Differentiable mappings in the Schoenflies theorem*, Comp. Math. **14** (1959), 83–151.

111. *Topologically nondegenerate functions on a compact n-manifold M*, J. Analyse Math. **7** (1959), 189–208.

112. *Fields of geodesics issuing from a point*, Proc. Nat. Acad. Sci. **46** (1960), 105–111.

113. *The existence of polar nondegenerate functions on differentiable manifolds*, Ann of Math. (2) **71** (1960), 352–383.

114. *A reduction of the Schoenflies extension problem*, Bull. Amer. Math. Soc. **66** (1960), 113–115.

115. *The existence of nondegenerate functions on a compact differentiable m-manifold M*, Ann. Mat. Pura Appl. **49** (1960), 117–128.

116. *Recurrent geodesics on a surface of negative curvature*, Trans. Amer. Math. Soc. **22** (1921), 84–110. (Multilithed with Historical Note, 1960.)

117. (With W. Huebsch) *An explicit solution of the Schoenflies extension problem*, J. Math. Soc. Japan **12** (1960), 271–289.

118. (With W. Huebsch) *A singularity in the Schoenflies extension*, Proc. Nat. Acad. Sci. **46** (1960), 1100–1102.

119. *On elevating manifold differentiability*, J. Indian Math. Soc. **24** (1960), 379–400.

120. (With W. Huebsch) *The dependence of the Schenflies extension on an accessory parameter*, J. Analyse Math. **8** (1960–1961), 209–271.

121. (With W. Huebsch) Abstracts: *A characterization of an analytic n-ball*, and *A. Schoenflies extension of a real analytic diffeomorphism of S into W*. (Multilithed April 17, 1961.)

122. *Boundary values of partial derivatives of Poisson integral*, Ann. Acad. Brasil. Ci. **33** (1961), 131–139.

123. (With W. Huebsch) *Conical singular points of diffeomorphisms*, Bull. Amer. Math. Soc. **67** (1961), 490–493.

124. (With W. Huebsch) *Schoenflies extensions of analytic families of diffeomorphisms*, Math. Ann. **144** (1961), 162–174.

125. (With W. Huebsch) *The Schoenflies extension in the analytic case*, Ann. Mat. Pura Appl. **54** (1961), 359–378.

126. *Schoenflies problems*, Fund. Math. **50** (1962), 319–332.

127. (With W. Huebsch) *Schoenflies extensions without interior differential singularities*, Ann. of Math. (2) **76** (1962), 18–54.

128. *Topological, differential, and analytic formulations of Schoenflies problems*, Rend. Circ. Mat. Palermo **21** (1962), 286–295. (Lecture delivered in Rome, April 1962.)

129. (With W. Huebsch) *Analytic diffeomorphisms approximating C^m-diffeomorphisms*, Rend. Circ. Mat. Palermo **11** (1962), 1–22.

130. (With W. Huebsch) *Diffeomorphisms of manifolds*, Rend. Circ. Mat. Palermo **11** (1962), 1–28.

131. *Schoenflies extensions and differentiable isotopies*, J. Mat. Pures Appl. **42** (1963), 29–41.

132. *An arbitrarily small analytic mapping into R_+ of a proper, regular, analytic r-manifold in E_m*, Rendi. Sci. Ist. Lombardo, Milano (A) **97** (1963), 650–660.

133. *Harmonic extensions*, Monatsh. Math. **67** (1963), 317–325.

134. (With W. Huebsch) *The dependence of the Shoenflies extension on an accessory parameter (the topological case)*, Proc. Nat. Acad. Sci. **50** (1963), 1036–1037.

135. (With W. Huebsch) *The bowl theorem and a model nondegenerate function*, Proc. Nat. Acad. Sci. **51** (1964), 49–51.

136. (With W. Huebsch) *Conditioned differentiable isotopies*, Differential Analysis Colloq., Bombay, India, 1964, pp. 1–25.

137. *The elimination of critical points of a nondegenerate function on a differentiable manifold*, J. Analyse Math. **13** (1964), 257–316.

138. *Bowls, f-fibre bundles and the alteration of critical values*, Ann. Acad. Brasil. Ci. **36** (1964), 245–259.

139. *Bowls of a nondegenerate function on a compact differentiable manifold*, Differential and Combinatorial Topology, Princeton Univ. Press, Princeton, N. J., 1965, pp. 81–103. (Symposium in honor of M. Morse.)

140. *Quadratic forms Θ and Θ-fibre bundles*, Ann. of Math. (2) **81** (1965), 303–340.

141. (With W. Huebsch), *A model nondegenerate function*, Revue Roumaine Math. Pures Appl. **10** (1965), 691–722.

142. *The reduction of a function near a nondegenerate critical point*, Proc. Nat. Acad. Sci. **54** (1965), 1759–1764.

143. *Projective methods (in photogrammetry)*, Photogrammetric Engineering, Sept. 1966, pp. 849–855.

144. (With J. Cantwell) *Diffeomorphism including automorphisms of $\pi_1(T_p)$*, Topology **4** (1966), 323–341.

145. *Nondegenerate functions on abstract differentiable manifolds M*, J. Analyse Math. **19** (1967), 231–272.

146. *Nondegenerate real-valued differentiable functions*, Proc. Nat. Acad. Sci. **57** (1967), 32–38.

147. *Focal sets of regular manifolds M_{n-1} in E_n*, J. Differential Geometry **1** (1967), 1–19.

148. *Bowls, f-fiber bundles and the alteration of critical values*, Proc. Nat. Acad. Sci. **60** (1968), 1156–1159.

149. (With S. S. Cairns) *Singular homology over **Z** on topological manifolds*, J. Differential Geometry **3** (1969), 257–288.

150. (With S. S. Cairns) *A setting for a theorem of Bott*, Proc. Nat. Acad. Sci. **65** (1970), 8–9.

151. *Mathematics in our culture*, The Spirit and the Uses of the Mathematical Sciences by T. L. Saaty and F. J. Weyl, McGraw-Hill, New York, 1969, pp. 105–120.

152. *Equilibrium points of harmonic potentials*, J. Analyse Math. **23** (1970), 281–296.

153. (With S. S. Cairns) *Elementary quotients of abelian groups, and singular homology on manifolds*, Nagoya Math. J. **39** (1970), 167–198.

154. *Subordinate quadratic forms and their complementary forms*, Proc. Nat. Acad. Sci. **68** (1971), 579.

155. *Model families of quadratic forms*, Proc. Nat. Acad. Sci. **68** (1971), 914–915.

156. (With S. S. Cairns) *Orientation of differentiable manifolds*, J. Differential Geometry **6** (1971), 1–31. (Dedicated to S. S. Chern and D. C. Spencer.)

157. *Subordinate quadratic forms and their complementary forms*, Rev. Roumaine Math. Pures Appl. **16** (1971), 559–569.

158. (With S. S. Cairns) *Singular homology on an untriangulated manifold*, J. Differential Geometry **7** (1972), 1–17.

159. *Axial presentations of regular arcs on M_n*, Proc. Nat. Acad. Sci. **69** (1972), 3504–3505.

160. *Singular quadratic functionals*, Math. Ann. **201** (1973), 315–340.

161. *F-deformations and F-tractions*, Proc. Nat. Acad. Sci. **70** (1973), 1634–1635.

162. *Fréchet curve classes*, J. Math. Pures Appl. **53** (1974), 291–298.

163. *Singleton critical values*, Bull. Inst. Math. Acad. Sinica **2** (1974), 317–333.

164. (With D. Landis) *Geodesic joins and Fréchet curve classes*, Rend. Mat. **8** (1975), 161–185.

165. *Connectivities R_i of Fréchet spaces in variational topology*, Proc. Nat. Acad. Sci. **72** (1975), 2069–2070.

166. *Topologically nondegenerate functions*, Fund. Math. **88** (1975), 17–52.

167. (With D. Landis) *Tractions in critical point theory*, Rocky Mountain J. Math. **5** (1975), 379–399.

168. *Fréchet numbers in global variational analysis*, Houston J. Math. **2** (1976), 387–403.

169. (With S. S. Cairns) *Fréchet numbers and geodesics on surfaces*, Bull. Inst. Math. Acad. Sinica **4** (1976), 7–34.

170. *Conjugate points on a limiting extremal*, Proc. Nat. Acad. Sci. **73** (1976), 1800–1801.

171. *Extremal limits of nondegenerate extremals*, Rend. Mat. **9** (1976), 621–632.

172. *Tubular presentations π of subsets of manifolds*, Proc. Nat. Acad. Sci. **74** (1977), 2209–2210.

173. *Nondegenerate point pairs in global variational analysis*, J. Differential Geometry **11** (1976), 617–632.

174. *Uses of the Fréchet numbers $R_i(M_n)$ of a smooth manifold*, Houston J. Math. **3** (1977), 503–513.

175. *Lacunary type number sequence in global analyses*, J. Math. Pures Appl. **57** (1978), 87–98.

176. *The Fréchet numbers of the differentiable product of two compact converted smooth manifolds*, Edited and completed by S. S. Cairns, Univ. of Illinois, for Marston Morse memorial issue of Bull. of Institute of Math. Acad. Sinica, Vol. VI, No. 2, October 1978.

Books

1. *Calculus of variations in the large*, Amer. Math. Soc. Colloq. Publ., no. 18, Amer. Math. Soc., Providence, R. I., 1934. (Address: P. O. Box 6248, Providence, R. I., 02904.)

2. *Functional topology and abstract variational theory*, Mémorial des Sciences Mathématiques 92, Gauthier-Villars, Paris, 1939.

3. *Topological methods in the theory of functions of a complex variable*, Princeton Univ. Press, Princeton, N. J., 1947. (Paperback available from Kraus Reprint Co., 16 E. 46th St., New York, N. Y. 10017.)

4. *Symbolic dynamics*, Lectures of 1938 with New Preface, 1966, 87 pp. (Notes by R. Oldenburger.) University Microfilms, 300 N. Zeeb Road, Ann Arbor, Mich. 48106.

5. (With S. S. Cairns) *Critical point theory in global analysis and differential topology*, Academic Press, New York, 1969.

6. *Variational analysis: Critical extremals and Sturmian extensions*, Wiley, New York, 1973.

7. *Global variational analysis: Weierstrass integrals on a Riemannian manifold*, Mathematical Notes, Princeton Univ. Press, Princeton, N. J., 1976.

Printed in the United States
By Bookmasters